Methods in Enzymology

Volume 208
PROTEIN–DNA INTERACTIONS

METHODS IN ENZYMOLOGY

EDITORS-IN-CHIEF

John N. Abelson   Melvin I. Simon

DIVISION OF BIOLOGY
CALIFORNIA INSTITUTE OF TECHNOLOGY
PASADENA, CALIFORNIA

FOUNDING EDITORS

Sidney P. Colowick and Nathan O. Kaplan

Methods in Enzymology

Volume 208

Protein–DNA Interactions

EDITED BY

Robert T. Sauer

DEPARTMENT OF BIOLOGY
MASSACHUSETTS INSTITUTE OF TECHNOLOGY
CAMBRIDGE, MASSACHUSETTS

ACADEMIC PRESS, INC.
Harcourt Brace Jovanovich, Publishers
San Diego   New York   Boston
London   Sydney   Tokyo   Toronto

Academic Press, Inc.
San Diego, California 92101

United Kingdom Edition published by
ACADEMIC PRESS LIMITED
24-28 Oval Road, London NW1 7DX

Library of Congress Catalog Card Number: 54-9110

ISBN   0-12-182109-9   (alk. paper)

PRINTED IN THE UNITED STATES OF AMERICA
91 92 93 94   9 8 7 6 5 4 3 2 1

Table of Contents

Section I. Purification and Characterization

Section II. DNA Binding and Bending

Section III. Biochemical Analysis of Protein–Nucleic DNA Interactions

Section IV. Genetic Analysis of Structure–Function Relationships

Contributors to Volume 208

Article numbers are in parentheses following the names of contributors.
Affiliations listed are current.

CHRISTOPHER R. AIKEN (21), *Infectious Diseases Laboratory, The Salk Institute, San Diego, California 92186*

BRUCE M. ALBERTS (3), *Department of Biochemistry and Biophysics, University of California, San Francisco, San Francisco, California 94143*

JACK BARRY (3), *Department of Biochemistry and Biophysics, University of California, San Francisco, San Francisco, California 94143*

JEREMY M. BERG (4), *Department of Biophysics and Biophysical Chemistry, The Johns Hopkins University, Baltimore, Maryland 21205*

JAMES U. BOWIE (27, 29), *Department of Chemistry and Biochemistry, University of California, Los Angeles, Los Angeles, California 90024*

RICHARD M. BREYER (27), *Laboratoire d'Immuno-Pharmacologie Moleculaire, Institut Cochin de Genetique Moleculaire, Paris, France*

THOMAS W. BRUCIE (20), *Department of Biological Chemistry and Department of Chemistry and Biochemistry and Molecular Biology Institute, University of California, Los Angeles, Los Angeles, California 90024*

HENRI BUC (14), *Unité de Physicochimie des Macromolécules Biologiques, Institut Pasteur, 75724 Paris Cedex 15, France*

MALCOLM BUCKLE (14), *Unité de Physicochimie des Macromolécules Biologiques, Institut Pasteur, 75724 Paris Cedex 15, France*

WLODZIMIERZ BUJALOWSKI (15), *Department of Human Biological Chemistry and Genetics, University of Texas Medical Branch, Galveston, Texas 77550*

RICHARD R. BURGESS (1), *McArdle Laboratory for Cancer Research, University of Wisconsin-Madison, Madison, Wisconsin 53706*

JANNETTE CAREY (8), *Department of Chemistry, Princeton University, Princeton, New Jersey 08544*

ARTEMIS E. CHAKERIAN (23), *Department of Biochemistry and Cell Biology, Rice University, Houston, Texas 77251*

CHING-HONG B. CHEN (20), *Department of Biological Chemistry and Department of Chemistry and Biochemistry and Molecular Biology Institute, University of California, Los Angeles, Los Angeles, California 90024*

DONALD M. CROTHERS (9), *Department of Chemistry, Yale University, New Haven, Connecticut 06511*

PETER B. DERVAN (24), *Arnold and Mable Beckman Laboratories of Chemical Synthesis, California Institute of Technology, Pasadena, California 91125*

WENDY J. DIXON (19), *Department of Biology, The Johns Hopkins University, Baltimore, Maryland 21218*

MARK DODSON (11), *Department of Biochemistry, Stanford University School of Medicine, Stanford, California 94305*

BETH A. DOMBROSKI (19), *Center for Medical Genetics, The Johns Hopkins Medical Institutions, Baltimore, Maryland 21205*

RICHARD H. EBRIGHT (30), *Department of Chemistry and Waksman Institute, Rutgers University, New Brunswick, New Jersey 08855*

HARRISON ECHOLS (11), *Department of Molecular and Cell Biology, University of California, Berkeley, Berkeley, California 94720*

ELISABETH M. EVERTSZ (13), *Institute of Molecular Biology, and Departments of Chemistry and Biology, University of Oregon, Eugene, Oregon 97403*

MATTHEW A. FISHER (16), *Department of Chemistry, Randolph-Macon College, Ashland, Virginia 23005*

TIM FORMOSA (3), *Department of Biochemistry, University of Utah, Salt Lake City, Utah 84132*

ALEXANDRE FRITSCH (14), *Unité de Physicochimie des Macromolécules Biologiques, Institut Pasteur, 75724 Paris Cedex 15, France*

JOSEPH A. GARDNER (23), *Department of Biochemistry and Cell Biology, Rice University, Houston, Texas 77251*

MARC R. GARTENBERG (9), *Department of Molecular Biology and Biophysics, Harvard University, Cambridge, Massachusetts 02138*

JOHANNES GEISELMANN (14), *Unité de Physicochimie des Macromolécules Biologiques, Institut Pasteur, 75724 Paris Cedex 15, France*

AMY L. GIBSON (31), *Department of Biological Chemistry, The University of Michigan, Ann Arbor, Michigan 48109*

STEPHEN P. GOFF (28), *Department of Biochemistry and Molecular Biophysics, Columbia University College of Physicians and Surgeons, New York, New York 10032*

JAY D. GRALLA (10), *Department of Chemistry and Biochemistry, and the Molecular Biology Institute, University of California, Los Angeles, Los Angeles, California 90024*

JACK GREENBLATT (3), *Banting and Best Department of Medical Research, University of Toronto, Toronto, Ontario, Canada*

RICHARD I. GUMPORT (21), *Department of Biochemistry, University of Illinois, College of Medicine, Urbana, Illinois 61801*

JEUNG-HOI HA (16), *Department of Biochemistry, University of Wisconsin—Madison, Madison, Wisconsin 53706*

JEFFREY J. HAYES (19), *Laboratory of Molecular Biology, National Institutes of Health, Bethesda, Maryland 20892*

WOLFGANG HILLEN (5, 18), *Institut für Mikrobiolgie und Biochemie, Friedrich-Alexander Universität Erlangen-Nürnberg, D-8520 Erlangen, Germany*

ANN HOCHSCHILD (17), *Department of Microbiology and Molecular Genetics, Harvard Medical School, Boston, Massachusetts 02115*

JOEL W. HOCKENSMITH (13), *Department of Biochemistry, University of Virginia School of Medicine, Charlottesville, Virginia 22908*

JAMES C. HU (27), *Department of Biology, Massachusetts Institute of Technology, Cambridge, Massachusetts 02139*

ANDRZEJ JOACHIMIAK (7), *Department of Molecular Biophysics and Biochemistry, Yale University, New Haven, Connecticut 06511*

JAMES T. KADONAGA (2), *Department of Biology and Center for Molecular Genetics, University of California, San Diego, La Jolla, California 92093*

RACHEL E. KLEVIT (6), *Department of Biochemistry, University of Washington, Seattle, Washington 98195*

KENDALL L. KNIGHT (27), *Department of Biochemistry and Molecular Biology, University of Massachusetts Medical Center, Worcester, Massachusetts 01655*

WILLIAM H. KONIGSBERG (25), *Department of Molecular Biophysics and Biochemistry, Yale University School of Medicine, New Haven, Connecticut 06510*

WILLIAM L. KUBASEK (13), *Department of Molecular Biology, Massachusetts General Hospital, Boston, Massachusetts 02114*

MICHIO D. KUWABARA (20), *Department of Biological Chemistry and Department of Chemistry and Biochemistry and Molecular Biology Institute, University of California, Los Angeles, Los Angeles, California 90024*

ARTHUR D. LANDER (12), *Department of Biology and Department of Brain and Cognitive Sciences, Massachusetts Institute of Technology, Cambridge, Massachusetts 02139*

JUDITH R. LEVIN (19), *Department of Chemistry, The Johns Hopkins University, Baltimore, Maryland 21218*

WENDELL A. LIM (12, 27), *Department of Biology, Massachusetts Institute of Technology, Cambridge, Massachusetts 02139*

TIMOTHY M. LOHMAN (15), *Department of Biochemistry and Molecular Biophysics, Washington University School of Medicine, St. Louis, Missouri 63110*

KATHLEEN S. MATTHEWS (23), *Department of Biochemistry and Cell Biology, Rice University, Houston, Texas 77251*

DENISE L. MERKLE (4), *Department of Chemistry, University of Florida, Gainesville, Florida 32611*

JEFFREY H. MILLER (26), *Department of Microbiology and Molecular Genetics, and the Molecular Biology Institute, University of California, Los Angeles, Los Angeles, California 90024*

MICHAEL C. MOSSING (27, 29), *Department of Biological Sciences, University of Notre Dame, South Bend, Indiana 46556*

DALE L. OXENDER (31), *Department of Biotechnology, Park-Davis Research Division of Warner-Lambert, Ann Arbor, Michigan 48106*

GRACE PÁRRAGA (6), *Department of Biochemistry, Biocenter of the University of Basel, Basel CH-4056, Switzerland*

DAWN A. PARSELL (27), *Howard Hughes Medical Institute, University of Chicago, Chicago, Illinois 60637*

VINAYAKA R. PRASAD (28), *Department of Microbiology and Immunology, Albert Einstein College of Medicine, Bronx, New York 10461*

M. THOMAS RECORD, JR. (16), *Department of Chemistry and Biochemistry, University of Wisconsin—Madison, Madison, Wisconsin 53706*

JOHN F. REIDHAAR-OLSON (27), *Department of Biochemistry and Biophysics, University of California, San Francisco, San Francisco, California 94143*

PASCAL ROUX (14), *Unité de Physicochimie des Macromolécules Biologiques, Institut Pasteur, 75724 Paris Cedex 15, France*

SELINA SASSE-DWIGHT (10), *Department of Genetics, Stanford University School of Medicine, Stanford, California 94305*

ROBERT T. SAUER (12, 27, 29), *Department of Biology, Massachusetts Institute of Technology, Cambridge, Massachusetts 02139*

KEVIN R. SHOEMAKER (27), *Department of Biology, Massachusetts Institute of Technology, Cambridge, Massachusetts 02139*

THOMAS E. SHRADER (9), *Department of Biology, Massachusetts Institute of Technology, Cambridge, Massachusetts 02139*

PAUL B. SIGLER (7), *Department of Molecular Biophysics and Biochemistry, Howard Hughes Medical Institute, Yale University, New Haven, Connecticut 06511*

DAVID S. SIGMAN (20), *Department of Biological Chemistry, Molecular Biology Institute, University of California, Los Angeles, Los Angeles, California 90024*

GARY D. STORMO (22), *Department of Molecular, Cellular and Developmental Biology, University of Colorado, Boulder, Colorado 80309*

KARLHEINZ TOVAR (5), *Institut für Mikrobiologie und Biochemie, Friedrich-Alexander Universität Erlangen-Nürnberg, D-6382 Erlangen, Germany*

THOMAS D. TULLIUS (19), *Department of Chemistry, The Johns Hopkins University, Baltimore, Maryland 21218*

PETER H. VON HIPPEL (13), *Institute of Molecular Biology, and Departments of Chemistry and Biology, University of Oregon, Eugene, Oregon 97403*

WILLIAM R. VORACHEK (13), *Department of Biochemistry, University of Virginia School of Medicine, Charlottesville, Virginia 22908*

MARGARET F. WEIDNER (19), *Seattle, Washington 98109*

KENNETH R. WILLIAMS (25), *Howard Hughes Medical Institute, Department of Molecular Biophysics and Biochemistry,* *Yale University School of Medicine, New Haven, Connecticut 06510*

ANDREAS WISSMANN (18), *BASF Bioresearch Corp., Cambridge, Massachusetts 02139*

Preface

DNA binding proteins are involved in a number of basic cellular processes including transcription, DNA replication, transposition, restriction, recombination, and DNA repair. The chapters in this volume cover a number of important methods, some old and some new, that are now widely used in the study of DNA binding proteins. These include purification and protein characterization, assays of protein–DNA binding and protein-induced DNA bending which can be used *in vitro* and *in vivo,* and biochemical and genetic methods for probing the structure, energy, and specificity of protein–DNA interactions. Previous volumes of this series that contain related methods include 65, 100, 130, 154, 155, and 170.

ROBERT T. SAUER

METHODS IN ENZYMOLOGY

VOLUME LV. Biomembranes (Part F: Bioenergetics)
Edited by SIDNEY FLEISCHER AND LESTER PACKER

VOLUME LVI. Biomembranes (Part G: Bioenergetics)
Edited by SIDNEY FLEISCHER AND LESTER PACKER

VOLUME LVII. Bioluminescence and Chemiluminescence
Edited by MARLENE A. DELUCA

VOLUME LVIII. Cell Culture
Edited by WILLIAM B. JAKOBY AND IRA PASTAN

VOLUME LIX. Nucleic Acids and Protein Synthesis (Part G)
Edited by KIVIE MOLDAVE AND LAWRENCE GROSSMAN

VOLUME LX. Nucleic Acids and Protein Synthesis (Part H)
Edited by KIVIE MOLDAVE AND LAWRENCE GROSSMAN

VOLUME 61. Enzyme Structure (Part H)
Edited by C. H. W. HIRS AND SERGE N. TIMASHEFF

VOLUME 62. Vitamins and Coenzymes (Part D)
Edited by DONALD B. McCORMICK AND LEMUEL D. WRIGHT

VOLUME 63. Enzyme Kinetics and Mechanism (Part A: Initial Rate and
Inhibitor Methods)
Edited by DANIEL L. PURICH

VOLUME 64. Enzyme Kinetics and Mechanism (Part B: Isotopic Probes
and Complex Enzyme Systems)
Edited by DANIEL L. PURICH

VOLUME 65. Nucleic Acids (Part I)
Edited by LAWRENCE GROSSMAN AND KIVIE MOLDAVE

VOLUME 66. Vitamins and Coenzymes (Part E)
Edited by DONALD B. McCORMICK AND LEMUEL D. WRIGHT

VOLUME 67. Vitamins and Coenzymes (Part F)
Edited by DONALD B. McCORMICK AND LEMUEL D. WRIGHT

VOLUME 81. Biomembranes (Part H: Visual Pigments and Purple Membranes, I)
Edited by LESTER PACKER

VOLUME 82. Structural and Contractile Proteins (Part A: Extracellular Matrix)
Edited by LEON W. CUNNINGHAM AND DIXIE W. FREDERIKSEN

VOLUME 83. Complex Carbohydrates (Part D)
Edited by VICTOR GINSBURG

VOLUME 84. Immunochemical Techniques (Part D: Selected Immunoassays)
Edited by JOHN J. LANGONE AND HELEN VAN VUNAKIS

VOLUME 85. Structural and Contractile Proteins (Part B: The Contractile Apparatus and the Cytoskeleton)
Edited by DIXIE W. FREDERIKSEN AND LEON W. CUNNINGHAM

VOLUME 86. Prostaglandins and Arachidonate Metabolites
Edited by WILLIAM E. M. LANDS AND WILLIAM L. SMITH

VOLUME 87. Enzyme Kinetics and Mechanism (Part C: Intermediates, Stereochemistry, and Rate Studies)
Edited by DANIEL L. PURICH

VOLUME 88. Biomembranes (Part I: Visual Pigments and Purple Membranes, II)
Edited by LESTER PACKER

VOLUME 89. Carbohydrate Metabolism (Part D)
Edited by WILLIS A. WOOD

VOLUME 90. Carbohydrate Metabolism (Part E)
Edited by WILLIS A. WOOD

VOLUME 91. Enzyme Structure (Part I)
Edited by C. H. W. HIRS AND SERGE N. TIMASHEFF

VOLUME 92. Immunochemical Techniques (Part E: Monoclonal Antibodies and General Immunoassay Methods)
Edited by JOHN J. LANGONE AND HELEN VAN VUNAKIS

VOLUME 154. Recombinant DNA (Part E)
Edited by RAY WU AND LAWRENCE GROSSMAN

VOLUME 155. Recombinant DNA (Part F)
Edited by RAY WU

VOLUME 156. Biomembranes (Part P: ATP-Driven Pumps and Related Transport: The Na,K-Pump)
Edited by SIDNEY FLEISCHER AND BECCA FLEISCHER

VOLUME 157. Biomembranes (Part Q: ATP-Driven Pumps and Related Transport: Calcium, Proton, and Potassium Pumps)
Edited by SIDNEY FLEISCHER AND BECCA FLEISCHER

VOLUME 158. Metalloproteins (Part A)
Edited by JAMES F. RIORDAN AND BERT L. VALLEE

VOLUME 159. Initiation and Termination of Cyclic Nucleotide Action
Edited by JACKIE D. CORBIN AND ROGER A. JOHNSON

VOLUME 160. Biomass (Part A: Cellulose and Hemicellulose)
Edited by WILLIS A. WOOD AND SCOTT T. KELLOGG

VOLUME 161. Biomass (Part B: Lignin, Pectin, and Chitin)
Edited by WILLIS A. WOOD AND SCOTT T. KELLOGG

VOLUME 162. Immunochemical Techniques (Part L: Chemotaxis and Inflammation)
Edited by GIOVANNI DI SABATO

VOLUME 163. Immunochemical Techniques (Part M: Chemotaxis and Inflammation)
Edited by GIOVANNI DI SABATO

VOLUME 164. Ribosomes
Edited by HARRY F. NOLLER, JR., AND KIVIE MOLDAVE

VOLUME 165. Microbial Toxins: Tools for Enzymology
Edited by SIDNEY HARSHMAN

VOLUME 166. Branched-Chain Amino Acids
Edited by ROBERT HARRIS AND JOHN R. SOKATCH

[1] Use of Polyethyleneimine in Purification of DNA-Binding Proteins

By Richard R. Burgess

One of the important early steps in purifying a DNA-binding protein is to separate the protein of interest from cellular nucleic acids. A large variety of methods have been used, including precipitation with streptomycin sulfate,[1] polyethylene glycol,[2,3] batch treatment of extracts with DEAE-cellulose,[4] and treatment with DNase I.[5] Perhaps the most effective method involves precipitation with polyethyleneimine.[6–8] This chapter will focus on the use of polyethyleneimine in early stages of the purification of DNA-binding proteins. It should be noted, however, that the use of polyethyleneimine in protein purification is not restricted to DNA-binding proteins and has been used successfully for proteins from both prokaryotic and eukaryotic sources. The most comprehensive summary of the applications of polyethyleneimine is found in a review by Jendrisak.[9]

Properties of Polyethyleneimine (PEI)

Polyethyleneimine (PEI) is a product of polymerization of ethyleneimine to yield a basic linear polymer with the structure

$$H_2N-\left(-CH_2CH_2-\overset{\displaystyle H}{\underset{\displaystyle |}{N}}-\right)_n-CH_2CH_2-NH_2$$

Typically, n equals 700–2000 to give a molecular weight range of 30,000–90,000. Since the pK_a value of the imino group is 10–11, PEI is a positively charged molecule in solutions of neutral pH. It has been produced for over 30 years in large quantities for uses such as a mordant for the dye industry by BASF (Charlotte, NC) under the trade name, Polymin P.

[1] M. J. Chamberlin and P. Berg, *Proc. Natl. Acad. Sci. U.S.A.* **48**, 81 (1962).
[2] C. Babinet, *Biochem. Biophys. Res. Commun.* **26**, 639 (1967).
[3] B. Alberts and G. Herrick, this series, Vol. 21, [11] (1971).
[4] W. F. Mangel, *Biochim. Biophys. Acta* **163**, 172 (1974).
[5] R. R. Burgess, *J. Biol. Chem.* **244**, 6160 (1969).
[6] W. Zillig, K. Zechel, and H. Halbwachs, *Hoppe-Seyler's Z. Physiol. Chem.* **351**, 221 (1970).
[7] R. R. Burgess and J. J. Jendrisak, *Biochemistry* **14**, 4634 (1975).
[8] J. J. Jendrisak and R. R. Burgess, *Biochemistry* **14**, 4639 (1975).
[9] J. J. Jendrisak, *in* "Protein Purification: Micro to Macro" (R. R. Burgess, ed.), p. 75, Alan R. Liss, New York, 1987.

For precipitation of nucleic acids and some proteins, it has the desirable properties of being inexpensive, working well at slightly alkaline pH's where proteins are usually stable, and rapidly forming finely divided precipitates which may be centrifuged and resuspended very easily.

Principles of PEI Precipitation

PEI causes precipitation of nucleic acids and acidic proteins at low ionic strength by forming charge neutralization complexes and cross-bridges between the complexes. This leads to flocculation.[10]

This precipitation is basically a titration. As PEI is added, it becomes intimately involved in the networked complexes and is thus part of the precipitate. PEI precipitation thus differs fundamentally from salt or solvent precipitation. In ammonium sulfate precipitation, for example, the precipitant sequesters H_2O, decreases H_2O available to solvate the protein, and thus decreases the solubility of the protein, without itself being substantially precipitated. To illustrate this difference, assume that a given extract requires 50% saturated ammonium sulfate or 0.2% (w/v) PEI to achieve 90% precipitation of a given enzyme. If the extract is diluted 10-fold, it will take *higher* ammonium sulfate (about 56% saturation) but *lower* PEI concentration (0.02%) to achieve a similar 90% precipitation of the enzyme.

The fraction of the total protein precipitated by PEI will depend on the pH (at higher pH, more proteins will carry a negative charge and be precipitated) and on the ionic strength (increasing ionic strength will weaken PEI–protein interactions). At low ionic strength (0–0.1 M NaCl), DNA and all acidic proteins will be precipitated. At higher ionic strength (0.1–1 M NaCl), only DNA and highly negatively charged proteins will precipitate. At greater than 1 M NaCl, only nucleic acids will precipitate. DNA–PEI precipitates dissolve at 1.4 to 1.6 M NaCl. There are thus three basic strategies for using PEI in the purification of DNA binding proteins.

Strategy A: Precipitate with PEI at high ionic strength to remove nucleic acid. Almost no proteins will bind to DNA or to PEI in 1 M NaCl. Therefore one can adjust the salt concentration of an extract to 1 M NaCl and titrate with PEI to achieve precipitation of nucleic acids, leaving the proteins in the supernatant. This titration can be conveniently followed by determining the UV absorption spectrum of the supernatants[9] and adding PEI until the supernatant shows an A_{280} value greater than the A_{260}, characteristic of protein solutions. While this strategy can be used, it is not common. One reason is that while nucleic acid is effectively removed,

[10] D. J. Bell, M. Hoare, and P. Dunhill, *in* "Advances in Biochemical Engineering/Biotechnology" (A. Fiechler, ed.), p. 1. Springer-Verlag, Berlin, 1982.

no purification from other proteins is achieved, unlike strategies B or C (below). Another problem is that if more PEI is added than is needed to precipitate the nucleic acids, excess PEI will be present in the supernatant with the proteins. When the supernatant is then diluted or dialyzed to lower ionic strength, PEI–acidic protein complexes will precipitate. This excess PEI can be removed by ammonium sulfate precipitation as described below.

Strategy B: Precipitate with PEI at low to medium ionic strength to remove nucleic acids and some proteins, leaving enzyme of interest in supernatant. This is like strategy A but, because it is carried out at lower ionic strength, a substantial number of proteins will be precipitated. The enzyme of interest will remain in the supernatant and will not only be nucleic acid free but also will be free of the precipitated acidic proteins. This strategy works well for neutral or slightly basic DNA-binding proteins that do not bind to DNA under the ionic conditions of the PEI precipitation.

Strategy C: Precipitate with PEI at lower ionic strength to precipitate nucleic acids and some proteins and then elute desired enzyme out of pellet with higher ionic strength. This strategy is by far the most commonly used. Generally one chooses a pH of 7.5–7.9 and an ionic strength of 0.1 to 0.2 M NaCl and then performs a PEI titration test to determine the minimal amount of PEI solution that causes complete precipitation of the enzyme of interest. The pellet, which contains nucleic acids and some proteins, is resuspended in a predetermined low ionic strength buffer to wash out any trapped proteins but not the desired enzyme. The washed pellet is then resuspended in a buffer of predetermined higher ionic strength to elute the enzyme, leaving the nucleic acids in the precipitate. The precise amounts of PEI added and the ionic strength of the wash and elution buffers are determined empirically as described below. The eluate contains the enzyme of interest, some other proteins, and a certain amount of PEI that was originally coprecipitated with the eluted proteins. The PEI is generally removed by ammonium sulfate precipitation of the protein, leaving the PEI in the ammonium sulfate supernatant. The ammonium sulfate pellet is dissolved in buffer and subsequent purification steps are performed.

Methods and Procedures

Preparing Extracts

Extracts may be prepared in a variety of ways from bacterial,[11] animal and plant[12] cells, or tissues. Generally, the crude extract is centrifuged at

[11] M. Cull and C. S. McHenry, this series, Vol. 182, [12] (1990).
[12] P. Gegenheimer, this series, Vol. 182, [14] (1990).

10,000 g for 10–30 min to remove unbroken cells and cell debris. The resulting supernatant is referred to as extract.

Preparing Neutralized 5% (w/v) PEI Stock Solution

Concentrated PEI solutions, which are usually (but not always!) 50% (w/v), can be obtained from several commercial suppliers, including ICN Biomedicals (Irvine, CA), Aldrich (Milwaukee, WI), and Sigma (St. Louis, MO). One hundred milliliters of 50% (w/v) PEI is dissolved in 800 ml of deionized water. The solution is adjusted to pH 7.9 by adding concentrated HCl (approximately 38–40 ml) with stirring. After diluting to 1000 ml, the neutralized solution is filtered through one layer of Miracloth, Calbiochem, San Diego, CA (to remove slight particulate matter), resulting in a clear, slightly bluish solution. This 5% (w/v) or 10% (v/v) stock solution is stored refrigerated and appears to be stable indefinitely.

While PEI solutions are sometimes dialyzed to remove low-molecular-weight components, this does not seem to be necessary.

Performing PEI Test Titration of Escherichia coli Extract (Strategy C)

Materials

PEI stock solution, 5% (w/v), pH 7.9 (see above)
E. coli extract (about 12 mg protein/ml and 0.2 M NaCl, pH 7.9) prepared as described[7]

Procedure

1. Two-milliliter portions of extract are placed in 11 small centrifuge tubes and 5% (w/v) PEI stock solution is added to give final PEI concentrations of from 0 to 0.5% (w/v) in 0.05% (w/v) intervals.

2. Tubes are rapidly mixed after PEI addition, incubated for 5 min at 0°, mixed again, and centrifuged for 5 min.

3. Supernatants are assayed for protein concentration and for the presence of the enzyme of interest. This may be done by enzymatic assay, by sodium dodecyl sulfate-polyacrylamide gel electrophoresis (SDS–PAGE) analysis, or by immunologic assay.

In the example given in Figs. 1A and 2A,[7] the E. coli RNA polymerase, measured by the presence of the high-molecular-weight β' and β subunits on Coomassie-stained SDS–PAGE, was found to be completely precipitated after addition of PEI to a final concentration of 0.175% (w/v) or 0.35 ml of PEI stock solution per 10 ml of extract. At this PEI concentration, all the nucleic acids and about 30% of the total protein precipitated.

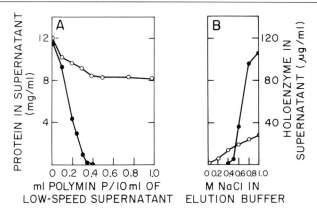

FIG. 1. PEI (Polymin P) precipitation curve and NaCl elution curve. (A) To 2-ml portions of the extract (low-speed supernatant), 5% (w/v) PEI was added in the indicated amounts and processed as described in Methods and Procedures. (B) An NaCl elution curve was prepared as described in Methods and Procedures. Supernatants were assayed for total protein (○) and for $\beta'\beta$ subunits of RNA polymerase on SDS–polyacrylamide gels shown in Fig. 2A and B (●). (Adapted from Ref. 7, with permission.)

Determining NaCl Elution Conditions (Strategy C)

Procedure

1. A 0.35-ml portion of 5% (w/v) PEI solution is added per 10 ml of extract as above, mixed, divided into 2-ml portions, centrifuged, and the supernatant discarded.

2. The identical pellets are resuspended for 10 min in 2 ml of TGED buffer (50 mM Tris-HCl, pH 7.9, 5% glycerol, 0.1 mM EDTA, 0.1 mM dithiothreitol) containing various concentrations of NaCl from 0 to 1 M and centrifuged.

3. Supernatants are again assayed for protein and enzyme.

As shown in Figs. 1B and 2B,[7] the *E. coli* RNA polymerase remained in the pellet during a 0.5 M NaCl wash but was completely eluted by 1 M NaCl. When a larger scale preparation was carried out with 0.175% (w/v) PEI precipitation, a 0.5 M NaCl wash, and a 1 M NaCl elution, a complete removal of nucleic acids, a 15-fold purification, a 29-fold increase in specific activity, and a 90% recovery of RNA polymerase was achieved.

Removal of PEI from Eluted Protein

When protein is eluted from the PEI pellet, PEI is also released into solution. If the eluate is dialyzed or diluted to lower ionic strength, the free protein and PEI will again interact and precipitate. Therefore, one

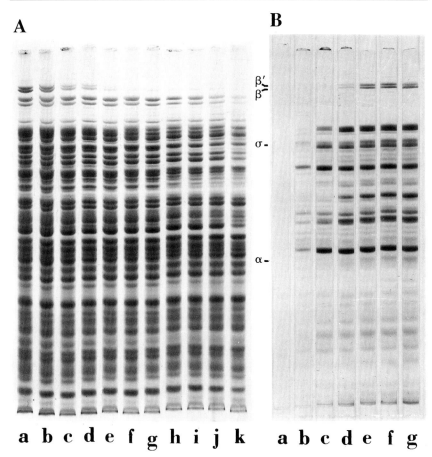

FIG. 2. SDS–polyacrylamide gel electrophoretic analysis of supernatants from the PEI precipitation curve and NaCl elution curve. Portions of supernatants were electrophoresed on an 8.75% acrylamide gel and stained with Coomassie Blue. The positions of the RNA polymerase subunits β', β, σ, and α (with apparent molecular weights of 165,000, 155,000, 87,000, and 39,000, respectively) are indicated. (A) Portions of supernatants of a PEI precipitation curve containing the following final PEI concentrations were analyzed: (a) 0%, (b) 0.05%, (c) 0.1%, (d) 0.125%, (e) 0.15%, (f) 0.175%, (g) 0.2%, (h) 0.25%, (i) 0.375%, (j) 0.5%, (k) 0.6%. (B) Portions of supernatants of an NaCl elution curve from Fig. 1B containing the following concentrations of NaCl: (a) 0.0 M, (b) 0.2 M, (c) 0.4 M, (d) 0.6 M, (e) 0.8 M, (f) 1.0 M, (g) 1.2 M. (Adapted from Ref. 7, with permission.)

must remove the free PEI. This is most commonly done by adding enough ammonium sulfate to the eluate to precipitate the protein. At this high ionic strength (often 40–60% saturated ammonium sulfate), PEI will not bind to protein and will remain in the supernatant after the ammonium sulfate precipitate is pelleted. This procedure removes the vast majority of the PEI and is sufficient for most proteins. If a protein is especially sensitive to residual traces of PEI, then a further step is required. The easiest method is to thoroughly resuspend the ammonium sulfate pellet in 40–60% saturated ammonium sulfate (sufficient to keep the protein insoluble) and recentrifuge. This way, any PEI physically trapped in the pellet will be removed.

Another method to remove PEI involves passing the eluate, or the dissolved ammonium sulfate precipitate, through a phosphocellulose column at 1 M NaCl. The protein will not bind to the column at this high ionic strength, but the PEI binds very tightly and is completely removed.[13]

Final Precautions and Comments

This basic approach for nucleic acid removal and protein purification has been widely employed (see references in Ref. 9). A few final comments, which might improve the success of this method, follow.

It is wise to carry out test titrations regularly, especially if the methods for preparing the extract change. The amount of PEI needed to achieve precipitation of the enzyme of interest will decrease if cell breakage is poor because the resulting extract will be more dilute.

Many proteins that bind to DNA are also acidic proteins that will precipitate with PEI even in the absence of nucleic acids. These are best treated with strategy C. Some DNA-binding proteins are neutral or basic and would only be expected to precipitate with PEI if they are bound to DNA and come down in the PEI–DNA precipitate. Therefore, the ionic strength at which PEI precipitation is done will determine whether one should use strategy B (high enough ionic strength to keep the protein of interest from binding to DNA and thus not precipitated with PEI) or strategy C (low enough ionic strength to allow the protein to bind DNA and be PEI precipitated). The optimal conditions must be determined empirically.

Precipitation with PEI occurs rapidly. There is a need to mix well during addition of PEI to prevent local overprecipitation. On large-scale (1–4 liters) quantities of extract this mixing is conveniently done in a 1-gal Waring blender.

[13] Z. F. Burton and D. Eisenberg, *Arch. Biochem. Biophys.* **205,** 478 (1980).

Occasionally, PEI is sold as a 25 (w/v) or 50% (v/v) solution diluted with an equal volume of H_2O to decrease viscosity and increase ease of dispensing. This can lead to an incorrect stock solution, so caution is advised. One way this can be detected is by observing the amount of HCl needed to titrate 100 ml of PEI concentrate to pH 7.9.

Acknowledgments

I would like to thank Mark Knuth for his helpful comments. This work was supported by NIH Grants CA07175, CA23076, and GM28575 and is dedicated to the outstanding research and training environment of McArdle Laboratory for Cancer Research on its 50th anniversary.

[2] Purification of Sequence-Specific Binding Proteins by DNA Affinity Chromatography

By James T. Kadonaga

Many biological processes, such as transcription, replication, and recombination, involve the action of sequence-specific DNA-binding proteins. To characterize the biochemical properties of these factors, it is necessary to purify them to homogeneity. In addition, the purified factors could be used to raise antibodies as well as to isolate the genes encoding the proteins. It has generally been difficult, however, to isolate these DNA-binding factors by conventional chromatography because they typically constitute less than 0.01% of the total cellular protein. Fortunately, it is now possible to purify sequence-specific DNA-binding proteins by DNA affinity chromatography.

DNA affinity chromatography was originally developed with nonspecific DNAs, such as calf thymus DNA, attached to cellulose[1] and agarose[2] supports. More recently, a variety of procedures have been described for DNA affinity purification of sequence-specific DNA-binding proteins. The different DNA affinity resins that have been employed include plasmid DNA adsorbed to cellulose,[3] biotinylated DNA fragments attached to agarose, cellulose, or magnetic beads by biotin–avidin or biotin–streptavi-

[1] B. Alberts and G. Herrick, this series, Vol. 21, p. 198.
[2] D. J. Arndt-Jovin, T. M. Jovin, W. Bähr, A.-M. Frischauf, and M. Marquardt, *Eur. J. Biochem.* **54**, 411 (1975).
[3] P. J. Rosenfeld and T. J. Kelly, *J. Biol. Chem.* **261**, 1398 (1986).

din interactions,[4-7] monomeric synthetic oligodeoxynucleotides covalently attached to agarose,[8,9] a Teflon-based support on which synthetic oligonucleotides are synthesized,[10] and multimeric synthetic oligodeoxynucleotides covalently coupled to agarose beads[11] or latex particles.[12] A sequence-specific DNA-binding protein was also purified by a preparative gel mobility shift procedure.[13] It is likely that many, if not all, of these different techniques are effective in the purification of sequence-specific DNA-binding proteins.

This chapter describes the preparation and use of a sequence-specific DNA affinity resin that consists of multimerized synthetic oligodeoxynucleotides covalently attached to an agarose support.[11] This particular procedure has been successfully employed in the purification of at least 50 different sequence-specific DNA-binding proteins, a few of which are listed in Table I. A brief description of the procedure is as follows. Complementary chemically synthesized oligodeoxynucleotides that contain a recognition site for a sequence-specific DNA-binding protein are annealed and ligated to give oligomers. The DNA is then covalently coupled to agarose beads with cyanogen bromide to yield the affinity resin. A partially purified protein fraction is first incubated with nonspecific competitor DNA to which the desired protein has very low affinity, and the resulting mixture is subsequently applied to the affinity resin. The desired sequence-specific DNA-binding protein binds to the recognition sites in the affinity resin rather than to the competitor DNA in solution, while other proteins, including nonspecific DNA-binding proteins, flow through the resin. The sequence-specific factor is then eluted from the affinity resin with a salt gradient. A typical protein can be purified 500- to 1000-fold with 30% yield by two sequential affinity chromatography steps. In addition, the use of tandem affinity columns containing different protein-binding sites allows the simultane-

[4] L. A. Chodosh, R. W. Carthew, and P. A. Sharp, *Mol. Cell. Biol.* **6,** 4723 (1986).

[5] M. S. Kasher, D. Pintel, and D. C. Ward, *Mol. Cell. Biol.* **6,** 3117 (1986).

[6] M. Leblond-Francillard, M. Dreyfus, and F. Rougeon, *Eur. J. Biochem.* **166,** 351 (1987).

[7] O. S. Gabrielsen, E. Hornes, L. Korsnes, A. Ruet, and T. B. Øyen, *Nucleic Acids Res.* **17,** 6253 (1989).

[8] C. Wu, S. Wilson, B. Walker, I. Dawid, T. Paisley, V. Zimarino, and H. Ueda, *Science* **238,** 1247 (1987).

[9] R. Blanks and L. W. McLaughlin, *Nucleic Acids Res.* **16,** 10283 (1988).

[10] C. H. Duncan and S. L. Cavalier, *Anal. Biochem.* **169,** 104 (1988).

[11] J. T. Kadonaga and R. Tjian, *Proc. Natl. Acad. Sci. U.S.A.* **83,** 5889 (1986).

[12] H. Kawaguchi, A. Asai, Y. Ohtsuka, H. Watanabe, T. Wada, and H. Handa, *Nucleic Acids Res.* **17,** 6229 (1989).

[13] I. Gander, R. Foeckler, L. Rogge, M. Meisterernst, R. Schneider, R. Mertz, F. Lottspeich, and E.-L. Winnacker, *Biochim. Biophys. Acta* **951,** 411 (1988).

TABLE I
SEQUENCE-SPECIFIC DNA-BINDING PROTEINS PURIFIED BY DNA AFFINITY TECHNIQUE

Factor	Source	Function	Refs.
Mammalian			
Sp1	HeLa cells	RNA polymerase II transcription	a
CTF/NFI	HeLa cells	RNA polymerase II transcription/ adenovirus DNA replication	b
AP-1	HeLa cells	RNA polymerase II transcription	c,d
AP-2	HeLa cells	RNA polymerase II transcription	e,f
NF-κB	Namalwa cells	RNA polymerase II transcription	g
	HeLa cells	RNA polymerase II transcription	h
	Bovine spleen	RNA polymerase II transcription	i
Pit-1/GHF-1/PUF-I	GC2 cells	RNA polymerase II transcription	j,k
	GH₃ cells	RNA polymerase II transcription	l
μEBP-E	Mouse plasmacytoma cells	RNA polymerase II transcription	m
ATP	HeLa cells	RNA polymerase II transcription	n
SRF/f-EBP/CBF	HeLa cells	RNA polymerase II transcription	o–q
TEF-1	HeLA cells	RNA polymerase II transcription	r
OBP	HSV-1-infected Vero cells	DNA replication	s
UBF1	HeLa cells	RNA polymerase I transcription	t
TFIIIC2	HeLa cells	RNA polymerase III transcription	u
Avian			
COUP factor	Chicken oviducts	RNA polymerase II transcription	v
PAL	Chicken erythrocytes	RNA polymerase II transcription	w
Drosophila melanogaster			
GAGA	Kc cells, embryos	RNA polymerase II transcription	x,y
DTF-1	Embryos	RNA polymerase II transcription	z
Zeste	Kc cells	RNA polymerase II transcription/ transvection	aa
Transposase	MTΔ2-3 cells	P element-mediated transposition	bb
Saccharomyces cerevisiae			
HSE-binding/HSTF	Strain BJ2168	RNA polymerase II transcription	cc
	Strain BJ926	RNA polymerase II transcription	dd
RAP-1/GRFI	Strain BJ2168	RNA polymerase II transcription/	ee
	Strain BJ926	DNA attachment to nuclear scaffold	ff
ABF-I/SBFI	Strain PEP4D	DNA replication	gg
CBP-1	Strain SK1	Centromere-binding protein	hh

[a] M. R. Briggs, J. T. Kadonaga, S. P. Bell, and R. Tjian, *Science* **234**, 47 (1986).

[b] K. A. Jones, J. T. Kadonaga, P. J. Rosenfeld, T. J. Kelly, and R. Tjian, *Cell* (*Cambridge, Mass.*) **48**, 79 (1987).

[c] W. Lee, P. Mitchell, and R. Tjian, *Cell* (*Cambridge, Mass.*) **49**, 741 (1987).

[d] P. Angel, M. Imagawa, R. Chiu, B. Stein, R. J. Imbra, H. J. Rahmsdorf, C. Jonat, P. Herrlich, and M. Karin, *Cell* (*Cambridge, Mass.*) **49**, 729 (1987).

[e] P. J. Mitchell, C. Wang, and R. Tjian, *Cell* (*Cambridge, Mass.*) **50**, 847 (1987).

ous purification of multiple DNA-binding proteins from a single protein fraction. This technique is not only effective, but it is also simple and straightforward to perform. Preparation of the DNA affinity resin can be easily accomplished within 2 days, and a DNA affinity chromatography experiment can be carried out in about 4 hr. Table I is a partial list of proteins that have been purified by using this technique. It may serve as a helpful guide for the purification of related proteins, and, in particular, the listed references may provide important information on

[f] M. Imagawa, R. Chiu, and M. Karin, *Cell (Cambridge, Mass.)* **51,** 251 (1987).

[g] K. Kawakami, C. Scheidereit, and R. G. Roeder, *Proc. Natl. Acad. U.S.A.* **85,** 4700 (1988).

[h] P. A. Baeuerle and D. Baltimore, *Genes Dev.* **3,** 1689 (1989).

[i] M. J. Lenardo, A. Kuang, A. Gifford, and D. Baltimore, *Proc. Natl. Acad. Sci. U.S.A.* **85,** 8825 (1988).

[j] H. J. Mangalam, V. R. Albert, H. A. Ingraham, M. Kapiloff, L. Wilson, C. Nelson, H. Elsholtz, and M. G. Rosenfeld, *Genes Dev.* **3,** 946 (1989).

[k] J.-L. Castrillo, M. Bodner, and M. Karin, *Science* **243,** 814 (1989).

[l] Z. Cao, E. A. Barron, and Z. D. Sharp, *Mol. Cell. Biol.* **8,** 5432 (1988).

[m] C. L. Peterson, S. Eaton, and K. Calame, *Mol. Cell. Biol.* **8,** 4972 (1988).

[n] T. Hai, F. Liu, E. A. Allegretto, Marin, and M. R. Green, *Genes Dev.* **2,** 1216 (1988).

[o] R. Treisman, *EMBO J.* **6,** 2711 (1987).

[p] R. Prywes and R. G. Roeder, *Mol. Cell. Biol.* **7,** 3482 (1987).

[q] L. M. Boxer, R. Prywes, R. G. Roeder, and L. Kedes, *Mol. Cell. Biol.* **9,** 515 (1989).

[r] I. Davidson, J. H. Xiao, R. Rosales, A. Staub, and P. Chambon, *Cell (Cambridge, Mass.)* **54,** 931 (1988).

[s] P. Elias and I. R. Lehman, *Proc. Natl. Acad. Sci. U.S.A.* **85,** 2959 (1988).

[t] S. P. Bell, R. M. Learned, H.-M. Jantzen, and R. Tjian, *Science* **241,** 1192 (1988).

[u] S. K. Yoshinaga, N. D. L'Etoile, and A. J. Berk, *J. Biol. Chem.* **264,** 10726 (1989).

[v] M. K. Bagchi, S. Y. Tsai, M.-J. Tsai, and B. W. O'Malley, *Mol. Cell. Biol.* **7,** 4151 (1987).

[w] B. M. Emerson, J. M. Nickol, and T. C. Fong, *Cell (Cambridge, Mass.)* **57,** 1189 (1989).

[x] M. D. Biggin and R. Tjian, *Cell (Cambridge, Mass.)* **53,** 699 (1988).

[y] D. S. Gilmour, G. H. Thomas, and S. C. R. Elgin, *Science* **245,** 1487 (1989).

[z] K. K. Perkins, G. M. Dailey, and R. Tjian, *Genes Dev.* **2,** 1615 (1988).

[aa] M. D. Biggin, S. Bickel, M. Genson, V. Pirrotta, and R. Tjian, *Cell (Cambridge, Mass.)* **53,** 713 (1988).

[bb] P. D. Kaufman, R. F. Doll, and D. C. Rio, *Cell (Cambridge, Mass.)* **59,** 359 (1989).

[cc] P. K. Sorger and H. R. B. Pelham, *EMBO J.* **6,** 3035 (1987).

[dd] G. Wiederrecht, D. J. Shuey, W. A. Kibbe, and C. S. Parker, *Cell (Cambridge, Mass.)* **48,** 507 (1987).

[ee] J. F.-X. Hofmann, T. Laroche, A. H. Brand, and S. M. Gasser, *Cell (Cambridge, Mass.)* **57,** 725 (1989).

[ff] A. R. Buchman, N. F. Lue, and R. D. Kornberg, *Mol. Cell. Biol.* **8,** 5086 (1988).

[gg] K. S. Sweder, P. R. Rhode, and J. L. Campbell, *J. Biol. Chem.* **263,** 17270 (1988).

[hh] M. Cai and R. W. Davis, *Mol. Cell. Biol.* **9,** 2544 (1989).

the preparation of partially purified protein fractions that are suitable for affinity chromatography.

General Approach to Purification of Sequence-Specific DNA-Binding Proteins

A strategy for the purification of sequence-specific DNA-binding proteins is as follows. First, the DNA sequence to which the factor binds is determined by techniques such as DNase I footprinting,[14] methidiumpropyl EDTA-Fe(II) footprinting,[15] and dimethyl sulfate methylation protection.[16] If possible, several different binding sites should be surveyed to identify a high-affinity recognition sequence, which would be useful in the construction of the DNA affinity resin. Second, optimal conditions are established for binding of the factor. The effect of temperature, Mg(II) concentration, pH, and monovalent ions, such as chloride versus glutamate,[17] should be examined. In addition, the sequence-specific binding of the factor should be tested in the presence of nonspecific DNAs, such as calf thymus DNA, poly(dI-dC), poly(dG-dC), and poly(dA-dT) (see Proper Use of Nonspecific Competitor DNAs, below). Third, the factor is partially purified to remove nucleases and other contaminants that might adversely affect the affinity resin (this will be discussed later). Finally, two or more DNA affinity resins that contain high-affinity recognition sites with different flanking DNA sequences are prepared. A control resin that does not contain any binding sites should also be made. If a protein is purified by passage through two different affinity resins, then contamination by proteins that bind fortuitously to flanking oligodeoxynucleotide sequences is minimized. The control resin is used to identify proteins that bind nonspecifically to DNA–agarose. The desired protein can be tentatively identified by using this approach because a species that is enriched by chromatography through the specific affinity resins, but is not enriched by chromatography through the control resin, is good candidate for either the sequence-specific DNA-binding protein or a molecule that is closely associated with the DNA-binding protein.

It is important to identify nonspecific DNA-binding proteins that may contaminate the protein preparations after DNA affinity chromatography. The use of a control DNA affinity resin that does not contain binding sites for the sequence-specific factor in parallel with the specific DNA affinity

[14] D. Galas and A. Schmitz, *Nucleic Acids Res.* **5,** 3157 (1978).

[15] M. W. Van Dyke and P. B. Dervan, *Nucleic Acids Res.* **11,** 5555 (1983).

[16] U. Siebenlist and W. Gilbert, *Proc. Natl. Acad. Sci. U.S.A.* **77,** 122 (1980).

[17] S. Leirmo, C. Harrison, D. S. Cayley, R. R. Burgess, and M. T. Record, Jr., *Biochemistry* **26,** 2095 (1987).

columns (using the same protein fraction as the starting material) should minimize misidentification of a nonspecific DNA-binding protein for the desired factor. In addition, the polypeptide that is responsible for the sequence-specific DNA-binding activity can be identified by recovery of protein from an SDS–PAGE gel and renaturation of the denatured polypeptide[18] followed by a DNA-binding assay (preferably a DNase I footprinting assay). In HeLa cells, there are two common nonspecific DNA-binding proteins that often contaminate preparations of affinity-purified factors: poly(ADP-ribose) polymerase (NAD$^+$ ADP-ribosyltransferase),[19,20] which has an M_r of 116,000, and the Ku antigen,[21] which consists of two polypeptides of M_r 70,000 and 80,000. Thus, when purifying DNA-binding factors from HeLa cells, it is prudent to be suspicious of polypeptides of M_r 70,000, 80,000, and 116,000.

Preparation of DNA Affinity Resin

Preliminary Considerations

A few practical considerations are worth noting before preparation of the DNA affinity resin (5-ml scale). First, oligodeoxynucleotides ranging in size from 14 to 51 bases have been used successfully. Since the coupling of the DNA to the CNBr-activated resin probably occurs via primary amino groups on unpaired bases, the oligodeoxynucleotides should have a 4-base overhang; I usually include 5'-GATC . . . at the 5' end of the oligodeoxynucleotides. Also, do not use 21- or 42-mers because, if the DNA sequence has a bend, the oligodeoxynucleotides may circularize on ligation. The amount of DNA that is coupled to the Sepharose is typically 80–90 µg/ml resin, which corresponds to a protein-binding capacity of 7 nmol/ml resin if there is one recognition site per 20 base pairs (bp). The affinity resins are stable in the storage buffer at 4° for at least 1 year and can be reused >30 times without any detectable decrease in the protein-binding capacity.

Materials and Reagents

Synthetic oligodeoxynucleotides, 250 µg each (500 µg total DNA) of complementary oligodeoxynucleotides that contain the recognition sequence of the desired protein. Note the recommendations given in

[18] D. A. Hager and R. R. Burgess, *Anal. Biochem.* **109**, 76 (1980).
[19] K. Ueda and O. Hayaishi, *Annu. Rev. Biochem.* **54**, 73 (1985).
[20] E. Slattery, J. D. Dignam, T. Matsui, and R. G. Roeder, *J. Biol. Chem.* **258**, 5955 (1983).
[21] T. Mimori, J. A. Hardin, and J. A. Steitz, *J. Biol. Chem.* **261**, 2774 (1986).

the previous section for the design of the oligodeoxynucleotides. The oligodeoxynucleotides should be purified by electrophoresis in a polyacrylamide–urea gel. To estimate concentration of single-stranded DNA, assume that 1 $A_{260\text{ nm}}$ unit = 40 μg/ml DNA

10× T4 polynucleotide kinase buffer: 500 mM Tris-HCl, pH 7.6, containing 100 mM MgCl$_2$, 50 mM dithiothreitol (DTT), 1 mM spermidine, and 1 mM EDTA. Store at 4°. Add extra 10 mM DTT just before use

T4 polynucleotide kinase (#201; New England Biolabs, Boston, MA)

10× Linker–kinase buffer: 660 mM Tris-HCl, pH 7.6, containing 100 mM MgCl$_2$, 150 mM DTT, and 10 mM spermidine. Store at 4°. Add extra 10 mM DTT just before use

T4 DNA ligase (#202; New England Biolabs)

Miscellaneous solutions for 5′ phosphorylation and ligation of the oligodeoxynucleotides for coupling to the Sepharose: TE buffer (10 mM Tris-HCl, pH 7.6, containing 0.1 mM EDTA); 10 mM ammonium acetate; 20 mM ATP (Na$^+$), pH 7.0; absolute ethanol; phenol equilibrated with TE; phenol–chloroform (1 : 1, v/v); chloroform–isoamyl alcohol (24 : 1, v/v); 3 M sodium acetate; 2-propanol

Sepharose CL-2B (17-0140-01; Pharmacia, Piscataway, NJ)

Cyanogen bromide (caution: *highly toxic*) (C9, 149-2; Aldrich)

Miscellaneous solutions required for coupling of the DNA to the Sepharose: 5 M NaOH; N,N-dimethylformamide (27,054-7; Aldrich); 10 mM potassium phosphate, pH 8.0 (500 ml); 1 M ethanolamine-HCl, pH 8.0 (200 ml); 1 M potassium phosphate, pH 8.0 (200 ml); 1 M KCl (200 ml)

Column storage buffer: 10 mM Tris-HCl, pH 7.6, containing 1 mM EDTA, 0.3 M NaCl, and 0.04% (w/v) sodium azide (freshly added)

Materials: Standard equipment for manipulation of DNA [microcentrifuges, water baths, Speedvac evaporator/concentrator (Savant, Hicksville, NY)], 60-ml coarse-sintered glass funnel; rotating wheel (multipurpose rotator, model 151; Scientific Industries, Inc.)

5′-Phosphorylation of Oligodeoxynucleotides

1. Combine 65 μl DNA (250 μg of *each* oligodeoxynucleotide in TE) with 10 μl of 10× T4 polynucleotide kinase buffer. Incubate at 88° for 2 min; 65° for 10 min; 37° for 10 min; and room temperature for 5 min.

2. Add 15 μl of 20 mM ATP, pH 7.0, containing about 5 μCi [γ-^{32}P]ATP, and 10 μl of T4 polynucleotide kinase (100 units) to a final volume of 100 μl. Incubate the solution at 37° for 2 hr.

3. Add 50 μl of 10 M ammonium acetate and 100 μl water to a final

volume of 250 μl. Heat at 65° for 15 min to inactivate the kinase. Let the sample cool to room temperature, and then add 750 μl ethanol. Mix by inversion. Spin (microcentrifuge) for 15 min at room temperature to pellet DNA. Discard supernatant.

4. Add 225 μl TE. Dissolve the pellet by vortexing. Then add 250 μl phenol–chloroform (1 : 1, v/v). Vortex for 1 min. Spin 5 min to separate phases. Transfer the upper layer to a new tube. Then add 250 μl chloroform–isoamyl alcohol (24 : 1, v/v). Vortex for 1 min. Spin 5 min to separate phases. Transfer the upper layer to a new tube.

5. Add 25 μl of 3 M sodium acetate. Mix by vortexing. Then add 750 μl ethanol. Mix by inversion. Spin 15 min to pellet DNA. Discard supernatant. Then add 800 μl of 75% ethanol. Mix by vortexing. Spin 5 min at room temperature. Discard supernatant. Dry pellet in Speedvac.

Ligation of Oligodeoxynucleotides

1. Dissolve the 5'-phosphorylated DNA pellet from above in 65 μl water. Then add 10 μl of 10× linker–kinase buffer to give a final volume of 75 μl. Vortex thoroughly to ensure that the pellet is completely dissolved.

2. Add 20 μl of 20 mM ATP, pH 7.0, and 5 μl of T4 DNA ligase (30 Weiss units) to a final volume of 100 μl. Incubate at 15° overnight.

Monitor the progress of ligation by agarose gel electrophoresis (use 0.5 μl of ligation reaction per gel lane). The average length of the ligated oligodeoxynucleotides should be at least 10-mers. Depending on the oligodeoxynucleotides used, the optimal temperature for ligation will vary from 4 to 30°. Short oligodeoxynucleotides ($\leq$15-mers) tend to ligate better at lower temperatures (4 to 15°), whereas oligodeoxynucleotides that have a moderate degree of palindromic symmetry, which have a tendency to self-anneal, ligate better at higher temperatures (15 to 30°).

Also, 5'-phosphorylated oligodeoxynucleotides often do not ligate on the first attempt. Thus, if ligation does not occur, extract the DNA once with 1 : 1 (v/v) phenol–chloroform, extract the DNA once with chloroform, and ethanol precipitate (using sodium acetate as salt). Dissolve the DNA in 225 μl TE, add 25 μl 3 M sodium acetate, and reprecipitate with ethanol (750 μl). Wash the 75% ethanol, dry in the Speedvac, and try the ligation again.

Preparation of Ligated DNA for Coupling to Sepharose CL-2B

1. Add 100 μl phenol to the 100 μl ligation reaction. Vortex 1 min. Spin 5 min at room temperature. Transfer upper layer to a new tube.

2. Add 100 μl of chloroform–isoamyl alcohol (24 : 1, v/v). Vortex 1 min. Spin 5 min at room temperature. Transfer upper layer to a new tube.

3. Add 33 μl 10 M ammonium acetate. Mix by vortexing.

4. Add 133 μl 2-propanol. Mix by inversion. Incubate at $-20°$ for 20 min. Spin 15 min to pellet DNA. Discard supernatant.

5. Add 225 μl TE. Vortex to dissolve pellet. Add 25 μl of 3 M sodium acetate. Mix by vortexing. Add 750 μl ethanol. Mix by inversion. Spin 15 min to pellet DNA. Discard supernatant.

6. Wash DNA two times with 75% ethanol. Dry pellet in Speedvac.

7. Dissolve the DNA in 50 μl glass-distilled water. Store at $-20°$. *Do not dissolve the DNA in TE; it will interfere with the coupling reaction.*

Coupling of DNA to Sepharose

The coupling of the DNA to the Sepharose CL-2B is similar to the procedure described by Arndt-Jovin *et al.*[2]

1. Wash 10 to 15 ml (settled bed volume) of Sepharose CL-2B extensively with glass-distilled water (500 ml) in a 60-ml coarse-sintered glass funnel.

2. Transfer the moist Sepharose to a 25-ml graduated cylinder and estimate a 10-ml settled bed volume of resin. Then add water to a 20-ml final volume. Transfer this slurry to a 150-ml glass beaker in a water bath equilibrated to 15° over a magnetic stirrer *in a fume hood.*

3. *In the fume hood,* measure 1.1 g of cyanogen bromide (*DANGER—TOXIC*) in a 25-ml Erlenmeyer flask with Parafilm covering the mouth of the flask. It is better to have slightly more than 1.1 g than slightly less than 1.1 g. Dissolve the cyanogen bromide in 2 ml of N,N-dimethylformamide. It will instantly dissolve. Add the cyanogen bromide solution dropwise over 1 min to the stirring slurry of Sepharose *in the fume hood.*

4. Immediately add NaOH as follows. Add 30 μl of 5 M NaOH every 10 sec to the stirring mixture (at 15°) for 10 min until the total volume of NaOH added is 1.8 ml.

5. Immediately add 100 ml of ice-cold water to the beaker and pour the mixture into a 60-ml coarse-sintered glass funnel. *At this point, it is very important that the resin is not suction filtered into a dry cake.* If the resin is accidentally dried during suction filtration, it will be necessary to start the CNBr activation procedure again from the beginning.

6. Wash the resin four times with 100 ml ice-cold water ($\leq$4°) each time, followed by two washes with 100 ml ice-cold 10 mM potassium phosphate, pH 8.0.

7. Immediately transfer one-half (5 ml) of the activated resin to a 15-ml polypropylene screw cap tube and add about 2 ml of 10 mM potassium phosphate, pH 8.0, until the resin has the consistency of a thick slurry.

[The remaining half (5 ml) of the activated resin could also be used for coupling another sample of DNA.]

8. Immediately add the DNA (in 50 μl water). Incubate on a rotating wheel overnight at room temperature.

9. Transfer the resin to a 60-ml coarse-sintered glass funnel, and wash twice with 100 ml water followed by a wash with 100 ml 1 M ethanolamine-hydrochloride, pH 8.0. Compare the level of radioactivity in the first few milliliters of the filtrate with the level of radioactivity in the washed resin to estimate the efficiency of incorporation of DNA to the resin. Usually, all detectable radioactivity is present only in the resin.

10. Transfer the resin to a 15-ml polypropylene screw cap tube and add 1 M ethanolamine hydrochloride, pH 8.0 until the mixture is a smooth slurry. Incubate the tube on a rotating wheel at room temperature for 4 to 6 hr (this step is to inactivate unreacted cyanogen bromide-activated Sepharose).

11. Wash the resin with the following solutions: 100 ml of 10 mM potassium phosphate, pH 8.0; 100 ml of 1 M potassium phosphate, pH 8.0; 100 ml of 1 M KCl; 100 ml water; and 100 ml of column storage buffer.

12. Store the resin at 4° (do not freeze the resin). The resin is stable for at least 1 year at 4°.

DNA Affinity Chromatography

Partial Purification of DNA-Binding Proteins before DNA Affinity Chromatography

The protein fraction that is to be subjected to affinity chromatography generally does not need to be extensively purified beyond the crude extract, although it is important that the fraction does not contain high levels of nucleases, which would destroy the affinity resin. The references listed in Table I may provide useful information for the preparation of crude extracts and partial purification of the DNA-binding factors. Typically, a crude nuclear extract is prepared, and the desired protein is partially purified by one column chromatography step and then applied to the affinity resins. For example, in the isolation of many sequence-specific transcription factors from HeLa cells, such as Sp1, CTF/NFI, AP-1, AP-2, AP-3, and AP-4 (for review, see Ref. 22), it is sufficient to prepare a standard nuclear extract[23] and then to subject the extract to Sephacryl S-300 gel filtration before affinity chromatography. Other chromatographic

[22] P. J. Mitchell and R. Tjian, *Science* **245**, 371 (1989).
[23] J. D. Dignam, R. M. Lebovitz, and R. G. Roeder, *Nucleic Acids Res.* **11**, 1475 (1983).

resins that are commonly used to carry out partial purification of DNA-binding proteins include heparin-agarose, DNA-cellulose,[1] phosphocellulose (P11; Whatman, Clifton, NJ), DEAE-cellulose (DE-52; Whatman), Mono S (Pharmacia), and Mono Q (Pharmacia) columns. It is also possible to purify partially the DNA-binding factor by chromatography with a specific DNA-Sepharose resin that is constructed with oligodeoxynucleotides that *do not* contain binding sites for the desired protein (for an example, see Ref. 24). This approach is effective if the desired factor elutes from this column at a low salt concentration while high-affinity nonspecific DNA-binding proteins elute from the resin at a high salt concentration. If such a separation of the desired protein and high-affinity nonspecific DNA-binding proteins can be achieved, then it would not be necessary to add competitor DNA during subsequent chromatography with a DNA affinity resin that contains binding sites for the desired factor. Finally, a variety of eukaryotic sequence-specific transcription factors have been found to contain multiple O-linked *N*-acetylglucosamine monosaccharide residues.[25] It has been demonstrated that three of these factors, Sp1, CTF/NFI, and HNF1, can be purified in high yield by wheat germ agglutinin affinity chromatography followed by sequence-specific DNA affinity chromatography.[26,27] Hence, it may be worthwhile to examine whether or not the desired DNA-binding factor can be partially purified by wheat germ agglutinin affinity chromatography.

Proper Use of Nonspecific Competitor DNAs

The proper use of nonspecific competitor DNAs is an important feature of successful purification of proteins. It is necessary to determine experimentally the appropriate amount and type of competitor DNA to add for each affinity chromatography experiment. DNA-binding assays (DNase I footprinting or gel mobility shift experiments) should be carried out with the starting material (the identical protein fraction that is to be applied to the affinity resin) in the presence of different types and amounts of competitor DNAs. Competitor DNAs that are commonly used include sonicated calf thymus DNA, poly(dI-dC), poly(dA-dT), and poly(dG-dC). Poly(dI-dC), poly(dA-dT), or poly(dG-dC) is dissolved to a final concentration of $10\ A_{260\ nm}$ units in TE containing 100 mM NaCl, heated to 90°, and slowly cooled over 30 to 60 min to room temperature before use. The average length of the competitor DNA should be around 1 kilobase (kb). It is

---

[24] P. D. Kaufman, R. F. Doll, and D. C. Rio, *Cell* (*Cambridge, Mass.*) **59**, 359 (1989).
[25] S. P. Jackson and R. Tjian, *Cell* (*Cambridge, Mass.*) **55**, 125 (1988).
[26] S. P. Jackson and R. Tjian, *Proc. Natl. Acad. Sci. U.S.A.* **86**, 1781 (1989).
[27] S. Lichtsteiner and U. Schibler, *Cell* (*Cambridge, Mass.*) **57**, 1179 (1989).

common for poly(dI-dC) to be sold in lengths greater than 9 kb. If the DNA is too long, it can be fragmented by sonication. Also, a mixture of different competitor DNAs can be used in a single experiment.

The appropriate amount and type of competitor DNA to use in affinity chromatography are estimated as follows. First, DNA-binding experiments are carried out using the partially purified protein fraction with varying levels of different types of competitor DNAs. In these studies, which can be viewed as small-scale DNA affinity experiments, the highest quantity of competitor DNA that does not adversely affect the binding of the protein is determined. The amounts of competitor DNA and protein fraction used in the DNA-binding studies are adjusted to full scale, and then the actual amount of competitor DNA that will be used for affinity chromatography is one-fifth of the quantity that was obtained by adjustment of the small-scale DNA-binding assay to full scale.

The method of estimation of competitor DNA is best explained by a simple example. For instance, the sequence-specific DNA binding of protein XX is strongly inhibited by either calf thymus DNA or poly(dA-dT), whereas it is weakly inhibited by poly(dI-dC). In a DNA-binding assay, the binding of XX (10 μl of a protein fraction containing XX) is affected as follows: (1) no detectable inhibition by 1 μg or less of poly(dI-dC); (2) weak inhibition by 2 μg of poly(dI-dC); (3) strong inhibition by 4 μg or more of poly(dI-dC). Thus, the highest level of competitor DNA that does not affect XX binding with 10 μl of the protein fraction is 1 μg of poly(dI-dC). If we have 10 ml of the protein fraction containing XX (which is 1000 × 10 μl), a direct unmodified scale-up of competitor DNA would be 1000 × 1 μg = 1000 μg poly(dI-dC). However, use one-fifth of the direct scale-up amount, which in this case would be 1000 μg × 1/5 = 200 μg poly(dI-dC).

Materials and Reagents

Buffer Z: 25 mM HEPES, K$^+$, pH 7.6, containing variable KCl (as indicated in the protocol), 12.5 mM MgCl$_2$, 1 mM DTT (freshly added just before use), 20% (v/v) glycerol, and 0.1% (v/v) Nonidet P-40 (NP-40). Adjust the pH of 1× buffer Z with KOH. Do not make a 10× buffer. Store at 4°. *Note:* Some preparations of HEPES buffer contain potent inhibitors of DNA-binding proteins. We typically use Sigma #H3375

Buffer Z^e: 25 mM HEPES, K$^+$, pH 7.6, containing variable KCl (as indicated), 1 mM DTT (freshly added just before use), 20% (v/v) glycerol, and 0.1% (v/v) Nonidet P-40. See comments on buffer Z

TM buffer: 50 mM Tris-HCl, pH 7.9, containing variable KCl (as indicated), 12.5 mM MgCl$_2$, 1 mM DTT, 20% (v/v) glycerol, and 0.1%

(v/v) NP-40. Store at 4°. This buffer can be used as a substitute for
buffer Z if inhibitors of DNA-binding proteins are present in the
HEPES buffer

Column regeneration buffer: 10 mM Tris-HCl, pH 7.8, containing 1 mM
EDTA, 2.5 M NaCl, and 1% (v/v) NP-40. Store at room temperature.
This solution will be cloudy and separate into two phases (NP-40 and
aqueous) on storage. Just before use, mix by swirling and shaking

Column storage buffer: 10 mM Tris-HCl, pH 7.6, containing 1 mM
EDTA, 0.3 M NaCl, and 0.04% (w/v) sodium azide (freshly added)

Chromatography columns: Bio-Rad (Richmond, CA) 2-ml disposable
Poly-Prep columns (#731-1550)

Competitor DNAs: Sonicated calf thymus DNA; poly(dI-dC);
poly(dA-dT); poly(dG-dC). The average length of the DNA should be
about 1 kb. If the average length is greater than 1 kb, the DNA should
be sonicated until it is sheared to the appropriate size

Affinity Chromatography

All operations are performed at 4°. This procedure is described with
buffer Z, but in some instances, factors have been found to bind to DNA
with higher affinity in buffer Z^c, which is identical to buffer Z except that
it does not contain MgCl$_2$.

1. Equilibrate the DNA affinity resin (1 ml settled bed volume) in a
Bio-Rad 2-ml Poly-Prep column with two washes of 10 ml of buffer Z +
0.1 M KCl.

2. The protein fraction (in TM + 0.1 M KCl, buffer Z + 0.1 M KCl,
or buffer Z^c + 0.1 M KCl) is combined with nonspecific competitor DNA.

3. The protein–DNA mixture is allowed to stand for 10 min. Then
centrifuge this solution at 10,000 rpm for 10 min in a SS-34 rotor to pellet
insoluble protein–DNA complexes.

4. The columns are loaded at gravity flow (15 ml/hr/column). When
purifying a large quantity of material, use multiple 1-ml columns.

5. After loading the starting material, wash each column four times
with 2 ml of buffer Z + 0.1 M KCl.

6. Elute the protein from each column by successive addition of 1-ml
portions of buffer Z containing 0.2 M KCl, 0.3 M KCl, 0.4 M KCl, 0.5 M
KCl, 0.6 M KCl, 0.7 M KCl, 0.8 M KCl, and 0.9 M KCl, and three times
with 1 ml of buffer Z containing 1 M KCl. Collect 1-ml fractions that
correspond to the addition of the 1-ml portions of buffer.

7. Save small aliquots of each fraction (20 μl), freeze all of the fractions
and aliquots in liquid nitrogen, and store at −80°.

8. Regenerate the affinity resin as follows. At *room temperature,* stop

the flow of the column, and then add 5 ml of regeneration buffer. Stir the resin with a narrow siliconized glass rod to mix the resin with the regeneration buffer. Let the buffer flow out of the column. Stop the flow of the column again, and repeat the addition and mixing of the regeneration buffer with the resin. Then, equilibrate the resin with two washes of 15 ml of column storage buffer, and store at 4°.

The protein fractions should be assayed for the sequence-specific DNA-binding activity. If further purification is desired, the fractions that contain the activity are combined. Then the pooled fractions are either diluted (with buffer Z without KCl) or dialyzed (against buffer Z plus 0.1 M KCl) to 0.1 M KCl final concentration. The protein fraction is subsequently combined with nonspecific competitor DNA (the amount and type of which need to be determined again experimentally) and reapplied to a DNA affinity resin.

Suggestions for Use and Handling of Affinity-Purified DNA-Binding Proteins

1. Quick-freeze the protein in liquid nitrogen.
2. Store the protein at $-70°$.
3. Thaw the protein quickly in water that is either cold or at room temperature. The thawed protein should then be maintained at 4°.
4. Since the protein concentration of affinity-purified proteins is typically low (in the range of 5 to 50 $\mu g/ml$), the buffers should contain 0.01 to 0.1% (v/v) NP-40 (a nonionic detergent) to minimize loss of protein by adsorption to plastic or glass.
5. Minimize handling of protein, especially with regard to exposure to plastic and glass, to which the protein might be irreversibly adsorbed. The use of siliconized plastic test tubes and pipette tips will minimize loss of protein by adsorption to plastic, but it is usually not necessary to use siliconized plasticware.
6. Typically, 2 to 10 ng of purified protein is sufficient for a complete footprint (with ≈10 fmol of DNA probe). Do not add competitor DNA to footprinting reactions with purified protein as it will severely inhibit the binding of the factors.

Acknowledgments

I would like to thank R. Tjian, B. Dynan, K. Jones, M. Biggin, M. Briggs, and D. Bohmann, as well as the many other colleagues who have contributed in many different ways to the development of the DNA affinity procedure. J.T.K. is a Lucille P. Markey Scholar in the Biomedical Sciences, and this work was supported in part by grants from the United States Public Health Service (GM 41249) and the Lucille P. Markey Charitable Trust.

[3] Using Protein Affinity Chromatography to Probe Structure of Protein Machines

By TIM FORMOSA, JACK BARRY, BRUCE M. ALBERTS,
and JACK GREENBLATT

Most proteins in cells function in large assemblies of proteins created by specific protein–protein interactions. Thus, for example, such diverse processes as DNA replication, RNA synthesis, transport vesicle budding and fusion, and the incorporation of specific proteins into membranes can be considered to be mediated by protein complexes that function as "protein machines." Many of the protein–protein interactions that play important roles in assembling and regulating these complex protein machines are too weak to allow copurification of the interacting species from cellular extracts. Such relatively weak binding is to be expected, since components that must bind to one another reversibly in the concentrated environment of the intracellular fluid will fall apart rapidly in a dilute extract. Since the study of processes mediated by protein machines requires the *in vitro* reassembly of active complexes, the subunit composition and enzymatic properties of which are the same as those found inside the cell, it is important to have methods for identifying and purifying all of the interacting proteins, starting with one member of a protein complex.

In this chapter we describe the use of protein affinity chromatography to characterize and isolate the interacting components of protein complexes. We summarize the techniques used to construct affinity matrices, the preparation of extracts from prokaryotic and eukaryotic sources, and methods for analyzing the binding fractions. This technique has been used successfully to study interactions between elements of the cytoskeleton,[1,2] protein transport machinery,[3] transcription complexes[4,5] and the DNA replication apparatus.[6,7] A chapter that focuses exclusively on microtubule

[1] D. G. Drubin, *Cell Motil. Cytoskeleton* **15**, 7 (1990).
[2] K. G. Miller, C. M. Field, B. M. Alberts, and D. R. Kellogg, this series, Vol. **196**, p. 303.
[3] S. Tajima, L. Lauffer, V. L. Rath, and P. Walter, *J. Cell Biol.* **103**, 1167 (1986).
[4] Z. F. Burton, M. Killeen, M. Sopta, L. G. Ortolan, and J. Greenblatt, *Mol. Cell. Biol.* **8**, 1602 (1988).
[5] R. J. Horwitz, J. Li, and J. Greenblatt, *Cell (Cambridge, Mass.)* **51**, 631 (1987).
[6] T. Formosa, R. L. Burke, and B. M. Alberts, *Proc. Natl. Acad. Sci. U.S.A.* **80**, 2442 (1983).
[7] B. M. Alberts, J. Barry, P. Bedinger, T. Formosa, C. V. Jongeneel, and K. Kreuzer, *Cold Spring Harbor Symp. Quant. Biol.* **47**, 655 (1983).

and actin filament affinity columns has been published previously in this series.[2] Given the widespread occurrence and importance of protein–protein interactions in biology, protein affinity chromatography should be useful for studying nearly all biological processes.

Preparing Affinity Columns

Selecting Activated Matrix

Protein affinity chromatography allows sensitive detection of protein–protein interactions because of the high concentration of binding sites presented by the matrix. To achieve optimal sensitivity, it is important to choose a matrix that will couple a maximal amount of protein in a minimal volume without introducing potentially denaturing multiple cross-links to individual protein molecules.

A second consideration in choosing a matrix is the minimization of interactions between proteins in a cellular extract and the column matrix itself. Agarose is a good neutral chromatographic matrix, but the process of activating it to accept a protein ligand can alter the agarose, creating unacceptably high levels of nonspecific protein binding. For example, the common method of activating agarose by treating it at high pH with cyanogen bromide leaves the matrix covered with a variety of charged residues.[8] These residues act as nonspecific ion exchangers, and sometimes bind such high levels of proteins themselves that weak interactions with the covalently bound protein substituent are obscured. Therefore, although results with cyanogen bromide-activated agarose have sometimes been acceptable,[9–11] the nonspecific binding by such matrices is often a problem.

We have had our most consistent success in optimizing for these considerations by coupling proteins to the N-hydroxysuccinimide-activated agarose matrix sold commercially by Bio-Rad (Richmond, CA), Affi-Gel 10 (we have had less favorable experiences with the similarly constructed but cationic matrix Affi-Gel 15). This matrix has not been cyanogen bromide treated, and it contains the active moiety at the end of a neutral spacer arm coupled to an agarose matrix. The active groups are present at a high enough concentration to couple proteins to the 20 mg/ml level, yet leave a neutral matrix with low background binding characteristics.

[8] M. Wilchek, T. Miron, and J. Kohn, this series, Vol. 104, p. 3.
[9] J. Greenblatt and J. Li, J. Mol. Biol. **147,** 11 (1981).
[10] J. Greenblatt and J. Li, Cell (Cambridge, Mass.) **24,** 421 (1981).
[11] M. Sopta, R. W. Carthrew, and J. Greenblatt, J. Biol. Chem. **260,** 10353 (1985).

In addition, the commercial availability allows good reproducibility and convenience.

Preparation of Protein Ligand

Ideally, the protein ligand that is used for affinity chromatography should be pure in order to ensure that every interacting protein that is detected is binding to the ligand rather than to a contaminant in the preparation. In practice, since true homogeneity of a protein preparation is not achievable, it is wise to develop independent criteria, either genetic[9,12,13] or biochemical,[2,4,11,14] that the proteins truly interact with each other. Optimally, one would like to be able to construct an appropriate control column that contains all of the contaminants, but none of the true ligand.[15] This can sometimes be achieved by performing parallel protein purifications from two strains or cell lines that differ only in their content of the protein of interest.

Coupling Proteins to Activated Matrix

To restrict the coupling reaction to the protein of interest, protein samples should be buffered with reagents lacking primary amines or sulfhydryls that will react with the activated Affi-Gel 10 matrix. Tris buffers usually are not suitable, but N-2-hydroxyethylpiperazine-N'-2-ethanesulfonic acid (HEPES), 4-morpholinopropanesulfonic acid (MOPS), and bicarbonate buffers are. The protein sample may contain glycerol, $MgCl_2$, NaCl and other reagents as long as no primary amines or sulfhydryls are present. (Some exceptions can be made if a protein couples efficiently and rapidly. In this case, the high concentration of active coupling groups ensures that sufficient sites are available for covalent binding of both the protein and the buffer component.) Our standard protocol is to dialyze about 0.5 to 3 mg of protein in a volume of 1–2 ml against several changes of coupling buffer [25 mM HEPES, pH 7.5–7.9, 5–20% (w/v) glycerol, 50–500 mM NaCl], at 4° for 5–15 hr. This protein sample is then coupled to an appropriate amount (often about 0.5 ml) of washed, activated matrix. Glycerol is useful for stabilizing nearly all proteins; it is therefore routinely used at a concentration of 10 to 20% in all buffers. High levels of NaCl have been used for coupling proteins that tend to aggregate at low ionic strengths.

[12] T. C. Huffaker, M. A. Hoyt, and D. Botstein, *Annu. Rev. Genet.* **21,** 259 (1987).
[13] D. G. Drubin, K. G. Miller, and D. Botstein, *J. Cell Biol.* **107,** 2551 (1988).
[14] A. E. Adams, D. Botstein, and D. G. Drubin, *Science* **243,** 231 (1989).
[15] K. F. Stringer, C. J. Ingles, and J. Greenblatt, *Nature (London)* **345,** 783 (1990).

The Affi-Gel 10 matrix is supplied as a slurry in 2-propanol and is stored at $-20°$. 2-Propanol must be removed before adding the protein sample. The washing and coupling can be done in a single tube; the matrix is quite sticky, so transferring should be minimized. We have found it convenient to wash activated matrix in 4–10 ml capped polycarbonate centrifuge tubes [e.g., Falcon 2059 or 2063 tubes], or in 1.5-ml microfuge tubes for smaller volumes. The matrix is equilibrated to $4°$, shaken to produce a slurry, then aliquots are removed using an Eppendorf pipette with the end of the disposable tip cut off to avoid clogging. Each aliquot is added directly to about 5 vol of cold, distilled H_2O, then the matrix is collected by centrifugation (1000 g) for 2 min. The supernatant is removed, the gel is washed twice with 5 vol of cold coupling buffer, and finally the matrix is suspended in the dialyzed protein sample. The coupling proceeds with gentle agitation on a rotator at $4°$ for several hours to overnight.

Using polycarbonate tubes and collecting the matrix by centrifugation decreases losses of material and drying out of the matrix compared to the use of filtration. However, the matrix can also be washed by gentle filtration, first with cold water and then with cold coupling buffer, usually six washes in all. The gel is then transferred into a plastic tube of appropriate size, weighed, and suspended in the dialyzed protein sample (2–5 ml of protein solution per gram of gel). We have found that the coupling is efficient even when the filtration process has made the gel somewhat difficult to resuspend. Most of the coupling occurs within a few hours, so the reaction can be terminated after 3–4 hr if desired. However, if short coupling times are used, the matrix should be blocked by treatment with neutralized ethanolamine (60–200 mM) for at least 2 hr.

Measuring Coupling Efficiency

The efficiency of coupling can be monitored in one of three ways. If the protein is pure, one can perform a Bradford assay for protein on a sample of the dialyzed protein used for the coupling and compare this with the supernatant from the reaction. Alternatively, the absorbance at 280 nm of both fractions can be measured. If the absorbance method is used, the sample and background controls should contain 0.1 M HCl to remove absorbance contributed by the unprotonated form of the N-hydroxysuccinimide moiety that is released during the coupling reaction. If the protein ligand is not pure, it is best to compare aliquots of the dialyzed protein and the postcoupling supernatant by sodium dodecyl sulfate-polyacrylamide gel electrophoresis (SDS–PAGE) followed by staining with Coomassie Brilliant Blue and densitometric determination of the amount of protein in the gel bands.

The Affi-Gel matrices contain about 15 μmol/ml of active esters. Assuming random distribution, these active groups are spaced about 5 nm apart, which is about equal to the diameter of a typical globular protein molecule of moderate size. It is therefore possible that multiple cross-links to the same protein molecule will form. If this acts to "freeze" proteins in denatured configurations, the columns will have lower than optimal binding capacities. In practice this has not usually been a problem. It can be detected and circumvented by allowing the activated matrix to hydrolyze in the presence of the aqueous coupling buffer for various amounts of time before adding the ligand protein. To prevent this problem, some researchers recommend using coupling conditions that allow not more than 80% of the ligand protein to be bound to the matrix[2]; controlling the linkage of microtubules to the agarose matrix in this way can produce columns that have a capacity that is three to four times greater than more heavily coupled columns.[9,16]

Constructing Affinity Column

The matrix is recovered by allowing it to settle for several minutes, removing a sample of the supernatant to check the coupling efficiency as described above, and then suspending the matrix in a suitable column buffer. Affinity matrices can be used to detect binding proteins using batch elution techniques; in this case, matrices are suspended in cell lysates and then collected by centrifugation, washed, and eluted by suspending the matrix in buffers of appropriate composition. However, greater resolution and more efficient washing can be obtained by continuous-flow column chromatography. A variety of column sizes, types, and geometries can be used. We describe below methods for preparing micro-, standard-, and preparative scale columns.

Microscale Columns. For small volumes, columns (20–100 μl) can be poured in capillary tubes. This is accomplished by heating a 200-μl capillary tube [for example, Clay Adams (Parsippany, NJ) Micropets, 12.6 cm long, 1.68-mm i.d.] in the middle over a flame and drawing it out. After cooling, the tube is scored with a diamond pencil and broken at the constricted region. This leaves one end of the tube with a narrow constriction and one end fully open. A slurry of glass beads [for example, VirTis Co. (Gardiner, NY) #16-220, 220-μm diameter] in water is drawn into the

---

[16] D. R. Kellogg, C. M. Field, and B. M. Alberts, *J. Cell Biol.* **109,** 2977 (1989).

open end of the tube using suction on the drawn-out end. The exact volume
of beads is not critical, at least in the 5- to 25-μl range, since beads have
not been observed to affect background binding by a column. The slurry
of beads, with more water added as needed, is then allowed to settle into
the narrow end. Excess liquid is allowed to drain out of the tube. A slurry
of the affinity matrix is then drawn into the open end of the tube up to the
bed of glass beads, the tube is set upright in a conical 1.5-ml centrifuge
tube, and the matrix is allowed to settle and drain by gravity. These
columns generally have a column bed that is about 1 cm high; they can be
loaded and washed by applying buffers with a 100-μl Hamilton syringe
and collecting the eluted fractions by periodically moving the columns to
fresh collection tubes. The flow rate in microcolumns depends on the
precise geometry of the drawn-out tube and is therefore somewhat variable, but should be in the range of 200–400 μl/hr.

Standard Scale Columns. Columns (0.2–3 ml) can be poured in plastic
sterile syringes cut to an appropriate length. The bottom of the syringe is
plugged with a polyethylene filter disk [for example, Ace Glass (Vineland,
NJ) 5848 filter disk support] or with siliconized glass wool (for example,
PhaseSep HGC166) and closed with a stopcock (for example, Bio-Rad
732-8107 three-way stopcocks). The matrix is suspended in about 5 vol of
column buffer, and then placed into the column with a 9-in. Pasteur pipette;
the tip of the pipette is inserted all the way to the bottom of the syringe to
avoid trapping bubbles in the barrel of the syringe. As the matrix settles
and the clear buffer drains from the bottom of the column, more slurry is
added so that the packing column bed is always submerged in the slurry.
It is important that the packing be continuous so that the column bed is
not disrupted by discontinuous layers that will cause nonuniform flow
characteristics. In order to test a column, a small amount of a solution
containing a visible dye can be applied to the column. The dye should
migrate through the column as a coherent disk.[2]

Once the bed is settled, it should be topped with a small amount of
column buffer, the stopcock closed, and the column cap applied. The
matrix bed should never be exposed to air, but rather should always have
a small layer of buffer covering it to avoid improper flow characteristics.
The column can be capped either with an appropriately sized silicone
stopper or with the tip of the plunger of the syringe after removing the
plastic handle. A needle is inserted into the stopper with the sharpened
end pointed out of the column and the base of the needle is twisted back
and forth until it breaks. This leaves the broken end of the needle inside
the column and the tip outside. Tubing from a peristaltic pump can be
attached to the tip of the needle and used to deliver solutions at uniform,

reproducible flow rates. The optimal flow rate depends on the cross-sectional area rather than the total bed volume; we typically use a rate of about 2 to 3 ml/hr for 1-ml syringes with 0.5- to 1-ml bed volumes.

Preparative Scale Columns. Preparative scale (10–50 ml) work can be done in standard chromatography columns (for example, disposable Bio-Rad Econo-Columns work quite well) run with a peristaltic pump. To assure good flow rates, it is best to avoid a column bed higher than 10 cm. We have used flow rates ranging from one to five bed volumes per hour, again depending on the geometry of the column; columns can generally be loaded with extract at higher flow rates than are appropriate for elution.

Control Columns. Control columns should be prepared in parallel to mimic as closely as possible the experimental matrix. However, we have found little difference between columns containing either no protein or an unrelated protein such as bovine serum albumin (BSA). Ideally, a protein with a similar isoelectric point should be used, although the issue of what constitutes an appropriate control for an affinity column is problematical. "Simple" ionic interactions cannot be discounted as irrelevant because most of the specific interactions that we have detected have an ionic component (see Table I). Would denatured protein be a reliable control, or might relevant domains for interactions reform on the matrix? Ultimately, any interactions detected by affinity chromatography should be validated as biologically relevant by genetic analysis or by other means such as the observation of colocalization of the two proteins *in vivo,* so the issue of an ideal control matrix is perhaps a secondary consideration.

Storing Matrices. If a stable protein is coupled to the matrix, the protein is not degraded by proteases in the extract, and the elution conditions used are mild, then affinity matrices can be equilibrated with buffer containing 50% glycerol, stored at −20°, and reused. High concentrations of glycerol are used to prevent the buffer from freezing, which both denatures proteins and destroys agarose matrices. For example, the T4 gene 32 protein can be coupled to an agarose matrix, then stored and reused for purifying the T4 DNA polymerase many times. Protein ligands on affinity columns are usually not harmed by high ionic strength elution buffers, but treatment with chaotropic agents such as high urea concentrations or SDS probably ruins these matrices permanently. Matrices that are to be reused should therefore not be treated in this way.

Preparing Extracts for Chromatography

Choosing a Method of Detecting Proteins

Detecting small amounts of protein normally involves polyacrylamide gel electrophoresis in the presence of sodium dodecyl sulfate (SDS), fol-

TABLE I
INTERACTIONS DETECTED BY PROTEIN AFFINITY CHROMATOGRAPHY

Ligand	Interacting protein	K_d (M) Apparent[a]	K_d (M) Actual	Condition for elution	Ref.
Phage λN protein	E. coli NusA	5×10^{-7}	3×10^{-7}	200–300 mM NaCl	9
E. coli NusA	E. coli RNA polymerase core component	10^{-7}	5×10^{-8}	200–300 mM NaCl	10
E. coli RNA polymerase	E. coli σ^{70}	ND	2×10^{-10}	500 mM NaCl	S. McCracken (unpublished data)
	E. coli NusA	ND	5×10^{-8}	500 mM NaCl	S. McCracken (unpublished data)
	E. coli NusG	ND	ND	500 mM NaCl	S. McCracken (unpublished data)
E. coli NusG	E. coli ρ factor	10^{-5}	ND	500 mM NaCl	J. Li (unpublished data)
E. coli ribosomal protein S10	E. coli NusB	3×10^{-6}	ND	1% SDS	S. Mason (unpublished data)
E. coli NusB	E. coli ribosomal protein S10	ND	ND	1% SDS	J. Li (unpublished data)
Mammalian RNA polymerase II	RAP38 (also known as SII or TFIIS)	10^{-7}	ND	200–300 mM NaCl	11
	RAP30/74 (also known as TFIIE or TFIIF)	2×10^{-8}	2×10^{-8}	200–300 mM NaCl	11
Yeast (S. cerevisiae) RNA polymerase II	yRAP37 (also known as P37)	10^{-7}	ND	200–300 mM NaCl	M. Sopta (submitted)
Activation domain of herpes simplex virus trans-activator VP16	Human or yeast TATA box factor TFIID	3×10^{-7}	ND	300–500 mM NaCl	15
Yeast (S. cerevisiae) TFIID	Human TFIIA	10^{-5}	ND	200–500 mM NaCl	J. Greenblatt and B. Honda (unpublished data)

[a] The apparent K_d is estimated as one-twentieth of the minimum concentration of ligand required to retain the interacting protein efficiently. ND, not determined.

lowed either by staining gels with silver or by using fluorography to detect radioactively labeled proteins. In the latter case $^{35}SO_4^{2-}$ or [^{35}S]methionine is usually used to label cells because of the low cost and high specific activity relative to ^{14}C. Radioactive labeling has the advantage of allowing quantitation and detection of fractions containing binding proteins by scintillation counting before polyacrylamide gel electrophoresis. However, radioactive labeling is relatively expensive and is not more sensitive than silver staining. In addition, the radioactive extracts must be handled more carefully, and fluorography is generally slower than silver staining. However, using radioactive extracts allows a distinction to be made between polypeptides that bind from the extract and proteolytic fragments of the ligand that might be released from the columns during a run, which would otherwise appear to be specific binding proteins. Depending on the ligand protein and the source of the extract, this can be a major consideration that favors the use of radioactive proteins.

Choosing Source of Starting Material

In some cases, a total cell extract is not the optimal starting point for affinity chromatography. For instance, if the ligand protein is known to act in the nucleus, a nuclear extract can provide a 10-fold enrichment of proteins that might interact, allowing much easier detection of the interacting species. In addition, for proteins present in cells in very low abundance, it might be necessary to use another chromatography matrix to fractionate an extract before screening various pools for interacting proteins. For example, a protein involved in DNA replication or transcription might interact with proteins that themselves bind to DNA; in this case, it could be beneficial to enrich for such proteins by DNA-cellulose chromatography to allow very minor interacting species to be detected. Such preliminary chromatographic steps can also assist in removing possible interfering substances from an extract.

Preparation of Whole-Cell Extracts

Whole-cell extracts are usually used as the source of interacting proteins. All procedures are performed at 0–4°.

Yeast cells can be broken by shaking or vortexing with 500-μm glass beads. We typically break the cells by suspending them in 1 vol of lysis buffer [1 ml/g of cells; 20 mM Tris-HCl (pH 7.5), 1 mM Na$_2$EDTA, 1 mM 2-mercaptoethanol, and 50 mM NaCl], adding 1 vol of glass beads and vortexing for about 10 min.

Fresh or frozen bacterial cells are most conveniently disrupted by sonication, but they can also be broken by grinding in a mortar with levigated alumina [Sigma (St. Louis, MO) type 305; 2 g/g of bacteria]. The bacteria are suspended (before sonication or after grinding) in 1–2 vol of lysis buffer (same as described above for yeast cells) or in 10 mM Tris-HCl, pH 7.9, 14 mM magnesium acetate, 1 mM dithiothreitol (DTT), and 60–300 mM KCl.

Mammalian cells grown in tissue culture are first washed with phosphate-buffered saline (PBS) and then either used fresh or kept frozen at $-70°$. We usually homogenize the cells, but have also had good success with sonication.[11] The cells are homogenized in 2 ml/g of 10 mM HEPES (pH 7.9), 1.5 mM MgCl$_2$, 10 mM KCl, and 0.5 mM dithiothreitol using 10 strokes of a Teflon-coated, motor-driven homogenizer. The nuclei in the extract are then disrupted by adding 1.5 ml/g of 50 mM HEPES (pH 7.9), 75% (v/v) glycerol, 1.26 M NaCl, 0.6 mM Na$_2$EDTA, 1.5 mM MgCl$_2$, 0.5 mM DDT, and homogenizing with 10 more strokes.

All extracts are cleared by centrifugation at high speed (1–3 hr at greater than 100,000 g average centrifugal force). It is crucial to remove particulate matter, which will otherwise cause high levels of nonspecific binding that can obscure specific interactions. Glycerol should be present in the fraction loaded onto the columns since this aids in maintaining the stability of many proteins and also reduces background nonspecific binding. [If glycerol is not already present in the extraction buffer, it should be added to at least 10% (w/v) after the high-speed centrifugation step.] We often lyse the cells in chromatography buffer and load the high-speed centrifugation supernatant without further treatment. However, if the extract is not in an appropriate buffer, it can be dialyzed, typically for 4–15 hr, against the affinity column buffer. Particulate matter that forms during dialysis must be removed by centrifugation just prior to loading onto the columns (at least 30 min at 10,000 g; 1 hr at 100,000 g is preferred).

Both the lysis buffer and the column buffers should contain protease inhibitors to minimize fragmentation of proteins in the extract and to protect the affinity ligand. While the recipe for an effective protease inhibitor cocktail will vary with the application, a good starting point is 0.2 mM phenylmethylsulfonyl fluoride (PMSF) (from a 100 mM stock in 2-propanol), 0.5 mM benzamidine-hydrochloride (from a 500 mM stock in H$_2$O), 0.5 μg/ml leupeptin (from a 0.5 mg/ml stock in H$_2$O), and 0.7 μg/ml pepstatin (from a 0.7 mg/ml stock in methanol) in a buffer containing 1 mM Na$_2$EDTA.

In some cases, a crude lysate will contain small molecules that will interfere with chromatography. This can often be remedied by first precipitating proteins by saturating the extract with ammonium sulfate, collecting the precipitate by centrifugation, and dissolving the precipitated protein in affinity column buffer followed by dialysis. Extracts prepared this way should still be centrifuged at high speed (at least 1 hr at greater than 100,000 g) to remove particulate matter that forms during the procedure.

Removing polymers such as DNA and RNA that could interfere with affinity chromatography is more problematical. Passing an extract containing 0.2 M NaCl through a DEAE-cellulose column will remove nucleic acids; this procedure is recommended provided that the proteins of interest do not bind to either nucleic acids or DEAE-cellulose under these conditions. Our experiences with enzymatic removal using nucleases have not been entirely positive; even relatively pure preparations of DNase I tend to show traces of protease contamination. Attempts to remove DNA from extracts with organic polymers such as polyethylene glycol or polyethyleneimine have often had undesirable side effects—such as the inability to resolubilize particular proteins under low ionic strength conditions, or a drastic increase in background binding, both of which are assumed to be due to the continued presence of the polymers in the extract even after extensive dialysis. If nucleic acids cause a problem, several approaches for removing them should be attempted, since no one method gives satisfactory results in every case.

Loading and Eluting Interacting Proteins

Loading Columns

Once suitable conditions for preparing an extract have been identified, the extract is loaded to the affinity matrix and the control matrix. For microscale columns run by gravity, the flow rate is determined by the geometry of the constricted end and the height of the buffer head. For a 20-μl column bed, it should be about 100 μl/15–30 min. For other column sizes, the natural maximal flow rate should be estimated by measuring the rate of flow of buffer while the open column is draining by gravity, then using a lower rate for subsequent operations. Usually an agarose matrix will support a flow rate of at least three column volumes per hour without adverse effects, although the geometry of the column will affect this parameter.

Our standard affinity chromatography loading buffer contains 10–20 mM Tris-HCl or HEPES (pH 7.5–7.9), 10–20% (w/v) glycerol, 1 mM Na$_2$EDTA, 1–2 mM 2-mercaptoethanol or DDT, and 50–100 mM NaCl. When the buffer contains 100 mM NaCl, there are very few proteins in a whole-cell extract that will bind to a control column and be eluted with salt.

Eluting Columns. After washing with at least 5, and preferably 10, column volumes of the low-salt loading buffer, bound proteins are eluted with increasing concentrations of salt. Since many interactions have at least a small ionic component, the elution of an interacting protein from a column can usually be accomplished with salt. We have observed many cases in which an interacting protein binds to the ligand in 100 mM NaCl and elutes with buffer containing 200–300 mM NaCl (see Table I). Increasing the ionic strength in steps is simple and limits the number of fractions that must be checked for binding proteins. However, information about the strength of interactions can be obtained more reliably by eluting with a gradient of increasing salt. It is also easier to distinguish specific from nonspecific binding using a gradient elution.

In pilot experiments, after a sufficiently harsh salt wash (usually 2 M NaCl), the matrix should be stripped by eluting at room temperature with 6 M urea containing 0.2% (w/v) SDS to detect any hydrophilic or extremely strong interactions. This of course sacrifices the column.

Strategic Considerations

Advantage can be taken of microcolumns to screen for the effects of a variety of conditions on the binding of proteins from extracts. An important variable to consider is the concentration of the immobilized protein ligand. There is often a critical optimum concentration that must be identified: concentrations of ligand that are too low do not retain enough protein from the extract, while concentrations that are too high can, with highly charged proteins, create ion exchangers that bind hundreds of proteins nonspecifically. An example of a ligand concentration curve using RNA polymerase II columns and mammalian whole-cell extract is shown in Fig. 1A.

The method of extract preparation, the chromatography buffer, and the volume of extract loaded onto the column are also important. It is worth keeping in mind that rare proteins constituting only $10^{-6}-10^{-5}$ of the total cell protein will only be detected by silver staining as gel bands containing 10–100 ng protein if the column is loaded with extract derived from at least 100 mg of cells (wet weight). Fortunately, we have generally found that, for rare proteins such as those involved in transcription, the signal-to-noise ratio improves as the extract load increases. This occurs because the background binding capacity of the matrix is saturated with a small amount of extract protein, while the relatively large amount of column ligand cannot be saturated by the small amounts of specific binding proteins in the extract. An example of this phenomenon, again with RNA polymerase II columns and mammalian whole-cell extract, is shown in Fig. 1B. It should be noted, however, that loading excessive amounts of protein can obscure weak interactions if proteins with stronger binding properties are present at high concentrations. The stronger interactions will effectively displace weaker ones if the binding sites overlap. This effect has been observed with actin filament columns.[2]

The binding capacity of a column for a particular protein can be determined by saturating the matrix with that protein, assuming that it is avail-

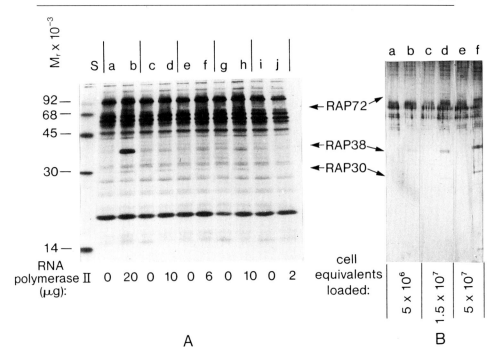

FIG. 1. Effects of ligand concentration and extract load on RNA polymerase II affinity chromatography. (A) Effect of ligand concentration. ^{35}S-Labeled whole-cell extract from murine erythroleukemia cells was loaded onto 20-μl microcolumns prepared with 0 μg (lanes a, c, e, g, i), 20 μg (lanes b, d, f), 10 μg (lane h), or 2 μg (lane j) of calf thymus RNA polymerase II. In lanes d and f the Affi-Gel 10 was inactivated in buffer for 30 or 60 min, respectively, before the addition of RNA polymerase II, resulting in the coupling of about 10 and 6 μg, respectively, to the matrix. Column eluates (0.2 M NaCl) were analyzed by SDS–PAGE and fluorography. S, Protein molecular weight standards. (B) Effect of extract load. Various amounts of murine erythroleukemia whole-cell extract were loaded onto control columns (lanes a, c, e) and RNA polymerase II columns (lanes b, d, f). Eluates (0.2 M NaCl) were analyzed by SDS–PAGE and silver staining. (Taken from Ref. 3.)

able in large quantity. Alternatively, after a column is eluted, the proteins that have flowed through the column are reapplied and the column is reeluted. If more of a given protein is retained in the second pass, then the amount bound in the first pass was probably the maximum that could have bound. If no more binds the second time, no conclusion can be drawn since the column may have been altered by the elution protocol; in this case the experiment can be repeated with two identical columns loaded in series and then eluted separately. If polyacrylamide gel electrophoresis (PAGE) reveals that components are completely missing from the flow-through fraction that were present in the load fraction, it can be concluded that the column was not saturated for those components. However, the presence of a given protein in the flow-through fraction does not mean the column was saturated, since some of the protein in question might have been denatured or otherwise unavailable for binding. If the binding rate is the problem, the column can be loaded two to four times more slowly.

The dissociation constants of the protein–protein interactions can be estimated from the results of microcolumn affinity chromatography experiments in which the ligand concentration is varied (Fig. 1A, for example). These columns are loaded with 3–10 column volume samples and washed with 10 column volumes of loading buffer before eluting the specifically bound proteins with salt or chaotropic agents. Proteins that remain bound after this extended washing must have dissociation constants at least 10-fold lower than the concentration of ligand coupled to the column. Therefore, the K_d of the interaction should be about one-twentieth of the lowest concentration of ligand that fully retains the binding protein, assuming that all of the ligand is available for binding.

Our experiences and those of others (J. Hosoda, personal communication, 1985) with T4 gene 32 protein columns indicate that about 10% of the 32 protein molecules are available for interacting with other proteins.[6] Estimates of dissociation constants derived from affinity chromatography experiments on the N–NusA,[9] NusA–RNA polymerase,[10] and RAP30/ 74–RNA polymerase II[11] interactions have been within factors of two of the dissociation constants measured in other ways (see Table I), suggesting that at least 50% of the N, NusA, and RNA polymerase II are available for binding. We do not know how typical these results are.

Analysis of Binding Fractions

Characteristics of Interacting Proteins

Proteins that interact with an affinity matrix can be characterized by PAGE or by assaying fractions for biochemical activities, assuming the conditions for elution did not alter the activity of the protein. Most proteins can be eluted with high concentrations of salt without affecting their activity, but the use of chaotropic agents such as urea or SDS will normally inactivate enzymes.

If extracts were radioactively labeled, the presence of proteins can be detected by assaying a sample of each fraction in a scintillation counter. Fractions containing protein are pooled—concentrated if necessary by precipitating with acetone or trichloroacetic acid, or by centrifuging against a dialysis membrane [for example, Amicon (Danvers, MA) Centri-

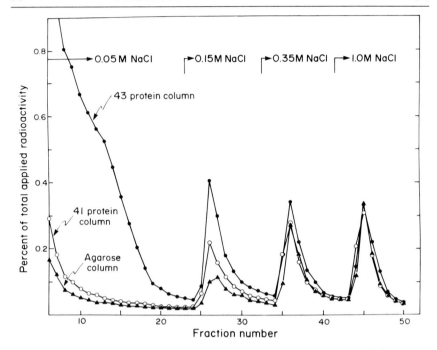

FIG. 2. The elution profile from a T4 bacteriophage DNA polymerase affinity column demonstrates specific interactions in the low-salt wash fractions. Affinity columns containing 1.2 mg/ml T4 bacteriophage gene 43 protein (DNA polymerase) or 0.28 mg/ml gene 41 protein were prepared and used to chromatograph extracts prepared from [^{35}S]methionine-labeled, T4-infected *E. coli* cells. Lysis and column buffers contained 20 mM Tris-Cl (pH 8.1), 1 mM Na$_2$EDTA, 1 mM 2-mercaptoethanol, 10% (w/v) glycerol, 10 mM MgCl$_2$, 0.5% (w/v) Triton X-100, and NaCl as noted. In this case, proteins bound specifically to the gene 43 protein column eluted during the initial 10 column volume wash with the lysis/loading buffer containing 50 mM NaCl. Examination of the eluted fractions by two-dimensional gel electrophoresis confirmed that the specifically bound proteins were known DNA polymerase accessory proteins.[7]

prep]—and then separated by PAGE in the presence of SDS. Proteins are detected by silver staining or fluorography. Weak but specific interactions are often revealed in the initial low-salt wash, so this region of the chromatogram must not be overlooked (see, for example, Fig. 2).

An alternative approach is to assay fractions for enzymatic activities. For instance, if the covalently bound ligand protein has an enzymatic activity, fractions can be assayed for the ability to alter this activity in the test tube. DNA polymerase accessory proteins bind specifically to a T4 DNA polymerase affinity column; these proteins enhance the rate, processivity, and accuracy of the polymerase and can be assayed by their effects on any of these properties.[7]

Still another approach is to assay the column flow-through for the loss of an enzymatic activity. For example, a HeLa cell nuclear transcription extract loses a protein that is essential for initiation by RNA polymerase II when it flows through an RNA polymerase II column.[17] The activity that binds to the column can be eluted and assayed.

The proteins that elute from an affinity column can next be chromatographed on standard matrices to obtain further purification. The same results also allow the identification of conditions that can be used to enrich for these species prior to future attempts to study the protein–protein interaction by affinity chromatography. If sufficient amounts of material are available, binding fractions can be used to generate antisera or to perform peptide sequencing. Either of these approaches can lead to the analysis, identification, or cloning of the genes encoding proteins detected by affinity chromatography.

Direct vs Indirect Interactions

If cell extracts were used for affinity chromatography, one must eventually determine whether an interacting protein is binding directly to the ligand or indirectly to the ligand via some other molecule. This can be accomplished, once the interacting protein has been purified, by rechromatographing the pure interacting protein on control and ligand columns.[10,15]

When a protein machine is held together solely by weak interactions, an affinity column made with any particular component of that machine may detect only interacting proteins that bind directly and sufficiently strongly to the ligand. For protein assemblies that are of low abundance in an extract, this occurs because the concentration of the primary interacting protein always remains too low on the column to bind secondary interacting proteins. However, such secondary interactions are detectable if the interacting protein has multiple subunits held together by strong forces.[4]

Possible Need for Additional Ligand

If a protein undergoes a change in its conformation when it binds to some ligand inside the cell and only then interacts with another protein, the method described above will fail to detect this interaction unless the needed ligand is present in the extract at a sufficient concentration. It is clear that this situation occurs in biological systems: for example, spectrin,

---

[17] Z. F. Burton, L. G. Ortolan, and J. Greenblatt, *EMBO. J.* **5**, 2923 (1986).

a cytoskeletal protein, binds to actin much more strongly if the spectrin has previously bound to band 4.1 protein.[18] Likewise, *Escherichia coli* RecA protein affinity columns have been found to bind LexA protein only in the presence of DNA.[19] We have encountered the same situation when analyzing the primosome of the T4 bacteriophage DNA replication apparatus.[20] Thus second generation protein affinity columns should be considered in which a preformed protein–protein or protein–nucleic acid complex is the immobilized affinity ligand, rather than a single protein alone.

Conclusion

Protein affinity chromatography has many potential uses for studying complex biochemical systems. Starting with a single element of a complex protein machine, it is possible in principle to reassemble the entire machine (see Fig. 3). The technique also has some limitations. For example, even though two proteins interact, they will not necessarily have the opportunity to do so on an affinity matrix. The T4 bacteriophage DNA polymerase binds tightly to a gene 32 protein affinity column, consistent with the specific interaction between these two proteins in solution (Fig. 4). However, although a T4 DNA polymerase column binds several polymerase accessory proteins, it does not bind gene 32 protein.[7] We assume that this lack of reciprocity reflects a nonrandom orientation of the polymerase molecules on the affinity matrix, with the 32 protein-binding domain in an inaccessible position. Perhaps coupling at a variety of pH values would reveal a condition that provides a more random attachment.

Protein affinity chromatography can detect interactions ranging in strength from K_d 10^{-5} to 10^{-10} M (see Table I). An interacting protein with a $K_d > 10^{-5}$ M will not remain bound to the column when the column is washed by the large volume of buffer that is necessary to lower the background binding, even if the immobilized protein concentration is 10 mg/ml. Alternatively, an interacting protein with a $K_d < 10^{-10}$ M might not be able to bind to the affinity matrix if it failed to dissociate rapidly enough from the endogenous ligand present in the extract. Such tight

[18] V. Ohanian, L. C. Wolfe, K. M. John, J. C. Pindu, S. E. Lux, and W. B. Gratzer, *Biochemistry* **23**, 4416 (1984).
[19] N. Freitag and K. McEntee, *Proc. Natl. Acad. Sci. U.S.A.* **86**, 8363 (1989).
[20] T.-A. Cha and B. M. Alberts, in "Cancer Cells 6; Eukaryotic DNA Replication." Cold Spring Harbor Laboratory, Cold Spring Harbor, New York, 1988.
[20a] J. Nodwell and J. Greenblatt, submitted.
[21] C. F. Morris, H. Hama-Inaba, D. Mace, N. K. Sinha, and B. Alberts, *J. Biol. Chem.* **254**, 6787 (1979).
[22] J. Rush and W. H. Konigsberg, *Prep. Biochem.* **19**, 329 (1989).

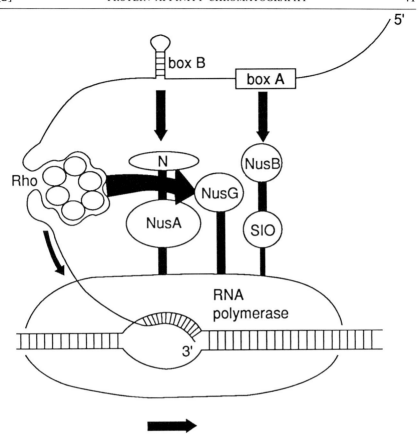

FIG. 3. Protein–protein and protein–nucleic acid interactions involved in the assembly of an elongation control particle containing the *N* gene transcriptional antitermination protein of phage λ. Bold lines are used to connect proteins that have been shown to interact by affinity chromatography (see Table I). Some of the proteins also interact with the box A and box B components of the RNA transcript of the *N* utilization site of the phage.[20a]

interactions might be more reliably detected by coimmunoprecipitation methods. In any case, failure to detect an interaction must be interpreted cautiously.

Finally, as noted above, detection of binding between two proteins *in vitro* cannot by itself be interpreted as a demonstration of a biologically significant protein–protein interaction. The two proteins must be shown to be involved in the same process *in vivo;* ideally the interaction should be demonstrated by a functional assay or by any of several genetic techniques that can detect protein–protein interactions.[12] Nevertheless, pro-

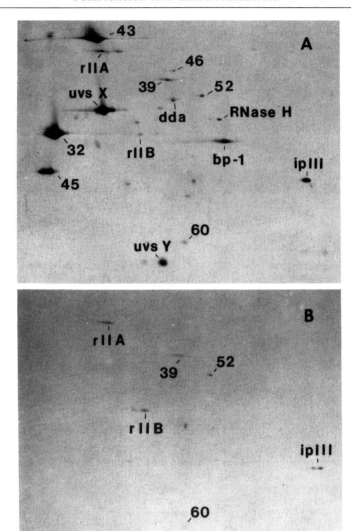

FIG. 4. Analysis of proteins bound to a T4 bacteriophage gene 32 protein affinity column by two-dimensional gel electrophoresis. Proteins from a radioactively labeled extract prepared from bacteriophage T4-infected *E. coli* cells were chromatographed on columns containing 8 mg/ml T4 gene 32 protein (A) or 12 mg/ml bovine serum albumin column (B). Samples from each of the eluted fractions were pooled, concentrated by precipitation, and fractionated by nonequilibrium two-dimensional gel electrophoresis. T4 bacteriophage proteins were detected by fluorography. Consistent with its role in all areas of T4 bacteriophage DNA metabolism, the gene 32 protein binds a large number of proteins known to be involved in DNA replication, genetic recombination, and DNA repair.[6]

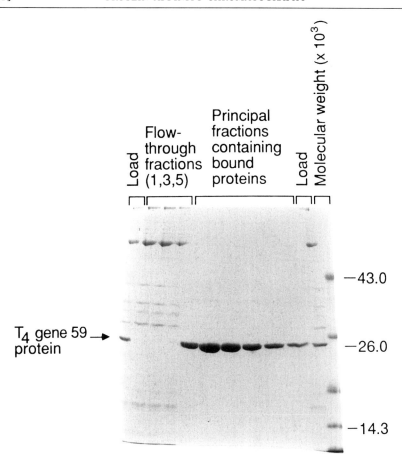

FIG. 5. SDS–polyacrylamide gel analysis of the purification of T4 gene 59 protein on an affinity column containing bound T4 gene 32 protein. Approximately 1 mg of T4 gene 59 protein, estimated to be 40% pure after DNA cellulose chromatography, was fractionated further by taking advantage of its affinity for the T4 gene 32 protein. The 59 protein was in 5 ml of a buffer consisting of 20 mM Tris-HCl (pH 8.1), 1 mM 2-mercaptoethanol, 1 mM Na$_2$EDTA, 3 mM MgCl$_2$, 10% (w/v) glycerol, and 400 mM NaCl. It was loaded at 1 ml/hr onto a 1-ml column of 32 protein-Affi-Gel 10 containing 3.8 mg of bound T4 gene 32 protein. After the column was rinsed with 4 ml of the buffer containing 400 mM NaCl, the bound proteins were eluted with the same buffer containing 2.4 M NaCl and 0.1% Triton X-100. One-milliliter fractions were collected throughout the run. The gel shows that only T4 gene 59 protein is able to bind quantitatively to the column at 400 mM NaCl and the eluted 59 protein is greater than 99% pure (fractions with bound proteins are overloaded on the gel, in order to provide a sensitive estimate of purity). The breadth of the 2.4 M NaCl eluted peak is unusual, and it attests to the strong interaction between T4 gene 59 protein and T4 gene 32 protein. This may also explain why the yield of 59 protein from this column is only about 50%.

FRACTION

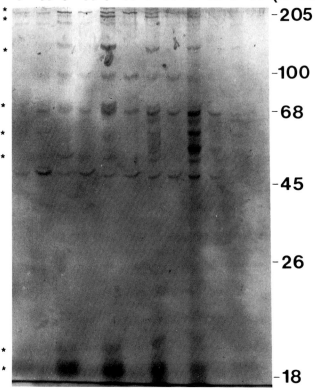

FIG. 6. Affinity chromatography with yeast DNA polymerase I. DNA Polymerase I was purified from a yeast strain overexpressing the *POL1* gene and approximately 400 mg was covalently attached to an Affi-Gel 10 agarose matrix. A control column containing a similar amount of human serum albumin was also constructed. These affinity matrices were used to chromatograph an extract prepared from whole yeast cells. Fourteen grams of a diploid strain lacking the major proteases PEP4 and PRB1 were harvested by centrifugation and stored frozen at $-70°$, thawed, and mixed with 10 ml of lysis buffer [20 mM Tris-HCl (pH 7.5), 1 mM Na$_2$EDTA, 1 mM 2-mercaptoethanol, 2 mM MgCl$_2$, 50 mM KCl, 0.5 mM PMSF, 0.5 μg/ml leupeptin, and 0.7 μg/ml pepstatin]. The cells were disrupted by vortexing for a total of 9 min with 11 ml of 500-μm glass beads. The lysate was cleared first by centrifuging for 10 min at 27,000 g, then for 2 hr at 160,000 g. The cleared lysate contained about 220 mg of protein in a volume of 12 ml. It was adjusted to 10% (w/v) glycerol and loaded directly to the affinity columns. The columns were eluted with a gradient of increasing KCl in lysis buffer (50–500 mM). Proteins were separated by polyacrylamide gel electrophoresis in the presence of SDS and stained with silver. Proteins binding to the polymerase (P) but not the human serum albumin (H) column are candidates for DNA polymerase accessory proteins (*).

TABLE II

T4 GENE 32 PROTEIN AFFINITY CHROMATOGRAPHY AS
FINAL STEP IN T4 DNA POLYMERASE PURIFICATION[a]

Parameter	Before ^{32}P affinity column	After ^{32}P affinity column
Endonuclease activity[b]	+	−
Amount of T4 gene 43 protein	20 mg	16.3 mg

[a] T4 gene 43 protein (the T4 DNA polymerase), purified by standard procedures,[21,22] including single-stranded DNA cellulose chromatography, often contains levels of endo-deoxyribonuclease that interfere with DNA enzymology. This nuclease activity can be easily removed by 32 protein affinity column chromatography. For example, 20 mg of 43 protein, estimated to be 98% pure (by SDS–polyacrylamide gel analysis) after standard purification steps[21,22] and still containing endonuclease, was further purified on a 20-ml column of 32 protein-Affi-gel 10 containing 68 mg of bound 32 protein (this is a molar ratio of 11:1 of bound 32 protein:43 protein). Flow rates throughout are one column volume per hour. The 43 protein was loaded onto the column in 33 ml of a buffer containing 20 mM Tris-HCl, pH 8.1, 1 mM Na$_2$EDTA, 2 mM 2-mercaptoethanol, 5 mM MgCl$_2$, 50 mM NaCl, and 10% (w/v) glycerol. Elution was with a 150-ml linear gradient from 50 to 400 mM NaCl in the same buffer; the 43 protein eluted (without endonuclease) in 30 ml with a peak concentration at 170 mM NaCl. The recovery was 82%.

[b] Conversion of RFI plasmid to RFII, as assayed by agarose gel electrophoresis.

tein affinity chromatography is very useful, both as an initial method for detecting the potential components of protein machines (Fig. 5) and for the large-scale purification of such components (Fig. 6 and Table II). For a biochemist, it represents an invaluable tool for exploring the complex protein machines that act in so many aspects of cellular metabolism.

[4] Metal Requirements for Nucleic Acid Binding Proteins

By DENISE L. MERKLE and JEREMY M. BERG

Introduction

In recent years, a number of families of nucleic acid binding and gene regulatory proteins have been found to require bound metal ions for full activity.[1] The largest and most well-characterized family is the so-called zinc finger protein family typified by *Xenopus* transcription factor IIIA (TFIIIA). This protein was initially purified as a protein that forms a complex with 5S RNA in *Xenopus* oocytes[2] and was subsequently shown to be required for the accurate initiation of transcription of 5S RNA genes.[3] TFIIIA was found to bind specifically to the internal control sequence found with such genes. In 1983, it was reported that the TFIIIA–5S RNA complex contained several equivalents of zinc as isolated.[4] Furthermore, removal of zinc by treatment with chelating agents resulted in loss of specific DNA-binding activity; this inactivation could be reversed by addition of zinc.[4] In 1984, the sequence of a TFIIIA cDNA clone was reported.[5] Subsequent analysis[6,7] revealed the presence of nine tandem sequences within the deduced primary structure that approximately matched the consensus (Tyr,Phe)-X-Cys-$X_{2,4}$-Cys-X_3-Phe-X_5-Leu-X_2-His-$X_{3,4}$-His-X_{2-6}, where X represents relatively variable amino acids. It was proposed that each of these sequences was capable of binding a zinc ion via the cysteine and histidine residues. Further investigation by a number of methods has revealed that the proposal is correct; these sequences do bind zinc ions via the cysteinate and histidine side chains to form structural domains that are essential for DNA binding.[8]

Since the characterization of TFIIIA, a large number of other proteins that have been implicated in nucleic acid binding or gene regulatory processes have been discovered to either require metal ions such as zinc(II)

[1] J. M. Berg, *J. Biol. Chem.* **265**, 6513 (1990).

[2] B. Picard and M. Wignez, *Proc. Natl. Acad. Sci. U.S.A.* **76**, 241 (1979).

[3] D. R. Engelke, S.-Y. Ng, B. S. Shastry, and R. G. Roader, *Cell (Cambridge, Mass.)* **19**, 717 (1980).

[4] J. S. Hanas, D. J. Hazuda, D. J. Bogenhagen, F. H.-Y. Wu, and C.-W. Wu, *J. Biol. Chem.* **258**, 14120 (1983).

[5] A. M. Ginsberg, B. O. King, and R. G. Roeder, *Cell (Cambridge, Mass.)* **39**, 479 (1984).

[6] J. Miller, A. D. McLachlan, and A. Klug, *EMBO J.* **4**, 1609 (1985).

[7] R. S. Brown, C. Sander, and P. Argos, *FEBS Lett.* **286**, 271 (1985).

[8] J. M. Berg, *Annu. Rev. Biophys. Biophys. Chem.* **19**, 405 (1990).

for full activity or to contain sequences that appear capable of forming metal-binding domains.[1] Some of these sequences are highly similar to the TFIIIA-type consensus whereas others are quite different but share the property of having sets of cysteine and/or histidine residues within short stretches of the primary structure that have the potential of forming metal-binding domains. It is important to realize that this large collection of proteins contains many distinct classes in which the three-dimensional structures of the metal-binding domains and the functions of the proteins as well as the role of the bound metal ion(s) are different.

It is the goal of this account to provide a description of some of the methods that can be used to determine whether a protein suspected of having metal-binding domains does, indeed, bind metal ions and if the presence of the bound metal ions affects the activity of the protein. The methods can be divided into three groups: (1) direct methods, (2) reconstitutive methods, and (3) functional methods. Two methods from each class will be discussed and the advantages and potential pitfalls will be noted. Examples of systems where the methods have been useful will be provided.

Direct Methods

The most direct method is the determination of the total amount of a given metal present in a sample of a purified protein. Atomic absorption spectroscopy provides one convenient and frequently utilized method. A solution of the protein is aspirated into a hot flame. A substantial proportion of the metal ions in the sample are reduced to the neutral atomic state. Atoms absorb light at characteristic frequencies with very small linewidths. This allows the determination of one element in the presence of many other elements. A light source that emits at the appropriate frequency is required for each element. The concentration of the element in the sample is determined by comparison with a standard curve. For zinc, concentrations of less than 1 μM can be easily determined using 1-ml sample sizes. The number of equivalents of metal can be evaluated if the protein concentration of the sample is known.

The main advantage of atomic absorption spectroscopy is its directness. The total amount of metal ion in the sample is evaluated without regard to its chemical form. Furthermore, atomic absorption spectroscopy and related methods (e.g., plasma emission spectroscopy) are reasonably sensitive and quite precise. There are, however, several disadvantages. First, the method requires a significant amount of purified protein and is destructive. For many DNA-binding proteins, the many micrograms of purified protein required are not available to burn. Second, the method measures the total amount of zinc in the sample, be it tightly bound in a metal-binding domain, adventitiously bound to the surface of the protein, or simply a contaminant of the buffer used. It is important always to check a sample of the buffer used in the final purification step (e.g., adjacent fractions from a chromatography column run) to be sure that it does not have the same metal ion content as does the protein sample in question. Finally, it is important to note that the result of a direct metal content determination may be quite dependent on the method of purification. Both false positive and false negative results are possible. Many proteins will bind ions such as zinc even if they do not play a role in the primary function of the protein. Thus, it has traditionally been the practice to dialyze protein samples against solutions containing low concentrations of chelating agents such as ethylenediaminetetraacetic acid (EDTA) prior to analysis. This ensures that the metal ions found are bound in at least a kinetically stable manner. False negative results are also possible if the protein sample is treated too harshly prior to analysis. For many of the DNA-binding proteins that contain metal-binding domains, the metal ions are bound tightly thermodynamically yet are relatively easily exchanged. Thus, dialysis against metal-free buffer or solutions of chelating agents could result of loss of metal ions that are required for function.

An example that illustrates many of these features is the case of TFIIA. In the initial report that this protein contained bound zinc,[4] samples of the TFIIIA–5S RNA complex that had been dialyzed against 5 mM EDTA were found by atomic absorption to contain 2–3 Eq of zinc per protein molecule. When the protein was freed from the RNA by nuclease treatment, dialysis against the same solution resulted in the loss of all zinc from the protein. Purification of the TFIIIA–5S RNA complex by another method that was designed to avoid potential chelating agents such as EDTA and dithiothreitol resulted in preparations of the complex that contained 7–11 zinc ions per protein by atomic absorption.[6] This number of zinc ions was very suggestive in light of the sequence analysis of the protein. The intensity of the X-ray fluorescence produced during extended X-ray absorption fine structure (EXAFS) studies of the TFIIIA–5S RNA complex provided independent evidence concerning the zinc stoichiometry.[9] The intensity observed was consistent with nine zinc ions per complex to within 10%. Since the EXAFS data could be well fit assuming only one type of zinc site with two cysteines and two histidines coordinated to the metal ion, these results strongly suggest that essentially all of the nine sequence repeats do, indeed, bind zinc ions. Quite different results would have been obtained if only two or three of the sequences bound zinc and

[9] G. P. Diakun, L. Fairall, and A. Klug, *Nature* (*London*) **324**, 698 (1986).

the remaining zinc present in the preparation was bound to different sorts of sites.

A second direct approach is the method of measurement of metal release. This is applicable to systems in which it is possible to chemically modify the protein in a manner that significantly weakens the protein–metal ion interaction. The metal released from the protein can then be detected in solution with the use of an appropriate metal-sensitive dye. The most generally used method[10] can be applied to proteins with metal–cysteinate coordination, a feature common to many metal-dependent nucleic acid-binding proteins. The modification agent is *p*-hydroxymercuribenzoate (PMPS), which reacts with cysteine to form a mercury–thiolate complex. This adduct absorbs at 250 nm so that the modification can be monitored spectrophotometrically. The dye used is 4-(2-pyridylazo)resorcinol (PAR), which forms a 2 : 1 complex with zinc that absorbs at 500 nm. The mercury modification can be reversed by treatment with excess thiol reagents. A disadvantage of this method is that it is dependent on having appropriately reactive ligands available for chemical modification. For example, for a zinc-binding site such as that in carbonic anhydrase, which has only histidine ligands, PMPS titrations would not be effective.

This method has been applied to several zinc-containing nucleic acid-binding proteins, including the bacteriophage gene 32 protein.[11] This protein binds single-stranded nucleic acids with high cooperativity. This protein was titrated with PMPS in the presence of PAR. Monitoring the absorbance at 250 nm revealed that 3.2 Eq of the adduct was formed with no additional change in absorbance on the addition of excess PMPS. Concomitant with this modification, an absorbance at 500 nm developed corresponding to the release of 1.12 zinc ions per protein. These observations are consistent with the binding of one zinc ion within the sequence **Cys**-Ser-Ser-Thr-**His**-Gly-Asp-Tyr-Ser-**Cys**-Pro-Val-**Cys,** which had been identified as a potential metal-binding site from a systematic computer-based sequence search.[12] Reversal of the modification with excess dithiothreitol produced the gene 32 apoprotein. Comparative studies of the native and apoproteins suggested that the presence of the bound zinc ion significantly affected the cooperativity of nucleic acid binding but only modestly reduced the binding affinity to a single-site oligonucleotide.[11] In addition, the metal ion appears to be important for protein stability as determined by susceptibility to proteolysis.[11]

[10] J. B. Hunt, S. H. Neece, H. K. Schachman, and A. Ginsberg, *J. Biol. Chem.* **259,** 14793 (1984).
[11] D. P. Giedroc, K. M. Keating, K. R. Williams, W. H. Konigsberg, and J. E. Coleman, *Proc. Natl. Acad. Sci. U.S.A.* **83,** 8452 (1986).
[12] J. M. Berg, *Science* **232,** 485 (1986).

Reconstitutive Methods

An alternative approach to the demonstration of metal-binding ability of a protein or peptide involves the ability of metal to bind to the metal-free form. The ion used in the reconstitution can be either the natural metal (either in stable or radioactive form) or a different metal with desired spectroscopic or chemical properties. These methods can be used both to demonstrate the presence of metal-binding activity and to prepare radioactively or spectroscopically labeled materials for other studies.

The first reconstitutive method we shall discuss involves cobalt substitution. Zinc(II) has a filled d shell and, therefore, is silent with regard to many spectroscopic probes. In contrast, cobalt(II) has many desirable spectroscopic features and it is structurally and chemically quite similar to zinc(II). Thus, it substitutes into many natural zinc sites to produce proteins that are structurally, and often functionally, similar to the native forms. However, the dissociation constants for the protein–cobalt(II) complexes are often several orders of magnitude larger than those for the zinc(II) complexes, so that care must be taken to ensure that adequate concentrations are used to allow incorporation. This represents a major disadvantage for the cobalt substitution method. Among the advantages is the fact that the position, shape, and intensity of the d-to-d transitions in the visible region provide a significant amount of information about the nature of the cobalt(II)-binding site.

This method has been applied to single zinc finger domains from a variety of proteins. The first such peptide to be studied corresponded to the second domain of TFIIIA.[13] This 30-amino acid peptide had been prepared by gene synthesis, expression of a fusion protein, and cleavage. Treatment of this peptide with solutions of cobalt(II) resulted in the production of an absorption spectrum with components in the ultraviolet and visible regions. A set of bands centered near 650 nm is due to d-to-d transitions. The relatively large extinction coefficients for these bands ($>500\ M^{-1}\ cm^{-1}$) indicated that the complex formed was distorted tetrahedral rather than five- or six-coordinated. The spectrum reached nearly its full intensity when 1 Eq of cobalt(II) had been added, indicative of a relatively stable 1:1 complex. Furthermore, the spectrum could be bleached by the addition of 1 Eq of zinc(II), providing evidence for the interaction between the peptide and the ion naturally found in TFIIA. Subsequent studies used these spectrophotometric methods to determine dissociation constants for the peptide–cobalt(II) and peptide–zinc(II) complexes.[14] The values determined indicated that the specificity for zinc(II)

[13] A. D. Frankel, J. M. Berg, and C. O. Pabo, *Proc. Natl. Acad. Sci. U.S.A.* **84**, 4841 (1987).
[14] J. M. Berg and D. L. Merkle, *J. Am. Chem. Soc.* **111**, 3759 (1989).

over other metal ions in reconstituting the intact protein was reflected in the ion affinities of a single zinc finger peptide.

The second reconstitutive approach is the method of zinc blotting.[15-17] In this method, proteins are separated from one another by sodium dodecyl sulfate-polyacryamide gel electrophoresis, reduced by soaking in an appropriate buffer, transferred to nitrocellulose filters, and then probed with radioactive zinc (^{65}Zn). After suitable washing, the presence of bound zinc can be detected by autoradiography. The conditions to be used at each step have been optimized with use of a series of known zinc-binding proteins together with a set of proteins known not to have a high affinity for zinc.[16,17] This method has the advantage that small, not necessarily homogeneous, samples can be easily analyzed. Detection of samples smaller than 100 pmol of protein has been demonstrated. Its disadvantages include difficulty in quantitating the number of zinc ions bound per protein as well as some potential for false positive or negative results due to binding at nonfunctionally important sites or the inability to restore the metal-binding activity during the renaturation steps.

This method has been applied to several important systems. These include poly(ADP-ribose)polymerase (NAD$^+$ ADP-ribosyltransferase),[17] a protein involved in damaged DNA recognition and repair, and retroviruses, in which case it was demonstrated that the nucleocapsid protein of avian myoblastosis virus had significant zinc-binding activity,[16] providing some of the earliest evidence that the Cys-X$_2$-Cys-X$_4$-His-X$_4$-Cys sequences in these proteins can, indeed, form metal-binding domains as had been proposed.[12]

Functional Methods

If a protein has a function that can be assayed in the laboratory, then studies of loss or gain of activity as a function of metal ion or chelator treatment can be extremely useful in providing evidence for the presence of bound metal ions and in determining their importance for biological function. In addition, if a specific protein sequence has been identified as a potential site for a metal ion interaction by any means, then studies of the effects of disrupting this site can provide support for or against such a hypothesis. Obviously, a wide variety of functional methods are possible, depending on the characteristics of a particular system. The two functional

[15] L. A. Schiff, M. L. Nibert, M. S. Co., E. G. Brown, and B. N. Fields, *Mol. Cell. Biol.* **8,** 273 (1988).

[16] L. A. Schiff, M. L. Nibert, and B. N. Fields, *Proc. Natl. Acad. Sci. U.S.A.* **85,** 4195 (1988).

[17] A. Mazen, G. Gradwohl, and G. de Murcia, *Anal. Biochem.* **172,** 39 (1988).

methods we shall discuss involve the use of chelator inactivation and metal ion reactivation on specific protein–DNA interactions and the use of site-directed metagenesis of proposed metal-binding residues in providing evidence for metal ion binding and identifying metal-binding domain functions.

As noted in the introduction, some of the first evidence that zinc played an important role in the function of TFIIIA came from the observation that EDTA treatment resulted in the loss of specific DNA-binding activity.[4] This approach can easily be applied to other proteins suspected of requiring metal ions for DNA-binding activity. The specific DNA-binding activity can be assayed by footprinting methods, such as the DNase I method used in the TFIIIA study, or by gel mobility shift methods. Additional, and stronger, evidence can be obtained if the specific DNA-binding activity can be restored by incubation of the chelator-inactivated preparation with metal salts. The chelators that can be used include EDTA and 1,10-phenanthroline with concentrations from 0.5 to 50 mM required to produce an effect. The chelated metal and free chelating agent can be removed from the protein by dialysis. A reducing agent such as dithiothreitol should be included in the dialysis buffers, as metal-free cysteine-rich proteins are often quite sensitive to air oxidation, which can cause loss of metal-restorable activity. Treatment of these samples with metal ion solutions can potentially produce reconstituted preparations that can be assayed for activity. A variety of metal ions can be tested, although zinc appears to be the most likely ion to be active. Various metal ion concentrations ranging from the micromolar to the millimolar range should be tested. Often an optimum is found as excess metal ion can inhibit some protein–DNA interactions.

This method has been applied to many different systems. For example, two partially purified factors that bind to the human histone H4 promotor were tested in this way using gel mobility shift methods to assay specific DNA binding.[18] Both H4TF-1 and H4TF-2 were found to be inactivated by 0.5 mM 1,10-phenanthroline and reactivated by zinc(II). One of the factors also appeared to be reactivated by iron(II). These relatively simple experiments provided the strong suggestion that these transcription factors contain zinc-binding domains of some sort that are directly or indirectly involved in their specific interactions with DNA. Analogous results were obtained with a 150-amino acid fragment corresponding to the DNA-binding domain of the glucocorticoid receptor that had been expressed in *Escherichia coli*.[19] Treatment of samples of this fragment that had been

[18] L. Dailey, S. B. Roberts, and N. Heintz, *Mol. Cell. Biol.* **7**, 4582 (1987).
[19] L. P. Freedman, B. F. Luisi, Z. R. Korzun, R. Basavappa, P. B. Sigler, and K. R. Yamamoto, *Nature (London)* **334**, 543 (1988).

made metal free with zinc(II) or cadmium(II) resulted in materials that specifically bound to glucocorticoid-responsive elements as assayed by gel mobility shift as well as footprinting methods. Maximal activity occurred at 250–500 μM zinc(II), but at 5 μM cadmium(II) levels.

The second functional method to be discussed involves the use of site-directed metagenesis in probing potential metal-binding domains. One of the reasons for the rapid development of our knowledge of the roles of metal-binding domains in nucleic acid-binding proteins is the fact that they are often recognizable at the deduced protein sequence level. Thus, members of the TFIIIA-like zinc finger protein class are easily identified by characteristic patterns of cysteine, histidine, and other residues whereas other classes of proteins have been proposed to contain metal-binding domains, prior to the availability of experimental data, based on other patterns of potential metal-binding residues. One approach to testing such hypotheses is to change one or more of the residues proposed to interact with the metal ion(s) and to examine the effects in a functional assay. Such an approach has the advantage that such sequence changes can be easily made using current site-directed mutagenesis protocols and that the method can be applied with any available assay. A disadvantage is that mutation of any residue, paricularly one that is known to be conserved in a family of proteins, can be expected to have functional effects regardless of the mechanistic basis for these effects. Thus, changing a cysteine may completely disrupt the activity of a protein even if the cysteine is not involved in metal binding.

This method has been applied to several retroviral nucleocapsid proteins[20–22] that contain potential metal-binding sequences that were discussed briefly above. Mutation of any of the three cysteine residues to serine in an infectious Moloney murine leukemia clone resulted in the production of virus particles that were not infectious.[20,21] Similar effects were seen when other conserved or conservatively substituted residues within the proposed metal-binding domain were mutated. Further characterization of the virions produced indicated that they lacked detectable amounts of viral RNA although some of the mutant virions contained a significant amount of cellular RNA species. The effects of the same sequence changes on the metal-binding properties of peptides corresponding to this region have been examined and in many cases the magnitude of the effects on metal-binding properties were found to correlate with the

[20] R. J. Gorelick, L. E. Henderson, J. P. Hanser, and A. Rein, *Proc. Natl. Acad. Sci. U.S.A.* **85**, 8420 (1988).

[21] C. Méric and S. P. Goff, *J. Virol.* **63**, 1558 (1989).

[22] J. E. Jentoft, L. M. Smith, F. Xiangdong, M. Johnson, and J. Leis, *Proc. Natl. Acad. Sci. U.S.A.* **85**, 7094 (1988).

strength of the effect on RNA packaging.[23] These mutagenesis results provide support that metal binding by the nucleocapsid protein occurs and is functionally important and revealed that the metal-binding domain might be involved with a specific interaction with part of the viral RNA required for encapsidation.

Conclusions

The methods discussed represent some of the most widely applicable techniques available for the identification and characterization of the roles of metal ions in nucleic acid-binding proteins. Given the large number of nucleic acid binding and gene regulatory proteins that have been found in recent years to contain metal ions or to require them for activity, initial tests for the presence of metal or for metal requirements should become part of the routine characterization of any newly discovered or purified factor. This continues to be a rapidly developing field and many new systems and new methods can be expected to be discovered in the upcoming years.

Acknowledgments

Work in our laboratory in this area has been supported by grants from The National Institutes of Health (GM-38230) and The National Science Foundation (DMB-8850069).

[23] L. M. Green and J. M. Berg, *Proc. Natl. Acad. Sci. U.S.A.* **87**, 6403 (1990).

[5] Large-Scale Preparation of DNA Fragments for Physical Studies of Protein Binding

By Karlheinz Tovar and Wolfgang Hillen

Introduction

The study of sequence-specific protein–DNA interactions profits greatly from the availability of pure proteins and DNA fragments containing the respective target sequences. Because the recognized sequence is only on the order of a few tens of base pairs or less, the DNA substrate is easily obtained in sufficient amounts by chemical synthesis. In many cases, however, cooperative binding distinguishes between the presence of one or more binding sites on the substrate DNA.[1,2] These cases require longer DNA fragments, which are difficult to obtain synthetically in sufficient amount and purity. The same holds true for studies of conformational changes of the DNA substrates on protein binding, which are gaining importance.[3–5] At present most of these studies are limited to methods that do not require large amounts of material. It has been our goal to optimize standard procedures of DNA fragment preparation to a state where large (tens of milligrams) amounts of homogeneous, small [approximately 100 base pairs (bp) DNA fragments from the class A through D *tet* regulatory regions[6,7] became available for physical studies, e.g., neutron scattering,[8] kinetic determinations using fluorescence spectroscopy,[9] electrooptics, and melting[10–12] of protein–DNA complexes, and circular dichroism (CD) studies of Tet repressor–*tet* operator interactions.[13] These methods may be used for efficient large-scale preparations of any desired fragment of DNA. They are based, in part, on previously described procedures.[14,15]

Outline of Procedure

The general strategy involves known procedures, which have been optimized for handling large amounts of material in a short period of time. In the first step the desired DNA fragment is cloned, possibly in multiple insertions, in a suitable plasmid vector, which should be amplifiable to a high copy number and exhibit genetic stability even without selection. These factors are more important than an easily selectable resistance marker for the cloning protocol. Particular care must be given to the selection of the proper restriction sites flanking the desired fragment be-

[1] A. Hochschild and M. Ptashne, *Cell* (*Cambridge, Mass.*) **44**, 681 (1986).
[2] J. D. Gralla, *Cell* (*Cambridge, Mass.*) **57**, 193 (1989).
[3] R. Schleif, *Science* **240**, 127 (1988).
[4] L. M. De Vargas, S. Kim, and A. Landy, *Science* **244**, 1457 (1989).
[5] O. Raibaud, *Mol. Microbiol.* **3**, 455 (1989).
[6] B. Mendez, C. Tachibana, and S. B. Levy, *Plasmid* **3**, 99 (1980).
[7] K. Tovar, A. Ernst, and W. Hillen, *Mol. Gen. Genet.* **215**, 76 (1988).
[8] H. Lederer, K. Tovar, G. Baer, R. P. May, W. Hillen, and H. Heumann, *EMBO J.* **8**, 1257 (1989).
[9] C. Kleinschmidt, K. Tovar, W. Hillen, and D. Pörschke, *Biochemistry* **27**, 1094 (1988).
[10] W. Hillen and B. Unger, *Nature* (*London*) **297**, 700 (1982).
[11] D. Pörschke, K. Tovar, and J. Antosiewicz, *Biochemistry* **27**, 4674 (1988).
[12] M. Wagenhöfer, D. Hansen, and W. Hillen, *Anal. Biochem.* **175**, 422 (1988).
[13] K. Tovar and W. Hillen, to be published.
[14] S. C. Hardies, R. K. Patient, R. D. Klein, F. Ho, W. S. Reznikoff, and R. D. Wells, *J. Biol. Chem.* **254**, 5527 (1979).
[15] W. Hillen, R. D. Klein, and R. D. Wells, *Biochemistry* **20**, 3748 (1981).

cause these sites must be digested on a large scale with a potentially costly amount of enzyme.

The cultivation of the transformed *Escherichia coli* strain is usually done in batches of 10–300 liters, depending on the facilities available. The amplification of plasmid DNA with chloramphenicol is advantageous, as it does not increase the final yield of DNA much, but rather simplifies the preparation procedure because the ratio of plasmid DNA to chromosomal DNA and to total protein is more favorable. Based on these considerations, amplification should not be omitted.

The protocol used for Brij deoxycholate (DOC) lysis of cells is quite crucial because the contamination with chromosomal DNA can be kept to a minimum at this step. Degradation of RNA and protein can be done in the supernatant of the lysed cells. Extensive degradation of proteins using proteinase K (obtained from Merck, Darmstadt, Germany) is very important for the expediency and efficiency of the following phenol extraction, because it seems to reduce the interface material and to increase the portion of DNA remaining in the aqueous phase.

Restriction of the plasmid DNA should be done with an inexpensive enzyme such as *Eco*RI, because it can be prepared in large amounts and excellent quality from overproducing *E. coli* strains.[16] However, *Bam*HI or other readily available enzymes may also be suitable. The results of a time course for *Eco*RI restriction on an analytical scale obtained for the four plasmids described below were reproduced in the preparative digestion. Because the preparative digestion is usually done for 24 hr or longer, it is advisable to split the total amount of plasmid DNA into several independent batches.

After digestion and removal of the restriction enzyme the bulk of the vector DNA is separated from the small insertion fragment by a fractionated polyethylene glycol (PEG) precipitation.[15,17] In the past, this procedure has been somewhat troublesome because it seemed only poorly reproducible. However, when done exactly as described here we have had repeated success with it. An enrichment of the desired DNA is essential because the expediency of the following HPLC purification depends a great deal on the amount of DNA to be separated.

The final purification of the desired DNA fragment is achieved by high-performance liquid chromatography (HPLC) on Nucleogen (obtained from Diagen, Düsseldorf, Germany). The products obtained this way have been sufficiently pure for physicochemical and enzymatic procedures. The following methods were used to purify *tet* regulatory fragments from four classes (A through D) of tetracycline resistance genes.[6] They may be applied, however, to any fragment of DNA.

[16] J. Bottermann and M. Zabeau, *Gene* **37,** 229 (1985).
[17] J. T. Lis and R. Schleif, *Nucleic Acids Res.* **2,** 383 (1975).

Methods and Results

Cloning of Desired DNA Fragments

Plasmid pWH802[18] is used as a vector for all four plasmids. It is a 3739-bp DNA derived from pVH51.[19] Its nucleotide sequence is known.[20] The *ori* fragment is derived from mini-Col[19] and its selectable marker is a colicin immunity.[14] For colicin E1 preparation[21] *E. coli* JF290 is grown on a rotary shaker to an OD_{650} of 1 at 37° in 1 liter of L broth (LB). Then 1 mg mitomycin is added and cultivation is continued at 37° for 6 hr. The cells are sedimented by centrifugation in a JA10 rotor at 6000 rpm for 30 min at 4°. To the supernatant 390 g of solid ammonium sulfate is added and the mixture incubated at 4° for 1 hr. The precipitate is spun down in a JA10 rotor at 10,000 rpm at 4°, dissolved in 10 ml of 0.1 M potassium phosphate, pH 7.0, dialyzed against 2 liters of 0.1 M potassium phosphate, pH 7.0, for 12 hr at 4°, lyophilized in aliquots of 500 μl, and stored at 4°. To determine the colicin activity of this preparation 200 μl of competent *E. coli* RR1 cells[14] are grown in 4 ml of LB for 90 min at 37° on a shaker and 200 μl of this culture is incubated without shaking at 37° for 30 min with varying amounts of a batch of colicin dissolved in 500 μl of LB. Then the entire mixture is spread on LB plates containing 10% (w/v) deoxycholate. The amount of colicin preventing growth in this assay is used for selection in all following transformations.[21] In our hands the colicin could be stored for months without loss of activity. The effort required to perform this somewhat elaborate selection procedure is outweighed by the excellent stability and high copy number of the plasmid, which has been found even with multiple insertions of the extremely strong phage T7 promoter A1.[22]

The *tet* regulatory DNA fragments are obtained by elution from polyacrylamide gels of a 76-bp *Fnu*4HI/*Tha*I fragment from pWH320[23] (class A), a 70-bp *Hinc*II/*Taq*I fragment from pWH106[24] (class B), a 94-bp *Ban*I/*Hae*III fragment from pBR322[25] (class C), and a 66-bp *Sau*3A/*Mnl*I fragment from pWH803[18] (class D). The protruding ends are filled in using DNA

[18] B. Unger, G. Klock, and W. Hillen, *Nucleic Acids Res.* **12,** 7693 (1984).

[19] V. Hershfield, H. W. Boyer, L. Chow, and D. R. Helinski, *J. Bacteriol.* **126,** 447 (1976).

[20] P. T. Chan, H. Ohmori, J.-I. Tomizawa, and J. Lebowitz, *J. Biol. Chem.* **260,** 8925 (1985).

[21] L. E. Maquat and W. S. Reznikoff, *J. Mol. Biol.* **125,** 467 (1978).

[22] H. Heumann, H. Lederer, W. Kammerer, P. Palm, W. Metzger, and G. Baer, *Biochim. Biophys. Acta* **909,** 126 (1987).

[23] G. Klock and W. Hillen, *J. Mol. Biol.* **189,** 633 (1986).

[24] W. Hillen, G. Klock, I. Kaffenberger, L. V. Wray, Jr., and W. S. Reznikoff, *J. Biol. Chem.* **257,** 6605 (1982).

[25] J. G. Sutcliff, *Cold Spring Harbor Symp. Quant. Biol.* **43,** 77 (1978).

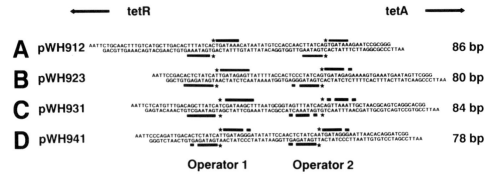

FIG. 1. Nucleotide sequences of the constructed restriction fragments containing the regulatory regions of the *tet* classes A to D. The classes and the respective plasmid designations are given on the left side and the length of the fragments (in bp) on the right. The operators are indicated by bars and stars at the center of their palindromic symmetry. The arrows at the top show the direction of *tet* gene transcription.

polymerase Klenow fragment[26] and an *Eco*RI linker (5′-GGAATTCC-3′) is ligated to the ends. The ligation mixtures are digested with *Eco*RI and the products eluted from a polyacrylamide gel and ligated in 5- through 25-fold molar excess with *Eco*RI-cleaved, dephosphorylated pWH802. The mixtures are transformed to *E. coli* RR1 and recombinants are screened for multiple insertions by their size. This results in plasmids pWH912, pWH923, pWH931, and pWH941. pWH912 carries a double insertion, pWH923 carries a triple insertion, and pWH931 as well as pWH941 contain single inserts of the class A to D *tet* fragments, respectively. Multiple insertions always have parallel orientations. The nucleotide sequences of the DNA fragments are shown in Fig. 1. We did not make further attempts to clone multiple insertions, because the yields turned out to be sufficient from these constructions (see below).

Large-Scale Plasmid Preparation

For plasmid isolation *E. coli* RR1 cells, transformed with the appropriate plasmid, are grown in 10- to 300-liter fermentors at 37°. M9 minimal medium (6 g/liter disodium hydrogen phosphate, 3 g/liter potassium dihydrogen phosphate, 0.5 g/liter sodium chloride, 1 g/liter ammonium chloride, adjusted to pH 7.4) supplemented with 0.4% glucose, 0.4% casein (acid hydrolyzed), 1 m*M* magnesium sulfate, 20 mg/liter thiamin dichloride is inoculated with 0.01 vol of a culture grown overnight in LB. During

[26] R. M. Wartell and W. S. Reznikoff, *Gene* **9,** 307 (1980).

cultivation the pH is observed and, if necessary, adjusted with ammonia. Cell growth is followed by measuring the optical density at 650 nm as a function of time. In the early log phase (OD_{650} 1.4) plasmid amplification is initiated by adding chloramphenicol to a final concentration of 150 mg/liter and cultivation is continued for 14 hr. Cells are harvested by centrifugation and the wet cell paste is stored at $-20°$. This procedure yields about 4 g of cells/liter of culture. Cells (200 g) are suspended in 1 liter of 25% sucrose, 50 mM Tris-HCl (pH 8.0), 1 mM EDTA. For lysis 16 ml of this cell suspension is mixed with 5 ml of 10 mg/ml lysozyme in 0.25 M EDTA, 50 mM Tris-HCl, adjusted to pH 8.0 in polycarbonate tubes for a Beckman (Palo Alto, CA) Ti55.2 rotor and kept on ice for 15 min. After adding 2 ml of an aqueous 2 : 1 (v/v) mixture of 10% Brij and 10% deoxycholate the tubes are gently inverted three times and kept on ice for 10 min. The suspension is centrifuged at 44,000 rpm in a Ti55.2 rotor at 4°. The supernatant is collected, RNase A from a stock solution (20 mg/ml in 1.5 M sodium acetate, boiled 10 min to inactivate DNases) is added to a final concentration of 50 μg/ml, and the mixture is kept on ice for 1 hr. After adjusting this solution to 1% sodium dodecyl sulfate (SDS) by adding the one-twentieth volume from a 20% stock solution the proteins are degraded with proteinase K at a final concentration of 50 μg/ml (added from a stock solution containing 20 mg/ml proteinase K in water). This mixture clears during incubation at 37° for 12 hr. The DNA is then precipitated with PEG by mixing the solution with 0.5 vol of 30% PEG 6000, 1.5 M sodium chloride, keeping it on ice for 30 min and spinning the precipitate down by centrifugation (JA10 rotor, 10,000 rpm, 4°). The DNA is dissolved in 500 ml of 10 mM Tris-HCl, pH 8.0, 0.1 mM EDTA/400 g wet cells, extracted three times with 1 vol of phenol, four times with 1 vol of ether, and again precipitated with PEG as described above. After dissolving this DNA in 500 ml of 10 mM Tris-HCl (pH 8.0), 0.1 mM EDTA the DNA concentration is determined by spectroscopy (1 OD_{260} = 50 μg DNA). The yields vary between 1.5 to 3.8 mg DNA/g of cells.

Large-Scale Restriction of Plasmid DNA

DNA is digested with *Eco*RI without any further purification in 100 mM Tris-HCl (pH 7.5), 50 mM sodium chloride, 10 mM magnesium chloride, 0.1 mg/ml bovine serum albumin at 37° for 24 hr. DNA concentrations vary between 1.5 and 2.5 mg/ml. The restriction enzyme stability and the required amount of enzyme for complete digestion are determined in analytical time courses. For that purpose 3 mg plasmid DNA is incubated as described above with 1000 to 5000 U *Eco*RI and tested for complete digestion after 2, 4, 6, 10, 24, and 48 hr. In preparative digestions half of

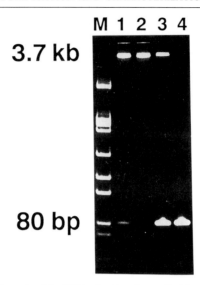

FIG. 2. Gel electrophoresis of the DNA after fractionated PEG precipitation. The analysis was done on a 5% polyacrylamide gel. Lane 1 contains 1 μg of restricted DNA, lane 2 contains 1 μg DNA of the PEG precipitate, lane 3 contains 1 μg DNA of supernatant, and lane 4 contains 1 μg DNA after final purification by HPLC. Lane M contains a molecular weight standard.

the necessary amount of enzyme is added to start the digestion and the other half after 5 hr. This yields a complete digestion after 24 hr of incubation at 37° with 400 to 1000 U *Eco*RI/mg plasmid DNA, depending on the number of cleavage sites in the respective plasmids. The reaction mixture is then extracted two times with 1 vol of phenol equilibrated with 10 mM Tris-HCl (pH 8.0), 0.1 mM EDTA, four times with 1 vol of ether, and residual ether is removed *in vacuo*.

Fractionated Precipitation of Restricted Plasmid DNA

In order to enrich the small restriction fragments with the *tet* regulatory regions the digested plasmid is selectively precipitated with PEG. This is done by mixing 1 vol 25% (w/v) PEG, 2.5 M sodium chloride with 4 vol of the restricted DNA, which are both preincubated at 37°. During an incubation time of 24 to 48 hr at 37° the precipitated vector DNA sediments to the bottom of the vessel. After this time the solution is decanted carefully without perturbing the vector DNA. The DNA in the supernatant contains nearly all of the small DNA fragment and is precipitated during a 12-hr incubation on ice, after adding 60 g/liter solid PEG. The precipitate

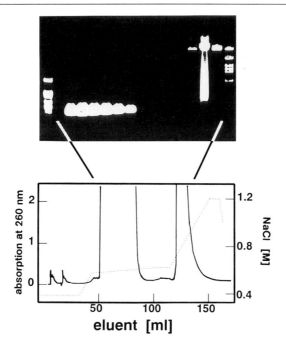

FIG. 3. Separation profile of the DNA on Nucleogen 500-10 IWC. The salt gradient is illustrated by the dashed line and the absorption of the eluent at 260 nm of a preparative run separating 26 mg of the DNA batch analyzed in lane 3 of Fig. 2 is indicated by the solid line. Five microliters of each fraction between the vertical lines was analyzed on the agarose gel shown above the elution profile.

is spun down at 10,000 rpm in a JA10 rotor for 30 min at 4°. A typical result of this procedure is shown in Fig. 2. The enrichment of the small restriction fragments by this procedure varies between 9-fold for the triple insertion plasmids and 80-fold for the single insertion plasmids.

HPLC Purification of Insert DNA

Final purification of the fragments from remaining vector DNA is achieved by HPLC on Nucleogen 500-10 IWC (125 × 10 mm, obtained from Diagen).[27,28] HPLC is performed on a low-pressure mixing liquid setup (LKB, Bromma, Sweden) equipped with a variable-wavelength UV detector set at 260 nm. All buffers are made from high-purity water (Millipore, Neu Isenburg, Germany) filtered through a 0.2-μm filter (Schleicher

[27] R. Hecker, M. Colpan, and D. Riesner, J. Chromatogr. **326**, 251 (1985).
[28] M. Colpan and D. Riesner, J. Chromatogr. **296**, 339 (1984).

TABLE I
CULTIVATION AND PURIFICATION STEPS

Parameter	tet resistance class			
	A	B	C	D
Transformed plasmid	pWH912	pWH923	pWH931	pWH941
Number and size of inserts (bp)	2 × 86	3 × 80	1 × 84	1 × 78
Total cultivation scale (liters)	200	100	200	200
Wet cell mass (g)	808	420	774	863
Total DNA after lysis (g)	2.2	1.0	3.0	1.3
DNA after PEG precipitation (mg)	110	135	88	44
Total yield of insert DNA (mg)	86	46	54	22
Relative yield of insert DNA (mg/liter of culture)	0.43	0.46	0.27	0.11

and Schuell, Dassel, Germany) and degassed *in vacuo*. Before use the column is washed with 6 vol of water, 4 vol of low-salt buffer (400 mM sodium chloride, 6 M urea, 30 mM potassium phosphate, pH 6.0), 2 vol of high-salt buffer (1.2 M sodium chloride, 5.5 M urea, 30 mM potassium phosphate, pH 6.0), and then equilibrated with low-salt buffer. After use the column is washed with 4 vol of water and stored after equilibration with methanol.

DNA is dissolved in low-salt buffer at concentrations varying between 2 and 4.5 mg/ml and insoluble particles are removed by centrifugation (JA20 rotor, 14,000 rpm, 30 min at 15°). The samples are injected using a loop injector (Rheodyne, Berkeley, CA) with a sample loop of 5-ml volume. In order to determine the optical conditions for the preparative run separating a maximum of 26 mg of DNA, analytical runs may be performed with 15 μg of the DNA samples. In these experiments the optimal slopes of the gradient for separating the insert from the vector DNA is determined. All chromatographic runs are done at a constant flow rate of 3 ml/min and linear gradients with varying slopes. A typical preparative separation profile is shown in Fig. 3. Including column equilibration and sample injection a complete chromatography is done in 80 min and can be very reproducibly repeated many times.

Collected fractions are analyzed by agarose gel electrophoresis and fractions containing only one species of DNA fragment are combined. The DNA is precipitated by adding 120 g/liter solid PEG and collected after a 12-hr incubation on ice by centrifugation (JA14 rotor, 10,000 rpm, 30 min at 4°).

Conclusion

A summary of the results obtained with these methods is presented in Table I. Between 22 and 86 mg of pure DNA fragments varying in size between 78 and 86 bp has been prepared to homogeneity. It may be noted that the triple insertion of the class B *tet* fragment does not lead to the anticipated higher yield compared to the double insert of the class A *tet* fragment, while the plasmids with the two single insertions show lower yields.

[6] Multidimensional Nuclear Magnetic Resonance Spectroscopy of DNA-Binding Proteins

By GRACE PÁRRAGA and RACHEL E. KLEVIT

Introduction

Multidimensional high-resolution nuclear magnetic resonance spectroscopy (2D NMR, 3D NMR, 4D NMR) is a relatively new method for studying macromolecular structure in aqueous and nonaqueous solutions. While the structural information obtained from an NMR analysis of a protein in solution is complementary to that from X-ray diffraction studies of protein crystals, kinetic, thermodynamic, dynamic, and conformational information can also be obtained via NMR spectroscopy.

X-Ray diffraction and high-resolution NMR spectroscopy have different sample and experimental requirements as well as different advantages and disadvantages. The first steps in an X-ray diffraction study involve obtaining crystals of the protein, preparing heavy atom derivatives, and solving the phase problem; these steps yield little structural information until the electron density map is calculated and analyzed. In contrast, a significant amount of information regarding, especially, the secondary structure of a protein is apparent from inspection of a two-dimensional nuclear Overhauser (NOESY) spectrum. For example, it was readily elucidated from the NOESY spectra of transcription factor IIIA (TFIIIA)-like zinc finger peptides ADR1a[1] and ARD1b[2] that the domains contained α-helical secondary structure. In contrast, analysis of the NOESY spectra

[1] G. Párraga, S. J. Horvath, A. Eisen, W. E. Taylor, L. Hood, E. T. Young, and R. E. Klevit, *Science* **241**, 1489 (1988).
[2] R. E. Klevit, J. R. Herriott, and S. J. Horvath, *Proteins: Struct. Funct. Genet.* **7**, 215 (1990).

for the human immunodeficiency virus (HIV) metal-binding domain,[3] thought to be structurally related to TFIIIA-like zinc fingers, indicated no α-helical secondary structure present. Thus, a qualitative analysis of the NOESY data revealed that the TFIIIA-like zinc finger domains and those in HIV (and by analogy similar sequences found in other viral proteins) were not structurally similar, even before a complete quantitative analysis and structure calculation was performed. Another advantage of NMR spectroscopy in structure–function studies is that once the NMR spectrum of one form of a protein (e.g., wild type) has been assigned, the assignment and structural analysis of altered forms (e.g., mutants, phosphorylated forms) can be quite straightforward.

A variety of DNA-binding proteins and domains have been studied by 2D NMR methods, and application of now well-established methods has led to ^{1}H assignments for *lac* repressor,[4,5] *cro* repressor,[6] zinc finger domains Xfin,[7] ADR1b[2] and human enhancer binding protein,[8] homeodomain *Antennapedia*,[9] Arc,[10,11] LexA,[12] glucocorticoid receptor[13] and estrogen receptor,[14] and incomplete assignments for the λ repressor,[15–18] *trp* repres-

[3] M. F. Summers, T. L. South, B. Kim, and D. R. Hare, *Biochemistry* **29,** 329 (1990).

[4] E. R. P. Zuiderweg, R. Kaptein, and K. Wuethrich, *Proc. Natl. Acad. Sci. U.S.A.* **80,** 5837 (1983).

[5] R. Kaptein, E. R. P. Zuiderweg, R. M. Scheek, R. Boelens, and W. F. van Gunsteren, *J. Mol. Biol.* **182,** 179 (1985).

[6] P. L. Weber, D. E. Wemmer, and B. R. Reid, *Biochemistry* **24,** 4553 (1985).

[7] M. S. Lee, G. P. Gippert, K. V. Soman, D. A. Case, and P. E. Wright, *Science* **245,** 645 (1989).

[8] J. G. Omichinski, G. M. Clore, E. Appella, K. Sakaguchi, and A. M. Gronenborn, *Biochemistry* **29,** 9324 (1990).

[9] G. Otting, Y.-Q. Qian, M. Mueller, M. Affolter, W. Gehring, and K. Wuethrich, *EMBO J.* **7,** 4305 (1988).

[10] J. N. Breg, R. Boelens, A. V. E. George, and R. Kaptein, *Biochemistry* **28,** 9803 (1989).

[11] M. G. Zagorski, J. U. Bowie, A. K. Vershon, R. T. Sauer, and D. J. Patel, *Biochemistry* **28,** 9813 (1989).

[12] R. M. J. N. Lamerichs, A. Padilla, R. Boelens, R. G. Kaptein, G. Ottleben, H. Ruterjans, J. Granger-Schnarr, P. Oertel, and M. Sharr, *Proc. Natl. Acad. Sci. U.S.A.* **86,** 6863 (1989).

[13] T. Hard, E. Kellenbach, R. Boelens, B. A. Maler, K. Dahlman, L. P. Freedman, J. Carlstedt-Duke, K. R. Yamamoto, J.-A. Gustafsson, and R. Kaptein, *Science* **249,** 157 (1990).

[14] J. W. R. Schwabe, D. Neuhaus, and D. Rhodes, *Nature (London)* **348,** 458 (1990).

[15] M. A. Weiss, R. T. Sauer, D. J. Patel, and M. Karplus, *Biochemistry* **23,** 5090 (1984).

[16] M. A. Weiss, A. Redfield, and R. H. Griffey, *Proc. Natl. Acad. Sci. U.S.A.* **83,** 1325 (1986).

[17] M. A. Weiss, M. Karplus, and R. T. Sauer, *Biochemistry* **26,** 890 (1987).

[18] M. A. Weiss, C. O. Pabo, M. Karplus, and R. T. Sauer, *Biochemistry* **25,** 897 (1987).

sor dimer,[19,20] 434 repressor,[21] and both the *lac* headpiece operator[22,23] and *Antennapedia*–DNA[24] complexes.

A number of excellent monographs[25,26] as well as Volumes 176 and 177 in this series describe the 2D NMR approach for protein structure in detail. Advances in higher dimension NMR experiments (3D NMR[27–29] and 4D NMR[30]) have been reviewed[31] and detailed elsewhere. This chapter will include a general description of the methodology and will outline the important considerations and limitations that are relevant for protein chemists and molecular biologists contemplating structural studies of DNA-binding proteins and protein–DNA complexes using multidimensional high-resolution NMR.

Some Background on Nuclear Magnetic Resonance

NMR spectroscopy is a powerful tool for the protein chemist and molecular biologist due to the significant amount of structural information offered even from simple one-dimensional nuclear magnetic resonance (1D NMR) experiments. A 1D ^{1}H NMR spectrum is a plot of intensity (or number of protons) versus resonance frequency or chemical shift (proton environment). In addition to the primary chemical shifts observed for protons in small organic molecules (dependent mostly on chemical bonding patterns), protein NMR spectra contain secondary chemical shifts that are a direct consequence of the *folded* three-dimensional structure of the

[19] C. H. Arrowsmith, L. Treat-Clemons, I. Szilagyi, R. Pachter, and O. Jardetsky, *Proc. 9th Colloque Ampere, Magn. Reson. Polymers* **7**, 10 (1989).

[20] C. H. Arrowsmith, J. Carey, L. Treat-Clemons, and O. Jardetsky, *Biochemistry* **28**, 3875 (1989).

[21] D. Neri, T. Szyperski, G. Otting, H. Senn, and K. Wuethrich, *Biochemistry* **28**, 7510 (1989).

[22] R. Boelens, R. M. Sheek, J. H. van Boom, and R. J. Kaptein, *Mol. Biol.* **193**, 213 (1987).

[23] R. J. J. N. Lamerichs, R. Boelens, G. A. van der Marel, J. H. van Boom, R. Kaptein, R. Buck, B. Fera, and H. Ruterjans, *Biochemistry* **28**, 2985 (1989).

[24] G. Otting, Y.-Q. Qian, M. Billeter, M. Mueller, M. Affolter, W. J. Gehring, and K. Wuethrich, *EMBO J.* **9**, 3085 (1990).

[25] K. Wuethrich, "NMR of Proteins and Nucleic Acids." Wiley, New York, 1986.

[26] K. Wuethrich, *Science* **243**, 45 (1989).

[27] H. Oschkinat, C. Griesinger, P. J. Kraulis, O. W. Sorenson, R. Ernst, A. M. Gronenborn, and G. M. Clore, *Nature (London)* **332**, 374 (1988).

[28] G. W. Vuister, R. Boelens, and R. Kaptein, *J. Magn. Reson.* **80**, 176 (1988).

[29] M. Ikura, L. E. Kay, and A. Bax, *Biochemistry* **29**, 4659 (1990).

[30] L. E. Kay, G. M. Clore, A. Bax, and A. M. Gronenborn, *Science* **249**, 411 (1990).

[31] A. M. Gronenborn and G. M. Clore, *Anal. Chem.* **62**, 2 (1990).

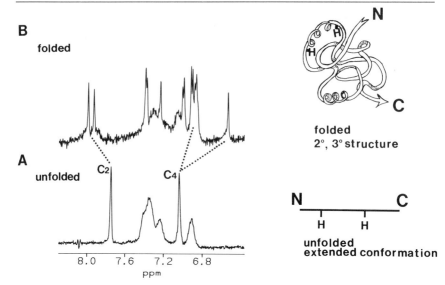

FIG. 1. Comparison of the aromatic regions of spectra of (A) unfolded protein and (B) folded protein. C-2 and C-4 imidazole proton resonances are labeled. N, N terminus; C, C terminus; H, histidine.

protein. A schematic representation of this is shown in Fig. 1. For proteins that are denatured or extended in conformation (Fig. 1A), amino acid side chains are exposed to roughly the same average solution environment so that specific protons of the same amino acid type (for example, the C-2 protons of two different His residues shown in Fig. 1A) resonate at the same chemical shift positions. When the protein assumes a stable folded structure (Fig. 1B), individual residues experience altered chemical environments, resulting in conformation-dependent chemical shift dispersion. As shown in Fig. 1B, C-2 and C-4 protons of the two different His residues experience different environments in the folded structure and now occupy new, unique chemical shift positions. Thus the NMR spectrum of a protein is a direct consequence of its folded secondary and tertiary structure, making NMR spectroscopy an extremely useful tool in structural analyses of protein conformation and conformational changes. The power of 1D NMR spectroscopy in revealing proton chemical shift changes and by inference protein conformational changes has been exploited in studies of another DNA-binding motif, the GCN4-DNA-binding domain.[32] On the basis of 1D NMR spectroscopy and circular dichroism (CD) spectroscopy,

[32] M. A. Weiss, T. Ellenberger, C. R. Wobbe, J. P. Lee, S. C. Harrison, and K. Struhl, Nature (London) 347, 575 (1990).

Weiss *et al.* have suggested that once the GCN4 basic region has bound to DNA it adopts a folded α-helical structure. Specific chemical shift changes of aromatic protons from unfolded or random coil positions (−DNA) to folded chemical shift positions (+DNA) in combination with increased α-helical content (CD spectroscopy) were offered as the basis for these suggestions.

The indication that NMR spectroscopy could be useful in solution structure determination[33] was made some time before the method became practicable. The main limitation arose from the necessity to assign each NMR resonance to its specific proton in the protein. However, even for a small protein, most of the peaks in the 1D NMR spectrum contain a contribution from more than one proton, thus making the proton resonance identification difficult if not impossible by 1D NMR. The incorporation of a second dimension in NMR spectra (2D NMR[34,35]) and recently a third (3D NMR[27–29]) and fourth dimension (4D NMR[30]) results in the distribution of resonances into two, three, or four dimensions. The extra dimensionality of these experiments serves two important functions. First, nuclear resonances can be distributed into additional frequency dimensions, substantially improving peak resolution. Second, the breakthrough development in 2D NMR was the ability to observe and identify the bonding and spatial relationships between protons. This greatly simplifies the process of resonance identification with respect both to amino acid residue type and to location in the sequence. Thus, it is both the improved resolution and the ability to observe the relationships between resonances that makes the identification of individual proton resonances possible using multidimensional NMR techniques. The structural information that had remained cryptic in protein 1D NMR spectra became accessible, opening the door to the determination of three-dimensional solution structures from NMR data.

The NMR literature is filled with an array of experiments with exotic names such as COSY, NOESY, ROESY, TOCSY/HOHAHA, and INADEQUATE. In this chapter, the spectroscopic details of these experiments will not be discussed. Rather, the type of *information* that is available from specific categories of experiments and the type of *protein sample* required for these experiments will be addressed.

NMR performed on biological molecules can be categorized according to (1) the type and number of nuclei that are observed or (2) the type of

[33] J. Jeener, Ampere International Summer School, Basko Polje, Yugoslavia, 1971, unpublished lecture.
[34] W. P. Aue, E. Bartholdi, and R. R. Ernst, *J. Chem. Phys.* **64,** 2229 (1976).
[35] J. Jeener, B. H. Meier, P. Bachmann, and R. R. Ernst, *J. Chem. Phys.* **71,** 4546 (1979).

information that is obtained. The first category consists of homonuclear and heteronuclear experiments. In homonuclear experiments, all the information obtained concerns a single kind of nucleus, most commonly ^{1}H for proteins. Since the ^{1}H nucleus is the most abundant naturally and is also the most sensitive in NMR spectroscopy, homonuclear ^{1}H experiments are by far the most widely used. In heteronuclear experiments, more than one type of nucleus is either observed or perturbed. For proteins, these other nuclei are usually ^{15}N and ^{13}C. However, because these isotopes are not abundant naturally, isotopic enrichment of the protein is usually required (discussed later in this section). Another way to categorize NMR experiments is according to the type of structural information that is obtained. Generally, NMR experiments can provide information about *bonding* relationships among nuclei (through bond) or pertaining to *spatial* relationships among nuclei (through space). The through bond-type experiments are also known as coherence transfer, *J* coupling, or scalar coupling experiments and include COSY (correlated spectroscopy), RELAY (relayed coherence transfer spectroscopy), TOCSY (total coherence spectroscopy), and HOHAHA (homonuclear Hartman–Hahn spectroscopy), HETCOR (heteronuclear correlated spectroscopy), HMBC (heteronuclear multiple bond correlation), and HMQC (heteronuclear multiple quantum correlation) experiments. In such experiments, the spectral information arises from scalar couplings between nuclei and therefore resonances of nuclei that are connected through a finite number of bonds are identified. In general (but not true in some three-dimensional triple-resonance heteronuclear experiments[29]), information is transferred among nuclei in each amino acid residue but not across the peptide bond. These experiments are therefore used to identify resonances in a spectrum that belong to a single "spin system" which, in the case of proteins, means a single amino acid residue. While this information is sufficient to categorize resonances according to their amino acid residue type (i.e., Gly versus Asp, etc.), it is not sufficient to identify the specific residue in the protein sequence from which that resonance is derived (i.e., Gly-52 versus Gly-68). For sequence-specific assignment of resonances, experiments are used in which the information is transferred through space, via dipolar coupling, regardless of the bonding relationships among the nuclei. The most commonly used experiment of this type is the NOESY (nuclear Overhauser effect spectroscopy), in which nuclei that are within about 5 Å of each other are identified. It is this distance or spatial information from which the details of the three-dimensional structure of the molecule will ultimately be obtained. The goal then is to identify as many peaks as possible in the NOESY spectrum with respect to the pairs of nuclei from which they originate. The greater the number of unambiguous peak assignments that can be made, the more precise and well defined the resulting experimentally determined three-dimensional structure will be.

An elegant new approach has been described for obtaining sequence-specific assignments using only through-bond information.[30] The method requires the production of a doubly labeled protein sample in which both ^{15}N and ^{13}C are uniformly incorporated. Advantage is taken of the strong scalar couplings that exist between the heteronuclei and identified spin systems (i.e., amino acid residues) are linked via the heteronuclear scalar coupling directly through the peptide bond. While this approach obviates the need to use through-space NOESY information to *assign* the spectrum, the NOESY information will still be the foundation of the structural determination itself.

General Sample Considerations

Prior to initiating NMR analysis of a DNA-binding protein or domain there are several important points to consider. Some of these considerations derive from limitations of NMR spectroscopy itself (inherently low sensitivity and spectral resolution). Both resolution and sensitivity are dependent on magnetic field strength; with the advent of higher field magnets, some of the current limitations may change. Whenever possible, the link between the sample considerations and the techniques themselves will be discussed so that the relevance of these limitations can be assessed with the development of newer NMR instrumentation.

Molecular Weight

A primary consideration in NMR studies of a macromolecule is molecular weight. If the goal is a detailed solution structure of the protein of interest, NMR analysis is presently limited to molecules of approximate molecular weight <30,000. The actual limit will depend on several properties of the systems of interest. For example, proteins that are mostly α helical tend to have poorly resolved spectra, so spectral overlap becomes a limiting factor at lower molecular weights than for proteins with β-sheet structure. Proteins or domains that form symmetric dimers have half as many resonances as monomeric proteins with the same effective molecular weight, but they will also have the linewidths that correspond to the dimer mass. NMR studies can be performed on larger proteins (M_r <100,000) but determination of a complete three-dimensional solution structure is not currently feasible. In such cases, incorporation of key isotopic labels may allow for selective observation of specific regions of the protein.

The molecular weight limitation stems from several aspects of the

NMR experiment itself. As the molecular weight of a molecule increases, its NMR peaks broaden. The net effect of line broadening is that resolution of individual resonances is decreased in both the one- and multidimensional NMR experiment. As well, broad lines often lead to decreased intensity in the cross-peaks of multidimensional experiments, leading to an effective decrease in obtainable signal to noise. In addition to the dependence of linewidth on molecular weight, with larger molecules there is greater spectral redundancy in the spectrum. For example, in most proteins, all methyl protons will resonate between about 2 and -1 ppm. As the number of methyl groups in the protein increases, the chance that more than one will resonate at any given frequency increases, leading to ambiguities even in the two-dimensional spectrum. This spectral overlap makes resonance identification difficult or impossible in 1D, 2D, and even 3D NMR spectra.

To illustrate the above points, the 1D ^{1}H NMR spectra of three different proteins are shown in Fig. 2. For the small 30-residue (molecular weight 3562) zinc finger peptide ADR1b, lines are fairly sharp and the resolution of one peak to the next throughout the spectrum is excellent. For comparison, the 1D NMR spectra of *Bacillus subtilis* His-containing protein (HPr; M_r 9100) and *Drosophila* calmodulin (M_r 16,700) are shown in Fig. 2B and 2C. The spectra are plotted so that the intensity of a single proton resonance is the same for all three spectra (see Fig. 2). The heights of other peaks become progressively greater and the widths of the peaks increase with molecular size, resulting in spectral overlap problems. The progressive broadening of peaks and loss of resolution are especially clear in the crowded methyl region (1–2 ppm) of the spectra.

An approach aimed at circumvention of the molecular weight limitation is to study well-defined structural and/or functional domains of a protein or protein–DNA complex by NMR. Because most domains of multidomain proteins consist of 200 residues or less, this is a practicable approach. There are a variety of choices for generating such domains and these will be addressed later in this chapter. Below are additional approaches that address the resolution problem that are useful both for small- to medium-sized proteins or domains and for the study of protein–DNA complexes.

Isotopic Labeling

Resonance overlap caused by either broad lines and/or resonance redundancies can be ameliorated and spectral resolution improved by isotopic labeling methods. There are a number of different approaches available, depending on the specific problem at hand. Isotopic labeling of proteins or domains is a viable course of action when working with bacte-

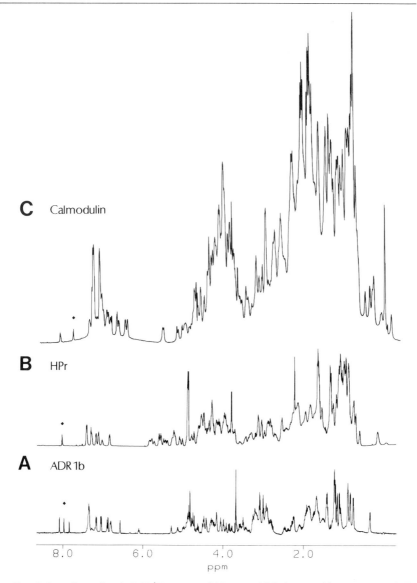

FIG. 2. One-dimensional NMR ^{1}H spectra of (A) yeast ADR1b (30 residues), (B) *B. subtilis* His-containing protein (HPr; 89 residues), and (C) *Drosophila* calmodulin (147 residues). ◆, Peak of single proton intensity.

rial expression systems as sources of protein. Some excellent reviews and articles describe in detail both selective and nonselective isotopic labeling techniques used in combination with multidimensional NMR to obtain spectral assignments for proteins up to 17 kDa[36-38] and the details will not be covered here.

There are two different isotopic labeling schemes that involve the incorporation of a nonobserved nucleus, deuterium, into the molecule. Because deuterons resonate at frequencies different from protons, they are "invisible" in an ^{1}H NMR spectrum. The first deuteration method involves uniform and random deuteration of the protein to about 30% enrichment. In this approach, all the resonances will still be observed in the spectrum, but due to the decrease in dipolar interactions between neighboring ^{1}H nuclei (now diluted by the incorporated deuterons), the linewidths are significantly reduced, thereby decreasing some of the overlap problems. A second approach is to label a selected group of amino acid residues in the protein with deuterated analogs. The result of this approach is spectral simplification: the spectrum of a deuterated analog is the original proton spectrum minus those peaks that now originate from deuterons. This technique was utilized in an attempt to determine sequential assignments for the *trp* repressor dimer of M_r 25K.[19] The scheme is actually better described as "selective protonation" because the bacteria are grown on a deuterated medium (an algal hydrolysate) in which the amino acids to be observed are supplemented in their natural, protonated forms. Thus, one can produce a protein in which only a limited subset of residues are observable. With five such selectively protonated analogs of the *trp* repressor, about 70% of the amide backbone assignments were determined. This approach has the added bonus that since most of the hydrogen atoms in the engineered molecules are deuterons, the resonances from the remaining protons have narrower linewidths, due to the decrease in dipolar coupling that results from the dilution of protons in the sample.

Another powerful approach to dealing with the resonance overlap problem involves isotopic labeling with ^{15}N and/or ^{13}C. As with the deuteration methods, both selective and random labelings are useful, depending on the particular problem to be addressed. Selective labeling is useful in the resonance assignment stage of NMR analysis. For example, by specifically incorporating ^{15}N-labeled Leu residues into a protein and using heteronuclear correlation experiments, the ^{1}H resonances derived from

[36] D. M. LeMaster, this series, Vol. 177, p. 23.

[37] D. C. Muchmore, L. P. McIntosh, C. B. Russell, D. E. Anderson, and F. W. Dahlquist, this series, Vol. 177, p. 44.

[38] D. W. Hibler, L. Harpold, M. Dell'Acqua, T. Pourmotabbed, J. A. Gerlt, J. A. Wide, and P. H. Bolton, this series, Vol. 177, p. 74.

Leu residues can be immediately identified. This approach may require the generation of several different selectively labeled samples and can therefore be rather expensive in both time and materials. However, a judicious choice of a subset of amino acid types can usually be made, for example, residues that are well dispersed throughout the sequence or potential DNA-binding residues. In a different kind of selective labeling approach, ^{13}C nuclei can be incorporated into specific positions of specific residues (controlled by the position of ^{13}C label in the glucose added to the growth medium). This scheme has been used to determine stereospecific assignments for the methyl groups in Val residues.[21] The specific incorporation of stable isotopes into side chains may also be a useful approach for observing specific intermolecular contacts, especially those observed in complexes formed between labeled DNA-binding proteins and their cognate DNA.

Uniform labeling with ^{15}N and/or ^{13}C allows one to perform a variety of 2D, 3D,[29,39] and 4D NMR[30] experiments that are capable of overcoming a significant amount of the overlap problem present in homonuclear 2D ^{1}H NMR spectra. These experiments allow for more unambiguous assignments of the information contained in the through-space (NOESY) spectra that are the basis for the three-dimensional structure determination.

Amount of Material Required

The question most often asked of an NMR spectroscopist engaged in the determination of protein structures concerns how much protein will be required for a complete NMR analysis. A *minimum* of one sample (natural isotopic abundance) with a concentration of at least 1–5 mM is required. In order to be expeditious with samples that are unstable or prone to aggregation and/or oxidation, it is preferable to have at least two such samples so that H_2O and D_2O experiments can be performed with a minimum of sample manipulation. As well, in our own experience we have noted that fresh protein samples tend to give higher quality spectra than older samples, so it is extremely useful to have a good supply of protein on hand in order to obtain the best spectra possible. It should be noted that NMR is a nondestructive technique, so samples can be reclaimed for other experiments. The required sample volume will vary from 350 to 700 μl, depending on the NMR instrument used. For a low-molecular-weight domain like the zinc finger from ADR1b (M_r 3562), a single 5 mM sample in 500 μl requires about 9 mg of material, whereas for a higher molecular weight protein like calmodulin (M_r 16,700) about 40 mg of purified protein

[39] E. R. P. Zuiderweg and S. W. Fesik, *Biochemistry* **28,** 2387 (1989).

is required for the same 5 mM sample. Of course, if heteronuclear experiments are to be performed, similar samples of isotopically labeled material will also be required.

Why is so much protein required for NMR spectroscopy when much less material is required for other types of spectroscopies? The inherent insensitivity of NMR spectroscopy derives from the small energy difference between the two states between which transitions are detected, resulting in a small net population difference between the two states. As the applied external magnetic field is increased (i.e., stronger magnets) the energy difference between states is increased and so is the population difference between states, resulting in a stronger signal and, hence, increased sensitivity. Such an increase has been realized with the availability of 600-MHz spectrometers. High-quality multidimensional NMR spectra of proteins at submillimolar concentrations have been obtained on these instruments.

Source of Proteins

Since NMR studies require large quantities of biological materials, the source and its cost and yields are important considerations. Several approachs are available for generating samples and their advantages and disadvantages are discussed below.

Protein Purification from Tissue Sources. Prior to advances in peptide synthesis and recombinant DNA technologies made in the last decade, the only sources for large amounts of protein for structural studies were tissues and organisms. In general, this source of biological material is less favored because of its inherent lack of flexibility toward important approaches such as isotopic labeling and mutagenesis. However, tissue sources remain an option when (1) the protein domain is too large to be synthesized chemically by solid-phase peptide synthesis, (2) a cDNA clone and/or expression system for the protein are not available, (3) overexpression of the protein in yeast or bacteria is lethal to the cells, (4) the bacterially expressed protein does not contain functionally and/or structurally critical posttranslational modifications found in the eukaryotically expressed protein.

Bacterial Expression of Cloned Genes. To date, most of the DNA-binding proteins and domains studied by NMR have been bacterial and phage repressor proteins (Lac,[4,5] Cro,[6] λ,[15-18] Trp,[19] Arc,[10,11] LexA[12]). Usually, these proteins have been purified from bacteria after transcriptional induction of the gene encoding the protein (overexpression using strong inducible promoters). Recombinant DNA technology makes these overexpression systems available for nonbacterial gene expression as well.

DNA-binding domains from the eukaryotic proteins *Antennapedia*,[40] the glucocorticoid receptor,[13] and the estrogen receptor[14] have been generated and studied by NMR.

There are many advantages to both the NMR spectroscopist and the molecular biologist in using an expression system as a source for protein. The most obvious advantage is the fact that with most expression systems there is a huge increase in protein synthesis. For DNA-binding proteins and transcription factors that are usually found in minute amounts in the cell, protein concentrations may be elevated to as high as 2–10% of the total cell protein. Some examples of DNA-binding proteins that have been purified in large amounts from bacterial expression systems are λ repressor,[41] *lac* repressor,[42] *cro* repressor,[43] *trp* repressor,[44,45] and *Antennapedia* homeodomain,[40] Arc,[46] and LexA.[47]

Another advantage to working with a bacterial expression system is that mutant and deletion forms of the protein can also be expressed and purified. As stated earlier, one of the greatest advantages of NMR in structural studies is the relative ease with which mutant forms of proteins can be studied once the spectrum of the wild-type protein has been assigned. Thus, an efficient bacterial expression system and multidimensional NMR are an extremely powerful combination of methods when applied to structure–function studies of wild-type and mutant molecules.

The use of bacterial expression systems also makes labeling of proteins with selective and nonselective isotopic labels quite straightforward. As described earlier, these approaches can increase the size of protein that can be studied by NMR. In general, then, the ability to label domains and proteins with different nuclei and the ease with which mutant proteins can be generated are probably the most advantageous or important reasons for using bacteria to express proteins of interest.

Chemically Synthesized Peptides and Domains. Recent advances in automated solid-phase peptide chemistry have opened a new era in peptide

[40] M. Mueller, M. Affolter, W. Leupin, G. Otting, K. Wuethrich, and W. J. Gehring, *EMBO J.* 4299 (1988).
[41] R. T. Sauer, K. Hehir, R. S. Stearman, M. A. Weiss, A. Jeitler-Nelsson, E. G. Suchanek, and C. O. Pabo, *Biochemistry* **25**, 5992 (1986).
[42] R. M. Sheek, E. R. P. Zuiderweg, K. J. M. Klappe, J. H. van Boom, R. Kaptein, H. Ruterjans, and K. Beyreuther, *Biochemistry* **22**, 228 (1983).
[43] P. L. Weber, D. E. Wemmer, and B. R. Reid, *Biochemistry* **24**, 4553 (1985).
[44] A. Joachimiak, R. L. Kelley, R. P. Gunsalus, P. Sigler, and C. Yanofsky, *Proc. Natl. Acad. Sci. U.S.A.* **80**, 668 (1983).
[45] J. L. Paluh and C. Yanofsky, *Nucleic Acids Res.* **14**, 7851 (1986).
[46] A. K. Vershon, J. Bowie, T. Karplus, and R. T. Sauer, *Proteins: Struct. Funct. Genet.* **1**, 302 (1986).
[47] M. Schnarr, J. Pouyet, M. Granger-Schnarr, and M. Daune, *Biochemistry* **24**, 2812 (1985).

research. Coincidental to the advances made in peptide synthetic techniques was the observation (via deletion and genetic analyses) that very short protein fragments or domains contained specific and discrete protein activities (phosphorylation sites, DNA-binding and activation domains). Thus, peptide synthesis has become an important source of substrates for structure–function studies of small domains and proteins.

In general, very short peptides (<30 residues) do not form stable secondary or tertiary structures in solution, with the exception of peptides that contain disulfide bridges. Thus, while it was generally acknowledged that some small domains like the helix–turn–helix domain of the *lac* repressor headpiece (60 residues) retained structure, it was thought that reconstitution of even smaller domains from synthetic peptides was not possible *in vitro*. There are now several examples of very small domains that possess stable structure in solution. One of the first examples of a short, nondisulfide-bridged peptide with stable structure was the 30-residue zinc finger domain. In their studies with a fragment from the transcription factor TFIIIA (TFIIIA-2), Frankel *et al.*[48] showed that the fragment could be folded *in vitro* on the addition of zinc. The zinc-bound peptide was also both proteolytically and thermally stable, supporting the notion that metal binding was essential for folding into a stable domain. More recently, an even smaller domain, the 17-residue viral zinc-binding domain, was also shown to possess a stable folded structure in the presence of zinc.[3] Thus, it would appear that in the absence of disulfide bonds, high-affinity metal binding is an efficient way to stabilize the structure of small polypeptides.

Molecules the size of either of the two aforementioned zinc-binding domains are quite amenable to either manual or automated chemical synthesis using solid-phase technique.[7,8,49,50] Currently, peptides of up to about 30 residues can be synthesized routinely in sufficient yield for structural studies. With special care, peptides of about 60 to 90 residues can be synthesized[51] but coupling must be *extremely* efficient to yield sufficient material for structural studies.

One advantage to peptide synthesis is the relative ease of purification. A one-step purification involving reversed-phase high-performance liquid chromatography (HPLC) is often sufficient for purifying synthetic pep-

[48] A. D. Frankel, J. Berg, and C. O. Pabo, *Proc. Natl. Acad. Sci. U.S.A.* **84**, 4841 (1987).
[49] D. Roise, S. J. Horvath, J. Tomich, H. Richards, and G. Schatz, *EMBO J.* **5**, 1327 (1986).
[50] M. F. Bruist, S. J. Horvath, L. Hood, T. A. Steitz, and M. I. Simon, *Science* **235**, 777 (1987).
[51] A. Wlodawer, M. Miller, M. Jaskolski, B. K. Sathyanarayana, E. Baldwin, I. T. Weaver, L. M. Slk, L. Clauson, J. Schneider, and S. B. H. Kent, *Science* **245**, 616 (1989).

tides. The caveat remains, however, that purification is straightforward only when the synthesis has proceeded cleanly and satisfactorily. HPLC purification becomes much more difficult when the HPLC profile of the synthesis product consists of many peaks rather than one single major product. This can occur with sequences containing duplicated Arg or Lys residues, commonly found in DNA-binding motifs. As well, Cys and His residues can be problematic. It is also advisable to avoid amino-terminal Met residues as these often oxidize irreversibly to Met sulfone.[52]

Many institutions have core peptide synthesis facilities with equipment for synthesis, sequencing, amino acid compositional analysis, and peptide purification. Alternatively, there are several companies that will produce purified peptides in sufficient quantities for structural analysis by NMR. The cost varies by several thousand dollars for about 30 mg of a purified 30-residue peptide.

Some disadvantages of chemical synthesis are as follow: (1) yield is finite and depends on how the synthesis proceeded, (2) mutant varieties of the peptide must be synthesized *de novo,* (3) isotopic labeling (^{2}H, ^{15}H, ^{13}C) during chemical synthesis can be an expensive enterprise,[36] (4) synthesis of peptides longer than 30 residues is difficult on most automated equipment and requires double and triple coupling of some residues, (5) reconstitution and solubility may be problematic (see Solubility and Reconstitution, below), (6) purification of mutant varieties or resynthesized batches may require different purification protocols. Given these limitations, though, we have found that chemically synthesized peptides can be useful substrates for NMR spectroscopic studies of small, stable DNA-binding domains.

Purity and Composition

The inherent insensitivity of NMR spectroscopy is almost an advantage when dealing with sample purity. Although it is necessary that samples be as pure as possible, as much as a 5% impurity (proteinaceous) in an NMR sample will not usually interfere with the 1D or 2D NMR spectrum. The most common source of impurities in samples is small organic molecules from HPLC columns, lyophilizers, and contaminated laboratory water. We have found it useful to perform all purification of NMR samples in three-times distilled and filtered H_2O [Millipore (Bedford, MA) Milli Q system]. It is also helpful to avoid lyophilizing voluminous amounts of different proteins and buffers simultaneously on the same lyophilizer.

---

[52] R. A. Houghten and C. H. Li, this series, Vol. 91, p. 549.

Another source of contamination is bacterial growth in samples causing proteolysis and degradation of the sample. The normal care taken with any protein sample during purification should be sufficient to avoid bacterial contamination and sample degradation. One must also be aware that paramagnetic ions are usually not compatible with NMR spectroscopy (Co^{2+}, Mn^{2+}, etc.) and their use should be avoided in the course of the sample purification and preparation. A low (0.1–0.5 mM) concentration of ethylenediaminetetraacetic acid (EDTA) can be added to the final NMR sample to chelate any contaminating paramagnetic ions to avoid the spectral broadening they might otherwise cause. Alternatively, distilled and purified water can be run over a Chelex columnn to rid the laboratory water of paramagnetics and other minerals and metals.

The final step in sample preparation is the removal of all protonated species (buffers, reducing reagents, and chelators) from the sample by exhaustive dialysis and/or gel filtration. This step is critical for optimal signal-to-noise and spectral quality. Even small amounts of protonated buffers will be observed in the NMR spectrum as sharp, intense lines and these contribute to the "dynamic range" problem in NMR. Dynamic range refers to the ability of the detection system of the instrument to measure and digitize signals from low concentration solutes (i.e., the protein) in the presence of other species present at higher concentration. In addition, an artifact known as t_1 streaking—strong vertical lines of noise in a multidimensional NMR spectrum—arises from the sharp signals of buffers. Deuterated buffers, acids, and bases are available and should be used whenever possible when preparing a sample either in D_2O or H_2O for an NMR experiment.

Accurate knowledge of the protein sequence, metal content, pH, and molecular weight are also important as sample heterogeneity or irreproducibility can complicate analysis of the resulting spectra. For samples that have been prepared via bacterial expression vectors, the expressed protein should be subjected to nondenaturing polyacrylamide gel electrophoresis (PAGE) and denaturing sodium dodecyl sulfate (SDS)–PAGE under reducing conditions. The protein sequence or at least an amino acid compositional analysis should be obtained to ensure that the composition and sequence is correct. For peptides that have been chemically synthesized, both the sequence and composition should be determined prior to structural studies. It is not advisable to attempt to sequence peptides directly using NMR, although in principle this can be done using the sequential assignment procedure.[53,54] Moreover, since deletion peptides or peptides

[53] K. Wuethrich, G. Wider, G. Wagner, and W. Braun, *J. Mol. Biol.* **155,** 3111 (1982).
[54] M. Billeter, W. Braun, and K. Wuethrich, *J. Mol. Biol.* **155,** 321 (1982).

that are incompletely deblocked (containing tosyl or other blocking groups on side chains) can sometimes be generated during synthesis and purification, the real molecular weight (M_r) of the peptide should be measured by time-of-flight, ion spray, or fast atom bombardment (Fab) mass spectrometry. Mass spectrometry should also clarify any sequence inconsistencies that have surfaced after sequence and amino acid compositional analysis.

Solubility and Reconstitution

Because highly concentrated samples are required for NMR analysis, solubility is a primary concern. Unfortunately, there is very little information in the literature on how such concentrated samples are dissolved or reconstituted into folded domains. Of course, each protein, domain, or peptide will have its own optimum conditions for solubility and stability, but there are some general considerations that must be kept in mind if high-quality spectra are to be obtained. In general, one would like to optimize solubility in order to gain sensitivity, but this must be achieved under conditions (pH, ionic strength) that do not promote aggregation. As well, as discussed below, there are certain pH limitations dictated by the NMR experiment itself (see pH and Stability, below).

Generally a lyophilized protein sample that has been treated by either gel filtration (Sephadex G-15 or G-25) or dialysis to rid the sample of protonated buffers and reagents is dissolved in 400 μl of D_2O and deuterated buffers. Initial characterization of the protein by NMR is undertaken under pH and buffer conditions shown optimal for the protein solubility and activity. When such information is not available, a program aimed at optimizing the desired properties must be undertaken. As an example, we offer our own experience with the ADR1 zinc finger peptides. For reconstitution of the peptides with $ZnCl_2$, we found (by trial and error) that Tris-HCl buffer at concentrations no greater than 50 mM with at least 10 mM acetate was ideal for peptide and $ZnCl_2$ solubility. We also found that heating this mixture to 70–80° for 5 min and slow cooling increased final solubility at room temperature. For domains or proteins that have been purified in the native state (for example, baculovirus expression), reconstitution of the folded state is not necessary. However, solubility/folding problems encountered on purification from bacterial inclusion bodies may be overcome by purification in guanidine hydrochloride or guanidine isothiocyanate.[55]

[55] K. Nagai, Y. Nakaseko, K. Nasmyth, and D. Rhodes, *Nature* (*London*) **332,** 284 (1988).

Aggregation State

It is important to know the aggregation state of the protein under the conditions to be used in the NMR experiments. This can be accomplished by analytical ultracentrifugation. Furthermore, evidence for oligomerization or aggregation can often be obtained from the 1D NMR spectrum of the protein. Proton linewidths are dependent on the effective molecular weight of the protein, as was shown in Fig. 2A–C. Some evidence for concentration-dependent aggregation in the submillimolar–millimolar range can be obtained by measuring linewidths as a function of protein concentration. For the zinc finger peptide ADR1a we monitored aromatic, $C\alpha$, and methyl proton peaks with respect to linewidths at 0.5 and 5 mM in the presence and absence of zinc in order to ensure that no concentration-dependent aggregation or dimerization was taking place. This was especially important for the initial studies on zinc finger domains, where the possibility existed for aggregation due to oxidation of Cys thiols or for metal coordination by more than one peptide.

pH and Stability

In order to determine structures by NMR, the backbone amide proton resonances must be observed in the NMR experiment. These amide protons are labile and often exchange rapidly with solvent, necessitating the collection of spectra in H_2O rather than the deuterated solvent. Even in H_2O, observation of labile protons is feasible only when the exchange with solvent is slow on the NMR time scale (exchange rate $< 1 \times 10^3$ min^{-1}). The rate of exchange is strongly pH dependent, as amide-solvent exchange is acid/base catalyzed.[56] Thus, it is desirable to work with samples that are at an acidic pH, where the solvent-exchange rate is slower. This explains the large number of protein NMR studies that are performed at pH values between 3 and 4.5.

Many proteins, however, are not structurally or functionally native at low pH values. As well, other properties of interest such as metal ion binding may also be pH dependent. The necessity to work at low pH is also problematic for proteins with p*I* values in the acidic range, as proteins tend to precipitate at pH values near their p*I*. Thus, NMR analysis must be a compromise between optimal conditions for the function and structure of the protein and optimal conditions for NMR analysis. In practice, we have found that nearly all amide proton resonances can be observed for *some* proteins at pH values as high as ~7.0. When "native" structural characteristics are of interest, it is essential to assess the dependence of

[56] K. Wuethrich and G. Wagner, *J. Mol. Biol.* **155**, 321 (1982).

the structure and/or function under the buffer and pH conditions where the NMR spectra are acquired. This can be done by investigating NMR chemical shift perturbations as pH or buffer conditions are altered to those conditions optimal for NMR experiments. Other than changes in the resonances from side chains that are ionizable over the pH range of interest, significant chemical shift changes should not be observed in the spectrum. This can be taken as evidence that the three-dimensional structure has not changed under the experimental conditions. To that end we monitored the 1D NMR spectra of the zinc finger peptides ADR1a and ADR1b as a function of pH in order to assess their stability and structure.[57] pH stability was an especially important concern for these molecules because Zn^{2+} binding by His imidazole nitrogens was proposed to be a driving force for domain folding and the normal pK_a of a His imidazole N is in the range pH 6–7. The pH dependence of the NMR spectra showed that a mixture of folded and unfolded species existed in the NMR tube below pH 5.5, indicating that the Zn^{2+}-stabilized tertiary structure begins to unfold below pH 5.5. Therefore, 2D NMR spectra for the determination of the zinc finger structure were acquired at pH 5.5 as a compromise between structural stability and amide exchange rates.

Protein–DNA Complexes

Ultimately, the application of multidimensional NMR techniques should allow determination of the three-dimensional structures of more DNA-binding proteins and protein–DNA complexes. To date, studies involving protein–DNA complexes have been reported in some detail for the *lac* repressor–operator system[22,23] and the *Antennapedia*–DNA complex.[24] Both studies have resulted in the determination of the approximate positioning or orientation of these domains onto the DNA. Both analyses involved homonuclear and heteronuclear 2D NMR experiments on 600-MHz spectrometers, using ^{15}N and ^{13}C isotopically labeled proteins that were overexpressed using bacterial expression vectors. The oligonucleotides utilized in both studies were chemically synthesized.

The addition of DNA resonances to already crowded protein NMR spectra further complicates 2D and even 3D and 4D NMR spectra. As described earlier, it is this spectral overlap that severely limits the interpretation/identification of observed NOEs and chemical shift perturbations. For studies of DNA–protein complexes, all the considerations described in this chapter will still be extant, with some additional ones as well. In order to restrict the molecular weight of the complex, the shortest func-

[57] G. Párraga, S. J. Horvath, E. T. Young, L. Hood, and R. E. Klevit, *Proc. Natl. Acad. Sci. U.S.A.* **87**, 137 (1990).

tional sequence of DNA is desirable. However, in order to avoid end-fraying effects on the DNA, an oligonucleotide slightly longer than the identified binding sequence is advisable. As well, binding affinities tend to decrease with decreasing DNA length for many DNA-binding proteins, presumably due to the loss of nonspecific protein–DNA backbone interactions. Since a stable complex is required, DNA length must be optimized versus complex stability. Synthetic oligonucleotides can be generated in high purity and yield on commercially available synthesizers. As well, protein–DNA complexes have been found to be of lower solubility,[58,59] than the individual components, presumably due to electrostatic neutralization. This problem will have to be addressed individually for each system.

It is anticipated that these considerations will be overcome for other protein–DNA complexes. The ever-growing list of multidimensional NMR techniques available to the spectroscopist, along with the many different biological molecules that can be generated by the molecular biologist and synthetic chemist, bode well for the elucidation of the structural details of these important molecular interactions.

[58] J. D. Baleja, Ph.D. Thesis, University of Alberta, Edmonton, Alberta, Canada.
[59] B. R. Reid, personal communication.

[7] Crystallization of Protein–DNA Complexes

By Andrzej Joachimiak and Paul B. Sigler

Introduction

X-ray crystallography is the most powerful tool with which to visualize the time- and space-averaged details of specific protein–nucleic acid interactions at the atomic level. Crystallographic analysis of protein–DNA complexes has been made possible by advances in the chemical synthesis of simulated DNA targets and the overexpression of the relevant proteins and protein domains from clones of their genes or cDNAs. A critical prerequisite of this method is the necessity of preparing well-ordered specific crystalline complexes that provide X-ray diffraction data of sufficient accuracy and resolution to assure a stereochemically useful model of the molecular interface. There is evidence that the difference between specific and nonspecific protein–DNA complexes can be very small,[1] therefore in order to understand the discriminatory features of a molecular

[1] B. Luisi, W.-X. Zu, Z. Otwinowski, L. Freedman, K. Yamamoto, and P. B. Sigler, *Nature* (*London*) in press (1991).

interface one needs accurate, high-resolution X-ray diffraction data. There is no hard and fast rule that relates crystal size, crystalline order, and accuracy of data acquisition to biochemical information; however, diffraction data to 2.5–2.8 Å of three-sigma accuracy will usually lead to a molecular model of the protein–nucleic acid complex that displays the chemistry of the interface in a useful way. Properly refined models, with data to 2.0-Å resolution or higher (especially if augmented with the repeated occurrence of the same chemical entity in the crystallographic asymmetric unit), should provide an unequivocal model that includes fixed solvent molecules and ions. In this chapter we identify considerations and recommend procedures that may enhance one's ability to produce X-ray quality crystals of specific protein–DNA complexes.

General Considerations

Physiological Relevance

Many crystalline specimens used in structural biology face the criticism that the nonphysiological conditions often required to produce X-ray quality crystals diminish if not negate the biological relevance of the study. This criticism is especially sharp where the stereochemistry of a specific protein–nucleic acid interface as seen in the crystal structure of the complex conflicts seriously with the expectations of most workers in the field. Although this concern is important to keep in mind in the design of the crystallization experiments, there are precious few cases in protein crystallography where a specific complex formed from sensibly chosen components has been rendered useless or misleading by the conditions of crystallization. This is especially true in the case of protein–DNA complexes where no challenge to the validity of the chemical relevance of a crystalline complex[2] has been substantiated by independent studies.

Special attention must be paid to designing and evaluating the DNA target used in cocrystallization. Obviously DNA targets must be chosen that include not only the recognition sequence (or an idealized version thereof) but the appropriate span of DNA with correctly identified boundaries. Careful attention to this issue paid handsome dividends to Phillips and co-workers in their study of the *met* repressor–operator complex that revealed tandem and overlapping binding involving an unexpected

[2] D. Staacke, B. Walter, B. Kisters-Woike, B. v. Wilcken-Bergmann, and B. Müller-Hill, *EMBO J.* **9**, 1963 (1990).

interface of the repressor.[3] To avoid mistakes the crystallographer and collaborators must experimentally establish the size and sequence of the optimal DNA segment with care (e.g., see Refs. 4 and 5). This includes the correct symmetry if any is suggested.[1] Reliance on others could be misleading, as evidenced by the flawed gel shift analysis of the *trp* repressor–operator complex by Staacke et al.[2] Last, consideration must be given to the possibility of crystallographic disorder resulting from the packing of helices in alternate directions. DNA duplexes that are nearly symmetrical and have chemically similar terminal base pairs can sometimes pack in two almost equivalent modes, leading to a confusing mixture of molecular images.

Plasticity and Crystalline Order

DNA-binding proteins and their DNA targets are usually flexible to facilitate their biological functions. This plasticity does not necessarily dim the prospects for producing well-ordered crystals, because a specific complex may often be more rigid than either of its components. Whereas the "flexible reading heads" of the uncomplexed *trp* repressor have a different orientation in various independent crystallographic representations, they are well fixed in a single specific orientation in all four independent representations of the *trp* repressor–operator complex.[6] DNA targets, especially the rather long ones, are structurally better ordered in cocrystals of regulatory complexes than in crystals of their uncomplexed counterparts.[6,7] Indeed, well-ordered crystals of duplexes longer than 12 bp are rare, whereas complexes of 14 up to 32 bp can be clearly visualized in cocrystalline complexes. The protein–DNA interface is more akin to the interior of the protein than to the solvent-exposed surface of the individual macromolecules.[6] Therefore, in addition to the fact that the macromolecular backbones are more restrained in the complex, the flexibility of the surface side chains of the protein is also significantly reduced when it forms an interface with its DNA partner.[6]

[3] S. E. V. Phillips, *Curr. Opinion Struct. Biol.* **1,** 89 (1991).
[4] B. K. Hurlburt and C. Yanofsky, *J. Biol. Chem.* **265,** 7853 (1990).
[5] T. Haran, A. Joachimiak, and P. B. Sigler, *Nature (London)* Submitted (1991).
[6] Z. Otwinowski, R. W. Schewitz, R.-G. Zhang, C. L. Lawson, A. Joachimiak, R. Marmorstein, B. F. Luisi, and P. B. Sigler, *Nature (London)* **335,** 321 (1988).
[7] R. Marmorstein, Ph.D. thesis, University of Chicago, Illinois, 1989.

Multiple Equilibria

Protein–DNA cocrystallizations are complex multicomponent systems whose behavior is described by more than just the dissociation constant of the protein–DNA complex. For example, there is likelihood of forming "hairpins" from single strands of a palindromic self-complementary duplex. Protein subunits that are in a monomer/dimer equilibrium may dimerize only when they bind to the DNA target, giving rise to the possibility that only one subunit of the protein will be bound to a symmetrical duplex if the latter is in excess. Recognition of such pitfalls is intrinsic to the strategy of growing crystals. Fortunately, because of the nearly millimolar concentration of macromolecules used in cocrystallization, a judicious choice of relative concentrations can help to drive the equilibrium toward homogeneity. There, concentrations, that are typically thousands times greater than the dissociation constant of the complex, can help overpower the potentially disruptive effects of precipitating agents (such as high salt and alcohols) used in cocrystallization.

Special Considerations

General Comments

The basic techniques used for the cocrystallization of DNA-binding proteins with their DNA targets are essentially the same as those used for any macromolecule or macromolecular complex. Thoroughly described vapor and liquid-diffusion techniques[8] that require only 1–20 µl of the macromolecular solution are used to conserve valuable materials while screening a wide panorama of crystallization conditions. The usual concern for purity and functional homogeneity in macromolecular crystallizations applies equally to the crystallization of protein and DNA in specific complexes. Special considerations that apply to protein–DNA cocrystals are discussed below.

Stoichiometry

The desired protein : DNA stoichiometry should be established in advance by solution experiments. Empirically it has been found that an excess of DNA is often, but not always, desirable. Most bacterial repressor–operator complexes, as well as the CAP-CAP site, have been crystallized best in systems where the concentration of DNA duplex was in excess

[8] See, for example chapters by A. McPherson [5] and G. N. Phillips [8] in "Methods in Enzymology," Vol. 114 (1985).

of the dimeric protein (1.2 : 1–1.6 : 1).[9–12] Several reasons are offered for this observation, all of which are based on the idea that the concentration of desired duplexes is actually less than expected. One cause is defective oligonucleotides, such as those damaged by acid depurination. Second is the fact that short DNA helices are more readily destabilized, especially symmetrical ones composed of self-complementary oligonucleotides that can form competing hydrogen-bonded hairpins. A third reason is the failure to supply equivalent amounts of each strand when forming asymmetric duplexes. This error is likely to be encountered unless care is taken to establish the extinction coefficient of each strand. Extinction coefficients can be easily calculated from the known absorbance of each base at a particular wavelength and the base sequence of the strand. The absorbance of an oligonucleotide solution must either be measured under conditions where there is no hypochromic effect (high temperature, complete hydrolysis) or corrected for the hypochromic effect. In the case of the cocrystallization of the glucocorticoid receptor DNA-binding domain with a modified version of its symmetrical DNA target, it was necessary to provide the protein and DNA in the exact stoichiometry of two domains per target since each subunit bound independently to each half-site and without positive cooperativity (see Ref. 1). An excess of DNA would lead to the formation of complexes in which only one protein subunit would be bound to one of the two half-sites of the symmetrical target.

The Optimal Protein

Segmentation. It is generally advisable to work with as compact a protein as possible consistent, of course, with the preservation of the function under study. This reflects an intuitive view that restricting the degrees of conformational freedom predisposes to well-ordered lattices. The wisdom of this approach is underscored by the success of those studying antibody structure and antibody–antigen interactions by crystallography. The Fab domain—a compact globular structure containing the antigen-binding site—has produced many well-ordered crystal structures whereas the entire immunoglobulin, composed of flexibly linked domains, does not crystallize readily in well-ordered lattices. To date, the stereochemical details of antigen–antibody interactions have been derived exclusively from complexes formed by the compact Fab fragment with its

[9] R. S. Jordan, T. V. Whitecomb, J. M. Berg, and C. O. Pabo, *Science* **230,** 1383 (1985).
[10] A. Joachimiak, R. Q. Marmorstein, R. W. Schevitz, W. Mandecki, J. L. Fox, and P. B. Sigler, *J. Biol. Chem.* **262,** 4917 (1987).
[11] H. C. Pace, P. Lu, and M. Lewis, *Proc. Natl. Acad. Sci. U.S.A.* **87,** 1870 (1990).
[12] S. C. Schultz, G. C. Shields, and T. A. Steitz, *J. Mol. Biol.* **213,** 159 (1991).

antigen and not from intact immunoglobulins. The utility of segmenting a multidomain protein either proteolytically or genetically as an aid to crystallization has been so well established that not to do so would be delinquent. One of the first of the bacterial DNA-binding regulatory proteins to be studied crystallographically was the N-terminal DNA-binding domain of the much larger and more elaborate λ repressor protein. It is also this domain that has given well-ordered crystals of the λ repressor–operator complex.[9] The complex of a template DNA and the Klenow fragment of *Escherichia coli* DNA polymerase I is another example of the utility of employing a proteolytically trimmed fragment that preserves important functions in crystallographic studies.[13]

Most eukaryotic DNA-binding genetic regulatory proteins are large polypeptide chains composed of several flexibly linked domains, each responsible for distinct functions. It can be a formidable task to cocrystallize such a multidomain, multifunctional, flexible protein with its cognate DNA sequence. Fortunately in most cases the DNA-binding domain constitues an independently stable substructure that is solely responsible for the specific affinity of the intact protein for a DNA-binding site. Such domains can be separated from the rest of the protein either proteolytically or genetically. Typical examples are the DNA complexes formed by homeodomains of transcriptional regulators of development,[15,16] the zinc finger-containing DNA-binding domains of the steroid receptors[1] and Zif 268.[17] Overexpressing bacterial clones that produce such fragments avoid the microheterogeneity of proteolytic digestion or posttranslational modifications and can be used to prepare pure protein in very large quantities.[18]

Thermophiles. Often a homolog of the protein of interest can be obtained from thermophilic organisms and may be more suitable for cocrystallization. Although there is neither systematic evidence nor rigorous logic behind this prejudice, there is a general feeling that such proteins (and presumably their DNA complexes) form better crystals. *Bacillus stearothermophilus*[19] and the strict thermophiles, *Thermus thermophilus*

[13] P. S. Freemont, J. M. Friedman, L. S. Beese, M. R. Sanderson, and T. A. Steitz, *Proc. Natl. Acad. Sci. U.S.A.* **85**, 8924 (1988).

[14] J. Anderson, M. Ptashne, and S. C. Harrison, *Proc. Natl. Acad. Sci. U.S.A.* **81**, 1307 (1984).

[15] C. Wolberger, C. O. Pabo, A. K. Vershon, and A. D. Johnson, *J. Mol. Biol.* **217**, 11 (1991).

[16] B. Liu, C. Kissinger, C. O. Pabo, E. Martin-Blanco, and T. B. Kornberg, *Biochem. Biophys. Res. Commun.* **171**, 257 (1990).

[17] N. Pavelatich and C. Pabo, personal communication (1991).

[18] F. W. Studier and B. A. Moffatt, *J. Mol. Biol.* **189**, 113 (1986).

[19] A. Yonath, F. Frolow, M. Shoham, J. Mussig, I. Makowski, C. Glotz, W. Jahn, S. Weinstein, and H. G. Wittmann, *J. Cryst. Growth* **90**, 231 (1988).

and *Thermus aquaticus,* have become common sources of bacterial proteins and can be expressed from overproducing *E. coli* strains.

Mutational Changes Involving Sulfur Atoms. Protein produced in bacteria (or, in principle, in yeast) offers the opportunity to exploit mutations that can aid crystallization and/or the structure determination. For example, complications caused by oxidation of cysteines can be circumvented by replacing or deleting the sensitive residues, provided the change does not affect function. Conversely, mutational changes of single Ser residues to Cys can be very helpful in obtaining heavy atom derivatives.[20] Replacement of Cys or Met with their selenium counterparts using Cys/Met auxotrophs is a gentle and elegant way to introduce a special scattering atom into a protein that, in conjunction with tunable synchrotron radiation, can provide phase information by the method of multiwavelength anomalous diffraction (MAD).[21] Such phases are, in principle, sufficient to determine the structure of a protein–DNA complex using only the selenium-substituted protein.

The Optimal DNA Fragment

The Sequence and Length of the DNA Target. The DNA fragment used for cocrystallization is selected on the basis of a careful analysis of genetic and biochemical data. The goal is to identify the DNA sequence elements that determine the specificity of protein binding and the segment of the DNA that is in contact with the protein. Whereas genetic experiments are valuable in determining the sequence determinants of specificity, biochemical and physicochemical studies are required to define the boundaries of the contact site. Methylation protection and footprinting, especially those conducted with small, "groove-specific" probes such as hydroxyl radicals, are helpful, but ethylation interference and primer extension analysis of gel retardation are clearly the most powerful methods to define the segment of DNA in contact with the protein. Excellent accounts of these latter techniques are given by Gartenberg and Crothers.[22]

The span of the DNA target is sometimes mistakenly equated with the span of the specificity determinants. This reflects a misunderstanding of how DNA binds to proteins. Most of the stabilizing contacts made at the specific interface between a protein and a nucleic acid involve the phosphate backbone. These phosphate interactions often involve nucleotides that extend beyond the bases that define the identity of the target. If one

---

[20] K. Nagai, C. Oubridge, T. H. Jenssen, J. Li, and P. R. Evans, *Nature (London)* **348,** 515 (1990).
[21] W. A. Hendrickson, J. R. Horton, and D. M. LeMaster, *EMBO J.* **9,** 1665 (1990).
[22] M. R. Gartenberg and D. M. Crothers, *Nature (London)* **333,** 824 (1988).

is to fully understand the chemistry of the specific complex one must cocrystallize a DNA fragment that spans the entire target, not just the segment bearing the identity elements.

The length of the DNA fragment in relation to the length of the contact site has significant effect on the nature of crystal. Harrison and co-workers initially cocrystallized relatively short 14-bp DNA fragments with 434 repressor[14] and 434 cro protein.[23] They found that the protein makes appropriate contacts with its operator, but in doing so it interacts with a neighboring duplex, causing a potential distortion at the junction between segments of DNA and complicating the interpretation of the protein–DNA interface.[23] In case of cocrystallization of the *engrailed* homeodomain with its DNA target, the DNA duplex was longer than the dimensions of the protein and therefore two proteins were seen to be bound to the DNA duplex. One protein molecule bound to the predicted site and the second bridged two adjacent DNA segments.[24]

Often DNA fragments are designed to produce a truly symmetrical target duplex. For such cases the "top" strand and "bottom" strand of the duplex are the same, therefore only one sequence defines the duplex. In cases where pseudopalindromic symmetry exists, it is not possible to use just one sequence. In such systems two different strands of the DNA target duplex must be designed to avoid the self-complementarity of individual strands. This problem is invariably encountered when the target has an odd number of base pairs and would be symmetrical except for a pseudo dyad that "skewers" a central base pair.

For experimental simplicity the sequence of the DNA fragment is often designed to contain idealized recognition elements, special flanking sequences, and symmetrical spacers that would be expected to aid cocrystallization. Experience indicates the need to check the binding behavior of such synthetic DNA elements before proceeding with cocrystallizations.

Optimizing the Crystal Packing. The design of DNA fragments for cocrystallization must consider the following issues. In all protein–DNA cocrystals except one (λ cro repressor–operator complex[25]) the DNA segments pack "end to end" to form a more or less distorted, extended helix. In order that the sequence repeats in register with the DNA helical repeat of 10.5 bp/turn, Anderson *et al.* suggested the 7n rule: that is, a sequence repeat every 7, 14, or 21 bp.[14] End-to-end packing would thereby produce an exact repeat every two (21 bp) or four (42 bp) helical turns.

[23] C. Wolberger, Y. Dong, M. Ptashne, and S. C. Harrison, *Nature (London)* **335**, 789 (1988).
[24] C. R. Kissinger, B. Liu, E. Martin-Blanco, T. B. Kornberg, and C. O. Pabo, *Cell* **63**, 579 (1990).
[25] R. G. Brennan, S. L. Roderick, Y. Takeda, and B. W. Matthews, *Proc. Natl. Acad. Sci. U.S.A.* **87**, 8165 (1990).

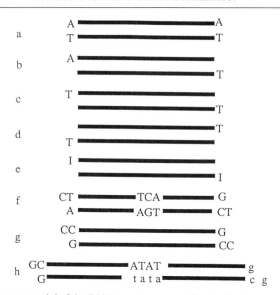

FIG. 1. Sequence termini of the DNA targets cocrystallized with the proteins listed in the Table I. Bars represent the span of a specificity determinant and the flanking regions. Top strand reads from 5' end on the left to 3' end on the right and the bottom strand reads in the opposite direction. Large space in h indicates a break in the strand.

For example, a 3_1 screw of a 14-bp sequence repeat will occur in exactly four helical repeats. This will produce a unit cell length along the 3_1 screw axis of 42 bp or about 143 Å. Although this is an attractive way to use an idealized DNA helix as a basis for crystallographic screw symmetry, this packing scheme has been observed only once. Clearly deviations from ideal B-DNA helices and end-to-end packing that is out of helical register subvert the application of this rule.

    Figure 1 shows the DNA termini that have been used successfully in cocrystals and thus serve as a guide in designing duplexes for cocrystallization: (1) blunt-ended double-helical units, either forming an extended helix register (7n) or without any end-to-end stacking, (2) complementary overhanging nucleosides such as a 5' A on the top strand and a corresponding 5' T on the bottom strand, (3) single overhanging nucleosides that are the same on the top and bottom strands, such as 5' T, 3' T, 5' C 5' I, etc.; although these obviously cannot form Watson–Crick base pairs, these identical nucleotides can base pair[6] or can stack to form a hooklike structure.[26] Identical overhanging bases provide the potential for symmetrical interactions such as those that relate the operator segments in the *trp*

[26] Y. C. Kim and J. M. Rosenberg, personal communication (1991).

repressor–operator cocrystals[6]; (4) overhanging bases can interact with a base pair at the end of the neighboring helix to form a "base triple"(Fig. 2). In this way adjacent helices can be connected end to end by two steps of a triple helix. An example is shown in Fig. 3, which also shows how different base triplets were used to form nearly isomorphous packing interactions.[1] A particularly successful example of DNA design for crystallization was described by Luisi et al.[1] These authors observed a triple helix mode of end-to-end packing in a complex between the DNA-binding domain of the glucocorticoid receptor and an artificially symmetrical DNA target. In order to grow isomorphous crystals containing the more natural, asymmetric DNA target, Luisi et al.[1] tailored the termini to form two different type of triple helices that were isostructural but ensured a unique polarity to the packing and hence to the orientation of the asymmetric complex in the lattice (Figs. 1, 2, and 3).

It seems quite clear that there is no general trend that specifies the optimal length and terminal structure of the oligonucleotides to be used in the cocrystallization of proteins and nucleic acids. In most cases the optimal DNA fragment must be determined experimentally by systematic variation of the internal sequence, length, and termini of the DNA duplex, guided by genetic and biochemical evidence.[9,27]

In order to scan a series of DNA targets more effectively (applying the guidelines described above) some investigators break the DNA segment into "modules" that are linked within the complex by "sticky ends." This combinatorial approach is a particularly effective strategy when dealing with long DNA targets. Fig. 1h shows the scheme by which Schultz et al.[12] successfully combined modules to produce X-ray quality crystals of the CAP-CAP site complex.

Halogenated Bases. The DNA component of a cocrystallized complex offers a simple and direct way to introduce atomic substitutions for solving the crystallographic phase problem. With increasing frequency 5-halogenated (brominated or iodinated) pyrimidines have been introduced into the DNA target. Typically 5-bromo-deoxyuridine and 5-iodo-deoxyuridine are introduced as an isomorphous substitutions for thymidine and 5-bromo-deoxycytosine and 5-iodo-deoxycytosine as nearly isomorphous substitutions for deoxycytosine. The bromine atom is a rather "light" heavy atom for phase determination by multiple isomorphous replacement (MIR) but it is an excellent anomalous scattering atom for the application of MAD phasing. Conversely the iodine atom, which possesses 53 electrons (44 more than the 5-methyl of T) is a suitable heavy atom for MIR phasing. Although the iodine anomalous signal is relatively strong for

[27] A. Joachimiak, J. Cryst. Growth 90, 201 (1988).

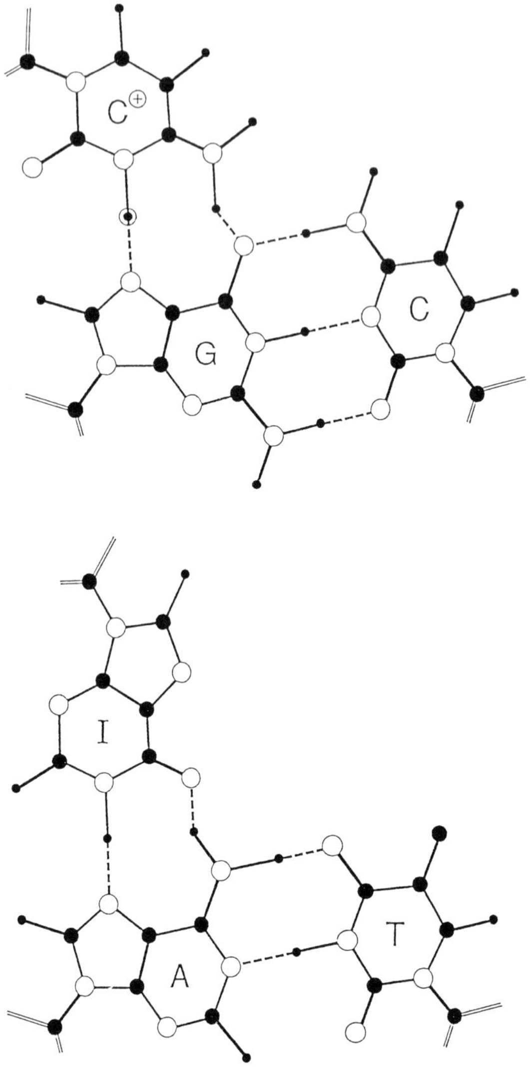

FIG. 2. Hoogsten-like interactions between a GC base pair and a protonated C base, and an inosine (I) interaction with an AT base pair, as observed in the cocrystals of the DNA-binding domain of the glucocorticoid receptor in a complex with its target DNA.[1]

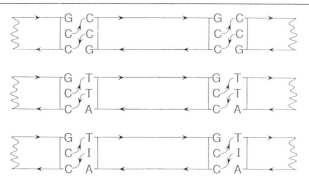

FIG. 3. *Top:* A schematic diagram of the end-to-end packing of the DNA helices observed in the crystal structure of the DNA-binding domain of the glucocorticoid receptor with DNA target showing symmetric C–G : C triple helix formation. *Middle* and *Bottom:* The proposed packing of the DNA duplexes using an asymmetric DNA target. Base triplets are formed by the interactions I–A : T T–A : T and C–G : C.

CuKα radiation (1.54 Å) from laboratory sources, it is not a good atom for application of MAD phasing with synchrotron radiation.

Preparation of DNA for Cocrystallization

Chromatographic Purification. DNA fragments of various length are usually synthesized with a commercially available DNA synthesizer on a 1 to 10 μmol scale. The polynucleotide is released from the column and the alkaline-labile base-protecting groups are removed in one step by treatment with a saturated ammonia solution at 55° for 8 hr. The oligonucle-otides (including halogenated bases) seem to survive this treatment intact. It is best to remove the by-products of ammonia treatment from the lyophilized ammonium salt with ethanol precipitation of the oligonucleo-tide (2.5 vol EtOH : 1.0 vol of a concentrated solution of oligonucleotides) from 0.3 M sodium acetate (pH 5.2).

DNA fragments can be purified in several alternative ways. The method we have found to be the most reliable and general, as well as least damag-ing, is preparative reversed-phase high-performance liquid chromatogra-phy (RP-HPLC) (C-4 or C-8 Vydac) at 50°. The procedure involves two chromatographic steps; the first is performed on the 5'-dimethoxytritylated (DMT) oligonucleotide. The oligonucleotide is then detritylated and sub-jected a second time to RP-HPLC on the same column matrix. The method depends on the retardation of the hydrophobic DMT group and hydropho-bic ion pairing in the presence of triethylammonium as counterion (0.1 M, pH 6.2). Both the triethylammonium acetate and acetonitrile used as eluant are volatile; however, the triethylammonium acetate used as the counter-ion/buffer can be damaging on lyophilization because a residue of less volatile acetic acid can form that will depurinate the oligonucleotide. This problem is best avoided by first exchanging the salt with triethylammonium bicarbonate. The elevated temperature serves to destabilize unwanted secondary structure. An alternative to the second column step that is sure to eliminate secondary structure is ion-exchange chromatography in 10 mM sodium hydroxide (pH 11.8) on a quaternary amine column [e.g., Pharmacia (Piscataway, NJ) Mono Q fast protein liquid chromatography (FPLC)]. Unfortunately, DNA fragments longer than 20 bp do not resolve well by this ion-exchange step and halogenated bases are at some risk.

Between the two RP-HPLC purification steps the dimethoxytrityl group, used to protect the 5′ terminus, is removed by acid treatment. This procedure, if carried too far, can depurinate DNA; therefore, this reaction is potentially harmful and should be strictly controlled. Greater than 90% detritylation is achieved by treatment with a 4:1 (v/v) mixture of glacial acetic acid and water at room temperature for 20 min. The reaction is stopped by neutralization on dry ice with the addition of equimolar amounts of cold triethylamine. The sample is dialyzed against 25 mM triethylammonium bicarbonate, then against water, and lyophilized. It is advisable to treat the detritylated oligonucleotide with saturated ammonia (55°, 12–18 hr). The ammonia treatment will split the chain at the sites of fortuitous depurination, thus facilitating their removal on the second chromatographic step. Properly done, this ammonia treatment should also remove all residual base-protecting groups. It is important not to overreact with ammonia when bromine- and particularly iodine-substituted oligo-nucleodides are involved. The halogenated oligonucleotides are also quite sensitive to photochemical reactions, therefore exposure to UV light (shorter than 350 nm) should be avoided.

Annealing Duplexes. Self-complementary or nearly self-complemen-tary oligonucleotides can base pair to form either palindromic duplexes or hairpins. The formation of duplexes is a second order reaction favored at high concentrations of oligonucleotide. The formation of hairpins is an intramolecular zero-order reaction that occurs equally well irrespective of concentration. One can drive the base pairing toward the duplex and minimize hairpinning of single strands by annealing at as high an oligonu-cleotide concentration as is practicable.

Self-complementary oligonucleotides are annealed as follows. The ly-ophilized sodium salt of DNA is dissolved in neutral 10 mM cacodylate and 50 mM sodium chloride at a concentration of 2–4 mM in duplex and heated for 5 min to a temperature much above the melting point. For the 18-bp self-complementary strands of the *trp* operator self-complementary nonadodecamer this temperature is 85°. The sample is then allowed to cool slowly to 4° in a water bath in the cold room at a rate of 1° every 4–6 min and is stored at 4°.

Surveying Crystallization Conditions

General Considerations. Screening conditions for the crystallization of protein–DNA complexes are carried out in much the same way as they are for other macromolecules and macromolecular assemblies. In addition to stoichiometry, there are special problems associated with the cocrystallization of specific protein–DNA complexes. Many DNA-binding proteins require high salt concentration (100–300 mM NaCl) to prevent their precipitation while in storage. This can necessitate the introduction of unwanted amounts of salt into the crystallization solution. For this purpose it is wise to use stock solutions of very high protein concentrations so as to minimize the volumes that are introduced into the drops. In some instances, high salt increases the solubility of the protein–DNA complex. One may take advantage of this property to grow cocrystals. One can lower the ionic strength in a vapor diffusion crystallization drop by using volatile salts such as ammonium acetate or ammonium bicarbonate.[1] Liquid diffusion offers the advantage that nonvolatile components can be readily removed from—or added to—the macromolecular solution and the system can be better maintained in a reducing environment. Unfortunately, liquid diffusion techniques are more cumbersome than vapor diffusion systems.

Crystallization conditions can be surveyed by vapor diffusion using droplets hanging from plastic coverslips over reservoirs sealed in Linbro boxes (Flow Labs, McLean, Va). This system allows one to use small droplets and thereby save expensive material. Initial concentrations of the complex in the droplet are generally chosen to be between 0.1 and 0.5 mM. Buffer, monovalent and divalent salts, and precipitating agents are usually at half the reservoir concentration. Droplets are equilibrated against one or more milliliters of the reservoir. In one instance, a crystalline complex was obtained by limited evaporation of the drop over an empty well.[28]

Precipitating Agents. Precipitating agents used for crystallization of protein–nucleic acid complexes fall into three categories: (1) high salt (typically ammonium sulfate and sodium chloride), (2) organic solvents [various alcohols such as ethanol, 2-methyl-2,4-pentanediol (MPD)], and (3) polyethylene glycols (PEG) of different molecular weights. It was once thought that high salt systems would not be useful for complex

---

[28] R. G. Brennan, Y. Takeda, J. Kim, W. F. Abderson, and B. Matthews, *J. Mol. Biol.* **188,** 115 (1986).

crystallization because a sufficiently high concentration of salt would ultimately dissociate even specific protein–nucleic acid complexes. This has proven not to be the case. Moderately well-ordered crystals of the 434 *cro* repressor–operator complex were obtained from concentrated solutions of ammonium sulfate,[14] the λ *cro* repressor–operator DNA complex was crystallized from sodium chloride,[28] and the *lac* repressor–operator DNA was crystallized from sodium acetate.[11] However, Table I shows that, so far, high-salt systems have not produced cocrystals that diffract to better than 3-Å resolution (in the best direction).

Organic solvents and PEG have been more effective for growing well-ordered crystals of protein–DNA complexes. The precipitating agents used include MPD, ethanol, and PEGs of average M_r 400, 3000, 4000, 6000, and 8000. PEG has the unique feature of being excluded from the crystal interstices, therefore this precipitating agent never intrudes on the macromolecular interface or crystal packing contacts.

As mentioned previously, protein–nucleic acid complexes are often much less soluble at low ionic strength. This fact has been exploited in the growth of crystals of the complex containing the DNA-binding domain of the glucocorticoid receptor and its DNA target by using a volatile salt such as ammonium bicarbonate, which can diffuse out of the droplet.

The hydrogen ion concentration seems to have a profound effect on the solubility, as well as the stability of the complex and the packing interactions that stabilize the crystal. The special effect of pH on the crystallization of protein–nucleic acid complexes results from the fact that both adenine and cytosine are partially titrated within two units of neutrality. Thus, stabilizing interactions in the complex of the DNA–binding domain of the glucocorticoid receptor–DNA target that involved a protonated "overhanging" cytidine (Figs. 1 and 3) are disrupted at a pH above 6.1. The profound effect of pH on solubility and lattice stability can be exploited to grow X-ray quality crystals. Crystals of MATα2–DNA complex, the *engrailed* homeodomain–DNA complex, and the glucocorticoid receptor DNA-binding domain–target DNA complex were obtained by forming a pH gradient between the drop and reservoir.[15,16]

Multivalent Ions. Multivalent cations play a dominant role in crystallizing specific protein–DNA complexes, presumably by forming stabilizing interactions with the phosphate oxygens of the DNA. The impact of such ions is reminiscent of the dominant influence of these ions on the crystallization of uncomplexed DNA. For example, in the *trp* repressor–operator system millimolar incremental changes in the concentration of Ca^{2+}, Mg^{2+}, Sr^{2+}, Ba^{2+}, $Co(NH_3)^{3+}$, Mn^{2+}, Cu^{2+}, and Fe^{2+} had dramatic effects on the solubility, crystal form, and degree of crystalline order.[10,27] Similarly, Schultz *et al.*,[12] in their study of the CAP-CAP site complex, showed

TABLE I
CRYSTALS OF PROTEIN–DNA COMPLEXES

Protein	Protein–DNA complex DNA	pH	Effector molecules	Precipitating agent	Diffraction limit (Å)	Ref.
*Eco*RI	TCGCGAATTCGCG GCGCTTAAGCGCT	7.4	EDTA	PEG 400	2.7	26
λ *cro* repressor	TATCACCGCGGGTGATA ATAGTGGCGCCCACTAT	6.9		NaCl	3.4	25
434 *cro* repressor	ACAATATATATTGT TGTTATATATAACA	8.0	Co(NH$_3$)$_6$Cl$_3$ MgCl$_2$	(NH$_4$)$_2$SO$_4$	3–5	23
λ repressor	TATATCACCGCCAGTGGTAT TATAGTGGCGGTCACCATAA	7.0		PEG 400	2.5	9
434 repressor	TATACAAGAAAGTTTGTACT TATGTTCTTTCAAACATGAA	7.0	MgCl$_2$, Spermine	PEG 3000	2.5–3.0	29
CAP	GCGAATGTGATATcacattg GCTTACACtatagtgtaagcg	5–6	cAMP, CaCl$_2$, Spermine	PEG 3350 or PEG 8000	3.0	12
trp repressor	TGTACTAGTTAACTAGTAC CATGATCAATTGATCATGT	7.2	CaCl$_2$ L-Trp	MPD	1.9	6
	CGTACTAGTTAACTAGTACG GCATGATCAATTGATCATGC	5.0 7.2	MgCl$_2$ L-Trp CaCl$_2$ L-Trp	MPD MPD	3.2 3.0	10 7
arg repressor	AAGTGAATAATTATTCACTTT TTTCACTTATTAATAAGTGAA	7.2	CaCl$_2$ L-Arg	PEG 3400	2.9	Joachimiak (unpublished)
DNase I	GCGATCgcgatcgc CGCTAGCGctagcg	7	EDTA	PEG 6000	2.0	30
lac repressor	TTGTGAGCGCTCACAA AACACTCGCGAGTGTT	6.5	2-Nitrophenyl-β-D-fucoside	CH$_3$COONa	6.5	11
engrailed homeodomain	ATTAGGTAATTACATGGCAA AATCCATTAATGTACCGTTT	6.7		pH gradient 8–9 to 6.7	2.0	16
MATα2	ACATGTAATTCATTTACACGC GTACATTAAGTAAATGTGCGT	7.5	CaCl$_2$	PEG 400; pH gradient 7.5 to 5	2.9	15
Glucocorticoid receptor	CCAGAACATCGATGTTCTG GTCTTGTAGCTACAAGACC	6.0	MgSO$_4$, Spermine ZnCl$_2$	MPD or Ethanol	2.5	1
	ICAGAACATCATGTTCTGA GTCTTGTAGCACAAGACTC	6.0		MPD	3.0	1
Zif268	TGCGGGTGCG CGCCCACGCT	6.5	NaCl	PEG 400	2.1	17

that Sr^{2+}, Ba^{2+}, spermine, and spermidine had significant effects on the type and quality of the crystals.

Calcium and magnesium ions are used most often to grow good quality crystals of the protein–nucleic complexes; cobalt hexamine chloride and polyamines are also useful. Schultz et al.[12] made the interesting observation that the presence of Ca^{2+}, spermine, spermidine, n-octyl-β-D-gluco-pyranoside, and MPD in the *stabilizing* medium can improve the limit of diffraction and reduce the amount of diffuse scattering and mosaicity. These observations suggest that it is necessary to search for optimal stabilization conditions as well as growth conditions if one is to achieve the highest resolution. The value of modifying the stabilizing supernatant can be extended one step further. Appropriate use of alcohols, diols, and the like can enable one to lower the temperature to as far as $-40°$. In favorable cases this causes a significant improvement in the degree of crystalline order and/or radiation resistance. This, in turn, can significantly enhance the resolution of the study.

Physiologically relevant cofactors are important for cocrystallization of specific complexes, presumably because they stabilize the correct stereochemistry. The corepressor ligand, L-tryptophan, was essential to grow the high-quality crystals of the *trp* repressor–operator complex[10] and *lac* repressor–operator DNA cocrystals have been grown in the presence of the anti-inducer 2-nitrophenyl-β-D-fucoside.[11] Metal ion cofactors have played a useful role in obtaining crystals of nuclease–DNA complexes. For example, depleting the system of magnesium ion cofactor enabled Rosenberg and co-workers to grow X-ray quality crystals of the *Eco*RI–target DNA complex.[26] Removing EDTA protection and supplying magnesium ions to the crystal causes *in situ* hydrolysis of the substrate. Similar observations have been made for cocrystals of the DNase I–DNA complex.[30]

The Future

We are at a very early point on the learning curve in our ability to grow X-ray quality crystals of specific protein–DNA complexes. Special problems have been recognized and strategies have been developed to deal with them. The case of the nucleosome has not been treated here because of its special architectural and biochemical features. As developments in overexpression continue to enhance our ability to obtain adequate

[29] A. K. Aggarwal, D. W. Rogers, M. Drottar, M. Ptashne, and S. C. Harrison, *Science* **242**, 899 (1988).

[30] D. Suck, A. Lahm, and C. Oefner, *Nature* (*London*) **332**, 464 (1988).

amounts of valuable proteins and with continued improvement in production and purification of oligonucleotides, the prospects for crystallizing increasingly complicated assemblies improve. With yearly advances in the application of synchrotron radiation and the development of new phasing techniques and data acquisition devices, the availability of crystalline specimens will remain the limiting experimental factor in developing an understanding of the stereochemistry of high-affinity interfaces between proteins and their DNA targets.

Section II

DNA Binding and Bending

[8] Gel Retardation

By JANNETTE CAREY

I. Introduction

Understanding gene regulatory mechanisms and other biological processes based on protein–nucleic acid interactions requires biochemical analysis of the binding reactions. While X-ray crystallography and nuclear magnetic resonance (NMR) are increasingly useful in revealing the *structural* details of such interactions, they cannot reveal the nature and magnitude of the *forces* that contribute to specific complex formation. The affinity, stoichiometry, and cooperativity of binding must also be analyzed to develop a complete picture of how the typically extraordinary specificity of these interactions is achieved. A fundamental difficulty in making such measurements for protein–nucleic acid interactions is the very high affinities that generally characterize these reactions. Such high affinities—often with subnanomolar dissociation constants—preclude direct observation of the binding equilibrium by spectroscopic methods, since few spectroscopic signals are detectable at concentrations comparable to the dissociation constants. This constraint has spurred development of numerous indirect and/or nonequilibrium methods, the most popular of which are gel retardation, filter binding, and footprinting.

Gel retardation is a powerful approach for both qualitative and quantitative analysis of protein–nucleic acid interactions. This method is based on the observation that the electrophoretic mobility of a nucleic acid through a polyacrylamide gel can be altered when a protein is bound to it. The current popularity of the technique dates from the 1981 reports of Garner and Revzin[1] and Fried and Crothers,[2] although the origins of the method can be traced to earlier literature.[3,4]

The gel retardation method has several advantages compared to other common assays for protein–nucleic acid interactions. First, a specific binding protein can be detected by its effect on mobility even when other binding proteins are present, as in a crude cell extract. Second, the specific nucleic acid target site of a given binding protein can be identified from a mixed population of fragments. Although both these applications can be

[1] M. M. Garner and A. Revzin, *Nucleic Acids Res.* **9**, 3047 (1981).
[2] M. G. Fried and D. M. Crothers, *Nucleic Acids Res.* **9**, 6505 (1981).
[3] B. K. Chelm and E. P. Geiduschek, *Nucleic Acids Res.* **7**, 1851 (1979).
[4] H. W. Schaup, M. Green, and C. G. Kurland, *Mol. Gen. Genet.* **109**, 193 (1970).

carried out with other methods, gel retardation is quick and simple, and the direct visualization of complexes is often informative. Third, because free protein is separated from complexes, only *active* protein will be present in the complex bands.[5] Fourth, complexes can often be resolved even if they differ only in their stoichiometries or in the physical arrangement of their components. These last two are the most important advantages of the gel method, and are unique to it.

Development of a gel retardation assay for a new binding protein can be approached systematically.[5] The virtues of systematic study include greater understanding, not only of the specific protein–nucleic acid system, but of the gel method itself. This chapter provides guidelines for such an approach, beginning with basic directions that work for a majority of proteins, and proceeding through troubleshooting steps to optimize the method for a new protein. These guidelines apply equally well to proteins that bind RNA or DNA, although in this chapter DNA binding is assumed for convenience. Presumably, these guidelines coupled with chemical intuition would work equally well for nonprotein DNA-binding molecules and/ or non-nucleic acid target molecules, although these applications have apparently not yet been demonstrated. Also presented here are methods for quantitative analysis of gel assay results, including determination of binding constants, stoichiometries, cooperativity, and specificity.

A phenomenological theory of gel retardation has been developed, which gives quite good agreement between theory and experiment for several simple systems.[6] In future, this approach may yield new practical insights as well as providing a rigorous foundation for analyzing gel retardation data.

II. Methods

A. *Pilot Conditions*

The choice of starting conditions for a gel retardation assay of a previously uncharacterized nucleic acid-binding protein is necessarily arbitrary. However, an exhaustive search for optimal conditions for the *trp* repressor of *Escherichia coli*[5] led to certain useful generalizations. The table summarizes starting conditions for the experimentally controllable variables that should work for many, but not all, protein–DNA systems.

140 × 140 × 1.5 mm gel (~35 ml)
Wells 0.5 to 1.0 cm wide with 3-mm spacing

[5] J. Carey, *Proc. Natl. Acad. Sci.* **85,** 975 (1988).
[6] J. R. Cann, *J. Biol. Chem.* **264,** 17032 (1989).

10% acrylamide (29 : 1 acrylamide : bis)
0.5× Tris-borate-EDTA buffer (TBE; Ref. 6), pH 8.3
Run in cold room (4°)
5% glycerol loading dye
100 V, 2 hr for 100-bp DNA

B. Detailed Protocol

1. Preparation of Gel. The optimal acrylamide concentration depends on the length of the DNA fragment used in the experiment. As the length of the DNA fragment increases, the acrylamide concentration should be reduced. A 10% gel works well for DNAs of up to 200 to 300 bp in length. The gel is poured as usual (see, e.g., Ref. 7), allowed to polymerize, mounted in the gel box, and equilibrated at 4° (preferably overnight). For reproducibility in quantitative work, the times between pouring, polymerization, and running should be kept constant. The gel box ideally should permit free circulation of coolant (air or water) on both surfaces of the gel plates to allow efficient heat dissipation. It should also allow for access to the wells so that the gel can be loaded while the current is turned on. Before loading, the gel is preelectrophoresed until the current drops to a constant value.

2. Reaction Mixes. If pure protein is available, a good first experiment is to titrate a small, constant amount of labeled DNA with increasing amounts of the protein. The concentration of DNA should be as low as is practical. High specific radioactivity DNAs labeled with ^{32}P by nick translation can be used at concentrations of 10 pM or less, as only a hundred or so counts per minute per band is adequate for detection.

A reaction mix is assembled containing the DNA plus necessary buffers, cofactors, salts, or other components if these requirements are known. If they are unknown, a reasonable choice of buffer to start with is 0.5× TBE with no added salt. The reaction mix need not have the same composition as the gel buffer. The reaction mix should also contain sucrose or glycerol at up to 5% final concentration, and loading dyes as desired, to facilitate loading. The mix is then aliquotted into reaction tubes. The protein is diluted over a wide range, say from the highest possible concentration down to a concentration equal to that of the DNA, and aliquots of protein are added to the reaction tubes. For a first experiment, 10-fold serial dilutions should be tested over this entire protein concentration range. One control sample should contain only the protein at the highest concentration tested, another only DNA at the chosen fixed concentration.

[7] T. Maniatis, E. F. Fritsch, and J. Sambrook, "Molecular Cloning: A Laboratory Manual," p. 174. Cold Spring Harbor Laboratory, Cold Spring Harbor, New York, 1982.

The samples are mixed and allowed to equilibrate. For unknown reaction rates, 15 min is a compromise length of time, and attainment of equilibrium can be verified by comparison with a much longer incubation time (1 to 2 hr). Sample volumes can range from 5 to 75 μl; the smaller surface-to-volume ratio of larger samples reduces reequilibration when sample and running buffers are different.[8]

3. Gel Loading. Immediately before loading, the wells of the prerun gel are cleaned out by flushing with reservoir buffer. The voltage is turned up to ~300 V (or more; see Ref. 9), and the samples are rapidly loaded. (*Caution:* Care must be exercised to avoid electrocution!) A glass capillary and mouth pipette, or a sequencing pipette tip and adjustable microliter pipettor, are convenient for loading. As soon as the last sample dyes have just entered the gel, the voltage is reduced to avoid heating, and the gel is run long enough for the free DNA to migrate about half the length of the gel. Longer running times may lead to excessive dissociation of complexes.

4. Detection. The gel is dismounted, and the lane containing only protein is cut off if it is to be stained with Coomassie Blue or silver. The rest of the gel is transferred to a sheet of filter paper for drying. A typical 1.5-mm gel requires about 1 hr under vacuum at 80° to dry completely. The dried gel is exposed to autoradiographic film. Overnight exposure is sufficient to detect several hundred ^{32}P counts per minute in a single gel band when an intensifying screen is used.

C. Troubleshooting

The conditions outlined above generally resolve complexes from free DNAs for many binding proteins. However, if the expected mobility shift is not observed, the protocol may require modification (as outlined in Section V). Troubleshooting can be carried out in a logical way by considering what occurs during the gel retardation process.

Before electrophoresis, the samples should contain an equilibrium mixture of free DNA, free protein, and complexes. When this mixture is loaded onto the running gel, all species begin to migrate under the influence of the electric field, each according to its own characteristic mobility, which is determined by its size, shape, and charge. Separation of bound and free components thus begins at the moment of loading, unless by chance two components have very similar electrophoretic mobilities.

Under most circumstances, free DNA has the highest mobility. As free DNA begins to enter the gel, the preexisting equilibrium may be perturbed by mass action in the direction of complex dissociation. On the other hand,

[8] D. Senear and M. Brenowitz, *J. Biol. Chem.* (in press).
[9] M. G. Fried, *Electrophoresis* **10**, 366 (1989).

the interface between the gel and the sample causes a discontinuity in the rate of electrophoresis, from the very fast rate in the sample to a much slower rate in the gel. Thus, focusing of the sample occurs at this interface, as is well illustrated by many reports in which samples initially occupying a volume of $1 \times 5 \times 10$ to 20 mm give rise after electrophoresis to free DNA bands of only $1 \times 5 \times 1$ to 5 mm. This focusing must concentrate the sample at the boundary, and could enhance complex formation by mass action, thereby opposing the effect of DNA depletion. Perturbations in either direction will depend on the relative rates of association and dissociation, and the clearance time for free DNA to fully enter the gel.

Once all the free DNA in the sample has entered the gel as a focused band, there can be no further increase in the amount of DNA in that band, provided it is the highest mobility species present. Even if *all* the complexes subsequently liberated their DNA by dissociation during electrophoresis, the DNA thus freed could not catch up with the initial free DNA band. Therefore, the free DNA band accurately reports on the equilibrium fraction of free DNA, except for perturbations during the clearance time. The clearance time for all free DNA to enter the gel is therefore the "dead time" of the method. Fortunately, this time is under experimental control via the variables of DNA length, acrylamide concentration, and loading voltage. In addition, clearance time is equal for all samples on a gel that is loaded with the current turned on.

The species with the next fastest gel mobility is ordinarily the protein–DNA complex. The remarkable fact that this complex migrates as a discrete band in gels was the founding observation of gel retardation. Bands corresponding to complexes are unexpected because the time required to run a gel can exceed the typical complex half-life by orders of magnitude. To explain their surprising persistence, the idea of "caging" was invoked.[2] The term caging has turned out to be a misunderstood, although not incorrect, way to describe what happens in the gel, because it is sometimes incorrectly interpreted as the cause of the apparently slow dissociation of complexes in the gels. In fact, all evidence suggests that the gel has *no* effect on dissociation rates. Rather, caging probably works by causing an increase in the effective concentration of the interacting components by slowing their electrophoretic escape after dissociation, leading to enhancement of the bimolecular (and thus concentration-dependent) reassociation reaction. It seems likely that the low water activity in polyacrylamide gels contributes to this effect as well. Thus, although complexes may dissociate many times during the gel run, reassociation is strongly favored, leading to persistence of complex bands even when the gel run time greatly exceeds the half-time for dissociation.

Mobility of the complex is related to mobility of the free protein.[5] The stained protein-only lane may thus provide a clue as to how to proceed in modifying the protocol if no shift is observed. If the protein forms a band in the gel, then it is negatively charged at the pH of the TBE buffer. If its mobility is too similar to that of the DNA, little or no retardation of the DNA will occur on binding.[5] In this case it is advantageous to run gels at lower pH values. Buffers should be chosen that have high buffering capacity at each pH value. Some buffers, such as dilute phosphate, are poor choices as they lose their buffering capacity during electrophoresis and must be recirculated. On the other hand, if the protein band is at the well rather than in the gel, the protein may either be very basic or highly aggregated. The following paragraphs give more details for troubleshooting (following the guide in Section V).

The pilot experiment described above may fail if the protein is too dilute or only partially active. Proteins can be concentrated by precipitation with ammonium sulfate, or by dialysis against solid Sephadex (Pharmacia, Piscataway, NJ) or Aquacide (Calbiochem, Los Angeles, CA), or by ultrafiltration (e.g., Amicon, Danvers, MA). Determination of protein activity is essential for accurate measurement of binding constants, and is discussed in Section III,D.

If there is no detectable free *or* bound DNA in the sample lanes, nuclease activity may be present as a contaminant of the protein preparation. On the other hand, this result could also be consistent with complex formation: if complexes are present initially, but dissociate slowly during electrophoresis, the liberated DNA can become so diffuse as to be undetectable as a band or even as a smear. In some cases the DNA may disappear at intermediate protein concentrations and reappear at higher concentrations, as high stoichiometry complexes form. Disappearance of the band corresponding to free DNA can be taken as evidence of complex formation,[1,5,10] and can be quantitated as described in Section III, if the possibility of nuclease activity is rigorously eliminated first. To accomplish this, a second DNA fragment lacking the binding site can be titrated simultaneously with the specific fragment (i.e., in the same reactions) if the two DNAs are of different lengths.[5]

Dissociation of complexes during electrophoresis is not, in and of itself, a problem for many purposes because the disappearance of free DNA can be quantitated instead of the appearance of complexes. However, dissociation may be minimized by increasing the concentration of acrylamide and/or bis in the gel, and reducing running times as much as possible. Dissociation of complexes during the dead time (while the DNA is entering the gel) is minimized by loading the gel while it is running at high

[10] A. Revzin, *BioTechniques* **7**, 346 (1989).

voltage[5,9] and/or by loading larger samples[11]; on the other hand, smaller samples may provide more focused bands.

Smeared, streaked, or uneven running of DNA can be due to several kinds of problems with the gel itself. If polymerization was too rapid it may have been uneven, or the gel may have been run too hot. The acrylamide concentration may be too low, or a stacking gel may be required to provide a second interface where the complexes can become further focused. The samples may have entered the gel too slowly, or a component of the loading dye may have interfered with migration (e.g., polyethylene glycol). All these potential sources of failure should be explored until the operator is confident of the technique.

If none of the suggested modifications succeeds, it may be necesary to explore a wider range of gel and/or solution conditions. If cofactors are required for binding, these may have to be incorporated into the gel when it is cast and included in the running buffer. Some compounds may become covalently incorporated into the gel by this method, forming, in effect, an affinity matrix for the protein; alternatively, cofactors may be introduced by electrophoresis or diffusion. The submarine horizontal gel system of Hendrickson and Schleif[12] allows for more efficient diffusion. Charged cofactors may become redistributed during electrophoresis. This can be minimized by recirculating the gel running buffer, or, in the case of titratable cofactors, by judicious choice of pH. In the case of *trp* repressor, the corepressor L-tryptophan is required in the gel to observe operator binding at 10% acrylamide, but at 20% acrylamide the corepressor is required only in the reaction mixes and not in the gel. This result implies that even small ligands may be caged to some extent by the polymeric gel matrix.

III. Quantitation

A. Affinity

Once adequate resolution is obtained of the free DNA from the complex band, and/or from any smear resulting upon its dissociation, it is possible to use the gel method to determine the quantitative parameters describing the binding reaction. The first point to establish is the protein concentration required to bind half the DNA. An experiment similar to the one described above can be used, with the protein concentration varied over the binding range such that at least two protein concentrations are tested in each log unit. The midpoint concentration can be estimated by eye from the

[11] M. Brenowitz, E. Jamison, A. Majumdar, and S. Adhya, *Biochemistry* **29**, 3374 (1990).
[12] W. Hendrickson and R. F. Schleif, *J. Mol. Biol.* **178**, 611 (1984).

autoradiogram, and gives an indication of the approximate magnitude of the dissociation constant, K_d. Visual estimation of K_d is generally accurate to within a factor of 10 or so, although precision is considerably better than this. This estimate of K_d is valid only if the DNA concentration is negligible compared to the protein concentration at the midpoint (at least 10-, and preferably 100-fold, lower).

To obtain a more accurate value for K_d, quantitation of the gel results can be done by film densitometry, direct scanning of gel radioactivity, or scintillation counting of excised gel bands. Smaller increments in protein concentration will improve the precision of the K_d value determined by these methods. The lane separation given in the protocol is essential to reliable quantitation by any of these methods. If the autoradiogram is to be scanned, it is important that the optical density of the bands be within the linear response range of the film. This is usually achieved through preflashing of the film to overcome the response threshold.[13]

The gel can also be sliced and counted in a scintillation spectrometer. For ^{32}P the quenching due to the acrylamide and filter paper is not generally a problem, but for lower energy isotopes it is necessary to slice the wet gel (no filter paper) and to dissolve the polyacrylamide by oxidation. This is accomplished using 17% $HClO_4$ and 21% H_2O_2 in tightly closed plastic scintillation vials at 60° overnight.[5,13] Chemiluminescence from peroxide reactions can be a problem, and can be minimized by altering the gel cross-linker so that oxidative dissolution is unnecessary.[14] Chemiluminescence is detectable as an instability of the cpm data with repeated measurement. An alternative treatment, therefore, is to allow chemical reactions to run their course, as determined by reproducibility of cpm values. Vials should be stored at least 24 hr in the dark, preferably in the scintillation counter, to reduce the background from photoluminescence as well. Direct scanning of the gels is much simpler because of its speed, linear response, and freedom from luminescence interference, although sensitivity and resolution may be less good than with film.

The fraction of DNA that is bound and free is determined after normalization, either using an internal standard[5] as described in Section II,C or by summing the total number of counts in each lane.[8] If not all the DNA is bound at very high protein concentrations, a correction should be made for this inactive population of DNA molecules. Partial activity is not generally found with restriction fragments, but can sometimes be a problem with synthetic DNA. Fractional activity is also a common, although poorly understood, problem with nucleic-acid binding proteins, and cor-

---

[13] R. A. Laskey and A. D. Mills, *Eur. J. Biochem.* **56**, 335 (1975).
[14] B. W. Fox, "Biochemical and Biophysical Methods," Vol. 5, Elsevier, New York, 1976.

rection must be applied for it. The correction factor is determined under quite different binding conditions, as described in Section III,D.

The data are plotted as fraction of free DNA vs protein concentration to determine K_d, which is approximately equal to the protein concentration at which half the free DNA has become bound, provided the DNA concentration is vanishingly small compared to the protein concentration at the midpoint. It may seem more logical to measure the appearance of complexes than the disappearance of free DNA, and in practice both should probably be plotted. The two resulting measurements of K_d will differ if the equilibrium is perturbed by the gel run. For good quantitation, it is often necessary to span six or more orders of magnitude in protein concentration in order to have well-determined baselines before and after the binding zone. The best way to plot such data is probably the so-called Bjerrum plot of fraction free DNA vs log of protein concentration. The principal advantage of this plot is that the log scale compresses the large range of protein concentrations into a convenient size. *All* binding data appear sigmoid on such plots, and the exact shape of the curve depends on the binding parameters, as described in Section III,D.

A very simple test can be applied to determine if the binding reaction is complicated by multiple binding sites or cooperativity. The experimental data are compared to a theoretical binding curve calculated from the binding expression for a model with independent and equivalent sites.[5] When good data do not fit this simple model, the information in the complex bands can be used to determine the correct binding isotherm,[8,11] as discussed in Section III,E.

Some discussion is in order about the validity of binding constants measured by gel retardation. As with any other method in which free and bound components are separated, the possibility exists of perturbing the equilibrium by mass action, as discussed above. Although this effect is minimized by analyzing the disappearance of free DNA rather than the appearance of complexes,[1,5–7] it may not be eliminated completely. For example, significant dissociation may occur during the dead time while free DNA is entering the gel. Although the dead time can be reduced by raising the loading voltage, it may still be long relative to the dissociation rate of the complexes or to the rate of reequilibration between sample buffer and loading buffer. Thus, it is essential to have a way to evaluate the assay before interpreting the absolute values of binding constants.

The best practical solution to this problem is to measure the binding constants by a second method that does not perturb the equilibrium. Ackers and co-workers[15] have shown that DNase I footprinting can be

[15] M. Brenowitz, D. F. Senear, M. A. Shea, and G. K. Ackers, this series, Vol. 130, p. 132.

used as a quantitative assay that meets this criterion for certain proteins. For uncharacterized proteins, it must be established independently that the footprinting conditions do not perturb the DNA binding equilibrium. This was shown to be the case for *trp* repressor,[5] and the DNase I assay in turn was used to validate the gel assay results. However, even when such tests are met, caution is advised in interpreting very large differences in binding constants determined by the gel method.

B. Kinetics

Rate measurements are useful for their role in studies of mechanism, and as a means of confirming a binding constant, as $K_d = k_{off}/k_{on}$. Although measurements of association and dissociation rates by gel retardation are straightforward, they are subject to the same perturbations as equilibrium measurements. Useful rate information can be obtained from gel retardation,[9,10] but it should not be the sole criterion used to validate a binding constant made by the same method.

Association rates are determined by mixing the components and loading samples at time intervals. Because the free DNA begins to be separated from complexes as soon as the sample is loaded onto the running gel, the loading process itself acts to quench the association reaction. For dissociation rates, complexes are preformed and allowed to reach equilibrium. The association reaction is quenched by either dilution or by addition of a quenching agent so that the dissociation reaction can be observed. Samples are loaded at time intervals after quenching. Typical quenching agents include unlabeled target DNA or nonspecific DNA, which are silent in autoradiograms of the gels. It is important to note that the quenching agent is competing with the labeled DNA for dissociated protein, which is usually in vast excess over the labeled DNA. Thus, the competitor must be in molar excess over the *protein* in order to observe competition. To verify that the quenching agent acts only as an inert sink for excess protein, k_{off} should be shown to be independent of competitor concentration.

C. Specificity

Use of a nontarget DNA as an internal control as described in Section II,C also allows simultaneous, independent observation of nonspecific binding. The nonspecific binding constant is evaluated identically, and it must be corrected for the statistical nature of nonspecific binding to DNA.[16] The ratio of affinity constants for specific and nonspecific binding is a measure of the specificity: the ability of a given protein to discriminate

[16] J. D. McGhee and P. H. von Hippel, *J. Mol. Biol.* **86**, 469 (1974).

between the true target site and nontarget sites. This ratio is expected to vary with solution conditions because the two kinds of complexes typically have differing contributions to their free energy of binding.[17] The gel retardation method is not the best way to determine specificity over the required wide range of conditions, however, because the gels themselves vary over these conditions. For example, as the salt concentration is raised, the gels run hotter, promoting dissociation as well as smearing of the bands. If it can be established that the gel method provides a valid measure of the equilibrium constant, then it may be adequate to vary conditions in the reaction mixes only, rather than in both the gel and the reactions. However, this approach is not recommended. It is preferable to use a true equilibrium method or footprinting under nonperturbing conditions when possible for these experiments.

D. Stoichiometry

The ability of the gel method to resolve complexes that differ only by their stoichiometries is a unique advantage of the technique, and can be exploited in a double-label experiment to determine the molar ratios of protein to DNA in each complex. Obviously, a requirement for this experiment is that discrete complex bands be detected in the gel. An experiment similar to that described in Section II,B above can be used with both labeled protein and labeled DNA.

The choice of isotopes for such an experiment is made according to the usual criterion for a double-label experiment[14]: the two isotopes should be quantifiable separately when present together, preferably with minimal corrections. All pairwise combinations of ^{32}P, ^{35}S, ^{14}C, and ^{3}H should work except for ^{35}S and ^{14}C, which have energy spectra that are too similar. However, quenching may shift the energy spectrum of an isotope[14]; therefore, counting efficiencies and spillover for the chosen isotope pair should be determined prior to the experiment under the exact conditions that will be used for scintillation counting. It is convenient if only one of the two isotopes is detectable by autoradiography, to facilitate preparing a template for fractionation of the gels,[5] but this is not essential.

^{32}P and ^{3}H are used as an example here for illustration, and they are a good choice: spillover corrections are not excessive, and counting is reasonably efficient. ^{32}P-Labeled DNAs are easily prepared in high yield and high specific radioactivity. The short half-life of this isotope confers a unique advantage: the complex correction factors sometimes required in double-label scintillation counting can be confirmed after a suitable

[17] M. T. Record, Jr., C. F. Anderson, and T. M. Lohman, Q. Rev. Biophys. 11, 103 (1978).

decay period and will give an internally consistent result if they were accurate.[5] ^{3}H-Labeled precursors for protein labeling *in vivo* or *in vitro* by biosynthesis or chemical modification are readily available at high specific radioactivities. Proteins labeled with ^{3}H should be stable for long periods when stored at 4° in dilute solution.

It is essential to have sufficient amounts of protein and DNA present in each complex band for accurate and precise quantitation. Because ^{3}H is somewhat inefficiently counted and sensitive to quenching, it is necessary to have well over 1000 dpm of ^{3}H/band. This amount of radioactivity is difficult to achieve in a single band under the conditions given in Section II,B, even with theoretical maximum specific radioactive labeling of the protein. The factor that limits the amount of protein in a complex band is the amount of *DNA* present. Thus, to improve detection of ^{3}H-labeled protein, the amount of DNA in the reaction mixes should be increased. Preliminary calculations can be carried out to determine the amount of DNA required to give a detectable amount of protein in the complex bands, using the estimated binding constant and quenching factors and the known specific radioactivity of the protein, and assuming a minimum stoichiometry of 1 : 1. The exact DNA concentration must be known to determine its specific radioactivity, and can be measured sensitively by any of several fluorescent methods.

After the gel is run, it is not dried but exposed to autoradiographic film while still wet. The pattern of ^{32}P-labeled DNA distribution is used as a template for excising bands from the wet gel.[5] The specific radioactivities of the protein and DNA are determined by excising empty regions of the gel and adding known chemical amounts of DNA or protein to each slice, and control samples are prepared to determine quenching and spillover correction factors.[14] Gel slices are oxidized as in Section II,A, and counted in scintillation fluid. The results are converted from counts per million to moles via the specific radioactivity and counting efficiency of each component, after correction for background and spillover.

In addition to excising individual complex bands, it is useful to count each entire lane of the gel to account for all the applied radioactivity and determine its distribution. This is especially important for the protein-only lane: the position of free protein must be known before a valid stoichiometry can be determined for the protein–DNA complexes. If the free protein migrates too closely to the complexes, the gel conditions (pH, acrylamide concentration, run time) can be altered to achieve better separation. If the protein is in large excess over the DNA, then the distribution of free protein will be approximately the same in the protein–DNA lanes as in the protein-only lane. In this case, the radioactivity in the protein-only slices can be subtracted from that in the equivalent complex slices.

A stoichiometry determination by gel retardation can also be used to establish if the protein is fully active for DNA binding. This application takes advantage of the fact that only *active* protein will be present in the complex bands; this is a unique attribute of the gel method compared to other DNA-binding assays. When the DNA concentration is at least 10- to 20-fold above the K_d, the input amount of protein required to bind half the DNA should be approximately equal to half the DNA concentration times the stoichiometry in the complex band. If more protein is required to achieve half saturation, this is an indication of partially active protein.

E. Cooperativity

Preliminary indications of cooperativity can be obtained simply by comparing the observed binding curve with the theoretical binding curve calculated from the expression for the binding constant, using a model of independent and equivalent sites. The presence of cooperativity is most easily assessed from the breadth of the transition on a semilog plot: for noncooperative interactions, an increase of 1.81 log units in protein concentration is required to increase the bound fraction from 10 to 90%. Positive cooperativity reduces this span, while negative cooperativity increases it. The degree of cooperativity is determined by fitting the data to theoretical curves calculated for various degrees of cooperativity. However, the presence of multiple classes of independent binding sites with different affinities cannot easily be distinguished from negative cooperativity unless the affinities differ by about 100-fold, and it may be necessary to physically separate the binding regions to decide between these two models.

A more sophisticated treatment of the data takes advantage of the unique ability of the gel to resolve individually liganded states. Binding constants for each ligation step can be extracted by fitting the data to an appropriate binding expression.[8] All complex bands as well as the free DNA band must be quantitated. High-quality data and discrete, well-resolved bands for each ligation state are required for this approach. Several reports have appeared illustrating its success in quantitative analysis of cooperativity, at least in simple cases.[8,11]

IV. Identification of New Binding Proteins

The gel method is ideal for purifying new proteins that bind to a defined target DNA. Two approaches can be taken: the gel method can be used to assay classical biochemical purification steps, or the protein can be purified directly from the complex in the gel. The most important element for success in the latter approach is an appropriate concentration of *DNA*.

As discussed in Section III,D above, the factor that limits the amount of binding protein in a complex is the amount of DNA present. Thus, raising the DNA concentration will increase the amount of bound protein, as long as the protein itself is not limiting.

In crude cell extracts, detection of binding activity toward a specific target DNA is complicated by the presence of other binding proteins. It is possible to detect a specific binding protein against this background of nonspecific binding activities because the specific and nonspecific proteins are competing for binding to the target DNA, but the specific complex has higher affinity under most conditions. The best way to exploit this difference in practice is to first do a trial experiment, like the one described in Section II,B, using two labeled fragments of different sizes simultaneously, one specific and one nonspecific, to determine the concentration of extract required to shift the DNAs out of their free positions. Then do a second experiment in which the reaction mix contains the labeled DNAs plus the chosen concentration of extract, and aliquots receive increasing amounts of an unlabeled competitor DNA. At some level of input competitor DNA, the nonspecific fragment will dissociate while the specific fragment remains bound. It may be necessary to examine a vast range of competitor concentrations, first in very coarse and then in very fine increments, to find the right concentration. Mixed sequence DNA is preferable to homopolymers as a competitor.

The binding protein can be partially purified from the complex band by denaturing gel electrophoresis. The complex band is excised from the wet retardation gel using the ^{32}P-labeled DNA pattern as a template. The gel slice is soaked briefly in a minimal volume of loading solution appropriate for the denaturing gel system that will be used; the gel slice is then stuffed into a well of the denaturing gel in contact with the bottom of the well.[5] An adjacent lane of the retardation gel should contain crude extract only. A slice should be excised from this protein-only lane at a position equivalent to that of the complex, and prepared for the denaturing gel in parallel with the complex band. A sample of the free labeled DNA should also be run alongside, preferably also in an excised gel slice. Localization of the protein can be by autoradiography for radioactively labeled proteins, silver staining in cases where enough protein is present, or by Western blotting if antibodies are available.[5] If protein cannot be detected in an ordinary Laemmli-type[18] gel, other gel systems can be tried (e.g., see Ref. 19 for proteins too small to be resolved by Laemmli gels).

[18] U. K. Laemmli, *Nature (London)* **227**, 680 (1970).
[19] A. L. Shapiro, E. Vinuela, and J. B. Maizel, *Biochem. Biophys. Res. Commun.* **28**, 815 (1967).

An innovation[20] has been to cross-link the protein to bromodeoxyuridine-containing DNA using UV irradiation before running the retardation gel, to enhance detection of the protein in the denaturing gel. The cross-linking step also works *in situ* in the retardation gel, whether agarose[21] or acrylamide (D. W. Ballard, personal communication, 1991).

V. Troubleshooting Guide

Problem	Possible causes
All DNA in free position at all protein concentrations	Inactive protein Too little protein Very acidic protein Gel pH too high
All DNA in bound position at all protein concentrations	Too much protein

No DNA in any position at all protein concentrations	Nuclease activity
	Disociation of complexes
	   Samples entered gel too slowly
	   Acrylamide concentration too low
	   Gel run too long

Some DNA remains unbound at all protein concentrations	Too little protein Partially inactive DNA
Complexes do not enter gel	Very large protein, aggregation, or cooperativity Very basic protein Gel pH too low Acrylamide concentration too high
No protein detectable in complex band	Too little DNA Very small protein

DNA smeared or unevenly streaked	Gel run too hot
	Uneven gel polymerization
	Dissociation of complexes
	   Acrylamide concentration too low
	   Samples entered gel too slowly
	   Gel run too long

Acknowledgments

I gratefully acknowledge valuable discussions with Drs. Michael Brenowitz, Michael Fried, and Werner Maas, and thank the many people who sent reprints and preprints.

[20] C. Wu, S. Wilson, B. Walker, I. Dawid, T. Paisley, V. Zimarino, and H. Ueda, *Science* **238**, 1247 (1987).
[21] D. W. Ballard, E. Böhnlein, J. W. Lowenthal, Y. Wano, B. R. Franza, and W. C. Greene, *Science* **241**, 1652 (1988).

[9] DNA Bending in Protein–DNA Complexes

By Donald M. Crothers, Marc R. Gartenberg,
and Thomas E. Shrader

Introduction

The realization that bound proteins may greatly distort nucleic acids is not new to molecular biology. The nucleosome, the earliest protein–DNA complex to be structurally characterized,[1] remains the most dramatic example of protein-induced DNA deformation. In more recent years noncrystallographic techniques have revealed that induced DNA bending is a common but not universal characteristic of regulatory proteins. The functional purpose of DNA flexure is in a simplistic sense obvious: the bend stabilizes a particular tertiary structure of a larger complex between DNA and one or more proteins. However, this facile answer sidesteps the issue of why DNA should be required to curve in such complexes when they might equally well be organized with proteins arrayed linearly along the DNA, especially in view of the significant energetic cost of DNA bending. It seems plausible that a compact shape facilitates the structural stabilization and regulatory signals contributed by interactions at protein–protein interfaces in multiprotein–DNA complexes. As experimental understanding of intrinsic and induced DNA bending increases, we can expect more sophisticated explanations for the phenomenon to emerge. This chapter is directed at experimental approaches to the problem that make use of standard molecular biological techniques, enabling widespread application to the numerous protein–DNA interaction systems that are now under study.

We divide the primary methods of interest into two categories: *comparative electrophoresis* and *sequence perturbation*. In principle both are capable of specifying the locus and direction of bends induced in DNA by proteins. The first method relies on the empirical fact that the electrophoretic mobility of DNA molecules depends on their shape, as well as on their contour length.[2] Even when a protein is bound, DNA shape still affects the mobility of the complex in nondenaturing gel electrophoresis,[2] as long as the DNA molecule is severalfold larger than the segment that interacts directly with the protein. The basic strategy of comparative

[1] T. J. Richmond, T. J. Finch, B. Rushton, D. Rhodes, and A. Klug, *Nature (London)* **311,** 532 (1984).
[2] H.-M. Wu and D. M. Crothers, *Nature (London)* **308,** 509 (1984).

electrophoresis is to alter DNA shape by changing the position of a bend in the molecule,[2] or by changing the helical phasing between two bends,[3] and to determine the relative electrophoretic mobility of the resulting constructs. Interpretation of experiments of this kind has a strongly empirical basis, requiring careful evaluation of alternative explanations as the technique is "bootstrapped" to new applications.

By sequence perturbation we mean systematic study of bending or binding strength of a DNA–protein complex as a function of the base sequence in a region where a DNA bend is suspected; this approach applies only outside the strongly conserved sequence that forms specific hydrogen bonds to the protein. The basis for interpretation of these experiments is provided by the rules developed for nucleosomes and *Escherichia coli* CAP protein, according to which A/T-rich sequences are preferred when DNA bends toward its minor groove, and G/C-rich segments are favored when the bend is toward the major groove.[4-7] The degree of bending is most sensitive to sequence at the locus where one of the grooves faces the protein,[6] corresponding to sites where roll between the base pairs causes the helix to bend. Hence, from the position of maximum sequence sensitivity, and the nature of the sequences that provide maximal bending, it is possible in principle with this relatively labor-intensive method to specify the bend locus and direction with high accuracy.

Comparative Electrophoresis

The three major properties that characterize a bend are (1) its locus, as described by the position of its center, (2) its direction, defined, for example, relative to the roll and tilt axes of a specific helix position such as the center of the bend, and (3) its magnitude, usually measured relative to a standard bend. Comparative electrophoresis is able to provide answers to all these questions, although different constructs are required for each of the three cases. However, the experiments share a common gel electrophoresis technology.

Methodology: Electrophoresis of Protein–DNA Complexes

To detect conformation-dependent changes in mobility, DNA molecules must migrate through the gel matrix by reptation, or snakelike motion, rather than by a seiving mechanism. The empirical fact is that bent

[3] S. S. Zinkel and D. M. Crothers, *Nature (London)* **328,** 178 (1987).
[4] H. R. Drew and A. A. Travers, *J. Mol. Biol.* **186,** 773 (1985).
[5] S. C. Satchwell, H. R. Drew, and A. A. Travers, *J. Mol. Biol.* **191,** 659 (1986).
[6] M. R. Gartenberg and D. M. Crothers, *Nature (London)* **333,** 824 (1988).
[7] T. E. Shrader and D. M. Crothers, *Proc. Natl. Acad. Sci. U.S.A.* **86,** 7418 (1989).

molecules migrate more slowly than linear ones in high-percentage acrylamide gels[8]; as the gel percentage is reduced, or if agarose gels are used, the anomaly in the mobility of the bent form lessens or disappears. This fact has been interpreted,[2] perhaps simplistically, as reflecting the predicted dependence of mobility on the mean square end-to-end distance for molecules of fixed contour length.[9,10] It is likely that motion through the pores of the gel acts to straighten the molecule out, since the entire chain must follow the path chosen by its leading segment; the energy needed to straighten bent molecules may be an important factor contributing to their relatively slow motion.[11] On the other hand, the gel retains some local elasticity, allowing the molecule to move toward the shape that is favored in free solution, thereby reducing its end-to-end distance. This tendency to return to solution shape may account for the qualitative success of theories that predict a correlation between mean square end-to-end distance and electrophoretic mobility.[9,10] However, the problem is a difficult one, and theory is not yet a quantitatively reliable guide to the electophoretic properties of DNA molecules of complex shape.

The protein–DNA complexes we have examined are quite stable and insensitive to moderate changes in electrophoresis conditions. Complexes are preformed in binding buffers appropriate for each protein, and commonly electrophoresed in nondenaturing polyacrylamide gels, of acrylamide percentage between 5 and 10%; the optimum acrylamide-to-bisacrylamide ratio has been found empirically[12] to be about 75 : 1. We prefer this ratio, used with the highest acrylamide percentage that gives adequate mobility to the complex, since we have found that these conditions yield optimally sharp bands for the complexes at a fixed migration distance.[12] This observation provides support for the view that high acrylamide concentration provides a "cage" around the complex that stabilizes it against dissociation during electrophoresis, although it is also possible that the phenomenon results from other factors, such as an influence of high gel percentage on the thermodynamic activity of water. (If bound water molecules are released from DNA or protein in forming the complex, then a lowered concentration of water in the gel will result in an increased binding affinity.)

The gels are generally 16 cm in length with a 1.5 mm by 14 cm cross-

[8] J. C. Martini, S. D. Levene, D. M. Crothers, and P. T. Englund, *Proc. Natl. Acad. Sci. U.S.A.* **79,** 7664 (1982).

[9] L. S. Lerman and H. L. Frisch, *Biopolymers* **21,** 995 (1982).

[10] O. J. Lumpkin and B. H. Zimm, *Biopolymers* **21,** 2315 (1982).

[11] S. Levene and B. H. Zimm, *Science* (in press).

[12] H.-N. Lui-Johnson, M. R. Gartenberg, and D. M. Crothers, *Cell (Cambridge, Mass.)* **47,** 995 (1986).

section. Loading dyes containing glycerol or sucrose may be added to binding reactions immediately before electrophoresis. To obtain sharp protein–DNA complex bands, the final concentration of glycerol should be kept to a minimum ($\leq$10%, v/v). We generally electrophorese the complexes at 420 V in 0.5× TBE buffer [45 mM Tris, 45 mM boric acid, 1 mM ethylenediaminetetraacetic acid (EDTA), pH 8.3], maintaining a temperature of 20–25° in the gel with a constant temperature gel apparatus (Hoefer, San Francisco, CA) and a circulating refrigerated water bath (Neslab, Portsmouth, NH). Electrophoresis at high voltage and low ionic strength also contributes to band sharpening, since those factors act to minimize complex dissociation during electrophoresis. Modifications to the electrophoresis buffer are tolerated and often necessary. EDTA may be eliminated if divalent cations such as Zn(II) are essential for protein binding. Buffers with alternate pH may be used, for example if the protein is small and has an acidic isoelectric point, causing it to confer little or no retardation on the mobility of the DNA to which it is attached.[13] DNAs are generally detected by ^{32}P autoradiography, β scanning, or ethidium fluorescence.

Mapping Bend Locus by Circular Permutation of Sequence

A DNA bend reduces the end-to-end distance of a DNA fragment and therefore its electrophoretic mobility if it is located near the center of the molecule; a bend near a molecular end influences neither appreciably (see Fig. 1). This simple determinant of gross molecular shape is the basis of the circular permutation assay used to map the locus of DNA bending.[2] A tandem dimer or multimer of a binding site fragment is cloned and cleaved with restriction enzymes that cut only once within the sequence, resulting in a set of circularly permuted DNA molecules that differ in the position of the binding site relative to the molecular ends. Figure 2 shows the variation in gel mobility for two series of circularly permuted *lac* promoters, one wild type (lanes 1–5) and one mutated (lanes 6–10), bound by *E. coli* CAP protein.[12] The slowest moving species (lanes 1 and 6) position the protein-binding site near the molecular center.

CAP-induced DNA bending and not the influence of bound protein is thought to be responsible for the observed position-dependent variation in gel mobility; complexes with *lac* repressor, another globular protein, migrate equivalently regardless of binding site position.[2] However, DNA-binding proteins with extended and negatively charged domains, as pro-

[13] J. Carey, *Proc. Natl. Acad. Sci. U.S.A.* **85**, 975 (1988).

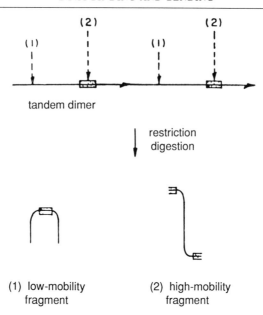

FIG. 1. Logic of the circular permutation experiment for determination of bend position.[2] A cloned tandem dimer or higher order repeat is cleaved with a series of restriction enzymes that cut once in the sequence of interest. Fragment (1), having a bend in the center, is highly bent and therefore of low gel mobility. Fragment (2) is nearly linear since the bend is near the end, and hence nearly normal in electrophoretic mobility.

posed for GCN4, may modulate electrophoretic mobility of protein–DNA complexes in a position-dependent manner without inducing DNA bends.[14]

For identification of the bend center one can use a plot such as that shown in Fig. 3, which is based on the data in Fig. 2. The mobilities of the complexes with circularly permuted DNAs are plotted against position of the DNA molecular end in the numbering system of the parent fragment. Distance moved down the gel reaches a maximum (corresponding to the minimum in the plot of Fig. 3) when the binding site, and hence the bend, is near the molecular end. Extrapolation of the lines on both sides of the minimum to their intersection enables identification of the bend center, often to within a few base pairs. If the DNA molecule has a more complex shape than provided by a single symmetric bend, the curve in Fig. 3 can be asymmetric, and the minimum need not correspond to the center of the local bend. For example, the minimum in Fig. 3 is displaced about half a

[14] M. R. Gartenberg, C. Ampe, T. R. Steitz, and D. M. Crothers, *Proc. Natl. Acad. Sci. U.S.A.* in press (1990).

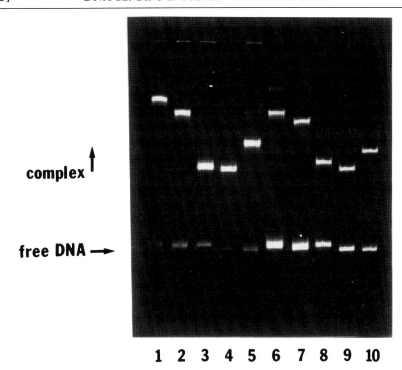

FIG. 2. Illustration of the circular permutation mobility assay.[12] Circularly permuted 203-bp *lac* promoter fragments complexed with CAP protein are shown in lanes 1–5, and a similar series is shown in lanes 6–10 for a mutated sequence (sy203). Electrophoresis is on a 10% polyacrylamide gel, 75 : 1 (w/w) acrylamide : bisacrylamide in TBE buffer.

helical turn downstream in the *lac* promoter from the center of the CAP binding site, an effect thought to be due to additional in-phase curvature in the promoter.[12]

The mobility of CAP protein complexed with the mutant DNA sequence in Figs. 2 and 3 is larger than that of the wild-type complex when the bend is near the center of the molecule, implying a smaller bend with the mutant sequence. In contrast, when the bend is near the end of the molecule, the mutant sequence provides a complex with lower gel mobility, an effect that was interpreted as a reflection of a bend in excess of 90°.[12]

Assigning DNA Bend Direction by Phase-Sensitive Detection

Static DNA bends are characterized by their direction relative to distal features, such as other DNA bends or bound proteins, as well as local helical parameters, including the major and minor grooves within the

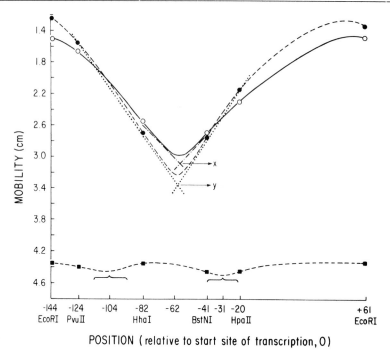

POSITION (relative to start site of transcription, 0)

FIG. 3. Mapping the bending locus.[12] Filled circles show the mobility of the CAP–wt203 complexes (Fig. 2); open circles indicate the CAP-sy203 complexes. Filled squares show the mobility of the naked DNAs, which run identically. The center of sequence symmetry is between −61 and −62. The bend center in the sy203 fragment, labeled x, is at about −58, and the estimated center of the wt203 bend (y) is at about −60. The regions of the naked DNA where a bend is suspected are indicated by braces.

binding site. Most protein-induced DNA bends examined are directed toward the protein; however, the structure of the DNase I–DNA cocrystal demonstrates that protein-induced bends need not be oriented in this fashion.[15] Structural variability extends further: at the center of some DNA-binding sites, such as those for γδ resolvase[16] and the nucleosome,[1] the major groove faces the protein, while at others, such as those for CAP[17] and typical dimeric prokaryotic repressors,[18–20] the minor groove faces the protein.

[15] D. Suck, A. Lahm, and C. Oefner, *Nature* (*London*) **332**, 464 (1988).

[16] G. F. Hatfull, S. Z. Noble, and N. D. F. Grindley, *Cell* (*Cambridge, Mass.*) **49**, 103 (1987).

[17] J. Waricker, B. P. Engelman, and T. A. Steitz, *Proteins* **2**, 283 (1988).

[18] C. Wohlberger, Y. Dong, M. Ptashne, and S. C. Harrison, *Nature* (*London*) **335**, 789 (1988).

[19] S. R. Jordan and C. O. Pabo, *Science* **242**, 893 (1988).

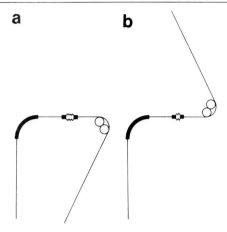

FIG. 4. Logic of the phase-sensitive experiment for determination of bend direction.[3] Shown are (a) cis and (b) trans isomers of DNA constructs containing both A tract (heavy line) and CAP-induced bends. Phasing between the two bends is controlled by the length of the variable linker region in the center of the molecule.

The relative directions of two DNA bends located in the same fragment may be determined from the electrophoretic mobilities of isomers that differ by the length of the intervening DNA; as the orientation of the bends varies, the mobility of the protein–DNA complexes should increase from a minimum when the end-to-end distance is short to a maximum when the end-to-end distance is long.[3,21] Figure 4 provides a schematic representation of the cis and trans isomeric structures on which this assay depends. In this example, a fragment containing a CAP binding site is fused to a fragment containing phased adenine (A) tracts. The spacing between the two loci is varied over a full turn of the double helix by the insertion of linkers of variable length. Adenine tracts provide an excellent standard because the direction[22] and magnitude[23] of the A tract bend have been determined. Exogenous bends are not required if only the relative orientation of multiple endogenous bends is of interest.[24,25]

[20] A. K. Aggarwal, D. W. Rodgers, M. Drottar, M. Ptashne, and S. C. Harrison, *Science* **242**, 899 (1988).
[21] J. J. Salvo and N. D. F. Grindley, *Nucleic Acids Res.* **15**, 9771 (1987).
[22] H.-S. Koo and D. M. Crothers, *Proc. Natl. Acad. Sci. U.S.A.* **85**, 1763 (1988).
[23] H.-S. Koo, J. Drak, J. A. Rice, and D. M. Crothers, *Biochemistry* **29**, 4227 (1990).
[24] J. J. Salvo and N. D. F. Grindley, *EMBO J.* **7**, 3609 (1988).
[25] U. K. Snyder, J. F. Thompson, and A. Landy, *Nature (London)* **341**, 255 (1989).

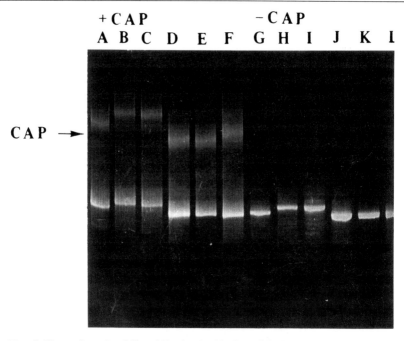

FIG. 5. Illustration of mobility shifts for the binding of CAP to the variable-linker constructs.[3] Lanes A–F, CAP–DNA complexes in which the distance between the centers of the CAP and A tract bends varies from 104 to 94 bp. Electrophoresis was in a 5% polyacrylamide gel, 39:1 acrylamide:bisacrylamide, in TBE buffer.

As shown in Figs. 5 and 6, the gel mobilities of CAP-bound spacer constructs vary sinusoidally with a period of approximately 10 bp in their dependence on linker length. The CAP-bound construct with an 18-bp insert (lane B, Fig. 5) migrates slowest, and is thereby identified as the cis isomer, in which the molecule is nearly planar and the two bends are in essentially the same direction (see Fig. 4). Determination of bend direction in the local coordinate frame of the nucleotides at the bend center requires that one know the number of helical turns between bend centers.

This experiment was originally designed to remove a residual ambiguity concerning the direction of curvature of A tracts, namely whether the bend is directed toward the major or the minor groove at the A tract center.[3] Given that DNA bends toward the *minor* groove at the center of the CAP binding site, only integral or half-integral numbers of helical turns need be considered between the positions of the CAP and A tract bends in the cis isomer. The centers are separated by 101.5 bp; with a helical repeat of 10.5 to 10.7 bp, the only acceptable answer is 9.5 helical turns.

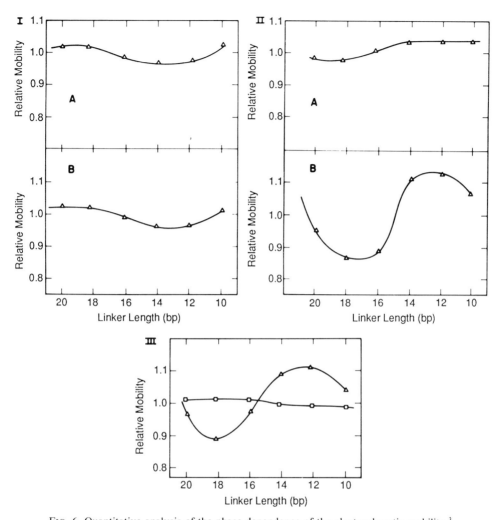

FIG. 6. Quantitative analysis of the phase dependence of the electrophoretic mobility.[3] (I) and (II) show relative mobility, normalized to the average for the set of values, for the *lac* repressor (I) and CAP protein (II) DNA complexes. In (I) and (II) the top graph (A) is for the free DNA, and the lower graph (B) for the protein–DNA complex. Note that for *lac* repressor, unlike the CAP–DNA complex, the fractional variation in gel mobility is very similar for free DNA and the complex. This is the behavior expected when a DNA fragment has some intrinsic curvature, which is unaltered by protein binding. (III) shows the ratio of the normalized mobilities for the protein–DNA complex, divided by the corresponding normalized free DNA mobility. Squares show values for CAP–DNA, and triangles are for repressor–DNA complexes. Note the strong periodic variation of relative mobility for the CAP–DNA complex normalized values. This is the behavior expected when a protein-induced bend is superimposed on intrinsic fragment curvature. From the mobility minimum, the construct with linker length 18 is the cis isomer. The lack of significant variation of normalized mobility in (III) confirms the lack of DNA bending by *lac* repressor.

Hence, with a half-integral number of turns, the coordinate frame for the A tract bend is half a turn displaced from that for the CAP bend, and the overall center of the A tract bend is directed toward the *major* groove at its center. (Individual A tracts bend toward the minor groove at their centers, but the center of the collective bend produced by four A tracts is toward the major groove in the central 5-bp segment that lies between A tracts.) As a result of this and related experiments, the A tract bend can now be used as a comparative standard in experiments designed to determine the direction of bending induced by proteins whose structure is less well understood than CAP. Almost no variation in mobility is observed for the same constructs when bound by *lac* repressor,[3] as expected for a nonbending protein.

The circular permutation and phase-sensitive detection assays are complementary in that the first maps bend loci irrespective of bend direction while the second assigns bend direction but provides insufficient information to map bend loci precisely.

Determination of Bend Magnitude by Comparison to Calibrated Standards

The magnitude of the protein-induced bends in a DNA fragment can be estimated by comparison of the electrophoretic mobility of the fragment with those of molecules containing well-characterized bends.[26] Specifically, a calibration curve is generated from the mobilities of DNA constructs in which the protein-binding site has been replaced by DNA sequences containing from three to nine adenine tracts, repeated with a period of 10.5 bp (Fig. 7). These "in-phase" bends add to produce molecules with increasing curvature, known to correspond to a deflection of the helix axis of about $18°/CA_6C$ repeat.[23] To ensure that the differences in mobility are caused predominantly by the shape of the DNA molecule, the total lengths of the DNA fragments are held nearly constant and a protein-binding site is included at the end of the calibration fragments (Fig. 7).

In Fig. 8, a typical calibration curve shows a plot of relative mobility versus the square of the number of resident A tracts. For molecules with less than six A tracts, the data are adequately fit by a second order polynomial. The mobility of the fragment with the centrally bound CAP is also shown. The interpolated bend magnitude is roughly 5.6 A tract equivalents or 101°, a value that is nearly invariant to changes in acrylamide concentration.[26]

[26] S. S. Zinkel and D. M. Crothers, *Biopolymers* **29**, 29 (1990).

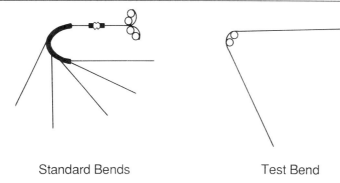

Standard Bends   Test Bend

FIG. 7. Logic of the experiment to determine the magnitude of the CAP-induced bend by comparison with A tract bends.[26] The standard bends consist of a set of DNA molecules containing runs of from three to nine A tracts phased at 10.5-bp intervals, located in the center of the molecule; the A tract bend in these sequences is known to be 18°/A tract.[23] A CAP-binding site at the end of the standard fragments corrects for gel retardation due to bound CAP protein, which is shown both cis and trans to the A tract bends; experiments showed that the relative orientation has little effect when the CAP bend is at the end of the molecule.[26] The magnitude of the CAP-induced bend is taken to be equal to that provided by A tracts in the standard-bend DNA molecule, the mobility of which (with CAP bound) matches the mobility of the test bend molecule with CAP bound.

An alternative approach[27] compares the ratio of the mobilities of molecules that have a bend of unknown magnitude in the center or at the end to the same ratio for molecules with standard bends.

Sequence Perturbation

Experimental evidence continues to accumulate showing that DNA flexibility and bending anisotropy play a role in protein recognition. DNA sequences isolated from chicken erythrocyte nucleosomal core particles are probably selected for favorable DNA binding and bending, and show a corresponding sequence signature in their nonrandom distribution of A · T- and G · C-rich sequences.[4,5] The observations include an increased probability for G · C-rich regions to occupy positions where the major groove is compressed as the DNA curves around the protein, whereas A · T-rich regions are favored in regions subjected to minor groove compression. These sequences preferentially occupy similar orientations in small, highly strained DNA circles,[4] suggesting that DNA flexibility may be a driving force in nucleosome placement. Similar rules have been deduced from the properties of mutants of the DNA-binding site for

[27] J. F. Thompson and A. Landy, *Nucleic Acids Res.* **16**, 9687 (1988).

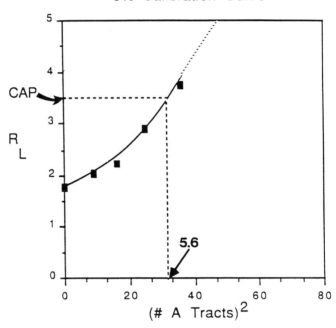

FIG. 8. Calibration curve for measurement of the magnitude of the CAP-induced bend.[26] The R_L value is the apparent number of base pairs in the CAP–DNA complex, determined by comparison with a ladder of normal DNA fragments, divided by the actual number of base pairs in the test fragment. Deviation of R_L from 1 when the number of A tracts is 0 reflects the gel retardation due to CAP protein bound at the end of the corresponding standard bend fragment. The plot of R_L versus the square of the number of A tracts was fitted to a quadratic polynomial for the purpose of interpolation to find the number of A tracts (5.6) that corresponds to the R_L value (3.49) measured for CAP bound to the center of the test bend fragment. Electrophoresis was on a 5% polyacryamide gel (acrylamide : bisacrylamide, 39 : 1).

E. coli CAP protein[6]; this system and the sequence preferences of nucleosomes are discussed more extensively below.

Another example is provided by the phage 434 repressor system, for which the protein-binding constant is affected by mutations in the central 4 bp of the binding site, shown by diffraction studies not to contact the protein.[28] These effects correlate with the stiffness of the central 4-bp segment, suggesting that the ability of this segment to bend and twist affects the binding constant by providing optimal protein–DNA contacts

[28] G. B. Koudelka, S. C. Harrison, and M. Ptashne, *Nature (London)* **326,** 886 (1987).

in more distal regions of the binding site.[29] However, effects due to bending may be of secondary importance relative to twisting in this case, since the mutated segment may have a critical effect on the helical phasing between the two DNA segments whose interactions dominate specific binding. We have reduced the importance of this factor in the case of CAP protein by focusing on bends that lie distal to the consensus recognition sequence, as measured from the central axis of the binding site. In the case of nucleosomes, it was possible to design sequences to test directly the importance of helical twist in determining relative binding constants.[30]

Our reported studies on CAP and nucleosomes were designed primarily for the purpose of elucidating the sequence dependence of DNA bending. In retrospect, it is clear that the results further refine our understanding of the nature and extent of the bent regions in the protein–DNA complexes. For example, the sites one helical turn and 1.5 helical turns from the dyad axis of the CAP–DNA complex showed maximal sensitivity in the dependence of bending on sequence.[6] However, no sequence-dependent modulation of bending could be detected at the locus two full turns removed (except when a permanent DNA bend resulting from an A tract was inserted). These results indicate that DNA is bent primarily by alternately compressing its major and minor grooves as it wraps around the protein, and that the bent region covers 1.5 helical turns on each side of the bending locus. In the case of nucleosomes, we were able to show that sequences that favor bending have little preference for any special site on the surface of the histone core, implying relatively uniform bending.[30] We expect that future studies of other complexes by these methods will be similarly informative about the position and direction of induced bends.

Escherichia coli CAP Protein

The CAP binding site consists of a consensus binding domain that interacts with the helix–turn–helix recognition motif of the protein and a distal binding domain that flanks the consensus and presumably interacts nonspecifically with a ramp of positive electrostatic potential that is present on the protein surface.[6,12,17,31] To examine the influence of DNA sequence on induced bending and binding, we generated an extensive library of CAP binding site mutants. We deliberately restricted the size of randomized regions in each mutant, so that the sequences examined can be regarded as local perturbations on the wild-type sequence. The extent

[29] M. E. Hogan and R. H. Austin, *Nature (London)* **329**, 263 (1987).
[30] T. E. Shrader and D. M. Crothers, *J. Mol. Biol.* in press.
[31] T. A. Steitz, *Q. Rev. Biophys.* **23**, 205 (1990).

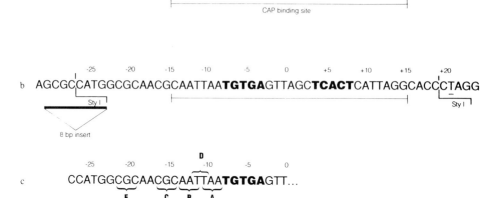

FIG. 9. Constructs for sequence-perturbation studies of CAP-induced bending.[6] (a) The sequence and features of the wild-type CAP-binding site in the *lac* promoter. The underlined region denotes the thermodynamically defined binding domain,[12] boldface type indicates nucleotides in the consensus sequence region, and an arrow identifies the pseudodyad axis of symmetry. Positions are numbered relative to the transcription start site. (b) The sequence changes introduced to create *Sty*I restriction sites; dinucleotide steps are numbered relative to the pseudodyad axis. A 211-bp *lac* promoter fragment containing this sequence was cloned into the *Eco*RI site of pGEM-2 (Promega). (c) The location of sites fully randomized by oligonucleotide replacement. Each site was mutagenized independently of the others.

of bending and stability of complexes of the mutant binding sites were compared to those of the wild type by nondenaturing gel electrophoresis.

Methods. Engineering of unique *Sty*I restriction sites flanking the CAP binding site in the *lac* promoter permitted derivatization of the natural sequence by replacement with synthetic oligonucleotides (Fig. 9a). Sequence changes were confined to the distal binding domains to avoid disruption of critical recognition contacts. The oligonucleotides contained regions of randomized sequence two or three consecutive base pairs in length, produced by degenerate synthesis. Each of the five sites A–E (Fig. 9b) were mutagenized independently of one another. Individual binding site clones were resolved from the pool of degenerate constructs by bacterial transformation and identified by sequencing of supercoiled miniprep DNA of individual colonies.[32]

Mutant binding sites, centrally located within 211-bp *lac* promoter fragments, were analyzed both for extent of induced DNA bending, taken

---

[32] Promega Biotech Technical Manual (1986).

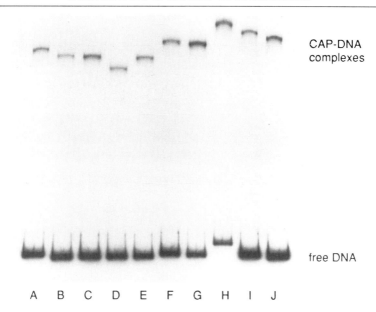

FIG. 10. Electrophoresis gel showing the variability of the mobility of CAP protein bound to DNA-binding site mutants.[6] Electrophoresis was on a 10% polyacrylamide gel, 75 : 1 acrylamide : bisacrylamide. The sample in lane H has an A tract bend centered at position −22, in phase with the CAP-induced bend, lane E contains an out-of-phase A tract bend centered at position −27, and lane F contains the wild-type sequence. The other lanes illustrate the electrophoretic properties of some of the mutants A–D in Fig. 9.

to be a simple function of the electrophoretic mobility anomaly, and relative CAP binding affinity. In order to calibrate the dependence of mobility on extent of bending, constructs were made in which A tracts were placed in phase (A tract center two helical turns from the central dyad axis) and out of phase (A tract 2.5 turns from the dyad). These constructs had mobilities about 20% smaller and larger, respectively, than the wild-type CAP complex. Figure 10 illustrates this analysis: lanes E, G, and H contain the wild-type binding sequences except that in E an out-of-phase A tract has been added in the region flanking the binding site, and in lane H the added A tract is in phase with the CAP bend; the fragment in lane G has no added A tract. (The lengths of the fragments in E, G, and H were held constant within 1 bp.) Since an A tract bends the helix by about 18°, we could, by extrapolating the nearly linear relationship between the three calibration points, assign a change in the CAP-induced bend angle to each of the mutants. The variation ranged from increased bending of about 10° to decreased bending by about 30°, the latter representing a substantial fraction of the total estimated bend angle of 101° for the CAP–DNA complex in solution.

Relative binding constants (K_{rel}) can be measured with high accuracy using a competition method and gel electrophoresis analysis.[33,34] A comparison sequence (usually the wild type) is incorporated in a shorter (<211 bp) DNA fragment, allowing the two free DNAs and their protein complexes to be separated on the basis of their electrophoretic mobilities. Standard nondenaturing gels with ^{32}P-labeled DNA fragments then yield lanes with variable amounts of radioactivity in the four bands corresponding to wild-type and mutant free DNA and their protein complexes. Analysis of such experiments is greatly facilitated by use of a two-dimensional β scanning device, which enables direct quantitation of the counts (β) in each band from the gel scan. We used a Betscope 603 Blot Analyzer (Betagen, Waltham, MA) for this purpose.

Calculation of K_{rel} from the count values β for each band uses the equation

$$K_{rel} = K_{mut}/K_{wt}$$
$$= \beta_{mut}(\text{complex}) \, \beta_{wt}(\text{free DNA})/\beta_{mut}(\text{free DNA}) \, \beta_{wt}(\text{complex})$$

where mut and wt refer to mutant and wild type, respectively. From the K_{rel} values, the difference in standard free energy ($\delta\Delta G°$) of binding between mutant and wild-type DNAs can be calculated from the thermodynamic equation $\delta\Delta G° = -RT \ln K_{rel}$.

Results. The data resulting from measurements of mobility changes and relative binding constants for a series of mutants provide a quantitative view of the influence of DNA sequence on these parameters. Figure 10 shows a polyacrylamide gel analysis of the mobility of a number of DNA binding site mutants complexed with DNA fragments mutated at the CAP binding site. We have described elsewhere how the results can be analyzed mathematically to rank particular dinucleotides for their contribution to bending and binding.[6,35]

A simpler, nonmathematical way to present the data is presented in Fig. 11, which shows the sequence of the dinucleotide changes at each position that yield the largest perturbations from the mobility of the wild-type complex. From the extremes, which are evident at positions one and 1.5 turns from the dyad axis, we deduce that bending is most sensitive to mutations at those positions, where the minor and major grooves turn successively to face the protein. Note that the mobility is highest (least

[33] M. G. Fried and D. M. Crothers, *Nucleic Acids Res.* **9,** 6505 (1981).
[34] M. M. Garner and A. Revzin, *Nucleic Acids Res.* **9,** 3047 (1981).
[35] M. R. Gartenberg, Ph.D. Thesis, Yale University, New Haven, Connecticut (1990).

bending) when G · C-rich dinucleotides are found at positions 10 and 11 nucleotides from the dyad, whereas mobility is lowest (greatest bending) when A · T-rich dinucleotides are found there. At sites 15 and 16, the rules are inverted, corresponding to the opposite direction of bending the DNA, toward the major groove in this case. (Although the diagram in Fig. 10 features only the dinucleotides at the extremes, we found that the properties of other A · T-rich and G · C-rich dinucleotides cluster near the values shown at the extremes.) Hence the data define the bending site and direction to base pair accuracy, and show that it does not extend to the locus two turns from the dyad axis.

Measurement of the relative binding constant for 56 DNA-binding sites mutated at sites A and D in Fig. 9 shows that there is substantial although imperfect correlation between binding strength and magnitude of bending.[35a] In general, A · T-rich dinucleotides at positions 10 and 11 from the dyad yield stronger binding, by as much as a factor of 50, than do G · C-rich sequences.

Nucleosomes

Nucleosomal core particles contain about 146 bp, or 14 turns of the DNA duplex, in close association with a protein octamer containing two each of the core histones H2A, H2B, H3, and H4. The DNA is wrapped around the outside of the octamer in about 1.75 superhelical turns, and bends toward the protein primarily by alternately compressing the major and minor grooves. As a result, the structure contains about 14 × 2 or 28 bending loci, which are phased at the helical repeat of DNA in the nucleosome. By use of synthetic DNA molecules one can investigate the relative preference when bulk DNA is replaced by special sequences having G · C- and A · T-rich segments that alternate every ~5 bp; we have called such sequences "anisotropically flexible" DNA because they are designed to conform to the rules for preferred bending alternately toward the major and minor grooves.[4–7]

The flexible sequences can be placed in various positions relative to the dyad axis of the core particle to determine whether the energetic demands on anisotropic bending are more pronounced in some regions than others. In addition, the sequence repeat can be varied to investigate the helical repeat of the nucleosome; one expects optimal incorporation of DNA molecules in which the anisotropically flexible regions are phased in exact compliance with the preferred helical repeat of DNA on the nucleosome. We describe here experiments illustrating the general

[35a] –. Dalma-Weiszhausz, M. R. Gartenberg, and D. M. Crothers, to be published.

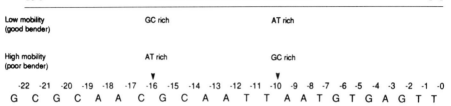

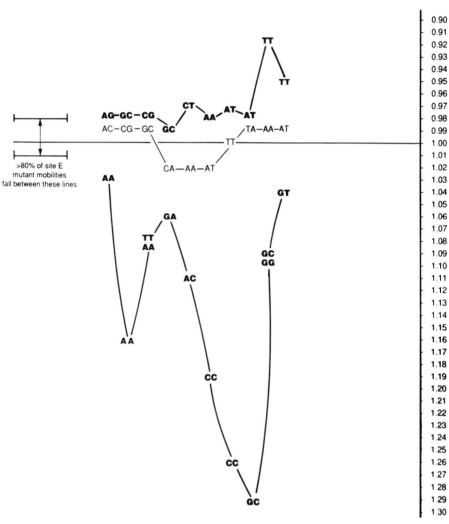

method, which involves competitive reconstitution of different DNA sequences into nucleosomal core particles, using a labeled DNA fragment to compete against a standard unlabeled sample. The extent of incorporation is determined from the percentage of a labeled DNA fragment that is converted to nucleosomal complex, which is separated from free DNA by electrophoresis on a polyacrylamide gel.

Methods

The artificial nucleosome positioning sequences used in this study are constructed by cloning multiple copies of synthesized oligonucleotides into a modified pGEM2 vector (Promega Biotech, Madison, WI). A unique, nonsymmetric AvaI restriction site is introduced into the original vector; AvaI cleavage of the site produces four-nucleotide, 5'-overhanging ends[36] that are complementary to each other but not self-complementary. Synthetic duplex oligonucleotides must have corresponding unique ends, each designed to hybridize with only one end of the cleaved vector, thus ensuring that all copies of the monomer assemble in a head-to-tail orientation rather than in a random manner.

In addition to forcing direct repeats on ligation, the oligonucleotides are constructed with a complete AvaI recognition site at only one end; incomplete ends contain the proper four-nucleotide overhang but not the terminal recognition base pair. Direct repeat of these sequences produces a single AvaI site at one end of the multimer. As a result of recreation of a single endonuclease site, different types of oligonucleotides may be combined in a defined order by successive cloning steps.

Oligonucleotides were synthesized on an Applied Biosystems (Foster City, CA) DNA synthesizer and purified by gel electrophoresis on 12% polyacrylamide gels in TBE buffer (45 mM Tris, 45 mM boric acid, 1 mM EDTA, pH 8.3) containing 50% (w/v) urea. This procedure removes most

---

[36] J. L. Hartley and T. J. Gregori, *Gene* **13,** 347 (1981).

FIG. 11. Sequence dependence of the mobility of CAP–DNA complexes. The dinucleotide estimated to confer the largest (bottom of the graph) and smallest (top of the graph) mobility on the CAP–DNA complex. The high-mobility sequences show extremes at positions − 10, − 11, and again at − 16, sites where the DNA minor and major grooves, respectively, face the protein. This observation implies maximum sensitivity of bending to sequence at the positions where the DNA bends alternately by roll into the minor and major grooves. Note that at positions − 10, − 11, the worst benders (greatest mobility) are the G/C-rich dinucleotides GC and CC, whereas the best benders are AT and AT. At the − 16 locus AA is the worst bender and GC is the best. The wild-type sequence is shown near the middle line in the figure.

of the longer and shorter oligonucleotides that represent ~10% of a crude DNA synthesis. Approximately 10 OD_{260} units of oligonucleotide is loaded per gel lane. The correct oligonucleotide band is identified by its UV shadow, excised from the gel, and isolated by soaking the crushed gel slice 8 hr at 65°. Acrylamide gel fragments are removed by two rounds of spinning the sample in an Eppendorf microfuge to pellet insoluble material. No further purification is required to anneal the single-stranded oligonucleotides and ligate the resulting duplexes into multimers.

Ten micrograms of the purified single strands is phosphorylated using 2–5 U of T4 polynucleotide kinase in T4 ligase buffer (25 mM Tris, pH 7.8, 10 mM $MgCl_2$, 4 mM 2-mercaptoethanol, 0.4 mM ATP). The strands are then mixed and annealed by slow cooling from 90 to 4°. Approximately 10 μg of the double-stranded oligonucleotide is polymerized by ligation for 15 min at room temperature (18–30°) and the ligation mixture is separated on a 6% polyacrylamide TBE gel. The correct multimer bands are visualized by staining in a 1 μg/ml ethidium bromide solution (exposure to ethidium bromide is kept as short as possible, ~5 min) and purified as above.

Multimers are cloned into the dephosphorylated asymmetric AvaI site of the pGEM2-Ava vector. One-half of the DNA isolated from each multimer band is used in the cloning step. Clones are screened first by restriction mapping and finally by dideoxy sequencing of DNA samples purified by the alkaline lysis method.[32] From the DNA isolated from 3 ml of an overnight growth of DH1 E. coli cells, 10% is required to restriction map and 40–50% is required for dideoxy sequencing. The sequences of the clones revealed that errors in the body of the multimer are extremely rare, while errors at the junction between the oligonucleotide multimer and the phosphatased vector occurred in 5–10% of the clones.

For the clones to be used for nucleosome reconstitution, DNA from 0.5 liter of cells grown 12–24 hr is isolated by an expanded alkaline lysis procedure (technical manual, Promega Biotech). The average yield from this procedure was about 1 mg plasmid DNA/liter of cells.

To purify a fragment for reconstitution, several different pairs of restriction enzymes are used. These enzymes remove the oligonucleotide-containing section surrounded by enough flanking vector DNA to create a fragment of the overall correct size. The EcoRI and HindIII sites are separated by 62 bp in pGEM2/Ava. Excision of a fragment constructed of five repeats of a 20-bp oligonucleotide with these enzymes would produce a 162-bp fragment, suitable for mononucleosome reconstitution, which could be labeled at both ends by the Klenow fragment of DNA polymerase I. To isolate fragments for footprinting studies, the pairs of enzymes EcoRI + PvuII or SacI + HindIII are used. These pairs of enzymes generate fragments that can be singly end labeled by Klenow fragment of DNA polymerase I and an appropriate choice of ^{32}P-labeled deoxynucleotide triphosphates. Digestions are made on ~50 μg of plasmid DNA. A 5% polyacrylamide TBE gel is used to separate the insert from the vector. Inserts are visualized by their UV shadows and purified as above.

Nucleosome reconstitution has recently been extensively reviewed.[7,30,37] The competitive reconstitution procedure that we and others have used is similar to previous salt exchange procedures with the addition of bulk competitor DNA to lower the total protein-to-DNA ratio.[38] The general approach (Fig. 12) is to mix chromatin (lacking its linker histones) or core nucleosomes with a labeled DNA fragment at a high salt concentration, where the histone octamers are free to exchange. After a suitable incubation time the exchange is stopped by lowering the ionic strength to a level where the histones are locked in place. Our recipe involves mixing ~5 μg of stripped chromatin[39] with 20 μg of bulk DNA and 0.1 μg of the labeled DNA fragment whose reconstitution is being tested. These components are incubated 30–45 min in 70 μl of buffer containing 1 M NaCl, 100 μg/ml albumin, and 0.1% Nonidet P-40. The ionic strength is lowered by three 210-μl additions of low salt buffer [1 mM Tris (pH 8.0), 0.1 mM EDTA] to bring the final NaCl concentration to 0.1 M.

The fraction of a DNA fragment that has been reconstituted into nucleosomes is determined by measuring the amount of radioactivity in the free and complexed band of a native polyacrylamide gel. This approach provides more information concerning possible degeneracy in the position of the histone octamer on the DNA fragment than agarose gels. We use a 5% gel [75 : 1 (w/w), acrylamide : N,N'-methylenebisacrylamide] to separate free from complexed DNA. In this system, as in others we have examined, a high acrylamide-to-bisacrylamide ratio produces the sharpest bands. The fraction of DNA reconstituted can be measured by using an autoradiogram of the dried gel to locate and excise the desired bands, which are then counted in a scintillation counter. Alternatively, the appropriate regions of the dried gels are counted directly on a Betascope 603 blot analyzer.

Control experiments from several laboratories suggest that the fraction of nucleosomes that are reconstituted reflects the equilibrium binding ratios at 1 M NaCl. First, neither lengthening the time of incubation in high salt buffer, nor the rate at which the ionic strength is lowered, changes the ratio of labeled fragment that is reconstituted into nucleosomes.[7] Sec-

[37] S. D. Jayasena and M. J. Behe, *J. Mol. Biol.* **208,** 320 (1986).
[38] N. Ramsay, *J. Mol. Biol.* **189,** 179 (1986).
[39] L. C. Lutter, *J. Mol. Biol.* **124,** 391 (1978).

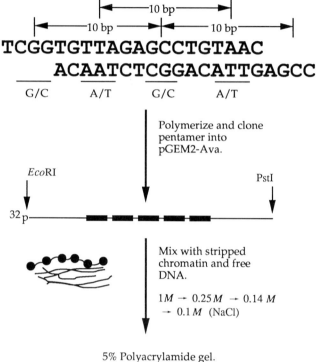

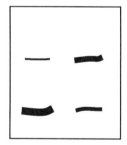

5% Polyacrylamide gel.

FIG. 12. Design and testing of anisotropically flexible DNA sequences for favorable incorporation into nucleosomes.[7] The anisotropically flexible sequence TG is shown at the top, with emphasis on its alternation of A/T and G/C-rich regions every 5 bp. Head-to-tail clones of this sequence are excised as restriction fragments of ~170 bp, and subjected to a competitive salt gradient procedure for chromatin reconstitution, followed by analysis of the extent of nucleosome formation on a polyacrylamide gel.

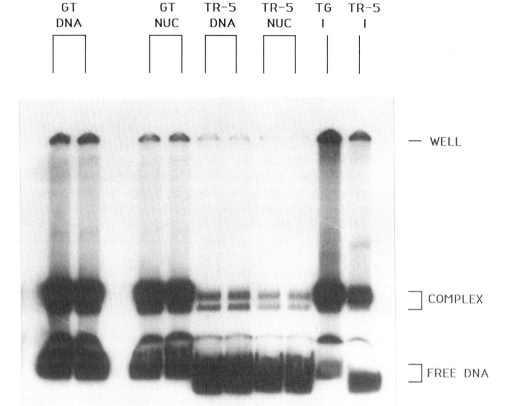

FIG. 13. Nucleosome reconstitution reflects equilibrium histone DNA binding. Fragments containing five copies of the GT (5'-TCGGGTTTAGAGCCTGTAAC-3') or the TR oligonucleotide (5'-TCGGAAGACTTGTCAACTGT-3') were reconstituted starting from free DNA on nucleosomal complexes. The fraction of a labeled fragment in the final nucleosomal complex is independent of its addition as free DNA (GT-DNA and TR-DNA) or as nearly completely complexed with histones (GT-NUC and TR-5-nuc). The nucleosomes used to add labeled fragments as preexisting complexes are shown in the two rightmost lanes. Additionally, it is clear that a very small fraction (3–5%) of the labeled DNA fragment exists in aggregrated forms that do not enter the gel.

ond, the same final labeled nucleosome-to-labeled free DNA ratio is reached regardless of whether the labeled fragment is introduced to the high salt incubation as free DNA or in nucleosomal complexes (Fig. 13). Finally, when a trace amount of labeled free nucleosomal DNA was incubated in high salt with otherwise identical unlabeled DNA and unlabeled nucleosomes at several protein to DNA ratios, the final fraction of labeled nucleosomes, after the ionic strength was lowered, reflected the percentage of free DNA in the original mixture.[37]

Results

To calibrate the size of the nucleosome positioning signal of our synthetic sequences, we determined the amount of anisotropically flexible DNA needed to mimic the histone binding affinities of some natural nucleosome positioning sequences. The fragments studied (Fig. 14) contain increasing amounts of artificial sequences roughly centered in fragments of nearly constant overall length. The molecule derived from two repeats of the 20-bp TG oligonucleotide in a 176-bp fragment binds histones as tightly as do the natural 5S RNA-encoding nucleosome-positioning DNA sequences.[40,41] Interestingly, the required 40-bp synthetic region is approximately the size of the region identified as essential for positioning in some natural fragments.[39,42]

In a second series of experiments we determined the optimal repeat of flexible A/T- and G/C-rich regions for histone binding.[30,43] We reasoned that this optimum should coincide with the most favored helical repeat of DNA on the nucleosome, whose value remains a point of controversy. Based on nuclease cleavage patterns and periodicities extracted from sequencing data, nucleosomal DNA has long been argued to be overwound.[4,5,44] However, more recent experiments involving the reconstitution on nucleosomes on small DNA circles argue for an unaltered value of the DNA twist.[45,46] Our system of repeated oligonucleotides, for which precisely defined changes of the phasing of flexible regions can be made, offered a new approach to this problem.

Figure 15 shows the series of oligonucleotides of differing lengths that were synthesized to study this problem. Nucleosomes were reconstituted onto DNA fragments containing five tandem repeats of each sequence; Fig. 15 shows the reconstitution energies, relative to a standard flexible sequence. The free energy minimum near a sequence repeat of 10.1 bp clearly supports the idea of DNA overwinding on the histone surface,

[40] R. T. Simpson and D. W. Stafford, *Proc. Natl. Acad. Sci. U.S.A.* **80,** 51 (1983).
[41] D. Rhodes, *EMBO J.* **4,** 3473 (1985).
[42] P. C. Fitzgerald and R. T. Simpson, *J. Biol. Chem.* **260,** 15318 (1985).
[43] T. E. Shrader, Ph.D. Thesis, Yale University, New Haven, Connecticut (1990).
[44] J. T. Finch, L. C. Lutter, D. Rhodes, R. S. Brown, B. Rushton, M. Levitt, and A. Klug, *Nature (London)* **269,** 29 (1977).
[45] I. Goulet, Y. Zivanovic, and A. Prunell, *J. Mol. Biol.* **200,** 253 (1988).
[46] Y. Zivanovic, I. Goulet, –. Revet, M. Le Bret, and A. Prunell, *J. Mol. Biol.* **200,** 267 (1988).

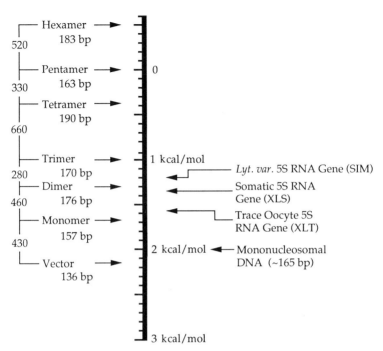

GT multimers
(total length)

FIG. 14. Nucleosome binding as a function of the length of anisotropically flexible DNA.[7] The free energies of reconstitution of three natural nucleosome-positioning sequences compared with designed sequences containing various lengths of flexible DNA. Sequences that reconstitute most favorably are at the top of the diagram. The binding of the best natural sequences can be mimicked with approximately 40 bp of repetitive DNA. Additionally, the binding free energy increment for each additional flexible region is quite constant. The total length of each multimer (vector DNA plus the indicated number of 20-bp GT oligonucleotides) is given below the multimer name. The difference in binding free energy between fragments is given at the extreme left. These free energies are all relative to the GT pentamer, which binds histones with essentially the same affinity as the TG pentamer.

since the optimal repeat is significantly shorter than the 10.5- to 10.6-bp helical repeat of B-DNA in solution.[47,48] This difference between optimal phasing for nucleosome formation and solution helical repeat is too large to be explained solely on the basis of the helical path of the DNA on the nucleosome surface. Consequently we believe that there is a true change

[47] D. Rhodes and A. Klug, *Nature* (*London*) **292**, 378 (1981).
[48] L. J. Peck and J. C. Wang, *Nature* (*London*) **292**, 375 (1981).

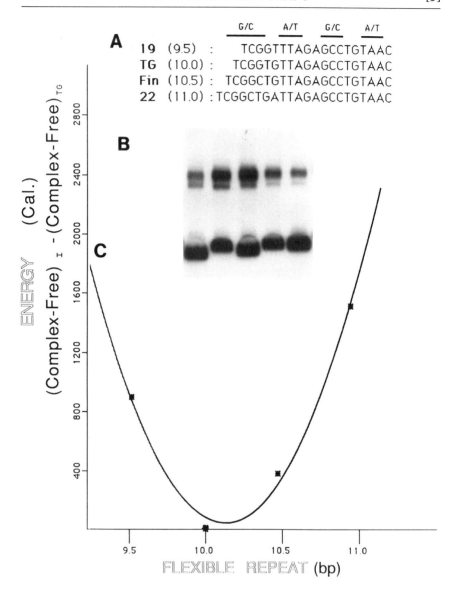

A

			G/C	A/T	G/C	A/T
19	(9.5)	:	TCGGTTTAGAGCCTGTAAC
TG	(10.0)	:	TCGGTGTTAGAGCCTGTAAC
Fin	(10.5)	:	TCGGCTGTTAGAGCCTGTAAC
22	(11.0)	:	TCGGCTGATTAGAGCCTGTAAC

B

C

ENERGY (Cal.)

$(\text{Complex-Free})_\text{I} - (\text{Complex-Free})_\text{TG}$

2800
2400
2000
1600
1200
800
400

9.5   10.0   10.5   11.0

FLEXIBLE REPEAT (bp)

M7.208-9-15

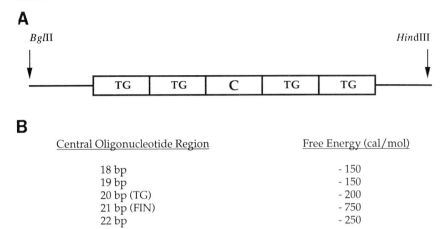

A

*Bgl*II   *Hind*III

| TG | TG | C | TG | TG |

B

Central Oligonucleotide Region	Free Energy (cal/mol)
18 bp	- 150
19 bp	- 150
20 bp (TG)	- 200
21 bp (FIN)	- 750
22 bp	- 250

FIG. 16. Helical repeat near the nucleosome dyad.[30] (A) Schematic sequence of fragments used to determine the helical repeat near the nucleosome dyad. Each *Bgl*II plus *Hind*III fragment contains DNA sequences derived from five oligonucleotides. The two TG oligonucleotides and vector DNA at either end of each fragment make up contant regions, common to all sequences. The central region is variable and is derived from one of several oligonucleotides. (B) Central oligonucleotides for the molecules tested. Also given are reconstitution free energies relative to a fragment containing five 20-bp TG oligonucleotides. For the nucleotide sequences of the oligonucleotides, see Fig. 11. Several of these sequences form nucleosomes more favorably than the reference sequence, including 20(TG), which contains the same oligonucleotide-derived region as the reference sequence. This is presumably due to the use of the pJC vector, rather than pGEMA2/*ava*, in the construction of these fragments.

FIG. 15. Optimal phasing of flexible regions.[30] (A) Oligonucleotides used to determine the optimal phasing of flexible (A/T or G/C rich) regions for nucleosome reconstitution. Only the top strand of each molecule is shown. All oligonucleotides were double-stranded with 4-bp asymmetric *Ava*T overhangs. (B) Polyacrylamide gel of competitive reconstitution of fragments constructed from five copies of the above oligonucleotides centered in a fragment of ~160- to 170-bp total length. From left to right, the fragments contain five copies of the oligonucleotides 19, TG, GT, FIN, and 22. The sequence of GT is identical to that of TG with inversion of the first TpG dinucleotide. These sequences (TG and GT) have essentially the same free energies of nucleosome reconstitution (≡0). (C) Free energy of nucleosome reconstitution plotted as a function of flexible repeat. Flexible repeat is defined as one-half the length of the oligonucleotide monomer that makes up the central 95- to 110-bp of each fragment. The curve represents a least-squares fit of a parabola to the data. Free energies are relative to the free energy for nucleosome reconstitution of a fragment containing five copies of the TG oligonucleotide. The minimum of the curve, at about 10.1 bp of the sequence repeat, should coincide with the preferred helical repeat of bending sites on DNA in nucleosomes.

in average DNA twist as it is packaged into nucleosomes in the correct direction to lessen the "linking number paradox."[49,50]

The degree of overwinding of DNA on the nucleosome surface determined from our experiments is similar to the values for this parameter measured in other studies. However, the appropriateness of a single average value for twist over the ~145 bp of nucleosomal DNA is unclear. Like the uneven bending of DNA seen in the nucleosomal crystal structure,[1] the overall DNA twist may be the average of disparate values from different regions of the structure. To explore the optimal repeat near the nucleosome dyad, the series of molecules depicted in Fig. 16 was constructed. These molecules all contain common vector sequences and two 40-bp "arms" derived from two copies of the TG oligonucleotide (10.0-bp sequence repeat), flanking a central region of variable length. The fragment containing the 21-bp central region reconstitutes most efficiently. The extra base pair, averaged over the 10 turns of DNA in the oligonucleotide-derived region, adjusts the overall helical repeat from 10.0 to 10.1 bp, leading to optimal alignment for the 40-bp arms. The fragments containing 19-, 20-, or 22-bp central regions reconstitute less well. This suggests that the local value for the helical repeat near the nucleosome dyad is not more than 10.5 bp/turn, close to the average of 10.1 bp.

[49] J. H. White and W. R. Bauer, *J. Mol. Biol.* **189**, 329 (1986).
[50] A. A. Travers and A. Klug, *Philos. Trans. R. Soc. London B* **317**, 537 (1987).

[10] Footprinting Protein–DNA Complexes *in Vivo*

By SELINA SASSE-DWIGHT and JAY D. GRALLA

This chapter describes the use of primer extension procedures to probe nucleoprotein complexes *in vivo* in *Escherichia coli* and *in vitro*. Included are an overview of the procedure (when it is appropriate and what materials are required), a description of the choice and preparation of materials, step-by-step protocols for primer extension probing with dimethyl sulfate and potassium permanganate, and a troubleshooting guide.

Overview of Primer Extension Probing

The technique of primer extension footprinting analysis[1] has been used for footprinting various regulatory regions *in vitro* on linear and su-

[1] J. D. Gralla, *Proc. Natl. Acad. Sci. U.S.A.* **82**, 3078 (1985).

percoiled DNA[1-3] and for examining protein–DNA interactions *in vivo* on both plasmid[2-8] and genomic DNA[9] in different bacteria.[10] It has been used to study viral transcription complexes in infected mammalian cells.[11-13] It has also been applied more generally to the study of specific modifications induced by various anticancer drugs.[14]

The protocol, which is illustrated in Fig. 1, involves three major steps. First, the DNA is covalently modified with an attacking reagent. Next, the modified DNA is isolated and, if necessary, broken at the sites of modification. Third, the sites of modification are detected by primer extension procedures. In the primer extension reaction a specific ^{32}P end-labeled synthetic oligonucleotide is hybridized to the modified template DNA and extended with a DNA polymerase. The resulting extension products, which all have the same ^{32}P end-labeled 5' end and a 3' end that varies according to the position of the template modification at which the primer extension is inhibited, are analyzed on a DNA sequencing gel.

If the original modification took place on protein-bound DNA, the attacking reagent may have had altered access to the DNA. In the simplest case this results in a "protected" band on the gel autoradiograph compared to a lane in which the protein was not present. The interpretation is just as in conventional footprinting techniques that involve using end-labeled template DNA rather than an end-labeled primer.[15,16]

The use of an end-labeled primer rather than end-labeled template DNA is advantageous in that it allows regulatory regions to be footprinted on circular templates either *in vivo* or *in vitro;* it avoids the isolation and end labeling of specific modified DNA fragments. The same end-labeled primer can be used to probe a given regulatory region either *in vivo* or *in vitro*. Another advantage is that various regions of the same DNA sample can be analyzed simply by splitting the sample and probing with different

[2] J. A. Borowiec, L. Zhang, S. Sasse-Dwight, and J. D. Gralla, *J. Mol. Biol.* **196,** 101 (1987).
[3] S. Sasse-Dwight and J. D. Gralla, *J. Biol. Chem.* **264,** 8074 (1989).
[4] J. A. Borowiec and J. D. Gralla, *Biochemistry* **25,** 5051 (1986).
[5] Y. Flashner and J. D. Gralla, *Proc. Natl. Acad. Sci. U.S.A.* **85,** 8968 (1988).
[6] Y. Flashner and J. D. Gralla, *Cell (Cambridge, Mass.)* **54,** 713 (1988).
[7] S. Sasse-Dwight and J. D. Gralla, *J. Mol. Biol.* **202,** 107 (1988).
[8] S. Sasse-Dwight and J. D. Gralla, *Proc. Natl. Acad. Sci. U.S.A.* **85,** 8934 (1988).
[9] S. Sasse-Dwight and J. D. Gralla, *Cell* **62,** 945 (1990).
[10] E. Morett and M. Buck, *Proc. Natl. Acad. Sci. U.S.A.* **85,** 9401 (1988).
[11] R. L. Buchanan and J. D. Gralla, *Mol. Cell. Biol.* **7,** 1554 (1987).
[12] L. Zhang and J. D. Gralla, *Nucleic Acids Res.* **18,** 1797 (1990).
[13] L. Zhang and J. D. Gralla, *Genes Dev.* **3,** 1814 (1989).
[14] J. D. Gralla, S. Sasse-Dwight, and L. Poljak, *Cancer Res.* **47,** 5092 (1987).
[15] D. Galas and A. Schmitz, *Nucleic Acids Res.* **5,** 3157 (1978).
[16] G. M. Church and W. Gilbert, *Proc. Natl. Acad. Sci. U.S.A.* **81,** 1991 (1984).

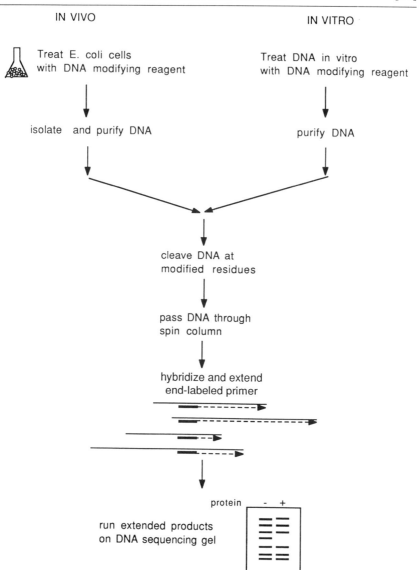

FIG. 1. Outline of the primer extension footprinting analysis technique *in vivo* and *in vitro*.

primers designed to read through different regions. Both strands can be probed on the same DNA template simply by designing two primers, each of which hybridizes to a different strand. In addition, reagents whose DNA modifications inhibit extension can be used without the necessity for DNA cleavage chemistry, theoretically increasing the diversity of reagents that can be employed for DNA footprinting. Last, primer extension footprinting is a relatively fast technique, as it eliminates the need for end-labeling each sample of modified template DNA.

Choice of Reagents

Since primer extension footprinting relies on extension blockage, the DNA modification reagents chosen must produce lesions that either directly impede extension or that can be specifically altered to serve as blocking sites. In addition, if the regulatory region is to be footprinted *in vivo*, the reagent must be able to penetrate the cell. Two chemicals have been used *in vivo*, dimethyl sulfate (DMS)[1,2,4] and potassium permanganate (KMnO$_4$).[3,8] The first has been useful for identifying residues protected and enhanced by protein binding while the second is most useful in identifying distorted DNA or DNA melted in transcription complexes. The procedure for using each of these reagents *in vitro* or *in vivo* on either plasmid or chromosomal DNA in bacterial cells will be described in the following sections. Although not described here, these techniques have also been applied in preliminary experiments in mammalian cells where the range of attacking reagents is broader due to more facile entry as compared with bacteria.[11–13] The technique has been used successfully with the enzymes DNase I and micrococcal nuclease.

Dimethyl sulfate (DMS), which has been employed for sequencing[17] and other footprinting protocols,[15,16] and potassium permanganate (KMnO$_4$) are two chemicals that have been used in primer extension probing. Both reagents are capable of penetrating the bacterial cell and have therefore been useful in studying protein–DNA interactions *in vivo*. In addition, both reagents meet the important requirement for primer extension analysis of modifying the DNA in such a way that the modifications can be made to serve as blocking sites for extension by a DNA-copying enzyme. Specifically, DMS modifies, among other sites, the N-7 position of guanine residues.[17] When treated appropriately with piperidine, the DNA strand becomes susceptible to cleavage at these modified guanine residues[17] and this break serves as a block to the DNA-copying enzyme during primer extension. Potassium permanganate, on the other hand,

[17] A. M. Maxam and W. Gilbert, this series, Vol. 65, p. 499.

modifies primarily T and to a lesser extent C residues.[18] This modification results in oxidation of the 5,6-double bonds to form thymine glycols.[18,19] The Klenow fragment of DNA polymerase I has been shown to stop extension after inserting a base across from a thymine glycol-modified residue.[19,20] Alternatively, treatment of the modified DNA with alkali results in the conversion of the thymine glycol residue to urea, which the polymerase is unable to copy.[19]

In addition to these two chemicals, several enzymes have been used in primer extension footprinting. These include DNase I[11-13] and micrococcal nuclease,[21] which have different modes of cleavage and thus give complementary information. They are not suitable for use in bacteria since they do not penetrate the cells, but have been used in primer extension protocols *in vitro* and in permeabilized mammalian cells.[11-13,21] Their use will not be described in detail here, since it is a matter of using standard digestions followed by the primer extension probing procedures that will be described below.

One advantage to using a variety of reagents is that each attacks the DNA differently and their joint use can give a very detailed picture of the DNA bound by the protein. For example, DNase I cleavage occurs by attack through the DNA minor groove, DMS attacks guanines in the DNA major groove, micrococcal nuclease attacks exposed positions on the DNA backbone, and permanganate attacks melted DNA preferentially (see references 4, 12, and 21). In principle, many chemicals and enzymes can be adapted for primer extension probing. The main limitation is that they must either break the DNA, modify it so as to allow breakage, or form lesions that block the progress of a copying enzyme. As an example, the following section describes the different information that can be obtained by using the two reagents DMS and $KMnO_4$.

Dimethyl Sulfate Footprinting

With the chemical dimethyl sulfate, we have used primer extension probing to reveal residues protected and enhanced by protein binding both *in vitro*[2,5,6] and *in vivo*.[2,4,7,8] For instance, DMS was used *in vitro* to determine the occupancy of the *lac* O_1 and O_3 operators on both linear and supercoiled DNA.[2] These results demonstrated that a single *lac* repressor molecule can bind simultaneously to these two operators only on supercoiled DNA. DMS was also used as a probe for *lac* repressor binding

[18] H. Hayatsu and T. Ukita, *Biochem. Biophys. Res. Commun.* **29,** 556 (1967).
[19] H. Ide, Y. W. Kow, and S. S. Wallace, *Nucleic Acids Res.* **13,** 8035 (1985).
[20] J. M. Clark and G. P. Beardsley, *Biochemistry* **26,** 5398 (1987).
[21] L. Zhang and J. D. Gralla, *Nucleic Acids Res.* **17,** 5017 (1989).

on the same plasmid DNA *in vivo* in experiments that demonstrated that this protein also binds cooperatively to the *lac* O_1 and O_3 operators *in vivo* and that this cooperativity requires DNA supercoiling.[7] These results point to some advantages associated with the primer extension footprinting protocols in probing supercoiled DNA and DNA in the living cell. In addition, in these and related experiments it was also important that the relative occupancy of the three *lac* operators on the same template be known reliably. The primer extension procedure allowed the very same sample to be split and probed with different primers. Such experiments demonstrated that DNA looping is an important control mechanism in regulating *lac* expression.

As described above, the primer extension technique is designed for facile probing of supercoiled DNA. However, the initial attack on DNA should not lead to a loss of supercoiling. For instance, DMS modifies purine residues without causing strand cleavage. These modified residues serve as sites for DNA cleavage by piperidine subsequent to DNA purification. Strictly speaking, if a DNA-breaking enzyme is used as a probe, the template should be broken at less than one site per template so that each attack occurs on a supercoiled template. This can be checked by agarose gel electrophoresis to show the incomplete loss of form I DNA. In practice, if supercoiling is required only for the formation but not the maintenance of the complex, this consideration is unimportant.

Thus, DMS serves as an effective probe for examining protein–DNA interactions *in vivo* by primer extension analysis. As has been observed previously for DMS, it is valuable in detecting specific contacts that the protein makes with individual guanine residues. In principle, other breakage protocols could be used to probe adenine and other modifications.[17]

Probing with Potassium Permanganate

In contrast to DMS, the reagent potassium permanganate preferentially attacks single-stranded DNA.[18] It serves as a probe for detecting sharply distorted DNA[2] and for DNA melted in an open complex with RNA polymerase.[3,8] This second property is illustrated in Fig. 2, which shows the pattern of protection obtained at the *lacUV5* promoter when potassium permanganate is used in a primer extension analysis. Lane 1 (Fig. 2) illustrates the control pattern where no proteins were present during modification. When RNA polymerase is allowed to bind to the *lac* promoter DNA prior to modification, the bands from -10 to $+4$ become very hyperreactive (lanes 2 and 3, Fig. 2). It was demonstrated that this hyperreactivity represents those bands that are melted in the open complex formed between RNA polymerase and *lac* promoter DNA.[3] Thus, perman-

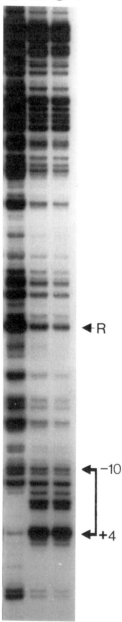

ganate serves as an efficient probe of open complex formation *in vitro* using primer extension footprinting analysis.

Figure 3 illustrates that permanganate can also detect open promoter complexes *in vivo*. When the same plasmid DNA used for the above *in vitro* experiment was probed under activating conditions for the *lac* promoter *in vivo*, a pattern essentially identical to that obtained *in vitro* was observed (lane 3). In this instance, rifampicin, which traps RNA polymerase in an open promoter complex *in vitro*[22–24] and *in vivo*,[4] was added to the cells just prior to permanganate addition in order to collect open complexes. This signal can be compared with that obtained under repression conditions in the absence of rifampicin (lane 1). These experiments and others led to the demonstration that permanganate can be used to determine the ratio of the rate of open complex formation to the rate of transcription initiation *in vivo*.[3]

More recently, it has also been observed that permanganate can be used as a footprinting reagent *in vitro* using end-labeled DNA rather than an end-labeled primer.[25] In these experiments it is necessary to cleave the DNA strand with piperidine at the sites of permanganate modification. This step is not required during primer extension analysis, but the results using the two techniques should differ only in unusual circumstances, where atypical permanganate lesions block extension but are not piperidine sensitive.

Probing on Chromosomal DNA

Thus far, these examples of primer extension analysis have been confined to footprints along multicopy plasmid DNA *in vitro* or *in vivo*. Because *E. coli* genomic DNA contains far more sequence complexity than

[22] A. Sippel and G. Hartmann, *Biochim. Biophys. Acta* **157**, 218 (1968).
[23] D. C. Hinkle, W. F. Mangel, and M. J. Chamberlin, *J. Mol. Biol.* **70**, 209 (1972).
[24] C. Bordier and J. Dubochet, *Eur. J. Biochem.* **44**, 617 (1974).
[25] K. R. Fox and G. W. Grigg, *Nucleic Acids Res.* **16**, 2063 (1988).

FIG. 2. *In vitro* potassium permanganate pattern obtained at the *lac*UV5 promoter along supercoiled plasmid DNA in the presence and absence of prebound RNA polymerase. Lane 1, *in vitro* pattern obtained in the absence of any proteins; lane 2, *in vitro* pattern obtained in the presence of RNA polymerase, which was allowed to bind the promoter DNA for 5 min before KMnO$_4$ was added; lane 3, *in vitro* pattern obtained with permanganate when the cAMP receptor protein (CRP) was allowed to bind in the presence of cAMP for 5 min followed by RNA polymerase binding for 5 min before treatment with permanganate. R refers to a reference band. (For a more detailed description of experimental procedures, see Fig. 1A of Ref. 3, from which this figure was reprinted with permission.)

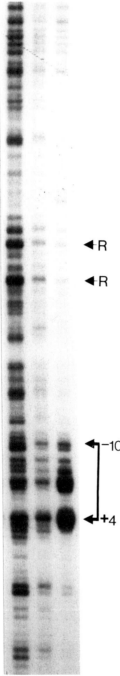

plasmid DNA and is present in only one copy per cell, it presents several problems for primer extension footprinting analysis. These include a far greater potential for nonspecific hybridization and a much weaker overall signal intensity. In order to surmount these problems, we have developed a modification of the polymerase chain reaction (PCR)[26] in which only one primer, the end-labeled primer, is used for extension. Because Taq polymerase is used, this permits both the hybridization and extension temperatures to be higher, lowering the probability of nonspecific hybridization. In addition, the multiple rounds of hybridization and extension result in a much stronger signal. These multiple rounds, although they do not exponentially amplify the signal as in a standard PCR reaction, lead to a theoretical doubling of the signal with every round. This allows a reasonable footprinting signal to be obtained on *E. coli* chromosomal DNA *in vivo* using primer extension analysis.

Results of such experiments using either DMS or $KMnO_4$ as the footprinting reagent are shown in Fig. 4. In these experiments, the *E. coli* glnAp2 promoter was probed under activating conditions. The DMS experiments demonstrate that protection of the promoter region (bands marked − 12 and − 24) by the RNA polymerase known to bind at this sequence *in vitro*[27,28] can be seen in the lane in which a wild-type strain was probed (lane 2) compared with a lane in which a strain deficient in the sigma factor (σ^{54}) of the polymerase was probed (lane 1). Similarly, lanes 4 and 3 show the permanganate patterns obtained at this promoter in the presence and absence of the required RNA polymerase. Strongly hyperreactive bands at the DNA transcriptional start site are observed only when the σ^{54} RNA polymerase is present in the cell. Thus, primer extension analysis of genomic DNA using a modification of the PCR can be used to detect specific protein–DNA contacts and open promoter complexes *in vivo* using DMS and potassium permanganate, respectively.

[26] K. B. Mullis and F. A. Faloona, this series, Vol. 155, p. 335.
[27] E. Garcia, S. Bancroft, S. G. Rhee, and S. Kustu, *Proc. Natl. Acad. Sci. U.S.A.* **74**, 1662 (1977).
[28] T. P. Hunt and B. Magasanik, *Proc. Natl. Acad. Sci. U.S.A.* **82**, 8453 (1985).

FIG. 3. *In vivo* potassium permanganate pattern obtained at the *lac*UV5 promoter along plasmid DNA under different metabolic conditions. Lane 1, permanganate pattern obtained under nonactivating conditions *in vivo;* lane 2, permanganate pattern obtained when $KMnO_4$ was added to the shaking cells just subsequent to induction of the *lac* operon with cAMP and IPTG; lane 3, permanganate pattern obtained when rifampicin was added to trap open complexes for 5 min following induction with cAMP and IPTG, isopropyl thiogalactoside. R refers to a reference band. (For a more detailed description of experimental procedure, see Fig. 3A of Ref. 3, from which this figure was reprinted with permission.)

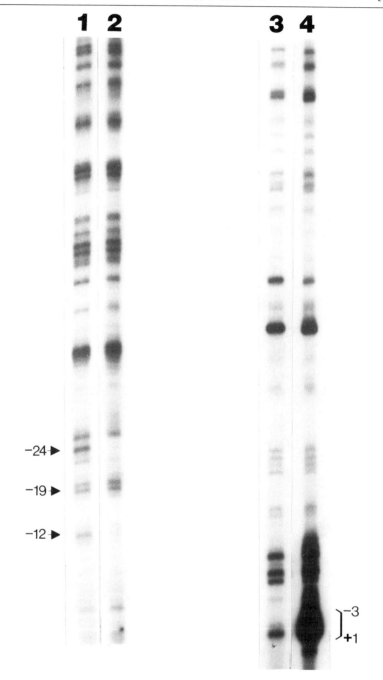

Preparation and Choice of Materials: General Considerations

DNA Preparation

The modified template DNA should be free of proteins and other contaminants that may induce artifactual stops during the primer extension. This requires that the DNA be carefully extracted with phenol in order to remove protein, preferably to the point where there is little or no contaminating interface. In all cases, either *in vitro* or *in vivo*, ultrapure or redistilled, neutralized phenol should be used in order to avoid introducing contaminants or DNA damage that might interfere with the extension reaction. In particular, the primer extension procedure is theoretically more sensitive to unwanted DNA damage since such modifications could block extension, thereby producing bands that, if not cleavable by piperidine, would not be observed on end-labeled DNA. In procedures involving piperidine (Aldrich, Milwaukee, WI), the reagent should be of the highest purity and retested periodically for induction of stop-inducing modifications in DNA.

In order to ensure removal of various ions and chemicals such as the reagent used to cleave the DNA at the modification sites, we strongly recommend passing the DNA through a spin column (Sephadex G-50–80), equilibrated in sterile, doubly distilled water, directly before the extension reaction. This removes contaminants such as magnesium, which can strongly inhibit extension at high concentrations. For this reason, if it is necessary to concentrate the final DNA stocks following column purification, precipitations should be carried out with 0.3 M sodium acetate rather than with magnesium salts.

Normally, 0.5–1.0 μg of whole-plasmid DNA is used per primer exten-

---

FIG. 4. DMS (lanes 1 and 2) and KMnO$_4$ (lanes 3 and 4) patterns obtained using a modification of the PCR when the chromosomal *glnAp2* promoter was footprinted under activating conditions *in vivo*. Lane 1, *in vivo* DMS pattern obtained in a strain that is deficient in the σ factor, which is required for binding at the *glnAp2* promoter, σ^{54}; lane 2, *in vivo* DMS pattern obtained for the wild-type strain in which σ^{54} is present. The -24 and -24 regions, representing the two major grooves to which the σ^{54} RNA polymerase binds, are marked by arrows, as is the -19 band, which serves as a reference band as it does not change in relative intensity when the protein is present. Lane 3, *in vivo* permanganate pattern at the chromosomal *glnAp2* promoter obtained in a strain that is deficient for the σ^{54} factor, which is required for binding at the *glnAp2* promoter, σ^{54}; lane 4, *in vivo* KMnO$_4$ pattern obtained in the wild-type strain in which σ^{54} is present. The positions of the melted residues are marked to the right of the figure. The growing cells were split into two samples just prior to modification so that the permanganate-treated cells (lanes 3 and 4) are identical to those treated with DMS (lanes 1 and 2, respectively).

sion analysis reaction. This gives a signal equivalent to approximately 2.5 ml of midlog *E. coli* cells when multicopy plasmid DNA is footprinted *in vivo*. In the case of chromosomal DNA, approximately 3 ml of midlog-phase cells will yield the desired amount of DNA for one sample of primer extension analysis using a modification of the polymerase chain reaction (PCR, see below).

Primer Preparation

The primer extension footprinting technique calls for the use of a 5'-^{32}P end-labeled primer. Although labeled dNTPs could theoretically also be used, we have found their use to be associated with high background signals. The primer is chosen based on location, end sequences, and its theoretical hybridization temperature. In general, oligonucleotides that are 17–20 bases in length and have a 3' end that hybridizes to a position at least 15–20 bp away from the sequence to be analyzed have been found to be most reliable. In addition, it is also best to design a primer such that its 5' and 3' ends are composed of either G or C residues. If the primer extension analysis is to be carried out on plasmid DNA, the hybridization temperature should be chosen to be approximately 50–55°, as determined by the formula

$$T_\mathrm{m} = 69.3 + 0.41(\mathrm{G} + \mathrm{C})\% - 650/L$$

where L is the length of the oligonucleotide in nucleotides and $(\mathrm{G} + \mathrm{C})\%$ represents the percentage G/C content.[29] In the case of chromosomal DNA, where a modification of the PCR is employed, we have found good results for primers with a T_m of around 57°.

Crude preparations of oligonucleotide should be purified on a poly-acrylamide gel. Precautions should be taken to remove acrylamide by passing the primer through a 0.2-μm Acrodisc (Gelman Sciences, Ann Arbor, MI). During purification of the primer, precipitation with ammonium ions should be avoided, as they strongly inhibit the subsequent kinasing reaction. Instead, primers should be precipitated with sodium acetate.

For the kinasing reaction, 20–30 pmol of oligonucleotide is end labeled with [γ-^{32}P]ATP.[29] The final reaction(50-μl volume) is then passed through Sephadex G-50–80 spin columns equilibrated in TE buffer pH 8.0 (10 mM Tris-HCl, 1 mM EDTA) in order to remove unincorporated label. This generally results in an end-labeled primer that is anywhere from 0.7 to 5 × 10^6 cpm/μl. As a rule, 0.3–0.5 × 10^6 cpm of ^{32}P end-labeled primer is added per primer extension reaction. As the primer decays, however,

[29] T. Maniatis, E. F. Fritsch, and J. Sambrook, eds., "Molecular Cloning: A Laboratory Manual." Cold Spring Harbor Laboratory, Cold Spring Harbor, New York, 1983.

the molar amount of primer required to give the desired amount of radioactivity increases. Since excessive molar amounts of unlabeled primer reduce hybridization specificity and compete with labeled primer, the labeled primers are generally discarded after they decay to a level of 0.3×10^6 cpm/μl or less. In general, a high specific activity for the end-labeled primer is advantageous.

Extension Enzyme

In the procedure discussed below we use the Klenow form of DNA polymerase I for extension. In order to minimize artifactual sequence-induced extension stops the reaction is done at 50–52°. We also describe a modified protocol in which Taq DNA polymerase is used. In principle, any DNA-copying enzyme can be used.

Detailed Protocols for Chemical Probing and Primer Extension

Note. A convenient way to prepare the Sephadex G-50–80 spin columns that are required for these experiments is to stopper the bottom of a 1-ml syringe with a polyethylene disk cut by a cork borer to be the diameter of the 1-ml syringe. Next, pour the Sephadex G-50–80, preequilibrated in the desired solution, into the syringe and let the Sephadex solution settle to 1 ml. Then pack the column to the 0.7-ml mark by spinning for 2 min in a clinical centrifuge at setting 4. Finally, load the sample carefully on top and spin again for 2.5 min at the same setting.

In Vitro Treatment with DMS or KMnO$_4$

DMS as the Modifying Reagent

1. Bring DNA samples to be treated with DMS (0.1–10 μg) to a final volume of 100 μl, where one-third of the volume is composed of 3× transcription salts (3×: 9 mM MgCl$_2$, 0.3M KCl, 0.6 mM DTT, 90 mM Tris-HCl, pH 8.0, 0.3 mM EDTA) or other salts required for the regulatory protein being examined. Prewarm the DNA for 2 min at 37° before adding the desired protein, also diluted to give 1× transcription salts or other desired salts.

2. While the protein–DNA mixture is incubating at 37°, dilute the DMS to 150 mM (from 10.6 M stock) with water and vortex well. Use care in handling DMS stock solutions, since the chemical is toxic.

3. Add 6.7 μl of 150 mM DMS to the protein–DNA sample and allow the modification to proceed for 5 min at 37°.

4. Then quench the reaction with 200 μl cold stop buffer (3 M ammo-

nium acetate, 1 M 2-mercaptoethanol, 250 μg/ml tRNA, 20 mM EDTA) and add 600 μl cold 95% (v/v) ethanol to precipitate the DNA for 10 min at $-70°$. Spin the samples for 10 min in a microfuge at 4°. The pellets are washed with 70% (v/v) ethanol and dried.

5. Following this, resuspend the samples in 100 μl 1 M redistilled piperidine, place black electrical tape over the microfuge tube top, close the lid, and place the tubes in a clamped holder for 30 min at 90°.

6. Place the piperidine-treated modified DNA on ice and then spin the sample through a 1-cm^3 Sephadex G-50–80 column preequilibrated in distilled water as described at the beginning of this section. Approximately 70 μl of cleaved DNA in water should be recovered and adjusted to give 0.5 μg/35 μl.

Potassium Permanganate Treatment

0.37 M KMnO$_4$ [formula weight (FW) of KMnO$_4$ = 158.04 g/mol, solubility limit = 60 g/liter or approximately 0.37 M]: Weigh out 12 g permanganate (since KMnO$_4$ is a strong oxidant, gloves and safety glasses should be worn when working with this reagent). Bring to just over 200 ml with distilled water and heat to boiling. Allow to boil for 3–5 min until the volume reaches 200 ml. Allow solution to cool and store in brown jar. This solution should be good for 1–2 months

1. Bring 0.5–2.0 μg DNA to 17.5 μl such that one-third of the volume is 3× salts (if a protein is to be added, bring it to only one-half this volume, 8.75 μl).

2. Prewarm the DNA samples 2 min at 37°. Add protein, diluted to give 1× transcription salts, and incubate for the desired time.

3. Add 2.5 μl 80 mM KMnO$_4$ (freshly diluted from 0.37 M stock) for exactly 2 min. Then quench the permanganate reaction with 2.0 μl 2-mercaptoethanol (14.7 M) and place on ice.

4. Add 6 μl 0.2 M EDTA and 27 μl water; extract samples by adding an equal volume of phenol and vortexing. Then place samples on ice for 2 min, heat for 3 min at 90°, and centrifuge. Remove DNA aqueous layer.

5. Spin the resulting aqueous sample through a Sephadex G-50–80 spin column preequilibrated in water as described at the beginning of this section. Dilute the resulting DNA, if necessary, such that its concentration is approximately 0.5 μg/35 μl.

In Vivo Treatment with DMS and KMnO$_4$

Note. Certain rich medias can quench permanganate. Therefore, minimal medium is generally used for KMnO$_4$ footprinting. It is advisable to check the medium itself to determine whether it quenches permanganate by adding $KMnO_4$ to the medium at the desired footprinting concentration. If the permanganate immediately turns brown, the medium is quenching this reagent. If the brown color takes a few minutes to develop, the medium can generally be used. When the experiment is done with medium containing cells, the cell pellet will often be brown and the supernatant will remain a purplish color. If the pellet remains white, this is another indication that the $KMnO_4$ has been quenched by the medium.

1. Cells to be footprinted are grown overnight in appropriate medium and diluted 1 : 100 the next morning.

2. *DMS:* At approximately $0.5–0.7$ OD_{660}, 10 ml of cells is treated with DMS for 5 min to give a final DMS concentration of $2–10$ mM (the optimal concentration may vary, so it is advisable to initially test the DMS concentration over this small range). The cells should remain shaking at $37°$ during DMS treatment.

$KMnO_4$: $KMnO_4$ is generally added to 10 ml of growing midlog-phase cells for 2 min to a final concentration of 10 mM from a 0.37 M stock. As with DMS, however, this concentration should be adjusted as needed. Since permanganate is useful in detecting open complexes, freshly made rifampicin (50 mg/ml in methanol), which traps the bacterial RNA polymerase in an open complex, can be added to a final concentration of 0.2 mg/ml to the shaking cells for 5 min just prior to permanganate addition when desired. This is necessary if the rate of open complex formation is much slower than the rate of initiation.[3]

3. Following chemical treatment of the cells with either DMS or $KMnO_4$, the samples are removed from the shaking $37°$ water bath and poured immediately into prechilled tubes. Following this, they are pelleted for 5 min at 5K rpm in a cold SS34 rotor and the supernatant is discarded.

4. DNA isolation:

Plasmid DNA. Plasmid DNA is normally isolated according to the method of Holmes and Quigley[30] with the following modifications: (1) the lysate is treated with RNase A (0.2 mg/ml for 30 min at $40°$) and proteinase K (0.2 mg/ml for 60 min at $52–57°$) before precipitation with 2-propanol; (2) following 2-propanol precipitation the pellets are resuspended in 550 μl of TE buffer, pH 8.0, and extracted once with phenol, three or four times with phenol : chloroform : isoamyl alcohol ($25 : 24 : 1$, v/v/v), and once with chloroform : isoamyl alcohol ($24 : 1$, v/v) (the phenol : chloroform : isoamyl alcohol extractions are performed by vortexing the extractions, heating the samples for 10–20 min at $50–55°$, cooling the samples, and spinning

[30] D. S. Holmes and M. Quigley, *Anal. Biochem.* **114**, 193 (1981).

them for 2 min to separate the interfaces); and (3) following extraction the samples are precipitated with 0.1 vol 3 M sodium acetate and 2.0 vol cold 95% (v/v) ethanol.

DMS: Samples are resuspended in 100 μl 1.0 M piperidine and cleaved at 90° for 30 min as described by Maxam and Gilbert.[17] Following piperidine cleavage they are spun through a 1-cm^3 Sephadex G-50–80 spin column preequilibrated in distilled water. For 10 ml of starting cells the final volume is diluted to 105 μl with distilled water.

KMnO$_4$: Samples are resuspended in 150–300 μl of distilled water (with heating at 50–55° when necessary). Up to 200 μl is spun on a 1-cm^3 Sephadex G-50–80 column preequilibrated in water. The resulting solution is ready for further analysis without dilution.

Chromosomal DNA. Escherichia coli genomic DNA is isolated according to the procedure of Owen and Borman[31] with the following modifications: (1) the samples are allowed to incubate overnight at 37–50° following proteinase K addition; (2) following the first chloroform extraction the samples are extracted two to four times more with chloroform, once with phenol, and once again with chloroform. This procedure is rapid and results in sufficiently pure DNA for primer extension analysis. The resulting pure pellets are treated as described above for the plasmid DNA pellets except that the final volume of DNA in water should be kept closer to 70 μl for DMS samples and 150 μl for KMnO$_4$ samples when 10 ml of cells is footprinted.

Primer Extension Analysis of DNA Treated in Vitro or in Vivo

At this point, all samples, irrespective of the DNA modification reagent or whether they are plasmid or chromosomal DNA, should be at an approximate concentration of 0.5 μg/35 μl in distilled water following passage through a spin column (see previous section). It is very important that the DNA be passed through this spin column in order to remove any contaminants that might interfere with the extension reaction. There are two different procedures that can be used for primer extension analysis of the modified DNA samples. The most common, alkaline denaturation, is recommended for *in vitro* or *in vivo* plasmid DNA samples. It involves denaturation of the modified DNA with alkali, neutralization, hybridization, and extension with Klenow fragment of DNA polymerase I. The second extension technique involves a modification of the PCR in which only one primer, the labeled primer, is used. Because this allows multiple rounds of denaturation, hybridization and extension, it has been developed

---

[31] R. J. Owen and P. Borman, *Nucleic Acids Res.* **15**, 3631 (1987).

for footprinting the more complex chromosomal DNA samples *in vivo*. In addition, the higher temperature of extension allowable with Taq polymerase mitigates the possibility of nonspecific hybridization.

The following describes the protocol for each of these extension techniques. In general, primer extension analysis should be performed for the first time on *in vitro* modified DNA. This is recommended as, although the overall strategy of primer extension is fast and straightforward, there are several steps that require attention to detail. In addition, we have found that adjustments sometimes need to be made for different primers.

Alkaline Denaturation Primer Extension (for Analysis of in Vitro or in Vivo Modified Plasmid DNA)

1. Begin with 35 μl of DNA in distilled water (approximately 0.5 μg). Add 1 μl ^{32}P end-labeled primer (diluted in distilled water, if necessary, to give 0.3–0.5 $\times$ 10^6 cpm/μl).

2. Add 1/9 vol (4 μl) 0.01 M NaOH to each extension reaction and mix well.

3. Heat samples for 2 min at 80° in order to denature template DNA; transfer directly to ice and allow to cool for at least 5 min.

4. Add 1/9 vol (5 μl) freshly made 10$\times$ TMD buffer [0.5 M Tris-HCl, pH 7.2, 0.1 M MgSO$_4$, 2 mM dithiothreitol (DTT)] to each sample and mix well.

5. Hybridize the end-labeled primer to the modified DNA template by heating the samples for 3 min at 45–50°; return the samples directly to ice. *Note:* the hybridization temperature may need to be altered slightly in accordance with the calculated T_m (see previous section) of the primer. It is best to keep the hybridization temperature at or just under the T_m of the primer in order to minimize nonspecific binding.

6. Add 1/9 vol (5 μl) of 4 $\times$ 5 mM dNTPs (a mixture containing the 4 dNTPs, each at a concentration of 5 mM). Tap samples well and return to ice.

7. Dilute the Klenow fragment of DNA polymerase I to 0.5–1.0 U/μl using Klenow diluent [50% glycerol (w/v), 50 mM KH$_2$PO$_4$, 1 mM DTT, 100 μg/μl bovine serum albumin (BSA); make 1 ml and store at -20°]. The enzyme should be placed at the bottom of a cold microcentrifuge tube without creating bubbles and the cold diluent slowly added. The enzyme can be mixed by gently tapping the bottom of the tube.

8. At time *t*, add 1 μl diluted Klenow fragment to a given sample and tap gently for 5–10 sec. Place the tube at 50° for exactly 10 min.

9. At time *t* plus 10 min, add 1/3 vol quench (4 M ammonium acetate, 20 mM EDTA) to the sample and mix thoroughly.

10. Precipitate the sample with 2.5–3.0 vol cold 95% ethanol and incu-

bate at $-70°$ for 20 min or overnight at $-20°$. Pellet the DNA in a microfuge for 10 min at $4°$ and wash the DNA pellet gently with 70% ethanol by rolling the microcentrifuge tube. Respin for 2 min, remove the liquid, and dry the pellet. Then proceed with the steps described under Polyacrylamide Gel Electrophoresis (below).

Extension by PCR Modification (for Analysis of Chromosomal DNA Modified in Vivo)

As stated above, this technique is often necessary when analyzing chromosomal DNA, as it magnifies the normally low signal by allowing multiple rounds of extension and permits the extension to occur at a higher temperature, thereby minimizing nonspecific hybridization. It is equivalent to the PCR except that only one primer is added to the reaction mix.

1. Prepare the samples by adding the following solutions in order:

Distilled water	49.5 μl
10× Taq reaction buffer (as provided with enzyme)	10.0 μl
4 × 5 mM dNTP mix (each of the 4 dNTPs present at 5 mM)	4.0 μl
^{32}P End-labeled primer (diluted to 0.5×10^6 cpm/μl with distilled water)	1.0 μl
DNA in distilled water	35.0 μl
Taq polymerase (2.5 U/μl)	0.5 μl

   2. Overlay samples with 100 μl mineral oil.
   3. Treat samples as follows:

A: 1.5 min at $94°$
B: 2 min at $57°$ (or T_m of primer)
C: 3 min at $72°$
D: 1 min at $94°$
E: Repeat steps B through D 15–20 times
F: Heat 2 min at $57°$ (or T_m of primer)
G: Heat 10 min at $72°$
H: Let samples cool to room temperature

   4. Separate the sample from the mineral oil by pipetting it up from the tube bottom and placing in a fresh tube.
   5. Extract samples once with chloroform : isoamyl alcohol (24 : 1).
   6. Precipitate DNA samples with 1/3 vol quench (4 mM ammonium acetate, 20 mM EDTA) and 2.5 vol cold 95% ethanol for 20 min at $-70°$ or overnight at $-20°$.
   7. Pellet DNA by spinning for 10 min at $4°$. Wash the pellet with 70%

ethanol by gently rolling the tube, respin for 2 min, remove the ethanol, and dry the pellet. The samples are now ready for polyacrylamide gel electrophoresis.

Polyacrylamide Gel Electrophoresis

The resulting primer extension pellets are now ready for resuspension in 4–6 μl of a loading dye composed of alkali/EDTA, formamide, urea, and marker dyes [0.075 g ultrapure urea, 100 μl deionized formamide, 8 μl 50 mM NaOH/1 mM EDTA, 8 μl 0.5% (w/v) of each xylene cyanol and bromphenol blue]. For consistent results, it is best to treat each sample in an identical fashion when resuspending. This is best accomplished by placing the samples with dye at 45–50° and removing each to tap forcefully for 30 sec. Following this, the samples should be vortexed in pairs for 30 sec. This technique ensures that most of the sample resuspends in the dye mix. This can be checked by making sure that most of the counts (over 90%) are now in the solution.

Following a brief spin in order to collect the dye, the samples are ready for loading. Following electrophoresis, the gel is dried and exposed on X-ray film using an intensifying screen, if necessary. Normally, a strong signal can be observed for the autoradiograph after 1 to 2 days.

Although approximately the same number of counts should be loaded per lane, the signal intensities observed on the autoradiograph can differ from one lane to another if different amounts of DNA template are present in the samples. This is not usually a problem *in vitro*, where exact DNA amounts are known, but can be a problem *in vivo*, where the efficiency of DNA isolation varies from sample to sample. This variability can often make comparison of lanes difficult. In order to reduce the difficulty in interpreting such footprinting patterns, it is advisable to pick one or more bands outside of the region where the protein binds and use this as a reference band. In this manner, taking the ratio of the intensity of a band whose signal changes in response to protein binding to the intensity of the reference band within the same lane allows internal normalization of signal intensity for a given lane. This ratio can then be compared with the same ratio for a different lane in order to make a fair comparison of band intensities between the two lanes. If a band is heavily protected by a protein during the modification reaction, this comparison can often be done by eye. However, when protein binding results in a subtle change in pattern or when quantitative information on binding is desired, the autoradiograph can be scanned in order to determine the desired ratio of the changing band to the reference band for each lane. This procedure was used to calculate the occupancies of the *lac* O_1 and O_3 operators *in vivo* over a range of effective repressor concentrations, making it possible to determine the relative binding affinities of the two operators for *lac* repressor *in vivo*.[7]

Troubleshooting

It is highly recommended that primer extension footprinting analysis first be performed on *in vitro*-modified plasmid DNA. This is best done on linearized DNA treated with a range of concentrations of the footprinting reagent. Once this procedure has been shown to work, it is ensured that the buffers are correct and that the primer design is acceptable. The following is a list of things that can go wrong with the extension analysis either *in vitro* or *in vivo* and suggestions on correcting them.

No Signal Observed

An autoradiograph in which no pattern is observed could indicate that the buffers are incorrect (such as the extension buffer or the dNTPs) or could occur due to problems with the primer. For instance, if the primer sequence is incorrect, hybridization will not occur. It is also important to check the predicted T_m of the primer (see General Considerations) and make sure that the hybridization temperature is just under or equal to, and not significantly above, this number. Another possibility concerns the purity of the DNA. If this DNA contains contaminants that interfere with the extension reaction, it could result in either a loss of signal or a signal that contains artifactual stops and/or smears. No signal can also result from either too little or too much $KMnO_4$. It is for this reason that we strongly recommend initial titration on naked DNA.

Smearing

Primer extension analysis should result in a very clean signal, especially when a multicopy plasmid serves as the template DNA. Although the chromosomal footprints also result in a relatively clean footprint, they can sometimes have a light background smear caused by the complexity of the genome, which may be unavoidable. However, a very bad, smeary background pattern that interferes or occludes the real signal may be preventable for either plasmid or chromosomal DNA *in vitro* or *in vivo*.

As stated above, extra bands in the primer extension pattern could result if there are contaminants that interfere with the extension reaction and result in random stopping, for instance high concentrations of magnesium ion. In addition, the primer could hybridize nonspecifically to the DNA. In this instance, it is advisable to see whether there are similar sequences along the template DNA at which the primer could hybridize. Extension along different sequences would give an indiscernible and often smeary signal. Smeary signals can also be avoided by choosing a primer with the recommended predicted T_m values and hybridizing the primer to the modified DNA at this temperature. Several temperatures can be tested to determine the optimal temperature for hybridization. However, we have found that some primers work better than others even when all the desired conditions have been met, so there are other unknown factors that may contribute to the quality of the primer with respect to primer extension analysis.

A smeary signal can sometimes result if the DNA cleavage chemistry employed to break the DNA strand at the modification sites also damages DNA in such a way that they now interfere with the extension reaction. This is an important consideration to take into account when designing new chemical probes. Although the piperidine cleavage used for DMS footprinting can result in some DNA damage the extent of damage is far less than the desired reaction, such that a smeary pattern should not be observed. A very common cause of extra bands is the use of less than ultrapure piperidine. We normally redistill the piperidine (Aldrich) and store for a few months and then discard. However, it is probably not necessary to redistill the piperidine and control lanes in which DMS or $KMnO_4$ were not added will reveal if there is a problem.

Nonspecific hybridization and therefore a smeary pattern can also result when the end-labeled primer decays to a value less than 0.3×10^6 cpm/μl (this value may be higher for chromosomal footprinting). This is due to the fact that the molar amount of primer that must be added to give the desired amount of radioactivity must be increased as the primer decays. It is therefore best to use the primer soon after the kinasing reaction.

Extra bands can also occur due to sequence-induced stopping of the copying enzyme. These can be prominent if the DNA template is supercoiled and will often lessen in intensity if the DNA is linearized before copying. As with DNA sequencing, such artifacts of DNA structure can be reduced by various means, including using different enzymes and modified deoxynucleotides. If such bands persist, they will appear falsely as unprotected nucleotides on footprinting.

Bottom- or Top-Heavy Signal throughout Lane

The pattern obtained within a lane should be relatively even with respect to signal intensity. However, it is sometimes observed that the extension looks like it does not proceed past a certain point, resulting in a pattern that is strongest near the bottom of the autoradiograph and very faint or lacking at the top of the autoradiograph. In some cases, it is possible that this is caused by contaminants that interfere with the extension reaction. However, the most likely possibility is that the DNA was overdigested during the modification reaction. This can be corrected by reducing the concentration of the modifying reagent and testing it over a given range until the pattern becomes more even.

In contrast, when the signal is most intense near the top of the autoradiograph, this is likely to be due to underdigestion of the template DNA during modification. Thus, the concentration of the DNA modifying reagent should be increased.

[11] Electron Microscopy of Protein–DNA Complexes

By MARK DODSON and HARRISON ECHOLS

Introduction

The electron microscope has proven to be an extremely valuable tool for studying protein–DNA interactions. The instrument provides a powerful means of obtaining both qualitative and quantitative information concerning the nature of specific and nonspecific binding of proteins to DNA. Although commonly used to corroborate conclusions about protein–DNA interactions obtained by other biochemical and biophysical techniques, electron microscopy often can provide critical information about protein–DNA interactions that is not obtained by conventional biochemical and biophysical techniques. Some examples are given in the final section.

There are numerous examples illustrating the use of electron microscopy in analyzing nucleoprotein structures. These include the visualization and mapping of the site-specific binding to DNA of RNA polymerases,[1-4] transcription factors,[5-12] site-specific recombinases,[13-16] and replication

[1] C. Bordier and J. Dubochet, *Eur. J. Biochem.* **44**, 617 (1974).
[2] R. Portman, J. M. Sogo, T. Koller, and W. Zillig, *FEBS Lett.* **45**, 64 (1974).
[3] J. Hirsch and R. Schleif, *J. Mol. Biol.* **108**, 471 (1976).
[4] R. C. Williams, *Proc. Natl. Acad. Sci. U.S.A.* **74**, 2311 (1977).
[5] F. Payvar, D. DeFranco, G. L. Firestone, B. Edgar, O. Wrange, S. Okret, J.-Å Gustafsson, and K. R. Yamamoto, *Cell (Cambridge, Mass.)* **35**, 381 (1983).
[6] A. M. Gronenborn, M. V. Nermut, P. Eason, and G. M. Clore, *J. Mol. Biol.* **179**, 751 (1984).
[7] J. Griffith, A. Hochschild, and M. Ptashne, *Nature (London)* **322**, 750 (1986).

initiator proteins.[17-25] Electron microscopy has shown that many site-specific DNA-binding proteins bind to several target sites on DNA and self-associate by protein–protein interactions to generate highly organized nucleoprotein structures. The DNA in many of these specialized nucleoprotein structures has been observed to assume some type of topological deformation in which the DNA appears to be bent, wrapped, looped, or unwound.[26-28] Moreover, electron microscopy has been used to thoroughly characterize the role of specialized nucleoprotein structures in the biochemical pathways that mediate site-specific recombination and DNA replication.[18,20,26-33] Immunoelectron microscopy has been used to eluci-

[8] H. Kramer, M. Niemoller, M. Amouyal, B. Revet, and B. von Wilcken-Bergmann, *EMBO J.* **6**, 1481 (1987).

[9] B. Theveny, A. Bailly, C. Rauch, M. Rauch, E. Delain, and E. Milgram, *Nature (London)* **329**, 79 (1987).

[10] M. Amouyal, L. Mortensen, H. Buc, and K. Hammer, *Cell (Cambridge, Mass.)* **58**, 545 (1989).

[11] N. Mandal, W. Su, R. Haber, S. Adhya, and H. Echols, *Genes Dev.* **4**, 410 (1990).

[12] W. Su, S. Porter, S. Kustu, and H. Echols, *Proc. Natl. Acad. Sci. U.S.A.* **87**, 5504 (1990).

[13] M. Better, C. Lu, R. Williams, and H. Echols, *Proc. Natl. Acad. Sci. U.S.A.* **79**, 5837 (1982).

[14] M. Better, S. Wickner, J. Aurebach, and H. Echols, *Cell (Cambridge, Mass.)* **32**, 161 (1983).

[15] H. W. Benjamin and N. R. Cozzarelli, *EMBO J.* **7**, 1897 (1988).

[16] K. A. Heichman and R. C. Johnson, *Science* **249**, 511 (1990).

[17] M. Salas, R. P. Mellado, and E. Vinuela, *J. Mol. Biol.* **119**, 269 (1978).

[18] R. S. Fuller, B. E. Funnel, and A. Kornberg, *Cell (Cambridge, Mass.)* **38**, 889 (1984).

[19] M. Dodson, J. Roberts, R. McMacken, and H. Echols, *Proc. Natl. Acad. Sci. U.S.A.* **82**, 4678 (1985).

[20] B. E. Funnel, T. A. Baker, and A. Kornberg, *J. Biol. Chem.* **262**, 10327 (1987).

[21] F. B. Dean, M. Dodson, H. Echols, and J. Hurwitz, *Proc. Natl. Acad. Sci. U.S.A.* **84**, 8981 (1987).

[22] D. K. Chattoraj, R. J. Mason, and S. Wickner, *Cell (Cambridge, Mass.)* **52**, 551 (1988).

[23] S. Mukherjee, H. Erickson, and D. Bastia, *Cell (Cambridge, Mass.)* **52**, 375 (1988).

[24] I. A. Mastrangello, P. V. C. Hough, J. S. Wall, M. Dodson, F. B. Dean, and J. Hurwitz, *Nature (London)* **338**, 658 (1989).

[25] M. Schnos, K. Zahn, F. Blattner, and R. B. Inman, *Virology* **168**, 370 (1989).

[26] H. Echols, *BioEssays* **1**, 148 (1984).

[27] H. Echols, *Science* **233**, 1050 (1986).

[28] H. Echols, *J. Biol. Chem.* **265**, 14697 (1990).

[29] S. Eisenberg, J. Griffith, and A. Kornberg, *Proc. Natl. Acad. Sci. U.S.A.* **74**, 3198 (1977).

[30] J. Sims, K. Koths, and D. Dressler, *Cold Spring Harbor Symp. Quant. Biol.* **43**, 349 (1978).

[31] M. Dodson, H. Echols, S. Wickner, C. Alfano, K. Mensa-Wilmot, B. Gomes, J. LeBowitz, J. D. Roberts, and R. McMacken, *Proc. Natl. Acad. Sci. U.S.A.* **83**, 7638 (1986).

[32] T. A. Baker, B. E. Funnel, and A. Kornberg, *J. Biol. Chem.* **262**, 6877 (1987).

[33] M. Dodson, F. B. Dean, P. Bullock, H. Echols, and J. Hurwitz, *Science* **238**, 964 (1987).

date the protein composition of nucleoprotein intermediates in DNA replication and site-specific recombination.[16,20,34-37]

The interactions of nonspecific DNA-binding proteins with DNA have also been very thoroughly characterized. In this area, electron microscopy has been crucial in contributing to our understanding of the function of RecA in homologous recombination.[38-40] The propensity for RecA to form thick filaments when bound to DNA has allowed the visualization of the overlying and underlying strands of knotted and catenated DNA; this technique has permitted critical topological insights into the pathway of strand exchange during site-specific recombination.[41] The interaction of single-stranded DNA-binding proteins with single-stranded DNA[42-45] and topoisomerase with double-stranded DNA[46] have been thoroughly characterized by electron microscopy. The development of a technique for spreading chromatin[47] has provided insight into the structural organization of chromatin and nucleosomes,[48-51] and has permitted the direct visualization of transcription and splicing.[52,53]

[34] S. I. Reed, J. Ferguson, R. W. Davis, and G. R. Stark, *Proc. Natl. Acad. Sci. U.S.A.* **72,** 1605 (1975).

[35] M. Wu and N. Davidson, *Nucleic Acids Res.* **5,** 2825 (1978).

[36] M. Dodson, R. McMacken, and H. Echols, *J. Biol. Chem.* **264,** 10719 (1989).

[37] B. D. Lavoie and G. Chaconas, *J. Biol. Chem.* **265,** 1623 (1990).

[38] A. Stasiak, E. DiCapua, and T. Koller, *Cold Spring Harbor Symp. Quant. Biol.* **47,** 811 (1983).

[39] A. Stasiak, A. Z. Stasiak, and T. Koller, *Cold Spring Harbor Symp. Quant. Biol.* **49,** 561 (1984).

[40] J. D. Griffith and L. Harris, *Crit. Rev. Biochem.* **23,** 543 (1988).

[41] S. A. Wasserman and N. R. Cozzarelli, *Proc. Natl. Acad. Sci. U.S.A.* **82,** 1079 (1985).

[42] H. Delius, N. J. Mantell, and B. Alberts, *J. Mol. Biol.* **67,** 341 (1972).

[43] P. C. van der Vliet, W. Keegstra, and H. S. Jansz, *Eur. J. Biochem.* **86,** 389 (1978).

[44] W. T. Ruyechan, *J. Virol.* **46,** 661 (1983).

[45] J. D. Griffith, L. D. Harris, and J. Register, *Cold Spring Harbor Symp. Quant. Biol.* **49,** 553 (1984).

[46] T. Kirkhausen, J. C. Wang, and S. C. Harrison, *Cell (Cambridge, Mass.)* **41,** 933 (1985).

[47] O. L. Miller, *Science* **164,** 955 (1969).

[48] J. D. Griffith and G. Christiansen, *Annu. Rev. Biophys. Bioeng.* **7,** 19 (1978).

[49] J. L. Cartwright, S. M. Abmayr, G. Fleischman, K. Lowenhaupt, S. C. R. Elgin, M. A. Keen, and G. C. Howard, *Crit. Rev. Biochem.* **13,** 1 (1982).

[50] I. Igo-Kemenes, W. Hörz, and H.-G. Zachau, *Annu. Rev. Biochem.* **51,** 89 (1982).

[51] H. Zentgraf and C.-T. Bock, and M. Schrenk, *in* "Electron Microscopy in Molecular Biology: A Practical Approach" (J. Sommerville and U. Scheer, eds.), p. 81. IRL Press, Washington D.C., 1987.

[52] M. F. Trendelenburg and F. Puvion-Dutilleul, *in* "Electron Microscopy in Molecular Biology: A Practical Approach" (J. Sommerville and U. Scheer, eds.), p. 101. IRL Press, Washington D.C., 1987.

[53] Y. N. Osheim and A. L. Beyer, this series, Vol. 180, p. 481.

The following sections describe the procedures that we have used to visualize and analyze nucleoprotein structures by electron microscopy. The protocols listed here are by no means the only ones available for preparing and examining such structures by electron microscopy. A more detailed review describing additional and specialized applications may be found in Ref. 54.

The method used for preparing nucleoprotein complexes for electron microscopy employs a protein free spreading technique. This method differs from the original Kleinschmidt method, in which a solution of DNA with a small basic protein (usually cytochrome *c*) is spread as a monolayer on a water surface.[55-58] DNA spread in this manner is covered by the positively charged protein and can be adsorbed to a plastic support film mounted on a grid; the protein-covered DNA is then stained with uranyl salts followed by rotary shadowing with Pt–Pd or tungsten. DNA treated by this procedure is easily observed in the electron microscope with single-stranded DNA being easily discernible from double-stranded DNA. However, this technique is generally unsatisfactory for visualizing the interaction of DNA-binding proteins with DNA because these interactions are obscured by the protein used for spreading and thickening the DNA. Consequently, several techniques that permit visualization of DNA-binding proteins bound to DNA were developed.[4,54,59-65] Two of these techniques will be presented in detail in this section. The techniques are easy to use, fast, reproducible, and require only nanogram quantities of DNA. A protocol for immunoelectron microscopy of nucleoprotein complexes is also presented. In the descriptions, we assume that the reader is familiar with the operation of both the electron microscope and the vacuum shad-

[54] L. Coggins, *in* "Electron Microscopy in Molecular Biology: A Practical Approach" (J. Sommerville and U. Scheer, eds.), p. 1. IRL Press, Washington D.C., 1987.

[55] K. A. Kleinschmidt, this series, Vol. 12B, p. 361.

[56] R. W. Davis, M. Simon, and N. Davidson, this series, Vol. 21, p. 413.

[57] H. J. Burkhardt and A. Puhler, *in* "Methods in Microbiology" (P. M. Bennet and J. Grinsted, eds.), Vol. 17, p. 133. Academic Press, New York, 1984.

[58] L. T. Chow and T. R. Broker, this series, Vol. 180, p. 239.

[59] J. Dubochet, M. Ducommun, M. Zollinger, and E. Kellenberger, *J. Ultrastruct. Res.* **35**, 147 (1971).

[60] J. D. Griffith, *in* "Methods in Cell Biology" (D. M. Prescott, ed.), Vol. 7, p. 129. Academic Press, New York, 1973.

[61] T. Koller, J. M. Sogo, and H. Bujard, *Biopolymers* **13**, 995 (1974).

[62] P. J. Highton and M. Whitfield, *J. Microsc. (Oxford)* **100**, 299 (1974).

[63] U. K. Laemli, *Proc. Natl. Acad. Sci. U.S.A.* **72**, 4288 (1975).

[64] H. J. Vollenweider, J. M. Sogo, and T. Koller, *Proc. Natl. Acad. Sci. U.S.A.* **72**, 83 (1975).

[65] P. Lahart and T. Koller, *Eur. J. Cell Biol.* **24**, 309 (1980).

owing device used for preparing and shadowing the electron microscope grids.

The preparation of nucleoprotein complexes for electron microscopy (EM) typically involves five steps. In the first step, the samples are treated with a reagent that cross-links functional groups of proteins. This "fixing" step serves to rigidify the bound protein structures and keep them trapped onto the DNA. However, certain DNA–protein complexes are destroyed by the cross-linking reagent, and so this step should be used with care. In the next step, unbound proteins are separated from the nucleoprotein complexes by gel filtration. This step removes excess protein that might obscure the nucleoprotein complex or confuse the interpretation. Unreacted cross-linking reagents are also removed so that any subsequent enzymatic steps, such as restriction enzyme digests, are not interfered with. In the third step, the nucleoprotein complexes are adsorbed to an electron microscope grid that is either covered with a plastic-supported carbon film or a carbon film alone. Prior to adsorption of the DNA, the grids are pretreated in a manner that renders the carbon surface hydrophilic so that the DNA will adhere. In the fourth step, excess sample is removed, and the grid is washed and stained with uranyl salts to enhance contrast. The washing procedure is necessary to prevent salts in reaction buffers from precipitating during staining or drying and interfering with the visualization of the nucleoprotein. Finally the specimen is rotary shadowed with metal and examined in the microscope.

Coating Grids with Formvar

Formvar is a plastic resin that is soluble in organic compounds such as dichloroethane. In this coating protocol, a thin layer of Formvar is formed on a water surface and then stretched over electron microscope grids. In the following protocol, the Formvar film is stabilized by coating it with a thin layer of carbon. Without the carbon overlay the Formvar would disintegrate in the electron beam of the microscope. The carbon-coated Formvar, transparent to the electron beam, serves as support for the specimens.

Required Materials

Large Pyrex dish or glass bowl either painted flat black or covered with black electrical tape

Round-bottom glass tube, 38 × 430 mm (Bellco, Vineland, NJ, glass pipette canister, round bottom)

Pasteur pipettes (9 in.) and rubber bulb

Solution of Formvar (1.5% in dichloroethane, Ted Pella, Redding, CA)

Electron microscope grids; Cat. No. 3HGC300, 3HGC400, or 3HGC500 (Ted Pella) are satisfactory

Fine-tipped forceps with a rubber O ring for keeping the jaws clamped together

Either an in-house vacuum line or small vacuum apparatus for picking up grids

Oven at 55°

Portable fluorescent lamp

Glass microscope slide

Dissecting microscope (optional, but highly recommended)

1. Wash out the glass dish with detergent, and then rinse well with high-quality distilled water. There must be no residual detergent after washing or the technique will not work.

2. Fill the bowl to about 1 or 2 in. from the top with high-quality distilled water at room temperature. If the water is too warm the technique will not work.

3. Wash off the round bottom of a 38 × 430 mm glass tube with detergent and rinse off with distilled water. Again, there must be no residual detergent. Dry the tube in the oven and then let it cool to room temperature. Do not touch the bottom so as to prevent contamination from finger grease.

4. Position the fluorescent lamp slightly to one side of the bowl so that it illuminates the surface of the water. The black coating on the bowl prevents glare and also contrasts with the silver-colored Formvar film that is to be spread on the water surface.

5. Fill the drawn-out portion of a 9-in. Pasteur pipette (approximately 200 μl) with Formvar solution.

6. This step is difficult, but crucial to the success of the technique. Hold the pipette so that the tip is about an inch or two above the surface of the water and to the side of the bowl. Next, quickly move the tip to the opposite side, keeping the tip the same distance above the water surface, and, while the tip is in motion, squirt the Formvar onto the water.

7. Look at the water surface. There should be a thin film on the surface. Look for a region that is silvery gray in color. Regions that are purple, blue, or gold may also appear. These regions are too thick for use and should be avoided. If a silvery gray area at least 0.5 by 0.5 in. cannot be found, then discard the water and begin again from step 1. The silvery gray area may appear somewhat streaked.

8. Lay about a dozen grids, shiny side down, onto a silvery gray patch. Make a tight pattern. Do not lay down more than 16 grids or the film will sink in the next step.

9. Hold a clean 38 × 430 mm glass tube vertically over the grids and *gently* bring the rounded bottom down onto the grids and submerge them about half an inch or less under water.

10. Maintain the glass tube in a vertical position with the grids submerged and *gently* move the tube to the side of the bowl.

11. Twist the remaining Formvar film up onto the sides of the tube.

12. Bring the tube back to the center of the bowl, tilt it steeply, and then remove it from the water. The grids should remain stuck to the bottom of the tube.

13. Examine the grids under the dissecting microscope. If there are drops of water on the grid surfaces, gently aspirate them off the grid with a finely drawn-out Pasteur pipette hooked up to a vacuum line.

14. Dry the grids for 2 min in an oven at 55°.

15. Using fine-tipped forceps, remove the grids from the glass tube, and lay them shiny side up on the end of a clean glass microscope slide. The grids are now ready for carbon coating.

Sometimes the grids consistently sink when attempting to adhere them to the glass tube in step 12. We are not sure what the cause of this is; but if this step does not work, then try picking the grids up with Parafilm. Do this by gently laying a 2-in. square of Parafilm over the grids and then carefully drape the remaining Formvar on top of the Parafilm. Gently remove the Parafilm and grids from the water surface, and then dry the grids overnight at room temperature. Once dry, the grids can be carbon-coated right on the Parafilm.

Carbon Coating Formvar-Covered Grids

Required Materials

Aluminum foil
Disk of Whatman (Clifton, NJ) filter paper
Graphite rods
Fine sand paper
Pencil sharpener

1. Sharpen the end of a graphite rod and insert into the spring-loaded holder of the vacuum shadower.

2. Flatten the end of another graphite rod and place it in the opposite holder so that the flattened surface of the rod contacts the sharpened rod.

3. Adjust the spring on the holder containing the sharpened rod so that the point continuously maintains pressure against the flattened surface of the adjacent carbon rod.

4. Fold a piece of aluminum foil over several times into a rectangle

that is about 1 in. wide × 5 in. long. Next, fold the rectangle into a ⌐ shape.

5. Cut a 1 × 1 in. square of filter paper and place it onto the foil so that only half of the paper is under the "roof" of the foil.

6. Place the glass slide with the Formvar-covered grids onto the foil at the end opposite to the "roof."

7. Place the foil with, the paper and grids on it directly under the carbon rods. The rods should be about 3 to 4 in. above the grids.

8. Evacuate the chamber to about 10^{-4} torr. Slowly apply current to the carbon electrodes until they just begin to spark, wait a few seconds, then stop the current flow.

9. Look at the filter paper. The portion underneath the foil roof should still be white and the area not protected by the roof should be a light brown. If it is not brown, repeat step 8 until the paper becomes a light brown, which indicates that the grids have been sufficiently coated with carbon.

10. Remove the grids from the vacuum shadower and store them in a glass Petri dish with a cover. Formvar-covered grids coated with carbon can be stored indefinitely.

Carbon Coating Grids without Use of Formvar Support

In this method the Formvar support film is dispensed with. Instead a thin layer of carbon film is directly applied onto the grids. The advantage of this method is that the contrast of specimens mounted on grids coated with carbon alone is better than that with grids coated with both Formvar and carbon. The disadvantage is that the carbon tends to easily tear without the Formvar support, which necessitates the use of finer meshed grids. This in turn reduces the visible area of view on the grid.

Required Materials

Büchner funnel with a length of plastic hose and a screw clamp near the end of the tubing. Attach the funnel to a heavy ringstand so that it does not wobble or vibrate when filled with water

Stainless steel mesh that has been bent into a level platform that stands about an inch high from the bottom of the Büchner funnel

Either 400- or 500-mesh tabbed electron microscope grids (3HGC400 or 3HGC500, Ted Pella)

Mica sheets (about 1 × 3 in., Ted Pella)

Fresh razor blade

Squares (1 in.) of Whatman filter paper

Clean glass microscope slides

1. Cleave several sheets of mica with a fresh razor blade. Lay the cleaved sides face up onto glass microscope slides. Use gloves or forceps to handle the sheets so as not to contaminate them with finger grease.

2. Place a square of filter paper between the mica and glass slide so that a portion of the paper sticks out from underneath the mica. The paper will be used to access the extent of carbon deposition on the mica.

3. Place a few of the mica sheets into the evaporator at varying distances from the carbon electrode. This arrangement will yield a gradient of differing carbon thicknesses at the carbon is evaporated onto the mica sheets.

4. Evacuate the chamber to at least 10^{-4} torr. Slowly apply current to a freshly sharpened carbon electrode just until sparks form and then stop the current about 5 sec later. Repeat this procedure until the paper becomes a light brown.

5. Clamp the end of the hose that is attached to the Büchner funnel and fill it with distilled water. Squeeze the tubing several times to remove air bubbles. (If there is any residual air in the tubing or funnel air bubbles may inadvertently travel up the tubing and displace the grids or carbon film.)

6. Using clean forceps, place about a dozen grids shiny side up onto the immersed mesh platform. When submerging the grids be sure to dislodge any air bubbles clinging to the grids or forceps, otherwise the grids may float or stick to the forceps.

7. Take a carbon-coated sheet of mica and gently breathe on it. This thin layer of moisture will assist in the separation of the carbon film from the mica. With the carbon side facing up, begin to submerge the mica at an angle of about 45°. As the carbon begins to separate from the mica and float, position it over the submerged grids.

8. Drain the water from the funnel. The carbon film should gently settle down onto the grids.

9. Allow the mesh and grids to drain for a few minutes. Place the mesh and grids into a desiccator with some Drierite and then evacuate and seal the desiccator. Allow the grids to dry overnight. Check the quality and contrast of a sample from each batch in the electron microscope.

Adsorbing Nucleoprotein Complexes to EM Grids

Two methods for applying nucleoprotein complexes to electron microscope grids are presented here. In the first,[4] the carbon-coated grids are made hydrophilic by a glow discharge, then a solution of polylysine is applied, and finally the nucleoprotein complexes are adsorbed to the treated grid. In the second method,[16,18,65,66] an organic dye is used to provide the grid surface with a positive charge, and the nucleoprotein complexes are then adsorbed to the dye-treated grid. The second method is useful if the vacuum shadower is not equipped with a glow discharge circuit.

Adsorbing Nucleoproteins to EM Grids Using Polylysine Technique

Required Materials

Formvar-covered grids coated with carbon
Lead buckshot that has been slightly flattened with a hammer
A small metal block
Pasteur pipette with a finely drawn-out tip about 0.5 mm in diameter; make by drawing out a 9-in. Pasteur pipette over a flame, score with a diamond-tipped scribe, and then break the tip off; hook the pipette up to a vacuum line and safety trap (clamp the modified Pasteur pipette to a ring stand and attach it to the vacuum via the safety trap)
Incandescent lamp
Dissecting microscope (optional, but highly recommended)
Fine-tipped forceps with a rubber O ring for holding the jaws shut
Parafilm
Two small glass Petri dishes with covers
0.1 and 0.01 M solutions of ammonium acetate
Polylysine, 1 μg/ml; prepare fresh from a 1 mg/ml stock solution of polylysine [poly(L-lysine), M_r 2000, Sigma (St. Louis, MO) #P0879] in distilled water; store 1-ml aliquots of 1 mg/ml stock polylysine solutions at $-20°$
Uranyl acetate solution ($\sim$5%); make by mixing 0.5 g uranyl acetate with 10 ml distilled water, vortex 3 min, remove undissolved salt by centrifugation for 5 min in a tabletop centrifuge at full speed, withdraw 8 ml of supernatant, and filter through a 0.45-μm filter; store refrigerated in the dark
DNA dilution buffer[4]: Tris, HEPES, phosphate, or triethanolamine buffers at pH values ranging from 7.5 to 8.0 at concentrations ranging from 10 to 50 mM plus Mg^{2+} at 3 to 10 mM are usually suitable for diluting and adsorbing DNA to grids. KCl or NaCl at $\leq$150 mM, formamide at $\leq$85% (v/v), glycerol $\leq$20% (v/v), or glyoxal $\leq$0.5 M may be present in the dilution buffer. If desired, EDTA and dithiothreitol may also be included at low concentrations as well ($\leq$1 mM)

[66] J. Sogo, A. Stasiak, W. DeBernardin, R. Losa, and T. Koller, in "Electron Microscopy in Molecular Biology: A Practical Approach" (J. Sommerville and U. Scheer, eds.), p. 61. IRL Press, Washington D.C., 1987.

Glow-Discharge Treatment of Grids. 1. Place grids onto a small metal block and cover the handles with the flattened buckshot. Covering the handles keeps them from becoming charged, which prevents fluid from running onto the jaws of the forceps in subsequent manipulations.

2. Place the block with the grids onto the center platform of the vacuum shadower, and insert the glow discharge electrode into its socket so that it is positioned about 4 to 5 in. above the grids. Be certain that the electrode does not come into contact with any other parts inside the chamber.

3. Evacuate the chamber to about 75-mtorr air pressure, then activate the glow discharge circuit. A pinkish-blue glow should be visible above the grids. After glow discharging for about 10 to 15 sec, switch off the glow discharge circuit.

4. Admit air to the chamber and remove the glow-discharged grids, which are now ready to be coated with polylysine. The grids will begin to lose their charge after several minutes, so it is important to coat them with polylysine as soon as possible.

Application of Polylysine to Glow Discharge-Treated Grids. 1. Pick up a glow-discharged grid by its handle with forceps, clamp the jaws shut with the O ring, and lay the forceps onto the benchtop so that the glow-discharged side is facing up.

2. Place a 7-μl drop of freshly prepared 1 μg/ml polylysine onto the grid.

3. About 30 sec later aspirate the polylysine drop off the grid by touching the edge of the grid to the drawn-out Pasteur pipette tip hooked up to a vacuum. Use a dissecting microscope and side illumination with the incandescent lamp to do this. If the grid has been properly glow discharged, it should take several seconds for the grid to dry as the fluid is aspirated off its surface, and interference colors should be visible at the trailing edge of the fluid.

Adsorption of Nucleoprotein Complexes to Polylysine-Treated Grids. 1. Dilute an aliquot of the nucleoprotein complexes so that the concentration of DNA is approximately 1 to 2 μg/ml.

2. Place a 5-μl droplet of the diluted nucleoprotein solution onto a polylysine-treated grid.

3. About 1 min later invert the grid and drag it several times over the surface of a solution of 0.1 M ammonium acetate contained in a small glass Petri dish. Do not submerge the back surface of the grid.

4. Touch the inverted grid to a 10-μl droplet of uranyl acetate (previously placed on the surface of a small square of Parafilm). The drop will stick to the grid as it is picked up.

5. Drag the grid two or three times over the surface of a solution of 0.01 M ammonium acetate contained in a small glass Petri dish. Again, do not submerge the back surface of the grid.

6. Aspirate the grid to dryness using the drawn-out Pasteur pipette hooked up to a vacuum line or blot excess fluid off the grid by touching the edge to a piece of filter paper. The grid is now ready to be rotary shadowed as described below.

Adsorbing Nucleoproteins to EM Grids Using the Alcian Blue Dye Technique

Required Materials

Fresh 0.2% solution of Alcian Blue 8GX[67] in 3% acetic acid
Formvar-covered grids coated with carbon
Fine-tipped forceps with a rubber O ring for holding the jaws shut
Parafilm
Uranyl acetate solution (same as above)
DNA dilution buffer: Same as for polylysine technique except that DNA
    will not stick if formamide is present[54]

Application of Alcian Blue to Grids and Adsorption of Nucleoprotein Complexes. 1. Dilute 0.2% Alcian Blue 8GX to 0.002% in doubly distilled water.

2. Pick up a carbon-coated grid by its handle with forceps and clamp the jaws shut with the O ring.

3. Place a 10-μl droplet of the diluted Alcian Blue solution onto the grid for 5 min.

4. Wash the grid by gently dragging it over the surface of double-distilled water in a Petri dish several times.

5. Remove any remaining water off the grid either by aspiration with a finely drawn Pasteur pipette or by touching a piece of filter paper to the side of the grid.

6. Adsorb DNA to grids as described above for grids that have been treated with polylysine.

Rotary Shadowing Nucleoprotein Complexes

We generally use a tungsten filament (0.02-in. diameter) for rotary shadowing. However, evaporation of a Pt–Pd wire wrapped around a tungsten filament works just as well.

1. Loosen both of the screws that clamp the tungsten filament onto the copper electrodes.

2. Loosen one of the copper electrodes from its retaining block by loosening the screw that holds the electrode in place.

[67] M. A. Hayat, *in* "Positive Staining for Electron Microscopy," p. 187. Van Nostrand-Reinhold, New York, 1975.

3. Slide the electrode out of its holder. Insert a 4.2-cm long tungsten filament into it and tighten the screw so that the tungsten filament is firmly clamped into the electrode.

4. Insert the electrode back into its holder.

5. Adjust the height of the electrodes so that the tungsten filament is at an 8° angle with respect to the center of the rotating platform, which will be 9.5 cm distant from the tungsten filament.

6. Insert the other end of the filament into the other electrode and tighten the screw so that it is held firmly in place.

7. With one hand squeeze the two vertical posts holding the electrode retaining blocks about a millimeter toward one another, and with the other hand tighten the screw that holds the electrode in place. This places tension on the filament and keeps it from bending as it heats up and expands when current is applied.

8. Place the grids to be shadowed onto the center of the rotating platform and move the platform so that the distance from its center to the tungsten filament is 9.5 cm.

9. Evacuate chamber to at least 10^{-5} torr.

10. Turn on the rotating platform and, if it is adjustable, set it to about 30 rpm.

11. Apply current to the filament and in one motion adjust the current to about 32 to 34 A. The filament should burn white hot for at least 1.5 to 2 min and then burn out. As the filament burns the current will decrease. Do not adjust current once it is set. If the filament either burns out too soon or burns too long, the shadowing will be too light. The optimal current setting required to achieve the correct filament burn time must be determined empirically for each vacuum shadowing apparatus.

12. Shut off the current and admit air to the chamber. The grids are now ready for examination in the electron microscope.

Preparing Nucleoprotein Complexes for Electron Microscopy

Required Materials

Glutaraldehyde, 8% aqueous, EM grade (Polysciences, Miles, IL)

Polyethylene microcentrifuge tubes (400 µl; Cat. #19002, Denville Scientific, Inc., Denville, NJ) for drip columns or 1-ml plastic syringes for spin columns

Spectra mesh, either polyethylene or nylon, 70- to 100-µm mesh size (Fisher Scientific, Pittsburgh, PA)

Sepharose 4B or CL-4B

Hypodermic needles, 25- or 22-gauge

Ring stand and thermometer clamp
Razor blad or scalpel
Hemostat
Parafilm
Safety glasses

Fixing Nucleoprotein Complexes with Glutaraldehyde

We generally use glutaraldehyde to fix proteins to DNA. Formaldehyde fixation followed by glutaraldehyde fixation can also be used.[7] The fixatives cross-link lysine residues with one another, and serve to cross-link and rigidify large multiprotein structures. Glutaraldehyde fixation does not cross-link protein to DNA. Sometimes aldehyde fixation fails to work, and may in fact alter a protein so that it dissociates from DNA. In this event it may be necessary to simply dilute the sample just prior to adsorbing it to the grid. The protein concentration of the diluted sample must be low enough so that it does not interfere with visual interpretation of the micrographs. Alternatively, other cross-linking reagents can be tried.

1. Open a fresh ampoule of 8% glutaraldehyde, dilute an aliquot into distilled water to yield a concentration of 1.5% solution, and store on ice.

2. DNA and proteins are mixed in reaction buffer so that the concentration of DNA is about 10 to 20 μg/ml in a volume of 15 to 25 μl. Try to avoid buffers such as Tris, which may react with the glutaraldehyde in the next step. If bovine serum albumin is to be included in the reaction (which usually is not necessary), be sure that its final concentration is less than 3 μg/ml when the sample is to be applied to a grid. After incubating at the desired temperature and time, add the freshly diluted glutaraldehyde to a final concentration of 0.1%, and continue to incubate an additional 10 to 15 min at the same temperature used for the reaction.

3. Remove excess unbound proteins and glutaraldehyde by gel filtration over a Sepharose 4B microcolumn.

Preparation of Sepharose 4B Microcolumns

Excess unbound protein can be removed either by passing the samples through drip columns[19] or by passing through spin columns.[20,37] Drip columns are slow, but if taken care of they can be used several times. Spin columns offer the advantage of speed, but on occasion excess protein may leak through.

Drip Column Gel Filtration

1. Grip a 25-gauge needle with a hemostat and cut off half of the plastic Luer (the hemostat prevents accidental injury to the fingers; wear safety glasses when cutting the Luer to prevent accidental eye injury).

2. Cut the tip off of a 400-μl microcentrifuge tube so that a hole about 2 mm in diameter is obtained.

3. Cut a piece of the flexible plastic mesh about 4 × 4 mm and plug the hole. A straightened paper clip is useful for this. Don't pack the mesh too tightly or the column will drain too slowly.

4. Fill the severed needle with distilled water and slip it onto the bottom of the microcentrifuge tube so that the Luer fits snugly.

5. Fill the tube with water and clamp it to a ringstand with a thermometer clamp. If the needle has been properly seated and the mesh is not too tightly packed into the hole, water should drain freely from the needle, and there should be no leakage at the Luer/tube joint.

6. Fill the microcolumn with a slurry of Sepharose 4B or CL-4B. A bed height of 3 cm is sufficient for most work. Do not become alarmed if the column runs dry, as there will still be enough water or buffer retained by capillary action to keep the bed wet for a couple of hours.

7. Make a tower for holding a reservoir of buffer by cutting the tip of another 400-μl microcentrifuge tube and connecting it to the top of the microcolumn.

8. The nucleoprotein complexes will be eluted in the void volume of the microcolumn. Determine the void volume as follows. Wrap an index card with a piece of Parafilm, which will be used to collect fractions from the microcolumn. Let the column drain and remove the reservoir tower. Layer 10 μl of a solution of blue dextran onto the surface of the drained bed, allow it to run in, and then chase it with 10 μl of reaction buffer. Fill the microcolumn with buffer, attach a reservoir tower, and then fill the tower with buffer. The first five fractions containing the blue dextran will be the fractions to pool when the nucleoprotein complexes are run over the microcolumn. Pass two column volumes of 4 M NaCl followed by two column volumes of distilled water to remove residual dye from the column. Radiolabeled DNA can also be used instead of blue dextran to determine the void volume.

9. Store the column until use by filling with distilled water, capping it, and then placing the needle tip into a 1.5-ml microcentrifuge tube filled with distilled water. Columns can be reused several times so long as they are washed with high salt and buffer or water after each use.

10. Equilibrate a calibrated Sepharose 4B microcolumn with two column volumes of buffer.

11. Drain the column and layer the solution containing the fixed nucleoprotein complexes onto the gel bed.

12. After the solution enters the bed, elute with buffer and pool the five peak fractions. After gel filtration the fixed nucleoproteins are usually stable for several hours at room temperature. At this stage the complexes are ready for electron microscopy, restriction enzyme digestion, or labeling with antibody. Gel filtration over a 2-cm high bed dilutes samples about two to threefold.

Spin Column Gel Filtration

1. Plug a 1-ml syringe with a swatch of Spectra Mesh.

2. Fill with a 1-ml slurry of Sepharose CL-4B in DNA dilution buffer.

3. Centrifuge at 2000 rpm in a tabletop centrifuge for 2 min, using a swinging bucket rotor.

4. Place a fresh microcentrifuge tube under the column, and load the fixed samples (50 μl) onto the column.

5. Recover the sample by centrifugation at 2000 rpm for 2 min.

Immunoelectron Microscopy of Nucleoprotein Complexes

Some nucleoprotein structures may consist of several proteins. So long as the appropriate antisera are available, the protein composition of a multiprotein complex can be qualitatively determined. The method is similar to other techniques, such as Western blotting, that probe with antibody and a visible secondary label. The major difficulty is optimizing the proper concentrations of both the antibody and the particular batch of protein A–colloidal gold (see below). The technique is extremely powerful once conditions have been optimized.

Required Materials

Purified IgG antibody

Protein A–colloidal gold, either 10- or 15-nm diameter gold particles, 5–10 μg/ml (E–Y Laboratories, San Mateo, CA)

Buffer A: 3-cm bed height Sepharose 4B microcolumn equilibrated with 10 mM Tris-HCl, pH 8.4, 150 mM NaCl, 10 mM MgCl$_2$

Buffer B: 3-cm bed height Sepharose 4B microcolumn equilibrated with 40 mM HEPES–KOH, pH 7.6, 100 mM KCl, 10 mM magnesium acetate

Buffer C: 4-cm bed height Sepharose 4B microcolumn equilibrated with 40 mM HEPES–KOH, pH 7.6, 100 mM KCl, 10 mM magnesium acetate. (*Note:* Spin columns may be substituted; however, the final column should be plugged with glass wool, which specifically traps excess colloidal gold.[37])

1. Apply glutaraldehyde-fixed nucleoprotein complexes to a 3-cm bed height Sepharose 4B microcolumn, elute with buffer A, and pool the five peak fractions.

2. Add 1 μl of purified antibody (0.5 to 2.5 μg) that has been diluted in equilibration buffer A. The dilution of IgG added to the nucleoprotein complexes to be probed is such that the complexes are not aggregated by the bound antibody. The correct dilution must be determined empirically for each purified IgG. Usually a 2- to 10-fold dilution of 5 to 25 mg/ml DEAE-Sephacel-purified IgG is adequate.

3. Incubate for 30 min at 37°.

4. Apply antibody-bound nucleoprotein complexes to a 3-cm bed height Sepharose 4B microcolumn, elute with buffer B, and pool the five peak fractions. This step removes unbound excess antibody.

5. Add 1 μl of protein A–colloidal gold conjugate particles and incubate for about an hour at room temperature.

6. Apply antibody-gold-bound nucleoprotein complexes to a 4-cm bed height Sepharose 4B microcolumn, elute with buffer C, and pool the five peak fractions. This step removes unbound excess colloidal gold. The samples are now ready for adsorption to grids.

Examination of Nucleoproteins in Electron Microscope and Analysis

We usually use an accelerating voltage of 80 kV. Lower acceleration voltages will enhance contrast but will reduce resolution. Magnifications ranging from ×20,000 to ×45,000 are the most useful for observing nucleoprotein complexes. Using the wobbler to focus does not always work well with older microscopes. We usually focus by eye at a magnification of ×35,000. If a molecule is to be photographed, it is best to take a through focus series of photographs, and then *immediately* develop the plates. If the photographs do not turn out well, it is easy to go back and rephotograph the molecule of interest. If possible, try to take photographs that depict several molecules in the same field of view. Presenting several similar molecules in a field of view is incontrovertible proof of the frequent occurrence of the structure. Always include a size bar on the micrographs as a means of giving the scale of magnification.

In order to measure the length of a segment of DNA an internal standard of known length must be included in the sample. The internal standard can be nicked circular, relaxed covalently closed, or linear DNA. The length of the internal standard should be as close as possible to that of the unknown but sufficiently different so that it can be unmistakably discerned.

Data that map the locations of protein–DNA binding sites and boundaries along a linear DNA segment are presented as a histogram. A homogeneous population will approximate a normal curve when plotted as a histogram in intervals of about half of a standard deviation. A heterogeneous population will have a large standard deviation, and the histogram may be skewed or bimodal. Measuring errors, such as inadvertently measuring molecules from micrographs of different magnification, will also exhibit these discrepancies. The data are displayed with the distances in base pairs or fractions of unit length from the end(s) of the DNA to either the edge(s) or the center of the protein moiety on the abscissa. The number of measurements in an interval is plotted on the ordinate.

If the numerical value of a quantity such as the distance from the end of a DNA segment to the edge of a nucleoprotein is to be stated, present the value as a mean plus or minus the sample standard deviation and include the number of measurements in the sample. Generally sample sizes ranging from 30 to 100 are satisfactory.

The frequency of a particular class of nucleoprotein can be determined by directly examining the specimen grids in the microscope so long as the structure of interest is easily discernable by eye or can be tagged with an identifiable marker such as colloidal gold. With a little practice in moving the stage with one hand and scoring molecules with a multikey hand tally with the other hand it is possible to scan several thousand molecules per hour. To scan a grid rapidly, reduce the magnification to a point where it is still possible to identify specific structures and distinguish separate DNA molecules. Score only nucleoprotein complexes that are completely in a field of view. Once all the interpretable molecules in a field of view have been scored move the stage just enough so that the previous field of view is not visible and score the next set of molecules. Follow a pattern so that the same area of the grid is scored only once. Score molecules in sets of 100. After accumulating at least 3 sets of 100 counts, a mean score and standard deviation for the frequency of each class of molecule scored can be calculated. For structures or classes not discernable by eye, it is necessary to take photographs and measure the size and frequencies from enlarged prints or projected images.

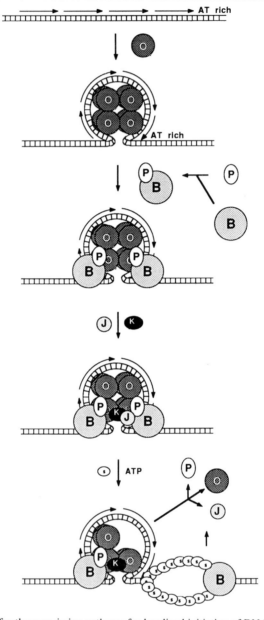

FIG. 1. Model for the prepriming pathway for localized initiation of DNA replication by phage λ. A sequential series of protein assembly and disassembly reactions at *ori* λ culminate in localized unwinding of the origin DNA, providing for the specific initiation of DNA replication. Electron micrographs depicting each of these stages are shown in Figs. 2–4. See text for details.

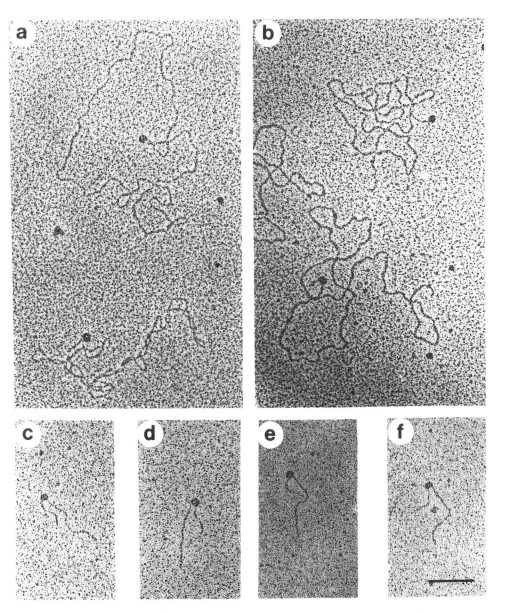

FIG. 2. Electron micrographs of phage λ O protein bound to *ori* λ DNA. (a) O protein bound to supercoiled and nicked circular plasmid DNA. (b) O protein bound to plasmid DNA linearized by restriction enzyme that cleaves adjacent to *ori* λ. (c–f) O protein bound to a restriction fragment containing *ori* λ. Note that the DNA enters and exits the O-some from the same side, which indicates that the DNA is either folded or wound about the protein core. The DNA-winding concept is also supported by measurements of the length of DNA associated with the nucleoprotein structure (bar: 0.1 μm). (Reprinted from Ref. 19.)

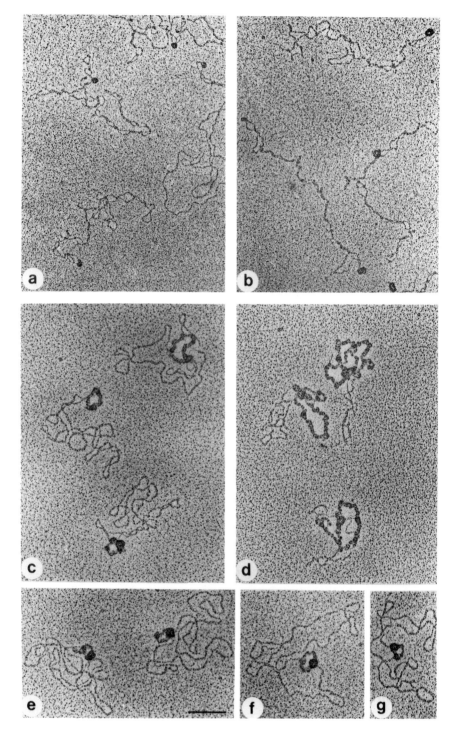

Some Examples of Exceptional Power of Electron Microscopy

The electron microscope is an enormously powerful tool for the analysis of the structure and dynamics of DNA–protein complexes. One special power of the technique is its ability to determine the path of the DNA in nucleoprotein complexes, information not available from other techniques of DNA–protein analysis (except for X-ray crystallography). Thus, electron microscopy has led to the realization that DNA-wound nucleoprotein structures are critical intermediates in the initiation of DNA replication and site-specific recombination. Electron microscopy has also been used to demonstrate self-association of transcriptional regulators bound to distant sites, and the association of a distant regulator with RNA polymerase (DNA-looping interactions). By the use of immunoelectron microscopy, the sequential assembly and disassembly of nucleoprotein complexes has been followed. To illustrate the special value of electron microscopy, we will present two examples from our work: the sequential pathway that initiates DNA replication by phage λ; and the DNA-looping interaction involved in transcriptional activation of the *glnA* promoter from the NtrC enhancer site.

Initiation of DNA Replication by Bacteriophage λ

We used two complementary electron microscopic approaches to show that a series of protein–DNA and protein–protein assembly/disassembly reactions precede the initiation of DNA replication at the bacteriophage λ origin of replication, *ori* λ. In the first approach, multiprotein structures were visualized by electron microscopy of rotary-shadowed nucleoprotein complexes. Using this technique, we visualized the successive assembly of three distinct nucleoprotein complexes that precede origin-specific initiation of DNA replication. In the second approach, antibody labeling of individual proteins with colloidal gold revealed the protein complement of these structures. The inferred pathway is shown in Fig. 1.

The assembly of the first nucleoprotein structure that catalyzes the cascade of assembly and disassembly reactions results from the binding

FIG. 3. Electron micrographs of nucleoprotein structures at *ori* λ. (a) O protein alone bound to *ori* λ plasmid DNA. (b) O, P, and DnaB proteins bound to *ori* λ plasmid DNA. (c) Unwound nucleoprotein structures generated by the interactions of DnaJ, DnaK, and single-stranded DNA-binding proteins and ATP with the O–P–DnaB–*ori* λ complex. (d) Highly unwound nucleoprotein complexes formed by the six-protein reaction of (c) in the presence of DNA gyrase. (e–g) *ori* λ DNA and the six-protein reaction of (c). Note the difference in size of the nucleoprotein structures at the two forks. The smaller structure is presumably DnaB because of its location at the unwinding fork (bar: 0.1 μm). (Reprinted from Ref. 31.)

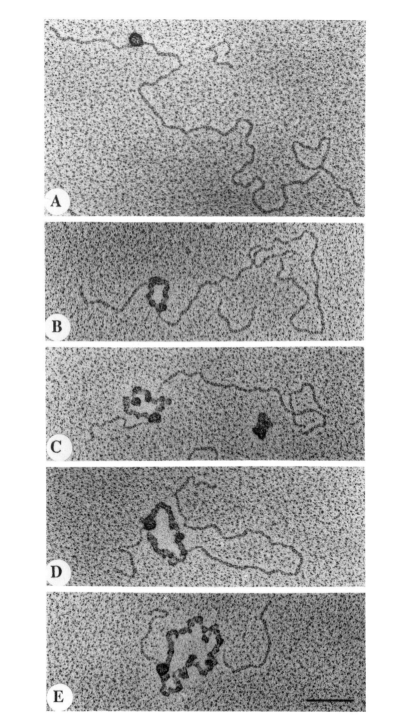

of a phage-encoded protein, λO, to the λ origin of replication, *ori* λ.[68] O protein binds to an inverted sequence that is repeated four times in *ori* λ.[69] Because each binding sequence is inverted, O is likely to bind as a dimer.[69] Thus a minimum of eight O monomers binds to linear *ori* λ DNA. However, electron microscopy showed that the functional unit is not a linear array, but a DNA-wound specialized nucleoprotein structure (snup) called the O-some[19] (Fig. 2).

In order for λ to initiate replication, the DnaB helicase must first be captured from the *Escherichia coli* host.[69] This is done by the λ-encoded P protein, which binds tightly to the DnaB helicase.[70] The second structure that is assembled in the pathway to initiating DNA replication arises from the interaction of the P–DnaB complex with the O-some. This second stage nucleoprotein intermediate is larger and more asymmetric than the O-some, and the structure includes all of the repeated O-binding sites plus an adjacent (A + T)-rich region[19,71] (Fig. 3b).

λ also requires several heat-shock proteins of the host to initiate replication.[69] In the presence of single-stranded DNA-binding protein and ATP, the heat-shock proteins DnaJ and DnaK interact with the second stage nucleoprotein to generate the key intermediate in initiating replication[31,71] (Fig. 3c–g). This structure consists of locally unwound DNA that is stabilized by single-stranded DNA-binding protein. Nearly complete unwinding occurs in the presence of a topoisomerase. The unwound entity is capable of associating with DnaG primase and DNA polymerase holoenzyme in

[68] T. Tsurimoto and K. Matsubara, *Nucleic Acids Res.* **9**, 1789 (1981).
[69] M. Furth and S. Wickner, *in* "Lambda II" (R. Hendrix, R. Weisberg, F. Stahl, and J. Roberts, eds.), p. 145. Cold Spring Harbor Laboratory, Cold Spring Harbor, New York, 1983.
[70] S. Wickner, *Cold Spring Harbor Symp. Quant. Biol.* **43**, 303 (1979).
[71] C. Alfano and R. McMacken, *J. Biol. Chem.* **264**, 10709 (1989).

FIG. 4. Electron micrographs of nucleoprotein structures generated on supercoiled *ori* λ DNA followed by cleavage with restriction enzyme. *ori* λ plasmid DNA was incubated with O, P, DnaB, DnaJ, DnaK, and single-stranded DNA-binding proteins and ATP, and then cut with a restriction enzyme that cleaves asymmetrically with respect to the origin. (A) Presumed O–P–DnaB–DnaJ–DnaK complex. (B) Partially unwound nucleoprotein structure formed by the six-protein unwinding reaction. (C–E) Highly unwound structures formed by the six-protein unwinding reaction in the presence of DNA gyrase. Note that unwinding is unidirectional from the left, and that a large nucleoprotein complex, presumably the incompletely disassembled initiation structure, is visible at the left fork. The second DnaB needed for bidirectional unwinding is presumably stalled in the initiation structure (bar = 0.1 μm). (Reprinted from Ref. 31.)

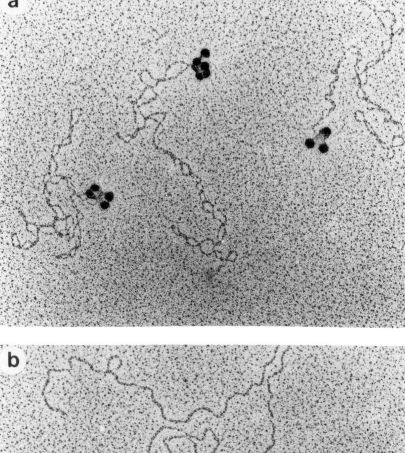

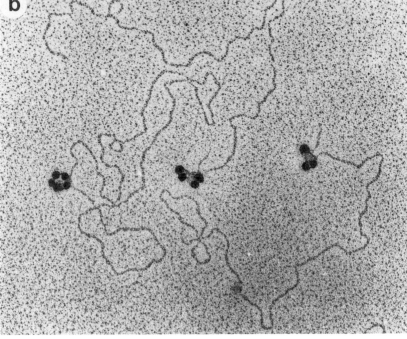

initiating DNA replication.[72,73] Unwinding and replication are unidirectional[31,72] (Fig. 4), which contrasts with the predominantly bidirectional mode of replication *in vivo*.[74] However, bidirectional unwinding and replication occur *in vitro* if a third heat-shock protein, GrpE, is included with the repertoire of components that yield the unidirectionally unwound structure.[75]

Simple rotary shadowing could not reveal the actual protein composition of these mutiprotein replication intermediates. Nor did it reveal the function of the heat-shock proteins in the pathway to unwinding and initiation of replication other than to show a requirement for their participation. Immunoelectron microscopy was used to address these problems.[36] Nucleoprotein structures were probed with specific antibody and secondarily labeled with colloidal gold–protein A conjugates (Fig. 5). With this powerful analytical tool, the protein composition of the nucleoprotein complexes comprising each stage of the pathway leading to the prepriming complex was discerned, and the function of the heat-shock proteins was determined (Fig. 1). Immunoelectron microscopy clearly showed that the second stage nucleoprotein structure is in fact composed of O, P, and DnaB proteins bound to *ori* λ. This technique also showed that the heat-shock proteins DnaJ and DnaK associate with the *ori* λ–O–P–DnaB structure, resulting in the assembly of a further intermediate stage. In the presence of ATP, the bound heat-shock proteins function in a disassembly reaction to eject P and some O from the initiation structure, allowing DnaB to proceed rightward in its unwinding reaction, with the newly unwound DNA stabilized by single-stranded DNA-binding protein. The bidirectional unwinding reaction presumably requires more extensive disassembly mediated by the participation of GrpE, so that a second DnaB helicase can proceed leftward.

As for origin-specific initiation of DNA replication, the precise regulation of promoter-localized initiation of transcription is mediated by snups. A DNA-looping model has been proposed as a mechanism for both positively and negatively regulating transcription (reviewed in Ref. 28). In this

[72] K. Mensa-Wilmot, R. Seaby, C. Alfano, M. S. Wold, B. Gomes, and R. McMacken, *J. Biol. Chem.* **264**, 2853 (1989).
[73] M. Zylicz, D. Ang, K. Liberek, and C. Georgopoulos, *EMBO J.* **8**, 1601 (1989).
[74] M. Schnos and R. Inman, *J. Mol. Biol.* **51**, 61 (1970).
[75] C. Vasilikiotis, D. Ang, C. Georgopoulos, and H. Echols, unpublished work (1991).

FIG. 5. Tagging of nucleoprotein structures with IgG antibody and *Staphylococcus* protein A–gold conjugate. (a) Gold-tagged O-somes on supercoiled *ori* λ plasmid DNA. (b) Gold-tagged O-somes on *ori* λ plasmid DNA linearized with restriction enzyme. (Reprinted from Ref. 36.)

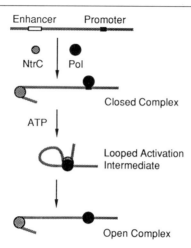

Fig. 6. Model for the action of NtrC from the enhancer site. NtrC protein binds to the enhancer and associates with promoter-bound RNA polymerase. The association facilitates the transition of RNA polymerase from an inactive closed complex into a productive open complex. Transcription of the *glnA* gene by polymerase ensues, and NtrC remains enhancer bound to initiate another round of activation. (Reprinted from Ref. 12.)

model, regulatory DNA sequences distant to the promoter exert their influence by first specifically binding to a regulatory protein. The DNA-bound regulatory protein then associates via direct protein–protein contacts with other regulatory proteins and/or with promoter-bound RNA polymerase to generate a snup. The snup then either represses or stimulates the initiation of transcription.

Transcriptional Activation of glnA Promoter from NtrC Enhancer Site

We used electron microscopy to directly visualize the central biochemical prediction of theDNA-looping model for the positive regulation of the prokaryotic *glnA* operon from its enhancer (Fig. 6).[12] The enhancer element of the *glnA* operon has two sites that tightly bind to the NtrC regulatory protein.[76] As with other enhancers, the *glnA* enhancer functions independently of distance and orientation to the promoter.[76] The NtrC regulatory protein facilitates the transition of promoter bound RNA polymerase from an inactive closed complex into an active open complex.[77] By electron microscopy, NtrC was seen bound at the enhancer, and RNA polymerase was seen bound at the promoter under conditions for open

[76] L. J. Reitzer and B. Magasanik, *Cell (Cambridge, Mass.)* **45**, 785 (1986).

[77] D. L. Popham, D. Szeto, J. Keener, and S. Kustu, *Science* **243**, 629 (1989).

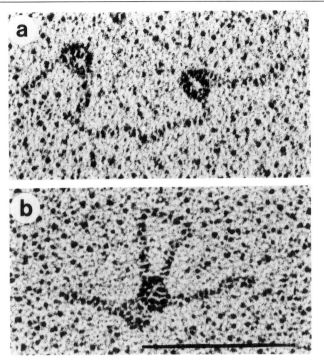

Fig. 7. Electron micrographs of NtrC bound to the enhancer and RNA polymerase bound to the promoter. (a) The product of the activation reaction in which NtrC is bound to the enhancer and RNA polymerase is stably bound to the promoter. (b) Loop formed by the association of enhancer-bound NtrC and promoter-bound RNA polymerase—the presumed activation intermediate. Note the higher electron density of the RNA polymerase, which presumably is shadowed more efficiently with tungsten than is NtrC (bar: 0.1 μm). (Reprinted from Ref. 12.)

complex formation (Fig. 7a). About 15% of these doubly bound DNA molecules exhibited a distinct looped structure in which the length of the loop was the engineered distance between the enhancer and promoter (Fig. 7b). Since these looped structures disappear when RNA polymerase is allowed to transcribe the DNA, it is likely that the structures are activation intermediates. The direct visualization of the heterologous interaction over distance between the DNA-bound NtrC regulatory protein and RNA polymerase shows that this enhancer systems operates through a snup-type mechanism. Other workers have used electron microscopy to visualize DNA-looping between homologous transcriptional regulatory proteins.[7-10]

Acknowledgments

We thank Chris Vasilikiotis and Wen Su for figures, Richard Eisner for editorial help, and Reid Johnson and Karen Heichman for expert advice. Research at Berkeley was supported by a grant from the National Institutes of Health (GM17078). M.D. was supported by Damon Runyon-Walter Winchell Cancer Fund Fellowship DRG-961.

[12] Analysis of DNA–Protein Interactions by Affinity Coelectrophoresis

By WENDELL A. LIM, ROBERT T. SAUER, and ARTHUR D. LANDER

Introduction

Gel retardation is a useful and convenient approach for analyzing protein–nucleic acid binding. Methods in current use have limitations, however, particularly when applied to the analysis of weak or rapidly dissociating interactions. We describe here a novel gel retardation method, referred to as affinity coelectrophoresis (ACE), which is generally applicable for studying protein–nucleic acid binding, and is particularly well suited for analyzing those interactions that are difficult to study using traditional gel retardation approaches. ACE is performed by electrophoresing a labeled nucleic acid fragment through an agarose gel with precast lanes containing protein at different concentrations. Unlike traditional gel retardation, binding is measured under equilibrium conditions, as the gel is running. The extent of binding can be measured from the extent of retardation in each lane.

In this chapter, we describe the theory and practice of ACE and its use in analyzing the binding of the N-terminal domain of the λ repressor protein to the O_L1 operator site. The advantages and disadvantage of ACE, as compared with other gel retardation methods, are also discussed.

Methods

ACE is performed using horizontal agarose gels that are cast with two types of wells: lanes, which are rectangular wells into which proteins mixed with low gelling temperature agarose are introduced and allowed to gel, and a slot, into which a nucleic acid sample is introduced just prior to electrophoresis. An apparatus for casting such gels may be assembled as shown in Fig. 1.

Briefly, a 75 × 100 mm Plexiglas casting tray ("t" in Fig. 1A) is

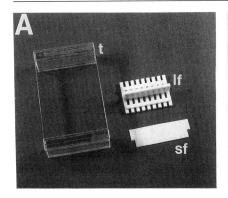
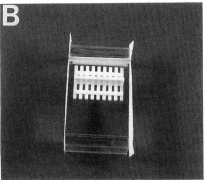

FIG. 1. Casting apparatus. (A) Gels are cast using a tray (t), a lane-forming comb (lf), and a slot-forming comb (sf), assembled as shown in (B). For details, see text.

fitted with a piece of GelBond film (FMC Bioproducts, Rockland, ME) and the sides are taped. Two types of combs are then placed into the tray. The lane-forming comb ("lf") is machined from a single piece of Teflon and consists of nine parallel bars, each $45 \times 4 \times 4$ mm, held together by cross-bracing with a spacing of 3 mm between each bar. This comb also contains a handle to permit its easy removal from formed gels. It is centered in the casting tray with the long sides of the rectangle parallel to the long sides of the tray.

The slot-forming comb is cut from a 25×75 mm rectangle of 1-mm thick teflon. Small 4.5×10 mm rectangles are removed from two corners to produce a comb with one 66-mm edge, as shown ("sf" in Fig. 1A). This comb is stood on its short edge, perpendicular to and at a distance of approximately 4 mm from the lane-forming comb (Fig. 1B). The slot-forming comb is stabilized in the upright position by pressing the tape used to seal the casting tray against the overhanging tabs of the comb.

One percent (w/v) low gelling temperature agarose (Bethesda Research Laboratories, Bethesda, MD) is prepared in electrophoresis buffer and heated until dissolved. In the experiments described below, a buffer of 50 mM potassium HEPES, pH 7.5, 1.5 mM CaCl$_2$, 0.1 mM EDTA, and 50, 100, or 200 mM KCl (as indicated) was used. Molten agarose (15–20 ml) is poured into the casting tray and allowed to cool. After the agarose has fully gelled, both combs are removed. Removing the lane-forming comb requires care in order to avoid tearing the lanes or slot. The best approach is to slide the comb a few millimeters forward and backward several times to break the seal between the Teflon and the agarose, then to pivot it about one end until it can be lifted easily out of the lanes. This process is made easier if detergent [0.1% Nonidet

P-40 (NP-40)] or nonspecific protein (0.1 mg/ml bovine serum albumin) is added to the molten agarose before pouring. Alternatively, flooding the top of the gel with electrophoresis buffer will help lubricate the lane-forming comb during its removal.

Next, proteins are prepared for embedding in the lanes. Protein samples are diluted in electrophoresis buffer to twice their desired concentrations (if detergent or nonspecific protein is used in pouring the gel, it should also be added here). Approximately 0.32 ml of each protein sample is mixed with an equal volume of 2% (w/v) low gelling temperature agarose prepared in the same buffer, previously melted and held at 37°. Immediately after mixing, each sample is pipetted into a lane, filling until level with the rest of the gel (to ensure that the mixture does not gel prematurely, protein samples should be warmed to room temperature, or if desired, to 37°, prior to addition of molten agarose).

For measurement of protein–nucleic acid affinities, it is convenient to introduce the same protein sample, at nine different concentrations, into each of the lanes. In the experiments described here, the protein used was the N-terminal domain (residues 1–102) of the bacteriophage λ repressor protein (for review, see Sauer et al.[1]), purified as described by Lim and Sauer.[2] Protein concentrations were determined using a calculated molar extinction coefficient at 276 nm of 7250 M^{-1} cm^{-1}. Protein samples were introduced into the gel lanes at final concentrations of 50, 100, 150, 200, 300, 400, 600, and 1200 nM.

After the samples in the lanes have solidified (typically 1–2 min), the gel is removed from the casting tray and submerged under buffer (minus detergent and protein) in the electrophoresis chamber. Gels are oriented with their long axis parallel to the direction of electrophoresis, with the slot end of the gel facing the cathode. Any appropriate-sized submarine gel box will do, provided that sufficient cooling and buffer recirculation can be provided. The experiments described here were performed using a Hoefer SuperSub (Hoefer Scientific Instruments, San Francisco, CA), which has built-in cooling and recirculation, and which can hold up to four ACE gels at once.

Radioactively labeled nucleic acid is then diluted into electrophoresis buffer containing a tracking dye (0.5% bromphenol blue) and 6% sucrose, and introduced into the slot by underlaying. Depending on the amount of agarose used in casting the gel, the slot will hold between 0.1 and 0.2 ml. In the experiments described here, a synthetic 29-bp double-stranded oligodeoxyribonucleotide was used. The sequence of the oligonucleotide was

[1] R. T. Sauer, S. R. Jordan, and C. O. Pabo, Adv. Protein Chem. **40**, 1 (1990).
[2] W. A. Lim and R. T. Sauer, J. Mol. Biol. **219**, 359 (1991).

```
5'–CGTA | TATCACCGCCAGTGGT | ATTGC
   GCAT | ATAGTGGCGGTCACCA | TAACGTTTT–5'
```

The boxed region corresponds to the O_L1 operator site of bacteriophage
λ. The duplex DNA was radiolabeled by filling in the four-nucleotide
overhang using $[\alpha\text{-}^{32}P]ATP$ and the enzyme Sequenase (U.S. Biochemical
Corporation, Cleveland, OH).

After introduction of the nucleic acid sample, electrophoresis is carried
out at 3–4 V/cm (90–120 V in the Hoefer SuperSub) for approximately
1.5–2 hr, or sufficient time for the nucleic acid to migrate most of the way
through the lanes (once the mobility of a nucleic acid sample is known
relative to bromphenol blue, electrophoresis times may be judged from
the migration of the dye front). Resistive heating during electrophoresis
will depend on buffer composition, but in buffers at or near physiological
ionic strengths, internal cooling within the gel box or buffer chambers is
necessary to avoid overheating. In the experiments described below, a
flow of cold tap water through the coolant ports of the electrophoresis box
was sufficient to maintain gel temperatures at approximately 22° throughout the runs.

After electrophoresis, gels are air dried and autoradiographed. Air
drying may be accomplished using forced warm air (2–3 hr in a Hoefer
Easy-Breeze), or a warm vacuum oven containing desiccant.

Results

Figure 2 illustrates the use of ACE to measure the affinity of a protein
for a nucleic acid. The protein sample is cast in the nine lanes at concentrations ranging from high (far left) to low (far right), and labeled nucleic acid
introduced into the slot (Fig. 2A). During electrophoresis, nucleic acid
migrates through each of the lanes, slowing down in lanes containing high
concentrations of binding protein. The amount by which nucleic acid
migration is retarded is not the same in each lane but varies with the
concentration of protein (Fig. 2B).

Figure 2C shows an example of this behavior. An amino-terminal
fragment of λ repressor was included in each of the lanes at concentrations
from 50 nM to 1.2 μM, and a radiolabeled oligonucleotide containing the
O_L1 operator site was introduced into the slot and electrophoresed. The
buffer, in this case, contained 50 mM KCl. As shown, the mobility of the
oligonucleotide was shifted progressively with increasing protein concentration, up to about 300 nM, above which no further retardation was
observed.

The pattern of progressive electrophoretic mobility retardation displayed in Fig. 2 can be understood as resulting from the progressive

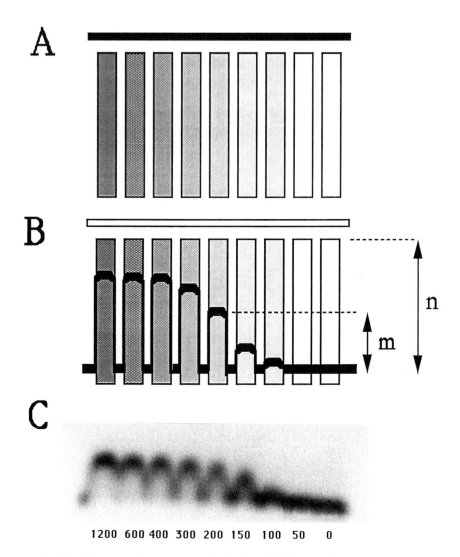

1200 600 400 300 200 150 100 50   0

Fig. 2. Mobility retardation patterns obtained with ACE. (A) Diagram of ACE gel prior to electrophoresis. Proteins have been cast into each of the lanes at decreasing concentrations from left to right, and labeled DNA has been introduced into the slot. (B) Diagram of the same gel following electrophoresis. The DNA front has moved toward the anode (at bottom), and has been progressively retarded with increasing protein concentration. m and n, the measurements used in determining the value of R, are shown. Note that it is preferable to measure n from the top of the lanes, not from the slot. (C) Autoradiogram of an ACE gel in which the protein was an amino-terminal fragment of the λ repressor and the DNA was a ^{32}P-labeled oligonucleotide containing the O_L1 operator site (see text). The protein concentrations in the lane are given in nanomolar units. In this experiment the electrophoresis buffer contained 50 mM KCl.

saturation of nucleic acid with protein. At high protein concentrations, the observed electrophoretic mobility presumably corresponds to that of the nucleic acid–protein complex. At intermediate protein concentrations, each nucleic acid molecule spends only a fraction of its time bound to protein. Provided that the kinetics of association and dissociation are reasonably rapid relative to electrophoresis times, a single intermediate mobility should be observed. If there are effectively only two electrophoretically different states of the nucleic acid (i.e., a single bound state and single free state), then the fraction of time that the nucleic acid spends in the bound state should be the same as the fraction by which its observed mobility is shifted from that of free nucleic acid to that of the bound complex.

This observation may be stated more precisely after defining a unitless number, R, the retardation coefficient, with which to quantify mobility shifts. In any lane of an ACE gel, R is found by measuring the difference between nucleic acid migration in that lane and the migration of free nucleic acid (i.e., its migration in a lane containing no protein), and normalizing that value to the migration of free nucleic acid (in Fig. 2B, $R = m/n$). If R_∞ is taken to represent the maximum value of R (i.e., when the protein concentration is saturating), then the above argument suggests that, in any lane, $R/R_\infty = \Theta$, where Θ represents the fractional saturation of nucleic acid with protein (i.e., bound DNA/total DNA) in that lane.

Because the dependence of R on protein concentration should parallel the dependence of Θ on protein concentration, data from ACE can be analyzed using several common graphic methods to model the binding reaction and measure the equilibrium binding constant. The simplest model for binding is a first order reaction:

$$P + D \xrightleftharpoons{K_d} PD \qquad K_d = [P][D]/[PD]$$

where P is protein and D is the DNA fragment. Since $\Theta = R/R_\infty$, the expression for binding can be written as

$$R = R_\infty/[1 + (K_d/[P])] \qquad (1)$$

In general, it is more convenient to plot values for the total rather than the free protein concentration in each lane. In this case, Eq. (1) becomes

$$R = R_\infty/\{1 + [K_d/([P_{tot}] - [PD])]\} \qquad (2)$$

which reduces to

$$R = R_\infty/[1 + (K_d/[P_{tot}])] \qquad (3)$$

whenever the concentration of nucleic acid is low ($<0.1\ K_d$). In practice,

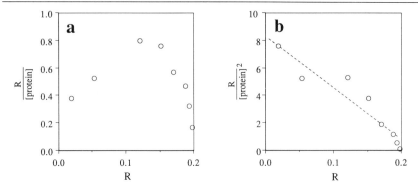

FIG. 3. Scatchard analysis. The data in Fig. 2C have been graphed on plots of R versus $R/[P_{tot}]$ (a) and R versus $R/[P_{tot}]^2$ (b). Only in (b) are the data points reasonably well fit by a straight line, suggesting that binding is effectively second order.

this condition is easy to meet when using radiolabeled nucleic acids, since picomolar amounts are usually sufficient to produce rapid autoradiographic results. In the experiment shown in Fig. 2C, the concentration of radiolabeled oligonucleotide was approximately 100 pM.

If the mechanism of binding is higher order, with more than one protein molecule binding in a strictly cooperative manner to each nucleic acid molecule, i.e.,

$$nP + D \rightleftharpoons P_nD \qquad K_d = [P]^n[D]/[P_nD]$$

then Eq. (3) can be generalized to

$$R = R_\infty/[1 + (K_d/[P_{tot}]^n)] \qquad (3')$$

where n equals the number of protein molecules binding to each molecule of nucleic acid.

Experimental data may be fit to Eq. (3') most accurately using nonlinear regression analysis, or transformed to the form of a Scatchard equation:

$$R/[P_{tot}]^n = -R/K_d + R_\infty/K_d$$

which permits the value of K_d to be calculated from the inverse slope of a plot of R versus $R/[P_{tot}]^n$.

Figure 3 presents the results of graphing the data from Fig. 2C on a plot of R versus $R/[P_{tot}]^n$ for $n = 1$ to model first order binding (Fig. 3A) and $n = 2$ to model second order binding (Fig. 3B). As shown, the data points in Fig. 3B are better fit by a straight line than those in Fig. 3A. This behavior is in agreement with the fact that the amino-terminal fragment of λ repressor, which is monomeric at the concentrations tested ($K_{dimer} =$

TABLE I
BEST FIT VALUES OF K_d AND R_∞ FOR THE
BINDING OF THE N-TERMINAL DOMAIN OF λ
REPRESSOR(1–102) TO O_L1 OPERATOR DNA[a]

[KCl] (mM)	K_d (M^2)	R_∞
50	1.9×10^{-14}	0.21
100	5.0×10^{-14}	0.17
200	9.8×10^{-14}	0.07

[a] As shown in Fig. 4, assuming second order binding [Eq. (3'), $n = 2$]. Under the conditions of these experiments, half-maximal binding is observed at a total protein concentration equal to $\sqrt{K_d}$.

$3 \times 10^{-4} M$; Weiss et al.[3]), binds strongly to the O_L1 oligonucleotide only as a dimer (for review, see Sauer et al.[1]).

Direct fits of the same data to Eq. (3') were also made by nonlinear least-squares analysis, and the derived curves for $n = 1$ and $n = 2$ are shown in Fig. 4A. Data were also analyzed from experiments in which the electrophoresis buffer contained either 100 or 200 mM KCl (Fig. 4B and C). It is clear that, in every case, the data are better fit by the second order than the first order model. The values of K_d derived from the best fit second order equations are given in Table I. Errors were calculated by assuming a 0.5-mm measurement error in the estimation of mobility shifts, and an insignificant error in the preparation of protein dilutions. Also presented in Table I are the derived values of R_∞, a measure of the maximum degree of mobility retardation under each of the buffer conditions tested.

From Table I it can be seen that the affinity of the N-terminal domain for O_L1 DNA decreases (K_d increases) with increasing salt concentration (the binding of intact λ repressor to operator shows a similar salt sensitivity[4]). At 50 mM KCl, the calculated value of K_d is $1.9 \times 10^{-14} M^2$. This value is very close to the value of $1.2 \times 10^{-14} M^2$ that was measured under similar buffer conditions by DNase I footprinting analysis.[5] It is also apparent that the value of R_∞ decreases with increasing salt concentration; the explanation for this phenomenon is not yet clear.

[3] M. A. Weiss, C. O. Pabo, M. Karplus, and R. T. Sauer, Biochemistry 26, 897 (1987).
[4] H. C. M. Nelson and R. T. Sauer, Cell (Cambridge, Mass.) 42, 549 (1985).
[5] J. F. Reidhaar-Olson, D. A. Parsell, and R. T. Sauer, Biochemistry 29, 7563 (1990).

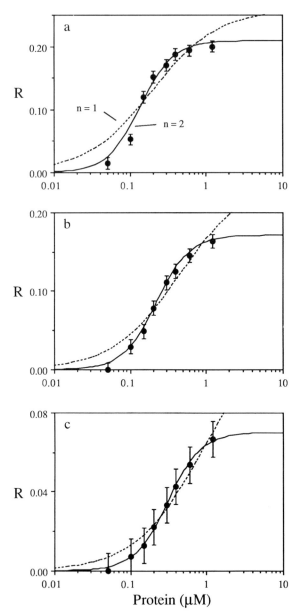

FIG. 4. Effect of KCl concentration on binding. R has been plotted as a function of protein concentration for the data shown in Fig. 2C (a), and for two other experiments in which the protein and DNA samples were identical, but the KCl concentration was increased from 50 mM (a) to 100 mM (b) and 200 mM (c). Nonlinear least-squares analysis was used to fit the data to first order ($n = 1$) or second order ($n = 2$) binding models, as described by Eq. (3'). Nonlinear analysis was performed using the NONLIN module of the statistical package SYSTAT (Systat, Inc., Evanston, IL).

Discussion

ACE is a simple electrophoretic method for analyzing protein–nucleic acid binding. In principle, ACE is a form of affinity electrophoresis,[6] and is generally applicable to the study of macromolecular binding. For example, ACE has been used to study the binding of heparin, an anionic oligosaccharide, to a variety of proteins.[7]

Differences between ACE and Other Gel Retardation Methods

For analyzing protein–DNA interactions, ACE differs from gel retardation methods that are in current use (referred to below as "traditional gel retardation") in several important ways.

Measurements are made under equilibrium conditions. Traditional gel retardation assays require separation of free DNA from DNA–protein complexes. When the dissociation of such complexes is too rapid, traditional gel retardation cannot be used (see [8], this volume). In contrast, the mobility shifts that occur in ACE develop without disturbing equilibrium; rapid dissociation therefore does not create difficulties. In practice, this means that ACE is well suited to measuring moderate to weak affinities. The data reported above illustrate this point. The amino-terminal fragment of the λ repressor exhibits a relatively weak affinity for DNA and its binding to operator is difficult to quantify by traditional gel retardation (W. A. Lim and R. T. Sauer, unpublished observations, 1991). In theory, there is no lower limit to the affinities that can be measured by ACE, although the ability to prepare lanes containing proteins at sufficiently high concentrations may be limited by the availability of material, or by protein solubility. Also, at very high protein concentrations (e.g., >10 mg/ml), the effects of protein on the electric field strength within the gel may become nonnegligible. So far, in the λ repressor system, values of K_d as low as $6 \times 10^{-9} M^2$ (indicating that half-maximal saturation occurs at 80 μM) have been measured by ACE (W. A. Lim, R. T. Sauer, and A. D. Lander, unpublished observations, 1991).

In ACE, data are derived from measurements of electrophoretic mobility, rather than film densitometry or radioactive counting. In traditional gel retardation assays, quantitative measurements of binding require determining the fraction of radioactivity in electrophoretic bands representing free or bound nucleic acid. If film densitometry is used for this purpose, errors can result from nonlinearity or film saturation. In the event that the

[6] V. Horejsi, *Anal. Biochem.* **112**, 1 (1981).
[7] M. K. Lee and A. D. Lander, *Proc. Natl. Acad. Sci. U.S.A.* **88**, 2768 (1991).

nucleic acid fragment used in a tradiational gel retardation assay contains a subpopulation of inactive molecules (apparently a common occurrence with synthetic oligonucleotides), corrections to the data must be made (cf. [8], this volume). In contrast, data from ACE are obtained from the electrophoretic pattern alone. Measurements can be made manually, and measurement errors estimated from the widths of the bands. Labeled, nonbinding molecules have no influence on the pattern of mobility shifts, but rather migrate independently as an unshifted front (W. A. Lim, R. T. Sauer, and A. D. Lander, unpublished observations, 1991).

ACE can measure binding over a wide range of conditions. Traditional gel retardation assays require the use of relatively sieving gels (e.g., 10% acrylamide) and high electric field strengths (at least during the time that samples enter the gel). In theory, the porosity of the gel matrix and the field strength can both affect binding. ACE is carried out in 1% agarose, a relatively nonsieving medium, and at low field strengths (e.g., 3–4 V/cm; experiments may also be carried out to test the effect of voltage on binding). Provided that an apparatus with adequate cooling is used, ACE may also be carried out in buffers of widely varying ionic strength (cf. Fig. 4).

ACE can fractionate oligonucleotides by affinity. If a mixture of oligonucleotides is electrophoresed through lanes containing a nucleic acid-binding protein, each oligonucleotide will produce a pattern of shifts reflecting its individual affinity. By recovering, amplifying, and identifying (sequencing) DNA molecules located at different positions within the lanes of an ACE gel, the relative affinities of multiple DNA sequences for DNA-binding proteins can be assessed simultaneously. By extension, this method can be used to identify, from a mixture of completely random oligonucleotides, those sequences that bind best to any DNA-binding protein (it has been reported that this type of approach can also be used with traditional gel retardation[8]). The need to be able to determine directly the DNA sequences to which proteins bind is likely to increase, as more DNA-binding proteins are identified simply on the basis of homology to other known DNA-binding proteins.

Limitations on the Uses of ACE

Although ACE should be generally applicable to the study of protein–nucleic acid interactions, there are several limitations.

The amount of protein used in ACE is likely to be greater than that

---

[8] T. K. Blackwell and H. Weintraub, *Science* **250**, 1104 (1990).

used in traditional gel retardation assays. In measuring affinity either by ACE or by traditional gel retardation, the same series of protein concentrations will be required, but significantly larger volumes will be needed for ACE, in order to fill the lanes. In the experiments described above, lanes of 45×4 mm were used and were filled to a height of approximately 3 mm (0.54 ml final volume). Protein samples may be conserved by reducing gel height (thereby reducing the amount of labeled DNA that can be loaded, and increasing the time required for autoradiography), or by using a lane-forming comb that produces shorter lanes and electrophoresing for a shorter time. The minimum lane length and electrophoresis time required for an ACE experiment depends on how substantial the mobility shifts are. In general, the greatest mobility shifts will occur when the oligonucleotides used are small and the proteins are large and/or basic. The data in Table I also suggest that mobility shifts are increased by decreasing the ionic strength. In the experiments described above, the protein was rather small ($M_r \sim 11,000$) and slightly acidic (Coomassie blue-stained ACE gels revealed a small degree of protein migration toward the anode), suggesting that many DNA-binding proteins could be expected to produce significantly greater mobility shifts. Thus for many ACE experiments, shorter lanes may be adequate. It should also be pointed out that, although maximal mobility shifts can be achieved by carrying out electrophoresis until all DNA has passed completely through each lane, Eq. (3′) is valid only when electrophoresis has been stopped while the free oligonucleotide front remains within the region of gel containing the proteins (a method is described below to correct for errors that arise when this condition is not met).

Protein–nucleic acid complexes that dissociate very slowly may fail to produce useful patterns in ACE gels. The assertion that ACE measures binding at equilibrium depends on the assumption that equilibrium is reasonably rapid compared with the time of electrophoresis. The effects of binding kinetics on affinity-electrophoretic separations in general have been modeled previously[9]; the results imply that, in ACE gels, slow kinetics will be manifested as spreading of the electrophoretic bands observed at intermediate R values. In general, significant spreading should occur when the value of $1/k_{off}$ is at least of the same order as the electrophoresis time. In practice, with electrophoresis times in the hours, only very high-affinity interactions would be expected to display this effect.

Exposure to molten low gelling temperature agarose might lead to denaturation of some proteins. Very heat-labile proteins, particularly at

[9] V. Matousek and V. Horejsi, *J. Chromatogr.* **245**, 271 (1982).

high concentrations, may be intolerant of spending even a short time at 37° (the temperature of the agarose that is mixed with protein samples prior to addition to the lanes of an ACE gel).

Graphical Analysis of ACE Data

As described in the previous section, by using Eqs. (3) or (3') or linear transformations of them, data from ACE experiments may be manipulated graphically so as to model binding and derive values for K_d and R_∞. The validity of these equations requires that several conditions be met, only some of which may be under the control of the investigator. These conditions are enumerated below. Where appropriate, corrections or alternate formulas are presented that may be used when these equations are not valid.

The nucleic acid concentration should be low compared with the value of K_d. As already described, the substitution of total for free protein concentration in the derivation of Eqs. (3) and (3') is valid only when the concentration of nucleic acid is significantly lower than the value of K_d. The effect of a higher DNA concentrations on the relationship between R and $[P_{tot}]$ may be illustrated by transforming Eq. (2) into Scatchard form:

$$\frac{R}{[P_{tot}]} = \frac{-R}{K_d + [D_{tot}](1 - R/R_\infty)} + \frac{R_\infty}{K_d + [D_{tot}](1 - R/R_\infty)}$$

When $[D_{tot}]$ is significant compared with K_d, a plot of R vs $R/[P_{tot}]$ will be concave downward, with a slope that depends on the values of K_d, $[D_{tot}]$, and R/R_∞.

Electrophoresis should be terminated before the nucleic acid front has migrated completely through any of the protein-containing zones. The retardation coefficient R is defined as the mobility shift normalized to the migration of the free oligonucleotide (m/n in Fig. 2B). Normalization assures that R should not depend on the length of time of electrophoresis. However, if electrophoresis is conducted long enough so that nucleic acid molecules pass completely through protein containing zones, the nucleic acid molecules will speed up again to their free mobility, and calculated values of R will fall with further running. Because nucleic acid fronts will exit the protein-containing regions at different times (the most retarded will exit last), values of R will be changed by different amounts for each lane. It therefore becomes necessary in the case of such "overruns" to correct R values to what they would have been at the time just before the emergence of the nucleic acid fronts. It should be pointed out that overruns are measured not by whether the free nucleic acid front has traveled beyond the boundaries of the cast *lane*, but by whether it has traveled past the position of the *protein*, which was initially cast within the lane, but which may have moved somewhat, either toward the anode or toward the cathode, during electrophoresis. Accordingly, to determine the exact amount by which to correct R, it is necessary to know how much the proteins in the lanes move during electrophoresis. In many cases, this can be determined easily by protein staining. Dried ACE gels may be stained by immersing in Coomassie Brilliant Blue [2.5 g/l in 45% (v/v) methanol, 5% (v/v) acetic acid] for 5 min, followed by destaining in 25% methanol. Protein migration can then be measured as the amount by which the stained rectangles are shifted from the lanes themselves (with anodal migration taken as positive and cathodal as negative). Corrected values of R may then be determined as follows:

$$\text{If } R < 1 - (L + p)/n, \quad R_{corr} = R[(n - p)/(L + m)]$$
$$\text{If } R \geq 1 - (L + p)/n, \quad R_{corr} = R$$

where L stands for the lane length, p for the measured amount of protein migration, R_{corr} for the corrected values of R, and m, n, and R are as defined previously. Thus, although it is simpler to obtain usable values of R when overruns are avoided, it is relatively easy to correct for the effects of overruns. Indeed, since overruns give the largest possible mobility shifts for any given lane length, it may be desirable when trying to conserve on protein to cast short lanes and deliberately overrun the nucleic acid front.

Nucleic acid and protein should combine to form only a single type of complex. Equating R/R_∞ with Θ presupposes that there are only two electrophoretic species in equilibrium, a bound species, which has one mobility and a free species, which has a greater mobility. This assumption can break down when more than one protein molecule can bind to a single nucleic acid molecule, or when more than one nucleic acid molecule can bind to a single protein. In such cases, each type of complex will possess its own electrophoretic mobility, and observed mobility shifts will be a weighted average of these mobilities. More precisely, if there are n such bound species, and θ_i represents the fractional saturation of the ith species, and r_i represents the mobility shift associated with the ith species (i.e., the mobility of the ith species subtracted from and normalized to the mobility of free nucleic acid), then it should be the case that

$$R = \sum_{i=1}^{n} \theta_i r_i$$

By substituting into appropriate equilibrium equations, one can determine the appropriate relationship between R and $[P_{tot}]$. It is important to note, however, that the relationship obtained will not necessarily match binding curves obtained by other techniques. For example, if a molecule of nucleic acid is capable of binding either one or two molecules of protein, i.e.,

$$2P + D \xrightleftharpoons{K_1} PD + P \xrightleftharpoons{K_2} P_2D \qquad K_1 = [P] \cdot [D]/[PD] \quad K_2 = [P] \cdot [PD]/[P_2D]$$

then the relationship of Θ to $[P_{tot}]$ is described by

$$\Theta = \frac{1 + \dfrac{K_2}{[P_{tot}]}}{1 + \dfrac{K_2}{[P_{tot}]} + \dfrac{K_1 K_2}{[P_{tot}]^2}} \tag{4}$$

where Θ is defined as bound DNA over total DNA. In contrast, in ACE, the relationship of R to $[P_{tot}]$ will be

$$R = r_1 \left(\frac{\dfrac{r_2}{r_1} + \dfrac{K_2}{[P_{tot}]}}{1 + \dfrac{K_2}{[P_{tot}]} + \dfrac{K_1 K_2}{[P_{tot}]^2}} \right) \tag{5}$$

The isotherms defined by Eqs. (5) and (4) will be equivalent only when $r_2/r_1 = 1$. In practice, the value of r_2/r_1 is likely to vary between 1 and 2, depending on the size of the DNA fragment and the properties of the protein being studied. For example, with a very large protein, binding of just a single protein molecule to the DNA is likely to result in nearly maximal retardation, and little further change should occur on binding of a second protein molecule; as a result, r_2/r_1 would be very close to 1. Conversely, with a very small protein, the mobility shift induced by the binding of a single molecule to DNA is likely to be such that a second molecule of protein shifts the complex almost as much as the first; as a result, r_2/r_1 should be around 2 [it is noteworthy that, when $r_2 r_1 = 2$, Eq. (5) generates an isotherm equivalent to what is obtained if Θ is expressed in terms of bound *protein* over total DNA, rather than bound *DNA* over total DNA].

Acknowledgments

This work was supported in part by NIH Grants NS-26862 (to A.D.L.) and AI-16892 (to R.T.S.). W.A.L. was supported by a predoctoral grant from the Howard Hughes Medical Institute.

[13] Laser Cross-Linking of Protein–Nucleic Acid Complexes

By Joel W. Hockensmith, William L. Kubasek,
William R. Vorachek, Elisabeth M. Evertsz, and
Peter H. von Hippel

Introduction

Ultraviolet (UV) light-induced photochemical cross-linking of proteins to nucleic acids has been demonstrated for a variety of nucleic acid-binding proteins.[1–16] Irradiation with UV light brings about the formation of a "zero-length" covalent bond between the protein and the nucleic acid to which it is bound, thereby "freezing" the interaction between the two biopolymers. Many model systems have been studied in attempting to understand the chemistry of this photochemical process.[17–25] These studies have shown that the mechanistic pathways involved are often complex,

[1] M. D. Shetlar, *Photochem. Photobiol. Rev.* **5**, 105 (1980).
[2] P. R. Paradiso, Y. Nakashima, and W. Konigsberg, *J. Biol. Chem.* **254**, 4739 (1979).
[3] P. R. Paradiso and W. Konigsberg, *J. Biol. Chem.* **257**, 1462 (1982).
[4] B. M. Merrill, K. R. Williams, J. W. Chase, and W. H. Konigsberg, *J. Biol. Chem.* **259**, 10850 (1984).
[5] G. F. Strniste and D. A. Smith, *Biochemistry* **13**, 485 (1974).
[6] Z. Hillel and C.-W. Wu, *Biochemistry* **17**, 2954 (1978).
[7] C. S. Park, Z. Hillel, and C.-W. Wu, *Nucleic Acids Res.* **8**, 5895 (1980).
[8] C. S. Park, Z. Hillel, and C.-W. Wu, *J. Biol. Chem.* **257**, 6944 (1982).
[9] A. Markovitz, *Biochim. Biophys. Acta* **281**, 522 (1972).
[10] J. Sperling and A. Havron, *Biochemistry* **15**, 1489 (1976).
[11] A. Havron and J. Sperling, *Biochemistry* **16**, 67 (1977).
[12] M. J. Modak and E. Gillerman-Cox, *J. Biol. Chem.* **257**, 15105 (1982).
[13] S. B. Biswas and A. Kornberg, *J. Biol. Chem.* **259**, 7990 (1984).
[14] S. Eriksson, I. W. Caras, and J. D. W. Martin, *Proc. Natl. Acad. Sci. U.S.A.* **79**, 81 (1982).
[15] H. Maruta and E. D. Korn, *J. Biol. Chem.* **256**, 499 (1981).
[16] Y. Shamoo, K. R. Williams, and W. H. Konigsberg, *Proteins: Struct. Funct. Genet.* **4**, 1 (1988).
[17] K. C. Smith, *Biochem. Biophys. Res. Commun.* **34**, 354 (1969).
[18] H. N. Schott and M. D. Shetlar, *Biochem. Biophys. Res. Commun.* **59**, 1112 (1974).
[19] M. Dizdaroglu and M. G. Simic, *Int. J. Radiat. Biol.* **47**, 63 (1985).
[20] A. E. Reeve and T. R. Hopkins, *Photochem. Photobiol.* **31**, 297 (1980).
[21] A. E. Reeve and T. R. Hopkins, *Photochem. Photobiol.* **31**, 413 (1980).
[22] M. D. Shetlar, K. Hom, J. Carbone, D. Moy, E. Steady, and M. Watanabe, *Photochem. Photobiol.* **39**, 135 (1984).
[23] M. D. Shetlar, J. Carbone, E. Steady, and K. Hom, *Photochem. Photobiol.* **39**, 141 (1984).
[24] K. C. Smith and D. H. C. Meun, *Biochemistry* **7**, 1033 (1968).
[25] M. D. Shetlar, J. Christensen, and K. Hom, *Photochem. Photobiol.* **39**, 125 (1984).

METHODS IN ENZYMOLOGY, VOL. 208

that reaction yields vary widely, and that competing processes may intervene. However, these studies have also shown that any amino acid residue can, in principle, be induced to form a UV-induced covalent cross-link with any nucleotide residue of DNA or RNA. Thus, despite the fact that reaction mechanisms are not fully understood, the introduction of covalent bonds between proteins and nucleic acids by UV irradiation represents a useful and reproducible technique that has been applied successfully *in vivo* and *in vitro*.[1-16,26-32]

Our primary motivation for developing a UV laser-induced version of this cross-linking methodology was to find a way to study dynamic aspects of rapid (millisecond) protein–nucleic acid interactions under conditions that freeze the "instantaneous" (on the time scale of the reaction) binding equilibria involved. Providing that the equilibria could be frozen, we also hoped to use such procedures to characterize thermodynamically and kinetically the sequential interactions that occur within protein–nucleic acid complexes of biological interest.

Our initial studies were carried out using purified components of the bacteriophage T4 DNA leading strand replication system.[33,34] This system is capable of incorporating several hundred nucleotide residues per second into new DNA.[35] Thus each template-dependent one-step reaction cycle, which involves a number of sequential reactions to insert a single nucleotide residue at the 3' end of the nascent DNA chain, must go to completion in milliseconds. Clearly, such processes are not amenable to analysis by conventional UV cross-linking methodology, which involves the use of fairly low-power, broad-band germicidal lamps and requires many minutes of UV irradiation to produce reasonable quantities of product.

Consequently we decided to develop a cross-linking procedure based on laser irradiation, which could, in principle, introduce many photons

[26] D. S. Gilmour and J. T. Lis, *Proc. Natl. Acad. Sci. U.S.A.* **81,** 4275 (1984).
[27] D. S. Gilmour and J. T. Lis, *Mol. Cell. Biol.* **5,** 2009 (1985).
[28] D. S. Gilmour and J. T. Lis, *Mol. Cell. Biol.* **7,** 3341 (1987).
[29] J. P. Schouten, *J. Biol. Chem.* **260,** 9916 (1985).
[30] D. Angelov, V. Y. Stefanovsky, S. I. Dimitrov, V. R. Russanova, E. Keskinova, and I. G. Pashev, *Nucleic Acids Res.* **16,** 4525 (1988).
[31] G. Dreyfuss, Y. D. Choi, and S. A. Adam, *Mol. Cell. Biol.* **4,** 1104 (1984).
[32] S. Piñol-Roma, S. A. Adam, Y. D. Choi, and G. Dreyfuss, this series, Vol. 180, p. 410.
[33] C. F. Morris, H. Hama-Inaba, D. Mace, N. K. Sinha, and B. M. Alberts, *J. Biol. Chem.* **254,** 6787 (1979).
[34] N. G. Nossal and B. M. Peterlin, *J. Biol. Chem.* **254,** 6032 (1979).
[35] B. Alberts, C. F. Morris, D. Mace, S. Sinha, M. Bittner, and L. Moran, *in* "DNA Synthesis and Its Regulation" (M. Goulian, P. Hanawalt, and C. F. Fox, eds.), p. 241. Benjamin, Menlo Park, California, 1975.

into the system in very short periods of time.[36,37] The laser cross-linking techniques presented in this chapter are fast and efficient, and can be used to link a wide variety of binding proteins (specific or nonspecific) to either single- or double-stranded nucleic acids. This technology has proven useful in exploring both static and dynamic protein–nucleic acid complexes. It can yield information about protein–nucleic acid binding constants and binding site sizes, and can be used to define actual points of contact between protein and nucleic acid species within complexes. Under some conditions this method can also be used to examine protein–protein interactions, protein–ligand interactions, and some aspects of nucleic acid structure. Procedures for designing experiments to yield such information are described below.

Experimental Procedures

Lasers and Irradiation Techniques

In principle a laser is merely another source of ultraviolet light, albeit more powerful than standard UV lamps. Its main advantage for our purposes is that it delivers this power very rapidly in monochromatic form. Although we have used a number of different lasers in our work, we find that a neodymium–yttrium–aluminum–garnet (Nd: YAG) laser (model DCR-3G, purchased from Spectra-Physics, Inc., Mountain View, CA) is the most practical. The DCR-3G laser emits light in a Gaussian mode, with the most intense light located at the center of the beam. We have also used a Spectra-Physics DCR-2 Nd: YAG laser successfully, but it should be noted that the DCR-2 emits light in a collimated output with a "hole" or "dot" directly in the middle of the beam; i.e., the beam is shaped like a doughnut. When using the DCR-2, the DCR-3D, or any laser with a "dot" profile, it is important to recognize that the most intense light is *not* located at the center of the beam and that samples must be positioned accordingly.

A schematic diagram of the laser setup we usually use is shown in Fig. 1. The power delivered is dependent on the number of pulses per unit time, and each Spectra-Physics Nd: YAG laser is optimized to operate at 1, 10, 20, or 30 pulses per second (Hz). The Nd: YAG lasers used for the studies presented here had repetition rates of 10 Hz. The Nd: YAG laser produces pulsed radiation at 1064 nm, with a pulse energy of 900 mJ (110-MW peak, DCR-3G). This high peak power permits efficient conver-

[36] C. A. Harrison, D. H. Turner, and D. C. Hinkle, *Nucleic Acids Res.* **10**, 2399 (1982).
[37] J. W. Hockensmith, W. L. Kubasek, W. R. Vorachek, and P. H. von Hippel, *J. Biol. Chem.* **261**, 3512 (1986).

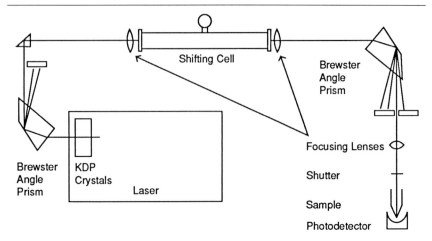

FIG. 1. Schematic diagram of the laser irradiation apparatus. The H_2-shifting cell is shown for completeness. It is not needed and can be omitted from the setup when protein–nucleic acid cross-linking is performed at an irradiation wavelength of 266 nm. See text for details.

sion of the 1064-nm radiation to 532-, 355-, and 266-nm photons, using a combination of nonlinear optical (potassium dihydrogen phosphate or potassium dideuterophosphate) "doubling" crystals that are generally provided as part of the apparatus. The average power output generated at these wavelengths is typically 360, 160, and 80 mJ/pulse, respectively. Pulses in the UV are typically about 5 nsec in duration.

Laser frequencies other than the "primary" ones listed above are generated by focusing one of the primary beams into a wavlength-shifting cell, which is filled with hydrogen gas at 10 atm.[38] The wavelengths of the emergent beams are shifted from that of the incident radiation to frequencies determined by the vibration spectrum of the hydrogen gas. This stimulated Raman-shifting process produces radiation that is displaced in both the Stokes (lower frequency) and anti-Stokes (higher frequency) directions relative to the incident radiation. These shifted beams carry significantly less power than does the incident irradiation. Wavelengths of 282.3, 252.7, and 228.7 nm are obtained using anti-Stokes shifts of the 532-nm band; wavelengths of 309.1, 273.9, 245.9, 223.1, and 204.2 nm result from anti-Stokes shifts of the 355-nm, band; and wavelengths of 239.5 and 217.8 nm are obtained from anti-Stokes shifts of the 266-nm band.

Radiation of the desired frequency is isolated using a dispersing quartz prism (see Fig. 1). We use Brewster angle prisms (also known as modified

[38] S. P. A. Fodor, R. P. Rava, R. A. Copeland, and T. G. Spiro, *J. Raman Spectrosc.* **17,** 471 (1986).

Pellin Broca prisms and obtainable from Continental Optical Corporation, Hauppauque, NY) to turn the beam by 90° while maintaining good dispersion. The beam is directed through a focusing lens, through a shutter, and into the sample. The sample itself is held by capillary action in the bottom of an open, horizontal Eppendorf tube. The desired incident radiation is focused just in front of the sample surface, so that the diverging circular beam strikes the full surface area of the solution (~3-mm diameter for a 10-μl sample).

This configuration was selected to minimize differences in focal position of the lens as a function of wavelength. Operating the laser in the constant pulsing mode produces the best pulse-to-pulse energy stability (±4% at 266 nm). Since the laser pulses continuously, it was necessary to devise a shutter apparatus that could withstand many pulses of irradiation, but could also operate rapidly enough to select and let through a single pulse from the incident "photon stream." We have found that a homemade shutter composed of a copper plate attached to a tubular solenoid (Guardian Electric Manufacturing Co., Chicago, IL, Cat. #LT4x12-C-12D) that is synchronized with the laser and controlled by a momentary switch is most efficient and cost effective. The copper plate is simply attached to the solenoid with epoxy and can be replaced as necessary.

The relative power at each irradiation wavelength was determined using a volume-absorbing disk calorimeter (Scientech, Inc., Boulder, CO, model 380103 or 380105) in conjunction with a power indicator (Scientech model 36-2002). At very low power levels we typically use an isoperibol enclosure (Scientech model 36-0203) around the calorimeter to protect it from ambient heat exchange due to air currents and incidental thermal input from the environment. In addition, we use a high-speed photodetector (Scientech model 301-020) in conjunction with a 520-MHz frequency counter (B&K Precision, Chicago, IL, model 1851) to count the number of pulses introduced into the sample.

While some of this sounds complicated, we emphasize that most laboratories that use lasers for spectroscopy will have much of this equipment. As a consequence molecular biologists interested in using this technique can often make arrangements to irradiate their samples in the laboratories of colleagues, and thus need not purchase all this apparatus. In typical applications only a few minutes of actual irradiation time are required to produce enough cross-linked material to occupy days or even weeks of subsequent analysis.

Proteins and Nucleic Acids

The proteins and nucleic acids used in the "calibrating" studies of this methodology that are described or mentioned here were prepared as follows. Bacteriophage T4 gene 43, 45, and 44/62 proteins were prepared by a modification of the method of Morris *et al.*[33] T4 gene 32 protein was prepared according to the procedures of Nossal and Bittner *et al.*[39,40] The proteolytic degradation products of gene 32 protein (G32P*I and G32P*III) were purified according to Lonberg *et al.*[41] *Escherichia coli* ρ protein was prepared according to Finger and Richardson[42]; *E. coli* RNA polymerase was prepared by the method of Burgess and Jendrisak,[43] and *E. coli* RecA protein was a gift of Dr. Stephen Kowalczykowski (Northwestern University Medical School, Chicago, IL). *Escherichia coli* DNA polymerase I was purchased from New England Biolabs (Beverly, MA) bacteriophage T4 polynucleotide kinase from New England Nuclear (Boston, MA) and horse heart myoglobin from Sigma (St. Louis, MO). Oligonucleotides were purchased from either Pharmacia (Piscataway, NJ) or New England Biolabs. The oligo(rC) and oligo(rU) series of oligonucleotides were a gift of Dr. James McSwiggen (USB, Cleveland, OH). Reagents for electrophoresis were purchased from International Biotechnologies (New Haven, CT) and [γ-^{32}P]ATP from New England Nuclear. Oligonucleotides were 5'-^{32}P end-labeled according to the method of Maxam and Gilbert.[44]

Analysis of Cross-Linking Experiments

Sample Volumes and Beam Attenuation by Other Adsorbing Materials. In our experiments samples have been irradiated at volumes ranging from 10 to 300 μl. Within this range the sample volume is not critical. However, the absolute number of "absorbing centers" (i.e., macromolecules with large extinction coefficients at the irradiation wavelength) within the sample *is* critical, since this directly affects the number of incident photons that are available to bring about cross-linking. This issue is discussed further below (see Fig. 11), but represents an important aspect of setting up an experiment and is here illustrated with an example. If one is irradiating a typical protein–nucleic acid sample of volume 10 μl, 0.25 μM nucleic acid and 0.4 μM protein at 266 nm, the presence of (e.g.) ~1 mM ATP will reduce the efficiency of cross-linking by 50% compared with a no-ATP-containing control, because (averaged over the sample volume) the ATP will absorb 50% of the incident photons. Thus dilution of the sample should not present a problem, but holding the macromolecule concentration con-

[39] N. G. Nossal, *J. Biol. Chem.* **249**, 5668 (1974).
[40] M. Bittner, R. L. Burke, and B. M. Alberts, *J. Biol. Chem.* **254**, 9565 (1979).
[41] N. Lonberg, S. C. Kowalczkowski, L. S. Paul, and P. H. von Hippel, *J. Mol. Biol.* **145**, 321 (1981).
[42] L. R. Finger and J. P. Richardson, *Biochemistry* **20**, 1640 (1981).
[43] R. R. Burgess and J. J. Jendrisak, *Biochemistry* **14**, 4634 (1975).
[44] A. M. Maxam and W. Gilbert, this series, Vol. 65, p. 499.

stant while increasing the volume may decrease the number of photons absorbed by other "absorbing centers" (e.g., ATP) present at constant concentration.

Separation and Identification of Cross-Linked Species. After irradiation, a variety of strategies may be used for separating cross-linked macromolecules, including filter binding, phenol extraction, column chromatography, and gel electrophoresis. We generally use relatively small oligonucleotides (single- or double-stranded) when we are trying to quantify cross-linking and establish the identity of a cross-linked protein subunit derived from an irradiated complex containing multiple protein species. This allows us to use the rate of migration of the cross-linked protein in a denaturing polyacrylamide gel to identify the protein species involved. Alternatively, nuclease digestion of larger, internally labeled oligonucleotide adducts prior to gel electrophoresis separation can also be used.

Typically, we work with 10-μl sample volumes, which are denatured by heating to 100° in 40 mM Tris-HCl, pH 6.8, 4% 2-mercaptoethanol, 1.6% (w/v) sodium dodecyl sulfate (SDS), and 10% (w/v) glycerol. Gel electrophoresis in the presence of SDS is performed essentially as described by Dreyfuss *et al.*,[45] using a 4% stacking gel and a 10% separating gel. The gels were run at a constant temperature of 37°, using a heater apparatus consisting of an aluminum plate, gasket, Plexiglas spacer, and Plexiglas sheet sandwiched such that water could be circulated between the Plexiglas sheet and the aluminum plate. The water temperature is controlled with a Lauda circulator and the aluminum surface of the heater plate is placed against one glass surface of the gel apparatus. This assembly reduces the background radioactivity in the gel and is critical for the analysis of some cross-linked complexes containing double-stranded DNA (SDS may promote renaturation or aggregation of certain DNAs). The gels are always prepared on a plastic support (GelBond) supplied by the FMC Corporation (Princeton, NJ). The plastic support allows the gels to be frozen and thawed without distintegrating and also preserves the size and shape of the gel during staining procedures, thereby allowing direct comparison to the autoradiographs. After electrophoresis, gels were covered with Saran wrap and autoradiography was performed at −80° using Kodak (Rochester, NY) XAR-5 film and Du Pont Cronex Lightning-Plus enhancing screens (X-Ray Corp., Portland, OR). Following autoradiography, the gels were stained with silver and dried.[46]

Silver Staining and Acid Lability of Cross-Links. Staining is always carried out *after* autoradiography, because the cross-linked adducts are

[45] G. Dreyfuss, S. A. Adam, and Y. D. Cho, *Mol. Cell. Biol.* **4**, 415 (1984).
[46] J. H. Morrissey, *Anal. Biochem.* **117**, 307 (1981).

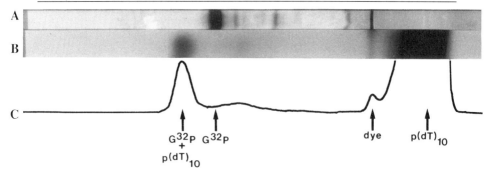

FIG. 2. Silver-stained polyacrylamide gel (A), autoradiogram (B), and densitometric scan of autoradiogram (C) for a typical photocross-linking experiment. Reaction conditions were 25 mM Tris–acetate, pH 7.5, 60 mM potassium acetate, 6 mM magnesium diacetate, 5 mM 2-mercaptoethanol, 3.9 μM T4 gene 32 protein, and 0.25 μM [5'-^{32}P](dT)$_{10}$ (in nucleotide residues). The sample was irradiated with 1.4 × 10^{17} photons at 245.9 nm.

not stable under the acidic conditions used to silver stain SDS–poly-acrylamide gels. Initially, we found that gels stained prior to autoradiography had reduced band intensities and higher backgrounds, suggesting that some of the cross-linked adduct was being released during the staining procedure. Earlier results of others had led to the suggestion that UV-induced cross-links between protein and nucleic acid might be acid labile.[2,4] We have shown, under conditions similar to those used for fixation and staining of a gel, that only 25% of the cross-linked adduct remains intact after a 30-min incubation in 10% (1.58 M) acetic acid and 50% (v/v) methanol. We have not investigated the mechanism of the adduct lability but suggest that the lability may be the result of depyrimidination.

Analysis. Band intensities from the autoradiographs and stained gels can be determined in a number of ways. Initially, we carried out densito-metric scanning and integration of the area under the peaks using a soft laser scanning densitometer (model SL-504-XL) from Biomed Instruments (Richmond, CA), coupled to an Apple IIe computer. A typical set of gel patterns and densitometer traces obtained in this way for an experiment in which bacteriophage T4 gene 32 protein was cross-linked to (dT)$_{10}$ is shown in Fig. 2. More recently, we have adopted a video-based scanning procedure, in which the autoradiographic images are digitized using a Dage video camera, an AT&T 6300 computer, and software supplied by the Computer Technology Center at the University of Virginia (Charlottesville, VA). An alternate procedure that we now also use avoids the use of films altogether by scanning the radioactive gels directly in real time using an Ambis radioactive gel scanning system (San Diego, CA).

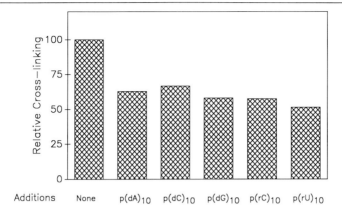

FIG. 3. Oligonucleotide competition experiments. The concentration of T4 gene 32 protein was 0.4 μM, and that of $[5'-^{32}P](dT)_{10}$ was 35 μM (in nucleotide residues). Prior to irradiation, the gene 32 protein–$(dT)_{10}$ complex was challenged with 35 μM $(dA)_{10}$, $(dC)_{10}$, $(dG)_{10}$, $(rC)_{10}$, or $(rU)_{10}$. Irradiation was with a single pulse of 5.9×10^{16} photons at 266 nm. Reaction conditions are as described in Fig. 2. Quantitation of data is described in the text. The data are normalized to 100% for the amount of photoadduct formed in the unchallenged reaction.

Calibration and Characterization of UV Laser Cross-Linking Reaction

T4-Coded Gene 32 Protein–Oligonucleotide Model System. We initially calibrated our procedures by monitoring the cross-linking of a variety of single-stranded 10-mer oligonucleotides to T4-coded gene 32 protein.[37] This set of protein–nucleic acid complexes had been previously characterized by standard solution thermodynamic methods, and the binding of the protein to these oligonucleotides has been shown to be independent of nucleotide composition.[47] As a consequence, we could interpret the efficiency of cross-linking of these oligonucleotide–gene 32 protein complexes as a function of laser power and wavelength directly in terms of the relative photoreactivities of the various oligonucleotides tested. This conclusion was directly confirmed with the cross-linking system by performing competition experiments in which gene 32 protein was mixed with $5'-^{32}P$-labeled $(dT)_{10}$ and then challenged with an equal concentration of an unlabeled competing oligomer. As shown in Fig. 3, all of the competing oligonucleotides reduce the yield of the gene 32 protein–$(dT)_{10}$ adduct in a proportionate fashion, thereby demonstrating directly that the binding affinity of gene 32 protein is indeed the same for all of the 10-mer oligonucleotides tested.

[47] S. C. Kowalczykowski, N. Lonberg, J. W. Newport, and P. H. von Hippel, *J. Mol. Biol.* **145**, 75 (1981).

TABLE I

UV CROSS-LINKING OF HOMOPOLYMER OLIGOMERS TO BACTERIOPHAGE
T4-CODED GENE 32 PROTEIN[a]

Wavelength (nm)	$(dT)_{10}$ cross-linked (%)	Cross-linking relative to $(dT)_{10}$			
		$(dT)_{10}$	$(dC)_{10}$	$(dG)_{10}$	$(dA)_{10}$
204.1	0.13	100	ND[b]	12.9	ND
217.8	1.8	100	1.2	ND	ND
223.0	0.76	100	ND	ND	ND
228.7	0.91	100	ND	ND	ND
239.5	2.2	100	3.1	ND	0.4
245.9	3.7	100	0.8	ND	ND
252.7	4.1	100	0.5	ND	ND
266.0	2.6	100	4.3	ND	2.7
273.9	3.6	100	1.0	ND	ND
282.3	3.1	100	0.6	ND	ND
309.1	0.06	100	ND	ND	ND

[a] Reaction conditions are described in Fig. 2. Oligonucleotide concentrations were held constant at 0.25 μM nucleotide residues. The gene 32 protein concentration was held constant at 0.4 μM. Irradiation was performed with a total flux of 1.4×10^{17} photons/sample at each wavelength.

[b] ND, No cross-linking detected, which at these levels of sensitivity corresponds to less than 0.005% of the input oligomer cross-linked.

Optimal Cross-Linking Wavelengths and Nucleotide Residues. To establish optimal wavelengths for cross-linking, as well as to define the nucleotide residues through which cross-linking can effectively occur, gene 32 protein was then mixed with $(dT)_{10}$, $(dC)_{10}$, $(dA)_{10}$, and $(dG)_{10}$, and irradiated at various wavelengths. The results are summarized in Table I (taken from Ref. 37).

As shown in Table I, cross-linking is most effective with the pyrimidine-containing oligonucleotides, $(dT)_{10}$ and $(dC)_{10}$. In Table I the $(dT)_{10}$ results are presented as the percentage of input oligomer that is cross-linked at each wavelength, while the results for the other oligonucleotides are normalized to the yield of $(dT)_{10}$–gene 32 protein cross-links obtained at each wavelength. Thymidine is, by far, the most efficient cross-linker at all wavelengths, and except at one or two isolated wavelengths effective cross-linking through purine nucleotides is not observed. At wavelengths greater than 245.9 nm, cross-linking efficiency follows the order: dT (1) $\gg$ dC (0.0084) > rU (0.0041) > rC, dA, dG (<0.001). The level of cross-linking, relative to dT, is shown for oligos $(dC)_{10}$, $(dG)_{10}$, and $(dA)_{10}$. The efficiency of cross-linking of gene 32 protein to the ribose-containing

TABLE II
CROSS-LINKING OF VARIOUS NUCLEIC ACID-BINDING PROTEINS
TO OLIGONUCLEOTIDES[a]

Protein (μM)	Cross-linking relative to gene 32 protein with $(dT)_{10}$		
	$(dT)_{10}$	$(rU)_{10}$	$(rC)_{10}$
Bacteriophage T4 gene 32 protein (0.4)	100	0.41	ND[b]
E. coli ρ protein (0.4)	1.53	0.21	0.82
E. coli RNA polymerase (0.4)			
σ subunit (0.4)	3.97	1.99	0.09
β and β' subunits (0.4)	1.88	0.84	ND
Rabbit muscle lactate dehydrogenase (5.0)	ND		

[a] Reaction conditions are described in Fig. 1. Oligonucleotide concentrations were held constant at 0.25 μM nucleotide residues. Irradiation was performed with a total flux of 1.4×10^{17} photons/sample using 245.9-nm light.
[b] ND, No cross-linking detected, <0.005% of oligomer cross-linked.

oligonucleotides $(rU)_{10}$ and $(rC)_{10}$ was also examined; the results are included in Table II.

It is clear from these findings that proteins can be cross-linked to either deoxyribose- or ribose-containing nucleic acids, but that the efficiency of cross-linking is very dependent on the nature of the nucleotide residue involved. Much of the following discussion applies equally to DNA- and to RNA-binding proteins, although for RNA cross-linking (through rU and rC residues), the specific activity of the ^{32}P-labeled RNA may need to be higher to yield a detectable signal.

Table II (taken from Ref. 37) shows that other proteins known to bind nucleic acids can also be cross-linked. The relative ratios of cross-linked products obtained for proteins with $(dT)_{10}$ and $(rC)_{10}$ vary from protein to protein, and may, in part, be related to differences in the relative binding constants of these proteins for the two oligonucleotides. This is clearly demonstrated in Table II for the E. coli ρ protein (an RNA-dependent ATPase), which can be efficiently cross-linked to $(rC)_{10}$. The binding affinity of this protein for $(rC)_{10}$ is more than 100 times greater than that of gene 32 protein for this oligomer,[48] and this tight binding is presumably responsible for the greater than 150-fold increase in cross-linking observed for the ρ–$(rC)_{10}$ complex, relative to the gene 32 protein–oligomer calibration system.

[48] J. A. McSwiggen, D. G. Bear, and P. H. von Hippel, J. Mol. Biol. 199, 609 (1988).

It is also clear, however, that relative binding affinities are not the only source of differences in cross-linking efficiency observed for different systems. The major factor in many cases may well be the relative "intimacy of contact" (in terms of the cross-linking mechanism) between the protein and a properly placed thymidine residue (for cross-linking with 266-nm radiation). An example of such an effect of detailed interaction geometry on the efficiency of cross-linking can be taken from our work on the cross-linking of site-specific binding proteins to DNA.[48a] Using single-pulse 266-nm radiation, we have shown that the saturating efficiency of cross-linking *E. coli* Cro protein to a specifically bound double-stranded oligonucleotide fragment carrying the canonical Cro-binding sequence, is approximately 12%. The efficiency of cross-linking Cro protein to a (nonspecifically bound) $(dT)_{20} \cdot (dA)_{20}$ fragment is <1%. In contrast, the observed efficiency of cross-linking *E. coli* cyclic AMP-binding protein (CAP) to its specific target sequence on a double-stranded DNA oligomer is ≪1% (we believe that this sequence contains no dT residues that are properly positioned to interact with the protein in the specific complex). Clearly, *both* binding affinity and the "intimacy" of appropriate (ideally involving dT residues if irradiating at 266 nm) protein–nucleic acid contacts are important in establishing cross-linking efficiency.

Cross-Linking to Define Protein–Nucleic Acid Contact Positions. Also shown in Table II are the relative cross-linking efficiencies for oligonucleotides with the α, β, and β' subunits of the *E. coli* RNA polymerase.[5–8,36] These RNA polymerase data show that laser cross-linking can be effective in discriminating contacts between nucleic acids and the various subunits of a protein complex. Similar results have now been obtained with more than 20 different nucleic acid-binding proteins (data not shown). In contrast to the nucleic acid-binding proteins, a protein such as lactate dehydrogenase, which has no intrinsic affinity for DNA, shows no adduct (cross-link) formation activity.

Cross-Linking Efficiency. Determination of the *yield* of cross-linked product is important in many types of analyses. As shown in Table I, the yield of gene 32 protein–$(dT)_{10}$ adduct with 252.7-nm irradiation corresponds to ~4% of the input oligonucleotides (in this experiment oligonucleotide is the limiting species, with protein present in considerable excess). Calculating from component concentrations and the known binding constant for this interaction ($K_{assoc} \approx 10^6 \, M^{-1}$),[47] we estimate that ~25% of the input oligomer is complexed with the gene 32 protein in these solutions. Therefore the *normalized* efficiency of cross-linking of $(dT)_{10}$ to

[48a] E. M. Evertsz, W. L. Kubasek, D. Kuninger, R. Brennan, and P. H. von Hippel, manuscript in preparation.

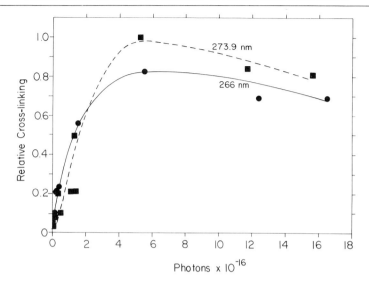

FIG. 4. Dose–response curve for the photocross-linking of T4 gene 32 protein to $(dT)_{10}$. Reaction conditions are as described in Fig. 2. The concentration of T4 gene 32 protein was 0.4 μM and that of $[5'-^{32}P](dT)_{10}$ was 72 nM (in nucleotides). Irradiation was with photons at either 266 nm (●) or 273.9 nm (■). A relative cross-linking yield of 1.0 corresponds to ~18% of the complex present in this set of experiments. The data are normalized to unity for the amount of photoproduct formed by 5.2 × 10^{16} photons at 273.9 nm.

gene 32 protein corresponds to an ~16% yield of the complex present during irradiation.

Quantum Yield for Cross-Linking. Making a few additional assumptions, we can also calculate the quantum yield (defined as the number of complexed oligomers that are cross-linked per absorbed photon) at each wavelength. For example, if we assume only a single dT residue per oligomer need be involved to obtain a cross-link (this has been demonstrated in Ref. 37), and that 25% of the oligomers are complexed in the reaction mix, we obtain a quantum yield (per nucleotide residue) of ~1 × 10^{-4} at 266 nm. This is comparable to quantum yields previously estimated with related model systems.[22]

In Fig. 4, we demonstrate that the amount of cross-linked product in a given experiment is a function of the total photon flux, as well as of the wavelength of the incident light. In general, at low levels of incident photons, the extent of cross-linking is a linear function of the total incident flux. At higher photon inputs the yield of cross-links reaches a maximum value, and then starts to decline. The data plotted in Fig. 4 were generated by using constant pulses of fixed power and increasing the total photon flux by increasing the total number of pulses. The same curves can be generated by using a fixed number of pulses and varying the power per pulse.[37] The data of Fig. 4 indicate that total photon fluxes exceeding 5×10^{16} photons do not significantly increase the yield of cross-linked product, and that higher levels may decrease the yield of adduct (see below).

Advantages of Single-Pulse Irradiation. The laser wavelengths used to generate the above data (with gene 32 protein and oligonucleotides of various composition) vary widely in their power per pulse; consequently, most samples were irradiated with more than one pulse so that the total *number* of incident photons could be kept constant in our quantum yield experiments. Having established that the highest quantum yields for cross-linking are associated with wavelengths between 245 and 280 nm, we have chosen to pursue single-pulse experiments using 266-nm light. Use of this wavelength provides ample power for single-pulse experiments, and also simplifies the experimental setup, since both the H_2 shifting cell and the second Pellin-Broca prism (see Fig. 1) are not needed. Our rationale for preferring single-pulse experiments involves consideration both of the quantum yields of the various changes that can be induced by UV irradiation of protein–nucleic acid complexes, and of the rates of association–dissociation reactions between, and conformational transition rates within, the macromolecular species.

Quantum Yields for Competing Photochemical Processes. We have measured a quantum yield of $\sim 10^{-4}$ for protein–(dT)$_{10}$ cross-link formation (per dT residue) with 266-nm radiation. Reported quantum yields for other UV-induced alterations of proteins and nucleic acids include (1) 4×10^{-2} for pyrimidine dimer formation, (2) 5×10^{-6} for backbone breakage of single-stranded DNA,[49] (3) 1×10^{-5} for backbone breakage of single-stranded RNA,[50] (4) 8×10^{-7} for DNA–DNA cross-link formation,[49] and (5) 7×10^{-6} for RNA–protein cross-link formation.[50] These data suggest that pyrimidine dimers will be introduced into DNA at approximately 400 times the rate at which protein–DNA cross-links are formed, while the other photochemical reactions mentioned above will be much less probable and not likely to complicate cross-linking studies of this type. However, pyrimidine-dimer formation can certainly

[49] G. B. Zavilgelsky, G. G. Gurzadyan, and D. N. Nikogosyan, *Photochem. Photobiophys.* **8**, 175 (1984).
[50] E. N. Dobrov, Z. H. Arbieva, I. S. Khromov, and S. V. Kust, *Photochem. Photobiol.* **43**, 493 (1986).

introduce problems into the interpretation of multipulse cross-linking experiments.[50a]

It is apparent that pyrimidine dimers will be introduced into DNA with high probability during the first pulse of a multiple-pulse laser cross-linking experiment, or during the initial phase of conventional ultraviolet irradiation. These pyrimidine dimers can cause significant distortion of the DNA to which the protein is bound. Since the rates of conformational transitions in macromolecular complexes are usually in the 100-μsec to 10-msec time range,[51] the geometries or affinities of protein–nucleic acid interactions would not be expected to be altered during the initial 5-nsec laser pulse. However, the ~100-msec interval *between* laser pulses certainly does provide ample time for conformations and contacts to be altered. Hence, the second incoming pulse could "freeze" an interaction that is substantially different from the interaction "seen" by the first pulse. This argues that the single-pulse irradiation mode should be used in experiments where protein–nucleic acid contacts might be altered by pyrimidine–dimer formation.

Estimated Rate of Protein–DNA Cross-Link Formation. Clearly photochemical *activation* of cross-link formation (by a single 5-nsec laser pulse) is fast compared to macromolecular association–dissociation and transconformation reactions. However, this does not prove that the subsequent chemical reaction that converts this photoactivation process into a protein–nucleic acid cross-link is also sufficiently fast to "freeze" labile macromolecular equilibria. Photochemical reactions can be quite slow (in the millisecond to second time range) and translational or rotational diffusion of macromolecular components occurring *after* the initial photochemical activation event could result in the appearance of significant nonspecific cross-linking artifacts.

To examine this question, we conducted the following experiment (details in Ref. 48a). Solutions of proteins that are not expected to interact significantly with DNA (e.g., horse heart myoglobin, alcohol dehydroge-

---

[50a] These comparisons of quantum yields with other competing photochemical processes include several that have been extensively studied only by low-intensity UV lamp irradiation procedures. The photon flux of a laser source is enormously greater than that achievable with a UV lamp. As a consequence quantum yields and photoproducts obtained by these methods may be very different. These issues are developed further in Evertsz *et al.*[48a] and have been reviewed in detail by D. N. Nikogosyan, *Int. J. Radiat. Biol.* **57**, 233 (1990). These potential differences between the products of laser and UV-lamp irradiation should also be considered in comparing the cross-linking results obtained here with those reported elsewhere in this volume [25].

[51] G. Careri, P. Fasella, and E. Gratton, *Crit. Rev. Biochem.* **3**, 141 (1975).

nase) were mixed with 5'-^{32}P-labeled (dT)$_{20}$, and subjected to single-pulse irradiation at 266 nm. We then looked for cross-link formation by our standard polyacrylamide gel electrophoresis procedures (see above) as a function of protein concentration. The results showed that no nonspecific cross-linking was observed at protein concentrations below approximately 100 μM. This is in contrast to the *specific* cross-linking in interactions of high affinity, which we observe at DNA-binding protein concentrations in the 1 μM concentration range in our "standard" cross-linking experiments.

Thus it appears that binding interactions involving association constants weaker than ~10^3 M^{-1} will not be detected by our cross-linking procedures. Furthermore, calculations of average mean free paths on the basis of estimated protein and nucleic acid translational diffusion constants suggest that the chemical part of the UV laser-induced protein–nucleic acid cross-link formation reaction goes to completion in less than 1 μsec. Thus photoactivated thymidine residues that are *not* in intimate contact with protein at the instant of activation will tend to be "quenched" by reaction with the solvent, and it appears that we can indeed "freeze" macromolecular binding equilibria and transconformation reactions of the sort we seek to study by this technique.

Analysis of Various Types of Protein–Nucleic Acid Interactions by UV Laser Cross-Linking Technique

The above discussion shows that single-pulse ultraviolet laser cross-linking can "freeze" protein–nucleic acid binding equilibria. As a further test of this proposition, we have examined a number of such equilibria that have previously been characterized by standard thermodynamic techniques. In each case we have measured binding isotherms and association constants comparable to those obtained by standard methods. In addition we have taken advantage of the ability of the laser cross-linking procedure to "freeze" a number of dynamic equilibria that are *not* easily studied by other techniques. Such results are briefly summarized in the following paragraphs to provide the reader with a further indication of the range and scope of problems that can be attacked by the UV laser cross-linking technique.

Binding Constant Measurements—Calibration with Gene 32 Protein. Using an association constant of 4 × 10^6 M^{-1} for the binding of T4-coded gene 32 protein to (dT)$_{10}$ (see Ref. 47), we can calculate the concentration of the protein–DNA complex that should be formed in the presence of varying concentrations of (dT)$_{10}$. Assuming that the efficiency of cross-linking is constant for all samples, the amount of complex produced should

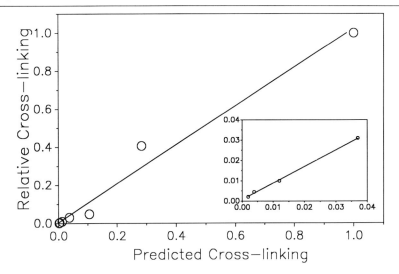

FIG. 5. Correlation between theoretical and actual cross-linking of $[5'-^{32}P](dT)_{10}$. T4 gene 32 protein (0.1 μM) was incubated with 1 μM $[5'-^{32}P](dT)_{10}$ under the reaction conditions described in Fig. 2. The complex was titrated with increasing amounts of $(dT)_{10}$ resulting in concentrations of $(dT)_{10}$ ranging from 1 to 100 μM (in nucleotides). Samples were irradiated with a single pulse of 4×10^{16} photons at 266 nm. The data were quantitated and normalized to unity for the amount of photoproduct formed at 1 μM $(dT)_{10}$. Using a binding constant of 4×10^6, the fraction of $(dT)_{10}$ bound to the gene 32 protein was "predicted" as a fraction of the total $(dT)_{10}$ present and normalized to unity for the 1 μM reaction. *Inset:* Expanded view of first four points.

be directly proportional to the amount of cross-linked adduct formed. Figure 5 demonstrates that the predicted cross-linking is exactly proportional to the actual cross-linking. Hence the complementary experiment, in which the concentration of $(dT)_{10}$ is held constant and the protein is varied, should also yield a binding curve that can be fit to a standard titration isotherm. Figure 6 shows this to be the case.

The general utility of determining binding constants by this laser cross-linking methodology is confirmed by the fact that the same binding constants are obtained over a 25-fold range of $(dT)_{10}$ concentration (10^{-8} to 2.5×10^{-7} M nucleotide). Clearly, the small reaction volume we use, coupled with the low concentrations of nucleic acid required, means that only very small quantities of material are needed to measure binding constants by this technique. (In addition, of course, these binding constants can be measured in impure solutions.)

The general applicability of such laser cross-linking determinations of protein-binding constants to nucleic acids, and more specifically for pro-

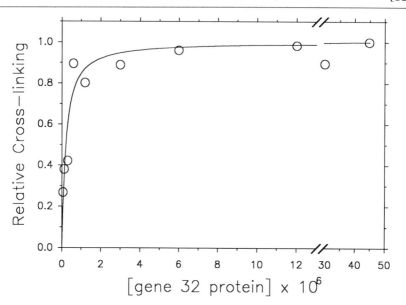

FIG. 6. T4 gene 32 protein titration of $(dT)_{10}$. Reaction conditions were 50 mM Na_2HPO_4, pH 7.7, 1 mM Na_2EDTA, 1 mM 2-mercaptoethanol, 50 nM $[5'-^{32}P](dT)_{10}$ (in nucleotides) and from 60 nM to 45 μM T4 gene 32 protein. Each sample was irradiated with a single pulse of 5.7×10^{16} photons at 266 nm.

teins that tend to translocate rather rapidly on the nucleic acid component (e.g., nucleases and polymerases) is demonstrated by our previously reported binding measurements of T4-coded DNA polymerase (gene 43 protein) to various DNA substrates.[52]

Salt Titrations. The binding of proteins to nucleic acids is supported by a variety of specific and nonspecific interactions. These interactions include the attraction of oppositely charged ionic groups; such attraction can, of course, be modulated by increasing salt concentrations. Figure 7 shows a salt concentration titration curve for the binding of gene 32 protein to $(dT)_{10}$, determined by the laser cross-linking technique. Again, standard methods confirm the profile that is generated.[47]

Nucleic Acid-Binding Protein Site Size Determinations. In most cases, the binding affinity of protein for nucleic acid should be maximal when the nucleic acid-binding domain of the protein is fully occupied. When the nucleic acid is too short to fill the site, favorable interactions will be

[52] J. W. Hockensmith, W. L. Kubasek, T. Cross, M. K. Dolejsi, W. R. Vorachek, and P. H. von Hippel, *in* "DNA Replication and Recombination" (R. McMacken and T. J. Kelly, eds.), Vol. 47, p. 111. Alan R. Liss, New York, 1987.

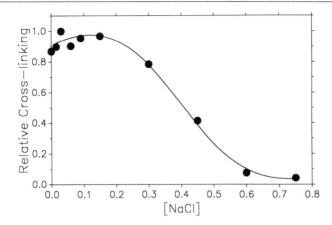

FIG. 7. Salt titration of T4 gene 32 protein–$(dT)_{10}$ interaction. Reaction conditions and irradiation are as described in Fig. 6, with the gene 32 protein concentration at 0.4 μM.

diminished; hence binding (and thus cross-linking) should be diminished. Nucleic acid moieties smaller than the actual nucleic acid-binding site may also be able to bind to the site in more than one way; measured binding constants in such cases will represent an average of the available binding modes. Consequently, if a nucleic acid-binding protein is bound to successively longer oligonucleotides, the most favorable binding should occur

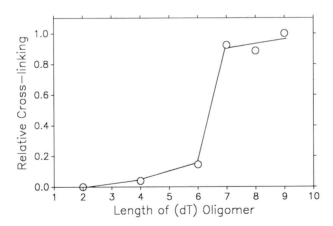

FIG. 8. Site size determination for the T4 gene 32*III protein. Reaction conditions are as described in Fig. 6, with gene 32*III concentration at 1.9 μM and $[5'-^{32}P]$oligo(dT) concentration at ~0.2 μM (in nucleotides). Irradiation was at 266 nm with a single pulse of 2.1 × 10^{16} photons.

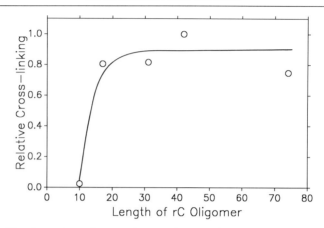

FIG. 9. Site size estimate for *E. coli* ρ protein. Reaction conditions are as described in Fig. 2. ρ protein concentration was 0.4 μ*M* and the [5'-³²P]oligo(rC) concentration was ~30 n*M* (in nucleotides). Irradiation was with 7.6 × 10¹⁶ photons at 245.9 nm.

when the site on the protein is filled. A dramatic example illustrating this point is presented in Figs. 8 and 9.

Figure 8 shows the cross-linking of the T4-coded gene 32 protein to oligo(dT) molecules of increasing length. There is a dramatic increase in cross-linking between $(dT)_6$ and $(dT)_7$, suggesting that an oligonucleotide of length seven exhibits the most thermodynamically favorable interaction with the nucleic acid-binding site of the protein. A binding site size of seven has been previously reported for the gene 32 protein binding cooperatively to single-stranded DNA.[53]

E. coli ρ *Protein RNA-Binding Site Size.* Cross-linking techniques can also be used to demonstrate RNA binding to proteins. Thus, we have cross-linked *E. coli* ρ protein to oligo(rC)$_n$; the results summarized in Fig. 9 suggest an RNA site size from 10 to 17 nucleotide residues/ρ protein monomers. This result is in good agreement with other data that suggest that the monomer site size for this protein ranges from 11 to 18 nucleotide residues, defined either as the site size needed to achieve full RNA binding, or the site size needed to achieve maximal ATPase activation.[48,54]

Bacteriophage T4-Coded Polymerase Accessory Proteins Interactions. While site size determination or "footprinting" by the laser cross-linking methodology provides information that can be obtained by other methods,

[53] D. E. Jensen, R. C. Kelly, and P. H. von Hippel, *J. Biol. Chem.* **251**, 7215 (1976).
[54] D. G. Bear, P. S. Hicks, K. W. Escudero, C. L. Andrews, J. A. McSwiggen, and P. H. von Hippel, *J. Mol. Biol.* **199**, 623 (1988).

TABLE III
T4 ACCESSORY PROTEIN–DNA CONTACTS[a]

Nucleic acid[a]	Addition	Relative cross-linking		
		G45P alone	G44/62P alone	G45P + G44/62P
$(dT)_{16}$	ATP	4	3/0	2, 2/170
$(dT)_{20}$	ATP	14	70/9	13, 6/310
$(dT)_{20} \cdot (dA)_{16}$	ATP	47	9/1	2, 0/24
$(dT)_{16}$	ADP	37	0/0	11, 5/0
$(dT)_{20}$	ADP	81	53/12	4, 0/0
$(dT)_{20} \cdot (dA)_{16}$	ADP	105	274/10	0, 0/0

[a] The number of molecules of (dT) oligomer was held constant for each of these experiments. Therefore the $(dT)_{20} \cdot (dA)_{16}$ experiment contained only one-fifth as much single-stranded DNA as the $(dT)_{20}$ experiment.

the cross-linking technique also has potential for providing information that is *not* easily obtained by other approaches.

For example, the bacteriophage T4 DNA replication system involves a DNA-dependent ATPase composed of genes 44, 62, and 45 proteins. When these three proteins are assembled on DNA in the presence of ATP, no footprint can be detected by neocarzinostatin or DNase I cleavage, suggesting that the hydrolysis of ATP greatly weakens the complex.[55] Laser cross-linking, however, is sufficiently fast to trap the transient complexes formed by the ATPase with the DNA in the presence of ATP (Table III, taken from Ref. 52). Experiments were performed using (1) gene 45 protein and DNA, (2) gene 44/62 proteins and DNA, or (3) a combination of gene 45 protein, gene 44/62 proteins, and DNA. Additionally, either ATP or ADP was added to the reaction. We have found that the gene 62 protein alone does not cross-link to $(dT)_{16}$ in the presence of ATP, but does cross-link to $(dT)_{20}$. Addition of the gene 45 protein results in a dramatic increase in cross-linking of the gene 62 protein to the three DNAs tested, but causes an apparent decrease in gene 44 protein cross-linking.

Cross-linking in the presence of ADP yields quite a different set of results. While the gene 62 protein still shows a preference for $(dT)_{20}$, the addition of gene 45 protein in the presence of ADP abolishes the gene 62 protein cross-linking. In addition, the gene 44 protein cross-links much better to $(dT)_{20}$ in the presence of ADP than in the presence of ATP. On

[55] H. E. Selick, J. Barry, T.-A. Cha, M. Munn, M. Nakanishi, M. L. Wong, and B. Alberts, *in* "DNA Replication and Recombination" (R. McMacken and T. J. Kelly, eds.), p. 183. Alan R. Liss, New York, 1987.

addition of the gene 45 protein, the gene 44 protein cross-linking is totally abolished in the presence of ADP.

These results demonstrate several things. First, the gene 62 protein cross-linking results show that an optimal length of DNA may be required to fill a binding site on the protein, thereby suggesting a site size between 16 and 20 nucleotide residues. Second, these results demonstrate that the laser cross-linking technique can be used to analyze multiple protein complexes and that proteins in such complexes may contact nucleic acids in a significantly different way than do the individual protein components in isolation. Finally, proteins that bind mononucleotides can have their polynucleotide contacts altered significantly by changing the nucleotide that is present.

In addition, as indicated above, these results show that the laser cross-linking technique can be used to "freeze" dynamic or rapidly interconverting protein–nucleic acid complexes. Such studies have been used to demonstrate rapid *changes* in protein–DNA contacts, and show that complexes such as the T4 DNA replication system cycle through altered protein conformations as proteins, substrates, and/or products (nucleotides) are added to or removed from the system.

Protein Photodegradation

Experimentation with the UV laser cross-linking procedure has also provided additional information that should be considered when attempting to cross-link proteins to nucleic acids by either conventional or laser ultraviolet irradiation. Photodegradation of proteins (protein chain breakage) after irradiation has been noted previously, although the reaction wavelengths and reaction mechanism have not been well characterized.[1] We have investigated photodegradation in more detail, since under certain circumstances it can significantly complicate UV laser cross-linking experiments.

Figure 10 presents an action spectrum for the photodestruction of protein. After laser irradiation at each wavelength, the protein was electrophoresed on an SDS–polyacrylamide gel and the protein remaining in the original protein band was quantitated. The photodegradation products are sufficiently random that they spread over the entire gel below the parent protein band, hence disappearing into the background. The decrease in the amount of stainable protein is related to both the wavelength and the input photon level, with maximum loss of stainable protein occurring at high photon flux in the region where peptide bonds absorb, a result that is certainly consistent with protein degradation.

Three sets of data are shown in Fig. 10. The high power data for

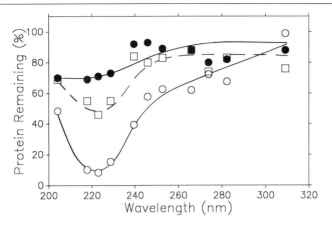

FIG. 10. Action spectrum for photodegradation. Reaction conditions are as described in Fig. 2. (○) T4 gene 32 protein $(0.4 \, \mu M)$ was mixed with $0.25 \, \mu M$ [$5'$-^{32}P](dT)$_{10}$ (in nucleotides) and irradiated with 1.4×10^{17} photons; (●) $3.9 \, \mu M$ T4 gene 32 protein was mixed with $0.22 \, \mu M$ [$5'$-^{32}P](dT)$_{10}$ and irradiated with 2.1×10^{16} photons; (□) $84 \, \mu g/ml$ T4 gene 45 protein was mixed with $0.22 \, \mu M$ [$5'$-^{32}P](dT)$_{10}$ and irradiated with 2.1×10^{16} photons. After irradiation samples were treated as described under Analysis of Cross-Linking Experiments.

the gene 32 protein shows considerable photodestruction, and we would suggest that this process accounts for the decreased cross-linking seen at the higher power levels of Fig. 4. Our data are consistent with a previous report that irradiation with $\sim 1.5 \times 10^{17}$ photons at 254 nm inactivates approximately one-third of the molecules in a gene 32 protein solution.[16] Alternatively the lower power level data for the gene 32 protein demonstrate that photodestruction can be minimized while maximizing the cross-linking (compare with Fig. 4). We have included photodegradation data obtained with gene 45 protein to illustrate that this phenomenon can occur with all proteins.

Photoreversal of protein–DNA cross-links has been suggested to occur at low wavelengths of irradiation.[2] We do not have any evidence for photoreversal of protein–nucleic acid cross-links, but should point out that photodegradation of complexes may mimic photoreversal. Consequently the optimization of laser cross-linking experiments should include experiments to find conditions that maximize cross-linking while minimizing protein photodegradation.

We note that photodegradation can be greatly minimized simply by avoiding the use of irradiating wavelengths below 245 nm. Users of low-pressure Hg fluorescent lamps should be aware that 14% of the emitted ultraviolet light is at wavelengths other than 254 nm, and high-pressure

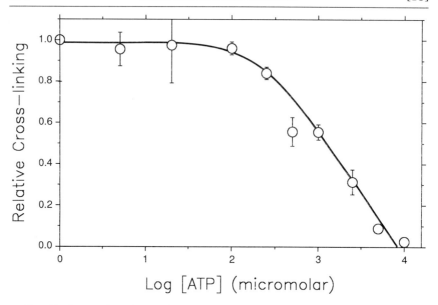

FIG. 11. Attenuation of cross-linking by ATP. Reaction conditions are as described in Fig. 2. Gene 32 protein (0.4 μM) was incubated with 0.25 μM [5'-^{32}P](dT)$_{10}$ and increasing concentrations of ATP prior to irradiation by a single pulse of 3.3×10^{16} photons at 266 nm.

Hg–Xe arc lamps have <1% of their ultraviolet output at 254 nm.[56] When using these light sources precautions should be taken to eliminate photodestructive wavelengths and effects.

Photoattenuation

The overall yield of cross-linked products may also be affected by the absorption of incoming photons by macromolecules other than those directly involved in the formation of protein–nucleic acid cross-links. Figure 11 shows the cross-linking of gene 32 protein to (dT)$_{10}$ in the presence of increasing amounts of ATP. At levels of ATP approaching 1 mM, the total cross-linking starts to diminish.

Viewing the depth of the sample to be irradiated as a series of very thin layers, we find that under normal conditions there is sufficient ultraviolet light to penetrate all the layers with some light remaining. When high levels of an absorbing center such as ATP are added, then most of the photons are removed by these absorbing centers within the first few layers below the surface. Layers located deeper in solution receive less light,

[56] T. P. Coohill, *Photochem. Photobiol.* **41**, 501 (1985).

and hence cross-linking is diminished in these layers, resulting in an overall decrease in cross-linking for the total solution.

While this attenuation is detrimental, in that it diminishes overall cross-linking, it has a beneficial effect in that it also reduces photodegradation of the protein (data not shown). In effect, the added adsorbing centers have resulted in fewer photons being available per mole of cross-linkable species, and consequently the experimental curve of Fig. 4 has been shifted so that the experiment is now being performed on the initial linear section of that curve. This situation may be "corrected" by merely increasing the input power. We suggest that when calibrating a system with respect to laser cross-linking it is useful to prepare a power curve such as that of Fig. 4.

Summary and Perspectives

This methodological chapter shows that ultraviolet laser cross-linking can be a very valuable tool in exploring protein–nucleic acid, protein–protein, and nucleic acid–nucleic acid interactions. The information obtained is often complementary to, and can be confirmed by, other techniques. However, in some instances this method will yield information that is difficult to obtain by other techniques.

The laser cross-linking technique is very simple and readily adaptable to a variety of systems. Typically laser cross-linking at 266 nm with a single pulse of $2–5 \times 10^{16}$ photons/sample works very well for protein concentrations ranging from 10^{-8} to 5×10^{-5} M, and for nucleic acid (in units of nucleotide residues) concentrations of 10^{-8} to 10^{-4} M. Higher concentrations of proteins and nucleic acids or nucleotide ligands can be used, but the input power level must be adjusted to take attenuation effects into account. Various other irradiation wavelengths can also be used, but should be evaluated for potential absorption by and photodegradation of the macromolecule of interest. Additionally, the technique permits examination of dynamic and static equilibria that have been perturbed in a variety of ways, such as by the addition of nucleotide cofactors and other complexing ligands.

We hope that this discussion of the UV laser cross-linking technique and its background will help others to devise additional variants of this powerful approach that may be better suited to other purposes and other experimental problems.

Acknowledgments

This work was supported in part by USPHS Grants GM-15792 and GM-29158 (to P.H.v.H.), by predoctoral traineeships on USPHS Institutional Training Grant GM-07759 (to W.L.K. and E.M.E.), by an USPHS Postdoctoral Fellowship GM-09353 (to J.W.H.), by a Jeffress Trust Award (to J.W.H.), by NSF Grant DMB-8704154 (to J.W.H.), and by an Institutional Grant to the University of Oregon by the Markey Charitable Trust. P.H.v.H. is an American Cancer Society Research Professor of Chemistry. We wish also to thank Professor Warner Peticolas and laboratory colleagues and the Laser Laboratory at the University of Oregon for access to lasers and for much help and advice in their use.

[14] Kinetic Studies on Promoter–RNA Polymerase Complexes

By Malcolm Buckle, Alexandre Fritsch, Pascal Roux, Johannes Geiselmann, and Henri Buc

Introduction

An understanding of the nature of the contacts between a protein and a nucleic acid existing in a given stereospecific complex at equilibrium is generally not enough to account for its function. In many instances knowledge of the corresponding pathway is a prerequisite for the elucidation of the mechanism by which a selected function is exerted. From a methodological point of view this is a much more difficult task.

A classical case is that of the interaction between RNA polymerase and its target site on the DNA, the promoter, where the overall rate of formation of the kinetically competent complex usually determines the frequency of initiaton of full-length transcripts or the strength of the promoter.[1-3] In this respect, it is necessary to use the time dimension to determine how protein–DNA contacts are sequentially established. A simple recording of changes in footprinting patterns on the DNA is not sufficient for a full characterization of the driving forces involved in the overall kinetic process, even when complemented with detailed genetic studies. Even though the protein component plays a predominant role, current technology is unfairly biased toward the DNA in that it is easier to investigate and interpret sequential changes in footprinting patterns than to follow modifications or subtle alterations in the protein occurring within the nucleoprotein complex.

This chapter will focus on two related aspects. First, an overview will be given of technological developments that allow the characterization of

[1] W. R. McClure, *Annu. Rev. Biochem.* **54,** 171 (1985).

[2] H. Bujard, M. Brunner, U. Deuschle, W. Kammerer, and R. Knaus, *in* "RNA Polymerase and the Regulation of Transcription. Proceedings of the 16th Steenbock Symposium" (W. S. Reznikoff, *et al.,* eds.), p. 95. Elsevier, New York, 1987.

[3] S. Leirmo and M. T. J. Record, *Nucleic Acids Mol. Biol.* **4,** in press (1990).

protein conformational changes taking place on discrete and separable nucleoprotein complexes representative of sequential changes in an overall reaction. The corresponding technology for the nucleic acid counterpart of the complex is rather well known and will not be reviewed in detail, except to stress the utility of a thorough characterization of the chemical reaction involved in the footprinting technique.

Second, methodologies used to characterize the change in the structure of the nucleic acid component, grouped here under the heading of footprinting techniques, may be adapted to obtain information about species only transiently present during the process of protein–nucleic acid recognition. This implies several technological advances (notably in the design of a rapid mixing device adapted to handle microvolumes and in the analysis of changes of chemical reactivity profiles as the reaction proceeds) that will be discussed in the second section of the chapter.

The interaction between *Escherichia coli* RNA polymerase (R) and the *lacUV5* promoter (P) provides a convenient system that illustrates the variability and versatility of these approaches. The process by which recognition and transcription of DNA into messenger RNA occurs may be schematically represented as, e.g.,[2,4]:

$$R + P \underset{K_B}{\rightleftharpoons} \underset{\substack{\text{closed} \\ \text{complex}}}{RP_c} \underset{k_2}{\overset{k_{-2}}{\rightleftharpoons}} \underset{\substack{\text{closed} \\ \text{intermediates}}}{(RP_i)} \underset{k_3}{\overset{k_{-3}}{\rightleftharpoons}} \underset{\substack{\text{open} \\ \text{complex}}}{RP_o} \underset{\substack{\uparrow \\ +XTP}}{\rightleftharpoons} \underset{\substack{\text{initiating} \\ \text{complex}}}{RP_{(init)}} \overset{-\sigma \nearrow}{\longrightarrow} \underset{\substack{\text{elongating} \\ \text{complex}}}{RP_{(el)}} \quad (1)$$

←—Recognition—→   ←—Isomerization—→   ←—Catalysis—→

It should be taken into consideration that Eq. (1) is derived primarily from kinetic studies that determine association constants (K_B) for promoter recognition (see Ref. 4) and forward rate constants (k_2, k_3, etc.) for the isomerization processes and from snapshot experiments conducted as the polymerase moves away from the promoter and into active transcription.[5-7] This minimal scheme will be challenged and further detailed. Characterization of the various intermediates requires a combination of three approaches: (1) the isolation of species appearing sequentially in the pathway after the open complex is formed, (2) changes in conditions that allow the isolation of certain intermediates prior to the appearance of RP_o, and (3) the recording of transient intermediates during the formation of RP_o.

At equilibrium, in the absence of ribonucleotides, at 37° and in the presence of 100 mM KCl at pH 8.0, the open complex, RP_o, is the predomi-

[4] H. Buc and W. R. McClure, *Biochemistry* **24**, 2713 (1985).
[5] A. Spassky, K. Kirkegaard, and H. Buc, *Biochemistry* **24**, 2723 (1985).
[6] A. Spassky, *J. Mol. Biol.* **188**, 99 (1986).
[7] J. R. Levin, B. Krummel, and M. J. Chamberlin, *J. Mol. Biol.* **196**, 85 (1987).

nant binary species. As detailed in Ref. 8 for the interaction between RNA polymerase and several variants of the T7A1 and *lac* promoters, various elongating complexes can also be isolated by the addition of a suitable choice of ribonucleotides. Hence a set of various elongating complexes, $RP_{(el)}$, where specified lengths of mRNA are present in the product site of the enzyme, can be obtained. The comparative studies of RP_o and of the subsequent $RP_{(el)}$ species are referred to below in the section Sequential Approach; when the enzyme is in the abortive mode $RP_{(init)}$, it is continuously synthesizing short transcripts; consequently, in this situation, only an average picture of the recycling process can be obtained.

The same methodology can be applied to the study of intermediates leading to the open complex, RP_o, if conditions can be found under which the intermediate of interest is either the predominant species or in rapid equilibrium with one of the intermediates on the pathway. Such is the case for a binary complex quenched at 16° at the *lacUV5* promoter $(RP_o)_{LT}$ (LT meaning low temperature), which undergoes a rapid isomerization with a postulated intermediate (RP_i) when the temperature is raised to 37°.[4] To a reasonable approximation, therefore, the "quenched" species RP_{LT} can be assumed to provide a picture of the short-lived species (RP_i). Note however that a careful kinetic study is required to justify this approach.

The RP_c binary complex, which, contrary to RP_i, has not yet been trapped, is the major species appearing transiently at 37°. At this temperature the kinetic scheme [Eq. (2)] may be reduced to

$$R + P \underset{K_B}{\rightleftharpoons} RP_c \underset{k_2}{\rightleftharpoons} RP_o \qquad (2)$$

Here again preliminary steady state kinetic studies allow prediction of the enzyme concentration [R] (in excess over [P]) and of the time scale required to study the decay of the RP_c binary complex into the final open complex under optimal conditions at 37° (Fig. 1; for more details see Ref. 5). The corresponding characteristic time is of the order of 0.1 sec^{-1}.

For most promoters, the complete formation of an open complex will require between 5 sec and 20 min. The classical static techniques will have to be adapted to a time scale ranging from 1 sec to 1 min, depending on the promoter considered.

Sequential Approach: Characterization of Stable Intermediates

Identification of Protein Components in Nucleoprotein Assemblies

The most commonly used technique to show an interaction between a protein and a nucleic acid is the band shift or gel retardation assay. The mobility of a radioactively labeled nucleic acid is altered by protein bind-

---

[8] B. Krummel and M. J. Chamberlin, *Biochemistry* **28**, 7829 (1989).

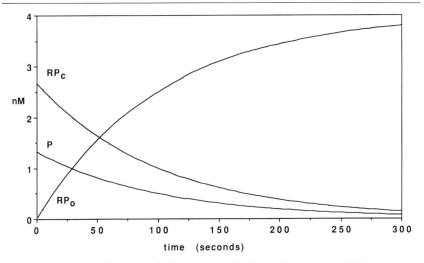

FIG. 1. Theoretical kinetic profile for open complex formation involving RNA polymerase and a weak promoter. The formation of an open complex is assumed to obey the simplified scheme given in Eq. (2). The relative concentrations of RNA polymerase (R), promoter (P), closed complex (RP_c), and open complex (RP_o) are calculated from the expressions $[P_o]_t = \lambda P_T e^{-t/\tau}$, $[RP_c]_t = (1 - \lambda)e^{-t/\tau}$, and $[RP_o]_t = P_T(1 - e^{-t/\tau})$, where PT represents the total promoter concentration, $\lambda = (1 + K_B[R])^{-1}$ and $\tau^{-1} = (1 - \lambda)k_2$. Values of $K_B = 10^7 M^{-1}$, $k_2 = 0.01 \text{ sec}^{-1}$, $[R] = 200$ nM, and $[P] = 4$ nM were used for this calculation.

ing. The main advantage of this technique is that it demonstrates a strong interaction between a given DNA fragment and a protein or protein assembly.[9,10] A disadvantage is that no details of the interactions involved are revealed. In its usual form identification of a nucleoprotein complex is carried out by visualization of a radioactively labeled DNA fragment following autoradiography. An innovation of this technique has been developed in this laboratory[11] that provides more detailed information from such assays. Nonlabeled DNA is used and after separation on native nondenaturing gels the complexes are electrotransferred to nitrocellulose sheets. Immunodecoration with antibodies and revelation with protein A labeled with ^{125}I allows visualization of the protein component of the complexes. The choice of ^{125}I allows quantification either by excision of the labeled bands and counting or by careful densitometric scanning of the autoradiogram. In order to illustrate this technique we will use the example of RNA polymerase in RP_o and $RP_{(el)}$ complexes.

Gel Electrophoresis. Polyacrylamide gels of approximately 1 mm thick-

[9] M. G. Fried and D. M. Crothers, *Nucleic Acids Res.* **9**, 6505 (1981).
[10] M. Garner and A. Revzin, *Nucleic Acids Res.* **9**, 3047 (1981).
[11] A. Kolb, M. Buckle, and H. Buc, *J. Cell. Biochem.* **12D**(Suppl.), 179 (1988).

ness and a length of 10 cm are prepared in TE buffer as described in Ref. 12. Complexes between 203-base pair (bp) linear fragments of *lac*UV5 DNA (20 nM) and RNA polymerase (30 nM) are formed at 37° as described in Ref. 12. Aliquots of 2 μl are taken into 2 μl of 22% Ficoll containing bromphenol blue and applied to a gel running in TBE (Tris–borate, EDTA) at a constant voltage of 100 V at room temperature. The length of migration and the percentage of acrylamide are empirically determined to allow visualization of retarded complexes as well as the free DNA fragment.

Electrotransfer. On completion of electrophoresis the gel is removed to Whatman (Clifton, NJ) filter paper and placed against a nitrocellulose sheet. Transfer is equally effective if carried out by placing the gel/nitrocellulose sandwich in solution between two electrodes, as classically used for Western blots,[13] or by arranging the gel/nitrocellulose between two graphite slabs, as in the so-called semidry blot.[14] Transfer is usually carried out by applying a potential of 7 V/cm distance between the electrodes for 1 hr independent of the technique of transfer used. Around pH 8.0, the efficiency of transfer greatly varies from one protein to the next. We found empirically that it can be maximized by varying first the pH of the transfer buffer containing a low amount of sodium dodecyl sulfate (SDS) (<0.01%) and then by increasing the SDS content. Immediately following electrotransfer, the protein component of the complexes reacts with dyes such as Ponceau Red or China ink as would a single protein. Such a coloration may be used to confirm the completion of transfer. Silver staining of the polyacrylamide gel subsequent to electrotransfer can also be used for this purpose.

Immunodecoration. The nitrocellulose sheets are incubated at 37° in a blocking solution such as phosphate-buffered saline containing either 5% nonfat powdered milk or a suitable nonionic detergent (0.05% Tween 20, rather than Triton X-100) and are then exposed to the specific antibody raised against one or more of the protein components of the complex. To date we have used polyclonal antibodies, although a powerful alternative would be the use of monoclonal antibodies whose efficiency of detection should reflect the accessibility of epitopes of the protein in various complexes with DNA.[15] If a quantification of the protein component within different complexes in the same gel is desired then subsequent detection with protein A labeled with [125]I[16] is favored, since this allows either efficient

[12] D. C. Straney and D. M. Crothers, *Cell* (*Cambridge, Mass.*) **43**, 449 (1985).
[13] H. Towbin, T. Staehelin, and J. Gordon, *Proc. Natl. Acad. Sci. U.S.A.* **76**, 4350 (1979).
[14] J. Kyshe-Anderson, *J. Biochem. Biophys. Methods* **10**, 203 (1984).
[15] X. M. Li and J. S. Krakow, *J. Biol. Chem.* **262**, 8383 (1987).
[16] S. W. Kessler, *J. Immunol.* **115**, 1617 (1975).

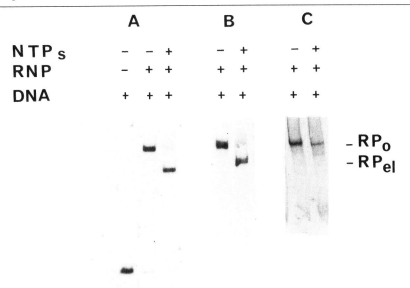

$$
\begin{array}{llllllll}
\textbf{A} & & & \textbf{B} & & \textbf{C} & \\
\end{array}
$$

	A			B		C	
NTP s	−	−	+	−	+	−	+
RNP	−	+	+	+	+	+	+
DNA	+	+	+	+	+	+	+

– RP$_0$
– RP$_{el}$

FIG. 2. Gel retardation assay of open and elongating complexes formed between RNA polymerase and the *lacUV5* promoter. Complexes between RNA polymerase (RNP) and the *lacUV5* promoter (DNA) in the presence or absence of nucleotide triphosphate substrates (NTPs) are formed as described in the text, and applied to nondenaturing 4% polyacrylamide gels. In (A), where ^{32}P-labeled DNA was used, the gel was autoradiographed; when nonlabeled DNA was present (B and C), the contents of the gel were electrotransferred to nitrocellulose sheets and incubated with either anti-α subunit antibodies (B) or anti-σ subunit antibodies (C) followed by ^{125}I-labeled protein A and autoradiography as discussed in the text. RP$_0$, open complex, no substrate; RP$_{el}$, elongating complex with substrate such that polymerase has synthesized a 10-mer ribonucleotide.

densitometric scanning of the resulting autoradiogram or counting of the excised bands. On the other hand, a second anti-Fc antibody, covalently labeled with biotin, may be incubated with the nitrocellulose sheet followed by a streptavidin–alkaline phosphatase complex, which in turn binds to the biotinylated anti-antibody.[17] A simple and rapid assay using 5-bromo-4-chloro-3-indolyl phosphate (PCIB) and nitroblue tetrazolium (NBT) gives a red-colored band on the nitrocellulose sheet.

Figure 2 shows how the presence or absence of several subunits of RNA polymerase can be seen at various stages of the reaction, following the reaction with protein A labeled with [125]I (cf. similar methods in Ref. 18).

[17] M. Wilchek and E. Bayer, *Anal. Biochem.* **171**, 1 (1988).
[18] D. C. Straney, S. B. Straney, and D. M. Crothers, *J. Mol. Biol.* **206**, 41 (1989).

Recording Protein Conformational Changes in Series of Consecutive Binary Complexes

At a higher level of complexity one would like to investigate important changes occurring in the protein itself. The initial rate of proteolytic digestion has been widely used for such a purpose (cf. Ref. 19 for an example). We will use here as an example the time course of digestion of the σ subunit (M_r 70,100) of *Escherichia coli* RNA polymerase with the protease V8 from *Staphylococcus aureus*. This protease has a specificity for D-G[19a] or E-X (where X is a polar residue) peptide bonds. On the σ subunit of RNA polymerase, V8 can theoretically attack two D-G sites as well as several E-X bonds.

Proteolytic Attack. The isolated σ subunit (final concentration in the reaction 0.5 mg/ml) prepared as described in Ref. 20 is incubated at 37° in 50 μl of a buffer containing 20 mM HEPES (pH 8.4), 3% (v/v) glycerol, 100 mM KCl, and 10 mM MgCl$_2$. Freshly prepared V8 protease in the same buffer is then added to a final concentration of 3 μg/ml [ratio of V8/σ, 1/160 (w/w)]. At time intervals, aliquots are removed to a quench solution containing 0.1 mM phenylmethylsulfonyl fluoride (PMSF) in SDS (0.5%, w/v). The time course of digestion is chosen as a function of the rate of disappearance, in this case, of the σ subunit. A typical profile of attack occurs over a 2-min period at 37°; however, in order to trap certain intermediates (RP$_i$) the temperature of the reaction must be decreased to around 16°. In comparing the digestion profile of these aptly named "frozen" complexes, the time scale of attack must necessarily be extended and clearly the comparisons with free subunit or in this case free RNA polymerase holoenzyme must be conducted at the same (16°) temperature. For example, the digestion of σ subunit in the holoenzyme at 37° took place over a period of 10 min under the conditions used for the more rapid digestion of isolated σ subunit. The digestion products are then separated by SDS–polyacrylamide gel electrophoresis (PAGE) and either stained with Coomassie Blue or electrotransferred to nitrocellulose and immunodecorated with anti-σ antibodies and ^{125}I-labeled protein A, (Fig. 3a). Quantitative densitometric scanning of these gels was carried out using a Bio-Rad (Richmond, CA) densitometer equipped with a Shimadzu (Columbia, MD) integrator. Normalization is then performed on a molar basis. It is not always possible to detect all the reaction products using SDS–PAGE. In this particular instance the three bands shown on the gel represent the larger fragments resulting from digestion.

[19] C. J. Lazdunski, J. Pages, and D. Louvard, *J. Mol. Biol.* **97**, 309 (1975).
[19a] D, Aspartic acid; E, glutamic acid; G, glycine.
[20] M. Gribskov and R. R. Burgess, *Gene* **26**, 109 (1983).

Quantitative Analysis of Proteolytic Products. It was first noted that the sum of the molar concentrations of the three major bands remains constant. Second, the time course of appearance and disappearance of the three major bands (Fig. 3b) indicates that the decay of the initial species follows a single exponential process, whereas two exponential phases are required to fit the two other curves. Within experimental error the characteristic time constants are equal, thus suggesting a scheme of the following type.

Hence for the σ subunit the rate of proteolysis by V8 can be characterized by three parameters, two of which (k_1 and k_2) represent conditions of initial attack, and the remaining one (k_3) a rate of interconversion between two products. It was verified that the same major products and the same scheme for their interconversion apply when σ was engaged in the RP_i, the RP_o, or various $RP_{(el)}$ species at *lac*UV5. The changes in the rate constants k_1, k_2, and k_3 are used to measure the conformational rearrangements in the σ subunit as elongation proceeds until σ is released.

Changes in Conformation of Cognate DNA-Binding Site as Probed by Chemical and Nucleolytic Reagents

The method of choice for the study of conformational changes in the nucleic acid component of nucleoprotein or nucleic acid–drug complexes, is to monitor the differential accessibility of base or sugar residues to chemical and nucleolytic probes. Ideally several conditions should be fulfilled.

1. The complex should adopt a *single* conformation so that the pattern of attack may be interpreted in simple terms (one pattern–one species). In default of this, one should search for conditions that favor the limiting species.

2. The reagent should attack the DNA and leave the protein intact. The differential rate of modification ($\pm$ protein) should reflect the DNA deformation in the *intact* binary complex. In particular, the reagent should neither displace nor perturb the complex that is probed.

3. In order to facilitate a structural interpretation one must be able to decide whether the reagent probes the major conformational state of the DNA or whether it reacts with another conformation in rapid equilibrium

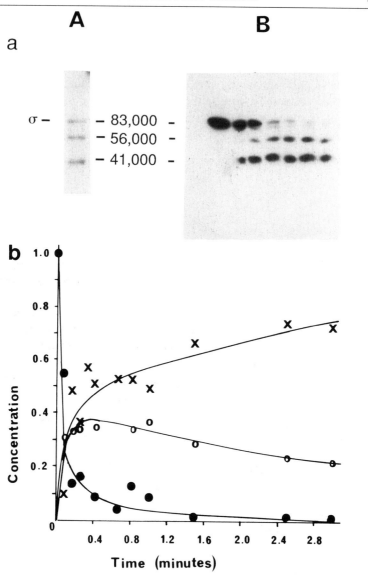

FIG. 3. V8 proteolytic digestion of the σ subunit of RNA polymerase. (a) Isolated σ subunit (A) or σ in the RNA polymerase holoenzyme (B) was digested with the protease V8 from *Staphylococcus aureus* over a limited time period as described in the text. The products were separated on SDS–PAGE and electrotransferred to nitrocellulose. Revelation was carried out using immunodecoration by anti-σ antibodies and ^{125}I-labeled protein A. (A) isolated σ alone following 30 sec of digestion with V8 protease at a protein-to-protease ratio of 160/1 (w/w) prior to quenching with 0.1 mM PMSF in 0.5% SDS. (B) Time course

with this species. An analysis of the rate of attack at each position allows one to distinguish between these possibilities.

In reality, none of these conditions is perfectly satisfied. Particular efforts to fulfill requirements (1) and (2) have been made and are described in other chapters of this volume (cf. also Refs. 21 and 22), but it should be stressed that even when binary complexes isolated from retardation gels are mildly attacked by a given probe, one is never certain that important rearrangements taking place during electrophoresis are not perturbing the initial complex. (A comparison of footprints before and after gel electrophoresis when a single retarded band is present should always be performed as a control.[23]

For quantitative analysis, conditions of single-hit kinetics should be respected (this means that for a 200-bp DNA fragment more than 90% of the DNA should remain intact); as a consequence the intensity of a given band will be independent of the position of the band with respect to the label (located at one end of a DNA strand). In many cases, however, a slight departure from single-hit kinetics can be accommodated by applying secondary corrections. A more crucial problem that cannot be easily circumvented in this way occurs in the case of a protein–DNA complex (cf. criterion (2) above). A first modification of the DNA can affect the reactivity at another position in the DNA. Single-hit conditions ensure that only *one* modification is present on each DNA molecule. The most obvious illustration of this situation is the attack of a complex formed on

[21] L. C. Lutter, *J. Mol. Biol.* **124,** 391 (1978).
[22] M. Brenowitz, D. F. Senear, M. A. Shea, and G. K. Ackers, this series, Vol. 130, p. 132.
[23] M. D. Kuwabara and D. S. Sigman, *Biochemistry* **26,** 7234 (1987).

---

of digestion of σ in the RNA polymerase holoenzyme. Conditions were as described in the text, each lane representing increments of 1 min of attack prior to quenching. (b) Kinetic profiles for the attack of V8 protease on isolated σ subunit. Conditions were carried out as described above. The relative proportions of each band were calculated from densitometric scans of the type of gels shown in (a). ●, Native σ subunit; ○, 56-kDa peptide; and ×, 41-kDa peptide. Curves were fitted according to the scheme shown in the text by these equations:

$$\sigma(t) = e^{-kt}$$

$$\mathbf{I}(t) = \frac{k_2}{k_3 - k} (e^{-kt} - e^{-k_3 t})$$

$$\mathbf{II}(t) = 1 + \frac{k_1 - k_3}{k_3 - k} e^{-kt} + \frac{k_2}{k_3 - k} e^{-k_3 t}$$

where $k = k_1 + k_2$.

a negatively supercoiled template by a nicking reagent.[24] Furthermore, single-hit considerations should apply to the attack on both the DNA and the protein when the reagent (such as diethyl pyrocarbonate or osmium tetroxide) can affect the two partners.[25] The initial rate of attack of DNA at any of the positions considered should be significantly larger than the rate of attack at a protein residue that, once modified, could affect the free energy of association and/or the conformation of the complex.

The third condition (3) is rarely taken into consideration. The simplest possible case, imino–proton exchange from a base pair in a free duplex,[26] serves to make the point: for an AT base pair, the reaction can be summarized as

$$\text{TH}\cdots\text{A} \underset{K_d}{\overset{k_{on}}{\rightleftharpoons}} \text{TH, A} \rightarrow \text{A} + \text{TH}^* \tag{3}$$

TH is a protonated thymine hydrogen bonded to a cognate adenine on the opposite strand. In the initial situation only the open state (TH, A) can react, the exchange on the thymine leading to TH* being catalyzed by the base components of the buffer.

A correct interpretation of the exchange time can only be made by following its dependence with respect to the catalyst concentration and comparing with the exchange time of a freely exposed base. At low catalyst concentration, many openings are required before exchange occurs and the exchange time is equal to that observed for the monomer measured under the same conditions of catalysis, divided by the base pair dissociation constant K_d. If the catalyst concentration is high enough, the exchange time is equal to the base pair lifetime, k_{on}^{-1}, since exchange occurs for each opening event.[26] In an analogous fashion one can carry out chemical modification of a base in a DNA molecule and measure the rate of attack as a function of the reagent concentration (C). We will examine two situations characterized by their respective rate profiles.

Hyperbolic Profile. When an overall profile is generated that shows a hyperbolic dependence of the rate on c then the same formalism developed for proton exchange will be used. At high reagent concentrations, an initial rate of attack that is independent of the concentration and nature of the reagent will be interpreted as the rate of access to a distorted, reactive intermediate [k_{on} in Eq. (3)]. At low reagent concentrations the initial rate of attack (now proportional to the reagent concentration) reflects the deformability of the complex. If, as in the case of imino–proton exchange, a model representation of the reactive species can be designed, a two-

[24] D. M. J. Lilley and B. Kemper, *Cell (Cambridge, Mass.)* **36**, 413 (1984).

[25] M. Buckle and H. Buc, *Biochemistry* **28**, 4388 (1989).

[26] J. L. Leroy, D. Broseta, and M. Guéron, *J. Mol. Biol.* **184**, 165 (1985).

step reaction scheme will be used to yield the corresponding equilibrium constant, K_d.

Linear Profile. Rapid reaction rates that are strictly first order with respect to the reagent at *all* reagent concentrations are classically interpreted as reflecting a property of the ground state. We define the ground state as being the major populated species that is the target for the reagent used at the initial equilibrium conditions. Such has been the case for a long time for DNase I attack on a double-stranded DNA fragment. This could turn out to be wrong, and to represent rather a minor population of conformations in rapid equilibrium with the ground state. As in the preceding case model systems representative of the putative reactive intermediate will be useful to clarify the problem. The practical problems that are then encountered in dealing with the comparison of reactivity profiles, obtained at various reagent concentrations and on several discrete binary complexes, arise from the quantitative scanning of gels of electrophoresis and the comparison of the corresponding optical density profiles (see below).

An example in which many of these considerations are examined in some detail is given in Ref. 25. In this study the reactivity of bases in the locally melted region of an open complex was examined using the reagents dimethyl sulfate (DMS) and diethyl pyrocarbonate (DEP). The reactivity of DMS with the N-7 of guanine residues in DNA occurs readily since no important distortion of the DNA is required to render the methylation site accessible. The kinetics of attack are always proportional to the concentration of the reagent. Conversely, the rate of DMS attack at position N-1 of cytosines that are not base paired presents a slower, hyperbolic profile. A similar situation is encountered for the reactivity of DEP with adenines in single-stranded DNA or in open complexes.[25] In both cases the reactive cytosine or adenine residues have to be in conformations that depart extensively from classical B-DNA structures. For bases in an open complex, the situation is, in part, analogous to that described in Eq. (3); the rate of modification measured at high concentrations of reagent reflects the k_{on} controlled formation of the distorted reactive intermediate. This deformation or isomerization step was shown to be a function of the position of a given base within the putative single-stranded region of the open complex.[25]

These considerations impose certain criteria governing the choice of a given reagent as a probe of DNA structures. Table I[27–40a] lists a selection of DNA-modifying or DNA-cleaving reagents that are currently in use.

[27] D. Suck and C. Oefner, *Nature (London)* **321**, 620 (1986).

[28] H. R. Drew and A. A. Travers, *J. Mol. Biol.* **186**, 773 (1985).

[29] D. S. Sigman, D. R. Graham, V. D'Aurora, and A. M. Stern, *J. Biol. Chem.* **254**, 12269 (1979).

[30] D. S. Sigman, A. Spassky, S. Rimsky, and H. Buc, *Biopolymers* **24**, 183 (1985).

TABLE I
MODIFYING AND CLEAVING REAGENTS AS PROBES

Reagent	Base	Site on base	Structure attacked	Ref.
DNase I	Nonspecific	—	Minor groove	27, 28, 29
OP_2Cu^+	Nonspecific	—	Minor groove	6, 30
Fe-EDTA	Nonspecific	—	Minor groove	31, 32
Dimethyl sulfate (DMS)	G	N-7, N-3, N-1	B-DNA	33
	C	N-3	Unpaired cytosines	25, 34
Diethyl pyrocarbonate (DEP)	A	N-7	Z-DNA	35
		(major groove)	Cruciform loops	36
			Hogsteen base pairing	37
			Open complexes	25
$KMnO_4$	T	C-5–C-6	Cruciforms	37a
		(double bond)	Open complexes	38
OsO_4	T	C-5–C-6	Supercoiled B-DNA	39
		(double bond)	B–Z junctions	40
			Open complexes	40a

The reagents are broadly classified into two general, nonmutually exclusive groups in a compilation that is by no means exhaustive. The division is based on the consideration that Fe-EDTA, $(OP)_2Cu^+$, DNase I, and finally DMS (for guanine residues) all represent reagents that, while not necessarily base nonspecific, probe a DNA structure in a fashion that reflects as closely as possible either the ground state of the DNA structure or a structure in rapid equilibrium with the ground state. One expects, therefore, a linear dependence on the concentration of the reagent, within the limits of single-hit conditions.

The reactivity profiles generated by the second group require an interpretation of the type afforded by Eq. (3). The plateau of reactivity observed for cytosines or adenines in an open complex may be interpreted as reflecting the rate of access to the reactive conformation of the base.

[31] T. D. Tullius, *Trends Biochem. Sci.* **12**, 297 (1987).

[32] T. D. Tullis, *Nature (London)* **332**, 663 (1988).

[33] W. Gilbert, A. Maxam, and A. Mirzabekov, in "Alfred Benzon Symposium XIII" (N. O. Kjeldgaard and O. Maaloe, eds.), p. 139. Munksgaard, Copenhagen, 1976.

[34] A. Spassky and D. Sigman, *Biochemistry* **24**, 8050 (1985).

[35] W. Herr, *Proc. Natl. Acad. Sci. U.S.A.* **82**, 8009 (1985).

[36] J. C. Furlong and D. M. J. Lilley, *Nucleic Acids Res.* **14**, 3995 (1986).

[37] D. Mendel and P. B. Dervan, *Proc. Natl. Acad. Sci. U.S.A.* **84**, 910 (1987).

[37a] D. M. J. Lilley and E. Palecek, *EMBO J.* **3**, 1187 (1984).

[38] S. Sasse-Dwight and J. D. Gralla, *J. Biochem. (Tokyo)* **264**, 8074 (1989).

[39] E. Lukasova, F. Jelen, and E. Palecek, *Gen. Physiol. Biophys.* **1**, 53 (1984).

[40] G. Galazka, E. Palecek, R. D. Wells, and J. Klysik, *J. Biol. Chem.* **261**, 7093 (1986).

[40a] M. Budde, personal communication.

Chemical Reactivity of Single-Stranded DNA. To illustrate the methodology involved in the kinetic approach, we will describe a series of experiments on putative single-stranded DNA regions with two reagents, diethyl pyrocarbonate (DEP) and osmium tetroxide (OsO_4). The single-stranded regions examined were either 57-mer oligomers or the same oligomers hybridized in complexes with complementary strands of DNA containing stretches of mismatches that produced fragments having "open" unpaired regions of varying sizes.

Attack by Osmium Tetroxide. DNA fragments (10 μg/ml) (5'-end-labeled by polynucleotide kinase with [γ-^{32}P]ATP) were incubated in a 20 mM HEPES buffer, pH 8.4, at 37° in the presence of 3% glycerol and 0.1 mg/ml sonicated calf thymus DNA. An aliquot of a stock solution (20 mM) of OsO_4 was added to a final concentration of 1 mM. At zero time the reaction was begun by the addition of pyridine to a final concentration of 0.1 M. Aliquots of the reaction were withdrawn at defined time intervals and quenched by precipitation of the DNA in ethanol. Samples were vacuum dried, resuspended in 1 M piperidine, heated at 90° for 30 min, lyophilized, and after denaturation in formamide at 90° for 3 min separated on 12% (w/v) polyacrylamide sequencing gels. Visualization of the bands was effected by classical autoradiography, quantification was carried out by scanning densitometry. A gel is shown in Fig. 4.

Attack by Diethyl Pyrocarbonate. DNA fragments as above were incubated in the HEPES buffer as described for the OsO_4 reaction. A stock solution of DEP, freshly prepared in ethanol, was diluted into water and added to the reaction mixture at zero time to give a final DEP concentration of 20 mM. Aliquots of the reaction mixture were removed at defined time intervals and quenched by the addition of 10 mM imidazole solution. The DNA was ethanol precipitated, vacuum dried, resuspended in 1 M piperidine heated at 90° for 30 min, and lyophilized. Polyacrylamide gel electrophoresis, autoradiography, and densitometry were carried out as described above.

The results confirmed that both DEP and OsO_4 react either with all bases in a single-stranded DNA, or exclusively with only those bases that lie within the mismatched region of the hybrids. The kinetic analysis demonstrated that single-hit conditions may be met only during a relatively short interval of time, but that in both cases the reaction was essentially second order. The reaction of adenines in an open complex with DEP has already been shown to be a first order process, limited by an isomerization that exposes the base.[25] A similar study of OsO_4 on thymines in RP$_0$ is currently under investigation. Furthermore, it has been shown[41] that the

[41] E. Palacek, A. Krejcova, M. Votjiskova, V. Podgorodnichenko, T. Ilynia, and A. Poverenny, *Gen. Physiol. Biophys.* **8**, 491 (1989).

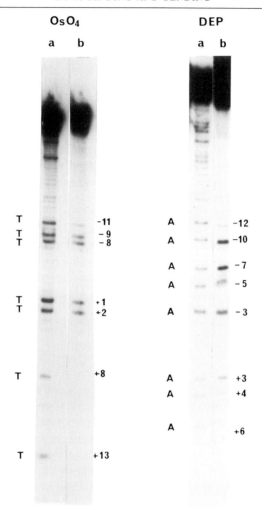

FIG. 4. Chemical cleavage of DNA by diethyl pyrocarbonate and osmium tetroxide. Single-stranded DNA (a) or duplex DNA (b) containing an internal 13-bp stretch of unpaired cytosines was modified by diethyl pyrocarbonate or osmium tetroxide as described in the text. Cleavage of the modified DNA at the sugar phosphate backbone was carried out by piperidine treatment at 90° for 30 min and the end-labeled fragments were separated on a 12% sequencing polyacrylamide gel. The 57-mer sequence was derived from a region of the *lac* promoter and the numeration on the gel refers to the position of a base with respect to the transcription start site at +1.

nature of the base, such as piperidine, is important in determining the reactivity of OsO_4 toward a given nucleotide.

Characterization of Transient Intermediates

When a given amount of RNA polymerase is added to a mixture containing a DNA fragment and the substrates necessary for abortive cycling of the enzyme at the corresponding promoter, a latency time, τ, is observed during the time course of product accumulation (Fig. 1). The rate profile for the formation of the kinetically competent binary complex RP_o can be traced from the results of the enzymatic assay using a simple derivation with respect to time (Fig. 1). In simple cases, based on that depicted in Eq. (3), it is expected to follow the relation

$$[RP_o]_t = [P]_T(1 - e^{-t/\tau}) \tag{4}$$

where $[P]_T$ is the concentration of the promoter containing DNA fragment. Analysis of the dependence of τ on RNA polymerase concentration in the enzymatic assay leads to an estimation of the constants K_B and k_2 for the two-step model given in Eq. (2), and thus allows the prediction, at each concentration of enzyme, of the concerted decay of the RP_c intermediate, as well as that of the free promoter (curves RP_c and P in Fig. 1; the two species being assumed to be in rapid equilibrium with respect to the time scale in Fig. 1).

Independent determinations of the deformations occurring in the DNA fragment and the enzyme during the same time interval will challenge and hence further define this model. Indeed, within this context, relatively rapid chemical attack on the DNA, at time intervals suitably spaced with respect to τ, can be used for this type of study. In practice, to date, this approach has been used only in a qualitative manner, using fairly weak promoters[5] where τ is generally longer than 3 min. We describe here improvements allowing a more rigorous and quantitative test of current models.

Analysis of Establishment of Chemical Reactivity Profiles

A typical footprinting experiment would yield a series of radioactively end-labeled DNA fragments generated by chemical attack for a short period (Δt) at various times after the mixing of a promoter and the enzyme RNA polymerase. Analysis of the cleavage pattern requires polyacrylamide gel electrophoresis and densitometric scanning. This approach is currently used to test and specify the model given in Eq. (2) and Fig. 1. For each concentration of RNA polymerase a single characteristic time τ will define the decay and appearance of the various species:

$$[P_o]_t = \lambda P_T e^{-t/\tau}$$
$$[RP_c]_t = (1 - \lambda)P_T e^{-t/\tau}$$
$$[RP_o]_t = P_T(1 - e^{-t/\tau})$$

where $\lambda = 1/(1 + K_B[R])$.

Besides checking the consistency of the two-step model, this kinetic footprinting experiment will generate a reactivity profile characteristic of the free promoter I_0, of the final complex I_2, and of the intermediate I_1 at each position of the sequence.

If the model is obeyed, then the intensity of a given band I_x should either stay constant or vary between the initial and the final state according to an exponential law:

$$I_x(t) - I_x(0) = [I_x(\infty) - I_x(0)]e^{-t/\tau} \qquad (5)$$

Errors can be estimated by a reconstitution experiment, as explained in Ref. 5. DNA attacked in the absence of enzyme or in the final complex are mixed in known proportions and applied to separate lanes of the gel. That the values of τ are consistent may be checked by examining all the sensitive bands corresponding to the kinetic system. Extrapolation to an infinite time yields $I_x(\infty)$, which corresponds to τ_2, the reactivity pattern of the open complex. Extrapolation to zero time yields $I_x(0)$, which is

$$I_x(0) = P_T[\tau_0\lambda + \tau_1(1 - \lambda)] \qquad (6)$$

Experiments performed at various RNA polymerase concentrations affect λ according to Eq. (4). I_0, I_1, and K_B can then be estimated from the change of $I_x(0)$ with respect to RNA polymerase concentration. Conversely, if the two-step model does not apply, then the minimal number of exponential components of the pattern will be derived using, for example, the Padé–Laplace approximants.[42] Eventually, the observations will yield the pattern of the reactivity profile of several consecutive intermediates. The following two sections describe technical details of signal acquisition and data analysis.

Rapid Mixing: A Microvolume Continuous/Stopped Flow Apparatus

In order to follow reactions that have a τ of around 10 sec, and to satisfy the requirements outlined above, a rapid mixing technique was required. A classical stopped-flow apparatus generally allows only one product sample per experiment to be assayed and requires a large amount of starting material in order to analyze several consecutive samples. For our purposes we are obliged to work with relatively small volumes (<100

---

[42] E. Yeramian and P. Claverie, *Nature (London)* **326**, 169 (1987).

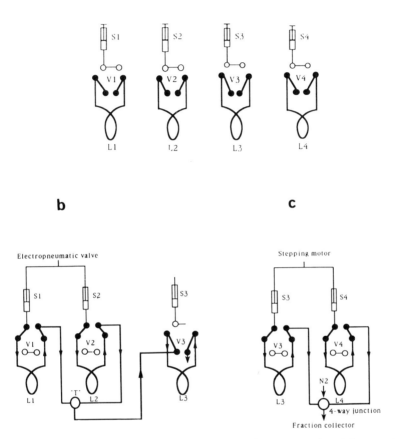

FIG. 5. Rapid mixing stopped-flow apparatus. Schematic representation of the operating principle. A detailed explanation is given in the text. S1 and S2 are electropneumatically driven water-filled syringes, S3 and S4 are water-filled syringes driven by a step motor. L1, L2, L3, and L4 are reaction loops. All valves labeled V are HPLC-type valves having negligible dead volumes.

μl) that should be used, however, to generate several data points on a time curve. In order to satisfy these requirements we designed the apparatus described in Fig. 5.

The system may be used either to follow a time course of, for example, a classical enzyme substrate reaction, or to study the effects of a reagent on a complex formed at $t = 0$ over a relatively short time course, as typified by a chemical attack on a complex between RNA polymerase and a promoter containing DNA fragment. The operating principle of the apparatus relies on the use of high-performance liquid chromatography (HPLC) valves having a negligible dead volume. For a simple rapid mixing and time course quench, the overall operation may be summarized as described below.

The loops L1, L2, and L4 (Fig. 5) are initially loaded with, for example, enzyme, radioactive substrate, and quenching solution, respectively. At time $t = 0$, the contents of loops L1 and L2 are propulsed via a mixing T junction into a third loop, L3, in which the enzyme reaction now takes place. The contents of loop L3 are then ejected as discrete fractions as a function of time into a second T junction, where they are mixed with an equal volume of quenching solution. The fractions are then carried by a continuous flow of an inert gas such as nitrogen into a waiting fraction collector. Prior to the reaction cycle all loops and connecting tubes are purged with air to prevent dilution due to lamellar flow. All reaction and mixing loops are immersed in a cryogenic fluid maintained at any given temperature. The volume of the loops L1 and L2 can be varied from 20 to 200 μl. To facilitate smooth flow through the tubing, bores of 0.3- to 0.8-mm diameter are used.

The force that propulses the contents of loops L1 and L2 through the T junction into loop L3 is provided via two water-filled syringes whose pistons are simultaneously driven by an electropneumatic valve. As soon as loop L3 is filled by the mixed reaction a second electropneumatic valve is activated. This operation places the loops L3, containing the reaction, and L4, containing the quench, under the control of a stepping motor. The controlled descent of this motor progressively injects small volumes of water into loops L3 and L4, which consequently expel the required amounts of reaction and quench into the exit four-way junction from where the stream of inert gas carries the mixture to the fraction collector. The inert gas flux also prevents premature mixing of the reaction with the quench solution. The stepping motor that drives the fraction collector is programmed to displace successive tubes in synchrony with the ejection of each sample.

The whole operation is commanded by an IBM PS2 microcomputer. The operating program allows selection of different volumes of the reaction to be removed to the quench and collector at various time points. The rapid mixing does not inactivate any of the enzymes examined to date nor does it induce shearing in linear DNA fragments.

In the second mode, RNA polymerase is placed in L1, DNA fragment in L2, and at $t = 0$ the two are propulsed into L3. Loop L4 contains a specified concentration of the modifying chemical reagent, which at given time intervals is mixed with the current nucleoprotein complex and ejected into the fraction collector. The awaiting collecting tubes can either contain a predetermined volume of a suitable quench or an aliquot of quench may be manually or mechanically added at the required times to the tubes in the fraction collector.

The advantages offered by this type of machine include short (millisecond) mixing times, negligible or nonexistent dead volumes (a logical consequence of using HPLC valves and of having a stream of inert gas that ejects all the samples into awaiting collecting tubes), no losses due to lamellar flow-induced mixing as solutions flow through narrow-bore tubes (avoiding here by having air present in all the mixing and connecting loops prior to reaction), and a reproducibility of time and volumes.

Quantitative Determination of Radioactivity Profile in Given Lane of Polyacrylamide Gel

The establishment of a reactivity profile assigned to the initial, transient, or final species (see above) implies quantitative determination of the bands resolved on a polyacrylamide gel. There are three principal techniques that are in current use. The first of these involves direct counting of excised radioactive bands and remains the most accurate and direct approach. It is necessary, however, that the sample loaded onto the gel is uniformly labeled (i.e., compare the pattern obtained from an end-labeled linear DNA fragment as opposed to that from a digest of a nonhomogenously reductively methylated ^{14}C-labeled protein) and that the bands are sufficiently resolved so as to permit discrete excision of each band. Both conditions are in practice rarely fulfilled. Consequently, the remaining two techniques, involving scanning of either an autoradiogram of the gel or the gel itself, must be used. The problems involved in densitometric scanning of autoradiograms have been treated elsewhere and the introduction of new devices is constantly improving the methodology.[43] We shall restrict ourselves here to four considerations.

1. It should be emphasized that it is the nature of the scanning device that dictates the type and quality of the electrophoretic gel, and not the converse. There are two types of apparatus that are in current use for densitometric scaning. For the sake of argument we will classify these into two categories, which we call line densitometers and band densitometers.

Line densitometers: This type of machine scans a narrow window (0.05 to 0.3 mm) down the length of a lane on a gel or autoradiogram. In practice the bands are aligned such that the slit passes through the center of each

---

[43] G. M. Smith and D. G. Thomas, *CABIOS* **6**, 93 (1990).

band. The scanner then displays or traces a series of peaks that correspond in surface area to the optical density profile across that portion of each band that is sampled. In general there is accompanying software that integrates the peaks thus obtained. The major specific reservation with this system is that the sampling of each band must provide a true representation of the distribution of material in the band. In order to be compatible with the relatively simple algorithms used in the integration process, the optical profile of a band should be as close to a Gaussian as possible and all the bands are assumed to possess identical profiles independent of their length of displacement through the gel.

Band densitometers: In this instance the whole gel is photographed and digitalized such that an associated computer can reconstruct an image in which the pixel density is proportional to the recorded optical density of the photograph. The experimenter then chooses, with the aid of appropriate software, the region that corresponds to a discrete peak that is then easily quantified for comparative purposes. This system thus eliminates the problem of representative sampling that flaws the line densitometers, since the whole of a band is actually sampled.

2. Independent of the type of scanning apparatus used, the peak signal obtained must be clearly differentiated from background noise in a uniform fashion that does not bias the integration procedure subsequently used to quantitate the area contained within the peak. This phenomenon is independent of the saturating capacity of the film.

3. Photographic films currently popular for autoradiography of sequencing type gels have a dynamic range of approximately 10^2; simply stated this means that the measured optical density should be linear over a range of 0.01 to 1.0 OD units (this is an approximate range and will vary among film types and autoradiographic conditions). Below the lower limit errors arise due to an increasingly subjective analysis of the signal-to-noise ratio on behalf of the operator. Above an OD of approximately 1.0 the response of the film is nonlinear and continued ignorance of this phenomenon induces large quantitative errors. Realistic scanning thus necessitates autoradiography of relatively short exposure times compared to the half-life of the isotope used and under constant conditions such that a series of exposure times allows the comparison of different bands on a gel (having ascertained that the response of a film to the radioactivity is constant with time).

4. The problems associated with densitometric analysis exemplified above, and the obviously powerful incentive to resolve them, have resulted in two relatively recent innovations in this field. The first approach retains the idea of scanning densitometry and addresses itself to the problem of the dynamic range of the photographic film. The answer adopted was to

use a special film that is an integral part of the scanner.[44] The film itself has a dynamic range of 10^5; furthermore, the bands are scanned in their entirety, thus eliminating sampling errors arising from anomolous migration. The film has the added advantage of being reusable. The second technique returns to a direct determination of the radioactivity in a gel. The polyacrylamide gel is placed just below a network of thin wires, the intersections of which record changes in electrical potential due to the presence of ionizing radiation. This grid is gradually displaced over a gel so as to map the distribution of radioactivity. Again, a computer equipped with the relevant algorithms reconstitutes an image where the pixel density reflects the counts acquired per unit area. The counts per unit area are used to carry out the comparative analysis between different regions (i.e., bands) on the gel. This system eliminates all the disadvantages of optical systems outlined above. However, use of this approach is restricted to the quantification of uniformly labeled samples. Furthermore, to date, the resolution between bands is no greater than 4 mm, i.e., insufficient for the separation of bands characteristic of a DNA sequencing gel.

Conclusion

We have attempted to provide an overview of how kinetic considerations may be merged with existing and emerging technologies to study conformational changes in nucleoprotein complexes. Reaction conditions can be adjusted so as to favor a given species at equilibrium. A variety of techniques can then be used for the characterization of this reaction intermediate. Both the protein and the nucleic acid component may be studied in considerable detail.

A kinetic study provides additional information concerning both the actual number of components in a given system as well as the rates of interconversion between the species. This is the only approach by which transient intermediates may be characterized. The most widely used techniques to date involve "footprinting." The time resolution is determined by the duration of signal acquisition (on the order of seconds, for most footprinting techniques).

A major advantage of footprinting techniques is that they allow the recording of changes occurring on all the bases of a nucleic acid target (~50 bp in the case of an RNA polymerase promoter interaction) and therefore provide a sample of data allowing quantitative testing of simple kinetic models. The most delicate part is probably the interpretation of changes in chemical reactivity patterns in terms of the deformation of the

[44] R. F. Johnston, S. C. Picket, and D. L. Barker, *Electrophoresis* **11**, 355 (1990).

nucleic acid template. Future technological advances will certainly include the development of equally powerful approaches to furnish details about the protein component responsible for these deformations.

Acknowledgments

We are extremely grateful to Dr. Annie Kolb and Dr. Kevin Gaston for general discussions and constructive criticisms of the text and to Nathalie Sassoon for technical expertise with the immunoblotting work. We are indebted to Mme. Odile Delpech for typing the manuscript.

This work was supported by grants from INSERM (No. 88-1003) and FRM. The CNRS has specifically financed the cost of development of the mixing device described in the section, Rapid Mixing. J.G. is a recipient of a long-term EMBO fellowship.

[15] Thermodynamic Methods for Model-Independent Determination of Equilibrium Binding Isotherms for Protein–DNA Interactions: Spectroscopic Approaches to Monitor Binding

By Timothy M. Lohman *and* Wlodzimierz Bujalowski

Introduction

A detailed understanding of macromolecular interactions, such as those involving proteins and nucleic acids, requires information about the energetics (thermodynamics) and kinetics of the interacting molecules as well as their structures. In particular, thermodynamic studies provide the information required to determine and quantify the forces that stabilize these interactions and any functionally important structural changes that occur within the complex. A quantitative study of the equilibrium binding of a protein to a nucleic acid is usually designed to determine the following information: (1) the stoichiometry of the interaction, (2) the existence and degree of cooperativity (positive or negative), (3) model-independent macroscopic binding constants and possibly model-dependent intrinsic binding constants, and (4) the dependence of the equilibrium binding parameters on solution variables (e.g., temperature, pH, and salt concentration), which can yield information about the molecular forces involved in the binding interaction.

In order to achieve these goals, equilibrium binding isotherms must be measured over a range of binding densities. An equilibrium isotherm for the binding of a ligand to a macromolecule is simply the relationship between the extent of ligand binding (ligand bound per macromolecule) and the free ligand concentration. True binding isotherms are model independent and reflect only this relationship. From such an isotherm one can obtain information about the macroscopic affinity of the ligand for the macromolecule and the stoichiometry of binding. However, for systems that involve the interaction of multiple ligands with a macromolecule, one can also analyze the binding isotherms to obtain physically meaningful *intrinsic* interaction constants that are related to the free energies of interaction. This is accomplished by comparing the experimental isotherms to theoretical predictions based on statistical thermodynamic models that incorporate the known molecular features of the interaction (cooperativity, overlap, etc.).

A number of techniques have been used to examine protein–nucleic acid binding equilibria. Sequence-specific binding has been examined quantitatively by filter binding,[1] footprinting,[2,3] and gel retardation techniques,[4,5] whereas nonspecific binding has been examined quantitatively using transport and chromatographic techniques.[6-11] Some of these techniques are thermodynamically rigorous, while others involve assumptions that must be verified for each system. Techniques such as equilibrium dialysis are thermodynamically rigorous and yield absolute binding isotherms without assumptions; however, this approach is time consuming and may not be applicable to studies of interactions between macromolecules.

More convenient methods become available if a change in a spectroscopic property of either the ligand or the macromolecule occurs on formation of a complex and these have been used to examine the binding of a variety of proteins, and other ligands, to both DNA and RNA[12-29]; how-

[1] A. D. Riggs, S. Bourgeois, and M. Cohn, *J. Mol. Biol.* **53**, 401 (1970).

[2] D. J. Galas and A. Schmitz, *Nucleic Acids Res.* **5**, 3157 (1978).

[3] M. Brenowitz, D. F. Senear, M. A. Shea, and G. K. Ackers, this series, Vol. 130, p. 132.

[4] M. M. Garner and A. Revzin, *Nucleic Acids Res.* **9**, 3047 (1981).

[5] M. Fried and D. M. Crothers, *Nucleic Acids Res.* **9**, 6505 (1981).

[6] K. R. Yamamoto and B. Alberts, *J. Biol. Chem.* **249**, 7076 (1974).

[7] P. L. deHaseth, C. A. Gross, R. R. Burgess, and M. T. Record, Jr., *Biochemistry* **16**, 4777 (1977).

[8] D. E. Jensen and P. H. von Hippel, *Anal. Biochem.* **80**, 267 (1977).

[9] A. Revzin and P. H. von Hippel, *Biochemistry* **16**, 4769 (1977).

[10] D. E. Draper and P. H. von Hippel, *Biochemistry* **18**, 753 (1979).

[11] T. M. Lohman, C. G. Wensley, J. Cina, R. R. Burgess, and M. T. Record, Jr., *Biochemistry* **19**, 3516 (1980).

[12] R. C. Kelly, D. E. Jensen, and P. H. von Hippel, *J. Biol. Chem.* **251**, 7240 (1976).

[13] D. E. Draper and P. H. von Hippel, *J. Mol. Biol.* **122**, 321 (1978).

[14] S. C. Kowalczykowski, N. Lonberg, J. W. Newport, and P. H. von Hippel, *J. Mol. Biol.* **145**, 75 (1981).

ever, these are indirect methods. If an "apparent" binding isotherm is obtained by indirect methods that involve assumptions, then the apparent isotherm and the interaction parameters that one obtains will be no more accurate than the assumptions. This may cause particular problems if an isotherm will be used to differentiate between alternative models for an interaction, since if one of the models does not reproduce the isotherm, this may be due either to the failure of the model or the failure of the assumptions on which the calculated isotherm is based. As a result, it is important to obtain a correct, model-independent binding isotherm, independent of assumptions.

In this chapter, we discuss the use of spectroscopic approaches to study binding phenomena and their rigorous thermodynamic analysis to obtain true, model-independent binding isotherms. In these approaches, one monitors a spectroscopic signal [e.g., absorbance, fluorescence, circular dichroism, nuclear magnetic resonance (NMR) linewidth, or chemical shift] from either the macromolecule or the ligand that changes on formation of the ligand–macromolecule complex. In order to use this signal change to calculate a binding isotherm, it is commonly *assumed* that a linear relationship exists between the signal change and the fractional saturation of the ligand or macromolecule. However, if a linear relationship does not hold, then the calculated apparent isotherm will be incorrect. In the general case, in which multiple ligands (e.g., proteins) bind to a macromolecule (e.g., DNA), the fractional signal change and the extent of binding may not (and often do not) have a simple linear relationship. This may occur if the ligand can bind in several modes, each possessing a different spectroscopic signal (e.g., if cooperativity exists). The extent of any deviation from the ideal linear behavior is usually unknown *a priori,*

---

[15] J. W. Newport, N. Lonberg, S. C. Kowalczykowski, and P. H. von Hippel, *J. Mol. Biol.* **145**, 105 (1981).

[16] F. Boschelli, *J. Mol. Biol.* **162**, 267 (1982).

[17] N. C. M. Alma, B. J. M. Harmsen, E. A. M. de Jong, J. V. D. Ven, and C. W. Hilbers, *J. Mol. Biol.* **163**, 47 (1983).

[18] D. Porschke and H. Rauh, *Biochemistry* **22**, 4737 (1983).

[19] W. Bujalowski and D. Porschke, *Nucleic Acids Res.* **12**, 1549 (1984).

[20] T. M. Lohman, L. B. Overman, and S. Datta, *J. Mol. Biol.* **187**, 603 (1986).

[21] W. Bujalowski and T. M. Lohman, *Biochemistry* **26**, 3099 (1987).

[22] W. Bujalowski and T. M. Lohman, *J. Mol. Biol.* **195**, 897 (1987).

[23] L. B. Overman, W. Bujalowski, and T. M. Lohman, *Biochemistry* **27**, 456 (1988).

[24] W. Bujalowski and T. M. Lohman, *J. Mol. Biol.* **207**, 249 (1989).

[25] W. Bujalowski and T. M. Lohman, *J. Mol. Biol.* **207**, 269 (1989).

[26] D. P. Mascotti and T. M. Lohman, *Proc. Natl. Acad. Sci. U.S.A.* **87**, 3142 (1990).

[27] P. L. deHaseth and O. C. Uhlenbeck, *Biochemistry* **19**, 6138 (1980).

[28] B. C. Sang and D. M. Gray, *J. Biomol. Struct. Dyn.* **7**, 693 (1989).

[29] T. M. Lohman and W. Bujalowski, *Biochemistry* **27**, 2260 (1988).

hence the degree of error introduced into the apparent isotherm and the resulting binding parameters is also unknown.

If an indirect spectroscopic approach is used, it is necessary to determine the relationship between the spectroscopic signal change and the degree of saturation of the ligand or the macromolecule. We will review thermodynamic approaches to analyzing such data that allow the determination of true binding isotherms in the absence of assumptions. The use of these methods allows one to use a spectroscopic signal to obtain a true binding isotherm, even when a direct proportionality does not exist between the fractional signal change and the degree of binding. Since the methods reviewed in this chapter are thermodynamically rigorous, they do not require calibration by other techniques. In some cases one can even determine spectroscopic parameters for a ligand interaction at individual binding sites.

The only constraint on the methods of analysis reviewed here is that they are not valid if the ligand or macromolecule undergoes a change in aggregation or self-assembly state within the experimental concentration range examined. Therefore, as in any case, knowledge of the self-assembly properties of the protein or nucleic acid under study is essential before a rigorous analysis of a binding phenomenon can be undertaken. However, these approaches are valid for binding systems that exhibit cooperativity of any type.

Previous discussions of the use of these rigorous approaches have appeared, both for the case in which the spectroscopic signal is from the macromolecule[24,25,29–32] as well as when the signal is from the ligand.[21,33–35] The methods of analysis differ for these two cases, as is discussed below. These methods of analysis are not constrained to the study of interactions that can be monitored by spectroscopic signals. Rather, they are general and can be used to analyze titrations that monitor any property that reflects binding (e.g., viscosity changes[31,36]).

Theory

In our discussions of binding, the term *macromolecule* refers to the species that binds multiple ligands. We first discuss the case in which the observed signal is from the macromolecule, since the results are fairly

[30] C. J. Halfman and T. Nishida, *Biochemistry* **11**, 3493 (1972).

[31] K. E. Reinert, *Biophys. Chem.* **13**, 1 (1981).

[32] H. Fritzsche and H. Berg, *Gazz. Chim. Ital.* **117**, 331 (1987).

[33] J. Bontemps and E. Fredericq, *Biophys. Chem.* **2**, 1 (1974).

[34] G. Schwarz, H. Gerke, V. Rizzo, and S. Stankowski, *Biophys. J.* **52**, 685 (1987).

[35] C. Gatti, C. Houssier, and E. Fredericq, *Biochim. Biophys. Acta* **407**, 308 (1975).

[36] B. M. J. Revet, M. Schmir, and J. Vinograd, *Nature (London) New Biol.* **229**, 10 (1971).

intuitive. We then discuss the case in which the observed signal is from the ligand, which requires a slightly different analysis. In each case, we derive the general binding density function (BDF) and show that the BDF is dependent on only the binding density distribution of ligand on the macromolecule, thus demonstrating that the BDF analysis is thermodynamically rigorous and without assumptions. The BDF equations are derived without specifying the actual signal that is monitored in order to emphasize their general character; however, specific cases are treated as examples.

In studies of protein–DNA and all ligand–macromolecule equilibria, two types of titrations can be performed to examine the binding equilibrium. In one case, the macromolecule is titrated with ligand and we refer to this as a "normal" titration, since the binding density, ν, (average moles of ligand bound per mole of macromolecule) increases during the progress of the titration. In the second case, the ligand is titrated with the macromolecule, which we refer to as a "reverse" titration, since the binding density decreases during the progress of the titration. Generally, the type of titration that is performed will depend on whether the signal being monitored is from the macromolecule (normal) or the ligand (reverse). Although this is not a strict requirement, the use of a normal titration when monitoring a signal from the ligand generally results in a titration curve that is less precise. The following sections discuss the rigorous methods that can be used to convert a titration curve, i.e., a change in signal as a function of titrant into a true, model-independent binding isotherm, which can then be analyzed to obtain binding and thermodynamic parameters for that interaction.

Thermodynamic Basis for Binding Density Function Method of Analysis

The methods of analysis reviewed here should be applied in cases in which an indirect signal is used to monitor binding and multiple equilibria may exist, i.e., multiple ligands can bind per macromolecule. The thermodynamic basis for these methods is that the distribution of ligands bound in different states i, symbolized by $\Sigma\nu_i$ (average number of ligands bound per macromolecule), is strictly determined at equilibrium by the free ligand concentration, L_F (more precisely the chemical potential of the free ligand).[37] Therefore, if the *free* ligand concentration is the same for two (or more) solutions at different total macromolecule concentrations (M_T), then the average ligand binding density distributions, $\Sigma\nu_i$, will also be the same

---

[37] T. Hill, "Cooperativity Theory in Physical Biochemistry." Springer-Verlag, New York, 1985.

in each solution. As a result, constant values of L_F and hence $\Sigma \nu_i$ will exist for a number of different combinations of *total* ligand and *total* macromolecule concentrations (L_{T_x}, M_{T_x}), that satisfy the mass conservation equation [Eq. (1)]:

$$L_T = (\Sigma \nu_i)M_T + L_F \tag{1}$$

Therefore, if the set of concentration pairs (L_{T_x}, M_{T_x}) can be found for which $\Sigma \nu_i$ and L_F are constant, then $\Sigma \nu_i$ and L_F can be determined from the slope and intercept, respectively, of a plot of L_T vs M_T, based on Eq. (1). The methods to determine the set of concentration pairs (L_{T_x}, M_{T_x}) that correspond to a constant L_F and hence $\Sigma \nu_i$ are described in the following sections and depend on whether the signal is from the ligand or the macromolecule.

Signal from Macromolecule

In this case, the signal reflects the average degree of saturation of the macromolecule and a normal titration (addition of ligand to a constant macromolecule concentration) is generally performed. Any physicochemical intensive property of the macromolecule (e.g., fluorescence, absorbance, circular dichroism, and viscosity) can be used to monitor binding if this property is affected by the state of ligation of the macromolecule.

Macromolecule Binding Density Function (MBDF) Expressions for Multiple Ligand Binding. We first consider the general relationship between the concentration of each macromolecular species with i ligands bound, M_i, and the experimentally observed signal from the macromolecule, S_{obs}. For a total macromolecule concentration, M_T, the equilibrium distribution of the macromolecule among its different ligation states, M_i, is determined by the free ligand concentration, L_F. Therefore, at each L_F, S_{obs} is the algebraic sum of the concentrations of the macromolecule in each state, M_i, each weighted by the value of the physicochemical property for that state, S_i. For example, if only one ligand can bind per macromolecule, then a two-state system as described in Eq. (2) applies,

$$S_{obs} = S_F M_F + S_B M_B \tag{2}$$

where the subscripts F and B refer to the free and bound states, respectively.

In general, a macromolecule will have the ability to bind r ligands, hence the general, multistate equation for the observed signal, S_{obs}, of a sample containing ligand at total concentration, L_T, and macromolecule at total concentration, M_T, is given by Eq. (3),

$$S_{obs} = S_F M_F + \sum_{i=1}^{r} S_i M_i \tag{3}$$

where S_F is the molar signal of the free macromolecule and S_i is the molar signal of the macromolecular complex with i bound ligands, M_i ($i = 1$ to r). Equation (3) is valid in the *absence* of macromolecular aggregation.

The mass conservation equation, which relates M_F and M_i to M_T, is given in Eq. (4):

$$M_T = M_F + \sum_{i=1}^{r} M_i \tag{4}$$

From the definition of the binding density, ν_i (i moles of ligand bound per mole of macromolecule), given in Eq. (5),

$$\nu_i = iM_i/M_T \tag{5}$$

one obtains the expression for M_i given in Eq. (6):

$$M_i = (\nu_i/i)M_T \tag{6}$$

On introducing Eqs. (4) and (6) into Eq. (3), one obtains Eq. (7), which can be rearranged to a general form as in Eq. (8):

$$S_{obs} = S_F M_T + \Sigma M_T (S_i - S_F)(\nu_i/i) \tag{7}$$
$$\Delta S_{obs} \equiv (S_{obs} - S_F M_T)/S_F M_T = \Sigma(\Delta S_i/i)(\nu_i) \tag{8}$$

In Eq. (8), $\Delta S_{obs} \equiv (S_{obs} - S_F M_T)/S_F M_T$ is the experimentally determined fractional molar signal change observed in the presence of total ligand and macromolecule concentrations, L_T and M_T, and $\Delta S_i/i$ is the average molecular signal change *per bound ligand* in the complex containing i ligands. Note that the quantity, $S_F M_T$, is simply the initial signal from the unliganded macromolecule, i.e., the signal at the start of the titration, before addition of ligand.

Equation (8) indicates that ΔS_{obs}, the macromolecular binding density function (MBDF), is a function *only* of the binding density distribution, $\Sigma \nu_i$, since $\Delta S_i/i$ is an intrinsic molecular property of the system. Therefore, the average degree of saturation of the macromolecule, $\Sigma \nu_i$, must be the same for any pair of total concentrations, L_T and M_T, for which ΔS_{obs} is constant. Since $\Sigma \nu_i$ is a unique function of the free ligand concentration, then the value of L_F at the same degree of saturation (constant MBDF) must also be constant. As a result, the MBDF is the appropriate function to be used to determine model-independent values of L_F and $\Sigma \nu_i$. The details for how this is accomplished are described below. The above derivation is rigorous and independent of any binding model and as such can be applied to any binding system, with or without cooperative interactions or overlapping binding sites (see below).

TABLE I
MACROMOLECULAR BINDING DENSITY FUNCTIONS FOR DIFFERENT SPECTROSCOPIC SIGNALS

Method	Macromolecular binding density function (MBDF)	Comments
Absorbance	$\Delta A_{obs} = \Sigma(\Delta\varepsilon_i/i)\nu_i$ where $\Delta A_{obs} = (A_{obs} - \varepsilon_F M_T)/\varepsilon_F M_T$ $\Delta\varepsilon_i = (\varepsilon_i - \varepsilon_F)/\varepsilon_F$	$\varepsilon_F M_T$ is initial absorbance of macromolecule solution in absence of ligand; ε_F and ε_i are extinction coefficients for free macromolecule and each state of bound macromolecule
Fluorescence	$Q_{obs} = \Sigma(Q_i/i)\nu_i$ where $Q_{obs} = (F_{obs} - F_F M_T)/F_F M_T$ $Q_i = (F_i - F_F)/F_F$	$F_F M_T$ is initial fluorescence of macromolecule solution in absence of ligand

In Table I, we give the expressions for the macromolecular binding density functions (MBDF) when the experimentally observed signal is due to ligand-dependent changes in the fluorescence or absorbance of the macromolecule; however, Eq. (8) is general and not restricted to spectroscopic changes.

Data Analysis When Signal Is from Macromolecule and "Normal" Titrations Are Performed. We will demonstrate the use of this method of analysis with data from studies of the binding of the oligonucleotide, $dT(pT)_{15}$, to the *Escherichia coli* single-strand binding (SSB) protein, which is a stable homotetramer under the conditions of the experiment.[24,25] The reader should consult the original papers for the experimental details of the fluorescence measurements, inner filter corrections, etc., as we will focus on the analysis of the experimental titrations once they have been obtained. In this case, the SSB tetramer is the macromolecule, since it binds multiple molecules of the ligand, $[dT(pT)_{15}]$.[24,25] The intrinsic SSB tryptophan fluorescence is substantially quenched on binding the oligonucleotide, hence this signal was used to monitor the interaction. Therefore, in this case, the MBDF is the observed degree of fluorescence quenching, Q_{obs}, and the appropriate relationship [from Eq. (8)] is given in Eq. (9) (see Table I),

$$Q_{obs} = \Sigma(Q_i/i)(\nu_i) \qquad (9)$$

where Q_i/i is the average degree of quenching per bound oligonucleotide, with i oligonucleotides bound per SSB tetramer.

Figure 1 shows the results of two titrations with $dT(pT)_{15}$ (ligand), at different SSB tetramer (macromolecule) concentrations, in which the values of Q_{obs} (MBDF) have been plotted as a function of total $dT(pT)_{15}$

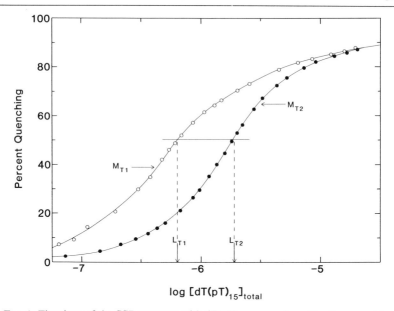

FIG. 1. Titrations of the SSB tetramer with dT(pT)$_{15}$, as monitored by the quenching of the intrinsic tryptophan fluorescence of the SSB protein in 20 mM KH$_2$PO$_4$, 0.2 M KCl (pH 7.4, 8°) at two different SSB tetramer concentrations: (○) 2.74 × 10^{-7} M and (●) 9.45 × 10^{-7} M. (Data from Bujalowski and Lohman.[24])

concentration (L_T). Also depicted in Fig. 1 is the process by which *one set* of values of $\Sigma \nu_i$ and L_F can be obtained from these data. A horizontal line intersecting both curves is drawn, defining *one* constant value of the MBDF (constant Q_{obs}). The point of intersection of this horizontal line with each titration curve defines the two sets of values of L_{T_x}, M_{T_x} (x = 1 and 2) for which L_F and $\Sigma \nu_i$ are constant, as discussed above. The example in Fig. 1 shows only two titrations, hence only two sets of values of L_{T_x}, M_{T_x} are obtained. For the analysis of n titrations, n sets of concentration pairs, L_{T_x}, M_{T_x}, are obtained (see below).

For the two titrations at macromolecular (SSB tetramer) concentrations, M_{T_1} and M_{T_2} ($M_{T_2} > M_{T_1}$), two mass conservation equations can be written in the form of Eq. (1) for each pair of concentrations, L_{T_x}, M_{T_x}. These two equations can be solved for $\Sigma \nu_i$ and L_F, with the results given in Eqs. (10a) and (10b):

$$\Sigma \nu_i = (L_{T_2} - L_{T_1})/(M_{T_1} - M_{T_2}) \qquad (10a)$$
$$L_F = L_T - \Sigma \nu_i (M_T) = (M_{T_1} L_{T_2} - M_{T_2} L_{T_1})/(M_{T_1} - M_{T_2}) \qquad (10b)$$

This process is then repeated for a series of m horizontal lines that can be

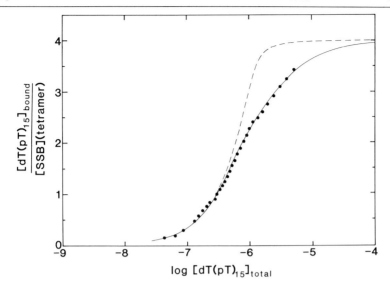

FIG. 2. The average number of dT(pT)$_{15}$ molecules bound per SSB tetramer as a function of the total concentration of dT(pT)$_{15}$, in 20 mM KH$_2$PO$_4$ (pH 7.4, 8°), 0.2 M KCl. This plot was constructed from an analysis of the two titration curves shown in Fig. 1, using Eqs. (10a) and (10b). The solid line is a theoretical curve based on the "square" model,[24] using the best-fit values of $K_{16} = 3.0 \times 10^7 M^{-1}$ and $\sigma_{16} = 0.17$. The tetramer concentration is 2.74×10^{-7} M. The dashed curve represents the theoretical isotherm for $K_{16} = 3.0 \times 10^7 M^{-1}$ and $\sigma_{16} = 1$. (Data from Bujalowski and Lohman.[24])

drawn to intersect both titration curves, each line yielding one set of values of L_F and $\Sigma \nu_i$. The analysis of the *series* of m horizontal lines will therefore yield a set of m values of L_F and $\Sigma \nu_i$, which can then be used to construct a *model-independent* binding isotherm. These data can also be used to determine the relationship between the signal change (Q_{obs}, in this case) and the average degree of ligand binding.

Figure 2 shows the model-independent binding isotherm ($\Sigma \nu_i$ vs L_T) constructed from an analysis of the two titration curves in Fig. 1, whereas Fig. 3 shows the relationship between the degree of fluorescence quenching (Q_{obs}) and the average degree of binding ($\Sigma \nu_i$). As can be seen from the deviation of the data from the dashed line in Fig. 3, Q_{obs} is *not* a linear function of the degree of binding, rather the average quenching per bound dT(pT)$_{15}$ decreases as the degree of binding increases.[24] Therefore, an incorrect apparent isotherm would have been calculated if a linear relationship had been assumed for this system. The solid line that describes the experimental isotherm in Fig. 2 is a theoretical isotherm based on a thermodynamic model that incorporates negative cooperativity among

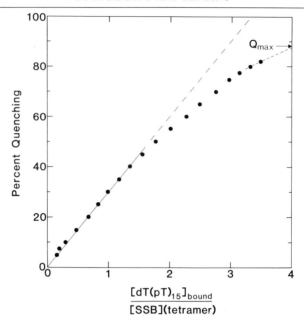

FIG. 3. The dependence of the quenching of the SSB protein fluorescence on the average number of dT(pT)$_{15}$ molecules bound per SSB tetramer in 20 mM KH$_2$PO$_4$, 0.2 M KCl (pH 7.4, 8°). The number of bound dT(pT)$_{15}$ molecules at each value of the fluorescence quenching was determined from the two titrations shown in Fig. 1, using Eqs. (10a) and (10b) in the text. The dashed line represents a linear extrapolation from the initial part of the curve. (Data from Bujalowski and Lohman.[24])

the four binding sites for dT(pT)$_{15}$ on the SSB tetramer, with intrinsic equilibrium constant $K_{16} = 3.0 (\pm 0.5) \times 10^7 \, M^{-1}$ and negative cooperativity parameter $\sigma_{16} = 0.17 \, (\pm 0.02)$.[24] The model-independent analysis of these binding data indicates that the first two molecules of dT(pT)$_{15}$ bind to two protomers of the SSB tetramer with approximately equal affinity, whereas the third and fourth molecules bind with lower affinity due to the negative cooperativity. A detailed discussion of this phenomenon and its implications for SSB binding to single-stranded (ss) DNA has been reported.[24,25]

In principle, as shown in Figs. 1 and 2, only two titration curves are needed to obtain values of $\Sigma \nu_i$ and L_F that cover the full range of binding densities, if one is able to cover the entire range of the signal change with each titration curve [this is not the case when a reverse titration is performed (see the section Signal from Ligand, below)]. This was possible for the data in Fig. 1 since the signal change (tryptophan fluorescence

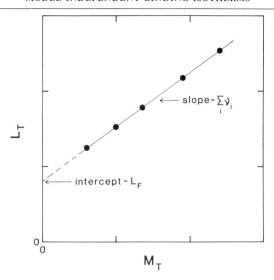

FIG. 4. An example of the graphical determination of the free ligand concentration, L_F, and the average binding density, $\Sigma \nu_i$, from a theoretical set of values (L_{T_x}, M_{T_x}) ($x = 1$ to 5) that would be determined from a binding density analysis of five titration curves.

quenching) for the SSB tetramer–dT(pT)$_{15}$ interaction is large and the affinity is fairly high under the conditions used. In this case, the data can be analyzed using Eqs. (10a) and (10b). However, more accurate values of $\Sigma \nu_i$ and L_F can be obtained if more than two titrations are performed and the data are analyzed graphically. For a case in which n titrations are performed, then n sets of values of $(L_{T_x}, M_{T_x}, x = 1$ to $n)$ will be obtained, one for each titration curve that is intersected by each horizontal line (constant MBDF). A plot of L_T vs M_T can then be constructed (see Fig. 4), which will result in a straight line, according to Eq. (1), from which the values of L_F and $\Sigma \nu_i$ can be obtained from the intercept and slope, respectively (see also Fig. 12). If a plot of L_T vs M_T determined in this manner is not linear, this may indicate one or more inconsistent sets of titration data or aggregation phenomena associated with the ligand or the macromolecule and these data should be viewed with caution.

We also note that this method can be used to analyze titrations for which one is not able to saturate the macromolecule with ligand (e.g., in cases of weak affinity or solubility limitations). This is due to the fact that the analysis does not require knowledge of the signal change at saturation. Of course, in such cases, data will be obtained only over a limited binding density range.

Signal from Ligand

In this case, some intensive property, S, of the ligand changes on binding to the macromolecule, hence the signal monitors the apparent degree of saturation of the ligand and a "reverse" titration (addition of macromolecule to a constant ligand concentration) is generally performed. A general discussion of the uses of this approach has been presented by Bujalowski and Lohman.[21]

Ligand Binding Density Function (LBDF) Expressions for Multiple Ligand Binding. Consider the equilibrium binding of a ligand [total concentration, L_T (moles per liter)] to a macromolecule [total concentration, M_T (moles per liter)] such that there can be r states of bound ligand ($i = 1$ to r), with each state possessing a different molar signal, S_i. The observed signal, S_{obs}, from the ligand solution in the presence of macromolecule has contributions from the free ligand and the ligand bound to the macromolecule in any of its r possible bound states and can be expressed by Eq. (11),

$$S_{obs} = S_F L_F + \sum_{i=1}^{r} S_i L_i \tag{11}$$

where S_F and L_F are the *molar* signal and concentration of free ligand, respectively, and S_i and L_i are the molar signal and concentration of the ligand bound in state i, respectively. Equation (11) is valid when the molar signal of each species is independent of concentration (i.e., in the absence of ligand aggregation). The concentrations of free and bound ligand are related to the total ligand concentration by conservation of mass [Eq. (12)],

$$L_T = L_F + \sum_{i=1}^{r} L_i \tag{12}$$

and furthermore

$$L_i = \nu_i M_T \tag{13}$$

where ν_i is the degree of ligand binding (moles of ligand bound per mole of macromolecule) for the ith state. Substituting Eqs. (12) and (13) into Eq. (11) yields Eq. (14):

$$S_{obs} - S_F L_T = M_T \Sigma (S_i - S_F) \nu_i \tag{14}$$

Dividing both sides by $S_F L_T$, the initial signal from the ligand before titration with macromolecule, and next multiplying by L_T/M_T yields Eq. (15),

$$[(S_{obs} - S_F L_T)/S_F L_T](L_T/M_T) = \Sigma[(S_i - S_F)/S_F]\nu_i \tag{15}$$

which can be rewritten as Eq. (16):

$$\Delta S_{obs}(L_T/M_T) = \sum_{i=1}^{r} (\Delta S)_i \nu_i \tag{16}$$

TABLE II
LIGAND BINDING DENSITY FUNCTIONS FOR DIFFERENT SPECTROSCOPIC SIGNALS

Method	Ligand binding density function (LBDF)	Comments
Absorbance	$\Delta A_{obs}(L_T/M_T) = \Sigma(\Delta\varepsilon_i)\nu_i$ where $\Delta A_{obs} = (A_{obs} - \varepsilon_F L_T)/\varepsilon_F L_T$ $\Delta\varepsilon_i = (\varepsilon_i - \varepsilon_F)/\varepsilon_F$	$\varepsilon_F L_T$ is initial absorbance of ligand solution in absence of the macromolecule; ε_F and ε_i are extinction coefficients for free ligand and each state of bound ligand
Fluorescence	$Q_{obs}(L_T M_T) = \Sigma(Q_i)\nu_i$ where $Q_{obs} = (F_{obs} - F_F L_T)/F_F L_T$ $Q_i = (F_i - F_F)/F_F$	$F_F L_T$ is initial fluorescence of ligand solution in absence of macromolecule

$\Delta S_{obs} \equiv (S_{obs} - S_F L_T)/S_F L_T$ is the experimentally observed fractional change in the signal from the ligand (normalized to the initial signal from the ligand, $S_F L_T$) in the presence of *total* ligand and *total* macromolecule concentrations, L_T and M_T, and $(\Delta S)_i = (S_i - S_F)/S_F$ is the intrinsic fractional signal change from the ligand when it is bound in state i. Note that the quantity, $S_F L_T$, is simply the initial signal from the free ligand in the absence of macromolecule, i.e., the signal at the start of the titration, before addition of macromolecule. If fluorescence of the ligand is being monitored, then ΔS_{obs} corresponds to the quenching of the ligand fluorescence. In Table II, we give the expressions for the ligand binding density functions (LBDF) for the cases in which the signal is due to fluorescence or absorbance changes; however, Eq. (16) is general and not restricted to only spectroscopic changes.

Equation (16) indicates that the quantity $\Delta S_{obs}(L_T/M_T)$ is equal to $\Sigma(\Delta S)_i\nu_i$, the sum of the binding densities for all i states of ligand binding, weighted by the intrinsic signal change for each bound state. Therefore, the quantity, $\Delta S_{obs}(L_T/M_T)$, is the ligand binding density function (LBDF). The weighting factor, $(\Delta S)_i$, is a molecular quantity that is constant for a particular binding state i, under a given set of experimental conditions. Therefore the quantity, $\Sigma(\Delta S)_i\nu_i$, and hence the LBDF, is constant for a given binding density distribution, $\Sigma\nu_i$. At equilibrium, the values of L_F and $\Sigma\nu_i$ (and each separate value of ν_i) are constant for a given value of $\Delta S_{obs}(L_T/M_T)$, independent of the macromolecule concentration, M_T. Hence, one can obtain model-independent estimates of $\Sigma\nu_i$ and L_F from plots of $\Delta S_{obs}(L_T/M_T)$ vs M_T for two or more titrations performed at different total ligand concentrations, under identical solution conditions. This is accomplished by obtaining the set of concentration pairs (L_{T_x}, M_{T_x}), from each titration, for which the binding density function is constant and solving for $\Sigma\nu_i$ and L_F, graphically as shown in Figs. 4 and 12, using the relationship given in Eq. (1). The procedure is therefore analogous to the case in which a signal from the macromolecule is monitored during the titration. The main difference is that a different formulation of the binding density function is required.

It is useful to compare Eqs. (8) and (16), which are the corresponding binding density functions for the macromolecule (MBDF) and ligand (LBDF), respectively. When the signal from the *macromolecule* is monitored, the MBDF is simply equal to ΔS_{obs}; however, when the signal from the *ligand* is monitored, the LBDF is equal to $\Delta S_{obs}(L_T/M_T)$. Therefore, the form of the BDF that is used in the analysis of binding data depends on whether the signal is from the ligand or the macromolecule. However, in both cases, the BDF is a function only of the molecular signal, which is constant for a given distribution of bound ligand, and the average binding density on the macromolecule.

Theoretical Behavior of LBDF When Data Are Collected by "Reverse" Titrations. It is informative to consider the general properties of the LBDF when experiments are performed as reverse titrations, since these differ considerably from the behavior of the MBDF when experiments are performed as normal titrations.[21] Figures 5 and 7 show the behavior of the LBDF, obtained from reverse titrations, and these should be compared to the behavior of the MBDF, obtained from normal titrations, shown in Fig. 1. Figure 5 represents a series of theoretical LBDF curves, when reverse titrations are performed, whereas Fig. 7 shows actual data for the binding of the *E. coli* SSB tetramer (ligand) to poly(U) (macromolecule), in which binding was monitored by the quenching of the intrinsic fluorescence of the SSB protein. In Fig. 5, the noncooperative, overlap model of McGhee and von Hippel[38] for large ligand binding to a linear nucleic acid [see Eq. (20) with $\omega = 1$] was used to generate the theoretical plots. The curves in Fig. 5 were generated for a single binding mode with the observed signal from the ligand with $(\Delta S)_i = 0.5$. However, the qualitative features of the LBDF plot are independent of this model and hold even for ligand binding to independent and identical sites on proteins (i.e., no overlap of potential binding sites) or if multiple binding modes or cooperative binding exists. Equation (16) indicates that the ordinate, $[\Delta S_{obs}(L_T/M_T)]$, of the plot in Fig. 5, is equal to the average binding density, ν, multiplied by $(\Delta S)_i$, hence the ordinate is a relative binding density scale. For a large ligand binding to a nucleic acid with site size, n, the maximum value of ν at saturation of the nucleic acid, ν_{max}, equals $1/n$ (0.1 in this case), hence the

[38] J. D. McGhee and P. H. von Hippel, *J. Mol. Biol.* **86**, 469 (1974); J. D. McGhee and P. H. von Hippel, *J. Mol. Biol.* **103**, 679 (1976).

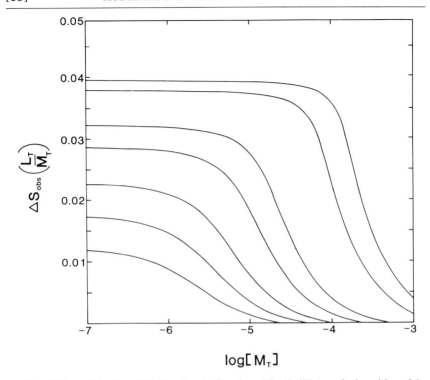

FIG. 5. Plots of the ligand binding density function, $\Delta S_{obs}(L_T/M_T)$, vs the logarithm of the total nucleic acid concentration, M_T, generated for the theoretical case of a "reverse" titration of a large ligand that binds noncooperatively to a homogeneous linear nucleic acid lattice. The noncooperative binding isotherm of McGhee and von Hippel[38] was used to generate these curves using the following parameters: ligand site size, $n = 10$ nucleotides, $K = 10^6\ M^{-1}$, and $\Delta S_{max} = 0.5$. The total ligand concentrations, L_T, are, from left to right: 5×10^{-8}, 10^{-7}, 2×10^{-7}, 5×10^{-7}, 10^{-6}, 5×10^{-6}, and $10^{-5}\ M$. (Taken from Bujalowski and Lohman.[21])

maximum possible value of the ordinate is 0.05, since we have specified $(\Delta S)_i = 0.5$ in the calculations for Fig. 5.

As is evident from a comparison of Fig. 1 with Figs. 5 and 7, one major difference between the behavior of the BDF when the data are collected by reverse titrations is that the curves for lower total ligand concentrations, L_T, do not cover the same range of binding densities as the curves for higher values of L_T. As L_T decreases, the largest value of ν attainable at that concentration of ligand also decreases, hence $\Delta S_{obs}(L_T/M_T)$, which equals $\nu(\Delta S)_i$, also decreases. This is a direct result of the use of reverse titrations and is a consequence of the fact that the binding density is solely determined by the free ligand concentration (for a given K), hence in the limit of low M_T, the value of ν is limited by the value of L_T, since L_F can never exceed L_T. As a result, a single binding density function curve obtained at one value of L_T usually cannot span the same range of binding densities as other curves obtained at different values of L_T. Only when $L_T \gg 1/K$ will the binding density function span the full range of binding densities. Hence, in order to obtain values of $\Sigma\nu_i$ and L_F that cover the full range of binding densities, a series of overlapping binding density function curves must be obtained. The behavior shown in Figs. 5 and 7 occurs even for ligands that bind with high cooperativity. We emphasize that this distinctive behavior results only when data are collected by a series of reverse titrations.

As discussed above and shown in Fig. 1, when normal titrations are performed and a signal from the macromolecule is monitored, all of the curves for different concentrations of macromolecule can span the full range of binding densities. Normal titrations of macromolecule with ligand can also be performed when the signal change is from the ligand and this may permit the full range of binding densities to be covered in each titration. However, in this case, each normal titration must be directly compared with a control titration in which buffer is titrated with ligand in order to calculate the binding density function, hence the relative accuracy of the data is generally reduced.

Data Analysis When Signal Change Is from Ligand and "Reverse" Titrations Are Performed. As an example of the use of the binding density function analysis when a signal from the ligand is monitored, we discuss a study of the interaction of *E. coli* SSB tetramer (ligand) with the single-stranded homopolynucleotide, poly(U) (macromolecule), in which the quenching of the intrinsic SSB tryptophan fluorescence was used to monitor binding.[21,22] Under the conditions of these experiments, the SSB tetramer binds to single-stranded nucleic acids at equilibrium in its $(SSB)_{65}$, binding mode, in which approximately 65 nucleotides of the nucleic acid wraps around the SSB tetramer, interacting with all 4 of the SSB protomers.[39–43] This method has also been used to analyze the binding of a series of oligopeptides containing lysine and tryptophan to single-stranded polynucleotides, with binding monitored by the fluorescence quenching of the peptide.[26]

A series of reverse titrations of *E. coli* SSB tetramer with poly(U) at

[39] T. M. Lohman and L. B. Overman, *J. Biol. Chem.* **260**, 3594 (1985).
[40] W. Bujalowski and T. M. Lohman, *Biochemistry* **25**, 7799 (1986).
[41] W. Bujalowski, L. B. Overman, and T. M. Lohman, *J. Biol. Chem.* **263**, 4629 (1988).
[42] S. Chrysogelos and J. Griffith, *Proc. Natl. Acad. Sci. U.S.A.* **79**, 5803 (1982).
[43] J. D. Griffith, L. D. Harris, and J. Register, *Cold Spring Harbor Symp. Quant. Biol.* **49**, 553 (1984).

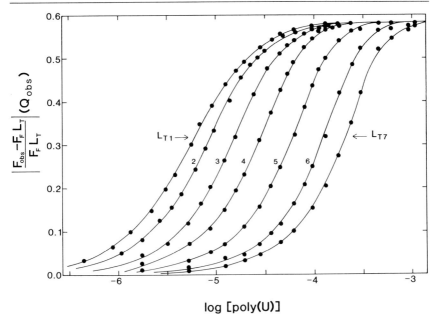

log [poly(U)]

FIG. 6. "Reverse" titrations (addition of nucleic acid to protein) of *E. coli* SSB protein with poly(U) at different SSB concentrations in 10 mM Tris (pH 8.1), 0.2 M NaCl, 25.0°. The extent of SSB fluorescence quenching is plotted vs the logarithm of the poly(U) nucleotide concentration. The SSB tetramer concentrations for each curve are (1) $5.69 \times 10^{-8}\ M$; (2) $1.22 \times 10^{-7}\ M$; (3) $2.44 \times 10^{-7}\ M$; (4) $5.28 \times 10^{-7}\ M$; (5) $1.30 \times 10^{-6}\ M$; (6) 2.84×10^{-6} M; (7) $5.20 \times 10^{-6}\ M$. (Data from Bujalowski and Lohman.[21])

different SSB tetramer concentrations is shown in Fig. 6. These data were used to construct the LBDF [$Q_{obs}(L_T/M_T)$] and the plot of $Q_{obs}(L_T/M_T)$ vs log[M_T] is shown in Fig. 7. As mentioned above, even though all of the reverse titrations shown in Fig. 5 span the same full range of SSB fluorescence quenching, not all of them span the full range of binding densities as seen from the LBDF plot in Fig. 7. As a result, multiple titrations at different values of L_T are required to obtain data over the full binding density range. At this point, the set of LBDF curves in Fig. 7 is analyzed in a similar manner to the analysis of the MBDF curves discussed above (see Fig. 1). A series of horizontal lines that intersect the LBDF curves are drawn, each line defining a constant value of the LBDF, $Q_{obs}(L_T/M_T)$. *Each* horizontal line will yield a value of L_F and its associated value of $\Sigma\nu_i$. The following describes the analysis for one such horizontal line. The points of intersection of one horizontal line with each binding density function curve determines the set of concentration pairs (M_{T_x},

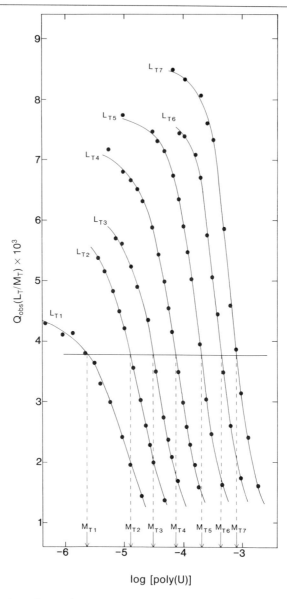

FIG. 7. The dependence of the ligand binding density function, $Q_{obs}(L_T/M_T)$ on the logarithm of the poly(U) concentration at different SSB protein concentrations in 10 mM Tris (pH 8.1) plus 0.2 M NaCl, 25.0°. The points of intersection with the separate curves of any horizontal line drawn at a constant value of the binding density function [e.g., $Q_{obs}(L_T/M_T) = 3.8 \times 10^{-3}$ in the figure], determine the set of total protein and macromolecule concentrations, L_{T_x}, M_{T_x}, for which the average binding density, $\Sigma \nu_i$ (and free protein

L_{T_x}) ($x = 1$ to 7) for which L_F and $\Sigma \nu_i$ are constant, as shown in Fig. 7 for one constant value of the LBDF. Based on Eq. (1), the average binding density, $\Sigma \nu_i$, and L_F, can then be determined, graphically, from the slope and intercept of a plot of L_T vs M_T. By repeating this procedure for a series of horizontal lines that span the range of values of $Q_{obs}(L_T/M_T)$, values of $\Sigma \nu_i$ as a function of L_F can be obtained and a model-independent binding isotherm can be constructed. A series of plots of L_T vs M_T, each generated from a different value of $Q_{obs}(L_T/M_T)$ from Fig. 7, is shown in Fig. 8.

Theoretically it is enough to analyze only two different protein concentrations to obtain two equations of the form shown in Eq. (1) (for L_{T_1}, M_{T_1} and L_{T_2}, M_{T_2}) from which the values of $\Sigma \nu_i$ and L_F can be determined using Eqs. (10a) and (10b). However, as mentioned above, since not all of the LBDF curves fully overlap, only a subset of them will intersect each horizontal line. Therefore, not all of the curves can be used to determine each value of $\Sigma \nu_i$ and L_F, especially at the higher binding densities. In addition, significantly more accurate determinations of L_F and $\Sigma \nu_i$ are obtained when titrations at several ligand concentrations (L_T) are performed. In the construction of a series of LBDF plots as shown in Fig. 7, one should cover as wide a range of ligand concentrations as possible; however, care should be taken to avoid large changes in total ligand concentration between two successive titrations, since this may bias the determination of L_F and $\Sigma \nu_i$. In our experience, six to eight titrations using successive total ligand concentrations that differ by a factor of two will generate an accurate set of data.

The model-independent binding isotherm constructed from the full analysis of the data in Fig. 7 for the binding of the *E. coli* SSB tetramer to poly(U) in 0.2 *M* NaCl, is plotted as a Scatchard plot[44] in Fig. 9. One can obtain intrinsic binding parameters by comparing this binding isotherm to theoretical isotherms generated by various models. The solid line in Fig. 9 is the best-fit theoretical isotherm, based on a model in which *E. coli* SSB tetramers bind to single-stranded nucleic acids with "limited" cooperativity to form a mixture of tetramers and octamers (dimers of tetramers)[22] [see Eq. (21)]. The best-fit theoretical isotherm for an "unlimited" cooperativity model[38] [see Eq. (20)] (dashed curve in Fig. 9), which as-

---

[44] G. Scatchard, *Ann. N.Y. Acad. Sci.* **51**, 660 (1949).

---

concentration, L_F) is constant. The SSB tetramer concentrations for each curve are as follow: (L_{T_1}) 5.69 × 10^{-8} *M*; (L_{T_2}) 1.22 × 10^{-7} *M*; (L_{T_3}) 2.44 × 10^{-7} *M*; (L_{T_4}) 5.28 × 10^{-7} *M*; (L_{T_5}) 1.30 × 10^{-6} *M*; (L_{T_6}) 2.84 × 10^{-6} *M*; (L_{T_7}) 5.20 × 10^{-6} *M*. (Data from Bujalowski and Lohman.[21])

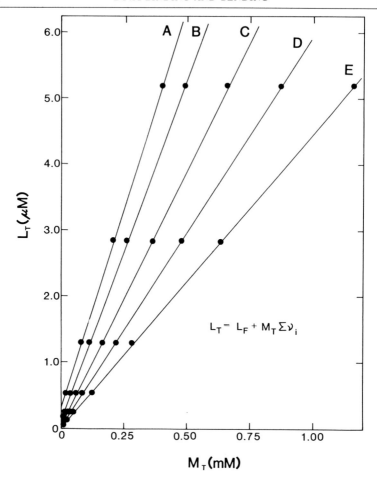

FIG. 8. Examples of the determination of the free SSB tetramer concentration (L_F) and the average binding density, $\Sigma\nu_i$ (SSB tetramers bound per nucleotide) from a plot of the total SSB tetramer concentration (L_T) vs the total poly(U) concentration (M_T) (nucleotides). Each straight line yields one set of values of L_F and $\Sigma\nu_i$ from the intercept and slope, respectively [see Eq. (1) in text]. The values of L_T and M_T for the five plots (A–E) shown were obtained from analysis of the ligand binding density function (LBDF) plots shown in Fig. 7. The values of the LBDF [$Q_{obs}(L_T/M_T)$] for each plot are (A) 6.4×10^{-3}; (B) 5.6×10^{-3}; (C) 4.4×10^{-3}; (D) 3.4×10^{-3}; (E) 2.6×10^{-3}. (Data from Bujalowski and Lohman.[21])

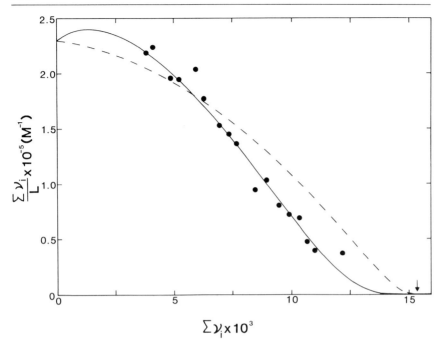

FIG. 9. Scatchard plot for *E. coli* SSB tetramer binding to poly(U) in its $(SSB)_{65}$ binding mode in 10 m*M* Tris (pH 8.1), 25°. The plot was constructed from data that were determined using the general method of analysis of all of the binding density function curves in Fig. 7. The solid line represents the best-fit isotherm for a model that describes the limited cooperative binding of SSB to single-stranded nucleic acids in its $(SSB)_{65}$ binding model (Bujalowski and Lohman[22]), with binding parameters $K = 1.15 \times 10^5 \, M^{-1}$; $\omega_{lim} = 420$, and $n = 65$ nucleotides per SSB tetramer. The arrow indicates the point of full saturation, $\nu = 1/n$. (Data from Bujalowski and Lohman.[22])

sumes that SSB tetramers can form a distribution of clusters, limited only by the length of the nucleic acid, does not describe the isotherm nearly as well.[22]

Correlation between Signal Change, ΔS_{obs}, and Average Binding Density. Measurements of the average binding density, $\Sigma \nu_i$, as a function of the free ligand concentration enable one to determine the relationship between the average signal change and the fraction of bound ligand. This is not necessary in order to obtain a binding isotherm as discussed above; however, if ΔS_{obs} is found to be directly proportional to the fraction of bound ligand (L_B/L_T), then binding isotherms can be constructed with much greater ease from a titration at a single ligand concentration (see below). Figure 10 shows the relationship between Q_{obs} and L_B/L_T [=

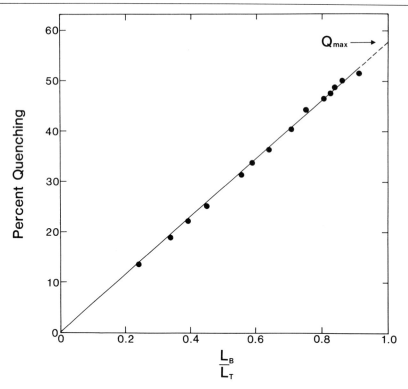

FIG. 10. The relation between the fluorescence quenching, Q_{obs}, and the fraction of protein bound L_B/L_T for SSB tetramer binding to poly(U) in 10 mM Tris (pH 8.1), 0.2 M NaCl, 25°; [SSB] = 2.44 × 10^{-7} M (tetramer) (curve L_{T_3} in Figs. 6 and 7). The concentration of bound protein, L_B, was calculated from $L_B = M_T(\Sigma\nu_i)$, where the average binding density, $\Sigma\nu_i$, was determined using the LBDF binding analysis method described in the text. The quenching of SSB fluorescence is observed to be directly proportional to the fraction of bound SSB, L_B/L_T. Therefore, for this case, $Q_{obs}/Q_{max} = L_B/L_T$. A linear extrapolation of the data to $L_B/L_T = 1$ (fully bound protein) yields a value of 57.7% for Q_{max}, under these solution conditions. (Data from Bujalowski and Lohman.[21])

$(\Sigma\nu_i)M_T/L_T]$, for the E. coli SSB tetramer–poly(U) data shown in Figs. 6–9. Under these conditions, Q_{obs} is directly proportional to the fraction of bound SSB tetramer over the range of L_B/L_T that was investigated. However, it is important to check the relationship between quenching and L_B/L_T over a wide range of binding densities, since the signal change can be dependent on binding density. In the case of E. coli SSB tetramer–single-stranded nucleic acid interactions in NaCl concentrations ≥0.2 M (pH 8.1), this direct proportionality holds for binding densities up to at least 0.012 (~78% saturation of the nucleic acid[21]). Therefore, under these

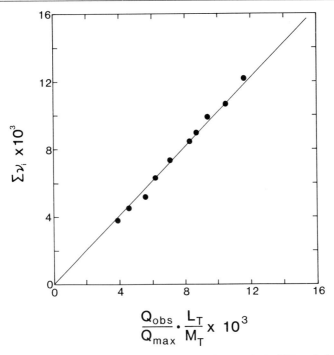

FIG. 11. The average binding density determined from a single SSB–poly(U) titration curve, $\nu_{app} = (Q_{obs}/Q_{max})(L_T/M_T)$ correlates well with the average binding density, $\Sigma\nu_i$, determined from the LBDF method of analysis. The slope of the line is 1.02. Data from curve 3 in Figs. 6 and 7 and a value of $Q_{max} = 57.7\%$ were used for the determination of ν_{app}. (Data from Bujalowski and Lohman.[21])

conditions, the fractional fluorescence quenching of the SSB tetramer is equal to the fraction of bound SSB, i.e., $Q_{obs}/Q_{max} = L_B/L_T$. If it can be demonstrated that Q_{obs} is directly proportional to L_B/L_T, one can determine the maximum extent of protein fluorescence quenching, Q_{max}, from a linear extrapolation of a plot of Q_{obs} vs L_B/L_T to $L_B/L_T = 1$ as shown in Fig. 10.

For a protein–nucleic acid interaction, it is possible that the relationship, $Q_{obs}/Q_{max} = L_B/L_T$ is valid only over a limited range of binding densities; hence one could miscalculate the extent of binding by assuming that a direct proportionality exists for all binding densities. A useful check of this relationship can be made over a wide range of binding densities by plotting $(Q_{obs}/Q_{max})(L_T/M_T)$ vs the true value of $\Sigma\nu_i$ obtained from the LBDF analysis. $(Q_{obs}/Q_{max})(L_T/M_T)$ is an *apparent* binding density, based on the assumption that $Q_{obs}/Q_{max} = L_B/L_T$. This type of plot is shown in Fig. 11 for the *E. coli* SSB–poly(U) data presented in Fig. 6. In this case

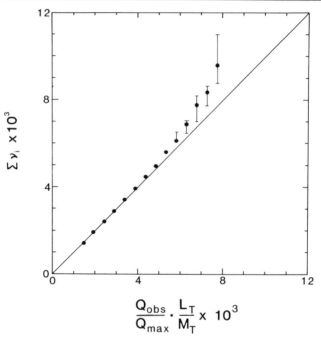

FIG. 12. Demonstration that Q_{obs}/Q_{max} is *not* equal to L_B/L_T for the SSB tetramer binding to poly(dT) in 2.0 M NaBr (10 mM Tris, pH 8.1, 25°). Data from five reverse titrations were analyzed using the LBDF method of analysis.[21] The apparent average binding density ν_{app} = $(Q_{obs}/Q_{max})(L_T/M_T)$, calculated assuming L_B/L_T = Q_{obs}/Q_{max}, with Q_{max} = 0.825, deviates from the true average binding density, $\Sigma\nu_i$, determined by the binding density function analysis. The line shows the expected relationship if Q_{obs}/Q_{max} = L_B/L_T. (Data from Overman *et al.*[23])

(0.2 M NaCl, pH 8.1, 25°), there is a direct correspondence between these two values over the range from 25 to 80% coverage of the nucleic acid, hence it can safely be concluded that Q_{obs}/Q_{max} = L_B/L_T in this range.

On the other hand, it would be unwise to conclude that Q_{obs}/Q_{max} = L_B/L_T for all solution conditions, based on an analysis under only one set of solution conditions (Fig. 11). In fact, we have found that Q_{obs} is *not* directly proportional to L_B/L_T for SSB–single-stranded nucleic acid binding in NaBr[23] as shown in Fig. 12. We have also observed a nonlinear relationship between Q_{obs} and L_B/L_T for the SSB–poly(C) interaction in 15 mM NaCl (T. Lohman, unpublished, 1991). The nonlinearity in the latter case reflects the binding density-dependent shift in SSB binding modes that has been previously observed,[41,43] since the $(SSB)_{35}$ mode exhibits a lower extent of quenching than either the $(SSB)_{56}$ or the $(SSB)_{65}$ binding modes.[29,39,40]

Generation of Binding Isotherms from Single Titration when $\Delta S_{obs}/$
$\Delta S_{max} = L_B/L_T$. The LBDF analysis allows one to rigorously determine a
model-independent binding isotherm and the relationship between the
ΔS_{obs} and the fraction of bound ligand, L_B/L_T. The LBDF method is time
consuming since six to eight titrations are required to construct a single
binding isotherm with good precision over a wide range of binding densities, although this is necessary if the relationship between ΔS_{obs} and L_B/L_T
is not known *a priori*. However, if it is determined from the LBDF analysis
that a linear relationship exists between ΔS_{obs} and L_B/L_T over a wide range
of binding densities, then one can use this relationship to determine the
average binding density and free ligand concentration from a single titration curve.[21] For this simple case, Eq. (16) reduces to Eq. (17),

$$\Delta S_{obs}/\Delta S_{max} = L_B/L_T \tag{17}$$

and it follows that

$$L_F = (1 - \Delta S_{obs}/\Delta S_{max})L_T \tag{18}$$

and

$$\nu = (\Delta S_{obs}/\Delta S_{max})(L_T/M_T) \tag{19}$$

Thus, a single titration can be used to obtain ν as a function of L_F.
However, we stress that one *should not assume* that the fractional signal
change is equal to the fraction of bound ligand, since this can lead to
significant errors if it is not true. On the other hand, if a direct proportionality does not exist between the signal change and the fraction of bound
ligand over a wide range of binding densities, the true binding isotherm
can still be constructed without any assumptions using the LBDF analysis.

Experimental Aspects

Since the methods of analysis reviewed here are applicable to a variety
of spectroscopic (and nonspectroscopic) approaches, we will not review
the details of data collection for each possible approach. However, there
are some general experimental details that we can address concerning the
collection of a set of titration curves to be used in a BDF analysis. We
discuss the case of an LBDF analysis obtained by reverse titrations as a
specific example. If the titrations require the addition of large volumes of
titrant, then one must be careful to account for the dilution of the total
ligand concentration, L_T, which will occur throughout the titration; i.e.,
the value of L_T will vary along the LBDF curve. Serious systematic
errors may result if this dilution is not considered in the calculation.
Alternatively, one can preadd ligand to the titrant (macromolecule) solu-

tion, at concentration L_T, in order to maintain the same total ligand concentration, L_T, throughout each titration. In this manner, no change in L_T occurs and the graphical analysis (see Fig. 7) is straightforward. The same procedure of "doping" the ligand titrant with macromolecule can be used to avoid changes in the total macromolecule concentration, M_T, throughout a normal titration (see Fig. 1).

Overlapping Binding Site Models for Nonspecific Binding of Protein to a Linear Nucleic Acid

The discussions presented above have focused on the generation of true, model-independent binding isotherms (i.e., the relationship between the average binding density and the free protein concentration). However, once a true binding isotherm has been obtained by a rigorous method, the next step is the analysis of this isotherm to obtain either model-independent macroscopic binding parameters or to use a statistical thermodynamic model to obtain intrinsic binding parameters. For site-specific binding of proteins to regulatory sites, this is a relatively straightforward process, although complications such as cooperativity among multiple binding sites within an operon may also need to be considered.[3,45] However, for nonspecific binding of proteins to a linear nucleic acid lattice, the analysis is more complex, since overlap of potential protein-binding sites must be considered.[38,46–49] Closed form expressions for the isotherms for noncooperative and unlimited nearest neighbor cooperativity have been derived for an infinite lattice[38,48] and for a finite lattice.[50] For the noncooperative case the binding is characterized by two parameters, the intrinsic equilibrium binding constant, K, and the site size, n, which is the number of nucleotides or base pairs occluded by the bound ligand. The unlimited cooperative case requires an additional parameter, ω, the nearest neighbor cooperativity parameter, which characterizes the interactions between two nearest neighbor proteins on the nucleic acid lattice.[38,47] In the original formulation of the unlimited cooperativity model, separate equations were obtained for the noncooperative vs cooperative cases[38]; however, a single equation for the binding isotherm, which is applicable for both cooperative and noncooperative binding has been reported.[51] The generalized isotherm

[45] M. Brenowitz, D. F. Senear, M. A. Shea, and G. K. Ackers, *Proc. Natl. Acad. Sci. U.S.A.* **83**, 8462 (1986).

[46] D. M. Crothers, *Biopolymers* **6**, 575 (1968).

[47] J. A. Schellman, *Isr. J. Chem.* **12**, 219 (1974).

[48] A. S. Zasedatelev, G. V. Gurskii, and M. V. Volkenshtein, *Mol. Biol.* **5**, 245 (1971).

[49] T. Tsuchiya and A. Szabo, *Biopolymers* **21**, 979 (1982).

[50] I. R. Epstein, *Biophys. Chem.* **8**, 327 (1978).

[51] W. Bujalowski, T. M. Lohman, and C. F. Anderson, *Biopolymers* **28**, 1637 (1989).

for the unlimited cooperativity model (applicable for all $\omega > 0$) is given in Eq. (20), in the Scatchard form:

$$\nu/L_F = K(1 - n\nu)\{[2\omega(1 - n\nu)]/[(2\omega - 1)(1 - n\nu) + \nu + R]\}^{n-1}$$
$$\times \{[1 - (n + 1)\nu + R]/[2(1 - n\nu)]\}^2 \qquad (20)$$

where

$$R = [((1 - (n + 1)\nu)^2 + 4\omega\nu(1 - n\nu))]^{1/2}$$

The analysis of experimental binding isotherms using this overlap binding model with nearest neighbor unlimited cooperativity has been reviewed.[52] An alternative overlap model considers a limited nearest neighbor cooperativity between bound proteins (i.e., the formation of clusters of dimers). The closed form expression for this limited cooperativity model[22] is given in Eq. (21):

$$\nu/L_F = K[q^2 - 2\nu q + (1 - \omega_{lim})\nu^2]^n/q^{(2n-1)} \qquad (21)$$

where $q = [1 - (n - 1)\nu] + \{[1 - (n - 1)\nu]^2 - \nu(1 - \omega_{lim})[2 - (2n - 1)\nu]\}^{1/2}$. This limited cooperativity model has been used to analyze the equilibrium cooperative binding of the *E. coli* SSB tetramer to long, single-stranded DNA in its $(SSB)_{65}$ binding mode to obtain the intrinsic equilibrium constant, K, and the limited cooperativity parameter, ω_{lim}, for the formation of dimers of SSB tetramers on the DNA[22] (see Fig. 9).

We emphasize that in all cases the *essential* starting point before embarking on such model-dependent analyses is a model-independent equilibrium binding isotherm. Only if this is available will the application of a specific statistical thermodynamic model, based on the known molecular aspects of the protein–nucleic acid system under study, provide meaningful interaction parameters.

Pitfalls in Analysis of Binding Data Obtained by Monitoring Spectroscopic Signals

Problems When the Assumption that $\Delta S_{obs}/\Delta S_{max} = L_B/L_T$ Is Not Valid: One Ligand Can Bind to a Nucleic Acid Lattice in Two Modes, Each Mode Possessing a Different Signal

We now discuss a theoretical case in order to illustrate the problems that one can encounter if a linear relationship between the fractional signal change and the fraction of bound ligand is incorrectly assumed to hold. We consider the nonspecific, noncooperative binding of a single protein to

[52] S. C. Kowalczykowski, L. S. Paul, N. Lonberg, J. W. Newport, J. A. McSwiggen, and P. H. von Hippel, *Biochemistry* **25**, 1226 (1986).

a linear nucleic acid (modeled as an infinite one-dimensional homogeneous lattice). However, the protein is able to bind in two different modes, each characterized by different site sizes (nucleotides covered per ligand), $n = 5$ and $k = 10$. Although the case treated here is hypothetical, there are several examples of proteins that behave in this manner under some conditions, e.g., the *E. coli* SSB protein[39,41,53] and the fd gene 5 protein.[54]

The theoretical treatment of this equilibrium binding problem requires a three-state lattice model. The treatment of the three-state lattice problem has been discussed,[51] using the sequence generating function approach.[55,56] The general secular equation for the three-state lattice model, in the absence of cooperativity, but considering overlap of potential binding sites, is given in Eq. (22):

$$x^{(n+k+1)} - x^{(n+k)} - K_2 L x^{(n+1)} - K_1 L x^{(k+1)} = 0 \tag{22}$$

where K_1 and K_2 are the intrinsic binding constants for the ligand in the first ($n = 5$) and second ($k = 10$) modes, respectively, and L is the free ligand concentration. The binding densities (ligand bound per nucleotide) in the two modes, ν_1 and ν_2, are then given in Eqs. (23) and (24):

$$\begin{aligned}\nu_1 &= \partial \ln x_1/\partial \ln(K_1 L) \\ &= K_1 L/[(n+k+1)x_1^{(n+k+1)} - (n+k)x_1^{(n+k)} - (n+1)K_2 L x_1^{(n+1)} \\ &\quad - (k+1)x_1^{(k+1)}K_1 L] \end{aligned} \tag{23}$$

$$\begin{aligned}\nu_2 &= \partial \ln x_1/\partial \ln(K_2 L) \\ &= K_2 L/[(n+k+1)x_1^{(n+k+1)} - (n+k)x_1^{(n+k)} - (n+1)K_2 L x_1^{(n+1)} \\ &\quad - (k+1)x_1^{(k+1)}K_1 L] \end{aligned} \tag{24}$$

where x_1 is the largest root of the secular equation.[51] Given values for n, k, K_1, K_2, and L, then ν_1 and ν_2 can be obtained as a function of the free ligand concentration, L.

We consider the hypothetical case in which the fluorescence of the ligand is partially quenched on binding to the nucleic acid and this signal is used to monitor binding. In this case, a reverse titration would be performed. Then, in the presence of the nucleic acid, Eq. (25) relates the observed quenching, Q_{obs}, to the average binding densities in the two modes.

$$Q_{obs}(L_T/M_T) = Q_1 \nu_1 + Q_2 \nu_2 \tag{25}$$

where Q_1 and Q_2 are the molecular fluorescence quenchings of the ligand

[53] T. M. Lohman, W. Bujalowski, and L. B. Overman, *Trends Biochem.* **13**, 250 (1988).
[54] J. W. Kansy, B. A. Clack, and D. M. Gray, *J. Biomol. Struct. Dyn.* **3**, 1079 (1986).
[55] S. Lifson, *J. Chem. Phys.* **40**, 3705 (1964).
[56] D. L. Bradley and S. Lifson, in "Molecular Associations in Biology" (B. Pullman, ed.), p. 261. Academic Press, New York, 1964.

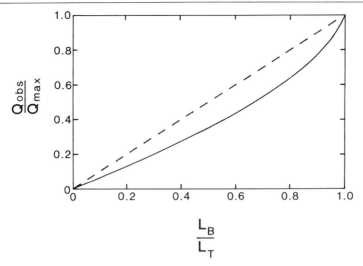

FIG. 13. The relationship between the fractional fluorescence quenching, Q_{obs}/Q_{max}, and the fraction of bound ligand, L_B/L_T, for the hypothetical case of a protein binding to a linear nuclear acid in two binding modes, possessing different site sizes (n and k) and degrees of fluorescence quenching (Q_1 and Q_2). For this case, the following parameters were used: mode 1, $n = 5$ nucleotides, $K_1 = 10^6\,M^{-1}$, $Q_1 = 0.2$; mode 2, $k = 10$ nucleotides, $K_2 = 3 \times 10^6$ M^{-1}, $Q_2 = 0.9$. The dashed line represents a direct proportionality between Q_{obs}/Q_{max} and L_B/L_T.

when bound in the two modes. Furthermore, the total ligand concentration, L_T, is given in Eq. (26):

$$L_T = L_F + (\nu_1 + \nu_2)M_T \qquad (26)$$

For the case in which $n = 5$ nucleotides, $K_1 = 1 \times 10^6\,M^{-1}$, $Q_1 = 0.2$, $k = 10$ nucleotides, $K_2 = 3 \times 10^6\,M^{-1}$, and $Q_2 = 0.9$, Fig. 13 indicates that the fractional change in ligand fluorescence quenching, Q_{obs}/Q_{max}, is *not* a linear function of the fraction of bound ligand, L_B/L_T. In Fig. 14 we have plotted the actual total binding density, $\Sigma\nu_i$ ($= \nu_1 + \nu_2$) vs the "apparent" binding density that is calculated if it is assumed that $Q_{obs}/Q_{max} = L_B/L_T$, i.e., "$\nu_{app}$" $= (Q_{obs}/Q_{max})(L_T/M_T)$. At low values of the actual binding density ($\Sigma\nu_i$), the apparent binding density is lower than the actual binding density; however, as the actual binding density increases, the apparent binding density decreases further due to a shift in the distribution of bound ligand toward the mode possessing the lower site size ($n = 5$), which also has a lower fluorescence quenching. It is clear for this case that the apparent binding isotherm, calculated by assuming a direct proportionality between Q_{obs}/Q_{max} and L_B/L_T, would be incorrect and

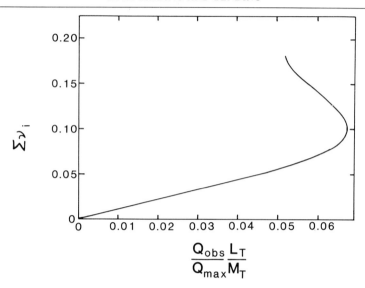

FIG. 14. The relationship of the actual average binding density, $\Sigma \nu_i$ (protein bound per nucleotide) to the apparent binding density, $\nu_{app} = (Q_{obs}/Q_{max})(L_T/M_T)$, calculated by assuming that $Q_{obs}/Q_{max} = L_B/L_T$. These theoretical data are based on a hypothetical model in which a protein can bind to a linear nucleic acid lattice in two binding modes, possessing different site sizes ($n = 5$ and $k = 10$ nucleotides), different extents of fluorescence quenching ($Q_1 = 0.2$, $Q_2 = 0.9$), and different binding constants ($K_1 = 10^6 \ M^{-1}$, $K_2 = 3 \times 10^6 \ M^{-1}$).

quite different than the true isotherm. Therefore, the analysis of this type of system would require the use of the binding density function analysis.

Potential Problems with the Use of Double-Reciprocal Plots to Determine Binding Parameters and Spectral Characteristics of Bound Ligand

A common procedure used to analyze titrations, especially those monitored by spectroscopic techniques, is to obtain the signal for fully bound ligand by performing experiments as a function of excess macromolecule concentration and plotting the signal vs $1/M_T$ and extrapolating to $1/M_T = 0$.[57] Correctly applied, the extrapolation is carried out in the range of low binding density and, as a result, this method will yield the maximal signal change *only* in the limit of low binding density. However, the signal change for the bound ligand at low binding density may not be the same as that for the bound ligand at intermediate or high binding densities, especially for a ligand that binds in a cooperative manner (i.e., there may

[57] H. H. Benesi and J. H. Hildebrand, *J. Am. Chem. Soc.* **71**, 2703 (1949).

be different signals associated with different binding modes that dominate at different binding densities). Therefore, the use of this approach to construct apparent binding isotherms always has these associated limitations and assumptions.

Detection of Binding Modes without Associated Signal Change

It is possible, for a complex macromolecular binding system, that not all of the ligand–macromolecule binding modes have an associated change in the physicochemical signal being monitored. In this case, the question arises as to whether these binding modes will be detected on monitoring that physicochemical signal. In fact, a true binding isotherm that incorporates all existing binding modes can still be obtained as long as at least one of the binding modes has an associated signal change. This results from the fact that the different binding modes are all thermodynamically coupled through the free ligand concentration, L_F, as shown in Eqs. (27) and (28):

$$L_F = L_T - (\Sigma \nu_i)M_T \qquad (27)$$
$$M_T = M_F + \Sigma M_i \qquad (28)$$

Therefore, even if there is no signal change on formation of a particular binding mode, the ligands or macromolecules bound in this "invisible" mode will still be detected since they affect L_F through mass conservation [Eq. (27)], hence the free ligand concentration, L_F, is linked to all binding modes, independent of whether they are accompanied by a signal change.

Analysis of Systems in Which Different Binding Modes Yield Signal Changes of Opposite Sign

An unlikely, although possible, situation is one in which different binding modes might have signal changes of opposite sign (i.e., fluorescence quenching vs enhancement). In such a case, a maximum or minimum in the observed signal would occur as a function of binding density and in the binding density function itself (i.e., $\partial S/\partial \nu = 0$). In such a case, the thermodynamic methods of analysis reviewed here can still be used to analyze a series of BDF curves; however, the analysis must be restricted to regions of each curve that have the same sign (e.g., $\partial S/\partial \nu > 0$). The analysis cannot be performed in the regions where $\partial S/\partial \nu = 0$, since no net signal change accompanies binding, although information can be obtained on either side of these regions.

Summary

The measurement of equilibrium binding constants for ligand–macromolecule interactions by monitoring a change in some spectral property of the ligand or the macromolecule is a common method used to study these interactions. This is due to the high sensitivity of the spectroscopic methods and general ease in applying these experimental procedures. In addition, binding can be monitored continuously, thus facilitating kinetic measurements. The main problem with these methods results from the fact that the spectroscopic signal is an indirect measure of binding, since the relationship between the change in the spectroscopic signal and the extent of binding is unknown, *a priori*. A common recourse is to assume a strict proportionality between the signal change and the fractional saturation of the ligand or macromolecule; however, it is often the case that such a direct proportionality does not hold. In this chapter we have reviewed the use of methods to analyze ligand–macromolecule equilibrium titrations that are monitored by indirect spectroscopic techniques. These methods of analysis yield thermodynamically rigorous, model-independent binding isotherms, hence assumptions concerning the relationship between the signal change and the extent of binding are not required. In fact, these methods can also be used to determine quantitatively the relationship between the signal change and the average degree of binding. In addition, the approaches discussed here are general and not limited to spectroscopic signals and therefore can be used with any intensive physicochemical property that reflects binding.

Acknowledgments

We thank Dave Mascotti, Nick Pace, Dave Giedroc, and Don Pettigrew for comments on the manuscript and Lisa Lohman for help in preparing the figures. The preparation of this chapter was supported by the National Institutes of Health (GM30498, GM39062), the Robert A. Welch Foundation (A-898), and the Texas Agricultural Experiment Station. T.M.L. is the recipient of an American Cancer Society Faculty Research Award (FRA-303).

[16] Analysis of Equilibrium and Kinetic Measurements to
Determine Thermodynamic Origins of Stability and
Specificity and Mechanism of Formation of Site-Specific
Complexes between Proteins and Helical DNA

By M. Thomas Record, Jr., Jeung-Hoi Ha,
and Matthew A. Fisher

Introduction

One of the central goals of molecular biology is to elucidate the basis
of the control of gene expression at the level of molecular interactions.
This requires understanding the origins of specificity and stability of protein–nucleic acid complexes. Specificity and stability are determined not
only by the sequences and functional groups on the macromolecular species but also by the physical environment (e.g., temperature, pH, and salt
concentration). Structural and equilibrium binding studies have been used
to probe the molecular and thermodynamic origins of stability and specificity. In some systems, specificity is kinetically determined, and mechanistic studies are required to understand its basis. In this chapter, thermodynamic and mechanistic conclusions derived from equilibrium and kinetic
studies on *Escherichia coli lac* repressor and RNA polymerase ($E\sigma^{70}$) are
used as examples of these analyses.

Harrison and Aggarwal[1] have reviewed the structural conclusions from
X-ray crystallographic investigations of complexes between DNA oligomers and site-specific binding proteins (434 repressor, 434 Cro, λ*cI* repressor, *E. coli trp* repressor), where the binding subdomain on the protein
contains the helix–turn–helix (HTH) motif. Sequence homology and NMR
data (for the *lac* repressor headpiece binding to an operator sequence)
suggest that the HTH motif is also used by *lac* repressor[2–5] and *E. coli*

[1] S. C. Harrison and A. K. Aggarwal, *Annu. Rev. Biochem.* **59**, 933 (1990).
[2] R. Kaptein, E. R. P. Zuiderweg, R. M. Scheek, R. Boelens, and W. F. van Gunsteren, *J. Mol. Biol.* **182**, 179 (1985).
[3] R. Boelens, R. M. Scheek, J. H. van Boom, and R. Kaptein, *J. Mol. Biol.* **193**, 213 (1987).
[4] R. Boelens, R. M. Scheek, R. M. J. N. Lamerichs, J. de Vlieg, J. H. van Boom, and R. Kaptein, *in* "DNA Ligand Interactions" (W. Guschlbauer and W. Saenger, eds.), p. 191. Plenum, New York, 1987.
[5] R. M. J. N. Lamerichs, R. Boelens, G. A. van der Marel, J. H. van Boom, R. Kaptein, F. Buck, B. Fera, and H. Rüterjans, *Biochemistry* **28**, 2985 (1989).

RNA polymerase[6,7] in site-specific binding. At the molecular level, the specificity of protein–DNA interactions may involve both (1) direct read-out of the DNA sequence, defined as the contribution to specific recognition from interactions between functional groups on the protein and functional groups on the base pairs that are accessible in the grooves of the DNA helix; and (2) indirect readout, defined as the contribution to specific recognition from the effects of nucleotide sequence on the conformation and physical properties of the specific DNA site.[1]

Direct readout is generally assumed to be the primary determinant of specificity. In the structures that have been solved to high resolution, direct readout is accomplished by hydrogen bonds and nonpolar contacts between amino acid side chains and functional groups on the base pairs that are accessible in the major groove. Factors that may contribute to indirect readout include sequence-dependent variations in the conformation of the sugar phosphate backbone, the twist angle of the base pairs, or the width of grooves, and/or differences in physical properties of DNA, including lateral and torsional stiffness and degree of static curvature.

Harrison and Aggarwal[1] point out that a recurring theme in these structures is the complementarity that exists between the proteins and both the sugar phosphate backbone and the base pairs of the DNA site. Record and collaborators[8–10] provide an analogous discussion from a thermodynamic perspective. To obtain a tightly bound complex, complementarity must exist at both the steric level and at the level of specific functional groups. Such complementarity is a hallmark of all noncovalent complexes, including those of enzymes with substrates and especially with transition states.[11] [The transition state stabilization theory of enzyme catalysis states that the surface of the enzyme catalytic site and of the substrate are sufficiently complementary to form a weakly bound enzyme–substrate complex, and that the chemical and physical changes in the substrate that convert it to the transition state (activated complex) on the path to product are driven in large part by the much greater degree of complementarity between the transition state and the enzyme catalytic site.[11]] In order to understand the origins of specificity and stability of

[6] M. Gribskov and R. R. Burgess, *Nucleic Acids Res.* **14,** 6745 (1986).

[7] D. A. Siegele, J. C. Hu, W. A. Walter, and C. A. Gross, *J. Mol. Biol.* **206,** 591 (1989).

[8] M. T. Record, Jr., and M. C. Mossing, *in* "RNA Polymerase and Regulation of Transcription" (W. Reznikoff *et al.*, eds.), p.61. Elsevier Science Publishing, New York, 1987.

[9] M. T. Record, Jr., and R. S. Spolar, *in* "Nonspecific DNA-Protein Interactions" (A. Revzin, ed.), p. 33. CRC Press, Boca Raton, Florida, 1990.

[10] M. T. Record, Jr., *in* "Unusual DNA Structures" (R. D. Wells and S. C. Harvey, eds.), p. 237. Springer-Verlag, New York, 1988.

[11] See for example the recent review by J. Kraut, *Science* **242,** 533 (1988).

noncovalent macromolecular associations, it will be important to quantify the extent of complementarity of a pair of recognition surfaces, the extent of local and global changes in conformation that accompany the attainment of this degree of complementarity, and the thermodynamic consequences thereof. Even in the case of enzyme–substrate complexes, this information is not yet available.

To date, no general molecular rules for specificity of protein–DNA complexes beyond the concept of mutual complementarity have emerged. Harrison and Aggarwal[1] review the complexity of site-specific recognition and conclude that "the variety of protein–base pair contacts is more evident than their uniformity." There is no evidence for a unique or even a degenerate "recognition code" involving amino acid side chains and base pairs. A single base pair may simultaneously interact with several different amino acids. For example, in the complex formed between 434 O_R1 and R(1–69) (the DNA-binding domain of 434 repressor), the AT at position 1 is contacted by Gln-17, Asn-36, and Gln-28.[1,12] A single amino acid may interact with different bases, may simultaneously interact with two or more base pairs, and/or may interact with different functional groups on a given type of base at different locations in the complex. For example, in the R(1–69)/O_R1 complex, different Gln residues contact A, G, C, and T.[1,12] Gln-28 makes two hydrogen bonds with A at position 1 and also makes a nonpolar contact with T at position -1. Gln-29 forms two hydrogen bonds with G at position 2 and also makes nonpolar contacts with T at positions 3 and 4. Gln-33 forms a hydrogen bond to T at position 4 and also makes a nonpolar contact with C at position 5. These structural observations imply that site-specific protein–DNA interactions will in at least some cases be strongly context dependent.[1,8,13]

In many discussions of specificity and stability of protein–DNA interactions based on structural data and/or genetic studies, it has been assumed (implicitly or explicitly) that the stability of protein–DNA complexes results from the same interactions that determine the specificity of recognition, and that the contributions of these interactions are "additive" (i.e., independent of context). Parallel measurements of equilibrium and kinetic parameters are required to gain insight into the thermodynamics ("energetics") and mechanisms of the process of forming these site-specific complexes, and to relate specificity and stability to the patterns of direct and indirect readout. To date no system has been extensively

[12] A. K. Aggarwal, D. W. Rodgers, M. Drottar, M. Ptashne, and S. C. Harrison, *Science* **242**, 899 (1988).
[13] M. C. Mossing and M. T. Record, Jr., *J. Mol. Biol.* **186**, 295 (1985).

investigated at *both* the structural and thermodynamic level, although a number of such studies are currently in progress.

Interpretations of structural data and of thermodynamic data obtained under a single set of solution conditions are complicated by the fact that functional groups on both proteins and DNA are capable of interacting not only with each other, but also with water and solutes. The latter interactions (including hydrogen bonds with water and Coulombic interactions of phosphates on the DNA polyelectrolyte with cations) may be intrinsically more or less favorable than the macromolecular interactions that replace them in the complex. The nature and consequences of small molecule–macromolecule interactions in solution cannot generally be ascertained from crystal structures of the complexed and uncomplexed species. The sites occupied by water, ions, and other solutes in the solid-state structure may not be relevant for the effects of these species in solution, and certainly provide little information about the strength of small molecule–macromolecule interactions. Such information is necessary in order to understand the thermodynamics of the process of pairing the mutually complementary macromolecular recognition surfaces, which must involve the rearrangement or loss of preexisting interactions with solutes and water. Interactions of proteins with nucleic acids are exchange reactions, and the stability and specificity of protein–DNA interactions are strong functions of the relevant environmental variables (temperature, pH, salt concentration, etc.) that affect these exchange processes. The exchange character of protein–DNA interactions must be considered in interpreting the dependence of equilibrium "constants" and rate "constants" on temperature and solution variables. In particular, characteristic effects of salt concentration and temperature are observed that differ at both a qualitative and a quantitative level from the effect of these variables on covalent reactions involving low-molecular-weight ionic solutes.

A well-known classical conceptual framework exists to interpret the effects of changes in salt concentration and temperature on reactions of low-molecular-weight ionic solutes. However, these reactions are not suitable model systems for the interactions of proteins with nucleic acids, or for other noncovalent interactions of biopolymers. For example, the composite variable "ionic strength" obtained from Debye–Hückel theory has been applied with some success to analyze effects of salt concentration on the thermodynamics and kinetics of reactions of small, charged solutes. However, "ionic strength" has been shown both theoretically and experimentally to be an inappropriate variable for analysis of ion effects on the interactions of oligocations and proteins with nucleic acid polyanions. As a second example, van't Hoff or calorimetric analysis of covalent reactions of small molecules typically yields values of ΔH° and ΔS° for the corresponding process that are relatively temperature independent. On the other hand, large heat capacity effects are observed in most noncovalent processes involving proteins. Consequently van't Hoff plots are highly nonlinear and $\Delta H°$ and $\Delta S°$ are strong functions of temperature.

To date only a few appropriate model systems for noncovalent interactions of proteins and nucleic acid have been characterized in detail. These include the transfer of small, nonpolar solutes from the pure state (liquid, solid, or vapor) to water and the binding of small oligocationic ligands to nucleic acid polyanions. Results with these model systems demonstrate that quantitative measurements of the effects of temperature and salt concentration on protein–DNA interactions provide detailed insight into the types of noncovalent exchange interactions involved and the magnitude of their contributions to the changes in standard-state thermodynamic functions for both nonspecific and site-specific binding. This review chapter provides a conceptual overview of the thermodynamic, mechanistic, and, in some cases, molecular information that can be deduced from the dependence of equilibrium and kinetic quantities (observed equilibrium "constants" and rate "constants") on temperature and salt concentration. For the thermodynamic discussion, we have chosen the interaction of *lac* repressor with its operator as our primary example. The kinetic studies reviewed here involve *lac* repressor and RNA polymerase. Although the examples that we discuss come from only a limited number of systems, this conceptual approach is generally applicable to all other site-specific and nonspecific protein–DNA and ligand–DNA interactions.

Equilibria and Thermodynamics of Site-Specific Protein–DNA Interactions

Thermodynamic Background

Specificity of Protein–DNA Interactions at the Thermodynamic Level. At a thermodynamic level, questions regarding the origins of stability and of specificity of site-specific protein–DNA interactions are interrelated. The stabilities of site-specific (**RS**) and nonspecific (**RD**) complexes are defined by the standard free energy changes ($\Delta G°$) for the appropriate processes of complex formation:

1. Protein + specific DNA site → specific complex:

$$\mathbf{R} + \mathbf{S} \xrightarrow{\Delta G°_{\mathbf{RS}}} \mathbf{RS} \tag{1}$$

2. Protein + nonspecific DNA site → nonspecific complex:

$$R + D \xrightarrow{\Delta G^{\circ}_{RD}} RD \tag{2}$$

We use the symbol R for protein, since both *lac* repressor and RNA polymerase are typically designated R. We use S for the specific DNA site, which differs from the standard abbreviation for operator (O) or promoter (P) DNA sites.] At a thermodynamic level, specificity is defined as the standard free energy change $\Delta G^{\circ}_{RD \rightarrow RS}$ for the process of transferring a protein from a nonspecific site (D) to a specific site (S):

$$\begin{aligned} \Delta G^{\circ}_{RD \rightarrow RS} &= \Delta G^{\circ}_{RS} - \Delta G^{\circ}_{RD} \\ &= -RT \ln(K_{RS}/K_{RD}) \\ &= (\overline{G}^{\circ}_{RS} - \overline{G}^{\circ}_{RD}) - (\overline{G}^{\circ}_{S} - \overline{G}^{\circ}_{D}) \end{aligned} \tag{3}$$

[In a biological sense, regulation (the consequence of specificity) may be under either thermodynamic or kinetic control.[8] In the latter case, the thermodynamic definition of specificity (K_{RS}/K_{RD}) is replaced by the appropriate ratio of rate constants.] For a DNA molecule with n_S specific sites and n_D nonspecific sites, the distribution of protein at low binding density is determined by the site-weighted specificity ratio $n_S K_{RS}/n_D K_{RD}$. The origins of stability of both site-specific and nonspecific protein–DNA complexes must be understood in order to deduce the specificity of the interactions as a function of environmental variables.

To understand the stability of protein–DNA complexes one must dissect at thermodynamic and molecular levels the contributions of various noncovalent interactions to the complicated process of bringing two macromolecular sites together and creating two complementary recognition surfaces. The process of juxtaposing these recognition surfaces involves disruption of preexisting interactions between individual functional groups on each biopolymer and water or solutes (e.g., ions), and replacement of these interactions by contacts between complementary functional groups on each macromolecule. Thus both nonspecific and specific binding are exchange processes involving both water and solutes. The stability of noncovalent complexes in aqueous solution is totally different from what would be predicted *in vacuo*, and is highly dependent on the details of the solution environment.

Binding Constants (K_{obs}) for Site-Specific Interactions. Thermodynamics provides information regarding processes, often gleaned from measurements of the corresponding equilibrium constants, and their dependence on physical variables. In the discussion that follows the process is the noncovalent association of protein with DNA. Typically, for the association of a protein and a DNA molecule containing one (or more) specific site(s), the extent of formation of the specific complexes at equilibrium is assayed as a function of total biopolymer concentrations in a titration of protein with DNA (or vice versa) under reversible binding conditions. (References in the following sections are representative but not inclusive; the bibliographies in these references should also be consulted.)

Quantitative equilibrium studies of site-specific binding have been performed with many prokaryotic proteins involved in regulation of gene expression, including the *E. coli* proteins *lac* repressor,[13-18] *trp* repressor,[19] *gal* repressor,[20] CAP,[21-25] RNA polymerases $(E\sigma^{70})^{26,27}$ and $(E\sigma^{32}),^{28,29}$ *Eco*RI endonuclease[14,30,31] and the bacteriophage proteins Mnt[32] and Arc[33] from P22 and *cI* repressor[34-36] and Cro[37,38] from phage λ. Commonly used assays include filter binding,[18,39] footprinting,[40,41] and the gel retardation assay.[21,22] The equilibrium extent of binding for a specified set of solution conditions is used to define a binding constant K_{obs} for the process:

[14] J.-H. Ha, R. S. Spolar, and M. T. Record, Jr., *J. Mol. Biol.* **209**, 801 (1989).
[15] M. D. Barkley, P. A. Lewis, and G. E. Sullivan, *Biochemistry* **20**, 3842 (1981).
[16] R. B. Winter and P. H. von Hippel, *Biochemistry* **20**, 6948 (1981).
[17] P. A. Whitson, J. S. Olson, and K. S. Matthews, *Biochemistry* **25**, 3852 (1986).
[18] A. D. Riggs, H. Suzuki, and S. Bourgeois, *J. Mol. Biol.* **48**, 67 (1970).
[19] J. Carey, *Proc. Natl. Acad. Sci. U.S.A.* **85**, 975 (1988).
[20] M. Brenowitz, E. Jamison, A. Majumdar, and S. Adhya, *Biochemistry* **29**, 3374 (1990).
[21] M. Fried and D. M. Crothers, *Nucleic Acids Res.* **9**, 6505 (1981).
[22] M. M. Garner and A. Revzin, *Nucleic Acids Res.* **9**, 3047 (1981).
[23] R. H. Ebright, Y. W. Ebright, and A. Gunasekera, *Nucleic Acids Res.* **17**, 10295 (1989).
[24] R. H. Ebright, A. Kolb, H. Buc, T. A. Kunkel, J. S. Krakow, and J. Beckwith, *Proc. Natl. Acad. Sci. U.S.A.* **84**, 6083 (1987).
[25] T. Heyduk and J. C. Lee, *Proc. Natl. Acad. Sci. U.S.A.* **87**, 1744 (1990).
[26] H. S. Strauss, R. R. Burgess, and M. T. Record, Jr., *Biochemistry* **19**, 3504 (1980).
[27] S. L. Shaner, P. Melancon, K. S. Lee, R. R. Burgess, and M. T. Record, Jr., *Cold Spring Harbor Symp. Quant. Biol.* **47**, 463 (1983).
[28] M. A. Fisher, Ph.D. Thesis, University of Wisconsin-Madison (1990).
[29] D. W. Cowing, J. Mecsas, M. T. Record, Jr., and C. A. Gross, *J. Mol. Biol.* **210**, 521 (1989).
[30] D. R. Lesser, M. R. Kurpiewski, and L. Jen-Jacobson, *Science* **250**, 776 (1990).
[31] B. J. Terry, W. E. Jack, R. A. Rubin, and P. Modrich, *J. Biol. Chem.* **258**, 9820 (1983).
[32] A. K. Vershon, S.-M. Liao, W. R. McClure, and R. T. Sauer, *J. Mol. Biol.* **195**, 311 (1987).
[33] A. K. Vershon, S.-M. Liao, W. R. McClure, and R. T. Sauer, *J. Mol. Biol.* **195**, 323 (1987).
[34] H. C. M. Nelson and R. T. Sauer, *Cell (Cambridge, Mass.)* **42**, 549 (1985).
[35] A. Sarai and Y. Takeda, *Proc. Natl. Acad. Sci. U.S.A.* **86**, 6513 (1989).
[36] D. F. Senear, M. Brenowitz, M. A. Shea, and G. K. Ackers, *Biochemistry* **25**, 7344 (1986).
[37] J. G. Kim, Y. Takeda, B. W. Matthews, and W. F. Anderson, *J. Mol. Biol.* **196**, 149 (1987).
[38] Y. Takeda, A. Sarai, and V. M. Rivera, *Proc. Natl. Acad. Sci. U.S.A.* **86**, 439 (1989).
[39] C. P. Woodbury, Jr., and P. H. von Hippel, *Biochemistry* **22**, 4730 (1983).
[40] R. T. Ogata and W. Gilbert, *J. Mol. Biol.* **132**, 709 (1979).
[41] M. Brenowitz, D. F. Senear, M. A. Shea, and G. K. Ackers, this series, Vol. 130, p. 132.

DNA site (S) + protein (R) → complex (RS) $K_{obs} \equiv [RS]/[R][S]$

$$(4)$$

This definition of K_{obs} requires knowledge of the equilibrium concentrations of both the complex and the free reactants. In practice, these quantities are often difficult to determine independently, and K_{obs} is more typically (but less appropriately) defined by neglecting all possible coupled or competitive equilibria that may be simultaneously occurring involving protein, DNA, and/or complex, and writing:

$$K_{obs} \equiv \frac{[RS]}{(R_T - [RS])(S_T - [RS])}$$

$$(5)$$

where R_T and S_T are total concentrations (on the molar scale) of protein and of specific DNA sites, respectively. If coupled or competitive equilibria (e.g., nonspecific binding, protein aggregation) are occurring and/or if significant concentrations of intermediate complexes are present at equilibrium, then $[R] \neq R_T - [RS]$ and/or $[S] \neq S_T - [RS]$, and the definitions of K_{obs} in Eqs. (4) and (5) are not equivalent. In these cases K_{obs} as defined by Eq. (5) may depend on the total concentration of protein and/or DNA.[42] Moreover, any dependence of K_{obs} on temperature, pH, salt concentration, or other variables (as discussed below) may contain contributions from the effects of these variables on the coupled process as well as on the process of protein–DNA association. While a significant amount is known about potential coupled equilibria for $E\sigma^{70}$ RNA polymerase,[27] the extent to which these aggregation and nonpromoter binding events contribute to the equilibria and kinetics of $E\sigma^{70}$–promoter interaction is not known, although *in vitro* experiments can be designed to minimize these contributions (cf. Kinetics and Mechanisms of Site-Specific Protein–DNA Interactions, below).

In most quantitative protein–DNA binding studies using the nitrocellulose filter assay or footprinting assays, the equilibrium concentration of complex has been determined as a function of R_T and S_T and the value of K_{obs} then calculated by Eq. (5). For RNA polymerase $E\sigma^{70}$, where multiple models of nonpromoter binding complicate the direct investigation of equilibrium binding at promoters, values of K_{obs} have been determined by competitive equilibrium[26,27] and kinetic[43] assays. Alternatively, K_{obs} has been evaluated from the ratio of the second-order association rate constant to the first-order dissociation rate constant, where the rate laws are ex-

[42] O. G. Berg, R. B. Winter, and P. H. von Hippel, *Biochemistry* **20**, 6929 (1981).
[43] T. R. Kadesh, S. Rosenberg, and M. J. Chamberlin, *J. Mol. Biol.* **155**, 1 (1982).

pressed in terms of the concentrations in Eq. (5)[44-48] (cf. Kinetics and Mechanisms of Site-Specific Protein–DNA Interactions, below). In some applications of the gel shift assay, the free DNA concentration has been determined instead of the concentration of the protein–DNA complex.[19,22] In other applications of the gel shift assay, concentrations of both free DNA and complex have been measured and used to define K_{obs}.[21,24,32,33] Because of these differences in the definition of K_{obs} and the possible dependence of K_{obs} on protein and/or DNA concentration, values of K_{obs} should be determined by several different methods and compared over a range of macromolecular concentrations at each specified set of environmental variables (temperature, pressure, pH, salt concentration, etc.).

Interpretation of Standard Free Energy Changes ΔG_{obs}°. The standard free energy change (ΔG_{obs}°) is evaluated from K_{obs} defined according to Eq. (4) [or, if necessary, Eq. (5)]:

$$\Delta G_{obs}^{\circ} = -RT \ln K_{obs} \qquad (6)$$

and provides the thermodynamic measure of stability of the protein–DNA complex relative to the reactants under the conditions of the experiment. When K_{obs} is defined according to Eq. (4), then ΔG_{obs}° is the free energy change for the process of converting reactants to products in a hypothetical mixture where the concentrations of reactant and product macromolecules are 1 M but the environment is that of a dilute solution, and where the temperature, pH, and concentrations of electrolyte ions and buffer components are those specified in determining K_{obs}.

ΔG_{obs}° contains information regarding the difference in molar free energies of the pure species and the differences in their interaction with water. Consider for simplicity the hypothetical case where no competitive multiple equilibria involving **R**, S, and/or **RS** are present, and where a thermodynamic equilibrium constant K_{eq} (expressed in terms of equilibrium activities on the molar scale and therefore a function of temperature and pressure only) and the corresponding standard free energy change ΔG° can be unambiguously determined for the process **R** + S → **RS**. For this case:

$$\Delta G^{\circ} = \bar{G}_{\mathbf{RS}}^{\circ} - (\bar{G}_{\mathbf{R}}^{\circ} + \bar{G}_{S}^{\circ}) = -RT \ln K_{eq} \qquad (7)$$

where $\bar{G}_{i}^{\circ}$ is the dilute solution standard state chemical potential (partial

[44] J.-H. Roe, R. R. Burgess, and M. T. Record, Jr., *J. Mol. Biol.* **176**, 495 (1984).
[45] J.-H. Roe, R. R. Burgess, and M. T. Record, Jr., *J. Mol. Biol.* **184**, 441 (1985).
[46] J.-H. Roe and M. T. Record, Jr., *Biochemistry* **24**, 4721 (1985).
[47] C. J. Dayton, D. E. Prosen, K. L. Parker, and C. L. Cech, *J. Biol. Chem.* **259**, 1616 (1984).
[48] G. Duval-Valentin and R. Ehrlich, *Nucleic Acids Res.* **15**, 575 (1987).

molar Gibbs free energy) of macromolecular species i (i = **R**, **S**, **RS**). Each $\bar{G}_i^\circ$ may be decomposed into contributions from the molar free energy of the pure liquid state of that species ($\bar{G}_i^\bullet$, obtained by extrapolation) and the contribution to the free energy of the solution per mole of added macromolecular solute resulting from its interaction with solvent ($RT \ln \gamma_i^{ds}$):

$$\bar{G}_i^\circ = \bar{G}_i^\bullet + RT \ln \gamma_i^{ds} + RT \ln \bar{V}_1^\bullet \tag{8}$$

where γ_i^{ds} is the composition-independent activity coefficient describing the interaction with solvent in the (ideal) dilute solution (relative to the ideal solution) and $\bar{V}_1^\bullet$ is the molar volume of the solvent.[14] The $RT \ln \bar{V}_1^\bullet$ term appears because a molar concentration scale is employed in the definitions of K_{eq} [Eq. (7)]. If a vapor phase reference state is chosen for $\bar{G}_i^\bullet$, then the interaction coefficient γ_i^{ds} is replaced by the Henry's law constant for that species.

Interpretation of ΔG_{obs}° defined by Eq. (6) differs from ΔG° of Eq. (7) because of the participation of electrolyte ions and buffer components in the interactions of proteins with nucleic acids in aqueous solution, as well as the presence of coupled macromolecular equilibria, if Eq. (5) is used to define K_{obs}. The stability of a noncovalent complex is determined by the difference in the quality and quantity of noncovalent interactions involving both the macromolecular recognition surfaces and solvent components (water, ions, etc.) in the complex (**RS**) and in the uncomplexed macromolecular species (**R**, **S**). The use of macromolecular concentrations in the definition of K_{obs} that do not take into account differences in the extent of association of ions, other small solutes, and solvent with the complex and with the reactants, as well as the neglect of activity coefficients describing nonideality arising from solute–solute interactions, results in a dependence of K_{obs} (and standard thermodynamic quantities derived from it) on solution variables (such as pH and ion concentrations) as well as on temperature and pressure. The classic biochemical study of this type is Alberty's[49] work on the effect of environmental variables (pH, pMg) on the thermodynamics of ATP hydrolysis. This dependence on solution conditions is indicated by the subscript "obs."

In summary, since stability is a free energy difference between complexed and uncomplexed species and depends to a major extent on solution conditions, it must be considered a relative rather than an absolute quantity. Stabilities of noncovalent complexes in solution cannot be deduced directly from structural data or molecular modeling exercises on the com-

---

[49] R. A. Alberty, *J. Biol. Chem.* **244**, 3290 (1969).

plex. Specificity [cf. Eq. (3)] is also a relative quantity that, in general, is highly sensitive to solution conditions.

Analyses of the effects of temperature and pressure on K_{obs} yield the standard enthalpy change ($\Delta H°_{obs}$) and the standard volume change ($\Delta V°_{obs}$), respectively. Determination of the dependence of K_{obs} on pH and salt concentration allows the evaluation of the stoichiometries of participation of protons (Δv_{H^+}) and electrolyte ions, respectively, in the association reaction. This thermodynamic information in turn provides insight into the molecular origins of stability of the site-specific protein–DNA complex. With the association reaction of *lac* repressor with the *lac* operator as our primary example, we discuss the analysis and interpretation of experimental data concerning the dependence of K_{obs} on temperature and on solution variables.

Van't Hoff Analysis of Effects of Temperature on K_{obs}

Information regarding the nature of the thermodynamic driving force for noncovalent macromolecular binding processes may be obtained from calorimetric determinations of $\Delta H°_{obs}$ and $\Delta C°_{p,obs}$ or from van't Hoff analysis of the temperature dependence of K_{obs}. Both approaches allow $\Delta G°_{obs}$ for the association process to be dissected into enthalpic and entropic contributions at the temperature of interest. Although enthalpies of interaction can in principle be measured directly and more accurately by isothermal mixing calorimetry than by the van't Hoff method, the calorimetric method has not yet been successfully applied to characterize site-specific protein–DNA interactions, presumably because its requirement for high (micromolar) concentrations of reactants may introduce competing interactions (e.g., aggregation of the protein and/or formation of nonspecific protein–DNA complexes).

In the van't Hoff method, applicable at any reactant concentration where product formation is accurately measurable, $\Delta H°_{obs}$ is calculated from the slope of a plot of ln K_{obs} versus $1/T$:

$$\left(\frac{\partial \ln K_{obs}}{\partial 1/T}\right)_P = -\frac{\Delta H°_{obs}}{R} \tag{9}$$

If ln K_{obs} is found to vary linearly with reciprocal absolute temperature, the van't Hoff $\Delta H°_{obs}$ is constant over the temperature range studied. This is the behavior exhibited by most covalent reactions over the relatively narrow range of absolute temperatures where protein–DNA interactions are usually examined. If $\Delta H°_{obs}$ is independent of temperature (i.e., if $\Delta C°_{p,obs} = 0$), then the standard entropy change $\Delta S°_{obs}$ will also be indepen-

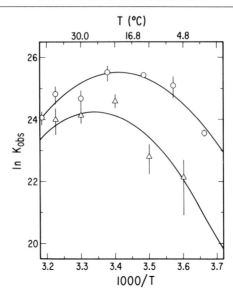

Fig. 1. Effects of temperature on site-specific binding of *lac* repressor to the O^{sym} operator (○) and *Eco*RI to its recognition sequence (△) in the form of van't Hoff plots. The curves are fits to the general relationship $\ln K_{obs} = (\Delta C^{\circ}_{P,obs}/R)[(T_H/T) - 1 - \ln(T_s/T)]$, useful when $\Delta C^{\circ}_{P,obs}$ is large in magnitude and independent of temperature. [From J.-H. Ha, R. S. Spolar, and M. T. Record, Jr., *J. Mol. Biol.* **209**, 801 (1989).]

dent of temperature and the only effect of temperature on ΔG°_{obs} will be the explicit linear dependence:

$$\Delta G^{\circ}_{obs} = \Delta H^{\circ}_{obs} - T\Delta S^{\circ}_{obs} \qquad (10)$$

Contrary to this simple behavior, all site-specific protein–DNA interactions investigated as a function of temperature to date exhibit highly nonlinear van't Hoff behavior.[14,20] Van't Hoff plots of equilibrium constants for site-specific binding of *lac* repressor and *Eco*RI are shown in Fig. 1. In all site-specific protein–DNA interactions examined to date, K_{obs} exhibits a relative maximum in the experimentally accessible temperature range. For the interaction of *lac* repressor with the symmetric *lac* operator (O^{sym}), the maximum in K_{obs} occurs at 20°, and for specific binding of *Eco*RI the maximum is at 26°.[14] For the association of RNA polymerase Eσ^{70} with the λP$_R$ promoter to form an open complex, the maximum is near 42°.[14,45] The maximum in K_{obs} defines a characteristic temperature T_H at which $\Delta H^{\circ}_{obs} = 0$. At lower temperature ($T < T_H$), ΔH°_{obs} is positive; at higher temperatures ($T > T_H$), ΔH°_{obs} is negative.

The strong temperature dependence of ΔH°_{obs} allows one to apply the

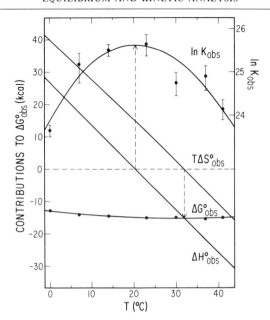

Fig. 2. The thermodynamics of the interaction of *lac* repressor with an isolated symmetric operator (O^{sym}) site. Values of ln K_{obs} and ΔG°_{obs} are plotted as a function of temperature. Enthalpic (ΔH°_{obs}) and entropic ($T \Delta S^{\circ}_{obs}$) contributions to ΔG°_{obs}, as well as theoretical fits to ln K_{obs} and ΔG°_{obs}, were obtained assuming a constant $\Delta C^{\circ}_{P,obs}$ of -1.3 kcal mol^{-1} K^{-1} over the temperature range investigated. [From J.-H. Ha, R. S. Spolar, and M. T. Record, Jr., *J. Mol. Biol.* **209**, 801 (1989).]

van't Hoff analysis to estimate the standard heat capacity change $\Delta C^{\circ}_{p,obs}$ for the process of complex formation, since

$$\Delta C^{\circ}_{p,obs} = \left(\frac{\partial \Delta H^{\circ}_{obs}}{\partial T} \right)_P \tag{11}$$

For all site-specific protein–DNA complexes investigated, the standard heat capacity change for the association process is a large negative quantity.[14,20] In fact $\Delta C^{\circ}_{p,obs}$ is much larger in magnitude than the entropy change (ΔS°_{obs}) throughout the experimentally accessible temperature range. As a result, enthalpy–entropy compensation is observed, leading to characteristic temperature dependences of the standard thermodynamic functions (see Fig. 2) with the following features: (1) Both ΔH°_{obs} and $T \Delta S^{\circ}_{obs}$ decrease rapidly with increasing temperature with nearly identical slopes, leaving ΔG°_{obs} relatively unchanged. Consequently there is a shift in the nature of the thermodynamic driving force with temperature. The site-specific binding of proteins to DNA is entropically driven at low temperature, and shifts toward enthalpically driven with increasing temperature; (2) at the characteristic temperature T_H where $\Delta H^\circ_{obs} = 0$, K_{obs} exhibits a relative maximum. At the characteristic temperature T_S where $\Delta S^\circ_{obs} = 0$, ΔG°_{obs} exhibits a relative minimum.

Sturtevant[50] has examined possible origins of $\Delta C^\circ_{p,obs}$ for a variety of processes involving proteins, including denaturation, oligomerization, and binding of small ligands. In general, $\Delta C^\circ_{p,obs}$ may include contributions from the hydrophobic effect (i.e., a change in the amount of water-accessible nonpolar surface area), from creation or neutralization of charged groups, from changes in internal vibrational modes, and/or from the existence of temperature-dependent changes in equilibria between two or more conformational states or states of aggregation of the protein.[50] For proteins, Sturtevant concluded that the contribution of the hydrophobic effect is approximately four times larger than that caused by changes in vibrational modes, and that these two factors are in general the dominant contributors to the $\Delta C^\circ_{p,obs}$ of noncovalent interactions involving proteins. Using Sturtevant's semiempirical relationships, Ha et al.[14] obtained similar estimates of the relative contributions to $\Delta C^\circ_{p,obs}$ from the hydrophobic effect and from stiffening of vibrational modes of protein–DNA interactions and proposed that the large negative $\Delta C^\circ_{p,obs}$ results primarily from the hydrophobic effect. Quantitative interpretation of the heat capacity change provides estimates both of the reduction in water-accessible nonpolar surface area on the protein and DNA that occur in formation of the specific complex, and of the contribution of the hydrophobic effect to the standard free energy change ΔG°_{obs}.[14]

To calibrate the heat capacity effect in terms of the reduction in water-accessible nonpolar surface, Spolar et al.[51] and Livingstone et al.[52] examined heat capacity changes for processes in which the change in water-accessible nonpolar surface area can be calculated from structural data. Heat capacity changes for the process of hydrocarbon transfer from the dilute aqueous solution standard state to the pure liquid phase and the process of protein folding in aqueous solution are large and negative. In both cases $\Delta C^\circ_{p,obs}$ is proportional to the reduction in water-accessible nonpolar surface area (ΔA_{np}); moreover the proportionality constant is the same within error for both hydrocarbon transfer and protein folding[51,52]:

$$\Delta C^\circ_{p,obs}/\Delta A_{np} \cong -0.3 \text{ cal K}^{-1} \text{ Å}^{-2} \tag{12}$$

[50] J. M. Sturtevant, *Proc. Natl. Acad. Sci. U.S.A.* **74**, 2236 (1977).

[51] R. S. Spolar, J.-H. Ha, and M. T. Record, Jr., *Proc. Natl. Acad. Sci. U.S.A.* **86**, 8382 (1989).

[52] J. R. Livingstone, R. S. Spolar, and M. T. Record, Jr., *Biochemistry* **30**, 4237 (1991).

Consequently the hydrophobic effect must be the dominant factor contributing to the $\Delta C^\circ_{p,obs}$ of protein denaturation.[50-52] Spolar et al.[51] proposed that this relationship can be applied to any noncovalent process involving a globular protein that exhibits a large $\Delta C^\circ_{p,obs}$. For example, Ha et al.[14] used Eq. (12) to estimate the reduction in nonpolar water-accessible surface area that occurs in the association of protein with DNA. For all protein–DNA interactions examined, the value of ΔA_{np} calculated from the experimental $\Delta C^\circ_{p,obs}$ is significantly larger than the estimated water-accessible nonpolar surface area of the complementary surfaces of the protein and the DNA. Consequently Ha et al.[14] proposed that conformational changes that remove additional nonpolar surface of the protein and/or DNA from water must occur as part of site-specific binding. Since the heat capacity effect accompanying DNA denaturation ($\Delta C^\circ_{p,obs} \approx 0.04$ to 0.06 cal g^{-1} K^{-1}) is substantially smaller on a weight basis than that observed in protein denaturation ($\Delta C^\circ_{p,obs} \approx 0.09$ to 0.15 cal g^{-1} K^{-1}),[50,53] it is reasonable to expect that changes in conformation of the protein are the primary origin of the effect.

For a process with a large $\Delta C^\circ_{p,obs}$, the contribution of the hydrophobic effect (ΔG°_{hyd}) to ΔG°_{obs} can be estimated using model compound transfer data. Baldwin[54] utilized thermodynamic data for the transfer of liquid hydrocarbons to water to estimate the contribution of the hydrophobic effect to the driving force for protein folding. Ha et al.[14] applied this approach to protein–DNA interactions. In the physiological temperature range, they demonstrated that

$$\Delta G^\circ_{hyd} \approx (8 \pm 1) \times 10^1 \, \Delta C^\circ_{p,obs} \tag{13}$$

which may be used to estimate ΔG°_{hyd} for a protein–DNA interaction from measurements of $\Delta C^\circ_{p,obs}$.

For the interaction of lac repressor with the symmetric lac operator DNA site, $\Delta C^\circ_{p,obs} = -1.3(\pm0.3)$ kcal mol^{-1} K^{-1}, leading to an estimate of ΔG°_{hyd} of approximately -10^2 kcal mol^{-1}. ΔG°_{hyd} is much larger in magnitude than ΔG°_{obs}, which is ~ -15 kcal mol^{-1} at the solution conditions examined. Therefore Ha et al.[14] proposed that the hydrophobic effect drives the thermodynamically unfavorable processes involved in forming the site-specific protein–DNA complex. These include the translational and rotational restrictions on the reactants in complexation[55,56] and various costs associated with any intrinsic lack of steric and/or functional group

[53] P. L. Privalov, O. B. Ptitsyn, and T. M. Birshtein, Biopolymers 8, 559 (1969).
[54] R. L. Baldwin, Proc. Natl. Acad. Sci. U.S.A. 83, 8069 (1986).
[55] M. I. Page and W. P. Jenks, Proc. Natl. Acad. Sci. U.S.A. 68, 1678 (1971).
[56] C. Chothia and J. Janin, Nature (London) 256, 705 (1975).

complementarity between the interacting surfaces, which may require deformations of the protein and/or DNA surfaces or net loss of favorable contacts on removal from water in the process of complexation.[8-10]

Effect of Solutes (Ligands) on K_{obs}

An equilibrium binding constant defined in terms of the equilibrium activities of all individual chemical species participating in the reaction [cf. K_{eq} in Eq. (7)] is a function of temperature and pressure only. However, the observed equilibrium constant K_{obs} [Eq. (4) or (5)] is formulated in terms of concentrations of reactant and/or product macromolecules, without regard to their extent of interaction with solutes or solvent. This experimentally convenient definition of K_{obs} neglects the possible participation of site-bound solutes (e.g., H^+, anions) in the binding process (and/or in coupled macromolecular equilibria) and neglects the effects of extreme macromolecular thermodynamic nonideality (e.g., the interaction of cations with nucleic acid polyanions). Therefore K_{obs} is generally a strong function of the corresponding compositional variables. In this and the following section, we examine effects on K_{obs} of the concentrations of uncharged solutes, protons, and electrolyte ions.

General Analysis of Effects of Solute (Ligand) Concentration on K_{obs}. When an uncharged solute acts as a ligand (L) and binds to different extents to complexed (AB) and free states of uncharged biopolymers (A, B), changes in concentration of the ligand will affect the equilibrium extent of association of A and B. The effect of such a ligand may be represented by the mass action of equation for the process:

$$A + B + \Delta v_L L + \Delta v_w H_2O \rightleftharpoons AB \tag{14}$$

If K_{obs} is defined in terms of the equilibrium concentrations of A, B, and AB, then it is readily shown that[57,58]

$$\left(\frac{\partial \ln K_{obs}}{\partial \ln[L]}\right)_{T,P} = \Delta v_L - \frac{[L]}{[H_2O]}\Delta v_w \tag{15}$$

where v_L and v_w are the average number of ligands and water molecules bound per biopolymer. [For both ligand ($x = L$) and water ($x = w$), $\Delta v_x \equiv v_{x,AB} - v_{x,A} - v_{x,B}$.] At low concentrations of the ligand ($<0.1\ M$), the term $\Delta v_w[L]/[H_2O]$ is usually negligible.

If the N sites for a ligand on a biopolymer (A, B, or AB) are independent and identical:

[57] J. Wyman, Jr., Adv. Protein Chem. 19, 223 (1964).
[58] C. Tanford, J. Mol. Biol. 39, 539 (1969).

$$v_L = \frac{NK_L[L]}{1 + K_L[L]} \tag{16}$$

where K_L is the intrinsic binding constant for association of L with an individual site. In this simplest possible situation, the concentration of ligand L will affect K_{obs} [cf. Eq. (15)] if K_L and/or N changes on complexation, since either of these will lead to a nonzero Δv_L. In general Δv_L is a function of ligand concentration and hence a plot of ln K_{obs} vs ln[L] will be nonlinear over an extended range of [L], with Δv_L obtained from the tangent to the curve. From Δv_L, information regarding changes in the intrinsic association constant (K_L) and/or in the number of ligand-binding sites (N) that accompany the process A + B → AB can be obtained. If binding of L is weak, then Δv_L is proportional to [L] and a plot of ln K_{obs} vs [L] may be the more suitable means of investigating changes in K_L and/or N.

Protons as Ligands: Differential Protonation Accompanying Binding. The dependence of K_{obs} on pH in the presence of excess electrolyte has been studied for a few site-specific protein–DNA interactions. Although in this case both the ligand and the macromolecules are charged, effects of pH on K_{obs} *in excess salt* may be analyzed using Eq. (15). Because $[H^+] \ll [H_2O]$, the net proton stoichiometry Δv_{H^+} is directly obtained from the pH dependence of K_{obs}:

$$\left(\frac{\partial \ln K_{obs}}{\partial \ln[H^+]}\right)_{T,P} = -\left(\frac{\partial \log K_{obs}}{\partial \text{ pH}}\right)_{T,P} = \Delta v_{H^+} \tag{17}$$

deHaseth et al.[59] and Lohman et al.[60] discuss two extreme scenarios to interpret Δv_{H^+}: (1) the requirement for protonation (possibly applicable to nonspecific binding of *lac* repressor) and (2) the shifted titration curve [applicable to $(Lys)_5$–DNA interactions]. If r independent and identical sites on the protein must be protonated as a prerequisite for binding (case 1), then:

$$\Delta v_{H^+} = \frac{r}{1 + K_H[H^+]} \tag{18}$$

where K_H is the intrinsic binding constant for a proton to an individual site. If protonation is required for complex formation, then at high pH where $K_H[H^+] \ll 1$, log K_{obs} is predicted to decrease linearly with increasing pH, with a slope of $-r$ (i.e., $\Delta v_{H^+} = r$). At sufficiently low pH, K_{obs} is insensitive to pH, because all ionizable groups are fully protonated

[59] P. L. deHaseth, T. M. Lohman, and M. T. Record, Jr., *Biochemistry* **16**, 4783 (1977).
[60] T. M. Lohman, P. L. deHaseth, and M. T. Record, Jr., *Biochemistry* **19**, 3522 (1980).

($\Delta v_{H^+} = 0$). At intermediate pH, Δv_{H^+} varies with pH, equaling $r/2$ at the pK_a of the protonating group.

As an example of the pH dependence of protein–DNA interactions, log K_{obs} for nonspecific binding of *lac* repressor to DNA at 0.13 M NaCl is found to be a linear function of pH over the pH range 7.7–8.4.[59] From the slope of this line [cf. Eq. (17)], Δv_{H^+} is determined to be 2.1 ± 0.2. The specific binding constant for formation of a repressor–operator complex is much less pH dependent (corresponding to $\Delta v_{H^+} < 1$ over the pH range 7.1–8.0, and $\Delta v_{H^+} \simeq 0$ in the pH range 8.0–8.4).[15] These results are consistent with a requirement for protonation at two sites ($pK < 7$) in nonspecific interactions, but a smaller requirement (if any) for protonation in the formation of the specific repressor–operator interaction.

Effects of Concentration and Nature of Electrolyte on K_{obs} of Protein–DNA Interactions

General Analysis of Effects of [MX] on K_{obs}. With exception, equilibrium constants K_{obs} for noncooperative site-specific and nonspecific association of proteins with double helical DNA are found to decrease *strongly* with increasing salt concentration [MX]. The dependence of K_{obs} on [MX] is significantly reduced by the presence of even low concentrations of competitive divalent (e.g., Mg^{2+}) or trivalent (e.g., spermidine) cations, and may of course be difficult to detect if the extent of binding is close to saturation. For both site-specific and nonspecific binding of proteins to double-helical DNA, plots of ln K_{obs} vs ln[MX] are typically linear over the experimentally accessible range, in the absence of competitive equilibria involving the protein, DNA, or the electrolyte. Many studies of site-specific binding of *lac* repressor and RNA polymerase have been carried out with mixtures of univalent and divalent cations (M^+, Mg^{2+}). At a fixed $[Mg^{2+}]$, K_{obs} is found to be relatively independent of $[M^+]$ at low $[M^+]$. With increasing $[M^+]$, the effect of $[M^+]$ on K_{obs} increases greatly. Only at very high $[M^+]$ does ln K_{obs} decrease linearly with increasing ln$[M^+]$. The nonlinearity of ln K_{obs} as a function of ln[MX] results from the competitive association of Mg^{2+} with DNA, the extent of which is itself a strong function of [MX].[9,26,27,60–62]

In the absence of Mg^{2+} or other competitors, the slopes (SK_{obs}) of plots of ln K_{obs} vs ln[MX] (where $SK_{obs} \equiv (\partial \ln K_{obs}/\partial \ln[MX])_{T,P}$), are typically large in magnitude, negative, and independent of salt concentration. For the various site-specific and nonspecific interactions of *lac* repressor and RNA polymerase with DNA, $-5 \geq SK_{obs} \geq -20$ (cf. reviews in Refs. 9, 10,

[61] M. T. Record, Jr., P. L. deHaseth, and T. M. Lohman, *Biochemistry* **16**, 4791 (1977).
[62] M. T. Record, Jr., C. F. Anderson, and T. M. Lohman, *Q. Rev. Biophys.* **11**, 102 (1978).

and 27). Consequently the salt acts as if it participated stoichiometrically as a product of the process of complex formation, with a large and relatively constant stoichiometric coefficient [cf. Eq. (15)]. Analysis of this behavior demonstrates that it results primarily from the Coulombic interactions of cations with the nucleic acid polyanion, the extent of which depends primarily on the axial charge density of the nucleic acid,[63-65] and is therefore reduced in processes that reduce the axial charge density.[9,59,62,64-67] This polyelectrolyte behavior, which is fundamentally different from site binding at both molecular and thermodynamic levels, nevertheless may be summarized as the release of a stoichiometric number of cations from the DNA when its axial charge density is reduced by binding of an oligocation (L^{z+}) or a protein with a positively charged DNA-binding domain. The large stoichiometry of cation release is usually the primary determinant of SK_{obs}. This contribution to SK_{obs} is denoted the "polyelectrolyte effect."[9,10,59,66]

More generally, extension of Eq. (15) to interpret effects of salt concentration on K_{obs} of protein–DNA interactions also requires consideration of anion–protein interactions and differential hydration effects. In forming the protein–DNA complex, site-bound or preferentially accumulated anions and water as well as cations may be displaced from the vicinity of the interacting surfaces. Neglecting differential protonation, a general stoichiometric description of protein–DNA interactions in an aqueous univalent (MX) salt solution is[9,62]

$$DNA + protein \rightleftharpoons complex + aM^+ + cX^- + (b + d)H_2O \qquad (19)$$
$$\scriptstyle (aM^+, bH_2O) \quad (cX^-, dH_2O)$$

The apparent stoichiometric coefficients a, b, c, and d are related to differences in the preferential interaction coefficients of the electrolyte ions and of H_2O with DNA, protein, and the protein–DNA complex. The dependence of K_{obs} on salt concentration is given by[9,62]

$$SK_{obs} \equiv \left(\frac{\partial \ln K_{obs}}{\partial \ln[MX]}\right)_{T,P} \approx -\left[a + c - (b + d)\frac{2[MX]}{[H_2O]}\right] \qquad (20)$$

The use of salt concentration rather than mean ionic activity as the independent variable has been justified.[62,66] Note that Eq. (20) differs in a

[63] G. S. Manning, *J. Chem. Phys.* **51**, 924 (1969).
[64] C. F. Anderson and M. T. Record, Jr., *Annu. Rev. Phys. Chem.* **33**, 191 (1982).
[65] M. T. Record, Jr., M. Olmsted, and C. F. Anderson, in "Theoretical Biochemistry and Molecular Biophysics" (D. L. Beveridge and R. Lavery, eds.), p. 285. Adenine Press, New York, 1990.
[66] M. T. Record, Jr., T. M. Lohman, and P. L. deHaseth, *J. Mol. Biol.* **107**, 145 (1976).
[67] G. S. Manning, *Q. Rev. Biophys.* **11**, 179 (1978).

nontrivial manner from Eq. (15) in that the *sum* of the stoichiometries of cations and anions is determined from the experimentally measurable value of SK_{obs}.

Protein–DNA interactions are typically investigated at low to moderate salt concentration ($[MX] = 0.05\text{–}0.20\,M$), where the magnitude of $2[MX]/[H_2O]$ is small and $-SK_{obs}$ is approximately equal to the sum of the stoichiometric coefficients of the electrolyte ions ($a + c$), where typically $a \gg c$ for interactions of proteins with double-stranded DNA. At higher salt concentration, where the contribution of water release to SK_{obs} is no longer negligible, SK_{obs} is expected to decrease in magnitude (i.e., become less negative) as a result of the linkage of water and salt activities required by the Gibbs–Duhem equation. (An increase in salt concentration reduces the water activity and makes the displacement of water to the bulk solution more favorable than it was at lower salt concentration.) Careful measurement and analysis of the behavior of SK_{obs} at high salt concentration in principle provides a useful route to determining the net stoichiometry of water participation ($b + d$) in the binding reaction. These experiments are difficult to perform because the binding constant K_{obs} is typically low, and are difficult to interpret because of uncertainties regarding effects of high salt concentration on the cation stoichiometry (a) and the anion stoichiometry (c).[68]

The extent of binding of some proteins to single-stranded nucleic acids has been shown to increase with increasing salt concentration. These apparent exceptions to the general behavior discussed in this section must result from one or more of the additional roles of salt concentration, including stabilizing specific conformations of the nucleic acid, favoring release of water, and/or in driving cooperative binding. For example, the extent of interaction of single-stranded oligonucleotides with site II of the *E. coli* ribosomal protein S1 increases with increasing salt concentration above 0.1 M NaCl,[69] and the extent of interaction of single-stranded oligonucleotides with the second site of *E. coli* single strand-binding (SSB) tetramers increases with increasing salt concentration below 0.1 M NaCl.[70] In the latter case, the origin of the effect is a decrease in anticooperativity of binding to the second site that occurs with increasing salt concentration. The intrinsic binding constant for interaction of the oligonucleotide with this site decreases with increasing salt concentration, but the salt concentration dependence of K_{obs} is dominated by the anticooperativity effect.

---

[68] J.-H. Ha, Ph.D. Thesis, University of Wisconsin-Madison (1990).
[69] D. E. Draper and P. H. von Hippel, *J. Mol. Biol.* **122**, 339 (1978).
[70] W. Bujalowski and T. M. Lohman, *J. Mol. Biol.* **207**, 269 (1989).

Cation–DNA Interactions and Effect of Electrolyte Concentration on Binding of Oligocationic Ligands to DNA (Polyelectrolyte Effect). Because DNA is a polyanion of high axial charge density, the extent of interaction of proteins with DNA is typically more sensitive to salt concentration than to the concentration of any other solution component, including the protein concentration! With two negatively charged phosphate groups per 3.4 Å, the axial charge density of B-DNA is so high as to cause the local accumulation of a high concentration of cations and the exclusion of anions from the vicinity of DNA.[63–65,71] At low salt concentration, these ion concentration gradients are predicted by all polyelectrolyte theories to be thermodynamically equivalent to the neutralization of approximately 88% of the structural charge on B-DNA.[64–66] As a consequence, B-DNA behaves like the weak electrolyte $(M_{0.88}^+ DNAP^-)_n$, where M^+ is a univalent cation and $DNAP^-$ is a phosphate charge of DNA.[9,64–66] As judged by cation nuclear magnetic resonance (NMR) experiments,[72] the molecular extent of neutralization of DNA phosphates by associated cations is relatively independent of the bulk univalent salt concentration ($[MX]$). Theoretical analyses predict that the thermodynamic degree of neutralization decreases with increasing salt concentration,[65,73] and with a decrease in the axial structural charge density of DNA.[62,64–66]

Binding of an oligocation (L^{z+}) to DNA in a univalent salt solution is a cation-exchange process in which the DNA structural charge density is reduced and cations (M^+) are released from the DNA to the bulk solution. Experimental determinations of the stoichiometry of univalent cation release [cf. Eqs. (19) and (20)] on binding of a variety of oligocations L^{z+} to double-helical DNA and RNA [including Mg^{2+}, polyamines, $(Lys)_N$, etc.] indicate that $a = (0.88 \pm 0.05)z$, independent to a first approximation of the bulk concentration of MX.[60,66,74–76] The anion stoichiometry in these cases is apparently negligible ($c \approx 0$). For the interaction of oligolysines with single-stranded polynucleotides, $a = (0.71 \pm 0.03)z$, as anticipated from the lower axial charge density of these polyanions, and the anion

[71] C. F. Anderson and M. T. Record, Jr., *Annu. Rev. Biophys. Biophys. Chem.* **19**, 423 (1990).
[72] S. Padmanabhan, M. Paulsen, C. F. Anderson, and M. T. Record, Jr., *in* "Monovalent Cations in Biological Systems" (C. Pasternak, ed.), p. 321. CRC Press, Boca Raton, Florida, 1990.
[73] P. Mills, C. F. Anderson, and M. T. Record, Jr., *J. Phys. Chem.* **90**, 6541 (1986).
[74] W. H. Braunlin, T. J. Strick, and M. T. Record, Jr., *Biopolymers* **21**, 1301 (1982).
[75] W. H. Braunlin, C. F. Anderson, and M. T. Record, Jr., *Biochemistry* **26**, 7724 (1987).
[76] G. E. Plum and V. A. Bloomfield, *Biopolymers* **29**, 13 (1990).

stoichiometry is negligibly small ($c = 0$).[77] Association of an oligocation L^{z+} with helical B-DNA may be represented by the cation-exchange equation:

$$L^{z+} + (M_{0.88}^+ DNAP^-)_n \rightleftharpoons complex + 0.88z\,M^+ \qquad (21)$$

Consequently ligand–DNA association is favored by low salt concentration, and K_{obs} increases strongly with decreasing salt concentration.

At low salt (MX) concentration, in the absence of specific anion effects, the stoichiometry of cation release ($0.88z$) on binding of L^{z+} to the helical B-DNA polyanion is obtained from the derivative [cf. Eq. (20)]:

$$SK_{obs} \equiv \left(\frac{\partial \ln K_{obs}}{\partial \ln[MX]}\right)_{T,P} = -0.88z \qquad (22)$$

For association of an oligocation (L^{z+}) with the B-DNA polyanion in excess univalent salt MX,

$$\ln K_{obs} = \ln K^\circ - 0.88z \ln[MX] \qquad (23)$$

and

$$\Delta G_{obs}^\circ = \Delta G_0^\circ + 0.88z RT \ln[MX] \qquad (24)$$

where $\ln K^\circ$ and the corresponding standard free energy change ΔG_0° are extrapolated values at the 1 M MX pseudostandard state. Since the extrapolated K° for binding of oligocations [e.g., $(Lys)_N$, polyamines] to nucleic acid polyanions is within an order of magnitude with unity, we conclude that the intrinsic preference of DNA for these oligocations (relative to M^+) is small and that these binding interactions are driven at low salt concentrations by M^+ release.[9,60,62,66,67,77] The resulting favorable free energy of dilution (an entropic effect) from counterion displacement into the bulk solution is called the polyelectrolyte effect.[9,10,59,66] Therefore, the contribution of the polyelectrolyte effect to the stability of an L^{z+}–B-DNA complex in an univalent salt solution (ΔG_{PE}°) may be estimated from the relationship:[9,10,66]

$$\Delta G_{PE}^\circ = 0.88z RT \ln[MX] \qquad (25)$$

In general, the interactions of oligocations [including polyamines with 2–4 positive charges, Mg^{2+}, $Co(NH_3)_6^{3+}$, and oligolysines with 3–10 positive charges] with nucleic acids behave as pure cation-exchange processes, without evidence of significant anion effects at low salt concentration.[60,66,74–77]

---

[77] D. P. Mascotti and T. M. Lohman, *Proc. Natl. Acad. Sci. U.S.A.* **87**, 3142 (1990).

If the binding of L^{z+} to DNA is investigated in a mixed salt buffer containing both univalent and di- or trivalent cations [e.g., binding of $(Lys)_5$ in an $NaCl/MgCl_2$ mixture], interpretation of the apparent stoichiometric coefficient a of univalent cation release is complicated by competition between the univalent and di- or trivalent cations.[60,61,74] These cations compete more effectively than monovalent cations for local accumulation near the surface of the DNA. Even low concentrations of a multivalent cation [e.g., Mg^{2+}, spermidine $(3+)$] will reduce both K_{obs} and $|SK_{obs}|$ for the binding of L^{z+} to DNA, especially at low salt ([MX]) concentration. Effects of Mg^{2+} concentration on K_{obs} and $|SK_{obs}|$ for the interaction of pentalysine with B-DNA decrease with increasing salt concentration.[60] This behavior can be accounted for with a simple competitive binding-polyelectrolyte model.[61,62]

Anion–Protein Interactions Involving Site-Binding or Preferential Interactions, Fundamentally Different from Interactions of Cations with Nucleic Acid Polyanions. *Both* the large number and high axial density of negative charges on the DNA polyanion are necessary to give rise to its characteristic polyelectrolyte behavior.[65,78] On the other hand, ligands like L^{z+} and the domains of proteins that interact with DNA are typically relatively short oligocations with a much smaller number of charges (and possibly a lower axial charge density) than that of polyanionic DNA. The thermodynamic extent of counterion association with a charged group on an oligocation or oligoanion is predicted to be far smaller than that characteristic of a polyion of the same axial charge density.[78] Experimental investigation of the interactions of oligolysines with single-stranded polynucleotides[77] demonstrates that Coulombic interactions of anions with oligocations (L^{z+}; $z \leq 10$) make only a minor contribution to the salt concentration dependence of K_{obs} for L^{z+}–DNA interactions [$c \approx 0$ in Eq. (20)].

For proteins, other relevant modes of interaction of anions with the DNA-binding domain may occur and lead to an effect on the anion stoichiometry c. Anions may bind to sites on the protein, generally with relatively small binding constants ($\leq 10^2 M^{-1}$).[62] Evidence for a more diffuse type of anion–protein interaction also exists in which a variety of anions are excluded (relative to water) from the vicinity of protein surfaces.[79,80] Both of these interactions become negligible at low salt concentrations. Although both weak binding and exclusion of anions may occur, effects of

[78] M. Olmsted, C. F. Anderson, and M. T. Record, Jr., *Proc. Natl. Acad. Sci. U.S.A.* **86,** 7766 (1989).
[79] T. Arakawa and S. N. Timasheff, *Biochemistry* **21,** 6545 (1982).
[80] T. Arakawa and S. N. Timasheff, *J. Biol. Chem.* **259,** 4979 (1985).

the nature of the anion on protein–DNA interactions are most simply modeled as resulting from competitive displacement of the anion from weak sites on the protein.[68,81–83] This competitive mode of action of anions will decrease K_{obs} and increase $|SK_{obs}|$ of the protein–DNA interaction relative to a situation in which no anions are displaced and SK_{obs} is determined only by the stoichiometry of cation release.[83] All the effects of anions are predicted to become more pronounced at higher anion concentration, and for anions that are at the strong-interacting end of the Hofmeister series.[83–85]

Effects of Salt Concentration on lac Repressor–Operator Interaction. The large effects of salt concentration on K_{obs} for interactions of *lac* repressor with restriction fragments containing either the wild-type primary *lac* operator (O$^+$) or a mutant *lac* operator sequence (O^{c}666, O^{c}640) are shown in Fig. 3. For O$^+$ and O^{c}666 in NaCl, $-SK_{obs} = 5 \pm 1$, whereas for O^{c}640, $-SK_{obs} = 10 \pm 1$.[13] For the interaction of *lac* repressor with nonoperator DNA, $-SK_{obs} = 11 \pm 1$.[86] Because these studies were conducted at relatively low salt concentration (0.1–0.2 M), differential hydration effects can be neglected so that $-SK_{obs}$ is equal to the sum of the stoichiometries of cation and anion release [$-SK_{obs} = a + c$; Eq. (20)].

Before proceeding to interpret these values of SK_{obs}, it is important to emphasize their practical consequences. First, since the sum of the stoichiometric coefficients of the cation and anion ($a + c = 5$–10) is far larger than that of *lac* repressor (with a stoichiometry of unity), errors in salt concentration (although typically much smaller than errors in protein concentration) are magnified and may contribute as much or more as uncertainties in protein concentration to the uncertainty in K_{obs}. Second, it is possible to obtain virtually any desired value of K_{obs} (if competing processes do not interfere) by appropriate choice of salt concentration. Therefore, values of K_{obs} (and of standard thermodynamic functions derived from K_{obs}) are only meaningful if all ion concentrations are accurately specified. Comparisons of values of K_{obs} for different DNA sequences or different systems generally must be made at the same concentrations of all ionic species. Extrapolations to a different salt concentration will be

[81] S. Leirmo, C. Harrison, D. S. Cayley, R. R. Burgess, and M. T. Record, Jr., *Biochemistry* **26**, 2095 (1987).
[82] L. B. Overman, W. Bujalowski, and T. M. Lohman, *Biochemistry* **27**, 456 (1988).
[83] S. Leirmo and M. T. Record, Jr., in "Nucleic Acids and Molecular Biology" (D. M. J. Lilley and F. Eckstein, eds.), Vol. 4, p. 123. Springer-Verlag, Berlin, 1990.
[84] P. H. von Hippel and T. Schleich, *Acc. Chem. Res.* **2**, 257 (1969).
[85] K. D. Collins and M. W. Washabaugh, *Q. Rev. Biophys.* **18**, 323 (1985).
[86] T. M. Lohman, C. G. Wensley, J. Cina, R. R. Burgess, and M. T. Record, Jr., *Biochemistry* **19**, 3516 (1980).

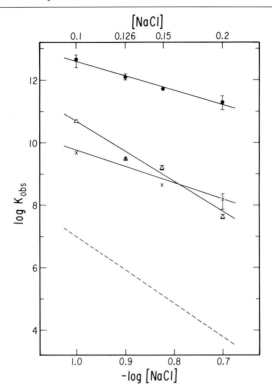

FIG. 3. Effects of NaCl concentration on the binding of *lac* repressor to specific and nonspecific DNA sequences. The logarithm of the specific binding constant K_{obs}^{RS} at (23°) is plotted vs the negative logarithm of the NaCl concentration for (●) O$^+$, (△) O^{c}640, and (X) O^{c}666. [Data from M. C. Mossing, and M. T. Record, Jr., *J. Mol. Biol.* **186**, 295 (1985).] Dashed line depicts the dependence of the logarithm of the nonspecific binding constant K_{obs}^{RD} on the negative logarithm of the NaCl concentration: $\log K_{obs}^{RD} = -(10.7 \pm 1.0)\log[\mathrm{Na}^+] - (3.7 \pm 1.0)$. [From T. M. Lohman, C. G. Wensley, J. Cina, R. R. Burgess, and M. T. Record, Jr., *Biochemistry* **19**, 3516 (1980).]

nonlinear (even on a log–log plot) if mixed salts are present (e.g., MgCl$_2$ and NaCl), and generally will be inaccurate. "Ionic strength" is demonstrably *not* a useful composite ionic variable for protein–DNA interactions, either in theory or in practice, and should not be used in these analyses.[9,46,61,62,87,88]

Analysis and interpretation of cation effects on K_{obs}: In order to decom-

[87] M. T. Record, Jr., S. J. Mazur, P. Melancon, J.-H. Roe., S. L. Shaner, and L. Unger, *Annu. Rev. Biochem.* **50**, 997 (1981).
[88] T. M. Lohman, *Crit. Rev. Biochem.* **19**, 191 (1985).

pose values of SK_{obs} in univalent salt (MX) solutions into contributions from the individual cation and anion stoichiometric coefficients a and c, it is useful to compare these values of SK_{obs} with those determined in the pure salt of a divalent cation and the same univalent anion X^- (e.g., MgX_2). A much lower range of MgX_2 concentrations yields the same experimentally accessible range of values of K_{obs} as investigated in MX. At the lower concentration of X^- in MgX_2, the anion stoichiometry c should be reduced to an extent consistent with a mass action model. Replacement of M^+ with Mg^{2+} is expected to reduce the cation stoichiometry to approximately one-half its value in M^+ (unless Mg^{2+} plays an additional specific role in complex formation).[59] For nonspecific binding of *lac* repressor in $MgCl_2$, $-SK_{obs} = 6 \pm 1$.[59] This reduction in SK_{obs} is consistent with the interpretation $-SK_{obs} = a$ and $c = 0$ for nonspecific binding of *lac* repressor. Therefore it appears that this process is well described by a pure cation-exchange model in Cl^- salts. For site-specific binding of *lac* repressor to the *lac* operon (containing two pseudooperators and the primary *lac* operator), Barkley *et al.*[15] observed that SK_{obs} was approximately halved when investigated in $MgCl_2$ as compared to NaCl, again consistent with an interpretation involving primarily cation release.

From a survey of structures of prokaryotic protein–DNA complexes solved to high resolution, Harrison and Aggarwal[1] conclude that "hydrogen bonds from positively charged side chains to DNA phosphates appear with only modest frequency. Coulombic interactions from less strongly anchored arginines and lysines do appear to be important, however, since in each structure a number of such residues lie near the DNA backbone." This observation suggests that neutralization of DNA phosphates by these proteins is generally not accomplished by classical salt bridges (ion pairs). However, the often large stoichiometry of cation displacement from DNA that accompanies binding of most proteins clearly indicates that the charge density of DNA is reduced on complexation. Presumably this is the consequence of the above-mentioned Coulombic interactions between DNA phosphates and positively charged groups on the protein, but the molecular details of these interactions in solution are unknown. Close juxtapositions of DNA phosphates and arginine and lysine residues also occur in the crystal structure of the eukaryotic *engrailed* homeo domain–DNA complex of Pabo and collaborators.[89]

If it is assumed that SK_{obs} for binding of *lac* repressor to these operator sequences is primarily an expression of the polyelectrolyte effect, then the thermodynamic contribution of cation release (ΔG°_{PE}) to the observed

[89] C. R. Kissinger, B. Liu, E. Martin-Bianco, T. B. Kornberg, and C. O. Pabo, *Cell* (*Cambridge, Mass.*) **63**, 579 (1990).

net standard free energy of association (ΔG°_{obs}) can be estimated from Eq. (25). At [NaCl] = 0.2 M, the polyelectrolyte effect contributes approximately -5 kcal to ΔG°_{obs}, which is approximately -15 kcal at this salt concentration.[13] The contribution of ΔG°_{PE} to ΔG°_{obs} varies widely for different specific and nonspecific protein–DNA interactions. For formation of the specific open complex between *E. coli* RNA polymerase $E\sigma^{70}$ and the λP_R promoter, the estimated ΔG°_{PE} is actually more favorable than ΔG°_{obs} at 0.2 M NaCl.[9,10]

Anion effects on K_{obs} and SK_{obs}: For both nonspecific and specific binding of *lac* repressor to the wild-type *lac* operon, replacement of Cl^- by acetate (Ac^-) leads to an approximately 30-fold increase in K_{obs} at constant salt concentration without a significant (<10–15%) change in SK_{obs}.[15,59] Detailed investigations of anion effects on K_{obs} and SK_{obs} for the interaction of *lac* repressor with the *lac* operon on λplac DNA[15] and with an isolated synthetic (strong-binding) *lac* operator[68] have been performed. For the interaction with λplac DNA at a fixed Na^+ concentration of 0.13 M, K_{obs} decreases by a factor of approximately 10^4 in the order $Ac^- \geq F^- > Cl^- > Br^- > NO_3^- > SCN^- > I^-$.[15] Comparisons of K_{obs} for the interaction of *lac* repressor with linear plasmid DNA carrying the symmetric strong-binding *lac* operator (O^{sym}) in K^+ salts [Cl^-, Ac^-, and glutamate (Glu^-)] demonstrate that K_{obs} decreases in the order $Glu^- > Ac^- > Cl^-$, and that the magnitude of the effect on K_{obs} due to the choice of anion increases with increasing salt concentration above 0.1 M.[68] Even though the nature of the anion has a detectable effect on SK_{obs}, the stoichiometry of cation release is in all cases the primary determinant of SK_{obs}.

No unique interpretation of these large anion effects is possible, primarily because the reference point of a noninteracting anion is not known. If it is assumed that all the effects are due to competitive anion binding, then perhaps Glu^- is a noninteracting anion [$c = 0$ in Eq. (20)]. Alternatively, it is possible that Cl^- does not interact with the DNA-binding site of *lac* repressor, as inferred from comparison of SK_{obs} in NaCl and $MgCl_2$. In this scenario, the effects of replacing Cl^- by Glu^- must be attributed to preferential hydration in Glu^- (i.e., preferential exclusion of Glu^-)[80] and/or to anomalous nonideality of glutamate salts.[90] Interpretation of the data as purely a preferential hydration effect leads to the deduction that as many as 200 mol of water may be displaced per mol of complex formed.[68]

In summary, effects of the nature of the anion on this (and other) protein–DNA interactions are *very* large, but their origin remains difficult

[90] D. Mascotti and T. M. Lohman, personal communication (1990).

to diagnose. These effects must be understood in order to refine the estimate of the contribution of the polyelectrolyte effect to stability and specificity, and to begin to ascertain the stability and specificity of protein–DNA complexes in the *in vivo* environment [a concentrated polyelectrolyte solution where the total concentration of electrolyte cations (primarily K^+ in *E. coli*) greatly exceeds that of electrolyte anions (primarily glutamate in *E. coli*) as a result of the high concentration of nucleic acid (RNA, DNA) polyanions].[91,92]

Kinetics and Mechanisms of Site-Specific Protein–DNA Interactions

Introduction

Relating Kinetics to Mechanism. Kinetic studies are required to determine the "mechanism," i.e., the path or sequence of elementary kinetic steps, by which a protein binds to a specific site on DNA to form a complex. From kinetic studies, information is obtained regarding the minimum number of significant steps in the mechanism, the number and (in some cases) thermodynamic or kinetic characteristics of kinetically significant intermediates, and the nature of the activation barriers between intermediates. For any reaction, these are difficult questions to address. Site-specific protein–DNA interactions are certainly no exception.

Rate Laws of Site-Specific Protein–DNA Interactions. Association typically exhibits second order (or pseudofirst order) kinetics: The kinetics of association of several proteins (including *lac* repressor and RNA polymerase of $E\sigma^{70}$) with specific DNA sites are observed to follow a second order rate law, with an experimentally determined second order rate constant k_{assoc} (M^{-1} sec^{-1}). (Our notation for rate constants is internally consistent in this section, although not always consistent with the notation in the cited work.) For the overall process $\mathbf{R} + S \rightarrow \mathbf{RS}$, the kinetics of association are found to conform to the rate equation:

$$\frac{d[\mathbf{RS}]}{dt} = k_{assoc}(\mathbf{R}_T - [\mathbf{RS}])(S_T - [\mathbf{RS}]) \qquad (26)$$

Note that it is by no means obvious *a priori* that this rate law must be applicable, except for the situation in which association occurs in a single elementary (i.e., diffusion–collision limited) bimolecular reaction step, and in the absence of any competing interactions that might significantly reduce the concentrations of the free protein and the free DNA site.

[91] B. Richey, S. Cayley, M. Mossing, C. Kolka, C. F. Anderson, T. C. Farrar, and M. T. Record, Jr., *J. Biol. Chem.* **262**, 7157 (1987).

[92] D. S. Cayley, B. A. Lewis, H. J. Guttman, and M. T. Record, Jr., *J. Mol. Biol.* in press (1991).

The mechanism of association of a protein with a specific DNA site embedded in a molecule containing nonspecific DNA sites will involve some or all of the following general classes of steps:

1. Changes in the state of aggregation of the protein, conformational changes in the protein or DNA, and/or nonspecific binding of the protein to DNA, prior to the elementary bimolecular association step
2. An elementary bimolecular step in which an initial complex at the specific DNA site is formed at the diffusion–collision rate or at a rate that is slower than the estimated maximum diffusion–collision rate because of orientation (entropic) effects
2'. An elementary bimolecular step in which an initial complex is formed at a distant nonspecific site at the (orientation-corrected) diffusion–collision rate, followed by a mechanistically distinct diffusional process in the domain of the DNA molecule to locate the specific site
3. Local and/or global conformational changes that occur in the initial complex at the specific site subsequent to the elementary bimolecular step, and result in formation of the functional specific complex

In the following sections, mechanisms involving these classes of steps are introduced, and interpretations of the experimentally determined second order association rate constant k_{assoc} in terms of these potential mechanisms are discussed. In general, the approach to the analysis of second order kinetic data in terms of mechanisms is the following: The rate constant k_{assoc} [Eq. (26)] is measured as a function of temperature and solution conditions (especially salt concentration, pH, and, where appropriate, solvent viscosity; DNA length is also an important variable if the process is facilitated as in class 2' above). If the magnitude of k_{assoc} is appropriate for a diffusion–collision reaction and k_{assoc} is only weakly dependent on temperature and salt concentration, then the interaction is probably diffusion limited. If k_{assoc} is only weakly dependent on temperature and salt concentration but its magnitude is significantly less than the diffusion limit, then a diffusion–collision mechanism with severe orientation restrictions may be appropriate. If k_{assoc} is smaller than the diffusion limit and strongly dependent on temperature and/or salt concentration (especially if k_{assoc} decreases with increasing temperature) then intermediates before and/or after the elementary diffusion–collision step must be considered. On the other hand, if k_{assoc} is significantly greater than the diffusion–collision limit, various facilitating mechanisms involving nonspecific DNA sites on the same molecule must be considered. The importance of a facilitating mechanism may be diagnosed from measurements of the dependence of k_{assoc} of DNA length and on salt concentration.

Dissociation typically first order (or pseudofirst order): Kinetics of dissociation of site-specific protein–DNA complexes follow the expected first order rate law:

$$-\frac{d[\mathbf{RS}]}{dt} = k_{\text{dissoc}}[\mathbf{RS}] \qquad (27)$$

where k_{dissoc} is the observed (generally composite) first order rate constant. The mechanism of dissociation must involve passage through the same steps (in the reverse direction) as in the association mechanism, according to the principle of microscopic reversibility. For example, if association is facilitated by the presence of contiguous nonspecific sites, these sites and facilitating mechanisms will also play a role in dissociation, and the equilibrium constant and thermodynamic quantities derived therefrom will be independent of this path-dependent effect. Observed deviations from this situation can usually be attributed to complications arising from differences in the way that the kinetics of association and dissociation are investigated.

In investigations of the kinetics of dissociation of site-specific complexes, a polyanionic competitor (e.g., heparin) is often added to sequester the free protein. In this case, it is important to examine whether k_{dissoc} is a true first order dissociation rate constant, or whether k_{dissoc} contains contributions from both dissociation (first order) and from competitor-induced displacement (typically pseudofirst order if competitor is in excess). The intrinsic contribution from dissociation may be separated from competitor-induced displacement by extrapolation to zero competitor concentration.[93]

Since cations may be considered fundamental and omnipresent competitors with proteins for the vicinity of the DNA polyanion, all elementary protein–DNA dissociation rate constants are actually pseudofirst order, because cations are reactants in the elementary step of dissociation of the protein from the DNA, and the cation concentration is in vast excess.[46,88,94] Hence these elementary rate constants generally vary as some large positive power of the cation concentration. This effect, which can serve as a diagnostic probe of mechanism, need not be (and indeed cannot be) extrapolated away, as long as measurements of association and dissociation kinetics and of equilibrium extents of binding are compared at identical concentrations of electrolyte ions.

Characterization of Intermediates. Two different and complementary approaches have been used to probe mechanisms of site-specific binding, as illustrated by studies on *E. coli* RNA polymerase. One approach has

[93] C. L. Cech and W. R. McClure, *Biochemistry* **19**, 2440 (1980).
[94] T. M. Lohman, P. L. deHaseth, and M. T. Record, Jr., *Biophys. Chem.* **8**, 281 (1978).

been to isolate and characterize intermediates using low-temperature or other "quenching" methods to block subsequent steps. This approach assumes that intermediates that can be isolated as a function of temperature (or other variables) bear a direct relationship to the transient intermediates that exist on the kinetic path for conversion of reactants to products at higher temperature. With RNA polymerase, structural methods, including electron microscopy[95] and various footprinting and chemical modification/chemical protection assays,[96] have been applied. Polymerase–promoter complexes investigated include those of σ^{70}–RNA polymerase ($E\sigma^{70}$) with the T7 A3,[97] $tetR$,[48] and $lacUV5$ promoters.[97,98] Similar studies have been carried out on intermediates in the interaction of σ^{32}–RNA polymerase ($E\sigma^{32}$) and the $groE$ promoter.[29] The complementary approach is to look for the characteristic dependencies of the individual elementary rate constants or overall composite rate constants on such variables as DNA length, sequence (of either protein or DNA), salt concentration, pH, and temperature. The analysis of effects of these variables in terms of mechanisms of site-specific interactions is discussed in the remainder of this chapter.

Use of Temperature and Salt Concentration to Investigate Mechanisms of Interactions of Proteins and DNA

Our focus in the following sections is on the interpretation of experimental rate constants of association and dissociation in terms of the mechanism of formation of the site-specific complex. In general, temperature, salt concentration, and pH are the important diagnostic probes of mechanism.

Investigations of the kinetics of site-specific protein–DNA interactions constitute a relatively young field, beginning with the elegant studies of Riggs, Bourgeois, and Cohn[99] on the lac repressor–operator interaction and of Hinkle and Chamberlin[100] on RNA polymerase $E\sigma^{70}$–promoter interactions. Analysis of kinetic data in terms of mechanisms of protein–DNA interactions is also a new area, in which apparent precedents provided by mechanistic studies on the covalent reactions between small ions or molecules are often not of direct relevance. This is particularly the case in analyzing the effects of temperature and salt concentration on the observed association and dissociation rate constants.

[95] R. C. Williams and M. J. Chamberlin, *Proc. Natl. Acad. Sci. U.S.A.* **74**, 3740 (1977).
[96] T. D. Tullius, *Annu. Rev. Biophys. Biophys. Chem.* **18**, 213 (1989).
[97] R. T. Kovacic, *J. Biol. Chem.* **262**, 13654 (1987).
[98] A. Spassky, K. Kirkegaard, and H. Buc, *Biochemistry* **24**, 2723 (1985).
[99] A. D. Riggs, S. Bourgeois, and M. Cohn, *J. Mol. Biol.* **53**, 401 (1970).
[100] D. Hinkle and M. Chamberlin, *J. Mol. Biol.* **70**, 187 (1972).

For both elementary and composite rate constants characterizing chemical reactions between low-molecular-weight solutes, the logarithm of the observed rate constant is typically a linear function of the reciprocal absolute temperature. The Arrhenius activation energy E is obtained from the slope. Interpretation of the magnitude and temperature dependence of an elementary rate constant for a covalent chemical reaction using the approximations of the Eyring quasithermodynamic model yields the activation parameters $\Delta H^{\circ\ddagger}$ (which is approximately equal to E) and (with further assumptions) $\Delta G^{\circ\ddagger}$ and $\Delta S^{\circ\ddagger}$ for formation of the transition state from the reactants. In the case of an elementary single-step reaction, E and $\Delta H^{\circ\ddagger}$ must be positive. Association processes in water where the rate-determining step is diffusion–collision should exhibit small activation energies of approximately 5 kcal, arising primarily from the temperature dependence of the diffusion coefficient ($D \propto T/\eta$, where η is the temperature-dependent solvent viscosity).

Rate constants (k) for second order association reactions between two low-molecular-weight ions (A^{Z_A}, B^{Z_B}) exhibit a modest dependence on the ionic strength (I) of the supporting electrolyte. For reactants of like charge, k increases with increasing ionic strength as a result of the more effective screening of the like charges of the reactants by the ionic atmospheres formed by the added salt. In the Brønsted–Bjerrum–Debye–Hückel theory, valid at low salt concentration (<0.01 M), log k is predicted to vary linearly with $I^{1/2}$: log $k/k_0 \simeq 1.02\,Z_A Z_B I^{1/2}$, where k_0 is the rate constant at $I = 0$.[101]

Differences in the conceptual framework and methods of analysis of thermodynamic data on covalent reactions of small ionic solutes and non-covalent interactions of large and highly charged solutes were discussed in Equilibria and Thermodynamics of site-specific Protein–DNA Interactions, above. In the analysis of effects of temperature and salt concentration on the kinetics of noncovalent interactions of proteins and nucleic acids, analogous conceptual and methodological differences occur. Arrhenius plots of the logarithms of association and dissociation rate constants versus reciprocal absolute temperature must be nonlinear if the van't Hoff plot of the logarithm of the corresponding equilibrium constant versus reciprocal absolute temperature is nonlinear. Interpretation of the resulting heat capacities of activation ($\Delta C_p^{\circ\ddagger}$) in terms of the hydrophobic effect (i.e., in terms of changes in the amount of water-accessible nonpolar surface in the various intermediates and/or transition states that determine the corresponding rate constant) provides molecular information about

[101] E. A. Moelwyn-Hughes, "Chemical Statics and Kinetics of Solutions." Academic Press, New York, 1971.

mechanism.[45] Likewise, effects of the concentration, valence, and species of electrolyte on the magnitude of the rate constant and on the log–log derivatives Sk_{assoc} and Sk_{dissoc} may be interpreted using polyelectrolyte theory (but *not* Brønsted–Bjerrum–Debye–Hückel theory) in terms of the participation of these ions in the various steps and transition states of the mechanisms.[46,88,94] These mechanistic analyses and molecular interpretations are discussed further in subsequent parts of this section.

Another fundamental issue, which remains to be resolved, regards the nature of a noncovalent transition state or activated complex, and more specifically the factors that determine its frequencies of decomposition into products and reactants. In the Eyring model, the frequency of decomposition of the transition state is the quantum mechanical frequency of vibration ν along the reaction coordinate ($\nu = k_B T/h$, where k_B is Boltzmann's constant and h is Planck's constant). Almost certainly this frequency factor is completely inappropriate for noncovalent transition states. In the absence of a theory to predict frequency factors for noncovalent transition states, one cannot obtain meaningful values of activation entropies ($\Delta S^{\circ\ddagger}$) or activation free energies $\Delta G^{\circ\ddagger}$ for noncovalent reaction steps from rate constants and their temperature dependences. If it is assumed that the frequency factors are invariant to changes in the recognition surfaces or in solution conditions, then relative values of activation entropies ($\Delta\Delta S^{\circ\ddagger}$) and activation free energies ($\Delta\Delta G^{\circ\ddagger}$) can be estimated.

A key similarity between attempts to relate kinetic data to mechanisms for noncovalent and covalent reactions is that no mechanism can be proven to be complete. Instead, one must proceed by demonstrating a minimal number and order of contributing steps, and by excluding possible alternative mechanisms on the basis of inconsistency with the kinetic data. A second similarity between analyses of mechanisms of noncovalent and covalent reaction is the reliance on mathematical approximations, such as the steady state assumption or rapid equilibrium approximations, in the analysis of a proposed mechanism. Even in the current era of powerful nonlinear fitting routines, such analytical approximations have their place, because of the insight provided by analytical expressions for the overall rate constants in terms of contributions from elementary rate constants. Strictly speaking, the assumptions involved in these approximations must be tested for all macromolecular sequences (i.e., site variants of the protein and/or DNA molecules) and environmental conditions (temperature, salt concentration, etc.) employed to be sure of their applicability. The nature and relative importance of individual kinetic steps are likely to be unique for each system investigated, and moreover may be different for different sequence variants of the recognition regions of the DNA and protein, and/or for different choices of environmental variables, since, in general, each step of the mechanism of formation of a specific complex displays a unique dependence on these variables.

Elementary Bimolecular Step in Protein–DNA Association

Magnitude of Diffusion–Collision Rate Constant. The elementary bimolecular diffusion–collision rate constant k_{dc}° for the reaction of two uncharged spheres, A and B, is given by the Smoluchowski equation:

$$k_{dc}^{\circ} = 4\pi(N/10^3)(D_A + D_B)(r_A + r_B) \, M^{-1} \, \text{sec}^{-1} \tag{28}$$

where the factor 4π is the spherical solid angle (indicating that all directions of approach of the spheres lead to reaction), D_A and D_B are diffusion coefficients (in $cm^2 \, sec^{-1}$), r_A and r_B are hydrodynamic radii (in cm), and N is Avogadro's number.

Proteins and nucleic acids are of course not adequately modeled as uncharged, uniformly reactive spheres. DNA is a highly charged locally cylindrical polyanion. Proteins may or may not be spherical, and more importantly are polyampholytes with an overall charge that is not uniformly distributed and is a function of pH. Only a small fraction of the molecular surface of either protein or DNA is "reactive." Long-range Coulombic interactions may increase or decrease the probability of collision. Introduction of these effects leads to an improved estimate of the diffusion–collision rate constant for noncovalent interactions of biopolymers (**R** and S)[88,102,103]:

$$k_{dc} = 4\pi\kappa f(N/10^3)(D_R + D_S)(r_R + r_S)M^{-1} \, \text{sec}^{-1} \tag{29}$$

where κ is the probability that the collision has the correct mutual orientation to lead to interaction, and f is a dimensionless factor that accounts for nonspherical geometry and long-range Coulombic effects ($f < 1$ for reactants of the same charge; $f > 1$ for reactants of opposite charge).[102] For the interaction of *lac* repressor (modeled as a sphere with a radius of 40 Å and an interaction surface that is approximately 20% of the total surface) with λ DNA (of negligible diffusion coefficient relative to *lac* repressor) containing the *lac* operator site (modeled as a cylinder with a radius of 10 Å and an interaction surface that is approximately 25% of the total surface of the operator site), von Hippel and Berg[103] estimate that $\kappa \approx 0.05$ and that the orientation-corrected diffusion-limited rate constant is on an order of magnitude of $10^8 \, M^{-1} \, \text{sec}^{-1}$ (neglecting geometrical and electrostatic corrections, which are expected to be relatively small). Since

[102] O. G. Berg and P. H. von Hippel, *Annu. Rev. Biophys. Biophys. Chem.* **14**, 131 (1985).
[103] P. H. von Hippel and O. G. Berg, *J. Biol. Chem.* **264**, 675 (1989).

the diffusion coefficients of spheres decrease slowly with increasing molecular weight ($D \propto M^{-1/3}$), this "order of magnitude" estimate should be generally applicable for site-specific interactions of proteins with macromolecular DNA; a somewhat higher estimate of the diffusion-limited rate constant will be obtained for the interactions of proteins with small DNA fragments, which diffuse more rapidly than the protein.

Rate constants for elementary bimolecular steps in mechanisms of noncovalent interactions may in principle be significantly less than those calculated as the diffusion–collision limit if the severity of orientation effects (an entropic activation barrier) has been underestimated in the use of Eq. (29). Enthalpic activation barriers in excess of that expected from the temperature dependence of the diffusion coefficient are inconsistent with an elementary diffusion-limited process. Observation of a large enthalpic barrier for a second order rate constant k_{assoc} indicates that it is not an elementary rate constant, and that additional steps, either preceding or following the bimolecular step, contribute to the second order rate constant (see below). For multistep mechanisms of association, if complex formation is investigated at sufficiently high dilution of reactants, if the reactants maintain the correct conformation and state of association upon dilution, and if any unimolecular steps subsequent to the bimolecular step are sufficiently fast in comparison to the rate of dissociation of the initial bimolecular complex, then (and only then) the overall rate of association should be determined by the orientation-corrected diffusion–collision rate of formation of the initial complex.

Effects of Temperature and Salt Concentration on Diffusion–Collision-Limited Reactions. The diffusion–collision rate constant k_{dc} [eq. (29)] is expected to be only weakly dependent on temperature, since the only enthalpic activation barrier to the process is that intrinsic to diffusion. Diffusion coefficients of solutes are proportional to the absolute temperature and inversely proportional to the temperature-dependent solvent viscosity η ($D \propto T/\eta$). In water, the temperature dependence of T/η leads to an estimated activation energy for a diffusion–collision interaction of approximately 5 kcal, corresponding to a 1.4-fold increase in a diffusion–collision-limited association rate constant between 25 and 37°. In addition to this relatively small dependence of k_{dc} on temperature, a related diagnostic is the inverse proportionality predicted to exist between k_{dc} and solvent viscosity. While changes in η have been used to examine whether reactions of small molecules are diffusion limited, additives such as sucrose or glycerol used to vary the viscosity of water over the range necessary to affect the rate constant to a significant extent also may affect the conformations and stabilities of interacting biopolymers as a result of weak, surface area-dependent preferential interactions.[104,105] Hence the effects of these solutes on the kinetics of association must be interpreted with caution.

Effects of long-range Coulombic interactions on the diffusion-limited rate constant for interactions of oligocations L^{z+} with the DNA polyanion have been examined by Lohman et al.[94] in the context of the two-state polyelectrolyte theory of Manning.[63,67] By analogy with the Brønsted–Bjerrum–Debye–Hückel theory of salt effects on reactions of low-molecular-weight charged species, Lohman et al. proposed that the electrolyte screens the interaction between DNA and L^{z+}, and hence reduces the rate of interaction of these oppositely charged reactants. For double-helical DNA, screening of phosphates is equivalent to the association of 0.12 univalent cation/phosphate.[66] The elementary diffusion–collision reaction is assumed to proceed via a low-entropy, a high-free energy "transition state" in which L^{z+} has penetrated the screening ion atmosphere surrounding the DNA and is localized in the vicinity of its surface, but in which the condensed cations have not yet been displaced. Assuming rapid equilibrium between transition state and reactants (as in covalent transition state theory), Lohman et al. predicted that screening effects should result in a modest linear dependence of the logarithm of k_{dc} on the logarithm of the univalent salt concentration[94]:

$$\frac{d \ln k_{dc}}{d \ln [\mathrm{MX}]} \equiv Sk_{dc} = -0.12z \qquad (30)$$

(This is probably a maximum estimate of the magnitude of Sk_{dc}, since use of a more realistic steady state analysis of the process of passing through the screening-controlled transition state will reduce the magnitude of Sk_{dc}.) As long as the ligand is an oligocation and not a polycation, screening effects are asymmetric and the consequences of screening of the charges on L^{z+} by MX are negligible by comparison.[78]

In the case of proteins like lac repressor, the situation is in principle more complicated because the net charge on the protein is negative, although the net charge on the DNA-binding domain is apparently positive. Goeddel et al.[106] found that the second order rate constant (k_{assoc}) for the interaction of lac repressor with a small synthetic operator (21 or 26 bp) was near the expected diffusion–collision limit (k_{assoc} ~1–2 × 10^9 M^{-1} $sec^{-1} \simeq k_{dc}$). [Their results exceeds the estimate of k_{dc} for the interaction

[104] J. C. Lee and S. N. Timasheff, J. Biol. Chem. 256, 7193 (1981).

[105] K. Gekko and S. N. Timasheff, Biochemistry 20, 4667 (1891).

[106] D. V. Goeddel, D. G. Yansura, and M. H. Caruthers, Proc. Natl. Acad. Sci. U.S.A. 74, 3292 (1977).

of *lac* repressor with λplac DNA ($\sim 10^8 \; M^{-1} \; \text{sec}^{-1}$) in part because of the higher diffusion coefficient of the small DNA fragment.] Goeddel *et al.* also found that k_{assoc} decreased slightly with increasing [KCl], with $Sk_{\text{assoc}} \simeq -0.5$ for wild-type repressor and $Sk_{\text{assoc}} \simeq -1$ for the tight-binding X86 repressor. These results are in semiquantitative agreement with the prediction of Eq. (30) since *lac* repressor behaves in site-specific binding as an oligocation with a valence z in the range $5 \leq z \leq 8$.[13–15,61,68] Both the magnitude and salt concentration dependence of k_{assoc} in this case appear consistent with a screening affected, diffusion–collision-limited rate constant. Furthermore it appears that the charge on the DNA-binding site of repressor is more important in determining Sk_{dc} than is the net charge on the protein.

Dependence of Elementary Rate Constant for Dissociation of L^{z+}*–DNA Complex on Salt Concentration.* If association of L^{z+} with DNA is a screening affected, diffusion–collision elementary process and if dissociation of the L^{z+}–DNA complex is also an *elementary* mechanistic step with rate constant k_{ed} (for elementary dissociation step) then the overall association–dissociation process can be represented as

$$L^{z+} + DNA \text{ ``site''} \underset{k_{\text{ed}}}{\overset{k_{\text{dc}}}{\rightleftharpoons}} \text{complex}$$

where $K_{\text{obs}} = k_{\text{dc}}/k_{\text{ed}}$. From Eq. (22) for SK_{obs} and Eq. (30) for Sk_{dc}, it follows that

$$Sk_{\text{ed}} = Sk_{\text{dc}} - SK_{\text{obs}} = 0.76z \qquad (31)$$

The dependence of k_{ed} on salt concentration [Eq. (31)] for this reaction results from the requirement that $0.76z$ univalent cations recondense on the DNA when L^{z+} dissociates. Hence cations are ''reactants'' in the dissociation process.[88,94] This is an example of what appears to be a general principle in analyzing effects of salt concentration on rate constants of ligand–DNA interactions. Rate constants for steps in which cations of the electrolyte associate with DNA exhibit a power dependence on salt concentration, where the exponent is the stoichiometry of participation of ions in the step. This principle even carries over to screening limited association, since the proposed origin of Sk_{dc} [Eq. (30)] is the reassociation of screening ions that accompanies dissociation of the ''transition state'' of that elementary step.[88,94] In mixtures of cations, especially of different valence, competitive interactions of the cations must be accounted for in estimating or analyzing salt concentration dependences of dissociation rate constants, as in thermodynamic analyses of SK_{obs}.[61,62]

Mechanisms Involving Intermediates

Calculated estimates of the magnitude of k_{dc} and of its relatively weak dependences on temperature and salt concentration provide convenient points of reference for comparisons of the behavior of experimentally determined second order rate constants k_{assoc} of site-specific protein–DNA associations. Generally one finds that $k_{assoc} < k_{dc}$, and that the magnitudes of the overall activation energy of association E_{assoc} and salt concentration dependence Sk_{assoc} are significantly larger than predicted for a diffusion–collision reaction. Here we consider some simple mechanisms involving intermediates that are consistent with second order kinetics. These demonstrate how temperature and salt concentration can be used to distinguish mechanistic alternatives.

Intermediates Subsequent to Elementary Bimolecular Step. Mechanism of open complex formation between RNA polymerase $E\sigma^{70}$ and promoter DNA: A moderately well-characterized example of a site-specific protein–DNA interaction in which conformational changes of the initial specific complex are mechanistically important is the association of *E. coli* RNA polymerase ($E\sigma^{70}$) with promoter DNA sequences. In excess RNA polymerase, the kinetics of association fit the form of Eq. (26) with $(\mathbf{R_T}-[\mathbf{RS}]) \simeq \mathbf{R_T}$, so that first order kinetics are observed:

$$\frac{d(S_T - [\mathbf{RS}])}{dt} = -k_{assoc}\mathbf{R_T}(S_T - [\mathbf{RS}]) \tag{32}$$

where $k_{assoc} \mathbf{R_T} \equiv k_{obs} = 1/\tau_{obs}$, and where τ_{obs} is a composite first order time constant. Kinetic studies demonstrate that τ_{obs} is a linear function of $\mathbf{R_T}$, with in general a positive (nonzero) intercept.[107,108] In other words, the kinetic order of association changes from second order (pseudofirst order) in the limit of low (excess) $\mathbf{R_T}$ to first order in the limit of very high $\mathbf{R_T}$.

Interpretation of this behavior and of the dependencies of the rate constants on temperature and salt concentration led to the proposal of a three-step minimal mechanism.[44–46,48,83,109,110] For association of $E\sigma^{70}$ with the λP_R promoter, the kinetically significant steps of the association mechanism under the conditions examined are found to be

$$\mathbf{R} + S \underset{k_{-1}}{\overset{k_1}{\rightleftharpoons}} I_1 \overset{k_2}{\rightarrow} I_2 \overset{k_3}{\rightarrow} \mathbf{RS} \tag{33}$$

[107] W. R. McClure, *Proc. Natl. Acad. Sci. U.S.A.* **77**, 5634 (1980).
[108] W. R. McClure, *Annu. Rev. Biochem.* **54**, 171 (1985).
[109] S. Rosenberg, T. R. Kadesch, and M. J. Chamberlin, *J. Mol. Biol.* **155**, 1 (1982).
[110] H. Buc and W. R. McClure, *Biochemistry* **24**, 2712 (1985).

where (to be consistent with the general notation of this chapter) $\mathbf{R}$ is the polymerase, S is the specific DNA site (promoter), $\mathbf{RS}$ is the functional ("open") complex (generally designated as RP_0) and I_1 and I_2 are intermediate closed complexes (generally designated RP_c and RP_i or RP_{c1} and RP_{c2}).

In excess $E\sigma^{70}$, the steady state solution of this mechanism (which assumes that the concentrations of the intermediate closed complexes are independent of time after an initial transient phase) is

$$k_{assoc} = \frac{k_{obs}}{\mathbf{R}_T} = \frac{k_1 k_2 k_3}{k_1 \mathbf{R}_T(k_2 + k_3) + k_3(k_{-1} + k_2)}$$
$$\tau_{obs} = \frac{k_{-1} + k_2}{k_1 k_2 \mathbf{R}_T} + \frac{1}{k_2} + \frac{1}{k_3} \tag{34}$$

In the limit of low concentration of $E\sigma^{70}$, k_{assoc} [Eq. (34)] approaches second order behavior:

$$\lim_{\mathbf{R}_T \to 0} k_{assoc} = \frac{k_1 k_2}{k_{-1} + k_2} \equiv k_a \tag{35}$$

where k_a is a composite second order rate constant.

In the limit of high concentration of $E\sigma^{70}$, from Eq. (34),

$$\lim_{\mathbf{R}_T \to \infty} (k_{assoc}\mathbf{R}_T) = \left(\frac{1}{k_2} + \frac{1}{k_3}\right)^{-1} \equiv k_i \tag{36}$$

and the association kinetics approach a true first order process, with the composite first order isomerization rate constant k_i. Expressed in terms of k_a and k_i, the time constant τ_{obs} [Eq. (34)] is

$$\tau_{obs} = \frac{1}{k_a \mathbf{R}_T} + \frac{1}{k_i} \tag{37}$$

This functional form is consistent with experimental observations on a wide variety of promoters.[108] [Mechanisms analogous to Eq. (33) with two or more steps yield the functional form of Eq. (37).]

Further information regarding intermediates on the pathway to formation of the functional specific complex may be deduced from measurement of the dissociation rate constant k_{dissoc}. For dissociation of $E\sigma^{70}$ from the λP_R promoter, the kinetically significant steps of dissociation under the conditions examined are

$$\mathbf{RS} \underset{k_3}{\overset{k_{-3}}{\rightleftharpoons}} I_2 \underset{k_2}{\overset{k_{-2}}{\rightleftharpoons}} I_1 \overset{k_{-1}}{\to} \mathbf{R} + S \tag{38}$$

Steady state analysis of mechanism (38), subject also to the assumption $k_3 \gg k_{-2}$ used to define the kinetically significant steps in the association mechanism (33), yields

$$k_{\text{dissoc}} = \frac{k_{-2}k_{-1}}{K_3(k_{-1} + k_2)} \tag{39}$$

where $K_3 \equiv k_3/k_{-3}$.

Consequently, from Eqs. (35) and (39),

$$\frac{k_a}{k_{\text{dissoc}}} = K_1K_2K_3 \cong K_{\text{obs}} \tag{40}$$

where $K_1 \equiv k_1/k_{-1}$, $K_2 \equiv k_2/k_{-2}$, and where K_{obs} is the observed equilibrium constant defined in Eq. (5). The approximate equality $K_1K_2K_3 \cong K_{\text{obs}}$ results from assuming that I_1 and I_2 are relatively unstable intermediates (in comparison to the open complex **RS**) that do not exist at significant concentrations at equilibrium.

Effects of temperature and salt concentration on composite rate constants for association and dissociation of $E\sigma^{70}$-promoter complexes:

1. Second Order Rate Constant k_a: The steady state solution for the second order rate constant k_a [Eq. (35)] for the $E\sigma^{70}$–promoter association process is in fact a general expression applicable to any situation in which one or more mechanistically important reversible bimolecular step(s) is followed by one or more mechanistically important irreversible unimolecular step(s) (e.g., Michaelis–Menten enzyme–substrate kinetics). Hence the following discussion is of more general applicability than to $E\sigma^{70}$–promoter interactions. The steady state expression for k_a encompasses a range of kinetic behavior, determined by the relative magnitudes of k_{-1} and k_2. The so-called rapid equilibrium limit of the steady state solution is obtained when $k_{-1} \gg k_2$, so that dissociation of the initial complex occurs frequently on the time scale of its isomerization. In this case,

$$\text{Rapid equilibrium } (k_{-1} \gg k_2): \quad k_a = \frac{k_1k_2}{k_{-1}} = K_1k_2 \tag{41}$$

The other limit is the sequential limit, in which $k_2 \gg k_{-1}$, so that formation of the initial complex is essentially irreversible on the time scale of its isomerization. In this case,

$$\text{Sequential } (k_2 \gg k_{-1}): \quad k_a = k_1 \tag{42}$$

Here the initial bimolecular step determines the association rate k_a. The steady state approximation is less accurate when the kinetics approach sequential behavior than when the rapid equilibrium limit is approached.

Temperature and salt concentration are useful variables to determine whether a second order rate constant k_a is closer to the rapid equilibrium or to the sequential limit of the steady state, and hence to determine relative values of k_{-1} and k_2. However, since k_{-1} and k_2 will generally exhibit different (and, in some cases, large) dependencies on these variables, one must recognize that both absolute and relative values of k_{-1} and k_2 may change over the ranges of temperatures and salt concentrations examined (see Activation Free-Energy Analysis of Association and Dissociation, below).

The activation energy E_a obtained from the temperature dependence of the steady state expression of k_a [Eq. (35)] is

$$E_a = RT^2 \frac{d \ln k_a}{dT} = \Delta H_1^\circ + \left(\frac{1}{k_{-1} + k_2}\right)(k_{-1}E_2 + k_2 E_{-1}) \qquad (43)$$

where ΔH_1° is the van't Hoff enthalpy change in the process of step 1 $[\Delta H_1^\circ = RT^2(d \ln K_1/dT)]$ and E_2 and E_{-1} are the Arrhenius activation energies of the corresponding elementary steps $[E_2 = RT^2(d \ln k_2/dT)$ and $E_{-1} = RT^2(d \ln k_{-1}/dT)]$. In the rapid equilibrium limit, $E_a = \Delta H_1^\circ + E_2$; in the sequential limit $E_a = E_1$.

Analogously, the salt concentration dependence of k_a, Sk_a, has the form

$$Sk_a = SK_1 + \left(\frac{1}{k_{-1} + k_2}\right)(k_{-1}Sk_2 + k_2 Sk_{-1}) \qquad (44)$$

In the rapid equilibrium limit, $Sk_a = SK_1 + Sk_2$; in the sequential limit $Sk_a = Sk_1$.

For the interaction of RNA polymerase with the λP_R promoter, $Sk_a = -12 \pm 1$,[46] independent of salt concentration over the range examined (cf. Fig. 4), but E_a is a function of temperature, ranging from 20 ± 5 kcal in the range 25–37° to 40 ± 15 kcal in the range 13–15° (cf. Fig. 5).[45] Since the magnitudes of k_a are less than the diffusion–collision limit over the range of conditions examined and since the magnitudes of Sk_a and E_a are much larger than expected for a diffusion-limited reaction, it is clear that the rapid equilibrium limit of the steady state is the appropriate one to consider, rather than the sequential limit. The decrease in E_a with increasing temperature could result from an intrinsic negative heat capacity of activation or from a large increase in the ratio k_2/k_{-1} with increasing temperature. In the latter possibility the kinetics would shift from near the

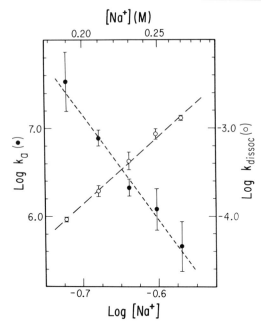

FIG. 4. Log–log plot of association (k_a) and dissociation (k_{dissoc}) rate constants for open complexes between Eσ^{70} RNA polymerase and the λP_R promoter as functions of NaCl concentration. The dashed lines are weighted linear least-squares fit to the data: log k_a = $-(1.2 \pm 0.7) - (11.9 \pm 1.1)$log[Na$^+$]; log k_{dissoc} = $(1.5 \pm 0.1) + (7.7 \pm 0.2)$log[Na$^+$]. [From J.-H. Roe and M. T. Record, Jr., *Biochemistry* **24**, 4721 (1985).]

rapid equilibrium limit at lower temperature to near the sequential limit at higher temperature. This possibility is inconsistent with the large magnitudes of Sk_a and E_a and with the analysis of dissociation kinetic data (below). Consequently, Roe et al.[45] concluded that the temperature dependence of E_a was the result of an intrinsic large negative $\Delta C_p^{\circ\ddagger}$, which they attributed to a protein conformational change that reduced the amount of nonpolar surface exposed to water in forming the transition state in isomerization of the initial closed complex ($I_1 \rightarrow I_2$). (ΔH_i° is relatively small and thought to be temperature independent.) The large magnitude of Sk_a appears to result from SK_1, which is the stoichiometry of ion release in forming the initial closed complex.[27,46,83]

2. First Order Dissociation Rate Constant k_{dissoc}: Analysis of k_a, Sk_a, and E_a for the interaction of Eσ^{70} with the λP_R promoter indicates that I_1 (the initial closed complex) is in rapid equilibrium with free Eσ^{70} and promoter DNA on the time scale of isomerization to I_2 ($k_{-1} \gg k_2$) over the range of conditions examined. In this limit,

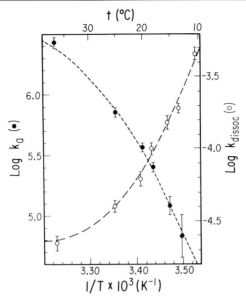

FIG. 5. Arrhenius plot of the temperature dependences of k_a and k_{dissoc} for the $E\sigma^{70}$–λP_R promoter interaction. Dashed curves are fits obtained using activation heat capacities $\Delta C_{p,a}^{\ddagger} = -1.13$ kcal K^{-1} and $\Delta C_{p,dissoc}^{\ddagger} = +1.28$ kcal K^{-1}, respectively. [From J.-H. Roe, R. R. Burgess, and M. T. Record, Jr., *J. Mol. Biol.* **184**, 441 (1985).]

$$k_{dissoc} = K_3^{-1} k_{-2}$$
$$E_{dissoc} = E_{-2} - \Delta H_3^{\circ} \qquad (45)$$
$$Sk_{dissoc} = Sk_{-2} - SK_3$$

Roe and Record found that $Sk_{dissoc} = 8 \pm 1$ (cf. Fig. 4)[46] and that E_{dissoc} was negative and highly temperature dependent, varying from -9 ± 4 kcal in the range 25–37° to -30 ± 10 kcal in the range 13–15 (cf. Fig. 5).[45] The negative E_{dissoc} is thought to result from the negative enthalpy change ($\Delta H_{-3}^{\circ} = -\Delta H_3^{\circ} < 0$) accompanying the conversion of the open complex (**RS**) to the intermediate closed complex (I_2), which is interpreted as the enthalpy of helix formation. (The fact that E_{dissoc} is negative demonstrates that the dissociation rate constant is a composite rate constant and not an elementary rate constant, since activation energies of elementary steps are positive.) The positive temperature dependence of E_{dissoc} appears to be an intrinsic positive heat capacity of activation, which is thought to result from the exposure of nonpolar surface in forming the transition state in the isomerization $I_2 \rightarrow I_1$. The large Sk_{dissoc} is thought to result primarily from $-SK_3$, which is interpreted as the stoichiometry of cation uptake that accompanies helix formation.

Intermediates Prior to Elementary Bimolecular Step. Conformational changes in the protein or DNA sites that occur prior to binding are consistent with overall second order kinetics of association, providing the conformational change is in rapid equilibrium on the time scale of association. Consider the hypothetical mechanism

$$\text{Rapid equilibrium } (k_{-1} \gg k_{dc}[\mathbf{R}],\ K_1 \equiv k_1/k_{-1}): \qquad \text{S} \underset{k_{-1}}{\overset{k_1}{\rightleftharpoons}} \text{S}' \tag{46}$$

$$\text{Diffusion-collision step:} \qquad \mathbf{R} + \text{S}' \overset{k_{dc}}{\longrightarrow} \mathbf{RS}$$

For this situation, the second order rate constant [cf. Eq. (26)] is

$$k_{assoc} = k_{dc}(1 + K_1^{-1})^{-1} \tag{47}$$

and therefore

$$E_{assoc} = E_{dc} - (1 + K_1)\Delta H_1^{\circ} \tag{48}$$

and

$$SK_{assoc} = \text{S}k_{dc} - (1 + K_1)^{-1}SK_1 \tag{49}$$

(A rapid conformational equilibrium in the protein would of course produce an analogous result. On the other hand, mechanisms involving intermediates in which the state of oligomerization of the protein changes prior to binding are generally inconsistent with overall second order kinetics of formation of $\mathbf{RS}$.)

In general, a conformational preequilibrium causes k_{assoc} to be significantly smaller than k_{dc}, and yields values of E_{assoc} and Sk_{assoc} that also should differ from the values expected for the diffusion–collision step alone. If intermediates also occur subsequent to the diffusion–collision step, clearly both will contribute to the behavior of k_{assoc} and its temperature and salt derivatives.

Nonspecific Binding as Competitor and/or as Facilitating Mechanism. All site-specific DNA-binding proteins exhibit a relatively weak and nonspecific binding mode drived principally by the polyelectrolyte effect. Equilibrium specificity ratios K_{RS}/K_{RD} for different binding proteins cover a wide range ($\sim 10^1$–10^8), and depend on ion concentrations, pH, temperature, and other environmental variables. Under conditions where nonspecific binding to a DNA molecule containing a specific site occurs to a significant extent, it may either retard or facilitate the rate of forming the site-specific complex.

Competition from nonspecific binding: Assume that nonspecific DNA sites D are present in excess, either on the same DNA molecule or on different DNA molecules from the specific sites, and that nonspecific binding equilibrates rapidly on the time scale of forming a specific complex. Then the competitive effect of nonspecific binding on the observed second order association rate constant for site-specific binding may be summarized by the mechanism

$$\text{Rapid equilibrium } (K_1 = k_1/k_{-1}): \quad \mathbf{R} + D \underset{k_{-1}}{\overset{k_1}{\rightleftharpoons}} \mathbf{RD} \tag{50}$$

$$\text{Elementary step:} \quad \mathbf{R} + S \xrightarrow{k_{dc}} \mathbf{RS}$$

(Conformational changes after formation of the initial site-specific complex may be included if they occur.)

If nonspecific sites are in excess, then $D_T = [D] + [\mathbf{RD}] \simeq [D]$ and the rapid equilibrium approximation for nonspecific binding yields the following expression for k_{assoc} [Eq. (26)]:

$$k_{assoc} = k_{dc}(1 + K_1 D_T)^{-1} \tag{51}$$

Competitive effects of DNA molecules not bearing the specific site are readily examined by determining the dependence of k_{assoc} on D_T.

Facilitation from nonspecific binding to a DNA molecule: Riggs et al.[99] made the striking observation that the second order rate constant k_{assoc} [Eq. (26)] for formation of a specific complex between *lac* repressor and the *lac* operator on phage λ DNA ($M_r \sim 3 \times 10^7$; 45 kbp) exceeded the calculated diffusion–collision limit ($\sim 10^8 \, M^{-1} \, \text{sec}^{-1}$) at all salt concentrations examined, and exceeded $10^{10} \, M^{-1} \, \text{sec}^{-1}$ at low salt concentration. Winter et al.[111] and Barkley[112] confirmed and extended these findings (cf. Fig. 6A). Near 0.1 M [KCl], k_{assoc} attains a maximum value of approximately $2 \times 10^{10} \, M^{-1} \, \text{sec}^{-1}$; above 0.1 M KCl, log k_a decreases linearly with log[K^+],[111] as predicted from analysis[94] of the kinetic data of Riggs et al.[99] in a mixed Mg^{2+}/K^+ buffer. Below 0.1 NaCl, k_a decreases with decreasing [K^+] to a low-salt plateau of approximately $3 \times 10^9 \, M^{-1} \, \text{sec}^{-1}$ for the repressor–operator interaction on λ plac DNA (Fig. 6A).[111]

To explain these observations, Richter and Eigen[113] and Berg, von Hippel, and co-workers[42,102,103,111] applied the general principle of Adam and Delbruck[114] that a reduction in dimensionality of the search will result in facilitation of a diffusional process. von Hippel and Berg[103] discuss the

[111] R. B. Winter, O. G. Berg, and P. H. von Hippel, *Biochemistry* **20,** 6961 (1981).
[112] M. D. Barkley, *Biochemistry* **20,** 3833 (1981).
[113] P. Richter and M. Eigen, *Biophys. Chem.* **2,** 255 (1974).
[114] G. Adam and M. Delbruck, *in* "Structural Chemistry and Molecular Biology" (A. Rich and N. Davidson, eds.), p. 198. Freeman, New York, 1968.

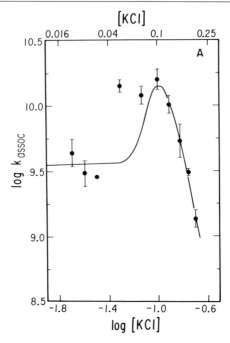

FIG. 6. Plot of log k_{assoc} vs log[KCl] for the interaction of *lac* repressor with the *lac* operator region contained on (A) λ_{plac} DNA (~45 kbp above) and (B) a 6.7-kbp *Eco*RI fragment (p. 337). The theoretical predictions of O. G. Berg, R. W. Winter, and P. H. von Hippel [*Biochemistry* **20**, 6929 (9181)] for a diffusional model of facilitation involving nonspecific binding (with the experimental value of SK_{obs}^{RD}) and sliding are shown by the solid lines. [Reprinted with permission from R. B. Winter, O. G. Berg, and P. H. von Hippel, *Biochemistry* **20**, 6961 (1981).]

possible random diffusional mechanisms by which nonspecific DNA sites on a molecule carrying a specific site may participate in the mechanism by which a protein locates the specific site. The domain of the DNA flexible coil provides a large target; the protein rapidly associates in a nonspecific manner with DNA sites in the domain by three-dimensional diffusion. Transfer within the domain to the specific site may involve one or more of the following diffusional processes: one-dimensional diffusion along the DNA chain ("sliding"), intersegment transfer (transient "looping"), and relatively long-lived inelastic collisions ("hopping"). None of these transfer processes is adequately expressed by a reaction mechanism with a small number of elementary steps.

For the situation in which sliding is the dominant transfer mechanism, Berg *et al.*[42] derive the result

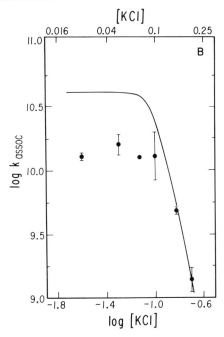

FIG. 6 *(continued)*

$$k_{assoc} = k_{dc}(1 + K_1 D_T)^{-1} F \qquad (52)$$

where F is the factor by which the competition-adjusted diffusion–collision rate constant [Eq. (51)] is increased by sliding. At high salt concentration, where nonspecific binding is relatively weak ($K_1 D_T \ll 1$), F is proportional to the square root of the nonspecific binding constant ($F \propto K_1^{0.5}$), because in this limit (where competitive effects of nonspecific binding are negligible) the difference between k_{assoc} and k_{dc} results from the effective extension of the operator site to include adjacent regions of nonspecific DNA from which the repressor may diffuse to the operator target before dissociating. In this limit, $Sk_{assoc} \simeq 0.5\ SK_1$. The predicted value of Sk_{assoc} is therefore -6 ± 1, based on $SK_1 = -12 \pm 2$. This compares well with the value $Sk_{assoc} = -5.3$ obtained by analysis[94] of the data of Riggs *et al.*,[99] and with that estimated from the data of Winter *et al.*[111] ($Sk_{assoc} \simeq -5$). At lower salt concentrations, the increasingly important competitive effect of nonspecific sites on a very large DNA molecule causes k_{assoc} to pass through a maximum (at the salt concentration where $K_1 D_T = 1$) and then decrease; this behavior is not observed on a 6.7-kbp DNA molecule, where the rate constant reaches a plateau at low salt concentration where the sliding range encompasses the entire molecule (cf. Fig. 6B).[111] At all salt concentrations (below 0.2 M) and all lengths of DNA molecules investigated, the facilitation factor F is much greater than unity; experimental rate constants k_{assoc} generally exceed $10^9 \, M^{-1} \, sec^{-1}$. The level of agreement between the experimental dependences of k_{assoc} on salt concentration and DNA length[111] and the predictions of the sliding facilitated mechanism of Berg *et al.*[42] provide strong support for this mechanism, although the ability of *lac* repressor to form stable looped complexes between distant pairs of specific DNA sites[115,116] suggests that the direct transfer (transient looping) mechanism may also play a role in facilitation of specific binding under some solution conditions.[42]

Whereas a value of k_{assoc} that is significantly greater than the calculated diffusion–collision limit (k_{dc}) provides strong evidence for a facilitating mechanism involving nonspecific sites, this does not imply that a value of k_{assoc} that is less than k_{dc} provides evidence against the existence of such mechanisms. For example, in a multistep mechanism such as Eq. (33), where an initial specific complex (I_1) rapidly equilibrates with free reactants on the time scale of subsequent conformational changes to form the functional specific complex **RS**, the magnitude and salt dependence of k_{assoc} provide no information regarding whether the formation of I_1 is facilitated. If formation of I_1 involves transfer from nonspecific sites, dissociation of I_1 will also be facilitated by this transfer step, and the equilibrium constant K_1 will be unaffected. In this case, facilitating mechanisms are not kinetically significant even if they do occur. For sequences or solution conditions where the association kinetics are described by the sequential limit of the steady state, then facilitating mechanisms will be kinetically significant if they do occur.

Activation Free Energy Analysis of Association and Dissociation

The energetics of the process of site-specific complex formation by a DNA-binding protein may be represented schematically on a "progress" diagram that plots the relative standard free energies of reactants and the various transition states and intermediates on the pathway. Standard free energy differences (ΔG_i°) between stable states (reactants, intermediates, products) are related to equilibrium constants (K_i) by $\Delta G_i^\circ = -RT \ln K_i$. Standard activation free energy differences ($\Delta G_i^{\circ\ddagger}$) between transi-

[115] H. Kramer, N. Niemöller, M. Amouyal, B. Revet, B. von Wilcken-Bergmann, and B. Müller-Hill, *EMBO J.* **6**, 1481 (1987).
[116] G. R. Bellomy and M. T. Record, Jr., *in* "Progress in Nucleic Acid Research and Molecular Biology" (W. Cohn and K. Moldave, eds.), Vol. 39, p. 81. Academic Press, New York, 1990.

tion states and initial (stable) states are related to the rate constants (k_i) of the corresponding elementary steps by an extension of the Eyring equation to noncovalent processes: $\Delta G_i^{o\ddagger} = -RT \ln(k_i/A_i)$, where A_i represents the various frequency and nonideality factors for formation and dissociation of the transition state of the ith reaction step.[117,118] Estimates of the magnitude of A_i require assumptions regarding these factors, and are available only for simple covalent transition states. Values of $\Delta G_i^{o\ddagger}$ cannot at present be calculated from rate constants of noncovalent processes. The temperature dependence of a rate constant provides relatively direct information about $\Delta H_i^{o\ddagger}$. However, $\Delta S_i^{o\ddagger}$ cannot be evaluated, for the same reasons as discussed above for $\Delta G_i^{o\ddagger}$. An additional complication in processes involving large changes in water-accessible nonpolar surface and/or vibrational modes is that the existence of a nonzero $\Delta C_{p,i}^{o\ddagger}$ results in $\Delta H_i^{o\ddagger}$ and $T\Delta S_i^{o\ddagger}$ being strong functions of temperature, while $\Delta G_i^{o\ddagger}$ is relatively temperature invariant.

Often one is interested in interpreting a composite rate constant in terms of cumulative free energy barriers along a reaction pathway. For example, in the multistep mechanism of Eq. (33) (and in classical Michaelis–Menten kinetics for a one-substrate reaction) the composite second order rate constant k_a [from Eq. (35)] is given by

$$k_a = \frac{k_1 k_2}{k_{-1} + k_2} = K_1 k_2 \left(1 + \frac{k_2}{k_{-1}} \right)^{-1} \tag{53}$$

Therefore

$$-RT \ln k_a = \Delta G_1^{o} + \Delta G_2^{o\ddagger} - RT \ln A_2 + RT \ln(1 + k_2/k_{-1})$$
$$\Delta G_a^{o\ddagger} \equiv \Delta G_1^{o} + \Delta G_2^{o\ddagger} = -RT \ln k_a/A_2 - RT \ln(1 + k_2/k_{-1}) \tag{54}$$

where $\Delta G_1^{o} = -RT \ln K_1$ and $\Delta G_2^{o\ddagger} = -RT \ln k_2/A_2$. Only in the rapid equilibrium limit ($k_{-1} \gg k_2$) is the composite activation free energy barrier $\Delta G_a^{o\ddagger}$ directly related to the composite experimental second order rate constant k_a, although the correction term $RT \ln(1 + k_2/k_{-1})$ is not large unless $k_2 \gg k_{-1}$ (the sequential limit). Current investigations of effects of protein and DNA sequence variants on kinetic parameters seek to interpret ratios of rate constants k_a in terms of differences in activation free energy barriers $\Delta\Delta G^{o\ddagger}$.[118-120] For these interpretations to be quantitatively valid, the terms A_2 and k_2/k_{-1} must be the same for all variants.

Figure 7A is a schematic diagram of the relative free energy levels of

[117] S. L. Leirmo, Ph.D. Thesis, University of Wisconsin-Madison (1989).
[118] B. Beutel, Ph.D. Thesis, University of Wisconsin-Madison (1990).
[119] A. Fersht, *Biochemistry* **26**, 8031 (1987).
[120] J. A. Wells, *Biochemistry* **29**, 8509 (1990).

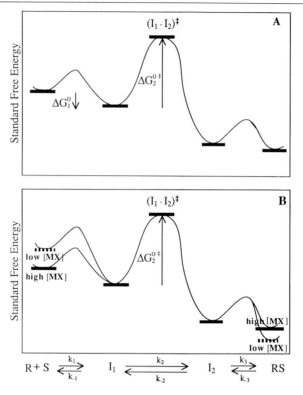

$$R + S \underset{k_{-1}}{\overset{k_1}{\rightleftharpoons}} I_1 \underset{k_{-2}}{\overset{k_2}{\rightleftharpoons}} I_2 \underset{k_{-3}}{\overset{k_3}{\rightleftharpoons}} RS$$

FIG. 7. Schematic free energy diagrams for the two-intermediate mechanism of interaction of $E\sigma^{70}$ (or $E\sigma^{32}$) RNA polymerase with a promoter to form an open complex. (A) The two-intermediate mechanism under conditions (e.g., high salt concentration, $T \simeq 37°$) where rapid equilibrium approximations ($k_{-1} \gg k_2; k_3 \gg k_{-2}$) are applicable in analysis of both k_{assoc} and k_{dissoc}. [An example of this behavior is the $E\sigma^{70}-\lambda P_R$ promoter interaction (see text).] ΔG_1° is the standard free energy difference between the initial closed complex (I_1) and reactants (**R**, S). $\Delta G_2^{\circ\ddagger}$ is the height of the activation barrier between I_1 and the transition state $(I_1 - I_2)^\ddagger$. This step $I_1 \rightarrow I_2$ is thought to involve a large conformational change in $E\sigma^{70}$. In the rapid equilibrium approximation, the composite second order rate constant k_a is related to the net activation barrier ($\Delta G_a^{\circ\ddagger} = \Delta G_1^\circ + \Delta G_2^{\circ\ddagger}$) between reactants (**R**, S) and transition state $(I_1 - I_2)^\ddagger$ by

$$k_a = A_2 e^{-\Delta G_a^{\ddagger}/RT}$$

where A_2 is the combination of frequency and nonideality factors involved in passing from intermediate I_1 to the transition state $(I_1 - I_2)^\ddagger$ [cf. Eq. (54)]. (B) Schematic demonstrating the effect of a reduction in univalent salt concentration [MX] on the standard free energies of promoter DNA and the open complex, and the resulting effect of [MX] on the rate constants k_{-1} and k_{-3} for the steps of the mechanism that involve cation uptake. The curve labeled "high [MX]" is the same as in Fig. 7A. At low [MX], the free energy barriers for steps requiring cation uptake ($\Delta G_{-1}^{\circ\ddagger}$, $\Delta G_{-3}^{\circ\ddagger}$) are increased, so the rate constants k_{-1} and k_{-3} are reduced from their high [MX] values. Consequently the kinetics of association at the various transition states, stable intermediates, reactants, and products for the prototypic mechanism of open complex formation by RNA polymerase at a promoter [Eq. (33)] at physiological temperature ($\sim37°$) and cation concentration (~0.2 M K^+). At this temperature and relatively high salt concentration, the highest activation free energy barrier in each direction appears to be that separating the two intermediate closed complexes.[44–46,83,117] Because the surrounding barriers are not as high, rapid equilibrium approximations ($k_{-1} \gg k_2$; $k_3 \gg k_{-2}$) appear appropriate to analyze both association and dissociation kinetic data (cf. Fig. 7A).

Changes in salt concentration have major effects on the relative heights of these barriers. As the salt concentration is reduced, the height of the free energy barrier to dissociation of the intermediate I_1 to free DNA increases, because of the additional free energy cost at low salt concentration of accumulating cations in the transition state by which I_1 dissociates (cf. Fig. 7B). At low salt concentration the free energy of the free DNA site also increases. The association kinetics are predicted to shift from rapid equilibrium toward sequential as the salt concentration is reduced, because the barrier for $I_1 \rightarrow \mathbf{R} + \mathbf{S}$ increases with decreasing salt concentration, whereas the barrier for isomerization ($I_1 \rightarrow I_2$) is thought to involve a conformational change in RNA polymerase and therefore be relatively salt concentration independent. The kinetics of association of $E\sigma^{32}$ RNA polymerase with the *groE* promoter on a 450-bp restriction fragment appear to shift in this manner when the [NaCl] is reduced.[28] At low salt concentration, the second order association rate constant k_a approaches the estimated diffusion–collision limit ($k_{dc} \simeq 10^8$ M^{-1} sec^{-1}) and the association process is sequential; at higher salt concentration k_a is drastically reduced because of the increase in the rate constant for dissociation of the initial closed complex (k_{-1}).[28] The kinetics of *lac* repressor–*lac* operator association may also exhibit a shift from rapid equilibrium toward sequential with decreasing salt concentration.[111] At the higher salt concentrations examined (>0.1 M KCl), the facilitating nonspecific binding step appears to be in rapid equilibrium on the time scale of specific complex formation. At lower salt concentration, the coupled processes of nonspecific association and transfer become more sequential, until in the low salt limit the lifetime of the nonspecific complex is sufficiently long that dissociation

---

sufficiently low salt concentration may no longer conform to the rapid equilibrium approximation, since k_{-1} may no longer be much larger than k_2. [This situation appears to apply to the interaction of $E\sigma^{32}$ with the *groE* promoter (see text).] Since the location of the progress diagram on the free energy axis is arbitrary, the diagram designated "low [MX]" has been normalized so that the [MX]-independent step $I_1 \rightarrow I_2$ is at the same free energy level as at "high [MX]."

apparently does not occur on the time scale of transfer to the specific
site.[111]

Summary

The concentration and nature of the electrolyte are key factors determining (1) the equilibrium extent of binding of oligocations or proteins
to DNA, (2) the distribution of bound protein between specific and nonspecific sites, and (3) the kinetics of association and dissociation of both
specific and nonspecific complexes. Salt concentration may therefore be
used to great advantage to probe the thermodynamic basis of stability and
specificity of protein–DNA complexes, and the mechanisms of association
and dissociation. Cation concentration serves as a thermodynamic probe
of the contributions to stability and specificity from neutralization of DNA
phosphate charges and/or reduction in phosphate charge density. Cation
concentration also serves as a mechanistic probe of the kinetically significant steps in association and dissociation that involve cation uptake. In
general, effects of electrolyte *concentration* on equilibrium constants
(quantified by SK_{obs}) and rate constants (quantified by Sk_{obs}) are primarily
cation effects that result from the cation-exchange character of the interactions of proteins and oligocations with polyanionic DNA. The competitive
effects of Mg^{2+} or polyamines on the equilibria and kinetics of protein–DNA interactions are interpretable in the context of the cation-exchange model. The nature of the anion often has a major effect on the
magnitude of the equilibrium constant (K_{obs}) and rate constant (k_{obs}) of
protein–DNA interactions, but a minor effect on SK_{obs} and Sk_{obs}, which
are dominated by the cation stoichiometry. The order of effects of different
anions generally follows the Hofmeister series and presumably reflects the
relative extent of preferential accumulation or exclusion of these anions
from the relevant surface regions of DNA-binding proteins. The question
of which anion is most inert (i.e., neither accumulated nor excluded from
the relevant regions of these proteins) remains unanswered.
The characteristic effects of temperature on equilibrium constants and
rate constants for protein–DNA interactions also serve as diagnostic
probes of the thermodynamic origins of stability and specificity and of the
mechanism of the interaction, since large changes in thermodynamic and
activation heat capacities accompany processes with large changes in the
amount of water-accessible nonpolar surface area. Although more work
remains to be done to verify the generality and the quantitative implications of these large heat capacity effects, one may anticipate that they will
serve to detect and quantify the contributions of removal of nonpolar
surface from water (the hydrophobic effect) to the steps of site-specific

binding, in much the same way that cation concentration effects have served to detect and quantify the contributions of neutralization of DNA phosphates and/or reduction in phosphate charge density (the polyelectrolyte effect).

Acknowledgments

Work from the authors' laboratory was supported by grants from the NIH and NSF. We thank Drs. T. M. Lohman, R. S. Spolar, and V. J. LiCata for discussions of aspects of this material and their comments on the manuscript, and thank Sheila Aiello for her assistance in its preparation.

[17] Detecting Cooperative Protein–DNA Interactions and DNA Loop Formation by Footprinting

By ANN HOCHSCHILD

Introduction

Weak interactions between adjacently and nonadjacently bound transcriptional regulators play an essential role in gene regulation in both prokaryotes and eukaryotes. In particular, many examples of "action at a distance" (i.e. regulatory events mediated by proteins bound far from the start point of transcription) involve long-range interactions between proteins bound at separated sites on the DNA. Such interactions induce the formation of loops in the DNA (for reviews, see Refs. 1 and 2).

Long-range protein–protein interactions were first implicated in the regulation of transcription by studies of the arabinose and galactose operons of *Escherichia coli*. The classical mechanism of transcriptional repression in prokaryotes is steric occlusion; the promoter and operator are overlapping sites and bound repressor prevents RNA polymerase from gaining access to the promoter. However, transcriptional repression of the arabinose operon was found to depend on an operator site located far upstream of the promoter.[3] This repression depends on an interaction between a molecule bound at the upstream site (located at position -280

---

[1] S. Adhya, *Annu. Rev. Genet.* **23**, 227 (1989).

[2] A. Hochschild, *in* "DNA Topology and Its Biological Effects" (N. R. Cozzarelli and J. C. Wang, eds.), p. 107. Cold Spring Harbor Laboratory, Cold Spring Harbor, New York, 1990.

[3] T. M. Dunn, S. Hahn, S. Ogden, and R. F. Schleif, *Proc. Natl. Acad. Sci. U.S.A.* **81**, 5017 (1984).

relative to the start point of transcription) and a second molecule bound near the start point of transcription.[4] Repression of the galactose operon also depends on an interaction between nonadjacently bound repressors, one bound upstream of the promoter (at position -61) and the other bound within the first structural gene (at position $+53$).[5-8] These are examples of two different mechanisms for transcriptional repression, both of which involve the formation of a DNA loop.

Transcriptional regulators of other prokaryotic operons also participate in long-range interactions,[9-15] as do DNA-binding proteins involved in other processes, such as site-specific recombination and DNA replication.[16-20] The function of these interactions differs from example to example. In some cases the interaction may serve simply to increase site occupancy. At the *lac* operon, for example, occupancy of the primary operator (which overlaps the promoter) is enhanced by the presence of two pseudo-operators, located 93 bp upstream and 401 bp downstream of the primary operator, respectively.[12] In other cases, the interaction may serve to impose a structural constraint on the intervening DNA. At the *gal* operon, for example, site occupancy alone is not sufficient to mediate repression; rather the looped structure contributes directly to repression.[8] Finally, long-range interactions may serve to bring together noncontiguous segments of DNA (such as in site-specific recombination) or to promote the formation of a well-defined higher order structure.

Several methodologies, including DNase footprinting, the gel-binding assay, and electron microscopy, have been exploited to demonstrate protein-mediated DNA loop formation *in vitro*. This chapter will focus on

[4] K. Martin, L. Huo, and R. F. Schleif, *Proc. Natl. Acad. Sci. U.S.A.* **83**, 3654 (1986).

[5] M. H. Irani, L. Orosz, and S. Adhya, *Cell (Cambridge, Mass.)* **32**, 783 (1983).

[6] H.-J. Fritz, H. Bicknase, B. Gleumes, C. Heibach, S. Rosahl, and R. Ehring, *EMBO J.* **2**, 2129 (1983).

[7] A. Majumdar and S. Adhya, *Proc. Natl. Acad. Sci. U.S.A.* **81**, 6100 (1984).

[8] R. Haber and S. Adhya, *Proc. Natl. Acad. Sci.U.S.A.* **85**, 9683 (1988).

[9] M. C. Mossing and M. T. Record, *Science* **233**, 889 (1986).

[10] H. Kramer, M. Niemoller, M. Amouyal, B. Revet, B. V. Wilcken-Bergmann, and B. Muller-Hill, *EMBO J.* **6**, 1481 (1987).

[11] S. Sasse-Dwight and J. D. Gralla, *J. Mol. Biol.* **202**, 107 (1988).

[12] S. Oehler, E. R. Eismann, H. Kramer, and B. Muller-Hill, *EMBO J.* **9**, 973 (1990).

[13] G. Dandanell and K. Hammer, *EMBO J.* **4**, 3333 (1985).

[14] L. J. Reitzer and B. Magasanik, *Cell (Cambridge, Mass.)* **45**, 785 (1986).

[15] A. J. Ninfa, L. J. Reitzer, and B. Magasanik, *Cell (Cambridge, Mass.)* **50**, 1039 (1987).

[16] H. Echols, *Science* **133**, 1050 (1986).

[17] M. Gellert and H. Nash, *Nature (London)* **325**, 401 (1987).

[18] N. L. Craig, *Annu. Rev. Genet.* **22**, 77 (1988).

[19] R. C. Johnson and M. I. Simon, *Trends Genet.* **3**, 262 (1987).

[20] S. Mukherjee, H. Erickson, and D. Bastia, *Cell (Cambridge, Mass.)* **52**, 375 (1988).

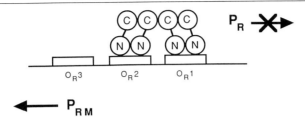

FIG. 1. Cooperative binding of λ repressor to adjacent sites in the right operator (O_R). In a λ lysogen repressor dimers bind cooperatively to O_R1 and O_R2, thereby repressing transcription from promoter P_R and activating transcription from promoter P_{RM}. O_R1 is a high-affinity site, whereas O_R2 and O_R3 are lower affinity sites. Thus, the dimer at O_R1 stabilizes the association of the second dimer with O_R2 and, at physiological concentrations of repressor, O_R3 remains mostly unoccupied because only two dimers cooperate at one time. Note that the carboxyl domain of repressor mediates the cooperative interaction.

DNase footprinting, as applied to complexes formed by the binding of λ repressor to artificially separated operator sites.[21,22]

Repressor Structure

λ repressor is a two-domain protein that binds its 17-bp operator as a dimer. Repressor dimers bind cooperatively to adjacent operator sites on the phage chromosome (see Fig. 1).[23] The amino domain of the repressor contacts the DNA, whereas its carboxyl domain mediates both dimerization and the cooperative interaction between adjacently bound dimers.[24,25]

Cooperative Binding

The DNase footprinting assay for protein-mediated DNA loop formation is based on the fact that an interaction between DNA-bound proteins will be reflected in cooperative DNA binding. This cooperativity should be detectable *in vitro* whether or not it is physiologically relevant *in vivo*. (Depending on the intracellular concentration of protein, site occupancy *in vivo* may or may not be enhanced appreciably by cooperative binding.)

[21] A. Hochschild and M. Ptashne, *Cell (Cambridge, Mass.)* **44**, 681 (1986).
[22] A. Hochschild and M. Ptashne, *Nature (London)* **336**, 353 (1988).
[23] A. D. Johnson, B. J. Meyer, and M. Ptashne, *Proc. Natl. Acad. Sci. U.S.A.* **76**, 5061 (1979).
[24] C. O. Pabo, R. T. Sauer, J. M. Sturtevant, and M. Ptashne, *Proc. Natl. Acad. Sci. U.S.A.* **76**, 1608 (1979).
[25] The amino domain itself dimerizes weakly and binds DNA even when separated from the carboxyl domain.

Cooperative binding is detected as follows. A template is prepared carrying two protein-binding sites (A and B). A second template is prepared carrying only one of the two sites (B, for example). Footprints are performed under equilibrium conditions to determine whether or not the affinity of the protein for site B is higher when site A is present on the same DNA fragment (see Fig. 2).

More specifically, we wish to determine the equilibrium dissociation constant for the binding of a repressor dimer to site B in the presence and absence of site A on the same DNA fragment. This equilibrium can be represented as follows

$$R_2O \rightleftharpoons R_2 + O \qquad (1)$$

where O represents free operator, R_2 represents free dimer, and R_2O represents bound operator.[26] The equilibrium dissociation constant for the repressor dimer/operator interaction is defined

$$K_D = [R_2][O]/[R_2O] \qquad (2)$$

At half-maximal site occupancy $[O] = [R_2O]$, and $K_D = [R_2]$. The quantity that we can actually measure in a footprint experiment is the total repressor concentration at which half-maximal site occupancy is achieved $[R_T]_{1/2}$. To determine K_D we must, therefore, be able to derive $[R_2]_{1/2}$ from $[R_T]_{1/2}$. This calculation is complicated by the fact that repressor exists as a mixture of monomers and dimers at equilibrium.[27] The monomer/dimer equilibrium

$$R_2 \rightleftharpoons R + R \qquad (3)$$

is governed by the equilibrium dissociation constant

$$K_1 = [R]^2/[R_2] = 2 \times 10^{-8} \, M \qquad (4)$$

If the operator concentration is low compared with the total repressor concentration at which half-maximal site occupancy is achieved, the presence of the operator-bearing DNA will not significantly perturb the monomer/dimer equilibrium, and we may calculate $[R_2]_{1/2}$ as follows. Using the stoichiometric relation

$$[R_T] = [R] + 2[R_2] + 2[R_2O] \qquad (5)$$

we replace $[R_2]$ in Eq. (4), obtaining

$$K_1 = 2[R]^2/([R_T] - [R] - 2[R_2O])$$

or

[26] A. D. Johnson, C. O. Pabo, and R. T. Sauer, this series, Vol. 65, p. 839.
[27] R. T. Sauer, Ph.D. Dissertation, Harvard University, Cambridge, Massachusetts (1979).

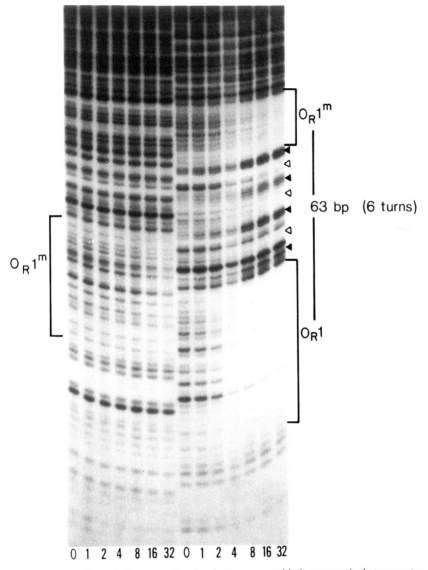

FIG. 2. DNase footprinting assay showing that λ repressor binds cooperatively to operator sites separated by six turns of the DNA helix. (Reprinted from Ref. 21; copyright held by Cell Press, Cambridge, MA.) The binding of repressor to operator site O_R1^m is measured in the absence (left panel) and presence (right panel) of a high-affinity site (O_R1) six turns away on the same DNA template. Reactions were incubated with increasing concentrations of repressor, a concentration of 1 corresponding to 13.5 nM repressor monomer. The experiment shows that between 16 and 32 relative units of repressor was required to occupy O_R1^m half-maximally on the single-site template, whereas only 4 relative units was required on the two-site template. Changes in the DNase I sensitivity of the region spanning the two sites are indicated by arrowheads; the black arrowheads indicate enhanced cleavages and the open arrowheads indicate diminished cleavages.

$$[R]^2 + (K_1/2)[R] = (K_1/2)Q \tag{6}$$

where

$$Q = [R_T] - 2[R_2O]$$

Solving Eq. (6) for [R], we find

$$[R] = -(K_1/4) + \sqrt{(K_1/4)^2 + (K_1/2)Q} \tag{7}$$

Since $[R_T] \gg [O_T]$, we may replace Q with $[R_T]$, giving

$$[R] = -(K_1/4) + \sqrt{(K_1/4)^2 + (K_1/2)[R_T]} \tag{8}$$

We thus obtain [R] as a function of the known quantities K_1 and $[R_T]_{1/2}$. We then use Eq. (5) to solve for $[R_2]$, i.e., K_D. Again, since $[R_T] \gg [O_T]$, we can simplify Eq. (5):

$$2[R_2] = [R_T] - [R] \tag{9}$$

Consider a numerical example. Assume that half-maximal occupancy of a given site occurs at a total repressor concentration of 10^{-8} M (expressed as repressor monomer) and that the concentration of operator containing DNA is $<10^{-10}$ M. Using Eq. (8), $[R] = 6 \times 10^{-9}$ M, and from Eq. (9) $[R_2] = 2 \times 10^{-9}$ M. Note that under these conditions we do not need to know the precise concentration of the operator containing DNA and, further, that small variations in operator concentration from experiment to experiment will not affect our determination of K_D.

The extra binding energy coming from the cooperative interaction will be reflected primarily in an increase in the affinity of the protein for the weaker of the two sites if they differ in strength (see Table I and legend for some sample values and derivation). This is because among the singly bound DNA molecules the strong site is more likely to be occupied than the weak, and the presence of the strong site will increase binding to the weak site more than the other way around. In practice, therefore, cooperativity is most easily detected by examining binding to a relatively weak site in the presence and absence of a strong site.

The Effect of Inter-operator Spacing on DNA Loop Formation

In our experiments with λ repressor we observed a periodic dependence of cooperativity on the spacing between the two binding sites. We detected cooperative binding when the center-to-center distance between the two operators was an integral number (n) of turns of the DNA helix (5, 6, or 7), assuming 10.5 bp/turn, but not when the inter-operator distance was close to $n + 1/2$ turns (4.6, 5.5, or 6.4). Since repressor binds primarily to one side of the DNA helix, the bound repressor molecules are located

TABLE I
THE EFFECT OF COOPERATIVITY ON SITE OCCUPANCY[a]

K_1 (M)	K_2 (M)	K_{coop}	K_2^* (M)	K_2/K_2^*
3.3×10^{-9}	3.3×10^{-9}	3.6×10^{-2}	6.3×10^{-10}	5
3.3×10^{-9}	5.0×10^{-8}	3.6×10^{-2}	3.5×10^{-9}	14
3.3×10^{-9}	5.0×10^{-7}	3.6×10^{-2}	2.1×10^{-8}	24
3.3×10^{-9}	5.0×10^{-6}	3.6×10^{-2}	1.8×10^{-7}	28

[a] Consider two operators, O_1 and O_2, located on the same DNA molecule. K_1 and K_2 are dissociation constants that describe the intrinsic affinities of O_1 and O_2 for repressor and K_{coop} is a constant that describes the cooperative term (defined below). The values chosen for K_{coop},[28] K_1 (all lines), and for K_2 (line 2) are based on the λ system; 3.3×10^{-9} and 5×10^{-8} M are the intrinsic dissociation constants for the binding of repressor to $O_R 1$ and $O_R 2$, respectively. K_2^* is a dissociation constant for the binding of repressor to O_2 when the binding is cooperative, and the ratio K_2/K_2^* is thus the factor by which binding to O_2 increases due to cooperativity. The table shows that for fixed values of K_1 and K_{coop}, the ratio K_2/K_2^* increases as O_2 is weakened. That is, the weaker O_2 is compared with O_1, the more the energy of cooperativity is reflected in an increase in the affinity of O_2 for repressor. For the given value of K_{coop}, the ratio K_2/K_2^* approaches a limiting value of approximately 28 when essentially all of the energy of cooperativity contributes to an increase in the affinity of O_2 for repressor.

To understand the meaning of K_{coop}, consider the following equilibrium relations: $K_1 = [R][O]/[RO_1]$ and $K_2 = [R][O]/[RO_2]$, where [R] denotes the concentration of free repressor, [O] the concentration of free DNA molecules, $[RO_1]$ the concentration of DNA molecules on which O_1 alone is occupied, and $[RO_2]$ the concentration of DNA molecules on which O_2 alone is occupied. If the binding to the two sites is noncooperative, then the dissociation of R_2O (a DNA molecule on which both O_1 and O_2 are occupied) to free repressor and free operator is described by the following equilibrium relation: $K_1 K_2 = [R]^2[O]/[R_2O]$. If binding to the two sites is cooperative, we define K_{coop} such that $[R]^2[O]/[R_2O] = K_1 K_2 K_{coop}$. K_{coop} can be determined experimentally by comparing the intrinsic affinity of each operator site for repressor with the affinity of that site when the binding is cooperative. That is, K_1 and K_1^* can be measured in separate experiments, as can K_2 and K_2^*. $K_{coop} = K_1^* K_2^*/K_1 K_2$.

The following calculation was performed to compute K_2/K_2^* for each value of K_2 (see Ackers et al.[28] for a detailed discussion of the λ system). Let fO_2 stand for the fractional occupancy of O_2. We can write:

$$fO_2 = \frac{[RO_2] + [R_2O]}{[O] + [RO_1] + [RO_2] + [R_2O]}$$

Dividing top and bottom by [O] and using the equilibrium relations, $K_1 = [R][O]/[RO_1]$, $K_2 = [R][O]/[RO_2]$, and $K_1 K_2 K_{coop} = [R]^2[O]/[R_2O]$, we can rewrite the equation:

$$fO_2 = \frac{[R]/K_2 + [R]^2/K_1 K_2 K_{coop}}{1 + [R]/K_1 + [R]/K_2 + [R]^2/K_1 K_2 K_{coop}}$$

We can now set $fO_2 = 0.5$ and solve for [R], which equals K_2^*. Performing the appropriate algebra gives the simplified equation:

$$[R]^2/K_1 K_2 K_{coop} + \{(K_1 - K_2)/K_1 K_2\}[R] - 1 = 0$$

on the same side of the helix when the operators are separated by an integral number of turns and on opposite sides when the sites are separated by $n + 1/2$ turns. To bring together two molecules that are bound on opposite sides of the DNA helix, bending of the DNA is insufficient; twisting and/or writhing is also required.

The amount of energy required to twist the DNA and bring sites that are on opposite sides of the helix into alignment for loop formation can be derived from the free energy of supercoiling, $\Delta G\tau$, which has been measured as a function of DNA length by Horowitz and Wang.[29] According to their data the cost of misalignment by half a turn is less than 0.1 kcal/mol for separations greater than ~2500 bp and ~3 kcal/mol for separations ~200 bp in length. With separations that are less than 100 bp, the cost of misalignment is predicted to be very high: ~7 kcal for 100 bp and ~13 kcal for 50 bp.[30] Therefore, the fact that we fail to observe cooperative binding to nonintegrally spaced sites separated by less than 100 bp is not surprising.

To strengthen the argument that the observed cooperative binding was mediated by a direct protein–protein interaction and that the apparent side-of-the-helix effect was due to the difficulty of bringing together molecules bound on opposite sides of the helix, we introduced a small gap into one strand of the DNA between two operators separated by a nonintegral number of turns (see Fig. 3A). Under these circumstances the DNA is

---

[28] G. K. Ackers, A. D. Johnson, and M. Shea, *Proc. Natl. Acad. Sci. U.S.A.* **79**, 1129 (1982).
[29] D. Horowitz and J. C. Wang, *J. Mol. Biol.* **173**, 75 (1984).
[30] J. C. Wang and G. N. Giaever, *Science* **240**, 300 (1988).

---

or

$$[R]^2 + (K_1 - K_2)K_{coop}[R] - K_1 K_2 K_{coop} = 0$$

Solving for [R], we find

$$[R] = \frac{(K_2 - K_1)K_{coop} + \sqrt{(K_1 - K_2)^2 K_{coop}^2 + 4K_1 K_2 K_{coop}}}{2}$$

Note that when $K_2 = K_1$, $[R] = K_1\sqrt{K_{coop}}$, and when $K_2 \gg K_1$ (O_2 significantly weaker than O_1),

$$[R] = \frac{K_2 K_{coop} + \sqrt{K_2^2 K_{coop}^2 + 4K_1 K_2 K_{coop}}}{2}$$

$$= \frac{K_2 K_{coop} + \sqrt{K_2 K_{coop}(K_2 K_{coop} + 4K_1)}}{2}$$

Finally, when K_2 is several orders of magnitude greater than K_1, $K_2 K_{coop} \gg 4K_1$, [R] approaches the limiting value $K_2 K_{coop}$, and the ratio K_2/K_2^* approaches $1/K_{coop}$.

expected to turn freely around the single-stranded region, and therefore the energetic barrier that would otherwise prevent cooperative binding should be removed. This prediction was borne out and repressor bound cooperatively to operators separated by 6.4 turns when a four-nucleotide gap was introduced into one of the DNA strands roughly midway between the operators (Fig. 3B).[21]

It should be noted that there may be proteins that bind cooperatively to nonadjacent operator sites irrespective of their angular alignment. For example, consider a hypothetical protein that binds its operator as a dimer and wraps around the DNA in such a way that the monomers lie on opposite sides of the DNA helix. Given such a configuration, one can imagine interactions involving either two or four monomers that would occur regardless of the spacing between the operators. It should also be noted that flexibility in the protein might reduce the amount of twisting that would be required to allow molecules that are not optimally aligned to interact. Finally, as the distance between binding sites is increased, any periodic effect of spacing is expected to diminish. Thus, although the observation of a periodic effect of spacing on cooperativity argues in favor of a direct protein–protein interaction, the absence of such an effect does not rule out the interaction.

Changes in the Susceptibility of Bent DNA to DNase Cleavage

The distortion of the DNA backbone that is induced by the cooperative binding of λ repressor to separated operators results in a change in the sensitivity of the DNA to DNase I cleavage. The binding of repressor to integrally spaced operators is accompanied by the appearance of an alternating set of enhanced and diminished cleavages affecting the DNA between the two operators (see Fig. 2). The enhanced and diminished cleavages are separated by roughly 5 bp; assuming that the interaction between the bound repressors induces a smooth bend in the DNA, the enhanced cleavages lie on the outside of the protein/DNA complex, whereas the diminished cleavages lie on the inside of this complex. A similar pattern of enhanced and diminished cleavages was previously observed with a small circular DNA molecule,[31] and it has also been observed in other cases of protein-mediated loop formation.[32-35]

[31] H. R. Drew and A. A. Travers, *J. Mol. Biol.* **186**, 773 (1985).

[32] H. Kramer, M. Niemoller, M. Amouyal, B. Revet, B. V. Wilcken-Bergmann, and B. Muller-Hill, *EMBO J.* **6**, 1481 (1987).

[33] J. J. Salvo and N. D. F. Grindley, *EMBO J.* **7**, 3609 (1988).

[34] D. Valenzuela and M. Ptashne, *EMBO J.* **8**, 4345 (1989).

[35] P. A. Beachy, unpublished observations with a Ubx protein from *Drosophila*.

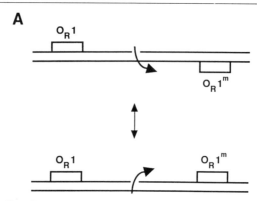

FIG. 3. The binding of repressor to a gapped duplex carrying nonintegrally spaced operator sites. (A) Schematic illustration of the gapped duplex, indicating free rotation around the single-stranded region. (B) DNase footprinting assay showing that repressor binds cooperatively to operator sites separated by 6.4 turns of the DNA helix on a gapped template. (Reprinted from Ref. 21; copyright held by Cell Press, Cambridge, MA.) The DNA template in the left six lanes is an intact duplex carrying O_R1 and O_R1^m separated by 6.4 turns of the DNA helix, and the DNA template in the right six lanes is the same except that it bears a four-nucleotide gap (on the unlabeled strand) in the region spanning the two operators. A repressor concentration of 1 corresponds to 13.5 nM repressor monomer. The experiment shows that ~16 relative units of repressor was required to occupy O_R1^m half-maximally on the intact duplex, whereas only about 2 relative units was required on the gapped duplex. Note that the presence of the gap changes the cleavage pattern between the two operators. However, the binding of repressor to the gapped template does not induce a pattern of enhanced and diminished cleavages, as is observed with the ungapped template.

These periodic changes in the susceptibility to DNase I cleavage of the DNA spanning two binding sites are a very sensitive indicator of loop formation. Under conditions where the stability of the looped and unlooped complexes are similar and cooperative binding may not be detectable, one may nevertheless detect changes in the cleavage pattern arising from a subpopulation of protein/DNA complexes that is looped.

The enhanced and diminished cleavages that are observed with bent DNA presumably reflect the narrowing and widening of the minor groove on the inside and outside of the bend, respectively. In addition, the DNA on the inside of the bent protein/DNA complex may be less accessible to DNase I. The presence of these anomalies in the cleavage pattern suggests that there is a single preferred conformation of the bent DNA in the looped complex. In fact, these anomalies are not seen when repressor binds to gapped DNA; presumably this is because the gapped DNA adopts a variety of conformations, or because the gapped DNA is kinked rather than bent smoothly. Likewise, when the protein-binding sites are located relatively far apart we would not necessarily expect to observe the anomalous cleav-

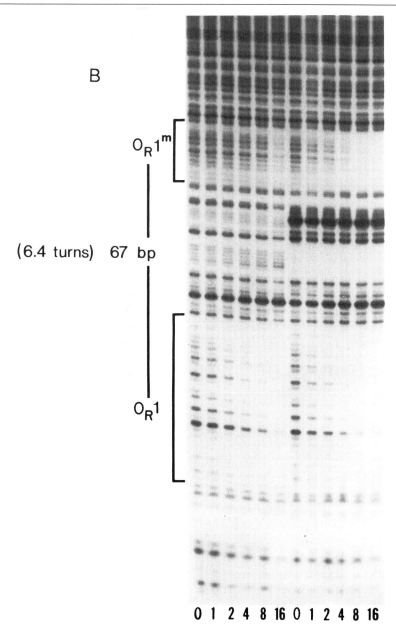

FIG. 3. (*continued*)

ages, because the conformation of the intervening DNA is apt to be more variable.

Protein Fragments and Fusion Proteins

In the case of λ repressor, cooperative binding to separated operator sites, like cooperative binding to adjacent operators, depends on the carboxyl domain; the purified amino domain binds noncooperatively to operators separated by both integral and nonintegral numbers of turns. Furthermore, a single amino acid substitution in the carboxyl domain of the repressor eliminates cooperative binding to both nonadjacent and adjacent operators.[22]

When carrying out experiments with protein fragments or with fusion proteins one may, of course, fail to observe cooperative binding even though the intact protein binds cooperatively. However, it is also possible that a fusion protein may bind cooperatively although the native protein does not. The foreign protein may, for example, mediate the formation of oligomers that can bind simultaneously to two binding sites. This situation would be analogous to that of *lac* repressor, which forms stable tetramers that can mediate loop formation by binding simultaneously to a pair of nonadjacent operators.[10,11]

Effect of Distance on Cooperative Binding

It would be useful to know how the probability of loop formation changes as a function of the distance separating the two protein-binding sites (see Refs. 2 and 30 for a more detailed discussion). When the sites are relatively close together, as are the λ operators on the constructs described here, this is a very complex problem. However, when the sites are sufficiently far apart on the DNA so that the size of the protein is negligible compared with the distance separating the binding sites, we can consider an analogous reaction, the cyclization of DNA fragments with complementary single-stranded ends. The cyclization probabilities, called *j* factors, have been measured previously as a function of fragment length.[36–38] The *j* value, which can be defined as the ratio of two equilibrium constants, K_c/K_a, where K_c is the cyclization constant and K_a is the bimolecular equilibrium association constant for joining two linear molecules, is a measure of the effective concentration of one end of the molecule

[36] J. C. Wang and N. R. Davidson, *J. Mol. Biol.* **19**, 469 (1966).

[37] J. E. Mertz and R. W. Davis, *Proc. Natl. Acad. Sci. U.S.A.* **69**, 3370 (1972).

[38] D. Shore, J. Langowski, and R. L. Baldwin, *Proc. Natl. Acad. Sci. U.S.A.* **78**, 4833 (1981).

in the vicinity of the other. Both theoretical and experimental analyses indicate that j reaches a maximum of $\sim 10^{-7}\ M$ for DNA fragments of ~ 700 bp. With decreasing fragment length, DNA flexibility becomes a critical parameter and the j value is determined primarily by an enthalpic term that describes the cost of deforming and bending the DNA. With increasing fragment length, on the other hand, the DNA more nearly resembles a flexible coil and the j value is determined primarily by an entropic term that describes the loss in configurational entropy of the DNA molecule when its ends are held in proximity to one another. This loss increases with fragment length; that is, the probability of one end finding the other decreases with increasing fragment length.

Based on this analogy with ring formation, the effective concentration of one DNA-bound protein in the vicinity of the other is not expected to exceed $10^{-7}\ M$, at least *in vitro*. We can therefore expect to observe loop formation over some range of intersite separations provided that the K_a for the bimolecular reaction of two DNA bound proteins exceeds $10^7\ M^{-1}$. This follows since K_c, the equilibrium constant for ring formation, equals $j(K_a)$, which is >1 if $K_a > 10^7\ M^{-1}$ (see Fig. 4). If we know the equilibrium association constant for the relevant protein–protein interaction in solution ($\sim 10^6\ M^{-1}$ for the association of free λ repressor dimers to form tetramers)[27] it is reasonable to assume that the K_a for the association of proteins bound to separate DNA molecules will be at least as strong. If there is an allosteric effect of DNA binding on the oligomerization reaction, the DNA-bound molecules may interact more strongly than free molecules in solution. It is also possible that an entropic effect may favor the interaction of DNA-bound molecules; for example, their immobilization on the DNA may serve to orient the interacting surfaces. In the case of λ repressor we do not observe loop formation at large separations, suggesting that the K_a for the bimolecular association of the DNA-bound dimers is less than $10^7\ M^{-1}$, or less than 10 times the value measured for free dimers ($10^6\ M^{-1}$).

Now let us consider relatively short separations. If the size of the protein is significant compared with the distance separating the two binding sites, the looped complex will not in fact resemble a DNA ring. In the case of interacting λ repressor molecules separated by about six turns of the helix the complex can be modeled as a semicircle bridged by the interacting protein molecules. Under these circumstances both DNA flexibility and protein flexibility are expected to be important parameters. Treating the protein as a rigid body one might expect that increasing the length of DNA over which the bend is distributed would facilitate complex formation. However, our observations with λ repressor are inconsistent with this expectation. The magnitude of the cooperative effect is relatively constant

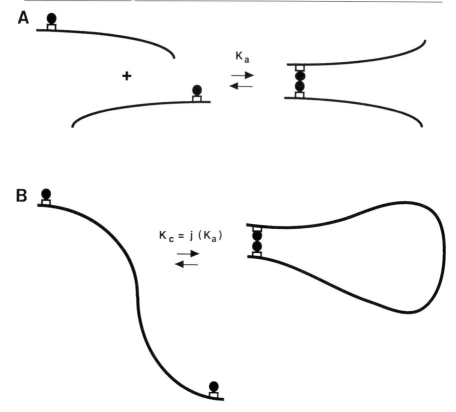

Fig. 4. Using the j value to estimate the probability of protein-induced loop formation. (A) The bimolecular association of two DNA-bound protein molecules. K_a is the bimolecular equilibrium constant for the association of two protein molecules that are prebound to separate DNA fragments. (B) The unimolecular association of two DNA-bound protein molecules. K_c is the unimolecular equilibrium constant for the association of two protein molecules that are prebound to a single DNA molecule. The j value is defined as the ratio K_c/K_a (see text), and so $K_c = j(K_a)$. Loop formation is favored provided that $K_c > 1$.

at 5, 6, 7, and 8 turns (within the sensitivity of our assay), and then diminishes gradually until 20 turns, at which point little or no cooperativity can be detected. These observations suggest that the detailed conformation of the looped protein–DNA complex may not be constant over this range of separations and that factors other than DNA flexibility are playing a significant role in determining the overall energetics of loop formation.

Construction of DNA Templates Carrying One or Two Operators

Several related templates were constructed carrying one or both of two repressor-binding sites, O_R1 and a mutant derivative of O_R1 (O_R1^m) to which repressor binds more weakly ($K_D O_R1^m = 8[K_D O_R1]$). To construct this series of templates we inserted a 45-bp DNA fragment carrying λ O_R1 at various positions within the polylinker of pEMBL8+, as outlined in Fig. 5.

To generate the gapped DNA template we prepared heteroduplexes by annealing an excess of single-stranded circles derived from the 6.4-turn plasmid to the 6.0-turn parent plasmid linearized at the BamHI site (see Fig. 5). The single-stranded circles were prepared by superinfecting plasmid-containing cells with phage F1 as described by Dente and Cortese.[39] The two DNA samples were mixed so that the single-stranded circles were present in an approximate fivefold molar excess over the double-stranded linears. The DNA in this mixture was denatured and renatured according to a procedure described by Kirkegaard and Wang[40] to produce the gapped circles.

Preparation of Labeled Restriction Fragments

The plasmid DNAs (~3 μg) were digested with HindIII and PvuII (see map) to generate an operator-containing fragment of ~250 bp in the case of the single-operator construction and fragments of ~300 bp in the case of the two-operator constructions. The gapped circles were treated in the same manner as the intact plasmids. The fragments were 3′ end labeled at the HindIII end with [α-32P]dATP and reverse transcriptase (purchased from Life Sciences, St. Petersburg, FL). On the gapped template this protocol results in the incorporation of the label at the end of the intact strand. The radiolabeled fragments were purified after electrophoresis through a 5% polyacrylamide gel. In each case the appropriate band was excised from the gel. The following protocol was used to elute the DNA: The gel slice is placed in a 15-ml tube and crushed with a glass rod. After the addition of 2.5 ml elution buffer (500 mM NH$_4$Ac, 10 mM MgAc, 1 mM EDTA, 0.1% SDS),[41] the tubes are rocked overnight at 37°. The samples are then centrifuged for several minutes in a clinical centrifuge to pellet the acrylamide. The supernatant containing the labeled DNA is poured into a clean tube. After the addition of ~15 μg of carrier tRNA, it is dripped through a 0.45-μm Millex filter (Millipore, Bedford, MA) into a 30-ml centrifuge tube. Ethanol (2.5 vol) is added and the labeled DNA is precipitated. The DNA pellet is redissolved in 400 μl of 0.3 M NaOAc, transferred to an Eppendorf tube, and reprecipitated.

To obtain an estimate of the specific activity of each DNA fragment, the amount of radioactivity in the dried DNA pellet is determined by

[39] L. Dente and R. Cortese, this series, Vol. 155, p. 111.
[40] K. Kirkegaard and J. C. Wang, *J. Mol. Biol.* **185**, 625 (1985).
[41] A. M. Maxam and W. Gilbert, this series, Vol. 65, p. 499.

A

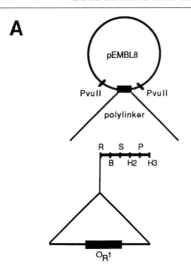

B

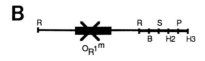

5.5-turn

6.0-turn

C

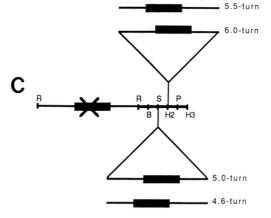

5.0-turn

4.6-turn

D

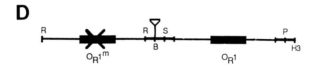

Cerenkov counting. For the purpose of these estimates, we assume 100% recovery of each gel-purified DNA fragment (so as not to underestimate the concentration of operator in the final reaction). Ideally it is desirable to load 10,000 cpm of labeled fragment in each lane so that a suitable footprint can be obtained after overnight exposure of the gel. In our case we wished to keep the concentration of DNA fragment below 0.2 nM, well below the repressor concentration required for half-maximal occupancy of our strongest site, ~20 nM. We generally obtain 1–2 × 10^6 cpm/labeled restriction fragment, or 20–40,000 cpm/2.4 × 10^{-14} mol fragment. The reaction volume is 200 μl (see below), and therefore we can normally use 20,000–40,000 cpm/reaction without exceeding a concentration of 0.12 nM in DNA fragment.

DNase Footprinting Reactions

Our footprinting protocol is essentially that of Johnson et al.[23] The reactions are performed in a buffer consisting of 10 mM Tris-HCl (pH 7.0), 2.5 mM MgCl$_2$, 1 mM CaCl$_2$, 0.1 mM EDTA, 200 mM KCl, bovine serum albumin at 100 μl/ml, and chick blood DNA at 2.5 μl/ml. After the addition of the labeled DNA fragment to an appropriate volume of protection buffer, the mixture is added to microcentrifuge tubes in 200-μl aliquots. Various amounts of repressor are added to these reaction tubes and the mixtures are allowed to come to equilibrium for 20 min at 37°. Fresh dilutions of the repressor stock solution are prepared for each experiment. The reactions are initiated by the addition of a predetermined amount of DNase I (see below) at 30-sec intervals. Each tube is quickly mixed by gentle vortexing after the addition of the DNase. The mixtures are incubated at 37° for 12 min, and then stopped by the addition of 50 μl of cold

---

FIG. 5. Construction of plasmid DNAs carrying one or two λ operators. The single-operator and two-operator constructions were generated as follows: (1) An *Eco*RI fragment carrying λ O$_R$1 was inserted into the unique *Eco*RI site of pEMBL8 + ; (2) a single base-pair substitution was introduced into O$_R$1 by site-directed mutagenesis to generate O$_R$1^m. The resulting plasmid was the source of the single-operator template; (3) the O$_R$1 fragment was reintroduced into either the *Sma*I site or the *Hin*dII site. O$_R$1 is not centered on the *Eco*RI fragment, and thus insertion of the fragment in one or the other orientation generates different spacing between O$_R$1 and O$_R$1^m. This set of plasmids was the source of four of the two-operator templates; (4) 4 bp was inserted between O$_R$1 and O$_R$1^m on the 6.0-turn construction by cutting the plasmid at the unique *Bam*HI site, filling out the 3' ends, and religating the blunt ends, thus generating the 6.4-turn construction. The gapped duplex was prepared by annealing single-stranded circular DNA carrying the 4-bp insertion at the *Bam*HI site to the 6.0-turn parent plasmid linearized at the *Bam*HI site. Restriction sites are abbreviated as follows: *Eco*RI, R; *Bam*HI, B; *Sma*I, S; *Hin*cII, H2; *Pst*I, P; and *Hin*dIII, H3.

8 M NH$_4$Ac containing 250 μg/ml tRNA and 1 ml of cold ethanol. The DNA is precipitated by immersing the reaction tubes into a dry ice–ethanol bath for ~5 min. The tubes are then centrifuged for 5 min in a microcentrifuge. The pellets should be clearly visible. The supernatant is aspirated; if care is taken to remove all visible droplets it is not necessary to rinse the pellets. After drying the pellets for several minutes under vacuum, the amount of radioactivity in each tube is again determined by Cerenkov counting.

In some cases it is necessary to subject the samples to a phenol/chloroform extraction prior to the precipitation (when using protein extracts, for example). This step was needed when examining the binding of mutant proteins that, because of a low affinity for operator DNA, had to be added at high concentrations. If the final DNA sample does not enter the gel properly, this may be an indication that the phenol/chloroform step is needed. In that case, the reactions are stopped by the addition of 200 μl of phenol/chloroform, the aqueous phase is transferred to a clean tube, and the DNA is precipitated as described above.

Finally the pellets are resuspended in 6–10 μl of loading dye consisting of 80% (v/v) deionized formamide/50 mM Tris-borate (pH 8.3)/1 mM EDTA/0.1% xylene cyanol/0.1% bromphenol blue. Prior to electrophoresis through an appropriate percentage acrylamide sequencing gel (in our case 8%), the samples are heated at 90° for 90 sec and then chilled on ice. If the samples are stored prior to electrophoresis it is preferable to freeze the dry pellets. However, we generally found that samples could be saved and rerun, if necessary, without significant fragment breakage.

Before carrying out a set of footprint experiments, a pilot experiment is performed to determine the optimum concentration of DNase I. The concentration should be adjusted so that each DNA molecule is cleaved only once between the labeled end and a point beyond the region of interest. That is, the probability of obtaining a second cut somewhere between the labeled end and a primary cut within the region of interest should be reduced as much as possible. Thus, the further the region of interest is from the labeled end, the lower the appropriate concentration of DNase I. The pilot experiment is performed over a range of DNase I concentrations to find a concentration at which the DNA is neither underreacted (too much uncleaved fragment and ladder too faint) nor overreacted (little or no uncleaved fragment and smaller fragments more abundant than larger fragments). For the DNA fragments decribed here we found that adding DNase to a final concentration of ~5 ng/ml gave a suitable amount of cleavage. We store our DNase I (purchased from Worthington, Freehold, NJ) in frozen aliquots at 1 mg/ml in H$_2$O. Fresh dilutions should be prepared for each experiment.

Before performing a series of experiments with a given protein preparation, the DNA mixture should be incubated with protein in the absence of DNase to ensure that there is no contaminating nuclease activity. In our experience, preparations of partially purified λ repressor have a significant amount of endogenous nuclease activity.

Experimental Reproducibility

To facilitate direct comparisons in these experiments, it is advisable to treat the single-site template as a standard and to measure binding to that template along with the two-site templates in any given experiment. If the K_D measurements are not consistent from experiment to experiment, possible sources of the variability include the following: (1) the protein preparation may be unstable; (2) the binding of some proteins (such as λ repressor) is quite temperature dependent and consequently small variations in the incubation temperature may have a significant effect on the K_D; (3) the concentration of DNA fragment may be too high so that the measured binding constant is not actually independent of DNA concentration. A good test of the latter possibility is to vary the DNA concentration intentionally over a small range and check that the K_D is not affected.

Section III

Biochemical Analysis of Protein–Nucleic DNA Interactions

[18] DNA Contacts Probed by Modification Protection and Interference Studies

By ANDREAS WISSMANN and WOLFGANG HILLEN

Introduction

Many methods for studying the sequence-specific recognition of DNA by proteins are available to date. One approach is to obtain cocrystals of the complex of interest and solve its X-ray structure at high resolution. This yields detailed information about the chemical nature of the contacts. To achieve this goal will usually take several years and the results reflect a static description of a dynamic complex. Another approach is to study protein–DNA complexes in solution using chemical methods.[1] This yields information about the complex much more quickly and is also very valuable for the confirmation and extension of crystallographic results to solution conditions.

The use of chemical methods relies on the fact that some chemicals react specifically with DNA and lead to breakage of the strand at the site of modification. Using mild reaction conditions only a single hit per molecule of DNA needs to be achieved, leading to a nucleotide-specific cleavage pattern. This may be influenced by the presence of a protein bound tightly to a specific sequence in the fragment. The result can either be protection of a nucleotide covered by the protein or enhanced reaction of another nucleotide in close proximity to the protein. This approach is outlined schematically in Fig. 1. Some examples of protection experiments are covered by other chapters in this volume. Here, modification protection protocols using dimethyl sulfate for purine N-7 methylation[2] and diethyl pyrocarbonate for carbethoxylation of purine N-7[3] will be presented.

Another particularly quick and informative chemical method to probe DNA–protein contacts is to determine sites where chemical modifications interfere with binding of the protein. In the case of methylation this is complementary to the protection experiment. The general approach of this experiment is outlined in Fig. 2. The crucial step is to separate modified DNA fragments according to their affinity for the binding protein. One portion should contain modifications only at positions where recognition

---

[1] U. Siebenlist, R. B. Simpson, and W. Gilbert, *Cell (Cambridge, Mass.)* **20,** 269 (1980).
[2] W. Hillen, K. Schollmeier, and C. Gatz, *J. Mol. Biol.* **172,** 185 (1984).
[3] L. Runkel and A. Nordheim, *J. Mol. Biol.* **189,** 487 (1986).

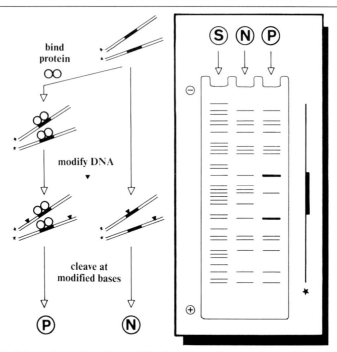

FIG. 1. Schematic outline of a modification–protection experiment. The left-hand side displays the experimental approach. DNA fragments are shown as bars with the protein-binding site in black. Asterisks denote a radioactive label at one end of a DNA strand. Protein is represented by a circle and is shown as a dimer. The modification is indicated by the black triangle. For the protection experiment the protein–DNA complex is modified and the DNA cleaved at the modified bases (P). In parallel, the same procedure is done for free DNA (N). At the right of the gel a schematic representation of the radioactively labeled DNA strand is given, the protein-binding site being indicated by the black box. (S) denotes a Maxam and Gilbert G + A reaction of the same DNA fragment as used in (P) and (N). In lane P residues protected by the protein can be identified as missing bands and residues with enhanced reactivity by increased intensities.

is not affected. The other portion contains modifications in sites preventing protein binding. This goal can either be achieved by polyacrylamide gel electrophoresis or by nitrocellulose filtration. In this chapter, interference experiments using dimethyl sulfate (purine N-7 methylation), N-ethyl-N-nitrosourea (phosphate ethylation), and diethyl pyrocarbonate (purine N-7 carbethoxylation) will be described.

Both experiments, modification protection (Fig. 1) and interference (Fig. 2), identify specific sites of the DNA that are located close to the bound protein. Experimental conditions for both approaches will be detailed and examples are presented for Z-DNA–antibody[3] and Tet repres-

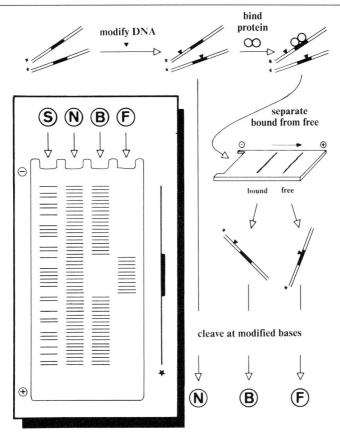

FIG. 2. Schematic outline of a modification interference experiment. The upper and right-hand side display the experimental approach. DNA, protein, and reagent are symbolized as in Fig. 1. The DNA is modified and a part of the sample is directly cleaved at the modified bases (N). The remaining portion of the modified DNA is incubated with the protein to allow binding. The bound and nonbound (free) fractions of the DNA are separated by gel electrophoresis, eluted, cleaved, and analyzed on sequencing gels (B, bound DNA; F, free DNA). On the left-hand side the anticipated results are presented. At the right of the gel a schematic representation of the radioactively labeled DNA strand is depicted with the protein-binding site indicated by a black box. (S) denotes a Maxam and Gilbert G + A reaction of the DNA fragment. The three other lanes show typical results of an ethylation interference experiment. Lane N shows all possible lengths of cleaved DNA fragments indicating the non-base-specific modification. DNA fragments analyzed in lane B define those sites of the DNA, where modification does not interfere with protein binding. The opposite holds true for sample F. Here, DNA fragments indicate that the modification resulted in interference with binding of the protein.

sor–*tet* operator complexes.[2,4] The technical aspects will be presented first, followed by a description of the results. Finally, the interpretation of the results will be discussed.

Modification of DNA

The DNA of interest needs to be labeled at one end of a single strand. This is achieved by published procedures using either [γ-^{32}P]dATP and polynucleotide kinase or [α-^{32}P]dATP and Klenow DNA polymerase. The labeled, redigested DNA fragment can finally be eluted from a polyacrylamide gel and purified by standard procedures.[5] The radioactivity needed at this point should be between 10^5 and 10^6 cpm; however, it is possible to obtain results with less.

Afterward the DNA is subjected to the modification reaction. It should be specific with respect to the attacked function on the DNA and, in particular for the interference analysis, rather small sizes of the introduced residues are desirable, because this allows a more specific interpretation in terms of protein–DNA contacts. The two mainly used chemicals for DNA modification fulfilling these requirements are dimethyl sulfate and *N*-ethyl-*N*-nitrosourea. Diethyl pyrocarbonate shows a structure-dependent reaction specificity[6] and has successfully been used to map protein binding to Z-DNA[3,7] while the application to B-DNA has been of only moderate success.[4] The products of their specific reactions with DNA are depicted in Fig. 3.

Methylation of DNA[5]

The most widely used modification to study protein–DNA interaction is the methylation with dimethyl sulfate, which attacks the N-7 of guanines (G) and the N-3 of adenines (A), as is well known from the chemical nucleotide sequencing method. With regard to the analysis of protein–DNA interactions the methylation of the guanine N-7 in double-stranded DNA is of greater practical importance. This reaction can either be performed as a protection or an interference experiment.

Between 0.1 and 1 pmol DNA fragment in 200 μl of 50 mM sodium cacodylate, pH 8.0, and 1 mM EDTA are mixed with 1 μl of a carrier DNA solution containing 1 mg/ml sonified pWH802 DNA (any DNA that does not contain the binding site of interest can be used here), 2 μl tRNA

[4] C. Heuer and W. Hillen, *J. Mol. Biol.* **202**, 407 (1988).
[5] A. M. Maxam and W. Gilbert, this series, Vol. 65, p. 499.
[6] W. Herr, *Proc. Natl. Acad. Sci. U.S.A.* **82**, 8009 (1985).
[7] B. Johnston and A. Rich, *Cell (Cambridge, Mass.)* **42**, 713 (1985).

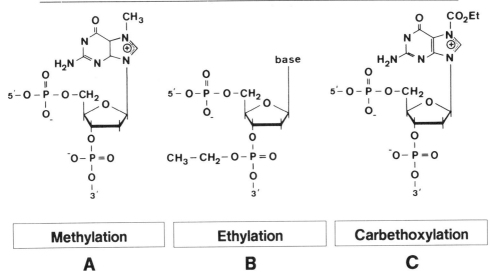

Methylation	**Ethylation**	**Carbethoxylation**
A	**B**	**C**

FIG. 3. Chemical structures of modified nucleotides. The reaction products of DNA and dimethyl sulfate (A), N-ethyl-N-nitrosourea (B), and diethyl pyrocarbonate (C) are shown. Dimethyl sulfate leads to the methylation of the N-7 in guanines and to a minor extent to methylation of the N-3 in adenines (not shown). N-ethyl-N-nitrosourea gives rise to ethylated phosphates and diethyl pyrocarbonate causes a carbethoxylation of the N-7 of purines (guanosine is shown as an example here).

solution containing 10 mg/ml yeast tRNA, and 0.5 μl dimethyl sulfate (New England Nuclear, Boston, MA) in that order and then incubated at ambient temperatures for 2 min. The methylation reaction is terminated by adding 50 μl of 1.5 M sodium acetate, 1 M 2-mercaptoethanol, 100 μg/ ml tRNA, followed by the addition of 750 μl of ethanol. The mixture is kept in liquid nitrogen for 2 min and centrifuged for 15 min at 13,000 rpm in a Biofuge A (Heraeus, Osterode, Germany). The pellet is dissolved in 250 μl of 0.3 M sodium acetate, 0.1 mM EDTA, 25 μg/ml tRNA and reprecipitated by adding 750 μl of ethanol and treatment as described above. The precipitate is dissolved in 50 μl of 0.4 M sodium chloride and reprecipitated with 1 ml of ethanol as described above. Finally, the pellet is washed three times with 0.1 ml ethanol each, dried for 2 min *in vacuo,* and stored for further use.

For the protection from methylation by dimethyl sulfate the solution of the DNA fragment is mixed with an equimolar or up to 10^4-fold excess of DNA-binding protein (Tet repressor in the example shown below) in 1 to 8 μl of protein storage buffer [5 mM potassium phosphate, pH 8.0, 4 mM 2-mercaptoethanol, 50% glycerol (v/v) in this case]. The complex is allowed to form for 10 min at ambient temperatures. Then the carrier DNA, tRNA, and dimethyl sulfate in the amounts described above are added. The following steps are performed as detailed above.

It should be noted that the amount of dimethyl sulfate, the reaction temperature, and time can be varied to yield the optimal methylation of one hit per fragment. Usually, however, this is not necessary due to the large excess of nonspecific competitor DNA. The strand cleavage reaction at the methylated positions is done exactly as described by Maxam and Gilbert.[5]

Ethylation of DNA[8,9]

Between 0.1 and 1 pmol DNA fragment in 100 μl of 50 mM sodium cacodylate, pH 7.0, 1 mM EDTA is mixed with 1 μl sonicated carrier DNA (1 mg/ml, pWH802 in the examples shown here, but any DNA may be used) and 100 μl of a freshly prepared saturated solution of N-ethyl-N-nitrosourea (Sigma, St. Louis, MO) in ethanol is added. The mixture is kept at 50° for 1 hr. After addition of 20 μl of 3 M sodium acetate the DNA is precipitated with 0.5 ml of ethanol as described above. The pellet is washed with ethanol, dried *in vacuo* for 2 min, and stored for further use.

For strand cleavage the ethylated DNA samples are dissolved in 30 μl of 10 mM sodium phosphate, pH 7.0, 1 mM EDTA. Five microliters of 1 M sodium hydroxide is added and the mixture is incubated at 90° for 30 min. Then 5 μl of 1 M HCl is added, followed by 1 μl of a glycogen solution (10 mg/ml) and 100 μl of ethanol. After precipitation for 5 min in liquid nitrogen and centrifugation the pellet is washed with ethanol, dried *in vacuo,* and redissolved in 98% formamide containing 0.1% bromphenol blue and xylene cyanol. The sample is incubated at 100° for 2 min, immediately placed in ice water, and applied to a sequencing gel.

Carbethoxylation of DNA[6,10]

DNA (0.1 to 1 pmol) in 200 μl of 50 mM sodium cacodylate, pH 7.0, 1 mM EDTA is mixed with 1 μl of carrier DNA solution (sonified pWH802 or any other DNA at 1 mg/ml) and 3 μl of ice-cold diethyl pyrocarbonate (obtained from Sigma). Since diethyl pyrocarbonate is not dissolved in water the mixture is vigorously vortexed for 15 min at ambient temperatures. The DNA is then precipitated with ethanol, washed with ethanol, and briefly dried *in vacuo.* Strand cleavage at the modified positions is done exactly as described.[6]

[8] U. Siebenlist and W. Gilbert, *Proc. Natl. Acad. Sci. U.S.A.* **77,** 122 (1980).
[9] W. Hendrickson and R. Schleif, *Proc. Natl. Acad. Sci. U.S.A.* **82,** 3129 (1985).
[10] D. A. Peattie and W. Gilbert, *Proc. Natl. Acad. Sci. U.S.A.* **77,** 4679 (1980).

For protection from ethoxylation[3] 1.5 μg of plasmid DNA is incubated with 10 μg monoclonal antibody (corresponding to a 90-fold molar excess) in 100 μl of 100 mM sodium chloride, 60 mM Tris-HCl, pH 7.4 for 20 min at 37°. The tubes are transferred to ice and 100 μl of cold, 50 mM sodium cacodylate, pH 7.1 is added followed by 3 μl of diethyl pyrocarbonate. After thorough vortexing the mixture is incubated at 20° for 10 min. Then the vortexing is repeated, followed by another 10-min incubation at 20°. The samples are further processed according to Herr[6] with a phenol extraction prior to the final precipitation. The DNA is then digested with a suitable restriction endonuclease, labeled at the 5' or 3'-end, and redigested with a second endonuclease. The products are separated on 5% (w/v) polyacrylamide gels, eluted, cleaved, and analyzed as described.[6]

Separation of Modified DNA Fragments According to Protein-Binding Capacity

The gel mobility shifts observed for most protein–DNA complexes provide a very convenient method to separate the bound fraction of modified DNA from those species in which the modification prevents DNA binding.[9] A second method is the filtration over nitrocellulose, which works well for complexes with high-affinity constants and a large protein portion, e.g., a strong promoter with RNA polymerase.[8] The experimental goal is, in any case, to obtain the most complete separation of these DNAs.

For the formation of Tet repressor–*tet* operator complexes the modified DNA is adjusted to a concentration of about 10^{-8} M in 10 to 30 μl of 10 mM Tris-HCl, pH 7.4, 10 mM potassium chloride, 10 mM magnesium chloride, 1 mM dithiothreitol, 0.1 mM EDTA, 5% glycerol (v/v), and 50 μg/ml bovine serum albumin. Then Tet repressor is added to reach a 4- to 10-fold molar excess over DNA and the mixture is incubated at 37° for 20 min. Immediately afterward the samples are applied to carefully rinsed slots of a 5% polyacrylamide gel (0.25% bisacrylamide) and electrophoresed at 40 V overnight in 60 mM Tris–borate, pH 8.3, 1 mM EDTA. The bands are visualized by autoradiography and eluted from the gel by a standard procedure.[5] The DNA is then cleaved and analyzed on sequencing gels.[5]

Modification Protection Analysis

Methylation Protection

Protection from methylation by dimethyl sulfate has been very widely used according to the protocol described above to determine protein–DNA contacts *in vitro*[1] and, in a few cases, *in vivo*.[11] It is not within the scope of this chapter to provide a list of references completely covering this topic; instead, only references to the original protocols are given.

The great popularity of this experiment is due to two main reasons. First, it is very easily done because it involves only the standard Maxam and Gilbert sequencing reaction for guanosine. It can also be used to study contacts to adenine (A); however, this requires more drastic reaction conditions, making it more difficult to observe A protection in addition to G protection. Second, dimethyl sulfate is a small reagent specific for the guanosine modification at position N-7 located in the major groove of the B-form structure of DNA (see Fig. 3). Thus, a protection can be interpreted as a result of a very close contact of the studied protein to the major groove of the bound DNA in the near vicinity or even at the N-7 of the guanine moiety. As shown schematically in Fig. 1, formation of a protein–DNA complex may also lead to an enhanced reactivity of residues. This is explained either by protein binding-induced structural alterations of the DNA leading to a more exposed reaction site, or by the formation of a so-called hydrophobic pocket by the protein on the DNA attracting the reagent and increasing its local concentration. In this interpretation the sites of increased reactivity may indicate the boundary of the protein on the DNA. A result of this experiment is not presented here. The data for Tet repressor–*tet* operator interaction discussed below are taken from Ref. 2.

Carbethoxylation Protection

Treatment of DNA with diethyl pyrocarbonate leads to carbethoxylation of purines at N-7 (see Fig. 3). It has been demonstrated that the efficiency of this reaction depends on the secondary structure of the DNA. The reactivity of purines in Z-DNA, e.g., is so much greater compared to B-DNA that modification by diethyl pyrocarbonate has been used as an indication for Z-DNA formation under conditions of negative superhelicity.[6,7] Consequently, the protection from this modification by binding of Z-DNA-specific monoclonal antibodies has been used to determine close contacts of the antibody to the DNA.[3] A typical result is displayed in Fig. 4. The optical impression of altered intensities are usually quantified by densitometry based on the intensity of a reference band located outside of the area of interest. In the case of antibody binding it was possible to combine the results of carbethoxylation protection with the global structure of antibodies known from X-ray analysis to propose a detailed model describing the orientation of the protein on the DNA.

---

[11] H. Nick and W. Gilbert, *Nature (London)* **313**, 795 (1985).

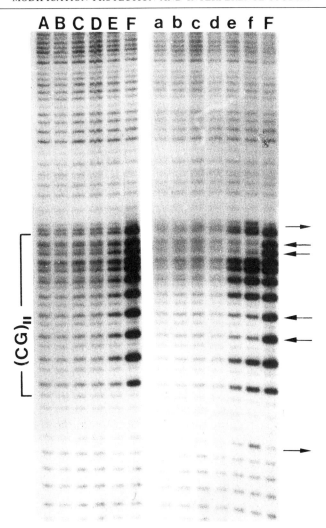

FIG. 4. Hyperreactivity of Z-DNA to diethyl pyrocarbonate and protection by an antibody. Preparations of a plasmid containing a $(C-G)_{11}$ repeat sequence were adjusted to six different negative superhelical densities, $-\sigma$ (A, 0; B, 0.012; C, 0.023; D, 0.035; E, 0.046; F, 0.058) and treated with diethyl pyrocarbonate in either the absence (lanes A to F) or presence (lanes a to f) of a monoclonal antibody. The bracket indicates the region of 22 bp of alternating d(C-G). Arrows indicate guanine residues with either a reduced (arrows pointing left) or enhanced (arrows pointing right) hyperreactivity toward diethyl pyrocarbonate on complex formation with a Z-DNA-specific monoclonal antibody. [Reproduced by permission of Academic Press, Inc., London (Runkel and Nordheim[3]).]

While carbethoxylation is particularly suitable for Z-DNA modification its use for protein-DNA complexes with B-type structures is rather limited. In the case of *tet* operator–Tet repressor recognition the modification of the pure *tet* regulatory sequence turned out to be quite inhomogeneous and the interpretation of modification interference was hampered by the lack of consistency within the palindromic *tet* operator sequences. However, although the interpretation of the results obtained for single base pairs was not straightforward, the points of interference are located in the center of the operator palindrome.[4] This is in good agreement with other protection[4] and genetic data[12] concerning Tet repressor–*tet* operator interaction.

Modification Interference Analysis

Separation of Bound from Nonbound Fraction of Modified DNA

For complexes with large protein portions showing complete binding to nitrocellulose the filtration technique is very suitable.[8] Repressor–operator complexes usually contain smaller proteins, which in some cases show only very poor retention on nitrocellulose.[13,14] For these cases in particular and for many other cases for the sake of simplicity the gel mobility shift technique is the method of choice.[15–17] An example is shown in Fig. 5 for the Tet repressor–*tet* operator complex with native and partially ethylated DNA. It is clearly visible that the ethylated DNA cannot be completely complexed by the repressor even though a sixfold molar excess of protein over DNA is used. The gel shown in Fig. 5 contains three regions of interest for the interference analysis. The first is the band corresponding to the complexed DNA. This should contain modifications only at those positions where the ethyl group does not interfere with repressor binding. In our hands, the analysis of this DNA always leads to clear results. The second band corresponds to the remaining free DNA. It contains modifications at sites where the ethyl residue prevents complex formation. The analysis of this band leads to results that are less clear. We believe that this fraction is always contaminated with some residual nonbound DNA, because binding may never be quantitative. This could lead to

[12] A. Wissmann, I. Meier, and W. Hillen, *J. Mol. Biol.* **202**, 397 (1988).
[13] A. Joachimiak, R. L. Kelley, R. P. Gunsalus, C. Yanofsky, and P. B. Sigler, *Proc. Natl. Acad. Sci. U.S.A.* **80**, 668 (1983).
[14] L. S. Klig, I. P. Crawford, and C. Yanofsky, *Nucleic Acids Res.* **15**, 5339 (1987).
[15] M. G. Fried and D. M. Crothers, *Nucleic Acids Res.* **9**, 6505 (1981).
[16] M. G. Fried and D. M. Crothers, *Nucleic Acids Res.* **11**, 141 (1983).
[17] A. Revzin, J. A. Ceglarek, and M. M. Garner, *Anal. Biochem.* **152**, 172 (1986).

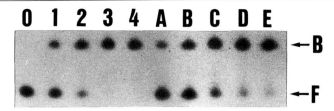

FIG. 5. Gel mobility of native and partially ethylated *tet* operator DNA in the presence of Tet repressor. A 64-bp *tet* operator containing DNA fragment was electrophoresed on a 5% polyacrylamide gel. Lanes 0 to 4 contain native DNA. In lane 0 no repressor was added to the DNA, whereas in lane 1 a 0.25-fold, in lane 2 a 0.75-fold, in lane 3 an equimolar, and in lane 4 a 4-fold molar excess of Tet repressor over *tet* operator was added. Lanes A to E contain the same DNA fragment after ethylation. The molar ratios of Tet repressor over *tet* operator added to the DNA in these lanes were as follows: lane A, 0.25-fold; lane B, 0.75-fold; lane C, equimolar; lane D, 4-fold; lane E, 6-fold. The arrows on the right denote the positions of the bound (B) and the free (F) DNA. [Reproduced by permission of Academic Press, Inc., London (Heuer and Hillen[4]).]

the increased background seen in those experiments. The third region is located between the two bands. It contains DNA from complexes that have undergone dissociation during electrophoresis and are, therefore, of reduced stability. An example of this analysis for methylation interference is discussed below.

Analysis of Interference

The electrophoretic analysis of cleavage products from the modified DNA fragments yields the positions where modification has interfered with protein binding. Ideally, these positions should be blank in the bound fraction of the modified DNA. When a gel separation is used this goal can generally be achieved for the positions showing the strongest interference for both methylated and ethylated DNA. An example from our own work on *tet* operator–Tet repressor recognition is shown in Fig. 6. The data indicate clearly that the interference signals are of different intensities. The methylation interference of the plus strand yields a single, strong signal for the guanosine at position 2, the same which is protected from methylation. For the minus strand, on the other hand, the corresponding strong G signal is accompanied by a weaker interference band for adenosine at position 3. However, since this is observed only on one strand of the two *tet* operators we are not sure that this signal indeed represents a contact. At operator O_1 an adenosine interference is found at position 4 of the minus strand (not shown). It has been shown that this methylation is less destructive for complex formation than the respective modification

OPERATOR 2

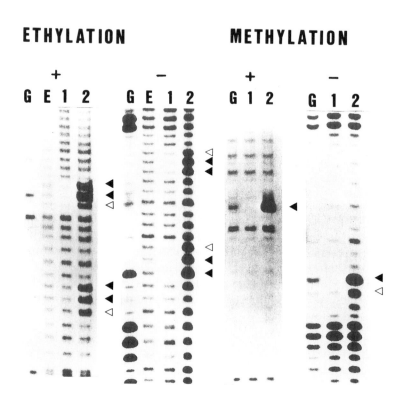

Fig. 6. Results of ethylation and methylation interference studies obtained for *tet* operator O_2. The + denotes the upper (5'-end-labeled) and the − the lower (3'-end-labeled) strand of the DNA sequence of *tet* operator O_2 shown at the bottom. In lanes G the products of the Maxam and Gilbert G reaction were loaded, whereas lanes E show the cleavage products of the ethylated DNA. Lanes 1 contain the bound and lanes 2 the nonbound fractions of the ethylated DNA. Stronger contacts are indicated by filled triangles, whereas weaker contacts are marked by open triangles. At the bottom the deduced contact sites are indicated in the sequence of *tet* operator O_2. The phosphate contacts are indicated as triangles as described

at the guanosine by comparing the signals obtained from the nonbound DNA with the ones from the region between the two bands on the polyacrylamide gel (result not shown). Very clearly, the G methylation prevents complex formation completely while the A methylation at position 4 yields a strong signal in the interband region. This has been interpreted as a weaker interference since the complex can be formed, but dissociates partially during gel electrophoresis.

The ethylation interference results in Fig. 6 reveal six positions on each strand of *tet* operator, where the modification reduces Tet repressor binding. They fall into two sets of three interference signals each, one of which is located close to the palindromic center of the *tet* operator, while the other is located at the outside. For the identification of the interference site it needs to be considered that strand cleavages in the G reaction and after ethylation lead to different products.[8] Since strand breakage can occur on both sides of the ethylated phosphate, the corresponding band are doublets, which are only resolved for short molecules. As a result, they trail the marker bands produced by G cleavage slightly. On the basis of the different intensities of these interference bands it is concluded that the modifications exhibit quantitatively different effects on Tet repressor binding. Inspection of the bound DNA reveals that the center positions are clearly blank while the outer positions show an increased background. This is interpreted as partial binding of the DNA modified at the outside while ethylation at the center prevents repressor binding completely under these conditions. It is, thus, concluded that the contacts at the inside of *tet* operator are more important for Tet repressor binding than the ones at the outside. Similar differences in intensity are found for the bands within each set of three contacted phosphates. A densitometric analysis of the data in Fig. 6 (not shown) indicates that the innermost interference signals of the inside contacts and the outermost signals of the outside contacts are somewhat weaker than the respective other two signals. It is, therefore concluded that the contacts of Tet repressor to these phosphates are less important for complex formation than the others. For this kind of analysis it is usually necessary to study a variety of modification conditions to yield the optimal undermodification of the sites on the DNA, and to vary the exposition times of the gels to yield the proper low intensities of bands for quantification. Usually, the experiments yielding the best pictures for publication are not the ones allowing the most reliable quantification.

---

above and contacts to the N-7 of G are marked by boxed nucleotides. The dots indicate the palindromic center of the operator. By definition this base pair is localized at position 0 of the operator. From here base pairs are numbered to the outside of the palindromic sequence (1 to 9). [Reproduced by permission of Academic Press, Inc., London (Heuer and Hillen[4]).]

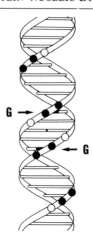

FIG. 7. Projection of the *tet* operator B-DNA structure showing Tet repressor contacts as deduced from modification protection and interference studies. The strong phosphate contacts of Tet repressor to *tet* operator DNA are indicated as filled circles, whereas the weaker contacts are marked by open circles. The G residues showing methylation protection and methylation interference are indicated by the filled bars at the respective base pairs. The palindromic center of the operator is shown as a black dot in the middle of the bar. [Reproduced by permission of Academic Press, Inc., London (Heuer and Hillen[4]).]

An important caveat relates to the interpretation of the results. An interference is observed, when the binding protein in the complex with DNA approaches the modified function so closely that the ethyl or methyl moiety cannot be accommodated. Furthermore, the energy required for a potential accommodation in the complex must reduce the association constant below the inverse concentrations of the binding experiment used to separate free from bound DNA. Thus, the interference results may depend on the methodology used. In addition, conclusions can only be derived in terms of contacts in the sense of distance and not in the sense of possible interactions that may or may not occur at the sites identified by interference analysis.

Interpretation of Results

The interpretation of protection and interference results is aided by either a three-dimensional computer drawing or a two-dimensional projection of the anticipated DNA structure in which the identified contacts are marked. For the examples described in this chapter, this has been done in the case of Z-DNA antibody binding to Z-DNA and Tet repressor binding to B-form *tet* operator DNA. In both cases the location of protein contacts on the surface of the DNA structure yields hints about the potential protein structures involved in complex formation. Assumptions about the structure of the Z-DNA-specific monoclonal antibodies have been derived from the crystal structures of other antibodies and indicate a good structural fit of the proteins to the contact sites on the DNA.

The DNA-binding domain of Tet repressor shows strong homologies to many other prokaryotic DNA-binding proteins having an α helix–turn–α helix supersecondary structure in this domain.[18] The ethylation interference results look indeed very similar to the ones obtained with these other proteins and their respective cognate DNAs.[19] However, the methylation protection and interference does not fit very well into this picture because it appears to be displaced from the center between the contacted phosphates. This interpretation is shown in Fig. 7. Furthermore, a quantitative saturating mutational analysis of the *tet* operator sequence for binding of Tet repressor has indicated that base pairs at positions 3 of the two half-sides are most important, followed by base pairs 2 and 4, while 1, 5, and 6 are less important.[20] Thus, the most important sequence as well as the protected guanosine are not located in the center between the ethylation contacts, as was proposed for other bacterial DNA-binding proteins.[19] Whether this pattern can be achieved by alterations in the position of the recognition α helix compared to other known structures or whether other parts of the Tet repressor contribute to *tet* operator binding cannot be distinguished by these results. The two examples presented here demonstrate clearly that DNA modification-protection and modification-interference analysis yields valuable information about the structure of protein–DNA complexes contributing one piece to the tertiary structure puzzle. In addition, the necessary experimental setup is commonly available and the experimental procedures invovled are not difficult. For these reasons they will most likely be even more extensively used in the future as they have been in the past.

---

[18] C. O. Pabo and R. T. Sauer, *Annu. Rev. Biochem.* **53,** 293 (1984).
[19] R. H. Ebright, *in* "Protein Structure, Folding and Design" (D. Oxender, ed.), p. 207. Alan R. Liss, New York, 1986.
[20] C. Sizemore, A. Wissmann, U. Gülland, and W. Hillen, *Nucleic Acids Res.* **18,** 2875 (1990).

[19] Hydroxyl Radical Footprinting

By WENDY J. DIXON, JEFFREY J. HAYES, JUDITH R. LEVIN,
MARGARET F. WEIDNER, BETH A. DOMBROSKI,
and THOMAS D. TULLIUS

Introduction

Advances in cloning and purification of biological molecules have made available for study an ever-increasing number of proteins that bind to DNA. With this large number of new proteins has come the recognition that there are several strategies in protein design that Nature has employed for producing sequence-specific protein–DNA complexes. The helix–turn–helix and the zinc finger are the best understood at the moment, but other protein structures are also known that mediate contact with DNA. Along with the variety in protein structural motifs that bind to DNA, one might expect to find a corresponding variety in the structures of protein–DNA complexes.

Footprinting[1] is one experimental approach that can provide detailed information on how a protein binds to DNA. Since footprinting was first devised, with its beginning in the use of the enzyme deoxyribonuclease I (DNase I) to digest a DNA–protein complex,[2–4] much progress has been made in employing other cleavage reagents that are capable of revealing higher resolution views of a protein–DNA complex. Many of these new reagents are small molecules, not enzymes. The smallest chemical species that has been used for footprinting, and therefore the one likely to provide the highest resolution structural information, is the hydroxyl radical (·OH).[5]

In this chapter we discuss the use of the hydroxyl radical to make footprints of protein–DNA complexes. In an earlier volume of "Methods in Enzymology" we presented our original experimental protocol for hydroxyl radical footprinting.[6] Since then we have refined the method, applied it to a wide variety of systems, and developed new experimental

---

[1] T. D. Tullius, *Annu. Rev. Biophys. Biophys. Chem.* **18**, 213 (1989).
[2] D. J. Galas and A. Schmitz, *Nucleic Acids Res.* **5**, 3157 (1978).
[3] A. D. Johnson, B. J. Meyer, and M. Ptashne, *Proc. Natl. Acad. Sci. U.S.A.* **176**, 5061 (1979).
[4] M. Noll, *Nucleic Acids Res.* **1**, 1573 (1974).
[5] T. D. Tullius, *Trends Biochem. Sci.* **12**, 297 (1987).
[6] T. D. Tullius, B. A. Dombroski, M. E. A. Churchill, and L. Kam, this series, Vol. 155, p. 537.

methods, based on the chemistry of the hydroxyl radical with DNA, that give additional information on the structure of a protein–DNA complex. The enhancements to the method that we will cover here include modifications to the composition of the cleavage reagent to give footprints in media containing scavengers of the hydroxyl radical, quantitative treatment of data from hydroxyl radical footprinting, the use of mobility-shift gel electrophoresis to "amplify" footprints, and a new method, the missing nucleoside experiment, which provides information on contacts with DNA that are important for stability in a protein–DNA complex.

Chemistry behind Hydroxyl Radical Footprinting

Three compounds are combined in the footprinting reaction mixture to produce the hydroxyl radical: $[Fe(EDTA)]^{2-}$, hydrogen peroxide, and sodium ascorbate. The reaction by which the hydroxyl radical is generated, called the Fenton reaction, is shown in Eq. (1):

$$[Fe(EDTA)]^{2-} + H_2O_2 \rightarrow [Fe(EDTA)]^{-} + OH^{-} + \cdot OH \qquad (1)$$

$$\underset{\text{ascorbate}}{\underline{}}$$

In this reaction an electron from iron(II)-EDTA serves to reduce and break the O–O bond in hydrogen peroxide, giving as products iron(III)-EDTA, the hydroxide ion, and the neutral hydroxyl radical. Sodium ascorbate is present to reduce the iron(III) product to iron(II), thereby establishing a catalytic cycle and permitting low (micromolar) concentrations of iron(II)-EDTA to be effective in cleaving DNA. A consequence of this scheme is that the concentrations of the three chemical species [iron(II)-EDTA, hydrogen peroxide, and sodium ascorbate] may be varied to optimize the generation of the hydroxyl radical under different solution conditions, e.g., to compensate for the presence of radical scavengers in the binding buffer of a protein–DNA complex. Specific examples of concentrations of each of the three reagents that are used in particular footprinting experiments are discussed in later sections.

The Fenton reaction [Eq. (1)] is not the only way to generate the hydroxyl radical for footprinting. We have published a protocol for making a footprint of the bacteriophage λ repressor using ionizing (γ) radiation to produce the hydroxyl radical in aqueous buffered solution.[7] This technique, which requires no additional chemical reagents, may be useful in some situations (for example, *in vivo*) where it is difficult to perform the Fenton reaction.

[7] J. J. Hayes, L. Kam, and T. D. Tullius, this series, Vol. 186, p. 545.

Updated Version of Basic Hydroxyl Radical Footprinting Technique

Reagents

The quality of the water used in buffers for hydroxyl radical foot-printing is critical. Organic contaminants in particular can interfere with the reaction, for example by reacting with iron(II), hydrogen peroxide, or the hydroxyl radical. For this reason all solutions used in experiments involving the hydroxyl radical are prepared with water purified by a Milli-Q system (Millipore, Bedford, MA), which includes cartridges for ion exchange and removal of organics, and a 0.44-μm filter for removal of bacteria.

To prevent nicking of the DNA by radiation products, radioactively end-labeled DNA is stored at an activity of 50,000 disintegrations per minute (dpm)/μl or lower in a dilute buffer, such as 10 mM Tris-Cl, 0.1 mM EDTA (pH 8.0) (which we refer to hereafter as TE buffer), at 4°.

An aqueous solution of iron(II) is prepared by dissolution of ferrous ammonium sulfate [(NH$_4$)$_2$Fe(SO$_4$)$_2$ · 6H$_2$O)] (99 + %; Aldrich, Milwaukee, WI) in water at a concentration of 10–100 mM. The complex of iron(II) with EDTA is prepared by mixing equal volumes of the solution of ferrous ammonium sulfate with a solution of EDTA (Gold Label; Aldrich) that is twice the concentration of the iron solution. These solutions are prepared in acid-washed glassware to prevent the introduction of extraneous iron. We segregate glassware used to prepare iron solutions from other glassware in the laboratory to obviate the possibility of iron contamination of DNA solutions. A solution of ascorbate ion is prepared by dissolving sodium ascorbate (Sigma, St. Louis, MO) in Milli-Q-purified water. Hydrogen peroxide solutions are made by dilution of a 30% (v/v) solution (J. T. Baker, Phillipsburg, NJ). The three reagents can be made up fresh prior to use. However, a more convenient approach is to freeze the ascorbate and [Fe(EDTA)]$^{2-}$ stock solutions in small aliquots. An alternative is to make stock solutions of iron(II) and EDTA separately. Iron(II) stock solutions can be stored as frozen aliquots, and EDTA solutions can be stored at 4°. These stock solutions can then be combined just prior to performing the experiment.

General Technique for Hydroxyl Radical Footprinting

A typical starting mixture for a hydroxyl radical footprinting experiment contains the binding buffer, 0 to 0.5 μg of nonspecific DNA, 50,000 to 200,000 dpm of singly end-labeled DNA containing the protein-binding site, and the desired amount of protein, in a volume of 70 μl. This solution should be incubated at the appropriate temperature to allow protein to bind to the DNA. Then 10 μl each of the three stock solutions (each at 10 times the final concentration desired in the reaction mixture) of ascorbate, hydrogen peroxide, and iron(II)-EDTA are placed as drops on the inside of the 1.5-ml Eppendorf reaction tube that contains the protein–DNA complex. To initiate the cleavage reaction, the reagents are mixed together on the side of the reaction tube and then added to the protein–DNA mixture. The reaction is allowed to proceed for 2 min at room temperature. The reaction can be performed at 4° if necessary, but this often requires modification of the reaction conditions to increase the rate of the cutting reaction. The reaction is quenched by adding 0.1 M thiourea (5 to 30 μl) (a hydroxyl radical scavenger), and 0.2 M EDTA (2 μl). To remove protein, samples are extracted with a volume of phenol equal to the volume of the reaction mixture, and then the phenol layer is back extracted with TE buffer. The aqueous solution is extracted with ether three times to remove any residual phenol. The DNA is then precipitated twice by addition of ethanol. The DNA pellet is rinsed with ethanol and dried in a SpeedVac concentrator (Savant, Farmingdale, NY).

Effect of Experimental Conditions on Hydroxyl Radical Cleavage of DNA

The concentrations of reagents in the hydroxyl radical footprinting method originally developed by Tullius and Dombroski[8] were 1 mM ascorbate, 0.03% hydrogen peroxide, 10 μM iron(II), and 20 μM EDTA. This mixture was found to generate sufficient hydroxyl radical for introduction of not more than one gap in the backbone of a DNA molecule.[9] These concentrations are adequate for DNA solutions that contain small amounts of scavengers of the hydroxyl radical, such as DNA in TE buffer. Such a cleavage reagent often is suitable for structural studies of free DNA molecules, and for some footprinting experiments. However, sometimes the reagents and buffers used for a protein–DNA complex reduce the rate of cleavage of DNA by the hydroxyl radical. The most avid scavenger of hydroxyl radical commonly encountered in protein–DNA systems is glycerol. We have found that a concentration of >0.5% (v/v) glycerol significantly inhibits cleavage of DNA by the hydroxyl radical. To a lesser degree, common buffers such as Tris and HEPES also reduce DNA cleavage. Thus DNA in a 10 mM Tris solution will be more efficiently cleaved than DNA in a 50 mM solution of Tris.[6] It is also possible that high concentrations of bovine serum albumin (BSA) or nonspecific DNA will

[8] T. D. Tullius and B. A. Dombroski, *Proc. Natl. Acad. Sci. U.S.A.* **83**, 5469 (1986).
[9] T. D. Tullius and B. A. Dombroski, *Science* **230**, 679 (1985).

decrease the hydroxyl radical cutting of the radioactively labeled DNA by competing for reaction with the hydroxyl radical.

There are two ways to increase the rate of cleavage of DNA. One is to lower the concentration of scavengers of the hydroxyl radical that are present in the reaction mixture. Since these reagents often are necessary for protein stability and binding to DNA, usually another means must be found for increasing DNA cleavage. This can be accomplished by increasing the concentrations of some or all of the reagents that generate the hydroxyl radical. Because some proteins are sensitive to hydrogen peroxide,[10] increasing its concentration is not advised. The best method for augmenting the cleavage reagent is to increase the concentrations of both $[Fe(EDTA)]^{2-}$ and ascorbate. Raising the concentration of $[Fe(EDTA)]^{2-}$ increases cutting by increasing the production of the hydroxyl radical by Fenton chemistry. Increasing the concentration of ascorbate is important because ascorbate reduces one of the products of the reaction, Fe(III)EDTA, to Fe(II)EDTA. Therefore it is preferable to have a concentration of ascorbate several times that of iron to maintain a catalytic cycle.

More cleavage is not always better, though. Reaction conditions are desired that produce no more than one cut per DNA molecule. Since the DNA is labeled at one end of one strand of the duplex, only the fragment resulting from the cut closest to the label will be seen on the autoradiograph of the gel. If the reaction conditions produce more than one cut per molecule, it is possible that an initial cut could occur far from the label and a second cut would be made between the label and the first cut. In this case, the fragment resulting from the first cut would be lost and only the smaller fragment containing the label would be observed.[11] Thus overcutting gives a biased picture of the cleavage frequency along the DNA molecule. To ensure that only one cut per molecule is produced, a simple Poisson calculation shows that the percentage of uncut DNA in the sample should be greater than 70%.[12] Conversely, a sufficient amount of cleavage is needed for a good signal-to-noise ratio in the experiment. This is why careful regulation of the length of the reaction and the concentrations of the cleavage reagents is necessary.

Effects of Hydroxyl Radical Cutting Reagents on Protein Binding

Previously it was demonstrated that DNase I footprinting can be used to test whether a particular component of the hydroxyl radical-generating system affects the ability of the protein of interest to bind to DNA.[6] This

[10] K. E. Vrana, M. E. A. Churchill, T. D. Tullius, and D. D. Brown, *Mol. Cell. Biol.* **8,** 1684 (1988).
[11] L. C. Lutter, *J. Mol. Biol.* **124,** 391 (1978).
[12] M. Brenowitz, D. F. Senear, M. A. Shea, and G. K. Ackers, this series, Vol. 130, p. 132.

control experiment is useful in cases where cleavage of DNA occurs in footprinting samples, but samples with protein give the same cleavage pattern as free DNA. In this procedure, each of the three components of the hydroxyl radical cleavage reaction ($[Fe(EDTA)]^{2-}$, ascorbate, and hydrogen peroxide) are added independently to samples of the DNA–protein complex. Then DNase I digestion is performed to determine whether protein remains bound in the presence of each reagent. Lack of a DNase I footprint in the presence of one of the reagents indicates sensitivity of the protein to that reagent.

So far the only one of the three reagents that we have found to which protein–DNA binding is sensitive is hydrogen peroxide. Both the zinc finger protein TFIIIA[6] and the "copper fist" protein CUP2[13] do not bind to DNA in the presence of the standard concentration of hydrogen peroxide, which may be due to the oxidation of sulfhydryl groups that are necessary for metal binding in these proteins. In order to footprint these proteins it was necessary to use a concentration of hydrogen peroxide of only 0.003%. To compensate, a concentration of 100 μM $[Fe(EDTA)]^{2-}$ was used for footprinting TFIIIA. For CUP2 footprinting, 1 mM $[Fe(EDTA)]^{2-}$ and 20 mM ascorbate were needed to provide sufficient cutting.

Sequencing Gel Electrophoresis and Gel Drying

The DNA pellet is dissolved in formamide–dye mixture, and denatured by heating to 90° for 3–5 min. Bromphenol blue is often omitted from the dye mixture for all but the marker lanes since it can interfere with the resolution of particular bands. Electrophoresis is performed on a denaturing polyacrylamide gel. Gels are made according to the procedure described by Maxam and Gilbert.[14] The gel and running buffer is 1 × TBE [100 mM Tris-Cl, 100 mM sodium borate, 2 mM EDTA (pH 8.3)]. Denaturing polyacrylamide gels are 50% by weight urea with a ratio of 19 : 1 acrylamide : bisacrylamide. For electrophoresis we use Hoefer (San Francisco, CA) Poker-face sequencing gel equipment. The products of a Maxam and Gilbert guanine-specific sequencing reaction[14] performed on the labeled DNA fragment serve as size markers for the gel.

The percentage of acrylamide used in the gel depends on the distance from the region of interest to the radioactively labeled end of the DNA. Table I lists the percentages of acrylamide required to resolve DNA fragments of particular lengths. Percentages between the values in Table I are used to optimize the resolution of a particular range of fragments. The

---

[13] C. Buchman, P. Skroch, W. Dixon, T. D. Tullis, and M. Karin, *Mol. Cell. Biol.* **10**, 4778 (1990).

[14] A. M. Maxam and W. Gilbert, this series, Vol. 65, p. 499.

TABLE I

PERCENTAGE OF ACRYLAMIDE FOR SEPARATING
DNA MOLECULES OF VARIOUS LENGTHS

Length of DNA fragment (bp)	Acrylamide (%)
3–14	25
5–30	20
10–50	15
15–120	10
50–200	5

ability to resolve the smallest fragments depends mostly on the salt content of the sample.

A larger number of fragments can be resolved on one gel by using wedge spacers, which are thin at the top of the gel and thick at the bottom. Gels are prerun for at least 1 hr at a constant power of 65–75 W, until the gel temperature is between 45 and 50°. Samples are loaded on the gel as quickly as possible. During electrophoresis the gel temperature should return to the original prerunning temperature and then remain at this temperature. This is generally achieved by electrophoresis at a constant power of 55–65 W.

After electrophoresis the gel is dried onto Whatman (Clifton, NJ) 3MM paper, with a Hoefer slab gel dryer (model SE 1160). Our technique for transfer of a sequencing gel to filter paper has been described.[15] This procedure is particularly well suited for transfer of high-percentage acrylamide gels, which are difficult to dry by the standard method.

Autoradiography

The dried gel is autoradiographed with preflashed Kodak (Rochester, NY) XAR-5 film, either at −70° with a Du Pont (Wilmington, DE) Cronex Lightning Plus intensifying screen, or at room temperature without a screen. A gel with a total of 100,000 dpm/lane should be exposed at −70° with an intensifying screen for 15 hr, or at room temperature without a screen for 6 days (150 hr). We generally make a first exposure at −70° to allow an early look at the gel and to determine the time needed for the room temperature exposure. Exposure at room temperature eliminates the parallax effect of both the screen and the gel contributing to darkening of the film, and gives sharper bands. To prevent exposure of the film by static electricity from gel cracking we place a piece of paper between the

[15] G. E. Shafer, M. A. Price, and T. D. Tullius, *Electrophoresis* **10**, 397 (1989).

gel and the film during the first exposure. If hydroxyl radical cutting is weak, exposing the film longer is advisable.

Data Analysis

Qualitative analysis of one-dimensional densitometer scans is adequate for determination of protections at the nucleotide level. A one-dimensional scan is produced by scanning the autoradiograph with a Joyce Loebl Chromoscan 3 densitometer at an aperture width of 0.05 cm. A full discussion of methods for quantitative analysis of footprinting data is presented in the next section.

Footprinting of NF I–DNA Complex with Hydroxyl Radical

As a specific example of a footprinting experiment, we present a protocol developed in this laboratory for footprinting nuclear factor I (NF I), a protein involved in initiation of replication by adenovirus (Ad). (Independent hydroxyl radical footprinting experiments on NF I have recently been published by another laboratory.[16])

A complication of this system is that the solution of protein and DNA found necessary to produce a saturated binding site contained a final glycerol concentration of 0.5%. We were unsuccessful in producing a footprint using the standard conditions for hydroxyl radical footprinting.[6,8] On performing the footprinting experiment with a series of concentrations of $[Fe(EDTA)]^{2-}$, we discovered that it was necessary to increase the concentration of $[Fe(EDTA)]^{2-}$ to 4.5 mM to produce enough cleavage of the DNA to see a footprint. In a volume of 25 μl, DNA samples were prepared by mixing 5 fmol of radioactively labeled Ad5 DNA (100,000 dpm) with 0.5 μl of 5 M NaCl in 2× NF I assay buffer [100 mM HEPES (pH 7.0), 50 mM MgCl$_2$, and 1 mM dithiothreitol (DTT)]. A 1 : 20 dilution of NF I from the stock solution [20 ng/μl NF I, 25 mM HEPES (pH 7.5), 20% glycerol, 0.01% Nonidet P-40 (NP-40), 100 mM NaCl, 1 mM EDTA, 1 mM DTT, and 0.1 mM phenylmethylsulfonyl fluoride (PMSF)] was prepared in NF I dilution buffer [10 mM HEPES (pH 7.0), 0.01% NP-40, 100 mM NaCl, 1 mM EDTA, and 1 mM DTT]. A 25-μl aliquot of the diluted protein solution, containing 25 ng of NF I, was added to the DNA mixture, for a final reaction volume of 50 μl. After the protein was added to the DNA solution, the sample was incubated for 20 min at room temperature to allow for protein binding. To DNA samples that did not contain protein, 25 μl of 1% glycerol in NF I dilution buffer was added to bring the final reaction volume to 50 μl.

[16] H. Zorbas, L. Rogge, M. Meisterernst, and E.-L. Winnaker, *Nucleic Acids Res.* **17,** 7735 (1989).

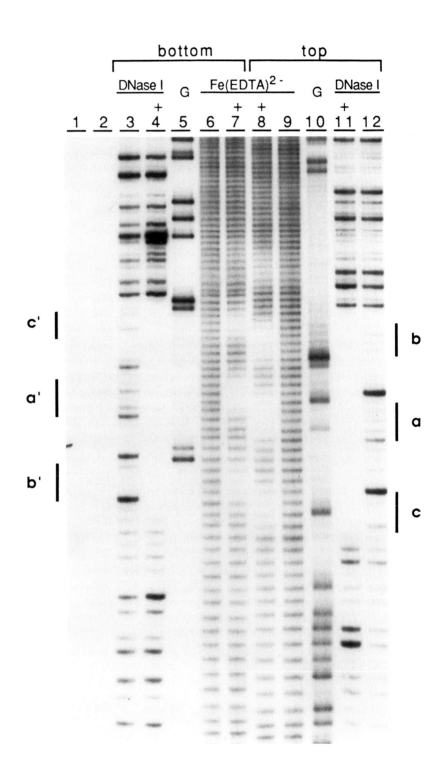

Immediately before the cutting reactions were initiated, equal volumes of Fe(II) (250 mM) and EDTA (500 mM) solutions were mixed. A solution of $[Fe(EDTA)]^{2-}$ that is 50 mM or greater in concentration is initially light green, but under these conditions Fe(II) is readily oxidized and the solution begins to turn rust in color. It is advised that, at these concentrations, a fresh solution of $[Fe(EDTA)]^{2-}$ be prepared for every three to four reaction samples.

The reaction was begun by the addition of 2 μl each of $[Fe(EDTA)]^{2-}$ [125 mM Fe(II), 250 mM EDTA], ascorbate (28 mM), and hydrogen peroxide (0.84%). After 2 min at room temperature the reaction was stopped with 150 μl of stop reagent [112 μl TE buffer, 10 μl 0.1 M thiourea, 25 μl 3 M sodium acetate, 2 μl 0.2 M EDTA, and 1 μl t-RNA (5 mg/ml)]. Protein was removed by phenol extraction. The DNA left in the aqueous solution was then subjected to two ether extractions, two ethanol precipitations, and rinsed and dried.[17]

An autoradiograph of the resulting NF I footprint is shown in Fig. 1. Three regions of protection are visible. The individual protections cover three to four nucleotides. The protected regions on each strand are offset by 3 base pairs (bp) in the 3' direction. This offset indicates equivalent protein interactions across the minor groove of DNA, since the backbone positions closest to each other across the minor groove are 3 bp apart in the sequence.[18] The individual protected regions are separated by 8.5 to 9.5 bp. In many aspects, the hydroxyl radical protection pattern of NF I resembles that of the prokaryotic bacteriophage λ repressor.[8] Using the protection results from λ repressor for comparison, it appears that NF I binds to one face of the Ad DNA helix as a dimer, with monomer units

[17] B. A. Dombroski, Ph.D. Dissertation, Johns Hopkins University, Baltimore, Maryland (1988).
[18] H. R. Drew and A. A. Travers, *Cell (Cambridge, Mass.)* **37**, 491 (1984).

FIG. 1. Hydroxyl radical footprints of nuclear factor I on both strands of the 326-bp Ad DNA fragment. Information for the bottom or the top strand was obtained by labeling the 5' or the 3' end of the *Hin*dIII/*Rsa*I fragment, respectively. Lane 1, untreated Ad DNA labeled on the top strand. Lane 2, Ad DNA, labeled on the top strand, subjected to the conditions of the hydroxyl radical cleavage reaction but with water substituted for the cutting reagents. Lanes 3 and 12, products of DNase I digestion of the Ad DNA fragment alone. Lanes 4 and 11, products of DNase I digestion of the nuclear factor I–Ad DNA complexes (25 ng NF I). Lanes 5 and 10, products from Maxam–Gilbert guanine-specific sequencing reactions. Lanes 6 and 9, products of hydroxyl radical digestion of free Ad DNA. Lanes 7 and 8, products of hydroxyl radical cleavage of the nuclear factor I–Ad DNA complex (25 ng NF I). The labels a, b, c, a', b', and c', and the lines associated with them identify the hydroxyl radical footprints.

contacting the DNA in adjacent major grooves, and contacting each other across the minor groove at the dyad axis of symmetry of the consensus sequence. Mutational analysis of the NF I consensus sequence is consistent with dimer binding. A mutation in one of the half-sites led to a decrease (but not complete loss) in affinity of NF I.[19]

Quantitative Treatment of Footprinting Data

As discussed in other sections of this chapter, the nucleotide-level resolution of hydroxyl radical footprinting allows the description of binding site topography at a level of detail afforded by few other techniques. While in many cases a qualitative treatment of footprinting data is sufficient to address the questions at hand, there are situations in which precise quantitation of reactivity at each nucleotide position is desirable. The relative lack of sequence specificity of hydroxyl radical cleavage of DNA makes it especially suitable for such applications. For example, the smoothly periodic pattern of hydroxyl radical cleavage observed with nucleosomal DNA, as opposed to the irregular pattern produced by digestion with DNase I, has allowed accurate calculation of the helical periodicity of specific DNA sequences when bound on the surface of a nucleosome.[20] The ability to examine structural periodicity at nucleotide resolution improves significantly over other techniques,[21,22] which detect the global conformation of the DNA but give no indication of local structural variations.

Another potential application of quantitative hydroxyl radical footprinting data is in the measurement of thermodynamic parameters of protein binding or DNA structural transitions. Ackers and co-workers have developed a footprint titration technique[12] in which the binding isotherm for the interaction of a protein with its specific recognition sequence on DNA is generated by quantitative analysis of DNase I footprints produced at various concentrations of protein. This technique, while powerful, is limited by the low resolution of DNase I as a probe of DNA structure. By using hydroxyl radical as the probe in the footprint titration experiment, it should be possible to examine both DNA structure and the thermodynamics of protein–DNA interactions at the nucleotide level.

Finally, quantitative analysis of hydroxyl radical reactivity at each position along the DNA backbone could give detailed structural informa-

---

[19] K. A. Jones, J. T. Kadonaga, P. J. Rosenfeld, T. J. Kelly, and R. Tijan, *Cell (Cambridge, Mass.)* **48**, 79 (1987).

[20] J. J. Hayes, T. D. Tullius, and A. Wolffe, *Proc. Natl. Acad. Sci. U.S.A.* **87**, 7405 (1990).

[21] J. C. Wang, *Proc. Natl. Acad. Sci. U.S.A.* **176**, 200 (1979).

[22] H.-M. Wu and D. M. Crothers, *Nature (London)* **308**, 509 (1984).

tion not easily obtainable by other methods. The measured rate of cleavage reflects the reactivity or accessibility of the target atom or bond in each nucleotide toward attack by the hydroxyl radical. This information could be used as an aid in modeling torsion angles in the DNA, and in positioning a protein domain on its binding site, in such a way that the observed pattern of reactivity is generated.

Since the mechanism of hydroxyl radical cleavage of DNA is as yet poorly understood, we do not know which atom(s) or bond(s) in a nucleotide to include in such a calculation. Instead, we have begun to approach this kind of quantitative application of hydroxyl radical footprinting from the reverse direction: that is, by comparing hydroxyl radical reactivity with calculated structural parameters from a protein–DNA complex for which the three-dimensional structure has been elucidated by X-ray crystallography. For this work we have studied the complex of λ repressor with the O_L1 operator site, for which the X-ray cocrystal structure was reported by Jordan and Pabo.[23] The goals of our work are twofold: in addition to evaluating the feasibility of deriving quantitative structural information from hydroxyl radical footprinting data, we hope to obtain important clues as to the mechanism of hydroxyl radical cleavage of DNA.

Methods

Autoradiographic data from footprinting experiments can be quantitated using either one-dimensional (1D) or 2D scanning densitometry. In the case of 1D densitometry the scanner beam is set to run down the length of a gel lane, detecting the optical density at each position along its path. This results in a series of peaks, each corresponding to a band on the gel (see Fig. 2). Visual inspection and comparison of such scans yields qualitative information about footprints; integration of the area under each peak is performed when quantitative data are desired. In theory, since electrophoresis is a 1D process, a 1D scan should accurately represent the data. However, 1D scans have inherent inaccuracies. For instance, the optical density across the width of the gel lane in a given band is not usually constant, and a scan with a narrow beam may not sample the same amount of all the bands. This problem is particularly obvious if the gel lane is not absolutely straight. Scanning with a beam as wide as the gel lane can reduce this problem. However, if the bands are not exactly perpendicular to the path of the scanning beam, resolution between peaks is decreased.

Because of the inherent tradeoff between accuracy and resolution in

[23] S. R. Jordan and C. O. Pabo, *Science* **242**, 893 (1988).

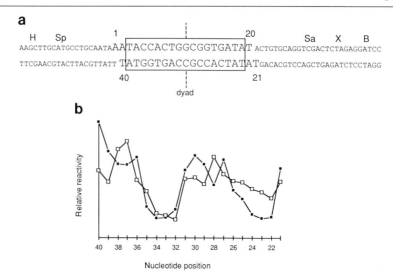

FIG. 2. Comparison of solvent-accessible surface area with hydroxyl radical cleavage frequency in the λ repressor–O_L1 complex. (a) DNA sequence of the insert region of plasmid pOL1-1. Larger letters indicate the sequence corresponding to the 20-mer oligonucleotide used in the cocrystal structure reported by Jordan and Pabo,[23] with numbering according to the scheme of Jordan and Pabo. The 17-bp binding site is boxed, with the pseudopalindrome dyad indicated. Restriction sites convenient for labeling are shown (H, *Hind*III; Sp, *Sph*I; Sa, *Sal*I; X, *Xba*I; B, *Bam*HI). (b) Relative hydroxyl radical cleavage frequencies (open squares) for the nucleotides on the bottom strand of the sequence shown in (a) are plotted along with the calculated accessible surface (probe radius 1.4 Å) for the sugar 3' hydrogens (closed squares). See text for experimental and theoretical details. The values have been adjusted to an arbitrary scale for purposes of comparison.

1D scanning, it is often preferable to scan in two dimensions. A 2D scan in effect transfers an image of the gel into a computer, for subsequent manipulation by image analysis methods. For the experiments described here, we used a photographic scanning system developed by the group of Professor Gary Ackers (Washington University, St. Louis, MO). The system consists of an Eikonix (Ektron Applied Imaging, Bedford, MA) model 1412 camera, interfaced to a Hewlett-Packard Vectra computer, which digitizes the optical density for each position on the autoradiograph and creates a computer file consisting of a series of pixels.[12] The digitized data can then be displayed on a computer and analyzed interactively using an image analysis program.[12] For these studies we used the program Image, which was written for the Macintosh II by W. Rasband (National Institutes of Health, Bethesda, MD). Using this software a box is drawn on the computer screen around each individual band to define the area for integral calculation. A pseudocolor feature, available in many image analysis programs, colors each pixel according to its optical density, and is useful in visualizing where the baseline between two adjacent bands occurs. Care must be taken to leave enough space between lanes when samples are loaded on the gel for the image analysis software to be able to determine an accurate background optical density level for the autoradiograph.[12]

We have begun to analyze gels with laser densitometers[24] that are capable of scanning autoradiographic films or imaging phosphor plates. The PhosphorImager[24] in particular allows for a much larger linear range for data acquisition than film-based systems, since imaging phosphor plates have a dynamic range of nearly 10^5.

With both 1D and 2D scans, there often is uncertainty as to where to place the border between adjacent bands. This can be caused by poor gel resolution or by electrophoretic artifacts, such as trailing of bands, or by the presence of minor bands between nucleotide positions, which are most likely side products of the hydroxyl radical cleavage reaction. Inaccurate division of bands can introduce error into the analysis of the data. In the case where quantitation is for the purpose of calculating helical periodicity,[20] this error can be greatly reduced by using a smoothing algorithm that averages the optical density of each band with those of the two adjacent bands. This three-band averaging results in a general dampening of the data, and so is useful only in cases where accurate quantitation of each band is not the goal, or where dramatic changes in optical density from one band to the next are not observed.

In most cases of footprinting proteins, three-band averaging would likely distort the footprint. Here, the best way to avoid error due to improper division between bands is to quantitate the footprints from several independent experiments and average the data. Comparison of data for the same strand of DNA labeled at the 5' end versus the 3' end can be a useful check for artifact. While the sequence itself will be inverted on the gel, and therefore contributions of electrophoretic anomaly to error should differ, the pattern of hydroxyl radical cleavage frequency should be independent of which end of the DNA is labeled.

Example: Quantitation of λ Repressor Footprint

Although hydroxyl radical footprinting of the λ repressor had already been reported by our laboratory,[8] the published experiments were not done under conditions designed for quantitative analysis of the footprinting data. We wished to repeat these experiments in a way that would optimize the resolution of the DNA cleavage products in the region of the repressor-

[24] Molecular Dynamics model 300 Computing Densitometer, and model 400 PhosphorImager, available from Molecular Dynamics, 240 Santa Ana Court, Sunnyvale, CA 94086.
[25] R. T. Sauer, C. O. Pabo, B. J. Meyer, M. Ptashne, and K. C. Backman, *Nature (London)* **279**, 396 (1979).

binding site, and which would generate data that would be directly comparable to the cocrystal structure.[23] Since the cocrystal contained an amino-
terminal fragment of λ repressor[25] bound to the O_L1-binding site, we set
out to duplicate this complex in our footprinting experiments.

A synthetic duplex oligonucleotide with the precise nucleotide sequence of the O_L1 operator, flanked by PstI-compatible 3' overhangs, is
cloned into the PstI site of plasmid pUC18 (Fig. 2a). Use of a cloning site
in the interior of the polylinker region provides us with multiple potential
labeling sites on either side of the repressor binding site, so that either
strand of DNA can be labeled at either the 5' or the 3' end, at distances
ranging from 13 to 28 nucleotides from the first nucleotide in the binding
site. Uniquely end-labeled DNA fragments are generated by digesting at
one of the polylinker restriction sites, labeling the 5' or 3' ends by standard
methods,[26] performing a second restriction digest with PvuII to generate
two labeled fragments, and isolating the O_L1-containing fragment from a
polyacrylamide gel.

Fragments labeled at several of the possible labeling sites are treated
with the hydroxyl radical and electrophoresed through sequencing gels
containing various concentrations of polyacrylamide to determine the optimum conditions for resolving the region of the O_L1-binding site. We obtain
the best results when the DNA is labeled approximately 20 nucleotides
from the first nucleotide in the binding site and the footprinting products
are run on a 15% (w/v) acrylamide sequencing gel (19 : 1 acrylamide : bisacrylamide) with electrophoresis continuing until the bromphenol blue
tracking dye has migrated 25–26 cm from the origin. These conditions
place the DNA fragment representing the first nucleotide in the O_L1-
binding site (i.e., the end of the site closest to the labeled end of the DNA)
very near the bottom of the gel, and array the 17 nucleotides comprising
the binding site over a distance of roughly 13 cm. While this combination
of labeling distance and electrophoresis conditions works well for these
experiments, other combinations likely exist that would yield comparable
results. Important considerations in designing this type of experiment are
to maximize gel resolution by labeling as close as possible to the region of
interest, keeping in mind that the closer the binding site is to the end of
the DNA, the more problems of inefficient ethanol precipitation of small
DNA fragments after hydroxyl radical cleavage, and potentially inefficient
binding of protein to a site near the end of the DNA, come into play.

After empirically determining the optimum concentrations of footprinting reagents to use in the presence of λ repressor-binding buffer

---

[26] T. Maniatis, E. F. Fritsch, and J. Sambrook, "Molecular Cloning: A Laboratory Manual."
Cold Spring Harbor Laboratory, Cold Spring Harbor, New York, 1982.

(see above), several footprinting experiments are carried out with the λ repressor 92-amino acid amino-terminal fragment used in the cocrystal study. (The repressor fragment was kindly provided by Dr. Carl Pabo of M.I.T.) Binding of the protein to the DNA is performed essentially as previously described, using a protein concentration that gives fully saturated footprints (i.e., complete protection) with DNase I in parallel reactions. Footprints are obtained for both DNA strands, and with the label at both ends of each strand. After gel drying and autoradiography, the films are scanned with the Eikonix camera system (courtesy of Dr. Gary Ackers) described above. The scanned image is imported into the Image program. The bands corresponding to the 17 nucleotides of the O_L1-binding site are individually integrated and the data are normalized to the integral for a band outside the binding site for which the cleavage frequency is known to be unaffected by the presence of λ repressor. Care is taken to ensure that the autoradiographic exposures are within the linear range of both the X-ray film and the densitometry equipment. Data from at least three independent footprinting experiments are averaged to provide a data set for comparison with crystallographic data.

Analysis of the atomic coordinates of the repressor cocrystal (provided by Dr. Carl Pabo) is performed with the QUANTA/CHARMm molecular modeling package from Polygen (Waltham, MA), run on a Silicon Graphics (Mountain View, CA) 4D/20 Personal Iris computer. The program uses the algorithm of Lee and Richards[27] to compute the solvent-accessible surface area for each atom of the DNA, using a variety of probe radii. A spreadsheet program (Microsoft Excel, run on a Macintosh II) is then used to sort the data by atom type and plot data for atoms of interest versus nucleotide position for comparison with hydroxyl radical cleavage frequency data.

For several of the deoxyribose hydrogens the pattern of surface accessibility to a probe sphere the size of a water molecule shows similarities to the hydroxyl radical footprinting pattern. As an example of such a comparison, the calculated surface accessibilities of the deoxyribose 3' hydrogens along the bottom strand of the O_L1-binding site are plotted with the relative hydroxyl radical cleavage rates for this sequence in Fig. 2b. Areas of strong protection from hydroxyl radical cleavage (e.g., positions 22–24 and 32–34) correspond to positions of decreased solvent accessibility, suggesting that reactivity of the sugars with the hydroxyl radical is determined at least in part by accessibility to the reagent. Furthermore, the correlation between reactivity and accessibility of the 3' hydrogen indicates that this hydrogen is one of the targets for abstraction by hy-

[27] B. Lee and F. M. Richards, *J. Mol. Biol.* **55,** 379 (1971).

droxyl radical. Other hydrogens may also be targets, since their accessibility patterns also resembled the footprinting pattern to varying degrees (data not shown), and since it is unlikely that solvent accessibility is the sole determinant of reactivity. Experiments in progress in our laboratory are aimed at determining the mechanism of DNA cleavage by the hydroxyl radical by replacing each sugar hydrogen with deuterium and measuring the resulting isotope effect on the rate of cleavage.

Correlation between chemical reactivity and calculated surface accessibility has been noted previously by other workers studying tRNA structure.[28–30] The observation that such a correlation exists in a protein–DNA complex demonstrates that it is possible to derive quantitative structural information from hydroxyl radical footprinting of macromolecular complexes. The sophistication of the derived structural information should increase as our understanding of the mechanism of hydroxyl radical cleavage of DNA improves.

Increasing Contrast of Hydroxyl Radical Footprint by Purifying Protein–DNA Complex on Mobility Shift Gel

Many proteins produce strong hydroxyl radical footprints when studied using the methods described above. However, proteins that bind to DNA weakly often produce very weak hydroxyl radical footprints, most likely due to a high level of signal from unbound DNA. Since the data from these weak footprints are hard to interpret, it is desirable to increase the footprint signal compared to the signal from unbound DNA. The footprint signal can be increased over the noise of the cleavage pattern of naked DNA by separating the protein–DNA complex from naked DNA by electrophoresis on a native (mobility shift[31,32]) polyacrylamide gel.

The protein is bound to the DNA, and the mixture is treated with the hydroxyl radical cleavage reagent as in a conventional footprinting experiment. After 2 min the reaction is stopped and the mixture is loaded immediately on a native polyacrylamide gel to separate the protein–DNA complex from the unbound DNA. The DNA from the band containing the protein–DNA complex is eluted from the gel, denatured by heating, and electrophoresed on a sequencing gel to produce the footprint pattern.

In this section, we apply our protocol for hydroxyl radical footprinting

[28] S. R. Holbrook and S.-H. Kim, *Biopolymers* **22**, 1145 (1983).
[29] D. Romby, D. Moras, M. Bergdoll, P. Dumas, V. V. Vlassov, E. Westhof, J. P. Ebel, and R. Giege, *J. Mol. Biol.* **184**, 455 (1985).
[30] J. A. Latham and T. R. Cech, *Science* **245**, 276 (1989).
[31] M. M. Garner and A. Revzin, *Nucleic Acids Res.* **9**, 3047 (1981).
[32] M. Fried and D. M. Crothers, *Nucleic Acids Res.* **9**, 6505 (1981).

with gel isolation of complexes to two proteins, the restriction endonuclease *Eco*RI and CUP2, a yeast transcriptional activator protein. While the footprint of *Eco*RI is clear in a standard footprinting experiment, a stronger footprint can be produced at a lower protein : DNA ratio by separating bound from free DNA on a mobility shift gel. With CUP2 we had difficulty observing any regions of decreased cutting using the standard procedure for hydroxyl radical footprinting. However, a clear CUP2 footprint was obtained by performing hydroxyl radical footprinting with gel isolation of the CUP2–DNA complex.

Hydroxyl Radical Footprinting of EcoRI–DNA Complex

Protocol for Footprinting without Mobility Shift Gel Electrophoresis. *Binding conditions for EcoRI–DNA complex:* The radioactively labeled DNA restriction fragment containing the *Eco*RI binding site (2 nM) is incubated with *Eco*RI endonuclease in binding buffer [0.1 M KCl, 10 mM Tris (pH 7.4), 1 mM EDTA] at room temperature for 20–30 min. [Since magnesium ion is necessary for cleavage of DNA by *Eco*RI, the lack of magnesium in this buffer permits binding of the enzyme but not cleavage of DNA.] The volume of the solution is 70 μl. Concentrations of *Eco*RI ranging from 3 to 80 nM are used for footprint titration experiments.

Hydroxyl radical cleavage reaction: Samples are subjected to hydroxyl radical cleavage by the addition of 10 μl of 2.5 mM [Fe(EDTA)]$^{2-}$, 10 μl of 10 mM sodium ascorbate, and 10 μl of 0.3% hydrogen peroxide. Samples are allowed to react for 2 min. The cleavage reaction is terminated by the addition of 90 μl of stop buffer (13 mM EDTA, 13 mM thiourea, 13 mg/ml tRNA, 0.1 M KCl). Samples are extracted with phenol and ether. The DNA is precipitated twice with ethanol. The pellet is rinsed with ethanol and dried. DNA samples are prepared for loading and electrophoresis on an 8% sequencing gel as described above.

Protocol for Footprinting with Mobility Shift Gel Electrophoresis. *Binding conditions for EcoRI–DNA complex:* Binding solutions contain binding buffer [20 mM KCl, 10 mM Tris (pH 7.4), 2.5 mM EDTA], and either 102 nM DNA [2 nM radioactively labeled restriction fragment, and 100 nM unlabeled self-complementary oligonucleotide, d(CGCGAATTCGCG), which contains the *Eco*RI recognition sequence] and 100 nM *Eco*RI, or 2 nM labeled DNA and 50 nM *Eco*RI. The total volume of the solution is 35 μl.

Hydroxyl radical cleavage reaction: Samples are subjected to hydroxyl radical cleavage in the same manner as described above, except that the reaction is quenched using 5 μl of 0.1 M thiourea. Native dye mixture [50% (v/v) glycerol, 2% (w/v) xylene cyanol] (10 μl) is added to samples prior to nondenaturing polyacrylamide gel electrophoresis, bringing the volume of the sample to 80 μl.

Nondenaturing gel electrophoresis: The EcoRI–DNA complex is separated from free DNA on a 16 cm × 19.5 cm × 1.5 mm gel consisting of 9.5% (w/v) acrylamide, 0.5% (w/v) bisacrylamide in 1 × TBE (pH 7.4). The gel is preelectrophoresed for 1 hr at 60 V. Samples are electrophoresed at 60 V at room temperature overnight. Autoradiography is used to identify radioactively labeled DNA in the gel. The gel slices containing radioactive DNA are crushed and incubated in 0.5 M ammonium acetate, 1 mM EDTA to elute the DNA. Small portions of the gel slices are crushed and eluted with 0.5 M ammonium acetate, 1 mM EDTA, and 10 mM MgCl$_2$ to test for DNA cleavage activity at the EcoRI sequence. Gel slices are incubated in elution buffer for 1.5 hr at 37°. The eluants are extracted with phenol and ether and made 0.3 M in sodium acetate. The DNA is precipitated twice with ethanol, rinsed, and dried. DNA is prepared for electrophoresis on an 8% sequencing gel as described above.

Results: We have produced hydroxyl radical footprints of EcoRI, bound to its recognition sequence GAATTC, with or without the additional step of separating the EcoRI–DNA complex from unbound DNA by mobility shift gel electrophoresis. The mobility shift gel shows that even at a ratio of 25 : 1 of EcoRI to its DNA recognition sequence a sizeable amount of unbound DNA is present in the reaction mixture (Fig. 3). From the autoradiographs and the densitometer scans of these footprints (Fig. 4) it is apparent that the protection patterns that result from the two procedures are quite similar. Both show a sharp decrease in hydroxyl radical cleavage two nucleotides prior to the G in the recognition sequence. This low level of cleavage is maintained throughout the recognition sequence. Past the C at the 3' end of the recognition sequence a gradual increase in cleavage occurs for about four nucleotides, at which point the cleavage frequency is again equal to that of the flanking sequences. The footprint pattern is stronger when unbound DNA has been separated by mobility shift gel electrophoresis. This is most apparent in the densitometer scans. Removal of free DNA by gel electrophoretic separation reveals that cleavage within the recognition sequence GAATTC is almost completely blocked by bound EcoRI. When the free DNA is not removed from the sample, weak bands are observed in the region of the recognition sequence.

In the X-ray crystal structure of the complex of EcoRI with its DNA-binding site,[33] the endonuclease is seen to interact with the DNA backbone

---

[33] J. A. McClarin, C. A. Frederick, B.-C. Wang, P. Greene, H. W. Boyer, J. Grable, and J. M. Rosenberg, *Science* **234**, 1526 (1986).

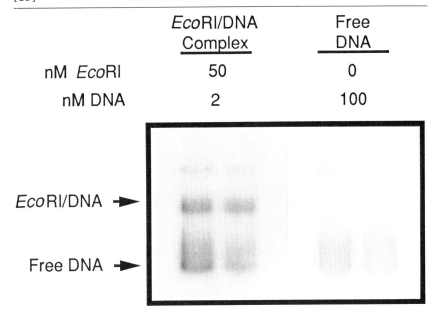

FIG. 3. Separation of the *Eco*RI–DNA complex from free DNA on a mobility shift gel. All samples contained a singly end-labeled 207-bp DNA restriction fragment containing the *Eco*RI recognition sequence GAATTC. Each sample was loaded in two lanes. Unbound DNA is labeled as free DNA, and migrates farther on the gel than DNA bound by protein.

of eight nucleotides, six that encompass the recognition sequence (GCGAATTC) and two nucleotides to the 5' side. The DNA backbone of those same eight nucleotides plus four others are protected from hydroxyl radical cleavage. The extra backbone protections are probably due to the presence of the protein at nearby nucleotides which results in a decrease in reaction with hydroxyl radical outside the recognition sequence. The hydroxyl radical footprint of the *Eco*RI–DNA complex corresponds in general to expectations based on the crystal structure.

Hydroxyl Radical Footprint of CUP2 Bound to Upstream Activation Site C of Yeast Metallothionein Gene

Binding Conditions for the CUP2–DNA Complex. Binding solutions contain 23.5 ng CUP2, 0.25 ng singly end-labeled DNA (500,000 dpm), 1 mg BSA, 1% poly(vinyl alcohol), 20 ng poly(dI) · poly(dC), 12.5 mM HEPES-NaOH buffer (pH 8.0), 50 mM KCl, 6.25 mM MgCl$_2$, 0.5 mM EDTA, and

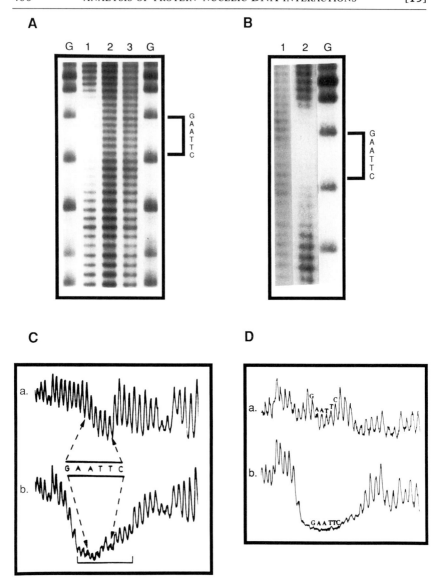

FIG. 4. Comparison of hydroxyl radical footprints of *Eco*RI produced with or without the additional step of gel isolation of the *Eco*RI–DNA complex. (A) and (C) show an autoradiograph and corresponding densitometer scan of the hydroxyl radical footprint of *Eco*RI produced by the standard footprinting technique (without gel isolation of the *Eco*RI–DNA complex). The autoradiograph and densitometer scan of a hydroxyl radical footprint of *Eco*RI produced by the standard footprinting technique followed by gel isolation of the *Eco*RI–DNA complex are shown in (B) and (D), respectively. In (A) are shown hydroxyl radical cleavage

0.5 mM DTT, in a total volume of 10 μl. To ensure that equilibrium is reached for binding of the protein to DNA, the solution is incubated on ice for 15 min.

Hydroxyl Radical Cleavage Reaction. After warming the CUP2–DNA mixture to room temperature for 1 min, the cutting reaction is initiated by mixing together on the inner wall of the Eppendorf tube 2 μl each of $[Fe(EDTA)]^{2-}$ [8 mM iron(II), 16 mM EDTA], 0.024% hydrogen peroxide, and 160 mM sodium ascorbate, and adding this reagent to the CUP2–DNA mixture. The final concentrations of the constituents of the cleavage reagent are 1 mM iron(II), 2 mM EDTA, 0.003% hydrogen peroxide, and 20 mM sodium ascorbate. After 2 min at room temperature the reaction is stopped by addition of 5 μl of 0.1 M thiourea.

Nondenaturing Gel Electrophoresis. To each sample, 4 μl of loading dye (0.25% bromphenol blue, 0.25% xylene cyanol, and 30% glycerol) is added. Samples are loaded on a 16 cm × 19.5 cm × 1.5 mm 6% native polyacrylamide gel (acrylamide : bisacrylamide, 80 : 1). Electrophoresis is performed at room temperature at 125 V for 3 hr in a buffer consisting of 25 mM Tris–borate–HCl (pH 8.3), 2.5 mM EDTA. Detection of radioactive DNA, excision and elution of DNA from the gel, and sequencing gel electrophoresis of DNA are performed in the same manner as described above for the *Eco*RI–DNA complexes.

Results. The CUP2 protein was bound to a restriction fragment containing the UASc- and UASd-binding sites.[13] Under the conditions used a single protein–DNA complex was observed via electrophoresis on a mobility shift gel. The best separation was achieved with the protein–DNA complex running slightly slower than the unbound DNA.

The resulting hydroxyl radical footprint of CUP2 is shown in Fig. 5. The footprint is highly symmetrical around the pseudodyad axis of symmetry of

products of a singly end-labeled 207-bp DNA restriction fragment (2 nM) in the absence (lane 3) and the presence (lane 2, 20 nM; lane 1, 80 nM) of *Eco*RI endonuclease. Lanes marked G are products of a Maxam–Gilbert guanine-specific sequencing reaction. The *Eco*RI recognition sequence, GAATTC, is indicated by a bracket. In (C), the upper scan (a) shows the cleavage of DNA in the absence of *Eco*RI and corresponds to lane 3 in (A). The lower scan (b) shows the cleavage pattern of DNA in the presence of *Eco*RI and corresponds to lane 1 in (A). The sequence reads 5' to 3', left to right, for each scan. The *Eco*RI recognition sequence is marked by arrows. In (B), lanes 1 and 2 show products of hydroxyl radical cleavage of a DNA restriction fragment that was isolated from the nondenaturing gel shown in Fig. 3. Lane 1, free DNA; lane 2, DNA isolated from the upper band on the mobility shift gel (the band containing the *Eco*RI–DNA complex). Lane marked G contains products of a Maxam–Gilbert guanine-specific sequencing reaction. The *Eco*RI recognition sequence, GAATTC, is indicated by a bracket. In (D), the upper scan (a) corresponds to lane 1 and the lower scan (b) corresponds to lane 2 in (B).

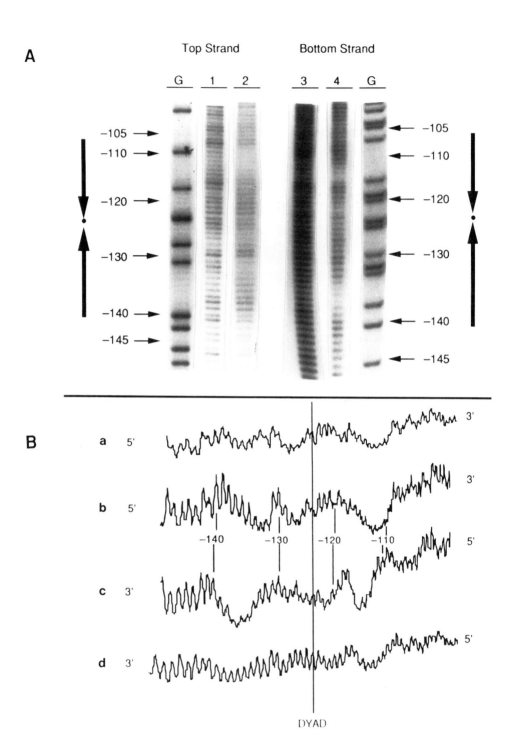

the binding site. CUP2 protects three regions of each strand. Each set of protections at the ends is offset across the DNA strands from each other by three bases in the 3' direction. The protected region on each strand in the center of the site is offset by six bases in the 5' direction from the corresponding protected region on the other strand.

These results indicate that in each half of the UASc, CUP2 crosses the minor groove at the ends of the binding site between the two major grooves where methylation interference detects contacts with guanines.[13] The protection at the center indicates that CUP2 interacts with the major groove at the center of the binding site.

Comments on Technique of Hydroxyl Radical Footprinting Followed by Mobility Shift Gel Electrophoresis

As discussed above, purification of a protein–DNA complex on a nondenaturing gel prior to sequencing gel electrophoresis can significantly enhance the hydroxyl radical footprint signal by elimination of bands that derive from cleavage of free DNA. However, to use this technique three technical pitfalls must be overcome. First, separation of a protein–DNA complex from free DNA by gel electrophoresis must be achieved. The conditions necessary to effect a separation depend on the protein, the DNA fragment length, and the number of binding sites. A general technique for separating free from bound DNA by mobility shift gel electrophoresis has been described.[34]

---

[34] W. Hendrickson and R. Schleif, *Proc. Natl. Acad. Sci. U.S.A.* **82**, 3129 (1985).

FIG. 5. Hydroxyl radical footprinting of the CUP2–UASc complex. (A) shows the autoradiograph of a gel containing the hydroxyl radical cleavage pattern for the UASc in the absence of protein (lanes 1 and 3), and with bound CUP2 (lanes 2 and 4). Lanes 1 and 2 are hydroxyl radical cleavage products derived from the top strand, and lanes 3 and 4 are cleavage products derived from the bottom strand. Lanes marked G are products of a Maxam–Gilbert G reaction performed on the respective labeled strands. Nucleotides are numbered relative to the start site of transcription. The vertical lines indicate the region of almost perfect dyad symmetry. The dots are placed at position − 124, the center of the pseudodyad. The corresponding densitometer scans are shown in (B). Scans a and b were made from lanes 1 and 2, and scans c and d were made from lanes 4 and 3, respectively. The vertical line marks the pseudodyad axis of symmetry. The footprint is clearly symmetrical from one strand to the other about the pseudodyad, with symmetry-related protections shaped identically. On the top strand the strongest protections occur at positions − 133, − 127, and − 112, while on the bottom strand the symmetry-related protections are at positions − 136, − 121, and − 115, respectively. From Buchman *et al.* (1990).[13]

Second, a substantial amount of DNA must be shifted into the protein–DNA complex and successfully eluted from the gel. Two means can be used to improve the amount of DNA that is shifted. One is to increase the amount of protein in the sample. The other is to omit the addition of marker dyes (bromphenol blue and xylene cyanol) to the protein–DNA sample. We usually realize a low recovery (about 50%) of DNA by elution from the gel. In order to have a sufficient amount of DNA to load on the sequencing gel, we have found it necessary to start the experiments with 5 to 10 times the desired amount.

Third, a protein–DNA complex must be stable during footprinting and loading onto the gel. Instability of the protein–DNA complex could lead to the following problems in interpreting the footprinting pattern of a protein–DNA complex that was gel isolated after the footprinting reaction. First, dissociation of the protein–DNA complex during gel loading or electrophoresis will diminish the protection signal, since some DNA that should have run in the band with complexed DNA instead runs as free DNA. Second, and more problematic, reassociation of protein with gapped DNA will be biased due to the effect of gaps on the affinity of protein for DNA, and will lead to a decreased cutting frequency observed at the nucleosides involved in specific contacts with the DNA. A gap at a critical place in the DNA sequence interferes with protein binding, as we have shown in the missing nucleoside technique[35] for studying protein–DNA complexes, which is described in the next section. In this situation, a decrease in signal in the data from the protein-bound DNA could therefore indicate either a protein protection or a protein contact.

The analysis becomes more complicated if there is more than one form of DNA–protein complex separated by the mobility shift gel, since a gap in the DNA will inhibit formation of the fully saturated complex, but may favor formation of one of the partially saturated complexes, resulting in an enhancement of signal for particular nucleosides in those complexes. Two control experiments may be performed to test for instability of the protein–DNA complex. One involves adding unlabeled DNA containing the binding site to the sample after hydroxyl radical treatment to see if exchange takes place. In the second control, comparison of the results of the experiment with conventional hydroxyl radical footprints and missing nucleoside data will show whether the pattern on the gel is more consistent with a footprint or a missing nucleoside signal.

---

[35] J. J. Hayes and T. D. Tullius, *Biochemistry* **28,** 9521 (1989).

Missing Nucleoside Experiment: Direct Information on Energetically
  Important Contacts of Protein with DNA

The set of all interactions between a DNA-binding protein and its
recognition site can be divided conceptually into two types based on the
architecture of DNA. One group of interactions is composed of the contacts between the protein and the sugar-phosphate backbone of DNA. The
other group includes the interactions between the base moieties of DNA
and the protein. The hydroxyl radical footprinting experiment yields high-
resolution structural information about protein–DNA contacts of the first
type.This technique shows where along a DNA molecule the sugar-phosphate backbone of DNA comes into close contact with the bound protein.
However, little is learned about the second type of interaction, that between the protein and the bases of DNA, from the footprinting experiment.
  We have described a new technique that can be used to obtain information about contacts between DNA and protein that fall into the second
group. This technique, called the missing nucleoside experiment,[35] can be
used to quickly determine at which positions in a DNA molecule the bases
are in contact with a sequence-specific DNA-binding protein. This method
is related to the missing contact experiment.[36] The missing nucleoside
experiment uses hydroxyl radical cleavage chemistry to randomly remove
nucleosides from DNA. The ability of the gapped DNA to bind to protein
is then tested by gel mobility shift. Bases that make important contacts to
protein are identified by sequencing gel electrophoresis of the bound and
unbound fractions of DNA recovered from the mobility shift gel. The
missing nucleoside experiment can be used to scan a DNA molecule for
specific contacts at single-nucleotide resolution in just one experiment.

Overview of Missing Nucleoside Method

The hydroxyl radical cleavage reaction is used to generate DNA fragments that contain, on average, fewer than one randomly placed single-
nucleoside gap per fragment. The gapped DNA molecules are then mixed
with the protein, and DNA bound to protein is separated from free DNA
by electrophoresis on a native polyacrylamide gel. Bands containing bound
and free DNA are excised from the native gel, and the DNA is eluted and
run on a sequencing electrophoresis gel for determination of the cleavage
patterns in the bound and free samples. Since the DNA strand is already
broken in the gapping reaction, no other treatment is necessary to develop
the pattern. Although this procedure might seem very similar to that
discussed in the previous section, for the missing nucleoside analysis

[36] A. Brunelle and R. F. Schleif, *Proc. Natl. Acad. Sci. U.S.A.* **84,** 6673 (1987).

naked DNA is treated by the hydroxyl radical before the protein is introduced. In contrast, for gel purification of DNA–protein complexes before sequencing gel electrophoresis, the hydroxyl radical reaction is performed on the DNA–protein complex in the usual way.

The missing nucleoside experiment detects gaps in the DNA molecule that interfere with binding enough that the protein and DNA no longer migrate as a complex on the gel. A nucleoside that is important to formation of the protein–DNA complex yields a weak or missing band on the sequencing gel in the lane containing DNA that was bound to protein, or a high-intensity band in the lane in which free DNA is run. Nucleosides that do not make energetically important contacts also give a positive experimental signal, but with the opposite pattern of band intensity. Complementary information is therefore obtained from the two lanes (containing bound and free DNA) that result from the original sample. In contrast, protection (footprinting) methods give a negative result for bases not involved in protein contact.

Advantages of Missing Nucleoside Experiment

Site-directed mutagenesis is a powerful adjunct to direct structural studies and can be used to investigate contacts with protein at each base pair in a recognition sequence. Many time-consuming manipulations are needed to scan even a small binding site using mutagenesis. For example, a saturation mutagenesis analysis of λ repressor binding to its operator required the synthesis of over 50 new versions of the DNA recognition site.[37] The binding affinity of repressor for each of the mutant operators was then determined. While the results from the mutation experiments and the missing nucleoside experiment generally agree,[35] the latter procedure took only a few days to complete—certainly a much shorter time than that required to complete the former experiment. A second advantage of the missing nucleoside experiment over mutational methods arises because often only one member of a base pair is involved in specific recognition.[23] Since the identities of both bases in a base pair are changed by mutagenesis, it is not easy to determine which is the important contact. The missing nucleoside experiment, on the other hand, can yield information about contacts to a particular base that is independent of the contribution of the base on the opposite strand.[35] Moreover, not all base substitutions at a particular position result in a reduction in binding affinity.[38,39]

Another experimental approach, methylation interference, involves

[37] A. Sarai and Y. Takeda, *Proc. Natl. Acad. Sci. U.S.A.* **86**, 6513 (1989).
[38] T. Pieler, J. Hamm, and R. G. Roeder, *Cell (Cambridge, Mass.)* **48**, 91 (1987).
[39] Y. Takeda, A. Sarai, and V. M. Rivera, *Proc. Natl. Acad. Sci. U.S.A.* **86**, 439 (1989).

assessing the effect of alkylation of the DNA bases on protein binding. Modified positions that interfere with protein binding are revealed on induction of backbone cleavage at the alkylated sites of the DNA molecules that were unable to bind to protein. However, information is available for only a subset of base-specific contacts and, unfortunately, only contacts with purines can generally be detected.

As mentioned above, the missing nucleoside experiment is related to the "missing contact" method of Brunelle and Schleif.[36] This experimental approach can also be used to assess the contribution to protein binding of each member of a base pair independently. However, in the missing contact method Maxam–Gilbert sequencing chemistry is used to remove the bases, so three separate experiments are needed to determine the bases important for protein binding. With the missing nucleoside technique a single chemical reaction, and thus a single lane on a sequencing gel, is all that is needed for the determination. The missing nucleoside experiment thus reduces the amount of work and material required for an analysis compared to the missing contact method, and since all of the signal appears in just one sequencing gel lane, the results are somewhat easier to interpret.

Methods

Proteins and DNA Fragments. Any protein–DNA complex can be analyzed by this method provided that the complexed DNA can be isolated from unbound DNA in some manner, and that enough sample is recovered to generate a signal on a sequencing gel. In our laboratory gel mobility shift is used to effect the separation, so the radioactively labeled DNA fragment must be of the appropriate size (see Troubleshooting, below). We also found that the ends of the fragment must be further than 15–20 bp from the binding site to be assayed, in order to prevent dissociation of short oligomers from the gapped DNA molecule.[40]

Hydroxyl Radical Cleavage. The labeled DNA molecule is randomly gapped by reaction with the hydroxyl radical. Typically, a relatively large quantity of labeled fragment, 250–500 fmol, enough for two or three missing nucleoside experiments, is dissolved in 70 μl TE buffer. For comparison, a typical hydroxyl radical footprinting experiment contains about 10–20 fmol of labeled fragment. The DNA is treated with a total of 30 μl of the cleavage reagent (10 μl of each of the individual components) in the manner described above. The cleavage reaction is terminated by the addition of 30 μl of 0.1 M thiourea, 10 μl of 0.2 M EDTA (pH 8.0), 20 μl of 3 M sodium acetate, and 40 μl of TE buffer, to bring the volume of the

[40] V. Rimphanitchayakit, G. F. Hatfull, and N. D. F. Grindley, *Nucleic Acids Res.* **17**, 1035 (1989).

reaction mixture to 200 μl. (These solutions can be added to the cleavage reaction together, as a cocktail.) The DNA is then precipitated by the addition of 500 μl of ethanol at $-20°$. The sample is reprecipitated by first dissolving the pellet in 200 μl of 0.3 M sodium acetate, 0.2 mM EDTA to remove traces of iron that may remain in the sample. The final pellet is rinsed with 70% ethanol ($-20°$), dried under vacuum, and then dissolved in a storage buffer consisting of 10 mM Tris-Cl, 0.1 mM EDTA (pH 8.0). In the case of proteins requiring metal cofactors, such as *Xenopus* transcription factor IIIA, EDTA is omitted from the final storage buffer.

Formation of Protein–DNA Complexes. Buffers for protein binding are prepared as in the usual procedure. Note that buffer systems that contain a high level of glycerol (or any other radical scavenger), which would be unsuitable for use in a hydroxyl radical footprinting experiment, are fully compatible with the missing nucleoside experiment. Protein is added to labeled, gapped DNA that is dissolved in binding buffer, and the system is allowed to come to equilibrium. A typical binding solution contains 125–250 fmol of gapped, labeled DNA. The amount of protein to be added is determined empirically and can be estimated by titrating small quantities of the labeled DNA with protein. (The amount of protein required to saturate a binding site on DNA in a footprinting experiment generally was found to be less than that required to saturate the DNA when measured by gel mobility shift.) The amount of protein or of unlabeled competitor DNA in the binding mixture is adjusted to give approximately 95% formation of complex as judged by the intensities of the bands on the mobility shift gel. This amount of protein generally results in good bound and unbound samples, but some proteins may require small adjustments in the amount of complex formed to optimize the signal observed on the sequencing gel (see Troubleshooting, below).

Fractionation and Analysis of Bound and Unbound DNA. DNA bound to protein is separated from free DNA by electrophoresis on a native polyacrylamide (mobility shift) gel as described.[41] For further comments on gel mobility shift techniques, see Troubleshooting, below. Radioactive bands containing bound and free DNA are excised from the gel and the DNA is eluted. DNA is precipitated by addition of ethanol, rinsed with ethanol, dried under vacuum, and dissolved in 10–20 μl of TE buffer. Note that typically in the unbound sample little or no radioactivity is detectable, while a majority of the original counts will appear in the bound sample. About one-fourth of each sample is then analyzed by sequencing gel electrophoresis. The amount of the bound or unbound sample that is loaded may be adjusted slightly on later sequencing gels to better match

[41] A. Wolffe, *EMBO J.* **7**, 1071 (1988).

the band intensities of the two samples, or to match the intensities of bands from corresponding footprinting sample.

Applications of Missing Nucleoside Experiment

The missing nucleoside method has been used to analyze two well-understood protein–DNA complexes, the complexes of the bacteriophage λ repressor and Cro proteins with the O_R1 operator site.[35] A detailed comparison of the missing nucleoside results with the protein–DNA contacts detected crystallographically and genetically has been presented.[35]

λ *Repressor–O_R1 Complex.* In the case of the λ repressor–operator complex, an example of the correlation between the cocrystal structure and the missing nucleoside experiment is most easily observed at position 2 in the operator. The crystal structure reveals an intricate network of contacts with the adenine member of the base pair at this position, while no contact is detected with the thymine. The missing nucleoside experiment[35] indicates exactly the same circumstance. A large contact signal is detected at the adenosine at position 2, while the signal corresponding to the thymidine indicates that this nucleoside is uncontacted by the protein.

Another interesting example of the data available from the missing nucleoside experiment involves the central 5 bp of the operator and the structural unit of the λ repressor that contacts them. The repressor has an amino-terminal arm that wraps around the center of the operator and contacts the backside of the DNA in the major groove.[42] The supposition that the arms contact the central bases of the operator is borne out by the cocrystal structure.[23] The missing nucleoside data[45] show that the arms make extensive and energetically important contacts with positions 7, 8, and the central dyad base. These contacts agree well with those detected by mutagenesis methods in the center of the operator.[37] The data also indicate that the structure of the amino-terminal arm is not the same in each half of the dimer. Specifically, the results suggest that the amino-terminal arm that extends from the repressor monomer that contacts the consensus half of the operator binds much more tightly to the DNA than does the arm that is associated with the nonconsensus half of the operator. This proposal correlates well with the cocrystal structure, which shows that the arm that contacts the consensus half-site is the more ordered.[23]

Cro–O_R1 Complex. We also have used the missing nucleoside technique to analyze a related system, the complex of the λ Cro protein with the O_R1 operator sequence. Cro and λ repressor bind to the same DNA sites, but with different affinities. The two proteins might thus be expected

[42] C. O. Pabo, W. Krovatin, A. Jeffrey, and R. T. Sauer, *Nature (London)* **298**, 441 (1982).

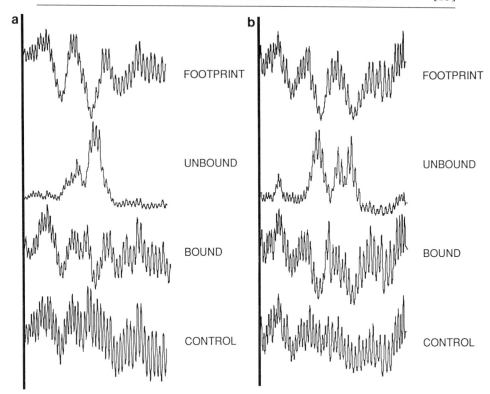

FIG. 6. Missing nucleoside analysis of the λ Cro–O_R1 complex. Shown are densitometer scans of the autoradiograph of the denaturing gel on which was separated DNA that was subjected to a standard hydroxyl radical footprinting reaction (footprint), DNA extracted from a mobility shift gel that was complexed with Cro (bound) and which ran as free DNA (unbound), and DNA that was subjected to cleavage by the hydroxyl radical in the absence of protein (control). Scans for the two DNA strands [(a) top strand, (b) bottom strand] are shown.

to make different sets of contacts with DNA. Indeed, despite marked similarities in the DNase I and hydroxyl radical footprints of Cro and repressor,[8] the missing nucleoside experiment yields different patterns for the two proteins.

One particular difference occurs at the center of the operator site. Cro protein does not have the same amino terminal arm that repressor has. Takeda et al.[39] demonstrated by mutagenesis experiments that the central 3 bp of O_R1 contribute little to the energetics of Cro binding. The missing nucleoside analysis shows that the center of the O_R1 site is largely devoid of contacts with Cro (Fig. 6). With λ repressor, by contrast, large contact signals are observed in this region,[35] which we attribute to the interaction of the arms of repressor with the center of the binding site.

Another Possible Use of the Method. The missing nucleoside experiment can be used to locate the DNA-binding site for a protein that is available only in a complicated extract, and for which the protein–DNA complex is observed only as a band on a mobility shift gel. Such an extract might be unsuitable for normal hydroxyl radical footprinting, because it contains high concentrations of radical scavengers (glycerol, for example). However, a sort of reverse footprint can be obtained by applying the missing nucleoside technique to this system. Protein-bound, mobility-shifted DNA fragments from such an experiment, even if the bound band represents only a small fraction of the total labeled DNA in the sample, should yield an obvious bound signal at the location of the recognition sequence.

Troubleshooting Missing Nucleoside Experiment

We have made several observations that might aid attempts to employ this technique with other systems. Each of these points may or may not apply to a particular protein–DNA complex. Some fine tuning of the parameters may be required before an optimal result is obtained. Most of the suggestions concern a process that is to date poorly understood: the separation of free DNA from protein–DNA complexes by mobility shift gel electrophoresis.[31,32]

It also should be noted that introducing gaps in a DNA molecule might affect the binding of a protein because of structural changes induced in the DNA backbone[43] as a result of the missing nucleoside. Comparison of our results for λ repressor with the cocrystal structure and with mutagenesis experiments showed that nearly all of the missing nucleoside signals are consistent with the loss of a contact with a nucleoside.[35] However, with other protein–DNA systems changes in DNA structure should also be considered.

1. *Gel.* Other gel systems than the one employed for this study have been used for mobility shift. We have obtained a quantitative shift for the TFIIIA–5S RNA gene complex using a 0.7% agarose gel with 0.5× TB buffer [50 mM Tris-Cl (pH 8.3), 50 mM boric acid] at room temperature. However, all of the proteins that we have investigated with the missing nucleoside technique worked well with the gel system described by Wolffe[41]: 4% acrylamide, 0.08% bisacrylamide, in a buffer consisting of

[43] G. B. Koudelka, P. Harbury, S. C. Harrison, and M. Ptashne, *Proc. Natl. Acad. Sci. U.S.A.* **85,** 4633 (1988).

20 mM HEPES (pH 8.3), 0.1 mM EDTA, 5% glycerol. The same solution without glycerol is used as the running buffer. The gel has the thickness of a sequencing gel, which conserves gel material and minimizes the volume of gel from which a sample must later be recovered. This thickness does, however, make it difficult to produce good wells and requires that the gel be loaded with a 0.2-mm diameter syringe. It is absolutely necessary for the large plate to be liberally siliconized before every gel is poured. Care must be taken when plastic wrap is put on and taken off since these gels are very fragile.

2. *Gel buffers*. Buffers such as 0.5× TB or TBE can be used in this system, depending on the effect of EDTA on protein binding. However, the HEPES buffer yielded superior results compared to any buffer containing Tris, especially in the experiments with TFIIIA. Although high concentrations of EDTA can interfere with the binding of metal-containing proteins, when a small amount of EDTA (0.1 mM) was included in the gel buffer the sharpness of bands was improved. This might be due to the scavenging of excess metal ions, such as free zinc that remains from TFIIIA–DNA complex formation. A low level of EDTA does not seem to affect the TFIIIA–DNA complex since the exchange rate of zinc in the protein–DNA complex is probably much slower than it is in the free protein.

3. *Gel running conditions*. Once a protein–DNA complex enters the gel, the complex seems to be very stable and migrates without dissociation. However, the equilibrium can be disturbed by the process of transferring the sample from a tube to the gel matrix. An interesting example involves λ repressor. With a gel run at room temperature, only a minute fraction of the labeled DNA sample can be made to run as a complex with repressor, regardless of protein concentration. Footprinting experiments show that this is true for concentrations of repressor well above that required to saturate the binding site *in vitro*. In contrast, Cro protein easily shifts the mobility of the DNA molecule under these conditions. However, when the repressor sample is cooled and run on a gel at 4°, a quantitative shift of DNA into the bound band results. The on/off rates for repressor might be much slower at 4°, essentially freezing the equilibrium of the sample and thus allowing the complex to enter the gel.

4. *Loading sample*. Artifacts can be introduced during the preparation and loading of the sample. For this reason the gel loading solution (1× HEPES/EDTA buffer, 40% (v/v) glycerol, 0.025% dye) is added just before loading. The sample is carefully layered into the well without mixing with the gel running buffer.

5. *Sample manipulations*. The strength of the contact signal observed can be improved by the above suggestions. In addition, certain manipula-

tions of the sample before loading, such as the addition of unlabeled competitor DNA, can help. A typical sample is prepared under conditions that maximize binding to a labeled DNA fragment. Then an excess of unlabeled, ungapped DNA, which usually contains the protein-binding site, is added. The amount of competitor DNA is adjusted so as to compete for only the most weakly bound protein. Alternatively, a large excess of competitor is added and the bound complexes are challenged for a specific length of time.[36] The time is controlled either by directly loading the sample on the gel or by cooling the sample to 0° on ice to slow the exchange process before loading.

6. *Miscellaneous.* The length of the DNA fragment is important, since a long fragment makes it difficult to separate bound from free DNA. As mentioned above, the binding site cannot be located too close to the end of the fragment, because dissociation of short oligonucleotides from the duplex after the gapping reaction will leave single-stranded regions in the binding site that are incapable of binding protein.[40] It might be necessary to do two separate experiments to get good bound and unbound samples. Since only a few bands occur in the unbound lane on the sequencing gel (and thus there is little radioactivity associated with unbound DNA), excellent unbound samples can be obtained by cutting out the place on the native gel where the free DNA is expected to run, even if there is no apparent radioactive band there.

Acknowledgments

We thank Carl Pabo for providing us with the oligonucleotide containing the O_L1 operator site and for samples of the λ Cro protein and the 1–92 fragment of λ repressor, Gary Ackers for supplying λ repressor and for making available to us the 2D gel scanning system, and Irina Russu for generous provision of *Eco*RI. We are grateful to Michael Karin for providing samples of *CUP2* and a clone containing the *UASc*, and Thomas J. Kelly for giving us samples of NF I and clones of Ad DNA. This work was supported by PHS Grant GM 41930. T.D.T. is a fellow of the Alfred P. Sloan Foundation, a Camille and Henry Dreyfus Teacher-Scholar, and the recipient of a Research Career Development Award (CA 01208) from the PHS. J.R.L. acknowledges the support of a postdoctoral fellowship from the Institute for Biophysical Research on Macromolecular Assemblies at Johns Hopkins University (an NSF Biological Center). W.J.D. is the recipient of a National Research Service Award from the PHS.

[20] Nuclease Activity of 1,10-Phenanthroline–Copper in Study of Protein–DNA Interactions

By DAVID S. SIGMAN, MICHIO D. KUWABARA, CHING-HONG B. CHEN, and THOMAS W. BRUICE

Introduction

The nuclease activity of the 1,10-phenanthroline–copper complex efficiently nicks DNA and RNA under physiological conditions.[1,2] Two reactants are essential for the cleavage reaction: the 1,10-phenanthroline–cuprous complex and hydrogen peroxide. The chemistry of B-DNA scission has been studied in detail, and proceeds by the kinetic mechanism summarized in Fig. 1.[3-9] The reaction is funneled through a noncovalent intermediate that is responsible for the structural and sequence-dependent variability of the nucleolytic activity. DNA scission is achieved by the oxidative species formed by the one-electron oxidation of the 1,10-phenanthroline–cuprous complex by hydrogen peroxide when the coordination complex is bound to the surface of DNA.

With the 2 : 1 1,10-phenanthroline–cuprous complex, reaction pathway **(a)** (Fig. 2) accounts for over 70–90% of the scission events. This pathway involves the C-1 hydrogen of the deoxyribose as the initial site of oxidative attack. Pathway **(b)** results from oxidative attack on the C-4 hydrogen.[5] Either reaction pathway requires that the initial sites of reaction be in the minor groove of B-DNA.

The nuclease activity of 1,10-phenanthroline–copper has been used in two distinct ways in the study of protein–nucleic acid interactions. In the first, copper complexes prepared from 1,10-phenanthroline (OP–Cu) or

[1] D. S. Sigman, D. R. Graham, V. D'Aurora, and A. M. Stern, *J. Biol. Chem.* **254,** 12269 (1979).

[2] D. S. Sigman, *Acc. Chem. Res.* **19,** 180 (1986).

[3] L. M. Pope, K. A. Reich, D. R. Graham, and D. S. Sigman, *J. Biol. Chem.* **257,** 12121 (1982).

[4] M. D. Kuwabara, C. Yoon, T. E. Goyne, T. B. Thederahn, and D. S. Sigman, *Biochemistry* **25,** 7401 (1986).

[5] T. E. Goyne and D. S. Sigman, *J. Am. Chem. Soc.* **109,** 2846 (1987).

[6] L. E. Pope and D. S. Sigman, *Proc. Natl. Acad. Sci. U.S.A.* **81,** 3 (1984).

[7] G. J. Murakawa, C.-H. B. Chen, M. D. Kuwabara, D. Nierlich, and D. S. Sigman, *Nucleic Acids Res.* **17,** 5361 (1989).

[8] D. S. Sigman, A. Spassky, S. Rimsky, and H. Buc, *Biopolymers* **24,** 183 (1985).

[9] T. B. Thederahn, M. D. Kuwabara, T. A. Larsen, and D. S. Sigman, *J. Am. Chem. Soc.* **111,** 4941 (1989).

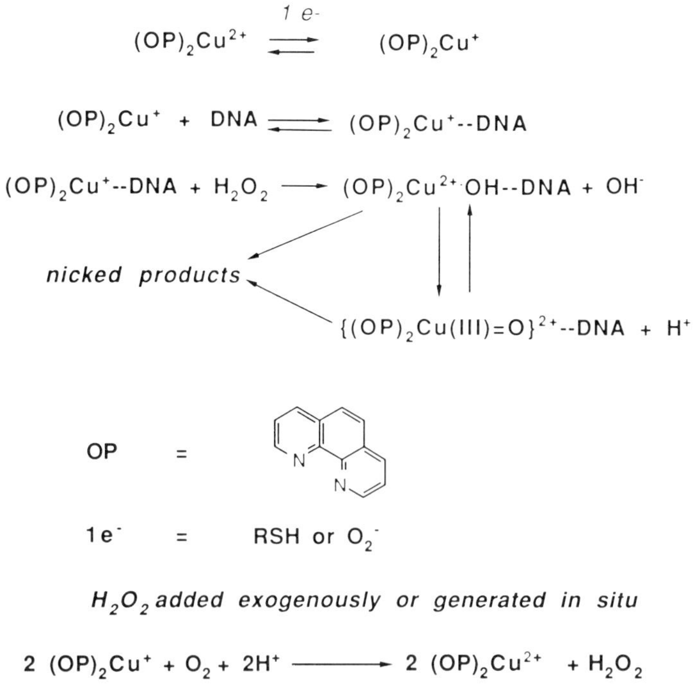

$$(OP)_2Cu^{2+} \xrightarrow[]{1\ e^-} (OP)_2Cu^+$$

$$(OP)_2Cu^+ + DNA \rightleftharpoons (OP)_2Cu^+\text{--}DNA$$

$$(OP)_2Cu^+\text{--}DNA + H_2O_2 \longrightarrow (OP)_2Cu^{2+}\cdot OH\text{--}DNA + OH^-$$

nicked products ↙

$$\{(OP)_2Cu(III){=}O\}^{2+}\text{--}DNA + H^+$$

OP   =   [structure of 1,10-phenanthroline]

$1\ e^-$ =   RSH or O_2^-

H_2O_2 added exogenously or generated in situ

$$2\ (OP)_2Cu^+ + O_2 + 2H^+ \longrightarrow 2\ (OP)_2Cu^{2+} + H_2O_2$$

FIG. 1. Kinetic mechanism of the nuclease activity of 1,10-phenanthroline–copper. The reaction pathway is obligatory because the 1,10-phenanthroline–cupric ion DNA complex is not reducible.

5-phenyl-1,10-phenanthroline (5-ϕ-OP–Cu) provides footprints of protein binding and can detect protein-induced structural changes in DNA.[10,11] In the second approach, 1,10-phenanthroline has been attached to DNA-binding proteins.[12] The orientation of the binding protein with respect to the DNA recognition sequence can be inferred by examining the positions of cleavage.

OP–Cu Footprinting for Studying Protein–DNA Interactions

The nuclease activity of 1,10-phenanthroline–copper makes single-strand nicks in double-stranded DNAs under physiological conditions. Piperidine treatment of the modified DNA is not necessary to cleave the

[10] A. Spassky and D. S. Sigman, *Biochemistry* **24**, 8050 (1985).
[11] M. D. Kuwabara and D. S. Sigman, *Biochemistry* **26**, 7234 (1987).
[12] C.-H. B. Chen and D. S. Sigman, *Science* **237**, 1197 (1987).

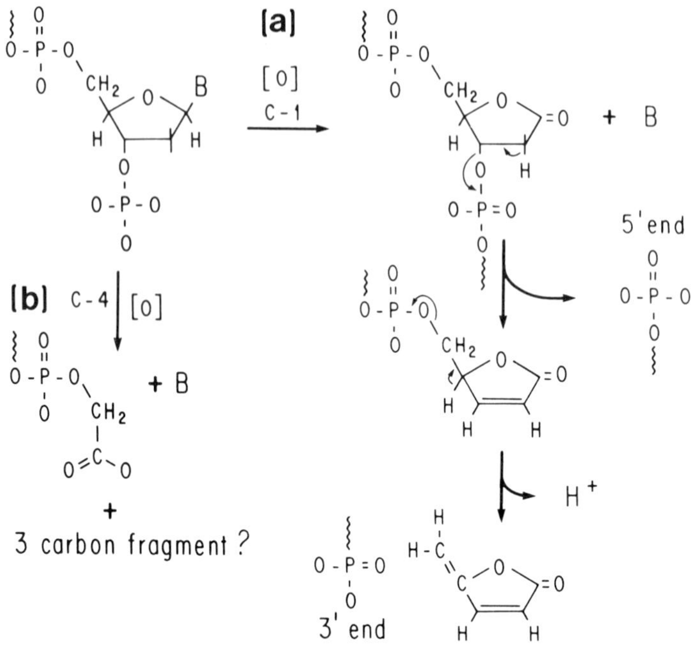

FIG. 2. Postulated reaction mechanism for the scission of DNA by the 2 : 1 1,10-phenan-throline–copper complex. Reaction pathway (a) accounts for greater than 70% of the scission events at any sequence position. The three-carbon fragment derived from ribose in pathway (b) has not been isolated although it is known that base propenals are not formed.

phosphodiester backbone. In contrast to DNase, which interacts with DNA both in the major and minor groove, the chemistry of the nuclease activity of OP–Cu is restricted to the minor groove. This aspect of the mechanism of OP–Cu is relevant to the interpretation of footprinting results. A ligand will protect scission at a given sequence position if it sterically blocks access to the C-1 hydrogen either directly or indirectly, or if it alters the minor groove geometry so that the tetrahedral coordination complex binds poorly. Conversely, if the ligand binds exclusively in the major groove without sterically blocking access to, or grossly perturbing, the dimensions of the minor groove of the recognition sequence, then no footprint will be observed. For example, EcoRI makes sequence-specific contacts in the major groove.[13,14] According to the X-ray structure of Rosenberg and colleagues, the minor groove is accessible to solvent even

[13] C. A. Frederick, J. Grable, M. Melia, C. Samudzi, L. Jen-Jacobson, B. C. Wang, P. Greene, H. W. Boyer, and J. M. Rosenberg, Nature (London) 309, 327 (1984).

[14] J. A. McClarin, C. A. Frederick, B. C. Wang, P. Greene, H. W. Boyer, and J. M. Rosenberg, Science 234, 1526 (1986).

though the DNA structure is slightly kinked. Although DNase I is blocked from cutting within the recognition sequence, the OP–Cu scission reaction is not inhibited.[4]

A unique feature of OP–Cu as a footprinting reagent is its ability to detect protein-induced changes in DNA.[10] This unanticipated feature of the reactivity of OP–Cu may arise from the affinity of this hydrophobic cation for protein surfaces and/or its intercalation into underwound DNA. For example, single-stranded DNA formed at the active site of *Escherichia coli* RNA polymerase of the transcriptionally competent open complex is efficiently cut on the template strand by OP–Cu and even more effectively cleaved by the copper complex of 5-phenyl-1,10-phenanthroline (5-φ-OP–Cu) (Fig. 3).[15] The same scission pattern is observed with the wild-type promoter in the presence of the cyclic AMP-binding protein as is observed in the strong cAMP-independent *lacUV5* promoter.[10] Presumably the DNA in both open complexes attains the same structures. Both reagents can be used to detect the series of intermediates formed during the initiation of transcription with the *lacUV5* promoter. In fact, 5-φ-OP–Cu can detect double-stranded DNA at the leading edge of a transcription bubble both in the open and elongation complexes (Fig. 3, lanes b and e).[15] The inference that DNA downstream of the positions of nucleotide triphosphate incorporation is double stranded relies on the observation of nicking on both strands. Scission sites on the template strand upstream of the site of nucleotide triphosphate incorporation are not associated with any cutting on the nontemplate strand. Magnesium ion is essential for the enzyme-induced melting of the double-stranded DNA at the active site of RNA polymerase.[11]

Detection of these strained intermediates appears to be a unique feature of the phenanthroline–copper complexes. Neither MPE nor DNase I can detect polymerase-induced changes at the start of the transcription. In addition to detecting single-stranded regions in the *lac* system, the OP–Cu reagent also can detect parallel sites in the *gal, mal, ara,* and *mer* operons.[16,17] 5-φ-OP–Cu also detects strain in the DNA induced by merR and RNA polymerase upstream of the *MerT* transcription start site.[17]

Footprinting within Gel Matrix

Experimentally, OP–Cu can be used as a footprinting reagent in a completely analogous fashion as DNase.[18] The protein–DNA complex is

[15] T. B. Thederahn, A. Spassky, M. D. Kuwabara, and D. S. Sigman, *Biochem. Biophys. Res. Commun.* **168,** 756 (1990).
[16] A. Spassky, S. Rimsky, H. Buc, and S. Busby, *EMBO J.* **7,** 1871 (1988).
[17] B. Frantz and T. V. O'Halloran, *Biochemistry* **29,** 4747 (1990).
[18] D. J. Galas and A. Schmitz, *Nucleic Acids Res.* **5,** 3157 (1978).

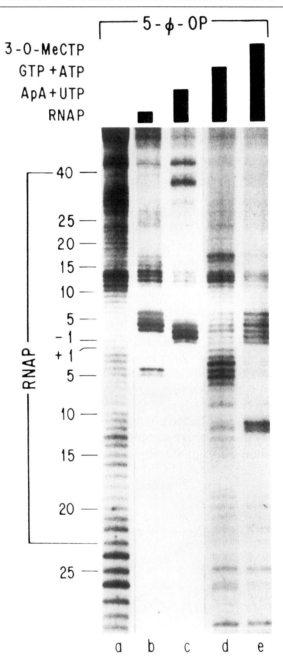

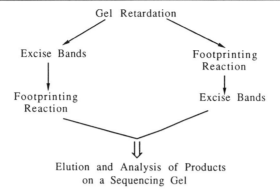

FIG. 4. Footprinting protein–DNA complexes using OP–Cu following separation by gel retardation. Two possible protocols to identify DNA sequences recognized by DNA-binding protein.

formed and attacked with the nucleolytic agents and the digested products analyzed on a sequencing gel. Alternatively, in procedures parallel to methylation interference assays, the DNA can be lightly digested either with DNase or OP–Cu (or ferrous EDTA or MPE), used to form a DNA–protein complex, and then separated from unbound DNA with the widely employed gel-retardation procedure.[19,20] The DNA bound to the protein will not be modified at sequence positions crucial for binding (see [19], this volume).

However, a novel footprinting procedure has been developed using OP–Cu that incorporates gel retardation *prior* to the nuclease reaction (Fig. 4).[11] In gel retardation assays, the binding of a protein to a labeled restriction fragment alters the migration of the DNA in a nondenaturing acrylamide gel.[19,20] Even in a complex mixture of proteins, separating a protein that binds to a labeled DNA is possible. To identify the sequence-

[19] D. M. Crothers, *Nature (London)* **325,** 464 (1987).
[20] M. M. Garner and A. Revzin, *Trends Biochem. Sci.* **11,** 395 (1986).

FIG. 3. Footprints by 5-ϕ-OP–Cu of the binding of RNA polymerase as a function of added nucleotides. The footprinting reagents used are indicated at the top. *lacUV5* fragment labeled on the 5′ end of the template strand was used for all experiments. Five different reaction conditions were used. The components of each reaction mixture are reflected by the size of the bar on top of each lane: (a) DNA scission pattern in absence of any addition; (b) in the presence of RNA polymerase; (c) ApA and UTP added; (d) nucleotides ApA, UTP, GTP, and ATP are added; (e) chain-terminating 3-O-methyl-CTP. Brackets indicate the limits of footprint observed with DNase I.

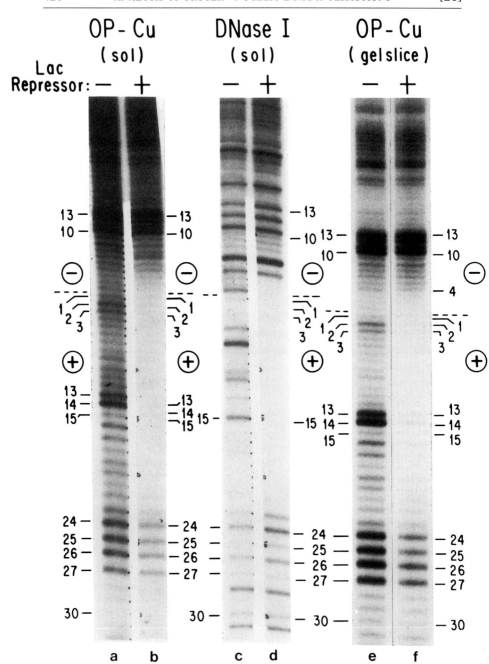

specific contacts between the protein and DNA, the footprinting reaction with OP–Cu can be carried out directly in the gel. The small size and ready diffusibility of the reactants of the nuclease activity—1,10-phenanthroline–copper ion, 3-mercaptopropionic acid, and hydrogen peroxide (molecular oxygen)—and the insensitivity of the reaction to inhibition by organic radical traps permit the reaction to proceed readily within the gel matrix. Deoxyoligonucleotide products are directly eluted from the gel and analyzed on sequencing gels. For initial calibration of this method, the binding of the *lac* repressor to the *lac* operator was examined. The same footprints were obtained within the gel as were observed in solution using OP–Cu. In turn, these patterns were consistent with those obtained in the solution using DNase I (Fig. 5).

In addition to convenience, combining these two methodologies has additional advantages. They ensure that the footprinting reaction is carried out on a discrete electrophoretic species. Moreover, they allow the footprinting of unstable complexes that are difficult to detect in solution for two reasons. First, unbound DNA is separated from the DNA–protein complex and, therefore, the background cutting of unbound DNA is greatly reduced. Second, the gel matrix inhibits the dissociation rate of the protein from its binding site on DNA. Unlike procedures in which the DNA is modified first and then used to form complex, there is no possibility that structural changes induced by the prior modification might influence the footprint obtained. Furthermore, the DNA–protein complex on which the footprints reaction is carried out is fully functional. As a result, molecules can be diffused in the gel matrix (e.g., ATP or inorganic ions) to interact with the protein–DNA complex prior to the footprinting reaction. It is then possible to determine their impact on the structure of the protein–DNA complex. For example, it was possible in this way to show that the magnesium ion was essential for forming single-stranded regions at the active site of RNA polymerase.[21]

---

[21] M. D. Kuwabara and D. S. Sigman, *Biochemistry* **26,** 7234 (1987).

FIG. 5. Comparison of the OP–Cu and DNase I footprints of the *lac* repressor–*lac* operator binding in solution with that obtained of the 1:1 complex isolated by gel retardation. The footprinting experiment was carried out in an acrylamide plug that had been excised from the gel. The 1:1 complex was isolated from an incubation mixture containing 9.4 n*M* L8-UV-5 DNA (2.5 × 10⁶ cpm) and 19 n*M lac* repressor. Lanes a and b, OP–Cu digestion in solution: (a) free DNA; (b) repressor–DNA complex; lanes c and d: (c) free DNA; (d) repressor–DNA complex; lanes e and f, OP–Cu digestion within the gel slice: (e) unbound DNA from gel retardation assay; (f) 1:1 repressor–DNA complex from gel retardation assay.

Experimental Procedures for Footprinting

Footprinting in Solution

5'-Labeled L8-UV5 186-bp DNA (10^5 cpm; 0.03 pmol) is incubated with 50 ng of *lac* repressor in 50 mM Tris-HCl, pH 8.0, 10 mM $MgCl_2$, and 0.2 mg/ml bovine serum albumin (BSA) in 10 μl for 10 min at 37°. For OP–Cu footprinting, 1 μl of 1 mM OP/0.23 mM $CuSO_4$ and 1 μl of 58 mM 3-mercaptopropionic acid (MPA) are added to the 10 μl mixture and incubated for 5.0 min at 37°. Digests are quenched by first adding 1 μl of 28 mM 2,9-dimethyl-OP and then adding 10 μl of the reaction mixture to the sucrose–urea mixture described in the paragraph below. The quenched solutions containing the sample are then heated at 90° for 5 min, and then loaded directly onto 10% sequencing gels that were run at 45 W for 3 hr. Gels are exposed to X-ray film at −20°.

The same footprinting experiment is carried out using DNase I in the following way. DNase I (1 μl of 40 U/ml) is added and incubated for 2.0 min at 37°. Digests are stopped by the addition of a 10-μl aliquot of the digestion solution to 10 μl of a mixture composed of the following: 300 mg of sucrose, 840 mg of urea, 40 μl of 0.5 M EDTA, 100 μl of 0.1% (w/v) bromphenol blue, and 100 μl of 0.1% (w/v) xylene cyanol. In order to obtain 10 ml of a homogeneous mixture, the latter is liquefied by heating to 90°.

In Situ OP–Cu Footprinting following Gel Retardation Assays

Slab gels are used for all gel retardation assays.[19,20]

Footprinting Reaction in the Acrylamide Matrix. The same four reaction components are necessary for the OP–Cu footprinting reaction in the acrylamide matrix as in solution: 1,10-phenanthroline (OP), 2,9-dimethyl-1,10-phenanthroline (2,9-dimethyl–OP), cupric sulfate, and MPA. Prepare the following solutions just prior to use: mix equal volumes of 40 mM OP (in 100% ethanol) with 9.0 mM $CuSO_4$ (in water); dilute 1/10 with water to 2.0 mM OP/0.45 mM $CuSO_4$ (solution A); dilute neat MPA 1/200 with water to 58 mM MPA (solution B).

Two methods of carrying out the footprinting reaction within the acrylamide matrix are used. In the first, the wet retardation gel is exposed to X-ray film for 30 min at room temperature until the retarded bands are visible. The bands of interest are then excised with a razor and immersed in 100 μl of 50 mM Tris-HCl, pH 8.0. Then, 10 μl of solution A is added followed by 10 μl of 28 mM 2,9-dimethyl-OP (in 100% ethanol) and 270 μl of a solution containing both 0.5 M ammonium acetate and 1 mM EDTA. The DNA is then eluted overnight at 37°. The eluted DNA is ethanol precipitated (carrier DNA is unnecessary), resuspended in 80% (v/v) formamide, 10 mM NaOH, 1.0 mM EDTA, 0.1% bromphenol blue, and 0.1% xylene cyanol, then loaded on a 10% sequencing gel.

An alternate method of carrying out the footprinting within the gel takes advantage of the ready availability of the reagents and involves immersion of the whole gel in a 1,10-phenanthroline–copper solution without the prior identification of the retarded bands. The first step of this alternative procedure is the immersion of the slab gel into 200 ml of Tris-HCl, pH 8.0 buffer. The gel is removed from its glass plate, then 20 ml of solution A and 20 ml of solution B are added; the digestion is allowed to proceed for 10 min at room temperature. To quench the reaction, 20 ml of 28 mM 2,9-dimethyl-OP is added, and the resulting solution is allowed to stand for 2.0 min. The gel is rinsed with distilled water and exposed to X-ray film for 30 min at room temperature. Bands of interest are cut from the gel and eluted overnight at 37° in 0.5 M ammonium acetate/1 mM EDTA as above. Eluted DNA is ethanol precipitated, resuspended in 80% (v/v) formamide, 10 mM NaOH, 1.0 mM EDTA, 0.1% bromphenol blue, and 0.1% xylene cyanol, and loaded on a 10% sequencing gel. The significant advantage of this procedure is that if a long exposure is essential to identify weak bands, the footprint will not be perturbed by protein denaturation. Reagents are not reusable because thiol is consumed and the 1,10-phenanthroline is oxidatively destroyed.

Reactivity Features of OP–Cu Relevant to Footprinting Applications

In this section, practical features of the reactivity of the nuclease activity of OP–Cu are reviewed.

Reducing Agent and Hydrogen Peroxide. Two species serve as coreactants of the nuclease activity: the 1,10-phenanthroline–cuprous complex and hydrogen peroxide. Typically the reaction has been carried out by reducing cupric ion *in situ* with thiol. The resulting cuprous complex is then oxidized by molecular oxygen via superoxide reforming the cupric complex and generating hydrogen peroxide through the spontaneous dismutation of superoxide.[22] Given the complexity of the system, it has not been possible to determine the equivalent of a K_m for H_2O_2. Although the addition of 3 mM H_2O_2 ensures that this coreactant will not be rate limiting, it has usually not been added because it causes background scission.

$$(OP)_2Cu^{2+} + RSH \rightarrow (OP)_2Cu^+ + \tfrac{1}{2}R\text{-}S\text{-}S\text{-}R \qquad (1)$$
$$(OP)_2Cu^+ + O_2 \rightarrow (OP)_2Cu^{2+} + O_2^- \qquad (2)$$

[22] D. R. Graham, L. E. Marshall, K. A. Reich, and D. S. Sigman, *J. Am. Chem. Soc.* **102**, 5419 (1980).

$$2O_2^- + 2H^+ \rightarrow H_2O_2 + O_2 \tag{3}$$

The thiol generally used to activate the chemistry is 3-mercaptopropionic acid. It has been chosen because it is not volatile and is a poorer chelating agent than mercaptoethanol, mercaptoethylamine, or dithiothreitol. Early studies with these thiols caused complexities that can be avoided with 3-mercaptopropionic acid.[23,24] NADH, in the presence of H_2O_2,[25] can also be used to activate the scission chemistry. However, the multistep pathway essential to the accomplishment of the one-electron reduction of cupric ion makes these reaction conditions very susceptible to inhibition. Ascorbic acid has also been used to activate the chemistry with apparently satisfactory results.[26]

The addition of thiol (or ascorbic acid) provides the most convenient way to activate the scission chemistry. Reaction mixtures containing 1,10-phenanthroline, cupric ion, and thiol should not be prepared and allowed to stand because thiol will be consumed [Eqs. (1)–(3)]. Activation of the scission chemistry by cupric ion addition is not advisable. Cupric ion is a ubiquitous contaminant, and will be present at sufficient concentrations to support measurable levels of reaction prior to the desired initiation time of the reaction.

Quenching Nuclease Activity. In footprinting applications, reactions must be carried out under single-hit conditions. As a result, the nuclease activity must be efficiently quenched to avoid overdigestion. This can be achieved with the ligand 2,9-dimethyl-1,10-phenanthroline, which is an analytical reagent (i.e., neocuproin) for copper. It forms a stable cuprous complex with a characteristic absorption maximum at 430 nm. It blocks the nuclease activity by sequestering all the available copper in an inert form. Because no other metal ion can substitute for copper, the reaction is efficiently inhibited.

The *o*-methyl substitutents stabilize the cuprous complex relative to the cupric complex through steric effects. In the tetrahedral cuprous complex, there is no interference between the methyl groups in the 2 : 1 complex. However, in the cupric complex a planar geometry is energetically disfavored due to steric hindrance of the methyl groups. In fact, a single ortho substitutent on a phenanthroline is sufficient to destroy the ability of its

[23] V. D'Aurora, A. M. Stern, and D. S. Sigman, *Biochem. Biophys. Res. Commun.* **78,** 170 (1977).

[24] V. D'Aurora, A. M. Stern, and D. S. Sigman, *Biochem. Biophys. Res. Commun.* **80,** 1025 (1978).

[25] K. A. Reich, L. E. Marshall, D. R. Graham, and D. S. Sigman, *J. Am. Chem. Soc.* **103,** 3582 (1981).

[26] Q. Guo, M. Lu, N. C. Seeman, and N. R. Kallenbach, *Biochemistry* **29,** 570 (1990).

copper complex to catalyze the oxidation of thiol and to nick DNA. The 1,10-phenanthroline–copper complex must be able to cycle between the oxidized and reduced forms in both of these reactions.

Sensitivity of Nuclease Activity to Reaction Conditions. Reaction mixtures used to study protein–DNA interactions often contain multiple components. The advantage of the nuclease activity of OP–Cu is its hardiness. It proceeds readily under conditions where biological catalysts are usually inactivated and other chemical nucleases can be inhibited.

Influence of Buffer Components

Glycerol/ethylene glycol: These frequently added components that are used to stabilize protein preparations have little effect on scission chemistry up to 1 M. Ferrous-EDTA, in contrast, is inhibited by these reagents in the millimolar range.[27]

EDTA: High concentrations of EDTA will inhibit the reaction by decreasing the copper concentration. The effect of EDTA will be decreased if magnesium ions are present.

Magnesium ions: 1,10-Phenanthroline is not a good ligand for magnesium ions (or other alkaline earth ions). The scission reaction will therefore not be inhibited by magnesium concentrations as high as 30 mM. However, a possible mechanism of inhibition at very high concentrations (0.1 M and above) is the displacement of the cupric ion by a magnesium ion. This can be overcome by the addition of cupric ions.

Chloride ions: At high concentrations (>0.1 M), chloride ions can sequester copper ions. The addition of cupric ions should counteract this problem.

Protein denaturants: A strong advantage of the nuclease activity is that it is functional even in the presence of high concentrations of protein denaturants (e.g., guanidine hydrochloride). Although these compounds may sequester copper ions or consume thiol, the reaction frequently will proceed if increased amounts of reactants are added. The addition of exogenous hydrogen peroxide (e.g., 7 mM) is particularly helpful if the reaction is sluggish.

Derivatization of DNA-Binding Proteins with 1,10-Phenanthroline–Copper

In addition to suggesting a method for the design of artificial restriction enzymes, the chemical modification of DNA-binding proteins with 1,10-phenanthroline provides a novel approach for studying the interaction

[27] T. D. Tullius and B. A. Dombroski, *Proc. Natl. Acad. Sci. U.S.A.* **83,** 5469 (1986).

of a protein to its recognition sequence. Presently, three proteins, the *Escherichia coli trp* repressor,[12] cAMP-binding protein,[28] and the bacteriophage λ Cro protein,[29] have been derivatized with 1,10-phenanthroline. In each case, the pattern of scission is consistent with modes of binding that have been inferred from crystallographic studies or other chemical approaches. An important component of all targeted cleavage reactions involving 1,10-phenanthroline–copper is that scission will occur only if the C-1 hydrogen of the deoxyribose in the minor groove is accessible to the scission reagent.

A general method involving lysine derivatization has been applied to the *E. coli trp* repressor. Iminothiolane reacts with primary amine groups to generate sulfhydryl groups that are then alkylated with 5-iodoacetyl-1,10-phenanthroline.[12] In the case of the *trp* repressor, four phenanthrolines are attached per *trp* repressor monomer.

$$
NH_3^+\text{—protein} + H_2C \underset{H_2C\text{—}CH_2}{\overset{\overset{\displaystyle\underset{H}{\overset{H}{N^+}}}{\underset{\|}{C}}}{\big\langle}} S \rightarrow HS\text{—}(CH_2)_3\text{—}\overset{O}{\underset{\|}{C}}\text{—}\overset{H\ \ H}{\underset{N^+}{}}\text{—protein} + H^+ \tag{4}
$$

$$
\text{(phenanthroline)}\ N\text{—}\overset{O}{\underset{\|}{C}}\text{—}CH_2\,I\ +\ HS\text{—}(CH_2)_3\text{—}\overset{O}{\underset{\|}{C}}\text{—}\overset{H\ \ H}{\underset{N^+}{}}\text{—protein} \rightarrow
$$

$$
\text{(phenanthroline)}\ N\text{—}\overset{O}{\underset{\|}{C}}\text{—}CH_2\text{—}S\text{—}(CH_2)_3\text{—}\overset{O}{\underset{\|}{C}}\text{—}\overset{H\ \ H}{\underset{N^+}{}}\text{—protein} \tag{5}
$$

A possible concern with any chemical modification procedure is the potential loss of sequence-specific binding. In the case of *trp* repressor, DNase

[28] R. H. Ebright, Y. W. Ebright, P. S. Pendergrast, and A. Gunasekera, *Proc. Natl. Acad. Sci. U.S.A.* **87**, 2882 (1990).
[29] T. W. Bruice, J. Wise, and D. S. Sigman, *Biochemistry* **29**, 2185 (1990).

footprinting indicated that this was not a problem; the modified protein bound with comparable affinity as the wild-type protein. Scission was achieved by the addition of both thiol and copper ion (Fig. 6). It is also dependent on the presence of the corepressor, L-tryptophan, which is essential for the sequence-specific binding of the protein.[30]

Two operators, *aroH* and *trpEDCBA*, were assayed as substrates for the OP-derivatized *trp* repressor. Both were nicked within the recognition sequence. The pattern of scission, however, with the two operators was distinct. These results are consistent with previous suggestions, based on DNase I footprinting and methylation studies, that the repressor binds to four successive major grooves on the *trpEDCBA* operator and three successive minor grooves on the *aroH* operator.[31] Thus, the OP–*trp* repressor provides chemical evidence that there is not a canonical orientation between a DNA-binding protein and a recognition sequence.

Ebright and colleagues have derivatized the *E. coli* cAMP-binding protein (CAP) at a unique cysteine residue using 5-iodoacetyl-1,10-phenanthroline.[28] This cysteine residue is at position 10 of the helix–turn–helix domain of this transcription activator. Scission on both strands at the center of the 22-nucleotide long symmetric recognition sequence was observed (Fig. 7), and was absolutely dependent on the addition of copper, thiol, and cAMP. Thus, both in the case of the *trp* repressor and CAP, the ligands essential for sequence-specific binding are also required for the scission chemistry. The observed sites of scission are fully consistent with the postulated model for CAP binding.

The helix–turn–helix format in DNA-binding proteins was first identified in the λ phage Cro protein by Matthews and colleagues.[32,33] In their pioneering studies, Matthews and co-workers suggested that the C-terminal end of the protein could extend into the minor groove of the recognition sequence and enhance the stability of the complex. Because the wild-type Cro protein contains no cysteine residue, a unique site of alkylation of the protein by 5-iodoacetamido-1,10-phenanthroline[33a] could be generated by changing the C-terminal alanine residue into a cysteine residue (mutant protein designated as Cro-Cys) using site-directed mutagenesis.

Gel retardation assays provided a convenient method to demonstrate that the site-directed mutagenesis of the Cro to generate Cro-Cys was achieved as planned and that Cro-Cys alkylated by 5-iodoacetamido-1,10-phenanthroline still could bind to the recognition sequence with high affin-

[30] D. N. Arvidson, C. Bruce, and R. P. Gunsalus, *J. Biol. Chem.* **261**, 238 (1986).

[31] A. A. Kumamoto, W. G. Miller, and R. P. Gunsalus, *Genes Dev.* **1**, 556 (1987).

[32] R. G. Brennan and B. W. Matthews, *Trends Biochem. Sci.* **13**, 286 (1989).

[33] R. G. Brennan and B. W. Matthews, *J. Biol. Chem.* **264**, 1903 (1989).

[33a] For synthesis of reagents, see end of chapter.

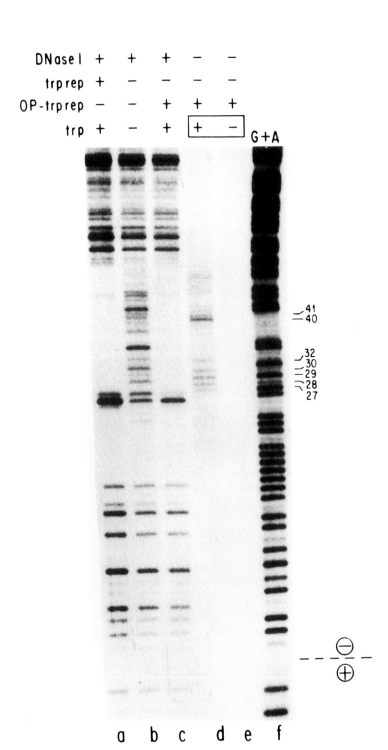

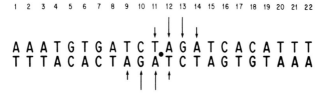

FIG. 7. Site-specific DNA scission of recognition sequence by cAMP-binding protein-alkylated cysteine-10 of the helix–turn–helix domain. (Adapted from Ebright et al.[28])

ity (Fig. 8A). Extracts of *E. coli* transformed with pUC119 containing the gene for Cro or Cro-Cys were used as the source of protein for the gel retardation assay. Chemical modification was carried out on the extracts directly.

To determine if sequence-specific scission could be achieved, copper ion and thiol were added to the gel slice containing the complex of OR-3 operator and Cro-Cys–OP (Fig. 8A). Scission is observed within the binding site identified by OP–Cu footprinting. These experiments provide direct chemical support for the model proposed by Matthews and colleagues. An important feature of the 1,10-phenanthroline-derivatized Cro protein is that it retains high affinity for the target sequence. The Cro protein, therefore, may serve as a convenient starting point for generating a family of semisynthetic restriction endonucleases. Possibly stable mutants of Cro-Cys can be generated in which the DNA-binding domain has been altered to bind other recognition sequences.

Methods for Preparation of 1,10-Phenanthroline-Modified Proteins

trp Repressor. The procedures used for derivatizing the *E. coli* trp repressor are summarized below. The protein is reacted with 2-iminothiolane (Pierce Chemical Company, Rockford, IL), according to a procedure adapted from a method first applied to the modification of ribosomal subunits.[34] From a freshly prepared 2-iminothiolane hydrochloride solution [0.5 M in 0.1 M (pH 8.0) phosphate buffer], 2.4 μl is added to 0.021

[34] R. Jue, J. M. Lambert, L. R. Pierce, and R. R. Traut, *Biochemistry* **17**, 5399 (1978).

FIG. 6. Binding and scission of the *aroH* operator with OP–*trp* repressor. Template strand 5' labeled. Lanes a to c: DNase footprints of the binding of the *trp* repressor and OP–*trp* repressor to the 5'-labeled template strand of *aroH*. Lane a: native *trp* repressor; lane b: control; lane c: OP–*trp* repressor; lane d: OP–*trp* repressor with L-tryptophan, Cu^{2+}, and 3-mercaptopropionic acid; lane e: same as lane d but lacking L-tryptophan; and lane f: G + A sequencing lane. Temperature = 25°.

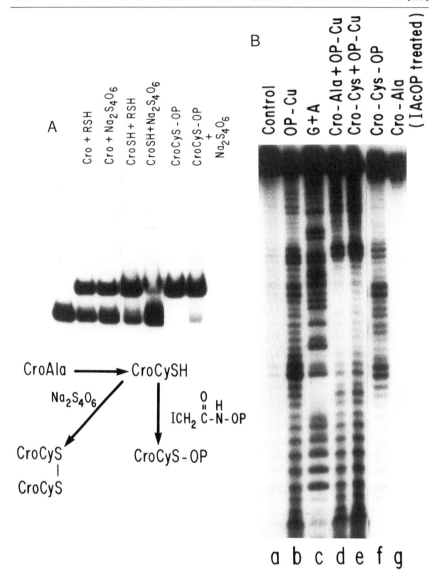

FIG. 8. Interaction of Cro protein with OR-3. (A). Gel retardation assay: (a) no protein added; (b) *E. coli* extract containing *cro* in the pUC119 plasmid. (B). Scission of retarded band within the gel. (a) Control DNA; (b) free DNA cleaved by OP–Cu; (c) G + A sequencing lane; (d) Cro–DNA complex cleaved by OP–Cu. e, f, g.

μmol of *trp* repressor in 250 μl of phosphate buffer containing 1% 2-mercaptoethanol and 1 mg of L-tryptophan at 4°. After overnight incubation at 4°, the mixture is passed through a Sephadex G-25 (coarse) spin column to remove the unreacted 2-iminothiolane. The spin column is equilibrated in 0.1 M phosphate buffer, pH 8.0, prior to use. Then, 1.6 mmol of 5-iodo[³H]acetamido-OP (specific activity, 3 mCi/mmol) in N,N-dimethylformamide (4 ml) is added to the eluant collected from the spin column. The alkylation reaction proceeds at 4° overnight. The product is isolated by passing the mixture through a Sephadex G-25 spin column.

The *aroH* operator region is used to assay binding and the scission activity of the modified protein. The operator derives from a 257-bp *Bam*HI–*Eco*RI fragment, which had been cloned into pBR327.[31] The control region of *trpEDCBA* is contained in a *Bam*HI–*Eco*RI restriction fragment that had also been cloned into pBR327. These fragments are 5' and 3' labeled. For the footprinting and cleavage reactions, a solution of labeled *aroH* operator fragment (100 nM, 10,000 cpm), *trp* repressor or OP–*trp* repressor (100 to 200 nM), and L-tryptophan (10 mM) is incubated at room temperature for 5 min in 10 μl of Tris-HCl (30 mM, pH 8.0), KCl (100 mM), and MgCl$_2$ (3 mM). Footprinting reactions are carried out with 0.04 U of DNase I for 5 min at room temperature. The reactions are quenched by the addition of "stop" solution [transfer RNA (200 mg/ml), 2 M sodium acetate, and 10 mM EDTA]. The DNA is then precipitated with ethanol, redissolved in Maxam–Gilbert loading buffer, and analyzed by the use of a 10% polyacrylamide gel that contains 8.34 M urea, and is cross-linked at a ratio of 19 : 1. The chemical cleavage reaction is initiated by adding 1 μl of 20 μM cupric sulfate and 1 μl of 58 mM mercaptopropionic acid. After 30 min at room temperature or at 37°, the reaction is quenched by adding 2,9-dimethyl-1,10-phenanthroline followed by 1 μl of stop solution. The DNA is then precipitated with ethanol, redissolved in Maxam–Gilbert loading buffer, and analyzed by 10% polyacrylamide (19 : 1 acrylamide :-bisacrylamide cross-linked gel) containing 8.34 M urea.

Cyclic AMP-Binding Protein. As described by Ebright and colleagues,[28] the cAMP-binding protein is derivatized with 5-iodoacetamido-1,10-phenanthroline in the following way. Reaction mixtures (50 μl) contain 2.7 nmol CAP, 210 nmol 5-iodoacetamido-1,10-phenanthroline, 20 mM Tris-HCl (pH 8.0), 200 mM KCl, 0.1 M EDTA, 5% glycerol, and 1.7% dimethylformamide. Reactions are carried out in the dark and proceeded for 3 hr at 23°, followed by 15 hr at 4°. The product is purified by chromatography on BioGel P-6DG (Bio-Rad, Richmond, CA), and stored at $-70°$ in 20 mM Tris-HCl (pH 8.0), 200 mM KCl, 5% glycerol. All solutions are treated with Chelex 100 (Bio-Rad) to remove trace metals.

Ebright and colleagues assay the scission reaction in a solution containing 1 nM ^{32}P-labeled target fragment, 80 nM OP-modified CAP, 7.5 μM cupric sulfate, 2.5 mM 3-mercaptopropionic acid, 0.2 mM cAMP, 10 mM MOPS–NaOH (pH 7.3), 20 mM NaCl, 50 mg/ml bovine serum albumin, and 2.2% ethanol. Reaction time is 1 hr at 37°; the reaction is quenched by the addition of 2,9-dimethyl-1,10-phenanthroline to 3.0 mM. Reaction mixtures are extracted twice with phenol and twice with chloroform prior to ethanol precipitation.

λ *Phage Cro Protein* Gel retardation techniques provide a powerful method for purifying DNA-binding proteins. It is therefore possible to derivatize DNA-binding proteins with 1,10-phenanthroline in crude reaction mixtures and then purify the derivatized protein by gel retardation. The scission reaction can then be activated within the gel matrix. This experimental protocol has been used in studies with Cro protein in which both the wild-type and Cro-Cys protein are expressed in low yield in the pUC119 plasmid.

In situ modification of proteins containing sulfhydryl groups can be carried out in extracts by first ensuring that the sulfhydryl group is reduced by adding a thiol (e.g., mercaptoethanol or 3-mercaptopropionic acid) to attain a concentration of 2.0 mM. Then 5-iodoacetamido-1,10-phenanthroline in dimethylformamide (DMF) is added to a final concentration of 5.0 mM in 10% DMF. The reaction mixture is incubated on ice for 60 min, then 3-mercapto-propionic acid (to 10 mM) is added to inactivate the 5-iodoacetyl derivative. The resulting preparation can then be used in gel retardation experiments. The protein is separated from excess 5-iodoacet-amido-1,10-phenanthroline by the gel retardation separation. This procedure allows an initial screen to determine if the modification procedure influences binding and whether a protein-targeted reaction can be achieved.

Synthesis of Reagents

5-Amino-1,10-phenanthroline. 5-Amino-1,10-phenanthroline is a key intermediate for derivatizing proteins and nucleic acids. A convenient high-yield synthesis of this known compound is presented below. Ten grams (0.044 mol) of 5-nitro-1,10-phenanthroline (available from G. F. Smith Co., Columbus, OH) is dissolved in 400 ml of absolute ethanol by boiling, and 475 ml of ammonium sulfide (~22%) is heated on a steam bath. The 5-nitro-1,10-phenanthroline solution is added dropwise to the ammonium sulfide with a continual flow of nitrogen. When the addition is completed, 250 ml more of ammonium sulfide is added and the reaction mixture is allowed to reflux for 1 hr. The solution is cooled to room temperature, and then is extracted with chloroform. The chloroform extracts are dried and distilled at atmospheric pressure. The residue is crystallized from ethanol/water and dried under vacuum. Yield: 5.5 g (63.5%); mp, 268–270°; thin-layer chromatography (TLC; silica, 5% ammonia in methanol) 1 spot, R_f 0.45, running slower than the starting material.

5-Iodoacetamido-1,10-phenanthroline. Five grams (0.024 mol) of dicyclohexylcarbodiimide is added to a solution of 10 g (0.054 mol) of iodoacetic acid in 150 ml of ethyl acetate. Dicyclohexylurea forms immediately, but stirring is continued for 2 hr. The urea is removed by filtration, and the solution is concentrated to dryness by rotatory evaporation.

5-Amino-1,10-phenanthroline (1.6 g, 0.0082 mol) is dissolved by heating in 200 ml of acetonitrile. The solution is allowed to cool at room temperature. Then the iodoacetic anhydride dissolved in 75 ml of acetonitrile is added. After the solution is stirred overnight at room temperature, a crystalline product is obtained. The product is washed with cold 5% sodium bicarbonate, washed with water, and finally dried under vacuum. Yield: 2 g (70%). Analytical data: TLC (silica, 5% ammonia in methanol) 1 spot; R_f 0.24. Elemental analysis: Calculated C 46.20%, H 2.75%, N 11.60%, I 35.00%; found C 46.41%, H 2.91%, N 11.74%, I 34.72%.

[21] Base Analogs in Study of Restriction Enzyme–DNA Interactions

By CHRISTOPHER R. AIKEN and RICHARD I. GUMPORT

Introduction

Enzymologists have long sought to discern the topography of the active sites of enzymes by examining substrate analogs for their ability to serve as reactants. Such investigations aim to contribute to our understanding of the kinetic and chemical mechanisms as well as the stereochemistry and stereoselectivity of a reaction. The interpretation of results from studies using analogs is based on the assumption that steric and electronic variations in the substrate will reveal contacts or interactions with the enzyme. The effects of substrate analogs are exerted on the enzyme–substrate complex as reflected in its equilibrium dissociation constant or the K_m value, or on catalysis itself as manifested by changes in the overall reaction rate (k_{cat}) or the rate of an individual step in the mechanism. Although the paradigm of enzyme and substrate as a rigid, lock-and-key-like system is incorrect because both partners are flexible to varying extents, the analog approach remains a valuable means to study the geometric and electronic interfaces between an enzyme and its substrate during a reaction. A complementary approach has more recently been developed through protein engineering techniques that create enzyme analogs in which one amino acid is substituted for another in order to monitor directly the role of specific amino acid side chains in the interactions. The combined use of mutant enzymes with substrate analogs should be a powerful means to elucidate mechanisms.

In this chapter we discuss the principles and assumptions of the substrate–analog approach, evaluate the merits, explain the limitations, and mention artifacts sometimes encountered when it is used to assess site-specific enzyme–DNA interactions. The aim of such studies with enzymes that react with specific sites on DNA has often been to determine the molecular basis of the "recognition" of the target nucleotide sequence. In this context, recognition means the capability of the enzyme to select the targeted sequence from relatively higher concentrations of similar sequences and to catalyze a specific reaction with the DNA. We will confine our detailed considerations to the type II restriction endonucleases. For a review of DNA recognition by restriction enzymes see Ref. 1.

Nucleic acid analogs have also been used with proteins that bind to specific sites on DNA but do not catalyze reactions with it, e.g., repressors. It should be noted that enzymes, in contrast to simple DNA-binding proteins, are afforded added opportunities to recognize their targets because chemistry occurs; covalent bonds are broken and formed. The natures of the substrates change during the course of the catalysis as a transition state develops from them and is tightly bound by the enzyme. Functional groups, possibly newly formed, on the developing transition state may interact with the enzyme. Such groups may be unavailable for contact with the enzyme in the initial enzyme–substrate complex. Thus, an enzyme may bind to DNA with some degree of site specificity yet fail to catalyze a reaction efficiently because prerequisites for the formation of the transition state are absent. With enzymes, some recognition can occur during the chemical steps as well as in the initial binding of the substrate.

The effects of analog substitutions can arise directly from lost interactions because of missing or added groups on the modified bases themselves

[1] S. P. Bennett and S. E. Halford, in "Current Topics in Cellular Regulation" (B. L. Horecker, E. R. Stadtman, P. B. Chock, and A. Levitzki, eds.), Vol. 30, p. 57. Academic Press, San Diego, California, 1989.

or from indirect effects as a result of conformational changes in the DNA introduced by the presence of the analogs. Whereas high-resolution structural analyses of specific DNA–protein complexes often definitively indicate the atomic interactions between the macromolecules, they cannot yet do so for the changing structures in an enzyme-catalyzed reaction with DNA. In addition, structural analyses of enzyme–substrate complexes are sometimes conducted in the absence of cofactors necessary for reaction; consequently, such structures may not represent the biologically active complexes. Kinetic studies using analogs and other methods that monitor the dynamics of a reaction under quasiphysiological conditions can sometimes uniquely reveal important interactions. For these reasons there is a continuing role for analog studies of protein–nucleic acid interactions despite the considerable ambiguities that exist in their interpretation (see below).

Analogs

DNA and oligodeoxyribonucleotides containing base analogs, i.e., bases different from Ade,[2] Cyt, Gua, and Thy, can be obtained in several ways. Natural base analogs exist in the genomes of some organisms. For example, methylated (Me) derivatives of Ade and Cyt occur in some bacterial geonomes,[3] and HmUra and glycosylated derivatives of HmCyt are found in the DNAs of some bacteriophages.[4] Analogs can also be formed in DNA and oligonucleotides through chemical modification reactions, e.g., 7-MeGua and 3-MeAde can be created by reaction with Me_2SO_4.[5] Such sources of analog-containing DNA are unsuitable for systematic studies because the variety of bases attainable is limited, and it is difficult to assure complete modification at a unique site.

Synthetic organic chemistry greatly enlarges the repertoire of structural variants available and, coupled with methods for their site-specific introduction into oligodeoxyribonucleotides, enables a large number of analogs to be examined. Variants at most of the sites available for potential interactions on each of the bases have been made. Tables I and II show some of the purine and pyrimidine compounds that have been incorporated into oligonucleotides, and list some of the restriction enzymes tested with

---

[2] Abbreviations follow the IUPAC suggestions in *Biochemistry* **9**, 4022 (1970), except that Pur stands for purine itself and not an unknown purine.

[3] H. A. Sober, ed., "CRC Handbook of Biochemistry, Selected Data for Molecular Biology," 2nd Ed., p. H-117. The Chemical Rubber Co., Cleveland, Ohio, 1970.

[4] H. A. Sober, ed., "CRC Handbook of Biochemistry, Selected data for Molecular Biology," 2nd Ed., p. H-4. The Chemical Rubber Co., Cleveland, Ohio, 1970.

[5] A. M. Maxam and W. Gilbert, this series, Vol. 65, p. 499.

TABLE I
PURINE ANALOGS

2,3-Positions	1,6-Positions	7-Position	Base	Enzymes examined
$H_2NC=N$	HN—CO	N	Guanine	
HC=N	HN—CO	N	Hypoxanthine	EcoRI,[a,b] EcoRV[c]
$H_2NC=N$	N=CH	N	2-Aminopurine	EcoRI,[a,b,d] EcoRV[c]
$H_2NC=N$	HN—CO	CH	7-Deazaguanine	EcoRI[d,e]
HC=N	HN—CO	CH	7-Deazahypoxanthine	HindII, SalI, TaqI[f]
OC—NH	HN—CO	N	Xanthine	None[g]
$H_2NC=N$	N=COCH$_3$	N	O^6-Methylguanine	HpaII, HhaI,[h] and others
CH$_3$NHC=N	HN—CO	N	N^2-Methylguanine	BglII, Sau3AI, MboI[i]
HC=N	N=CNH$_2$	N	Adenine	
HC=N	N=CH	N	Purine	EcoRI,[b] EcoRV[c]
$H_2NC=N$	N=CNH$_2$	N	2,6-Diaminopurine	EcoRI[a,b]
HC=N	N=CNHOCH$_3$	N	N^6-Methoxyadenine	None[j]
HC=N	N=CNHCH$_3$	N	N^6-Methyladenine	EcoRI,[a,d] EcoRV,[k] Sau3AI, MboI, MflI[l]
HC=CH	N=CNH$_2$	N	3-Deazaadenine	EcoRV,[m,n] BglIII, Sau3AI, MboI[i]
HC=N	N=CNH$_2$	CH	7-Deazaadenine	EcoRI,[o] EcoRV,[k,m] HindII, SalI, TaqI,[f] BglII, Sau3AI[p]
HC=N	N=CH	CH	7-Deazapurine	HindII, SalI, TaqI[f]

[a] C. A. Brennan, M. D. Van Cleve, and R. I. Gumport, *J. Biol. Chem.* **261**, 7270 (1986).

[b] L. W. McLaughlin, F. Benseler, E. Graeser, N. Piel, and S. Scholtissek, *Biochemistry* **26**, 7238 (1987).

[c] J. Mazzarelli, S. Scholtissek, and L. W. McLaughlin, *Biochemistry* **28**, 4616 (1989).

[d] D. R. Lesser, M. R. Kurpiewski, and L. Jen-Jacobson, *Science* **250**, 776 (1990).

[e] F. Seela and H. Driller, *Nucleic Acids Res.* **14**, 2319 (1986).

[f] J. Jiricny, S. G. Wood, D. Martin, and A. Ubasawa, *Nucleic Acids Res.* **14**, 6579 (1986).

[g] R. Eritja, D. M. Horowitz, P. A. Walker, J. P. Ziehler-Martin, M. S. Boosalis, M. F. Goodman, K. Itakura, and B. E. Kaplan, *Nucleic Acids Res.* **14**, 8135 (1986).

[h] J. M. Voigt and M. D. Topal, *Biochemistry* **29**, 1632 (1990).

[i] A. Ono and T. Ueda, *Nucleic Acids Res.* **15**, 3059 (1987).

[j] H. Nishino, A. Ono, A. Matsuda, and T. Ueda, *Nucleic Acids Res. Symp. Ser.* No. **21**, 123 (1989).

[k] A. Fliess, H. Wolfes, F. Seela, and A. Pingoud, *Nucleic Acids Res.* **16**, 11781 (1988).

[l] A. Ono and T. Ueda, *Nucleic Acids Res.* **15**, 219 (1987).

[m] P. C. Newman, V. U. Nwosu, D. M. Williams, R. Cosstick, F. Seela, and B. A. Connolly, *Biochemistry* **29**, 9891 (1990).

[n] R. Cosstick, X. Li, D. K. Tuli, D. M. Williams, B. A. Connolly, and P. C. Newman, *Nucleic Acids Res.* **18**, 4771 (1990).

[o] F. Seela and A. Kehne, *Biochemistry* **26**, 2232 (1987).

[p] A. Ono, M. Sato, Y. Ohtani, and T. Ueda, *Nucleic Acids Res.* **12**, 8939.

TABLE II

PYRIMIDINE ANALOGS

2-Position	3,4-Positions	5-Position	Base	Enzymes examined
O	HN—CO	CH$_3$	Thymine	
O	HN—CO	H	Uracil	EcoRI,[a,b] EcoRV[c,d]
O	HN—CS	CH$_3$	4-Thiothymine	EcoRV[d,e]
S	HN—CO	CH$_3$	2-Thiothymine	EcoRV[d]
O	N=CH	CH$_3$	5-Methyl-2-pyrimidinone	EcoRV[d,f]
O	HN—CO	(H)CH$_3$	5,6-Dihydrothymine[g]	None[h]
O	HN—CO	F	5-Fluorouracil	MvaI[i]
O	HN—CO	Br	5-Bromouracil	EcoRI,[a] EcoRII,[j] EcoRV,[k] BglII, Sau3AI, MboI[l]
O	HN—CO	CN	5-Cyanouracil	BglII, Sau3AI, MboI[l]
O	HN—CO	C$_2$H$_5$	5-Ethyluracil	BglII, Sau3AI, MboI[l]
O	N=CNH$_2$	H	Cytosine	
O	N=CNH$_2$	CH$_3$	5-Methylcytosine	EcoRI,[a,b] EcoRII,[j] EcoRV,[c,m] MvaI[i]
O	N=CNH$_2$	Br	5-Bromocytosine	EcoRV[m]
O	N=CNHCH$_3$	H	N^4-Methylcytosine	Sau3AI, MboI, MflI[n]

[a] C. A. Brennan, M. D. Van Cleve, and R. I. Gumport, *J. Biol. Chem.* **261**, 7270 (1986).

[b] L. W. McLaughlin, F. Benseler, E. Graeser, N. Piel, and S. Scholtissek, *Biochemistry* **26**, 7238 (1987).

[c] J. Mazzarelli, S. Scholtissek, and L. W. McLaughlin, *Biochemistry* **28**, 4616 (1989).

[d] P. C. Newman, V. U. Nwosu, D. M. Williams, R. Cosstick, F. Seela, and B. A. Connolly, *Biochemistry* **29**, 9891 (1990).

[e] B. A. Connolly and P. C. Newman, *Nucleic Acids Res.* **17**, 4957 (1989).

[f] H. P. Rappaport, *Nucleic Acids Res.* **16**, 7253 (1988).

[g] The 5,6-double bond is reduced in dihydrothymine and there is an additional hydrogen at position 6.

[h] D. Molko, A. M. Delort, and R. Teoule, *Biochemie* **67**, 801 (1985).

[i] E. A. Kubareva, C.-D. Pein, E. S. Gromova, S. A. Kuznezova, V. N. Tashlitzki, D. Cech, and Z. A. Shabarova, *Eur. J. Biochem.* **175**, 615 (1988).

[j] A. A. Yolov, M. N. Vinogradova, E. S. Gromova, A. Rosenthal, D. Cech, V. P. Veiko, V. G. Metelev, V. G. Kosykh, Y. I. Buryanov, A. A. Bayev, and Z. A. Shabarova, *Nucleic Acids Res.* **13**, 8983 (1985).

[k] A. Fliess, H. Wolfes, A. Rosenthal, K. Schwellnus, H. Blocker, R. Frank, and A. Pingoud, *Nucleic Acids Res.* **14**, 3463 (1986).

[l] T. Hayakawa, A. Ono and T. Ueda, *Nucleic Acids Res.* **16**, 4761 (1988).

[m] A. Fliess, H. Wolfes, F. Seela, and A. Pingoud, *Nucleic Acids Res.* **16**, 11781 (1988).

[n] A. Ono and T. Ueda, *Nucleic Acids Res.* **5**, 219 (1987).

them. These lists are not exhaustive with respect to either analogs or enzymes. In this chapter, we have selected representative studies with analogs that illustrate useful approaches or methods.

For an ideal analysis, one would remove or alter each of the functional groups of the heterocyclic base that has the potential to interact specifically with the enzyme, namely, the O or N atoms on accessible surfaces of the DNA that could act as hydrogen bond donors or acceptors and the methyl group of Thy that could be involved in hydrophobic or van der Waals interactions.[6] The perfect analog would have only one of its specificity determinants removed without added new atoms that might introduce steric blockages or electronic incompatibilities that interfered with any other interactions near the site of modification. Examples of analogs approaching this ideal are purines (Pur[2]) for Ade, where the exocyclic amino group of Ade is replaced with a hydrogen atom,[7] or Ura, where the methyl group of Thy is similarly replaced.[7,8] In practice, one functional group has been substituted for another, for example, bromine replaced a methyl group when BrUra was substituted for Thy.[8] Sometimes a single, larger group has been added in place of a hydrogen atom, e.g., when 5-MeCyt replaces Cyt.[8] The relative positions of the amino and carbonyl groups on a base pair have also been interchanged, as when a 2,6-$(NH_2)_2$Pur : Ura base pair was substituted for a Gua : Cyt base pair.[7]

Ideally, the introduction of the analog should not decrease the thermal stability of the duplex under the conditions of the assay so that the substrate is significantly denatured. The analog should also not cause conformational changes in the DNA structure that could affect nearby interactions with the normal base pairs by altering their relative locations. Adverse conformational effects could exert themselves on the direct readout of adjacent bases or the indirect readout of sequence-specific contacts with the ribose-phosphodiester backbone. Since some site-specific DNA-binding proteins significantly distort the DNA on binding, an additional, related requirement is that the analog should not alter the deformability of the DNA. Since all these qualities are unlikely to be present with any analog, the extent to which a particular structural variant lacks them will determine the degree to which the interpretation of the results attained with it will be compromised. These difficulties often necessitate ancillary information for a more complete understanding of the molecular effects of the substitutions. For example, the effects of several analogs on the bend-

[6] N. C. Seeman, J. M. Rosenberg, and A. Rich, *Proc. Natl. Acad. Sci. U.S.A.* **73,** 804 (1976).

[7] L. W. McLaughlin, F. Benseler, E. Graeser, N. Piel, and S. Scholtissek, *Biochemistry* **26,** 7238 (1987).

[8] C. A. Brennan, M. D. van Cleve, and R. I. Gumport, *J. Biol. Chem.* **261,** 7270 (1986).

ing of the *Eco*RI recognition sequence have been correlated with their effects on the activity of the enzyme.[9] The location of the enzyme with respect to the major and minor grooves of the bound DNA target, as determined by X-ray crystallography,[10] has also been useful in explaining the effects of analog substitution experiments.[7,8] Just as with enzyme or protein analogs produced by mutagenesis procedures, the more structural information available, the easier it is to design and interpret substrate–analog experiments.

Preparation of Substrates for Type II Restriction Enzymes

If properly protected and activated nucleotide analogs are available, an oligodeoxyribonucleotide can be synthesized by current automated chemical methods. Analog synthons containing 7-deazaAde, N^6-MeAde, 5-BrCyt, 5-IoCyt, 5-MeCyt, 7-deazaGua, O^6-MeGua, hypoxanthine (Hpx), Pur, 2,6-$(NH_2)_2$Pur, Ura, 5-BrUra, 5-FlUra, and 5-IoUra are currently available commercially. The preparations of many other analog-containing oligonucleotides have been described (see Tables I and II for references) but the protected, activated nucleotides are not commercially available. Alternatively, T4 RNA ligase[11] may be used to prepare sufficient quantities of modified oligonucleotides for the kinetic analyses used in most studies of restriction endonucleases. If the 2'-deoxyribonucleoside of the analog is available, it is converted to the 2'-deoxyribonucleoside 3',5'-bisphosphate derivative by reaction with pyrophosphoryl chloride.[12] The deoxyribonucleoside bisphosphate serves as a substrate for RNA ligase,[13] and is joined by the enzyme to the 3'-hydroxyl of an oligodeoxyribonucleotide at least three nucleotides long. After removal of the 3'phosphate from this initial product, the extended oligonucleotide bearing the analog is similarly added to a second oligomer to form the analog-containing product.[11] When annealed to a complementary strand, a duplex containing an analog at a unique site is formed. Because chemical synthesis provides more material and thus facilitates the characterization and structural studies that can aid in interpreting results, it is the preferred method.

Oligodeoxyribonucleotides, eight or more base pairs long, serve as

[9] S. Diekmann and L. W. McLaughlin, *J. Mol. Biol.* **202,** 823 (1988).
[10] J. A. McClarin, C. A. Frederick, B.-C. Wang, P. Greene, H. W. Boyer, J. Grable, and J. M. Rosenberg, *Science* **234,** 1526 (1986).
[11] C. A. Brennan and R. I. Gumport, *Nucleic Acids Res.* **13,** 8665 (1985).
[12] J. R. Barrio, M. D. Barrio, N. J. Leonard, T. E. England, and O. C. Uhlenbeck, *Biochemistry* **10,** 2077 (1987).
[13] O. C. Uhlenbeck and R. I. Gumport, *in* "The Enzymes" (P. D. Boyer, ed.), Vol. 15, p. 31. Academic Press, New York, 1982.

substrates for many restriction enzymes.[1,14] Since the sequence recognized by most type II enzymes is palindromic, the easiest synthetic approach is to make a self-complementary oligonucleotide. Self-complementary oligonucleotides can sometimes form intramolecular hairpin structures that complicate their use as restriction enzyme substrates.[15] Duplexes formed by self-complementary oligomers that contain an analog will have symmetrically disposed substitutions in each strand. If the effect of an analog in one strand only is to be tested, then complementary oligonucleotides with unique sequences must be synthesized. For enzymes recognizing palindromic sequences, this requires synthesizing two oligonucleotides with sufficiently long, nonself-complementary flanking sequences to suppress intramolecular hairpin formation by either strand.

At a minimum, the melting temperature (T_m) of the duplex under the conditions of the assays must be determined to ensure that the substrate is a duplex. Usually, this is impossible to measure directly because of the low concentrations of oligomer used in the assays. As a result, the T_m must be measured at several higher concentrations, and the value interpolated to the lower concentration.[11,16] Other information, e.g., structural analyses of oligomers containing analogs, can facilitate the interpretation of the results obtained.[9,17–19] Careful consideration should also be given to the base pairs adjacent to the recognition sequence, because they can affect the activity of the enzyme. Flanking base pairs influence the stability and structure of the oligonucleotide,[20] and with *Eco*RI, by contacting the enzyme,[21] can directly affect its activity.[22–24]

One must also ascertain whether the decreased ability of a substrate analog to be cleaved is due to the modified base or to the presence of an adventitious enzyme inhibitor. Careful purification and characterization

[14] P. J. Greene, M. S. Poonian, A. L. Nussbaum, L. Tobias, D. E. Garfin, H. W. Boyer, and H. M. Goodman, *J. Mol. Biol.* **99**, 237 (1975).

[15] F. Seela and A. Kehne, *Biochemistry* **26**, 2232 (1987).

[16] O. C. Uhlenbeck, P. N. Borer, B. Dengler, and I. Tinoco, *J. Mol. Biol.* **73**, 483 (1973).

[17] A.-M. Delort, J. M. Neumann, D. Molko, M. Herve, R. Teoule, and S. T. Dinh, *Nucleic Acids Res.* **13**, 3343 (1985).

[18] G. M. Clore, H. Oschkinat, L. W. McLaughlin, F. Benseler, C. S. Happ, E. Happ, and A. M. Gronenborn, *Biochemistry* **27**, 4185 (1988).

[19] P. C. Newman, V. U. Nwosu, D. M. Williams, R. Cosstick, F. Seela, and B. A. Connolly, *Biochemistry* **29**, 9891 (1990).

[20] B. A. Connolly and F. Eckstein, *Biochemistry* **23**, 5523 (1984).

[21] A.-L. Lu, W. E. Jack, and P. Modrich, *J. Biol. Chem.* **256**, 13200 (1981).

[22] M. Thomas and R. W. Davis, *J. Mol. Biol.* **91**, 315 (1975).

[23] J. Alves, A. Pingoud, W. Haupt, J. Langowski, F. Peters, G. Maass, and C. Wolff, *Eur. J. Biochem.* **140**, 83 (1984).

[24] M. D. Van Cleve, Ph.D. Thesis, University of Illinois, Urbana, Illinois (1989).

of the oligonucleotide reduces the likelihood of this potential artifact. Experiments in which the effect of the modified oligonucleotide on the cleavage of DNA or another oligonucleotide containing the unmodified recognition sequence are useful for differentiating these phenomena.[7,8,25] Inhibition due to binding of the modified, duplex oligonucleotide to the enzyme can be clearly distinguished from nonspecific general inhibition as follows.[7] If the T_m of a modified oligonucleotide is lower than that of an unmodified substrate (and both T_m values are within the range of temperatures at which the enzyme is active), the reaction is performed at a temperature between the two T_m values. If the modified substrate is denatured while the normal substrate remains a duplex and the inhibition of cleavage of the unmodified substrate disappears, it can be concluded that the analog itself is binding to the enzyme and that there is no thermally stable, general inhibitor present.[7] This method depends on the endonuclease using only duplex DNA as a substrate. Analyses of the mode of inhibition by analog-containing oligonucleotides have revealed examples of competitive inhibition and allowed the determination of the dissociation constant (K_I) of the enzyme–analog complex.[7,26]

Assays of Type II Restriction Endonucleases with Analog-Containing Substrates

Early studies with analog-containing substrates often reported only qualitative results using preparations of enzymes that were not homogeneous. Some studies simply reported whether the DNA was cleaved, while others determined the rate of cleavage at a single concentration of substrate with excess enzyme. Quantitative interpretations of the effects of analog substitutions require the determination of the K_m and k_{cat} values from steady state studies or of individual rate constants from single-turnover experiments and of equilibrium association constants (K_A).

Analog-containing substrates are usually synthesized with 5′-hydroxyl groups that are subsequently labeled with T4 polynucleotide kinase and [γ-^{32}P]ATP. Cleavage can be followed by separating the products from the substrates by thin-layer chromatography[27] or gel electrophoresis.[28] Cleavage has also been followed by elution of the products from a DEAE-cellulose filter and quantifying the disappearance of substrate.[7] For the

[25] J. Mazzarelli, S. Scholtissek, and L. W. McLaughlin, *Biochemistry* **28,** 4616 (1989).

[26] C. R. Aiken, L. W. McLaughlin, and R. I. Gumport, *J. Biol. Chem.* **266,** (1991), in press.

[26a] C. R. Aiken, E. W. Fisher, and R. I. Gumport, *J. Biol. Chem.* **266,** (1991), in press.

[27] E. Jay, R. Bambara, R. Padmanabhan, and R. Wu, *Nucleic Acids Res.* **1,** 331 (1974).

[28] A. A. Yolov, M. N. Vinogradova, E. S. Gromova, A. Rosenthal, D. Cech, V. P. Veiko, V. G. Metelev, V. G. Kosykh, Y. I. Buryanov, A. A. Bayev, and Z. A. Shabarova, *Nucleic Acids Res.* **13,** 8983 (1985).

first two methods, the locations of substrate and product are determined by autoradiography, and the extent of reaction determined by excising the labeled components and quantifying them by liquid scintillation counting. The filter-binding method measures only the amount of substrate remaining after hydrolysis, and is less reliable for determining small amounts of product. Alternatively, the hydrolysis reaction of unlabeled oligonucleotides can be monitored by UV absorbance after high-performance liquid chromatography[29] (HPLC) to follow the disappearance of substrate or appearance of product.[7,19,30,31] HPLC assays do not readily lend themselves to the analysis of reaction mixtures containing low concentrations of oligonucleotides because they rely on UV absorbance for analyte detection. Since no continuous assay has yet been developed for oligonucleotide substrates, aliquots of a reaction mixture must be analyzed individually.

For steady state analyses, the assays are manipulated so that the kinetics fit the Michaelis–Menten equation. The Mg^{2+} cofactor must be saturating so that the measured velocities are a function of only the oligonucleotide concentrations. The following conditions must be met for the assays to be valid. The initial velocity (v_0) of the reaction must be linear with respect to both the times of reaction and the enzyme concentrations used. The DNA concentration should be sufficiently greater (usually by at least 10-fold) than that of the enzyme, and the assay times and enzyme concentrations chosen so that enzyme binding or the reaction itself does not consume so much (>5 to 10%) of the oligonucleotide that the v_0 changes as a consequence of free substrate depletion. With very poor substrates, it is sometimes difficult to measure the reaction without using relatively high concentrations of enzyme.[30] For the best estimates of K_m, the substrate concentrations should range from about three to eight times below and above the ultimately determined K_m value, and curve-fitting techniques (see Methods of Procedure, below) rather than plots of the various linear forms of the Michaelis–Menten equation should be used to determine the kinetic values and their associated errors of measurement. In practice, the data obtained tend to be erratic, and several measurements at each substrate concentration are required to obtain reliable values of k_{cat} and K_m. In addition, one should attempt to ensure that reaction components other than the DNA do not influence the reaction. For example, the K_m values for Mg^{2+} with an analog-containing substrate and the unsubstituted control were different with EcoRV.[30]

[29] L. W. McLaughlin, Chem. Rev. **89**, 309 (1989).

[30] P. C. Newman, D. Williams, R. Cosstick, F. Seela, and B. A. Connolly, Biochemistry **29**, 9902 (1990).

[31] J. Alves, A. Pingoud, W. Haupt, J. Langowski, F. Peters, G. Maass, and C. Wolff, Eur. J. Biochem. **140**, 83 (1984).

For reactions conducted under single-turnover conditions, two criteria need be met to ensure that all the substrate is enzyme bound. The concentration of enzyme should be greater than that of the substrate, and the concentrations of enzyme and substrate should each be greater than the K_A value for the formation of the enzyme–substrate complex. The enzyme excess is proved by showing that the observed initial velocity is not increased by the further addition of enzyme.[32] Additional confirmation that all the substrate is enzyme bound by a physical technique such as mobility retardation during gel electrophoresis has also been reported.[32] The assay conditions should be those that favor cleavage at canonical sites in spite of high enzyme concentrations, which often favor "star" activity.[33] This consideration also applies to steady state studies, i.e., avoidance of cleavage at noncanonical sites by maintenance of high-specificity conditions. The reactions are usually initiated by the addition of Mg^{2+}, and the progress of the reaction is monitored by the withdrawal of aliquots with time. The rate of product release from the enzyme does not influence the results, for it is eliminated as a consideration by the strongly denaturing conditions used to quench the reaction.[32] The substrates can be designed so that the cleavage products of the two strands have different sizes or, alternatively, either strand can be labeled individually so that hydrolysis of each can be monitored independently. Equations that describe the appearance of the radiolabeled cleavage products from each strand of a nonself-complementary oligonucleotide duplex in terms of the four first-order rate constants that together describe the complete cleavage reaction of an asymmetrically located EcoRI site have been reported.[32]

Equilibrium association constants (K_A) for specific binding to a target sequence have been determined by nitrocellulose filter-binding assays.[32,34,35] These measurements are carried out in the absence of Mg^{2+} to prevent hydrolysis of the substrates.[34] Direct determinations under equilibrium conditions[34,35] or indirect assays using equilibrium–competition techniques[34,36] have been reported. Equations for the solution of equilibrium–competition data from experiments using oligonucleotides with noncanonical sites or mismatched base pairs in the recognition sequence of EcoRI that are applicable to analog-containing oligonucleotides

[32] D. R. Lesser, M. R. Kurpiewski, and L. Jen-Jacobson, *Science* **250,** 776 (1990).
[33] B. Polisky, P. Greene, D. Garfin, B. McCarthy, H. Goodman, and H. Boyer, *Proc. Natl. Acad. Sci. U.S.A.* **72,** 3310 (1975).
[34] B. J. Terry, W. E. Jack, R. A. Rubin, and P. Modrich, *J. Biol. Chem.* **258,** 9820 (1983).
[35] L. Jen-Jacobson, M. Kurpiewski, D. Lesser, J. Grable, H. W. Boyer, J. M. Rosenberg, and P. J. Greene, *J. Biol. Chem.* **258,** 14638 (1983).
[36] L. Jen-Jacobson, D. Lesser, and M. Kurpiewski, *Cell (Cambridge, Mass.)* **45,** 619 (1986).

have been published.[37] For assays under equilibrium conditions, the K_A values determined with excess enzyme should equal those obtained with excess oligonucleotide. Efforts were made with EcoRI to show that the absence of Mg^{2+} does not yield a K_A value different from the association constant under conditions allowing hydrolysis.[32] Some restriction endonucleases, e.g., EcoRV and TaqI, do not appear to bind specifically to their target sequences in the absence of Mg^{2+}.[38,39]

Since the endonucleases have affinities for DNA lacking their recognition sequences, i.e., they bind nonspecifically to DNA,[34] the effects of the concentration of potential nonspecific sites on the oligonucleotide should be considered in calculating the K_A value for specific binding. If the ratios of K_A (specific) to K_A (nonspecific) are similar to those for EcoRI binding to plasmid DNA ($\sim 10^4$ to 10^5),[34] the correction for nonspecific binding on a short oligonucleotide would be negligible because there are relatively few nonspecific sites. The effects on the binding resulting from the proximity of termini to the target sequence should also be kept in mind. In addition to the structure of DNA at duplex ends being different than that located within the helix,[20] the enzymes probably contact more base pairs than the canonical recognition sequence[10,21] and specific targets near a terminus may lack one or more of these nucleotides. K_m values are higher with short oligonucleotides than with polymeric DNA, supporting the view that the flanking sequences contribute to the interactions between enzyme and substrate.[1,14]

Interpretation of Results

The effects of the analog substitutions are commonly assessed by determining the kinetics of the hydrolysis reaction under either steady state or single-turnover conditions. The most easily interpretable result is the lack of an effect on the reaction metrics. In this case, one can safely conclude that the determinant under consideration does not interact with the enzyme and, additionally, that the analog substitution does not alter the DNA conformation such that any nearby interactions are affected. Usually, however, an analog substitution affects the reaction in at least one of its kinetic or thermodynamic constants. In a study of EcoRI with a battery of analog substrates, this finding has been interpreted to indicate a close interfacial contact between the enzyme and substrate.[8]

For analog substitutions that have kinetic consequences, the straight-

[37] V. Thielking, J. Alves, A. Fliess, G. Maass, and A. Pingoud, Biochemistry 29, 4682 (1990).
[38] J. D. Taylor, I. G. Badcoe, A. R. Clarke, and S. E. Halford, Biochemistry 30, (1991) in press.
[39] J. Zebala and F. Barany, personal communication.

forward interpretation is that the effect is due to the loss of an interaction between the enzyme and a group that has been deleted, or to a steric interference between the enzyme and the region of the base to which a new group was added. The direct inference is that the enzyme and the substrate normally "touch" one another at this site at some time during the reaction. Such an unambiguous conclusion is seldom warranted in the absence of other information that eliminates the indirect causal effects previously mentioned.[8,19,25,30] Given this general caveat, analog experiments have been interpreted as follows.

Effects on both the initial velocity of the reaction and substrate binding are determined. The specificity constant, k_{cat}/K_m, is calculated from steady state results.[40] For reactions measuring single turnovers, the counterpart of the specificity constant, $k_1 K_A$, is calculated.[32] In this expression, k_1 is the first-order rate constant for the irreversible hydrolysis of the first phosphodiester bond and K_A is the equilibrium association constant of the enzyme and its recognition sequence in the absence of Mg^{2+}.[32] Both the k_{cat}/K_m and $k_1 K_A$ values reflect the recognition of the substrate by the enzyme. For example, for two substrates A and B under steady state conditions, the ratio of the velocities $v_A/v_B = \{[A](k_{cat}/K_m)_A\}/\{[B](k_{cat}/K_m)_B\}$ expresses the relative selectivity of the enzyme for the two substrates.[40] For equimolar concentrations of A and B, the ratio of the specificity constants directly reflects the selectivity of the enzyme for one substrate with respect to the other. Because the values are derived from steady state experiments, the specificity constant, k_{cat}/K_m, measures the selectivity of the enzyme as a function of the appearance of the free, cleaved products of the reaction. If the "nicked" product after cleavage of the first strand of the duplex dissociates from the enzyme, steady state studies can measure the rate of nicking rather than of double-strand cleavage.

Analogously, the expression $v_A/v_B = \{[A](k_1 K_A)_A\}/\{[B](k_1 K_A)_B\}$ provides the relative selectivity of an enzyme under single-turnover conditions. The $k_1 K_A$ ratio reflects the selectivity as a function of the hydrolysis of the first bond cleaved.[32] The specificity constant ratio may be more physiologically relevant than the ratio of $k_1 K_A$ values because DNA that has been cleaved in both strands at a target sequence may have more of a biological consequence than does nicked DNA. Conversely, the value $(k_1 K_A)$ is valuable for determining the energetics of the atomic interactions of recognition because it is a more direct measure of the events at a discrete step in the reaction—the first bond hydrolysis.[32] This value is not subject to ambiguities introduced by such mechanistic considerations as whether

[40] A. Fersht, "Enzyme Structure and Mechanism." Freeman, New York, 1985.

the rate-limiting step of the reaction under steady state conditions occurs after bond hydrolysis.

In steady state studies, effects predominantly on k_{cat} have been regarded as indicative of alterations in interactions between the enzyme and the transition state.[3,30,40] Since the rate-limiting step with polymeric DNA substrates is related to product release, i.e., is at some point after the initial phosphodiester bond hydrolysis with EcoRI[41] and EcoRV,[42] the interpretation of the steady state values may be complicated because the "recognition" occurs before the rate-determining step of the reaction. However, with some short oligonucleotide substrates, k_{cat} values approach those seen for the bond hydrolysis step determined under single-turnover conditions with DNA.[8,30,43] This finding suggests that product release may not be rate limiting with these oligodeoxyribonucleotides and that k_{cat} may approximate k_1, the first-order rate constant for bond hydrolysis. This hypothesis has not been confirmed.[30] Although K_m is likely not a true equilibrium dissociation constant (K_S) for the enzyme–substrate complex because of the complexity of the mechanisms of the endonucleases, changes in this value have been interpreted as arising primarily from effects on macroscopic binding phenomena involving all of the enzyme-bound forms of the substrate.[40] Keeping in mind that $K_m \neq K_S$, K_m values have been considered to be indicative of total binding effects.

Both k_{cat}/K_m and $k_1 K_A$ ratios can be used to estimate the binding energy provided by a single group on the base in the transition state given appropriate circumstances. If the determinant under consideration is not chemically involved in the bond-making and bond-breaking reactions and there are no inductive effects, the ratios are the binding energy ratios.[40,44–46] By comparing the canonical sequence with the analog-containing sequence using EcoRV, the general equation $\Delta G_{app} = RT \ln[(k_{cat}/K_m)_{analog}/(k_{cat}/K_m)_{canonic}]$ has been used to estimate the contribution to the interaction energy of a group that has been deleted in the analog.[30] Likewise, with EcoRI, $\Delta\Delta G_I^{\ddagger} = -RT \ln[(k_1 K_A)_{noncanonic\ or\ analog}/(k_1 K_A)_{canonic}]$ has been used to estimate the energy cost in reaching the transition state of deleting a group.[32] The former method depends on product appearance under steady state conditions, whereas the latter measures the first-order rate of the initial bond hydrolysis under single-turnover conditions. In spite of their

[41] P. Modrich and D. Zabel, J. Biol. Chem. 251, 5866 (1976).
[42] S. E. Halford and A. J. Goodall, Biochemistry 27, 1771 (1988).
[43] B. J. Terry, W. E. Jack, and P. Modrich, in "Gene Amplification and Analysis" (J. G. Chirikjian, ed.), Vol. 5, p. 103. Elsevier Science Publishing, New York, 1987.
[44] A. R. Fersht, Trends Biochem. Sci. 12, 301 (1987).
[45] A. R. Fersht, Biochemistry 26, 8031 (1987).
[46] A. R. Fersht, Biochemistry 27, 1577 (1988).

differences, both methods have yielded anticipated values for the energies of some interactions.[44,47] Unanticipated values have required interpretations that invoke indirect effects such as steric clashes and impaired conformational adjustments of enzyme or substrate to account for the energy discrepancies observed.[30,32] These elegant studies represent the first attempts to quantify the energetics of the interactions between a submolecular determinant on the oligonucleotide in the transition state and the amino acid side chains of restriction endonucleases.

Methods of Procedure

Oligodeoxyribonucleotides containing modified bases are now most commonly synthesized using automated methods with β-cyanoethyl phosphoramidite synthons (see references in Tables I and II). After synthesis, the deprotected oligonucleotides are dried *in vacuo*, and purified by polyacrylamide gel electrophoresis in the presence of 7 M urea or by reversed-phase HPLC. We have found that HPLC is convenient for the isolation of oligonucleotides shorter than 20 bases.

Procedure 1. Purification of Oligonucleotides by Reversed-Phase HPLC

Prepare 1 liter of each of the following solutions using HPLC-grade methanol:

Buffer A: 50 mM potassium phosphate, pH 5.9
Buffer B: 50 mM potassium phosphate, pH 5.9,
    50% (v/v) methanol

Equilibrate an octylsilica (C_8) reversed-phase HPLC column (4.6 × 150 or 250 mm) at 1 ml/min with buffer A. Inject 5 to 10 optical density (OD) units (measured at 260 nm) of oligomer in 200 μl of buffer A, and separate with a linear gradient of increasing buffer B. Perform a trial run using 0.1 OD unit of material to determine the conditions for optimal separation. In our hands, a decanucleotide eluted at a concentration of approximately 60% buffer B (~30% methanol). Collect the product peak, as detected by absorbance at 254 nm, and reinject a small amount (0.05 OD units) with the detector set at high sensitivity to analyze for purity. Oligonucleotides do not necessarily elute in order of their increasing lengths during this procedure, so it can be difficult to identify the desired product unless it is the major product of the synthesis. An analysis using polyacrylamide gel

[47] J. A. Wells, *Biochemistry* **29**, 8509 (1990).

electrophoresis may be performed to check the length of the product. It is necessary to equilibrate the column completely using at least 20 column volumes of solvent before each injection to ensure reproducible results. Dry the pooled fractions containing the product *in vacuo*, dissolve in H_2O, and desalt by passage over a Sephadex G-25 column (NAP-5 Sephadex from Pharmacia, Inc., Piscataway, NJ) and elution with H_2O. Store the product frozen.

Procedure 2. Purification of Oligonucleotides by Polyacrylamide Gel Electrophoresis

Polyacrylamide gel electrophoresis is the method of choice for purification of longer oligonucleotides (>20 bases), but it can also be employed in the purification of shorter products.

Pour a 20% (w/v) polyacrylamide–7 M urea sequencing gel.[5] Dry the oligonucleotide *in vacuo* and dissolve it in the loading buffer [10 mM NaOH, 80% (v/v) formamide, 1 mM EDTA, 0.1% (w/v) xylene cyanol, 0.1% (w/v) bromphenol blue]. Load approximately 0.5 OD units (A_{260}) of oligomer in 5 μl/cm of well width. Separate the mixture at 1000 V until the dyes have migrated a sufficient distance to indicate movement of the product to approximately half the length of the gel. Remove the gel and transfer it to a polyethylene film (Saran wrap). Remove one glass plate, overlay the gel with the plastic wrap, invert the plate, and tease the gel away from the glass with a spatula while simultaneously pulling on the plastic wrap.

Visualize the bands under reflected, short-wave UV light by placing the plastic-wrapped gel on a support that fluoresces on UV illumination, e.g., an X-ray film intensifying screen or a thin-layer chromatography plate impregnated with a fluorophore. Excise the desired products from the gel, and isolate the oligonucleotides by electroelution or by a crush-and-soak method.[48] Our laboratory routinely uses the EpiGene device (EpiGene, Baltimore, MD) for electroeluting the oligonucleotides onto columns of DEAE-cellulose (5 mm diameter × 3 mm) in the bicarbonate form. The DNA is recovered (~50% yield) by washing and eluting the columns with 0.5 ml 2 M triethylammonium bicarbonate (TEABC) [prepared by neutralization with CO_2 (gas) of triethylamine that has been distilled in the presence of chlorosulfonic acid to eliminate primary and secondary amines and to remove nonvolatile impurities]. Remove the TEABC *in vacuo* by repeated coevaporation with 50% (v/v) methanol. Dissolve the product in H_2O.

[48] T. Maniatis, E. F. Fritsch, and J. Sambrook, "Molecular Cloning: A Laboratory Manual," 2nd Ed. Cold Spring Harbor Laboratory, Cold Spring Harbor, New York, 1989.

Procedure 3. Nucleoside Analysis of Oligodeoxyribonucleotides Containing Modified Bases[11]

After purification, the presence of the modified base is confirmed by HPLC analysis of the products of snake venom phosphodiesterase and alkaline phosphatase digestion of the oligomer.

Prepare a reaction mixture (50 μl) containing:

Purified oligonucleotide: 0.2 OD units (A_{260})
Snake venom phosphodiesterase (5 mg/ml): 5 μl
Bacterial alkaline phosphatase (1 mg/ml): 5 μl
Tris-HCl (0.1 M, pH 7.5), 20 mM MgCl$_2$: 25 μl
Sufficient H$_2$O to bring to a final volume of 50 μl

Combine the solutions and incubate at 37° for 2 hr. Inject a sample of the digested oligonucleotide onto a reversed-phase column (C$_8$ or C$_{18}$) equilibrated in buffer A as described in procedure 1. We advise the use of a guard column to protect the primary column from contamination. Elute the column with a linear gradient of 0–30% buffer B in 20 min, followed by a 30 to 50% elution over 10 min. Wash the column for 3 min in buffer B before reequilibrating it. The deoxyribonucleosides are detected at 254 nm. The areas of the peaks are determined with an integrator that is connected to the spectrophotometer of the HPLC apparatus. Deoxynucleoside standards, including the analog deoxynucleoside, are individually injected to determine their retention times and relative recoveries. The mole fraction of each nucleoside is determined by dividing the area of each peak by the extinction coefficient at 254 nm of the corresponding deoxyribonucleoside, and dividing the area of the normalized peak by the lowest value obtained.

Procedure 4. Partial Snake Venom Digestion of Oligonucleotides to Determine Length

Prepare the reaction as in procedure 3 but use a 5′-^{32}P-labeled oligonucleotide (10^5 to 10^6 cpm) and omit the alkaline phosphatase. Start the reaction by adding the phosphodiesterase. Withdraw 5-μl aliquots at intervals of 1, 2, 5, 10, 20, 30, 40, 50, and 60 min and spot them onto a plastic-backed, DEAE-cellulose thin-layer plate (20 cm long). Develop the plate as described in procedure 7. After autoradiography, the number of components in the nested set of products will correspond to the length of the oligonucleotide.

If desired, the oligonucleotides can be sequenced by either the wandering-spot[27] or the Maxam–Gilbert[5,49] methods. In practice, this is usually unnecessary if the length is correct, and the expected 5'-nucleoside and the modified base are present. The 5'-terminal nucleoside is determined by complete digestion of the 5'-^{32}P-labeled oligonucleotide as described previously followed by thin-layer chromatography of the deoxynucleotides and standards on cellulose TLC plates using saturated $(NH_2)_2SO_4/1$ M sodium acetate/2-propanol [80/18/2 (v/v/v)] as the eluant.

Assay of Type II Restriction Endonucleases with Oligodeoxyribonucleotides

Assays of type II restriction endonucleases using oligonucleotides containing base analogs as substrates are most commonly performed using radiolabeled oligomers followed by the separation of substrates and products by homochromatography[27] or polyacrylamide gel electrophoresis.[28]

Procedure 5. Labeling of Oligonucleotides Using Polynucleotide Kinase and [γ-^{32}P]ATP

Evaporate to dryness 1 nmol of purified oligonucleotide in a siliconized microfuge tube. Redissolve in 12.5 μl of a solution containing:

Tris-HCl, pH 7.5, 100 mM
MgCl$_2$, 20 mM
Dithiothreitol, 20 mM
Bovine serum albumin, 0.1 mg/ml
Spermidine (neutralized to pH 7.5), 2 mM

Add 9.5 μl of distilled water, and 2 μl of [γ-^{32}P]ATP (6000 Ci/mmol, 140 mCi/ml). Add 1 μl (10 units) of T4 polynucleotide kinase, and incubate at 37° for 30 min. If possible, the ATP concentration should be >1 μM to allow its efficient utilization by polynucleotide kinase. Add 1 μl of 1 mM ATP and an additional 1 μl of the kinase, and return the reaction to 37° for 30 min. The second addition of unlabeled ATP and enzyme ensures that all the oligonucleotide is phosphorylated. In endonuclease assays using short oligomers, it is necessary for all the molecules to be phosphorylated to ensure that the enzyme has only one possible substrate. In addition, the concentration of the substrate may be more accurately estimated if only one species is present. An alternative labeling procedure involves decreasing the specific radioactivity of the commercial [γ-^{32}P]ATP with ATP, and using a molar excess over the oligonucleotide of total ATP

---

[49] A. M. Banaszuk, K. V. Deugau, J. Sherwood, M. Michalak, and B. R. Glick, *Anal. Biochem.* **128**, 281 (1983).

with a lower specific radioactivity in the phosphorylation reaction. This procedure allows one to control more easily the specific radioactivity of the products.

Purification of Labeled Oligonucleotides and Determination of Concentrations

After labeling, it is sometimes necessary to purify the oligonucleotides before using them in the enzyme assays. In assays of the *Rsr*I and *Eco*RI endonucleases using decanucleotides, contaminating [γ-^{32}P]ATP, remaining from the polynucleotide kinase reaction, comigrated with the trimer product of hydrolysis during homochromatography. Chromatography of the labeled oligonucleotide mixtures using Nensorb 20 (Du Pont, Inc., Wilmington, DE) eliminated virtually all of the unreacted ATP. The columns were used as described by the manufacturer. However, the oligonucleotides obtained were not free of UV-absorbing impurities. Therefore, we employed chromatography using NAP-5 Sephadex columns (see procedure 1) as a final purification step. The resulting products were free of enzyme inhibitors and ATP, and yielded spectra typical of pure oligonucleotides. The oligonucleotide concentrations were determined spectrophotometrically using extinction coefficients calculated by summing the published values for the individual mononucleotides.[50] A correction in the molar extinction coefficients to account for hypochromicity can be introduced by measuring the increase in absorbance of a sample of the *duplex* oligonucleotide on digestion to completion with snake venom phosphodiesterase and alkaline phosphatase.[19]

Procedure 6. Kinetic Assays of Restriction Endonucleases Using Labeled Oligonucleotides as Substrates

Dispense aliquots containing the desired amounts of the labeled oligonucleotides necessary to form 50-μl reactions into siliconized microfuge tubes. Evaporate to dryness *in vacuo*. Redissolve in 25 μl restriction endonuclease assay buffer. The buffer should be one that gives maximal activity and high recognition specificity (no star activity) with the enzyme under study. Heat to 65° when using decanucleotides,[7] or to higher temperatures for longer oligonucleotides, for 5 min, and cool to room temperature. This protocol ensures the formation of a duplex oligomer. Equilibrate the tubes at the desired temperature (20° for decanucleotides). The temperature is chosen to be low enough to

[50] B. Janik, "Physicochemical Characteristics of Oligonucleotides." IFI/Plenum, New York, 1971.

maintain the oligonucleotides in the duplex form and high enough to allow enzyme activity. Incubate the mixture for at least 5 min to dissolve the oligonucleotide. Start the reactions by adding 25 μl of the assay buffer containing an appropriate dilution of the restriction endonuclease. Withdraw 5-μl aliquots at intervals, and stop the reaction by either spotting directly onto DEAE-cellulose thin-layer plates (for homochromatography) or by pipetting the aliquot into 5 μl of a solution containing a fivefold molar excess of EDTA over the Mg^{2+} in the reaction mixture (for gel electrophoresis). We have found that aliquots can be withdrawn every 2 min from six separate reactions, if the starting times are staggered. Normally, five aliquots are taken from each reaction and analyzed for product formation. The assays must be replicated enough times to provide precise results.

Procedure 7. Homochromatography for Separation of Substrates and Products

Prepare homochromatography solvent (Homomix 6).[27] Mix together 80 ml H_2O and 20 g yeast RNA (Sigma Chemical Co., St. Louis, MO). With cooling and stirring, add 20 ml 5 M KOH. Incubate at 37° with gentle agitation for 48 hr. Remove, add 100 ml H_2O, and neutralize with 1 M HCl. Add 420 g urea and H_2O to bring to a volume of 1000 ml. Stir the suspension until the urea is dissolved. Filter through Whatman (Clifton, NJ) #1 paper, and store at 4°. This protocol prepares Homomix 6, which is useful for separating oligonucleotides up to 12 nucleotides long. For longer oligonucleotides, other homomixes can be prepared by decreasing the amount of KOH or the incubation time, or both.[27] In a thin-layer chromatography tank, place 50–100 ml of Homomix (enough to cover the bottom to the depth of 0.5 cm). Place the entire tank in a 65° oven or incubator, and allow the tank to equilibrate. Spray the bottom 3 cm of the 20-cm long DEAE thin-layer plate onto which reaction aliquots were spotted with a fine mist of water from an atomizer. The chromatogram should be wet but not dripping. Without allowing the plate to dry, place it in the chromatography tank and develop the solvent to the top of the chromatogram. Remove the plate, allow it to dry, and mark the corners with radioactive ink. The Homomix solvent may be stored in a sealed container and reused. Cover with polyethylene film and subject the chromatogram to autoradiography on X-ray film. Develop the film and mark the positions of the spots corresponding to the substrate and product on the chromatogram. Excise these regions of the chromatogram, and determine their radioactivity by Cerenkov or liquid scintillation counting. For each time point, the extent of reaction is determined using the following equation:

$$\text{Product (\%)} = \frac{\text{radioactivity in product}}{\text{radioactivity in product} + \text{radioactivity in substrate}}$$

The initial rate of each reaction is determined by plotting the product formation, expressed as moles, as a function of time and determining the slope of the straight line by regression analysis.

Procedure 8. Gel Electrophoresis for Separation of Substrates and Products

If the oligonucleotides are too long for the products to be separated from the substrate by homochromatography, electrophoresis in polyacrylamide gels as described in procedure 2 is used. Dry each aliquot, dissolve it in gel-loading buffer, and load the sample onto a 20% polyacrylamide gel containing 7 M urea.[5,28] The electrophoresis is performed as described above, and the gel dried and subjected to autoradiography. Excise the product and substrate bands, quantify their radioactivities, and determine the extent of the reaction as described in procedure 7.

Procedure 9. Analysis of Data and Calculation of Kinetic Constants

The kinetic assays should be performed so that their initial velocities are directly proportional to the times of reaction and the concentrations of enzyme used. After measuring the initial rates of reaction for various concentrations of modified and unmodified substrates, the kinetic constants may be calculated using one of the standard linearization equations (Lineweaver–Burk[7] or Eadie–Hofstee[8]); or, preferably, they are determined by computer-assisted direct fitting to the Michaelis–Menten equation. Our laboratory has employed a Fortran version of the kinetics program[51] of W. W. Cleland (courtesy of B. V. Plapp, University of Iowa) to calculate k_{cat} and K_m values from the kinetic data. The computer-assisted analysis has the advantage of providing an estimate of the precision of the measurements. In practice, the data obtained using the procedures given here require several measurements of the reaction rate at each concentration of substrate to give reliable values.

[51] W. W. Cleland, this series, Vol. 63, p. 103.

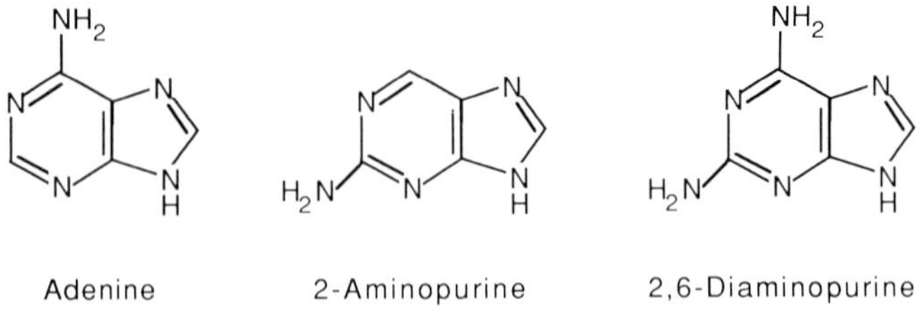

FIG. 1. Adenine and two analogs used in studies of the EcoRI and RsrI endonucleases. In 2-aminopurine, the amino group of adenine, which is found in the major groove of B-form DNA, has been removed and placed at C-2, where it lies in the minor groove. The analog 2,6-diaminopurine contains amino groups in both grooves.

Examples with EcoRI and RsrI Endonucleases

RsrI and EcoRI endonucleases catalyze the identical reaction of cleaving the duplex sequence d(GAATTC) at the phosphodiester bonds between the G and A residues.[52–54] Both enzymes have been purified to homogeneity and their genes sequenced.[41,53–57] The gene sequences reveal that the proteins have >50% amino acid identity, and that the amino acids postulated to be involved in DNA recognition and catalysis as determined by crystallographic and mutational analysis of EcoRI are present in corresponding positions in both amino acid sequences.[54,55] The enzymes have similar molecular weights and both function as homodimers.[41,53] They display similar turnover numbers, Michaelis constants, and specificity constants when assayed on plasmid DNA substrates. Both enzymes cleave DNA at noncanonical sites under appropriate reaction conditions.[33,53] The two enzymes differ with respect to their isoelectric points, optimal reaction conditions, and sensitivity to N-ethylmaleimide.[53] Unlike EcoRI, RsrI endonuclease is essentially inactive at 37° and has reduced activity at NaCl concentrations above 10 mM.[53,54]

An experiment was performed with EcoRI endonuclease using the duplex octadeoxyribonucleotide d(GGAATTCC)[8] with 2-NH$_2$Pur or 2,6-(NH$_2$)$_2$Pur (2NH$_2$-Ade) (Fig. 1) at the second adenine position. The results

[52] J. Hedgepeth, H. M. Goodman, and H. W. Boyer, Proc. Natl. Acad. Sci. U.S.A. 69, 3448 (1972).

[53] C. Aiken and R. I. Gumport, Nucleic Acids Res. 16, 7901 (1988).

[54] P. J. Greene, B. T. Ballard, F. Stephenson, W. J. Kohr, H. Rodriguez, J. M. Rosenberg, and H. W. Boyer, Gene 68, 43 (1988).

[55] F. H. Stephenson, B. T. Ballard, H. W. Boyer, J. M. Rosenberg, and P. J. Greene, Gene 85, 1 (1989).

[56] P. J. Greene, M. Gupta, H. W. Boyer, W. E. Brown, and J. M. Rosenberg, J. Biol. Chem. 256, 2143 (1981).

[57] A. K. Newman, R. A. Rubin, S.-H. Kim, and P. Modrich, J. Biol. Chem. 256, 2131 (1981).

of this experiment indicate the importance of having ancillary information to aid in interpreting results. The experiment was performed to determine if there was a contact between the enzyme and the C-6 amino group of the central Ade base. The analog 2-NH$_2$Pur lacks this amino group in the major groove of B-form DNA but has an added amino group in the minor groove that forms a hydrogen bond with the paired Thy base in the opposite strand. With this analog, the amino group on the purine has been switched from the major to the minor groove of the DNA. When measured under steady state conditions, the specificity constant, k_{cat}/K_m, of the oligomer containing the 2-NH$_2$Pur base was decreased 11-fold with respect to the unmodified octamer.[8] This result suggested that the amino group in the major groove of the DNA made an important contact with the enzyme.

The analog 2,6-(NH$_2$)$_2$Pur was used as a control to test for the effect of the addition of an amino group to the minor groove of the DNA. When this analog, containing 2,6-(NH$_2$)$_2$Pur, was tested, the specificity constant remained depressed ninefold.[8] The result with this substrate indicated that the major effect caused by the analog substitutions was due to the insertion of the amino group into the minor groove and not to its deletion from the major groove. In the absence of any other information, one might have concluded that the addition of the amino group in the minor groove interfered with some interaction with the enzyme in that groove of the DNA. Fortunately, an X-ray structural of the cocrystal of the enzyme and oligonucleotide containing the recognition site was available[10] and showed that protein was absent from the minor groove. The structure also revealed that the DNA in the complex was not in a normal B form, i.e., it was kinked with the minor groove compressed. Thus, we were able to conclude that the effect of the addition of the amino group to the minor groove was probably due to its interference with the kinking of the DNA observed when it is bound to the enzyme.[10,58] A subsequent experiment, in which the C-6 amino group was simply deleted by substitution of Pur for Ade at the central position of the sequence, indicated that the removal of the amino group decreased the specificity constant twofold.[7] Taken together, these results suggest that although there is likely a contact between the amino group and the enzyme, its loss is not as disruptive as is the introduction of an amino group into the minor groove. Thus, these experiments were more revealing of the role of DNA conformational factors than of amino acid–base interactions. In general it has been observed that the removal of a functional group from a base is less disruptive of the interaction between the enzyme and DNA than is addition of a new group.[7,8]

[58] Y. Kim, J. C. Grable, R. Love, P. J. Greene, and J. M. Rosenberg, *Science* **249**, 1307 (1990).

Another example from studies with *Eco*RI illustrates the value of experiments with analogs in the absence of confirmatory structural information. Removal of the methyl groups from either of the Thy bases in the GAATTC recognition sequence yielded decreased specificity constants with *Eco*RI, with the effect being larger when the second Thy residue was converted to Ura.[7,8] The activities of the substrates were restored on the substitution of BrUra for the Ura.[8] These results suggested that the methyl groups of these two pyrimidines were involved in interactions with the enzyme, although the structure of the enzyme–DNA complex failed to reveal electron density in this region and recognition was assigned to interactions solely with the purines.[10] Genetic studies[59] have implicated, and the determination of the *Eco*RI–DNA structure in several new crystal forms[58] now accommodate, a role for interactions of the enzyme with the pyrimidines.

The analog 2-NH$_2$Pur has also been useful in a comparative study of the *Rsr*I and *Eco*RI endonucleases.[26] Because the two enzymes catalyze the identical reaction and have a high degree of amino acid identity,[55] we anticipated that they would interact with their duplex target sequence, d(GAATTC), similarly. They appear to do so in the complex that is formed during site-specific binding in the absence of Mg^{2+}. If the enzyme is bound site specifically to covalently closed circular DNA and the complex relaxed with a type I topoisomerase, both enzymes induce the same unwinding of the helix ($\sim$25°) as determined by the change in the linking number of the DNA.[26a,60] Similarly, when both enzymes are tested for site-specific binding to DNA fragments in which the target site is located at different distances from the ends of the DNA, they bend the DNA to the same extent ($\sim$50°) as determined by gel mobility shift assays.[26a,61] Finally, the site-specific and nonspecific equilibrium thermodynamic binding constants for the two enzymes are similar and they thus show nearly identical ratios of specific to nonspecific binding ($\sim$10^5) despite the different solvent and temperatures at which the values were determined.[26a] All these results suggest similar modes of interaction between the enzymes and the target sequence.

However, when experiments with analog-containing substrates were performed in order to study the cleavage reactions, in contrast to the binding events, substantial differences between the enzymes were observed. Decameric oligodeoxyribonucleotides containing 2-NH$_2$Pur at the central Ade position of the recognition sequence (GAATTC) were assayed

[59] J. Heitman and P. Model, *Proteins* **2**, 185 (1990).
[60] R. Kim, P. Modrich, and S.-H. Kim, *Nucleic Acids Res.* **12**, 7285 (1984).
[61] J. Thompson and A. Landy, *Nucleic Acids Res.* **16**, 9687 (1989).

with both enzymes. *Eco*RI cut the substrate with a k_{cat}/K_m ratio ninefold lower than that with the unmodified control as a result of a relatively unchanged k_{cat} value and an increased K_m value.[7,8,26] In contrast, *Rsr*I failed to cut the analog-containing sequence, and the analog was not an inhibitor of the cleavage of the unmodified sequence, suggesting that the analog did not bind to the enzyme.[26] Thus, *Eco*RI binds and cleaves the substrate containing 2-NH$_2$Pur whereas *Rsr*I fails even to bind the compound.

In order to insert an amino group into the minor groove, 2,6-(NH$_2$)$_2$Pur was substituted at this position for Ade, and the two enzymes compared. *Eco*RI cleaved the substrate well, whereas *Rsr*I hydrolyzed it at a rate just above the level of detection.[26] For *Rsr*I, the analog containing 2,6-(NH$_2$)$_2$Pur was a strong competitive inhibitor ($K_I \sim 10^{-9} M$) of the cleavage of the unmodified sequence.[26] Thus, in spite of the high degree of amino acid identity and the similar interactions with the target sequence in the absence of Mg^{2+}, the two enzymes react with the same substrate analogs differently, i.e., they exhibit selectivity or recognition in the chemistry of the cleavage and not just in the binding. Because this difference in reactivity occurs under conditions *in vitro* in which each enzyme exhibits high-specificity cleavage, it is likely that it is due to a property of the enzymes themselves rather than to the reaction conditions.

Studies using substrate analogs and restriction endonucleases have provided information about interactions between the enzymes and their DNA substrates and, additionally, between the enzymes and transition states of the reactants. The ability to monitor interactions between the enzyme and DNA at stages of the hydrolysis reaction beyond the initial ground-state, enzyme–substrate complex is a valuable attribute of this approach. Such studies have directly indicated that recognition may occur during the chemistry of the reaction as well as at the level of substrate binding. Studies with substrate analogs should continue to serve as useful partners with genetic and structural approaches in our attempts to understand the molecular nature of sequence-specific protein–DNA interactions.

Acknowledgments

We thank the many colleagues who supplied us with their results prior to publication, and ask their forbearance if we failed to cite them. We are grateful to N. J. Leonard and R. L. Switzer for reading the manuscript, and offering comments. The experiments reported from our laboratory were supported in part by a grant (NIH GM 25621) from the National Institutes of Health.

[22] Probing Information Content of DNA-Binding Sites

By Gary D. Stormo

Introduction

Information content is a convenient way to describe the specificity of a DNA-binding protein.[1] In its simplest form information content is equivalent to the common description of a restriction enzyme as a "6-base cutter" or a "4-base cutter." That is, information content is a measure of the amount of specificity required for recognition, independent of the mechanism of recognition. Equivalently, it is a measure of the frequency of sites expected in a random sequence. We usually measure information content in *bits*, rather than bases. Since the choice of 1 base requires two bits, a "6-base cutter" has an information content of 12 bits. A restriction enzyme with multiple recognition sites, such as *Hinc*II, which cuts at GTYRAC (Y = C or T, R, = A or G),[2] recognizes 10 bits, or 5 bases, of information, even though its recognition site is 6 bases wide, because of the ambiguity in some of the positions.

The utility of information content as a measure of specificity is that it is applicable even to cases of high ambiguity, which is a typical characteristic of regulatory sites and which distinguishes them from restriction sites. Even though restriction sites can contain ambiguity they have essentially an all-or-none activity. That is, *Hinc*II cuts at every site that matches the sequence GTYRAC, and only such sites are cut. Regulatory sites, on the other hand, are usually similar to a "consensus sequence," but may differ at almost any position and at more than one position in any particular example. Furthermore, some positions within the consensus sequence may be more variable than others. The most studied example is *Escherichia coli* promoter sequences. They contain a two-part consensus sequence: a "−35" region of TTGACA and a "−10" region of TATAAT. In a compilation of over 200 promoters each of the 4 bases has been found at least once at each position of the consensus sequence, and none of the examples is a perfect match to it.[3] But promoters are not random, because any particular promoter will likely match at most of the consensus positions, and some of the positions are highly, although not absolutely, conserved.

[1] T. D. Schneider, G. D. Stormo, L. Gold, and A. Ehrenfeucht, *J. Mol. Biol.* **188,** 415 (1986).
[2] G. D. Stormo, *Annu. Rev. Biophy. Biophy. Chem.* **17,** 241 (1988).
[3] C. B. Harley and R. P. Reynolds, *Nucleic Acids Res.* **15,** 2343 (1987).

A

A	9	214	63	142	118	8
C	22	7	26	31	52	13
G	18	2	29	38	29	5
T	193	19	124	31	43	216

B

A	0.04	0.88	0.26	0.59	0.49	0.03
C	0.09	0.03	0.11	0.13	0.22	0.05
G	0.07	0.01	0.12	0.16	0.12	0.02
T	0.80	0.08	0.51	0.13	0.18	0.89

C

A	-2.76	1.82	0.06	1.23	0.96	-2.92
C	-1.46	-3.11	-1.22	-1.00	-0.22	-2.21
G	-1.76	-5.00	-1.06	-0.67	-1.06	-3.58
T	1.67	-1.66	1.04	-1.00	-0.49	1.84

D

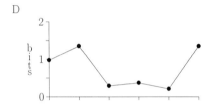

Fig. 1. Matrices from *E. coli* − 10 promoter sequences.[3] (A) The number of occurrences of each base at each position of the 242 nonmutant promoters. (B) The fraction of each base at each position, from the above data. (C) The logarithms (to the base 2) of those fractions divided by 0.25, the genomic frequency of each base in *E. coli*. (D) The information content[1] of each position. This is the dot product of the vectors in each column of the matrices in (B) and (C), as calculated in Eq. (1). (Reprinted with permission.[4])

The simplistic notion of specificity used above for restriction sites is not easily adapted to the highly variable sequences in regulatory sites, but information content is. Figure 1[4] illustrates the calculation of information content for the − 10 region of *E. coli* promoters. The basic formula for the information content at any position in a binding site is

$$I_{seq} = \sum_{b=A}^{T} f_b \log_2(f_b/p_b) \tag{1}$$

where f_b is the fraction of each base b in the aligned sites, and p_b is the genomic proportion of each base.[1,2,4] When the genomic proportions of

[4] G. D. Stormo, this series, Vol. 183, p. 211.

each base are equal ($p_b = 0.25$ for all b) the formula reduces to $I_{seq} = 2 + \Sigma f_b \log_2 f_b$, which can vary from 0, for a random position, to 2, for a position that is absolutely conserved.[1] The DNA-binding sites for several regulatory proteins have been examined.[1,5] The information content of each is very close to that expected based on the number of sites in the genome. For highly ambiguous sites information content determined assuming independent positions, as in Eq. (1), is a lower bound on the value that would be determined using higher order analyses.[6] Therefore, the result of near equality between expected and observed frequencies[1] is not inconsistent with the idea that regulatory sites should be "overspecified."[7,8] That result does allow one to estimate the frequency expected for a particular type of site, and to gauge whether or not the pattern obtained for the specificity of a particular protein is appropriate. For example, the large discrepancy we originally observed for the T7 promoters led to the experiments described below, from which we determined that there were additional constraints on phage promoters than just transcription initiation activity.[5]

Experimental Determination of Information Content

While the information content determined from several examples of the binding sites of a protein can be a valuable tool in understanding its specificity, often a more detailed analysis is needed. For example, the known binding sites may have other constraints that contribute to their information content, but which are not required for function by the protein. It is also likely that not enough natural sites are known to have a reliable picture of the specificity of the protein from those data alone.

A variety of experimental procedures can also be used to determine the information content of a protein. The basic strategy is to randomize the binding sites and to select from the entire population those that retain functional activity. The definition of a functional site may be qualitative; either a particular site works or it does not, or it may involve a quantitative assay that lets one use the relative affinity of the protein for different sites in the calculation of information content. Such experiments can either be carried out *in vitro* or by putting the randomized sites back into cells and determining function *in vivo*.

[5] T. D. Schneider and G. D. Stormo, *Nucleic Acids Res.* **17**, 659 (1989).
[6] G. Z. Hertz and G. D. Stormo, submitted.
[7] P. H. von Hippel, *in* "Biological Regulation and Development" (R. F. Goldberger, ed.), p. 279. Plenum, New York, 1979.
[8] O. G. Berg and P. H. von Hippel, *J. Mol. Biol.* **193**, 723 (1987).

In Vivo Experiments

In our early studies of information content in regulatory sites the T7 promoters were an exception to the observation that information content is approximately equal to that expected from the frequency of sites.[1] In fact, T7 promoters had about twice the expected information content. We hypothesized then that perhaps only part of the information was required for promoter function, polymerase binding and transcription initiation, and that the remaining information was conserved for other, unknown, reasons. In order to test this hypothesis we synthesized randomized variants of a 27-bp region that contained the promoter consensus sequence.[5] At each position of this synthesized promoter we used a mixture of nucleotides designed to give the consensus base 85% of the time and each of the other bases 5% of the time. In so doing we expected to get the complete consensus sequence about 1% of the time, assuring us that we would have a positive control of promoters that had to function. We also expected that there would be an average of about four changes per promoter. These randomized promoter sequences were then cloned into a plasmid that would serve to screen for functional promoters in cells that could be induced to make T7 RNA polymerase. The screen was that functional promoters killed the cells when induced, whereas nonfunctional promoters did not. Comparing the sequences that were recovered by the screen to analyses done by Chapman and Burgess,[9] we concluded that promoters had to be within two- or threefold the activity of the consensus promoter to pass the screen.

Two hundred clones were screened and 58 (29%) of them were found to be functional. Five had multiple inserts, leaving 53 (including 3 with the consensus sequence) that were functional promoters. Figure 2 shows the composition by position of the functional promoters and of the T7 phage promoters for comparison. In determining the information content from these data two factors are important. One is that the population proportions of the bases, p_b, are not equimolar, but rather they have been biased to be 85% the consensus base, which varies from position to position. When these numbers are used in Eq. (1), the information content comes out at about 2.9 bits, which is not too far removed from the 1.8 bits expected from the 29% functional promoters. But the other important consideration to remember is that this is the information relative to the biased synthetic promoters. What we really want to know is the amount of information relative to a T7 infection, in which case the genomic proportions of the bases are approximately equal. We could have obtained that

[9] K. A. Chapman and R. R. Burgess, *Nucleic Acids Res.* **15,** 5413 (1987).

L	φ10	Experiment T7-W				Phage Promoters			
		A	C	G	T	A	C	G	T
-21	a	43	5	4	1	8	3	4	2
-20	a	49	2	1	1	12	0	2	3
-19	a	41	0	12	0	10	1	3	3
-18	t	3	2	2	46	3	1	3	10
-17	t	3	1	1	48	0	3	0	14
-16	a	48	0	3	2	16	0	0	1
-15	a	51	1	0	1	17	0	0	0
-14	t	2	0	0	51	0	0	0	17
-13	a	47	1	1	4	14	1	0	2
-12	c	0	53	0	0	0	16	1	0
-11	g	7	0	46	0	2	0	15	0
-10	a	47	2	1	3	17	0	0	0
-9	c	0	53	0	0	0	17	0	0
-8	t	0	0	0	53	0	0	0	17
-7	c	0	53	0	0	0	17	0	0
-6	a	48	1	2	2	17	0	0	0
-5	c	0	53	0	0	0	16	1	0
-4	t	1	3	1	48	0	0	0	17
-3	a	42	4	4	3	17	0	0	0
-2	t	2	0	0	51	5	0	0	12
-1	a	42	4	5	2	16	0	0	1
0	g	2	0	50	1	2	0	15	0
1	g	1	1	50	1	2	0	15	0
2	g	2	4	46	1	5	0	12	0
3	a	45	4	2	1	10	0	7	0
4	g	3	4	42	4	5	1	11	0
5	a	47	3	1	2	13	2	0	2

FIG. 2. Tabulated results for strong T7 promoters. The numbers of each kind of base at each position (L) are given for both the functional promoters recovered by the screen and for sequences at promoters in wild-type T7 phage. The column labeled φ10 refers to a particular T7 promoter, the sequence of which is identical to the phage consensus. A symmetry element is indicated by boxed letters. (Reprinted with permission.[5])

number by completely randomizing the promoters instead of leaving a bias toward the consensus sequence. However, since we had only a screen, not a selection, for activity, that would have required screening many thousands of clones to get a few that were functional. Instead, we can use the data we collected and "normalize" it to see what we would have obtained had we done the complete randomization. The assumption is that the data obtained accurately reflect the selectivity of the polymerase. We assume that the ratios of f_b/p_b are equal to the ratios of the activities for each variant base.[2,4,5,8] That is equivalent to assuming that the data reflect the differences in binding energies to each possible base, and that those energy differences are independent of the base proportions in the population of possible binding sites.

The data we have are of f_b and p_b values. The normalization procedure is to find the f_b' values for new values of p_b'.[5] The constraints that must be satisfied are that the sums of f_b and p_b must equal one, and that the ratios of all the f_b/p_b values must be unchanged. For the case of $p_b' = 0.25$ this is straightforward, because then $\sum f_b'/p_b' = 4$. The fraction of each base calculated to be in binding sites if the nonsite bases are equimolar is

$$f_b^s = (f_b/p_b)/(\sum f_b/p_b) \tag{2}$$

We consider this situation, of equimolar nonsite bases, to be a "standard state," and the binding constant determined at this state, $K_b^s = 4f_b^s$, to be the specific binding constant. It can vary between 0 and 4, with a value of 1 for nonspecific binding. The formula for information content then becomes

$$I_{seq} = \sum_{b=A}^{T} f_b^s \log_2 K_b^s \tag{3}$$

which is easily seen to be proportional to the average specific binding energy.[2,4,13]

When our randomized T7 promoter data are treated in this manner, the resulting information content at each position is as shown in Fig. 3. Also shown is the information content as calculated from the phage promoters.[1,5] The most important result is that much of the information within the phage promoters is not required for promoter activity. In fact, only about half of the total conserved information is required, placing T7 promoters into the same class as other regulatory sites where the information content observed is about as expected from the frequency of sites.[1] Careful examination of the data in Figs. 2 and 3 reveals several interesting features. First, there are some positions, such as -7, -8, and -9, that are absolutely conserved in both the phage and the functional promoters. There are several other positions, such as -3, -4, -6, and -10, that are absolutely conserved in the phage promoters, but at which any base can exist in a functional promoter. This finding alone confirms our hypothesis that there are constraints at T7 promoters other than those required by the polymerase for transcription activity. At several other positions there is information in the functional promoters, but less than in the phage promoters. The results at positions -5 and -12, where it looks like functional promoters contain more information than phage promoters, are not significant, but due to the small sample size. If we had seen even one nonconsensus base at those positions the information content would

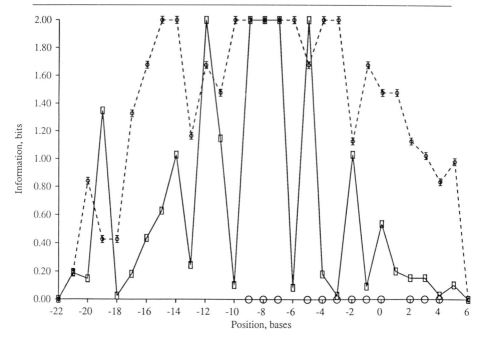

FIG. 3. Information at T7 promoters. The abscissa is the position along the sequence, in bases, while the ordinate shows the amount of information at each position in bits. (Φ——Φ), Information from the phage promoters; (□——□), normalized information in strong T7 promoters recovered by the screen. (O), Symmetry elements on the abscissa. (Reprinted with permission.[5])

have been lower than the phage value. That is, given the bias in the randomized promoters, the phage data lead us to expect between zero and one nonconsensus base in a sample size of 53, and the observation of zero cannot be considered significantly different than that expectation. At position − 19 the result is significant and indicates that a G at that position may lead to more promoter activity than the consensus A, which must be conserved for other reasons. Overall the information content analysis of *in vivo* functioning T7 promoters gives us a very different picture of the sequence requirements than is obtained from the phage examples.

If we had quantitative measurements of how strong each promoter is we could have incorporated those data into the analysis. For example, several years ago we reanalyzed data about the context effects on nonsense suppression in *E. coli*.[10] The efficiency of amber suppression by *su2* had been determined for many different contexts, and we analyzed the data

[10] G. D. Stormo, T. D. Schneider, and L. Gold, *Nucleic Acids Res.* **14,** 6661 (1986).

by multiple regression using several alternative models. As shown in Fig. 4, the simple model that the context effect is determined by the 6 bases surrounding the amber codon, three on each side, and that the effects of each base are independent and additive does quite well in describing the data. Figure 4b shows the fit by this model between the predicted and observed values, which has a correlation coefficient of about 0.93. Figure 4a shows the values of relative ln K_b that were determined by the regression, in which we had arbitrarily set $K_T = 1$ at each position. (There are only three independent parameters per position, so we set one and determine the others relative to that one.)[10] We can normalize those values to a standard activity constant, as described above, and then calculate the information content for suppression contexts, as in Eq. (3). The result is a total of 0.70 bits, with 0.58 coming from the two positions 3' to the amber codon, positions 3 and 4. While it was clear in our original paper that these were the two most important positions for determining suppression efficiency, this information content analysis lets us quantitate the relative importance of each position.

In Vitro Experiments

One can also determine information content from *in vitro* binding experiments. In fact, these make it easy to use quantitative assays of binding affinity in the determination. One method is to perform extensive binding analysis on mutant binding sites, as has been done for the phage λ proteins Cro and Repressor.[11,12] In these experiments every single base change was made to the consensus λ operator and the binding constant was determined for each protein to each variant operator. The results were presented as $\Delta\Delta G$ values, which can easily be converted to specific binding constants, as described above. The information content determined by these calculations is about 15.5 and 14 bits for Cro and Repressor, respectively. These numbers are low compared to that determined from the known binding site sequences or from the frequency of known operators.[1] These calculations, based on quantitative measurements of independent changes, tend to underestimate the complete specificity of the proteins.[6] Nonetheless, these values are consistent with many experiments of determining the affinity to sites with multiple changes, and serve to constrain the models of how the proteins obtain their required specificity.[11,12]

Information content can also be determined from experiments in which the protein is bound to many different sites at once, providing there is a

[11] Y. Takeda, A. Sarai, and V. M. Rivera, *Proc. Natl. Acad. Sci. U.S.A.* **86**, 439 (1989).
[12] A. Sarai and Y. Takeda, *Proc. Natl. Acad. Sci. U.S.A.* **86**, 6513 (1989).

a

position:		-3	-2	-1	O	+1	+2	+3	+4	+5
	A	+0.02	-0.20	+0.30	a			+1.72	-1.15	-0.22
	C	+0.35	+0.28	+0.16				+0.54	-0.98	-0.45
base	G	-0.03	+0.07	-0.53			g	+1.78	-1.55	-0.30
	T	0.00	0.00	0.00	t			0.00	0.00	0.00

+ 1.85   =   ln(suppression activity)

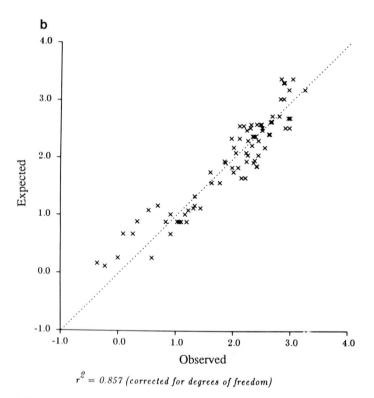

$r^2 = 0.857$ *(corrected for degrees of freedom)*

FIG. 4. Regression results for six mononucleotides vs suppression activity. (a) Matrix of relative ln(suppression activity) constants. (b) Plot of the observed ln(activities) vs those predicted from the matrix of (a). (Reprinted with permission.[10])

means of determining the partitioning of each site between bound and unbound fractions. We have been developing such a method for the Mnt repressor of phage P22.[13] Mnt is a repressor that binds as a tetramer to a symmetric operator sequence to regulate expression from the P_{ant} promoter.[14] We synthesized an operator with one randomized position, which could be cut with a different restriction enzyme for each different base in the variable position: . . . AGGTCCACGGTGGA*n*CTagt . . . The uppercase letters are the wild-type operator sequence, and the *n* is the randomized position. Depending on whether the varible base is an A, C, G, or T, the oligonucleotide can be cut by *Spe*I, *Sau*96I, *Alu*I, or *Mbo*I, respectively. Mnt repressor was added to the randomized synthetic operator in sufficient quantity to bind approximately half of the DNA, and the bound and unbound fractions were separated in a gel mobility shift assay.[15] Both bands were eluted from the gel and the proportion of each base in each fraction was determined by cutting with the restriction enzymes. The results of this experiment were that the specific binding constants, K_b^s, were 0.71, 2.36, 0.28, and 0.64 for A, C, G, and T, respectively. The information content for this position is 0.41 bits.

The partitioning of each site into the two fractions could be determined by quantitative sequencing of each fraction. Then an equivalent experiment could be done at any position, regardless of whether an appropriate set of restriction enzymes could be found. It would also be possible to randomize several positions at once and determine the information content of each in a single experiment. Then only a few binding experiments might be sufficient to determine the complete specificity of the protein. So as to minimize the effects of nonindependent binding interactions, such an experiment should probably be done with nonadjacent positions randomized in a particular experiment.

Summary

An information content analysis of protein-binding sites gives a quantitative description of the specificity of the protein, independent of the mechanism of specificity. It gives useful information about the total specificity of the protein and about the individual positions within the binding sites. Information content is consistent with both thermodynamic and statistical analyses of specificity.[2,4] When applied to a collection of known binding sites, the description provided may be limited by the sample size

[13] G. D. Stormo and M. Yoshioka, *Proc. Natl. Acad. Sci.* **88,** 5699 (1991).
[14] A. K. Vershon, S.-M. Liao, W. R. McClure, and R. T. Sauer, *J. Mol. Biol.* **195,** 311 (1987).
[15] K. L. Knight and R. T. Sauer, *Proc. Natl. Acad. Sci. U.S.A.* **86,** 797 (1989).

or by unknown constraints on those sites. Experimental procedures to determine the information content can give much more reliable measures. A large number of functional sites can be obtained from a much larger pool of randomized potential sites. Quantitative assays for the activity of different sites can be easily incorporated into the analysis, thereby increasing its sensitivity. Both *in vitro* and *in vivo* experiments are amenable to information content analysis.

Acknowledgments

The experiments on the Mnt-binding specificity were performed using protein kindly provided by Ken Knight and Robert Sauer. The work described in this chapter has been supported by NIH Grant GM28755.

[23] Protein Chemical Modification as Probe of Structure–Function Relationships

By KATHLEEN S. MATTHEWS, ARTEMIS E. CHAKERIAN, and JOSEPH A. GARDNER

Chemical modification is a useful tool for examining the participation of specific amino acid side chains in the structure and function of proteins. In addition, chemical modification provides the potential for cross-linking and, in the case of enzymes, for design of reagents that are activated by the enzymatic mechanism. A number of reviews and books are available for consultation in designing an experimental approach for chemical modification of DNA-binding proteins.[1-7] Figure 1 provides a flow chart for developing an experimental design to modify a protein. In the ideal situation, a specific reagent will react with only one type of side chain; in

[1] R. L. Lundblad and C. M. Noyes, "Chemical Reagents for Protein Modification, Volumes I and II." CRC Press, Boca Raton, Florida, 1984.
[2] C. H. W. Hirs, ed., this series, Vol. 11.
[3] C. H. W. Hirs and S. N. Timasheff, eds., this series, Vol. 25.
[4] C. H. W. Hirs and S. N. Timasheff, eds., this series, Vol. 47.
[5] A. N. Glazer, R. J. DeLange, and D. S. Sigman, "Chemical Modification of Proteins." American Elsevier, New York, 1975.
[6] G. E. Means and R. E. Feeney, "Chemical Modification of Proteins." Holden-Day, San Francisco, California, 1971.
[7] R. E. Feeney, *Int. J. Pept. Protein Res.* **29**, 145 (1987).

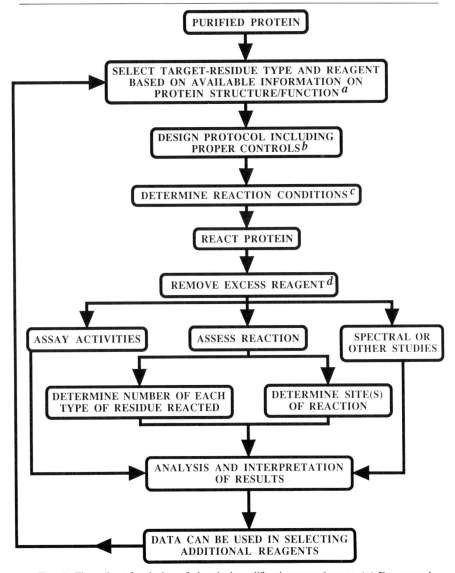

FIG. 1. Flow chart for design of chemical modification experiments. (a) For example, DNA-binding proteins may have lysine or arginine residues in or near the binding site. (b) Controls may include no reagent, preblocked reagent, with/without ligands, etc. (c) Reaction conditions to be considered include buffer, ionic strength, pH, temperature, reaction time, reagent concentration (ratio or absolute), presence or absence of sulfhydryl reagents, metal ions, reaction stop procedure, etc. (d) Excess reagent may be removed by dialysis, chromatography, etc.

reality, however, multiple residues are frequently affected, and the effects of modification of each type must be assessed carefully to interpret the results obtained. By adjusting the reaction conditions (e.g., molar ratio of reagent to protomer, pH, salt concentration, temperature, and time of reaction), it may be possible to elicit greater selectivity for residue type. It should be noted that in some cases it is the ratio between reagent and protein that determines the extent of modification, while in others, the absolute concentration of the reagent is the determining factor. Thus, care in selecting reaction conditions as well as reagent and protein concentrations is essential. Once modification has been achieved, identification of the sites of reaction and assessment of the effects on the structural and functional properties of the protein must be accomplished.

Determination of the effects of modification on activity is specific for each protein under consideration; all the activities must be evaluated (DNA binding, RNA binding, modulator ligand binding, enzymatic activity, oligomer formation, etc.). Where possible, kinetic as well as equilibrium parameters should be measured. Modification of specific amino acid residues can affect the conformation of the protein and thereby exert an indirect effect on binding or enzymatic properties; reaction can also alter residues that are directly involved in these functions. Distinguishing conformational from direct effects can be difficult; detailed analyses of the reaction pattern and functional properties under a variety of experimental conditions may provide sufficient information to make the distinction. In addition, other methods to determine conformational effects (e.g., circular dichroism spectra, proteolytic susceptibility, and NMR studies) can be used to detect alterations in structure consequent to modification. The ability of ligands/substrates to protect residues against modification is frequently a useful method to confirm the participation of specific side chains in the binding of these molecules. However, it is essential to confirm that the ligands (particularly DNA and RNA) do not themselves react with the reagent and do not interfere in nonpassive ways with protein reaction.

The purity of the protein is a consideration in designing chemical modification experiments, since the relative reactivity of different proteins varies, and a contaminant may alter the concentration of reagent. Selection of control conditions is exceptionally important in order to have a standard for comparison; in some cases, exposure to reaction conditions alone can cause alterations in functional capacity of the protein. Thus, it is essential to include protein processed identically, but without reagent, for comparison to the corresponding sample actually exposed to reagent. In some cases, a "stop" compound is added to prevent further reaction (e.g., dithiothreitol to sulfhydryl reagents, lysine to reagents specific for the ε-amino group); where this type of protocol is followed, the control sample includes the "stop" compound added to modifying reagent prior to exposure to the protein. Thoughtful design of the experimental approach and inclusion of multiple controls can allow unambiguous interpretation of results and prevent unnecessary experimental repetition.

Analysis of the modified protein to determine all residues affected by the reaction is essential. In some cases, assessment can be made by spectrophotometric or fluorometric methods, while in others a secondary reaction is required. These will be detailed below for each of the residue types and different reagents. Extent of modification of a residue type provides useful information, but identification of the specific amino acid(s) modified is of greatest utility in interpreting the effects of a reaction on the structure and the functional capacity of a protein. This analysis can be complex and sometimes requires careful experimental design; a few specific examples will be provided in the following sections. The microtechniques now available for amino acid analysis, peptide separation, and peptide sequencing diminish significantly the quantities necessary for this type of study and make the task of determining the site(s) of reaction within the structure more feasible on proteins that are available only in small amounts.

Selectivity of a reagent is a major consideration, both with respect to residue type and to specific reaction at given sites within the protein. Reaction is determined by a combination of the inherent reactivity of the side chain, the influence of the surrounding protein environment on reaction (inhibitory or facilitative), and the conditions selected for the reaction. An excellent example of increased reactivity and specificity is Lys-7 reaction in ribonuclease with reagents that have an acidic moiety that mimicks the phosphate on RNA.[8] The selectivity of a particular modification can be increased in some cases by conducting the reaction at low molar ratios of reagent to protomer or by limited time exposure of the protein to the reagent. In most cases, experimental assessment is required to determine the optimal reaction conditions for a specific protein with a given reagent. However, some guidelines and examples of previous modification reactions with DNA-binding proteins will be provided below for each reagent.

Two methods that have been widely used to determine the number of a specific side-chain type that are essential for protein function are based on the kinetics of the reaction[9] and on statistical analysis.[10,11] For the

[8] G. E. Means and R. E. Feeney, *J. Biol. Chem.* **246**, 5532 (1971).

[9] D. E. Koshland, Jr., W. J. Ray, Jr., and M. J. Erwin, *Fed. Proc.* **17**, 1145 (1958).

[10] C.-L. Tsou, *Sci. Sin., Ser. B (Engl. Ed.)* **11**, 1535 (1962).

[11] R. B. Yamasaki, A. Vega, and R. E. Feeney, *Anal. Biochem.* **109**, 32 (1980).

kinetic method, the pseudo-first order rate constant for the reaction of a given side-chain type and the associated loss of functional activity are compared to yield data on the number of essential residues. In the statistical approach (or graphical method), the number of essential residues is determined from measurement of the number of residues modified and the remaining functional capacity in samples varying in total extent of modification. Although useful in estimating the number of essential residues, these methods assume reactivity of only one residue type and are compromised by the problem of whether previous modification affects subsequent reactivity. In general, utilization of multiple reagents for a given side chain combined with careful analysis of the protein at low levels of modification provide more interpretable data.

Classification of reagents with respect to the primary amino acid residue modified provides a means to select specific side chains and reagents for a particular study. Based on the reagent, a set of conditions can be identified for initiating an experimental program to determine the effects of modification on structure and function of a DNA-binding protein (Fig. 1 and Tables I–III). It should be noted that selection of both side chain and reagent is important, and previous efforts using specific reagents may be helpful in making this choice. Useful reference material can be found in Refs. 1–7 for selecting reagent(s) as well as the initial set of conditions to be explored and precautions to be exercised. For example, some reagents and products are photolabile or thermolabile or unstable in water; thus, familiarity with the requirements specific to working with a given reagent will prevent expensive and potentially time-consuming errors in experimental design. The sections that follow describe reagents for modification of sulfhydryl residues (cysteine), amino groups (lysine), guanidino groups (arginine), phenols (tyrosine), indoles (tryptophan), and imidazoles (histidine). By far, the most specific reactions can be obtained with sulfhydryl residues, and these will be discussed first and in the most detail.

Cysteine Residues

Cysteine side chains are generally the most reactive in proteins, and a wide variety of reagents are available with reactivities that are influenced significantly by the relative surface exposure and surrounding environment of the residues. For modification of cysteine residues, it is essential that all potentially reactive agents (e.g., dithiothreitol and 2-mercaptoethanol) be removed from the sample and that the cysteine residues be protected from oxidation. The most straightforward way to achieve this is dialysis or gel filtration into the buffer to be utilized for modification. The buffer must be purged of oxygen to prevent sulfhydryl oxidation using either an inert gas or nitrogen (10–20 min of bubbling gas through the buffer solution from the bottom using a gas dispersion tube is normally sufficient). In evaluating the extent of reaction and effects on functional activities, it is necessary to handle the modified protein in such a manner as to ensure that the product is not exposed to conditions that might reverse the reaction.

There are several important categories of cysteine reagents listed in Table I with examples of application to DNA-binding proteins and an indication of the conditions used in these experiments. A brief summary of the different reactions and more information on specific examples follow.

α-Ketoalkyl Halides

α-Ketoalkyl halides (e.g., iodoacetic acid and iodoacetamide) have been used to modify cysteines in a variety of proteins, and derivatives of these reagents provide an even greater range of modification. T7 RNA polymerase modification with iodoacetamide with incorporation of ~1.2 mol/mol protein led to inactivation of the protein.[12] The reagent excess utilized was 10- to 20-fold over protein at pH 7.7 in 20 mM sodium phosphate, 100 mM NaCl, 1 mM EDTA or 100 mM NH$_4$HCO$_3$, 1 mM EDTA. The reaction was stopped by addition of 2-mercaptoethanol, and peptide analysis indicated the sites of reaction. Although the highest reactivity is with cysteine residues and careful control of conditions can ensure selective reaction, these reagents will also react with histidine and lysine moieties, particularly at higher reagent concentrations. Reaction occurs primarily with exposed residues for the parent compounds, but less available sulfhydryls can be modified with derivatives substituted with apolar moieties. The apolar moieties apparently facilitate reaction at hydrophobic sites that are removed from the surface of the protein. For example, in the native tetrameric lactose repressor protein, none of the three cysteines in each monomer is reactive to iodoacetic acid or iodoacetamide, but at least two of these side chains can be modified with 2-bromoacetamido-4-nitrophenol (BNP) and one with at least partial selectivity with N-[(iodoacetyl)aminoethyl]-5-naphthylamine 1-sulfonate (IAEDANS).[13–15]

General reaction conditions include protein at approximately 1–10 μM with reagent in excess at a 10- to 200-fold molar ratio. High pH can facilitate the reaction, and a usual buffer would be Tris-HCl with the pH adjusted to ~8–9. The reaction can be stopped with excess sulfhydryl-

[12] G. C. King, C. T. Martin, T. T. Pham, and J. E. Coleman, *Biochemistry* **25**, 36 (1986).
[13] C. F. Sams, B. E. Friedman, A. A. Burgum, D. S. Yang, and K. S. Matthews, *J. Biol. Chem.* **252**, 3153 (1977).
[14] J. M. Schneider, C. I. Barrett, and S. S. York, *Biochemistry* **23**, 2221 (1984).
[15] J. A. Gardner and K. S. Matthews, *Biochemistry* **30**, 2707 (1991).

TABLE I
REACTION OF CYSTEINE RESIDUES

Reagent	Protein	[Protein]	[Reagent]	Buffer	Additives	Ref.
α-Ketoalkyl halides						
Iodoacetamide	T7 RNA polymerase	1–10 μM	10–20 M excess	20 mM sodium phosphate, pH 7.7	100 mM NaCl/1 mM EDTA	a
				100 mM NH$_4$HCO$_3$, pH 7.7	1 mM EDTA	a
BNP	lac repressor	0.2–0.7 mg/ml	10–200 M excess	0.1 M Tris-HCl, pH 9.0	1.0 M NaCl	b
IAEDANS	lac repressor	1–3 mg/ml	~8 mM	0.2 M Tris-HCl, pH 8.0–8.6	0.5 M KCl/0.1 mM EDTA	c, d
Mercurials						
p-CMB	Yeast RNA polymerase I	0.04–0.06 mg/ml	1–20 μM	50 mM Tris-HCl, pH 8	0.1 M KCl/0.5 mM EDTA/10% glycerol	e
CMNP	lac repressor	0.1–0.5 mg/ml	1–5 M excess	1.0 M Tris-HCl, pH 9.0		f
FMA	lac repressor	0.5 mg/ml	0–5 M excess	0.2 M Tris-HCl, pH 8.0		g
Disulfide exchangers						
DTNB	RecA	6–20 μM	0.5–1 mM	0.01 M potassium phosphate, pH variable	1.0 mM EDTA/0.5 M KCl	h, i
	CAP	0.2 mg/ml	5–40 μM	0.1 M sodium phosphate, pH 8.0	0.1 M NaCl	j
	Histone H3	50 μM	78 μM	2.4 mM sodium phosphate, pH 5.6	Varied	k
	Glucocorticoid receptor	Cytosol extract	0.25 mM	50 mM HEPES, pH 7.9–8.0		l
MMTS	Glucocorticoid receptor	Cytosol extract	0.25 mM	50 mM HEPES, pH 7.9–8.0		l
DFDS	lac repressor	12.5 μM	1–100 M excess	0.1 M Tris-HCl, pH 7.6		m
	Histone H3	~33 mg/ml	~3 mg/ml	10 mM sodium phosphate, pH 7.5		n
Maleimides						
N-Ethylmaleimide	lac repressor	20–50 μM	1–100 M excess	0.24 M potassium phosphate, pH 7.0	5% glycerol	o
	Integrase protein from λ	40 μg/ml	5 mM	50 mM Tris-HCl, pH 7.4	0.9 M KCl/0.1 mM EDTA/1 mM mercapto-ethanol	p

	RNA polymerase σ	5–20 mM	0.5–0.6 mg/ml	50 mM TEA-sulfate, pH 7		q
	Glucocorticoid receptor	5 mM	Crude extract, 20 mg/ml	10 mM Tris-HCl, pH 7.5	0.25 M sucrose	r
	TFIIIA	10 mM		10 mM HEPES, pH 7.5	65 mM KCl/5 mM $MgCl_2$/0.5 mM DTT/10% glycerol	s
N-Pyrenemaleimide	lac repressor	1–25 M excess	20 μM	0.24 M potassium phosphate, pH 7.0	5% glycerol	o
o-PDM	lac repressor	1–10 M excess	0.2–1.0 mg/ml	50–100 mM HEPES, pH 8		t
Oxidation						
NBS	lac repressor	1–5 M excess	10–70 μM	0.1 M Tris-HCl, pH 7.8	5% glucose	u
	CAP			0.24 M potassium phosphate, pH 7.0		u

[a] G. C. King, C. T. Martin, T. T. Pham, and J. E. Coleman. *Biochemistry* **25**, 36 (1986).
[b] C. F. Sams, B. E. Friedman, A. A. Burgum, D. S. Yang, and K. S. Matthews, *J. Biol. Chem.* **252**, 3153 (1977).
[c] J. M. Schneider, C. I. Barrett, and S. S. York, *Biochemistry* **23**, 2221 (1984).
[d] D. E. Kelsey, T. C. Rounds, and S. S. York, *Proc. Natl. Acad. Sci. U.S.A.* **76**, 2649 (1979).
[e] P. Bull, U. Wyneken, and P. Valenzuela, *Nucleic Acids Res.* **10**, 5149 (1982).
[f] D. S. Yang, A. A. Burgum, and K. S. Matthews, *Biochim. Biophys. Acta* **493**, 24 (1977).
[g] A. A. Burgum and K. S. Matthews, *J. Biol. Chem.* **253**, 4279 (1978).
[h] S. Kuramitsu, K. Hamaguchi, T. Ogawa, and H. Ogawa, *J. Biochem. (Tokyo)* **90**, 1033 (1981).
[i] S. Kuramitsu, K. Hamaguchi, H. Tachibana, T. Horii, T. Ogawa, and H. Ogawa, *Biochemistry* **23**, 2363 (1984).
[j] E. Eilen and J. S. Krakow, *J. Mol. Biol.* **114**, 47 (1977).
[k] J. Palau and J. R. Daban, *Arch. Biochem. Biophys.* **191**, 82 (1978).
[l] J. E. Bodwell, N. J. Holbrook, and A. Munck, *Biochemistry* **23**, 1392 (1984).
[m] T. J. Daly, J. S. Olson, and K. S. Matthews, *Biochemistry* **25**, 5468 (1986).
[n] E. Wingender and A. Arellano, *Anal. Biochem.* **127**, 351 (1982).
[o] R. D. Brown and K. S. Matthews, *J. Biol. Chem.* **254**, 5128 (1979).
[p] W. Ross and A. Landy, *Proc. Natl. Acad. Sci. U.S.A.* **79**, 7724 (1982).
[q] C. S. Narayanan and J. S. Krakow, *Nucleic Acids Res.* **11**, 2701 (1983).
[r] M. Kalimi and K. Love, *J. Biol. Chem.* **255**, 4687 (1980).
[s] J. J. Bieker and R. G. Roeder, *J. Biol. Chem.* **259**, 6158 (1984).
[t] C. Pampeno and J. S. Krakow, *Biochemistry* **18**, 1519 (1979).
[u] S. P. Manly and K. S. Matthews, *J. Biol. Chem.* **254**, 3341 (1979).

containing compounds (e.g., dithiothreitol, 2-mercaptoethanol), and remaining reagent is removed by dialysis or gel filtration. The protein can then be analyzed spectrophotometrically or by peptide mapping techniques. One advantage of many iodoacetamide derivatives is the presence of a spectroscopically detectable moiety, using either standard spectrophotometry in the visible range or fluorometry. These probes can be used for spectral studies and yield information regarding the effects of modification. A number of probes with a variety of fluorophores and chromophores are available (e.g., 2-bromoacetamido-4-nitrophenol, IAEDANS).

Mercurials

Several different types of mercurials are available, including *p*-chloromercuribenzoate (*p*-CMB), fluorescein mercuric acetate, and 2-chloromercuri-4-nitrophenol. Reaction results in altered spectral characteristics of the reagent in most cases and thus can be monitored spectrophotometrically. These modifications are normally executed at pH 8–9 in Tris-HCl buffers with very low molar ratios of reagent, as reaction is nearly stoichiometric.[16–18] Titration using small aliquots of reagent is an effective means to determine the total number of reactive cysteines in the protein. An advantage of the mercurial reagents and the disulfide exchange reagents to be discussed next is the ability to reverse the reaction using excess sulfhydryl-containing compounds (2-mercaptoethanol or dithiothreitol); while this reversibility is useful in determining the effects of reaction and assuring that no irreversible changes have occurred, it may complicate mapping procedures and require development of specific strategies for identifying sites of reaction.

Disulfide Exchangers

Free sulfhydryls will exchange with disulfide-containing reagents to produce a mixed disulfide at the site of reaction. Several different reagents are available, including 5,5'-dithiobis(2-nitrobenzoic acid) (DTNB), methyl methane thiosulfonate (MMTS), and difluoroscein disulfide. DTNB reaction with sulfhydryl residues results in the release of TNB anion, which has maximum absorption at ~410 nm, so that reaction can be followed spectrophotometrically. Reaction of exposed sulfhydryl residues with disulfide exchange reagents can be stoichiometric, but buried residues or those in an unusual environment may require higher molar ratios of

[16] P. Bull, U. Wyneken, and P. Valenzuela, *Nucleic Acids Res.* **10**, 5149 (1982).
[17] D. S. Yang, A. A. Burgum, and K. S. Matthews, *Biochim. Biophys. Acta* **493**, 24 (1977).
[18] A. A. Burgum and K. S. Matthews, *J. Biol. Chem.* **253**, 4279 (1978).

reagent to achieve reaction. All three cysteine residues in the *Escherichia coli* RecA protein are reactive with DTNB; however, in the presence of ATP or ADP, only one cysteine residue reacts.[19,20] Binding of single-stranded (ss) DNA to the RecA protein does not affect the reactivity with DTNB.[20] For these studies, potassium phosphate buffer was utilized at pH 7.9 at varying salt concentration with 1.0 mM DTNB and 6 μM protein. With the cAMP-binding protein (CAP) from *E. coli*, it has been observed that two cysteines will react in the absence of cAMP, whereas in its presence, the formation of a disulfide linkage occurs between the two sulfhydryls.[21] This result was interpreted as indicating the close proximity of two antiparallel β-sheet structures when the CAP protein is in its DNA-binding conformation.

Local environment may be an important factor in reactivity. For example, it has been shown that reaction of DTNB with histones can be facilitated by electrostatic effects,[22] but this phenomenon is suppressed by the presence of high salt or guanidine hydrochloride. In eukaryotic systems, DTNB has been used to treat cellular extracts to determine the effects on glucocorticoid receptor complexes; the results indicate that sulfhydryl modification inhibits the binding of activated dexamethasone–receptor complexes to DNA cellulose.[23] The conditions for these modifications were pH 8 with a concentration of 0.25 mM reagent.

MMTS does not contain a chromophoric probe to allow spectrophotometric assessment, but its small size and uncharged -S-Me addition to a free sulfhydryl make it a very effective reagent for assessment of the role of cysteine residues in protein function. Reaction of MMTS with exposed and partially buried sulfhydryl residues appears to be stoichiometric; however, in the lactose repressor protein this reagent can also react at high molar ratios with residues that are involved in a subunit interface and are unavailable to other sulfhydryl reagents.[24] Generally, MMTS reaction is carried out at pH 7.5 to 8.0 in Tris-HCl or HEPES buffer; following modification,detection of modified residues requires irreversible reaction with a second reagent to provide a chromophoric signal for unreacted cysteine sites.[25] In this fashion, the cysteines that were not modified

[19] S. Kuramitsu, K. Hamaguchi, T. Ogawa, and H. Ogawa, *J. Biochem. (Tokyo)* **90**, 1033 (1981).
[20] S. Kuramitsu, K. Hamaguchi, H. Tachibana, T. Horii, T. Ogawa, and H. Ogawa, *Biochemistry* **23**, 2363 (1984).
[21] E. Eilen and J. S. Krakow, *J. Mol. Biol.* **114**, 47 (1977).
[22] J. Palau and J. R. Daban, *Arch. Biochem. Biophys.* **191**, 82 (1978).
[23] J. E. Bodwell, N. J. Holbrook, and A. Munck, *Biochemistry* **23**, 1392 (1984).
[24] T. J. Daly, J. S. Olson, and K. S. Matthews, *Biochemistry* **25**, 5468 (1986).
[25] J. A. Gardner and K. S. Matthews, *Anal. Biochem.* **167**, 140 (1987).

by MMTS can be identified by conventional peptide mapping methods. Because of its small size, this reagent is very useful for determining the significance of an ionizing sulfhydryl moiety to function, and the effects are readily reversible by addition of sulfhydryl reagents.

Difluorescein disulfide has been utilized to examine cysteine residues in histone H3; this fluorophore allows assessment of conformational changes that influence the fluorescence parameters associated with the probe.[26] Reaction occurred at pH 7.5 in a sodium phosphate buffer; in this case, high concentrations of histone were utilized (~33 mg/ml) with ~3 mg/ml reagent. Prior to reaction, a low level of dithiothreitol was utilized to reduce the sulfide linkage between the two cysteines in histone H3; reagent was then added in excess over the combined protein and added dithiothreitol concentrations. This protocol illustrates the ability to modify cysteine residues that are normally linked by disulfide bonds in the protein structure.

Maleimide Derivatives

N-Ethylmaleimide and other N-substituted maleimides have been used widely to modify cysteine residues. This reagent is generally cysteine specific and requires high molar ratios for reaction; with the lactose repressor, N-substituted maleimide molar ratios of 1- to 100-fold were utilized in a potassium phosphate buffer at pH 7 with protein concentrations of 20 to 50 μM.[27] For N-ethylmaleimide, there is no chromophore for assessment of extent of reaction; for other N-substituted maleimides (e.g., N-pyrenemaleimide), spectrophotometric or fluorometric methods can be used to assess the extent of reaction. Spectral probes introduced by this reaction can be used to assess environmental changes consequent to ligand binding.[28] The maleimide reaction is irreversible, a feature that facilitates mapping procedures to identify specific sites of modification. Differential effects on functional properties have been observed in some studies. For example, for the integrase protein (int) from bacteriophage λ, reaction with N-ethylmaleimide (40 μg/ml protein in 0.9 M KCl, 50 mM Tris-HCl, pH 7.4, 1 mM 2-mercaptoethanol, 0.1 mM EDTA, 2 mg/ml bovine serum albumin) resulted in selective loss of int interaction with one of the consensus recognition sequences for the protein[29]; this result was consistent with previous reports that this modification abolished the ability of int to execute recombination and to form a heparin-resistant complex with DNA

[26] E. Wingender and A. Arellano, *Anal. Biochem.* **127,** 351 (1982).
[27] R. D. Brown and K. S. Matthews, *J. Biol. Chem.* **254,** 5128 (1979).
[28] R. D. Brown and K. S. Matthews, *J. Biol. Chem.* **254,** 5135 (1979).
[29] W. Ross and A. Landy, *Proc. Natl. Acad. Sci. U.S.A.* **79,** 7724 (1982).

without effect on its topoisomerase activity.[30] This type of differential effect on separate activities can be utilized to decipher roles of various subdomains of protein sequence/structure in these functional properties.

RNA polymerase modification with N-ethylmaleimide resulted in reaction of three cysteine residues with accompanying loss of σ subunit activity[31]; kinetic analysis of the inactivation data indicated that a single cysteine was essential for activity. Reaction was carried out in 50 mM triethylamine-sulfate at pH 7 with 5–20 mM N-ethylmaleimide and 0.6 mg/ml σ subunit. The reaction can also be carried out in crude fractions; for example, the cytosol fraction from rat liver homogenates was modified in 10 mM Tris-HCl, 0.25 M sucrose, pH 7.5 at a total protein concentration of 20 mg/ml and 5 mM reagent concentration.[32] The results of this study indicated that N-ethylmaleimide treatment yielded a glucocorticoid receptor complex that was frozen in its intact, unactivated form. Similarly, reaction of transcription factor IIIA (TFIIIA) with 10 mM N-ethylmaleimide (15 min, 30°) either in the absence or presence of 5S RNA or the 5S RNA gene results in abolition of its ability to promote transcription.[33]

An example of using specific chemical modification for cross-linking is the use of o-phenylenedimaleimide to cross-link the subunits of the cAMP-binding protein (CAP). The cross-linking reaction is modulated by the presence of cAMP, and the cross-linked protein loses DNA-binding activity.[34] A 10-fold molar excess of reagent was sufficient to generate the cross-linked species in HEPES at pH 8; addition of excess dithiothreitol was used to stop the reaction prior to analysis.

Oxidation

Controlled oxidation of cysteine residues can also be of utility in deciphering the role specific sulfhydryls play in the activity of proteins. Although most oxidizing agents affect a broader range of side chains, careful control of reaction conditions can result in specific cysteine modification. N-Bromosuccinimide has been used with the lactose repressor protein, and the effects of cysteine and methionine oxidation were distinguished by ligand protection measurements.[35] Reaction with N-bromosuccinimide can be carried out between pH 7 and 8 in either phosphate or Tris-HCl

[30] Y. Kikuchi and H. A. Nash, *Proc. Natl. Acad. Sci. U.S.A.* **76**, 3760 (1979).
[31] C. S. Narayanan and J. S. Krakow, *Nucleic Acids Res.* **11**, 2701 (1983).
[32] M. Kalimi and K. Love, *J. Biol. Chem.* **255**, 4687 (1980).
[33] J. J. Bieker and R. G. Roeder, *J. Biol. Chem.* **259**, 6158 (1984).
[34] C. Pampeno and J. S. Krakow, *Biochemistry* **18**, 1519 (1979).
[35] S. P. Manly and K. S. Matthews, *J. Biol. Chem.* **254**, 3341 (1979).

buffer; the reagent can be inactivated by the addition of excess dithiothreitol. By varying the molar ratio, it may be possible to control the extent of cysteine oxidation. Other protocols and reagents can be utilized for these types of experiments, and the choice of oxidizing reagents is broad. Care must be exerted to determine all residues affected by the reaction and to assess, where possible, whether the product is sulfenic (-SOH), sulfinic (-SO$_2$H), or sulfonic (-SO$_3$H) acid at the cysteine sites. The sulfonic acid derivative is not affected by addition of sulfhydryl reagents, while the sulfenic and sulfinic acid products can revert at least partially to free sulfhydryl under reducing conditions. In addition, disulfide bonds can form under oxidizing conditions.

Lysine Residues

The α- and ε-amino groups in a protein are targets for reactions with a variety of reagents with relatively high specificity. A list of reagents with examples of DNA-binding proteins examined and conditions utilized is given in Table II. Care must be taken to evaluate the reaction of cysteine residues in a protein when lysine side chains are the targeted amino acid due to the high reactivity of the sulfhydryl group. The unprotonated form of the amino group is usually the reactive species; thus, elevated pH (i.e., above 8) to generate a significant proportion of the unprotonated species is generally required for reaction to occur at a reasonable rate. It should be noted that many buffer systems can be used for reacting lysine residues, but in general Tris and other compounds which contain a primary or secondary amine should be avoided, as they will react with reagent and lower the concentration available for protein modification. As with other amino acid side chains, the specific reactivity of a residue is influenced by the surrounding environment. In DNA-binding proteins, lysine and arginine residues are involved frequently in nonspecific ionic interactions with the sugar phosphate backbone, and it is important to assess the degree to which changes in specific affinity may be correlated with alterations in the nonspecific binding properties of the modified protein.

Aryl and Sulfonyl Halides

These classes of compounds have been useful in a number of contexts, particularly the introduction of spectroscopic probes into protein structure (e.g., dansyl chloride, fluorodinitrobenzene). Low levels of modification with concomitant introduction of chromophore/fluorophore frequently can be achieved with minimal effects on the functional properties of the protein to generate a species that can be used in a variety of experimental designs to measure physical and functional parameters. Dansyl chloride modification of the lactose repressor protein resulted in loss of specific DNA binding at low molar ratios of reagent without effect on inducer-binding properties.[36] The presence of nonspecific DNA results in protection of two lysine residues and concomitant protection of operator DNA-binding activity. The dansyl group can be used for fluorescence measurements in addition to determining lysine participation in functional activity.

Trinitrobenzene Sulfonate (TNBS)

This reagent reacts with relatively high specificity with amino groups, and the reaction can be followed spectrophotometrically.[37] Reaction of the σ subunit of RNA polymerase with TNBS required modification of five lysine residues for complete inactivation; using kinetic analysis, one lysine was implicated as critical for function.[38] The modified σ subunit was able to form holoenzyme with binding affinity similar to the unmodified protein; however, the modification compromised the ability to form a tight complex with promoter and to stimulate RNA chain initiation. It is noteworthy that proteolytic digestion analysis indicated a conformational change in the σ subunit following trinitrophenylation; this example illustrates the importance of complete and careful analysis of the reaction product.

Lactose repressor protein modified with TNBS exhibited differential effects on its three binding activities: inducer binding was increased, operator and nonspecific DNA binding were decreased, although to differing degrees.[39] The presence of ligands influenced the extent of reaction and the degree of binding perturbation caused by the modification. The trinitrophenyl chromophore is useful for mapping the sites of reaction, as it absorbs in the visible region; this absorbance can also be used to estimate the extent of reaction of TNBS with protein, although ancillary analysis is required to confirm these results.

Pyridoxal Phosphate

This reagent selectively produces a Schiff's base with α- and ε-amino groups in proteins; in addition, pyridoxal phosphate (PLP) reaction introduces a chromophore, thereby facilitating analysis, and the reaction is reversible (although reduction can be utilized to generate an irreversible

---

[36] W.-T. Hsieh and K. S. Matthews, *Biochemistry* **24**, 3043 (1985).
[37] A. F. S. A. Habeeb, *Anal. Biochem.* **14**, 328 (1966).
[38] C. S. Narayanan and J. S. Krakow, *Biochemistry* **24**, 6103 (1982).
[39] P. A. Whitson, A. A. Burgum, and K. S. Matthews, *Biochemistry* **23**, 6046 (1984).

TABLE II
REACTION OF LYSINE RESIDUES

Reagent	Protein	[Reagent]	[Protein]	Buffer	Additives	Ref.
Sulfonyl halides and sulfonates						
Dansyl chloride	*lac* repressor	1–32 *M* excess	1 mg/ml	0.24 *M* potassium phosphate, pH 8.4	5% glucose	a
TNBS	RNA polymerase σ	1–10 m*M*	0.3–1.0 mg/ml	80 m*M* Tris, pH 9.0	80 m*M* borate/2.5 m*M* EDTA	b
	lac repressor	2–16 *M* excess	0.6–2.0 mg/ml	2.5% triethanolamine, pH 9.0		c
Pyridoxal phosphate	DNA polymerase I	1 m*M*	1 μ*M*	50 m*M* HEPES, pH 8.0	10 m*M* NaCl/20% glycerol/5 m*M* Mg(OAc)₂/1 m*M* DTT	d
	DNA polymerase β	0–500 μ*M*	200 μg/ml	50 m*M* HEPES, pH 7.8	1 m*M* MnCl₂/10 m*M* NaCl/20% glycerol	e
Acid anhydrides						
Dimethylmaleic anhydride	Chromatin	0–0.5 mg/ml	0.19 mg DNA/ml	10 m*M* HEPES, pH 8.2		f
	Nucleosomes	0.5–0.7 mg/ml	0.14–0.35 mg DNA/ml	10 m*M* HEPES, pH 8.2	5 m*M* EDTA 0.1 *M* PMSF	g
Reductive alkylators						
Sodium cyanoborohydride } Formaldehyde	fd gene 5 protein	10 m*M* 10-fold [Lys]	0.5–1.0 mg/ml	20 m*M* HEPES, pH 7.5		h
Formaldehyde } Sodium borohydride	CAP	6 m*M* ~0.4 mg/ml	1.6 mg/ml	45 m*M* potassium phosphate, pH 8.0	4.5 m*M* EDTA/10 μ*M* cAMP/10% glycerol	i

Imido esters

Reagent	Protein			Additive	Buffer	Ref.
Ethyl acetimidate	Pf1 ssDNA-binding protein	6.6 mg/mg protein	6 μM		50 mM NaHCO$_3$, pH 8.4	[j]
Ethyl acetimidate	fd gene 5 protein	0.13 mg/ml	2 mg/ml		0.1 M triethanolamine-HCl, pH 8.3	[k]
Sulfosuccinimidyl 6-(biotinamido)-hexanoate	CAP	0.6 mM	1.6 mg/ml	1 mM EDTA/5% glycerol	20 mM HEPES, pH 8	[l]
2-Iminothiolane	trp repressor	5 mM	85 μM	1% mercaptoethanol/4 mg/ml tryptophan	0.1 M phosphate, pH 8	[m]

[a] W.-T. Hsieh and K. S. Matthews, *Biochemistry* **24**, 3043 (1985).
[b] C. S. Narayanan and J. S. Krakow, *Biochemistry* **24**, 6103 (1982).
[c] P. A. Whitson, A. A. Burgum, and K. S. Matthews, *Biochemistry* **23**, 6046 (1984).
[d] A. K. Hazra, S. Detera-Wadleigh, and S. H. Wilson, *Biochemistry* **23**, 2073 (1984).
[e] A. Basu, P. Kedar, S. H. Wilson, and M. J. Modak, *Biochemistry* **28**, 6305 (1989).
[f] E. Palacián, A. López-Rivas, J. A. Pintor-Toro, and F. Hernández, *Mol. Cell. Biochem.* **36**, 163 (1981).
[g] J. Jordano, M. A. Nieto, and E. Palacián, *J. Biol. Chem.* **260**, 9382 (1985).
[h] L. R. Dick, A. D. Sherry, M. M. Newkirk, and D. M. Gray, *J. Biol. Chem.* **263**, 18864 (1988).
[i] M. G. Fried and D. M. Crothers, *Nucleic Acids Res.* **11**, 141 (1983).
[j] A. Tsugita and G. G. Kneale, *Biochem. J.* **228**, 193 (1985).
[k] S. Bayne and I. Rasched, *Biosci. Rep.* **3**, 469 (1983).
[l] A. M. Brown and D. M. Crothers, *Proc. Natl. Acad. Sci. U.S.A.* **86**, 7387 (1989).
[m] C.-H. B. Chen and D. S. Sigman, *Science* **237**, 1197 (1987).

product). Inhibition of modified *E. coli* DNA polymerase I large fragment with PLP required two molecules of PLP, although reduction of the modified enzyme demonstrated incorporation of 3 mol PLP/mol of enzyme.[40] Interestingly, the presence of dNTPs during modification resulted in a decrease of 1 mol PLP incorporated/mol enzyme with a product that exhibited DNA polymerase activity but was compromised in the elongation function. The site of PLP attachment is Lys-758, and the inactivation is dependent on the presence of divalent metal ions.[41] Similar results have been observed with mammalian DNA polymerase β, although the complexity of the reaction appears greater with this enzyme.[42] The site protected by dNTPs in DNA polymerase β is Lys-71, presumed to participate in the nucleotide substrate-binding pocket.

Carboxylic Acid Anhydrides

A variety of anhydrides will react with amino groups in proteins at alkaline pH, although modification of tyrosine can complicate interpretation of the results with this type of modification. Charge neutralization or reversal of charge at reacted lysines can be achieved with this reagent class (e.g., acetylation or succinylation). Chromatin dissociation has been obtained using dimethylmaleic anhydride (citraconic anhydride).[43] Incubation at pH 6 reverses the reaction, and the proteins released from chromatin by this method are able to reform nucleosome-like structures. Similar results have been obtained with nucleosomal particles and with isolated core-histone octamers; dimethylmaleic anhydride elicits a biphasic dissociation of H2A : H2B dimers, while acetic anhydride does not have this effect.[44,45] Use of both reagents at higher molar excesses yields dissociation of nucleosomal particles, a result that was interpreted to signify the participation of lysine groups in binding of histones to the DNA. These studies illustrate the specific effects that can be elicited by modification and the necessity for careful planning and exploration of multiple reagents for a specific purpose.

Reductive Alkylation

This reaction has the advantage of preserving the charge characteristic of the lysyl side chain, although side reactions can occur. Either mono- or disubstituted products are observed, depending on the reagent and

[40] A. K. Hazra, S. Detera-Wadleigh, and S. H. Wilson, *Biochemistry* **23**, 2073 (1984).
[41] A. Basu and M. J. Modak, *Biochemistry* **26**, 1704 (1987).
[42] A. Basu, P. Kedar, S. H. Wilson, and M. J. Modak, *Biochemistry* **28**, 6305 (1989).
[43] E. Palacián, A. López-Rivas, J. A. Pintor-Toro, and F. Hernández, *Mol. Cell. Biochem.* **36**, 163 (1981).
[44] J. Jordano, M. A. Nieto, and E. Palacián, *J. Biol. Chem.* **260**, 9382 (1985).
[45] M. A. Nieto and E. Palacián, *Biochemistry* **27**, 5635 (1988).

conditions. Although larger alkyl groups can be introduced, methyl is most commonly used in proteins to minimize perturbation of structure. Formaldehyde is utilized as the source of the methyl group, and sodium cyanoborohydride serves as the reducing agent; the use of the latter is important, as it is stable in aqueous solution at neutral pH, is specific for Schiff's base reduction, and does not affect aldehydes or disulfide bonds. Several authors have noted the necessity of recrystallizing the sodium cyanoborohydride immediately prior to use and problems with interference from other components in the reaction solution (e.g., sulfhydryl reagents, metal ions, ammonium ion).[46,47] This method has found application in the area of DNA-binding proteins in the introduction of ^{13}C or ^{14}C into the structure of proteins. For example, the lysine residues in gene 5 protein from fd bacteriophage, a ssDNA-binding protein, have been modified using sodium cyanoborohydride (10 mM) and formaldehyde (10-fold lysyl side-chain concentration) in a 20 mM HEPES buffer at pH 7.5.[48] Of the seven lysyl residues, six react in the free protein, while only three are available in the presence of ssDNA sequences; identification of the three protected residues was possible by using radiolabeled formaldehyde for the reaction, and the results suggested that these lysine side chains may be involved in binding to DNA. Reductive methylation has also been utilized to radiolabel the CAP protein for determination of protein–DNA stoichiometry.[49] The reaction with CAP protein utilized sodium borohydride as the reductant following reaction with formaldehyde.

Imido Esters

This class of compounds shares the advantage of reductive alkylation in maintaining the charge on the lysine residues. Cross-linking reactions are a specific application of imido ester modification, and a broad variety of reagents, some cleavable, to react amino functions is available (e.g., dimethyl suberimidate). Many types of imido esters have been utilized in protein modification, although relatively minimal application to DNA-binding proteins has occurred. One reason may be marginal effects on the functional properties of DNA-binding proteins due to the maintenance of charge in the reaction product; this situation has been observed using reaction of several imido esters with the lactose repressor.[50] ssDNA-binding proteins have been examined using ethyl acetimidate, and the

[46] N. Jentoft and D. G. Dearborn, *J. Biol. Chem.* **254,** 4359 (1979).
[47] N. Jentoft and D. G. Dearborn, *Anal. Biochem.* **106,** 186 (1980).
[48] L. R. Dick, A. D. Sherry, M. M. Newkirk, and D. M. Gray, *J. Biol. Chem.* **263,** 18864 (1988).
[49] M. G. Fried and D. M. Crothers, *Nucleic Acids Res.* **11,** 141 (1983).
[50] K. S. Matthews and J. Rex, unpublished data (1980).

data indicate that modification of lysines may alter functional properties and that the presence of ssDNA protects specific residues from acet-imidation.[51,52] An example of a specialized imido ester reaction is the biotinylation of CAP protein from *E. coli* using sulfosuccinimidyl 6-(biotinamido)hexanoate; the biotin introduced by this reaction was uti-lized in identifying and separating subunits involved in exchange reactions with unmodified CAP dimer.[53] Modification was carried out on a DNA cellulose substrate to protect binding activity in a 20 mM HEPES buffer at pH 8 with 0.6 mM reagent; reaction was stopped with a Tris-HCl buffer at similar pH.

Iminothiolane (an imido thioester) has been used to convert the amino function to a sulfhydryl group in the *trp* repressor from *E. coli* using a phosphate buffer, pH 8, ~5 mM reagent, ~0.1 mM protein.[54] The resulting sulfhydryl was alkylated by 5-iodoacetamido-1,10-phenanthroline to gen-erate a site-specific endonuclease based on the affinity of the *trp* repressor for its cognate operator sequences. Such combined reactions can be used to manufacture altered proteins with unusual properties and with desired specificities.

Other Reagents

A broad range of additional reagents is utilized for lysine modification, but these have found limited application in the study of DNA-binding proteins. As with all of the amino acids, careful investigation of the large number of available reagents is required to determine the best choice for a particular application.

Arginine

A limited number of reagents is available for arginine modification, in part due to the high pK of the guanidino moiety. Those widely used are phenylglyoxal (and its derivatives), 2,3-butanedione, and 1,2-cyclohex-anedione. Examples of reactions with DNA-binding proteins and condi-tions used are summarized in Table III(A).

Phenylglyoxal

Phenylglyoxal will react with amino groups, particularly the α-amino group, and this side reaction must be assessed in determining the effects of modification. Takahashi[55] has demonstrated that two phenylglyoxal

[51] A. Tsugita and G. G. Kneale, *Biochem. J.* **228**, 193 (1985).
[52] S. Bayne and I. Rasched, *Biosci. Rep.* **3**, 469 (1983).
[53] A. M. Brown and D. M. Crothers, *Proc. Natl. Acad. Sci. U.S.A.* **86**, 7387 (1989).
[54] C.-H. B. Chen and D. S. Sigman, *Science* **237**, 1197 (1987).
[55] K. Takahashi, *J. Biol. Chem.* **243**, 6171 (1968).

molecules react with a single guanidino group. RNA polymerase holoenzyme from *E. coli* has been modified using this reagent, and kinetic analysis of the inactivation data indicates that reaction of a single arginine is responsible for the loss in activity.[56] Inhibition of DNA polymerases from eukaryotic, prokaryotic, and RNA tumor viruses with phenylglyoxal has been demonstrated; the modification appears to interfere with template binding of the DNA polymerases and suggests arginine interaction with the template as well as a common mechanism for these enzymes.[57] The presence of template during reaction protected against activity loss. The reaction was carried out in a HEPES buffer at pH 7.8, with phenylglyoxal concentrations up to 0.5 mM. Phenylglyoxal reaction with the lactose repressor protein did not affect inducer binding, but diminished operator and nonspecific DNA binding with modification of one to two equivalents of arginine per monomer.[58] Protection of activity was observed with reaction in the presence of operator DNA concomitant with diminished reactivity of arginine residues. Reaction was carried out in this study using sodium bicarbonate buffer, pH 8.3, with protein concentration $\sim$25 μM and reagent concentrations ranging to 2.5 mM. Bicarbonate/carbonate buffers have been shown to increase the rate of reaction with arginine and to minimize reaction with the α- and ε-amino groups.[59] It is possible to obtain (Amersham, Arlington Heights, IL) or synthesize radiolabeled phenylglyoxal to facilitate determination of the extent of reaction and sites of modification. Several derivatives of phenylglyoxal (*p*-nitro-, *p*-hydroxy-, and *p*-azido-) are available, but utilization of these reagents with DNA-binding proteins has been minimal.

2,3-Butanedione

Borate stabilizes the reaction product between butanedione and arginine, and this buffer should be utilized for experiments using butanedione. Optimal pH and borate concentration for reaction should be determined for each protein. Removal of borate or the presence of excess free arginine can reverse the reaction and regenerate the side chain in the protein. In addition, this reagent can polymerize, and the polymeric species display diminished reactivity relative to the monomer[60]; thus, distillation of the reagent on a routine basis is requisite as well as storage in the dark to prevent photoreactions.[60,61] Side reactions have been observed with

[56] V. W. Armstrong, H. Sternbach, and F. Eckstein, *FEBS Lett.* **70**, 48 (1976).
[57] A. Srivastava and M. J. Modak, *J. Biol. Chem.* **255**, 917 (1980).
[58] P. A. Whitson and K. S. Matthews, *Biochemistry* **26**, 6502 (1987).
[59] S.-T. Cheung and M. L. Fonda, *Biochem. Biophys. Res. Commun.* **90**, 940 (1979).
[60] J. F. Riordan, *Biochemistry* **12**, 3915 (1973).
[61] D. Petz, H.-G. Löffler, and F. Schneider, *Z. Naturforsch. C: Biosci.* **34**, 742 (1979).

TABLE III

REACTION OF OTHER RESIDUES

Reagent	Protein	[Reagent]	[Protein]	Buffer	Additives	Ref.
A. Reaction of arginine residues						
Phenylglyoxal	RNA polymerase holoenzyme	0.36 mM	0.45 mg/ml	100 mM bicine, pH 8.0	10 mM MgCl$_2$	a
	DNA and RNA polymerases	Up to 0.5 mM		50–80 mM HEPES, pH 7.8	Variable	b
2,3-Butanedione	lac repressor	0–2.5 mM	2.7 × 10^{-5} M	0.1 M NaHCO$_3$, pH 8.3		c
	lac repressor	0–25 mM	2.4 × 10^{-5} M	50 mM borate, pH 8.0		c
	BamHI methylase	0–40 mM	6 units	20–50 mM borate, pH 8.0		d
	BamHI endonuclease	0–30 mM	50 units	0–200 mM borate, pH 8.0		e
	BS-NS	10 mM	0.77 mg/ml	100 mM HEPES, pH 8.0	50 mM borate	f
B. Reaction of histidine residues						
Diethyl pyrocarbonate	RNA polymerase	0.11–1.0 mM	1–4 µM	0.1 M phosphate, pH 6.0		g
	lac repressor	0–3 mM	0.5 mg/ml	0.24 M potassium phosphate, pH 7.5	5% glycerol	h
C. Reaction of tyrosine residues						
TNM	fd gene 5 protein	64 M excess	0.8 mM	0.05 M Tris-HCl, pH 8	0.15 M NaCl	i
	lac repressor	2–50 M excess	0.1–1.0 mg/ml	0.1 M Tris-HCl, pH 8.0		j
	lac repressor	2–50 M excess	0.1–1.0 mg/ml	0.24 M potassium phosphate, pH 8.0	5% glucose	j
	Topoisomerase I and II	1–3 mM	10–35 µg/ml	50 mM Tris-HCl, pH 8.0		k

488

I₂/KI	*lac* repressor	0–60 *M* excess I₂ / 0–240 *M* excess KI	0.6–0.8 mg/ml	0.04 *M* Tris-HCl, pH 7.5	0.01 *M* Mg(OAc)₂/0.2 *M* KCl/5% glycerol	*l*

D. Reaction of tryptophan residues

NBS	*lac* repressor	1–20 *M* excess	25–100 μ*M*	1.0 *M* Tris-HCl, pH 7.8		*m*
	lac repressor	1–20 *M* excess	25–100 μ*M*	0.2 *M* potassium phosphate, pH 7.9		*m*
	RNA polymerase	25–30 *M* excess		pH 7.6		*n*
Oxidation						
Trichloroethanol	T4 gene 32 protein	0.1 *M* trichloroethanol/UV irradiation	~1–10 μ*M*	10 mM cacodylate, pH 7.6	10 mM NaCl/0.2 mM EDTA	*o*
UV irradiation	*lac* repressor	UV irradiation	0.15 mg/ml	0.1% NH₄HCO₃, pH 8.7		*p*

[a] V. W. Armstrong, H. Sternbach, and F. Eckstein, *FEBS Lett.* **70**, 48 (1976).
[b] A. Srivastava and M. J. Modak, *J. Biol. Chem.* **255**, 917 (1980).
[c] P. A. Whitson and K. S. Matthews, *Biochemistry* **26**, 6502 (1987).
[d] G. Nardone, J. George, and J. G. Chirikjian, *J. Biol. Chem.* **259**, 10357 (1984).
[e] J. George, G. Nardone, and J. G. Chirikjian, *J. Biol. Chem.* **260**, 14387 (1985).
[f] M. Lammi, M. Paci, and C. O. Gualerzi, *FEBS Lett.* **170**, 99 (1984).
[g] A. W. Abdulwajid and F. Y.-H. Wu, *Biochemistry* **25**, 8167 (1986).
[h] C. F. Sams and K. S. Matthews, *Biochemistry* **27**, 2277 (1988).
[i] R. A. Anderson, Y. Nakashima, and J. E. Coleman, *Biochemistry* **14**, 907 (1975).
[j] W.-T. Hsieh and K. S. Matthews, *J. Biol. Chem.* **256**, 4856 (1981).
[k] L. Klevan and Y.-C. Tse, *Biochim. Biophys. Acta* **745**, 175 (1983).
[l] T. G. Fanning, *Biochemistry* **14**, 2512 (1975).
[m] R. B. O'Gorman and K. S. Matthews, *J. Biol. Chem.* **252**, 3565 (1977).
[n] B. Wasylyk and A. D. B. Malcolm, *Biochem. Soc. Trans.* **3**, 654 (1975).
[o] J.-J. Toulmé, T. LeDoan, and C. Hélène, *Biochemistry* **23**, 1195 (1984).
[p] M. Spodheim-Maurizot, M. Charlier, and C. Hélène, *Photochem. Photobiol.* **42**, 353 (1985).

histidine.[61] Reaction of 2,3-butanedione with the lactose repressor protein resulted in loss of arginine with no evident modification of lysine residues.[58] Reaction of one to two arginine residues was sufficient to inhibit operator DNA binding, but no effect was observed on inducer binding; these data are consistent with the results obtained using phenylglyoxal to modify the arginines in the lactose repressor protein.

BamHI methylase is inhibited by treatment with 2,3-butanedione with protection against inactivation by the presence of DNA but not by S-adenosylmethionine. Borate at 50 mM was used in these studies with 20–40 mM butanedione; the extent of enzyme inactivation was maximal at 50 mM borate.[62] In like manner, BamHI endonuclease is inhibited by reaction with 2,3-butanedione, with the dinucleotide pdGpdG providing maximum protection for the enzyme against inactivation by this reagent; however, in this case the optimum concentration of borate was 100 mM.[63] The effects of arginine modification by 2,3-butanedione on DNA binding of the prokaryotic DNA-scaffolding protein BS-NS from *Bacillus stearothermophilis* has indicated that at least one arginine residue in this protein may be required for binding to DNA.[64] The buffer used in this study was 100 mM HEPES, pH 8.0, 50 mM borate with 10 mM 2,3-butanedione. It should be emphasized again that the presence of borate stabilizes the adduct formed with arginine and 2,3-butanedione,[65] and the optimal concentration for each system must be determined.

1,2-Cyclohexanedione has also been used to modify arginine, and an adduct is stabilized by borate; however, there are side reactions that are irreversible as well. This and other complications have presumably prevented widespread use of this reagent with DNA-binding proteins.

Histidine

Although a number of reagents will modify the imidazole moiety of histidine, specificity of reaction is obtained with only a few of these compounds. When attempting to modify histidine residues, particular care must be exercised to assess reactions with all other potential targets of reaction.

[62] G. Nardone, J. George, and J. G. Chirikjian, *J. Biol. Chem.* **259**, 10357 (1984).
[63] J. George, G. Nardone, and J. G. Chirikjian, *J. Biol. Chem.* **260**, 14387 (1985).
[64] M. Lammi, M. Paci, and C. O. Gualerzi, *FEBS Lett.* **170**, 99 (1984).
[65] J. F. Riordan, *Mol. Cell. Biochem.* **26**, 71 (1979).

General Reagents

Early studies of DNase II from porcine spleen utilized iodoacetamide reaction to demonstrate alkylation at a single histidine to inactivate the protein.[66] Substrate DNA partially protected the enzyme against this inactivation using 0.2 M sodium acetate buffer, pH 4.6, with up to 0.2 mM iodoacetate. Photooxidation was another method used to target histidine, but difficulties with specificity have precluded widespread application with DNA-binding proteins. While a number of reagents have been applied to other proteins, none has achieved the specificity observed with diethyl pyrocarbonate [Table III(B)].

Diethyl Pyrocarbonate

Diethyl pyrocarbonate (ethoxyformic anhydride) reacts with high specificity with histidine in the pH range from 5.5 to 7.5, although side reactions with lysine and tyrosine, and in some cases cysteine, must be monitored carefully. Diethyl pyrocarbonate reaction with histidines is a function of the concentration of the reagent rather than the molar ratio to the protein in solution. The reagent itself is sensitive to hydrolysis in aqueous solutions, with a half-life of approximately 10 min at pH 7 and shorter times at higher pH.[67] Stock solutions of reagent can be prepared in anhydrous ethanol and standardized by reaction with imidazole following the reaction spectrophotometrically. Monosubstitution of the imidazole ring can be reversed at alkaline or acidic pH or at neutral pH with hydroxylamine, while dicarbethoxylated imidazole is stable. Reaction is monitored by the increase in absorbance at 240 nm ($\Delta\varepsilon$ ~3200 M^{-1} cm^{-1}),[68] and reaction with tyrosine can be spectrophotometrically determined by a decrease in absorbance at 280 nm.

RNA polymerase undergoes rapid inactivation on exposure to diethyl pyrocarbonate with formation of carbethoxyhistidine and no tyrosine, lysine, or cysteine reaction.[69] A large number of histidines were modified (six to nine), although substrate ATP (but not template DNA) provided protection against inactivation. Using the statistical method of analysis, it was determined that a single histidine was critical for enzyme activity with a reaction rate seven times greater than other histidines. Phosphate buffer, pH 6, was utilized for the reaction with 0.1 to 1.0 mM reagent. Reaction

[66] R. G. Oshima and P. A. Price, *J. Biol. Chem.* **248**, 7522 (1973).
[67] E. W. Miles, this series, Vol. 47, p. 431.
[68] J. Ovadi, S. Libor, and P. Elodi, *Acta Biochim. Biophys. Acad. Sci. Hung.* **2**, 455 (1967).
[69] A. W. Abdulwajid and F. Y.-H. Wu, *Biochemistry* **25**, 8167 (1986).

of lactose repressor protein with diethyl pyrocarbonate resulted in diminished inducer, operator, and nonspecific DNA binding.[70] A maximum of three histidines per subunit were modified along with a single lysine residue; the loss of DNA binding was correlated with histidine modification, and the binding could be restored by the addition of hydroxylamine. The presence of inducer sugars had no effect on histidine modification or loss of DNA-binding activity; however, inducer protected against lysine reaction and prevented loss of inducer binding. An interesting aspect of this study was the absence of evidence for a difference in incorporation of radiolabel with/without inducer present during reaction; diethyl pyrocarbonate can apparently function as a catalyst to cross-linking between carboxylate and amino functions, and the reagent may catalyze internal linkages as well as cross-links between molecules (this type of reaction may occur more readily at lower pHs).[67]

Tyrosine

The modification of the phenolic side chain can be achieved with a measure of specificity using several different types of reagents [Table III(C)]. Reaction in some cases alters the UV/visible absorption properties of the tyrosine and results in a chromophore useful for spectral studies.

Tetranitromethane (TNM)

Nitration of tyrosine in proteins significantly alters the pK of the phenolic group, and reduction to aminotyrosine further alters the pK. Spectral properties are also influenced by this reaction, with nitrotyrosine absorbing in the visible range (~380–430 nm). Alkaline pH favors tyrosine nitration, and it is essential to monitor oxidation of sulfhydryl groups under these conditions. Although rare, histidine, methionine, and tryptophan modification have been observed on occasion. Another concern with tetranitromethane modification is cross-linking between tyrosine residues (both intra- and intermolecular cross-links have been observed); this reaction tends to occur more readily at low pH. For this reason, low concentrations of protein are desirable to minimize the possibility of intermolecular reactions. Determination of cross-linking is required to interpret the effects of tetranitromethane reaction on functional activities. The stability of nitrotyrosine to hydrolysis enables ready assessment of the extent of reaction by amino acid analysis of the product. Reduction of nitrotyrosine to aminotyrosine can be accomplished using $Na_2S_2O_4$ at slightly elevated

[70] C. F. Sams and K. S. Matthews, *Biochemistry* **27**, 2277 (1988).

pH. Tertiary reactions can be targeted at the p-amino group. Tetranitromethane is soluble in apolar solvents, and reaction cannot be assumed to relate to solvent exposure, as "buried" groups may react due to concentration of the reagent in an apolar region of the molecule.

Gene 5 protein of fd phage has been reacted with tetranitromethane with modification of three of the five tyrosines in the protein.[71] Reaction results in ~100-fold reduced binding affinity for fd DNA. The presence of fd DNA during the reaction protects all tyrosyl residues from nitration. Similarly, nitration of gene 32 protein from bacteriophage T4 results in modification of five of the eight tyrosines with complete loss of DNA-binding activity and protection from nitration in the presence of DNA.[72] The effects of reaction on cysteine residues were not determined in these studies.

In studies of the reaction of the lactose repressor protein, the effects of nitration were obscured by oxidation of cysteine residues.[73] Reaction of cysteine residues with N-ethylmaleimide (which does not affect the functional properties of the lactose repressor) followed by tetranitromethane treatment resulted in loss of DNA-binding activities with modification of primarily two tyrosine residues.[74] Cross-linking observed in this system was minimized by reducing the concentration of the protein. Reduction of the nitrotyrosine to aminotyrosine with $Na_2S_2O_4$ restored nonspecific DNA binding and partially restored operator DNA binding. The presence of nonspecific DNA during the modification with tetranitromethane reduced nitration and protected DNA-binding capabilities. Topoisomerases I and II from *Micrococcus luteus* are inactivated by treatment with tetranitromethane, and activity is protected by the presence of DNA in the reaction mixture.[75] Unfortunately, this study did not determine effects on other amino acid side chains.

Iodination

A variety of methods are available for iodination of proteins, although side reactions may occur due to the oxidizing capacity of the reagents. This reaction is particularly useful for the introduction of radiolabel at specific sites in proteins. Iodine/iodide solutions at alkaline pH are nor-

[71] R. A. Anderson, Y. Nakashima, and J. E. Coleman, *Biochemistry* **14**, 907 (1975).

[72] J. E. Coleman, K. R. Williams, G. C. King, R. V. Prigodich, Y. Shamoo, and W. H. Konigsberg, *J. Cell. Biochem.* **32**, 305 (1986).

[73] M. E. Alexander, A. A. Burgum, R. A. Noall, M. D. Shaw, and K. S. Matthews, *Biochim. Biophys. Acta* **493**, 367 (1977).

[74] W.-T. Hsieh and K. S. Matthews, *J. Biol. Chem.* **256**, 4856 (1981).

[75] L. Klevan and Y.-C. Tse, *Biochim. Biophys. Acta* **745**, 175 (1983).

mally utilized; it is important to remove sulfhydryl reagents from the buffer prior to reaction. It is also possible to iodinate using iodine monochloride (ICl) at slightly alkaline pH or by peroxidase/hydrogen peroxide/sodium iodide. Cross-linking has been observed in iodination reactions, and this by-product should be monitored carefully in the analysis of the reaction product. The lactose repressor protein has been iodinated with I_2/KI in Tris-HCl at pH 7.5; this reaction resulted in loss of operator-binding activity with minimal effect on inducer binding.[76] The presence of operator DNA during reaction provided partial protection of binding activity and diminished modification; it should be noted that in this case oxidation of cysteine residues was observed, an occurrence that complicates interpretation of the effects of tyrosine iodination on binding activity.

Tryptophan

Tryptophan is sensitive to oxidation by a variety of agents, including N-bromosuccinimide (NBS), UV irradiation, and photosensitized oxidation [Table III(D)]. Unfortunately, most of these methods are not specific, and other amino acids are affected. Thus, it is essential to determine the effect of modification on residues other than tryptophan (e.g., cysteine, methionine, tyrosine). Assessment of tryptophan modification is complicated by the instability of this amino acid to many protocols for acid hydrolysis preparatory to amino acid analysis.

N-Bromosuccinimide

Modification with NBS requires recrystallization of the reagent from water immediately prior to each use, and halide ions should be avoided in the buffer system. Low molar ratios of reagent to tryptophan are sufficient to achieve modification, and greater specificity for tryptophan is observed at acid pH values. Reaction can be followed by observing the decrease in absorbance at 280 nm; although the extent of modification can be estimated spectrophotometrically, the decrease in absorbance can undergo reversal in the presence of excess reagent and other products of the reaction. Thus, it is essential to evaluate the extent of tryptophan oxidation by alternative means, e.g., amino acid analysis of protein hydrolyzed under conditions to preserve tryptophan. The reaction of NBS with lactose repressor protein illustrates the problems encountered at neutral to basic pH (required for stability of this protein); cysteine is as reactive as tryptophan, and methionine and tyrosine modification was also observed at molar ratios up to 20-

[76] T. G. Fanning, *Biochemistry* **14**, 2512 (1975).

fold.[77] Absorption and fluorescence spectra of modified protein can provide information on the contribution of specific tryptophans to the spectral properties of the parent protein.[78] Similarly, RNA polymerase is completely inactivated by modification with NBS (25-fold molar ratio, pH 7.9), even when the sulfhydryl residues in the protein are protected by DTNB.[79] Spectral measurements indicated oxidation of approximately two tryptophan residues, and the affinity for ATP was decreased fourfold by the reaction. However, protection against inactivation by NBS was not observed in the presence of DNA/ATP/GTP. Circular dichroism spectra indicated no differences between modified and unmodified RNA polymerase. These two studies indicate the significant problems in achieving specificity for this reaction, and any efforts to modify tryptophan must include careful attention to potential side reactions.

Oxidation

Direct photooxidation of tryptophan, sensitized photochemical reaction of this side chain, and selective free-radical oxidation have all been utilized to examine the role of tryptophan in DNA-binding proteins. Trichloroethanol has been used as a sensitizing agent for oxidizing gene 32 protein from phage T4 (gp32); a decrease in absorbance at 280 nm was observed, with an accompanying increase at wavelengths less than 260 nm and greater than 290 nm.[80] Only one type of photoproduct appears to be produced in this reaction.[81] Three tryptophan residues were modified, and the two remaining intact tryptophan residues exhibit minimal fluorescence emission (presumably due to energy transfer to the photoproduct). Irradiation in the presence of trichloroethanol resulted in diminished affinity for DNA; protection of tryptophan by the presence of DNA was not investigated in this study. Similar experiments on gp32 have been executed using oxidation by selective free-radical anions at slightly acidic pH. The radical anions I_2^-, Br_2^-, and SCN_2^- were produced by steady state γ radiolysis.[82] Low irradiation doses were sufficient to decrease binding to DNA; in addition to tryptophan oxidation, cysteine was also affected by this reaction. As cysteine oxidation also partially inhibits DNA-binding activity, differentiation of the effect of modification of sulfhydryl and

[77] R. B. O'Gorman and K. S. Matthews, *J. Biol. Chem.* **252**, 3565 (1977).
[78] R. B. O'Gorman and K. S. Matthews, *J. Biol. Chem.* **252**, 3572 (1977).
[79] B. Wasylyk and A. D. B. Malcolm, *Biochem. Soc. Trans.* **3**, 654 (1975).
[80] J.-J. Toulmé, in "Progress in Tryptophan and Serotonin Research," p. 853. de Gruyter, Berlin, 1984.
[81] J.-J. Toulmé, T. LeDoan, and C. Hélène, *Biochemistry* **23**, 1195 (1984).
[82] J. R. Casas-Finet, J.-J. Toulmé, C. Cazenave, and R. Santus, *Biochemistry* **23**, 1208 (1984).

indole groups is complex. However, the presence of ssDNA protects the tryptophans and one of the sulfhydryl residues from oxidation, consistent with the interpretation that the tryptophans are required for gp32 activity.

Ultraviolet irradiation of the lactose repressor protein (without sensitizer) results in photooxidation of 1 Eq of the two tryptophan residues per monomer with the formation of N-formylkynurenine.[83] The formation of this photoproduct results in quenching of fluorescence of the remaining tryptophan by energy transfer and precludes photoreaction at the second site. The photooxidation occurs with equal frequency at each of the two tryptophan residues without ligand present, but in the presence of inducer, one tryptophan is protected while the other is slightly more vulnerable to oxidation.

Conclusion

Chemical modification is an effective tool for evaluating the participation of specific types of side chains in the functional properties of a DNA-binding protein. Examples of a variety of reactions have been presented to provide an overview of the range of reagents and their specificity. Careful selection of reagent and reaction conditions as well as controls utilized is essential for interpretation of results. In addition, it is imperative that assessment of reaction products include not only the targeted amino acid based on apparent reagent specificity, but also all potential by-products of reactions with other amino acids. Finally, the effects of reaction on all activities and on the structural integrity of the protein must be determined; coupled with this measurement, data on ligand protection from reaction may be useful in interpreting the role of specific side chains in the functional properties of the protein. Application of this technique without thorough preparation and careful experimental design can result in misinterpretation of results that in turn misleads other investigators. Despite the limits inherent in chemical modification of proteins and the pitfalls of crafting the appropriate protocol for a given protein, this method coupled with other approaches, genetic studies and physical measurements in particular, has yielded significant information about the contribution of specific amino acids to the structure and function of DNA-binding proteins and has a place in the experimental repertoire applied to investigations of this class of proteins.

Acknowledgments

Support by NIH GM22441 and Welch C-576 grants for our work reported here is gratefully acknowledged.

[83] M. Spodheim-Maurizot, M. Charlier, and C. Hélène, *Photochem. Photobiol.* **42,** 353 (1985).

[24] Characterization of Protein–DNA Complexes by Affinity Cleaving

By Peter B. Dervan

High-resolution crystallographic views of protein–DNA complexes reveal the structural complexity of protein–DNA interactions.[1-5] The combination of direct protein–DNA contacts mediated by multiple hydrogen bonds and sequence-dependent DNA conformational effects limits our ability to make detailed structural predictions, even if a new DNA-binding protein can be assigned to a structural class such as helix–turn–helix,[1-4] double-barreled helix,[5] zinc-binding finger,[6] or scissor grip-leucine zipper.[7] In the absence of high-resolution crystallographic and nuclear magnetic resonance (NMR) data, solution methods such as affinity cleaving can be used to characterize the topology of protein–DNA complexes and correlate sequence similarities with known structural classes.

The conversion of a sequence-specific DNA-binding protein into a sequence-specific DNA-cleaving protein by covalent attachment of the iron chelator, ethylenediaminetetraacetic acid (EDTA), to a specific amino acid residue creates a class of hybrid affinity-cleaving proteins that are available through chemical synthesis. Moreover, a structural domain consisting of naturally occurring amino acids that binds transition metals and oxidatively cleaves DNA extends this method to recombinant methods for protein synthesis.

[1] A. K. Aggarwal, D. W. Rodgers, M. Drottar, M. Ptashne, and S. C. Harrison, *Science* **242,** 899 (1988).

[2] J. E. Anderson, M. Ptashne, and S. C. Harrison, *Science* **326,** 846 (1987).

[3] Z. Otwinowski, R. W. Schevitz, R. G. Zhang, C. L. Lawson, A. Joahimiak, R. Q. Marmorstein, B. F. Luisi, and P. B. Sigler, *Nature (London)* **335,** 321 (1988).

[4] S. R. Jordan and C. O. Pabo, *Science* **242,** 893 (1988).

[5] J. A. McClarin, C. A. Frederick, B. C. Wang, P. Greene, H. W. Boyer, J. Grable, and J. M. Rosenberg, *Science* **234,** 1526 (1986).

[6] J. Miller, A. D. McLachlan, and A. Klug, *EMBO J.* **4,** 1609 (1985); J. M. Berg, *Proc. Natl. Acad. Sci. U.S.A.* **85,** 99 (1988); G. Párraga, S. J. Horvath, A. Eisen, W. E. Taylor, L. E. Hood, E. T. Young, and R. E. Klevit, *Science* **241,** 1489 (1988); M. S. Lea, G. P. Gippert, R. V. Saman, D. A. Case, and P. Wright, *Science* **245,** 635 (1989).

[7] W. H. Landschulz, P. F. Johnson, and S. L. McKnight, *Science* **240,** 1759 (1988); C. R. Vinson, P. B. Sigler, and S. L. McKnight, *Science* **246,** 911 (1989).

Affinity Cleaving

Attachment of EDTA · Fe to a DNA-binding moiety creates a DNA-cleaving molecule that functions under physiologically relevant pH, temperature, and salt conditions.[8] The cleavage reaction can be initiated by addition of a reducing agent such as dithiothreitol or sodium ascorbate.[8] If the DNA-binding molecule is sequence specific, the EDTA · Fe cleaves at highly localized sites on DNA restriction fragments and plasmids.[9–16] Because the EDTA · Fe-cleaving moiety is not sequence specific, the cleavage specificity is determined only by the binding specificity of the molecule being investigated. EDTA · Fe-equipped DNA-binding molecules cleave DNA by oxidation of the deoxyribose backbone via a diffusible oxidant, presumably hydroxyl radical.[9–16] Cleavage of both DNA strands is observed and typically extends over 4 to 6 bp.[9–16] Due to the right-handed nature of double-helical DNA, the groove in which the EDTA · Fe is located can be identified by cleavage pattern analysis.

Cleavage Pattern Analysis

Affinity cleaving has been used to study the sequence-specific recognition of double-helical DNA by naturally occurring DNA-binding antibiotics[9,10], peptide analogs that bind in the minor groove,[11] and oligonucleotide-

---

[8] R. P. Hertzberg and P. B. Dervan, *J. Am. Chem. Soc.* **104**, 313 (1982); R. P. Hertzberg and P. B. Dervan, *Biochemistry* **23**, 3934 (1984).

[9] P. G. Schultz, J. S. Taylor, and P. B. Dervan, *J. Am. Chem. Soc.* **104**, 6861 (1982); J. S. Taylor, P. G. Schultz, and P. B. Dervan, *Tetrahedron* **40**, 457 (1984); P. G. Schultz and P. B. Dervan, *J. Am. Chem. Soc.* **105**, 7748 (1983).

[10] P. B. Dervan, *Science* **232**, 464 (1986).

[11] R. S. Youngquist and P. B. Dervan, *Proc. Natl. Acad. Sci. U.S.A.* **82**, 2565 (1985); R. S. Youngquist and P. B. Dervan, *J. Am. Chem. Soc.* **107**, 5528 (1985); J. H. Griffin and P. B. Dervan, *J. Am. Chem. Soc.* **108**, 5008 (1986); J. P. Sluka and P. B. Dervan, *in* "New Synthetic Methodology and Functionally Interesting Compounds, Proceedings of the 3rd International Kyoto Conference on New Aspects of Organic Chemistry" (Z. I. Yoshida, ed.), Vol. 25, p. 307. Elsevier, New York, 1986; R. S. Youngquist and P. B. Dervan, *J. Am. Chem. Soc.* **109**, 7564 (1987); W. S. Wade and P. B. Dervan, *J. Am. Chem. Soc.* **109**, 1574 (1987); J. H. Griffin and P. B. Dervan, *J. Am. Chem. Soc.* **109**, 6840 (1987).

[12] H. E. Moser and P. B. Dervan, *Science* **238**, 645 (1987); L. C. Griffin and P. B. Dervan, *Science* **245**, 967 (1989).

[13] J. P. Sluka, S. J. Horvath, M. F. Bruist, M. I. Simon, and P. B. Dervan, *Science* **238**, 1129 (1987); J. P. Sluka, S. J. Horvath, A. C. Glasgow, M. I. Simon, and P. B. Dervan, *Biochemistry* **29**, 6551 (1990); D. P. Mack, J. A. Shin, J. H. Griffin, M. I. Simon, and P. B. Dervan, *Biochemistry* **29**, 6561 (1990).

[14] M. G. Oakley and P. B. Dervan, *Science* **248**, 847 (1990).

[15] K. Graham and P. B. Dervan, *J. Biol. Chem.* **265**, 16534 (1990).

[16] J. P. Sluka, J. H. Griffin, D. P. Mack, and P. B. Dervan, *J. Am. Chem. Soc.* **112**, 6369 (1990).

directed recognition of the major groove by triple-strand formation.[12] From these studies, which involve less complicated DNA-binding motifs, numerous cleavage patterns caused by a diffusible oxidant produced in either the minor or the major groove of duplex DNA have been analyzed. EDTA · Fe located in the minor groove generates an asymmetric cleavage pattern with maximal cleavage loci shifted to the 3' side on opposite strands. When the EDTA · Fe is located in the major groove, the maximal cleavage loci are 5' shifted; in addition, cleavage of lower efficiency occurs on the distal strands of the adjacent minor grooves.[13] This results in an overall pattern of a *pair* of 3'-shifted asymmetric cleavage loci of *unequal* intensity on opposite strands. These patterns can be explained if the diffusible radical generated from the localized EDTA · Fe reacts with the major and minor grooves of DNA with unequal rates and preferentially (although not necessarily exclusively) in the minor groove.[17] This database allows interpretation of the affinity-cleaving results from conformationally more complex proteins (Fig. 1).

Characterization of Protein–DNA Structures by Affinity Cleaving

Incorporation of EDTA · Fe at discrete amino acid residues within a protein allows the position and groove location of the modified residues to be mapped to nucleotide resolution.[13-16] The secondary and tertiary structures of DNA–protein complexes can be analyzed by affinity cleaving using two approaches. First, the amino acid location of the EDTA on a polypeptide chain of constant length may be varied in order to reveal key topological features of the protein–DNA complex (e.g., location of NH_2 versus COOH terminus).[13-15] Moreover, the length of the polypeptide may be incrementally changed while the EDTA is kept at the same terminus to examine the influence of substructures on the binding affinity of the protein and the conformational flexibility of a specific peptide fragment.[13,14]

Sequence Specificity Mapping

Protein–EDTA · Fe can cause sequence-specific double-strand breaks on DNA.[13] The locations of site-specific protein binding on large DNA such as plasmids can be mapped by nondenaturing agarose gel electrophoresis. This provides information about DNA sites of similar sequence that

[17] T. D. Tullius and B. A. Dombrowski, *Science* **230**, 679 (1985); T. D. Tullius and B. A. Dombrowski, *Proc. Natl. Acad. Sci. U.S.A.* **83**, 5469 (1986); T. D. Tullius, B. A. Dombroski, M. E. A. Churchhill, and L. Kam, this series. Vol. 155, p. 537.

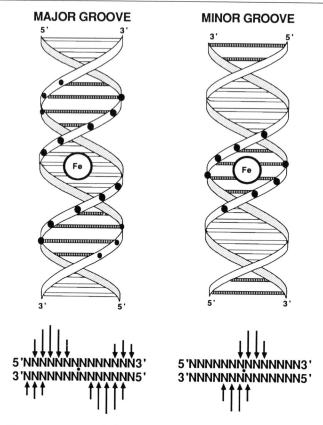

Fɪɢ. 1. Cleavage pattern analysis. Filled circles represent points of cleavage along the phosphodiester backbone. Sizes of circles represent extent of cleavage.

have affinity for the protein, which bears on the assignment of DNA base pair–protein interactions critical for sequence-specific recognition.

Methods

Synthetic procedures for the introduction of the metal chelator, EDTA, at unique amino acid positions of proteins by solid-phase methods have been described.[16] Two protected derivatives of EDTA compatible with Merrifield solid-phase protein synthesis employing *N-tert*-butyloxycarbonyl (Boc)-protected amino acids were developed. The first reagent is a dipeptide with three of four carboxyl groups of EDTA protected as benzyl

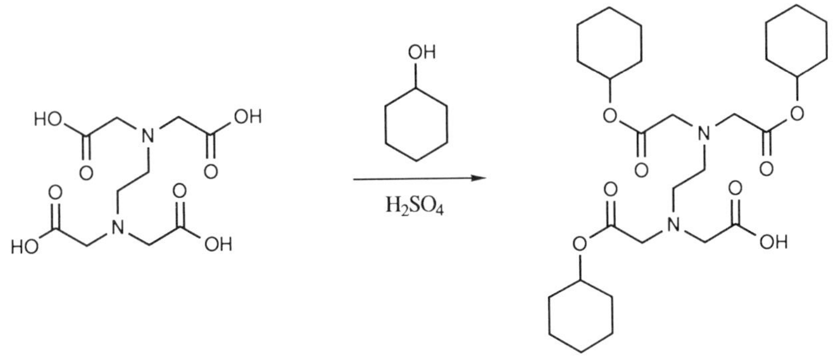

FIG. 2. Synthesis of tricyclohexyl ester of EDTA (TCE).[16]

esters and the fourth coupled to a γ-aminobutanoic acid linker, referred to as tribenzyl-EDTA–GABA (BEG).[13,16] A second reagent is the tricyclohexyl ester of EDTA, TCE.[16] BEG and TCE allow the modification of the NH_2 terminus and/or lysine side chains of resin-bound peptides and proteins. Due to the ease of synthesis, stability, coupling efficiency, and flexibility in choice of linker, TCE is the likely reagent for most applications. On deprotection and cleavage from the resin, a protein is produced with EDTA at a defined amino acid position. The availability of protein–EDTA conjugates extends the affinity-cleaving method to the study of protein–DNA complexes in solution.

Synthesis of TCE[16]

A stirred slurry of EDTA (10 g, 34 mmol) in cyclohexanol (70 ml) containing concentrated sulfuric acid (2 ml) is heated at 115° for 10 hr, at which time nearly all of the EDTA has dissolved. The mixture is cooled to room temperature, poured into saturated aqueous sodium bicarbonate solution (500 ml), and extracted with ether (twice with 200 ml). The combined extracts are dried (Na_2SO_4) and concentrated under reduced pressure and then *in vacuo* to afford a thick gum. The product is purified by flash chromatogrpahy [3% (v/v) CH_3OH in CH_2Cl_2 followed by 30% CH_3OH in CH_2Cl_2] to afford TCE as a brittle foam (3.9 g, 21%). ^{1}H NMR (deuterium-labeled dimethylsulfoxide, Me_2SO-d_6): δ 4.68 (m, 3H), 3.50 (s, 4H), 3.49 (s, 2H), 3.37 (s, 2H), 2.71 (s, 4H), 1.78–1.21 (m, 30H) ppm; infrared (IR; film) 3448, 2937, 2858, 1735, 1634, 1452, 1198 cm^{-1} (see Fig. 2).

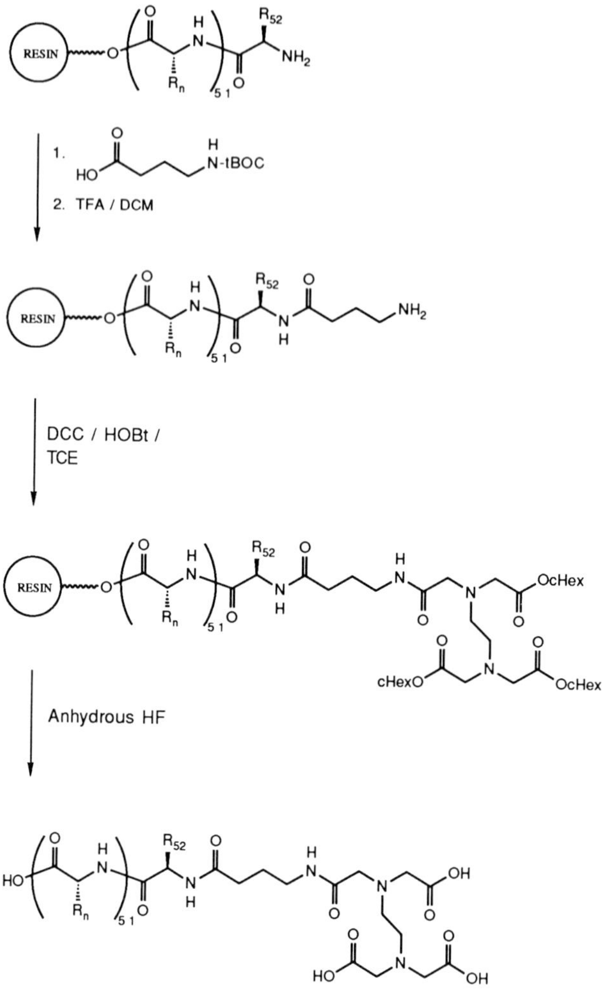

FIG. 3. Scheme for the attachment and deprotection of TCE to the NH₂ terminus of a protein using solid-phase methods.[16]

Peptide Syntheses[16]

Peptides and proteins are prepared manually or by automated methods employing *N-tert*-butyloxycarbonyl (Boc) amino acid derivatives for Merrifield solid-phase synthesis.[18] *N-α*-Boc-L-amino acids are used with the following side-chain protecting groups: Arg(Tos), Asp(OBzl), Glu(OBzl), His-dinitrophenol (DNP), Lys(Cl-Z), Ser(Bzl), Thr(Bzl), Trp-carbohydrate (CHO), and Tyr(Br-Z). Coupling yields are determined by quantitative ninhydrin monitoring, with acceptable values being ≥99.7% in the beginning of the synthesis and gradually decreasing to 99% near the end.

TCE Coupling to NH_2 Terminus of Resin-Bound Peptides (Fig. 3)

A typical procedure for the coupling of TCE to the NH_2 terminus of resin-bound peptide is as follows.[16] An ~100-mg sample of peptide/resin (ca. 100 μmol/g, total peptide ca. 10 μmol) is placed in a 12 × 80 mm reactor, and the resin is swollen in CH_2Cl_2 for 15 min. The resin is washed five times with CH_2Cl_2, the terminal Boc-protecting group is removed, and the resin neutralized using standard procedures. TCE (125 mg, 194 μmol) and HOBt (50 mg, 330 μmol) are dissolved in dimethylformamide (DMF, 2 ml) and DCC (42 mg, 203 μmol) is added. The solution is stirred 30 min at room temperature and added to the peptide/resin along with sufficient DMF to fill the reactor two-thirds full. Ninhydrin analysis indicates ≥98% reaction yield after 2–3 hr. The peptide/resin is washed four times with DMF and four times with CH_2Cl_2.

TCE Coupling to ε-NH_2 Side Chain of Lysine Residue (Fig. 4). N-α-Boc-N-ε-Fmoc-L-lysine is coupled to the growing peptide as an HOBt ester.[16] The resin is washed five times with CH_2Cl_2, followed by five washes with DMF. The N-ε-Fmoc group is selectively removed using 20% piperidine in DMF (a 1-min reaction step followed by a 10-min step) and the resin is washed five times with DMF. TCE (1.08 g, 2 mmol, 4 Eq based on a 0.5-mmol synthesis) and HOBt (0.46 g, 3.4 mmol, 6.8 Eq) are dissolved in DMF (2 ml) and DCC (0.41 g, 2 mmol, 4 Eq) is added. The solution is stirred 30 min at room temperature and added to the peptide resin along with sufficient DMF to fill the reactor vessel two-thirds full. Ninhydrin analysis indicates 99.9% reaction after 1 hr. The peptide resin is washed five times with DMF and five times with CH_2Cl_2. The N-α-Boc is removed as usual with trifluoroacetic acid (TFA) and the synthesis continued with the standard protocols.

Deprotection and Purification of Protein–EDTA

Deprotection of the resin-bound peptide is carried out as described by Kent.[19] The histidine-protecting group, DNP, is removed at 25° using 20% (v/v) 2-mercaptoethanol and 10% (v/v) DIEA in DMF; this treatment is repeated twice (30 min each). After removal of the N-α-Boc group with TFA and drying of the resin, all other side chain-protecting groups as well as the peptide–resin bond are cleaved using anhydrous HF, in the presence

---

[18] R. B. Merrifield, *Adv. Enzymol.* **32,** 221 (1969).
[19] S. B. H. Kent, *Annu. Rev. Biochem.* **57,** 957 (1988).

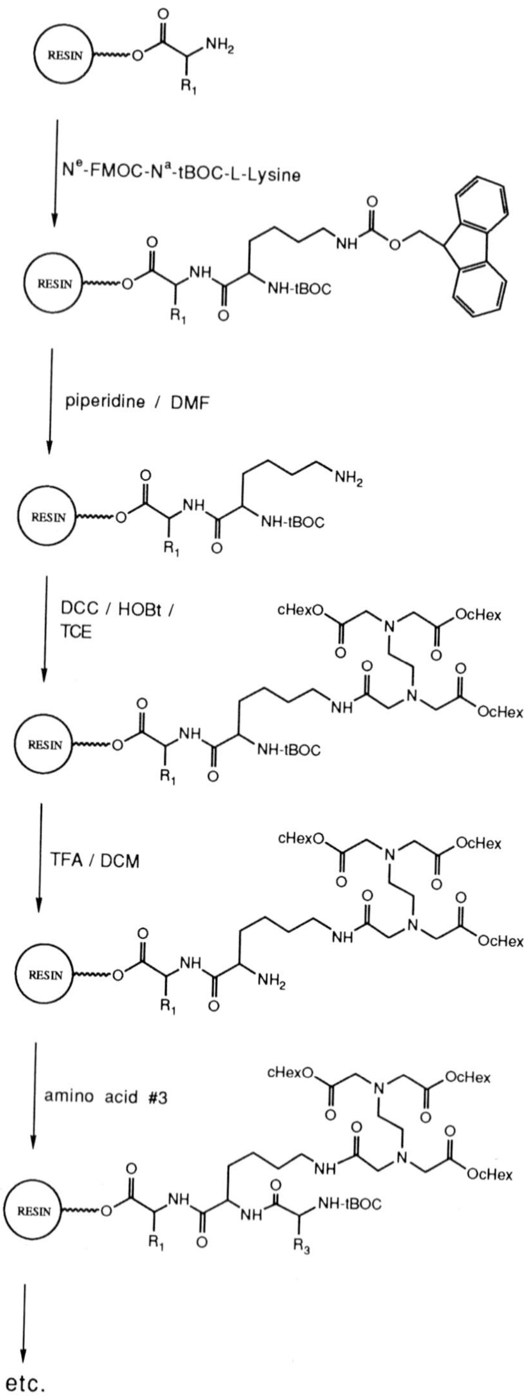

RESIN
NH2
R1
Ne-FMOC-Na-tBOC-L-Lysine
NH-tBOC
piperidine / DMF
DCC / HOBt / TCE
cHexO
OcHex
TFA / DCM
amino acid #3
R3
etc.

of p-cresol and p-thiocresol as scavengers, for 60 min at 0°. For peptides containing a Trp residue, the scavengers used are anisole and 1,4-butanedithiol. The HF is removed under vacuum. The crude protein is precipitated with diethyl ether, collected on a fritted funnel, dissolved in water, and filtered through, leaving the resin on the frit. A small sample is then removed, microfiltered, and subjected to analytical high-performance liquid chromatography (HPLC) [Vydac, 25 × 4.6 mm C_4 column (Separations Group, Hesperia, CA), 0 to 60% acetonitrile/0.1% TFA over 60 min]. The remaining solution is frozen and lyophilized. Residual DNP groups are removed from the crude protein by treatment in 4 M guanidine hydrochloride, 50 mM Tris, pH 8.5, and 20% 2-mercaptoethanol for 1 hr at 50°.[20] This solution is injected onto a semipreparative C_4 HPLC column (25 × 1 cm) and developed with H_2O/0.1% TFA until the guanidine and 2-mercaptoethanol have eluted. A gradient of 0–60% acetonitrile/0.1% TFA is then run over 240 min. Fractions are collected and assayed for the presence of the desired protein by analytical HPLC.

Cleavage Reactions by Protein–EDTA · Fe on DNA Restriction Fragments

Protein–EDTA · Fe cleavage reactions on DNA restriction fragments labeled at the 5′ (and 3′) ends are run in a total volume at 15 μl. Final concentrations are 20 mM phosphate, pH 7.5, 20 mM NaCl, 100 μM in base pairs of calf thymus DNA, ~15,000 cpm of ^{32}P-end-labeled DNA restriction fragment, 5 mM dithiothreitol (DTT), and 5 μM protein–EDTA · Fe. The protein–EDTA · Fe is allowed to equilibrate with the DNA for 10 min at 25°; cleavage is then initiated by the addition of DTT and allowed to proceed for 60 min at 25°. The reactions are terminated by ethanol precipitation, dried, and resuspended in 5 ml of 100 mM Tris–borate–EDTA, 80% formamide solution. The ^{32}P-labeled DNA products are analyzed by denaturing polyacrylamide gel electrophoresis followed by autoradiography. Densitometric analysis of the gel autoradiogram and comparison of individual lanes with sequence marker lanes allows assignment of DNA cleavage to nucleotide resolution.

[20] N. Camerman, A. Camerman, and B. Sarkar, J. Chem. 110, 7572 (1988).

FIG. 4. Synthetic scheme for the attachment of TCE to the ε-NH_2 group of a lysine in the second amino acid position on a resin-bound peptide. [D. P. Mack, J. A. Shin, J. H. Griffin, M. I. Simon, and P. B. Dervan, Biochemistry 29, 6561 (1990).]

Double-Strand Cleavage of Plasmid DNA by Protein–EDTA · Fe

Protein–EDTA · Fe cleavage reactions on linearized plasmid DNA ^{32}P-labeled at one (or both) ends are run in a total volume of 45 μl. Final concentrations are 20 mM phosphate, pH 7.5, 20 mM NaCl, 100 μM in base pairs of calf thymus DNA, ~30,000 cpm of ^{32}P-end-labeled linearized plasmid DNA, 5 mM DTT, and 2.5 μM protein–EDTA · Fe. The protein–EDTA · Fe is allowed to equilibrate with the DNA for 10 min at 25°; cleavage is then initiated by the addition of DTT and allowed to proceed for 60 min at 25°. The reaction is terminated by the addition of 5 μl of 10× Ficoll solution (25% in water, type 400; Sigma, St. Louis, MO). The ^{32}P-labeled DNA products are analyzed on a 4 mm × 20 cm 1% (w/v) agarose gel. Densitometric analysis of the gel autoradiogram and comparison of individual lanes with molecular weight markers allow assignment of cleavage loci to within 20 bp.

Cleavage Reactions by Ni · GGH–Protein on DNA
Restriction Fragments

Nickel-mediated Gly-Gly-His (GGH)–protein cleavage reactions on DNA restriction fragments labeled with ^{32}P at the 5' (and 3') ends are run in a total volume of 20 μl. Final concentrations are 20 mM phosphate, pH 7.5, 20 mM NaCl, 100 μM in base pairs of calf thymus DNA, ~15,000 dpm of ^{32}P-end-labeled DNA restriction fragment, 5 μM monoperoxyphthalic acid, magnesium salt, and 5 μM Ni · GGH–protein. The Ni · GGH–protein is allowed to equilibrate with the DNA for 10 min at 25°; cleavage is then initiated by the addition of monoperoxyphthalic acid and allowed to proceed for 15 min at 25°. The reactions are terminated by ethanol precipitation, dried, and resuspended in 50 μl of 0.1 M *n*-butylamine and heated to 90° for 30 min. The reactions are then dried by lyophilization and resuspended in 5 μl of 100 mM Tris–borate–EDTA, 80% formamide solution. The ^{32}P-labeled products are analyzed by denaturing polyacrylamide gel electrophoresis, followed by autoradiography. Densitometric analysis of the gel autoradiogram and comparison of individual lanes with sequence marker lanes allow assignment of DNA cleavage to nucleotide resolution.

Discussion

Synthetic Procedure and Experimental Design

Protecting groups for the EDTA carboxylic acids are chosen based on standard side chain-protection methods used for glutamic and aspartic acids in Merrifield solid-phase protein synthesis employing *N-tert*-butyloxycarbonyl (Boc) amino acid derivatives. In BEG, three of the four carboxyl groups of EDTA are protected as benzyl esters. The fourth carboxyl is coupled via an amide bond to a γ-aminobutanoic acid (GABA) linker designed to minimize disruptions of protein structure by the attached EDTA $\cdot$ Fe chelates (Fig. 2). The protected dipeptide, tribenzyl-EDTA-GABA (BEG), can be synthesized in four steps.[16] The utility of BEG was demonstrated by the introduction of EDTA at the NH_2 terminus of the DNA-binding domain from *Hin* recombinase, residues 139–190.[13]

The tricyclohexyl ester of EDTA (TCE) has been developed as an improved alternative to BEG.[16] TCE may be prepared on a larger scale than BEG, is more stable than BEG, and offers more flexibility than BEG in that a variety of amino acid linkers (or no linker at all) may be used. TCE was synthesized in a single step by incomplete esterification of EDTA with cyclohexanol (Fig. 3). TCE couples with high efficiency to the NH_2 terminus or ε-NH_2 group on the side chain of a lysine residue at the COOH terminus of resin-bound peptides (>98%) in the presence of DCC and HOBt in DMF.

Although BEG and TCE are formally amino acids, they are suitable only for capping an amino terminus or lysine side chain since they do not have a primary or secondary amino group for further chain extension. Current methodology only allows attachment of EDTA to the NH_2 residues of resin-bound protected peptides. In addition, the approach of transient protection of the ε-NH_2 side chain of lysine with Fmoc, for subsequent coupling with TCE, may be limited to the first few residues at the COOH terminus at the beginning of the peptide chain synthesis. EDTA modification of internal ε-lysine side chains of peptides in the *tert*-Boc-based synthesis has not yet been documented. Thus, BEG and TCE represent only the first steps in the development of a class of modified amino acids compatible with Merrifield synthesis for the construction of hybrid proteins with novel functions.

Cleavage Pattern Analysis of Helix–Turn–Helix Motif Protein–DNA Complexes

This methodology has been used to investigate the tertiary structures of two putative helix–turn–helix protein–DNA complexes and a leucine zipper–DNA complex. The DNA-binding domains of *Hin* recombinase and $\gamma\delta$ resolvase are thought to bind DNA by a helix–turn–helix motif. A synthetic 52-residue protein based on the sequence-specific DNA-binding domain of *Hin* recombinase (residues 139–190) with EDTA $\cdot$ Fe at the

[21] K. Kim, S. G. Rhee, and E. R. Stadtman, *J. Biol. Chem.* **260,** 15394 (1985).

NH$_2$ terminus, Fe · EDTA–Hin(139–190), produced two cleavage patterns on the symmetry axis side of the 13-bp inverted half-sites (see Fig. 5).[13] The cleavage patterns on opposite strands of DNA for Fe · EDTA–Hin(139–190) were shifted to the 3' side at all binding sites. This asymmetry reveals that the NH$_2$ terminus of Hin(139–190) is bound in the minor groove of DNA near the symmetry axis of Hin-binding sites.[13] A binding model put forward for Hin(139–190) includes a helix–turn–helix–turn–helix structure in the major groove with residues at the NH$_2$ terminus extending across the DNA phosphodiester backbone, making specific contacts to the adjacent minor groove.[13] Attachment of EDTA · Fe to a lysine side chain (Ser-183 → Lys-183) at the COOH terminus of Hin(139–184) affords an asymmetric cleavage pattern on opposite strands that is also shifted to the 3' side. Within the context of the helix–turn–helix model, these data are consistent with the putative recognition helix oriented toward the symmetry axis of the site (Fig. 6).[13]

From sequence comparison, the 43-residue DNA-binding domain of γδ resolvase is also thought to bind DNA by a helix–turn–helix motif. Incorporation of EDTA · Fe at the NH$_2$ and COOH termini of γδ(141–183), respectively, affords cleavage patterns similar to Hin(139–190). We have assigned the NH$_2$ terminus bound in the minor groove of DNA near the symmetry axis of γδ-binding sites and the recognition helix oriented toward the symmetry axis of the site.[15]

Cleavage Pattern Analysis of Y-Shape Motif Protein–DNA Complexes

The DNA-binding domain of the transcriptional activator GCN4 is thought to bind DNA by a dimeric Y-shaped leucine zipper motif.[7] EDTA was attached to the NH$_2$ terminus of the DNA-binding domain of GCN4 (residues 222–281).[14] Because both halves of the Fe · EDTA 60-mer dimers contain Fe · EDTA at the NH$_2$ terminus, the affinity-cleavage pattern should result from the sum of two localized Fe · EDTA moieties. The cleavage pattern observed consists of three cleavage loci of unequal intensity. With regard to the cleavage model in Fig. 1, the simplest interpretation of this cleavage pattern is that it results from the superimposition of two adjacent major groove cleavage patterns along one face of the DNA. Cleavage occurs on both strands in three adjacent minor grooves, with the most efficient cleavage in the central minor groove proximal to both Fe · EDTA moieties. These data strongly suggest that the two Fe · EDTA moieties are located in adjacent major grooves. The positions of the Fe · EDTA moieties may be assigned by assuming that they lie in the center of the two 5'-shifted patterns. The Fe · EDTA moieties, and hence the NH$_2$ termini of the dimer, are 9 to 10 bp apart and are located 4 to 5 bp on

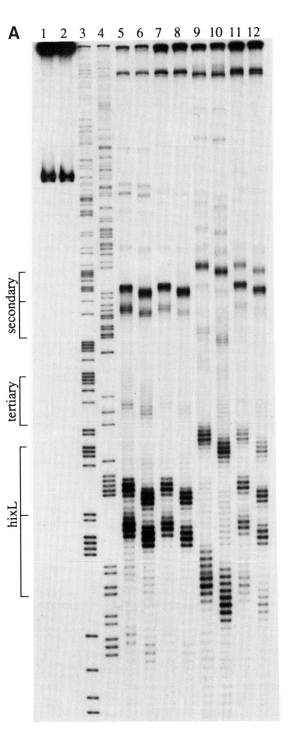

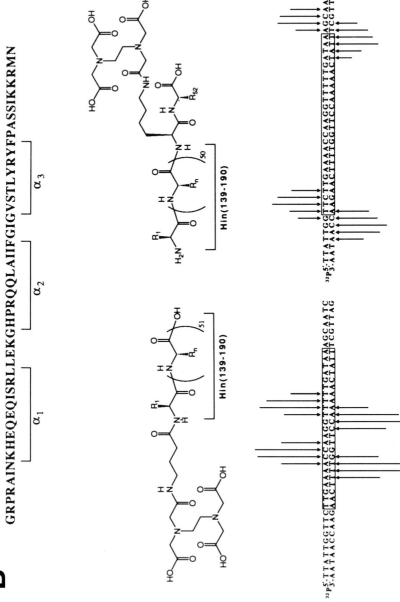

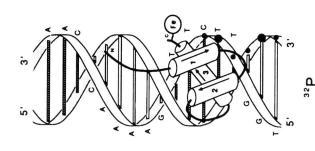

³²P

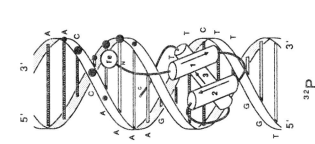

³²P

FIG. 5. (A) Autoradiogram of a high-resolution denaturing polyacrylamide gel of affinity-cleaving reactions on a ³²P-end-labeled restriction fragment containing five binding sites for *Hin*(139–190). [D. P. Mack, J. A. Shin, J. H. Griffin, M. I. Simon, and P. B. Dervan, *Biochemistry* **29**, 6561 (1990).] Odd- and even-numbered lanes contain 5' and 3' end-labeled DNA, respectively. Lanes 1 and 2 are DNA controls and 3 and 4 are A-specific sequencing lanes. Lanes 5 and 6 contain 5 μM Fe · EDTA–*Hin*(139–190), lanes 7 and 8 contain 10 μM Fe · EDTA–*Hin*(139–184), lanes 9 and 10 contain 10 μM *Hin*(139–184)–EDTA · Fe, and lanes 11 and 12 contain 10 μM Fe · EDTA–*Hin*(139–184)–EDTA · Fe. [D. P. Mack, J. A. Shin, J. H. Griffin, M. I. Simon, and P. B. Dervan, *Biochemistry* **29**, 6561 (1990).] HixL is 26 bp long and has nearly 2-fold symmetry. *Hin* binds to HixL and secondary sites as a dimer. Tertiary site binds *hin* weakly. (B) Sequence is the DNA-binding domain of *Hin* recombinase (residues 139–190). Putative α helices are indicated by brackets. Below, left: Fe · EDTA–*Hin*(139–190) mapping the location of the NH₂ terminus of the DNA-binding domain of *Hin* in the minor groove toward the symmetry axis site of the 13-bp *Hin* half-site. Below, right: *Hin*(139–186)–EDTA · Fe maps the location of the COOH terminus of the DNA-binding domain of *Hin* recombinase in the major groove. The putative recognition helix is oriented (N → C) toward the symmetry axis of the *Hin*-binding site. Filled circles represent points of cleavage along the phosphodiester backbone. Sizes of circles represent extent of cleavage. (See Ref. 13.)

PESSDPAALKRARNTEAARRSRARKLQRMK
Basic Region

QLEDKVEELLSKNYHLENEVARLKKLVGER
Leucine Zipper Region

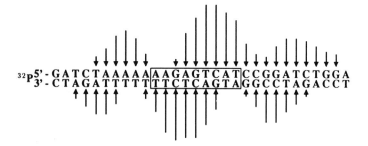

GCN4(222-281)

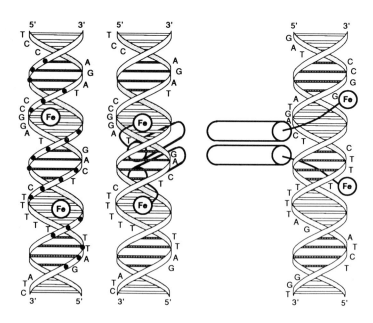

either side of the central C in the binding sites 5'-CTGACTAAT-3' and 5'-ATGACTCTT-3'[14] (see Fig. 6). Although not a proof, this is consistent with a Y-shaped model for a class of DNA-binding proteins important in the regulation of gene expression.

Design of Sequence-Specific DNA-Cleaving Proteins Using Naturally Occurring α-Amino Acids

From X-ray diffraction analysis, the tripeptide, GGH, is known to bind Cu(II) in a square planar complex with coordination from an imidazole nitrogen, two deprotonated peptide nitrogens, and the terminal amino group.[20] Although a crystal structure of GGH · Ni(II) is not available, the Ni(II) complex of GGH has been studied by other techniques.[22] Crystal structures of tetraglycine with Cu(II) or Ni(II) indicate that the metal ions are bound by peptide ligands in a similar fashion.[23]

GGH was attached to the NH_2 terminus of the DNA-binding domain of *Hin* recombinase (residues 139–190) to afford a new 55-residue protein, GGH[*Hin*(139–190)].[24] GGH[*Hin*(139–190)] at 1.0 μM concentrations (pH 7.5, 25°) in the presence of 1 Eq of Ni(OAc)$_2$ and monoperoxyphthalic acid cleaves double-helical DNA at four 13-bp sites (termed *hixL* and secondary) on base workup (see Fig. 7[25]). The cleavage patterns observed at the *hixL*-binding sites are strong and occur predominantly on one strand of each DNA site with single base specificity, while the cleavage at the secondary sites is modest and covers one to two base positions on both DNA strands. Maximal cleavage on opposite strands of the DNA is asymmetric to the 3' side, consistent with the known location of the NH_2 terminus of *Hin*(139–190) in the minor groove. On base workup, the

[22] F. P. Bossu and D. W. Margerum, *Inorg. Chem.* **16,** 1210 (1977); C. E. Bannister, J. M. T. Raycheba, and D. W. Margerum, *Inorg. Chem.* **21,** 1106 (1982); T. Sakurai and A. Nakahara, *Inorg. Chim. Acta* **34,** L243 (1979).

[23] H. C. Freeman and M. R. Taylor, *Acta Crystallogr.* **18,** 939 (1965); H. C. Freeman, J. M. Guss, and R. L. Sinclair, *Chem. Commun.* (1968); F. P. Bossu, E. B. Paniago, D. W. Margerum, S. T. Kirsey, and J. L. Kurtz, *Inorg. Chem.* **17,** 1034 (1978).

[24] D. P. Mack, B. L. Iverson, and P. B. Dervan, *J. Am. Chem. Soc.* **110,** 7572 (1988).

[25] D. P. Mack and P. B. Dervan, *J. Am. Chem. Soc.* **112,** 4604 (1990).

FIG. 6. Sequence is the DNA-binding domain of GCN4 (residues 222–281). Putative dimerization domain and DNA contact basic region are indicated by brackets. Fe · EDTA-GCN4(222–281) reveals that the location of the NH_2 termini of a dimer of this DNA-binding protein are separated by 9 to 10 bp in the major groove on the same face of the DNA. (See Oakley and Dervan.[14])

GGHGRPRAINKHEQEQISRLLEKGHPRQQLAIIFGIGVSTLYRYFPASSIKKRMN

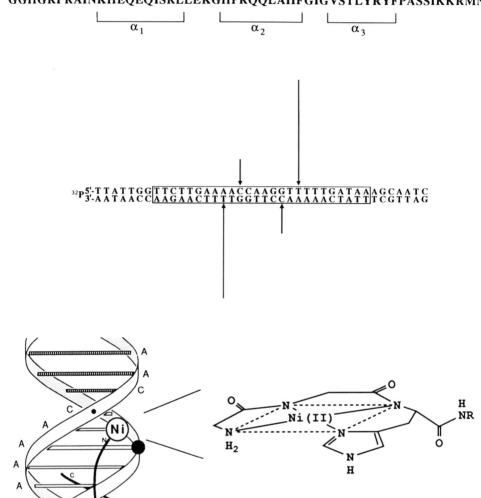

DNA termini at the cleavage site are 3'- and 5'-phosphate consistent with oxidative degradation of the deoxyribose backbone.

The small number of deoxyribose cleavage positions within each binding site implies that a nondiffusible oxidant is generated by the Ni(II) · GGH[(*Hin*(139–190))]–peracid reactions. One explanation for the difference in reactivity observed for the sites is that the structure of the DNA is different at each site, and the nondiffusible reactive moiety on the metalloprotein is sensitive to these sequence-dependent conformational differences. The highly reactive nature of the oxidizing species, the precision of the cleavage reaction, and the identity of the DNA termini created on treatment with butylamine imply that the species may be a high-valent nickel-bound oxygen that abstracts a specific hydrogen atom on the deoxyribose backbone.[25]

Future Applications: Protein Affinity Cleaving

Because the reactive oxidant generated from EDTA · Fe can cause cleavage of proteins,[21,26] protein–EDTA · Fe molecules can be used for affinity-cleaving studies of protein structure in the presence and absence of DNA. In principle, analyses of protein fragments caused by protein autocleavage of protein–EDTA · Fe complexes should reveal the proximal location of the amino acid equipped with EDTA · Fe and its neighbor amino acid sequences in three-dimensional space.

Acknowledgments

This work was supported by grants from the DARPA University Research Initiative Program and the National Foundation for Cancer Research. I am indebted to my students and co-workers who made this work possible; in particular, J. P. Sluka, D. P. Mack, J. H. Griffin, K. Graham, J. A. Shin, and M. G. Oakley. We are grateful for helpful discussions and collaborations with the M. J. Simon research group and for generous technical advice from S. B. H. Kent and S. J. Horvath.

[26] T. M. Rana and C. F. Meares, *J. Am. Chem. Soc.* **112**, 2457 (1990).

FIG. 7. Sequence of designed protein with GGH at the NH$_2$ termini of the DNA-binding domain of *Hin* recombinase. Nickel-mediated oxidative cleavage occurs in the minor groove predominantly on one strand of the *Hin*-binding site by a nondiffusible oxidant (see Mack *et al.*[24]; Mack and Dervan,[25]). Circle represents point of cleavage.

[25] Identification of Amino Acid Residues at Interface of Protein–Nucleic Acid Complexes by Photochemical Cross-Linking

By KENNETH R. WILLIAMS and WILLIAM H. KONIGSBERG

Introduction

This chapter will focus on providing a general approach for identifying amino acid residue(s) that have been cross-linked to nucleic acids using light. Ultraviolet light can be employed to detect the presence of specific protein–nucleic acid complexes. Because irradiation of these complexes with ultraviolet light will produce covalent linkages between amino acid residues and nucleic acid bases with minimal perturbation of the system being studied, this technique can be used to identify residues at the interface between these species. The reaction is thought to occur via a free radical mechanism where the radical is generated by photoexcitation of a nucleic acid base followed by abstraction of a hydrogen atom from a favorably positioned amino acid residue to produce a purinyl or, much more likely, a pyrimidinyl radical along with a radical on the side chain of the amino acid residue that serves as the hydrogen atom donor. The formation of the covalent bond results from the geminate recombination of the purinyl or pyrimidinyl radical with the corresponding radical on the proximate amino acid residue.[1] For this reason, ultraviolet photolysis produces zero-length cross-links, in contrast to chemical cross-linking agents, which interpose spacers of varying distance between the protein and the nucleic acid. Since model studies suggest that virtually any amino acid residue can participate in photolysis, the promiscuous nature of the ultraviolet induced reaction makes it very useful for probing the interface of these complexes (for reviews see Refs. 1 and 1a). In addition to the generality of its application (it can be used *in vitro* or *in vivo*) the simplicity of the procedure is also an advantage. For example, the only equipment required is an inexpensive germicidal lamp that is available commercially.

Photochemical cross-linking has also been employed to detect protein bound to specific sites on double-stranded DNA by a modification of the method that permits transfer of ^{32}P from specific phosphodiester bonds at

---

[1] M. D. Shetlar, *Photochem. Photobiol. Rev.* **5**, 105 (1980).

[1a] J. W. Hockensmith, W. L. Kubasek, W. R. Vorachek, E. M. Evertsz, and P. H. Von Hippel, this volume [13].

the binding site, to the protein.[1b] This approach allows proteins that are present in only small amounts to be identified. By using ^{32}P-labeled site-specific probes, it may provide a valuable way of estimating pool sizes of specific DNA-binding proteins that cannot be done by DNase I, free radical footprinting, or electrophoretic mobility shift assays.[1b]

While any nucleotide residue in DNA or RNA might, in principle, take part in the photo-cross-linking reaction, it appears that, in those systems that have been most extensively studied, thymine is by far the most reactive base. Hence, at 252.7 nm, a wavelength that is close to the maximum emission (253.4 nm) for a germicidal lamp, no cross-linking of $(dG)_{10}$ or $(dA)_{10}$ to the bacteriophage gene 32 single-stranded DNA-binding protein (gp32) could be observed while the ratio of $(dT)_{10}$ to $(dC)_{10}$ cross-linking to gp32 was 200 : 1.[1a]

To test the potential of any protein to undergo photo-cross-linking, the simplest approach involves the use of an oligonucleotide such as $(dT)_8$ that has been 5'-end labeled with $[^{32}P]ATP$. After photolysis most of the non-cross-linked $(dT)_8$ can be removed by trichloroacetic acid precipitation prior to digestion of the remaining mixture of free and cross-linked protein with trypsin. By taking advantage of the large negative charge on tryptic peptides that are cross-linked to $(dT)_8$, it is possible to use anion-exchange high-performance liquid chromatography (HPLC) to rapidly separate cross-linked from non-cross-linked tryptic peptides. If there are multiple cross-linking sites on the protein then reversed-phase ion-pairing HPLC can be used for the final step in purification. Surprisingly, recoveries of oligonucleotide–peptide adducts from reversed-phase supports are very poor unless an ion pairing reagent such as triethylammonium acetate is present.[2,3] The amino acid residue that has been cross-linked can then be identified by amino acid sequencing. If a solid-phase sequencer is used, then the site of attachment can be identified by scintillation counting of an aliquot of each of the resulting phenylthiohydantoin amino acid derivatives. If a gas or liquid phase instrument is used, the phenylthiazolinone derivative of the $[^{32}P](dT)_{10}$-labeled amino acid will not be extracted from the Polybrene-coated support disk and so a direct identification cannot be made on the basis of scintillation counting. However, there will be a "hole" at that position in the sequence that, if the primary structure of the protein is known, can be used to assign the amino acid residue involved in cross-linking.

[1b] B. Safer, R. B. Cohen, S. Garfinkel, and J. A. Thompson, *Mol. Cell. Biol.* **8,** 105 (1980).
[2] B. M. Merrill, K. R. Williams, J. W. Chase, and W. H. Konigsberg, *J. Biol. Chem.* **259,** 10850 (1984).
[3] B. M. Merrill, K. L. Stone, F. Cobianchi, S. H. Wilson, and K. R. Williams, *J. Biol. Chem.* **263,** 3307 (1988).

An alternative approach to that just described involves the use of a long polynucleotide or naturally occurring nucleic acid such as bacteriophage fd DNA in place of an oligonucleotide of defined length. In this instance sedimentation[4] or gel filtration can be used to separate the free and cross-linked nucleic acid from the non-cross-linked protein. Following extensive micrococcal nuclease digestion, the cross-linked protein–oligonucleotide complex can then be treated with trypsin and the cross-linked peptide–oligonucleotide complex isolated by electrophoresis at low pH[4] or by anion-exchange and reversed-phase HPLC in the presence of an ion pairing reagent. The nucleic acid can either be uniformly labeled *in vivo* prior to cross-linking or, alternatively, it can be 5′-end labeled with bacteriophage T4 polynucleotide kinase and $[\gamma\text{-}^{32}P]ATP$ after the nuclease digestion. While the use of a high-molecular-weight nucleic acid permits the facile separation of free and cross-linked protein, it introduces a significant level of heterogeneity that complicates the task of purifying the final peptide–oligonucleotide complex. Unless the micrococcal nuclease digestion is complete, leaving only a single nucleotide residue linked to the protein, there will be a population of cross-linked protein (and ultimately peptide) molecules that will be covalently linked to a series of oligonucleotides of different lengths and base composition with the potential to separate on anion-exchange and reversed-phase ion-pairing HPLC. For example, despite extensive micrococcal nuclease digestion of the bacteriophage fd gene 5 protein–fd single-stranded (ss) DNA cross-linked complex, the cross-linked gene 5 protein still migrated on sodium dodecyl sulfate (SDS)-polyacrylamide gel electrophoresis as a diffuse band corresponding to a population of gene 5 monomers linked to a series of oligonucleotides ranging in length from 3 to 10.[4] This heterogeneity substantially decreased the yield of any single oligonucleotide–peptide species. If, on the other hand, conditions could be found to obtain complete nuclease digestion this would substantially decrease the ability of the anion-exchange step to separate the cross-linked peptide from all of the other non-cross-linked tryptic peptides.

Experimental Procedures

Optimization of Photocross-Linking Conditions

To use photochemical cross-linking for identifying amino acid residues at the interface of protein–nucleic acid complexes sufficient amounts of the purified oligonucleotide–peptide covalent complex must be isolated to

---

[4] P. R. Paradiso and W. H. Konigsberg, *J. Biol. Chem.* **257**, 1462 (1982).

permit detailed chemical characterization. Since the yield of the cross-linked product is dependent on the extent of cross-linking that can be achieved, it is necessary to determine the dose response. This information can be easily obtained using a short oligonucleotide such as $(dT)_8$ that has been $5'$-^{32}P end labeled using polynucleotide kinase as described by Maniatis et al.[5] The resulting $[^{32}P](dT)_8$ can then be separated from unincorporated $[\gamma$-$^{32}P]$ATP using a Du Pont (Newton, CT) Nensorb 20 column following the manufacturer's recommended procedures and buffers. The $[^{32}P](dT)_8$ and the protein of interest are then mixed under conditions (low salt, high protein concentration, and, if necessary, excess oligonucleotide) that are known to result in near stoichiometric binding based on the amount of protein present. The irradiation is usually carried out in a cold room to minimize sample heating using a 15-W germicidal lamp (General Electric Co.). The sample (20–150 μl) is placed on a Parafilm sheet at a distance of 5–15 cm away from the light source. At a distance of 5 cm, irradiation times typically range from 1 to 30 min and the actual photon flux can be measured with a dosimeter (S. I. Schumberger Co.). Maximal extent of cross-linking is usually reached with an exposure that will provide photon fluxes in the range of 0.2 to 2.0 $\times$ 10^5 ergs/mm^2 (see Refs. 2–4, 6). The amount of protein that has been cross-linked to the labeled oligonucleotide can then be estimated by SDS–polyacrylamide gel electrophoresis followed by autoradiography. More quantitative measurements of the extent to which the protein has been cross-linked can be made by using HPLC gel filtration in the presence of a denaturant such as 6 M guanidine hydrochloride[2] or, if the cross-linked complex is acid stable, by a simple filter assay.[6] In the latter case, irradiated samples can be blotted onto Whatman (Clifton, NJ) gf/c41 filter paper that has been presoaked in 15% trichloroacetic acid (TCA) and then dried prior to use. After spotting several samples onto a grid that has been penciled onto the gf/c41 filter paper, it is placed on a Büchner funnel, rinsed once under vacuum with 50 ml cold 10% TCA, three times with 50 ml cold H_2O, and three times with 20 ml cold ethanol. The filters are then baked at 67° for 15 min prior to immersing in 10 ml Hydrofluor scintillation fluid. After scintillation counting, the maximum extent of protein cross-linking can then be determined based on the specific activity of the $[^{32}P](dT)_8$ and the initial protein concentration.

In the case of three single-stranded nucleic acid-binding proteins whose cross-linking sites have been determined, using the procedure outlined

[5] T. Maniatis, E. F. Fritsch, and J. Sambrook, in "Molecular Cloning: A Laboratory Manual," p. 125. Cold Spring Harbor Laboratory, Cold Spring Harbor, New York, 1982.
[6] Y. Shamoo, K. R. Williams, and W. H. Konigsberg, Proteins: Struct. Funct. Genet. 4, 1 (1988).

TABLE I

PHOTOINACTIVATION OF PROTEINS UNDER CONDITIONS USED
FOR IDENTIFYING CROSS-LINKING SITES

Nucleic acid-binding protein	Dose (ergs/mm^2)	Inactivation[a] (%)	Ref.
Bacteriophage fd gene 5	1.5×10^5	0	7
E. coli SSB	1.3×10^5	50	2
Bacteriophage T4 gene 32	0.9×10^5	33	6
HeLa A1 hnRNP	0.6×10^5	0	3

[a] As determined either by stoichiometric fluorescence titrations with nucleic acid before and after irradiation, or in the case of the fd gene 5 protein by comparing cross-linking efficiencies before and after irradiation.

above (which will be described in more detail below)[2,3,6] the percentage cross-linking ranged from 11.8% for the *Escherichia coli* SSB ssDNA-binding protein[2] to 33% for the A1 hnRNP protein.[3] The amount of protein used in these studies ranged from 39 nmol for the bacteriophage T4 ssDNA-binding protein encoded by gene 32 (gp32, Ref. 6) to 250 nmol for *E. coli* SSB.[2] Based on these three studies[2,3,6] it is reasonable to expect an overall average recovery of about 50% for the cross-linked peptide(s) following tryptic digestion and anion-exchange HPLC.

Controls Required to Determine Specificity of Labeling

Prior to identifying a site of photochemical cross-linking it is essential to show that cross-linking is limited to the nucleic acid-binding site of the native protein. Since exposure to UV light can lead to protein denaturation it is important to determine the extent of denaturation that is occurring during photolysis and to demonstrate that only the native protein can be cross-linked to the nucleic acid. As shown in Table I,[7] the extent of protein denaturation that has been presumed to occur under the conditions used for identifying cross-linking sites in four nucleic acid-binding proteins ranges from 0 to 50%. Since the data in Table I show that similar doses of UV light lead to 50% inactivation of *E. coli* SSB[2] compared to no detectable inactivation of the fd gene 5 protein,[7] it is apparent that there is considerable variation in the photosensitivity of proteins. Thus, an essential control requires determination of the extent of protein nucleic acid cross-linking that can be obtained before and after the protein alone has been subjected

[7] P. R. Paradiso, Y. Nakashima, and W. H. Konigsberg, *J. Biol Chem.* **254**, 4739 (1979).

to the same photon flux as that used in the cross-linking reaction. Hence Merrill *et al.*[2] demonstrated that prior irradiation of SSB with 1.3×10^5 ergs/mm^2 of UV light in the absence of [^{32}P](dT)$_8$ led to a 38% *decrease* in the extent of cross-linking compared to an identical sample of SSB that had not been preirradiated.[2] These data argue strongly that it is the native rather then the photoinactivated SSB that is involved in cross-linking. Another important control is the demonstration that disruption of the complex, for example by adding a sufficiently high NaCl concentration to result in at least a 50% dissociation of the protein–nucleic acid complex, directly correlated with a decreased extent of cross-linking.[2,3,6] Finally, advantage has also been taken of the finding[1,7] that thymine is the most photoreactive base. Hence, Merrill *et al.*[3] found that a 20 : 1 molar ratio of (dA)$_8$ to [^{32}P](dT)$_8$ completely prevented cross-linking as estimated by SDS polyacrylamide gel electrophoresis followed either by Coomassie Blue staining or autoradiography. These data are best explained by assuming that [^{32}P](dT)$_8$ can cross-link only at the A1 oligonucleotide-binding site(s) and, hence, when the [^{32}P](dT)$_8$ is effectively removed from the binding site by (dA)$_8$ competition, the extent of A1 : [^{32}P](dT)$_8$ cross-linking is decreased. The final proof that cross-linking is nonrandom comes from actually identifying the site of cross-linking, as will be described below.

Measurement of Extent of Photodegradation of the Cross-Linked Protein–Nucleic Acid Complex

In addition to photoinactivation of the protein[2,3,6] and photolysis reactions involving the oligonucleotide,[1] another factor that accounts for the inability to obtain 100% cross-linking of the protein is reversal or photodegradation of the cross-linked protein–nucleic acid complex. In the case of the bacteriophage fd gene 5 protein, it appears that the cross-linking reaction can be reversed. Thus, a maximum of 22% cross-linking of the gene 5 protein–poly(dT) complex was observed when either a mixture of the individual components or the covalently linked gene 5–poly(dT) complex itself was exposed to 1.5×10^5 ergs/mm^2 of UV light.[7] To obtain this result with the covalent gene 5–poly(dT) complex, the complex was first isolated by gel filtration in 1.0 *M* NaCl, and then, after desalting, it was exposed to the same dosage of UV light that was used to form the complex initially. Gel filtration in SDS of the complex after reexposure to UV light indicated that reirradiation resulted in breakage of nearly 80% of the covalent bonds in the gene 5 protein–poly(dT) complex.[7] Most importantly, it was also shown that the free gene 5 protein and poly(dT), which had failed to cross-link during the initial UV exposure, and which had been separated by gel filtration in 1.0 *M* NaCl, gave the same cross-linking efficiency of about

22% when desalted, recombined, and exposed again to the same dosage of UV light.[7] From these results it was concluded that in the case of gene 5 protein, no detectable photodenaturation occurred on exposure of the protein to 1.5×10^5 ergs/mm^2 of UV light.[7] Paradiso et al.[7] demonstrated a significant difference in the action spectrum for the forward and reverse photochemical reaction. They found that the action spectrum for covalent bond formation between the gene 5 protein and [32P](dT)$_4$ was similar to the UV absorption spectrum of thymine, which has a maximum near 265 nm. This result suggested that absorption of UV light by thymine results in its excitation and that this is the primary event leading to covalent bond formation with the gene 5 protein.[7] In contrast, the extent of bond breakage in the reverse reaction was inversely related to the incident wavelength (at a constant dosage of 1.5×10^5 ergs/mm^2) down to 230 nm, which was the lowest wavelength tested.[7] It appeared therefore that in the gene 5–[32P](dT)$_4$ system, the nature of the absorbing species in the product is different from that in the reactant. In addition to the gene 5–[32P](dT)$_4$ system, photodissociation of the cross-linked protein–nucleic acid complex has also been demonstrated in the case of the E. coli SSB protein–(dT)$_8$ complex.[2] In this instance, exposure of the purified SSB–[32P](dT)$_8$ cross-linked complex to 1.3×10^5 ergs/mm^2 of UV light at 257 nm resulted in bond breakage of about 40% of the covalent complex.[2]

Chemical Stability of Photochemically Cross-Linked Protein–Nucleic Acid Complexes

Because of the numerous possible cross-linking reactions that may occur when protein–nucleic acid complexes are exposed to UV light, the resulting covalent complexes may differ substantially in their stability to acids, bases, or denaturants, depending on the protein being studied. In all instances examined so far in detail, the resulting complexes are stable to high concentrations of salts and denaturants, such as $6\ M$ guanidine hydrochloride and $8\ M$ urea, as well as to boiling in SDS at pH 8. Similarly, the covalent complex formed with the gene 5 protein is stable to 7% perchloric acid[7] and those formed with bacteriophage T4 gp32,[6] E. coli SSB,[2] and the A1 hnRNP protein[3] are resistant to exposure to 5–10% trichloroacetic acid at 25°. In contrast, the covalent gene 5 protein–DNA complex is disrupted by 50% (v/v) formic acid[4] or by treatment with 1.0 M acetic or $6\ N$ HCl for 15 min at 25°.[7] Similarly, Hockensmith et al.[1a] found that a 30-min incubation at room temperature in 10% (v/v) acetic acid (1.6 M) and 50% (v/v) methanol led to 75% disruption of the covalent gp32–DNA complex that had been formed using a laser UV light source. These conditions are similar to those routinely used for fixing and Coomassie Blue staining of SDS–polyacrylamide gels. While the *E. coli* SSB–DNA complex is stable at 25° to anhydrous trifluoroacetic acid (TFA) it is disrupted in this acid if the temperature is raised to 45°.[2] Although the covalent gene 5 protein–DNA complex has been shown to be stable when 15-hr incubations were carried out in 0.1 N NaOH or in concentrated NH₄OH at 25°,[7] the alkali stability of the other protein–nucleic acid complexes mentioned above has not yet been studied.

Isolation of Peptides Cross-Linked to ³²P-Labeled Oligonucleotides or Nucleic Acids

The approach that has most often been used to identify nucleic acid cross-linking sites in proteins is to cross-link the protein to a ^{32}P-labeled oligonucleotide such as $(dT)_8$, digest the cross-linked complex with trypsin, and then use anion-exchange HPLC to rapidly isolate the corresponding $[^{32}P](dT)_8$-labeled tryptic peptide. Using this approach the non-cross-linked $[^{32}P](dT)_8$ can be removed by a trichloroacetic acid precipitation (TCA) step that immediately follows the photolysis.[2,3,6] After washing the TCA pellet three times with cold acetone, the mixture of cross-linked and non-cross-linked protein is then suspended in a volume of 8 *M* urea, 0.4 *M* NH₄HCO₃ to give a final protein concentration in the range of 0.1 to 5 mg/ml. If the protein contains disulfide bonds, the sample can be reduced and carboxamidomethylated, to ensure complete denaturation, and digested as described previously.[8,9] Briefly, the protein is reduced by adding dithiothreitol to a final concentration of 4.5 m*M* and incubating it at 50° for 15 min. After cooling to room temperature, iodoacetamide is added to a final concentration of 10 m*M* and the sample is incubated for an additional 15 min at room temperature. Prior to adding a 1 : 25 (w/w) ratio of trypsin, the sample is diluted with water to give a final urea concentration of 2 *M*. After 24 hr at 37° the resulting peptide mixture can be directly injected onto an AX300 (SynChrom, Lafayette, IN) anion-exchange HPLC column that had been equilibrated at a flow rate of 1.0 ml/min with buffer A [20 m*M* KH₂PO₄, pH 6.6, 4.8% (v/v) ethanol, 40 m*M* KCl]. The column is then eluted over a period of 100 min or more with a linear gradient extending to 1.0 *M* KCl in buffer A.[2,3,6] Peptides are detected by their absorbance

[8] K. L. Stone, M. B. LoPresti, N. D. Williams, J. Myron Crawford, R. DeAngelis, and K. R. Williams, *in* "Techniques in Protein Chemistry" (T. E. Hugli, ed.), p. 377. Academic Press, New York, 1989.

[9] K. L. Stone, M. B. LoPresti, J. Myron Crawford, R. DeAngelis, and K. R. Williams, *in* "A Practical Guide to Protein and Peptide Purification for Microsequencing" (P. T. Matsudaira, ed.), p. 33. Academic Press, New York, 1989.

at 220 nm[2] or 254 nm[3,6] and by Cerenkov counting of the resulting fractions, which can conveniently be collected in 1.5-ml Eppendorf tubes that are then capped prior to storage. At this point individual peaks can either be desalted and then analyzed and sequenced or can be further purified by directly injecting them onto a reversed-phase HPLC column in the presence of an ion-pairing reagent. A Du Pont Nensorb 20 nucleic acid purification cartridge can be used to desalt cross-linked peptides rapidly that have been isolated by anion-exchange HPLC. As described previously,[3,6] the Nensorb 20 cartridge is equilibrated with 5 mM triethylammonium acetate at pH 7.0 prior to sample loading. After washing the cartridge with the same buffer, the cross-linked peptide(s) can be eluted with 1 ml 75% methanol and then reduced in volume using a Savant (Farmingdale, NY) Speed-Vac. Fractions selected for further purification can be injected directly onto a Vydac (Hesperia, CA; 300-Å pore size, 5-μm particle size) or similar C$_{18}$ column that had been equilibrated at a flow rate of 1 ml/min with buffer B [10% (v/v) acetonitrile, 90% (v/v) 0.1 M triethylammonium acetate at pH 6.8]. After washing the column with the same solvent, peptides can be eluted with a linear gradient extending over about 100 min to 80% (v/v) acetonitrile, 20% (v/v) 0.1 M triethylammonium acetate. Peptides are detected as previously[2,3] by their absorbance at 254 nm and by Cerenkov counting of the collected fraction. If necessary, they can be individually desalted as described above prior to amino acid analysis or sequencing.

If the protein of interest has been cross-linked to a high-molecular-weight nucleic acid such as fd ssDNA[4,7] then the non-cross-linked protein can be removed by high-speed centrifugation[4] or gel filtration in the presence of a high concentration of salt. At this point it may be possible to carry out an extensive chemical and/or proteolytic degradation of the protein followed by high-speed centrifugation or gel filtration, now in the presence of a low concentration of salt, to isolate the cross-linked peptide(s). Alternatively, the mixture of cross-linked and non-cross-linked DNA may be digested with micrococcal nuclease as previously described[4] and the free oligonucleotides removed by dialysis. If necessary, the cross-linked protein : oligonucleotide complex can be 5'-end labeled using [γ-^{32}P]ATP and bacteriophage T4 polynucleotide kinase.[4] The peptide–oligonucleotide complex, which because of the expected variability in the length and base composition of the resulting oligonucleotide may be heterogeneous, can then be purified by anion-exchange and reversed-phase HPLC as described above or by paper electrophoresis at pH 1.9.[4,7]

Identification of Amino Acid Residues that Participate
in Cross-Linked Adduct

Providing that the complete primary structure of the protein being studied is known, it may be possible to identify the cross-linked peptide by amino acid analysis following acid hydrolysis. Having done this, Paradiso *et al.*[7] then used extensive aminopeptidase M and pronase digestion to release all non-cross-linked amino acids from the cross-linked oligonucleotide. Following performic acid oxidation and hydrolysis, the cross-linked amino acid was then identified as cysteine, based on amino acid analysis.[7] This may not represent a general approach, however, since it depends on the ability of acid hydrolysis to regenerate the free amino acid from the amino acid–nucleotide adduct. Differences in the extent of regeneration of the original amino acid may be expected, depending on the exact chemical nature of the cross-link. In most instances, the identification of the cross-linked peptide and amino acid has been made on the basis of amino acid sequencing. In the case of the bacteriophage T4 gene 32, *E. coli* SSB[2] and A1 hnRNP[3] proteins, the site of cross-linking was inferred from the absence of an identifiable phenylthiohydantoin derivative at a particular position in the peptide sequence. In all three of these proteins only the amino acid side chain (as opposed to the peptide bond) appeared to be involved in cross-linking because in each case amino acid sequencing proceeded as expected before and after the cross-linked residue was reached. In the case of *E. coli* SSB, the cross-linked peptide was subjected to solid-phase sequencing so that a sufficiently polar solvent, such as trifluoroacetic acid, could be used that would extract the oligonucleotide-labeled phenylthiazolinone derivative of the cross-linked amino acid.[2] Since the oligonucleotide had been ^{32}P-labeled it was possible using this solid-phase sequencing approach to directly confirm the cycle position corresponding to the site of cross-linking.

Results

Relative Efficiency and Extent of Cross-Linking of Several Single-
Stranded Binding Proteins to Nucleic Acids

In terms of nucleotide base specificity, results obtained using a germicidal lamp source[3,7] seem to parallel similar experiments carried out with laser cross-linking.[1a] In both cases thymine appears to be the most reactive base. As shown in Table II for the bacteriophage fd gene 5 protein, no cross-linking was detected with either poly(dA) or poly(dG) and poly(dT)

TABLE II
PHOTOCROSS-LINKING OF BACTERIOPHAGE fd
GENE 5 PROTEIN TO NUCLEIC ACIDS[a]

Nucleic acid	Gene 5 protein cross-linked (%)
Poly(dT)	21.4
Poly(dA-dT)	11.4
fd ssDNA	7.0
Poly(dC)	4.4
Poly(dA)	0
Poly(dG)	0

[a] Data are taken from Ref. 7 and were determined by BioGel A-5 M chromatography (Bio-Rad, Richmond, CA) that was carried out in 0.5% SDS following exposure of mixtures of ^{4}C-labeled gene 5 protein and the above nucleic acids to 1.5×10^{5} ergs/mm^{2} ultraviolet light. The irradiation was carried out at 4° in 10 mM Tris-HCl, pH 7.2, 1 mM EDTA, and 0.1% 2-mercaptoethanol.

was five times more reactive in this regard than poly(dC). When poly(dA-dT) and fd ssDNA was used, there was a clear correlation between the content of thymine and the extent of cross-linking. Thus, poly(dA-dT) had 50% efficiency and fd ssDNA, which contains 24% thymine, had 33% cross-linking efficiency relative to poly(dT).[7] Although an extensive study was not carried out, the A1 hnRNP protein appears similar to the gene 5 protein in that A1 cross-links with a maximum efficiency of about 35% to oligo(dT)$_8$ (Fig. 1) compared with no detectable cross-linking to (dA)$_8$.

The maximum extent of cross-linking depends not only on the specific protein as well as the UV dose but also on the fraction of the protein that is actually complexed to the nucleic acid under the conditions of irradiation. Table III gives relative cross-linking efficiencies for a number of single-stranded nucleic acid-binding proteins. Although 1.3×10^{5} ergs/mm^{2} appeared to provide near maximum cross-linking for F plasmid SSB (data not shown), the data in Table III indicate that with the homologous E. coli SSB protein, optimum cross-linking requires at least twice this dose. With the calf thymus UP1 protein, which is actually a proteolytic fragment corresponding to the NH$_2$-terminal two-thirds of A1,[10,11] binding

[10] A. Kumar, K. R. Williams, and V. Szer, J. Biol. Chem. 261, 11266 (1986).
[11] S. Riva, C. Morandi, P. Tsoulfas, M. Pandolfo, G. Biamonti, B. Merrill, K. R. Williams, G. Multhaup, and K. Beyereuther, EMBO J. 5, 2267 (1986).

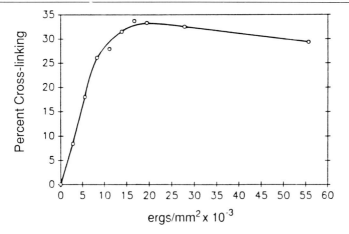

FIG. 1. Dose–response curve for the photocross-linking of A1 to $[^{32}P](dT)_8$. A1 was mixed with $[^{32}P](dT)_8$ in a 1 : 40 ratio of protein mononmer to phosphate, exposed for varying lengths of time to UV light, then chromatographed on an HPLC gel-filtration column equilibrated in 6 M guanidine hydrochloride as described in Ref. 3. The results indicate that the maximum amount of A1 cross-linked is approximately 33%, on exposure of the A1-oligodeoxynucleotide complex to 16,700 ergs/mm² of UV light.

TABLE III

EFFICIENCY OF CROSS-LINKING OF VARIOUS SINGLE-STRANDED NUCLEIC ACID-BINDING PROTEINS TO $[^{32}P](dT)_8{}^{a}$

Protein	Monomer protein concentration (μM)	Molar ratio of olignucleotide to protein monomer	Dose (ergs/mm²)	Protein cross-linked (%)
Bacteriophage T4 gene 32 protein	20	2.5 : 1	2.0×10^5	20.0
F plasmid SSB	20	2.5 : 1	1.3×10^5	27.3
E. coli SSB	20	2.5 : 1	1.3×10^5	11.8
	20	2.5 : 1	2.0×10^5	22.0
Rat liver lactate dehydrogenease	20	2.5 : 1	2.0×10^5	3.8
Calf thymus UP1	5	2.5 : 1	2.0×10^5	37.8
	5	8.0 : 1	2.0×10^5	85.6

^{a} Data are taken from Ref. 2 and were determined by HPLC gel filtration on a Waters (Milford, MA) I-125 column that had been equilibrated with 6 M guanidine hydrochloride. The $[^{32}P](dT)_8$: protein mixtures were exposed to the indicated dose of ultraviolet light at 4° in 0.1 ml of 10 mM Tris-HCl, pH 8.0, 1 mM 2-mercaptoethanol, 10% glycerol, 310 mM NaCl.

to $(dT)_8$ in 310 mM NaCl is sufficiently weak that even a 2.5-fold molar excess of this oligonucleotide was insufficient to drive most of the protein into the complex. Hence, the molar ratio of oligonucleotide to protein monomer must be increased further, to about 8 : 1, to reach a near maximal extent of UP1 cross-linking (Table III). The unusually large percentage of UP1 cross-linking may result partly from the existence of four different cross-linking sites in this protein.[3] Similarly, the seemingly low percentage of cross-linking of the rat liver lactate dehydrogenase (LDH) may result from its relatively low affinity for ssDNA. Although this enzyme binds sufficiently tightly to ssDNA that it was mistaken for a new rat liver helix-destabilizing protein, it binds less tightly than the other proteins listed in Table III to $(dT)_8$. It is likely therefore that the percentage cross-linking efficiency for LDH might be considerably improved by increasing the oligonucleotide to protein ratio above 2.5, the value that was used and reported in Table III.

Cross-Linked Amino Acid Residues Identified in Several ssDNA- and RNA-Binding Proteins

Identification of the phenylalanine residue in the *E. coli* SSB protein photo-cross-linked to $[^{32}P](dT)_8$ provides a good example of how the approach outlined above can be applied to a specific protein. As shown in Fig. 2, anion-exchange HPLC readily separates even the unusually acidic T-14,15 tryptic peptide, which has a net negative charge of -5, from the cross-linked peptide, which elutes at about 45 min and has a net negative charge of at least -7. Under the conditions used in Fig. 2 all other SSB tryptic peptides elute in the void volume of the column and a small amount of residual $[^{32}P](dT)_8$ elutes as shown at about 66 min. All four A1 tryptic peptides that were cross-linked to $[^{32}P](dT)_8$ also eluted from an anion-exchange HPLC column prior to free $[^{32}P](dT)_8$[3] whereas, in the case of the T4 gene 32 protein, the cross-linked tryptic peptide actually coeluted with free $[^{32}P](dT)_8$.[6] As shown in Fig. 3 further purification of the $[^{32}P]dT_8$ cross-linked, SSB tryptic peptide was achieved by reversed phase HPLC in the presence of an ion-pairing reagent. Solid-phase amino acid sequencing of the two major peaks in Fig. 3 indicated that both contained the same 6-residue tryptic peptide that corresponds to residues 57 to 62 in *E. coli* SSB. Since only a single SSB tryptic peptide was found cross-linked to $[^{32}P](dT)_8$, the multiple peaks resolved on ion-pairing HPLC (Fig. 3) may result from photochemical changes such as thymine dimer formation occurring within the oligonucleotide. As was also the case with the bacteriophage T4 gene 32 and the A1 hnRNP protein,[3] the actual site of SSB cross-linking was indirectly identified by the absence of an expected phe-

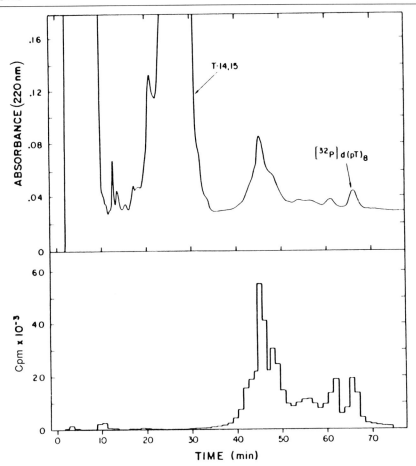

FIG. 2. Anion-exchange HPLC of [^{32}P](dT)$_8$ cross-linked tryptic peptides from SSB. SSB (250 nmol of monomer) and [^{32}P](dT)$_8$ (2 mmol of phosphate) were mixed together, irradiated with 1.3×10^5 ergs/mm^2 of UV light, precipitated with trichloroacetic acid, and then digested with trypsin as described in Ref. 2. The resulting peptide mixture was then injected onto a Synchropak AX300 anion-exchange HPLC column equilibrated with 20 mM potassium phosphate, pH 6.6, 4.8% ethanol, 40 mM KCl. The column was eluted at a flow rate of 1.0 ml min^{-1} with linear gradients of 0.8 M KCl in 20 mM potassium phosphate, 4.8% ethanol (buffer B) into 20 mM potassium phosphate, 48% ethanol (buffer A), 0–30 min (5–50% buffer B), 30–63 min (50–70% buffer B), and 63–99 min (70–100% buffer B). The elution positions of the SSB carboxyl-terminal tryptic peptid (T-14,15, residues 155–177) and of free [^{32}P](dT)$_8$ are indicated above.

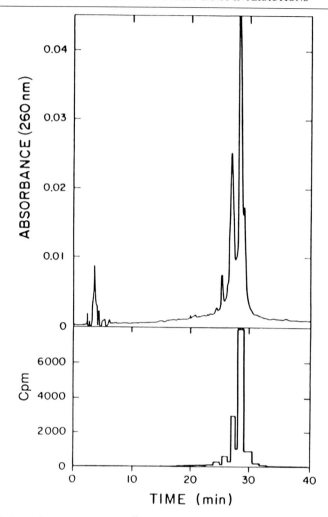

FIG. 3. Ion-pairing HPLC of the [^{32}P](dT)$_8$ cross-linked tryptic peptide from SSB. The cross-linked peptide isolated via anion-exchange HPLC (see Fig. 2) was further purified as described in Ref. 2 by ion-pairing HPLC on a Waters C$_{18}$ column equilibrated in 90% 0.1 M triethylammonium acetate (buffer A) and 10% acetonitrile (buffer B). The column was eluted with a linear gradient from 10 to 25% buffer B over 40 min at a flow rate of 1 ml/min. Both major peaks eluted contain the same cross-linked peptide, which corresponds to SSB tryptic peptide T-8 (residues 57–62).

nylthiohydantoin derivative of a particular amino acid residue following sequencing of the cross-linked peptide. As shown in Fig. 4, in the case of the *E. coli* SSB cross-linked peptide, the phenylthiohydantoin derivative of Phe-60, which corresponds to cycle 4 in Fig. 4, was missing. Further evidence suggesting Phe-60 as the site of cross-linking was provided by subjecting a portion of the trifluoroacetic acid wash from each sequence cycle to scintillation counting. As expected, most of the ^{32}P label was released at cycle 4.[2]

As summarized in Table IV,[12–14] at least four different amino acids (cysteine, phenylalanine, tyrosine, and methionine) have been identified as photochemical cross-linking sites in nucleic acid-binding proteins. In half of the examples cited in Table IV, oligo(dT)$_8$ was the nucleic acid whereas in the other half a naturally occurring, high-molecular-weight nucleic acid was cross-linked. Although the A1 hnRNP protein binds RNA *in vivo*, a deoxyoligonucleotide was used for the cross-linking experiment because of its high efficiency of cross-linking. In general, single-stranded RNA versus single-stranded DNA-binding proteins have only about an order of magnitude difference in specificity with respect to sugar type[15] and hence it seems reasonable in these instances to take advantage of whichever oligonucleotide has the higher cross-linking efficiency. An appropriate control would be to determine whether an oligonucleotide such as (U)$_8$ could effectively compete in a [^{32}P](dT)$_8$ photo-cross-linking assay. Of the proteins listed in Table IV the A1 hnRNP was the oniy one in which multiple cross-linking sites were found. Phe-16 is the major A1 cross-linking site since it accounted for nearly 65% of the bound [^{32}P](dT)$_8$. Approximately 10–15% cross-linking occurred at each of the remaining three sites in A1, which together actually account for two analogous pairs of cross-linking sites. That is, A1 contains an approximately 90-amino acid internal region of sequence similarity such that when residues 3–93 are aligned with residues 94–194, 32% of the amino acid residues in these two regions are identical.[16] Hence, Phe-16 and Phe-107 as well as Phe-58 and Phe-149 occupy analogous positions in terms of this 90-amino acid internal region of sequence similarity. The finding that both of these 90-amino acid domains can be cross-linked to [^{32}P](dT)$_8$ greatly strengthened the notion that each of these domains corresponds to an independently folded nucleic acid-binding domain.[3] In the case of the L4 and S7 ribosomal proteins the

[12] P. Maly, J. Rinke, E. Ulmer, C. Zwieb, and R. Brimacombe, *Biochemistry* **19,** 4179. (1980).
[13] K. Moller, C. Zwieb, and R. Brimacombe, *J. Mol. Biol.* **129,** 489 (1978).
[14] C. Zwieb and R. Brimacombe, *Nucleic Acids Res.* **6,** 1775 (1979).
[15] J. Chase and K. R. Williams, *Annu. Rev. Biochem.* **55,** 103 (1986).
[16] B. M. Merrill, M. B. LoPresti, K. L. Stone, and K. R. Williams, *J. Biol. Chem.* **261,** 878 (1986).

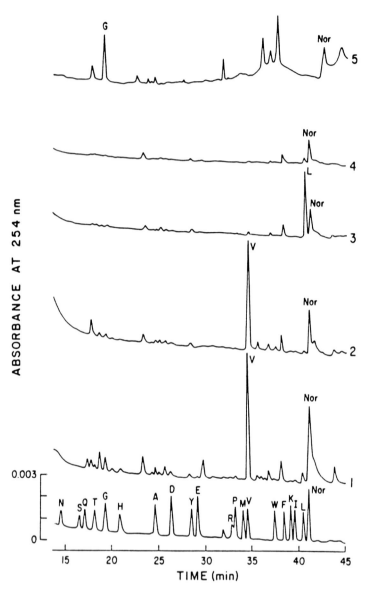

FIG. 4. HPLC traces of phenylthiohydantoin derivatives from cycles 1–5 of the solid-phase sequencing of the SSB–$[^{32}P](dT)_8$ cross-linked peptide coupled to an aminopolystyrene resin. An internal standard of 200 pmol norleucine (Nor) was added to each sequenator cycle. Samples were run on a Waters HPLC system as described in Ref. 2. The sequence is indicated by the one-letter amino acid symbol adjacent to the corresponding peak. The bottom tracing represents 100 pmol of a standard phenylthiohydantoin derivative mixture. The norleucine (Nor) peak in cycle 5 elutes slightly later than those in cycles 1–4 because cycle 5 was run on the HPLC on a different date.

TABLE IV
IDENTIFICATION OF PHOTOCHEMICAL CROSS-LINKING SITES
IN NUCLEIC ACID-BINDING PROTEINS

Protein	Nucleic acid	Cross-linked amino acid	Ref.
Bacteriophage fd gene 5	fd ssDNA	Cys-33	4,7
Bacteriophage T4 gene 32	$(dT)_8$	Phe-183	6
E. coli SSB	$(dT)_8$	Phe-60	2
Human A1 hnRNP	$(dT)_8$	Phe-16	3
		Phe-58	
		Phe-107	
		Phe-149	
E. coli 50S ribosomal protein S7	23S RNA	Tyr-35	12
E. coli 30S ribosomal protein S7	16S RNA	Met-114	13,14

exact site of cross-linking was located both with respect to the protein as well as the nucleic acid. In both cases a single uridine residue was involved that was contained within the sequence A-A-U-G-G in 16S RNA[13,14] and within the sequence A-A-U-A-G- in 23S RNA.[12] Despite the observation that with the T4 gene 32 protein $(rU)_{10}$ has less than 0.5% of the laser cross-linking efficiency of $(dT)_{10}$,[1a] uridine does provide a high degree of cross-linking in the case of the L4 and S7 ribosomal proteins.

Use of Nucleotide Analogs to Increase Photocross-Linking Efficiency of Nucleic Acid–Protein Complexes

Incorporation of photosensitive analogs into nucleic acids may greatly increase their cross-linking efficiency, thus enabling the study of protein–nucleic acid complexes that would not normally form cross-links. An added benefit of this approach is that the photon flux required to obtain cross-linking is greatly reduced and, hence, problems due to photodenaturation of the protein are minimized. In the case of the bacteriophage T4 gene 32 protein, 20% cross-linking to $(dT)_8$ required 9×10^4 ergs/mm^2, which resulted in inactivating nearly 33% of the free protein that was initially present.[6] In contrast, by using poly(adenylic-8-azidoadenylic) acid, which had been prepared with 8-azidoadenosine 5'-diphosphate and polynucleotide phosphorylase, nearly 100% cross-linking efficiency was achieved[17] with a dose $(9 \times 10^2$ ergs/mm$^2)$ that was equal to only 1% of the photon flux previously used for $(dT)_8$.[6] Similarly, 50% cross-linking of the *lac* repressor to bromodeoxyuridine-substituted *lac* operator required

[17] R. L. Karpel, V. Y. Levin, and B. E. Haley, *J. Biol. Chem.* **262,** 9359 (1987).

8×10^3 ergs/mm^2 (Ref. 18), which is considerably less than the dosages that are typically used (Table III) in the absence of a photosensitive nucleotide analog. Other photosensitive analogs that have been used include 5-azido-2'-deoxyuridine 5'-triphosphate, which can be incorporated into DNA by DNA polymerase I,[19] and 4-thio-UTP.[20] In the latter case the 4-thio-UTP was incorporated into RNA transcripts by RNA polymerase III transcription complexes that could then be photo-cross-linked to the nascent RNA transcripts by irradiating with ultraviolet light above 280 nm.[20]

Azido derivatives of nucleotides have also been cross-linked to proteins in high yield. In the case of the complex formed between *E. coli* RecA protein and 8-azido-ATP, a maximum of about 15% cross-linking was achieved[21] and a single tyrosine residue was found to be labeled.[22] Another system that has been examined in detail is 8-azido-dATP cross-linking to the Klenow fragment of DNA polymerase. A maximum of 50% cross-linking was achieved and covalent bonds were formed at five adjacent chemical groups centered around Tyr-766[23]: that is, the peptide carbonyl of Ile-765, the α-carbon of Tyr-766, and the peptide amide, β-carbon, and γ-carbon of Tyr-766. Cross-linking to either the peptide carbonyl or α-carbon was manifested by an inability to sequence past Leu-764 and Ile-765, respectively.[23] One potential disadvantage of the use of the azido photoaffinity label is that, as in the case of the Klenow fragment, the high reactivity of the azido group may result in cross-linking to multiple sites around a single amino acid residue, which then may result in heterogeneity on reversed-phase HPLC peptide mapping. Thus, HPLC fractionation of a tryptic digest of the 8-azido-[^{32}P]dATP-labelled Klenow fragment resulted in five different radioactive peaks (corresponding to each of the five proposed cross-linking sites around Tyr-766), all of which separated and contained the same tryptic peptide.[23]

Another approach that has been used to decrease the extent of photoinactivation while cross-linking proteins to nucleotides is the use of a photosensitizer such as acetone. Whereas irradiation of ribonuclease with ultraviolet light of wavelength equal to 260 nm or above resulted in 80% inactivation after 1 hr, similar irradiation with ultraviolet light of wavelength equal to 300 nm or above in the presence of 5% acetone resulted in

[18] S.-Y. Lin and A. D. Riggs, *Proc. Natl. Acad. Sci. U.S.A.* **71**, 947 (1974).
[19] R. K. Evans, J. D. Johnson, and B. E. Haley, *Proc. Natl. Acad. Sci. U.S.A.* **83**, 5382 (1986).
[20] B. Bartholomew, C. F. Meares, and M.E. Dahmus, *J. Biol. Chem.* **265**, 3731 (1990).
[21] K. L. Knight and K. McEntee, *J. Biol. Chem.* **260**, 867 (1985).
[22] K. L. Knight and K. McEntee, *J. Biol. Chem.* **260**, 10185 (1985).
[23] J. Rush and W. H. Konigsberg, *J. Biol. Chem.* **265**, 4821 (1990).

less than 60% inactivation.[24] Under the latter conditions the extent of cross-linking of ribonuclease to pUp was about 15%.[24] The primary sites of pUp cross-linking to ribonuclease appeared to be Ser-80 and Thr-82.[25]

DNA Affinity Labeling of Proteins Binding to Specific Sequences in Double-Stranded DNA

Numerous sequence elements in double-stranded DNA have been identified that exhibit specifity for the binding of certain proteins. Usually these sequence specific DNA-binding proteins are involved in the regulation of gene expression. To aid in the identification and characterization of these proteins, Safer et al.[1b] have developed a method that allows transfer of ^{32}P from specifically labeled phosphodiester bonds at or near the center of the binding site to the protein by ultraviolet irradiation of a mixture of proteins (usually present in a cellular extract or in an eluate from chromatographic columns used to fractionate the extract) with a double-stranded deoxynucleotide probe that has been labeled with ^{32}P in one phosphodiester bond in both the coding and in the noncoding strands. Safer et al.[1b] used three different methods to introduce the ^{32}P label at single internal phosphodiester bonds on each DNA strand (the Safer et al. reference should be consulted for the experimental details of probe labeling). Incubations with the probe were carried out at 24° for 10 min with about 2 μl of concentrated protein fraction in a total of 20 μl of a buffer [containing 40 mM(NH$_4$)$_2$SO$_4$, 1 mM EDTA, 20 mM Tris-HCl (pH 7.9), and 1 mM dithiothreitol]. Generally 10 fmol of the ^{32}P-labeled DNA probe was incubated with a solution that contained between 1 to 10 μg of total protein. Photolysis was performed with UV light (260 to 280 nm) at 4° with an average intensity of 600 μW/cm^2. After a specified time of irradiation, CaCl$_2$ was added to a final concentration of 10 mM followed by addition of 50 units of micrococcal nuclease. After a 30-min digestion at 37° a portion of each sample was subjected to SDS–PAGE followed by autoradiography.

In this situation described here, sequence-specific proteins bind stably at the site into which ^{32}P-labeled phosphodiester bonds have been introduced while nonspecific protein–DNA interactions are randomly distributed across the fragment and are weaker. The extensive nuclease digestion leaves only small oligonucleotides covalently linked to the proteins. Some of these oligonucleotides contain the ^{32}P-labeled specific binding sequence. The SDS–PAGE separation followed by autoradiography allows the visualization of only those protein(s) that have been photo-cross-linked to the

[24] J. Sperling and A. Havron, Biochemistry 15, 1489 (1976).
[25] A. Havron and J. Sperling, Biochemistry 16, 5631 (1977).

labeled region of the probe. It also permits the estimation of the apparent molecular weight of the protein, since the small oligonucleotide remaining attached to the protein would not significantly alter its mobility. Using this technique Safer et al.[1b] identified two proteins having molecular weights of 116,000 and 45,000 from a HeLa whole-cell extract that could be covalently linked by this procedure to ^{32}P-labeled probes containing the adenovirus type 2 major late upstream promoter sequence (UPS). One of these probes was a 63-base pair (bp) fragment and another contained 357 bp. Unlabeled competitor DNA fragments containing the same internal UPS-binding sequence were used as competitors and, as expected, the extent of UV cross-links diminished with increasing concentration of the competitor DNA. A DNA fragment without the UPS failed to diminish the extent of ^{32}P-labeled cross-link product when the competitor fragment was incubated at high molar excess in the presence of the ^{32}P-labeled probe containing the UPS. When a shorter, 38-bp probe containing within it the UPS ^{32}P labeled at position 19 was used, the specificity of labeling was reduced because of the formation of covalent links with end-binding proteins that extend far enough along the DNA oligomers to allow them to be cross-linked. This problem could be overcome by labeling in the presence of higher concentrations of nonspecific unlabeled DNA competitors that increase the signal-to-noise ratio. One of the requirements for this method to succeed is the presence of ^{32}P phosphodiester bonds adjacent to pyrimidines within the binding site. Most often the sequence of these binding sites are such that this requirement is met. The fact that the probes contain only normal nucleotides rather than analogs such as bromodeoxyuridine, which has been used in other cross-linking studies, is an advantage since UV-induced breaks at bromodeoxyuridine sites in the DNA probes are avoided and thus the background of nonspecific end binding proteins that might otherwise be visualized in this procedure is minimized. Finally Safer et al.[1b] point out that if an excess of the labeled probe is used and the duration and/or intensity of the photolysis is determined experimentally by the methods described, then this technique may be useful for estimating the pool size of these DNA-binding proteins in a crude extract. They state that, because of its sensitivity, it is possible to detect proteins that would not ordinarily be found with a DNase I footprint assay.

Discussion

Factors that Influence Rate and Extent of Photochemical Cross-Linking

The extent of protein–nucleic acid cross-linking that can be achieved is a complex function of several factors that relate both to the experimental conditions as well as to the intrinsic nature of the nucleic acid, the protein, and the three-dimensional structure of the resulting complex. An important variable often overlooked is the fraction of the protein in the reaction mixture that is actually complexed with the nucleic acid. That is, is the final salt concentration low enough and the final protein and nucleic acid (binding site) concentration high enough to ensure that most of the protein is actually in a protein–nucleic acid complex as opposed to being free in solution? If the K_{app} for the complex is not known then preliminary experiments should be carried out where the extent of protein cross-linking is measured as a function of an increasing concentration of nucleic acid. The value of this approach is illustrated in Table III, which shows that increasing the molar ratio of oligonucleotide to UPI from 2.5 to 8 resulted in increasing the extent of cross-linking from about 40% to nearly 86%. Obviously, at the UPI concentration used in this experiment (Table III), the 2.5 molar ratio of oligonucleotide/UPI was not sufficiently high to drive most of the protein into a complex. Another important variable is the dose of ultraviolet light that is used. While it is best to generate a dose–response curve similar to that in Fig. 1 for each system that is being studied, a total dose of about 2.0×10^5 ergs/mm^2, based on the data in Table III, has so far proved to be near optimum. With a General Electric 15-W germicidal lamp this does corresponds to about a 20-min exposure at a distance of about 7.5 cm. Once an optimum dose has been found for the system under study then it is important to determine the extent of photoinactivation of the protein that is occurring with this dose and, using one or more of the methods discussed in the text, to demonstrate that only the native, as opposed to the denatured protein, can be cross-linked and then only when it is complexed with nucleic acid. By carrying out a full dose–response study it is possible to determine rapidly the maximum dose of ultraviolet light that is consistent with minimal photoinactivation of the protein and photodegradation of the resulting amino acid–base adduct.

While there appears to be agreement that thymine is the most photoreactive deoxynucleotide and uridine the ribonucleotide base yielding the greatest extent of photochemical cross-linking to proteins, no data are currently available on the relative tendencies of the 20 different amino acids (when they are in proteins as opposed to being free in solution) to participate in adduct formation with a photochemically excited base. In general, model studies carried out on amino acid–nucleotide mixtures seem to be of little use in this regard. For example, while Schott and Shetlar[26] failed to observe any phenylalanine–thymine cross-linking (whereas they observed significant thymine cross-linking to cysteine, lysine, and arginine), phenylalanine accounts for six of the nine cross-linking

[26] H. N. Schott and M. D. Shetlar, *Biochem. Biophsy. Res. Commun.* **59**, 1112 (1974).

sites that have been identified in the six nucleic acid-binding proteins listed in Table IV. The high propensity for phenylalanine to be cross-linked may reflect an unusually high reactivity for this amino acid when it is held in a certain configuration in proteins or it may simply reflect the essential role that this amino acid may play in the overall mechanism of binding of proteins to single-stranded nucleic acids. There is considerable biochemical and physicochemical data that suggest that the bacteriophage T4 gene 32, the *E. coli* SSB, and the A1 hnRNP proteins (which together account for all of the phenylalanine cross-linking sites identified in Table IV) share a common mechanism of binding involving the close approach of aromatic amino acid side chains with the bases of the single-stranded polynucleotide (see Ref. 15 for a review). In the absence of any relevant data, it seems prudent to assume that any of the 20 different amino acids found in proteins can in principal be photo-cross-linked to nucleic acids.

Because ultraviolet light is a "zero-length" cross-linker, an amino acid residue would have to be in extremely close proximity to cross-link to an excited base. In addition, studies on the bacteriophage fd gene 5 protein suggest that topology must also play an essential role in determining the ability of amino acids to cross-link to nucleic acids. Hence, even though ^{1}H nuclear magnetic resonance (NMR) data requires that Tyr-26 and Phe-73 form part of the DNA-binding domain of the bacteriophage fd gene 5 protein[27] and the data in Table IV demonstrate that both of these amino acids can be cross-linked to nucleic acids, neither of these residues can in fact be cross-linked in the gene 5–ssDNA complex.[4,7] Despite these studies on the gene 5 protein, which suggest that a relatively specific topological arrangement between an amino acid and a nucleotide base must be reached in order to achieve photochemical cross-linking, the fact remains that all five of the single-stranded nucleic acid-binding proteins (bacteriophage T4 gp32, *E. coli* and F plasmid SSB, calf thymus UPI, and in Refs. 4 and 7 the bacteriophage fd gene 5 protein) that have been subjected to UV photolysis in the presence of oligonucleotides can in fact be cross-linked to oligo(dT) or poly(dT) with an efficiency above 20%.

Photochemical Cross-Linking Limited to Those Amino Acid Residues
in Native Protein that Are at Interface of Protein–Nucleic
Acid Complex

Without exception, all of the numerous controls that have been used, including prior denaturation of the protein and the use of competing nucleic acids and high salt concentrations, confirm that only the *native* protein

---

[27] G. C. King and J. E. Coleman, *Biochemistry* **27**, 6947 (1988).

can be cross-linked and only when it is complexed with nucleic acid. That cross-linking is restricted to amino acid residues that are at the interface of the protein–nucleic acid complex is suggested by studies on four proteins whose three-dimensional structures have been solved and whose cross-linking sites have been identified. In the case of the bacteriophage fd gene 5 protein, the cysteine residue that was cross-linked to fd DNA (Cys-33) is located within a predicted DNA-binding groove that is composed of a three-stranded antiparallel β sheet encompassing approximately residues 11–40.[7,28] The three amino acids that are cross-linked in the ribonuclease A–(pUp) complex (Ser-80, Ile-81, and Thr-82) are known to constitute the bottom of the binding site for the pyrimidine ring.[25,29] Based on the recently published crystal structure of the U1 small ribonucloprotein A, which contains a eukaryotic RNA binding motif that is homologous to the two, 90-residue domains found in the A1 hnRNP protein, the two phenylalanine residues that were cross-linked in each of the A1 RNA binding domains (3) are located exactly next to each other on adjacent β-strands that form the proposed RNA binding surface.[30] Similarly, 8-azido-dATP cross-linking to Tyr-766 in the Klenow fragment of DNA polymerase I is consistent with the location of the dNTP-binding site as determined by crystallography, chemical modification, footprinting, and NMR spectroscopy (see Ref. 23 for a summary of these studies). Although a three-dimensional structure is not yet available for the *E. coli* SSB ssDNA-binding protein, the involvement of Phe-60 [which was the residue found cross-linked to (dT)$_8$ (Ref. 2)] in binding has been confirmed by site-directed mutagenesis. Substitution of Phe-60 by several other amino acid residues revealed a direct correlation between the hydrophobicity of the amino acid at this position and the relative *E. coli* SSB affinity for single-stranded nucleic acids.[31] Since NMR spectroscopy revealed little or no conformational change as a result of these mutations, the conclusion was reached that Phe-60 in SSB is involved with a hydrophobic interaction with ssDNA.[31]

In summary, photo-cross-linking provides a useful approach for identifying one or more amino acid residues at the interface of protein–nucleic acid complexes. This technique requires minimal instrumentation, is equally applicable *in vivo* and *in vitro*, is highly specific, and proceeds with minimal perturbation of the system being studied.

[28] G. D. Brayer and A. McPherson, *J. Mol. Biol.* **169** 565 (1983).
[29] F. M. Richards and H. W. Wyckoff, *in* "Atlas of Molecular Structures in Biology" (F. M. Richards and H. W. Wyckoff, eds.), p. 1. Oxford Univ. Press (Clarendon), London and New York, 1973.
[30] K. Ngai, C. Oubridge, T. Jessen, J. Li, and P. Evans, *Nature (London)* **348,** 515 (1990).
[31] I. Bayer, A. Fliess, J. Greipel, C. Urbanke, and G. Maass, *Eur. J. Biochem.* **179,** 399 (1989).

Section IV

Genetic Analysis of Structure–Function Relationships

[26] Use of Nonsense Suppression to Generate Altered Proteins

By Jeffrey H. Miller

Introduction

The study of protein structure–function relationships relies heavily on the ability to generate altered proteins with known sequence changes. The first investigations of altered proteins involved variants that were recognized in the population.[1,2] Later, mutagens were employed to create altered proteins (e.g., Refs. 3–5). These *in vivo* techniques were responsible for much of the knowledge of the effects of altered proteins, until the onset of the recombinant DNA era in the mid-1970s. At that point, a variety of methods became available that allowed the investigator to program specific changes into cloned genes and create the exact altered protein that was desired. These procedures, some of which are reviewed in this volume, include site-directed mutagenesis,[6] gene synthesis,[7] Kunkel mutagenesis,[8] cassette mutagenesis,[9] and combinatorial cassette mutagenesis.[10] Each technique can be used to great advantage in the appropriate situation. This chapter describes an additional method for generating altered proteins that is based on the suppression of nonsense mutations. This latter method greatly simplifies the task of examining a large number of amino acid replacements in the protein under study.

[1] V. M. Ingram, *Nature (London)* **180**, 326 (1957).
[2] M. F. Perutz and H. Lehmann, *Nature (London)* **219**, 902 (1968).
[3] C. Yanofsky, B. C. Carlton, J. R. Guest, D. R. Helinski, and U. Henning, *Proc. Natl. Acad. Sci. U.S.A.* **51**, 266 (1966).
[4] G. Streisinger, F. Mukai, W. J. Dreyer, B. Miller, and S. Horiuchi, *Cold Spring Harbor Symp. Quant. Biol.* **26**, 25 (1961).
[5] G. Streisinger, Y. Okada, J. Emrich, J. Newton, A. Tsugita, E. Terzaghi, and M. Inouye, *Cold Spring Harbor Symp. Quant. Biol.* **31**, 77 (1966).
[6] M. Smith, *Annu. Rev. Genet.* **19**, 423 (1985).
[7] H. G. Khorana, *Science* **203**, 614 (1979).
[8] T. A. Kunkel, *Proc. Natl. Acad. Sci. U.S.A.* **82**, 488 (1985).
[9] J. A. Wells, M. Vasser, and D. B. Powers, *Gene* **34**, 315 (1985).
[10] J. F. Reidhaar-Olson and R. T. Sauer, *Science* **241**, 53 (1988).

The Principle of the Method

Nonsense mutations signal chain termination during protein synthesis (see reviews by Eggertsson and Soll[11] and Gorini[12]). In bacteria, and in most other organisms, the triplets UAG, UAA, and UGA serve as nonsense codons. In the absence of tRNAs that recognize these codons (the normal situation), protein release factors mediate chain termination and release from the ribosome. In *Escherichia coli,* there are two release factors, RF1, which recognizes UAG and UAA, and RF2, which recognizes UAA and UGA. The effects of nonsense mutations can be reversed by suppressor mutations, which allow some translation through the nonsense codon. The nonsense codons have alternate names. UAG is termed "amber," UAA "ochre," and UGA "opal." The initial suppressors were found to result in altered tRNAs, most of which carried a change at the anticodon of a dispensible copy of a tRNA, now enabling it to read one of the three nonsense codons. Each suppressor tRNA inserted a specific amino acid, namely the same amino acid inserted by the unaltered tRNA. Subsequent work, however, showed that at least one suppressor tRNA inserted a different amino acid than the original tRNA. When the tRNA^{Trp} is converted to a UAG suppressor by a change at the anticodon, glutamine is inserted at a UAG site 95% of the time.[13,14] For a summary of many of the suppressors found *in vivo,* see the reviews by Gorini[12] and by Eggertsson and Söll.[11]

The initial set of suppressors allowed the insertion of up to five amino acids at a UAG codon. These included glutamine, tyrosine, serine, and leucine, at efficiencies greater than 10%, and lysine at efficiencies from 1 to 2%. Lysine, glutamine, and tyrosine could also be inserted at UAA codons, and tryptophan could be inserted at a UGA codon. In 1979, these suppressors were used on a collection of 90 nonsense mutations,[15] most of which were UAG mutations, in the *lacI* gene of *E. coli,* which encodes the *lac* repressor protein. The approximately 300 altered repressors that were generated by this technique, each with a known amino acid replacement, did represent an extensive collection of mutant proteins with interesting properties. However, the limitations of this procedure were evident. Namely, only certain amino acids could be inserted, because of the very small set of nonsense suppressors available. Also, nonsense mutations

[11] G. Eggertsson and D. Söll, *Microbiol. Rev.* **52,** 354 (1988).
[12] L. Gorini, *Annu. Rev. Genet.* **4,** 107 (1970).
[13] M. Yaniv, W. R. Folk, P. Berg, and L. Soll, *J. Mol. Biol.* **86,** 245 (1974).
[14] J. Celis, C. Coulondre, and J. H. Miller, *J. Mol. Biol.* **104,** 729 (1976).
[15] J. H. Miller, C. Coulondre, M. Hofer, U. Schmeissner, H. Sommer, A. Schmitz, and P. Lu, *J. Mol. Biol.* **131,** 191 (1979).

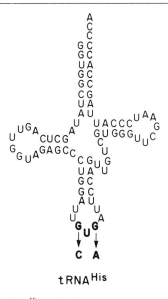

tRNA^{His}

FIG. 1. The structure of tRNA^{His}, with the anticodon altered to read the UAG codon.

could, for the most part, be obtained only by single base changes *in vivo*. Thus, only a fraction of the residues in a protein could be exchanged by nonsense suppression. Both of these limitations could be overcome with the aid of recombinant DNA techniques, as described below.

Construction of Synthetic Nonsense Suppressor Genes

The use of site-directed mutagenesis makes possible the creation of a nonsense mutation at any point in a gene. Also, new suppressor genes can be designed and constructed *in vitro* by annealing synthetic oligonucleotides. With this in mind, the author and co-workers have constructed a set of new suppressor tRNA genes, engineered to read the amber (UAG) codon.[16–18] This involved changing the anticodon of each resulting tRNA to 5'-CUA-3', as given in the example in Fig. 1. The synthetic genes were

[16] J. Normanly, J.-M. Masson, L. G. Kleina, J. Abelson, and J. H. Miller, *Proc.Natl. Acad. Sci. U.S.A.* **83**, 6548 (1986).

[17] L. G. Kleina, J.-M. Masson, J. Normanly, J. Abelson, and J. H. Miller, *J. Mol. Biol.* **213**, 705 (1990).

[18] J. Normanly, L. G. Kleina, J.-M. Masson, J. Abelson, and J. H. Miller, *J. Mol. Biol.* **213**, 719 (1990).

constructed as a set of four to six oligonucleotides, as shown in Fig. 2. When the synthetic genes were reintroduced into *E. coli* cells, suppressor activity was monitored by measuring the level of β-galactosidase produced by the *lacZ* gene carrying a UAG mutation. Figure 3 summarizes the methodology. The nature of the amino acid inserted was determined by purifying dihydrofolate reductase produced in a strain carrying a UAG mutation early in the plasmid-encoded *fol* gene, which encodes dihydrofolate reductase, depicted in Fig. 4. The amino-terminal end of the suppressed protein was then sequenced.

Some of the newly constructed suppressors operated at high efficiency, whereas others did not. Therefore attempts were made to improve the level of suppression.[17] In some cases, changing the bases directly following the anticodon to -A-A- improved the efficiency of suppression (Fig. 5), as expected from previous work.[19,20] In one case, a hybrid tRNA consisting of the anticodon stem and loop of the phenylalanine tRNA UAG suppressor and the remainder of the molecule from the proline tRNA was constructed (Fig. 6) and found to operate at high efficiency,[17] in contrast to the parent proline tRNA suppressor, which did not function as a suppressor at all. In a second case, McClain and Foss[21] used a variant of a hybrid tRNA^{Phe}/tRNA^{Arg} suppressor to construct an arginine-inserting tRNA amber suppressor.

The determinants for correct recognition of tRNAs by their cognate synthetases are under intense study (see review by Schulman and Abelson[22]). At least two classes of tRNAs can be identified. Some tRNAs do not appear to use the anticodon as part of the recognition, whereas others rely heavily on the anticodon for synthase recognition. The first category can be converted into nonsense suppressors and still retain recognition by the correct synthase, but the second cannot, at least not without additional constructions. The sequencing results of the suppressed dihydrofolate reductases bear out these predictions, since in some cases the amino acid inserted by the suppressor is the correct one, and in other cases it is either a mixture of amino acids, or else is predominantly or exclusively either glutamine or lysine.

Table I summarizes the information on the efficiency and nature of the amino acid inserted for a series of suppressors, including both the naturally occurring suppressors and those derived by *in vitro* constructions.

[19] M. Yarus, *Science* **218,** 646 (1982).
[20] D. Bradley, J. V. Park, and L. Soll, *J. Bacteriol.* **145,** 704 (1981).
[21] W. H. McClain and K. Foss, *Science* **241,** 1804 (1988).
[22] L. H. Schulman and J. Abelson, *Science* **240,** 1591 (1988).

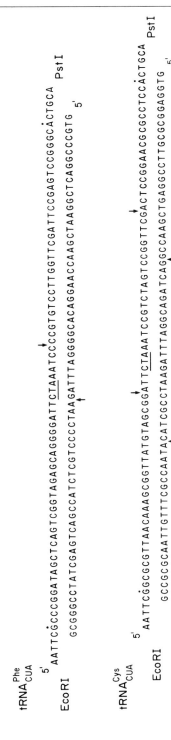

Fig. 2. Synthetic genes for tRNA[Phe] and tRNA[Cys]. The first and last nucleotides encoding each tRNA (indicated by a dot) are immediately flanked by the cohesive ends of *Eco*RI and *Pst*I restriction endonuclease sites, respectively. Arrows indicate the oligonucleotide junctions, and the altered anticodon is underlined. Note that the tRNA[Phe] gene does not encode the 3'-terminal C and A residues. These residues are added *in vivo* by the tRNA nucleotidyltransferase [M. P. Deutscher, *Crit. Rev. Biochem.* **17**, 45 (1984)].

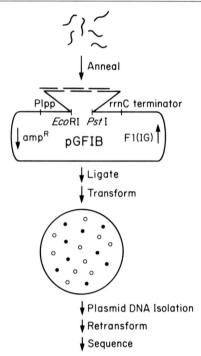

FIG. 3. The steps involved in constructing a synthetic tRNA suppressor gene that suppresses a *lacZ* amber mutation (see text).

Advantages and Disadvantages

Let us examine some of the advantages and disadvantages of using nonsense suppressors. The strong point of the technique is that it makes possible the generation of a large number of variants with a relatively small amount of work. If mutations are being generated by site-directed mutagenesis, then by constructing an amber (UAG) mutation at a given site, 12–13 different mutants are created simply by employing different suppressors. It becomes a feasible project to examine 200 altered proteins with known sequence changes simply by constructing 16 site-directed mutations. As will be seen below, systematic studies of amino acid exchanges can be carried out that would not otherwise be attempted. This method thus serves as an important extension to simple site-directed mutagenesis, since it offers a 12- to 13-fold payoff for each amber mutation constructed by site-directed mutagenesis. In addition, in cases where it is important to control for unwanted secondary mutations resulting from site-directed mutagenesis, only the amber construct needs to be sequenced in its entirety, rather than each individual missense construct.

```
                2                                                      12
DHFR wt:    Ile Ser Leu Ile Ala Ala Leu Ala Val Asp Arg

            ATC AGT CTG ATT GCG GCG TTA GCG GTA GAT CGC

DHFR am:    ...  ...  ...  ...  ...  ...  ...  ...  am Asn ...

            ...  ...  ...  ...  ...  ...  ...  ...  TAG AAT ...
```

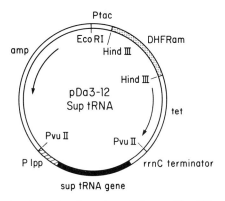

FIG. 4. Plasmid used to determine the amino acid inserted by different amber suppressors. The *E. coli fol* gene was mutated to an amber at the tenth coding position, and the following codon was also changed from GAT to AAT. [For details of the construction see J. Normanly, J.-M. Masson, L. G. Kleina, J. Abelson, and J. H. Miller, *Proc. Natl. Acad. Sci. U.S.A.* **83,** 6548 (1986).]

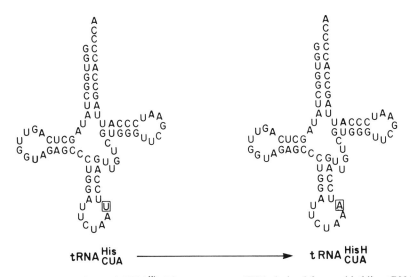

FIG. 5. Alterations of tRNA^{His}. The suppressor tRNA derived from a histidine tRNA, and resulting from a synthetic tRNA gene is shown, as is the change made near the anticodon that improves suppression efficiency.

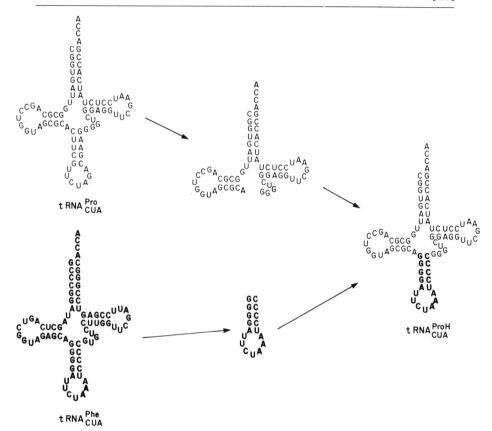

FIG. 6. Construction of tRNA^ProH. The sequences used to produce a hybrid tRNA via gene synthesis, derived from sequences of tRNA^Phe and tRNA^Pro, are shown.

There are potential pitfalls, however, some of which can be overcome with additional experiments. First, there are still seven amino acids that are not inserted efficiently via nonsense suppression. If one of these residues (valine, isoleucine, aspartic acid, asparagine, methionine, tryptophan, or threonine) is to be examined, then additional mutations must be generated. It is anticipated that at least some of the remaining seven amino acids will also be able to be inserted via amber suppression, since efforts to convert the altered recognition by tRNA synthases are underway in several laboratories. However, as will be seen below, the consequences of 12 or 13 amino acid replacements reveal an enormous amount of information regarding the tolerance to substitutions of a specific amino acid residue. Second, the efficiency varies for each suppressor, and for each amber site. Although

TABLE I
EFFICIENCY OF SUPPRESSION AND SPECIFICITY OF INSERTION OF tRNA SUPPRESSORS[a]

Suppressor	Gene	Amino acid inserted	Efficiency of suppression at UAG codons (%)
Su1	supD	Serine	6–54
Su2–89	supE	Glutamine	32–60
Su3	supF	Tyrosine	11–100
Su5	supG	Lysine	0.6–3; 5–29[b]
Su6	supP	Leucine	30–100
tRNA$_{CUA}^{Ala}$	Synthetic	Alanine	8–83
tRNA$_{CUA}^{Cys}$	Synthetic	Cysteine	17–54
tRNA$_{CUA}^{GluA}$	Synthetic	Glutamic acid (80%), glutamine (20%)	8–100
tRNA$_{CUA}^{GlyI}$	Synthetic	Glycine	24–100
tRNA$_{CUA}^{HisA}$	Synthetic	Histidine	16–100
tRNA$_{CUA}^{Lys}$	Synthetic	Lysine	9–29
tRNA$_{CUA}^{Phe}$	Synthetic	Phenylalanine	48–100
tRNA$_{CUA}^{ProH}$	Synthetic	Proline	9–60
FTOIΔ26	Synthetic	Arginine	4–28; 4–47[b]

[a] Taken from L. G. Kleina, J.-M. Masson, J. Normanly, J. Abelson, and J. H. Miller, *J. Mol. Biol.* **213,** 705 (1990).

[b] Assayed in a strain carrying the *mar* mutation (S. M. Ryden and L. A. Isaksson, *Mol. Gen. Genet.* **193,** 38 (1984).

for the most part the efficiency range is only severalfold, there are cases where a greater range is exhibited. If threshold levels of a protein need to be achieved for activity, or if the actual specific activity of a suppressed protein is being determined, the variation of levels is not a significant factor. Third, there is the possibility that certain suppressors add a second amino acid at undetectable (less than 2–3%) levels. This could complicate certain analyses. The best use of the method is to screen a large number of altered proteins to find the few very interesting ones, and then to reconstruct these interesting mutants by designing missense mutations. This would solve both the variation in amount problem, and also the potential problem from minute amounts of a second amino acid being inserted at an amber site. Finally, there are potential complications arising from fragments resulting from the chain-terminating codon. These fragments can occur by two different mechanisms. Since nonsense suppression is rarely 100%, there will always be some level of the polypeptide containing the amino-terminal portion of the protein up until the point corresponding to the amber fragment. If this fragment is not degraded and can interfere with the wild-type protein (negative complementation), then difficulties arise with the nonsense suppression method. Also, sometimes reinitiation of protein synthesis after chain termination can occur at internal reinitiation sites.[23–25] In the *lacI* gene, this occurs at several positions in the beginning of the mRNA, resulting in fragments that negatively complement wild-type repressor.[26] Since reinitiation is rarely efficient, and since only the unsuppressed nonsense codons are subject to subsequent reinitiation, this problem can be overcome by very efficient suppression, or by converting the reinitiation codons to other codons still specifying the same amino acid.[27]

Types and Sources of Suppressors Available

In order to employ nonsense suppressors, it is necessary to be able to place each suppressor in combination with each nonsense mutation in the same strain. If the constructions for achieving this are very laborious, then it defeats the purpose of using nonsense suppression. Suppressors are currently available on the *E. coli*[11] and *Salmonella typhimurium* chromosome,[28] on different plasmids, and on specialized transducing phage such as λ and ϕ80. Although the simplest situation is to have the suppressor on the chromosome, this is not always possible. Also, some of the synthetic suppressors have greatly reduced efficiency when encoded by single-copy genes on the chromosome. On the other hand, greatly increased copy number of some of the suppressor tRNAs inhibits cell growth. Thus, several tRNA suppressors are virtually lethal when cloned onto the very high copy number pUC series of vectors. This effect varies with the suppressor. The serine-inserting Su1 suppressor is very efficient when cloned onto pBR322 and also allows normal growth, whereas the tyrosine-inserting Su3 suppressor causes slow growth and operates inefficiently on the same type of plasmid. Since many genes that will be subjected to suppression of nonsense mutations are carried on plasmids, it is important to employ compatible plasmids for the cloned gene and the cloned suppressor tRNA genes. The synthetic suppressors operate efficiently on a pBR322-based plasmid derived from pEMBL, pGFIB,[29] shown in Fig.

[23] A. Sarabhai and S. Brenner, *J. Mol. Biol.* **27,** 145 (1967).
[24] T. Grodzicker and D. Zipser, *J. Mol. Biol.* **38,** 305 (1968).
[25] T. Platt, K. Weber, D. Ganem, and J. H. Miller, *Proc. Natl. Acad. Sci. U.S.A.* **69,** 897 (1972).
[26] J. G. Files, K. Weber, and J. H. Miller, *Proc. Natl. Acad. Sci. U.S.A.* **71,** 667 (1974).
[27] L. G. Kleina and J. H. Miller, *J. Mol. Biol.* **212,** 295 (1990).
[28] F. Winston, D. Botstein, and J. H. Miller, *J. Bacteriol.* **137,** 443 (1979).
[29] J.-M. Masson and J. H. Miller, *Gene* **47,** 179 (1986).

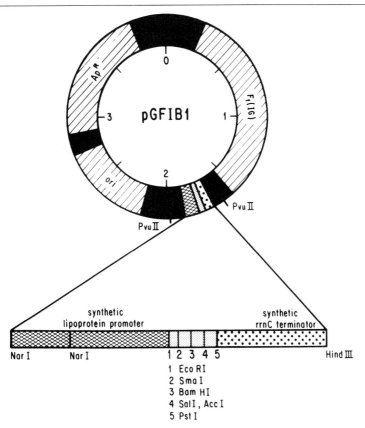

FIG. 7. Plasmid pGFIB-1. This vector [J.-M. Masson and J. H. Miller, *Gene* **47,** 179 (1986)] is a derivative of pEMBL-8 + [L. Dente, G. Cesareni, and R. Cortese, *Nucleic Acids Res.* **11,** 1645 (1983)]. The *lac* promoter was replaced with a synthetically constructed promoter based on the promoter sequence of the *E. coli* lipoprotein gene *lpp* [K. Nakamura and M. Inouye, *Cell (Cambridge, Mass.)* **18,** 1109 (1979)].

7, which would necessitate using a plasmid such as pACYC for the cloned gene carrying the amber mutations. However, different pACYC derivatives of most of the synthetic suppressors have been constructed (R. Crowl, M. Michaels, and C. W. Kim, unpublished, 1991), and although the efficiency of the suppressors is reduced severalfold, strains carrying these plasmids grow much better than those carrying the suppressors on the pEMBL plasmid. The use of pACYC plasmids carrying the synthetic tRNA suppressor genes creates the opportunity to employ pBR322-based plasmids for the cloned gene carrying the amber mutations.

TABLE II

SUPPRESSOR EFFICIENCY IN DIFFERENT CONTEXTS

tRNA suppressor	β-Galactosidase activity (% wild type)[a]			
	A26[b]/AUG	A16/ACA	O17/CGC	O13/CUG
Su1 (Ser)	25.4	21.1	6.3	26.0
Su2 (Gln)	17.8	9.6	0.8	11.3
Su3 (Tyr)	100	42.6	10.6	46.4
Su6 (Leu)	100	61.7	30.4	57.8
tRNA$_{CUA}^{His}$	23.7	17.5	0.8	11.7
tRNA$_{CUA}^{Phe}$	100	75.2	48.0	80.8

[a] β-Galactosidase activity was determined in a *lacI-Z* fusion system as described by J. H. Miller and A. M. Albertini, *J. Mol. Biol.* **164**, 59 (1983). See also L. G. Kleina, J.-M. Masson, J. Normanly, J. Abelson, and J. H. Miller, *J. Mol. Biol.* **213**, 705 (1990).

[b] Site/following codon.

Technical Factors: Procedures and Problems

Context Effects on the Efficiency of Suppression

The efficiency of amber (as well as other nonsense) suppression is affected by the sequence surrounding the amber triplet in the mRNA.[17,30–32] Studies using *lac* fusions indicate that the most important factor in the mRNA sequence is the two bases following the UAG.[31,32] The codons fit into several patterns. In general, UAG triplets followed by codons beginning with A are suppressed efficiently, those followed by G are moderately well suppressed, and those followed by U and C are poorly suppressed. One exception is the CUX family of codons, which allows very efficient suppression when following an amber codon. Also, the AUX family of codons allows the most efficient suppression of all, when following an amber codon. Table II gives some examples of different suppressors operating on UAG codons followed by different sequences. Although this information offers the possibility of constructing each nonsense mutation with the most favorable sequence context, in most cases the amino acid following the one corresponding to the UAG codon would be substituted, a situation one clearly wishes to avoid. Therefore, in order to maintain the same surrounding amino acid sequence, one is limited to completely silent changes. The three amino acids encoded by six different triplets (leucine,

[30] L. Bossi and J. R. Roth, *Nature (London)* **286**, 123 (1980).

[31] L. Bossi, *J. Mol. Biol.* **164**, 73 (1983).

[32] J. H. Miller and A. M. Albertini, *J. Mol. Biol.* **164**, 59 (1983).

TABLE III
EFFECT ON SUPPRESSION EFFICIENCY OF
CODONS FOLLOWING UAG TRIPLET

Codon following UAG triplet	Effect on UAG suppression
Leucine	
CUX	Favorable
UUA, UUG	Unfavorable
Arginine	
CGX	Unfavorable
AGA, AGG	Favorable
Serine	
UCX	Unfavorable
AGU, AGC	Favorable

serine, and arginine) offer the only possibilities for improving the context without changing the amino acid sequence. Thus, when an amber mutation is followed by a Leu, Ser, or Arg codon, it is feasible to change that to a better codon. Table III shows the preferred codons for each of the groups of triplets that comprise the sixfold degenerate codons. In each case one set of triplets is favorable and the other unfavorable. For instance, the CGX family of arginine codons is extremely unfavorable for suppression, whereas the AGA and AGG family is favorable. Likewise, the UUA and UUG codons for leucine are unfavorable for suppression when they follow the amber codon, but the CUX family of triplets is very favorable for suppression.

Instability of Suppressors

Maintaining tRNA suppressor genes on moderate or high-copy-number plasmids requires constant selection to prevent cells that have lost the suppressor from overtaking the population, which they can do quite readily, since they can grow more rapidly. Selection to maintain the plasmid itself is usually carried out by using relevant antibiotic resistance markers. For instance, the pGFIB-based constructs are maintained in the presence of ampicillin, and the pACYC constructs are maintained in the presence of chloramphenicol. However, the suppressor tRNA gene insert can be lost from the plasmid, so it is necessary to have a suppressible amber mutation in the strain. It is convenient to use amber mutations in required genes. The original strains carrying the plasmid-encoded suppressor tRNA genes[16,17] also carry an *argE* amber mutation that requires at least 2–3% suppression to allow normal growth in the absence of arginine.

Therefore, cells are usually grown and stored on minimal medium with ampicillin (or chloramphenicol). They can grow on rich medium (with antibiotic) overnight if inoculated from these minimal plates without suffering a severe loss of the plasmid insert.

Applications

Let us examine some applications of the amber suppressors in generating altered proteins. Two studies serve to illustrate the typical uses of this technology.

Studies with the lac Repressor

In the *lacI* gene of *E. coli*, nonsense mutations have been constructed at over 300 of the 360 coding positions in the gene (P. Markiewicz, L. G. Kleina, and J. H. Miller, unpublished, 1991). The initial 90 of these were obtained by *in vivo* mutagenesis,[15] and the remainder added by site-directed mutagenesis *in vitro* (Kleina and Miller[27], P. Markiewicz and J. H. Miller, unpublished, 1991). In designing the UAG mutations, unfavorable leucine, arginine, or serine codons that followed the UAG triplet were converted to more favorable ones if the need arose. This is illustrated in Fig. 8. In the case of the *lacI* gene, the amber mutations were carried on an F' *lac proAB* eipsome that could be transferred into each suppressor strain by conjugation. In practice, 50 amber mutants were gridded onto nutrient agar in a Petri dish and replica mated into the suppressor strains. The newly created suppressor derivatives were then further replicated onto different indicator media to allow examination of the different LacI phenotypes. The fortunate use of a conjugative plasmid and the availability of sensitive indicator plates permit the examination of a very large number of altered *lac* repressors.

Figure 9 represents the effects of more than 3300 amino acid replacements created by suppressing nonsense mutations at 270 sites in the *lac* repressor (see also Kleina and Miller[27]). This quasi-systematic study of amino acid replacements reveals much about the structure–function relationships of the protein. The distribution of residues that are sensitive to replacement is certainly not even. In the amino-terminal end (the first 59 residues), which has been shown to bind to DNA and operator,[33–36] many

[33] T. Platt, J. G. Files, and K. Weber, *J. Biol. Chem.* **248**, 110 (1973).
[34] B. Mueller-Hill, *Prog. Biophys. Mol. Biol.* **30**, 227 (1975).
[35] N. Geisler and K. Weber, *Biochemistry* **16**, 938 (1977).
[36] R. Ogata and W. Gilbert, *Proc. Natl. Acad. Sci. U.S.A.* **75**, 5851 (1978).

Amino Acid Position of Amber	Sequence			Amino Acid Position of Amber	Sequence		
4	PRO	VAL	THR	20	THR	VAL	SER
	CCA	GTA	ACG		ACC	GTT	TCC
		↓↓↓				↓↓↓	↓↓
	CCA	TAG	ACG		ACC	TAG	AGC
5	VAL	THR	LEU	21	VAL	SER	ARG
	GTA	ACG	TTA		GTT	TCC	CGC
		↓↓	↓ ↓			↓↓	↓ ↓
	GTA	TAG	CTG		GTT	TAG	AGA
8	TYR	ASP	VAL	22	SER	ARG	VAL
	TAC	GAT	GTC		TCC	CGC	GTG
		↓ ↓				↓↓↓	↓
	TAC	TAG	GTC		TCC	TAG	GTC
9	ASP	VAL	ALA	23	ARG	VAL	VAL
	GAT	GTC	GCA		CGC	GTG	GTG
		↓↓↓				↓↓	↓
	GAT	TAG	GCA		CGC	TAG	GTC
10	VAL	ALA	GLU	24	VAL	VAL	ASN
	GTC	GCA	GAG		GTG	GTG	AAC
		↓↓↓			↓	↓↓	
	GTC	TAG	GAG		GTC	TAG	AAC
14	ALA	GLY	VAL	25	VAL	ASN	GLN
	GCC	GGT	GTC		GTG	AAC	CAG
		↓↓↓			↓	↓ ↓	
	GCC	TAG	GTC		GTC	TAG	CAG
15	GLY	VAL	SER	28	ALA	SER	HIS
	GGT	GTC	TCT		GCC	AGC	CAC
		↓↓↓	↓↓			↓↓↓	
	GGT	TAG	AGT		GCC	TAG	CAC
16	VAL	SER	TYR	29	SER	HIS	VAL
	GTC	TCT	TAT		AGC	CAC	GTT
		↓↓				↓ ↓	
	GTC	TAG	TAT		AGC	TAG	GTT
19	GLN	THR	VAL	30	HIS	VAL	SER
	CAG	ACC	GTT		CAC	GTT	TCT
		↓↓↓				↓↓↓	↓↓
	CAG	TAG	GTT		CAC	TAG	AGT

Fɪɢ. 8. Construction of new amber sites in the *lacI* gene constructed *in vitro*. The number given is that of the amino acid specified by the codon converted to amber. Arrows indicate the sequence changes made from the wild type to create the mutant. In some cases the bases in the following coding position were altered so as to allow more efficient suppression without changing the following amino acid. [See also L. G. Kleina and J. H. Miller, *J. Mol. Biol.* **212**, 295 (1990).]

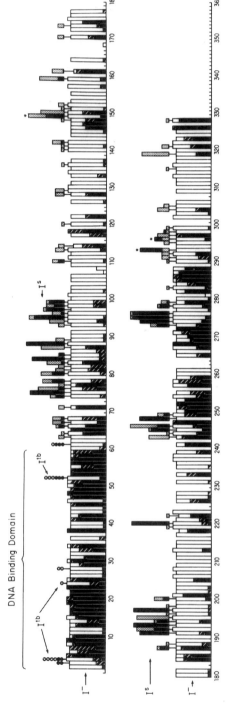

FIG. 9. Effects of amino acid replacements in the *lac* repressor. The consequences of over 3300 single amino acid replacements are depicted here. The repressor protein is shown as an array of 360 sites (amino acids). At each position where amino acids can be exchanged through nonsense suppression a bar has been drawn on the linear axis of the repressor polypeptide. The height of the bar represents the number of exchanges made (12 or 13 in the case of UAG sites, 3 to 4 in the case of UAA sites, and 1 in the case of UGA sites). Each bar is divided into invisible segments, the number of segments corresponding to the number of replacements made. When a replacement generates the I⁻ phenotype, a corresponding portion of the bar is filled in. If the replacement does not alter the I phenotype, then the segment is left open or blank. Intermediate effects are represented by a diagonal stripe. Thus, if all of the replacements destroy repressor function, then the whole bar is filled in (see position 22 or 38). If none of the replacements creates the I⁻ phenotype, then the whole bar is left open (see positions 153 and 189). Above the first set of bars the replacements that create Iˢ repressors are shown. Here the bars indicate only those replacements causing this change in phenotype, the height of the bar indicating the number of replacements that generate the Iˢ repressor. The black segments are strong effects, and the open or dotted segments are weaker effects. The Iˢ phenotype can sometimes result from tighter binding to operator, as occurs for certain replacements in the amino-terminal portion (residues 1 to 61 of the repressor). These repressors are depicted as Iᵗᵇ (tight binding) and are represented by circles above the main set of bars. Filled-in circles indicate strong effects, and open circles weaker effects (P. Markiewicz, L. G. Kleina, and J. H. Miller, unpublished, 1991).

residues are sensitive to substitutions. This is in contrast to the remaining amino acids, many of which can be freely substituted, as is evident in the bottom portion of Fig. 9. Even so, many of these later residues do not tolerate proline, which shows up in this figure as the solitary black segment in an otherwise open box (see the legend to Fig. 9 for more details). There are, of course, residues in the latter part of the protein that are essential, and an inspection of Fig. 9 reveals that the segment between amino acids 239 and 287 has a high density of residues sensitive to substitution.

The upper part of Fig. 9 displays the position of resides involved in inducer binding and/or the allosteric change in response to inducer binding, which causes the repressor to lose affinity for operator DNA. The resulting I^s phenotype is easy to score. While some type of I^s effect results from replacing many of the residues from positions 70–100, in the latter half of the protein I^s repressors are generated by substitutions at five almost evenly spaced clusters. These can be seen in Fig. 9 at residues centered around positions 195, 220, 246, 274, and 296. This spacing has prompted speculation that some of these clusters may represent the ends of β turns in the repressor structure.[37]

The circles shown above the line in the amino-terminal portion of the repressor represent repressors that bind operator and DNA more tightly. Several examples of such tight-binding repressors have been detected before,[38–40] resulting from both missense and suppressed nonsense mutations.

The diagram in Fig. 9 compiles much detailed information into a single perspective. Many of the individual altered repressors have been analyzed more quantitatively.[27] The use of nonsense mutations facilitates spectroscopic studies of proteins, as has been shown for both fluorescence and NMR studies of the repressor.[41–43] Together with the emerging three-dimensional structure of the repressor,[44] the data represented in Fig. 9

[37] J. H. Miller, *J. Mol. Biol.* **131**, 249 (1979).
[38] A. Jobe and S. Bourgeois, *J. Mol. Biol.* **72**, 139 (1972).
[39] A. Schmitz, C. Coulondre, and J. H. Miller, *J. Mol. Biol.* **123**, 431 (1978).
[40] J. L. Betz, *J. Mol. Biol.* **195**, 495 (1987).
[41] H. Sommer, P. Lu, and J. H. Miller, *J. Biol. Chem.* **251**, 3774 (1976).
[42] M. A. C. Jarema, K. T. Arndt, M. Savage, P. Lu, and J. H. Miller, *J. Biol. Chem.* **256**, 6544 (1981).
[43] M. A. C. Jarema, P. Lu, and J. H. Miller, *Proc. Natl. Acad. Sci. U.S.A.* **78**, 2702 (1981).
[44] H. C. Pace, P. Lu, and M. Lewis, *Proc. Natl. Acad. Sci. U.S.A.* **87**, 1870 (1990).

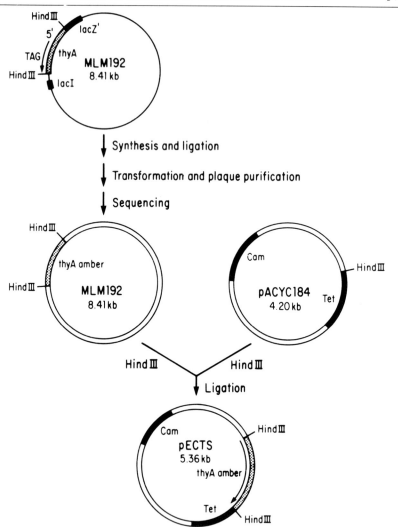

FIG. 10. Construction of pECTS amber mutants. The *thyA*-containing *Hin*dIII fragment from plasmid pATAH was subcloned into M13mp19 to create MLM192. Oligonucleotides containing amber stop codons were annealed to the template and extended and ligated with T4 DNA polymerase and ligase. Mutants were identified by sequencing and were subcloned into pACYC 184 to create the pECTS amber mutant series. [For further details see M. L. Michaels, C. W. Kim, D. A. Matthews, and J. H. Miller, *Proc. Natl. Acad. Sci. U.S.A.* **87,** 3957 (1990).]

TABLE IV
SUPPRESSION PATTERNS OF THYMIDYLATE SYNTHASE[a]

Residue	Nonpolar					Polar				Basic			Acidic
	Gly	Ala	Pro	Phe	Leu	Cys	Ser	Tyr	Gln	His	Lys	Arg	Glu
Glu-14	+	+	+	+	+	+	+	+	+	+	+	+	+
Arg-21	+/−	−/+	+/−	−	−	−	+/−	−	−	+/−	−	+	−
Phe-30	−	−	−	+	+	−	−	+	−	−	−	−	−
Gln-33	+	+	+	+	+	+	+	+	+	+	+	+	+
Arg-35	+	+	+	+	+	+	+	+	+	+	+	+	+/−
Trp-80	−/+	−/+	−	+/−	−	−	−	−	−	−/+	−	−	−/+
Asp-81	+	+	+	+	+	+	+	+	+	+	+	+	+
Asp-105	+	+	+	+	+	+	+	+	+	+	+	+	+
Asp-110	−	−	−	−	−	+/−	+/−	+/−	−	−	−	−	−
Asn-121	+	+	+	+	+	+	+	+	+	+	+	+	+
Arg-126	−/+	−/+	+/−	−	−	−	+/−	+	−	+	+	+	−
Arg-127	+	+	+	+	+	+	+	+	+	+	+	+	+
Cys-146	−	−	−	−	−	+	+/−	−	−	−	−	−	−
His-147	+	+	+	−	+/−	+	+	−	+/−	+	−	+	−
Gln-151	−	−	−/+	−	−	−/+	−	−/+	+	+	−	−	+
Arg-166	−/+	−	−	−	−	−	−	−	−	−/+	−	+	−
Asp-169	−	−	−	−	−	+/−	−	−	−	−	−	−	−
Asn-177	+/−	−	−	−	+/−	−	+/−	−	+/−	−	−	−/+	−
Gly-204	+	+	−	−	−	−	+/−	−	−	−	−	−	−
Glu-223	+	+	+	+	+	+	+	+	+	+	+	+	+

[a] Purified colonies of 20 *thyA* amber mutations transformed into 13 amber suppressor strains were streaked onto media with or without thymidine and growth of individual colonies was compared at 37°: +, 50–1005; +/−, 10–50%; −/+, less than 10%; −, no individual colonies observed. [From M. L. Michaels, C. W. Kim, D. A. Matthews, and J. H. Miller, *Proc. Natl. Acad. Sci. U.S.A.* **87,** 3957 (1990).]

should allow a detailed understanding of the structure–function relationships in the repressor.

Studies with Thymidylate Synthase

Escherichia coli thymidylate synthase is required for the biosynthesis of dTMP (thymidine monophosphate) from dUMP (deoxyuridine monophosphate), in the presence of the cofactor 5,10-methylenetetrahydrofolate.[45] We used site-directed mutagenesis to introduce 20 different amber mutations into the *E. coli thyA* gene.[46] The sites were chosen based on the

[45] D. V. Santi and P. V. Danenberg, *in* "Folates and Pterines" (R. L. Blakely and S. J. Benkovic, eds.), Vol. 1, p. 345. Wiley, New York, 1984.
[46] M. L. Michaels, C. W. Kim, D. A. Matthews, and J. H. Miller, *Proc. Natl. Acad. Sci. U.S.A.* **87,** 3957 (1990).

TABLE V
ACTIVITIES OF SUPPRESSED THYMIDYLATE SYNTHASE (TS) VARIANTS

Wild-type amino acid	Amino acid substituted	Growth phenotype	Proposed role	Total activity (per milligram extract protein per hour)	$[TS]_{\mu g/ml}$ in the extract	Specific activity (per milligram TS protein per hour)	Percent of wild type
Phe-30	Ala-30	−	Crystal structure suggests an important role in tertiary structure of the binding pocket	0.80 (±0.73)	5.12 (±1.75)	0.16 (±0.20)	0.32
	Phe-30	+		973.52 (±142.38)	32.51 (±1.80)	23.36 (±4.71)	46.46
	Ser-30	−		0.37 (±0.13)	4.96 (±1.11)	0.08 (±0.05)	0.16
	Tyr-30	+		77.03 (±17.26)	1.88 (±0.59)	42.58 (±22.90)	84.69
	Leu-30	+		246.82 (±18.46)	3.94 (±0.73)	61.51 (±16.00)	122.33
Gln-33	Ala-33	+	Conserved amino acid among all known TS's.	38.28 (±5.74)	1.16 (±0.04)	26.09 (±4.81)	51.89
	Phe-33	+		409.02 (±169.86)	6.06 (±2.22)	36.35 (±28.41)	72.30
	Ser-33	+		519.72 (±39.88)	7.12 (±0.71)	49.72 (±8.78)	98.89
	Tyr-33	+	Crystal structure suggests no role in catalysis	485.25 (±16.25)	16.34 (±10.44)	48.67 (±32.72)	96.80
	Leu-33	+		675.62 (±63.32)	31.30 (±8.92)	22.26 (±8.43)	49.27
	Gln-33	+		602.52 (±54.22)	16.15 (±0.91)	49.25 (±7.21)	97.95
Arg-127	Ala-127	+	Crystal structure suggests an important role in dUMP binding	27.56 (±14.12)	0.69 (±0.02)	28.78 (±15.63)	57.24
	Phe-127	+		54.10 (±1.56)	10.51 (±2.37)	4.61 (±1.17)	9.17
	Ser-127	+		54.94 (±5.27)	4.22 (±0.35)	10.60 (±1.90)	21.08
	Try-127	+		48.93 (±8.42)	8.76 (±1.19)	8.48 (±2.61)	16.86
	Leu-127	+		550.26 (±77.75)	27.74 (±5.58)	20.05 (±6.87)	39.88
	Arg-127	+		361.54 (±21.40)	13.24 (±2.29)	31.82 (±7.39)	63.29
His-147	Ala-147	+	Crystal structure suggests an important role during catalysis	93.01 (±11.82)	38.43 (±3.31)	3.45 (±0.74)	6.86
	Phe-147	−		0.73 (±0.03)	49.30 (±3.11)	0.03 (±0.01)	0.06
	Ser-147	+		21.19 (±1.67)	57.18 (±3.51)	0.63 (±0.09)	1.25
	Tyr-147	−		1.02 (±1.70)	9.96 (±3.26)	0.12 (±0.20)	0.24
	Leu-147	±		5.71 (±1.97)	19.14 (±2.96)	0.49 (±0.24)	0.97
	His-147	+		1436.61 (±23.89)	46.69 (±4.40)	43.08 (±4.78)	85.68
Wild-type plasmid	None	+		790.05 (±151.35)	26.40 (±7.22)	50.28 (±23.38)	100.00

three-dimensional structure of the enzyme.[47] In this case, we employed a plasmid that expressed the *thyA* gene in a strain that was ThyA⁻. Figure 10 depicts the use of this plasmid to construct the mutations. The *thyA*-encoded plasmid was derived from pACYC,[48,49] which is compatible with the pBR322-based plasmids carrying the synthetic superrepressor genes. Although the activity of thymidylate synthase can be roughly estimated by scoring for Thy⁺ and Thy⁻, as shown in Table IV, it is more informative to assay directly for thymidylate synthase activity in crude extracts, and then to calculate specific activities after determining the concentration of thymidylate synthase by quantitative Western blotting. Table V depicts a sample of the results obtained.

Summary

    The use of suppressed nonsense mutations to generate altered proteins can greatly simplify studies in which a large number of defined mutant proteins are sought. If site-directed mutagenesis is used to generate specific mutations, than for every amber (UAG) mutation constructed, as many as 13 different amino acids can be inserted at the corresponding site in the protein. This allows a rapid screening of many altered proteins for those with interesting properties. Once identified, the interesting substitutions can be regenerated by missense changes, to avoid some of the potential problems of the method. Nonsense suppression has been used to generate more than 3300 amino acid replacements in the *E. coli lac* repressor, and close to 250 amino acid substitutions in *E. coli* thymidylate synthase.

Acknowledgment

    The work described here was partly supported by a grant from the National Science Foundation (DMB-8417353).

---

[47] D. A. Matthews, K. Appelt, and S. J. Oatley, *J. Mol. Biol.* **205,** 449 (1989).
[48] A. Y. C. Chang and S. N. Cohen, *J. Bacteriol.* **148,** 1141 (1978).
[49] M. Belfort and J. Pederson-Lane, *J. Bacteriol.* **160,** 371 (1984).

[27] Random Mutagenesis of Protein Sequences Using Oligonucleotide Cassettes

By John F. Reidhaar-Olson, James U. Bowie,
Richard M. Breyer, James C. Hu, Kendall L. Knight,
Wendell A. Lim, Michael C. Mossing, Dawn A. Parsell,
Kevin R. Shoemaker, and Robert T. Sauer

Introduction

Investigations of protein structure and function often rely on the analysis of mutant proteins. With the advent of methods for rapid and economical chemical synthesis of DNA, there has been a steadily increasing use of oligonucleotide-directed mutagenesis[1] and cassette mutagenesis[2,3] to create specific mutations at particular sites. These synthetic methods can also be used to create random mutations at one or more codons in a gene. In this chapter, we describe and evaluate several methods of cassette mutagenesis that allow random mutations to be generated, and discuss the application of these methods to the analysis of the structure and function of DNA-binding proteins. In addition to the techniques described here, other methods are available for the efficient generation of protein variants. These include approaches based on the suppression of amber mutations[4] and oligonucleotide-directed methods that generate either random changes[5,6] or specific changes.[7]

A simple form of cassette-mediated random mutagenesis involves mutating a single codon to encode all 20 naturally occurring amino acids. Individual clones are then isolated from the resulting population of mutants, sequenced, and screened for neutral, conditional, or defective phenotypes. This method can be somewhat labor intensive, but because changes are limited to a single position, the interpretation of the resulting phenotypes is relatively straightforward. As described below, this ap-

---

[1] M. J. Zoller and M. Smith, this series, Vol. 100, p. 468.
[2] S. J. Eisenbeis, M. S. Nasoff, S. A. Noble, L. P. Bracco, D. R. Dodds, and M. H. Caruthers, *Proc. Natl. Acad. Sci. U.S.A.* **82**, 1084 (1985).
[3] J. A. Wells, M. Vasser, and D. B. Powers, *Gene* **34**, 315 (1985).
[4] J. H. Miller, this volume [26].
[5] D. D. Loeb, R. Swanstrom, L. Everitt, M. Manchester, S. E. Stamper, and C. A. I. Hutchison, *Nature (London)* **340**, 397 (1989).
[6] J. D. Hermes, S. C. Blacklow, and J. R. Knowles, *Proc. Natl. Acad. Sci. U.S.A.* **87**, 696 (1990).
[7] B. C. Cunningham and J. A. Wells, *Science* **244**, 1081 (1989).

proach is easily extended to the simultaneous mutagenesis of several codons, either in contiguous blocks or in noncontiguous residue positions. In either of these multiple mutagenesis experiments, a biological selection is generally applied to identify active clones, which are then sequenced. Such experiments generate lists of functional sequences, from which one can determine the spectrum of substitutions that are tolerated. An analysis of the resulting substitution patterns is then used to determine the importance of the mutagenized positions and to check for possible combinatorial effects.

All cassette mutagenesis procedures involve the synthesis of a small, double-stranded DNA molecule that can be ligated into a larger vector fragment to reconstruct the gene of interest (Fig. 1). The cassette is created by chemical synthesis or a combination of chemical and enzymatic synthesis. The backbone molecule is generated by digestion of plasmid or viral DNA with restriction enzymes. To introduce a cassette into a particular region of a gene requires that appropriate restriction sites occur in the gene sequence. If such sites are not present in the wild-type gene, it is usually possible to modify the sequence in such a way that restriction sites are introduced every 30–40 bases without altering the encoded protein sequence.[8,9]

In the following sections, we discuss mutagenic strategies and issues that arise in the interpretation of results from random mutagenesis experiments. We then describe techniques for preparation of oligonucleotide cassettes, for preparation of plasmid backbones, for transformation, and for rapid sequencing of mutant genes. Finally, we discuss problems that can arise in cassette mutagenesis experiments.

Mutagenic Strategies

Genetic Selections and Screens. Efficient functional selections or screens are extremely important in most randomization experiments, especially when multiple codons are mutagenized. For sequence-specific DNA-binding proteins that can be expressed in *Escherichia coli,* it is often possible to develop selections *in vivo* based on antibiotic resistance.[10,11]

[8] M. Nassal, T. Mogi, S. S. Karnik, and H. G. Khorana, *J. Biol. Chem.* **262,** 9264 (1987).

[9] A number of computer programs are available that identify sites in gene sequences at which recognition sites for restriction endonucleases may be introduced without altering the protein sequence. One such program, SeqSearcher for the Apple Macintosh computer, is available from Jim Hu, Dept. of Biology, Room 16-833, Massachusetts Institute of Technology, Cambridge, MA 02139.

[10] M. C. Mossing, J. U. Bowie, and R. T. Sauer, this volume [29].

[11] S. J. Elledge, P. Sugiono, L. Guarente, and R. W. Davis, *Proc. Natl. Acad. Sci. U.S.A.* **86,** 3689 (1989).

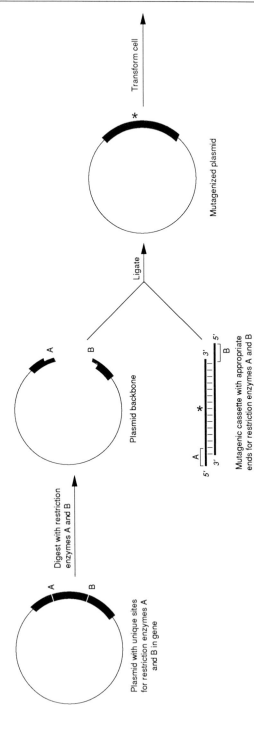

FIG. 1. Cassette mutagenesis procedure. A plasmid or viral vector is digested with restriction enzymes A and B to generate a backbone fragment. An oligonucleotide cassette corresponding to the region between sites A and B is synthesized with mutations at one or more positions. The cassette is then ligated to the backbone fragment to generate the mutagenized vector, which is transformed into the appropriate cells.

	T	C	A	G
T	TTC Phe TTG Leu	TCC TCG Ser	TAC Tyr TAG Stop	TGC Cys TGG Trp
C	CTC CTG Leu	CCC CCG Pro	CAC His CAG Gln	CGC CGG Arg
A	ATC Ile ATG Met	ACC ACG Thr	AAC Asn AAG Lys	AGC Ser AGG Arg
G	GTC GTG Val	GCC GCG Ala	GAC Asp GAG Glu	GGC GGG Gly

FIG. 2. Genetic code with third codon position restricted to G or C. All of the codons shown are accessible in an NN$_C^G$ randomization. Further restrictions are also possible. For example, an NT$_C^G$ randomization would restrict the available codons to the first column.

Colony screens based on repression of *lacZ* or *galK* expression are also straightforward in many cases. In interpreting phenotypes based on any selection or screen, it is important to know what level of activity is required. If the stringency of the selection or screen is very high, then small reductions in activity or level may result in a conditional or even defective phenotype, and fewer residue substitutions will be tolerated. If the stringency is lower, then more substitutions would be expected to be classified as functionally neutral. In our experience, a selection that requires approximately 10% of wild-type activity provides a useful metric that allows the importance of residue positions to be established.

Codon Restrictions. In codon randomization experiments, the position(s) of interest can be changed to either a limited set or a complete set of the 20 naturally occurring amino acids. Complete randomization at the codon level can be achieved by an NNN randomization (where N indicates a mixture of A, C, G, and T). However, as shown in Fig. 2, the third codon position can be restricted to G or C without eliminating any of the amino acids, and thus an NN$_C^G$ randomization is sufficient to achieve complete randomization with respect to possible amino acid sequences. This restriction is useful both because it reduces the overall DNA sequence complexity and because it reduces the coding discrepancy between residues like Met and Trp, which have a single codon, and residues like Leu, Arg, and Ser, which have six codons in an NNN randomization but only three codons in an NN$_C^G$ randomization. Randomizations can also be restricted to specific regions of the genetic code. For example, a randomization using

NT_C^G would be restricted to the five hydrophobic amino acids Phe, Leu, Ile, Met, and Val, whereas an NA_C^G randomization would be restricted to the seven relatively hydrophilic side chains Tyr, His, Gln, Asn, Lys, Asp, and Glu. In either complete or limited randomizations of this type, equal quantities of each desired base are generally included. Thus in unselected populations, one would expect to recover each residue at frequencies that were roughly proportional to the number of codons for that residue. However, random mutagenesis can also be performed using unequal quantities of each base,[12,13] in order to bias recoveries toward a specific residue, which is often the wild type. For example, a "biased" randomization might include 70% of the wild-type base at each codon position and 10% of each of the other bases.

Single-Codon Randomizations. Mutagenesis of single-codon positions is the most basic kind of cassette randomization experiment. In general, one uses an unbiased NN_C^G randomization to generate all of the possible amino acids and then analyzes the sequences and phenotypes of clones from the resulting population of variants. A good mixture of different sequences can generally be obtained simply by sequencing unselected clones. The phenotypes conferred by each different sequence can then be determined. Alternatively, it may be desirable in some cases to screen clones for phenotype first and then to sequence several candidates from each phenotypic class.

How many unselected clones from an NN_C^G randomization need to be sequenced to recover a given number of different amino acids? Monte Carlo simulations (assuming a random distribution at the nucleotide level) show that sequencing 30 candidates generally results in the identification of 14–16 residues, whereas sequencing almost 100 candidates is necessary to have a greater than 50% chance of recovering all 20 amino acids.[14] For most studies, there is no need to analyze all possible substitutions; enough different changes are recovered by sequencing 20–30 candidates to evaluate the importance of the residue being studied. When recovery of a complete set of substitutions is desirable, it is often easiest to sequence approximately 50 clones, and then make the remaining 1 or 2 substitutions by conventional site-directed mutagenesis using a cassette.

Randomization of Multiple Codons. Simultaneous randomization of more than one codon is useful because many positions can be examined in a single experiment. Moreover, potential interactions among residues can be tested. However, in randomization experiments involving multiple

[12] A. R. Oliphant, A. L. Nussbaum, and K. Struhl, *Gene* **44,** 177 (1986).

[13] D. E. Hill, A. R. Oliphant, and K. Struhl, this series, Vol. 155, p. 558.

[14] J. F. Reidhaar-Olson and R. T. Sauer, *Proteins: Struct. Funct. Genet.* **7,** 306 (1990).

codons, it is rarely desirable to try to recover or analyze all of the resulting sequences. This is true both because of numerical complexity and because analysis of multiply mutant sequences that confer a defective phenotype is generally uninformative. As a consequence, the analysis is usually limited to active sequences that are identified by using a biological selection for activity. Screening among unselected candidates can also be used to find active clones, but can be tedious if only a small number of the randomly generated sequences are active.

In multiple codon randomizations, it is often desirable to limit the number of codons being mutated or the extent of randomization to ensure that some active sequences are recovered. If an NN_C^G randomization is applied to 5 codons, then there are 20^5 or 3,200,000 different amino acid sequences generated. The complexity at the level of nucleic acid sequences is even greater. In principle, with a sufficiently powerful selection, one could recover any active sequence (including wild type). In practice, however, the total library of transformants may be smaller (e.g., 10^5–10^6) than the sequence complexity and thus there is no guarantee that any active sequences will be recovered. By contrast, if the NN_C^G randomization were restricted to 3 codons, then the amino acid sequence complexity would be only 8000, and there would be an excellent chance of recovering all active sequences from the mutagenized library. If the goal of a study is a comprehensive survey of all sequences, then it is probably best to limit complete randomizations to three or at most four codons.

Sequence complexity in multiple codon randomizations can also be limited by restricting the randomization to a subset of the amino acids or by biasing the randomization toward the wild-type residue (see section on codon restriction). Restricting the randomization is useful if the subset of residues introduced at each position (these subsets can be different for different codons) provide useful information. For example, in one study of hydrophobic packing in the core of λ repressor, a randomization of 3 positions was restricted to 5 hydrophobic residues. allowing a comprehensive survey of the 125 possible residue combinations.[15] In biased randomizations, the extent of mutagenesis can be controlled by adjusting the level of mutant bases. For example, in studies of the Arc repressor,[16] blocks of approximately 10 adjacent codons were mutagenized by contaminating the wild-type nucleotide at each position with 7.5% of each of the other 3 bases. At this level of mutagenesis, each codon has about a 60% chance of encoding the wild-type residue and about a 40% chance of encoding a mutant amino acid. Hence, the mutagenized population should contain

[15] W. A. Lim and R. T. Sauer, *J. Mol. Biol.* **219**, 359 (1991).
[16] J. U. Bowie and R. T. Sauer, *Proc. Natl. Acad. Sci. U.S.A.* **86**, 2152 (1989).

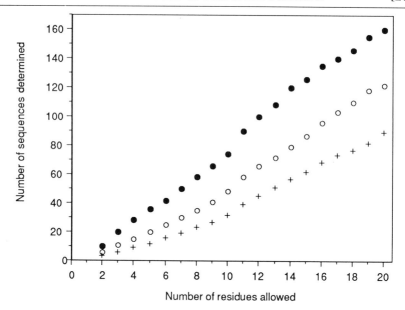

FIG. 3. Number of sequences required to observe all allowed residues. Monte Carlo simulations of NN$_C^G$ random mutagenesis experiments were performed to find the number of sequences that need to be determined to observe all allowed residues 50% of the time (+), 80% of the time (○), and 95% of the time (●). For example, if one determines 50 sequences and observes 10 different residues at a randomized position, then one knows with about 80% confidence that all allowed residues have been recovered. With 10 observed residues, about 75 sequences would be required to increase the confidence level to 95%.

occasional wild-type sequences (at a frequency of approximately 0.6%) but most genes should contain an average of four amino acid substitutions in the mutagenized region. Substitutions resulting from "biased" randomizations will be somewhat limited because single base changes are more probable than double changes, and triple changes are extremely rare. In practice, however, enough substitutions can generally be recovered to assess the general requirements for function.

In analyzing functional variants, how does one decide that enough sequences have been determined to be confident that all tolerated substitutions have been identified? As shown in Fig. 3, the number of sequences that need to be determined depends both on the number of residues allowed and the desired confidence limits. Nevertheless, several points are clear. For example, if 20 sequences are determined and only 1 or 2 residues are recovered at a given position, then it is quite likely that other residues would be nonfunctional. However, if in the set of 20 sequences, 10 different residues are recovered, then sequencing more candidates is likely to result in additional allowed residues. In general, if a large number of different side chains are observed to be tolerated at a given position, then it is probably dangerous to assume that the residues that have not been recovered are nonfunctional. However, it is generally not necessary to know with certainty that all allowed changes have been recovered. The simple fact that a large number of chemically dissimilar side chains allow function suggests that the position under study cannot be critical for function.

In any experiment where more than one codon is randomized, it is possible that some of the functional substitutions recovered at one position depend on simultaneous changes at other positions in the sequence. In such cases, the observed substitutions might not be tolerated as single substitutions in otherwise wild-type backgrounds. Such examples are most common in the case of interacting residues. For example, when three or four interacting residues in the hydrophobic core of the N-terminal domain of λ repressor were randomized together, residue substitutions were found to be allowed that were known not to be tolerated as single substitutions.[17]

Applications and Analysis

Identification of Allowed Substitutions. Cassette randomization experiments are extremely useful in identifying the importance of residues in a protein sequence. Because the method of mutagenesis ensures that a complete ensemble of sequences is present in the initial randomized population, many different substitutions should be allowed if the chemical identity of the side chain is not important for function. In contrast, if only the wild-type residue or a small set of related side chains is recovered among the functional sequences, then it is reasonable to infer that most other substitutions would result in a defective phenotype.

Figure 4 shows residue substitutions that are functionally allowed in the helix 1 region of the N-terminal domain of λ repressor.[14] Some positions tolerate many chemically dissimilar side chains, suggesting that these positions play little or no role in either structure or function. Other positions are restricted to a single residue or a small set of chemically similar side chains. In the crystal structure of the N-terminal domain bound to operator DNA,[18] the ε-amino group of Lys-19 and the hydroxyl of Tyr-22 make direct contacts with the phosphate backbone of the operator, indicating a direct role of these side chains in DNA binding. In addition to its role in DNA binding, the ring portion of Tyr-22 forms part of the hydrophobic core of the protein and thus serves a structural role as well. Ala-15,

[17] W. A. Lim and R. T. Sauer, *Nature (London)* **339**, 31 (1989).
[18] S. R. Jordan and C. O. Pabo, *Science* **242**, 893 (1988).

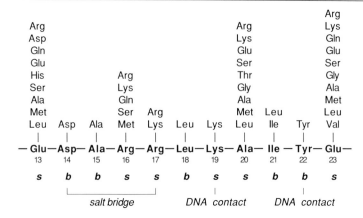

FIG. 4. Allowed substitutions identified from random mutagenesis experiments. The allowed residues in a helical region of the N-terminal domain of λ repressor are shown above the wild-type sequence. The numbers below the wild-type sequence indicate residue numbers. An *s* below the sequence indicates that the wild-type residue is on the surface of the protein; a *b* indicates that the residue is buried (0–25% fractional side chain accessibility). The side chains of Asp-14 and Arg-17 interact through a salt bridge. Lys-19 and Tyr-22 make contacts with the operator DNA. In general, residues that are either structurally or functionally important are conserved.

Leu-18, and Ile-21 also form part of the hydrophobic core, while Asp-14 and Arg-17 form a charge stabilized hydrogen bond. Hence, these residues are likely to be important for maintaining the folded structure of the protein.

Discriminating between Structural and Functional Effects. The examples discussed above show that allowed substitutions can be restricted because the residue mediates contact with the DNA or is important for protein folding or stability. In the absence of structural information, these effects can usually be distinguished by biochemical studies of purified proteins.[19,20] However, it is sometimes possible to eliminate mutations that cause structural defects if more than one phenotype that depends on structure can be monitored. For example, in studies of the positive control function of λ repressor, randomized candidates were first selected for their ability to mediate repression (thereby ensuring that the mutant proteins could fold and bind DNA) and then screened for a positive control phenotype.[21] This allowed the identification of a single residue that was critical

[19] M. H. Hecht, J. M. Sturtevant, and R. T. Sauer, *Proc. Natl. Acad. Sci. U.S.A.* **81,** 5685 (1984).
[20] H. C. M. Nelson and R. T. Sauer, *J. Mol. Biol.* **192,** 27 (1986).
[21] F. D. Bushman, C. Shang, and M. Ptashne, *Cell (Cambridge, Mass.)* **58,** 1163 (1989).

for positive control. As discussed below, strategies involving antibody screens or intracellular proteolysis can also be used to distinguish structural and functional effects.

Proteolysis and Structural Defects. Thermally unstable variants of many DNA-binding proteins, including the N-terminal domain of λ repressor, λ Cro, and P22 Arc, are subject to rapid degradation in *E. coli.*[22-25] As a result, variants bearing destabilizing substitutions are generally present at intracellular levels lower than wild type, and the presence of normal levels of a mutant protein *in vivo* is often indicative of thermal stability. For example, in studies of the Arc repressor, blocks of approximately 10 codons at a time were subject to a "biased" randomization and unselected colonies were screened by sodium dodecyl sulfate (SDS) gel electrophoresis for Arc protein levels.[16] The *arc* genes from colonies displaying moderate to high levels of protein were sequenced, and the corresponding proteins were purified and studied by circular dichroism to confirm that they were stably folded. These studies resulted in a list of "structurally" tolerated substitutions. By comparing this list with that of functionally tolerated substitutions, it was possible to distinguish positions likely to be directly involved in DNA recognition from those likely to be involved in stabilization of structure.

Although proteolysis screens can be used in a preliminary way to identify structurally stable proteins, the absence of a mutant protein in a cell lysate does not always indicate structural instability. For example, in a study of the N-terminal domain of λ repressor, the five C-terminal codons of the gene were subjected to a complete NN_C^G randomization and colonies were screened by SDS gels for protein levels.[26] Candidates found to have extremely low levels of the N-terminal domain (and to be rapidly degraded had hydrophobic C-terminal pentapeptides. By contrast, candidates with high levels of protein had hydrophilic C-terminal pentapeptides. While these results provided clear evidence that hydrophobic residues at the C terminus resulted in proteolytic instability, the T_m of one of the most rapidly degraded variants was found to be identical to wild type. Hence, proteolytic instability and thermal instability are not always correlated. This example illustrates that purification and biochemical analysis of mutant proteins are ultimately required to understand the effects of particular mutations.

[22] D. A. Parsell and R. T. Sauer, *J. Biol. Chem.* **264,** 7590 (1989).

[23] J. U. Bowie and R. T. Sauer, *J. Biol. Chem.* **264,** 7596 (1989).

[24] A. A. Pakula, V. B. Young, and R. T. Sauer, *Proc. Natl. Acad. Sci. U.S.A.* **83,** 8829 (1986).

[25] A. A. Pakula and R. T. Sauer, *Proteins: Struct. Funct. Genet.* **5,** 202 (1989).

[26] D. A. Parsell, K. R. Silber, and R. T. Sauer, *Genes Dev.* **4,** 277 (1990).

Observed Densities of Restricted Sites in DNA-Binding Proteins. In the 92-residue N-terminal domain of λ repressor, 60 residues have been studied by random mutagenesis.[14,17,21,27–30] Of these residues, roughly one-half exhibit highly restricted substitution patterns, and thus play important roles of some kind in repressor structure and function. Many of the important residues are buried in the hydrophobic core or the dimer interface of the protein. Of the remaining restricted positions, some are involved in hydrogen bonds or salt bridges, some are directly required for DNA binding, and one is essential in protecting the protein from intracellular proteolysis.

In the P22 Arc repressor, the entire 53 residues of the protein have been characterized in a "biased" randomization study.[16] Here, the identities of approximately one-third of the residues were found to be functionally important, and one-half were structurally important. The remaining residues could be freely substituted and thus are unimportant for either structure or function.

Interplay of Single-Codon and Multiple-Codon Randomizations. It is often useful to apply single-codon and multiple-codon randomizations in a sequential fashion, as the two methods can provide different types of information. For example, in mapping antibody-binding epitopes on the surface of λ repressor, a region of the gene was subjected to a multiple-codon "biased" randomization and candidates displaying repressor activity and binding to a conformation-specific monoclonal antibody were identified.[28] Several surface positions were found to be invariant in the antibody-reactive clones, suggesting that the side chains at these positions played important roles in antibody binding. This was confirmed by performing single-codon NN_C^G randomizations on each of these positions, sorting the resulting mutants into reactive and nonreactive classes, and then purifying representative mutants and measuring antibody affinities directly. In this case, the multiple-codon randomization provided a rapid way to identify important positions, while the single-codon randomizations provided detailed information about the sequence requirements at each of these positions.

Multiple-codon randomizations can also be performed after single-codon studies, as a test of the additivity of allowed substitutions. For example, in studies of leucine zipper function, single-codon randomizations were first performed to identify allowed substitutions at four of the

[27] J. F. Reidhaar-Olson and R. T. Sauer, *Science* **241**, 53 (1988).
[28] R. M. Breyer and R. T. Sauer, *J. Biol. Chem.* **264**, 13355 (1989).
[29] J. F. Reidhaar-Olson, D. A. Parsell, and R. T. Sauer, *Biochemistry* **29**, 7563 (1990).
[30] N. D. Clarke and C. O. Pabo, manuscript in preparation.

conserved leucine positions.[31] In these experiments, other hydrophobic residues such as Val, Ile, and Met were found to be tolerated at each of the four positions. However, when all four of these leucine positions were randomized simultaneously, it was found that each of the functional sequences contained at least two and usually three leucines, suggesting that many of the substitutions that were tolerated singly could not be tolerated together. In this case, the multiple-codon randomization revealed combinatorial effects that could not be identified in the single-codon randomizations.

Random Mutagenesis Techniques

Synthesis of First Strand of Cassette

Oligonucleotides can be conveniently prepared using any of a number of commercially available DNA synthesizers. On some synthesizers, the machine can be programmed to deliver variable amounts of any of the four bases during any coupling step. If this is not the case, then it is generally necessary to have additional bottles with mixtures of bases (e.g., equal amounts of A, C, G, and T). These bottles may then be attached to extra delivery ports on the synthesizer, if available. If the machine has capacity for only four bottles of nucleotide solutions, it will be necessary to interrupt the synthesis and replace one of the bottles with the base mixture at the appropriate step of the synthesis. Following removal of the final product from the solid support, the oligonucleotide may be of sufficient purity to use directly. However, gel purification on a 12–20% (w/v) polyacrylamide/urea gel[32] is often desirable in order to remove incomplete DNA fragments that accumulate during the course of the synthesis. Following electrophoresis, the appropriate band is located by UV shadowing or staining with ethidium bromide, and excised. The oligonucleotide is then eluted from the gel slice and can be further purified (e.g., by passage over a reversed-phase C_{18} Sep-Pak column from Waters Associates, Milford, MA).

Chemical Synthesis of the Complementary Strand

Figure 5A and B shows two possibilities for chemical synthesis of the second strand of the cassette. In the first, the same NN_C^G random mixture is included in both strands at the codon or codons being randomized (Fig. 5A). In theory, mismatches generated at the randomized codon could lead

[31] J. C. Hu, E. K. O'Shea, P. S. Kim, and R. T. Sauer, Science 250, 1400 (1990).
[32] F. M. Ausubel, R. Brent, R. E. Kingston, D. D. Moore, J. G. Seidman, J. A. Smith, and K. Struhl, "Current Protocols in Molecular Biology." Wiley, New York, 1989.

A. Random bases on both strands

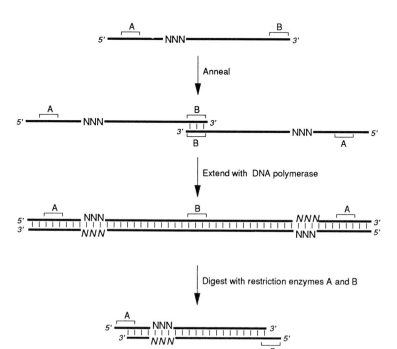

B. Inosine pairing

C. Enzymatic second-strand synthesis

Anneal

Extend with DNA polymerase

Digest with restriction enzymes A and B

to inefficient pairing or problems with heterogeneity following transformation. However, in practice this method appears to be completely satisfactory, in both efficiency and randomness, when codons are randomized individually. We have not tested the technique extensively with multiple randomizations. An alternative method (Fig. 5B) uses inosine in the second strand opposite the randomized positions in the first strand. Since inosine is able to base pair with each of the four standard bases,[33] such cassettes are able to anneal and ligate with high efficiency. As discussed later, this procedure occasionally introduces some bias toward particular bases, but appears to be sufficiently random in most instances. This technique has been used successfully to randomize as many as three residue positions simultaneously.[27]

Procedure. Following synthesis and purification, each strand of the cassette is diluted to a concentration of 0.5 μM in 40 μl kinase buffer [50 mM Tris-HCl (pH 7.5), 10 mM MgCl$_2$, 5 mM dithiothreitol, 0.1 mM spermidine, 0.1 mM EDTA, 2 μM ATP], and phosphorylated using 20 units of T4 polynucleotide kinase (New England Biolabs, Beverly, MA). Phosphorylation is carried out for 1.5 hr at 37°, followed by 10 min at 65° to inactivate the kinase. (As discussed later, phosphorylation of the oligonucleotides is particularly important when using the inosine method.) The two oligonucleotides are then annealed by mixing at 0.2 μM in 40 μl annealing buffer [10 mM Tris-HCl (pH 8.0), 10 mM MgCl$_2$], heating at 80° for 10 min, and cooling slowly to room temperature.

Enzymatic Second Strand Synthesis

When both strands of the mutagenic cassette are prepared by chemical synthesis, there is always the potential problem of noncomplementarity due to mismatch formation during the annealing step. This can be avoided

[33] F. H. Martin, M. M. Castro, F. Aboul-ela, and I. Tinoco, Jr., *Nucleic Acids Res.* **13**, 8927 (1985).

FIG. 5. Three strategies for construction of the complementary strand of a mutagenic cassette. (A) Random bases are included on both strands of the cassette during synthesis. (B) Inosines are included on the complementary strand opposite the randomized positions on the first strand. (C) The complementary strand is synthesized enzymatically to prevent mismatches in the cassette. In this case, the first strand is synthesized with a self-complementary region at its 3' end corresponding to restriction site B. Following extension with DNA polymerase, the double-stranded molecule is digested with restriction enzymes A and B to yield the final cassette. An italicized *N* indicates an enzymatically-inserted base that is complementary to the random base on the opposite strand. Throughout this figure, randomized codons are indicated as NNN for simplicity; in practice, complete randomization may be accomplished by using NN$_C^G$.

by performing enzymatic second strand synthesis, using the first, random-ized strand as a template as shown in Fig. 5C.[12,13] The first strand is synthesized with a self-complementary 3' end that contains the recognition sequence for one of the two restriction enzymes. The recognition sequence for the other restriction enzyme is included near the 5' end. Extension of the oligonucleotide is performed using the Klenow fragment of DNA polymerase I. The result of the self-primed DNA synthesis is a double-stranded oligonucleotide that contains no mismatches. Digestion with the two restriction enzymes yields a cassette with the appropriate ends for insertion into the plasmid backbone. Efficient digestion requires the pres-ence of several additional base pairs beyond the restriction sites at the ends of the cassette; for most enzymes, three additional bases at the 5' end of the oligonucleotide appear to be sufficient. The enzymatic method is more involved than either of the chemical methods for second strand synthesis, and can lead to higher levels of unwanted mutations at cassette positions that have not been mutagenized (see Bonus Mutations, below). However, the enzymatic method does not appear to skew randomizations toward particular bases, does not introduce potential problems due to mismatched bases, and is almost certainly the best method to use in "biased" randomizations.

Figure 6 shows a variation on the enzymatic approach that allows the synthesis of larger mutagenic cassettes.[15] In this case, two oligonucleotides are synthesized, each covering half the distance between the two restric-tion sites, with a complementary 9-base overlap at their 3' ends. Each oligonucleotide contains one of the two appropriate restriction sites plus several additional bases at its 5' end. The oligonucleotides are annealed and filled in, each priming second strand synthesis of the other. Digestion with the two enzymes yields the final randomized cassette with the appro-priate ends for cloning into the plasmid backbone. Any of the codons between the two sites may be randomized. This method has been used to randomize four codons combinatorially, the most distant of which were 100 bp apart in the sequence.[17]

Procedure. For enzymatic second strand synthesis, we use approxi-mately 5 μg of the template oligonucleotide. Following annealing, exten-sion is performed in 50 μl of a solution containing 10 mM Tris-HCl (pH 7.5), 10 mM MgCl$_2$, 100 μg/ml bovine serum albumin, 50 mM NaCl, and 250 μM dNTPs using 10 units of the Klenow fragment of DNA polymerase I (Boehringer-Mannheim, Mannheim, Germany) or Sequenase (United States Biochemical, Cleveland, OH). The Sequenase enzyme has higher processivity but also exhibits a higher misincorporation rate (see below). After 1 hr at 37°, 5 μl of 2.5 mM dNTPs, 0.5 μl of 0.5 M dithiothreitol,

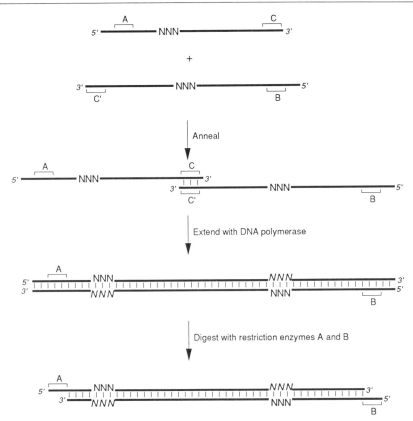

FIG. 6. A strategy for randomizing multiple codons distant in the sequence. Two oligonu-cleotides are synthesized that together encode the region between restriction sites A and B. The first oligonucleotide corresponds to the top strand, the second oligonucleotide corre-sponds to the bottom strand. The two oligonucleotides are synthesized with a complementary 9-base overlap at their 3' ends. Each molecule serves as a template for extension of the other by DNA polymerase. Digestion with restriction enzymes A and B yields the final, full-length cassette.

and 10 units of enzyme are added, and the reaction is allowed to proceed at 37° for an additional hour. The DNA is then ethanol precipitated, and approximately 10 μg of the double-stranded oligonucleotide is digested with 40 units of each restriction enzyme at 37° overnight. Extension and digestion reactions are monitored by running the DNA on 6% (w/v) dena-turing polyacrylamide gels. The final extended and digested cassettes are purified on 8% nondenaturing gels.

Preparation of Plasmid Backbone

If plasmids containing the wild-type gene (or any active gene) are used to prepare a backbone fragment, then the main consideration is preventing significant contamination of the backbone fragment by uncut or singly cut plasmid molecules. Such contamination will give rise to an unwanted background of unrandomized clones. This can be an especially serious problem in multiple codon randomizations, where only a small fraction of the randomized sequences may be active. A simple strategy that allows contamination due to uncut or singly cut molecules to be detected is to introduce a silent mutation at one of the nonrandomized positions in the cassette to a degenerate codon. When candidates are sequenced, contaminants can be readily identified because they will have the wild-type DNA sequence.

Contamination with uncut or singly cut molecules can be reduced by overdigestion with restriction enzymes or by careful gel purification of the backbone fragment, but these precautions are not always sufficient. The best solution to this problem is to purify the backbone from a plasmid that contains a large "stuffer" fragment cloned between the restriction sites to be used for insertion of the mutagenic cassette (for an example of the use of "stuffer" fragments, see Ref. 27). The stuffer fragment serves two purposes. First, it disrupts the coding sequence, which should result in inactivation of the gene. This ensures that any functional genes recovered following random mutagenesis result from the insertion of oligonucleotide cassettes rather than from backbone reclosure. Second, if the stuffer fragment is fairly large, its excision following digestion with restriction endonucleases leads to a significant change in mobility during gel electrophoresis. Consequently, the plasmid backbone can be readily purified away from uncut or singly cut plasmid DNA.

Procedure. Plasmid DNA that has been purified by CsCl gradient centrifugation is the most reliable for digestion with restriction enzymes, although DNA purified by minipreparation procedures[32] is often of sufficient purity. In typical experiments, 5 μg of plasmid DNA is digested with 5–10 units of each restriction enzyme for 1 hr at 37°. An additional 5–10 units of each enzyme is added, and the incubation is continued for 1 hr. Restriction fragments are separated by electrophoresis on low melting point agarose gels. DNA fragments are isolated from gel slices by any of several methods, including the Qiagen Cartridge (Qiagen, Inc., Studio City, CA), Elutip (Schleicher & Schuell, Heane, NH), and Gene Clean (Bio 101, Inc., La Jolla, CA) protocols. Alternatively, the slice may be melted, phenol extracted, and the DNA purified by ethanol precipitation.[32]

Double-stranded oligonucleotide cassettes are ligated to plasmid backbone overnight at 4–14° in ligation buffer [50 mM Tris-HCl (pH 7.5), 10 mM MgCl$_2$, 10 mM dithiothreitol (DTT), 1 mM spermidine, 1 mM ATP, 100 μg/ml bovine serum albumin]. Ligation reactions are performed at roughly equimolar concentrations of insert and plasmid backbone.

Transformation

Since the goal of a random mutagenesis experiment is usually to sample as much of sequence space as possible, the number of residue positions that can be examined in a single experiment rapidly becomes limited by the efficiency of each step in the procedure, and especially by the efficiency of transformation. We have found, using the transformation protocol of Hanahan,[34] that it is possible to obtain 10^5–10^6 transformants from 100 ng of DNA. This is enough to ensure that most of sequence space is sampled in experiments in which one to three residue positions are randomized at a time. However, randomizing more than three positions generally requires scaling up the transformation procedure. An alternative is to use electroporation,[32] which gives transformation efficiencies about 100-fold greater than Hanahan transformation.

Sequencing

Since random mutagenesis generates a large number of mutant genes, the ability to rapidly sequence these genes is essential. We have found it convenient to perform mutagenesis using plasmids bearing an M13 origin of replication to allow production of single-stranded plasmid DNA.[35] This DNA is then sequenced using the dideoxy method.[36] An alternative is to isolate and sequence the double-stranded plasmid DNA.[37,38]

We typically perform 48, 72, or 96 sets of dideoxy-sequencing reactions at a time. Sequencing this many candidates at once is facilitated by the use of microtiter dishes for all of the sequencing reactions. Sequencing reagents may be rapidly dispensed into the microtiter wells using a repeating pipette. The labeling and extension reaction is performed in one row of the dish, and then aliquots of each reaction are transferred using a multichannel pipette to separate wells for the A, C, G, and T termination reactions. Sequencing kits are available (e.g., from Amersham, Arlington Heights, IL) with reagents already dispensed into microtiter wells. The

[34] D. Hanahan, *J. Mol. Biol.* **166,** 557 (1983).
[35] R. J. Zagursky and M. L. Berman, *Gene* **27,** 183 (1984).
[36] F. Sanger, S. Nicklen, and A. R. Coulson, *Proc. Natl. Acad. Sci. U.S.A.* **74,** 5463 (1977).
[37] E. Y. Chen and P. H. Seeburg, *DNA* **4,** 165 (1985).
[38] D. Seto, *Nucleic Acids. Res.* **18,** 5905 (1990).

use of double-fine sharkstooth combs on 34-cm wide gels allows 24 sets of sequencing reactions to be loaded per sequencing gel.

Procedure. To prepare single-stranded DNA for sequencing, a 1.5-ml aliquot of $2\times$ YT[39] containing 1.5×10^7 M13 RV-1 helper phage[40] is inoculated with 30 μl of a fresh overnight culture. The cells are grown aerobically in a roller drum and infection is allowed to proceed for 6 hr at 37°. The cultures are then transferred to 1.5-ml Eppendorf tubes and centrifuged for 10 min. The full 10-min centrifugation is necessary to pellet cell debris that can otherwise interfere with the DNA sequencing. A 1.2-ml portion of the supernatant, containing the phage, is transferred to an Eppendorf tube containing 300 μl of 2.5 M NaCl in 20% polyethylene glycol (M_r 8000). After 30 min at room temperature, this solution is centrifuged for 10 min at 4° to pellet the phage. The supernatant is poured off, the tubes are briefly centrifuged again, and the residual supernatant is removed by aspiration. It is crucial that all the polyethylene glycol be removed, since it will inhibit the sequencing reactions. The phage pellet is suspended in 100 μl TES [20 mM Tris-HCl (pH 7.5), 10 mM NaCl, 0.1 mM EDTA], a 50-μl aliquot of phenol is added, and the tubes are vortexed for 30 sec. After 5 min at room temperature, the tubes are again vortexed for 30 sec, and then centrifuged for 10 min at 4°. An 80-μl portion of the upper, aqueous supernatant is removed and added to 4 μl of 3 M sodium acetate. The single-stranded DNA is precipitated by addition of 200 μl ethanol. The DNA is pelleted by centrifugation for 15 min at 4° and rinsed with 200 μl 70% ethanol. Following another 5-min centrifugation, the DNA is dried in a Savant (Farmingdale, NY) Speed-Vac and dissolved in 25 μl TES. Eight microliters of this DNA is used for sequencing according to the Sequenase (United States Biochemical) protocol. Using this amount of DNA, 200–300 bases can be sequenced. However, when sequencing very close to the sequencing primer (<50 bases), better results are obtained if the purified single-stranded DNA is dissolved in 12 μl of TES, and an 8-μl aliquot is used for sequencing.

To prepare double-stranded plasmid DNA for sequencing, 20 μl of a fresh overnight culture is added to 2 ml of L broth (LB),[39] and the cells are grown at 37° for 5 hr (it is important to use a freshly saturated culture). A 1.5 ml portion of culture is transferred to a 1.5-ml Eppendorf tube, and the cells are pelleted by centrifugation for 30 sec. The cell pellet is suspended in 300 μl STET [8% sucrose, 0.5% (v/v) Triton X-100, 50 mM EDTA, 50 mM Tris-HCl (pH 8)], and a 20-μl aliquot of 10 mg/ml lysozyme

---

[39] J. Miller, "Experiments in Molecular Genetics." Cold Spring Harbor Laboratory, Cold Spring Harbor, New York, 1972.
[40] A. Levinson, D. Silver, and B. Seed, *J. Mol. Appl. Genet.* **2**, 507 (1984).

is added. After 5 min at room temperature, the solution is heated in a boiling water bath for 2 min and then centrifuged for 5 min at room temperature. The pellet is removed with a toothpick, and 200 μl of 2.5 M ammonium acetate/75% 2-propanol is added. After 5 min at room temperature, the solution is centrifuged for 5 min at room temperature, and the pellet is washed with 200 μl 70% ethanol. Following another 5-min centrifugation, the pellet is dried in a Savant Speed-Vac, and then resuspended in 20 μl TE [10 mM Tris-HCl (pH 8), 1 mM EDTA].

To sequence using the double-stranded DNA as a template, the entire sample prepared as described above is used. To the DNA are added 4 μl of 16 ng/μl sequencing primer and 6 μl of 2 M NaOH. The solution is incubated at 37° for 15 min. The DNA is precipitated by addition of 10 μl of 3 M sodium acetate (pH 5.2) and 200 μl of cold ethanol, and is pelleted by centrifugation for 15 min at 4°. The pellet is washed with 200 μl of 70% ethanol. Following a 5-min centrifugation, the supernatant is removed and the pellet is dried in a Savant Speed-Vac. The dried pellet is resuspended in 12 μl of sequencing buffer [prepared by mixing 2 μl of 5× Sequenase buffer (United States Biochemical), 1 μl of 0.1 M DTT, and 9 μl of water] and the sequencing is continued according to the Sequenase protocol, beginning with the labeling step.

Methodological Considerations

Sources of Nonrandomness. In using random mutagenesis techniques to study protein structure and function, it is important that the distribution of amino acids at the mutagenized positions be as near to truly random as possible. The nature of the genetic code dictates that the distribution cannot be perfectly random, since different amino acids are encoded by different numbers of codons. However, at the nucleotide level, a random distribution should be possible. In practice, there are several potential sources of nonrandomness to consider. Some of these sources are difficult or impossible to control experimentally. For example, the stability or translation of an mRNA may be sensitive to the presence of particular codons at some positions, causing certain protein sequences to be underrepresented among the functional sequences. Deviations from a random distribution of bases may also arise in the construction of the cassette. If the coupling efficiency during oligonucleotide synthesis is not the same for each of the four bases, there will be an unequal distribution of bases in the final cassette. Bias may also be introduced in the annealing of the two oligonucleotide strands. For example, in the inosine pairing method, randomized sequences rich in C may pair better than other sequences

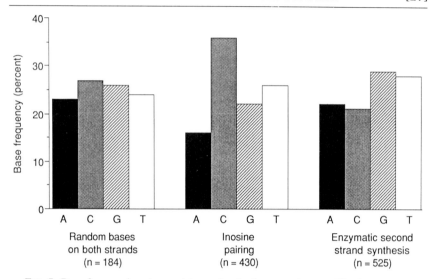

Fig. 7. Base frequencies observed in randomization experiments. The frequency with which each base was observed on the randomized strand is shown, using three different strategies for construction of the complementary strand. In each case, sequences from unselected populations were used, and only data from fully randomized positions were included (i.e., the third position in NN^G_C randomizations was omitted). The total number of base sequences determined for each strategy is indicated.

to the inosine-containing strand.[33] As a result, such sequences may be overrepresented in the mutagenized pool.

   Evaluations of Base Distributions. To estimate the nucleotide distribution within pools of mutagenized plasmids generated by different methods, we sequenced a number of randomized genes from transformants that were not subjected to a functional selection. Figure 7 shows the frequency of each nucleotide observed at randomized positions, mutated using (1) random bases on both strands of the cassette, (2) the inosine pairing method, or (3) enzymatic second strand synthesis. This figure shows data from fully randomized bases only; that is, the third base in NN^G_C randomizations is omitted. All three methods give fairly even distributions of the four bases, although the inosine-pairing method overrepresents C and underrepresents A. The bias toward C in the inosine method probably reflects a slight base-pairing preference.[33] If the 5' ends of the oligonucleotide cassette are not phosphorylated prior to ligating into the plasmid backbone (thereby leaving a nick), then this bias toward C can be extreme at positions near the end of the cassette. For example, in one such experiment, C was recovered 100% of the time at the terminal codon, 95% of the time at the penultimate codon, and approximately 70% of the time at the next two codons. This may result from exonuclease digestion of the randomized bases near the nick, followed by preferential insertion of C opposite the inosines in the gapped duplex during repair synthesis *in vivo*.

Bonus Mutations. Another problem that may be encountered during construction of the mutagenic cassette is the incorporation of the wrong base at nonrandomized positions. When randomizing positions by including NN_C^G on both strands or by using the inosine-pairing method, "bonus" mutations are observed at a frequency of about 0.03%/base throughout the cassette region. As a consequence, for cassettes of 30–60 bp, roughly 1–2% of the sequences are found to contain mutations at nonrandomized positions. These mutations may arise from misincorporation of nucleotide bases during synthesis of the oligonucleotides on the DNA synthesizer. Alternatively, they could reflect chemical modification of bases or failure to deprotect bases completely, with subsequent errors arising during replication in the cell. Another problem is the occurrence of single-base deletions in the cassette region. Such deletions appear in roughly 10% of the sequences from unselected populations.

A somewhat higher frequency of bonus mutations is observed when the second strand of the cassette is synthesized enzymatically. In one set of experiments in which the Sequenase enzyme was used to extend the oligonucleotides as shown in Fig. 5C, 30% of the candidates analyzed contained additional mutations (including deletions) at nonrandomized positions. Since the cassettes in these experiments were roughly 100 bp in length, this frequency represents a 0.3% misincorporation rate at each base. Many of these errors probably arise during the extension reaction, both as a consequence of the lack of an editing function in the Sequenase enzyme and because the extension reaction is usually performed at high nucleotide concentrations. Although the Sequenase enzyme is more error prone than the Klenow fragment of DNA polymerase I, it can be used to extend some oligonucleotides that cannot be extended with the Klenow fragment. Hence, there is a trade-off of efficiency versus fidelity when choosing between these two enzymes. Native T7 polymerase has many desirable properties that may help to overcome some of these problems,[41] although we have not had extensive experience with this enzyme.

Heterogeneity. In cassette randomization experiments, there is always heterogeneity among the plasmid molecules, prior to transformation into recipient cells. Single plasmids will encode different sequences as a consequence of the deliberate introduction of random bases during construction of the plasmids, and some or even all of the plasmids may bear mismatches as a consequence of the method of second strand synthesis of the muta-

[41] K. Bebenek and T. A. Kunkel, *Nucleic Acids Res.* **17,** 5408 (1989).

genic cassette. Heterogeneity may persist after transformation if a single cell is transformed with plasmids bearing different sequences or if DNA replication of mismatched strands yields more than one sequence. Fortunately, such examples are rare in our experience. Presumably, multiple transformation events are uncommon and repair *in vivo* corrects most mismatches before replication can occur. As a consequence, it is possible to use primary transformants for selections, screens, and sequencing in most cases. Nevertheless, the possibility of mixed populations in cells must be kept in mind. It is good practice to restreak candidates to single colonies before single-stranded template DNA is prepared for sequencing, and to retest activity phenotypes after colony purification. Heterogeneity may be indicated as a potential problem if phenotypes are not stable, if independent isolates of the same sequence appear to confer different phenotypes, or if sequencing suggests the presence of more than one base at a given position. In these cases, plasmid DNA should be purified and used to retransform cells before further experiments.

[28] Linker Insertion Mutagenesis as Probe of Structure–Function Relationships

By Stephen P. Goff and Vinayaka R. Prasad

The *in vitro* modification of cloned genes is one of the most powerful methods available for the localization of functional domains of a given gene product. The analysis of the function of mutant gene products, and correlation of the position of mutations with their effects, can quickly permit a determination of the essential regions encoding that function. A variety of mutagenesis techniques, variously generating substitutions, insertions, and deletions, can be used to reveal the structural organization of domains in proteins. Amino acid substitution, readily achieved with oligonucleotides, is perhaps the most commonly used form of mutagenesis. When the structure of a gene is only poorly understood, however, it is difficult and expensive to use this method to make a large library of mutants with changes scattered across the gene. In contrast, linker insertion mutagenesis offers a rapid means of structure–function analysis in the absence of clues suggesting specific site-directed mutations.

In linker insertion mutagenesis, short palindromic oligonucleotides containing the recognition sequence for a particular restriction enzyme are inserted at known locations throughout the length of the gene, and the effects of these insertions on protein function and stability are studied.

Insight into the localization of functional domains of several proteins has been obtained by the procedure (for examples, see Refs. 1–9). Mutagenesis by linker insertion is attractive because the insertions are genetically stable, the mutations are easily mapped and monitored, they facilitate subsequent manipulations (such as the generation of terminal or internal deletions) employing the new restriction sites inserted, and the effects of the mutations on protein function are significant but often restricted to a relatively local part of the protein. In a surprising number of cases, linker insertion mutations have been found to confer a temperature-sensitive phenotype on the target gene.[10–15] This property is one of the most powerful features of the method.

We have been interested in understanding the structure and function of reverse transcriptase (RT), a key enzyme in the replication of retroviruses. It is a multifunctional protein, displaying RNA-directed DNA polymerase, DNA-directed DNA polymerase, and ribonuclease H activities. Understanding the manner in which these enzymatic activities, residing on the same polypeptide, are coordinated during replication requires knowledge of the spatial organization of domains encoding them. To this end, we have used linker mutagenesis for mapping the two major domains of retroviral reverse transcriptases. The availability of bacterially expressed functional porteins allowed us to carry out analysis of a series of linker insertion and deletion mutants of two different viral RTs.[4,9] We describe below linker mutagenesis methods, listing the advantages and disadvantages for each method, and briefly summarize the results of the analysis of RTs as examples of applications.

[1] J. C. Stone, T. Atkinson, M. Smith, and T. Pawson, *Cell (Cambridge, Mass.)* **37,** 549 (1984).

[2] M. Ng and M. L. Privalsky, *J. Virol.* **58,** 542 (1986).

[3] S. D. Lyman and L. R. Rohrschneider, *Mol. Cell. Biol.* **7,** 3287 (1987).

[4] N. Tanese and S. P. Goff, *Proc. Natl. Acad. Sci. U.S.A.* **85,** 1777 (1988).

[5] L. A. Donehower, *J. Virol.* **62,** 3958 (1988).

[6] W. Z. Cai, S. Person, C. DebRoy, and B. H. Gu, *J. Mol. Biol.* **201,** 575 (1988).

[7] M. Chen and M. S. Horwitz, *Proc. Natl. Acad. Sci. U.S.A.* **86,** 6116 (1989).

[8] J. Y. Zhu and C. N. Cole, *J. Virol.* **63,** 4777 (1989).

[9] V. R. Prasad and S. P. Goff, *Proc. Natl. Acad. Sci. U.S.A.* **86,** 3104 (1989).

[10] J. Colicelli, L. I. Lobel, and S. P. Goff, *Mol. Gen. Genet.* **199,** 537 (1985).

[11] D. DiMaio and J. Settleman, *EMBO J.* **7,** 1197 (1988).

[12] E. T. Kipreos, G. J. Lee, and J. Y. J. Wang, *Proc. Natl. Acad. Sci. U.S.A.* **84,** 1345 (1987).

[13] G. E. Tullis, P. L. Labieniec, K. E. Clemens, and D. Pintel, *J. Virol.* **62,** 2736 (1988).

[14] J. E. DeClue and G. S. Martin, *J. Virol.* **63,** 542 (1989).

[15] N. Tanese, M. Roth, H. Epstein, and S. P. Goff, *Virology* **170,** 378 (1989).

Generation of Conventional Linker Insertion Libraries

It is straightforward to generate a single insertion, or a small number of insertions, at specific locations in a cloned gene. Briefly, a circular plasmid DNA is cleaved by an enzyme that cuts only once, linkers are joined to the ends of the linear DNA, and the DNA is recircularized. But the real power of the method is in its capability of generating larger libraries of mutants, members of which contain insertions at single sites scattered along the entire coding region. The procedure for creating insertions at multiple sites involves the generation of a pool of permuted linear plasmid DNA molecules obtained by cleavage at one of a series of positions along the length of the gene, adding linkers to the ends of the linear DNAs, recircularizing the DNA and transforming bacteria, and screening the colonies for the presence of the newly created restriction site in their plasmids (Fig. 1). We will describe each step in turn.

Generation of Permuted Linear DNA Molecules

Several different methods can be used to generate a random pool of linear molecules derived from the cleavage of supercoiled plasmid containing the gene of interest. One can either use controlled digestion of the plasmid DNA with DNase I to generate double-stranded breaks,[16,17] or partial digestion with a restriction enzyme that cuts the gene of interest at multiple sites. Restriction enzymes are preferable, because the precise site of insertion and the altered amino acid sequence can usually be deduced from limited mapping data and a known nucleotide sequence. When DNase I is used to create sites for insertion, the resulting mutations must generally be examined very closely—often by DNA sequence analysis—to precisely map the position and structure of the mutations. DNase I does not often make blunt ends, and the resulting insertions are often "dirty," i.e., associated with loss or duplication of parental nucleotides.

In choosing a restriction enzyme for linearizing the plasmid, it is best to use enzymes with the maximum number of sites in the sequence of interest, and the minimum number of sites in the vector. In general this requirement demands that the vector be as small as possible; the pUC series of plasmids are eminently suitable. Simplest to use are enzymes that produce blunt ends; our preferred enzymes include *Hae*III, *Alu*I, *Fnu*DII, *Rsa*I, and *Nla*IV. Enzymes that produce staggered cuts yield cohesive termini that must be blunted with the Klenow fragment of DNA

[16] F. Heffron, M. So, and B. J. McCarthy, *Proc. Natl. Acad. Sci. U.S.A.* **75**, 6012 (1978).
[17] K. Tatchell, K. A. Nasmyth, B. D. Hall, C. Astell, and M. Smith, *Cell (Cambridge, Mass.)* **27**, 25 (1981).

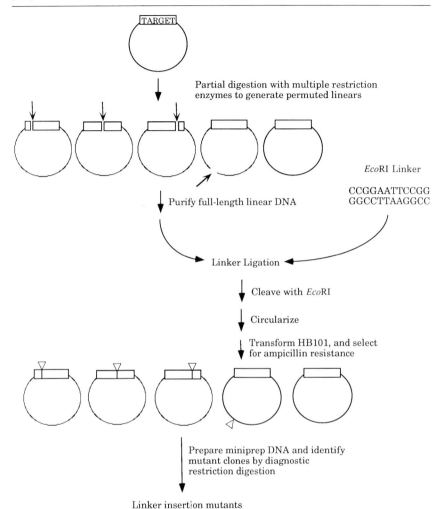

FIG. 1. Construction of a library of linker insertion mutants. A plasmid DNA carrying the target gene is exposed to limited digestion by any one of a variety of restriction enzymes to produce a collection of permuted linear molecules. The full-length linear DNA is purified, and synthetic oligonucleotides containing a restriction enzyme recognition site are added to the termini. The DNA is treated with the cognate enzyme (in this example, *Eco*RI) and recircularized by ligation at low DNA concentrations. The DNA is used to transform bacteria, and individual clones are screened for the presence and position of insertions.

polymerase I (see below) before linkers are added; suitable examples of this class are *Dde*I, *Hinf*I, *Sau*3AI, and *Sau*96I.

A partial digestion of the plasmid DNA containing the target gene is carried out in one of the following ways.

Time–Course Method. This is the most common method for producing partial digests of plasmid DNA. One sets up a digestion of about 5–20 μg of DNA with 1–2 units of the restriction enzyme of choice, and aliquots of 2–5 μg are withdrawn into an equal volume of 20 m*M* EDTA at various times. Digestion times are often very short, in the range of 1–10 min. With some enzymes, digestion must be limited by carrying out the reaction at reduced temperatures of 20° or even 0°. The products of digestion are displayed by electrophoresis on a 0.8% (w/v) agarose gel to determine the time point at which most DNA is in the full-length, linear form. Ideally, the products of partial digestion should contain very few molecules that are uncut or multiply cut and should contain the maximum amount of full-length linears. Conditions of electrophoresis that maximize the separation of full-length linears from contaminating form I and form II DNAs should be used whenever possible.

While this method is generally easy to use, it is often hard to hit on the time point that yields an appropriate level of digestion due to the fact that some enzymes act too rapidly. With some enzymes there can be serious difficulties with "hot spots," i.e., preferential cleavage at one or a few sites.

Ethidium Bromide Method. Slowing the rate of restriction enzyme reaction by including inhibitors increases the chances of obtaining an appropriate level of digestion. If the inhibitor acts specifically or selectively on those DNAs that have already been cleaved, the effect is to increase the yield of singly cut molecules. One way of achieving this is to include ethidium bromide in the reaction mixture.[18] Ethidium will bind to a limited extent in supercoiled DNA, inhibiting digestion to a similarly limited extent; once the DNA has been cleaved (or nicked), much more ethidium will intercalate and so inhibit further digestion. In our laboratory, we use the following method.

Set up digestions as follows:

Supercoiled DNA	2–5 μg
Recommended buffer (10×)	10 μl
Ethidium bromide (0.5 mg/ml)	0, 5, 10, 20, or 50 μl
Water	to 100-μl total volume
Restriction enzyme	25 units (U)

---

[18] R. C. Parker, R. M. Watson, and J. Vinograd, *Proc. Natl. Acad. Sci. U.S.A.* **74,** 851 (1977).

The digestion is allowed to proceed for an extended period (1–3 hr, depending on the restriction enzyme being used). At the end of the reaction, a small aliquot of the reaction is electrophoresed on a 0.8% agarose gel alongside an equal quantity of uncut plasmid DNA to determine the extent of digestion. If an appropriate level of digestion has been obtained in one of them, then the rest of that reaction is fractionated on a preparative 0.8% agarose gel.

In all cases, once a set of permuted linear DNAs has been generated, the full-length DNA is isolated after electrophoresis. If the termini produced by the chosen restriction enzyme are not blunt, or if DNase I is used to linearize the DNA, the ends must be blunted with the Klenow fragment of DNA polymerase I. A typical incubation includes

DNA	~1 μg
Tris-HCl, pH7.5	50 mM
MgCl₂	10 mM
DTT	0.1 mM
dXTP	100 μM each
Water	to 50-μl total volume
Klenow	2–5 U

Reactions are carried out for 15–30 min at 20°; the DNA is recovered after phenol extraction and precipitation with ethanol.

Addition of Linkers to DNA Termini and Circularization

First, a linker of the correct size and sequence must be chosen. With phosphorylated linkers, the sequence must include a recognition site for a restriction enzyme that does not cut the parental DNA. Linkers containing recognition sites for a large array of enzymes are now available, and usually a suitable linker is readily identified. There are some additional considerations in choosing a linker. The sequence should generally not include a terminator triplet (TAG, TAA, or TGA) in any of the three possible reading frames; thus, *Spe*I linkers (sequence GACTAGTC) or *Bcl*I linkers (sequence CTGATCAG) would be poor choices. In addition, the insertion of the linker into a particular restriction site should not create a terminator triplet. Thus, the insertion of some *Cla*I linkers (those of sequence GATCGATC) into *Rsa*I sites (sequence GT'AC) would be ill-advised. There is not a great deal of information concerning the effects of insertions of different sequence. Because random insertions of a given linker occur in all three translational reading frames of the target gene, three different sets of inserted amino acids are generated; different novel condons are also created at the joints between linker and site. Presumably some inserted amino acids have much more serious consequences than others, but no systematic studies of these ideas have been made.

We have had most experience with linkers 8, 10, or 12 nucleotides (nt) in length. Although 6-nt linkers can be used,[19] these shorter linkers are generally added very inefficiently; thus, if it is desirable to add a 6-mer, a 12-mer consisting of a tandem repeat of a 6-mer may be easier. For example, in the case of *Eco*RI, a linker of the sequence GAATTCGAATTC is first added; the eventual digestion with *Eco*RI and religation will reduce the size of the final insertion to only 6 bp (see below). There is some evidence that short insertions (say, 6 bp or less) are more often silent than larger insertions, but studies of this notion are limited. It is probably also true that 3- and 6-bp insertions generate temperature-sensitive phenotypes more often than larger insertions.

There are two ways to add linkers to DNA termini.

Phosphorylated Linkers. The chemistry used in the synthesis of linkers generally produces nonphosphorylated 5' termini. Many commercial firms now provide prephosphorylated linkers, and we have used these preparations with good success. If phosphorylated linkers are not available, linkers can be efficiently and easily phosphorylated with polynucleotide kinase and ATP:

Linker	100–1000 ng
10× Buffer L (500 mM Tris-HCl, pH 7.6; 100 MgCl$_2$; 100 mM dithiothreitol (DTT); 5 mM ATP)	1 μl
Water	to 10-μl total volume
T4 polynucleotide kinase	2–10 U

Incubation is for 30 min at 37°.

In adding linkers to DNA, we have found that the use of a 40- to 50-fold molar excess of linkers over DNA termini is sufficient to alow ligation of more than one linker per end. It is helpful to keep the reaction volume small to promote addition. A ligation reaction is set up as follows:

Permuted linear plasmid DNA	100 ng
Phosphorylated *Eco*RI linkers	40- to 50-fold molar excess (100 ng, depending on the DNA length)
10× Buffer L (see above)	1 μl
Water	to 10-μl total volume
T4 DNA ligase	2–10 U

[19] F. Barany, *Proc. Natl. Acad. Sci. U.S.A.* **82**, 4202 (1985).

The reaction is allowed to proceed for 12–18 hr at 20°. It is helpful to replenish the ATP once after 8 hr. Add 0.5 μl of 10 mM ATP and continue the incubation.

At the conclusion of the reaction, the ligase is inactivated by heating the ligation mix at 65° for 5 min and the DNA is digested with the restriction enzyme whose recognition sequence is present on the linker. This digestion removes the majority of the linkers appended to each terminus, and leaves on each end only one-half of the total duplex linker; each terminus now displays a cohesive end. It is necessary to use a large excess of the enzyme in order to obtain complete digestion of all the linker molecules that are ligated, in the presence of a large excess of free linkers. Alternatively, one can first electrophoretically separate the unligated linkers from the DNA containing the linkers, and then digest it.

In the final step, the digested DNA is recircularized in a simple reaction.

DNA	50–100 ng
10 × buffer L	5 μl
Water	to 50-μl total volume
T4 DNA ligase	2–10 U

The optimal DNA concentration for promoting self-ligation is low, in the range of 1–10 μg/ml. The reaction is allowed to proceed for 10–20 hr at 15°.

Nonphosphorylated Linkers. Occasionally, one needs to insert linkers containing a restriction site that is also present on the gene being mutated. This will lead to deletions when the linkers on the DNA termini are subjected to digestion. In such an instance, the use of nonphosphorylated linkers will obviate the need for digestion with the restriction enzyme during the cloning step. The procedure is essentially similar to the one described above except that the linkers are not phosphorylated. The ligation of the 3'-OH ends of the double-stranded linker to the PO_4 at the 5' ends of the linear DNA proceeds as expected, while the 5'-OH ends of the linker cannot be ligated to the 3'-OH ends of the linear DNA molecule. After one linker is added to the DNA, other linkers cannot be added to the modified termini. Thus, there is no need to remove oligomerized linkers by cleavage with the cognate restriction enzyme. In these reactions, it must be noted that there is often a significant background of circularization of the DNA without linkers.

At the end of the ligation, the DNA is heated at 65° to melt the unligated linker strand from the DNA, and then the linear DNA is separated from circles and free linkers by electrophoresis. After recovery, the DNA can generally be used directly to transform bacteria; the presence of the single-stranded linker tails on the termini is usually sufficient to induce circularization by annealing. The nicks in the DNA are repaired by the bacteria *in vivo*. Alternatively, the linear DNA can be phosphorylated with fresh ATP and polynucleotide kinase, and circularized *in vitro* using ligase reactions set up as described above.

Recovery of Clones and Mapping Insertions

After circular molecules containing inserts are generated, the DNA is used to transform bacteria. Competent cells can generally be transformed directly with a portion of the ligation mix without any purification of the DNA. The resulting colonies should each contain plasmid molecules with a single linker inserted at a particular position.

In the next step, the insertion mutations in individual clones must be localized. Insertions in the target gene being analyzed as well as in nonessential regions of the plasmid vector will be recovered. The most straightforward approach is to prepare minipreparations of DNAs and to digest each preparation with a pair of restriction enzymes: the enzyme whose recognition sequence is on the linker, and an enzyme cutting once on the parental plasmid. Such a digest will give the distance from the marker site to the linker, but it will leave symmetry ambiguities. Subsequent double digestion with an enzyme recognizing a different marker site can then be used to resolve these ambiguities.

In many cases it will be important to determine the exact site of insertion of a linker. When there are many potential target sites (i.e., when the sites for cleavage by the linearizing restriction enzyme are closely spaced) fine mapping is required. We usually examine each mutant DNA by digestion with a frequent-cutting enzyme, with or without the linker enzyme, and examine the digests by electrophoresis on polyacrylamide gels. Very occasionally it will be necessary to determine the DNA sequence in the area of the insertion to resolve ambiguities.

It is worth noting that sometimes short deletions are generated in the procedure. These result from joining of linkers to linear DNAs that have been cleaved not once but twice by the partial digestion; if two cuts are very close together the resulting linear will be nearly full length and will not be resolved from the singly cut linears. Even more rarely, Klenow DNA polymerase I used to fill in the cohesive ends can remove bases unexpectedly before linkers are added. It should also be noted that sometimes larger insertions than expected can be formed. These often arise when tandem insertions of two or more copies of the linker are formed at a given site; usually the cause is a failure by the linker restriction enzyme to remove the extra linkers added to the termini. In all these cases, frameshift and other more drastic mutations can be generated. It is very helpful to have an independent check that the mutants still encode a full-length protein before firmly assigning a particular phenotype to a given, supposedly simple, insertion mutation.

Mutagenesis of Human Immunodeficiency Virus Reverse Transcriptase

We have applied the simple linker insertion method described above to probe the structure of the reverse transcriptase enzyme (RT) encoded by the human immunodeficiency virus (HIV-1).[9] This enzyme is a hetero-dimeric protein in the virion particle, consisting of 66- and 51-kDa subunits that are closely related; the smaller subunit is a proteolytic fragment of the larger. The two share the same N-terminal sequence. For mutagenesis of HIV RT sequences, we employed an expression construct termed pHRTRX2 that encodes the p66 version of HIV RT as a fusion protein with a portion of the bacterial TrpE protein.[20]

We cleaved pHRTRX2 DNA with a variety of enzymes, controlling the digestions with time, isolated full-length linears, and when necessary blunted the termini with Klenow. We added phosphorylated *Eco*RI linkers followed by digestion with *Eco*RI. The inserts were mapped, and were found to be broadly scattered along the gene (Fig. 2). In virtually every case, the mutant protein was still synthesized and was stably retained in the cell. The analysis of the mutants for the RNA-dependent DNA polymerase activity showed that the polymerase activity was localized to the N-terminal two-thirds of the RT molecule; most mutants with insertions in the N-terminal region were inactive, while most in the C-terminal region were still active (Fig. 2). To define the polymerase domain more firmly, we then generated directional deletions from each end, employing the *Eco*RI sites in the inserted linkers. This deletion analysis helped further localize the minimal domain encoding the polymerase function to the N-terminal p51 region. Interestingly, rare insertions in HIV RT were highly disruptive of polymerase activity even though they mapped in the C terminus. For example, insertion mutant A5 in the C-terminal region affects the polymerase function even though many deletions that span this mutation do not (e.g., mutant R8). This insertion, therefore, may interfere with the proper folding of the molecule.

It should be added that this type of domain mapping would not yield simple results for a protein without discrete functional domains. For example, a linker insertion analysis of adenovirus DNA polymerase similar to that described above for RT showed that functional residues of that polymerase were distributed across the length of the molecule.[7]

[20] N. Tanese, V. R. Prasad, and S. P. Goff, *DNA* 7, 407 (1988).

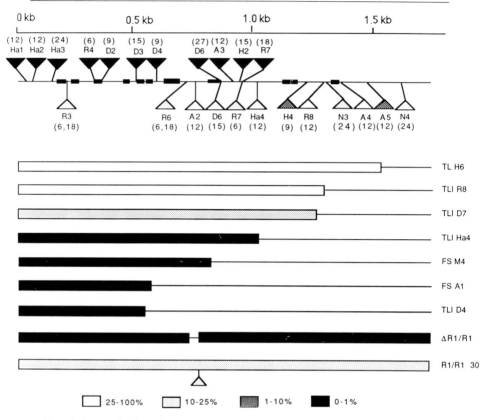

FIG. 2. An application of linker insertion mutagenesis: analysis of HIV reverse transcriptase. A library of mutants was generated, and members of the library were assayed for the levels of DNA polymerase activity in the encoded protein. Top line: Triangles indicate the position of each insertion; the numbers in parentheses indicate the size of the insertion (in bp). The enzyme activity of each mutant is indicated by the shading from black to white (see key at bottom). Lower lines: a series of C-terminally truncated enzymes were similarly assayed. (From Ref. 9.)

Cassette-Mediated Linker Insertions

In some cases the addition of linkers to linearized DNA is inefficient, as when the linearizing restrictions enzyme leaves cohesive or heterogeneous termini; large numbers of colonies need to be screened to find those DNAs containing linkers. There are methods that help eliminate the burden of screening a large number of colonies in search of the rare mutant clones.

Such methods rely on the use of cassettes that contain selectable or screenable genetic markers, symmetrically flanked by multiple restriction sites. Markers used in cassette-mediated linker insertions include genes for resistance to kanamycin[21] and streptomycin/spectinomycin[22] or a gene for suppressor tRNA, *supF*, in combination with *lacZam* host bacteria.[23] The general idea is a two-step procedure: first the cassette is inserted at random positions along the target gene; then the bulk of the cassette is excised, leaving only a few nucleotides at the site of the original insertion. The final result is similar to that achieved with conventional linker insertion.

The method differs in several ways from the conventional method described above. We will outline the procedure for the *supF* cassette ("sup-link"), discussing each difference from the simple linker method in turn.

Design and Preparation of Linker Insertion Cassette

The cassette consists of a small fragment of DNA, carrying an autonomous marker; for example, we built a cassette beginning with a 220-bp fragment containing the *supF* gene from the plasmid πVX.[23] Restriction sites are first engineered to flank the gene on both termini. The *supF* gene is flanked by *Eco*RI sites and was recovered by simple cleavage with *Eco*RI. The choice of the enzyme site proximally flanking the marker gene is important because this is the site that will ultimately be present in the insertion mutation in the target gene, and in particular the site must not appear in the target plasmid. A second site is then appended to the fragment, usually in the form of linkers. In the *supF* case, the termini were filled in by treatment with Klenow polymerase I, and the *Eco*RI sites were reconstructed by addition of *Pvu*II linkers to the blunted termini (Fig. 3). The resulting DNA thus consists of the *supF* gene, flanked first by *Eco*RI sites, and then in turn flanked by *Pvu*II sites. Cleavage with *Pvu*II generates the final cassette.

The cassette used for insertional mutagenesis is generally built only once and cloned into a suitable vector. To clone the *supF* cassette, we simply inserted the fragment into the single *Pvu*II site of pBR328. Once formed, the cassette can be prepared whenever needed simply by excision of the cassette from a stock of plasmid DNA.

[21] M. L. Smith and G. F. Crouse, *Gene* **84,** 159 (1989).
[22] P. Prentki and H. M. Krisch, *Gene* **29,** 303 (1984).
[23] L. I. Lobel and S. P. Goff, *Proc. Natl. Acad. Sci. U.S.A.* **81,** 4149 (1984).

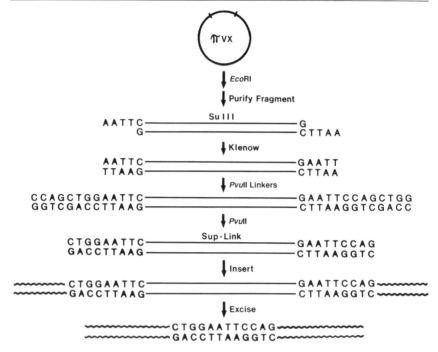

FIG. 3. Construction and use of the suppressor–linker cassette. A DNA fragment carrying the *supF* marker was isolated after *Eco*RI cleavage, the overhanging termini were filled in, and *Pvu*II linkers were added. Cleavage with *Pvu*II generates the "sup-link" DNA, a blunt-ended DNA cassette suitable for insertional mutagenesis. Successive insertion of the cassette onto a blunt-ended linear target DNA, followed by excision with *Eco*RI, generates a net 12-bp insertion (bottom line). (From Ref. 23.)

Insertion of Cassette into Target Gene

The target DNA is linearized as before and the termini are blunted if necessary. The full-length linear DNA is simply ligated to the cassette (produced by cleavage with *Pvu*II) in a small volume; we generally use a large molar excess of the cassette to drive recovery of mutant clones. Typically 0.1–0.5 μg of target DNA is joined to 100 ng of the cassette, in a total volume of perhaps 10 μl. This ligation mix is then used directly to transform bacteria to drug resistance. In our experiments, we use the host strain CC114 [Δ(*ara leu*)7697 LacZ⁻ *Y14*(Am) *galU galK* HsdR⁻ *hsdM*⁺ *strAr rifr argE*(Am) *srl*::Tn*10*, *recA1*] as best able to stably propagate many plasmids and to serve as an indicator of the *supF* gene. Recipient cells can be directly plated on Maconkey lactose plates containing ampicillin, and the presence of a *supF* insertion is directly visible as a red colony.

The overall efficiency of the insertion process may vary, but the proportion of the selected (red) colonies that contain an insertion mutant is very high (essentially 100%).

In some cases, it is preferable to pick individual red colonies and map the insertions in the resident plasmids of each clone. Desirable clones can be chosen and used in the step to follow. Alternatively, it is often preferable to pool large numbers of red colonies and prepare bulk DNA from the pooled colonies. This pool of DNA can then be used in the next step.

Excision of Cassette: Generation of Final Insertion

In a second step, the bulk of the cassette is excised. Here the design of the cassette makes this trivial. The plasmid DNA is simply cut with the enzyme whose sites proximally flank the selectable marker of the cassette; this cut removes the selectable marker and leaves only the outer bases of the original cassette appended to the termini of the linear target DNA (Fig. 3). The linear DNA is then recircularized by ligation at low concentrations, and the resulting DNA is used to transform bacteria. Clones from which the cassette has been successfully excised can readily be detected by screening or selecting for loss of the selectable marker. If a single mapped cassette insertion was subjected to the excision procedure, the position of the resulting small insertion is of course known. If a pool of cassette insertion clones was originally made, individual mutants can now be recovered and mapped as for the conventional linker insertion method.

Insertions in Moloney Murine Leukemia Virus Reverse Transcriptase

We made use of the *supF* cassette in our own experiments to generate insertions in reverse transcriptase of the Moloney murine leukemia virus.[4] The parental plasmid for the procedure was pSH1, expressing the monomeric 80-kDa enzyme as a TrpE fusion protein.[24] The enzyme exhibits both activities normally associated with RTs: DNA polymerase activity and RNase H. Our goal was to localize both activities if possible, and determine whether they could be separated by mutation.

We used a variety of restriction enzymes to generate full-length linears of the target DNA, joined the linears to the *supF* cassette, and recovered a large number of clones carrying the suppressor tRNA marker. The efficiency was high even when rather poor enzymes like *Mnl*I were used for the partial digestion. We excised the marker gene from the clones with *Eco*RI and recovered the final mutants (Fig. 4). The inserts ranged in size from 9 to 15 bp.

[24] N. Tanese, M. Roth, and S. P. Goff, *Proc. Natl. Acad. Sci. U.S.A.* **82,** 4944 (1985).

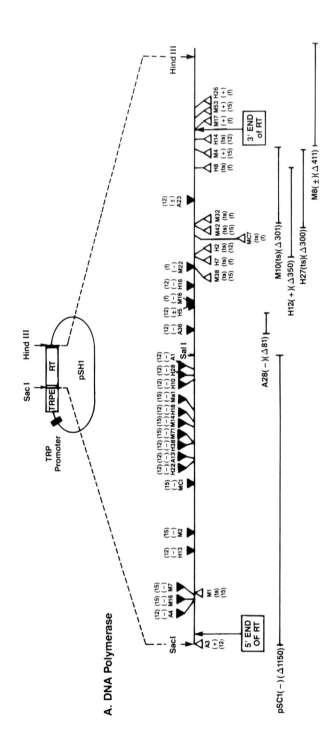

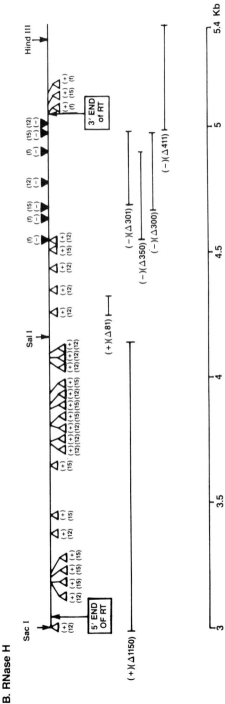

Fig. 4. An application of cassette-mediated insertional mutagenesis: analysis of Moloney murine leukemia virus reverse transcriptase. A plasmid encoding an enzymatically active fusion protein was used to generate a library of insertion mutants with the *supF* cassette, and individual mutants were assayed for both DNA polymerase and RNase H activity. The parental plasmid is drawn at the top; expanded views of the target gene are shown below. (A) DNA polymerase activity; (B) RNase H activity. The position of each linker insertion is indicated by the position of a triangle. Solid triangles denote little (±) or no (−) activity, and open triangles denote wild-type (+) or substantial levels of temperature-sensitive (ts) activity. Numbers in parentheses indicate the number of base pairs inserted at each site. Frameshift mutations are indicated by (f). A number of deletion mutants are included; the position and extent of the deleted sequences are indicated by the bars below each map. (From Ref. 4.)

Each bacterial clone was induced to express the enzyme, and lysates were tested for both the RNA-dependent DNA polymerase activity and RNase H activity characteristic of the murine viral enzyme. In the case of this enzyme the results were extremely clear: insertions in the N-terminal two-thirds of the protein abolished the polymerase activity with little or no effect on the RNase H activity; insertions in the C-terminal one-third abolished the RNase H activity with little or no effect on the polymerase. The conclusion was inescapable that the enzyme was a two-domain protein, with an N-terminal polymerase and a C-terminal RNase H.

The presence of the *Eco*RI site at each mutation made it very simple to build more variants using the insertion mutants as parents. Because the boundary between the two domains was apparent from the analysis, we generated new clones that separately expressed either the polymerase or RNase H domains alone. The assays of these constructs confirmed the existence of two domains. Fully separable, nonoverlapping sequences of amino acids were able to express either one or the other of the two activities.

Short Insertions Created without Linkers

While the insertion of short sequences via linker insertion is broadly useful, there are other, less frequently employed approaches for the formation of small insertions.

Insertion by Fill-In Reaction

This is a simple method of insertion mutagenesis, especially useful in introducing one-codon insertions into a gene, and therefore one-residue insertions into the gene product.[12,25] A target plasmid DNA is treated with a frequently cutting restriction enzyme that generates a 3-base, 5'-protruding end; digestion is limited such that full-length linears are produced. A number of such enzymes are available, namely, *Ava*II, *Dde*I, *Hinf*I and *Sau*96I. After the linear molecules are generated, the single-stranded projections are filled in by treatment with the Klenow fragment of DNA polymerase I, and the flush-ended, linear molecules are recircularized in a simple ligation reaction at low DNA concentration. The result is the net insertion of 3 bp at the original site of cleavage. In general no new restriction site is created; the mutation must be detected as the loss of a restriction site at that position.

[25] J. D. Boeke, *Mol. Gen. Genet.* **181**, 288 (1981).

Although the method of insertion per se is easy, confirming the presence of the insertion and mapping it can be time consuming. As the process of insertion by filling in destroys the existing restriction site and does not necessarily create new sites, no further manipulations at that site can be carried out without resorting to subsequent oligonucleotide-directed mutagenesis.

Insertion by Oligonucleotide-Directed Mutagenesis

All the methods described above are dependent on the presence of a restriction site at the position of the insertion. For initial random mutagenesis, no single site is a particularly important target for mutagenesis. But as information about a protein is acquired, it may be desirable to place an insertion at a very critical site, and because the probability of finding a restriction site at any given position is not high, no restriction enzyme site may be available in the vicinity. In such cases one must employ oligonucleotide-directed linker insertion mutagenesis. An oligonucleotide is synthesized with a short insertion relative to the wild-type sequence, and is used to prime synthesis on single-stranded clones in M13. The procedure allows insertion of a desired sequences (encoding any desired amino acids) at any site.[26]

Conclusions

Linker insertion is a broadly useful procedure for the generation of libraries of mutations scattered along the coding region of a target gene. The mutations often provide the most rapid definition of the regions of a protein required for a given enzymatic activity.

Acknowledgments

The authors were supported by Grant R01 CA 30488. V.R.P. is a scholar of the American Foundation for AIDS Research.

---

[26] T. A. Kunkel, *Proc. Natl. Acad. Sci. U.S.A.* **82,** 488 (1985).

[29] A Streptomycin Selection for DNA-Binding Activity

By Michael C. Mossing, James U. Bowie, and Robert T. Sauer

Introduction

Any DNA-binding protein that can bind to a specific target site and block access of RNA polymerase or accessory proteins to nearby promoter sites can regulate gene expression. With the appropriate configuration of promoters, binding sites, and reporter genes, it is possible to monitor the activity of DNA-binding proteins in *Escherichia coli* and also to require such activity for survival of the bacterium. The chief advantage of performing such screens or selections in *E. coli* is that many candidate clones can be processed in parallel and thus rare active genes can be identified. This is true whether the candidate genes are members of a cDNA library and the goal is to identify the source of a particular binding activity, or whether the candidates are heavily mutagenized variants of a well-characterized protein and the goal is to probe the range of amino acid substitutions that are compatible with function.

Screens have been described for the activities of particular DNA-binding proteins based on transcriptional fusions of heterologous promoter–operator elements to easily assayed reporter genes such as *lac*Z,[1] *gal*K,[2] *trp*E,[3] and chloramphenicol acetyltransferase.[4] Biological selections for the activities of specific repressor proteins based on nutritional pathways[5] or phage resistance[6,7] have also been reported. Several more general selection schemes have also been developed.[8–10] Each of these

[1] K. Bertrand, K. Postle, L. V. Wray, and W. S. Reznikoff, *J. Bacteriol.* **158,** 910 (1978).

[2] K. McKenney, H. Shimatake, D. Court, U. Schmeissner, C. Brady, and M. Rosenberg, "Gene Amplification and Analysis" (J. G. Chirikjian and T. S. Papas, eds.), pp. 383–415. Elsevier/North-Holland, New York, 1981.

[3] K.-O. Cho and C. Yanofsky, *J. Mol. Biol.* **204,** 41 (1988).

[4] A. K. Vershon, J. U. Bowie, T. M. Karplus, and R. T. Sauer, *Proteins: Struct. Funct. Genet.* **1,** 302 (1986).

[5] J. H. Miller, "Experiments in Molecular Genetics." Cold Spring Harbor Laboratory, Cold Spring Harbor, New York, 1972.

[6] M. H. Hecht and R. T. Sauer, *J. Mol. Biol.* **186,** 53 (1985).

[7] P. Youderian, A. Vershon, S. Bouvier, R. T. Sauer, and M. M. Susskind, *Cell (Cambridge, Mass.)* **35,** 777 (1983).

[8] N. P. Benson, P. Sugiono, D. N. Arvidson, R. P. Gunsalus, and P. Youderian, *Genetics* **114,** 1 (1987).

methods involves the construction of hybrid promoter–operator regions that directly or indirectly control a gene whose expression can be conditionally required for cell survival. In this chapter, we describe the use of a positive selection involving streptomycin resistance to select for the DNA-binding activities of Lac repressor, λ Cro, and P22 Arc repressor. We also discuss several applications of this selection, including the isolation of neutral mutations, the isolation of second site revertants, and attempts to select proteins with altered DNA-binding specificity.

In *E. coli*, the wild-type ribosomal protein S12 is the target of the antibiotic streptomycin. Streptomycin-resistant alleles of the S12 gene are recessive to the wild-type, sensitive allele.[11] Thus, a diploid expressing both the resistant and sensitive genes is killed by the antibiotic. In a strain in which the streptomycin-resistant gene is expressed constitutively from its natural promoter, and the dominant, streptomycin-sensitive gene is expressed from a promoter that is repressible by a DNA-binding activity of interest, resistance to streptomycin is dependent on repressor activity.[10] This is illustrated schematically in Fig. 1. The stringency of a selection of this type can be altered in a number of ways. First, the concentration of streptomycin to which resistance is required can be increased or decreased. Second, expression of the DNA-binding protein can be regulated so that the level of activity subject to selection can be varied. Finally, the expression level of the S12 gene can be adjusted by varying gene dosage, promoter strength, operator sequences, or translational signals.

Methods

In the following sections, we describe the construction of vectors that bring the S12 gene under control of the P22 Arc, Lac, or λ Cro repressors. In some cases, these vectors are described in some detail, since they may prove useful as starting points for construction of vectors that could be used in positive selections for other DNA-binding proteins. We also describe control experiments that address potential problems in using the streptomycin selection.

To bring the wild-type S12 gene under Arc control,[10] the S12 gene was separated from its promoter and translational signals by restriction cleavage with *Hpa*I, which cuts after the fourth codon of the S12 gene (see

[9] S. J. Elledge, P. Sugiono, L. Guarente, and R. W. Davis, *Proc. Natl. Acad. Sci. U.S.A.* **86,** 3689 (1989).
[10] J. U. Bowie and R. T. Sauer, *J. Biol. Chem.* **264,** 7596 (1989).
[11] J. Lederberg, *J. Bacteriol.* **61,** 549 (1951).

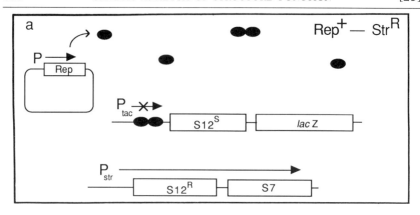

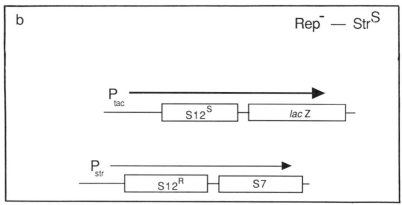

FIG. 1. Principle of the streptomycin selection for DNA-binding activity. Both cells contain a constitutively expressed copy of a steptomycin-resistance gene (S12^R) and a copy of a dominant, streptomycin-sensitive gene (S12^s) under control of a *tac* promoter. The cell shown in (a) also contains a plasmid that expresses a DNA-binding protein (Rep), which represses transcription of the S12^s gene initiated at P$_{tac}$. As a consequence, cell a is resistant to streptomycin whereas cell b (which contains no DNA-binding protein) is sensitive to streptomycin. Note that a cell that expressed an inactive form of Rep would also be strepto-mycin sensitive.

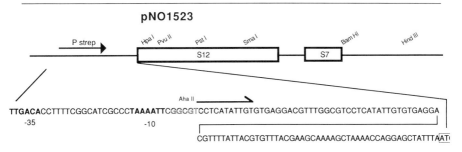

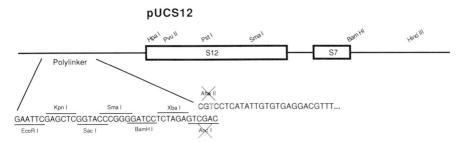

FIG. 2. S12 subcloning. Plasmid pNO1523[12] contains the wild-type S12 gene of *E. coli*. The nucleotide sequence of the S12 promoter and leader mRNA region are shown. The transcription start point is indicated by an arrow and the translation initiation site is boxed. The *Aha*II restriction site overlaps the transcription start site. Plasmid pUCS12 contains the *Aha*II–*Hin*dIII fragment of pNO1523, cloned into the *Acc*I–*Hin*dIII backbone of plasmid pUC118 (J. Vieira and J. Messing, this series, Vol. 153, p. 3). The *Acc*I–*Aha*II junction destroys the sites for both restriction enzymes.

Fig. 2).[12] The 5' end of the gene was then replaced with a synthetic cassette that recreated the S12-coding sequence and ribosome binding site and introduced a unique *Bst*EII restriction site upstream of the gene. This *Bst*EII site was used to introduce synthetic sequences containing the wild-type *arc* operator and the Arc-repressible promoter P_{ant} (bearing a mild down mutation T→A at position −7 of the promoter). The sequence of the promoter–operator region is shown in Fig. 3A. The P_{ant}–S12 fusion was then cloned into a pBR322-derived plasmid containing the wild-type *arc* gene (under *tac* promoter control) to generate plasmid pUS405. Otherwise identical plasmids containing defective alleles of the *arc* gene were also constructed. *Escherichia coli* strain UA2F is normally streptomycin resistant because it contains the *str*A allele of the S12 gene (see Table I

[12] D. Dean, *Gene* **15**, 99 (1981).

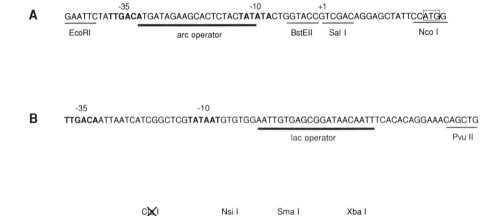

FIG. 3. Sequences of promoter-operator regions. (A) Sequence of the *ant* promoter and *arc* operator. The operator is positioned between the − 10 and − 35 regions of the promoter. The *Nco*I site overlaps the start point of translation of the wild-type S12 gene. (B) A portion of the *Eco*RI–*Pvu*II fragment, containing the *tac* promoter and *lac* operator from plasmid ptac12.[13] The *Eco*RI site is approximately 200 bases to the left of the − 35 region. The translational start of the S12 gene is approximately 115 bases to the right of the *Pvu*II site. (C) A portion of the *Eco*RI–*Xba*I fragment containing a *tac* promoter and the λ1 operator. The *Eco*RI site is approximately 200 bases to the left of the − 35 region. The translational start of the S12 gcne is approximately 100 bases to the right of the *Xba*I site.

for complete genotype). When this strain was transformed with pUS405, transformants were found to be streptomycin resistant if the plasmid contained an active *arc* gene and streptomycin sensitive if the plasmid contained a defective *arc* gene.

In selections for Arc activity, strain UA2F was transformed with pUS405 or related plasmids bearing defective or highly mutagenized *arc* genes. Following the heat shock step of the transformation protocol, cells were allowed to grow at 37° in 5 ml of LB broth plus 20 mM MgSO$_4$ for 1 hr. The cells were then collected by centrifugation and resuspended in 0.5 ml of LB broth containing 20 mM MgSO$_4$ and 100 μg/ml ampicillin. Selections for Arc$^+$ transformants were performed by plating 100-μl portions of cells on LB plates supplemented with 100 μg/ml ampicillin, 50 μg/ml streptomycin, and 2 mg/ml IPTG (isopropyl-β-D-thiogalactopyrano-side). In strain UA2F/pUS405, Arc synthesis from the *tac* promoter is repressed by Lac repressor. The inducer IPTG is included in the selection plates because it inactivates Lac repressor, thereby resulting in higher

TABLE I
BACTERIAL STRAINS, PHAGES, AND PLASMIDS

	Comments	Source/Ref.
Strains		
GW5100	Δ(*lac pro*) *thi str*A *Sup*E *end*A *sbc*B with F' *tra*D36 *Pro*AB *lac*I^Q *lac*Z ΔM15	G. Walker (MIT)
GW5180	*rec*A::cml derivative of GW5100	G. Walker (MIT)
GW5100F⁻	F⁻ derivative of GW5100	G. Walker (MIT)
GW5180F⁻	F⁻ derivative of GW5180	G. Walker (MIT)
UA2F	*str*A *thi*⁻ *his*⁻ *lac*Z⁻ *lac*Y⁺ *sup*° *rec*A⁻ (λAC201) F' *lac*I^Q *lac*Z::Tn5(kanR) *pro*⁺	(4)
Phages		
λ112	*imm*²¹ λP$_R$–*trp*A–*lac*Z fusion	(a)
λtac-S12	*imm*²¹ P$_{tac}$–*lac* operator–S12 fusion	This study
λtac/λ1-S12[b]	*imm*²¹ P$_{tac}$–λ1 operator–S12 fusion	This study
λAC201	*imm*²¹ P$_{ant}$–*cat*–*lac*Z fusion	(4)
Plasmids		
pNO1523	Contains wild-type S12 gene (see Fig. 2)	Pharmacia[12]
pAP119	pBR322 derivative, P$_{tac}$–λ Cro fusion	(c)
pI^Q	pACYC184 derivative, contains *lac*I^Q gene on ~1-kb fragment inserted at *Eco*RI site	J. Wang (Harvard)
pUCroRS	pUC18 derivative, contains P$_{tac}$–λ Cro fusion from pAP119 inserted in polylinker	This study
pUCS12	pUC118 derivative, contains S12 sequences from pNO1523 (see Fig. 2)	This study
pUCtacS12	pUCS12 derivative, P$_{tac}$–*lac* operator–S12 fusion (see Fig. 3)	This study
pUCtac/λ1	pUC118 derivative, contains P$_{tac}$–λ operator construct (see Fig. 3)	This study
pUS405	pBR322 derivative, P$_{ant}$–*arc* operator–S12 fusion, P$_{tac}$–arc fusion	(10)

[a] R. Maurer, B. J. Meyer, and M. Ptashne, J. Mol. Biol. **139**, 147 (1980).

[b] Phage with the λ2, λ3, and λ4 operators are designated λtac/λ2-S12, etc.

[c] A. A. Pakula and R. T. Sauer, *Proteins: Struct. Funct. Genet.* **5**, 202 (1989).

levels of Arc expression. The streptomycin selection for Arc activity was generally found to be reliable as long as it was performed on freshly transformed cells. However, to ensure that streptomycin-resistant candidates were indeed Arc⁺, an independent screen for Arc activity was performed. This screen relies on the fact that strain UA2F contains a gene for chloramphenicol acetyltransferase under control of an Arc-repressible promoter.[4] As a result, a cell that is Arc⁺, such as UA2F/pUS405, is both streptomycin resistant and chloramphenicol sensitive.

In constructing the P$_{ant}$–S12 fusion used for the Arc selection, 5′ sequences apparently required for stability or high-level translation of the

S12 mRNA were removed. While expression of the S12 gene from the multicopy plasmid pUS405 was sufficient to confer a streptomycin-sensitive phenotype, we worried that this reduced expression might preclude the construction of single-copy selection vectors. To facilitate the construction of selection strains in which a single copy S12 gene would be expressed at high levels, the S12 gene in plasmid pNO1523[12] was separated from its natural promoter near the start of transcription by restriction cleavage with AhaII (see Fig. 2). The resulting S12 fragment, which includes all of the natural leader sequences, was cloned into the AccI site of the polylinker in plasmid pUC118 to generate pUCS12, as shown in Fig. 2. The S12 gene was then brought under lac operator control by cloning the EcoRI–PvuII restriction fragment bearing the tac promoter and lac operator from plasmid ptac12[13] between the EcoRI and SmaI sites in the polylinker of pUCS12. We refer to the resulting fusion as the tac–S12 fusion and the plasmid as pUCtac–S12. Plasmid pUCtac–S12 was viable in a lacI^Q strain, which overproduces Lac repressor but could not be propagated, even in the absence of streptomycin, in the presence of IPTG or in strains producing low levels of Lac repressor. Presumably, the combination of the strong tac promoter, the natural leader sequences of the S12 gene, and expression from a multicopy plasmid results in a lethal level of S12 expression.

To allow the tac–S12 fusion to be maintained in a single copy per genome, it was cloned into a λimm[21] phage. A phage bearing the S12 fusion was constructed from the left and right arms of λ112,[14] as shown in Fig. 4. This construction also places the lacZ gene under control of the tac promoter, which allows the presence of a promoter insert to be detected by the formation of blue plaques on X-gal (5-bromo-4-chloro-3-indolyl-β-D-galactopyranoside) plates. Following ligation of the gel-purified fragments, the resulting phage were packaged and infected into GW5100, a streptomycin-resistant, lacZ^- strain (see Table I for genotype). Phage forming turbid blue plaques on X-gal plates were purified by two rounds of replating. Lysogens—designated GW5100(λtac–S12)–were then isolated by streaking cells from the centers of the turbid plaques onto LB plates containing X-gal, and restreaking candidates until all progeny displayed a homogeneous blue color. To ensure that single lysogens were selected, β-galactosidase assays[5] were performed on lysogens obtained from four independent plaques and the strain with the lowest level was selected. At most, one or two of the lysogenic strains showed levels of activity that were multiples of the most common single lysogen level.

[13] E. Amann, J. Brosius, and M. Ptashne, Gene 25, 167 (1983).
[14] R. Maurer, B. J. Meyer, and M. Ptashne, J. Mol. Biol. 139, 147 (1980).

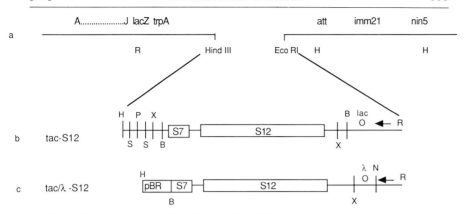

FIG. 4. Prophage maps. (a) The left arm of λ112[14] contains λ genes A to J, the *lacZ* gene, and the *trpA* gene. Transcription of *lacZ* and *trpA* is dependent on a leftward-directed promoter cloned between the *Hind*III and *Eco*RI sites. The right arm of λ112 contains the phage attachment site, the immunity region from phage 21, and the nin5 deletion. (b) A phage containing the tac–S12 fusion was constructed in a three-part ligation containing the left and right arms of λ112 and the *Eco*RI–*Hind*III fragment from plasmid pUCtac–S12 (see Methods). (c) Phages containing the tac/λ-S12 fusions were constructed in a four-part ligation of the *Eco*RI arm of λ112, an *Eco*RI–*Xba*I fragment containing the *tac* promoter and λ operator, the *Xba*I–*Hind*III fragment of pUCS12, and the left arm of λ112 Restriction enzyme sites are indicated by single letters: R, *Eco*RI; H, *Hind*III; B, *Bam*HI; X, *Xba*I; N, *Nsi*I; S, *Sac*I; P, *Pst*I; S, *Sma*I.

Similar lysogenic strains were constructed to select for λ Cro DNA-binding activity. In this case, the *tac* promoter was again used to drive wild-type S12 gene expression and a λ operator site was positioned several bases downstream of the − 10 region of the *tac* promoter (see Fig. 3C). This hybrid tac/λ promoter–operator was constructed from an *Eco*RI–*Cla*I fragment (containing the − 35 region of the promoter) from plasmid pEA300[13] and a synthetic oligonucleotide that provided the − 10 promoter sequence and an operator site for the λ Cro protein. The 17-bp operator sequence synthesized was

5'-**TATCACC**C(G/C)G**GGTGATA**
3'-**ATAGTGG**G(C/G)C**CCACTAT**

The outer 7 bp in each operator half-site correspond to a consensus λ operator site.[15] The central base was synthesized using an equal mixture of C and G. This yielded two different operators with central *Sma*I restriction sites (CCCGGG) shifted by 1 bp. We refer to the operator with a

---

[15] M. Ptashne, "A Genetic Switch: Gene Control and Phage λ." Cell Press and Blackwell Scientific Publ., Cambridge, Massachusetts and Palo Alto, California, 1986.

central G on the top strand as λ1 and the operator with a central C on the top strand as λ2. Additional operators in which a single base pair had been inserted (λ3) or deleted (λ4) at the center of the site were constructed by cleaving the λ1 and λ2 operators and swapping their right halves (SmaI–XbaI fragments). The fusion of the tac/λ1 promoter–operator region to the S12 gene was made by ligation of the EcoRI–XbaI fragment containing the promoter–operator to the XbaI–HindIII fragment from pUCS12 containing the S12 gene. As described in the legend to Fig. 4, this occurred as part of a four-part ligation used to construct a λimm²¹ phage bearing the tac/λ1–S12 fusion. Phage DNA was packaged in vitro and transfected into GW5100. Single lysogens—designated GW5100(λtac/λ1-S12)—were isolated as described above. Strains GW5100(λtac/λ2-S12), GW5100(λtac/λ3-S12), and GW5100(λtac/λ4-S12) were constructed in a similar manner.

Lysogens containing the streptomycin-sensitive S12 gene under Lac repressor or λ Cro repressor control were repressible and thus streptomycin resistant in the presence of a wild-type repressor gene. Strain GW5100(λtac-S12) expresses elevated levels of Lac repressor (from the lacIᑫ gene on the F episome) and was found to be as resistant to streptomycin (>600 μg/ml) as GW5100, the nonlysogenic parent strain. However, when the Lac repressor was inactivated by addition of the inducer IPTG, the GW5100(λtac-S12) cells were killed in the presence of 50 μg/ml of streptomycin. Similarly, strain GW5100(λtac/λ1-S12) was transformed with the Cro-overproducing plasmid pAP119[16] to determine if λ Cro could repress expression of S12. Even at low levels of expression, Cro was able to express S12 synthesis and render the cells resistant to 50 μg/ml streptomycin.

To correlate streptomycin resistance with the level of S12-transcription, Lac repressor activity was varied by growth of strain GW5100(λtac-S12) F⁻ containing the plasmid pIᑫ in media containing different amounts of IPTG[16a] and cells were assayed for growth on LB-amp plates in the presence of 50, 200, and 600 μg/ml streptomycin. In parallel experiments, lysates of the same cells grown in LB broth were assayed for β-galactosidase activity to monitor expression from the tac promoter. As shown in Table II, cells in which β-gal expression and thus tac promoter activity is low are resistant to streptomycin and vice versa.

[16] A. A. Pakula and R. T. Sauer, Proteins: Struct. Funct. Genet. 5, 202 (1989).
[16a] It is important when attempting to vary Lac repressor activity smoothly over a wide range with IPTG that the host strain be lacY⁻, as the Y gene product is both induced by and is a transporter of IPTG. This makes the internal concentration of IPTG in a lacY⁺ cell increase very rapidly once the amount of IPTG in the media exceeds a critical threshold where passive diffusion into the cell leads to induction.

TABLE II

CORRELATION BETWEEN β-GALACTOSIDASE (β-gal) ACTIVITY AND
STREPTOMYCIN RESISTANCE[a]

IPTG (μM)	β-gal units	Growth on streptomycin		
		50 μg/ml	200 μg/ml	600 μg/ml
0	300	+	+	+
1	200	+	+	+
3	300	+	+	+
10	350	+	+	−
30	800	+	−	−
60	1100	+	−	−
100	1600	−	−	−
300	1800	−	−	−
Control	2000	−	−	−

[a] Lac repressor activity was varied by titration with IPTG in strain GW5100(λtac-S12) F⁻ containing plasmid pI^Q. β-Galactosidase activity of liquid cultures grown in LB broth was assayed by the method of Miller.[5] Growth on LB plates containing 200 μg/ml of streptomycin and 100 μg/ml ampicillin and 15 μg/ml tetracycline is indicated by a +. The control strain is GW5100(λtac-S12) F⁻, which contains no Lac repressor.

In a selection experiment, there are several ways other than acquisition of repressor activity that a cell could become streptomycin resistant. These include alteration of the streptomycin-sensitive gene by prophage loss, mutation, or recombination. To assess the probability of these events, we measured the frequency at which colonies lacking a functional Lac repressor acquired a streptomycin-resistant phenotype spontaneously. In either the RecA⁺ strain GW5100(λtac-S12) F⁻ or the RecA⁻ strain GW5180(λtac-S12) F⁻, streptomycin-resistant colonies arose at frequencies of approximately 10^{-5}. Roughly 80% of these resistant colonies were immune to λimm[21] and were lacZ⁺, indicating that the cells are still lysogenic and the tac promoter is still functional. As a result, resistance appears to arise mainly via mutation of the streptomycin-sensitive S12 gene. We presume that most of these S12 mutations simply result in inactivation of the gene. It is straightforward, however, to distinguish between bona fide candidates that have acquired DNA-binding activity and those with spontaneous mutations to streptomycin resistance because plasmids isolated from the latter cells would not yield streptomycin-resistant colonies at high frequency following retransformation of a naive strain.

Another potential problem in using the streptomycin selection is that of phenotypic lag. Imagine that the promoter–operator–S12 fusion is present in a cell devoid of endogenous repressor. This cell is then transformed with a population of plasmids, only some of which encode functional protein, and transformants are selected for survival in the presence of streptomycin. A freshly transformed cell will contain a large population of streptomycin-sensitive ribosomes, and the cell might remain sensitive to killing for some time even if the synthesis of the sensitive S12 gene were repressed immediately. To test this possibility, we varied the outgrowth time between transformation of strain GW5100(λtac/λ1-S12) with the Cro-producing plasmid pAP119 and plating of the transformants on plates containing ampicillin (100 μg/ml) and streptomycin (200 μg/ml). The ratio of ampicillin-resistant to streptomycin-resistant transformants was found to remain constant from 30 min to 4 hr after transformation, indicating that once the cells stop synthesis of the streptomycin-sensitive S12 protein, they rapidly acquire resistance to the antibiotic.

Applications

Isolation of Neutral Mutations. Selections for activity can be combined with random mutagenesis to investigate the tolerance of different residue position to amino acid substitutions. If a given residue position tolerates chemically diverse side-chain substitutions without affecting protein function, then this position cannot be a critical structural or functional determinant. On the other hand, if a residue position is intolerant of substitutions or accepts only conservative substitutions, then it is likely that the side chain plays an important role in the structure, function, or intracellular stability of the protein.

The Arc repressor of phage P22 is a 53-residue DNA-binding protein (for review, see Ref. 17). To examine which Arc residues were tolerant of substitutions, the *arc* gene in plasmid pUS405 was subjected to "biased" cassette mutagenesis procedure in blocks corresponding to approximately 10 residues (Ref. 18; for description of technique, see Reidhaar-Olson *et al.* [27], this volume). Following transformation of strain UA2F, active *arc* genes were isolated using the streptomycin selection and sequenced.[18] Sixty-one different, functionally neutral amino acid substitutions were identified at 24 residue positions. At many of these positions, nonconservative residue substitutions were allowed. Moreover, at many of the positions where functionally neutral substitutions were not recovered, it was

[17] K. L. Knight, J. U. Bowie, A. K. Vershon, R. D. Kelley, and R. T. Sauer, *J. Biol. Chem.* **264,** 3639 (1989).
[18] J. U. Bowie and R. T. Sauer, *Proc. Natl. Acad. Sci. U.S.A.* **86,** 2152 (1989).

shown that residue substitutions did not disrupt the folded structure of Arc. By comparing the sets of functionally neutral and structurally neutral substitutions, it was determined that 15 Arc residues had the properties expected for functionally important positions.[18]

A second example of the use of the streptomycin selection in probing structure–function relationships is illustrated by studies of λ Cro residues 52–58, which form the dimer interface in the crystal structure of the protein.[19] This region of the *cro* gene was mutagenized using a "biased" cassette procedure and streptomycin-resistant colonies were selected.[20] When the selection was performed under conditions of low Cro expression (transcription of the *cro* gene repressed by Lac repressor), non-conservative residue changes were tolerated at position 53, but the remaining positions tolerated no changes or only conservative changes, such as Lys-56→Arg and Phe-58→Tyr (see Fig. 5, left side). When the stringency of the selection was reduced by higher-level Cro expression, a greater degree of residue tolerance was evident (see Fig. 5, right side). Now, only position 58 is seen to be strictly conserved, whereas the remaining positions accept a variety of amino acid substitutions. These results show how hierarchies of residue importance can be established. Phe-58 appears to be the most important side chain in this region of λ Cro; Pro-57, Lys-56, Val-55, Glu-54, and Ala-52 seem to play lesser roles in activity; and Glu-53 appears to be relatively unimportant for Cro dimerization.

Isolation of Revertants. Intragenic, second site revertants can provide information about interactions that correct mutant defects or improve protein function is some other way. In DNA-binding proteins, for example, reverting mutations have included sequence changes that enhance DNA binding (Ref. 21; Oxender and Gibson [31], this volume) and improve thermal stability.[16]

In P22 Arc, a large number of missense mutations result in loss of Arc function.[4] Arc gene fragments containing one of three single mutations (Met-4→Ile, Val-18→Gly, and Asp-20→Asn) were subjected to hydroxylamine mutagenesis or random primer mutagenesis *in vitro*, the fragments were recloned into plasmid pUS405, and the streptomycin selection for Arc activity was imposed following transformation of strain UA2F.[10] The *arc* genes from colonies that were both streptomycin resistant and chloramphenicol sensitive (see Methods) were sequenced. Second site revertants were isolated for each of the primary mutations. One revertant

[19] W. F. Anderson, D. H. Ohlendorf, Y. Takeda, and B. W. Matthews, *Nature (London)* **290**, 754 (1981).

[20] M. C. Mossing and R. T. Sauer, *Science* **250**, 1712 (1990).

[21] H. C. M. Nelson and R. T. Sauer *Cell (Cambridge, Mass.)* **42**, 549 (1985).

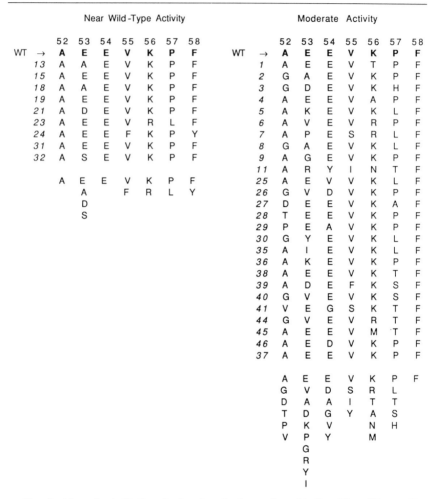

Near Wild-Type Activity

		52	53	54	55	56	57	58
WT	→	A	E	E	V	K	P	F
13		A	A	E	V	K	P	F
15		A	E	E	V	K	P	F
18		A	A	E	V	K	P	F
19		A	E	E	V	K	P	F
21		A	D	E	V	K	P	F
23		A	E	E	V	R	L	F
24		A	E	E	F	K	P	Y
31		A	E	E	V	K	P	F
32		A	S	E	V	K	P	F
		A	E	E	V	K	P	F
			A			F	R	L	Y
			D						
			S						

Moderate Activity

		52	53	54	55	56	57	58
WT	→	A	E	E	V	K	P	F
1		A	E	E	V	T	P	F
2		G	A	E	V	K	P	F
3		G	D	E	V	K	H	F
4		A	E	E	V	A	P	F
5		A	K	E	V	K	L	F
6		A	V	E	V	R	P	F
7		A	P	E	S	R	L	F
8		G	A	E	V	K	L	F
9		A	G	E	V	K	P	F
11		A	R	Y	I	N	T	F
25		A	E	V	V	K	L	F
26		G	V	D	V	K	P	F
27		D	E	E	V	K	A	F
28		T	E	E	V	K	P	F
29		P	E	A	V	K	P	F
30		G	Y	E	V	K	L	F
35		A	I	E	V	K	L	F
36		A	K	E	V	K	P	F
38		A	E	E	V	K	T	F
39		A	D	E	F	K	S	F
40		G	V	E	V	K	S	F
41		V	E	G	S	K	T	F
44		G	V	E	V	R	T	F
45		A	E	E	V	M	T	F
46		A	E	D	V	K	P	F
37		A	E	E	V	K	P	F

52	53	54	55	56	57	58
A	E	E	V	K	P	F
G	V	D	S	R	L	
D	A	A	I	T	T	
T	D	G	Y	A	S	
P	K	V		N	H	
V	P	Y		M		
	G					
	R					
	Y					
	I					

FIG. 5. Allowed substitutions in the dimerization region of λ Cro. The wild-type Cro residue is indicated in the one-letter code at the top of each column. Below are listed the sequences from active *cro* genes, isolated following cassette mutagenesis. Sequences listed on the left side were isolated in strain GW5100(λtac/λ1-S12), in which Lac repressor limits Cro expression. Sequences listed on the right side were isolated in strain GW5100(λtac/λ1-S12) F⁻, where Cro expression is unrepressed. With the pUCroRS vector used for these studies (see below), unrepressed expression of wild-type Cro is lethal; as a result genes with wild-type or near wild-type activity are probably not represented among the moderate activity class. To mutagenize residues 52–58, the corresponding codons of one DNA strand of a cassette were synthesized using a mixture of 62.5% of the wild-type base and 12.5% of each of the other three bases. The double-stranded DNA cassettes for both the randomization and monomer construction were generated by the method of A. K. Oliphant, A. L. Nussbaum, and K. Struhl [*Gene* 44, 177 (1986)] and ligated into pUCroRS. This plasmid consists of the *Eco*RI to *Sma*I fragment of pAP119[16] cloned between the *Eco*RI and *Hinc*II sites of pUC18 [C. Yanisch-Perron, J. Vieira, and J. Messing, *Gene* 33, 103 (1985)].

contained a secondary Gly-52→Asp mutation in addition to the primary mutant. Surprisingly, however, the remaining revertants contained either −1 or +1 frameshift mutations near the 3' end of the *arc* gene. These mutations resulted in bypass of the normal termination codon and resulted in C-terminal additions of 26 residues or 8 residues to the wild-type Arc sequence. These C-terminal extension or "tail" sequences act by slowing the intracellular degradation of the mutant proteins.[10] This occurs because the tails alter the susceptibility of the proteins to an *E. coli* protease that recognizes C-terminal sequences.[22] This case demonstrates how the isolation of revertants can reveal unexpected ways of compensating for mutant defects.

The 26-residue long tail sequence described above did not interfere with the DNA-binding activity or gene regulatory properties of the wild-type Arc protein.[10] We wished to make similar C-terminal fusions to the λ Cro protein, as a means of generating an active Cro molecule that could be easily distinguishable from the wild-type protein in gel shift experiments. However, although the C terminus of Cro is disordered in the crystal structure, these residues had been proposed to contribute to DNA binding.[19] As a result, we worried that addition of the tail might perturb the structure or function of Cro. To maximize our chances of obtaining functional Cro variants; we generated proteins with random fusion endpoints, and then identified active molecules by using the streptomycin selection. A plasmid was constructed containing a tandem arrangement of the *cro* gene with its normal TAA termination codon, a 21-bp sequence containing a unique *Pst*I restriction site, and the coding sequences for a 35-residue tail sequence (see Fig. 6; the first 9 residues of the tail sequence correspond to λ repressor residues 94–102 and the next 26 residues correspond to the long tail C-terminal extension discussed above). The plasmid DNA was cleaved at the *Pst*I site and digested with the *Bal*31 exonuclease. At several times aliquots were removed, the exonuclease reaction was stopped, and the linearized plasmids were religated. Following transformation into GW5100(λtac/λ1-S12), colonies resistant to 200 μg/ml streptomycin were selected. Cro expression in resistant candidates was induced with IPTG and lysates were electrophoresed on sodium dodecyl sulfate (SDS) polyacrylamide gels to determine the size of the fusion product. Two candidates containing Cro variants of increased size were selected and the corresponding *cro* genes were sequenced to determine the nature of the junction. Figure 6 shows the deletion endpoints. The Δ2 deletion results in a fusion protein that contains all of the wild-type residues in Cro followed by the sequence SLRSEYERKVEAP-

[22] D. A. Parsell, K. R. Silber, and R. T. Sauer, *Genes Dev.* **4**, 277 (1990).

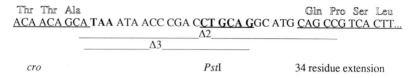

FIG. 6. Strategy for isolating Cro-fusion proteins. The C-terminal codons of the *cro* gene are shown on the left, followed by a 21-bp sequence containing a unique *Pst*I site, followed by the coding sequence for a 34-residue extension. The Δ2 and Δ3 deletions were generated by *Bal*31 digestion from the *Pst*I site and result in active Cro variants with C-terminal extensions 33 and 19 residues, respectively (see text).

TAVTVRASVVSKSLEKNQHE. The Δ3 deletion results in a fusion protein that also contains all of the normal Cro residues, followed by a C-terminal extension with the 19-residue sequence ACSRHLDLSM-SVKLRPQRP.

Attempts to Isolate Cro Variants that Bind +1 or −1 Operators. As described in the Methods section, we constructed variants of the tac/λ-S12 fusion in which the λ operator was altered by insertion or deletion of a single base pair from the center of the operator. Wild-type Cro did not bind specifically to these +1 or −1 operators in gel shift experiments and gave no streptomycin resistant colonies under selection conditions. We then used the library of residue 52–58 randomized sequences to ask if the Cro dimerization interface could be altered by substitution to allow recognition of the +1 or −1 operators. However, no streptomycin-resistant clones were obtained for either mutant operator. Hence, variability in the spacing of operator half-sites cannot be readily accommodated by changes in the sequence of the Cro dimer interface. Recent results, however, suggest that Cro variants with residue insertions in the dimer interface region can repress the +1 operator (M. C. Mossing, unpublished).

Discussion

We have described examples in which the streptomycin selection has been used to probe for DNA-binding activity of three different repressor proteins. The streptomycin selection should, in principle, be applicable to any site-specific DNA-binding protein that can be stably expressed in *E. coli*. The main requirement in modifying the selection for a new DNA-binding protein is that a binding site for the protein be placed in a position where it can repress expression of the wild-type gene for ribosomal protein S12. In the Lac repressor and λ Cro cases, the center to center distances between the −10 region of the *tac* promoter and the operator sites were

20 and 15 bp, respectively (see Fig. 3). Hence, both the *lac* and λ operator sites overlap the startpoint of transcription. To set up the selection for a new DNA-binding protein, it would probably be best to place the DNA-binding site in an analogous position relative to the promoter. In cases where the wild-type gene encoding a DNA-binding protein is available for manipulation, the system can obviously be calibrated with the wild-type protein expressed at various expression levels, and then used to select active variants. The streptomycin selection may also prove useful in cloning genes for DNA-binding proteins whose recognition sites are known, although in this case, it will not always be clear how to position the operator with respect to the promoter or whether bound protein will, in fact, prevent transcription initiation by RNA polymerase.

The efficiency of the streptomycin selection is limited, in some sense, by the frequency at which streptomycin-resistant colonies arise in the absence of the appropriate DNA-binding activity. In the cases where the promoter–operator–S12 fusion is present at the level of a single copy per cell as a lysogen, the background frequency of streptomycin-resistant colonies was about 1 in 100,000. It should, however, be possible to select active proteins that arise less frequently than this by employing a two-step procedure. For example, the selection could be applied, plasmid DNA purified from pooled streptomycin-resistant colonies, and the selection could be reapplied following retransformation of a naive strain. Such two-step procedures would not be effective in cases like that of Arc, where the promoter–operator–S12 fusion is carried on the same plasmid as the gene for the DNA-binding protein. However, in these cases, the effectiveness of the selection could be extended by recloning the gene for the DNA-binding protein between sequential applications of the selection.

Acknowledgment

This work was supported in part by NIH Grant AI-16982.

[30] Identification of Amino Acid–Base Pair Contacts by Genetic Methods

By RICHARD H. EBRIGHT

Introduction

This chapter discusses the use of genetic methods, i.e., the construction and analysis of single amino acid substitutions, to identify contacts between individual amino acids of a DNA-binding protein and individual base pairs of the DNA site in the protein–DNA complex. The experimental techniques are straightforward and do not require specialized equipment: (1) site-directed mutagenesis of the gene encoding the DNA-binding protein, followed by analysis of the DNA sequence-recognition properties of the substituted protein, and (2) random mutagenesis of the gene encoding the DNA-binding protein, followed by genetic selection and/or screening, DNA sequence determination, and analysis of the DNA sequence-recognition properties of the substituted protein.

The construction and analysis of single amino acid substitutions is an effective method to identify amino acid–base pair contacts. In addition, the construction and analysis of single amino acid substitutions is the sole method to determine the apparent binding free energy contribution and the apparent specificity free energy contribution of an amino acid–base pair contact (see Ref. 1), and the sole method to determine the alternative specificities that arise when an amino acid involved in a specificity-determining contact is replaced by 1 of the 19 other amino acids.

Other methods to identify amino acid–base pair contacts include X-ray diffraction and two-dimensional nuclear magnetic resonance (2D NMR) spectroscopic analysis of the specific protein–DNA complex.[2–5]

[1] R. Ebright, A. Kolb, H. Buc, T. Kunkel, J. Krakow, and J. Beckwith, *Proc. Natl. Acad. Sci. U.S.A.* **84,** 6083 (1987).
[2] J. Anderson, M. Ptashne, and S. Harrison, *Nature (London)* **326,** 846 (1987).
[3] A. Aggarwal, D. Rodgers, M. Drottar, M. Ptashne, and S. Harrison, *Science* **242,** 899 (1988).
[4] C. Wolberger, Y. Dong, M. Ptashne, and S. Harrison, *Nature (London)* **335,** 789 (1988).
[5] S. Jordan and C. Pabo, *Science* **242,** 893 (1988).

Loss-of-Contact Approach

The standard genetic method to identify amino acid–base pair contacts is the "loss-of-contact" approach.[6,7]

Logic

The logic of the loss-of-contact approach is outlined in Fig. 1. The basis of this approach is to eliminate the ability of the amino acid under study to make a contact to DNA, and then to determine at which base pair, if any, specificity is eliminated. To eliminate the ability of the amino acid under study to make a contact to DNA, the amino acid is substituted by an amino acid having a shorter side chain (e.g., Gly or Ala; see below).

A loss-of-contact substitution will result in four identifiable characteristics:

1. The substituted protein will lack the ability to discriminate between the consensus base pair and nonconsensus base pairs at one position, i, within the DNA site, corresponding to the position that in the wild-type protein–DNA complex is contacted by the amino acid under study. (In the case of an inverted-repeat or direct-repeat DNA site, the substituted protein will lack the ability to discriminate between the consensus base pair and nonconsensus base pairs at one position, i, in each repeat.) The substituted protein will exhibit approximately equal affinities for the consensus DNA site and for derivatives of the consensus DNA site having a nonconsensus base pair at position i: $K_{D,\text{noncons},i}/K_{D,\text{cons}} \approx 1$.

2. The substituted protein will retain the ability to discriminate between the consensus base pair and nonconsensus base pairs at each other position, j, within the DNA site. The substituted protein will exhibit a higher affinity for the consensus DNA site than for derivatives of the consensus DNA site having a nonconsensus base pair at position j: $K_{D,\text{noncons},j}/K_{D,\text{cons}} \gg 1$.

3. As compared to the wild-type protein, the substituted protein will exhibit a decreased affinity for the consensus DNA site. This is due to the presence in the wild-type protein–DNA complex, but absence in the substituted protein–DNA complex, of an energetically favorable interaction between the amino acid under study and the consensus base pair at position i.

4. As compared to the wild-type protein, the substituted protein will exhibit an unchanged affinity (Fig. 1-I) or an increased affinity (Fig. 1-II) for derivatives of the consensus DNA site having a nonconsensus base pair

---

[6] R. Ebright, *J. Biomol. Struct. Dyn.* **3**, 281 (1985).
[7] R. Ebright, *Proc. Natl. Acad. Sci. U.S.A.* **83**, 303 (1986).

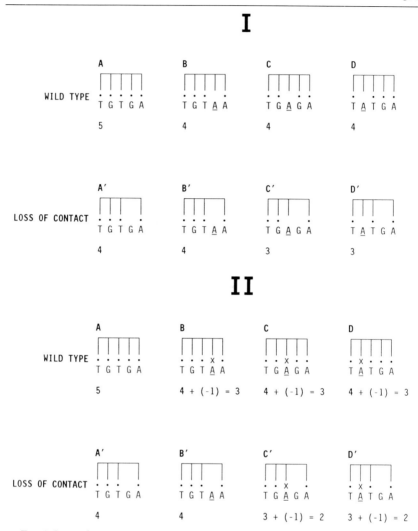

FIG. 1. Loss-of-contact approach: rationale. The basis of the approach is to eliminate the ability of the amino acid under study to make a contact with DNA, and then to determine at which base pair of the DNA site specificity is eliminated. A loss-of-contact substituted DNA binding protein will: (1) lack the ability to discriminate between the consensus base pair and nonconsensus base pairs at one position, i, within the DNA site, corresponding to the position that in the wild-type protein–DNA complex is contacted by the amino acid under study, but (2) retain the ability to discriminate between the consensus base pair and nonconsensus base pairs at each other position, j, within the DNA site. Two cases are presented: case I, which considers only energetically favorable contacts, and case II, which considers both energetically favorable and energetically unfavorable contacts. *Case I.* (A) Interaction of the wild-type DNA-binding protein with five base pairs of the consensus DNA

at position *i*. This is due to the absence in both the wild-type protein–DNA complex and the substituted protein–DNA complex of an energetically favorable interaction between the amino acid under study and the nonconsensus base pair at position *i* (Fig. 1-I, 1-II); and to the possible presence in the wild-type protein–DNA complex, but absence in the substituted protein–DNA complex, of an energetically unfavorable interaction between the amino acid under study and the nonconsensus base pair at position *i* (Fig. 1-I).

Characteristics 1 and 2 in the list above are the critical, defining characteristics of a loss-of-contact substitution. Conclusions regarding the identification of amino acid–base pair contacts can be drawn only if both characteristics 1 and 2 are documented. Characteristics 1 and 2 are specificity characteristics. It is important not to confuse specificity and affinity. Specificity is the factor by which DNA-binding protein *X* prefers to interact with one DNA site vs another DNA site—for example, to interact with the consensus DNA site vs a nonconsensus DNA site. The ratio $K_{D,noncons}/K_{D,cons}$ is a measure of specificity.

---

site (five favorable contacts). The residue of the DNA-binding protein participating in each contact is indicated as a vertical line; the presence of a favorable contact is indicated as a dot. (B–D) Interaction of the wild-type DNA-binding protein with derivatives of the consensus DNA site, each having a nonconsensus base pair at one position (four favorable contacts). (A′) Interaction of the loss-of-contact substituted DNA-binding protein with the consensus DNA site (four favorable contacts). (B′) Interaction of the loss-of-contact substituted DNA-binding protein with the derivative of the consensus DNA site having a nonconsensus base pair at the position, *i*, that in the wild-type protein–DNA complex is contacted by the amino acid under study (four favorable contacts; identical to the number with the wild-type DNA site). (C′ and D′) Interaction of the loss-of-contact substituted DNA-binding protein with derivatives of the consensus DNA site, each having a nonconsensus base pair at one other position, *j* (three favorable contacts). *Case II*. (A) Interaction of the wild-type DNA-binding protein with five base pairs of the consensus DNA site (five favorable contacts). The residue of the DNA-binding protein participating in each contact is indicated as a vertical line; the presence of a favorable contact is indicated as a dot. (B–D) Interaction of the wild-type DNA-binding protein with derivatives of the consensus DNA site, each having a nonconsensus base pair at one position (four favorable contacts, one unfavorable contact; net three favorable contacts). The presence of an unfavorable contact is indicated as an ''x.'' (A′) Interaction of the loss-of-contact substituted DNA-binding protein with the consensus DNA site (four favorable contacts). (B′) Interaction of the loss-of-contact substituted DNA-binding protein with the derivative of the consensus DNA site having a nonconsensus base pair at the position, *i*, that in the wild-type protein–DNA complex is contacted by the amino acid under study (four favorable contacts; identical to the number with the wild-type DNA site). (C′ and D′) Interaction of the loss-of-contact substituted DNA-binding protein with derivatives of the consensus DNA site, each having a nonconsensus base pair at one other position, *j* (three favorable contacts, one unfavorable contact; net two favorable contacts).

Experimental Design

The optimal experimental design for use of the loss-of-contact approach depends on the amount of information available. Examples are presented here for three cases.

Case One: High-Resolution Structure or Detailed Model of the Protein–DNA Complex Exists. In this case, the structure or model indicates which amino acids are in close proximity to which DNA base pairs in the protein–DNA complex. The objective for use of the loss-of-contact approach is to verify the predicted amino acid–base pair contacts: i.e., to establish that the predicted contacts in fact occur and make energetically significant contributions to DNA binding and DNA sequence specificity.

The optimal experimental design is as follows: Site-directed mutagenesis[8,9] is used to substitute the amino acid under study by Gly and by Ala. The specificities of the wild-type protein, the Gly-substituted protein, and the Ala-substituted protein are analyzed at the position within the DNA site predicted to be contacted by the amino acid under study (position i), and, as a control, at one other position within the DNA site (position j). The ratio $K_{D,noncons,i}/K_{D,cons}$ is determined for each of the three possible derivatives of the consensus DNA site having a nonconsensus base pair at position i. (For example, in the case where the consensus base pair at position i is G : C, the ratio $K_{D,noncons,i}/K_{D,cons}$ is determined for the derivatives of the consensus DNA site having A : T, C : G, and T : A at position i.) The ratio $K_{D,noncons,j}/K_{D,cons}$ is determined for each of the three possible derivatives of the consensus DNA site having a nonconsensus base pair at position j.

Other amino acid substitutions may be used in this approach[1,10]; however, in general, the Gly and Ala substitutions are preferred. Gly has no side chain. The Gly substitution has the highest likelihood of eliminating the contact by the amino acid under study, and of not introducing a new energetically significant interaction. Ala has a one-carbon side chain. The Ala substitution has the second highest likelihood of eliminating the contact by the amino acid under study, and of not introducing a new, energetically significant interaction. As compared to the Gly substitution, the Ala substitution has a higher likelihood of not introducing complications due to altered protein conformation.

This experimental design has been used to verify two predicted amino

---

[8] T. Kunkel, *Proc. Natl. Acad. Sci. U.S.A.* **82**, 488 (1985).

[9] J. Wells, M. Vasser, and D. Powers, *Gene* **34**, 315 (1985).

[10] R. Ebright, A. Gunasekera, X. Zhang, T. Kunkel, and J. Krakow, *Nucleic Acids Res.* **18**, 1457 (1990).

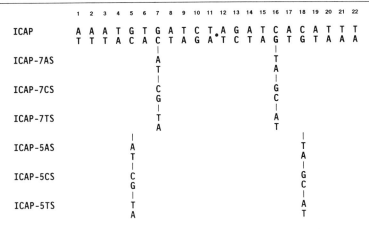

FIG. 2. DNA sites used to verify a predicted amino acid–base pair contact in the CAP–DNA complex (A. Gunasekera, X. Zhang, M. Smith, T. Kunkel, and R. Ebright, in preparation). ICAP is the consensus DNA site for CAP. The site is 22 bp in length and has perfect twofold sequence symmetry. The twofold axis is located between positions 11 and 12. ICAP-7AS, ICAP-7CS, and ICAP-7TS are derivatives of the consensus DNA site substituted at position 7 of the DNA half-site. ICAP-5AS, ICAP-5CS, and ICAP-5TS are derivatives of the consensus DNA site substituted at position 5 of the DNA half-site.

TABLE I

DNA SEQUENCE-RECOGNITION PROPERTIES OF WILD-TYPE CAP, [Gly-181]CAP, AND [Ala-181]CAP[a,b]

DNA site[c]	$K_D/K_{D,cons}$		
	Wild-type CAP	[Gly-181]CAP	[Ala-181]CAP
ICAP (consensus site)	[1]	[1]	[1]
ICAP-7AS	>7000	0.6	2
ICAP-7CS	>7000	1	2
ICAP-7TS	3000	2	3
ICAP-5AS	2000	80	60
ICAP-5CS	>7000	300	>800
ICAP-5TS	700	30	50

[a] A. Gunasekera, X. Zhang, M. Smith, T. Kunkel, and R. Ebright (in preparation).
[b] Data are from nitrocellulose filter-binding assays.
[c] Sequences of the DNA sites are in Fig. 2.

acid–base pair contacts in the CAP–DNA complex.[11,12] Glu-181 of CAP is known to contact position 7 of the DNA site in the CAP–DNA complex[1,13,14] (see Experimental Design, Case Three, below). In recent work, Glu-181 of CAP has been substituted by Gly and by Ala.[11] The data in Fig. 2 and Table I show that [Gly-181]CAP and [Ala-181]CAP lack the ability to discriminate between the consensus base pair G : C and the nonconsensus base pairs A : T, C : G, and T : A at position 7 of the DNA site ($K_{D,noncons,7}/K_{D,cons} \approx 1$). In contrast, [Gly-181]CAP and [Ala-181]CAP retain the ability to discriminate between the consensus base pair G : C and the nonconsensus base pairs A : T, C : G, and T : A at position 5 of the DNA site ($K_{D,noncons,5}/K_{D,cons} \gg 1$). These data are consistent with the known contact. The model for the structure of the CAP–DNA complex[10,14,15] predicts that Arg-180 of CAP contacts position 5 of the DNA site in the CAP–DNA complex. In recent work, Arg-180 of CAP has been substituted by Gly and by Ala.[12] The data show that [Gly-180]CAP and [Ala-180]CAP lack the ability to discriminate between the consensus base pair G : C and the nonconsensus base pairs A : T, C : G, and T : A at position 5 of the DNA site ($K_{D,noncons,5}/K_{D,cons} \approx 1$). In contrast, [Gly-180]CAP and [Ala-180]CAP retain the ability to discriminate between the consensus base pair G : C and the nonconsensus base pairs A : T, C : G, and T : A at position 7 of the DNA site ($K_{D,noncons,7}/K_{D,cons} \gg 1$). These data are consistent with the predicted contact. The identification of two amino acid–base pair contacts in the CAP–DNA complex provides information sufficient to define the orientation of the helix-turn-helix DNA-binding motif of CAP with respect to DNA in the CAP–DNA complex.[12]

Case Two: No High-Resolution Structure or Detailed Model of the Protein–DNA Complex Exists; However, Information Suggests Which Amino Acids Make Contacts. In this case, amino acid sequence similarities to a well-characterized DNA-binding protein, or previous biochemical or genetic results, suggest which amino acids of the DNA-binding protein under study are candidates to make amino acid–base pair contacts in the protein–DNA complex. The objectives for use of the loss-of-contact approach are: (1) to determine whether a candidate amino acid makes an amino acid–base pair contact, and, if so, (2) to determine which base pair it contacts.

The optimal experimental design is as follows: Site-directed mutagene-

[11] A. Gunasekera, X. Zhang, M. Smith, T. Kunkel, and R. Ebright, in preparation.
[12] X. Zhang and R. Ebright, *Proc. Natl. Acad. Sci. U.S.A.* **87**, 4717 (1990).
[13] R. Ebright, P. Cossart, B. Gicquel-Sanzey, and J. Beckwith, *Nature (London)* **311**, 232 (1984).
[14] R. Ebright, P. Cossart, B. Gicquel-Sanzey, and J. Beckwith, *Proc. Natl. Acad. Sci. U.S.A.* **81**, 7274 (1984).
[15] I. Weber and T. Steitz, *Proc. Natl. Acad. Sci. U.S.A.* **81**, 3973 (1984).

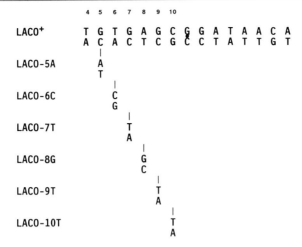

FIG. 3. DNA sites used to identify an amino acid–base pair contact in the Lac repressor–DNA complex [R. Ebright, *J. Biomol. Struct. Dyn.* **3**, 281 (1985); R. Ebright, *Proc. Natl. Acad. Sci. U.S.A.* **83**, 303 (1986); R. Ebright and X. Zhang, unpublished data]. LACO⁺ is the wild-type *lac* operator. The site is 21 bp in length and has approximate twofold sequence symmetry. The twofold axis is located at position 11. (Base positions are numbered with respect to the start point of *lac* transcription; using this convention, the leftmost position in the *lac* operator is designated as position 4.) LACO-5A, LACO-6C, LACO-7T, LACO-8G, LACO-9T, and LACO-10T are derivatives of the *lac* operator substituted at positions 5, 6, 7, 8, 9, and 10 of the DNA half-site.

TABLE II

DNA-SEQUENCE RECOGNITION PROPERTIES OF WILD-TYPE LAC REPRESSOR,
[Gly-18]LAC REPRESSOR, AND [Ala-18]LAC REPRESSOR[a,b,c]

DNA site[d]	$K_D/K_{D,O^+}$		
	Wild-type Lac repressor	[Gly-18]Lac repressor	[Ala-18]Lac repressor
LACO⁺ (wild-type *lac* operator)	[1]	[1]	[1]
LACO-5A	100	20	100
LACO-6C	90	20	60
LACO-7T	50	*0.9*	*0.3*
LACO-8G	70	10	50
LACO-9T	20	20	40
LACO-10T	90	20	200

[a] R. Ebright, *J. Biomol. Struct. Dyn.* **3**, 281 (1985).
[b] R. Ebright, *Proc. Natl. Acad. Sci. U.S.A.* **83**, 303 (1986).
[c] R. Ebright and X. Zhang (unpublished data).
[d] Sequences of the DNA sites are in Fig. 3.

sis[8,9] is used to substitute the amino acid under study by Gly and by Ala (see Experimental Design, Case One, above, for the rationale for choice of Gly and Ala substitutions). The specificities of the wild-type protein, the Gly-substituted protein, and the Ala-substituted protein are analyzed at *each* position of the DNA site. For each position of the DNA site, the ratio $K_{D,noncons}/K_{D,cons}$ is determined for one derivative of the consensus DNA site having a nonconsensus base pair at the position. The position that in the wild-type protein–DNA complex is contacted by the amino acid under study is identified as the position, i, within the DNA site at which $K_{D,noncons}/K_{D,cons} \approx 1$.

This experimental design has been used to identify an amino acid–base pair contact in the Lac repressor–DNA complex.[6,7,16] Amino acid sequence similarities to well-characterized helix-turn-helix motif sequence-specific DNA-binding proteins,[17,18] and previous genetic results,[19] had suggested that Gln-18 of Lac repressor was a strong candidate to make an amino acid–base pair contact in the Lac repressor–DNA complex. Gln-18 of Lac repressor has been substituted by Gly and by Ala.[6,7,16] The specificities of wild-type Lac repressor and of the two substituted Lac repressor variants have been assessed with respect to positions 5, 6, 7, 8, 9, and 10 of the DNA site.[6,7,16] The data in Fig. 3 and Table II show that [Gly-18]Lac repressor and [Ala-18]Lac repressor lack the ability to discriminate between the consensus base pair G : C, and the nonconsensus base pair T : A at position 7 of the DNA site ($K_{D,noncons,7}/K_{D,cons} \approx 1$). In contrast, [Gly-18]Lac repressor and [Ala-18]Lac repressor retain the ability to discriminate between the consensus and nonconsensus base pairs and at each other position of the DNA site ($K_{D,noncons,j}/K_{D,cons} \gg 1$). These data suggest that Gln-18 of Lac repressor contacts position 7 of the DNA site in the Lac repressor–DNA complex.

This experimental design also has been used to identify amino acid–base pair contacts in the λ repressor–DNA complex,[20,21] in the λ cro–DNA complex,[20,22] in the σ^H–DNA complex,[23] and in the AraC–DNA complex.[24]

[16] R. Ebright and X. Zhang, unpublished data.
[17] C. Pabo and R. Sauer, *Annu. Rev. Biochem.* **53,** 293 (1984).
[18] Y. Takeda, D. Ohlendorf, W. Anderson, and B. Matthews, *Science* **221,** 1020 (1983).
[19] J. Miller, *J. Mol. Biol.* **180,** 205 (1984).
[20] A. Hochschild and M. Ptashne, *Cell (Cambridge, Mass.)* **44,** 925 (1986).
[21] A. Brunelle and R. Schleif, *Proc. Natl. Acad. Sci. U.S.A.* **84,** 6673 (1987).
[22] Y. Takeda, A. Sarai, and V. Rivera, *Proc. Natl. Acad. Sci. U.S.A.* **86,** 439 (1989).
[23] P. Zuber, J. Healy, H. L. Carter, S. Cutting, C. Moran, and R. Losick, *J. Mol. Biol.* **206,** 605 (1989).
[24] A. Brunelle and R. Schleif, *J. Mol. Biol.* **209,** 607 (1989).

Case Three: No High-Resolution Structure or Detailed Model of the Protein–DNA Complex Exists; No Information Suggests Which Amino Acids Make Contacts. In this case, no predictions are available. The objectives for use of the loss-of-contact approach are: (1) to identify an amino acid making an amino acid–base pair contact, and (2) to determine which base pair it contacts.

The optimal experimental design involves random mutagenesis of the gene encoding the DNA-binding protein, followed by genetic selection and/or screening, followed by DNA sequence determination.

The genetic selection and/or screening procedure takes advantage of characteristics 3 and 4 of a loss-of-contact substitution (decreased affinity for the consensus DNA site, unchanged or increased affinity for derivatives of the consensus DNA site having a nonconsensus base pair at one position, i, within the DNA site; see Logic, above). The genetic selection and/or screening procedure has two steps.[13] In step 1, a selection or screen is performed in order to isolate a pool of mutant clones, each having a substituted DNA-binding protein with decreased affinity for the consensus DNA site. This is most efficiently done using a selection or screen for decreased fractional occupancy, θ_{cons}, of the consensus DNA site (see Experimental Methods, *In Vivo* Assay, below). In step two, within the resulting pool of mutant clones, a selection or screen is performed in order to isolate rare mutant clones, each having a substituted DNA-binding protein with unchanged or increased affinity for a derivative of the consensus DNA site having a nonconsensus base pair at one position, i, within the DNA site. This is most efficiently done using a selection or screen for unchanged or increased fractional occupancy, $\theta_{noncons,i}$, of a derivative of the DNA site having a nonconsensus base pair at position i. It is possible to reverse the order of steps one and two.[25,26] In ideal cases, it is possible to perform steps 1 and 2 on a single agar plate.[13]

For each mutant clone isolated in the genetic selection and/or screening procedure, it is imperative to confirm that the substituted DNA-binding protein in the mutant clone exhibits characteristics 1 and 2 of a loss-of-contact substitution (loss of the ability to discriminate between the consensus base pair and nonconsensus base pairs at one position, i, within the DNA site; retention of the ability to discriminate between the consensus base pair and nonconsensus base pairs at each other position, j, within the DNA site; see Logic, above). The specificities of the wild-type DNA-binding protein and of the substituted DNA-binding protein in the mutant

[25] D. Siegele, J. Hu, W. Walter, and C. Gross, *J. Mol. Biol.* **206,** 591 (1989).
[26] T. Gardella, H. Moyle, and M. Susskind, *J. Mol. Biol.* **206,** 579 (1989).

clone are analyzed at each position of the DNA site. For each position of the DNA site, the ratio $K_{D,noncons}/K_{D,cons}$ is determined for one derivative of the consensus DNA site having a nonconsensus base pair at the position.

For each verified mutant clone, the DNA nucleotide sequence of the gene encoding the DNA-binding protein is then determined. This permits the identification of the amino acid substitution responsible for the loss-of-contact phenotype.

This experimental design has been used to identify an amino acid–base pair contact in the CAP–DNA complex.[1,13,14] At the time the experiments were initiated (1982), there were conflicting models for the structure of the CAP–DNA complex,[27–30] and there was no experimental information to suggest which amino acids of CAP make amino acid–base pair contacts. The gene encoding CAP was subjected to random mutagenesis.[13] A selection was performed in order to isolate mutant clones, each having a substituted CAP variant with decreased affinity for the wild-type *lac* DNA site for CAP; within the resulting pool of mutant clones, a screen was performed in order to isolate rare mutant clones, each having a substituted CAP variant with increased affinity for the derivative of the *lac* DNA site for CAP having A : T at position 7.[13] The selection and the screen were performed on a single agar plate. Three mutant clones were isolated. DNA nucleotide sequence determination demonstrated that in all three mutant clones the identical amino acid of CAP—amino acid 181—was substituted; the substitutions were Val, Leu, and Lys. It was concluded that Glu-181 contacts base pair 7 of the DNA site in the CAP–DNA complex.[1,13,14]

This experimental design also has been used to identify amino acid–base pair contacts in the σ^{70}–DNA complex.[25,26]

Experimental Methods

In Vivo Assay of $K_{D,noncons}/K_{D,cons}$. If the DNA-binding protein under study is a transcriptional repressor or a transcriptional activator protein, it is possible to use an *in vivo* assay to determine the ratio $K_{D,noncons}/K_{D,cons}$. This can be efficient and convenient, since it eliminates the requirement to purify the wild-type protein and each substituted protein.

The method employs a set of isogenic "tester" strains. In each tester strain, the identical reporter promoter is placed under the control of the

[27] D. McKay and T. Steitz, *Nature* (*London*) **290**, 744 (1981).
[28] F. R. Salemme, *Proc. Natl. Acad. Sci. U.S.A.* **79**, 5263 (1982).
[29] R. Ebright, *in* "Molecular Structure and Biological Activity" (J. Griffen and R. Duax, eds.), p. 91. Elsevier, New York, 1982.
[30] C. Pabo and M. Lewis, *Nature* (*London*) **298**, 443 (1982).

DNA site under study. In one tester strain, the reporter promoter is placed under the control of the consensus DNA site. In the other tester strains, the reporter promoter is placed under the control of derivatives of the consensus DNA site having a nonconsensus base pair at one or more positions within the DNA site.

In cases where the DNA site is an inverted repeat or a direct repeat, equivalent qualitative results are obtained using derivatives of the consensus DNA site substituted in each repeat, or derivatives of the consensus DNA site substituted in only one repeat.[10] Derivatives of the consensus DNA site substituted in each repeat normally are preferred. These derivatives exhibit larger values for the ratio $K_{D,noncons}/K_{D,cons}$, reducing the significance of experimental error. (In most cases, the ratio $K_{D,noncons}/K_{D,cons}$ for a derivative of the consensus DNA site substituted in each of two repeats is the square of the ratio $K_{D,noncons}/K_{D,cons}$ for the analogous derivative of the consensus DNA site substituted in only one repeat.)

In order to analyze the profile of specificity of DNA-binding protein X, the plasmid encoding DNA-binding protein X is introduced into each of the tester strains. Note that each of the resulting plasmid-bearing strains is identical except for the sequence of the DNA site. One important implication is that in each of the resulting plasmid-bearing strains the intracellular concentration of DNA-binding protein X is equal. In the data reduction, all calculations are performed using only data from strains having the plasmid encoding the identical DNA-binding protein X. Therefore, the analysis does not incur complications due to either (1) differing rates of synthesis of wild-type vs substituted proteins, or (2) differing stabilities of wild-type vs substituted proteins.

Data collection and data reduction are performed as follows: For each strain, the expression of the reporter promoter is determined. This is accomplished by measurement of the differential rate of synthesis of the reporter mRNA, or by measurement of the differential rate of synthesis of the reporter protein. The fractional occupancy of the reporter promoter, θ, is calculated from the experimental data using Eq. (1) or Eq. (2). Ratios of equilibrium dissociation constants, $K_{D,1}/K_{D,2}$, are calculated from the values for θ using Eq. (3).

In the case of a repressor, the fractional occupancy of the reporter promoter, θ, is calculated as follows[6,7]:

$$\theta = 1 - (E/E_M) \tag{1}$$

where E denotes the expression of the reporter promoter in the presence of repressor, and E_M denotes the expression of the reporter promoter in the absence of repressor. E_M is determined by measurement of the expression of the reporter promoter in the presence of a chemical that inactivates the repressor (e.g., isopropyl-thio-β-D-galactoside in the case of Lac repressor).[6,7] Alternatively, E_M can be determined by measurement of the expression of the reporter promoter in a derivative of the tester strain having a control plasmid that does not encode the repressor.

In the case of an activator protein, the fractional occupancy of the reporter promoter, θ, is calculated as follows[10]:

$$\theta = E/E_M \tag{2}$$

where E denotes the expression of the reporter promoter in the presence of the activator protein, and E_M denotes the expression of the reporter promoter in the presence of a saturating concentration of the activator protein. E_M is determined by calculation using comparison *in vitro* data.[10] Alternatively, E_M can be determined by measurement of the expression of the reporter promoter in a derivative of the tester strain having a plasmid construct that results in a saturating concentration of the activator protein. The values used for E and E_M should be corrected for activator-independent expression of the reporter promoter.

Ratios of equilibrium dissociation constants, $K_{D,1}/K_{D,2}$, are calculated as follows[6,7,10]:

$$K_{D,1}/K_{D,2} = [\theta_2(1 - \theta_1)]/[\theta_1(1 - \theta_2)] \tag{3}$$

Equation (3) applies for data from any two strains 1 and 2 in which the intracellular concentration of unbound DNA-binding protein X in strain 1 is equal to the intracellular concentration of unbound DNA-binding protein X in strain 2 (e.g., any two tester strains having a plasmid encoding the identical DNA-binding protein X; see above).

In Vitro Assay of $K_{D,noncons}/K_{D,cons}$: Method One. This method is the simplest and most reliable method. In this method, equilibrium DNA binding experiments are performed using purified wild-type DNA binding protein and purified substituted DNA binding protein, and each of a set of double-stranded DNA fragments containing the DNA site. One DNA fragment in the set contains the consensus DNA site. The other DNA fragments in the set contain derivatives of the consensus DNA site having a nonconsensus base pair at one or more positions within the DNA site.

In cases where the DNA site is an inverted repeat or a direct repeat, equivalent qualitative results are obtained using derivatives of the consensus DNA site substituted in each repeat, or derivatives of the consensus DNA site substituted in only one repeat.[1] Derivatives of the consensus DNA site substituted in each repeat normally are preferred (see Experimental Methods, *In Vivo* Assay, above).

Equilibrium dissociation constants, K_D, are measured using the nitrocellulose filter-binding assay,[31] the gel retardation assay,[32,33] or the quantitative nuclease footprinting assay.[34]

In Vitro Assay of $K_{D,noncons}/K_{D,cons}$: *Method Two.* A chemical modification interference assay is performed as follows[35,36]: A sample of end-labeled double-stranded DNA fragment containing the DNA site of interest is randomly chemically modified, using reaction conditions such that at most one chemical modification occurs per DNA fragment. The sample of modified DNA fragment then is incubated with excess DNA-binding protein until the establishment of equilibrium. The DNA molecules that remain unbound at equilibrium are separated from the protein-bound DNA molecules. The DNA molecules that remain unbound comprise those having a chemical modification positioned so as to interfere with binding of the protein; the locations of the interfering modifications are identified by thermal- and/or alkali-mediated DNA cleavage at the modified nucleotide, followed by denaturing polyacrylamide gel electrophoresis and autoradiography. There are two relevant types of chemical modification interference assays: methylation interference assays,[35,36] which probe the effects of methylation of guanine N-7 atoms (and, under certain conditions, also of adenine N-3 atoms), and depurination/depyrimidation interference assays,[21,35] which probe the effects of removal of purines or pyrimidines.

Hochschild and Ptashne[20] and Brunelle and Schleif[21,24] have shown that a chemical modification interference assay can be a rapid, efficient method to obtain a rough estimate of the ratio $K_{D,noncons}/K_{D,cons}$ at each of several positions within the DNA site. The methylation interference assay permits the estimation of the ratio $K_{D,\text{N-7-methyl-G}}/K_{D,cons}$ at each guanine of the DNA site. The depurination/depyrimidation interference assay permits the estimation of the ratio $K_{D,\text{abasic site}}/K_{D,cons}$ at each base of the DNA site. The procedure is as follows: The methylation interference and/or depurination/depyrimidation interference patterns exhibited by the wild-type protein and a substituted protein are determined and compared. The position that in the wild-type protein–DNA complex is contacted by the amino acid under study is identified as the position, i, at which the wild-

[31] A. Riggs, H. Suzuki, and S. Bourgeois, *J. Mol. Biol.* **48,** 67 (1970).

[32] M. Garner and A. Revzin, *Nucleic Acids Res.* **9,** 3047 (1981).

[33] M. Fried and D. Crothers, *Nucleic Acids Res.* **9,** 6505 (1981).

[34] M. Brenowitz, D. F. Senear, M. A. Shea, and G. K. Ackers, this series, Vol. 130, p. 132.

[35] J. Majors, Ph.D. Thesis, Harvard University, Cambridge, Massachusetts (1977).

[36] U. Siebenlist, R. Simpson, and W. Gilbert, *Cell (Cambridge, Mass.)* **20,** 269 (1980).

type protein exhibits an interference (i.e., $K_{D,noncons,i}/K_{D,cons} \gg 1$), but at which the substituted protein does not exhibit an interference (i.e., $K_{D,noncons,i}/K_{D,cons} \approx 1$).

Using this procedure, all positions within the DNA site susceptible to the chemical modification are assessed *in a single experiment*. This procedure is useful in the case where no high-resolution structure or detailed model of the protein–DNA complex exists, but information suggests which amino acids make contacts (see Experimental Design, Case Two, above).

It is important to note that the methylation interference assay involves the introduction of a methyl group, and the introduction of a positive charge, at the guanine N-7 atom. The depurination/depyrimidation assay involves the introduction of an abasic site. These changes can have significant effects on local DNA conformation, complicating the interpretation of results. Important results obtained using the methylation interference or depurination/depyrimidation assays should be confirmed using one of the two preceding procedures.[24]

New-Specificity Approach

The logic of the new-specificity approach[37] is outlined in Fig. 4. The approach is to substitute the amino acid under study by an amino acid that: (1) makes no energetically favorable interaction with the consensus base pair at the position within the DNA site that in the wild-type protein–DNA complex is contacted by the amino acid under study, but (2) makes an energetically favorable interaction with one nonconsensus base pair at the position within the DNA site that in the wild-type protein–DNA complex is contacted by the amino acid under study. A new-specificity substitution will result in four identifiable characteristics:

1. The substituted protein will prefer a particular nonconsensus base pair to the consensus base pair at one position, i, within the DNA site, corresponding to the position that in the wild-type protein–DNA complex is contacted by the amino acid under study. (In the case of an inverted-repeat or direct-repeat DNA site, the substituted protein will prefer a particular nonconsensus base pair to the consensus base pair at one position, i, in each repeat.) The substituted protein will exhibit a lower affinity for the consensus DNA site than for one derivative of the consensus DNA site having a nonconsensus base pair at position i: $K_{D,noncons,i}/K_{D,cons} \ll 1$.

2. The substituted protein will retain the wild-type ability to discriminate between the consensus base pair and nonconsensus base pairs at each

[37] P. Youderian, A. Vershon, S. Bouvier, R. Sauer, and M. Susskind, *Cell* (*Cambridge, Mass.*) **35**, 777 (1983).

other position, j, within the DNA site. The substituted protein will exhibit a higher affinity for the consensus DNA site than for derivatives of the consensus DNA site having a nonconsensus base pair at position j: $K_{D,noncons,j}/K_{d,cons} \gg 1$.

3. As compared to the wild-type protein, the substituted protein will exhibit a decreased affinity for the consensus DNA site. This is due to the presence in the wild-type protein–DNA complex, but absence in the substituted protein–DNA complex, of an energetically favorable interaction between the amino acid under study and the consensus base pair at position i (Fig. 4-I, 4-II); and to the possible presence in the substituted protein–DNA complex of an energetically unfavorable interaction between the amino acid under study and the consensus base pair at position i (Fig. 4-II).

4. As compared to the wild-type protein, the substituted protein will exhibit an increased affinity for one derivative of the consensus DNA site having a nonconsensus base pair at position i. This is due to the absence in the wild-type protein–DNA complex, but presence in the substituted protein–DNA complex, of an energetically favorable interaction between the amino acid under study and the nonconsensus base pair at position i (Fig. 4-I, 4-II); and to the possible presence in the wild-type protein–DNA complex, but absence in the substituted protein–DNA complex, of an energetically unfavorable interaction between the amino acid under study and the nonconsensus base pair at position i (Fig. 4-II).

Examples of new-specificity substitutions have been obtained in the following ways: (1) in the course of construction of a large number of substitutions (up to all 19 possible substitutions) of an amino acid known or predicted to make an amino acid–base pair contact,[11,38–40] (2) by interchanging, or "swapping," one amino acid within a pair of highly homologous DNA-binding proteins having DNA sites that differ at only one base pair[41] (see also Refs. 42–46), and (3) by random mutagenesis of the gene

[38] R. Wharton and M. Ptashne, *Nature (London)* **326,** 888 (1987).

[39] N. Lehming, J. Sartorius, S. Oehler, B. von Wilcken-Bergmann, and B. Müller-Hill, *Proc. Natl. Acad. Sci. U.S.A.* **85,** 7947 (1988).

[40] S. Bass, V. Sorrelis, and P. Youderian, *Science* **242,** 240 (1988).

[41] L. Altschmied, R. Baumeister, K. Pfleiderer, and W. Hillen, *EMBO J.* **7,** 4011 (1988).

[42] N. Lehming, J. Sartorius, M. Niemöller, G. Genenger, B. von Wilcken-Bergmann, and B. Müller-Hill, *EMBO J.* **6,** 3145 (1987).

[43] S. Hanes and R. Brent, *Cell (Cambridge, Mass.)* **57,** 1275 (1989).

[44] J. Treisman, P. Gonczy, M. Vashishtha, E. Harris, and C. Desplan, *Cell (Cambridge, Mass.)* **59,** 553 (1989).

[45] M. Danielson, L. Hinck, and G. Ringold, *Cell (Cambridge, Mass.)* **57,** 1131 (1989).

[46] K. Umesono and R. Evans, *Cell (Cambridge, Mass.)* **57,** 1139 (1989).

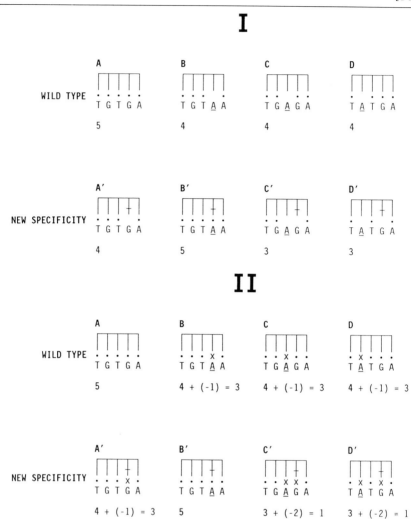

FIG. 4. New-specificity approach: rationale. A new-specificity substituted DNA-binding protein will: (1) prefer a particular nonconsensus base pair to the consensus base pair at one position, *i*, within the DNA site, corresponding to the position that in the wild-type protein–DNA complex is contacted by the amino acid under study, but (2) retain the wild-type ability to discriminate between the consensus base pair and nonconsensus base pairs at each other position, *j*, within the DNA site. The figure presents two cases: case I, which considers only energetically favorable contacts, and case II, which considers both energetically favorable and energetically unfavorable contacts. *Case I.* (A) Interaction of the wild-type DNA-binding protein with five base pairs of the consensus DNA site (five favorable contacts). The residue of the DNA-binding protein participating in each contact is indicated as a vertical line; the presence of a favorable contact is indicated as a dot. (B–D) Interaction of the wild-type DNA-binding protein with derivatives of the consensus DNA site, each

encoding the DNA-binding protein, followed by genetic selection and/or screening for decreased affinity for the consensus DNA site and for increased affinity for one nonconsensus derivative of the consensus DNA site.[23,37]

Empirically, new-specificity substitutions are relatively rare. This is not surprising, since a new-specificity substitution must not only eliminate an interaction, but also introduce a new interaction. Glu-181 of CAP is known to contact position 7 of the DNA site in the CAP–DNA complex[1,13,14] (see Experimental Design, Case One and Experimental Design, Case Three, above). In recent work, all 19 possible amino acid substitutions at residue 181 of CAP have been constructed; all 19 substituted proteins have been purified; and the DNA sequence-recognition properties of all 19 substituted proteins have been assessed with respect to positions 5, 6, and 7 of the DNA site.[11] Fully 15 of 19 amino acid substitutions at residue 181 of CAP result in the loss-of-contact phenotype (Gly, Ala, Val, Leu, Ile, Pro, Ser, Thr, Cys, Met, Asn, Gln, His, Lys, and Arg). Only 1 of 19 amino acid substitutions at residue 181 of CAP results in the new-specificity phenotype (Asp). Empirically, new-specificity substitutions also are difficult to predict. The sole substitution at residue 181 of CAP that results in the new-specificity phenotype was not predicted.[11]

---

having a nonconsensus base pair at one position (four favorable contacts). (A') Interaction of the new-specificity substituted DNA-binding protein with the consensus DNA site (four favorable contacts). (B') Interaction of the new-specificity substituted DNA-binding protein with one derivative of the consensus DNA site having a nonconsensus base pair at the position, i, that in the wild-type protein–DNA complex is contacted by the amino acid under study (five favorable contacts). (C' and D'). Interaction of the new-specificity substituted DNA-binding protein with derivatives of the consensus DNA site, each having a nonconsensus base pair at one other position, j (three favorable contacts). *Case II*. (A) Interaction of the wild-type DNA-binding protein with five base pairs of the consensus DNA site (five favorable contacts). The residue of the DNA-binding protein participating in each contact is indicated as a vertical line; the presence of a favorable contact is indicated as a dot. (B–D) Interaction of the wild-type DNA-binding protein with derivatives of the consensus DNA site, each having a nonconsensus base pair at one position (four favorable contacts, one unfavorable contact; net three favorable contacts). The presence of an unfavorable contact is indicated as an "x." (A') Interaction of the new-specificity substituted DNA-binding protein with the consensus DNA site (four favorable contacts, one unfavorable contact; net three favorable contacts). (B') Interaction of the new-specificity substituted DNA-binding protein with one derivative of the consensus DNA site having a nonconsensus base pair at the position, i, that in the wild-type protein–DNA complex is contacted by the amino acid under study (five favorable contacts). (C' and D') Interaction of the new-specificity substituted DNA-binding protein with derivatives of the consensus DNA site, each having a nonconsensus base pair at one other position, j (three favorable contacts, two unfavorable contacts; net one favorable contact).

New-specificity substitutions, when obtained, can permit the identification of amino acid–base pair contacts (cf. Refs. 2, 3, and 38). However, since new-specificity substitutions are rare and difficult to predict, in most instances it is not efficient to seek new-specificity substitutions in order to identify an amino acid–base pair contact.

Altered Methylation Protection Approach

Methylation protection experiments identify guanine N-7 and adenine N-3 atoms of the DNA site that exhibit reduced reactivities to dimethylsulfate on formation of the protein–DNA complex. Methylation protection conventionally is interpreted as indicating that amino acids of the protein are close to, or in contact with, the protected guanine N-7 or adenine N-3 atoms. Pabo et al.[47] and Knight and Sauer[48] have shown that deletion of, or substitution of, amino acids of the DNA-binding protein can, at least in certain cases, produce position-specific alterations in the methylation protection pattern. Such position-specific alterations in the methylation protection pattern can permit the identification of amino acid–base pair contacts.

Pabo et al.[47] have used the altered methylation protection approach to identify amino acid–base pair contacts in the λ repressor–DNA complex. Wild-type λ repressor protects positions 4, 6, 7, 8, and 9 of the consensus DNA site; in contrast, a derivative of λ repressor, lacking amino acids 1 to 3, protects only positions 4, 6, and 7 of the consensus DNA site. These results suggest that amino acids 1 to 3 of λ repressor are near to, or in contact with, positions 8 and 9 of the consensus DNA site in the λ repressor–DNA complex. This suggestion has been confirmed in two ways: (1) using an approach related to the loss-of-contact approach,[49] and (2) by determination of the three-dimensional structure of the specific λ repressor–DNA complex.[5] Knight and Sauer[48] have used the altered methylation protection approach to identify amino acid–base pair contacts in the Mnt repressor–DNA complex. Wild-type Mnt repressor protects positions 4, 5, 7, 8, 10, and 11 of the consensus DNA site; in contrast, a derivative of Mnt repressor having the Arg → Lys substitution at amino acid 2 protects only positions 4, 5, 7, and 8 of the DNA site. These results suggest that Arg-2 of Mnt repressor is near to, or in contact with, positions 10 and 11 of the consensus DNA site in the Mnt repressor–DNA complex.

In the two examples cited, the altered methylation protection approach

[47] C. Pabo, W. Krovatin, A. Jeffrey, and R. Sauer, Nature (London) **298,** 441 (1982).
[48] K. Knight and R. Sauer, J. Biol. Chem. **264,** 13706 (1989).
[49] J. Eliason, M. Weiss, and M. Ptashne, Proc. Natl. Acad. Sci. U.S.A. **82,** 2339 (1985).

has been applied to an extended N-terminal "arm" segment.[47,48] The success of the approach may have been a function of the relative flexibility[50] of the arm segment. It is not known whether an amino acid substitution not in an arm segment would be sufficient to permit access of dimethyl sulfate to a previously protected guanine N-7 or adenine N-3 atom.

Caveats

Protein Conformation

When an amino acid substitution is constructed in a protein, the substitution can result in changes in protein conformation. This complicates the interpretation of results. It is necessary to distinguish between results that are due to first-order effects (i.e., the removal and/or addition of amino acid side-chain atoms), and results that are due to second-order effects (i.e., the alteration of protein conformation).

In light of these considerations, to identify an amino acid–base pair contact using the approaches discussed in this chapter, the evidence must meet the following standards:

1. The effects must have large magnitudes. To demonstrate the retention of specificity at a position within the DNA site, or to demonstrate a new specificity at a position within the DNA site, the ratio $K_{D,1}/K_{D,2}$ must be ≥ 5 (≈ 1 kcal/mol). In the case of an inverted-repeat or direct-repeat DNA site altered in each repeat, the ratio $K_{D,1}/K_{D,2}$ must be ≥ 20 (≈ 2 kcal/mol).

2. The effects must be position specific. Substitution of the amino acid under study must affect specificity or methylation protection only at one position, or at two adjacent positions, within the DNA site.

3. At least two different substitutions of the amino acid under study must result in the loss-of-contact phenotype, the new-specificity phenotype, or the altered methylation protection phenotype.

4. The inferred amino acid–base pair contact must be consistent with the structures, if known, of the protein and of the specific protein–DNA complex.

Protein Concentration: In Vivo Assays

When the wild-type DNA-binding protein and a substituted DNA-binding protein are expressed *in vivo* under identical conditions, the intracellular concentration of the unbound wild-type DNA-binding protein can

[50] M. Weiss, R. Sauer, D. Patel, and M. Karplus, *Biochemistry* **23,** 5090 (1984).

be different from the intracellular concentration of the unbound substituted DNA-binding protein. This is due to several factors: i.e., differences in rates of synthesis, differences in stability, and differences in nonspecific DNA binding affinity.

The section Experimental Methods (see above) describes quantitative *in vivo* assays in which a reporter promoter is placed under the control of the DNA site of interest. In interpreting the results of such assays, one must not compare the levels of expression of the reporter promoter for two different DNA-binding proteins (e.g., for the wild-type DNA-binding protein vs a substituted DNA-binding protein). This comparison is not controlled for differences in the *in vivo* concentrations of the unbound DNA-binding proteins.

It is imperative to restrict all comparisons and calculations to data for a single DNA-binding protein. Failure to observe this requirement has resulted in the misinterpretation of results.[51–54]

Acknowledgments

I thank Dr. Jon Beckwith for helpful discussions. Preparation of this chapter was supported by National Institutes of Health Grant GM41376.

---

[51] In Refs. 13 and 14, the Val-181, Leu-181, and Lys-181 substitutions of CAP were misidentified as new-specificity substitutions. In fact, these are loss-of-contact substitutions.[1]

[52] In Ref. 40, the Phe-79, Thr-80, and Ser-81 substitutions of Trp repressor are misidentified as new-specificity substitutions. In fact, these are neither new-specificity nor loss-of-contact substitutions; these substitutions result in essentially unchanged, wild-type specificity (Table I of Ref. 40).

[53] J. Sartorius, N. Lehming, B. Kisters, B. von Wilcken-Bergmann, and B. Müller-Hill, *EMBO J.* **8**, 1265 (1989).

[54] In Ref. 53, the [Ser-17;Ala-18], [Ala-17;Ala-18], [Ser-17;Thr-18], [Pro-17;Thr-18], [Ile-17;Thr-18], [His-17;Val-18], and [His-17;Ile-18] substitutions are misidentified as new-specificity substitutions. In fact, these are loss-of-contact substitutions (Table 1 of Ref. 53).

[31] Second-Site Reversion as Means of Enhancing DNA-Binding Affinity

By DALE L. OXENDER and AMY L. GIBSON

Introduction

As a first approach to the study of protein structure, random mutagenesis techniques are useful for identifying key residues and domains of the protein that play important roles in structure or function.[1,2] Second-site reversion studies of key mutations can be a powerful genetic tool to identify additional important residues and to elucidate complex interactions between specific amino acid residues important for the structure and function of a protein.[3–6] A specific application for the selection of second-site revertants illustrated in this chapter is the production of an altered DNA-binding protein that shows increased binding activity.

Studies by Yanofsky and colleagues demonstrated the range of information available from reversion analysis in classic genetic studies of the α subunit of tryptophan synthase from *Escherichia coli*.[7,8] Using random mutagenesis techniques, the investigators found that 8 of the 268 residues of the enzyme were essential for function. A missense mutation resulting in replacement of one of these eight amino acid residues inactivated the enzyme. Subsequent mutagenesis and selection for revertants showed that four of these positions could be mutated to one or, in some cases, several new amino acid residues, which restored significant activity. These results are summarized in Fig. 1. For example, replacement of Thr-183 by isoleucine resulted in reduced enzymatic activity. Activity was restored by the substitution of either aspartic acid or serine at position 183. At position 211, seven different primary mutations were reversed by one of eight substitutions.

Table I summarizes the effects on α-tryptophan synthase activity of a variety of reversions, second-site reversions, and double mutations produced by molecular splicing methods. An interesting set of results involves

[1] C. Auerbach and J. M. Robson, *Nature (London)* **157,** 302 (1946).
[2] R. L. Kelley and C. Yanofsky, *Proc. Natl. Acad. Sci. U.S.A.* **82,** 483 (1985).
[3] D. R. Helinski and C. Yanofsky, *J. Biol. Chem.* **238,** 1043 (1963).
[4] M. H. Hecht and R. T. Sauer, *J. Mol. Biol.* **186,** 53 (1985).
[5] H. C. M. Nelson and R. T. Sauer, *Cell (Cambridge, Mass.)* **42,** 549 (1985).
[6] L. S. Klig, D. L. Oxender, and C. Yanofsky, *Genetics* **120,** 651 (1988).
[7] E. J. Murgola and C. Yanofsky, *J. Mol. Biol.* **86,** 775 (1974).
[8] C. Yanofsky, *JAMA* **218,** 1026 (1971).

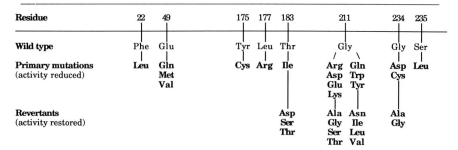

Residue	22	49		175	177	183		211		234	235
Wild type	Phe	Glu		Tyr	Leu	Thr		Gly		Gly	Ser
Primary mutations (activity reduced)	Leu	Gln Met Val		Cys	Arg	Ile		Arg Asp Glu Lys	Gln Trp Tyr	Asp Cys	Leu
Revertants (activity restored)						Asp Ser Thr		Ala Gly Ser Thr	Asn Ile Leu Val	Ala Gly	

Fig. 1. Summary of mutations in the α subunit of tryptophan synthase. Amino acid positions are given, with the original residue, primary mutations, and revertants listed. [Data are from E. J. Murgola and C. Yanofsky, *J. Mol. Biol.* **86**, 775 (1974).]

two missense mutations: substitution of glutamic acid for Gly-211 and replacement of Tyr-175 with cysteine. Either mutation alone results in inactive enzyme, but a double mutant with both changes is active. Yanofsky suggested that these two residues must be close together in the folded protein.

The three-dimensional structure of the $\alpha_2\beta_2$-tryptophan synthase complex with a substrate analog bound in the α subunit active site was deter-

TABLE I

TRYPTOPHAN SYNTHASE α SUBUNIT: REVERSION AND SECOND-SITE REVERSION[a]

Gene source	Amino acid sequence				Activity
	175	177	211	213	
Wild type	-Tyr-	Leu-	Leu-...-	Gly-Phe-Gly-	+
Primary mutation	-Tyr-	Leu-	Leu-...-	Glu-Phe-Gly-	−
Phenotypic revertant	-Tyr-	Leu-	Leu-...-	V̄al-Phe-Gly-	+
Double mutant	-Tyr-	Leu-	Leu-...-	V̄al-Phe-Val-	−
Phenotypic revertant	-Tyr-	Leu-	Leu-...-	Val-Phe-Āla-	+
Second-site revertant	-Tyr-	Leu-	Leu-...-	Gly-Phe-V̄al-	+
Second-site revertant	-Tyr-	Leu-	Leu-...-	Āla-Phe-Val-	+
Second-site revertant	-Tyr-	Leu-	Ārg-...-	Val-Phe-Val-	+
Recombinant	-Tyr-	Leu-	Ārg-...-	Gly-Phe-Gly-	−
Second-site revertant	-C̄ys-	Leu-	Leu-...-	Glu-Phe-Gly-	+
Recombinant	-C̄ys-	Leu-	Leu-...-	Gly-Phe-Gly-	−

[a] Results of sequential mutagenesis of tryptophan synthase α subunit. The wild-type amino acid sequence at positions 175 to 177 and 211 to 213 are shown. Sequential amino acid replacements are underlined. Effects on enzymatic activity are given by + and −. Two mutations obtained by second-site reversion were placed in an otherwise wild-type protein by gene splicing methods. Data are from C. Yanofsky, *JAMA* **218**, 1026 (1971).

mined.[9] The structural model for the enzyme confirms predictions made by Yanofsky based on results of the reversion studies. Based on the crystallographic results, at least one of the eight key residues identified in Fig. 1 (Glu-49) is directly involved in catalysis. Six of the remaining seven (absenting Thr-183) participate in the substrate-binding site. As predicted from the study cited in Table I, Tyr-175 and Gly-211 lie in close spatial proximity in the substrate-binding site. Computer graphics model building[10] suggests that either cysteine at position 175 or glutamic acid at position 211 would disrupt the geometry of the binding site, but that the paired changes described in Yanofsky's studies could restore the proper geometry. The close correspondence of results obtained by the two methods illustrates the power of fine-structure genetic analysis involving second-site reversion.

Tryptophan Repressor

Reversion analysis of the tryptophan repressor, TrpR, demonstrated that the method can also be used to obtain second-site mutations that enhance DNA binding by the repressor.[6] This result may prove to be broadly applicable to many DNA-binding proteins and could provide an additional degree of freedom in applications involving "engineered" regulation of gene expression. Second-site reversion provides an explicit search strategy for finding amino acid substitutions that (introduced singly into wild-type proteins) cause increased DNA binding.

Expression of the tryptophan biosynthetic genes of *E. coli* is regulated by binding of a tryptophan-activated (K_D 10^{-5} M[11,12]) specific repressor at the promoter/operator region of the operon, preventing initiation of transcription.[13,14] The repressor is encoded by the *trpR* gene and is subject to autoregulation. TrpR also regulates expression of a third operon, *aroH*, involved in biosynthesis of aromatic amino acids.

The repressor has been purified[15] and crystallographic studies have

[9] C. C. Hyde, S. A. Ahmed, E. A. Padlan, E. W. Miles, and D. R. Davies, *J. Biol. Chem.* **263,** 17857 (1988).
[10] S. Nagata, C. C. Hyde, and E. W. Miles, *J. Biol. Chem.* **264,** 6288 (1989).
[11] D. N. Arvidson, C. Bruce, and R. P. Gunsalus, *J. Biol. Chem.* **261,** 238 (1986).
[12] R. Q. Marmorstein, A. Joachimiak, M. Sprinzl, and P. B. Sigler, *J. Biol. Chem.* **262,** 4922 (1987).
[13] J. K. Rose, C. L. Squires, C. Yanofsky, H.-L. Yang, and G. Zubay, *Nature (London) New Biol.* **245,** 133 (1973).
[14] C. L. Squires, F. D. Lee, and C. Yanofsky, *J. Mol. Biol.* **92,** 93 (1975).
[15] A. Joachimiak, R. L. Kelley, R. P. Gunsalus, C. Yanofsky, and P. B. Sigler, *Proc. Natl. Acad. Sci. U.S.A.* **80,** 668 (1983).

revealed the structures of the inactive aporepressor,[16] the active holore-pressor,[17] and a repressor/operator complex containing an 18-mer syn-thetic oligonucleotide duplex.[18] The current high-resolution model for tryp-tophan repressor shows a dimer of unusual architecture. The majority (72%) of the residues participate in one of six helices (designated A–F). There is a solid "central core" formed by the interlocked amino-terminal halves of the two subunits (helices A, B, and C). The central core is symmetrically flanked by two flexible "DNA-reading heads," each formed from the carboxyl-terminal half of one monomer. The D and E helices (residues 68–92) of each monomer form a classic bihelical DNA-binding feature, related by sequence homology to portions of the repressors from λ and coliphage 434, and catabolite repressor protein and P22 Cro protein.[2]

Surface pockets that bind corepressor are wedged between the solid core and the flexible DNA-reading heads. Side chains of two arginine residues at positions 54 and 84 form a sandwich around the indole moiety of the corepressor, tryptophan. The hydroxyl of Thr-44 and the guanidino group of Arg-84 interact with the carboxyl group of the ligand.

An unusually large DNA contact surface was identified and operator recognition appears to rely on sequence-determined DNA backbone con-formation. TrpR functional regions were mapped by random chemical mutagenesis methods.[2] Mutations were distributed throughout the *trpR* gene, although a cluster of mutations was identified in the helix–turn–helix region of the protein. Primary mutations chosen for second-site reversion analysis were drawn from both classes.

Methods

Choice of Primary Mutations and Mutagens

The pool of TrpR mutants included proteins with altered tryptophan-binding sites (TM44), altered DNA-binding features (GS78) or single muta-tions affecting both regions (TM81, RH84, and GR85).[19] These primary

[16] R. G. Zhang, A. Joachimiak, C. L. Lawson, R. W. Schevitz, Z. Otwinowski, and P. B. Sigler, *Nature* (*London*) **327**, 591 (1987).

[17] R. W. Schevitz, Z. Otwinowski, A. Joachimiak, C. L. Lawson, and P. B. Sigler, *Nature* (*London*) **317**, 782 (1985).

[18] Z. Otwinowski, R. W. Schevitz, R.-G. Zhang, C. L. Lawson, A. Joachimiak, R. Q. Marmorstein, B. F. Luisi, and P. B. Sigler, *Nature* (*London*) **335**, 321 (1988).

[19] Specific missense mutations are designated by the one-letter amino acid codes of the original residue, followed by the replacement residue, and the number of the amino acid that has been changed. T, Threonine; M, methionine; G, glycine; S, serine; R, arginine; H, histidine; E, glutamic acid; K, lysine; P, proline; L, leucine; D, aspartic acid; N, asparagine; A, alanine; V, valine.

mutations were produced by *in vitro* exposure to hydroxylamine, which causes primarily C-T transitions. The same unidirectional mutagen was chosen for obtaining second-site revertants to minimize the appearance of true revertants. A second advantage of the use of hydroxylamine is that it is possible to obtain a level of mutagenesis that approaches one mutation/10^3 bases. Second-site mutation candidate genes should always be completely sequenced to verify that only one new change has been incorporated.

Hydroxylamine Mutagenesis

Mutagenesis was performed on plasmids derived from pRLK13.[2] Plasmid DNA was purified by CsCl banding and mutagenized by exposure to hydroxylamine according to the protocol of Davis *et al.*[20] A solution of 1 *M* hydroxylamine (pH 6.0) was made immediately before use by mixing 0.56 ml 4 *M* NaOH and 0.35 g NH_2OH in sterile water to a final volume of 5.0 ml. Four volumes of hydroxylamine solution was added to 1 vol of DNA (0.8 *M* final hydroxylamine concentration) and the mixture was incubated for 36 hr at 37°. The mutagen was removed by overnight dialysis against TE buffer (10 m*M* Tris-HCl, pH 7.4, 1 m*M* EDTA). DNA was ethanol precipitated and resuspended in TE buffer.

Isolation of Revertants

Mutagenized plasmid DNA was introduced into *E. coli* strain CY15075, a λ lysogen carrying a *lacZ* gene fusion producing β-galactosidase under control of the *trp* promoter/operator region.[21] The strain has a frameshift mutation in the chromosomal *trpR* gene and produces no endogenous repressor. Plasmid-containing colonies were selected for growth on chloramphenicol-containing plates. Cells were replicated to media with and without tryptophan to compare repressor activation. Inclusion of X-Gal permitted screening by color for colonies with reduced β-galactosidase activity. Up to 100,000 transformants were screened from each starting plasmid.

Characterization of Second-Site Mutations

DNA Sequencing. Plasmid DNA was isolated from second-site mutation candidates and the *trpR* genes subcloned to M13 replicative-form DNA for Sanger sequencing.[22]

[20] R. W. Davis, D. Botstein, and J. R. Roth, eds., "Advances in Bacterial Genetics, a Manual for Genetic Engineering," p. 94. Cold Spring Harbor Laboratory, Cold Spring Harbor, New York, 1980.

[21] C. Yanofsky and V. Horn, *J. Bacteriol.* **145,** 1334 (1981).

[22] F. Sanger, S. Nicklen, and A. R. Coulson, *Proc. Natl. Acad. Sci. U.S.A.* **74,** 5463 (1977).

Repressor Activity Measured by β-Galactosidase Assay. The relative activities of mutant and wild-type repressor molecules were assayed *in vivo* by determining β-galactosidase levels[23] in the expression indicator strain with and without added tryptophan. Overnight cultures of strains bearing double-mutant plasmids were diluted with 5 ml minimal medium containing 1 mM isopropyl-β-D-thiogalactoside (IPTG), and subcultured at 37° to OD_{600} of 0.3 to 0.7. Cultures were chilled on ice and 10 to 100 μl was mixed with 0.8 ml Z buffer (see below). Cells were permeabilized by adding 2 drops chloroform and 1 drop 0.1% (w/v) sodium dodecyl sulfate (SDS), then vortexed 15 sec. Extracts were warmed to 30° in a water bath and 200 μl substrate [o-nitrophenyl-β-D-galactoside (β-ONPG)] was added. The reaction was timed, stopped by adding 0.5 ml 1 M Na$_2$CO$_3$, and left on ice. Tubes were centrifuged 5 min to remove cell debris and the OD_{420} recorded. Activity was calculated according to the formula:

$$\text{Units} = (1000)(OD_{420})/(\text{time in minutes})(\text{culture volume})(OD_{600})$$

Z-buffer salts: Na$_2$HPO$_4$ · 7H$_2$O (8.0 g), NaH$_2$PO$_4$ · H$_2$O (2.8 g), KCl (0.38 g), MgSO$_4$ · 7H$_2$O (0.13 g)

Salts were dissolved in water to a volume of 480 ml and the pH was adjusted to 7.0; 13.5 ml 2-mercaptoethanol was added and the final volume adjusted to 500 ml.

Results

Activity levels of the wild-type repressor and single and double mutants are shown in Table II. As expected, compensatory mutations increased the activity of repressors with primary mutations. In the case of TrpR, no double mutant had activity exceeding that of wild-type repressor or of the corresponding superrepressor mutation alone. Second-site reversions that restored activity to four of the five primary mutations are summarized in Table III. A single compensatory mutation was found for TM44 and GS78. Three different second-site reversions were found for both RH84 and GR85. Most of these second-site mutations were represented by several independent isolates. Three additional double mutants, including one that restored activity to TM81, were constructed by subcloning (see Discussion, below). The majority of second-site reversions (18 of 27 isolates) consisted of previously known superrepressor mutations. The remaining nine mutations involved changes at positions 45 or 46. Many of the same second-site changes were obtained regardless of which primary mutation

[23] T. Platt, B. Müller-Hill, and J. Miller, *in* "Experiments in Molecular Genetics" (J. Miller, ed.), p. 352. Cold Spring Harbor Laboratory, Cold Spring Harbor, New York, 1972.

TABLE II

REPRESSOR ACTIVITY OF *trpR* MUTANTS WITH SECOND-SITE CHANGES AND
SUPERREPRESSOR MUTANTS[a]

Mutant plasmid[b]	Second-site change	Units of β-galactosidase activity[c]		Superrepressor plasmid[b]	Units of β-galactosidase activity[c]	
		− Trp	+ Trp		− Trp	+ Trp
TM44		9,000	5,900	(Wild type[b])	500	10
TM44	EK18	3,700	1,600	No plasmid[d]	12,000	12,000
TM44	EK49	2,100	400	EK18	179	4
GS78		6,000	2,300	EK49	68	4
GS78	EK18	3,500	500	DN46	167	12
TM81		8,300	4,000	AV77	188	7
TM81	EK18	7,800	2,400			
RH84		10,000	5,300			
RH84	EK18	7,500	1,200			
RH84	PL45	7,100	1,000			
RH84	DN46	5,300	400			
RH84	EK49	2,100	500			
GR85		600	2,000			
GR85	EK18	1,800	500			
GR85	EK49	400	400			
GR85	AV77		200			

[a] Repressor activity of TrpR proteins containing single and double mutations are compared to activity of wild-type TrpR. Units of β-galactosidase activity (with and without added tryptophan) are shown for mutants with impaired activity, second site revertants of those mutants, and TrpR superrepressors. Reprinted by permission of the Genetics Society of America.

[b] Each *trpR* gene was present in pACYC184, in the repression indicator strain, CY15075. This strain contains the *trpR2* allele, a frameshift allele with no repressor activity.

[c] Cultures were grown in the presence or absence of exogenous tryptophan and assayed for β-galactosidase activity.

[d] Recipient strain. When this strain contained the parental plasmid pACYC184, identical values were obtained.

was involved. No compensatory changes were observed at the codon responsible for the primary mutation.

Construction of Specific Double Mutations

One mutation (EK18), known to compensate for two distinct primary mutations (GS78 and GR85), was introduced by subcloning into plasmids containing distant mutations (TM44, TM81, and RH84). This procedure produced a new class of double mutants with moderate superrepressor

TABLE III
AMINO ACID CHANGES IN PRIMARY MUTANTS AND SECOND-SITE
REVERTANTS AND CONSTRUCTS[a]

Name	Primary mutation	Compensatory change	Number of isolates
TM44	Thr-44→Met		
TM44–EK18		Glu-18→Lys	C[b]
TM44–EK49		Glu-49→Lys	3
GS78	Gly-78→Ser		
GS78–EK18		Glu-18→Lys	2
TM81	Thr-81→Met		
TM81–EK18		Glu-18→Lys	C
RH84	Arg-84→His		
RH84–EK18		Glu-18→Lys	C
RH84–PL45		Pro-45→Leu	1
RH84–DN46		Asp-46→Asn	8
RH84–EK49		Glu-49→Lys	6
GR85	Gly-85→Arg		
GR85–EK18		Glu-18→Lys	3
GR85–EK49		Glu-49→Lys	2
GR85–AV77		Ala-77→Val	2

[a] Summary of amino acid changes in TrpR primary mutations used in this study, second-site reversions obtained, and selected double mutants constructed by gene splicing. Data are from L. Klig, D. L. Oxender, and C. Yanofsky, *Genetics* **120,** 651 (1988).
[b] Double mutant constructed as described in the text.

activity that would not have been detected by the colony-color screening method involving X-Gal plates.

Search for Additional Superrepressors

Superrepressor mutations were originally isolated by screening for the appearance of tryptophan auxotrophy in a tryptophan bradytroph strain transformed with mutagenized pRLK13 plasmid containing the wild-type *trpR* gene. When second-site mutations occurred at positions 45 and 46, a second search for superrepressors was carried out. One of the two new second-site mutations (DN46, isolated as a compensatory mutation for RH84) was recovered, demonstrating that this mutation was also of the superrepressor type.

Discussion

Second-site reversion mutations that restored DNA-binding activity to TrpR were either clustered in the helix–turn–helix region (residues 45, 46, and 49) or scattered elsewhere in the linear amino acid sequence (residues

18 and 77) at positions near the DNA-binding region of the folded protein. The TrpR superrepressors obtained by second-site reversion were essentially equivalent in activity to the wild-type molecule in the presence of tryptophan, but were also active in the absence of corepressor. Three of these mutations involve the loss of a negatively charged residue (glutamic acid or aspartic acid). In the case of DN46, the charged residue is replaced with neutral asparagine. In two superrepressors (EK18 and EK49) glutamic acid was replaced by a positively charged lysine residue. This recurring theme of alterations in surface charges suggests that electrostatic repulsion plays a role in the suboptimal binding activity characteristic of wild-type TrpR. The most common superrepressors (EK18 and EK49) restored activity to a variety of mutant repressors, suggesting that this class of mutations (net charge change of four per dimer) act globally, rather than in a site-specific fashion. EK49 has the highest repressor activity in the absence of tryptophan seen to date. This protein has been purified and its activity has been examined in an *in vitro* transcription assay.[24] The superrepressor was shown to have a 10-fold higher affinity for operator DNA due to a decrease in its rate of dissociation from DNA. It is likely that mutations of the EK class may increase activation by providing more favorable electrostatic interactions between the repressor and operator DNA.

The effects of the PL45 compensatory mutation in an otherwise wild-type background have not been observed. This substitution does not affect the charged surface of the protein, but it has been suggested that loss of the proline at that position could result in recruitment of additional residues into helix C, bringing new portions of the solvent-exposed protein surface into contact with DNA.[6]

An alanine residue at position 77 is found in the turn portion of the bihelical DNA-binding motif. The side chain of the residue at position 77 must be small, like that of alanine, to permit a conformational change necessary to go from the active form (with bound tryptophan) to the inactive aporepressor. Substitution of valine at this position (AV77), with its more bulky side chain, keeps the *trp* repressor locked in its active form whether the corepressor is present or absent. Interestingly, several other DNA-binding proteins that employ the bihelical motiff have a valine at the position analogous to position 77 in the *trp* repressor. It is interesting that one of the second-site revertants found for GR85 was replacement of alanine with valine (AV77), even though GR85 is unable to bind tryptophan.

The classes of TrpR superrepressors described illustrate three possible molecular mechanisms for enhancing operator binding: (1) Changes in

[24] L. S. Klig and C. Yanofsky, *J. Biol. Chem.* **263**, 243 (1988).

surface electrostatic forces (from negative to neutral or positively charged: EK18, DN46, EK49) may act globally to slow dissociation from DNA; (2) changes in the size of the DNA contact surface (e.g., PL45) may alter the hydrophobic effect or intermolecular hydrogen bonding to increase repressor–operator affinity; (3) Steric constraints preventing conformational change (e.g., AV77) may circumvent allosteric control mechanisms.

Using a combination of second-site reversion and subcloning approaches, it was possible to successfully explore the repertoire of changes available to enhance DNA binding by TrpR, and to classify the effects of those changes based on the structure of the protein.

Generalization of Method

Direct genetic selection is the preferred method for detecting mutant proteins with enhanced DNA-binding affinity whenever possible, but for a variety of reasons direct selection may not be available for a given protein. The preferred strategy usually requires establishment of bacterial growth conditions where tighter DNA binding would prevent accumulation or utilization of a toxic metabolite, providing a selective advantage to cells containing a superrepressor-type mutation. An easy to use media-based direct selection strategy may not be available for a particular DNA-binding protein. Development of a direct selection method may also require extensive strain construction and gene splicing. Finally, a wild-type DNA-binding protein may already exhibit near-optimal binding affinity, making the task of detecting incremental improvements even by direct selection more difficult. Direct selection has been used successfully, however, to isolate TrpR superrepressor mutations.[2] Contributing to the success of the selection strategy were (1) comparatively low operator–repressor binding affinity. (2) the existence of an extensive collection of required background strains, and (3) existence of a substrate analog to permit superrepressor selection (5-methyl anthranilate, which can be converted to a toxic analog, 5-methyltryptophan, in cells in which the *trpEBCD* operon is expressed). TrpR superrepressors obtained using this strategy included EK18 and EK49.

In cases where direct genetic selection methods are difficult or not possible, the second-site reversion approach described above provides a technique to amplify the differential in binding affinities between wild-type and mutant repressors. Mutagenesis of a primary mutant with impaired DNA binding allows screening to detect restored function. When second-site mutations are identified, the specific amino acid changes can be introduced into otherwise wild-type proteins and their effects assayed. The approach efficiently permits the selection of mutations that cause global superrepressor activities.

Acknowledgments

The authors gratefully acknowledge the research contributions of Lisa S. Klig and Charles Yanofsky and the comments of Mark Adams. These studies were supported by grants from the National Science Foundation (DBM 8703685) and the American Heart Association (69-015). D.L.O. was an American Cancer Society Scholar at the time of these studies.

Author Index

Numbers in parentheses are footnote reference numbers and indicate that an author's work is referred to although the name is not cited in the text.

Bossu, F. P., 513
Botstein, D., 26, 41(12), 552, 645
Bottermann, J., 56
Bourgeois, S., 259, 297, 321, 335(99), 559, 633
Bouvier, S., 604, 634
Bowie, J. U., 64, 74(11), 96(11), 569, 573, 574(16), 604, 605, 609(4), 614, 615(18), 617(10), 618
Bowie, J., 75
Boxer, L. M., 13
Boyer, H. W., 57, 398, 416, 439, 443, 444(10, 14), 454, 455(10), 456(10, 55), 497
Boyer, H., 443
Bracco, L. P., 564
Bradley, D. L., 286
Bradley, D., 546
Brady, C., 604
Brand, A. H., 13
Bratzer, W. B., 40
Braun, W., 78
Braunlin, W. H., 311, 313(74)
Brayer, G. D., 539
Breg, J. N., 64, 74(10), 96(10)
Brennan, A., 436, 449(11)
Brennan, C. A., 437, 438, 439, 441(8), 444(8), 445(8), 446(8), 455(8), 456(8), 457(8)
Brennan, R. G., 89, 95, 96(28), 427
Brenner, S., 552
Brenowitz, M., 106, 109, 110(8), 111, 115(8, 11), 245, 259, 284, 297, 303(20), 384, 390(12), 392(12), 393(12)
Brent, R., 575, 580(32), 581(32), 635
Breyer, R. M., 574
Briggs, M. R., 12
Brimacombe, R., 531, 533(12, 13, 14)
Broker, T. R., 171
Broseta, D., 246
Brosius, J., 610, 611(13)
Brown, A. M., 483, 486
Brown, D. D., 384
Brown, E. G., 51
Brown, R. D., 475, 478
Brown, R. S., 46, 142
Brown, W. E., 454
Bruce, C., 427, 643
Bruice, T. W., 426
Bruist, M. F., 498, 499(13), 501(13), 507(13), 508(13)

Brunelle, A., 405, 411(38), 628, 629, 633(21, 24), 634(24)
Brunner, M., 236, 237(2)
Buc, H., 169, 195(10), 237, 238(4), 239, 246, 247, 248(25, 30), 249(25), 297, 299(24), 321, 328, 414, 417, 620, 625(1), 630(1), 632(1)
Buchanan, R. L., 147, 149(11), 150(11)
Buchman, A. R., 13
Buchman, C., 401
Buchman, 403
Buck, F., 291
Buck, M., 147
Buck, R., 65, 81(23)
Buckle, M., 239, 246, 247(25), 248(25), 249(25)
Bueron, M., 246
Bujalowski, W., 260, 261(21, 24, 25, 29), 265(24, 25), 266(24), 267(24, 25), 270(21), 272(21), 273(21), 274, 275(21), 277(22), 278(21), 279(21, 22), 280(21), 281(21), 282(21, 23, 29, 40), 283(21), 285, 286, 310, 314
Bujard, H., 171, 236, 237(2)
Bull, P., 475, 476
Bullock, P., 169
Burgess, R. R., 3, 6(7), 7(7), 8(7), 14, 15, 216, 242, 259, 292, 297, 298(26), 299, 302(45), 314, 323(45), 328(44), 329(45), 332(27), 333(45), 341(44, 46), 461
Burgum, A. A., 473, 475, 476, 481, 483, 493
Burke, R. L., 24, 37(6), 42(6), 216
Burkhardt, H. J., 171
Burton, Z. F., 9, 24, 26(4), 39
Buryanov, Y. I., 437, 441, 450(28), 453(28)
Busby, S., 417
Bushman, F. D., 572, 574(21)

C

Cai, M., 13
Calame, K., 13
Camerman, A., 505, 513(20)
Camerman, N., 505, 513(20)
Campbell, J. L., 13
Cann, J. R., 104, 111(6)
Cao, Z., 13
Caras, I. W., 211, 212(14)
Carbone, J., 211, 223(22)

H

Subject Index

addition of nonphosphorylated linkers
    to, 593–594
addition of phosphorylated linkers to,
    592–593
Drosophila melanogaster
calmodulin, 1D ^{1}H NMR spectrum of,
    70–71
DTF-1, purification by DNA affinity
    chromatography, 12
GAGA, purification by DNA affinity
    chromatography, 12
transposase, purification by DNA affin-
    ity chromatography, 12
Zeste, purification by DNA affinity
    chromatography, 12
DTNB. *See* 5,5′ Dithiobis(2-nitrobenzoic
    acid)
Duplexes, formation of, 94

E

*Eco*RI, hydroxyl radical footprinting, with
    gel isolation of *Eco*RI–DNA complex,
    397–401
*Eco*RI–DNA complex
cocrystallization, cofactors for, 98
crystals, 97
gel isolation of, in hydroxyl radical
    footprinting of technique, 397–401
X-ray crystal structure of, 398–399
*Eco*RI endonuclease
properties of, 454
site-specific binding, quantitative equilib-
    rium studies, 297
studies with substrate analogs, 454–456
EDTA, tricyclohexyl ester of. *See* TCE
Electron microscopy
applications of, 189
carbon coating of grids without Formvar
    supports, 175–176
Formvar-coated grids, carbon coating of,
    174–175
Formvar coating of grids for, 172–174
of initiation of DNA replication by
    bacteriophage λ, 184–185, 187–194
nucleoprotein complexes for
adsorption to grids, 176–179
Alcian Blue dye technique, 177, 179
polylysine technique, 176–179
glutaraldehyde fixation, 181

preparation of, 171–172, 180–181
removal of excess unbound protein
    from, by gel filtration, 172, 181
drip column gel filtration, 182–183
preparation of Sepharose 4B micro-
    columns for, 181
spin column gel filtration, 183
rotary shadowing, 179–180
of nucleoprotein structures, 171–195
examination of nucleoproteins in
    electron microscope and analysis,
    185–188
of protein–DNA complexes, 168–195
applications of, 168–169
of transcriptional activation of *glnA*
    promoter from NtrC enhancer site,
    194–195
engrailed homeodomain–DNA complex,
    crystallization, 96–97, 316
Escherichia coli
CAP protein. *See* CAP protein
chromosomal DNA, primer extension
    probing, 155–157
DNA-binding proteins, screens and
    selections of, 604
DNA polymerase I
lysine residues, chemical modification
    of, 482, 484
preparation of, 216
NusA–RNA polymerase interactions,
    31, 37
NusB–ribosomal protein S10 interac-
    tions, 31
NusG–ρ factor interactions, 31
promoter sequences, information content
    of, 458–459
Rec A protein, cysteine residues, chemi-
    cal modification of, 474, 477
release factors, 544
ρ protein, preparation of, 216
ρ protein–oligonucleotide system, laser
    cross-linking, 219–221
nucleic acid-binding protein site size
    determinations, 230
ribosomal protein S10–NusB interac-
    tions, 31
RNA polymerase
5-phenyl-1,10-phenanthroline foot-
    printing of, 417–419
preparation of, 216

N

Namalwa cells, NF-κB, purification by DNA affinity chromatography, 12

NBS. *See* N-Bromosuccinimide

Nitrogen-15, protein labeling with, for NMR studies, 72–73

NOESY, 68–69
two-dimensional, information obtained from, 63–64

Nonsense codons, 544

Nonsense suppression
advantages and disadvantages of, 548–552
in *E. coli*, context effects on, 464–466
in generation of altered proteins, 543–563
principle of, 544–545

Nonsense suppressor genes, synthetic, construction of, 545–551

Nonsense suppressors. *See also* tRNA suppressors
sources of, 552–553
types available, 552–553

NtrC enhancer site, *glnA* promoter from, electron microscopy of transcriptional activation of, 194–195

Nuclear factor I–DNA complex, hydroxyl radical footprinting of, 387–390

Nuclear magnetic resonance spectroscopy, 65–69, 103
on biological molecules
information obtained by, 67–68
type and number of nuclei observed, 67–68
^{13}C, 68
coherence transfer, 68
COSY (correlated spectroscopy) experiments, 68
of DNA-binding proteins
samples for, 69–81
amount required, 73–74
composition of, 77–79
deuteration, 72
isotopic labeling, 70–73
molecular weight, 69–71
purity of, 77–78
sources of proteins for, 74–77
studies of structural and/or functional domains, 70

dynamic range in, 78
of folded versus unfolded protein, 65–66
four-dimensional (4D), 67
^{1}H, 65–66, 68
HETCOR (heteronuclear correlated spectroscopy), 68
heteronuclear experiments, 68
HMBC (heteronuclear multiple bond correlation), 68
HMQC (heteronuclear multiple quantum correlation), 68
HOHAHA (homonuclear Hartman–Hahn spectroscopy), 68
homonuclear experiments, 68
J coupling, 68
multidimensional
of DNA-binding proteins, 63–82
information obtained from, 63–64
of protein–DNA complexes, 81–82
^{15}N, 68
one-dimensional (1D), 65–67
RELAY (relayed coherence transfer spectroscopy) experiments, 68
samples for
aggregation state of, 80
pH, 80–81
reconstitution of, 79
solubility, 79
stability of, 80–81
scalar coupling experiments, 68
sequence-specific assignments with, 68–69
in solution structure determination, 67, 69
spectral simplification, with deuterated analogs of proteins, 72
three-dimensional (3D), 67
through bond experiments, 68–69
through space experiments, 68
TOCSY (total coherence spectroscopy) experiments, 68
two-dimensional (2D), 67, 620

Nuclear Overhauser effect spectroscopy. *See* NOESY

Nuclease–DNA complex, cocrystallization, cofactors for, 98

Nucleic acid binding proteins
metal content of, determination of, by metal release, 49
metal requirements, 46–54

Y

Yeast
DNA polymerase I, affinity chromatography with, 44
RNA polymerase I, cysteine residues, chemical modification of, 474
RNA polymerase II–yRAP37 interactions, 31
TFIID–human TFIIA interactions, 31

Z

Z-DNA
carbethoxylation of, modification-protection analysis, 372–374
modification protection and interference studies with, 366–368
interpretation of results, 378–379
Z-DNA–antibody, modification protection

and interference studies with, 366–368, 378–379
Zif 268–DNA complex, crystals, 97
Zinc-binding domains, synthetic, stability of, 76
Zinc blotting, in determination of metal-binding ability of protein, 51
Zinc-containing nucleic acid-binding proteins
metal-binding ability, determination of, by cobalt substitution, 50–51
metal content of, determination of
by atomic absorption mass spectrometry, 47–49
by metal release, 49
Zinc finger domains
aggregation state of, and NMR studies, 80
2D NMR studies of, 64
Zinc fingers, TFIIIA-like, 63–64

Arthur N. Popper
Richard R. Fay
Editors

Hearing by Bats

With 138 Illustrations

Springer-Verlag
New York Berlin Heidelberg London Paris
Tokyo Hong Kong Barcelona Budapest

Arthur N. Popper
Department of Zoology
University of Maryland
College Park, MD 20742, USA

Richard R. Fay
Parmly Hearing Institute and
Department of Psychology
Loyola University of Chicago
Chicago, IL 60626, USA

Series Editors: Richard R. Fay and Arthur N. Popper

Cover illustration: A big brown bat, *Eptesicus fuscus,* is about to capture a mealworm hanging on a fine filament. Photograph by S. P. Dear and P. A. Saillant. This figure appears on p. 148 of the text.

Library of Congress Cataloging in Publication Data
Hearing by bats / Arthur N. Popper, Richard R. Fay, editors.
     p.   cm. — (Springer handbook of auditory research : v. 5)
     Includes bibliographical rcferences and index.
     ISBN 0-387-97844-5.
     1. Bats—Physiology.   2. Echolocation (Physiology)   3. Hearing—
Behavior.   I. Popper, Arthur N.   II. Fay, Richard R.   III. Series.
QL737.C5H435   1995
599.4'041825—dc20   94-41860
                                                         CIP

Printed on acid-free paper.

© 1995 Springer-Verlag New York, Inc.

Production managed by Terry Kornak; manufacturing supervised by Jeffrey Taub.
Typeset by TechType Inc., Upper Saddle River, NJ.
Printed and bound by Braun-Brumfield, Ann Arbor, MI.
Printed in the United States of America.

9 8 7 6 5 4 3 2 1

ISBN 0-387-97844-5 Springer-Verlag New York Berlin Heidelberg

Series Preface

The *Springer Handbook of Auditory Research* presents a series of comprehensive and synthetic reviews of the fundamental topics in modern auditory research. It is aimed at all individuals with interests in hearing research including advanced graduate students, postdoctoral researchers, and clinical investigators. The volumes will introduce new investigators to important aspects of hearing science and will help established investigators to better understand the fundamental theories and data in fields of hearing that they may not normally follow closely.

Each volume is intended to present a particular topic comprehensively, and each chapter will serve as a synthetic overview and guide to the literature. As such, the chapters present neither exhaustive data reviews nor original research that has not yet appeared in peer-reviewed journals. The series focuses on topics that have developed a solid data and conceptual foundation rather than on those for which a literature is only beginning to develop. New research areas will be covered on a timely basis in the series as they begin to mature.

Each volume in the series consists of five to eight substantial chapters on a particular topic. In some cases, the topics will be ones of traditional interest for which there is a solid body of data and theory, such as auditory neuroanatomy (Vol. 1) and neurophysiology (Vol. 2). Other volumes in the series will deal with topics which have begun to mature more recently, such as development, plasticity, and computational models of neural processing. In many cases, the series editors will be joined by a co-editor having special expertise in the topic of the volume.

Richard R. Fay
Arthur N. Popper

Preface

Of all vertebrate groups, the echolocating mammals have probably provided the most intrigue to auditory researchers. The biological significance of echolocation sounds is obvious, yet the sounds are generally inaudible to humans and are used in ways that seem alien to our auditory experience.

Of the echolocators, bats are the most widely and intensively studied, and are the best understood. In fact, we probably know and understand more about the acoustic neuroethology of bats than of any other vertebrate group.

Our decision to include a volume on bats in the *Springer Handbook of Auditory Research* was motivated by several considerations. First, to our knowledge, there has been no modern volume that addresses hearing and the neuroethology of echolocation in bats in a systematic and comprehensive way. Thus, this volume provides detailed insights into the current state of our knowledge on these fascinating species, as well as a guide to areas for future research. Second, and as pointed out by George Pollak, Jeffery Winer, and William O'Neill in Chapter 10, bats are exceptionally interesting and useful model systems for the study of mammalian hearing in general, and as such, this volume provides investigators interested in other species with detailed insights into the auditory mechanisms and capabilities of this model system. Finally, as pointed out elegantly by Alan Grinnell in Chapter 1, our understanding of hearing and echolocation by bats is " . . . one of the triumphs of neuroethology . . . " and, as such, bats provide one of the few comprehensive "stories" about the uses of sound in the auditory literature.

Grinnell (Chapter 1) provides an overview of what is known about hearing and echolocation by bats, and provides a valuable perspective on the questions and controversies of greatest interest. In Chapter 2, Fenton discusses the natural history and evolution of echolocation and provides an introduction to the myriad species that are discussed in the rest of this volume. The behavioral capabilities of bats are discussed in Chapter 3 by Moss and Schnitzler, and discrimination is treated in detail by Simmons and his colleagues in Chapter 4. The structure and function of the auditory system is dealt with in detail in Chapters 5 through 9. In Chapter 5, Kössl and Vater describe the inner ear. The brainstem and lower auditory pathways are discussed by Covey and Casseday in Chapter 6, while Pollak and Park describe the inferior colliculus in Chapter 7. The thalamus is treated in Chapter 8 by Wenstrup and the auditory cortex in Chapter 9 by O'Neill. Finally, Pollak, Winer, and O'Neill give an overview of the mammalian auditory system and provide compelling arguments for the use of bats as a model system for mammalian hearing.

In inviting the authors to write chapters for this volume we asked them to not only consider the species upon which they work, but also to provide a comparative overview of the auditory system of the wide variety of bat species. Thus, the volume not only deals with bat hearing per se, but also provides a perspective that shows the wide range of variation in detection and processing mechanisms even within this one amazing group of mammals.

The editors would like to express their gratitude to George Pollak for his guidance, insight, and help in developing this volume. Dr. Pollak has a unique and invaluable perspective on bats, and he shared his thoughts with us freely and generously when we invited him to contribute chapters to this volume.

Arthur N. Popper
Richard R. Fay

Contents

Contributors

John H. Casseday
Department of Neurobiology, Duke University Medical Center, Durham, NC 27710, USA

Ellen Covey
Department of Neurobiology, Duke University Medical Center, Durham, NC 27710, USA

Steven P. Dear
Department of Neuroscience, Brown University, Providence, RI 02912, USA

M. Brock Fenton
Department of Biology, York University, North York, Ontario, Canada M3J 1P3

Michael J. Ferragamo
Department of Neuroscience, Brown University, Providence, RI 02912, USA

Alan D. Grinnell
Department of Physiology, UCLA School of Medicine, Jerry Lewis Neuromuscular Research Center, Los Angeles, CA 90024, USA

Tim Haresign
Department of Neuroscience, Brown University, Providence, RI 02912, USA

Manfred Kössl
Zoologisches Institut, Universität Munchen, D8000 Munchen 2, Germany

David N. Lee
University of Edinburgh, Edinburgh, Scotland

Cynthia F. Moss
Department of Psychology and Program in Neuroscience, Harvard University, Cambridge, MA 02138, USA

William E. O'Neill
Department of Physiology, University of Rochester School of Medicine and Dentistry, Rochester, NY 14642, USA

Thomas J. Park
Department of Zoology, The University of Texas at Austin, Austin, TX 78712, USA

George D. Pollak
Department of Zoology, The University of Texas at Austin, Austin, TX 78712, USA

Prestor A. Saillant
Department of Neuroscience, Brown University, Providence, RI 02912, USA

Hans-Ulrich Schnitzler
Department of Animal Physiology, University of Tübingen, Tübingen, Germany

James A. Simmons
Department of Neuroscience, Brown University, Providence, RI 02912, USA

Marianne Vater
Universität Regensburg, Fachbeireich Biologie, Zoologisches Institut, 93040 Regensburg, Germany

Jeffrey J. Wenstrup
Department of Neurobiology, Northeastern Ohio Universities College of Medicine, Rootstown, Ohio 44272, USA

Jeffery A. Winer
Department of Molecular and Cell Biology, Division of Neurobiology, University of California, Berkeley, CA 94720-2097, USA

Janine M. Wotton
Department of Neuroscience, Brown University, Providence, RI 02912, USA

1

Hearing in Bats: An Overview

Alan D. Grinnell

1. Introduction

As predominantly visual animals, we have great difficulty imagining how sound can be used to orient precisely in a complex natural environment. Indeed, most animals, even nocturnal animals, share our dependence on vision. Comparatively recently evolved, hearing has reached a high degree of sophistication in birds and mammals. No wonder, then, that bats, with their abilities to orient, find food, and lead active lives in the dark, have long fascinated humans. This volume summarizes the current understanding of hearing in bats — the behavioral skill with which bats obtain information about their environment and the adaptations of the mammalian auditory nervous system that make this possible.

Our understanding of how bats use hearing began in the last years of the eighteenth century, when the great Italian scientist, L. Spallanzani, noted that blinded bats could fly, avoid obstacles, land on walls and ceiling, and survive in nature as well as bats with sight. He and his Swiss counterpart, C. Jurine, established that hearing was the sense bats used in orientation (see Galambos 1942a; Griffin 1958; and Dijkgraff, 1946 for accounts of these experiments). However, it was not until 140 years later that technological advances enabled Donald R. Griffin, then an undergraduate at Harvard, to demonstrate that bats emitted trains of high-frequency sounds during flight and could use the echoes of these sounds to detect objects and orient in a complex environment. In the next few years, Griffin and his collaborators showed that bats use "echolocation" to do almost everything a bird such as a swift or flycatcher can do with vision (Griffin 1958; see also Dijkgraaf 1946 for accounts of independent experiments leading to similar conclusions). At the same time, Griffin's friend and colleague, Robert Galambos, recorded cochlear microphonics from bat ears to at least 98 kHz (Galambos 1942b).

From the beginning, the study of hearing in bats has been characterized by a synergistic partnership between behavioral and physiological approaches. The results have been nothing short of spectacular, as this

1

volume, written by leaders in the field, testifies. Echolocation, for all of its unanswered questions, is one of the triumphs of neuroethology—better understood, perhaps, than any other complex mammalian behavior, certainly any auditory behavior.

The reasons are straightforward. It is not that bats are intrinsically easier to work with or that their nervous systems are more simple. On the contrary, the size of bats, the difficulty of obtaining and maintaining them in captivity, the inaccessibility of their native habitats, and their nocturnal activity in a territory that covers square miles all combine to make the study of bats daunting. What has made bats such superb subjects for the study of acoustic behavior and neural processing is that they rely almost exclusively on hearing for the information they need about their environments, at least while flying. We can construct test situations in which we know the information they need as well as the signals they use to obtain that information. Moreover, these signals tend to be simple in structure and are employed in reproducible ways under similar conditions. Although bats use much more complex sounds to communicate with one another, and in many cases take advantage of prey-generated sounds, the overwhelming emphasis is on extracting information from echoes of their relatively simple and stereotyped echolocation signals. Moreover, echolocation sounds are consistent within a given species and differ in species-specific ways. Comparative studies show that these differences correlate with preferred habitats and hunting strategies. The differences also help establish the information-gathering value of different components of the echolocation sounds. Finally, the behavioral skills of bats are so remarkable, even unbelievable, that investigators find them fascinating subjects.

2. The Natural History of Echolocating Bats

Bats evolved early in the history of mammals (Jepsen 1970). Enlargements in fossil bat skulls from the Eocene reflect enlarged auditory neural structures. Furthermore, still older fossils of noctuid moths—a favorite prey of bats—demonstrate the existence of tympanic organs, suggesting that the animals that preyed on them (presumably bats) used echolocation. As Fenton points out in Chapter 2 echolocating bats have radiated widely into seemingly all conceivable niches except the polar cap. There are more than 800 species in the suborder Microchiroptera, all of which apparently can echolocate. (One of the puzzles of the bat world is the existence of another suborder of Chiroptera, the Megachiroptera, restricted to the Old World, all with excellent night vision and, with the exception of one genus, all lacking echolocation. That exception, *Rousettus*, has independently evolved a primitive form of echolocation using tongue clicks similar to those used by some cave-dwelling birds [Griffin 1958]). Little study has been devoted to hearing in nonecholocating bats, which probably are like most other small animals in their hearing capabilities (Grinnell and Hagiwara 1972b). This volume is devoted almost exclusively to echolocating bats and their orientation skills. In this we also reflect the field's general neglect of social communication sounds and their processing. A good introduction to that subject can be obtained from Fenton (1985).

Echolocating bats range in size from 2 g to more than a kilogram, with wing spans from 8 in. to more than 2 m. Different species feed exclusively or in various combinations on pollen, nectar, fruit, blood, on flying, crawling, or floating insects or other arthropods, and on frogs, fish, and small birds and mammals. All bats must be able to avoid obstacles such as tree branches, wires, buildings, cave walls, and other flying bats. Typically, they land by approaching a potential landing site and, at the last moment, flipping around and grabbing a surface irregularity with their hind claws. Many drink by swooping down over a body of water and dipping their lower jaw. These maneuvers require not only superb flight skills but a precise knowledge of where objects are located. In most cases, this information is obtained from echoes of emitted sounds. It is not that other cues are unavailable. Many bats probably have much better eyesight than is typically assumed, and some find their food by prey-generated sounds or olfaction. In most species, however, echolocation appears to be the necessary and adequate source of information.

2.1 Echolocation Signal Characteristics

2.1.1 FM Bats

As mentioned, bats have evolved a variety of echolocation sounds that can, to some degree, be correlated with different hunting strategies and mechanisms of information processing. Most families of bats use relatively short, frequency-modulated (FM) sounds that sweep through about an octave ("FM bats"). One or more harmonics increase the bandwidth of these signals. With few exceptions, the frequencies employed are ultrasonic (i.e., above the human range of hearing). Typically these sounds last several milliseconds and are emitted 2–10 times per second while a bat searches for prey. After the bat detects a target of interest, it increases the repetition rate and decreases the duration of its signals. As it nears the target, the bat produces a terminal "buzz," emitting pulses as rapidly as 100–200/sec (Fig. 1.1A). At this time, the intensity and the starting frequency of the sweeps typically decrease. Representative sonograms of a number of species are shown in Fenton, Chapter 2, Figure 2.2. Depending on the target size, a bat first detects an object at distances as close as a few centimeters to as distant as 3–5 m. With these echoes, the bats determine the direction and distance of targets and discriminate the nature of the target.

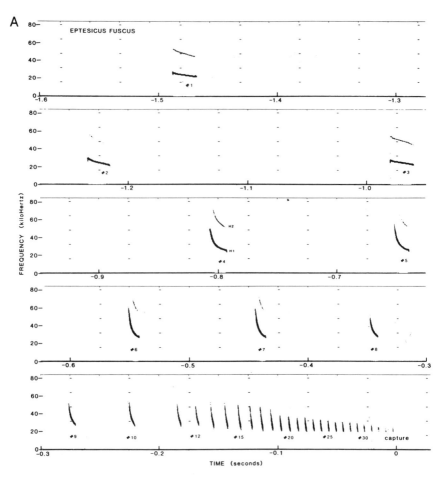

FIGURE 1.1. (A) Continuous spectrogram record of echolocation sounds emitted by a big brown bat, *Eptesicus fuscus*, during pursuit of an insect in the field. Time is shown with respect to the moment of capture. Note the very shallow sweep of the initial search pulses. Courtesy of J. Simmons based on data from O.W. Henson, Jr.

2.1.2 Long CF/FM Bats and Doppler Shift Compensation

A much smaller number of species belonging to three different families have evolved a different type of signal, with a 10- to 100-msec-long, constant-frequency (CF) component preceding an FM sweep ("long CF/FM bats"). In the two best-studied species, the CF component is approximately 83 kHz in the horseshoe bat, *Rhinolophus ferrumequinum*, and 60–62 kHz in the mustached bat, *Pteronotus parnellii*. Individual bats exhibit slight differences, but these signals are remarkably consistent in any given bat.

Long CF/FM bats are strikingly specialized for detailed analysis of sounds in the range of the emitted CF. A large fraction of the inner ear and

B

FIGURE 1.1. (B) (cont'd.) Continuous record of pulse emission pattern in a long constant-frequency/frequency-modulated (CF/FM) bat, *Rhinolophus ferrumequinum*, during pursuit and capture of a moth (shown) in [B]). Numbers below the sonograms indicate the time of each image. Capture occurred between images 3 and 4. The CF frequency was 82 kHz; the vertical calibration is 10 kHz. (From Neuweiler, Bruns, and Schuller, 1980.)

most of the neurons in all auditory neural centers are devoted to this narrow range of frequencies. This neural configuration has been termed an "acoustic fovea" (Schuller and Pollak 1979). The importance of this narrow frequency band to the bats is demonstrated by their ability to compensate for Doppler shifts of returning echoes by lowering the emitted CF just enough to restrict the echo CF within the narrow range of frequencies where their sensitivity, and change in sensitivity with frequency, are greatest (see Pollak and Park, Chapter 7, Fig. 7.14). This frequency regulation is accurate to within 50 Hz in 83 kHz, or within 0.06% (Schuller, Beuter, and Schnitzler 1974). Given the concentration of the signal energy within this frequency range, these adaptations optimize the bat's detection of echoes that are slightly displaced in frequency from background echoes, as would result from a difference in the relative velocity of target and bat. The Doppler shift changes systematically with angle relative to the bat's flight direction. But near the axis of flight, or to a bat emitting CF/FM signals while perched on a branch, the echo of an approaching insect would stand out clearly. Indeed, the bats appear to be exquisitely sensitive to the small frequency and amplitude oscillations caused by insect wingbeats and able to use these oscillations not only to detect but also to discriminate prey (Schnitzler and Flieger 1983).

CF/FM bats tend to hunt in cluttered environments where prey detection is probably harder for bats that use exclusively FM signals. CF/FM bats also emit FM signal components, however. During pursuit of prey, these bats progressively shorten the CF portion of their pulses and increase the repetition rate until, in some species, the pulses in the final buzz are composed of almost pure FM signals (Fig. 1.1B). In all bats, it is likely that broadband, sharply timed FM components are used for distance detection and target localization.

2.1.3 Short CF/FM Bats

More numerous than the long CF/FM bats are bats that employ an intermediate pulse design, with pulses containing a short CF component (up to 8–10 msec) and terminating in an FM sweep. These species have been comparatively little studied. Short CF/FM bats may use Doppler shift information to some degree, but they are less specialized for analysis of the CF frequency band. Instead, they probably use the CF component to enhance target detection, but then obtain most information about the target from the FM sweep. Some bats that emit pure FM pulses when close to vegetation prolong the pulse and reduce the amount of sweep in uncluttered environments. Presumably this modification helps them detect faint echoes from relatively distant targets (see Fig. 1A, and Chapter 4, Fig. 2, this volume).

Thus there is considerable facultative flexibility in the design of echolocation pulses. The duration, amount of CF and FM sweep, overall frequency range, intensity, and repetition rate can be varied in ways that optimize the type and amount of information obtainable in a given echolocation situation. The major aims of research in echolocation have been directed at determining the accuracy with which bats obtain information through echolocation and the probable mechanisms that underlie this behavior. Most of this work is discussed in detail in the chapters by Moss and Schnitzler (Chapter 3) and by Simmons and his colleagues (Chapter 4) (this volume). I mention a few of the major findings and issues here.

3. Echolocation Abilities

The ability of bats to detect and capture small insects close to shrubbery or among tree branches involves remarkable echolocation skills. However, our understanding of these skills has grown only gradually as the technology of recording and analyzing ultrasonic sounds has improved and as ever more ingenious ways of explaining their capabilities have been developed. The most important initial studies were done by Griffin and his collaborators. In careful laboratory experiments, they documented the changes in orientation sound characteristics and emission patterns as bats detected and avoided arrays of fine wires. In these studies, the distance at which the pulse emission rate changed were used as the criterion for the distance of detection (Griffin 1958). Parallel field observations established that bats detect and capture flying insects in the wild with a similar pattern of search, pursuit or approach, and terminal buzz pulses. In the late 1950s, Griffin and his colleagues, in collaboration with Fred Webster, succeeded in moving insect capture into the lab (a large Quonset hut) where small FM bats, *Myotis lucifigus*, were strobe photographed and recorded simultaneously as they caught small insects or mealworms thrown into the air by a "mealworm gun." Not only could these bats localize airborne targets accurately enough to catch them with the tail membrane or the tip of a purposely outstretched wing, but they also identified and caught mealworms in the presence of a number of other targets of similar size but different shape (e.g., disks or spheres) that contributed interfering signals (clutter) (Webster 1967; Webster and Brazier 1965; also see Chapter 4 by Simmons et al. [this volume] for a detailed description of these and similar recent experiments).

In short, bats can detect and identify individual small targets in the presence of other targets that return similar echoes that overlap in time; they can determine the distance and direction of each target, and they can avoid unwanted targets while catching the "insect" (see Simmons et al., Chapter 4, Fig. 4.8). It is difficult to escape the conclusion that bats are able to form an acoustic image of the world in front of them. It appears that the echoes of each pulse, returning from different directions and delays corresponding to the distance of different targets, are processed by the brain to provide a three-dimensional map. This map is comparable to the visual image produced by the flash of a strobe light and tells the bats simultaneously about the location of potential prey and all other objects within hearing range, such as tree branches, wires, buildings, and the ground. Indeed, one of the most inexplicable capabilities of bats is that of memorizing a home territory of several square miles, within which they fly along regular flight paths, use the same night roosts, and always find their way back to the colony site. Bats also migrate long distances (e.g., from summer colony sites to overwintering caves), a related skill that is even less well understood (Griffin 1958).

How do bats construct this acoustic image, and how accurately can they do it? Moss and Schnitzler (Moss and Schnitzler, Chapter 3) discuss the experimental results that help answer, or at least define, these questions. The most illuminating experiments have been done with the behavioral conditioning techniques introduced by James Simmons. Simmons showed that bats could readily be trained to discriminate accurately between targets presented simultaneously or sequentially in an automatic forced choice (AFC) or yes/no paradigm. The bats indicate their choice by crawling to one side or the other of a holding platform. Real targets can be used, or "phantom" targets produced by recording the emitted pulse with a microphone near the bat and playing back either a faithful or altered "echo" at the desired delay and angle from a loudspeaker (see Moss and Schnitzler, Chapter 3, Fig. 3.2). This technique makes it is possible to test how sensitively a bat can detect echoes, measure echo delays and discriminate various targets at different distances, resolve target directions in both horizontal and vertical axes, discriminate target shapes, dimensions, and surface textures, and make all these judgments in the presence of different amounts of acoustic clutter.

3.1 Detection

In a laboratory setting, the detectability of an echo is subject to forward and backward masking by echoes from platform, microphone, loudspeaker, walls, and floor. Consequently, no adequate measure of absolute threshold of echo detection has been made. There is no reason to believe, however, that bats are absolutely more sensitive than any other mammal at its best frequency. Behavioral (and neurophysiological) audiograms show that bats tend to be sensitive at the range of their emitted orientation pulses, a frequency range far above that used by most other mammals. FM bats tend to be broadly sensitive throughout a frequency range that often covers several harmonics. CF/FM bats have a sharp peak in sensitivity near the most prominent CF frequency in their emitted sounds (which is usually the second harmonic), with lesser peaks at other harmonics of this frequency (Fig 1.2) (see also Kössl and Vater, Chapter 5, Figs 5.2, 5.3, and Moss and Schnitzler, Chapter 3, Fig. 3.1).

The CF signal used by CF/FM bats is thought to increase their ability to detect distant targets by concentrating signal energy into a narrow frequency band. This is premised on the assumption that information returning at that frequency can be integrated for the duration of the CF component, up to 100 msec. Because any given frequency in an FM signal typically lasts only a fraction of a millisecond, an equivalent ability to use FM signals for detection would require that the auditory system act as a matched filter receiver, programmed by the outgoing signal to sum all the energy in returning sounds with the same frequency–time structure. Although this capability cannot be ruled out (and evidence from ranging experiments suggests that such an operation does occur for detection of pulse–echo delay), two observations argue against detection involving such matched filter performance. First, field observations show that FM bats which hunt in open space tend to elongate their search pulses and reduce

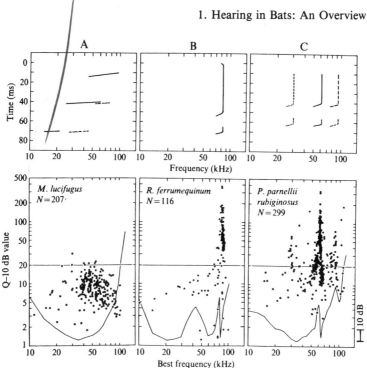

FIGURE 1.2. Characteristic frequency–time plots (sonograms) of the emitted echolocation sounds of a typical FM bat, *Myotis lucifugus*, and the two most-studied long CF/FM bats, *Rhinolophus ferrumequinum* and *Pteronotus parnellii* (*top panels*) as well as behavioral or neurophysiological audiograms of each species (*solid lines* in *lower panels*, with 10 dB calibration on right) and the $Q_{10\ dB}$ values of single auditory nerve fibers (*dots*). (From Suga and Jen 1977, including data from Suga 1973 (A), Suga, Neuweiler, and Möller, 1976 (B), and Long and Schnitzler 1975 [audiogram in B].)

their frequency sweep, concentrating most of the signal energy into a narrow frequency band. Second, FM bats are almost equally capable of detecting artificial "echoes" that sweep upward instead of downward (Møhl 1986; Masters and Jacobs 1989). Further research is needed to determine whether the integration time differs for different echo frequencies and how bats alter their emitted signals to optimize target detection under different echolocation conditions.

The effective limit to the range of echolocation may be that imposed by the emission of another orientation sound. In most cases, pulse duration appears to be reduced progressively during the approach to a target so the emitted sound and its returning echo never overlap. On the other hand, with FM pulses and sharp frequency resolution, it is unclear why some overlap should not be tolerated. And there is no convincing evidence that the echoes of one pulse, returning to a bat after a subsequent emission, are not useful, as might be the case for echoes from large obstacles in the environment or from the ground. Moreover, the acoustic reaction to a target, whether pursuit or avoidance, appears to be a facultative response. Bats show such a response when they are pursuing a target or avoiding a particular obstacle or landing, but they still obtain useful information about other objects in the environment — branches, other insects, etc. — that they do not respond to with pursuit sequences and buzzes.

3.2 Temporal Resolution

3.2.1 Ranging

Sound travels at 34.4 cm/msec in air; consequently, the interval between an emitted sound and a returning echo is an accurate measure of target distance. To capture a flying insect, even in its tail or wing membranes, a bat must be able to localize it to within about 1 cm at a distance of 10 cm or more (a minimal reaction time at a bat's typical flight speed). Many behavioral measurements have been made of the ability of different bats to discriminate target distance in the laboratory: identification of the closer of two real or phantom targets presented simultaneously or sequentially, or a yes/no decision whether a target is at a given trained absolute distance. Interestingly, all such tests using any of the experimental paradigms outlined here, including recognition of absolute distance, have yielded a maximum accuracy of about 10–11 mm (50–55 μsec time separation) for all species studied (see Moss and Schnitzler, Chapter 3, Fig. 3.4).

3.2.2 "Jitter" Resolution

Initially this 50- to 55-μsec value was accepted as a realistic limit because it could explain the bats' behavioral ability to localize a target in space. The fact that similar values were obtained when bats were tested at long or short target distances was bothersome. Then it was realized that as a bat moves around on its holding platform, it shifts the position of its head and body from one side to the other as it emits pulses, and the head movement might well introduce this much variability in the distance being measured. Hence this type of experiment was supplanted by one in which there are two phantom targets, one of which returns "echoes" at a constant delay, while the other shifts back and forth in time, from one pulse to the next, by variable amounts (Simmons 1979). In such jitter experiments, the bat holds its head in one position as it scans one target with its pulses, then holds its head relatively constant in a different position as it directs pulses toward the other. An intensity comparator is used to restrict "echo" production to only one target at a time (jittering or nonjittering), the one toward which the bat is directing its emission. The results of these experiments have been so dramatic that they have dominated behavioral research in the field for the last several years.

Initial results with *Eptesicus* yielded a jitter threshold of less than 5 μsec, a value that was soon lowered to about 1 μsec (Simmons 1979). Further training with the same bats eventually pushed the value down to the incredible value of 10 nsec (corresponding to a range sensitivity of less than 2 μm) (see Moss and Schnitzler, Chapter 3, Fig. 3.5) (Simmons et al. 1990b).!! This finding has been received with understandable skepticism (Pollak 1993; Simmons 1993; also see Moss and Schnitzler, Chapter 3; Simmons et al., Chapter 4). Besides the difficulty of explaining how the nervous system could encode an event with 10-ns accuracy, two major problems argue against interpretation of the jitter experiments as showing micrometer range resolution.

First, even in the jitter situation, the bat moves its head. Not as much as a centimeter, surely, but even while facing in one direction, a bat's entire body jerks with each emission. Its head movement is likely to vary by at least a fraction of a millimeter between pulses (and the jitter is being measured between pulses). Variations in head positions could be minimized if the bat emits its pulses and hears the echoes with its head at approximately the same phase of each pulse-induced vibration. Its seems inevitable, however, that residual changes in the bat's head position (not to mention the complications of ear movements) would be much greater than the resolution implied by 10-nsec threshold jitter detection. Perhaps even more troubling is that bats consistently perform in behavioral experiments as if they interpret a 1-dB reduction in echo intensity as a 13- to 17-μsec increase in delay (distance) (Simmons et al. 1990a) (see Chapter 3, Fig. 3.6). A 1-dB difference in emitted sound or reflected echo strength of two successive (jittered) echoes would yield an intensity/latency shift about 1000-fold greater than their apparent jitter resolution.

Nevertheless, reproduction of the jitter experiment in a different lab, with instrumentation that allowed jitter down to 400 nsec, showed no decrement of performance (Menne et al. 1989, Moss and Schnitzler 1989). Moreover, variants on these experiments showed that bats are sensitive to 90°–180° phase shifts in phantom echoes, indicating that they can resolve the phase of sounds at these frequencies. So despite reluctance to believe the 10-nsec finding and the widespread feeling that there must be some qualitative difference between the jittering and nonjittering targets, no one has identified an artifactual origin or a flaw in the experimental procedure that could yield this result. In fact, one is tempted to suspend disbelief in this result for the simple reason that temporal resolution of almost this magnitude seems to be necessary for bats to resolve one target from others nearby (as Griffin and Webster and their colleagues showed so clearly) and to discriminate the dimensions of individual targets (see Simmons et al., Chapter 4).

3.2.3 Multidimensional Targets and Two-Wavefront Range Discrimination Resolution

Targets of the size caught by most insectivorous bats reflect echoes that consist of two or more "glints," echoes from different parts of the target that reflect stronger echo components than other parts. Each glint is an echo of the emitted pulse, displaced in arrival time by a few tens of microseconds at most. Hence, the glints overlap for most of their duration. Interference and reinforcement between overlapping echoes produces troughs and peaks in echo intensity during the frequency sweep. To simulate such natural sounds, echoes that contain two wavefronts a few microseconds apart have been used to test discriminability of such overlapping glints. For example, two different hole depths in targets can be detected with thresholds of 0.6–0.9 mm, corresponding to 4-Δ to 6-μsec time resolution. Initially, these experiments were interpreted as showing detection of spectral cues, but in fact they can be interpreted equally well as a time difference (Mogdans, Schnitzler, and Oswald 1993, and see Moss and Schnitzler, Chapter 3). Indeed, in a particularly intriguing experiment, Simmons and his collaborators (1990a) reported that such overlapping echoes can be mimicked by echoes that jitter back and forth by the same Δt but are always heard separately. In summary, although the specific cues that bats use are hotly debated, they clearly are exquisitely sensitive to time or spectral cues that result from multiple glints from targets. Furthermore, they may be capable of range discrimination on the order of a millimeter or less. Is this an exciting field or what?!

3.2.4 Some Other Puzzles in Ranging

The ranging, jitter, and two-wavefront discrimination experiments pose an interesting dilemma. The accuracy of resolution of range of two different targets, presented either simultaneously or sequentially on either side of the animal or sequentially from the same direction, is always about 1 cm (50 μsec) or more. In contrast, jitter or two-wavefront discrimination appears to be well under 1 mm. Why this difference? The answer is not clear. However, both figures could be correct. Greater head movement in two-target range resolution experiments might explain the difference, but that explanation is not particularly compelling. If the bats can process all available and useful information, then almost inevitably echoes from other objects in the environment (the microphone or loudspeaker, for example) at a constant distance would serve as an accurate reference against which target distance could be compared, independent of head position. Also, experiments with two real targets yield the same 50- to 60-μsec Δt value when echoes from bats would be heard overlapping in time (but from different angles), where again head position should have been unimportant.

Thus, the difference in range thresholds for the two experimental situations very likely is real. One possible explanation for this might be that echoes returning from different targets, at significantly different angles, are kept separated in the brain (a separation that seems necessary for an effective auditory "map" of space, which bats appear to enjoy). Different populations of neurons would process the information from different targets, or at least those targets separated sufficiently in space. In contrast, glints (or jittered echoes) from the same target would be analyzed by the same population of neurons. It may be that temporal comparisons (or spectral notches) are measured much more accurately within one population than between different populations. (The 50-μsec ranging limit for sequential presentation of targets at different Δt were done with a minimal time separation of several seconds between trials at different apparent distance. This measurement might be no more accurate than that between different populations.)

Another interesting question is whether bats can detect range information from single pulse–echo pairs. This ability seems necessary if bats are to adjust their flight path to intercept fast-moving insects, as some bats do (see Simmons et al., Chapter 4). On the other hand, it is possible that range determination requires the integration of several echoes. This is suggested by behavioral evidence from the lesser fishing bat, *Noctilio albiventris*, a short CF/FM bat. When these bats were required to perform an easy range discrimination task in the presence of loud white noise interrupted by brief gaps of silence, they could do so only if at least two pulse–echo pairs fell within a single silent gap. Difficult discriminations required as many as seven pulse–echo pairs. Simple detection required only a single pulse–echo pair (Roverud and Grinnell 1985a). This might be explained simply as forward and backward masking, but related experiments show that free-running loud CF/FM pulses simulating the bat's own sounds could also interfere with range discrimination. This interference occurred even when the sounds appeared well after or before a natural pulse–echo pair, if one such artificial pulse fell between any two natural pulse–echo pairs. If two emitted sounds and their echoes occurred without such an artificial pulse, the bats could discriminate range (Roverud and Grinnell 1985a).

These experiments also support the idea that an echo (or a sound simulating the bat's emission) is interpreted as bearing range information if it occurs during a restricted time window following each emitted pulse. In *Noctilio albiventris* a loud artificial FM sweep disrupted range discrimination for about 27 msec after the onset of the emitted pulse (corresponding to a target distance of approximately 5 m). Outside that window, the FM sweep by itself was ineffective; only a more complete CF/FM pulse could degrade the bat's performance (Roverud and Grinnell 1985b). These findings suggest that a loud CF signal at the right frequency resets the system, activating a time window of approximately 25–30 msec during which a returning FM sweep is interpreted as carrying distance information. If the artificial FM sweep is sufficiently loud during this time window, it can interfere with range discrimination. Outside this time window, it is effective only if the gate is again opened by a loud CF component.

Other laboratory experiments with *Noctilio* have been aimed at discovering which features of the FM sweep are critical to range determination. *Noctilio* can be induced to increase the repetition rate of its pulses as it would to an approaching target if echoes of its emitted pulses are returned at progressively shorter delays. The acoustic situation mimics that of an object moving toward the bat, and clearly gets the bat's attention. By manipulating the properties of returning echoes, R. Roverud (personal communication) has shown that the direction of the FM sweep is important for this response (in contrast to detection). An upward sweeping signal does not induce the response. With this technique, Roverud has also shown that a series of brief CF tone components, electronically merged seamlessly in phase with each other, can replace the FM setup if there are more than 90 steps and none is closer together than about 100 Hz (Roverud 1993). A 10-kHz series of such steps (50-μsec duration and 100-Hz separation) is as effective as a 40-kHz sweep (50-μsec duration, 400-Hz separation), and the tonal steps can be as short as 20 μsec without degrading the recognition of that component as one of the steps. This finding not only implies extremely rapid and accurate discrimination of different frequencies in an echo, but also that the FM component of an echo is recognized by the sequential activation of a required number of separate channels of information.

3.3 Angular Resolution and Target Localization

Bats appear able to determine the direction of an echo's source to within ±2°–5°. This conclusion reflects several lines of evidence: their ability to capture a flying insect by spreading their tail membrane like a catcher's mitt or reaching out to haul it in with a tip of a wing membrane, their ability to point their head directly at an insect with that degree of accuracy even if the flight path is on a different line (Masters, Moffat, and Simmons 1985), and their ability to resolve and discriminate one target from several others in the same approximate direction (see Simmons et al., Chapter 4). Laboratory behavioral experiments confirm this degree of accuracy. In the azimuth, *Eptesicus* can discriminate targets at a separation as small as 1.5° (Simmons et al. 1983). Most mammals are much less accurate in resolving the direction of a sound source in the vertical plane. In bats, however, forced-choice behavior experiments show vertical localization to be almost as accurate, about 3° (Lawrence and Simmons 1982).

How is this done? More importantly, how are many targets at different directions and distances localized simultaneously with this accuracy, which seems necessary to explain the ability to catch an insect in the midst of branches and foliage? To a remarkable degree, bats act as if this were no problem. Indeed, as mentioned earlier, multiple targets seem not to be interpreted as interfering signals, but instead are analyzed accurately in their own right to construct an acoustic map, or image, of the world in front of the bats. How they construct this map is a source of some disagreement, as the chapters in this volume attest.

The available cues for target localization are interaural differences in intensity and arrival time, either of the envelope of the sound or of the phase of each cycle in it. Localization is almost certainly not accomplished by directing a narrow beam of sound in various directions like a flashlight. In vespertilionids, with which much of the behavioral work has been done, the emitted sounds are not beamed very directionally (Hartley and Suthers 1989). Even if they were, information must be obtained about many objects, in many directions, with each echo. Because of the combined directionality of sound emission and hearing, either ear is almost equally sensitive to targets anywhere between straight ahead and 30°–40° to the side of that ear. On the other side, sensitivity decreases by 20–30 dB within 30° (see Simmons et al., Chapter 4, Fig. 4.13). The head and external ears (the latter in bats tend to be extremely large and complicated by the presence of a conspicuous tragus in front of the pinna) create strong sound shadows at ultrasonic frequencies. A target 10°–20° to one side of the axis of a bat's head returns an echo that may be 30–40 dB louder at one ear than at the other. Binaural inhibitory interactions further exaggerate interaural differences within the nervous system.

Vertical localization is as important to bats as horizontal localization. In other mammals it depends, at least in part, on the timing of multiple reflections from the pinna into the ear canal (Middlebrooks 1992). The huge pinna and prominent tragus characteristic of bat ears make them good candidates for performing a similar function. The importance of the pinna and tragus to vertical localization has been demonstrated behaviorally (Lawrence and Simmons 1982). Moreover, long CF/FM bats rapidly flick one pinna forward and the other backward with each pulse emission; perhaps this rapid alternation permits correlation of ear position with echo direction. External ear shape and movements introduce differences in intensity as well as timing of reflections. Until it has been determined whether a bat can locate targets in the vertical dimension monaurally, the possibility of binaural comparisons cannot be excluded. Indeed, because hearing is directional in the vertical as well as horizontal dimension, a comparison of interaural intensity differences at three or more different frequencies in an echo (or the whole FM sweep) could, in principle, pinpoint the direction of a target in both planes (Grinnell and Grinnell 1965; Fuzessery and Pollak 1984).

The other major cue to direction is interaural arrival time (or phase) differences. Echoes arrive slightly earlier at the closer ear. In most bats, these time differences are 50 μsec or less. For directions within 15° of the midline, the interaural time differences would be 10 μsec or less. Because there is a trade-off between intensity and response latency, and a 1-dB intensity difference between the two ears is equivalent to approximately a 15-μsec difference in arrival time (Simmons et al. 1990a), it would seem that interaural intensity differences are much the more prominent cues for signal direction. Moreover, there is an orderly array of neurons in the inferior colliculus that respond preferentially to different interaural intensity differences, corresponding to different angles in the azimuth (see Pollak and Park, Chapter 7, Fig. 7.12; and Wenstrup, Fuzessery, and Pollak 1986; Park and Pollak 1993).

On the other hand, the situation becomes enormously more complicated with additional targets. For example, when two targets are at approximately the same distance on either side of the head (e.g., two wires the bat must avoid), the echoes, one slightly louder than the other, arrive at the two ears displaced by a few microseconds. Because time and intensity are interchangeable once they enter the nervous system, cues that are not subject to these transformations would be preferable. Indeed, Simmons and his colleagues (Simmons et al., Chapter 4) argue that intensity and spectral cues are useless for localization in the usual situation in which there are multiple echoes from different directions overlapping in time as they return to the bat. Because the "integration time" of either ear is 300–400 μsec, that is, for this length of time the ear integrates all the information it receives at a given frequency and assigns it a single amplitude, a simple measurement of interaural intensity difference when there are multiple echoes would result in an intermediate and false target localization. Instead, Simmons et al. argue that correct localization can be done only by using the spectral cues generated at each ear by the interference patterns caused by overlapping echoes from different targets (or glints from parts of targets). Arriving at each ear, the overlapping echoes cause peaks and troughs in the echo intensity at different frequencies, the location of which (spectral cues) is the result of differences in arrival time (phase). The height and depth of the peaks and troughs reflect the relative intensity. The spectral peaks and troughs in the composite echo can in principle be deconvoluted to determine the arrival times of each echo at each ear, which, given the ability to resolve interaural time differences as short as 1 μsec, can then be compared to determine the direction of each echo source. This is not a trivial neural computation, and it remains to be demonstrated that bats actually use these cues for localization.

3.4 Overcoming Clutter

Most remarkably, bats can detect and determine the direction and distance of targets and discriminate between them in the presence of an enormous amount of acoustic clutter, the loud outgoing sounds and echoes of multiple targets at different distances and directions (see Simmons et al., Chapter 4, Figs. 4.8, 4.10). This clutter does interfere with fine discrimination abilities; both forward masking (e.g., by the emitted sound and closer objects) and backward masking (e.g., by branches or a wall behind the target) degrade performance, but much less than might be expected (Simmons et al. 1988; Hartley 1992).

To overcome the deafening effect of their own emitted sounds, bats, like other mammals, reduce the sensitivity of their hearing during the emission by contracting the middle ear muscles. They must, of course, be as sensitive as possible to returning echoes. Henson (1965) showed that FM bats (e.g., the Mexican free-tailed bat, *Tadarida*) begin contracting the middle ear muscles a few milliseconds before pulse emission. Contraction was maximal at approximately the onset of emission, and relaxation began immediately thereafter. Sensitivity at the level of cochlear microphonics was reduced by 20–30 dB during pulse emission and was gradually restored to maximum levels within 5–8 msec after the pulse. This mechanism is especially important for the relatively long search pulses emitted at low repetition rates. As the interval between pulses decreases, however, the contractions become progressively more shallow and sustained.

3.4.1 Automatic Gain Control

Contraction of the middle ear muscles strongly attenuates sensitivity to the emitted signal, weakly attenuates response to echoes from nearby targets, and leaves the auditory system maximally sensitive to echoes from distant objects. At the same time, because of the spatial dispersion of both signal and echo energy, echo strength falls sharply with distance. The result is what is now recognized as a form of automatic gain control (Simmons and Kick 1984). The decrease in echo strength with distance is compensated for by the increase in auditory sensitivity associated with pulse–echo interval. The effective echo strength stays relatively constant as a bat approaches a target. Whether this stabilization of echo loudness independent of distance is of major importance remains to be demonstrated, but this is plausible. It should be noted, however, that bats often decrease the intensity of their emitted pulses as they approach a target.

Moreover, there is another form of suppression of response to the emitted sound. A neural inhibition of responsiveness occurs at the level of the lateral lemniscus (Suga and Shimozawa 1974). Like contractions of the middle ear muscles, this inhibition is triggered by the motor program of pulse production. The time course of the inhibition and recovery from it are unclear, but the mechanism may reduce the perceived intensity of echoes in ways that sharply alter the gain control associated with middle ear muscle function that has been documented with cochlear microphonics.

Bats employing long CF/FM signals have a similar but seemingly more severe problem: they are still emitting a loud sound during most of the time the echo is returning. In these bats, the middle ear muscles contract during most of the period of pulse emission, reducing sensitivity to the emitted sounds (Henson and Henson 1972). Surprisingly, their returning echoes are much less suppressed. The principal reason is the extraordinarily sharp tuning of their auditory system. The emitted CF, in a flying bat, is at a frequency of much reduced sensitivity, while Doppler-shifted echoes fall in a region of high sensitivity, with little inhibition from the lower emitted frequencies (Suga, Neuweiler, and Möller 1976; and see Kössl and Vater,

Chapter 5, Fig. 5.5 for a characteristic single-cell tuning curve illustrating the sharpness of tuning near the CF_2).

3.5 Target Discrimination

Some of the earliest experiments with flying bats established that they are remarkably skilled at discriminating one target from another (Webster 1967). Under natural conditions, bats appear to be able to distinguish insects from other objects and preferred insect prey species from undesirable insects. In the laboratory, a bat trained to catch mealworms propelled into the air can identify and capture the worm despite the presence of decoy disks or spheres (Webster 1967; Simmons and Chen 1989).

In long CF/FM bats, this ability depends at least in part on relative movement of the target and the surroundings. Doppler shifts will distinguish a flying insect from nearby stationary objects. Moreover, the wingbeat frequency and degree of modulation of the CF echo differ for different insects, and the CF signal is long enough to encompass one or more complete insect wingbeat cycles. Given the exquisite ability of long CF/FM bats to discriminate frequencies around their CF, frequency and amplitude modulations can be used to distinguish differences in wingbeat frequency as small as 4%–9% (Schnitzler 1970; Goldman and Henson 1977; Emde and Menne 1989; Roverud, Nitsche, and Neuweiler 1991).

The emitted pulses of most FM bats are too short to detect significant change in echo strength resulting from movement of an insect's wings, and FM signals typically are swept in frequency much too quickly to take advantage of (or be misled by) Doppler-shift information. Yet FM bats can discriminate fluttering targets only slightly less well than long CF bats (Sum and Menne 1988; Roverud, Nitsche, and Neuweiler 1991). What cues do the FM bats use?

The phenomenal ability of bats to resolve "glints" or multiple overlapping wavefronts has already been described. Perhaps their temporal resolution is good enough to get an instantaneous image of an insect's shape based on all the glints returning from its different body parts. Alternatively, bats may use differences between the Doppler-shifted echo from moving wings and the (relatively) unshifted echoes of the body (Sum and Menne 1988). These factors are discussed in detail by Moss and Schnitzler in Chapter 3.

In the context of insect identification and capture, there is a fascinating subplot to the story (see Fenton, Chapter 2 for the more complete version). Many insects have developed simple but effective ears to detect bats and to allow evasion. Arctiid moths go one step further, producing loud ultrasonic clicks of their own when they hear bat cries. These prey-generated sounds cause the bats to veer away, abandoning the chase. As arctiid moths appear to be distasteful to bats in captivity, it is probable that the signal is telling the bat, "You don't want me; I taste bad." Alternatively, and less plausibly, the moth-generated sounds may disorient the bat.

4. Specializations of the Auditory System for Echolocation

The bat auditory nervous system is constructed with the same basic elements as that of other mammals: the same identifiable nuclei, cell types, synaptic connections, and pharmacology. However, echolocation poses special requirements: for extraordinarily precise temporal resolution, for accurate sound localization in three-dimensional space, and for formation of an acoustic image with each echo. The circuits and processes responsible probably represent selection and refinement of general mammalian mechanisms rather than evolution of novel pathways or mechanisms.

Most species of bats have large and complex pinnae, a tragus in front of each air canal, and, in many, elaborate nose leaves and other nasal structures associated with sound emission. The inner ear and auditory nervous system are no less dramatically specialized for the use of sound. At each level of the auditory system there are specializations in structure, organization, and physiological processing that are critical to echolocation. Much of this volume is devoted to reviews of the current understanding of these specializations and the mechanisms of information processing used in echolocation. Most of this work has been done with the FM bats *Eptesicus* and *Myotis* and the CF/FM bats *Rhinolophus* and *Pteronotus parnellii*. Not surprisingly, the auditory nervous system of these two groups differs in fascinating ways, reflecting their different signals and echolocation behavior.

4.1 Cochlea

4.1.1 Morphology

The cochlea, while conforming to the general mammalian model, is clearly specialized for the use of high frequencies and, in CF/FM bats, for hyperacuity around the CF (see Kössl and Vater, Chapter 5). Readers wishing to gain more perspective on specializations of the mammalian cochlea are advised to consult the excellent chapter by Echteler, Fay, and Popper (1994) in a companion volume of this series. The basilar membrane is appropriately narrow, with unusual thickenings especially at the basal end. The inner hair cells (IHCs) entirely overlie the bony spiral lamina. The basilar membrane is longer, relative to body weight, than in other mammals, especially in CF/FM bats. In most FM bats, the basilar membrane increases in width systematically and decreases in thickness along its length in typical mammalian fashion, forming a logarithmic topographic map of frequency. In *Trachops*, a frog-eating bat, and perhaps in others taking advantage of prey-generated sounds, the basilar membrane shows special adaptations near the apical end that extend hearing to comparatively low frequencies (Bruns, Burda, and Ryan 1989).

It is in long CF/FM bats, however, that the cochlear specializations become most dramatic. In both *Rhinolophus* (Bruns 1976a,b) and *Pteronotus parnellii* (Henson 1978), much of the basal half of the basilar membrane is approximately the same width and thickness, appropriate to analyzing a narrow band of frequencies around the CF_2 (see Kössl and Vater, Chapter 5, Fig. 5.13); hence the term, "acoustic fovea." Distal to the expanded CF_2 region, there is an abrupt widening and decrease in thickness associated with analysis of lower frequencies. In *Pteronotus*, the cochlear microphonics (CMs) show a pronounced resonance at the frequency of the Doppler-shifted echo CF (60 kHz) that persists after an acoustic stimulus anywhere near that frequency (Henson, Jenkins, and Henson 1982). Curiously, *Rhinolophus*, with even more dramatic discontinuities in basilar membrane stiffness, does not show such resonance, but somehow this species also achieves sharp frequency selectivity for the dominant CF component. The innervation pattern of the hair cells also conforms to the general mammalian pattern, although the few studies that have been done show some interesting quantitative specializations. In *Rhinolophus*, for example, 10%–20% of the afferent fibers innervate outer hair cells (OHCs), compared with about 5% in other mammals (Bruns and Schmieszek 1980). In bats using CF components, there is dense innervation of the basilar membrane in regions corresponding to the representations of the second and third harmonics, with sparse innervation in between (Zook and Leake 1989). The density of innervation at these sites is close to the maximum observed for other mammals (25–35:1); in the FM bat *Myotis*, interestingly, the density is about twice that high (70:1) (Ramprashad, Money, Landolt and Laufer 1978).

The efferent innervation of the cochlea in bats also shows some interesting specializations. As in other mammals, there are both medial and lateral olivocochlear efferent systems. Unlike other mammals, however, the lateral system contains only an ipsilateral component, not one from the contralateral side (see Kössl and Vater, Chapter 5, Fig. 5.19). *Rhinolophus* is unique among the species studied in having no medial olivocochlear system and lacking efferent synapses on the OHCs (Bishop and Henson 1987). *Pteronotus* is unusual in a different way. Each OHC receives only one efferent ending, and each efferent fiber forms terminals on only a few OHCs, in contrast to most other mammals. Because the overall function of the efferents is so poorly understood, it is not easy to interpret the differences among mammals or between bats and other mammals, but an understanding of their significance in bats may provide critical insights into the role of cochlea efferents in general.

Kössl and Vater (Chapter 5) also discuss the implications of these specializations in cochlear structure and innervation pattern for the important "second filter" mechanism of achieving sharply tuned low-threshold responses. *Pteronotus* exhibits particularly prominent otoacoustic emissions (OAEs) at approximately the frequency of its second harmonic CF (60–62 kHz), about 100-fold louder than in other animals. There is no compelling evidence, however, that active movement of OHCs is responsible. On the other hand, active micromechanical processes in the OHCs are likely responsible for the sensitivity peaks at high frequency, with specialization in shape and thickness of the basilar and tectorial membranes shaping the tuning curves.

4.1.2 Physiology

The first physiological study of bat hearing was the recording by Galambos (1942b) of CMs in the little brown bat, *Myotis lucifugus*, which he obtained up to 98 kHz. This finding supported the idea that bats could use frequencies that high, but presented a severe and persistent problem for models of transduction in the cochlea. In fact, most bats utilize frequencies in the 25- to 100-kHz range for echolocation, and some emit and analyze principal components as high as 150 kHz. Although frequencies this high offer advantages for echolocation acuity (and disadvantages for detection of distant targets), they require considerable specialization of the middle and inner ears. In FM bats, recordings of CMs, auditory nerve single units, and the synchronized onset response of the auditory nerve (N_1) all indicate broad sensitivity throughout the range of emitted frequencies. With the exception of cetaceans, this is well beyond the range of high sensitivity in other mammals. Single units show best frequencies (BFs) to more than 100 kHz, and tuning curves similar to those of other mammals, but with $Q_{10\ dB}$ values (the BF divided by the bandwidth of the tuning curve 10 dB above threshold at the BF) up to 30. This tuning is two- to threefold sharper than in most other mammals.

Responses in long CF/FM bats are much more specialized, reflecting the dramatic adaptations for analysis of frequencies around the CF_2 at early stages of sound processing. CMs and N_1 show peak sensitivity slightly above the frequency of the CF second harmonic (CF_2) and sometimes near CF_3 as well (Pollak, Henson, and Novick 1972). On both sides of these maxima, especially the low-frequency side, thresholds rise sharply. Single auditory nerve fibers in this frequency range show $Q_{10\ dB}$ values as high as 400! (see Fig. 1.2; and Kössl and Vater, Chapter 5, Fig. 5.3). Single units tuned to other frequencies are much less numerous than those in the CF region and have $Q_{10\ dB}$ values like those in FM bats. Associated with the steep tuning near the CF_2, especially in *Pteronotus*, is a distinct long-lasting resonance in the CM just above the frequency of the CF (Suga, Simmons, and Jen 1975). Whatever is giving rise to this resonance is presumably also responsible for the sharp tuning of single units and the phenomenally accurate resolution of frequency around the CF_2.

4.2 Brainstem Auditory Nuclei and Specializations in Processing of Information for Echolocation

The mammalian auditory system is a series of parallel frequency-tuned pathways, beginning with the tonotopic distribution of eighth nerve endings along the basilar membrane and maintained as a tonotopic map in (almost) every division of every acoustic nucleus en route to the cortex. In addition, beginning with the input to the cochlea nuclei, the information in each frequency-tuned pathway is duplicated and sent through multiple channels that process it in different ways. Some are specialized for frequency and intensity analysis, others for precise registration of the timing of signals. In bats, this information is integrated at various levels to provide the range of targets, their localization in space, and the dimensions, movement, and surface texture of each target. It seems probable that all this information is eventually combined somehow to produce an acoustic image of the world around the bat.

Figure 2 in Chapter 7 shows diagrammatically the principal nuclei of the auditory system. The earliest recordings were from the inferior colliculi (Grinnell 1963a; Suga 1964), but much has now been done at all neural levels through the cortex, revealing a great deal about mechanisms of information processing for echolocation and the circuitry involved. This work is summarized in chapters by Covey and Casseday (Chapter 6), Pollak and Park (Chapter 7), Wenstrup (Chapter 8), and O'Neill (Chapter 9).

Information from the auditory nerve is sent to three different divisions of the cochlea nucleus (CN), which then project in parallel to different subsets of nuclei in the brainstem, some ipsilaterally, most contralaterally. In many cases, from the superior olivary nuclei centrally, inputs converge from the two ears. These may both be excitatory (EE), but more often one is excitatory, the other inhibitory (EI).

A separation of time and intensity information is seen initially in the medial superior olivary nuclei (MSO) and lateral superior olivary nuclei (LSO). Responses at the MSO tend to be broadly tuned and either sustained or phasic with only one or two spikes to the onset of a sound. In most mammals, the MSO receives approximately equal excitatory input from both cochlea nuclei. The relative timing of these binaural inputs determines the activity of different populations, signaling target azimuth. In contrast, the MSO of bats receives predominantly contralateral input, with analysis of binaural timing apparently shifted to higher neural levels (Covey, Vater, and Casseday 1991). Responses in the LSO, in bats as in other mammals, emphasize frequency and intensity information. Neurons have V-shaped tuning curves much as at the CN, and respond with sustained firing that reflects a balance of excitatory input from the ipsilateral CN and inhibitory input for the contralateral CN.

The nuclei of the lateral lemniscus are complex and much hypertrophied in bats, and clearly are an important part of the temporal analysis pathway. Parts receive binaural input, others are monaural. Of particular interest are neurons in the dorsal nucleus (DNLL) that show strong facilitation to the second of two identical sounds at a specific delay. These "best delay" (BD) neurons are all selective for intervals within the range needed in echolocation (Covey 1993; see also Grinnell 1963b; Suga and Schlegel 1973; Feng, Simmons, and Kick 1978). These neurons probably are responsible for the finding—in the first study of auditory neurophysiology in bats, using the evoked potential (N_4) representing input to the inferior colliculus—that there is strong facilitation of response to the second of a pair of sounds (Grinnell 1963b). The rapidity of full recovery of N_4 (2–3 msec) and exaggeration of recovery at ascending neural levels from auditory nerve to lateral lemniscus is characteristic only of echolocating bats (again, with the exception of cetaceans [Bullock et al. 1968]).

The most prominent of the nuclei of the lateral lemniscus is the columnar division of the ventral nucleus ($VNLL_c$). This nucleus is especially conspicuous in echolocating bats, consisting of several densely packed, highly organized columns of monaurally driven small neurons separated by fiber tracts. These neurons differ from others in the auditory system to this point in that they show broad tuning, presumably by convergence of inputs from cells with widely separated BFs, and they are highly phasic in response, usually firing just one spike at the onset of a sound. The dominant input is from cochlear nucleus axons that form large calyx-like endings on the cell body. Like other cells that have been recorded in the inferior colliculus (Bodenhamer, Pollak, and Marsh 1979), the response latency of these cells is remarkably invariant for any given cell, independent of changes in frequency or intensity (Covey, Vater, and Casseday 1991). Thus they are ideal for detecting wavefront arrival time.

Other parts of the nuclei of the lateral lemniscus encode time in another way, showing sustained, nonadapting responses firing throughout the duration of a signal. These are robust responders that can follow rapid frequency and amplitude modulations and can be viewed as intensity/ duration encoders. In Chapter 6, Covey and Casseday help sort out all the complexities of these brainstem nuclei.

4.3 Inferior Colliculus

Information coursing through the brainstem takes many alternative pathways, but all converge at the inferior colliculus. In echolocating bats, the inferior colliculus (IC) is a relatively huge nucleus, readily accessible near the dorsal surface of the skull in most species and thus intensively studied. These studies, and our current understanding of the organization and function of the IC, are summarized in Chapter 7 by Pollak and Park.

The central nucleus of the IC (IC_c) receives input from at least 10 lower auditory nuclei as well as descending input from higher centers. This includes excitatory input from the ipsilateral CN, MSO, intermediate nucleus of the lateral lemniscus (INLL), and contralateral LSO; inhibitory input comes from other parts of the nucleus of the lateral lemniscus and from the ipsilateral LSO. While some neurons receive monaural excitatory input, most are either excited by both ears or excited by inputs from one ear and inhibited by the opposite side. In addition, there are many inhibitory interneurons within the colliculus and large tracts promoting interaction between the colliculi. The result of all this convergence and synaptic interaction is a conspicuous transformation of information.

In the frequency–intensity domain, one of the conspicuous transformations is the sharpening of single-unit tuning curves. Instead of the mostly V-shaped or broad tuning curves of lower centers, a large percentage of collicular neurons have narrower tuning curves with nearly vertical cutoffs on either side of the BF. This has the important consequence that these cells respond at approximately the same time independent of the intensity of the sound, for example, a narrow band within an FM sweep. Moreover, many neurons have "closed" tuning curves, that is, they respond well to faint sounds (e.g., echoes), but are inhibited by high-intensity sounds (e.g., emitted sounds), even at their BFs.

The IC of long CF/FM bats, like all other auditory nuclei, reflects the existence of an acoustic fovea near the bat's CF_2 (Pollak and Bodenhamer 1981). In *Pteronotus*, about a third of the neurons in IC are tuned to 60–62 kHz. As at more peripheral levels, the tuning is exquisitely sharp, with Q_{10} dB values reaching well over 300. Indeed, the tuning of these units is so sharp that many of them are responsive to a Doppler-shifted echo while being almost totally unresponsive to the slightly lower emitted CF_2. These neurons (and their equivalents at the cortex) are also highly sensitive to the slight frequency and amplitude modulations introduced into echoes by the wingbeats of insects (Schuller 1979; Bodenhamer and Pollak 1983). Even frequency modulations as small as 10 Hz on a carrier of 63 kHz evoke clear responses (see Pollak and Park, Chapter 7, Fig. 7.23).

Because it is so large, the slab of IC dealing with CF_2 signals in CF/FM bats has proved useful for discerning the organization of inputs within a single isofrequency contour. Populations of cells receiving monaural input, binaural excitatory input (EE), and binaural excitatory/inhibitory (EI) input are arranged topographically, each receiving a different subset of inputs from brain stem nuclei. Some of the inhibitory inputs are mediated by the neurotransmitter γ-aminobutyric acid (GABA), others by glycine. As in other bats, a large fraction of the EI-driven neurons show inhibition of response at high intensities and even "closed" tuning curves. The EI neurons in *Pteronotus*, and probably in FM bats as well, show a further important form of organization. They are excited via the contralateral ear, inhibited by sounds to the ipsilateral ear, and highly sensitive to interaural intensity differences. Pollak and his colleagues have shown that there is a systematic topographic distribution of the relative effectiveness of inputs from the two sides: at one extreme, the ipsilateral input needed to be 25 dB louder than the contralateral input to effect a 50% reduction in response; at the other extreme, the ipsilateral input caused equivalent inhibition even when it was 15 dB fainter (Wenstrup, Fuzessery, and Pollak 1986). This suggests that the neurons are arranged in an orderly sequence with respect to the angle of incidence of a sound at which ipsilateral inhibition becomes effective.

The IC also marks an important transformation in the processing of temporal information. About one-third of the neurons respond with sustained activity throughout a signal; the rest are strongly phasic. The latter normally respond to an effective stimulus with just one spike. The minimum latency of that response is about 5 msec, but in most cells it is considerably longer. In fact, the response latencies for different IC neurons cover a much wider range than at any earlier nucleus: from 5 to more than 30 msec, compared with a range of 3–6 msec at the lateral lemniscus (Casseday and Covey 1992, Park and Pollak 1993). Response latencies are also topographically ordered. Most of the wide spread of latencies occurs near the dorsal surface of the IC. With increasing depth, the minimum latency and the range of latency both decrease. The mean latency is about 12–17 msec, depending on the intensity and depth of recording.

Unlike many of the neurons of the nucleus of the lateral lemniscus, which show striking specialization for rapid recovery and even facilitation of response to the second of two signals, IC neurons tend to recover relatively slowly. However, the information about echoes must be getting beyond the IC. The explanation is thought to lie in those neurons that respond with latencies so long that echoes from objects up to several meters distance return before the response to the initial sound. In these cases, if the excitatory input from an echo can reach a population of target cells at the same time as the delayed response from these long-delay IC neurons, the coincidence might be strongly excitatory (Park and Pollak 1993; Sullivan 1982b; Jen and Schlegel 1982; Dear, Simmons, and Fritz 1993; and see Pollak and Park, Chapter 7, Fig. 7.46). The specific subpopulation of long-delay IC neurons that provides an input coincident with that from a particular echo potentially allows the nervous system to determine the echo delay. We shall see later that this is in fact the case in the medial geniculate. Park and Pollak (1993) have shown that the wide range of latencies in these collicular neurons is caused, at least in part, by GABA-mediated inhibitory synapses, because locally applied bicuculline greatly shortens the response latency. It is possible that cortical feedback also contributes to the long latency responses.

4.4 Medial Geniculate Body (MGB)

The medial geniculate body is a composite of several structurally and functionally distinct divisions, each a part of different parallel pathways to the cortex. It is the least thoroughly studied of the principal auditory nuclei in bats. Indeed, it has mainly been studied in the CF/FM bat *Pteronotus* with the motivation of understanding the origin of the beautifully organized combination-sensitive neurons of the auditory cortex (see following).

The MGB in bats receives its input mainly from the central nucleus of the ipsilateral IC (in the cat, there is a strong contralateral input as well). While some parts of the MGB, especially the ventral division, are tonotopically organized with sharply tuned neurons sensitive to sound direction, the main interest has been in the populations of neurons in the dorsal and medial divisions that do not show clear tonotopic organization but are instead sensitive to combinations of inputs at different frequencies. As Wenstrup (Wenstrup, Chapter 8) points out, these neurons, which tend to display long latencies and broad tuning, are the first to show strongly facilitated responses to combinations of first and second or first and third harmonic spectral components. This is a remarkable response specificity. Where most investigators expected to find specialized populations of neurons re- sponding to the same narrow band of frequencies in the emitted sound and echo at intervals down to a fraction of a millisecond (which may be the case in FM bats, but see following), the long CF/FM bats—or *Pteronotus* at least—are doing something much more clever. At their CF_2, which provides critical information for target detection and discrimination, they employ a receiver so sharply tuned that there is little response to the emitted signal, but high sensitivity, from cochlear to cortical levels, to Doppler-shifted echoes. At the MGB (and even more so at the auditory cortex), however, such sharp tuning is not enough. Many of these sharply tuned neurons will not respond to the Doppler-shifted CF_2 component unless they receive input at the CF_1 at the same time. The CF_1 component of the emitted sound is relatively faint, but to the bat emitting it, it is adequate to excite input to the MGB combination-sensitive neurons. The combination of CF_1 and CF_2 elicits response up to 50 fold greater than that to either harmonic alone.

Of the neurons sensitive to CF combinations, about 70% are specific to CF_1/CF_2 and 30% to CF_1/CF_3. Interestingly, some of these appear specialized for echolocation when the bat is at rest. They are sharply tuned close to the resting emitted CF_1, but are excited only when the CF_2 (or CF_3) component is delayed with respect to the CF_1; they are strongly inhibited when the two begin simultaneously. Others are specialized for echolocation in flight. They are excited by CF_1 only when it is shifted downward to compensate for Doppler shift (Olsen and Suga 1991a).

There are also large numbers of neurons selective for different FM–FM combinations, almost equal numbers preferring FM_1-FM_2 and FM_1-FM_3, and even a few selected for FM_1-FM_4. These FM_1-FM_n populations are beautifully adapted for encoding target distance. They respond better when the FM_1 signal is louder than the FM_n (echo) component, and each is maximally facilitated within a narrow range of FM_n delays. The best delays range from 0 to 23 msec in *Pteronotus*, with most between 1 and 10 msec, corresponding to the distances of greatest importance in echolocation (Olsen and Suga 1991b; and see Wenstrup, Chapter 8, Fig. 8.13).

The CF_1/CF_n and FM_1–FM_n combination-sensitive neurons are segregated in the MGB, but a finer degree of organization is not clear. In the cortex, on the other hand, the CF/CF neurons are topographically arranged according to amount of Doppler frequency shift and FM/FM neurons are arranged by best delay (see following). Wenstrup (Chapter 8) discusses in detail the circuitry and pharmacology of the combination-sensitive pathways.

4.5 Auditory Cortex

Studies of the auditory cortex of echolocating bats have yielded the clearest pictures of how these animals extract the necessary information from echoes. Four species have been studied in detail: two FM bats (*Myotis lucifugus* and *Eptesicus fuscus*) and two CF/FM bats (*Pteronotus parnellii* and *Rhinolophus ferrumequinum*). The organizational principles for these two groups are so different that it is not easy to generalize about the bat cortex. O'Neill describes this work in Chapter 9.

4.5.1 Long CF/FM Bats

Of particular importance has been the series of elegant studies by Suga and his colleagues of the cortex of the CF/FM mustached bat, *Pteronotus*. The spectacular organization of the auditory cortex in this species (see O'Neill, Chapter 9, Fig. 9.2) illustrates well the advantage of working with an animal for which one knows the signals used and the information needed.

As in all mammals, a major portion of the *Pteronotus* auditory cortex (the AI equivalent) is arranged in tonotopical order. However, within this sequence, the narrow band of frequencies just above the emitted CF_2 (61–63 kHz) is even more overrepresented than at lower auditory centers. It comes to occupy fully one-third of the tonotopic portion of the cortex. In the large "foveal" portion of the tonotopic map dealing with Doppler-shifted, constant-frequency signals (the DSCF area), neurons are extremely sharply tuned, even at high intensity, and neurons with BFs from 61 to 63 kHz are further organized tonotopically in concentric rings. Those tuned to 61 kHz are at the center with an orderly progression to 63 kHz at the periphery. Superimposed on this concentric high-resolution frequency map is a map of "best amplitudes" (at which the response is maximal), arranged radially in an orderly sequence from about 10 dB to 90 dB. Finally, the populations preferring low-amplitude signals tend to receive EE input from both ears; those preferring louder sounds receive EI input and are better adapted for target localization. It is clear from this organization that CF_2 components of echoes will excite different populations in the DSCF area depending on the amplitude, the relative velocity of bat and target, and the direction of the echo.

Lying outside the main tonotopic area are at least seven secondary fields that are not organized tonotopically. In these secondary fields, responses are fascinatingly adapted for various features of echolocation. Neurons respond well only to combinations of sounds, similar to the combination neurons described for the MGB but more dependent on the combination of inputs. One of these nontonotopic fields is an area selective for combinations of CF_1 with CF_2 or, in an immediately adjacent strip, with CF_3. Within these strips, there is an orderly sequence of preference for small to large Doppler shifts of the CF_2 and CF_3 components.

Just dorsal to this area are a series of cortical strips selective for combinations of FM_1 with FM_2, FM_3, or FM_4 (see O'Neill, Chapter 9, Fig. 9.2). These cells are arranged, from one end to the other, according to the most effective delay of the second signal with respect to FM_1; that is, the cells are "delay tuned" (see O'Neill, Chapter 9, Figs. 9.6, 9.7, 9.10). The best delays range from 0.4 to 18 msec, covering most of the range of echo delays of importance to the bat (Suga, O'Neill, and Manabe 1978). These fields, thus, map echo delay or target distance. Several other small outlying regions of the auditory cortex also code echo delay, mostly over restricted parts of the full range of echo delays. The segregation of function between these various secondary nuclei is not clear in all cases, but the ability of the nervous system to construct computational maps of auditory space is evident and has reoriented much of the research on cortical organization in other animals as well.

The other principal long CF/FM bat, *Rhinolophus*, has been studied less extensively but shows qualitatively similar organization and combination-sensitive stimulus requirements. The organization is not so distinct, however, and facilitation by harmonically related combinations of signals is not as clear-cut. These differences are described in detail in Chapter 9 by O'Neill.

4.5.2 Cortical Organization in FM Bats

The auditory cortex of FM bats also has both tonotopically and nontono-topically organized regions, but they are quite different than in CF/FM bats. The tonotopic map covers the range of frequencies heard, with overrepresentation of the frequencies in the FM sweep, but there is no equivalent of the DSCF region. Most cells are more sensitive to an FM sweep than to any pure tone falling within the sweep, and between 10% and 20% respond only to FM sweeps. A much smaller fraction respond only to CF signals. These CF-specific neurons are segregated at one end of the tonotopic map and are tuned to frequencies at the low end of the FM sweep.

Many of the FM-sensitive cells are delay tuned; some have short best delays and fixed response latencies but are strongly facilitated by echoes in a narrow range of delays, and others have long best delays and response latencies that are time locked to the echo (O'Neill and Suga 1982; Sullivan 1982a, b; Dear et al. 1993). Moreover, in contrast to the requirement for different harmonic combinations seen in CF/FM bats, facilitation of response in echo-sensitive units requires instead a pair of FM sounds differing in amplitude (Sullivan 1982a). In *Eptesicus*, best delays extend to more than 30 msec, with cells preferring long delays typically found in the nontonotopic areas while short delays are preferred in the tonotopic region. How delay tuning can be achieved has already been discussed in the context of responses in the IC and MGB.

In general, the models depend on coincidence of inputs via two different pathways with different delay lines (or different amounts of inhibition). One has a high-amplitude signal pathway with long delay lines (or prolonged inhibition), the other a low-amplitude signal pathway with relatively short delay lines (or little or no inhibition). Interestingly, the convergent inputs that elicit the strongest response to pairs of FM sounds are not necessarily to the same frequency component of the two FM signals. Usually, in *Myotis*, best facilitation occurred when the pulse was a frequency about 8 kHz higher than the echo (Berkowitz and Suga 1989). Perhaps this is one step in the direction of the harmonic combination specificity of CF/FM bat neurons. Dear, Simmons, and Fritz (1993) postulated that populations of neurons like these in the IC and cortex, which fire with latencies up to 30 msec, allow the nervous system to process information from targets at all distances simultaneously, helping build a true acoustic image with each pulse.

Another important feature of many cortical neurons in FM bats is that they are tuned to two different frequency ranges, typically harmonically related in the relationship 1:3, with the lower frequency near the low end of the FM sweep (Dear et al. 1993). This may be exactly the property needed for analysis of the spectral peaks and notches generated by overlapping echo "glints" that are postulated to produce the information necessary for bats to resolve the dimensions of targets.

Unfortunately, despite beautiful studies of the directional sensitivity of neurons at cortical and lower auditory levels, especially by Pollak and his colleagues, the mechanisms of target localization in three-dimensional space are not clear.

4.6 Summary and Comments

Field observations and behavioral experiments with echolocating bats document remarkable, sometimes seemingly impossible, skills at detecting, localizing, and discriminating the nature of targets by echoes of emitted sounds. Using either CF/FM or FM sounds in ways adapted to different echolocation conditions, and using different mechanisms of information processing, bats behave as if they can construct a full acoustic image of nearby objects in space by the echoes of each pulse. Ingenious behavioral experiments have led to a number of interesting and controversial models of how they might be obtaining the necessary information.

Neurophysiological experiments with both CF/FM-and FM-emitting species have revealed neural organization and functional specializations associated with each echolocation strategy. Research continues to be driven to a large extent by behavioral experiments. The major emphasis at present is on mechanisms of preserving and analyzing microsecond and submicrosecond differences in arrival time of echoes from different targets and glints from a given target. There is much interest in the possibility that small time differences become expanded at the level of the colliculus and higher, but how are these differences encoded to that point, and how might they be magnified?

The auditory nervous system is extremely complex. Often there seems a surfeit of possible pathways or mechanisms to help explain a given behavioral ability. Based on response properties, one can make educated guesses about the function of different cell populations, such as different divisions of the auditory cortex. To test these guesses, however, it is important to attempt to alter behavior in predictable (and reversible) ways by localized injections of pharmacological agonists or blocking agents, much as Suga and his colleagues (Riquimaroux, Gaioni, and Suga 1992) have done to verify the role of the DSCF area for frequency discrimination around CF_2 in the mustached bat.

The ultimate aim of those studying echolocation is to understand how a bat can form the equivalent of a visual image with the multiple echoes of a single emitted sound, and do the same for each successive sound. This requires independently registering the directions (as well as distances) of all objects returning echoes. (There is a real need for behavioral experiments that begin to test the accuracy of information that can be obtained simultaneously about multiple targets.) It is difficult enough to understand the basis for accurate range discrimination and localization of a single target in the vertical and horizontal directions. To be able to do this simultaneously for many different targets at different directions, and at the same time analyze the fine structure of each echo to identify the dimensions and surface features of the reflecting object, and then put together successive images to determine the rate and direction of movement of each target as well—these abilities frankly boggle the mind.

Several aspects of hearing in bats have been largely neglected to date and deserve much more attention. To what extent is echolocation hard wired, for example, and what role does experience play in the perfection of echolocation abilities? How do neurophysiological adaptations develop ontogenetically? The little work that has been done on development, especially in CF/FM bats, shows that there are coordinated changes in the frequencies of maximal hearing sensitivity and of CF vocalization (Grinnell and Hagiwara 1972a; Rübsamen 1987; Rübsamen, Neuweiler, and Marimuthu 1989; Rübsamen and Schäfer 1990a). Is this primarily the result of growth, or can vocalizations be greatly altered to compensate for changes in hearing capabilities? Can neural plasticity, during development or in mature animals, compensate for abnormalities such as experimental removal of a tragus, change in pinna shape, or partial deafening (see, e.g., Rübsamen and Schäfer 1990b)? Also important are the questions: In what ways do the emitted pulse, or the motor commands for pulse emission, play roles in "priming" the auditory system for echo analysis (see Metzner 1993)? How is information obtained in the sensory pathways translated into flight maneuvers? How is spatial memory for a home territory encoded?

There is also a general need for more comparative research. Much of what we understand about echolocation is the result of natural experiments: evolution of different pulse design and adaptation of different species to different habitats and hunting strategies. Only a small number of species have been studied in the lab or in the field. Study of natural echolocation behavior of many more species is sure to reveal new and important adaptations. These will demand neural correlates and eventually help us not only to understand echolocation behavior, but to gain additional valuable insights into hearing mechanisms in man.

Finally, many bats are gregarious, highly social animals, with most interactions occurring in the dark. Almost certainly the auditory brains that have been revealing such remarkable adaptations for analysis of echolocation sounds are also highly adapted for complex acoustic communication (see Fenton 1985 for a good review of the literature). Study of the wide repertoire of communication sounds, their uses, and their analysis is likely to yield rich rewards.

This seems a long list of unknowns. It is, and it is just a beginning. However, no one would dispute that enormous progress has been made in explaining the remarkable behavior skills of echolocating bats at the neurophysiological level, and that bats constitute a striking example of the adaptability of the mammalian brain.

Acknowledgements It is a privilege to have been asked to write an overview chapter on this exciting subject. Most of the data and ideas are, of course, those of my colleagues in the field, but I would like especially to acknowledge the many happy hours spent discussing bats and echolocation with Drs. D.R. Griffin, H.-U. Schnitzler, J. A. Simmons, R. Roverud, N. Suga, G. Pollak, and G. Neuweiler. I also thank Dr. S.A. Kick for valuable help in refining the manuscript.

References

Berkowitz A, Suga N (1989) Neural mechanisms of ranging are different in two species of bats. Hear Res 41:255–264.

Bishop AL, Henson OW Jr (1987) The efferent cochlear projections of the superior olivary complex in the mustached bat. Hear Res 31:175–182.

Bodenhamer RD, Pollak GD (1983) Response characteristics of single units in the inferior colliculus of mustache bats to sinusoidally frequency modulated signals. J Comp Physiol 153:67-79.

Bodenhamer RD, Pollak GD, Marsh D S (1979) Coding of fine frequency information by echoranging neurons in the inferior colliculus of the Mexican free-tailed bat. Brain Res 171:530-535.

Bullock TH, Grinnell AD, Ikezono E, Kameda K, Katsuki Y, Nomoto M, Sato O, Suga N, Yanagisawa K (1968) Electrophysiological studies of central auditory mechanisms in cetaceans. Z Vgl Physiol 59:117-156.

Bruns V (1976a) Peripheral auditory tuning for fine frequency analysis by the CF-FM bat, *Rhinolophus ferrumequinum*. I. Mechanical specializations of the cochlea. J Comp Physiol 106:77-86.

Bruns V (1976b) Peripheral auditory tuning for fine frequency analysis by the CF-FM bat, *Rhinolophus ferrumequinum*. II, Frequency mapping in the cochlea. J Comp Physiol 106:87-97.

Bruns V, Schmieszek E (1980) Cochlear innervation in the greater horseshoe bat: demonstration of an acoustic fovea. Hear Res 3:27-43.

Bruns V, Burda H, Ryan MJ (1989) Ear morphology of the frog-eating bat (Trachops cirrhosus, Family: Phyllostomidae): apparent specializations for low-frequency hearing. J Morphol 199:103-118.

Casseday JH, Covey E (1992) Frequency tuning properties of neurons in the inferior colliculus of an FM bat. J Comp Neurol 319:34-50.

Covey E (1993) Response properties of single units in the dorsal nucleus of the lateral lemniscus and paralemniscal zone of an echolocating bat. J Neurophysiol 69:842-859.

Covey E, Vater M, Casseday JH (1991) Binaural properties of single units in the superior olivary complex of the mustached bat. J Neurophysiol 66:1080-1093.

Dear SP, Simmons JA, Fritz J (1993) A possible neuronal basis for representation of acoustic scenes in auditory cortex of the big brown bat. Nature 364:620-623.

Dear SP, Fritz J, Haresign T, Ferragamo M, Simmons JA (1993) Tonotopic and functional organization in the auditory cortex of the big brown bat, Eptesicus fuscus. J Neurophysiol 70:1988-2009.

Dijkgraaf S (1946) Die Sinneswelt der Fledermäuse. Experientia (Basel) 2:438-448.

Echteler SM, Fay RR, Popper AN (1994) Structure of the mammalian cochlea. In: Fay RR, Popper AN (eds) Comparative Hearing: Mammals. New York, Springer-Verlag, pp. 134-171.

Emde GVD, Menne D (1989) Discrimination of insect wingbeat-frequencies by the bat Rhinolophus ferrumequinum. J Comp Physiol A 164:663-671.

Feng AS, Simmons JA, Kick SA (1978) Echo-detection and target ranging neurons in the auditory system of the bat, Eptesicus fuscus. Science 202:645-648.

Fenton MB (1985) Communication in the Chiroptera. Bloomington: Indiana University Press.

Fuzessery ZM, Pollak GD (1984) Neural mechanisms of sound localization in an echolocating bat. Science 225:725-728.

Galambos R (1942a) The avoidance of obstacles by bats: Spallanzani's ideas (1794) and later theories. Isis 34:132-140.

Galambos R (1942b) Cochlear potentials elicited from bats by supersonic sounds. J Acoust Soc Am 14:41-49.

Goldman LJ, Henson OW (1977) Prey recognition and selection by the constant frequency bat, *Pteronotus p. parnellii*. Behav Ecol Sociobiol 2:411-419.

Griffin DR (1958) Listening in the Dark. New Haven: Yale University Press.
Grinnell AD (1963a) The neurophysiology of audition in bats: intensity and frequency parameters. J Physiol 167:38–66.
Grinnell AD (1963b) The neurophysiology of audition bats: temporal parameters. J Physiol 167:67–96.
Grinnell AD, Grinnell VS (1965) Neural correlates of vertical localization by echolocating bats. J Physiol 181:830–851.
Grinnell AD, Hagiwara S (1972a) Adaptations of the auditory nervous system for echolocation. Studies of New Guinea bats. Z Vgl Physiol 76:41–81.
Grinnell AD, Hagiwara S (1972b) Studies of auditory neurophysiology in nonecholocating bats, and adaptations for echolocation in one genus, *Rousettus*. Z Vgl Physiol 76:82–96.
Hartley DJ (1992) Stabilization of perceived echo amplitudes in echolocating bats. I. Echo detection and automatic gain control in the big brown bat, *Eptesicus fuscus*, and the fishing bat, *Noctilio leporinus*. J Acoust Soc Am 91:1120–1132.
Hartley DJ, Suthers RA (1989) The sound emission pattern of the echolocating bat, *Eptesicus fuscus*. J Acoust Soc Am 85:1348–1351.
Hanson MM (1978) The basilar membrane of the bat *Pteronotus p. parnellii*. Am J Anat 153:143–159.
Henson MM, Jenkins DB, Henson OW Jr (1982) The cells of Boettcher in the bat, *Pteribitus parnellii*. 7:91–103.
Henson OW Jr (1965) The activity and function of the middle ear muscles in echolocating bats. J Physiol Lond 180:871–887.
Henson OW Jr, Henson MM (1972) Middle ear muscle contractions and their relations to pulse and echo-evoked potentials in the bat, *Chilonycteris parnellii*. In: AIBS-NATO Symposium on Animal Orientation Navigation, Wallops Station, VA, pp. 355–363. Galler, S.R, Schmidt-Koenig, K., Jacobs, G.J., & Belleville, R.E (eds) Publ: Scientific and Technical Information Office, NASA, Washington, D.C.
Jen PH-S, Schlegel PA (1982) Auditory physiological properties of the neurons in the inferior colliculus of the big brown bat, *Eptesicus fuscus*. J Comp Physiol A 147:351–363.
Jepsen GL (1970) Bat origins and evolution. In: Wimsatt WA (ed) Biology of Bats, Vol. 1. New York: Academic Press, pp. 1–64.
Lawrence BD, Simmons JA (1982) Echolocation in bats: the external ear and perception of the vertical positions of targets. Science 218:481–483.
Long GR, Schnitzler H-U (1975) Behavioral audiograms from the bat, *Rhinolophus ferrumequinum*. J. Comp. Physiol. 100:211–219.
Masters WM. Jacobs SC (1989) Target detection and range resolution by the big brown bat (*Eptesicus fuscus*) using normal and time-reversed model echoes. J Comp Physiol A 166:65–73.
Masters WM, Moffat AJM, Simmons JA (1985) Sonar tracking of horizontally moving targets by the big brown bat, *Eptesicus fuscus*. Science 228:1331–1333.
Menne D, Kaipf, I, Wagner J, Ostwald J, Schnitzler, H-U, (1989) Range estimation by echolocation in the bat *Eptesicus fuscus*: trading of phase versus time cues. J Acoust Soc Am 85:2642–2650.
Metzner W (1993) An audio-vocal interface in echolocating horseshoe bats. J Neurosci 13:1899–1915.
Middlebrooks JC (1992) Narrow-band sound localization related to external ear acoustics. J Acoust Soc Am 92:2607–2624.

Mogdans J, Schnitzler H-U, Ostwald J (1993) Discrimination of 2-wavefront echoes by the big brown bat, *Eptesicus fuscus* – Behavioral experiments and receiver simulations. J Comp Physiol A 172:309-323.

Møhl B (1986) Detection by a pipistrellus bat of normal and reversed replica of its sonar pulses. Acoustica 61:75-82.

Moss CF, Schnitzler H-U (1989) Accuracy of target ranging in echolocating bats: Acoustic information processing. J Comp Physiol A 165:383-393.

Neuweiler G, Bruns V, Schuller G (1980) Ears adapted for the detection of motion, or how echolocating bats have exploited the capacities of the mammalian auditory system. J Acoust Soc Am 68:741-753.

Olsen JF, Suga N (1991a) Combination sensitive neurons in the medial geniculate body of the mustached bat: encoding of relative velocity information. J Neurophysiol 65:1254-1274.

Olsen JF, Suga N (1991b) Combination sensitive neurons in the medial geniculate body of the mustached bat: encoding of target range information. J Neurophysiol 65:1275-1296.

O'Neill WE, Suga N (1982) Encoding of target range and its representation in the auditory cortex of the mustached bat. J Neurosci 2:17-31.

Park TJ, Pollak GD (1993) GABA shapes a topographic organization of response latency in the mustache bat's inferior colliculus. J Neurosci 13:5172-5187.

Pollak GD (1993) Some comments on the proposed perception of phase and nanosecond time disparities by echolocating bats. J Comp Physiol A 172:523-531.

Pollak GD, Bodenhamer RD (1981) Specialized characteristics of single units in the inferior colliculus of mustache bats: frequency representation, tuning and discharge patterns. J Neurophysiol 46:605-620.

Pollak GD, Henson OW Jr, Novick A (1972) Cochlear microphonic audiograms in the 'pure tone' bat, *Chilonycteris parnellii parnellii*. Science 176:66-68.

Ramprashad F, Money KE, Landolt JP, Laufer J (1978) A neuroanatomical study of the little brown bat (*Myotis lucifugus*). J Comp Neurol 178:347-363.

Riquimaroux H, Gaioni SJ, Suga N (1992) Inactivation of the DSCF area of the auditory cortex with muscimol disrupts frequency discrimination in the mustached bat. J Neurophysiol 68:1613-1623.

Roverud RC (1993) Neural computations for sound pattern recognition: evidence for summation of an array of frequency filters in an echolocating bat. J Neurosci 13:2306-2312.

Roverud RC, Grinnell AD (1985a) Discrimination performance and echolocation signal integration requirements for target detection and distance determination in the CF/FM bat, *Noctilio albiventris*. J Comp Physiol 156:447-456.

Roverud RC, Grinnell AD (1985b) Echolocation sound features processed to provide distance information in the CF/FM bat, *Noctilio albiventris*: evidence for a gated time window utilizing both CF and FM components. J Comp Physiol 156:457-469.

Roverud RC, Nitsche V, Neuweiler G (1991) Discrimination of wingbeat motion by bats correlated with echolocation sound pattern. J Comp Physiol A 168:259-263.

Rübsamen R (1987) Ontogenesis of the echolocation system in the rufous horseshoe bat, *Rhinolophus rouxi*, (Audition and vocalization in early postnatal development.) J Comp Physiol A 161:899-913.

Rübsamen R, Schäfer M (1990a) Ontogenesis of auditory fovea representation in the inferior colliculus of the Sri Lakan rufous horseshoe bat, *Rhinolophus rouxi*. J Comp Physiol A 167:757-769.

Rübsamen R, Schäfer M (1990b) Audiovocal interactions during development? Vocalisation in deafened young horseshoe bats vs. audition in vocalisation-impaired bats. J Comp Physiol A 167:771–784.

Rübsamen R, Neuweiler G, Marimuthu G (1989) Ontogenesis of tonotopy in inferior colliculus of a hipposiderid bat reveals postnatal shift in frequency-place code. J Comp Physiol A 165:755–769.

Schuller G (1979) Coding of small sinusoidal frequency and amplitude modulations in the inferior colliculus of the CF-FM bat, *Rhinolophus ferrumequinum*. Exp Brain Res 34:117–132.

Schuller G, Pollak GD (1979) Disproportionate frequency representation in the inferior colliculus of horseshoe bats: evidence for an "acoustic fovea" J Comp Physiol 132:47–54.

Schuller G, Beuter K, Schnitzler H-U (1974) Response to frequency-shifted artificial echoes in the bat, *Rhinolophus ferrumequinum*. J Comp Physiol 89:275–286.

Schnitzler H-U (1970) Comparison of echolocation behavior in *Rhinolophus ferrumequinum* and *Chilonycteris rubiginosa*. Bijdr Dierkd 40:77–80.

Schnitzler H-U, Flieger E (1983) Detection of oscillating target movements by echolocation in the greater horseshoe bat. J Comp Physiol 153:385–391.

Simmons JA (1979) Perception of echo phase in bat sonar. Science 204:1336–1338.

Simmons JA (1987) Acoustic images of target range in bat sonar. Naval Res Rev 39:11–26.

Simmons JA (1993) Evidence for perception of fine echo delay and phase by the FM bat, *Eptesicus fuscus*. J Comp Physiol A 172:533–547.

Simmons JA, Chen L (1989) The acoustic basis for target discrimination by FM echolocating bats. J Acoust Soc Am 86:1333–1350.

Simmons JA, Kick SA (1984) Physiological mechanisms for spatial filtering and image enhancement in the sonar of bats. Annu Rev Physiol 46:599–614.

Simmons JA, Moss CF, Ferragamo M (1990a) Convergence of temporal and spectral information into acoustic images of complex sonar targets perceived by the echolocation bat, *Eptesicus fuscus*. J Comp Physiol A 166:449–470.

Simmons JA, Ferragamo M, Moss CF, Stevenson SB. Altes RA (1990b) Discrimination of jittered sonar echoes by the echolocating bat, *Eptesicus fuscus*: the shape of target images in echolocation. J Comp Physiol A 167:589–616.

Simmons JA, Kick AS, Moffat AJM, Masters WM, Kon D (1988) Clutter interference along the target range axis in the echolocating bat, *Eptesicus fuscus*. J Acoust Soc Am 84:551–559.

Simmons JA, Kick SA, Lawrence BD, Hale C, Bard C, Escudie' B (1983) Acuity of horizontal angle discrimination by the echolocating bat, *Eptesicus fuscus*. J Comp Physiol 153:321–330.

Suga N (1964) Recovery cycles and responses to frequency-modulated tone pulses in auditory neurons of echolocating bats. J Physiol 175:50–80.

Suga N (1973) Feature extraction in the auditory system of bats. In: Møller AR (ed) Academic Press, pp. 675–744. New in auditory neurons of echolocating bats. J Physiol 175:50–80.

Suga N, Schlegel P (1973) Coding and processing in the auditory systems of FM-signal producing bats. J Acoust Soc Am 54:174–190.

Suga N, Shimozawa T (1974) Site of neural attenuation of responses to self-vocalized sounds in echolocating bats. Science 183:1221–1213.

Suga N, Jen PH-S (1977) Further studies on the peripheral auditory system of

'CF-FM' bats specialized for fine frequency analysis of Doppler-shifted echoes. J Exp Biol 69:207-232.

Suga N, Simmons JA, Jen PH-S (1975) Peripheral specialization for fine analysis of Doppler shifted echoes in the auditory system of the CF-FM bat *Pteronotus parnellii*. J Exp Biol 63:161-192.

Suga N, Neuweiler G, Möller J (1976) Peripheral auditory tuning for fine frequency analysis by the CF-FM bat, *Rhinolophus ferrumequinum*. IV. Properties of peripheral auditory neurons. J Comp Physiol 106:111-125.

Suga N, O'Neill WE, Manabe T (1978) Cortical neurons sensitive to combinations of information-bearing elements of biosonar signals in the mustached bat. Science 200:778-781.

Sullivan WE (1982a) Neural representation of target distance in auditory cortex of the echolocating bat *Myotis lucifigus*. J Neurophysiol 48:1011-1032.

Sullivan WE (1982b) Possible neural mechanisms of target distance coding in auditory system of the echolocating bat, Myotis lucifugus. J Neurophysiol 48:1033-1047.

Sum YW, Menne D (1988) Discrimination of fluttering targets by the FM bat *Pipistrellus stenopterus*? J Comp Physiol A 163:349-354.

Webster FA (1967) Performance of echolocating bats in the presence of interference. In: Busnel R-G (ed) Animal Sonar Systems: Biology and Bionics. Jouy-en-Josas-78, France: Laboratoire de Physiologie Acoustique, pp. 673-713.

Webster FA, Brazier O G (1965) Experimental studies on target detection, evaluation, and interception by echolocating bats. TDR No. AMRL-TR-65-172, Aerospace Medical Division USAF Systems Command, Tucson, AZ.

Wenstrup JJ, Fuzessery ZM, Pollak GD (1986) Binaural response organization within a frequency-band representation of the inferior colliculus: implications for sound localization. J Neurosci 6:692-973.

Wenstrup JJ, Fuzessery ZM, Pollak GD (1988) Binaural neurons in the mustache bat's inferior colliculus: responses of 60-kHz EI units to dichotic sound stimulation. J Neurophysiol 60:1369-1383.

Zook JM, Leake PA (1989) Connections and frequency representation in the auditory brainstem of the mustache bat, *Pteronotus parnellii*. J Comp Neurol 290:243-261.

2

Natural History and Biosonar Signals

M. Brock Fenton

1. Introduction

The diversity of bats is reflected in many aspects of their appearance and behavior. There are approximately 900 species of extant bats with most occurring in the tropics and subtropics. From continental settings to islands, including remote oceanic ones such as Hawaii, bats often are prominent members of the mammal fauna. In almost any country in subsaharan Africa, for example, bats are the most diverse group of mammals, with more species than rodents (Smithers 1983). Whether one considers bats by diet or roosting habits, they present an impressive array of approaches to living. This diversity is clearly reflected in their echolocation behaviors.

Echolocation is an active form of orientation that permits animals to be independent of lighting conditions. It is polyphyletic in vertebrates, having evolved in both birds and mammals (classes Aves and Mammalia, respectively; Fenton 1984). Furthermore, echolocation has appeared more than once in both of these classes, in the avian orders Caprimulgiformes and Apodiformes and in the mammalian orders Insectivora, Cetacea, and Chiroptera (Fenton 1984).

Although echolocation often is assumed for all bats, not all bats echolocate, and not all echolocating bats use the same echolocation signals or use echolocation for the same purpose. Differences in echolocation behavior can raise basic questions about the evolution of bats, manifested by arguments about their classification. Traditionally, bats are placed in the mammalian order Chiroptera with living species arranged in two suborders, the Megachiroptera and the Microchiroptera (Table 2.1). While some families of bats include living and fossil representatives (Megachiroptera or Microchiroptera), one is known only as fossils (Table 2.1). The Eocene and Oligocene fossils are typically different genera or species from living taxa.

TABLE 2.1. The classification, fossil, worldwide, and dietary diversity of bats.

Suborder/superfamily, family	First fossils		Worldwide distribution	Variation in diet	Number of living species
	Geological age	Years ($\times 10^6$)			
Palaeochiropterygoidea					
Palaeochiropterygidae (the "old bats")	Eocene	50	Europe and North America	Insects	0
MEGACHIROPTERA					
Pteropodidae (the Old World fruit bats; flying foxes)	Oligocene	30	European fossils;today, Old World tropics	Fruit, nectar, and pollen	150
MICROCHIROPTERA					
Emballonuroidea					
Rhinopomatidae (the rat-tailed bats)	Not known	—	North Africa to Southern Asia and Borneo	Insects	2
Craseonycteridae(hog-nosed bats)	Not known	—	Thailand	Insects	1
Emballonuridae (sheath-tailed bats)	Eocene	50	Tropics	Insects	44
Rhinolophoidea					
Megadermatidae (false vampire bats)	Eocene	50	European fossils; today, Old World tropics	Animals, from insects to vertebrates	5
Nycteridae (slit-faced bats)	Not known	—	Africa to Java and Sumatra	Animals, from insects to vertebrates	13
Rhinolophidae (horseshoe bats)	Eocene	50	Old World	Insects	69
Hipposideridae (Old World leaf-nose bats)	Eocene	50	Old World tropics	Insects	56
Phyllostomoidea					
Noctilionidae (bulldog bats)	Not known	—	New World tropics	Insects and fish	2
Mormoopidae (mustached bats)	Not known	—	New World tropics	Insects	8
Mystacinidae (short-tailed bats)	Not known	—	New Zealand	Insects, fruit, nectar, carrion	1
Phyllostomidae (New World leaf-nose bats)	mid-Miocene	22	New World tropics	Insects, fruit and pollen, vertebrate blood	123

Vespertilionoidea

Natalidae (funnel-eared bats)	Not known	—	New World tropics	Insects	4
Furipteridae (thumbless bats)	Not known	—	New World tropics	Insects	2
Thyropteridae (New World sucker-footed bats)	Not known	—	New World tropics	Insects	2
Vespertilionidae (plain-nosed bats)	Eocene	50	Worldwide	Insects, fish, and other vertebrates	283
Myzopodidae (Old World sucker-footed bats)	Not known	—	Madagascar	Insects	1
Molossidae (freetailed bats)	Not known	—	Tropical	Insects	82

The family names of bats are shown bold print. The geological name of the time of the first fossils in the family are shown with the approximate are in millions of years.

39

1.1 Phylogeny, Evolution, and Classification

Classifying bats in one order, the Chiroptera, implies that they are monophyletic, with all living and fossil bats sharing an immediate common ancestor. This position infers that the Megachiroptera and the Microchiroptera are more closely related to one another than either is to any other group of mammals. The alternative view, that bats are diphyletic, has been proposed by various workers during the past 100 years with the specific suggestion that the Megachiroptera and the order Primates share an immediate common ancestor, while the Microchiroptera are more closely related to mammals in the order Insectivora, the shrews and their allies (e.g., Pettigrew et al. 1989).

In one sense, the question of bat phylogeny revolves around the relative importance of similarities versus differences. Although the Megachiroptera share many characteristics with the Microchiroptera, the two groups also differ substantially. Although the wings of bats are generally similar (Thewissen and Babcock 1991), the Megachiroptera and Microchiroptera have quite different cervical vertebrae, with those of megachiropterans resembling the typical mammalian condition. The Microchiroptera, however, have extremely flexible necks and their cervical vertebrae differ strikingly from those of other mammals (Fenton and Crerar 1984). Similarly, the visual pathways of megachiropteran bats are more like those of primates, while those of microchiropterans resemble other mammals (Pettigrew 1991). Several recent papers have explored the similarities and differences between the two suborders in search of the "correct" interpretation (e.g., Pettigrew et al. 1989; Baker, Novacek, and Simmons 1991; Pettigrew 1991; Simmons, Novacek, and Baker 1991) and two schools of thought remain with no clear prevailing view or agreement about "truth" (e.g., Rayner 1991; Thewissen and Babcock 1991, 1992). A recent study of the structure of the epsilon-globin gene supports the view that bats are monophyletic (Bailey, Slightom, and Goodman 1992).

The study of echolocation is affected by the debate about the phylogeny of bats because most of the 150 or so species of Megachiroptera do not echolocate (Hill and Smith 1984; Fenton 1992). Furthermore, the echolocating megachiropteran, *Rousettus aegyptiacus* (the Appendix provides more information about each species of bat mentioned in this chapter), uses a completely different approach to signal production than do species in the Microchiroptera. While *R. aegyptiacus* makes echolocation sounds by clicking its tongue, the microchiropterans studied to date all use sounds produced in the larynx (vocalizations) as echolocation signals (Suthers 1988). The monophyletic position poses two alternatives: (1) that echolocation is an ancestral trait (e.g., Fenton 1974) that has been lost in the Megachiroptera and reacquired in one species; or (2) that echolocation is not an ancestral trait, but has evolved independently in both suborders. The diphyletic position necessitates the independent evolution of echolocation in the two major groups of bats.

The fossil record does not help to resolve the question of bat phylogeny. The first fossil bats occur in Eocene strata, a time when at least eight families were represented. Although there are no fossils of animals that appear to be intermediate between bats and something else, most mammalogists suspect that bats developed from a small, arboreal, insectivorous mammal with long arms and fingers. Flight is presumed to have evolved when the proto-bat glided from one tree to another. The "top-down" theory proposes that the ancestor of bats captured insects along the trunks and branches of trees, working its way up to the top before gliding down to begin again on another tree (Norberg 1990). In a variation on the top-down theory, Jepsen (1970) proposed that the ancestor of bats used membranes between the fingers and body as insect nets so the incipient wing membranes simultaneously may have served at least two functions. The diversity of bats in the Eocene (see Table 2.1) makes it easy to presume that the group has a long history, perhaps originating in the early Palaeocene or late Cretaceous.

In spite of their early appearance and diversity in the fossil record, there are relatively few fossil bats. One currently accepted phylogeny of the Chiroptera (Fenton 1992) reflects the presumed relationships among the living families and superfamilies. Together with the data in Table 2.1, the phylogeny (Fig. 2.1) reflects the diversification of living bats in terms of both diet and geographic distribution. Although the general relationships among the families and superfamilies appear clear, as noted there is still no general agreement about whether the Megachiroptera are the closest living relatives of the Microchiroptera.

There are other questions about relationships among the Chiroptera. In Figure 2.1, the New Zealand Mystacinidae are grouped with New World bats that constitute the superfamily Phyllostomoidea, reflecting a number of morphological and molecular traits (Pierson et al. 1986). Traditionally, the Mystacinidae have been placed in the Vespertilionoidea (e.g., Koopman and Jones 1970). In this way, the arrangement of living Microchiroptera in different families changes with the appearance of new evidence or the reanalysis of older data.

Other points of contention in bat classification concern the numbers of families, with some biologists arguing for fewer families and others for more. In the past, the three living species of vampire bats have been treated as a distinct family, the Desmodontidae, largely because of their specializations for feeding on blood. In their basic structure, however, including skeletal features and sperm morphology, genetics, and biochemical features, the vampires closely resemble bats in the family Phyllostomidae, and most biologists now place them there as a subfamily (Desmodontinae; Koopman 1988). The opposite trend involves some Old World bats. While some workers combine the Hipposideridae and the Rhinolophidae into one family (Rhinolophidae; e.g., Vaughan 1986), others stress the differences between these bats and treat them as distinct families (e.g., Fenton 1992). As recently as 1974 a new family of living bats was described (Hill 1974), suggesting that we may not yet appreciate the full diversity of bats.

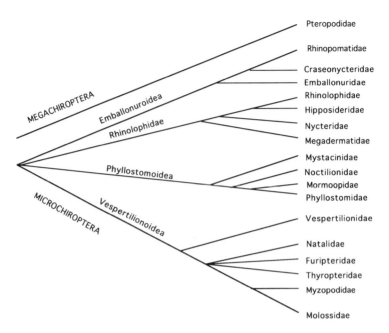

FIGURE 2.1. In this phylogeny of the families of living bats, the suborders Megachiroptera and the Microchiroptera do not share a common ancestor. Furthermore, the Mystacinidae are included among the superfamily Phyllostomoidea rather than with the Vespertilionoidea. The arrangement may or may not reflect reality (after Fenton 1992), and recent molecular evidence (Bailey, Slightom, and Goodman 1992) suggests that both suborders share an immediate common ancestor.

At another level, changes in the classification of bats affect the names by which they are known in the literature. For the student of echolocation it is vital to realize that *Chilonycteris parnellii* is now called *Pteronotus parnellii*, following Smith's (1972) revision of the family Mormoopidae. In some cases the morphological diversity of bats may have been underestimated by earlier classifications. For example, the tribe Plecotini constitutes a distinct group of vespertilionid bats, characterized by enormous ears and, presumably, distinctive foraging behavior. Although there have been various assessments of the systematic status of the bats in this group, the most recent study suggests that the North American forms are distinct from those in Eurasia (Tumlison and Douglas 1992) and should be placed in separate genera (*Corynorhinus*, and *Idionycteris* versus *Plecotus*). This situation also may be reflected by differences in echolocation and foraging behavior (e.g., Simmons and O'Farrell 1977; Anderson and Racey 1991).

1.2 The Diversity of Bats

There are some obvious generalizations about megachiropteran and microchiropteran bats (Hill and Smith 1984; Fenton 1992). Megachiropteran bats occur in the Old World tropics and feed mainly on fruit or nectar and pollen. Their teeth are highly specialized for these diets, and they depend upon flight only to get from one place to another. Their pectoral girdles are relatively unspecialized. The available data suggest that the echolocating species of Megachiroptera uses this form of orientation to find its way in the darkness of its roosts, which are usually located in dark hollows, often in caves.

Microchiropteran bats, by comparison, occur everywhere there are bats. They fill different trophic roles, feeding on a variety of animals, fruit, nectar and pollen, and blood. With the exception of blood feeders, the teeth of Microchiroptera are relatively unspecialized. Many species, however, use flight when feeding and their pectoral girdles are often highly specialized, reflecting aerial agility and maneuverability. Many microchiropterans use echolocation to detect and evaluate targets they treat as prey (see Schnitzler, Simmons, and O'Neill, Chapters 3, 4, and 9, respectively). In various aspects of the brain, reproductive systems, cervical vertebrae, and other features, the Megachiroptera differ from the Microchiroptera, but in general appearance, particularly in wing structure, they are very similar (Hill and Smith 1984; Thewissen and Babcock 1992).

Bats range in size from adult body masses of 2 g to 1500 g, with most species weighing less than 50 g. Length of forearm, a general indication of body size in bats, ranges from about 25 mm to about 230 mm, corresponding to wingspans of about 15 cm and 200 cm, respectively. While most species of bats (see Table 2.1) feed mainly on insects, others eat animals ranging from frogs to fish, from scorpions to millipedes, and from other bats to birds. Larger species tend to take larger animal prey, including small vertebrates, in their diets (Norberg and Fenton 1988). In the Megachiroptera (Family Pteropodidae), most species are thought to feed mainly on fruit, but a few are specialized for nectar and pollen (Hill and Smith 1984; Fenton 1992). This dependence on plant material as food is paralleled in the Neotropics by various subfamilies within the microchiropteran Family Phyllostomidae, the New World leaf-nosed bats. Other phyllostomids feed on animals, and three species, the vampires, depend entirely on the blood of other vertebrates. Fish-eating is known from microchiropteran bats in at least four families (Noctilionidae, Megadermatidae, Nycteridae, and Vespertilionidae), but details of fishing behavior have been studied only in the Noctilionidae (Suthers 1965; Hill and Smith 1984; Fenton 1992).

While some species of bats, perhaps most of them, live relatively solitary lives and roost in foliage, others congregate in huge numbers in roosts such as caves and some artificial structures. A few species live in small social units, often harems (e.g., Bradbury 1977), which appear to be well organized and involve a high level of individual recognition of group members. This organization is most evident in common vampire bats (*Desmodus rotundus*), among whom individuals will regurgitate blood for roost mates that are unsuccessful foragers (Wilkinson 1985). Little is known about the social lives of most of the 900 or so species of bats, but the available data suggest diversity, generally reflecting that shown in dietary habits and morphological structures (Hill and Smith 1984; Fenton 1992).

1.3 Ontogeny

In bats studied to date, as in many other mammals, hearing develops after birth even though neonates begin to vocalize almost immediately (Brown 1976; Brown and Grinnell 1980; Rubsamen, Neuweiler, and Marimuthu 1989; Rubsamen and Schafer 1990a). In some species, echolocation calls appear in the first 10 days or so, coinciding with the development of hearing (Brown 1976). This pattern prevails both in bats such as *Antrozous pallidus*, with typically mammalian patterns of hearing sensitivity (audiograms; Brown 1976), and in species such as *Hipposideros speoris* and *Rhinolophus rouxi*, which have specialized patterns of hearing coinciding with a strikingly different approach to echolocation (see Section 2.2; and Simmons; Kössl and Vater; Casseday and Covey; Pollak and Parks; Wenstrup; and O'Neill and Winer, Chapters 4–10, respectively) apparently specialized for detecting fluttering targets (Rubsamen, Neuweiler, and Marimuthu 1989; Rubsamen and Schafer 1990a). In *Rhinolphus rouxi* (Rubsamen, Neuweiler and Marimuthu 1989) there are developmental shifts in the cochlear frequency-place code associated with specialized hearing, and in *Hipposideros speoris* the development of hearing specialization cannot be explained by a single developmental parameter (Rubsamen, Neuweiler, and Marimuthu 1989). By deafening young animals, Rubsamen and Schafer (1990b) demonstrated that in *Rhinolophus rouxi* echolocation pulse production is under auditory feedback control. The emergence and development of the specialized hearing curve (the so-called auditory fovea) is an innate process that occurs through shifts in frequency tuning.

The development of vocalizations and hearing suggests that in a variety of bats (e.g., *Antrozous pallidus* [Brown 1976], *Myotis lucifugus* [Buchler 1980], *Rhinolophus rouxi* [Rubsamen, Neuweiler, and Marimuthu 1989], and *Hipposideros speoris* [Rubsamen and Schafer 1990a]), young animals can echolocate at least a week to 10 days before they start to fly. The clumsiness of bats' first flights could reflect the mechanical challenges of learning to fly and the difficulties of using echolocation during flight (e.g., Buchler 1980).

2. Biosonar Signals of Bats

Abbreviations are commonly used when discussing bats and their approaches to echolocation, and the literature abounds with terms such as CF (constant-frequency) and FM (frequency-modulated), describing different bat signals (e.g., Busnel and Fish 1980; Hill and Smith 1984; Nachtigall and Moore 1988). In the same volumes, and in other literature, these modifiers often are extended to the bats themselves, giving us CF bats, CF-FM bats, and FM bats (Fig. 2.2). While FM clearly describes a signal (or component) showing frequency modulation over time, CF is more ambiguous because it is used for signals with bandwidths of less than 1 kHz as well as signals with bandwidths of several kilohertz. As noted elsewhere (e.g., Neuweiler and Fenton 1988), it would be more prudent to recognize that the shorthand describes the signals and their components, but not necessarily the bats or their use of them.

This point is particularly important in the context of a bat's ability to use echolocation to detect a fluttering target, usually an insect flapping its wings. Traditionally, flutter detection has been associated with bats that produce CF signals most of the time, "high duty cycle" species (Fig. 2.3) that actually produce signals at least 50% and usually 80% or more of the time they are echolocating (e.g., Neuweiler and Fenton 1988). The high duty cycle bats are horseshoe bats (Rhinolophidae), Old World leaf-nosed bats (Hipposideridae), and the mormoopid *Pteronotus parnellii*. Other micro-chiropterans are "low duty cycle" species that produce calls less than 20% of the time they are echolocating.

Flutter-detection behavior is not limited to high duty cycle bats using narrowband signals. While Roverud, Nitsche, and Neuweiler (1991) showed that two CF species were better at flutter detection than one low duty cycle FM species (*Rhinolophus rouxi* and *Hipposideros lankadiva*, versus *Eptesicus fuscus*), Summe and Menne (1988) found that the FM bat *Pipistrellus stenopterus* used its echolocation to detect fluttering targets. Furthermore, Casseday and Covey (1991) found "filter" neurons in the inferior colliculus of *Eptsicus fuscus* that were as sharply tuned as those in some high duty cycle CF bats (see Covey and Casseday, Chapter 6).

The overlap between species and intraspecific flexibility in echolocation calls parallels the situation in wing morphology (e.g., Norberg and Rayner 1987), communities (e.g., Aldridge and Rautenbach 1987), or general morphology (e.g., Findley and Wilson 1982), and together these suggest broad overlap between species.

The following sections document the ways that bats vary different aspects of their echolocation signals, demonstrating the danger of generalizing broadly about different approaches to echolocation. Evidence supports two basic approaches, one based on signal production and the other on duty cycle. Beyond these fundamental differences in signal production, abbreviations used to describe signals and their components are most valuable when the broader picture of variation is clearly presented.

The Chiroptera studied to date appear to depend mainly on two kinds of signals for echolocation, the tongue clicks of the echolocating Megachiroptera (the pteropodid *Rousettus aegyptiacus*) and the vocalizations of the

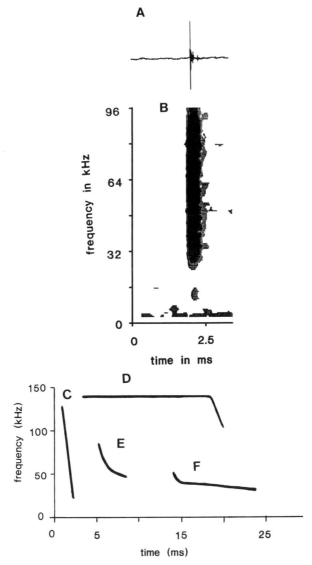

FIGURE 2.2A–F. While the pteropodid *Rousettus aegyptiacus* produces tongue clicks as echolocation calls (A, time-amplitude plot; B, sonogram), the microchiropteran bats use tonal calls that show structured change in frequency over time (C–F). Search-phase bat echolocation calls, include those that would be considered frequency modulated (FM; C, E, F) and constant frequency (CF) ending in an FM sweep (D), reflect low and high duty cycles, respectively. The short, broadband call (C) is typically of low to medium intensity and common among species of *Myotis* that take prey from surfaces (= glean; e.g., *Myotis auriculus* [Fenton and Bell 1979]; *Myotis emarginatus* [Schumm, Krull, and Neuweiler 1991]). The long call dominated by a narrowband (CF) component (D)

FIGURE 2.3A,B. Two time-amplitude plots compare a sequence of echolocation calls produced at low duty cycle (A: 13.6%, *Lasiurus borealis*) and high duty cycle (B: 56.8%; *Rhinolophus landeri*).

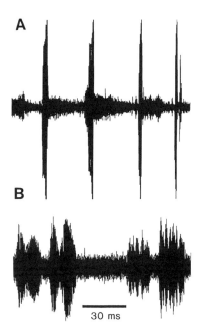

30 ms

Microchiroptera (Suthers 1988). Echolocating Microchiroptera use a variety of call designs (see Fig. 2.2) to collect information about their surroundings, coinciding with different approaches to echolocation. Bats also produce other vocalizations, apparently for communication with other bats. Many putative social communication calls are much longer in duration than echolocation calls (Fenton 1985), and this, combined with the speed of sound in air, atmospheric attenuation of high-frequency sound, and the flight speeds of bats, probably precludes the use of longer social calls in echolocation. Quite simply, long-duration signals from a flying bat could mask their echoes rebounding from nearby targets. A possible exception is provided by the long, low-frequency calls given by Hawaiian hoary bats (*Lasiurus cinereus*) during intraspecific chases at feeding grounds (Belwood and Fullard 1984). A comparison of some features of bat echolocation calls is presented in Table 2.2.

←

FIGURE 2.2A–F. (*continued*) is typical of high duty cycle bats, rhinolophids, hipposiderids, and the mormoopid *Pteronotus parnellii*, with different species showing different CF frequencies (e.g., Heller and von Helverson 1989). One call (E) combines narrowband and broadband components and is typical of many species that hunt airborne targets in along the edges of woodlands, for example, while the longer call dominated by a narrowband component (F) is commonly associated with species that hunt in more open settings.

TABLE 2.2. The sizes and echolocation features of extant bats.

Suborder/superfamily	Size FA (nm)	Echolocate (yes/no)	Signal type	Duty cycle %	Intensity	Duration (msec)	Bandwidth
MEGACHIROPTERA							
Pteropodidae	40–230	No	—	—	—	—	—
Rousettus aegyptiacus		Yes	Click	≤20	Low	1–2	Broad
MICROCHIROPTERA							
Emballonuroidea							
Rhinopomatidae	55–70	Yes	Vocal	≤20	High	10	Narrow
Craseonycteridae	22–26	Yes	Vocal	≤20	High	5	Broad
Emballonuridae	32–80	Yes	Vocal	≤20	High	10–20	Narrow
Rhinolophoidea							
Megadermatidae	50–115	Yes	Vocal	≤20	Low	1	Broad
Nycteridae	36–60	Yes	Vocal	≤20	Low	1	Broad
Rhinolophidae	30–75	Yes	Vocal	≥80	High	>20	Narrow
Hipposideridae	30–110	Yes	Vocal	≥80	High	10	Narrow
Phyllostomoidea							
Noctilionidae	70–92	Yes	Vocal	≤20	High	10	Broad
Mormoopidae	35–65	Yes	Vocal	≤20	High	5–10	Broad
Pteronotus parnellii		Yes	Vocal	≥80	High	10	Narrow
Mystacinidae	44–48						
Phyllostomidae	30–105	Yes	Vocal	≤20	Low	1–4	Broad
Vespertilionoidea							
Natalidae	27–41	Yes	Vocal	≤20	Low	1–5	Broad
Furipteridae	30–40	Yes	Vocal	≤20	?	?	?
Thyropteridae	27–38	Yes	Vocal	≤20	?	?	?
Vespertilionidae	22–75	Yes	Vocal	≤20	High	5	Broad
		Yes	Vocal	≤20	High	10–20	Narrow
		Yes	Vocal	≤20	Medium	1–5	Broad
Myzopodidae	44–48	(Yes)	(Vocal)	(≤20)	?	?	?
Molossidae	27–85	Yes	Vocal	≤20	High	10–20	Broad

Size is shown as forearm (FA) length in millimeters; signal type is either tongue click (click) or vocalization (vocal).

2.1 Intensity

It has been known for some time that the echolocation calls of bats vary in intensity (measured 10 cm in front of the bat) from low (<60 dB SPL [sound pressure level]) to high (>110 dB SPL) (Griffin 1958). Among echolocating bats there is an obvious gradient in call intensity that includes many species producing calls of intermediate strength. The situations in which bats operate appear to influence call intensity. For example, bats searching for airborne targets produce intense echolocation calls, while those searching for prey on surfaces depend more on quieter calls. Call intensity is consistent in some families; for example, the Phyllostomidae use low-intensity calls and the Molossidae high-intensity calls (Table 2.2). In other families, such as the Vespertilionidae, some species produce high-intensity calls and others low-intensity ones. At least one species, the vespertilionid *Myotis emarginatus*, adjusts the intensity of its call according to the situation in which it is hunting (Schuum, Krull, and Neuweiler 1991).

Differences in call intensity pose important challenges to people studying the echolocation of bats. Species using high-intensity calls are easily monitored with microphones (bat detectors) sensitive to the frequencies in the calls, while those using low-intensity calls are virtually undetectable even by very sensitive bat detectors (e.g., Fenton and Bell 1981; Fenton et al. 1992).

2.2 Durations, Duty Cycles, and Pulse Repetition Rates

Two fundamentally different approaches to microchiropteran echolocation are reflected in the duty cycles of their calls, or the percentage of time that signals are being produced. As noted earlier, most echolocating bats have duty cycles of less than 20% (Table 2.2; Fig. 2.3), and these species appear unable to tolerate overlap between pulse and echo (Schnitzler 1987). In the other approach, species in the families Hipposideridae and Rhinolophidae, and the mormoopid *Pteronotus parnellii* (Table 2.2), duty cycles regularly exceed 80% and these species can tolerate overlap between pulses and their echoes. The auditory systems of high duty cycle bats are highly specialized for exploiting Doppler-shifted echoes generated by fluttering targets (see Moss and Schnitzler, Simmons et al., Kössl and Vater, Covey and Casseday, Pollak and Park, Wenstrup, and O'Neill, Chapters 3–9, respectively).

Low duty cycle bats adjust the durations of individual calls according to the situations in which they are operating. During an attack on an airborne target, species such as *Lasiurus cinereus* shorten their calls from 20 msec to about 1 msec, as they detect, approach, and close in on a target (Barclay 1986; Obrist 1989). Shorter echolocation calls are essential in the final stages of the attack to ensure that the outgoing pulse does not mask the fainter returning echo. Other species, for example, *Myotis daubentoni*

(Kalko and Schnitzler 1989), use shorter echolocation calls (≤ 5 msec) when searching for targets and calls of 1 msec or less when closing with prey.

High and low duty cycle bats that hunt airborne targets increase their pulse repetition rates as individual calls shorten through an attack sequence. When searching for targets, low duty cycle bats produce calls at intervals ranging from 50 to 300 ms, while during attacks (feeding buzzes; Griffin, Webster, and Michael 1960), intercall intervals are closer to 5 msec. Even during feeding buzzes, the duty cycles of these bats rarely exceed 20% (Obrist 1989).

When low duty cycle bats take prey from surfaces (glean) they usually produce short (< 1-msec) echolocation calls (see Table 2.2) during their final approaches, but show no evidence of high pulse repetition rates (feeding buzzes) during actual attacks on targets (e.g., *Myotis auriculus* [Fenton and Bell 1979]; *Trachops cirrhosus* [Barclay et al. 1981]; *Nycteris grandis* and *Nycteris thebaica* [Fenton, Gaudet, and Leonard 1983]; *Myotis evotis* [Faure and Barclay 1992]). At least two species that take prey from surfaces also pursue airborne targets, situations in which they produce feeding buzzes (*Myotis emarginatus* [Schumm, Krull, and Neuweiler 1991]; *Antrozous pallidus* [Krull 1992]). This pattern of behavior probably also occurs in other gleaners.

The high duty cycle bats (rhinolophids, hipposiderids, and *P. parnellii*) produce calls that vary in duration from around 10 msec in hipposiderids to more than 50 msec in rhinolophids. During attacks on airborne targets, these bats shorten the durations of individual calls while maintaining high duty cycles (e.g., Griffin and Simmons 1974; Neuweiler et al. 1987).

2.3 Patterns of Frequency Change over Time

In these patterns, there are two basic kinds of echolocation pulses. The first are short, broadband clicks typical of *Rousettus aegyptiacus* (see Fig. 2.2b) and other echolocators such as birds, shrews, and cetaceans. The second are those of the Microchiroptera, tonal sounds showing structured changes in frequency over time (see Fig. 2.2). There is no evidence that broadband clicks offer any special advantage or disadvantage in echolocation relative to broadband tonal sounds (Buchler and Mitz 1980). It is possible that tonal echolocation sounds originated as communication signals (Fenton 1984).

An array of sonograms of bat echolocation calls (see Fig. 2.2) illustrates variations in patterns of frequency change over time as well as different patterns of bandwidth and duration. Short, broadband calls typically show sharp changes in frequency over time, while longer calls of narrower bandwidth have more gradual changes. While hunting, some species using short calls always show rapid changes in frequency over time (e.g., *Myotis daubentoni*, as noted previously), but others such as *Lasiurus cinereus* use rapid rates of frequency change mainly in feeding buzzes.

2.4 Variations in Echolocation Calls

The various combinations of intensity, duration, duty cycle, and patterns of frequency change over time (see Table 2.2) illustrate the diversity of echolocation call design in bats. Some workers have proposed using variation in echolocation calls to address questions about the evolution of bats (e.g., Simmons and Stein 1980). The recurring use of short, broadband signals or long, narrowband signals by species in several families (Table 2.2), however, weakens the argument that echolocation call design reflects evolution and phylogeny.

Flexiblity is particularly evident in the case of *Myotis emarginatus* (Schumm, Krull, and Neuweiler 1991), which dramatically alters the intensity, durations and patterns of frequency change over time in its calls according to the situation in which it is hunting. When searching for prey on surfaces, these bats use short broadband signals of medium intensity but when the targets are airborne, they use long signals of high intensity that are dominated by narrowband components. There are hints of comparable behavior in other species (e.g., *Megaderma spasma* [Tyrrell 1988]; *Myotis septentrionalis* [Faure, Fullard, and Dawson 1993, and unpublished observations]). Together, the data on flexibility in call design suggest a strong environmental component, with bats adjusting their call design to provide best operation in the specific settings where they are hunting. If this is true, then individuals hunting in the open may be expected to use different call design and pulse repetition rates (narrowband and low pulse repetition rates) than when they are hunting in cluttered situations (broadband and higher pulse repetition rates).

There is also a geographical component to variation in call design as illustrated by the search-phase echolocation calls of *Eptesicus fuscus*. In eastern North America, for example, around Toronto (Canada), *E. fuscus* typically produces five 10-msec-long broadband echolocation calls with energy distributed across a broad bandwidth (Brigham, Cebek, and Hickey 1989). In south central British Columbia (Canada), however, this species produces ten 15-msec-long calls dominated by narrow bandwith components (Obrist 1989). Such variation may reflect more situation-specific behavior; the differences between hunting in open versus in cluttered habitats.

An initial indication of intraspecific variation in echolocation calls was provided by observations of the frequencies dominating the high duty cycle vocalizations of rhinolophid and hipposiderid bats (summarized in Novick 1977; see also Heller and von Helverson 1989). The frequencies of the narrowband (CF) calls of some rhinolphids and hipposiderids varied geographically (e.g., Fenton 1986) as do those in lower duty cycle, broadband calls (e.g., Thomas, Bell, and Fenton 1987; Brigham, Cebek, and Hickey 1989).

More recently, Cebek (1992) has found statistical evidence supporting the preliminary observations of Brigham, Cebek, and Hickey (1989) that *Eptesicus fuscus* uses colony-specific echolocation calls. In both studies, statistical analysis of different features of echolocation calls (frequency with maximum energy, lowest frequency) allowed a computer to correctly assign bats to some colonies on the basis of features of echolocation calls. J.S. Wilkinson (personal communication, February 1993) has similar data for *Nycticeius humeralis*. In *E. fuscus*, colony-specific echolocation calls do not coincide with genetic differences (Cebek 1992) and may reflect the communicative nature of echolocation calls (see Section 6.1).

Other studies have produced evidence of intraspecific variation in echolocation calls according to the setting in which bats were operating. In a rhinopomatid (Habersetzer 1981), some emballonurids (Barclay 1983), and at least one molossid (Zbinden 1989), bats changed the frequencies of their echolocation calls when flying with other (conspecific) bats. In a more detailed, longer-term study of marked individuals, Obrist (1989) observed that four species of vespertilionids might change the frequencies dominating their calls, call strengths, and pulse repetition rates when flying with other bats. Variations in calls in these settings could reflect jamming, avoidance behavior, or some communicative role for echolocation calls.

Although it is clear that bats vary their echolocation calls, these vocalizations often are significantly less variable in some parameters than their social calls (Fenton, 1994).

2.5 Species-Specific Echolocation Calls

In spite of a growing appreciation of the variations in the echolocation calls of microchiropteran bats just noted (e.g., Thomas, Bell, and Fenton 1987), many species produce quite distinctive echolocation calls (e.g., Ahlén 1981; Fenton and Bell 1981; Thomas and West 1989; Fullard et al. 1991). It is clear that in any location the value of identifying bat species by their echolocation calls will depend upon the local bat fauna and on the measurement instrumentation. For example, at many locations in southern Ontario (Canada), the echolocation calls of some of the eight local species of bats are readily distinguished from one another using a narrowband bat detector (Fig. 2.4). At 20 kHz, the long, chirplike signals of *Lasiurus cinereus* are distinctive, while similar sounds at 40 kHz are usually the calls of *Lasiurus borealis*. Tuned to 40 kHz, however, the same bat detector will pick up echolocation calls of *Eptesicus fuscus*, *Lasionycteris noctivagans*, *Pipistrellus subflavus*, and three species of *Myotis*, complicating the situation. As noted in the original papers about using echolocation calls to identify bats, however, monitoring the calls of individual bats marked with active tags can maximize the chances of correctly identifying the signaller (Bell 1980; Fenton and Bell 1981).

2.6 The Operational Range of Echolocation

Echolocation in air is a short range operation (Griffin 1958). In 1982, Kick reported the results of behavioral experiments which demonstrated that an

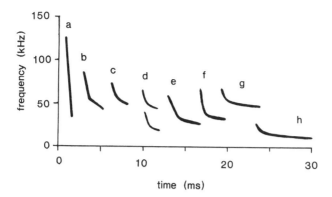

FIGURE 2.4. Eight species of bats that occur in southeastern Ontario produce different search-phase echolocation calls. Shown here are the calls of *Myotis septentrionalis* (*a*), *Myotis lucifugus* (*b*), *Myotis leibii* (*c*), *Pipistrellus subflavus* (*d*), *Eptesicus fuscus* (*e*), *Lasionycteris noctivagans* (*f*), *Lasiurus borealis* (*g*), and *Lasiurus cinereus* (*h*). The calls of *Pipistrellus subflavus* show a distinct second harmonic that is conspicuous when heard over a narrowband bat detector (strongest signals at 20 and 40 kHz). Most of these data are from Fenton and Bell (1979, 1981), Barclay (1986), and Obrist (1989), those for *P. subflavus* from MacDonald et al. (1994).

echolocating *Eptesicus fuscus* first detected a 19-mm-diameter sphere at a range of 5 m, the equivalent of 40 nosetip-to-tailtip lengths. These data confirmed that echolocation in air is a short-range operation, an impression remaining from observation of the distances at which bats reacted to targets and obstacles (e.g., Griffin 1958).

Interpulse intervals probably give an impression of the operational ranges of echolocating bats using low duty cycle calls. These bats are presumed to be intolerant of overlap between the outgoing pulse and the returning echo (Schnitzler 1987; see also Schnitzler, Chapter 3, and Simmons, Chapter 4). When interpulse intervals are considered with respect to the speed of sound in air, the low duty cycle echolocating bats could have operational ranges from 2.4 m to just over 62 m, corresponding to interpulse intervals of 20 msec and 365 msec for *Nycteris grandis* and *Euderma maculatum*, respectively [Fenton 1990]). Interpulse intervals, however, may provide no indication of the range at which bats are actively searching for prey.

3. Echolocation and the Hunting Strategies of Animal-Eating Bats

Microchiropteran bats may or may not use echolocation to detect, track, and assess targets. Species hunting airborne prey, usually insects, appear to depend more on echolocation than do species taking prey from surfaces such as the ground or foliage. Airborne targets are tracked by bats using low duty cycle, broadband or narrowband signals as well as high duty cycle calls dominated by narrowband components (see Table 2.2). While short, broadband, frequency-modulated sweeps provide bats with detailed information about target detail (e.g., Schmidt 1988), particularly at short range, other call designs may serve different functions. A convincing demonstration of this is Roverud's (1987) work showing that the narrowband component in the echolocation calls of *Noctilio albiventris* primes the auditory system for processing echoes by opening a window for analysis. In behavioral experiments, Roverud (1987) could effectively jam these bats' echolocation systems by presenting them with appropriate narrowband pulses of sound. Roverud (1987) suggested similar processes in *Rhinolophus ferrumequinum*.

There remains no clear indication of the function of the broadband FM sweeps that dominate the terminal parts of the echolocation calls of rhinolophid and hipposiderid bats and *Pteronotus parnellii* (e.g., Fig. 2.2b). Evidence that these call components provide details about the target is provided by some studies but not others. Support for this function comes from observations of *Hipposideros caffer* approaching (to within about 15 cm) but not attacking the fluttering target presented by a small electric motor with rotating flaps of masking tape (Bell and Fenton 1984). *Pteronotus parnellii*, however, attacks and grabs a similar fluttering target (Goldman and Henson 1977). While *H. caffer* clearly made an abort decision before contacting the target, *P. parnellii* only rejected the target after contact, suggesting differences in echolocation behavior.

3.1 Airborne Targets

Echolocating bats may hunt airborne targets while in flight or from a perch. While there have been many studies of the echolocation signals and behavior of bats that hunt in flight, less is known about the ones hunting from perches. Furthermore, the approach to foraging may change over time, with young animals hunting from perches and adults from flight (Buchler 1980). Several species switch from one approach to the other (e.g., Vaughan 1976; Fenton and Rautenbach 1986; Neuweiler et al. 1987), particularly when food is scarce (Fenton et al. 1990). Cost of hunting is one important difference between the two strategies, because flying bats consume energy at 7 to 25 times the rate of sitting bats (Thomas 1987). The two strategies also can result in bats encountering suitable prey at different rates.

As noted earlier, bats change the design and pattern of their echolocation calls as they detect, follow, and close with an airborne target, culminating in a feeding buzz (Fig. 2.5). Traditionally, the different stages in the process have been identified, from "search" through "approach" and "terminal" (e.g., Griffin, Webster, and Michael 1960; Barclay 1986; Kalko and Schnitzler 1989), reflecting times when bats are collecting different kinds of

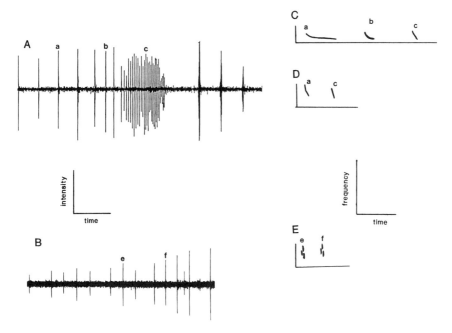

FIGURE 2.5A–E. Bats attacking airborne targets typically produce feeding buzzes (A) as they close with their prey (500-msec-long sequence), while species taking prey from surfaces (B) do not (500-msec-long sequence). While some species dramatically change the durations and patterns of frequency change over time of their calls over a feeding buzz (e.g., *Lasiurus cinereus*; C), others show less striking changes (e.g., *Myotis lucifugus*; D). Species such as *Macrotus waterhousii* show virtually no change in call durations or patterns of frequency change over time during an approach to a target on the ground (E). Calls indentified by lowercase letters in the attack sequences (A, B) correspond to sonograms with the same letters (C, D, E). *Lasiurus cinereus* calls are 15 msec long (*a*), 5 msec long (*b*), and 3 msec long (*c*); *Myotis lucifugus* calls are 3 msec (*a*) and 1 msec (*c*) long; and *Macrotus waterhousii* calls are 0.75 msec long (*e, f*).

information (detection versus tracking and evaluation). Kick and Simmons (1984) proposed "track" as a fourth stage between search–approach and terminal, a position not widely accepted by others (e.g., Kalko and Schnitzler 1989). The situation is complicated because different species show different patterns of behavior. For example, low duty cycle bats using long narrowband calls when searching for targets dramatically change both pulse repetition rates and call design during an attack sequence. However, low duty cycle bats, using shorter broadband signals and higher pulse repetition rates, make more gradual changes in call design and pulse repetition rate during an attack sequence (see Fig. 2.5). In some species, therefore, there are marked changes in call features and call rates over an attack sequence, while in others the changes in call features are less striking.

Apparent changes in echolocation calls over the course of an attack sequence may reflect differences in a bat's position relative to a microphone rather than actual changes in the calls. Photographs of flying *Myotis lucifugus* and *Lasiurus borealis* attacking targets (Griffin, Webster, and Michael 1960) demonstrated the drastic changes in body posture and head position associated with bats detecting and responding to airborne targets. We still lack details about the changes in head position made by foraging bats as they search for airborne targets, although Masters, Moffat, and Simmons (1985) documented changes in head position for *Eptesicus fuscus* using echolocation to track a target while sitting on a platform.

Flying bats actually catch airborne targets in different ways, either in the interfemoral membrane or on parts of the wing (Hill and Smith 1984). Work with captive *Rhinolophus ferrumequinum* revealed that some individuals invariably caught flour-coated mealworms in the wing pocket near their wrists (Trappe and Schnitzler 1982). The available data, much of it anecdotal and descriptive, leaves a strong impression that hunting bats show striking behavioral flexibility, particularly during the final pursuit. For example, *Lasiurus borealis* and *Lasiurus cinereus* dive to within 1 m of the ground in pursuit of a moth, and once a *L. cinereus* made an Immelman turn with a radius of about 1.5 m as it turned back to chase a stone that had been tossed into the air (Fenton, unpublished data). The bat began its feeding buzz within a second of making the turn and contacted the stone, which it then dropped. In other situations, high duty cycle bats have been observed taking prey from surfaces such as the ground, trunks of trees, or foliage (e.g., Bell and Fenton 1984; Link, Marimuthu, and Neuweiler 1986).

It is obvious that bats chasing airborne targets use echolocation to detect and track prey (Campbell and Suthers 1988). Bat attacks on stones and other small objects thrown into the air (e.g., *Lasiurus borealis* and *Lasiurus cinereus* [Hickey and Fenton 1990; Acharya and Fenton 1992]; *Myotis lucifugus* and *Myotis yumanensis* [R.M.R. Barclay and R.M. Brigham, personal communication, October 1992]) cast some doubt on how much prey evaluation is done by echolocation. Temporal and spectral cues are vital to these operations and may play important roles in each stage of the process. A large body of evidence demonstrates the value of temporal information to bats, partly through illustrations of their ability to finely measure time (e.g., Simmons, Saillant, and Dear 1992). Other work highlights the importance of spectral cues to echolocating bats (e.g., Schmidt 1988). More recently, descriptions of the interactions between attacking bats and their arctiid moth prey (Section 6.2.2) have shown that echolocating *Lasiurus borealis* may continue to use temporal and spectral information after they enter the feeding buzz stage of their attacks (Acharya and Fenton 1992).

Narrowband echolocation calls are presumed to provide a greater effective range than broadband calls because more energy is focused in a smaller range of frequencies. Signal theory suggests, however, that calls of this design provide fewer details about targets than signals of broader bandwidth (Simmons and Stein 1980). As noted previously, in low duty cycle bats interpulse interval may provide an indication of functional range, and concentrating of energy means that long, narrowband calls should contribute to greater operational range. Bats with low pulse repetition rates usually use long, narrowband calls, making it difficult to separate cause from effect in this situation.

Signal theory proposes that broadband calls provide better resolution of target detail (Simmons and Stein 1980), but atmospheric attenuation means that calls with a preponderance of higher frequency sounds have shorter functional ranges (Griffin 1971). Bats using short, broadband calls at low duty cycle typically use relatively high pulse repetition rates, presumably reflecting shorter operational ranges.

3.2 Long-Range and Short-Range Bats

Two distinct patterns of hunting behavior associated with narrowband and broadband calls are exemplified by *Lasiurus cinereus* and *Lasionycteris noctivagans* (Barclay 1986). Low pulse repetition rates and long narrowband echolocation calls are typical of *Lasiurus cinereus* searching for prey. These bats typically make one attack per pass through a concentration of insects (for example, around a light) and appear to fix on targets at ranges of at least 5–10 m. Higher pulse repetition rates and shorter broadband echolocation calls are typical of *Lasionycteris noctivagans* searching for prey. These bats often make several attacks as they move through a concentration of insects and appear to react to targets at ranges of 5 m or less. There is a wealth of comparable data from other species of bats that have been studied in the field (reviewed in Neuweiler and Fenton 1988). This fundamental distinction has been recognized for some time, and Brosset (1966) spoke of insectivorous bats that react to targets at close range (hunting "aux bout de leur nez") and distinguished them from others that worked at longer distances.

A 1-cm-diameter stone tossed into the air where bats are hunting can provide a graphic demonstration of these two basic approaches to echolocation and foraging. A bat such as *L. cinereus,* using long, narrowband echolocation calls, will attack an airborne stone and actually capture it, while another like *L. noctivagans* will turn toward the stone, but turn away without making physical contact with it. Geographical variation in *Eptesicus fuscus* behavior, however, emphasizes that there is a continuum between these two extremes. As noted in Section 2.4, this bat uses shorter broadband echolocation calls in eastern North America and longer narrowband calls in the west. The difference in call structure appears to coincide with differences in behavior, supporting the long- versus short-range distinction noted earlier (Obrist 1989). A more striking difference is provided by the variation in echolocation calls and feeding behavior of *Myotis emarginatus* (Schumm, Krull, and Neuweiler 1991).

With further evidence of the behavioral flexibility associated with echolocation and foraging, it will becokme increasingly difficult to label most bats by their echolocation calls.

3.3 Nonairborne Targets

Depending upon the setting, a flying insect is a hard target on a soft background, meaning that it offers an echolocating bat a different perceptual challenge than an insect sitting on a surface, which is a hard target on a hard background. While some bats can use echolocation to find hard targets on hard backgrounds (Suthers 1965; Fiedler 1979; Bell 1982), and some bats use the same signals in either situation (Neuweiler 1989), more and more studies show that echolocating bats use prey-generated sounds or vision to find hard targets on hard backgrounds (e.g., Barclay et al. 1981; Fenton, Gaudet, and Leonard 1983; Ryan and Tuttle 1987; Faure and Barclay 1992). As noted in Section 2.2, low duty cycle bats attacking targets on surfaces usually do not produce feeding buzzes. High duty cycle, flutter-detecting bats readily use echolocation to find fluttering targets on surfaces (e.g., *Hipposideros caffer* [Bell and Fenton 1984]).

As in the situation with airborne targets, some bats hunt nonairborne targets from flight (e.g., *Trachops cirrhosus* [Tuttle and Ryan 1981]; *Myotis myotis* [Audet 1990]; *Antrozous pallidus* [Krull 1992]) and others hunt from perches (e.g., *Cardioderma cor* [Vaughan 1976]; *Nycteris grandis, N. thebgaica* [Fenton, Gaudet, and Leonard 1983; Aldridge et al. 1990]; *Cardioderma cor* [Ryan and Tuttle 1987]). At least some species alternate (e.g., *Nycteris* species [Aldridge et al. 1990; Fenton et al. 1990]) between the two foraging modes, perhaps because of the costs of flight (Thomas 1987).

Bats locate nonairborne targets using a variet of cues. When the target is moving, some bats clearly use echolocation to detect and track prey. *Noctilio leporinus*, for example, uses echolocation to detect and track prey moving across the surface of water (Campbell and Suthers 1988). In other situations, bats use other cues. Sounds of movement may be the most common, whether they are the footfalls of mice (*Megaderma lyra* [Fiedler 1979]), the rustling sounds of insects (*Nycteris grandis* or *Nycteris thebaica* [Fenton, Gaudet, and Leonard 1983]; *Macrotus californicus* [Bell 1985]; *Cardioderma cor* [Ryan and Tuttle 1987]; *Plecotus auritus* [Anderson and Racey 1991]), or the fluttering of wings (*Antrozous pallidus* [Bell 1982]; *Myotis evotis* [Faure and Barclay 1992]). Frog calls (*Trachops cirrhosus* [Tuttle and Ryan 1981]) and the calls of some orthopterans (phyllostomine bats [Tuttle, Ryan, and Belwood 1985; Belwood and Morris 1987]) also are important cues for some species, and at least one, *Macrotus californicus*, sometimes hunts by vision (Bell 1985).

Just how much these bats depend on echolocation to find and assess their prey remains unknown. Many of them that obviously take a lot of their prey from surfaces have large, conspicuous ears that effectively amplify the low-frequency sounds associated with movement (Obrist, Fenton, Eger, and Schlegel 1993). The echolocation calls of these species tend to be short (≤ 1 msec), broadband, and low in intensity. This combination of features means that the calls are barely detectable even by very sensitive bat detectors. It is difficult, therefore, to monitor the production of these echolocation calls, a factor contributing to our ignorance about the role echolocation plays in the lives of these bats. Some *Megaderma lyra* use echolocation to find frogs whose heads protrude from the water, but relatively few individual *Megaderma lyra* exhibit this behavior (G. Marimuthu, personal communication). When taking prey from surfaces, several bats are known to grab prey in their mouths (e.g., *Myotis auriculus* [Fenton and Bell 1979]; *Trachops cirrhosus* [Tuttle and Ryan 1981]; *Megaderma lyra* [Fiedler 1979]), while others use their wings to envelop their prey (*Nycteris grandis* and *Nycteris thebaica* [Fenton, Gaudet, and Leonard 1983]). When taking prey from surfaces, bats may use echolocation to assess the background from which they are taking their victims.

4. Echolocation and the Foraging of Bats That Eat Plant Material and Blood

The available evidence, which is not extensive, suggests that *Rousettus aegyptiacus* uses echolocation to gain access to dark roosting sites such as caves. We still have no clear indication of the role that echolocation plays in the lives of the phyllostomid bats that feed on fruit, nectar and pollen, or blood. These bats produce short (≤ 1 msec), broadband, low-intensity echolocation calls (e.g., *Carollia perspicillata* [Hartley and Suthers 1987]), but appear to rely heavily on other cues such as vision and olfaction for orientation and for finding and evaluating food. In the blood-feeding *Desmodus rotundus*, the inferior colliculus contains neurons tuned to the sounds associated with the heavy breathing of sleeping mammals (Schmidt et al. 1991); there are no clear indications of the use these bats make of echolocation. Obviously any of these species could use echolocation to gain access to dark roosts such as caves or tree hollows, but many of them roost in foliage, situations where echolocation should make little difference to roosting (e.g., stenodermine phyllostomids [Morrison 1980]).

Echolocation behavior may reflect a bat's dependence on different food material. Concentrations of protein, for example, are often very low in fruit and nectar, making it an important factor for bats that eat plant material (Thomas 1984). Howell (1974) found that phyllostomids more specialized for flower feeding obtained their protein from pollen, while less specialized taxa supplemented their diets with insects. She related the differences in echolocation acuity to the move from insects to pollen as a source of protein, presumably reflecting a decreased role for echolocation in finding and assessing food.

5. Echolocation and Foraging Ecology

It is well known, through studies of foraging behavior, that echolocation directly affects the ecology of animal-eating bats (e.g., Aldridge and Rautenbach 1987; Norberg and Rayner 1987). Lack of information about the role that echolocation plays in the lives of bats eating fruit, nectar and pollen, or blood precludes drawing conclusions about how echolocation affects their ecology.

By affecting a bat's ability to detect potential obstacles and food in its surroundings, echolocation influences patterns of habitat use. Bats using broadband signals at relatively high pulse repetition rates appear to collect detailed information about their surroundings, permitting them to work in more closed habitats. Bats using broadband signals at higher pulse repetition rates are better suited to dealing with clutter (echoes returning from objects other than the target of interest) than species using narrowband calls at lower pulse repetition rates (Simmons and Stein 1980). For working in thick vegetation such as the closed crowns of trees and shrubs, short, broadband calls and high pulse repetition rates are essential. Many bats working in these settings, however, depend on other cues (e.g., sounds of prey or sight) to detect and locate their prey. Intraspecific variability in echolocation behavior complicates defending broad generalizations about the association between echolocation call design and habitat use. Even so, many bat communities are composed of very different species. For example, *Lasiurus cinereus* uses long narrowband calls and low pulse repetition rates while *Myotis septentrionalis* usually relies on short broadband calls and high pulse repetition rates. Both species eat insects, but while *Lasiurus cinereus* seems to depend entirely on echolocation to find its prey, *Myotis septentrionalis* often uses prey-generated sounds to find its victims (see Fig. 2.4). Although different species like these are sympatric, they may forage in different settings (see also Neuweiler 1989).

Among high duty cycle echolocators in the Rhinolophidae and Hipposideridae, sympatric species use echolocation calls dominated by different narrowband frequencies (Heller and von Helverson 1989). The frequencies are more different than expected by chance, and there also is a relationship between call frequency and bat size, with smaller bats using higher-frequency calls (Heller and von Helverson 1989). The ecological consequences of these differences remain to be determined. Many smaller hipposiderid and rhinolophid species, however, have echolocation calls with most energy at frequencies beyond the range at which moth ears are most sensitive (Fenton and Fullard 1979; see Section 6.2.2). Heller and von

Helverson (1989) also noted that differences in echolocation call frequencies could be associated with communication (see Section 6).

The various approaches to foraging and echolocating outlined here coincide with striking differences in the auditory systems of bats. In the best illustration of this to date, Neuweiler, Singh, and Sripathi (1984) documented differences in hearing, echolocation calls and foraging behavior in a community of sympatric bats in southern India (see Chapters 3, 4, and 5). Hearing and echolocation is one axis along which bat communities may be organized (Neuweiler 1989).

Different foraging strategies and differences in echolocation behavior can influence the prey available to animal-eating bats. By taking nonairborne prey, bats immediately have access to larger animals, including both insects and arthropods, and small vertebrates (Norberg and Fenton 1988). Hunting nonairborne prey may also provide access to more prey because there often are more insects on the ground than in the air (Rautenbach, Kemp, and Scholtz 1988). The energetic implications of differences in prey availability were demonstrated by Barclay (1991) working on the eastern slopes of the Rocky Mountains. He found that *Myotis evotis*, which commonly takes nonairborne prey (Faure, Fullard, and Barclay 1990; Faure and Barclay 1992), bears young in his study area while two sympatric species (*Myotis lucifugus* and *Myotis volans*) that eat airborne targets do not.

Access to prey also is influenced by actual feeding behavior. Comparing the numbers of different-sized noctuid and sphingid moths in the diets of *Lasiurus cinereus* and *Nycteris grandis* (Fig. 2.6) illustrates this point. Both species weigh about 30 g, but while *L. cinereus* hunts and eats on the wing, taking only airborne targets (e.g., Barclay 1986; Hickey 1993), *N. grandis* sometimes hunts on the wing but other times from a perch, taking a mixture of airborne and nonairborne targets (Fenton et al. 1990). Furthermore, *N. grandis* appears to consistently perch while eating, perhaps enhancing its capacity to handle large prey. The differences in behavior coincide with striking differences in prey consumed (insects alone versus a varied diet of insects, other arthropods, and small vertebrates [Hickey 1993; Fenton et al. 1990]), even when the comparison is restricted to two families of moths.

6. Communication Role of Biosonar Signals

The signals one bat produces to collect information about its surroundings also are available to other animals, making information leakage a reality of echolocation. While many intraspecific interactions mediated by echolocation calls clearly meet most definitions of "communication" (e.g., Slater 1983), other interactions do not. The magnitude of information leakage is determined partly by the strength of the echolocation signals and partly by the hearing properties of the eavesdroppers. Using bat detectors to monitor the activity and behavior of echolocating bats (see Section 2) takes

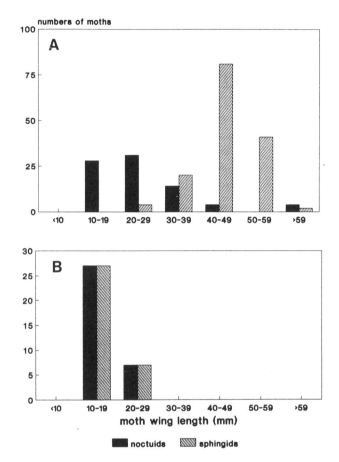

FIGURE 2.6A,B. A comparison of the numbers of noctuid (*black bars*) and sphingid (*shaded bars*) moths of different sizes taken by *Nycteris grandis* (A) and *Lasiurus cinereus* (B). Both bats weigh about 30 g as adults, but while *Nycteris grandis* takes prey from the ground and in flight, *Lasiurus cinereus* attacks only airborne prey. Differences in foraging strategy affect the prey available to a hunting bat. The data for *N. grandis* are from Fenton et al. (1993), and those for *L. cinereus* are from Hickey (1993).

advantage of information leakage. While the calls of some bats are readily detectable over dozens of meters, others are picked up only when the bats fly within 1 or 2 m of the microphone, reflecting differences in call intensity (see Section 2.1).

6.1 Intraspecific Interactions

It is evident that some bats respond to the echolocation calls of conspecifics, but the nature of the response varies by species. Möhres (1967) observed that some captive *Rhinolophus ferrumequinum* appeared to use the echolocation calls of conspecifics to find the locations of preferred roostmates. Using playback presentations, Barclay (1982) demonstrated that free-flying *Myotis lucifugus* flew toward speakers presenting the echolocation calls of conspecifics or sounds of similar bandwidth and duration. These bats responded significantly more to presentations of echolocation calls than to control signals generated by playing the recorded calls backward. These bats are quite gregarious at roost sites and in feeding areas, places where Barclay (1982) found a positive response. He proposed that in either situation the individuals whose calls attracted conspecifics suffered no penalty.

A contrasting picture emerged from similar experiments with *Euderma maculatum*, a species that forages alone over swamps and in open woodland. These bats showed two patterns of response to playback presentations (Leonard and Fenton 1984). While they sometimes flew directly at the speaker, other times they flew away from it, supporting the suggestion that echolocation calls serve to space individuals in foraging areas.

The sight of one bat chasing another often has been interpreted as evidence of territorial behavior (e.g., Rydell 1986), especially when the interactions are accompanied by vocalizations other than echolocation calls (e.g., Belwood and Fullard 1984; Racey and Swift 1985). But things are not always what they seem. Griffin (1958) proposed that in *Lasiurus borealis* chases occurred when one or more bats used the feeding buzzes of another to cue on an available (vulnerable?) prey. Hickey and Fenton (1990) showed that intraspecific chases in *L. borealis* did not correspond to exclusive use of rich patches of food and that their incidence was not related to prey abundance. Balcombe and Fenton (1988) used playback presentations to demonstrate that in this species feeding buzzes were cues that precipitated chases. The data from individually marked bats and from playback experiments (Balcombe and Fenton 1988; Hickey and Fenton 1990) supported Griffin's original (1958) interpretation. The chases in *L. borealis* did not reflect territorial behavior, but rather the presence of the medium-sized moths that form the bulk of the bats' diet at the study site in southwestern Ontario.

A variety of evidence, including analysis of signals and the results of playback presentations, suggests that echolocation calls mediate some aspects of interactions between mother bats and their young. Thomson, Barclay, and Fenton (1985) showed that infant *Myotis lucifugus* were more attentive to the echolocation calls of their mothers than to those of other females, and Balcombe and McCracken (1992) found evidence that in *Tadarida brasiliensis* echolocation calls were one factor in reunions between mothers and their own offspring.

While echolocation calls by themselves can mediate communication, Suthers (1965) demonstrated that when two *Noctilio leporinus* were on a collision course, one or both bats would alter their echolocation calls by dropping the terminal FM sweep by an additional octave. These "honks" apparently alerted the bats to a possible collision and were associated with one or both animals changing their course. Similar "honks" have been reported from a variety of other bats, including several species of *Myotis* (Fenton and Bell 1979).

6.2 Interspecific Interactions

There are two obvious interspecific audiences of bats: other species of bats, and other kinds of animals. These usually involve predator–prey interactions where bats may be the predators or the prey.

6.2.1 Bat–Bat and Bat–Bird Interactions

Just as chases between bats may not represent territorial behavior, chases involving bats and other animals may reflect other phenomena. Shields and Bildstein (1979) proposed that chases of common nighthawks (*Chordeiles minor*: Aves) by big brown bats (*Eptesicus fuscus*) reflected efforts by the bats to exclude the birds from a rich food source. Alternatively, the chases could reflect missed communication signals because the birds would not have heard the ultrasonic honks of bats on collision courses with them. In this situation, chases reflect the hazard of missed warning signals.

Interspecific communication among bats can involve echolocation calls. Both Barclay (1982) and Balcombe and Fenton (1988) demonstrated that responses to echolocation calls often involved species other than the one whose calls were presented in playback situations. Furthermore, the statistical association that Bell (1980) found between some species responding to rich patches of prey could also result from a communicative role of echolocation calls, perhaps the specific availability of prey as indicated by feeding buzzes.

In parts of South America, Africa, Asia, and Australia, some bats regularly or occasionally eat other bats (Norberg and Fenton 1988). These predators are found in three families: the Phyllostomidae (*Vampyrum spectrum*, *Chrotopterus auritus*), the Nycteridae (*Nycteris grandis*), and the Megadermatidae (*Cardioderma cor*, *Megaderma lyra*, *Macroderma gigas*). Although it is tempting to think that bat-eating bats depend on other species' echolocation calls to locate and identify their prey, this hypothesis remains unproven. Fenton, Gaudet, and Leonard (1983) found that captive *Nycteris grandis* sometimes responded to the vocalizations of other bats, as well as to their wing flutterings. Vaughan's (1976) observations of *Cardioderma cor* attacking, killing and eating some *Pipistrellus* species could have involved echolocation calls as a cue.

Other predators of bats, such as owls or raptors such as bat hawks (*Macheirhamphus alcinus* [Fenton 1992]), are unlikely candidates as eavesdroppers on bat echolocation calls because birds cannot hear ultrasonic

(>20 kHz) sound (Welty 1975). Predators such as small mammals with keen ultrasonic hearing (Bench, Pye, and Pye 1975) are more obvious candidates for using echolocation calls to find and identify bat as prey.

6.2.2 Bat–Insect

The story of moths with ears that detect the echolocation calls of hunting bats (Roeder 1967) is certainly one of the "eye-opening" discoveries in animal behavior (Griffin 1976). We know that insects such as moths, lacewings, and various orthopterans have bat-detecting ears (for reviews, see Fullard 1987; Surlykke 1988), and recently mantids have been added to the list (Yager, May, and Fenton 1990). Insect bat detectors may be located on wings (lacewings), face, thorax or abdomen (moths), or mid venter on the thorax (mantids). While insect ears sensitive to echolocation calls tend to occur in pairs, providing information about direction, mantids have only one ear and show no directionality in their response to the calls of an approaching bat (Yager, May, and Fenton 1990).

The actual response of a moth, lacewing, cricket, or mantid to an approaching bat appears to be mediated by the strength of the bat's echolocation calls, although some species appear to react to changes in pulse repetition rates. The echolocation calls of a distant bat, perceived as quiet calls, produce negative phonotactic behavior as the insect flies away from the bat. A close bat, perceived as loud calls, may cause erratic behavior ranging from diving to the ground to complex spiral and zigzag patterns. Some species of tiger moths (Arctiidae) produce clicks in response to loud bat calls, and these often affect the bats' attacks (Fullard 1987; Surlykke 1988).

Bat-detecting ears in insects range from simple structures with one sensory neuron to more complex ones involving more neurons. Moth ears are not equally sensitive to all frequencies in the bandwidths used by echolocating bats, and the frequencies of their echolocation calls make some bats much less conspicuous to moths than others (Fullard 1987; Surlykke 1988). Intensity also affects the conspicuousness of bat calls to moths and other insects (Faure, Fullard, and Barclay 1990). Although it appears that the "average" moth with ears has 40% less chance of being taken by a bat than a deaf moth (Roeder and Treat 1960), ear structure and sensitivity together combine with characteristics of bat echolocation calls to influence this situation. Fenton and Fullard (1979) demonstrated the impact of bat call frequency and intensity on the conspicuousness of bats to moths. They found that an "average" bat, one using high-intensity echolocation calls with energy between 35 and 60 kHz, was detected by the "average" moth at about 40 m. With the operational range of bat echolocation restricted to a few meters (see Section 2.6), the potential advantage to the moth is considerable.

Moths, including species with bat-detecting ears, form a large part of the diets of some echolocating bats. Sometimes this can be explained by the inability of the moths to hear the bats' echolocation calls. Thus, the small African hipposiderid *Cloeotis percivali* eats many moths (Whitaker and Black 1976), and most of the energy in its echolocation calls is at about 212 kHz (Fenton and Bell 1981), well above the frequencies at which sympatric moths can hear (Fenton and Fullard 1979). In the same way, the North American vespertilionid *Euderma maculatum* sometimes eats moths (Ross 1967), and its echolocation calls range in frequency from 9 to 15 kHz, with most energy at 10–12 kHz (Leonard and Fenton 1984; Obrist 1989), well below the lower frequency threshold of most sympartric moths (Fullard, Fenton, and Fulonger 1983).

Many species of moths with bat-detecting ears, however, are taken by bats using intense echolocation calls with most energy in the range of the moths' best hearing. In the laboratory, L.A. Miller (personal communication) found that some individual *Pipistrellus pipistrellus* learned to thwart the auditory-based defensive maneuvers of lacewings. In the field, Hickey (1993) and Hickey and Fenton (1990) have observed that several *Lasiurus cinereus* and *Lasiurus borealis* often forage together in places where flying moths are abundant. L. Acharya (personal communication) has observed that in these situations moths with ears usually react to attacking bats by flight maneuvers, but the individual that successfully evades one bat may quickly (within a second) be caught by another. This means that although individual *L. cinereus* and *L. borealis* typically succeed in only 40%–50% of their attacks on moths (Hickey 1993; Hickey and Fenton 1990), the vulnerability of the moths is complicated by the presence of several bats. In the final analysis, the hearing-based defenses of moths, like other defensive systems (Edmunds 1974), do not provide absolute immunity from predators.

Dunning and Roeder (1965) noted that the clicks of some arctiid moths affected the behavior of *Myotis lucifugus* trained to take mealworms tossed into the air. They found that if arctiid clicks were presented just as the bats closed with their targets, they veered away, aborting the attack. Three hypotheses have been used to explain the bats' behavior: (1) moth clicks could interfere with or "jam" the bats' echolocation, (2) moth clicks could startle the bats, and (3) moth clicks may alert the bats to bad-tasting arctiids. A number of studies have presented different sets of data on this topic (e.g., Dunning 1968; Fullard, Fenton, and Simmons 1979; Surlykke and Miller 1985; Stoneman and Fenton 1988; Bates and Fenton 1990; Miller 1991; Acharya and Fenton 1992; Dunning et al. 1992). Some laboratory work contradicts the first hypothesis, that moth clicks jam bat echolocation (e.g., Surlykke and Miller 1985; Stoneman and Fenton 1988), while other work supports it (Miller 1991). Some captive animals clearly reject arctiids, and in the field, free-flying bats take arctiids less often than expected by their general incidence in moth populations (Dunning et al. 1992).

Presentations of muted free-flying arctiids (*Hypoprepia fucosa*) to hunting *L. borealis* showed that the moths use their clicks to warn bats of their bad taste (Acharya and Fenton 1992). Free-flying *L. borealis* approached but did not attack intact *H. fucosa*, but attacked, caught, and then rejected these moths that had been muted.

Arctiid moth clicks affect the behavior of bats. Bates and Fenton (1990) used captive *Eptesicus fuscus* to demonstrate that a bat's prior experience determined its response to moth clicks. Bats that have learned to associate moth clicks with bad taste avoid prey that click, while naive bats ignore clicks after an initial period of being startled. Field studies reveal that the clicks of some arctiids act as aposematic signals for bats that feed heavily on moths, namely *L. borealis* and *L. cinereus* (Acharya and Fenton 1992). Further work will certainly reveal variations on these themes, including the likelihood of Batesian mimics depending on acoustic displays (cf. Dunning 1968). It remains to be seen if the sounds of other insects also influence the prey taken by bats (e.g., peacock butterfly clicks [Møhl and Miller 1976]; defense stridulations [Masters 1979]).

7. Biosonar and the Lives of Bats

Although we do not always know just what role echolocation plays in the lives of bats, especially in species producing weak calls, several lines of evidence suggest that this mode of orientation has profoundly affected bats. Echolocation can be expensive, particularly when the cost is measured in information leakage (see Section 6, earlier in this chapter). By making themselves more conspicuous, echolocating bats can affect their chances of avoiding or attracting conspecifics, catching prey, or being taken by predators.

Other costs of echolocation could be in the actual production and processing of sounds. Speakman and Racey (1991), however, used two species of echolocating bats (*Pipistrellus pipistrellus* and *Plecotus auritus*) to demonstrate that the costs of echolocation were not additive to those of flight. They concluded that echolocation was relatively inexpensive for animals already expending so much energy in flying. The situation could be different for bats that hunt from perches.

The short effective range of bat echolocation is probably one factor contributing to bats the small size of (Barclay and Brigham 1991). Being restricted to collecting information from just in front of it may be an acceptable limitation for a small maneuverable animal, but not a large unmaneuverable one. In this context, it is not surprising that faster-flying bats have longer interpulse intervals (= effective range; see Sections 2.2, 2.6) than slow-flying species. Although the fossil record reveals that a much larger vampire bat (30% bigger than the living species, which weigh about 40 g) once lived in South America (Ray, Linares, and Morgan 1988), there are no other records of "giant" bats. It is evident that being small is a predominant feature of most insectivorous bats (Fig. 2.7).

The facial features of bats (Fig. 2.8), including their ears, provide one of the most graphic demonstrations of the impact of echolocation on the physiognomy of bats. Obrist et al. (1993) demonstrated three general trends in bats' ears that are associated with different orientation behaviors. Ear (pinna) size was directly correlated with echolocation call frequency in bats that used high duty cycles (rhinolophids, hipposiderids, and the mormoopid *Pteronotus parnellii*). In low duty cycle species, the relationship between pinna size and frequency varied from being relatively strong in species using narrowband signals to weak in species using broadband calls. Bats that listen to sounds coming from their prey have large ears that amplify low-frequency sounds. These bats occur in several evolutionary lines; the Nycteridae and Megadermatidae, the phyllostomine Phyllostomidae, and some vespertilionids.

Leaflike structures are conspicuous on the faces of bats in the families Hipposideridae, Rhinolophidae, Megadermatidae, and Phyllostomidae, while less marked leaflike structures occur in the Rhinopomatidae, some Vespertilionidae, and Mormoopidae (Fenton 1992). In the Nycteridae, a conspicuous slit dominates the animals' faces. Hartley and Suthers (1987) demonstrated how noseleaf position in the phyllostomid *Carollia perspicillata* affects the pattern of sound radiation from the bat's face and thus its acoustic perception of its surroundings. Still and motion photographs of the

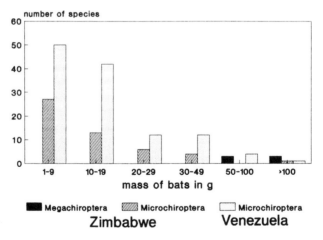

FIGURE 2.7. The small size of bats is illustrated here as the number of species of different adult body mass from Zimbabwe and Venezuela. The data for Zimbabwe come from Smithers (1983) and those for Venezuela from Linares (1986). *Black bars*, Megachiroptera, *shaded bars*, Microchiroptera, Zimbabwe; *clear bars*, Microchiroptera, Venezuela.

A

B

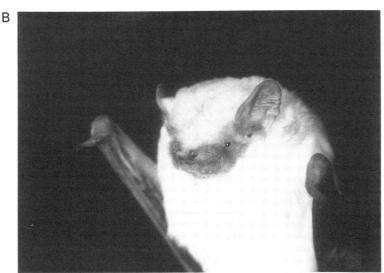

FIGURE 2.8A-H. The faces and ears of bats of bats appear to reflect their echocation behavior. While some species producing low duty cycle, broadband, frequency-modulated (FM) calls have plain faces (e.g., A: *Balantiopteryx plicata*, Emballonuridae; B: *Chalinolobus variegatus*, Vespertilionidae), species using high duty cycle, narrowband calls tend to have facial ornaments (e.g., C: *Rhinolophus hildebrandti*, Rhinolophidae; D: *Hipposideros caffer*, Hipposideridae). Other species with facial ornaments produce low duty cycle, low-intensity, broadband echolocation calls (E: *Nycteris grandis*, Nycteridae; F: *Megaderma lyra*, Megadermatidae; G: *Chrotopterus auritus*; H: *Tonatia brasiliensis*, Phyllostomidae).

C

D

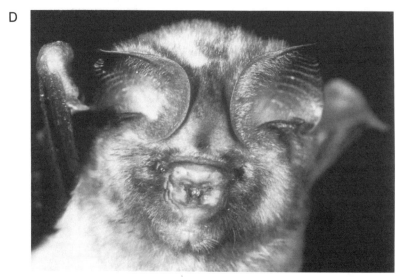

FIGURE 2.8 *(continued)*

phyllostomid *Trachops cirrhosus* approaching frogs suggest changes in noseleaf position in these situations (Tuttle and Ryan 1981), indicating that the noseleaf may serve a dynamic role in orientation.

As noted in Section 4, our lack of knowledge about the role that echolocation plays in the lives of the low-intensity bats precludes strong statements about the relationship between facial features and echolocation. In Hipposideridae and Rhinolophidae, independent ear movements and the presence of a noseleaf coincide with a high duty cycle approach to

E

FIGURE 2.8 (*continued*)

echolocation, presumably affecting the bats' sonar prowess (Pye and Roberts 1970). The temptation to associate facial features and noseleafs is tempered by two facts: first, the absence of such structures in the mormoopid *Pteronotus parnellii*, another high duty cycle echolocator, and second, their presence in nycterids, megadermatids, and phyllostomids, which use low duty cycle echolocation calls that are broadband and low intensity. The situation is ripe for further investigation.

8. Summary

Echolocation typifies species in the suborder Microchiroptera but is exceptional among Megachiroptera. *Rousettus aegyptiacus*, the echolocating megachiropteran, produces short, broadband pulses of sound by clicking its tongue, while the echolocation calls of microchiropterans are vocalizations, sounds produced in the larynx. The echolocation calls of microchiropteran bats vary considerably in different acoustic parameters such as intensity,

Figure 2.8 (continued)

duration, bandwidth, duty cycles, and patterns of frequency change over time. Each species appears to produce a distinctive echolocation call. Species using echolocation to detect, track, and assess airborne targets generally use low duty cycle ($\leq 20\%$) calls and marked increases in pulse repetition rate during attacks on targets. At some stages in the attack process, these bats use short (≤ 1-msec) echolocation calls dominated by frequency-modulated (FM) components. Species in the families Rhinolophidae and Hipposideridae and the mormoopid *Pteronotus parnellii* use high duty cycle ($\geq 80\%$) calls dominated by narrow bandwidth components. These bats, known as constant-frequency (CF) species, appear specialized for flutter detection.

Among the Microchiroptera, many species hunt nonairborne targets they take from the ground, foliage, or other surfaces. These species, from the families Nycteridae, Megadermatidae, Phyllostomidae, and Vespertilionidae, use short (≤ 1-msec), broadband echolocation calls of low intensity but often depend on prey-generated sound cues or vision to find their targets. Bats in the family Phyllostomidae that feed on blood, fruit, nectar, and pollen also produce short, low-intensity, broadband echolocation calls but

G

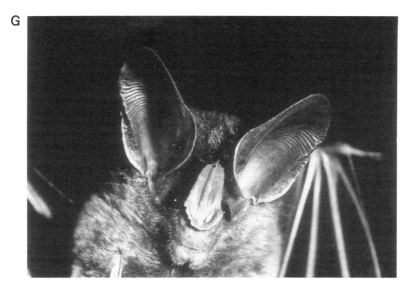

H

FIGURE 2.8 (*continued*)

as yet the role that echolocation plays in their lives is not clear. The echolocation calls of many bats serve a communication function, and playback presentations elicit various responses to the calls of conspecifics. Echolocation calls also mediate interspecific interactions in which bats may be prey or predators. The bat-detecting ears of many insects are among the most notable of interspecific interactions mediated by echolocation calls. Echolocation has profoundly affected the lives of microchiropteran bats, from their body size to the appearances of their faces.

Acknowledgments. I thank Lalita Acharya, Doris Audet, Robert M. R. Barclay, James H. Fullard, David Johnston, Jennifer Long, Cathy Merriman, Martin Obrist, David Pearl, Stuart Perlmeter, and Daphne Syme for reading earlier drafts of this manuscript and making helpful suggestions for improving it. I am particularly grateful to J. E. Cebek, P. A. Faure, J. H. Fullard, M. B. C. Hickey, G. Marimuthu, and G. S. Wilkinson for sharing some of their unpublished results with me and to R. A. Suthers for providing the images of Rousettus sounds. My research on bats has been supported by operating and equipment grants from the Natural Sciences and Engineering Research Council of Canada.

References

Acharya L, Fenton MB (1992) Echolocation behaviour of vespertilionid bats (*Lasiurus cinereus* and *Lasiurus borealis*) attacking airborne targets, including arctiid moths. Can J Zool 70:1292–1298.

Ahlén I (1981) Field identification of bats and survey methods based on sounds. Myotis 18–19:128–136.

Aldridge HDJN, Rautenbach IL (1987) Morphology, echolocation and resource partitioning in insectivorous bats. J Anim Ecol 56:763–778.

Aldridge HDJN, Obrist M, Merriam H G, Fenton M B (1990) Social calls and habitat use: roosting and foraging by the African bat *Nycteris thebaica*. J Mammal 71:242–246.

Anderson MB, Racey PA (1991) Feeding behaviour of captive brown long-eared bats, *Plecotus auritus*. Anim Behav 42:489–493.

Arroyo-Cabrales J, Jones JK Jr (1988) *Balantiopteryx plicata*. Mammal Species 301:1–4.

Avery M I (1991) Pipistrelle. In: Corbet G B, Harris S (eds) The Handbook of British Mammals, 3d Ed. Oxford: Blackwell Scientific, pp. 124–128.

Audet D (1990) Foraging behavior and habitat use by a gleaning bat, *Myotis myotis*. J Mammal 71:420–427.

Audet D, Krull D, Marimuthu G, Sumithran S, Singh JB (1991) Foraging behavior of the Indian false vampire bat, *Megaderma lyra* (Chiroptera: Megadermatidae). Biotropica 23:63–67.

Bailey WJ, Slightom JL, Goodman M (1992) Rejection of the "flying primate" hypothesis by phylogenetic evidence from the epsilon-globin gene. Science 256:86–89.

Balcombe JP, Fenton MB (1988) Eavesdropping by bats: the influence of echolocation call design and foraging strategy. Ethology 79:158–166.

Balcombe JP, McCracken GF (1992) Vocal recognition in Mexican free-tailed bats: do pups recognize their mothers? Anim Behav 43:79–88.

Baker RJ, Novacek MJ, Simmons NB (1991) On the monophyly of bats. Syst Zool 40:216–231.

Barclay RMR (1982) Interindividual use of echolocation calls: eavesdropping by bats. Behav Ecol Sociobiol 10:271–275.

Barclay RMR (1983) Echolocation calls of emballonurid bats from Panama. J Comp Physiol 151:515–520.

Barclay RMR (1986) The echolocation calls of hoary (*Lasiurus cinereus*) and silver-haired (*Lasionycteris noctivagans*) bats as adaptations for long- versus short-range strategies and the consequences for prey selection. Can J Zool 64:2700–2705.

Barclay RMR (1991) Population structure of temperate zone insectivorous bats in relation to foraging behaviour and energy demand. J Anim Ecol 60:165–178.

Barclay RMR, Brigham RM (1991) Prey detection, dietary niche breadth, and body size in bats: why are aerial insectivorous bats so small? Am Nat 137:693–703.

Barclay RMR, Fenton MB, Tuttle MD, Ryan MJ (1981) Echolocation calls produced by *Trachops cirrhosus* (Chiroptera: Phyllostomatidae) while hunting for frogs. Can J Zool 59:750–753.

Bates DL, Fenton MB (1990) Aposematism or startle? Predators learn their responses to prey. Can J Zool 68:49–52.

Bell GP (1980) Habitat use and responses to patches of prey by desert insectivorous bats. Can J Zool 58:1876–1883.

Bell GP (1982) Behavioral and ecological aspects of gleaning by a desert insectivorous bat, *Antrozous pallidus* (Chiroptera: Vespertilionidae). Behav Ecol Sociobiol 10:217–223.

Bell GP (1985) The sensory basis of prey location by the California leaf-nosed bat *Macrotus californicus* (Chiroptera: Phyllostomidae). Behav Ecol Sociobiol 16:343–347.

Bell GP, Fenton MB (1984) The use of Doppler-shifted echoes as a flutter detection and clutter rejection system: the echolocation and feeding behavior of *Hipposideros ruber* (Chiroptera: Hipposideridae). Behav Ecol Sociobiol 15:109–114.

Belwood JJ, Fullard JH (1984) Echolocation and foraging behaviour in the Hawaiian hoary bat, *Lasiurus cinereus semotus*. Can J Zool 62:21130–2120.

Belwood JJ, Morris GK (1987) Bat predation and its influence on calling behavior in neotropical katydids. Science 238:64–67.

Bench RJ, Pye A, Pye JD (eds) (1975) Sound Reception in Mammals (Symposium No. 37). London: Zoological Society of London.

Bradbury JW (1977) Social organization and communication. In: Wimsatt WA (ed) Biology of Bats, Vol. 3. New York: Academic, pp. 2–72.

Brigham RM, Cebek JE, Hickey MBC (1989) Intraspecific variation in the echolocation calls of two species of insectivorous bats. J Mammal 70:426–428.

Brosset A (1966) La biologie des chiroptères. Paris: Masson et Cie.

Brown P (1976) Vocal communication in the pallid bat, *Antrozous pallidus*. Z Tierpsychol 41:34–54.

Brown PE, Grinnell AD (1980) Echolocation ontogeny in bats. In: Busnel R-G, Fish J (eds) Animal Sonar Systems. New York: Plenum, pp. 355–380.

Brown PE, Brown TW, Grinnell AD (1983) Echolocation development and vocal communication in the lesser bulldog bat, *Noctilio albiventris*. Behav Ecol Sociobiol 13:287–298.

Buchler ER (1980) The development of flight, foraging, and echolocation in the little brown bat (*Myotis lucifugus*). Behav Ecol Sociobiol 6:211–218.

Buchler ER, Mitz AR (1980) Similarities in design features of orientation sounds used by simpler, nonaquatic echolocators. In: Busnel R-G, Fish JF (eds) Animal Sonar Systems. (Nato Advanced Study Institutes, Vol. A28.) New York: Plenum, pp. 871–874.

Busnel R-G, Fish JF (eds) (1980). Animal Sonar Systems. (Nato Advanced Study Institutes, Series A28.) New York: Plenum.

Campbell KA, Suthers RA (1988) Predictive tracking of horizontally moving targets by the fishing bat, *Noctilio leporinus*. In: Nachtigall PE, Moore PWB (eds)

Animal Sonar: Processes and Performance, Vol. A156. New York: Plenum, pp. 501–506.

Casseday J H, Covey E (1991) Frequency tuning properties of neurons in the inferior colliculus of an F M bat. J Comp Neurobiol 319:34–50.

Cebek J E (1992) Social and genetic correlates of female philopatry in the temperate zone bat, *Eptesicus fuscus*. Ph.D. Dissertation, Department of Biology, York University, North York, Ontario, Canada.

Dunning DC (1968) Warning sounds of moths. Z Tierpyschol 25:129–138.

Dunning DC, Roeder KD (1965) Moth sounds and the insect-catching behavior of bats. Science 147:173–174.

Dunning DC, Acharya L, Merriman C, Dal Ferro L (1992) Interactions between bats and arctiid moths. Can J Zool 70:2218–2223.

Edmunds M (1974) Defence in Animals. London: Longman.

Faure PA, Barclay RMR (1992) The sensory basis of prey detection by the long-eared bat, *Myotis evotis*, and the consequences for prey selection. Anim Behav 44:31–39.

Faure PA, Fullard JH, Barclay RMR (1990) The response of tympanate moths to the echolocation calls of a substrate gleaning bat, *Myotis evotis*. J Comp Physiol A 166:843–849.

Faure PA, Fullard JH, Dawson JW (1993) The gleaning attacks of northern long-eared bats, *Myotis septentrionalis*, are relatively inaudible to moths. J Exp Biol. 178:173–189.

Fenton MB (1974) The role of echolocation in the evolution of bats. Am Nat 108:386–388.

Fenton MB (1984) Echolocation: implications for the ecology and evolution of bats. Rev Biol 59:33–53.

Fenton MB (1985) Communication in the Chiroptera. Bloomington: Indiana University Press.

Fenton MB (1986) *Hipposideros caffer* (Chiroptera: Hipposideridae) in Zimbabwe: morphology and echolocation calls. J Zool (Lond) 210:347–353.

Fenton MB (1990) The foraging behaviour and ecology of animal-eating bats. Can J Zool 68:411–422.

Fenton MB (1992) Bats. New York: Facts on File.

Fenton MB (1994) Assessing signal variability and reliability: 'to thine ownself be true'. Anim Behav 47:757–764.

Fenton MB, Bell GP (1979) Echolocation and feeding behaviour in four species of *Myotis* (Chiroptera: Vespertilionidae). Can J Zool 57:1271–1277.

Fenton MB, Bell GP (1981) Recognition of species of insectivorous bats by their echolocation calls. J Mammal 62:233–243.

Fenton MB, Crerar LM (1984) Cervical vertebrae in relation to roosting posture in bats. J Mammal 65:395–403.

Fenton MB, Fullard JH (1979) The influence of moth hearing on bat echolocation strategies. J Comp Physiol A 132:77–86.

Fenton MB, Rautenbach IL (1986) A comparison of the roosting and foraging behaviour of three species of African insectivorous bats. Can J Zool 64:2860–2867.

Fenton MB, Gaudet CL, Leonard ML (1983) Feeding behaviour of the bats *Nycteris grandis* and *Nycteris thebaica* (Nycteridae) in captivity. J Zool (Lond) 200:347–354.

Fenton MB, Swanepoel CM, Brigham RM, Cebek JE, Hickey MBC (1990) Foraging behavior and prey selection by large slit-faced bats (*Nycteris grandis*). Biotropica 22:2-8.

Fenton MB, Rautenbach IL, Chipese D, Cumming MB, Musgrave MK, Taylor JS, Volpers T (1993) Variation in foraging behaviour, habitat use and diet of large slit-faced bats (*Nycteris grandis*). Z Saeugetierk 58:65-74.

Fenton MB, Acharya L, Audet D, Hickey MBC, Merriman C, Obrist MK, Syme DM, Adkins B (1992) Phyllostomid bats (Chiroptera: Phyllostomidae) as indicators of habitat disruption in the Neotropics. Biotropica 24:440-446.

Fiedler J (1979) Prey catching with and without echolocation in the Indian false vampire bat (*Megaderma lyra*). Behav Ecol Sociobiol 6:155-160.

Findley JS, Wilson DE (1982) Ecological significance of chiropteran morphology. In: Kunz TH (ed) Ecology of Bats. New York: Plenum, pp. 243-260.

Flemming TH (1988) The short-tailed fruit bat. Chicago: University of Chicago Press.

Fujita MS, Kunz TH (1984) *Pipistrellus subflavus*. Mammal Species 228:1-8.

Fullard JH (1987) Sensory ecology and neuroethology of moths and bats: interactions in a global perspective. In: Fenton MB, Racey PA, Rayner JMV (eds) Recent Advances in the Study of Bats. Cambridge: Cambridge University Press, pp. 244-272.

Fullard JH, Fenton MB, Fulonger CL (1983) Sensory relationships of moths and bats sampled from two nearctic sites. Can J Zool 61:1752-1757.

Fullard JH, Fenton MB, Simmons JA (1979) Jamming bat echolocation: the clicks of arctiid moths. Can J Zool 57:647-649.

Fullard JH, Koehler C, Surlykke A, McKenzie NL (1991) Echolocation ecology and flight morphology of insectivorous bats (Chiroptera) from south-western Australia. Aust J Zool 39:427-438.

Goldman LJ, Henson OW Jr (1977) Prey recognition and selection by constant frequency bat, *Pteronotus parnellii parnellii*. Behav Ecol Sociobiol 2:411-420.

Greenhall AM, Schmitd U (eds) (1988) Natural History of Vampire Bats. Boca Raton: CRC Press.

Griffin DR (1958) Listening in the Dark. New Haven: Yale University Press.

Griffin DR (1971) The importance of atmospheric attenuation for the echolocation of bats (Chiroptera). Anim Behav 19:55-61.

Griffin DR (1976) The Question of Animal Awareness. New York: Rockefeller University Press.

Griffin DR, Simmons JA (1974) Echolocation of insects by horseshoe bats. Nature 250:730-731.

Griffin DR, Webster FA, Michael CR (1960) The echolocation of flying insects by bats. Anim Behav 8:151-154.

Guppy A, Coles RB (1988) Acoustical and neural aspects of hearing in the Australian gleaning bats, *Macroderma gigas* and *Nyctophilus gouldii*. J Comp Physiol A 162:653-668.

Habersetzer J (1981) Adaptive echolocation sounds in the bat *Rhinopoma hardwickei*: a field study. J Comp Physiol A 144:559-566.

Hartley D J, Suthers R A (1987) The sound emission pattern and the acoustical role of the noseleaf in the echolocating bat, *Carollia perspicillata*. J Acoust Soc Am 82:1892-1900.

Heller K-G, von Helverson O (1989) Resource partitioning of sonar frequency bands in rhinolophoid bats. Oecologia 80:178-186.

Herd RM (1983) *Pteronotus parnellii*. Mammal Species 209:1-5.

Herd RM, Fenton MB (1983) An electrophoretic, morphological, and ecological investigation of a putative hybrid zone between *Myotis lucifugus* and *Myotis yumanensis* (Chiroptera: Vespertilionidae). Can J Zool 61:2029-2050.

Hickey MBC (1993) Thermoregulatory and foraging behaviour of hoary bats, *Lasiurus cinereus*. Ph.D. Dissertation, Department of Biology, York University, North York, Ontario, Canada.

Hickey MBC, Fenton MB (1990) Foraging by red bats (*Lasiurus borealis*): do intraspecific chases mean territoriality? Can J Zool 68:2477-2482.

Hill JE (1974) A new family, genus and species of bat (Mammalia, Chiroptera) from Thailand. Bull Br Mus (Nat Hist) Zool 27:303-336.

Hill JE, Smith JD (1984) Bats: A Natural History. London: British Museum of Natural History.

Howell DJ (1974) Acoustic behavior and feeding in glossophagine bats. J Mammal 55:293-308.

Hudson WS, Wilson DE (1986) Macroderma gigas. Mammalian Species 260:1-4.

Jepsen GL (1970) Bat origins and evolution. In: Wimsatt WA (ed) Biology of Bats, Vol. 1. New York: Academic, pp. 1-64.

Kalko EKV, Schnitzler H-U (1989) The echolocation and hunting behavior of Daubenton's bat, *Myotis daubentoni*. Behav Ecol Sociobiol 24:225-238.

Kick SA (1982) Target-detection by the echolocating bat, *Eptesicus fuscus*. J Comp Physiol A145:431-435.

Kick SA, Simmons JA (1984) Automatic gain control in the bat's sonar receiver and the neuroethology of echolocation. J Neurosci 4:2725-2737.

Koopman KF (1988) Systematics and distribution. In: Greenhall AM, Schmidt U (eds) Natural History of Vampire Bats. Boca Raton: CRC Press, pp. 7-19.

Koopman KF, Jones JK Jr (1970) Classification of bats. In: Slaughter BH, Walton DW (eds) About Bats. Dallas: Southern Methodist University Press, pp. 29-50.

Krull D (1992) Jagdverhalten und Echoortung bei *Antrozous pallidus* (Chiroptera: Vespertilionidae). Ph.D. Dissertation, Fakultat fur Biologie der Ludwig-Maximilians Universitat Munchen, Germany.

Kurta A, Baker RH (1990) *Eptesicus fuscus*. Mammal Species 356:1-10.

Leonard ML, Fenton MB (1984) Echolocation calls of *Euderma maculatum* (Chiroptera: Vespertilionidae): use in orientation and communication. J Mammal 65:122-126.

Linarés O (1986) Murcielagos de Venezuela. Caracas: Cuadernos Lagoven.

Link A, Marimuthu G, Neuweiler G (1986) Movement as a specific stimulus for prey catching behavior in rhinolophid and hipposiderid bats. J Comp Physiol 159:403-414.

MacDonald K, Matsui E, Stevens R, Fenton MB (1994) Echolocation calls of *Pipistrellus subflavus* (Chiroptera: Vespertilionidae), the eastern pipistrelle. J Mammal 75:462-465.

Masters MW (1979) Insect disturbance stridulation: its defensive role. Behav Ecol Sociobiol 5:187-200.

Masters MW, Moffat AJM, Simmons JA (1985) Sonar tracking of horizontally moving targets by the big brown bat *Eptesicus fuscus*. Science 228:2332.

Medellin RA (1989) *Chrotopterus auritus*. Mammal Species 343:1-5.

Medellin RA, Artia HT (1989) *Tonatia evotis* and *Tonatia sylvicola*. Mammal Species 334:1-5.

Miller LH (1991) Arctiid moth clicks can degrade the accuracy of range difference discrimination in echolocating big brown bats, *Eptesicus fuscus*. J Comp Physiol A 199:571-579.

Møhl B, Miller LA (1976) Ultrasonic clicks produced by the peacock butterfly: a possible bat repellant. J Exp Biol 64:639-644.

Möhres FP (1967) Communicative characters of sonar signals in bats. In: Busnel R-G (ed) Animal Sonar Systems, Vol. 2. (NATO Advanced Study Institutes.) New York: Plenum, pp. 939-945.

Morrison DW (1980) Foraging and day-roosting dynamics of canopy fruit bats in Panama. J Mammal 61:20-29.

Nachtigall PE, Moore PWB (eds) (1988) Animal sonar: processes and performance. (NATO Advanced Study Institutes 156.) New York: Plenum.

Navarro L-D, Wilson D E (1982) *Vampyrum spectrum*. Mammal Species 184:1-4.

Neuweiler G (1989) Foraging ecology and audition in bats. Trends Ecol Evol 4:160-166.

Neuweiler G, Fenton MB (1988) Behavior and foraging ecology of echolocating bats. In: Nachtigall PE, Moore PWB (eds) Animal Sonar Systems: Processes and performance. (NATO Advanced Study Institutes 156.) New York: Plenum, pp. 535-550.

Neuweiler G, Singh S, Sripathi K (1984) Audiograms of a south Indian bat community. J Comp Physiol A154:133-142.

Neuweiler G, Metzner W, Heilman U, Rubsamen R, Eckrich M, Costa HH (1987) Foraging behavior and echolocation in the rufus horseshoe bat, *Rhinolophus rouxi*. Behav Ecol Sociobiol 20:53-67.

Norberg UM (1990) Vertebrate flight. Zoophysiology 27:1-291.

Norberg UM, Fenton MB (1988) Carnivorous bats? Biol J Linn Soc 33:383-394.

Norberg UM, Rayner JMV (1987) Ecological morphology and flight in bats (Mammalia, Chiroptera): wing adaptations, flight performance, foraging strategy and echolocation. Philos Trans R Soc Lond B Biol Sci 316:335-427.

Novick A (1977) Acoustic orientation. In: Wimsatt WA (ed) Biology of Bats, Vol. 3. New York: Academic, pp. 73-289.

Obrist KM (1989) Individuelle Variabilitat der Echoortung: Vergleichende Freilanduntersuchungen an Vier Vespertilioniiniden Fledermausarten Kanadas. Ph.D. Dissertation, Faculty of Science, Ludwig-Maximilians-Universitat Munchen, Germany.

Obrist M, Aldridge HDJN, Fenton MB (1989) Roosting and echolocation behavior of the African bat, *Chalinolobus variegatus*. J Mammal 70:828-833.

Obrist KM, Fenton MB, Eger JL, Schlegel P (1993). What ears do for bats: a comparative study of pinna sound pressure transformation in Chiroptera. J Exp Biol 180:119-152.

Pettigrew JD (1991) Wings or brain? convergent evolution in the origin of bats. Syst Zool 40:199-216.

Pettigrew JD, Jamieson BGM, Robson SK, Hall LS, McNally KI, Cooper HM (1989) Phylogenetic relations between microbats, megabats and primates (Mammalia: Chiroptera and Primates). Philos Trans R Soc Lond B Biol Sci 325:489-559.

Pierson ED, Sarich VM, Lowenstein JM, Daniel MJ, Rainey WE (1986) A molecular link between the bats of New Zealand and South America. Nature 323:60-63.

Pye JD, Roberts LH (1970) Ear movements in a hipposiderid bat. Nature 225:285-286.

Racey PA, Swift SM (1985) Feeding ecology of *Pipistrellus pipistrellus* (Chiroptera: Vespertilionidae) during pregnancy and lactation. 1. Foraging behaviour. J Anim Ecol 54:205–215.

Ransome RD (1991) Greater horseshoe bat. In: Corbet GB, Harris S (eds) The Handbook of British Mammals, 3d Ed. Oxford: Blackwell Scientific, pp. 88–94.

Rautenbach IL, Kemp A C, Scholtz C H (1988) Fluctuations in availability of arthropods correlated with microchiropteran and avian predator activities. Koedoe 31:77–90.

Ray CE, Linares OJ, Morgan GS (1988) Paleontology. In: Greenhall AM, Schmidt U (eds) Natural History of Vampire Bats. Boca Raton: CRC Press, pp. 19–30.

Rayner JMV (1991) Complexity and a coupled system: flight, echolocation and evolution in bats. In: Schmidt-Kittler N, Vogel K (eds) Constructional Morphology and Evolution. Berlin: Springer-Verlag, pp. 173–191.

Roeder KD (1967) Nerve cells and insect behavior, revised edition. Cambridge: Harvard University Press.

Roeder KD, Treat AE (1960) The acoustic detection of bats by moths. In: XI International Entomological Congress, Vienna.

Ross A (1967) Ecological aspects of the food habits of insectivorous bats. Proc West Found Vertebr Zool 1:205–263.

Roverud R C (1987) The processing of echolocation sound elements in bats: a behavioural approach. In: Fenton MB, Racey P A, Rayner JMV (eds) Recent Advances in the Study of Bats. Cambridge: Cambridge University Press, pp. 152–170.

Roverud RC, Nitsche V, Neuweiler G (1991) Discrimination of wingbeat motion by bats, correlated with echolocation sound pattern. J Comp Physiol A168: 259–263.

Rubsamen R, Schafer M (1990a) Ontogenesis of auditory fovea representation in the inferior colliculus of the Sri Lankan rufous horseshoe bat, *Rhinolophus rouxi*. J Comp Physiol A167:757–769.

Rubsamen R, Schafer M (1990b) Audiovocal interactions during development? Vocalisation in deafened young horseshoe bats vs. audition in vocalisation-impaired bats. J Comp Physiol A167:771–784.

Rubsamen R, Neuweiler G, Marimuthu G (1989) Ontogenesis of tonotopy in inferior colliculus of a hipposiderid bat reveals postnatal shift in frequency-place code. J Comp Physiol A165:755–769.

Ryan MJ, Tuttle MD (1987) The role of prey-generated sounds, vision and echolocation in prey location by the African bat, *Cardioderma cor* (Megadermatidae). J Comp Physiol A161:59–66.

Rydell J (1986) Feeding territoriality in female northern bats, *Eptesicus nilssoni*. Ethology 72:329–337.

Schmidt S (1988) Evidence for a spectral basis of texture perception in bat sonar. Nature 331:617–619.

Schmidt U, Schlegel P, Schweizer H, Neuweiler G (1991) Audition in vampire bats, *Desmodus rotundus*. J Comp Physiol A168:45–51.

Schnitzler H-U (1987) Echoes of fluttering insects: information for echolocating bats. In: Fenton MB, Racey PA, Rayner JMV (eds) Recent Advances in the Study of Bats. Cambridge: Cambridge University Press, pp. 226–243.

Schumm A, Krull D, Neuweiler G (1991) Echolocation in the notch-eared bat, *Myotis emarginatus*. Behav Ecol Sociobiol 28:255–261.

Shields WM, Bildstein KL (1979) Birds versus bats: behavioral interactions at a localized food source. Ecology 60:468–474.

Simmons JA, O'Farrell MJ (1977) Echolocation in the long-eared bat, *Plecotus phyllotus*. J Comp Physiol A122:201–214.

Simmons JA, Stein RA (1980) Acoustic imaging in bat sonar: echolocation signals and the evolution of echolocation. J Comp. Physiol A135:61–84.

Simmons JA, Saillant PA, Dear SP (1992) Through a bat's ear. IEEE Spectrum, 29 March, pp. 46–48.

Simmons NB, Novacek M J, Baker R J (1991) Approaches, methods and the future of the chiropteran monophyly controversy: a reply to J.D. Pettigrew. Syst Zool 40:239–243.

Slater PJB (1983) The study of communication. In: Halliday TR, Slater PJB (eds) Animal Behaviour 2. Communication. New York: Freeman, pp. 9–81.

Smith JD (1972) Systematics of the chiropteran family Mormoopidae. Univ Kansas Mus Nat Hist Misc Publ 56:1–132.

Smithers RHN (1983) Mammals of the Southern African Subregion. Pretoria: University of Pretoria Press.

Speakman JR, Racey PA (1991) No cost of echolocation for bats in flight. Nature 350:421–423.

Stoneman MG, Fenton MB (1988) Disrupting foraging bats: the clicks of arctiid moths. In: Nachtigall P E, Moore PWB (eds) Animal Sonar: Processes and performance. (NATO Advanced Study Institutes 156.) New York: Plenum, pp. 635–638.

Summe YW, Menne D (1988) Discrimination of fluttering targets by the FM bat *Pipistrellus stenopterus*. J Comp Physiol A163:349–354.

Surylkke AM (1988) Interactions between echolocating bats and their prey. In: Nachtigall PE, Moore PWB (eds) Animal Sonar: Processes and Performance. (NATO Advanced Study Institutes 156.) New York: Plenum, pp. 635–638.

Surlykke AM, Miller LH (1985) The influence of arctiid clicks on bat echolocation: jamming or warning? J Comp Physiol A156:831–843.

Suthers RA (1965) Acoustic orientation by fish-catching bats. J Exp Zool 158:253–258.

Suthers RA (1988) The production of echolocation signals by bats and birds. In: Nachtigall P E, Moore PWB (eds) Animal Sonar Systems: Processes and Performance. (NATO Advanced Study Institutes 156.) New York: Plenum, pp. 23–46.

Thewissen JGM, Babcock SK (1991) Distinctive cranial and cervical innervation of wing muscles: new evidence for the monophyly of bats. Science 251:934–936.

Thewissen JGM, Babcock S K (1992) The origin of flight in bats. Bioscience 42:340–345.

Thomas DW (1984) Fruit intake and energy budgets of frugivorous bats. Physiol Zool 57:457–462.

Thomas DW, West SD (1989) Sampling methods for bats. Gen. Tech. Rep. PNW-GTR-243. Portland, Oregon: U.S. Dept. of Agriculture, Forest Service, Pacific NW Research Station.

Thomas DW, Bell GP, Fenton MB (1987) Variation in echolocation call frequencies in North American vespertilionid bats: a cautionary note. J Mammal 68:842–847.

Thomas SP (1987) The physiology of flight. In: Fenton MB, Racey PA, Rayner J M V (eds) Recent Advances in the study of bats. Cambridge: Cambridge University Press, pp. 75–99.

Thomson CE, Barclay RMR, Fenton MB (1985) The role of infant isolation calls in mother-infant reunions in the little brown bat (*Myotis lucifugus*). Can J Zool 63:1982-1988.

Trappe M, Schnitzler H-U (1982) Doppler shift compensation in insect-catching horseshoe bats. Naturwissenschaften 69:193-194.

Tumlison R, Douglas M E (1992) Parsimony analysis and the phylogeny of the plecotine bats. J Mammal 73:276-285.

Tuttle MD, Ryan MJ (1981) Bat predation and the evolution of frog vocalizations in the neotropics. Science 214:677-678.

Tuttle MD, Ryan MJ, Belwood JJ (1985) Acoustic resource partitioning by two species of phyllostomid bats (*Trachops cirrhosus* and *Tonatia sylvicola*). Anim Behav 33:1369-1371.

Tyrrell K (1988) The use of prey-generated sounds in flycatcher-style foraging by *Megaderma spasma*. Bat Res News 29:51.

Vaughan TA (1976) Nocturnal behavior of the African false vampire bat (*Cardioderma cor*). J Mammal 57:227-248.

Vaughan TA (1977) Foraging behaviour of the giant leaf-nosed bat (*Hipposideros commersoni*). J East Afr Wildl 15:237-249.

Vaughan TA (1986) Mammalogy, 3d Ed. Philadelphia: Saunders.

Warner RM (1982) *Myotis auriculus*. Mammal Species 191:1-3.

Welty JC (1975) The Life of Birds, 2d Ed. Philadelphia: Saunders.

Whitaker JO Jr, Black H L (1976) Food habits of cave bats from Zambia. J Mammal 57:199-204.

Wilkinson JS (1985) The social organization of the common vampire bat. I. Pattern of cause and association. Behav Ecol Sociobiol 17:111-121.

Wilkinson JS (1992) Information transfer at evening bat colonies. Anim Behav 44:501-518.

Yager DD, May ML, Fenton MB (1990) Ultrasound-triggered, flight-gated evasive maneuvers in the praying mantis *Parasphendale agrionina*. 1. Free flight. J Exp Biol 152:17-39.

Zbinden K (1989) Field observations on the flexibility of the acoustic behaviour of the European bat *Nyctalus noctula* (Schreber, 1774). Rev Suisse Zool 96:335-343.

Appendix

The scientific and common names of bats referred to in the text are shown with forearm lengths (in millimeters) and family name given in parenthesis, followed by information about distribution, diet, and foraging behavior. The asterisk (*) identifies species illustrated in Figure 2.8.

Antrozous pallidus, the pallid bat (53-60; Vespertilionidae), is widespread in western North America. It often takes arthropod prey from surfaces, having located it by passive cues (Bell 1982). *Antrozous* also use echolocation to locate flying targets (Krull 1992).

Balantiopteryx plicata, the least sac-winged Bat (38-44; Emballonuridae), is an insectivorous species of Central America (Arroyo-Cabrales and Jones 1988).

Cardioderma cor, the heart-nosed bat (54–59; Megadermatidae), occursin east Africa and takes animal prey, from insects to bats (Vaughan 1976). This species often catches its food on the ground having located it by passive cues (Ryan and Tuttle 1987).

Carollia perspicillata, the short-tailed fruit bat (40–45; Phyllostomidae), is common and widespread in the neotropics. This bat eats a variety of species of fruit and, occasionally, insects (Fleming 1988).

*Chalinolobus variegatus, the butterfly bat (41–45; Vespertilionidae), is a widespread insectivorous species in the African savannah (Smithers 1983). Its echolocation calls have been described by Obrist et al. (1989).

*Chrotopterus auritus, the woolly false vampire bat (78–87; Phyllostomidae), is a large "carnivorous" bat that eats large insects, lizards, and small bats. It is widespread in South and Central America, but few details are known about its biology and behavior (Medellin 1989).

Cloeotis percivali, the short-eared trident bat (30–35; Hipposideridae), is a small insectivorous species from southeastern Africa. Known to feed heavily on moths (Whitaker and Black 1976), its echolocation calls are dominated by constant frequency components of more than 200 kHz (Fenton and Bell 1981).

Desmodus rotundus, the common vampire Bat (52–63; Phyllostomidae), is the most abundant and widespread of the blood-feeding vampire bats. Like the others, it occurs in South and Central America. It is more common and better known than the other vampire bats (Greenhall and Schmidt 1988; Schmidt et al. 1991).

Eptesicus fuscus, the big brown bat (41–52; Vespertilionidae), a widespread insectivorous species of North America, also occurs in the West Indies and in the northern part of South America. This has been "the bat" of many echolocation studies (e.g., Kick 1982; Masters, Moffat, and Simmons 1985; Miller 1991), and its biology is well known (Kurta and Baker 1990).

Euderma maculatum, the spotted bat (48–51; Vespertilionidae), is a spectacular black-and-white insectivorous bat of western North America. The bat has enormous ears and uses echolocation calls that are clearly audible to most human observers (Leonard and Fenton 1984).

*Hipposideros caffer, Sundevall's leaf-nosed bat (40–48; Hipposideridae), is widespread in savannah regions of subsaharan Africa. It commonly roosts in large numbers in caves and old mines and often is mistaken for the similar *Hipposideros ruber* (e.g., Bell and Fenton 1984; Fenton 1986), which is more common in rain forest situations (Smithers 1983). Moths are among the insects commonly taken by this species.

Hipposideros lankadiva, Kelaart's leaf-nosed bat (48–55; Hipposideridae), occurs in Sri Lanka and southern India, and the ontogeny of its echolocation has been reported by Rubsmen and colleagues (Rubsamen and Schafer 1990a, b; Rubsamen, Neuweiler, and Marimuthu, 1989).

Hipposideros speoris, Schneider's round-leaf bat (48–52; Hipposideridae), occurs in peninsular India and Sri Lanka. Insectivorous, its foraging behavior and echolocation have been studied in southern India (Neuweiler 1989; Neuweiler, Singh, and Sripathi 1984).

Lasionycteris noctivagans, the silver-haired bat (36–45; Vespertilionidae), is widespread in North America. This migratory species feeds on airborne insects and roosts in hollows and crevices around trees (Barclay 1986).

Lasiurus borealis, the red bat (36–42; Vespertilionidae), is widespread in North and Central America. This migratory species roosts in foliage and feeds on airborne prey (Hickey and Fenton 1990).

Lasiurus cinereus, the hoary bat (54–58; Vespertilionidae), occurs from Hawaii to the Galapagos and throughout much of North, Central, and South America. Another migratory species, *Lasiurus cinereus*, roosts in foliage and feeds on airborne insects (Barclay 1986).

Macroderma gigas, the ghost bat (105–115; Megadermatidae), occurs over much of northern Australia. This large bat feeds on prey ranging from large insects to vertebrates such as birds and mice, which it locates mainly by listening to their sounds of movement (Hudson and Wilson 1986).

Macrotus californicus, the California leaf-nosed bat (45–58; Phyllostomidae), occurs in the southwestern United State and adjacent Mexico. It takes prey from surfaces, having located them by prey-generated sounds or acute vision (Bell 1985).

*Megaderma lyra, the Indian false vampire bat (65–70; Megadermatidae), is a widespread species extensively studied in the laboratory (Fiedler 1979; Schmidt 1988; Neuweiler 1989) and field (Audet et al. 1991). This bat takes a range of prey from large arthropods to frogs, mice, and other bats and tends to rely on prey-generated sounds to locate its targets.

Megaderma spasma, the lesser false vampire bat (54–61; Megadermatidae), occurs in southeast Asia and feeds mainly on medium-sized to large insects. In captivity it takes airborne prey and prey sitting on surfaces (Tyrrell 1988).

Myotis auriculus, the Mexican long-eared myotis (37–40; Vespertilionidae), occurs in the American Southwest (Warner 1982) and takes prey from surfaces (Fenton and Bell 1979). In its behavior, *M. auriculus* resembles *Myotis emarginatus*, *Myotis evotis*, and *Myotis septentrionlis*.

Myotis daubentoni, Daubenton's bat (35–39; Vespertilionidae), appears to be the Eurasian counterpart of *Myotis lucifugus*, a small mouse-eared bat that hunts airborne targets or grabs insects from the surface of the water (Kalko and Schnitzler 1989).

Myotis emarginatus, the notch-eared Bat (37–43; Vespertilionidae), of Europe, is a versatile species sometimes taking prey from surfaces, sometimes in the air. It produces a range of echolocation calls (Schumm, Krull, and Neuweiler 1991).

Myotis evotis, the long-eared bat (36–41; Vespertilionidae), is a large-eared *Myotis* of western North America that commonly takes prey from surfaces, having located it by prey-generated sounds (Faure and Barclay 1992; Faure, Fullard, and Barclay 1990).

Myotis lucifugus, the little brown bat (34–40; Vespertilionidae), is a widespread bat of North America that commonly roosts in buildings and hibernates in caves and old mines. This species hunts airborne targets but also takes prey from the surface of water (Fenton and Bell 1979; Buchler 1980; Thomson, Barclay, and Fenton 1985).

Myotis septentrionalis, the northern long-eared bat (35–40; Vespertilionidae), is widespread in North America. Like other large-eared *Myotis*, this species commonly takes prey from surfaces (Fenton 1992).

Myotis volans, the long-legged myotis (36–44; Vespertilionidae), occurs in western United States and Canada where it feeds on airborne targets hunted in the open. For a *Myotis*, this species uses a relatively narrowband echolocation call (Fenton and Bell 1979).

Myotis yumanensis, the Yuma myotis (33–37; Vespertilionidae), occurs in western North America from British Columbia to Mexico. The bat hunts airborne targets and feeds over water (Herd and Fenton 1982).

Noctilio albiventris, the lesser bulldog bat (60–68; Noctilionidae), occurs in the Neotropics where it commonly feeds over water. Its echolocation has been studied in the laboratory (Roverud 1987) and in the field (Brown, Brown, and Grinnell 1983).

Noctilio leporinus, the greater bulldog bat (81–88; Noctilionidae), is another neotropical species that feeds on fish and other animals taken from the surface of water or from land. Its echolocation has been studied in a variety of situations (Suthers 1965; Campbell and Suthers 1988).

*Nycteris grandis, the large slit-faced bat (57–66; Nycteridae), occurs mainly in central Africa and uses low-frequency sounds of movement to locate prey ranging from insects to frogs, bats, birds, and fish (Fenton et al. 1990).

Nycteris thebaica, the Egyptian slit-faced Bat (42–52; Nycteridae), is widespread in Africa from Egypt to South Africa. This species feeds on insects ranging from orthoperans to moths, often taking prey from surfaces (Fenton, Gaudet, and Leonard 1983; Aldridge et al. 1990).

Nycticeius humeralis, the evening bat (34–39; Vespertilionidae), is common in the east-central United States where it roosts in buildings and takes airborne prey (Wilkinson 1992).

Pipistrellus pipistrellus, the pipistrelle (28–35; Vespertilionidae), is one of the more common Eurasian bats that has been well studied in many parts of Europe (e.g., Racey and Swift 1985; Avery 1991).

Pipistrellus stenopterus, the narrow-winged pipistrelle (38–42; Vespertilionidae), is known from Malaysia and Sarawak. Its echolocation behavior has been studied by Summe and Menne (1988).

Pipistrellus subflavus, the eastern pipistrelle (32–36; Vespertilionidae), is a common and widespread species of eastern North America. It appears to feed on airborne insects, but there are few data about its echolocation behavior (Fujita and Kunz 1984).

Pteronotus parnellii, Parnell's mustached bat (50–60; Mormoopidae), is widespread neotropical species noted for its specializations for detecting fluttering targets. As the New World high duty cycle bat, itts echolocation behavior has been widely studied (e.g., Goldman and Henson 1977), but there is less information about its behavior in the wild (Herd 1983).

Rhinolophus ferrumequinum, the greater horseshoe bat (50–59; Rhinolophidae), was once common and widespread in Europe and Asia but has disappeared from large parts of its original range. The topic of many studies of echolocation (e.g., Griffin and Simmons 1974; Trappe and Schnitzler 1982), there is a growing body of information about its behavior in the field (e.g., Ransome 1991).

*Rhinolophus hildebrdandti, Hildebrandt's horseshoe bat (62–66; Rhinolophidae), is the largest of the horseshoe bats and alternates between hunting from a perch and from continuous flight. This insectivorous species is widespread in eastern Africa, where it often forages in riverine forest (Fenton and Rautenbach 1986).

Rhinolophus landeri, Lander's horseshoe bat (42–45; Rhinolophidae), occurs widely in subsaharan Africa and commonly roosts in caves or other hollows (Smithers 1983). This insectivorous species produces echolocation calls dominated by 120 kHz CF components (Fenton and Bell 1981).

Rhinolophus rouxi, the rufous horseshoe bat (45–50; Rhinolophidae), occurs from Sri Lanka to Vietnam. The foraging behavior and echolocation have been studied in Sri Lanka (Neuweiler et al. 1987; Rubsamen and Schafer 1990a).

Rousettus aegyptiaca, the Egyptian fruit bat (90–105; Pteropodidae), is the species of Pteropodidae known to echolocate. This species is commonly exhibited in zoos and has been widely studied in the field and in captivity (Smithers 1983).

Tadarida brasiliensis, the Mexican free-tailed bat (38–45; Molossidae), roosts in huge numbers in some caves. It is widespread from the southern United States into Central and South America, and there are excellent studies of mother–young interactions in this species (e.g., Balcombe and McCracken 1992).

*Tonatia brasiliensis, the pygmy round-eared bat (32–36; Phyllostomidae), is a little-known insectivorous species of South and Central America (Medellin and Arita 1989).

Trachops cirrhosus, the fringe-lipped bat (58–64; Phyllostomidae), is best known for its use of male songs to locate and identify the frogs on which it feeds (Tuttle and Ryan 1981). This bat also eats other prey and although it is not well known, occurs widely in the neotropics (Tuttle, Ryan, and Belwood 1985).

Vampyrum spectrum, Linnaeus' false vampire bat (105–115; Phyllostomidae), is the largest of the New World bats and occurs widely in the Neotropics. It takes prey from birds to bats and other small vertebrates, but has been little studied in the wild (Navarro and Wilson 1982).

3

Behavioral Studies of Auditory Information Processing

Cynthia F. Moss and Hans-Ulrich Schnitzler

1. Introduction

Echolocating bats are nocturnal animals that rely largely on auditory information to orient in the environment and intercept prey. Bats produce high-frequency vocal signals and perceive their surroundings by listening to the features of the echoes reflecting off targets in the path of the sound beam (Griffin 1958). Computations performed on these echoes by the auditory system allow the bat to extract fine spatial information about its world through acoustic channels. This chapter attempts a comprehensive review and synthesis of data on auditory information processing and perception by sonar in echolocating bats.

The information available to a bat's acoustic imaging system is constrained by the characteristics of its species-specific sonar emissions. Species diversity in echolocation signal design therefore gives rise to differences in auditory information processing in bats (see Fenton, Chapter 2). This chapter includes a discussion of psychophysical studies of auditory information processing in species that use both constant-frequency (CF) and frequency-modulated (FM) signals. In this chapter, research on target echo detection, range estimation, horizontal and vertical localization, and movement discrimination are examined. Although some early behavioral work in the field is covered, the emphasis of this chapter is on research conducted since 1979. For a detailed review of echolocation behavior in bats before this period, see Schnitzler and Henson (1980).

2. Detection

Sonar sounds used by bats may exceed 100 dB sound pressure level (SPL) at a distance of 10 cm (Griffin 1958), and yet spherical spreading losses and excess attenuation (Lawrence and Simmons 1982a; Hartley 1989) result in echoes from small targets (e.g., insects) that may be 90 dB weaker than the transmitted signal at a distance of 1.6 m (Kober and Schnitzler 1990). Given the bat's success at capturing small prey, we must infer that it can detect very weak echoes. Indeed, a bat can only process acoustic information about the spatial dimensions of a sonar target if the signal level exceeds the animal's threshold for detection.

Target detection occurs when a bat accurately determines that an echo arriving at its ears has reflected from an object in the path of its sonar transmission. There are many parameters that influence the detection process, including the structure of the emitted signal, the characteristics of the auditory receiver, the reflective properties and the distance of the target, and the acoustic environment (see Møhl 1988). Here, each of these parameters is considered, with particular emphasis on the characteristics of the auditory receiver and the acoustic environment.

Early studies of echo detection relied on methods of obstacle avoidance and behavioral responses to airborne targets to estimate the operating range of bat sonar. A most impressive result of wire avoidance studies is that many species of bat can detect wires with diameters much less than the wavelength of their sonar sounds (summarized in Schnitzler and Henson 1980). From field studies of bats chasing airborne targets, Griffin (1953) found that the big brown bat, *Eptesicus fuscus*, first reacts to a 1-cm-diameter target at a distance of 200 cm, and in the laboratory Webster and Brazier (1965) reported that the little brown bat, *Myotis lucifugus*, reacts to a 2.1-mm-target at a distance of 60 cm and a 4.2-mm-diameter sphere at about 120–135 cm. A basic premise of this work is that the bat's first behavioral reaction to a target serves as a reliable indicator of its detection of that target. Using estimates of the level of the bat's sonar emissions, its detection range, and the reflective properties of the target, echo detection threshold estimates were 23–28 dB SPL for *Myotis lucifugus* (Griffin 1958), 25–34 dB SPL for the greater horseshoe bat *Rhinolophus ferrumequinum* (Sokolov 1972), and 9.2–21.6 dB SPL for *Myotis oxygnathus* (Airapetianz and Konstantinov 1974), all performing in wire avoidance tasks. Threshold estimates were 17 dB SPL for *Eptescius fuscus* reacting to a sphere in the field (Griffin 1958) and 15–30 dB SPL for *Myotis lucifugus* hunting fruitflies in the laboratory (Griffin, Webster, and Michael 1960). This early work and more recent field studies on the behavioral responses to airborne targets are discussed elsewhere in this volume (see Fenton, Chapter 2; Simmons et al., Chapter 4).

Changes in the bat's sonar signals as it approaches obstacles or attempts to intercept prey serve to describe the animal's vocal-motor control of acoustic information guiding spatial perception by sonar. In fact, the very features of the sonar sounds used during different phases of insect pursuit suggest the acoustic information a bat requires at different distances from its prey (Griffin 1958). However, the bat's motor response to an obstacle or target may be delayed with respect to absolute detection, and estimates of the sound level of echolocation cries produced by flying bats can be distorted by the directional properties of the transmitter and microphone. Together, these factors limit the information one can gather on target detection and the operating range of sonar in flying bats. While laboratory studies do not draw upon the richness of the bat's behavior observed under natural conditions, they present the opportunity to carefully control and measure the bat's acoustic environment. Indeed, psychophysical experiments on target detection complement data gathered in the field, and results from these different methods can be assembled to develop a unified description of target detection in bat sonar.

Behavioral studies of the bat sonar receiver have employed diverse methods. Traditional psychoacoustic measures of passive hearing with pure tone stimuli have generated data on absolute sensitivity and frequency selectivity. Experiments requiring the bat to actively interrogate its environment using sonar have yielded complementary data on the animal's detection of targets by echolocation. Collectively these studies show that the bat sonar receiver is well suited for the detection of species-specific echolocation sounds.

2.1 Absolute Sensitivity

The most extensive data on receiver characteristics important for signal detection in echolocating bats are found in behavioral audiograms. Behavioral audiograms obtained in different species show that echolocating bats hear over a broad frequency range, often spanning several octaves. Minimum thresholds are comparable to other mammalian species (see Fay 1988), typically in the range of -5 to $+10$ dB SPL. Figure 3.1 displays behavioral audiograms for the following species: the big brown bat, *Eptesicus fuscus* (Dalland 1965; Poussin and Simmons 1982), the little brown bat, *Myotis lucifugus* (Dalland 1965), the horsehoe bat, *Rhinolophus ferrumequinum* (Long and Schnitzler 1975), the fish-catching bat, *Noctilio leporinus* (Wenstrup 1984), the Indian false vampire bat, *Megaderma lyra* (Schmidt, Türke, and Vogler 1983), and the Egyptian tomb bat, *Rousettus aegyptiacus* (Suthers and Summers 1980).

The audiogram for *Rhinolophus ferrumequinum* was measured using classical conditioning of the heart rate response, and all other audiograms were obtained using operant conditioning procedures. Above the abscissa of each audiogram in Figure 3.1 is a horizontal bar to indicate the frequency band of the harmonic components of the echolocation sounds produced by that species. The thickness of each bar denotes the relative strength of the individual harmonic components. For all microchiropteran species, the frequency range of the bat's echolocation sound corresponds closely to a frequency region of maximum auditory sensitivity. *Rhinolophus ferrumequinum* shows the lowest threshold to a narrow frequency band, flanked by frequency regions of relative insensitivity. This narrow frequency region of high auditory sensitivity corresponds closely to the constant frequency

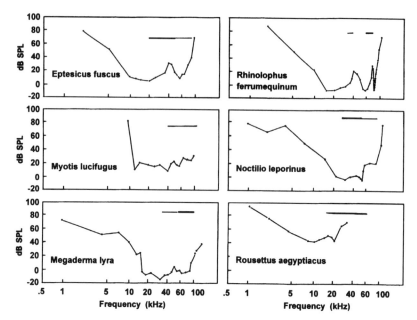

FIGURE 3.1. Behavioral audiograms of six different bat species: *Eptesicus fuscus* (Dalland 1965; Poussin and Simmons 1982), *Myotis lucifugus* (Dalland 1965), *Rhinolophus ferrumequinum* (Long and Schnitzler 1975), *Megaderma lyra* (Schmidt, Türke, and Vogler 1983), *Noctilio leporinus* (Wenstrup 1984), and *Rousettus aegyptiacus* (Suthers and Summers 1980). The horizontal bars above each audiogram show the frequency band of the echolocation sounds of each species. SPL, sound pressure level.

component of its echolocation signal. The species that shows the least sensitivity to pure tones across its audible range is *Rousettus aegyptiacus*, a megachiropteran fruit-eating bat that uses clicks for echolocation (Suthers and Summers 1980).

While most echolocation signals produced by bats are in the range of 25 to 100 kHz, many of the behavioral audiograms show thresholds well below 50 dB SPL at 10 kHz; in fact, the audible range of hearing for *Eptesicus fuscus* and *Megaderma lyra* extends down to about 1 kHz (Poussin and Simmons 1982; Schmidt, Turke, and Volger 1983). Hearing at frequencies below those used for echolocation may play an important role in social communication and detection of prey and predators through passive listening.

2.2 Frequency Selectivity

While it is of interest to establish the bat's absolute sensitivity to sounds in quiet, animals that rely on echolocation must detect signals against background noise. Wind, noise associated with flight, the communication sounds of other nocturnal animals, and jamming signals from other bats and possibly insects may all create acoustic interference for echo detection. For this reason, masked auditory thresholds provide a useful description of an animal's resistance to noise. If a masking stimulus is broadband white noise, only a portion of the noise band actually contributes to the masking of a pure tone stimulus. This was originally demonstrated by Fletcher (1940), who introduced the concept of the critical band, the frequency region about a pure tone that is effective in masking that tone. A large critical band indicates that the noise must be summed over a wide frequency band to mask the signal and therefore indicates relatively poor frequency resolution of the auditory system. By contrast, a small critical band indicates relatively high frequency resolution. Thus, the critical band indexes the animal's resistance to noise as a function of signal frequency (see Long 1994).

Fletcher included in the concept of the critical band a hypothesis proposing that the power of the noise integrated over the critical band equals the power of the pure tone signal at threshold. This implies that a critical band can be determined indirectly by measuring the detection of a pure tone against broadband masking noise, rather than directly by measuring the threshold against a variety of noise bandwidths. If one knows the level of the tone at threshold and the spectrum level of the noise, the ratio of the two provides the necessary information to determine the critical bandwidth based on Fletcher's assumptions. This ratio has been termed the critical ratio (Zwicker, Flottorp, and Stevens 1957).

Critical bands and critical ratios generally show parallel and systematic increases with signal frequency in vertebrates; however, there are noteworthy exceptions, among them the horseshoe bat, *Rhinolophus ferrumequinum*. This bat shows a sharp decline in critical ratio (i.e., a marked increase in frequency resolution) at 83 kHz, relative to neighboring frequencies (Long 1977). This specialization for frequency resolution at 83 kHz parallels that observed for absolute hearing sensitivity in this bat (see Fig. 3.1) and reflects specialization in the horseshoe bat's cochlea (e.g., Bruns 1980; Neuweiler, Bruns, and Schuller 1980). Critical ratios have also been estimated in the echolocating megachiropteran bat *Rousettus* (Suthers and Summers 1980). The shape of the critical ratio function in this species differs from most other mammals, showing an unusual elevation of critical ratios at frequencies below 8 kHz.

Critical bands have been measured for the false vampire bat, *Megaderma lyra*, a species that uses a broadband FM signal for echolocation (Schmidt 1993). As in other mammals, the width of the auditory filters in *Megaderma* increases with center frequency. However, unlike the auditory filters measured for nonecholocating mammals, the slopes are steeper and the filter shape around the center masker frequency remained symmetrical at high masker levels. Schmidt (1993) proposed that this departure from the standard mammalian critical band filter may reflect adaptations for the perception of broadband stimuli in noisy and echo-cluttered environments.

2.3 Directionality

Directional sensitivity of the bat's hearing also plays a role in echo detection. Behavioral measurements of directional sensitivity as a function of signal frequency have been conducted only in the CF bat *Rhinolophus ferrumequinum* (Grinnell and Schnitzler 1977). This species exhibits a ventral sidelobe of relative insensitivity, yet when these behavioral measurements are combined with acoustic measurements of sonar emission directionality (Schnitzler and Grinnell 1977), the complete echolocation system of *Rhinolophus* shows radially symmetrical sensitivity, with a −6 dB bandwidth of about 15°. The directionality of emission patterns has been studied in detail in other species, including the FM bat *Eptesicus fuscus* (Hartley and Suthers 1989), the grey bat *Myotis grisescens* (Shimozawa et al. 1974), and the leaf-nosed bat *Carollia perspicillata* (Hartley and Suthers 1987); however, behavioral sensitivity measures of directionality in these animals are currently lacking.

2.4 Psychophysical Studies of Echo Detection

In the past decade, psychophysical studies on echo detection in bats have been carried out in laboratory settings where the distance between the bat and the target has been controlled. It is important to note that estimates of echo detection threshold depend on several parameters: the features of the bat's sonar signals, the conditions of the environment, and the characteristics of the bat's sonar receiver.

Sonar signal structure in detection experiments is under the bat's active control and can be adapted to improve target detection under different conditions. Increasing the duration and level of the sonar signal and decreasing the bandwidth all serve to improve conditions for echo detection. It is therefore essential that signal structure be continuously monitored in a detection experiment.

The acoustic environment influences the extent to which masking of the target echo occurs, and there are several potential sources of masking: the bat's own sonar transmission, clutter echoes surrounding the target of interest, noise generated in the environment, and noise internal to the bat's sonar receiver. With the exception of internal noise, all other sources of noise can be monitored and controlled in behavioral studies of detection.

Also of importance to measures of echo detection is the sensitivity of the bat's auditory receiver following each sonar transmission. The bat's hearing sensitivity is reduced during and after signal emission by contractions of the middle ear muscles (Henson 1965; Suga and Jen 1975) and by neural suppression in the central auditory system (Suga and Schlegel 1972; Suga and Shimozawa 1974). There is a gradual recovery of hearing sensitivity following each sonar emission, and therefore changes in echo detection depend on the delay (distance) of the target echo.

2.4.1 Detection of Target Echoes in Quiet

Estimates of echo detection thresholds obtained from psychophysical studies are summarized in Table 3.1. The first behavioral laboratory study to examine detection of targets at controlled distances was conducted by Kick (1982), who trained the FM bat *Eptesicus fuscus* in a two-alternative forced-choice (2-AFC) procedure to respond to the presence of a ball (to the left or right of its observing position) by crawling toward it on a Y-shaped platform. In a 2-AFC procedure, the subject is required to make a response in every two-choice trial, and its performance level should directly reflect the relative difficulty of the task. In this particular experiment the bat's performance was nearly 100% correct when the target was positioned close to the animal and the target echoes were strong. As the target was placed further from the bat, the echoes were weaker, and the bat's performance level dropped off. When the bat was no longer responding to the presence of the target, its mean performance was close to that expected by chance (50% correct). In this study, as well as most others implementing a 2-AFC procedure, the criterion for threshold was the SPL producing 75% correct responses, that is, halfway between 100% correct and chance performance.

A schematic of the experimental conditions in Kick's study is presented in Figure 3.2B. The solid circle represents the target, which was presented either to the bat's left or right on a given trial. The bottom trace in this figure illustrates the bat's emission followed by a target echo; the time separation between these sounds depended on the distance between the bat and the target (see Section 3, Ranging). This particular study minimized echoes from other objects in the room, and hence no clutter echo is shown. The bat's task was to crawl down one arm of the Y-shaped platform, toward the side on which the target was presented. The distance of this target was increased over days of testing, resulting in successively weaker echoes returning to the bat's ears at increasing delay times. Detection of two sizes of spheres was studied (4.8 mm and 19.1 mm in diameter), and the detection range for the two spheres was 2.9 and 5.1 m, respectively. As noted previously, the maximum effective range of bat sonar was previously inferred from changes in the animal's echolocation behavior in obstacle avoidance and target interception tasks, and these data represent the bat's *distance of reaction*. Kick's data represent the bat's *distance of detection*, and they are the first to demonstrate the operation of the bat's sonar receiver for distances beyond 3 m. The threshold for echo detection in Kick's experiment was estimated at 30 and 60 kHz to be +5 and −12 dB SPL, respectively, for the 4.8-mm sphere and −2 dB and −21 dB SPL, respectively, for the 19.1-mm sphere. Because the echo level is stronger at 30 kHz than at 60 kHz, Kick suggested that the bat uses this component to detect the presence of the target. Thus, she concluded that the echo detection threshold of *Eptesicus* is approximately 0 dB SPL (see data plotted for the two sizes of spheres in Fig. 3.3A).

TABLE 3.1 Summary of behavioral data on detection by sonar in echolocating bats.

Species	Procedure	Target distance (cm delay/msec)	Speaker angular position	Speaker position relative to target echo	Threshold echo sound level	Author(s) (year)
Eptesicus fuscus	2-AFC simultaneous method of limits	To 510 cm (29.6 msec) real targets	40°	n/a	O dB SPL	Kick (1982)
Eptesicus fuscus	2-AFC simultaneous method of limits	17 cm/1.0 msec; 110 cm/6.4 msec; sonar sound playback	40°	25 cm/1.45 msec 2.1 m/12.2 msec after target echo	36 dB SPL 8 dB SPL	Kick and Simmons (1984)
Eptesicus fuscus	Yes/no sequential staircase	46 cm/2.7 msec; sonar sound playback	Single channel	68 cm/3.9 msec after target echo	37–38 dB SPL	Møhl and Surlykke (1989)
Eptesicus fuscus	2-AFC simultaneous method of limits	80 cm/4.6 msec; canned signal playback	40°	15 cm/0.87 msec before target echo	26 dB SPL	Master and Jacobs (1989)
Eptesicus fuscus and *Noctilio leporinus*	Yes/no sequential staircase	100 cm/5.8 msec; sonar sound playback	Single channel	100 cm/5.8 msec after target echo	40 dB SPL 40 dB SPL	Hartley (1992a)
Eptesicus fuscus	2-AFC simultaneous method of limits	54 cm/3.1 msec 40 cm/2.3 msec; 80 cm/4.6 msec; 160 cm/9.3 msec; all with sonar sound playback	40° 40°	100 cm/5.8 msec after target echo; 12 cm/0.7 msec before target echo;	33 dB SPL 21 dB SPL 11 dB SPL 20 dB SPL	Simmons et al (1988)
Eptesicus fuscus	2-AFC simultaneous method of limits	40 cm/2.3 msec; 80 cm/4.6 msec; sonar sound playback	40°	12 cm/0.7 msec before target echo	23 dB SPL 12 dB SPL	Simmons, Moffat, and Masters (1992)
Eptesicus fuscus	2-AFC simultaneous method of limits	57 cm/3.3 msec; phantom target sonar sound playback	40°	20 cm/1.16 msec before target echo	55–59 dB SPL	Moss and Simmons (1993)
Pipistrellus pipistrellus	Yes/no sequential staircase	138 cm/8.0 msec; triggered canned signal playback	Single channel	85 cm/4.9 msec before target echo	35 dB peSPL	Møhl (1986)
Eptesicus serotinus	Yes/no sequential staircase	55 cm/3.2 msec; sonar sound playback	Single channel	88 cm/5.1 msec after target echo	40–48 dB peSPL	Troest and Møhl (1986)

2-AFC, two-alternative forced-choice detection experiments; peSPL, peak equivalent sound pressure level SPL, sound pressure level; n/a, not available. level.

DETECTION

A. yes-no real target	B. 2-AFC real target	C. yes-no phantom target	D. yes-no phantom target	E. 2-AFC phantom target	F. 2-AFC phantom target
decision platform	decision platform	decision platform	decision platform	decision platform	decision platform
bat echo	bat echo	bat clutter echo	bat clutter echo clutter	bat clutter echo	bat clutter echo clutter

RANGING

G. yes-no real target	H. 2-AFC real targets	I. yes-no phantom target	J. yes-no phantom target	K. 2-AFC phantom target	L. 2-AFC phantom target
decision platform	decision platform	decision platform	decision platform	decision platform	decision platform
bat echo	bat echo	bat clutter echo	bat clutter echo clutter	bat clutter echo	bat clutter echo clutter

FIGURE 3.2A–L. Schematic of behavioral experiments for target detection (above, A–F) and target range difference discrimination (below, G–L). For all the experiments, the bat was trained to echolocate from the base of a decision platform and to indicate a response by crawling (or flying) to the left or right of its start position (see text). *Filled circles* represent real targets; *open ovals* represent phantom targets; *m*, microphone, illustrated with concentric rings; *s*, speaker, illustrated with concentric rings. Below each schematic is an illustration of the temporal sequence of sounds the bat hears under given experimental conditions: *bat*, bat's echolocation cry; *clutter*, echoes from speaker and microphone; *echo*, target echo or playback signal. *Detection*: In the yes/no detection experiments, the bat was rewarded for reporting whether the target was present or absent. It crawled down one arm of the platform to indicate "yes" (target present) and down the other arm of the platform to indicate "no" (target absent). In the two-alternative forced-choice (2-AFC) detection experiments, the bat was rewarded for reporting whether the target was presented to the left or right of its observing position. It learned to crawl down the arm of the platform on which it detected the target. *Ranging*: In the yes/no ranging experiments, the bat was rewarded for reporting whether the target was presented at a fixed distance (d_1) or at a greater distance (d_2). It crawled down one arm of the platform to indicate "yes" (d_1 presented) and down the other arm of the platform to indicate "no" (d_1 not presented). The *arrows* shown by the targets in G and I illustrate the change in target distance from trial to trial. In the 2-AFC ranging experiments, the bat was rewarded for crawling toward the closer target, which was presented to the its left or right following a pseudorandom schedule.

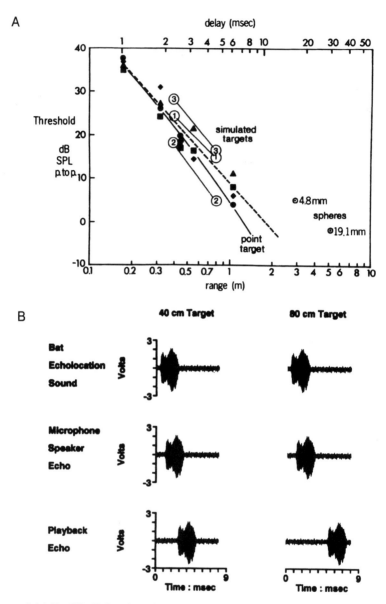

FIGURE 3.3A,B. (A). Echo detection thresholds for *Eptesicus fuscus* as a function of target distance or echo delay. Thresholds were all based on 75% correct detection performance. *Open circles* (4.8- and 19.1-mm spheres) show data taken from Kick (1982; real targets, see Fig. 3.2B); *solid symbols (triangles, squares, diamonds, circles)* represent data for individual animals taken from Kick and Simmons (1984; phantom target echoes; see Fig. 3.2F) *Open circles* marked *1, 2, 3,* are data points from individual animals taken from Simmons, Moffat, and Masters. (1992; phantom target echoes; see Fig. 3.2E). *Dashed line* shows the regression line fit to the solid data points; the *solid line* shows the amplitude of

Subsequent psychophysical studies of echo detection in bats have often electronically simulated echoes from targets. In phantom target experiments, the bat is trained to rest on a platform and emit sonar cries. Broadcast back to the bat through loudspeakers are either delayed playback replicas of its sonar emissions or triggered playbacks of a digitally stored sonar signal. In both instances, the bat is trained to respond to the presence or absence of a sonar signal playback that is delayed to represent an echo from a target positioned at ranges typically from 40 to 120 cm (see Fig. 2C–F).

No psychophysical data are available for bats that use CF FM sounds (e.g., *Rhinolophus ferrumequinum*, or the mustached bat, *Pteronotus parnelli*); however, *Noctilio leporinus* uses short quasi-CF signals (i.e., very shallow FM) in its repertoire (see Wenstrup 1984; Wenstrup and Suthers 1984). The most extensive work on detection in bats has employed the FM bat *Eptesicus fuscus*. Perhaps the most striking comparison shown in Table 3.1 is the difference in echo detection threshold estimates for *Eptesicus* reported by Kick (1982) using real targets and others using target simulation methods. In fact, the detection threshold reported by Kick ranges from 8 to 59 dB lower than estimates obtained with echo playback systems.

Why are Kick's detection threshold estimates so much lower than those obtained using virtual targets? And why are the detection threshold estimates obtained using virtual targets so variable? Most important is the influence of clutter echoes in virtual target experiments. Unavoidable in playback experiments are echoes from objects in the apparatus, such as microphones and speakers, that are used to generate virtual target echoes (in Fig. 3.2, compare A and B with C–F). These clutter echoes may serve as either forward or backward maskers, depending on their temporal relation to the virtual target echoes. For example, clutter echoes from the microphone-speaker pair shown in Figures 3.2C and 3.2E may serve as forward maskers of the phantom target echoes, while clutter echoes from the speakers shown in Figures 3.2D and 3.2F may serve as backward maskers (see also Fig. 3.3B).

2.4.2 Clutter Interference

Several experiments have shown that clutter echoes interfere with the detection of playback echoes, even if these sounds are temporally isolated

---

FIGURE 3.3 (*continued*) echoes from an ideal point target at different distances. (B) Schematic of the temporal relationship between sounds produced by bats in a detection experiment and the resulting clutter echoes and phantom target playback echoes for each sonar emission, based on the 40-cm and 80-cm target echoes studied by Simmons, Moffat, and Masters (1992; Fig. 3.2E). Note that the conditions for forward masking differ between the 40- and 80-cm target distances.

from other signals. For example, Troest and Møhl (1986) showed that the detection threshold of *Eptesicus serotinus* for a phantom target in a single-channel yes/no experiment dropped by 20 dB when the speaker diameter in the playback apparatus was reduced from 75 to 15 mm. Simmons et al. (1988) reported that the zone of clutter interference depends on the absolute range of the target. For a target distance of 54 cm, surrounding objects produce a threshold elevation if separated from the target by less than 15 cm. Hartley (1992a) also demonstrated the effect of clutter on target detection in *Eptesicus fuscus* and *Noctilio leporinus* by placing rings around the loudspeaker to increase the clutter echoes by measured amounts. These experiments showed backward masking of virtual target echoes by clutter. There are also data to suggest that forward masking by clutter can influence echo detection thresholds as well (e.g., Simmons et al. 1988; Moss and Simmons 1993). Thus, the large differences between estimates of echo detection thresholds in bats may be largely the result of the masking effects of clutter echoes. In addition, changes in the bat's sonar emission level across trials and days, directly influencing the level of sonar playback echoes, contribute error to estimates of echo detection thresholds, which in many studies are based on very few measures of the SPL of the bat's sonar transmissions.

2.4.3 Detection of Target Echoes in Noise

In addition to the forward and backward masking effects of clutter, the presence of simultaneous and interrupted noise can disturb target echo detection by sonar in bats. This finding has behavioral relevance to echolocation performance of bats foraging under natural conditions where environmental noise may influence target detection. Roverud and Grinnell (1985a) demonstrated this phenomenon in the CF-FM bat *Noctilio albiventris* under conditions of continuous and pulsed noise (schematic of behavioral testing shown in Fig. 3.2A). Animals were unable to detect the presence of a 19-mm sphere at a distance of 35 cm when white noise (20–80 kHz) was broadcast continuously at levels above 54 dB SPL. When the white noise was pulsed, target detection performance remained above chance levels for noise duty cycles below 80%. While the bats performing in this task increased the repetition rate of their sonar sounds in the presence of noise, there was no evidence of an increase in sound duration. No information was provided in this report on the estimated echo levels required for detection in quiet or the signal-to-noise ratio (S:N) that interferes with echo detection performance. These critical aspects of target detection and masking were carefully considered by Troest and Møhl (1986), who used a single-channel target simulator to measure simultaneous masking in *Eptesicus serotinus* (a schematic of the apparatus is shown in Fig. 3.2D). In this study, the bat's detection threshold in quiet was 40–48 dB peSPL (peak equivalent sound pressure level; see Stapells, Picton, and

Smith 1982) and elevated by 7–8 dB in the presence of continuous broadband noise. Under simultaneous masking conditions, the ratio of signal energy to noise spectrum level ranged from 36 to 49 dB. The loudspeaker used in these masking experiments was positioned 88 cm from the bat and had a 75-mm diameter, producing a strong clutter echo that limited the bat's detection performance (see earlier).

2.4.4 Aural Integration Time

Also using a single-channel target simulator, Møhl and Surlykke (1989) compared simultaneous and backward masking in *Eptesicus fuscus*, but in this experiment used a smaller diameter speaker (15 cm in diameter at a distance of 68 cm) to reduce the effect of clutter. In this experiment, they found that echo detection in quiet was about 37.5 dB SPL. When continuous broadband noise was presented, the detection thresholds were elevated by about 15 dB relative to those measured in quiet, and the signal energy to noise spectrum level ratio was about 35 dB. The effect of masking was relatively stable over masker durations from 2 to 37.5 msec, and backward masking was evident only for echo–noise delays as long as 2 msec. The duration of the bats' sounds was 1.7–2.4 msec, suggesting that masking only occurred for noise pulses that overlapped the playback echoes. The results of this experiment suggest that aural integration time in the bat is tied to the duration of the bat's sonar cries. This estimate of integration time is an order of magnitude lower than that reported in the human psychophysical literature (e.g., de Boer 1985) and suggests an auditory specialization for echolocation in bats.

2.4.5 Echo Gain Control

As a bat flies toward a stationary target, the echoes returning to its ears increase in amplitude, because the distance over which spherical spreading losses occur is decreased. For each twofold reduction in target range, there is a 12-dB increase in echo amplitude. This increase in echo amplitude could produce a systematic change in the discharge latency of auditory nerve fibers in response to echoes, because neural response latency decreases with increasing stimulus level (see Simmons, Moss, and Ferragamo 1990). In turn, a decrease in response latency could disturb the bat's accurate estimate of echo delay, its perceptual cue for target distance (e.g., Simmons 1973; Simmons et al. 1990; see also Section 3, Ranging). Thus, the increase in echo sound level that occurs when a bat approaches a target could potentially distort its perception of target range.

Kick and Simmons (1984) proposed that there is little or no change in the sensation level of echoes over a target distance of about 1.5 m, largely because of middle ear muscle modulation of hearing sensitivity. When a bat emits sonar pulses, the middle ear muscles contract, at least in part to protect the inner ear from damage caused by the intense sound levels of its vocalizations. These middle ear muscle contractions serve to reduce hearing sensitivity, producing a maximum rise in threshold around the time of vocal emission (Henson 1965; Suga and Jen 1975). As the middle ear muscles relax during the several milliseconds following each sonar emission, hearing sensitivity increases. In addition, there is evidence for central attenuation of neural responses to emitted sounds, which could affect sensitivity to echoes occurring within a restricted time window (Suga and Schlegel 1972). Thus, the bat should be more sensitive to echoes returning from more distant targets at longer delays than to closer targets at shorter delays. Kick and Simmons (1984) suggested that the net effect is an automatic gain control mechanism whereby the bat's sensitivity to echoes increases by about 11–12 dB for each doubling of target range, effectively offsetting the effect of spherical spreading losses on signal detectability over a distance of 1.5 m.

The gain control hypothesis developed from data on the hearing sensitivity of *Eptesicus fuscus* to phantom target echoes at simulated distances ranging from 17 to 110 cm. Kick and Simmons (1984) trained bats in a 2-AFC procedure to echolocate from the base of a Y-shaped platform and to indicate the presence of a target echo to the left or right of its observing position. The bat's echolocation sounds were picked up by two condenser microphones placed beyond the arms of the platform approximately 10 cm from the observing position (schematic shown in Fig. 3.2F). These sounds were played back to the bat on either the left or right side through an electrostatic speaker whose distance determined the total delay (or perceived distance) of the phantom target echo. The total delay of the phantom target echoes was set by the travel time of the bat's sound to the microphone and that from the speaker back to the bat. The sound level of the playback echo was determined by the bat's own emission level, and echo level could be modified by attenuating the signal output to the speaker. The peak-to-peak SPL producing 75% correct detection performance was estimated from recordings of sounds produced by bats during the task. A measure of 105–110 dB was taken and used to assign thresholds for each delay or simulated range studied. The results of this study show that the threshold for detecting a phantom target echo decreased by about 11 dB for each doubling of target distance, a change that would approximately cancel the spherical spreading losses in echo level of approximately 12 dB for each doubling of distance (data plotted as solid symbols in Fig. 3.3A).

As described previously, the detection of a target echo is strongly influenced by the presence of clutter echoes from objects present in the behavioral testing setup. Using phantom targets necessitates the presence of a microphone and a loudspeaker whose echoes can serve as both forward and backward maskers, depending on their spatial relationship with respect to the playback echo. In addition, the sonar transmission itself can function as a forward masker if the echo delay is short with respect to the duration of the bat's echolocation sound. In the Kick and Simmons study, a phantom target echo was presented at a shorter distance, or delay, than the clutter echo returning from the loudspeaker (see Fig. 3.2F). As the phantom target was presented at shorter distances, the speaker was moved closer to the bat, thus reducing the time interval between the playback echo and the clutter echo. The time between the bat's sonar transmission and echo also shortened, increasing the effectiveness of the bat's own echolocation sound as a forward masker. Moreover, as the speaker–bat distance decreased, the amplitude of the speaker echo increased. Both the reduction in temporal separation between the target and clutter echoes and the increase in clutter echo level could serve to increase the magnitude of backward masking effects on target detection at shorter distances. Without measures of the contribution of clutter to the echo detection thresholds in this study, the change in threshold with target distance is difficult to evaluate.

In an attempt to minimize backward masking effects in estimating the slope relating hearing sensitivity and target distance, Simmons, Moffat, and Masters (1992) used a modified experimental setup. The speakers in this second 2-AFC experiment remained fixed at a distance of 22 cm from the bat, and a variable electronic delay of sound playbacks simulated target echoes at two different distances, 40 and 80 cm from the position of the bat (see Fig. 3.2E). The data from this study also showed that the bat's detection threshold decreased by about 11 dB for a doubling of target range (see open symbols numbered 1, 2, and 3 in Fig. 3.3A). However, this experiment was not free of masking effects either. In this study, the playback echo delays were 2.32 and 4.64 msec for the 40- and 80-cm target distances, respectively. The microphone and speaker pair, positioned at a distance of 22 cm, produced clutter echoes at a delay of 1.27 msec. Echolocation sound durations recorded in this study were 2–4 msec, and thus there was considerable overlap between echoes from the speaker-microphone pairs and the phantom target echoes, particularly for the 40-cm target (see Fig. 3.3B). In addition, each of the bat's sonar transmissions may have interfered with detection of target echoes, especially at the lesser of the two distances. Therefore, in this experiment, forward masking may have influenced detection thresholds, and presumably such masking was stronger for the 40-cm target than that for the 80-cm target. Here again, the slope relating detection threshold to target distance presumably includes some contribution of masking effects.

Hartley (1992a), having noted the potential contribution of backward masking to the data reported by Kick and Simmons (1984), used a different experimental paradigm to study gain control in both *Eptesicus fuscus* and *Noctilio leporinus*. With a one-channel phantom target playback apparatus, Hartley trained bats to report which of two successive echoes sounded louder. He presented pairs of sounds to the bats, simulating echoes from targets at distances ranging from 0.28 to 11.48 m; the second echo playback in the pair arrived at approximately double the delay of the first. The speaker-microphone pair that picked up and played back the bat's sounds was positioned approximately 10 cm from the animal's head and produced a clutter echo of approximately 60 dB SPL. The echo playback level of the pairs of sounds was modified on line by attenuators adjusted to present suprathreshold echoes in the range of 70–80 dB SPL.

Hartley's data indicate that bats perceive two echoes to be equally loud when their delay (target distance) differs by a factor of two and their SPL differs by about 6.7 dB for *Eptesicus fuscus* and 7.2 dB for *Noctilio leporinus*. From these results, Hartley inferred that the slope of the echo gain control function is about 6–7 dB per halving of target distance, rather than the 11–12 dB reported by Simmons and colleagues. Furthermore, Hartley measured the sonar emission levels of *Eptesicus* tracking an approaching target (maximum range 1 m) and found that the bat reduced pulse intensity by 6 dB per halving of target range (Hartley 1992b). This distance-dependent intensity compensation of sonar emissions, together with the automatic gain control, would serve to stabilize the sensation level of echoes over a target range of about 1–2 m. Thus, two very different approaches to the question of echo gain control in bats generated conflicting results, but congruent conclusions.

How do we reconcile the results and interpretation of these studies? First, it may be that the slope of the gain control function is level dependent: it may be steeper at echo levels close to threshold for detection (Simmons, Moffat, and Masters 1992) than at suprathreshold levels used for loudness discrimination (Hartley 1992a). A complete test of this idea would require estimating the slope of the gain control function at several different echo levels relative to threshold. There is also evidence to suggest that the level of a bat's sonar emissions changes with target size and distance (Reetz and Schnitzler 1992; Hartley 1992a,b), a factor that was not considered in using a fixed estimate of sonar emission level to determine detection thresholds for targets at different distances (Kick and Simmons 1984; Simmons, Moffat, and Masters 1992). Using measures of bat sonar transmission levels for each of the target distances tested might change estimates of echo detection thresholds and possibly alter the slope of the gain control function.

As noted, it is also possible that measures of the gain control slope based on echo detection thresholds are influenced by forward masking by the sonar transmission and by clutter echoes from microphones and speakers, but the magnitude of this effect was not estimated in the study by Simmons, Moffat, and Masters (1992). Measures of the signal to clutter ratios in detection experiments (see for example, Møhl 1986; Troest and Møhl 1986; Masters and Jacobs 1989; Hartley 1992a) at different distances could help to clarify the contribution of backward and forward masking to estimates of echo gain control.

2.5 Summary

In conclusion, estimates of echo detection thresholds in bats yield data differing by more than 50 dB in the same species. Echo detection in bats reflects the combined properties of the signal, environment, target, and receiver, all of which varied across the studies reviewed here. Moreover, the level of the sonar transmissions used by the bats may have differed across individual animals, conditions, and experimental trials within a single study (Hartley 1992a,b; Reetz and Schnitzler 1992; Denzinger and Schnitzler 1994), contributing error to estimates of echo detection calculated from a few measures of sonar transmission level. Indeed, the critical parameters determining the limits of echo detection were not measured in many of the studies cited. For future work in this area, we would like to emphasize the importance of careful control and measurement of these parameters.

3. Ranging

Insectivorous bats require an accuracy of about 1–3 cm to successfully intercept prey (Webster and Griffin 1962; Trappe 1982); however, some behavioral studies of the bat's target ranging performance in the laboratory demonstrate distance discrimination accuracy several orders of magnitude higher. Psychophysical estimates of ranging accuracy depend on the bat's perceptual task, the presentation of the stimuli, and the behavioral methods employed. Most important is the perceptual task, and in this section the discussion is organized around three types of tasks employed in the study of target ranging in bats: range difference discrimination, range jitter discrimination, and range resolution/two-wavefront discrimination.

3.1 Range Difference Discrimination

Echolocating bats use the time delay between sonar emission and returning echo to determine the distance to a target (Hartridge 1945). This was first demonstrated experimentally by Simmons (1973), who trained bats in a 2-AFC procedure to discriminate the distance between two targets positioned at different ranges. He first estimated the bat's distance discrimination threshold using two spheres, positioned to the left and right of the animal's observing position. The bat was trained to scan the two targets from the base of a Y-shaped platform and crawl down one arm of the platform toward the closer of the targets (schematic of the apparatus is shown in Fig. 3.2H). The closer target's left or right position was randomized from trial to trial, and the difference in distance between the two was gradually reduced until the bat's performance fell to chance (see Fig. 3.4A). Simmons then replaced the spheres with playback replicas of the bat's sonar sounds whose delays were adjusted to simulate target echoes at different ranges (schematic shown in Fig. 3.2L). (The two-way travel time of the bat's sonar emission and the returning echo results in an echo delay of about 58 μsec for each centimeter of target distance.) Simmons found that the performance curves and range difference thresholds using real targets and virtual targets were similar, suggesting that the bat experienced the playback

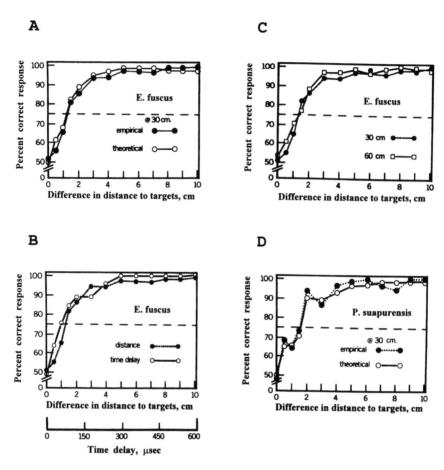

FIGURE 3.4A–D. Psychometric functions on range difference discrimination for *Eptesicus fuscus* (A–C) and *Pteronotus suapurensis* (D) using a two-alternative forced choice (2-AFC) procedure. (A) Empirical and theoretical curves for *Eptesicus* at 30 cm. (B) Discrimination performance using real and phantom targets is compared for *Eptesicus* at 30 cm. (C) Behavioral performance curves for *Eptesicus* using real targets at 30 and 60 cm. (D) Empirical and theoretical curves for *Pteronotus* using real targets at a distance of 30 cm. Theoretical curves are based on the envelope of the autocorrelation function of the species' sonar transmissions, corrected for head movements in the two-choice task, (From Simmons 1973.)

echoes at different delays as targets at different ranges (Fig. 3.4B). The behavioral comparison in performance using real and phantom targets was conducted only for the species *Eptesicus fuscus*.

Simmons (1973) reported that *Eptesicus* can discriminate the range difference of two targets separated by about 12 mm (70-μsec echo–delay difference), and that this threshold is relatively stable over test distances between 30 and 240 cm (e.g., see the performance data at 30 and 60 cm shown in Fig. 3.4C). He also measured range difference thresholds using real targets in the horseshoe bat *Rhinolophus ferrumequinum*, the naked-backed bat *Pteronotus suapurensis*, and the spear-nosed bat *Phyllostomus hastus*. Both *Pteronotus suapurensis* (see Fig. 3.4D) and *Phyllostomus hastus* were tested with targets at different distances, and, like *Eptesicus*, showed that range difference threshold is relatively independent of absolute distance. The stability of range difference thresholds over a distance of 30 to 240 cm has important implications for evaluating receiver models, and this is discussed in Section 4.

During the past 20 years, range difference (or echo delay difference) thresholds have been measured in a variety of bat species. These data are summarized in Table 3.2. While estimates of echo detection thresholds in bats differ widely (see Table 3.1), behavioral measures of range difference thresholds are remarkably similar, even across different experimental methods and species. (For schematic diagrams of different methods used to study range difference discrimination, see Fig. 3.2G–L.) This finding suggests that range difference discrimination may be less sensitive to the effects of clutter and therefore more stable under varying experimental conditions. Most species studied show range difference thresholds between about 8 and 15 mm, corresponding to echo-delay differences of 46 and 87 μsec. The two striking exceptions are *Rhinolophus ferrumequinum* (Simmons 1973; Airapetianz and Konstatinov 1974), tested in a 2-AFC simultaneous discrimination task (see Fig. 3.2H), and the lesser bulldog bat, *Noctilio albiventris* (Roverud and Grinnell 1985a), tested in a single-target yes/no sequential discrimination (see Fig. 3.2G), both of which show range difference discrimination thresholds of about 30–41mm (174–238 μsec of echo–delay difference).

The higher discrimination thresholds obtained for *Rhinolophus ferrume-quinum* may directly reflect limitations of the information carried in its sonar signals about target range. It is widely accepted that the CF-FM bat uses the FM component of its echolocation sound for target distance estimation, and the bandwidth of the FM signal influences the bat's ranging performance (e.g., Simmons 1973; Schnitzler and Henson 1980). Surlykke (1992) reported data consistent with this view. She demonstrated that the range difference threshold in *Eptesicus fuscus* was elevated by a factor of 2–3 when it listened to 1-msec triggered-playback echoes of its FM sounds that were either 40 kHz lowpass or 40 kHz highpass filtered. Summarized in Table 3.2 is the approximate bandwidth of the FM components used by different species of bats performing in ranging tasks. The bandwidth of the FM component of *Rhinolophus*'s sound is narrower than that of other species showing lower range difference discrimination thresholds. The role of bandwidth on echo ranging performance is discussed further in Section 4, Receiver Models.

The elevated threshold of *Noctilio albiventris* in the single-target sequen-

TABLE 3.2 Summary of behavioral data on range difference discrimination in bats.

Species	Procedure	Target Distance (cm delay/msec)	Target Angular Separation	Signal Bandwidth (strongest harmonic, kHz)	Threshold (mm/delay/μsec)	Author(s) (year)
Eptesicus fuscus	2-AFC simultaneous method of limits	30 cm/1.74 msec; 60 cm/3.48 msec; 240cm/13.9 msec; real targets	40°	25-35	12 mm/70 μsec 13 mm/75 μsec 14 mm/81 μsec	Simmons (1973)
Eptesicus fuscus	2-AFC simultaneous method of limits	30 cm/1.74 msec; sonar sound playback	40°	25-35	10 mm/58 μsec	Simmons (1973)
Eptesicus fuscus	2-AFC simultaneous method of limits	80 cm/4.6 msec; canned signal playback	40°	25-35	7 mm/41 μsec– 17 mm/99 μsec	Masters and Jacobs (1989)
Eptesicus fuscus	Yes/no sequential staircase	54-74 cm, 3.1-4.3 msec; canned signal playback	Single channel	25-35	6 mm/35 μsec– 11 mm/64 μsec	Miller (1991)
Eptesicus fuscus	Yes/no sequential staircase	40-46 cm, 2.32-2.67 msec; canned signal playback	Single channel	25-35	10 mm/58 μsec– 20 mm/116 μsec	Surlykke (1992)
Eptesicus fuscus	2-AFC simultaneous method of limits	30 cm/1.74 msec; sonar sound playback	40°	25- 35	13.8 mm/80 μsec	Denzinger and Schnitzler (1994)
Phyllostomus hastus	2-AFC simultaneous method of limits	60 cm/3.48 msec; 120 cm/6.9 msec; real targets	40°	35	12 mm/70 μsec 12 mm/70 μsec	Simmons (1973)
Pteronotus suapurensis	2-AFC simultaneous method of limits	30 cm/1.74 msec; 60 cm/3.48 msec; real targets	40°	22.5	15 mm/87 μsec 17 mm/99 μsec	Simmons (1973)
Rhinolophus ferrumequinum	2-AFC simultaneous methods of limits	30 cm/1.74 msec; real targets	40°	15	30 mm/174 μsec	Simmons (1973)
Rhinolophus ferrumequinum	2-AFC simultaneous method of limits	100 cm/5.8 msec; real targets	13°	15	41 mm/238 μsec	Airapetianz and Konstantinov (1974)

Myotis oxygnathus	2-AFC simultaneous method of limits	100 cm/5.8 msec; real targets	13°	85	8 mm/46 μsec– 12 mm/70 μsec	Airapetianz and Konstantinov (1974)
Pipistrellus pipistrellus	2-AFC simultaneous method of limits	24 cm/1.39 msec; real targets	75°	65	15 mm/87 μsec	Surlykke and Miller (1985)
Noctilio albiventris	2-AFC simultaneous method of limits	35 cm/2.03 msec; real targets	40°	20	13 mm/75 μsec	Roverud and Grinnell (1985a)
Noctilio albiventris	Yes/no sequential method of limits	35 cm/2.03 msec; real targets	Single target	20	30mm/174 μsec	Roverud and Grinnell (1985a)

tial discrimination task compared to that obtained in the 2-AFC simultaneous discrimination are the only data showing a large difference in the bat's ranging performance with changes in the behavioral paradigm. In the yes/no sequential task, the bat was trained to recognize a sphere at 35 cm and respond differentially between this target range and a more distant one (see Fig. 3.2G). Thus, the bat learned to compare a target's distance with a stored representation of 35-cm range, and it is perhaps not surprising that its performance under these conditions was poorer than in the simultaneous discrimination task. However, other researchers have used the single-target sequential discrimination method and reported range difference thresholds of about 6–15 mm in *Eptesicus fuscus* (e.g., Miller 1991; Surlykke 1992), which compare well with estimates obtained using a modified method of limits and a 2-AFC procedure in the same species (e.g., Simmons 1973; Masters and Jacobs 1989; Denzinger and Schnitzer 1994). The congruence of the threshold estimates using different methods with *Eptesicus fuscus* prompt us to further evaluate the conflicting range difference data in *Noctilio albiventris*.

The experimental conditions that apparently differentiate the single-target sequential discrimination study by Roverud and Grinnell (1985a) and others cited (Miller 1991; Surlykke 1992) include both the use of real targets and the psychophysical procedure. It may be that the real targets impose difficulties for the bat discriminating range differences from successive stimulus presentations, because any variation in the animal's start position on the platform will influence its estimate of target distance. In a single-channel phantom target ranging task, however, the bat's start position on the platform would not influence its estimate of target distance if it learns to use the echo from the speaker or microphone as a fixed reference from which to measure target echo delay (Miller 1991; Surlykke 1992), and this might explain the lower threshold estimates using virtual targets in single-channel range discrimination tasks.

The psychophysical procedure employed by Roverud and Grinnell (1985a) might have also contributed to a higher threshold estimate in the yes/no task. Roverud and Grinnell used a modified method of limits, in which all of the trials on a given test day contained the sphere at a range of 35 cm and one alternative at a greater distance; on succeeding test days, the range difference between the two stimuli was decreased. By contrast, others employing a single-target sequential presentation of targets used a staircase method, in which the range difference between the stimuli was decreased by a fixed step size for each correct response and increased for an incorrect response. The staircase method allows the bat many more trials close to threshold, and changes in the range of the alternative (more distant stimulus) perhaps provide the bat with information that facilitates its discrimination performance. This factor may be of importance in the design of future behavioral studies on echolocation in bats. The modified method of limits has been widely used in 2-AFC simultaneous experiments, and it appears to yield data that are reliable across species and tasks. By contrast, the bat may show its best performance in single-target sequential discrimination tasks when presented with many comparison stimuli around some reference, as in a staircase procedure.

3.1.1 Range Difference Discrimination in the Presence of Interfering Signals

When echolocating bats forage, sounds from the environment can disrupt ranging accuracy, interfering with estimates of target echo arrival time. This environmental noise might take the form of sonar cries from neighboring bats, echoes from nearby vegetation, or ultrasound clicks from insects. The influence of interference signals on range estimation in bats has been investigated in several species. Roverud and Grinnell (1985a) reported for *Noctilio albiventris* that range difference discrimination of two spheres separated by 5 cm deteriorated in the presence of white noise, presumably because of decreased S:N. With continuous white noise at 93.5 dB SPL, the bat's discrimination performance fell to chance. When white noise at this high level was presented in pulses, the bat's distance discrimination performance depended on the range separation of the targets and the duty cycle of the noise pulses. To reliably discriminate range differences of 5–10 cm, the bat required a minimum silent interval of 300 msec. Shorter intervals disrupted range discrimination, which may be explained by backward or forward masking of the echoes by the noise pulses.

In another set of experiments, *Noctilio*'s discrimination of two spheres separated by 5 cm deteriorated in the presence of high-amplitude interference signals that contained both CF and FM components similar to those found in species-specific sonar sounds (Roverud and Grinnell 1985b). As with white noise, the magnitude of this effect depended on the intensity and duty cycle of the artificial sonar pulses. Isolated CF and FM interference signals presented randomly with respect to the target echo did not disrupt range discrimination of 5 cm. However, FM signals that were time locked to the target echo interfered with distance discrimination if they occurred within 8–27 msec after the onset of the FM component of the bat's own pulse and within a correspondingly shorter interval of the FM component of the target echo. Roverud and Grinnell (1985b) interpreted these results to demonstrate that the CF component of the echo activates a gating mechanism that sets up a time window for processing distance information contained in the FM sweep. The width of the postulated time window depends, however, on the SPL of the interfering signal, suggesting that this hypothesis may require modification.

Roverud (1989a) conducted a set of distance discrimination experiments with the roufous horseshoe bat, *Rhinolophus rouxi*, that were similar to those of Roverud and Grinnell (1985b). *Rhinolophus rouxi*'s discrimination of two spheres positioned at 35 and 43 cm was disrupted by free-running CF-FM pulses with no regular temporal relationship to the target echoes when the interference signals were presented at frequencies near the first or second harmonic of this species' echolocation sounds (Roverud 1989a). When a free-running, 2-msec CF sound preceded an FM sound by about 10–65 msec, range discrimination performance also deteriorated. FM pulses alone only disrupted the bat's ranging behavior when triggered by the bat's own sonar emissions and played back after a delay of approximately 45–65 msec. The triggered playback FM sounds arrived close in time to the FM echoes from the spheres. Roverud (1989b) proposed a gating mechanism for the processing of distance information in *Rhinolophus* similar to that in *Noctilio*, and suggested that the temporal requirements of this gating mechanism depend on the duration of the CF component of the individual species' echolocation signals. For *Rhinolophus rouxi*, he did not test whether the width of the time window depends on the SPL of the interfering signal, leaving open the interpretation of this result.

Surlykke and Miller (1985) measured range difference thresholds in *Pipistrellus pipistrellus* in the presence of broadband clicks that occurred randomly with respect to echoes from real targets. The purpose of this study was to test the hypothesis that clicks, such as those produced by arctiid moths in response to bat echolocation sounds (see Fullard and Fenton 1977; Miller 1983), might interfere with the bat's ability to estimate distance. The clicks were playbacks of sounds produced by noxious arctiid moths (the ruby tiger, *Phragmatobia fuliginosa*) recorded in response to computer-generated bat sonar cries. These clicks were broadcast from electrostatic speakers built into two balls used for the range difference discrimination experiment. Using a 2-AFC procedure, Surlykke and Miller (1985) reported that *Pipistrellus pipistrellus* discriminated a range difference of 1.5 cm in a baseline experiment and of 1.0 cm in the presence of clicks. Thus, in this experiment, the bat's range difference discrimination threshold was not disturbed by random presentation of ultrasound clicks.

When the presentation of interfering clicks is synchronized to target echoes, a different pattern of results is observed. Miller (1991) studied range difference discrimination of phantom targets in *Eptesicus fuscus* in experiments in which arctiid moth click trains were time locked to the playback echoes. In this study, he found that the bat's range difference threshold increased by a factor of almost 40 when the interfering click train began about 760 μsec before the phantom target echo.

A general conclusion that can be drawn from the experiments on range difference discrimination in the presence of interfering signals is that the bat's performance is disturbed largely when such signals coincide close in time with the arrival of the target echo. Future experiments should be directed at better understanding the mechanisms involved in this process.

3.2 Range Jitter Discrimination

Range difference thresholds in bats are no greater than 40 mm for any species studied, which is close to the accuracy estimated for bats catching insects (Webster and Griffin 1962). Using other measures of ranging performance, however, the psychophysical and insect capture data diverge. In particular, behavioral data using a range jitter discrimination task developed by Simmons (1979) suggests that the bat can detect changes in target distance many orders of magnitude smaller than appear biologically relevant or neurophysiologically possible (but see Simmons et al. in press).

In Simmons's (1979) first range jitter discrimination task, the bat was trained to rest on the base of a Y-shaped platform and emit echolocation sounds into two microphones, positioned at the end of each arm of the platform. The bat's sounds were picked up by the microphones, delayed, and played back to the bat through two loudspeakers positioned 40° apart to the left and right of the animal's observing position (Fig. 3.5A). A voltage comparator determined the microphone receiving the stronger signal (the side toward which the bat's head was aimed) and activated the loudspeaker on that side. Thus, as the bat scanned its frontal field, it received delayed replicas of its sonar transmissions, delivered sequentially from the loudspeakers. Through one loudspeaker, the bat's sounds were returned at a fixed delay, simulating echoes from a stationary target at a distance of about 50 cm. Through the other loudspeaker, the bat's sounds were returned at a delay that alternated between two time values, simulating echoes from a jittering target, also at a distance of about 50 cm (see Fig. 3.5B). The delay of the playback echo was selected to minimize temporal overlap between the bat's outgoing sonar transmissions and clutter echoes from speakers and microphones with the phantom target echoes.

The side on which the jittering target appeared was randomized from trial to trial, and the bat's task was to crawl down the arm of the platform toward the jittering target. For each of the bat's sonar transmissions, it received only one playback echo, and the level, duration, bandwidth, and sweep rate were all controlled by the bat. Simmons found that the bats performing in this task could discriminate jitter in echo delay of 1–2 μsec (based on a 75% correct criterion taken from Fig. 1; Simmons 1979). This astonishing result stimulated great controversy, prompting further investigation (e.g., Menne et al. 1989; Moss and Schnitzler 1989) using digital playback systems that would not produce spectral artifacts associated with delay that are inherent to the analog system used in Simmons's 1979 study.

Range jitter discrimination performance has since been studied in *Eptesicus fuscus* by several investigators, using both single-channel yes/no (Moss and Schnitzler 1989) and two-channel 2-AFC (Menne et al. 1989; Simmons et al. 1990; Moss and Simmons 1993) psychophysical procedures. The echo jitter values tested in these studies were typically between 0.4 and

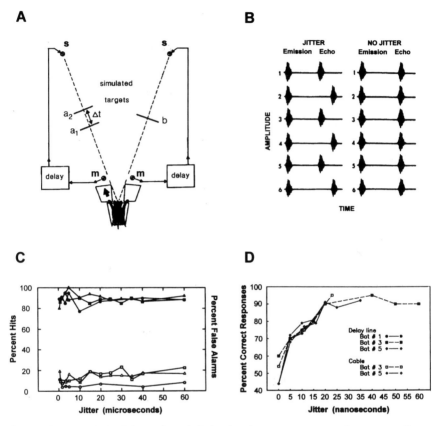

FIGURE 3.5A–D. (A) Schematic of behavioral test apparatus for range jitter discrimination (from Simmons 1989). (B) Schematic of echo time waveforms under conditions with and without jitter in playback delay. For each sonar emission, the bat receives a single playback echo (adapted from Moss and Schnitzler 1989). (C) Jitter discrimination performance of three individual bats for jitter values between 0.4 and 60 μsec (adapted from Moss and Schnitzler 1989). In this experiment, the bat performed in a yes/no discrimination task; *solid symbols* plot percentage hits and *open symbols* plot percentage false alarms. (D) Jitter discrimination performance for three individual bats tested in a 2-AFC task with temporal jitter between 0 and 60 nsec. Bat 3 (*squares*) and bat 5 (*diamonds*) were tested with both analog delay lines (*solid symbols*) and coaxial cables differing in length (*open symbols*). Bat 1 was tested using only the analog delay line (*solid circle*) (from Simmons, et al. 1990).

100 μsec (range jitter, 0.07–17 mm). In addition, Simmons et al. (1990) tested sensitivity to jitter in echo delay in the nanosecond range.

Using a playback apparatus that could digitally generate changes in echo playback delay as small as 0.4 μsec, Moss and Schnitzler (1989; see Fig. 3.5C) and Menne et al. (1989) found that *Eptesicus fuscus* showed stable discrimination performance (>85% correct) down to the smallest jitter values that could be tested, which were smaller than those reported by Simmons (1979). Moss and Schnitzler examined the role of echo bandwidth in range jitter discrimination and found that *Eptesicus's* performance was not disturbed when the playback echoes were low-pass filtered at cutoff frequencies as low as 40 kHz (filtering out the upper echo frequencies of the first harmonic and all energy in the higher harmonics of the bat's sonar signals). The effect of low-pass filtering in this study was, however, inconclusive, because performance could not be assessed at jitter values around threshold for the unfiltered echo condition. High-pass filtering of playback echoes at 40 kHz disrupted the bat's behavior, and the animals in the study refused to perform the jitter discrimination task under these echo conditions, suggesting that the lower frequencies (22–40 kHz) of *Eptesicus's* sonar sound are important for carrying information about target range. This frequency band also contains the most energy in this species' echolocation sound.

Menne et al. (1989) were the first to demonstrate that *Eptesicus fuscus* is sensitive to the phase of echoes relative to emissions, an idea first proposed by Simmons (1979). Bats performed in a baseline jitter discrimination experiment and in an experiment in which the phase of the jittering echo alternated between ±45° (90° phase change between echoes of the pair). In the baseline experiment, sensitivity to echo jitter was studied for values between 0.4 and 120 μsec, echo–delay alternations that the bats consistently discriminated at levels well above chance. In the phase jitter experiment, these time delay values were combined with 90° phase alternations between echoes of the jittering target, and again the bat successfully discriminated all jitter/phase-shift pairs. They also found that the bat was sensitive to a phase alternation alone (jitter = 0 μsec), a result which has since been replicated (Simmons et al. 1990; Moss and Simmons 1993). The study by Menne et al. also bears importance to the assessment of receiver models (see section 4 below).

While Moss and Schnitzler (1989) and Menne et al. (1989) presented data consistent with Simmons's original (1979) report that *Eptesicus* is sensitive to changes in echo delay of about 1 μsec, some have proposed that the bat does not use the timing of echo arrival to perform the jitter discrimination task. For example, Pollak (1993) has proposed that the bat may use a spectral cue for echo jitter discrimination. He argues that clutter echoes from the loudspeakers and microphones of the target simulator might overlap the playback replicas of the bat's sonar emissions, resulting in spectral interference patterns whose features would vary with changes in echo delay (or jitter). Thus, the jittering target would have a spectral color that alternates between echo playbacks and the nonjittering target would have a spectral color that remains constant. Overlap between playback echoes and clutter echoes occurs when the bat's sounds exceed a particular duration, which would depend on the spatial arrangement of the loudspeakers and microphones relative to the bat. The spectral information arising from interference between playback and clutter echoes that might potentially characterize jitter in echo delay of only a few microseconds is, however, well outside the range of hearing in the bat, and theoretical arguments suggest the use of spectral cues in jitter discrimination to be highly improbable (Menne et al. 1989; Moss and Schnitzler 1989). In addition, Simmons et al. (1990) presented experimental data to support the view that bats use echo delay alone to discriminate target jitter and that spectral artifacts do not contribute to the bat's performance, at least in the microsecond range.

Simmons et al. (1990) exploited the phenomenon of amplitude-latency trading to determine whether echo delay is the bat's perceptual cue in performing the jitter discrimination task. The logic behind this study is as follows: If the bat's cue for jitter discrimination is the relative timing of the playback replicas of its sonar emissions, its performance should be susceptible to amplitude-induced latency shifts in the auditory system. For stimulus levels approximately 15 dB above threshold, brainstem auditory evoked responses in *Eptesicus fuscus* show a latency change of -13 to -18 μsec for each decibel increase in the amplitude of a simulated biosonar sound (Burkard and Moss 1994; Simmons, Moss, and Ferragamo 1990) (Fig. 3.6A). Thus, one would predict that an amplitude change in the playback echoes in the jitter experiment would influence the bat's estimate of echo arrival time and disturb its perception of target distance. This is in fact what Simmons et al. (1990) reported. In a jitter discrimination study, they conducted two control experiments in which the amplitude of the echoes of the jittering target pair differed by 1 dB. In one experiment, the amplitude of the echo with the longer delay was increased by 1 dB relative to the echo with the shorter delay. In another experiment, the amplitude of the echo with the shorter delay was decreased by 1 dB relative to the echo with the longer delay (see Fig. 3.6B). For each playback echo, the absolute amplitude was controlled by the bat: The stronger its sonar emissions, the stronger the playback echoes.

Variation in sonar emissions and playback echoes presumably occurs and could disturb the bat's performance in the jitter discrimination task. The bat, however, appears able to overcome this potential difficulty, perhaps by

---

$\longrightarrow$

FIGURE 3.6A–C. (A) Evoked potentials recorded from the inferior colliculus of the bat (*inset*). Stimuli were FM sweeps, ranging in level between 34 and 79 dB SPL (peak to peak). Latency of N_1 response as a function of stimulus level relative to threshold. At about 15 dB SL, the slope of the regression line fit to the data is about -13 μsec/dB (from Simmons, Moss, and Ferragamo 1990). (B) Schematic of jitter apparatus when one of the echoes from the jitter–echo pair is adjusted in amplitude by 1 dB (a_1, -1 dB; a_2, $+1$ dB) to produce an amplitude-induced latency shift. (C) Percentage errors in jitter discrimination of a single bat under conditions when the amplitude of $a_1 = a_2$ (0 dB), when a_1 is reduced by 1 dB, and when a_2 is increased by 1 dB (from Simmons et al. 1990).

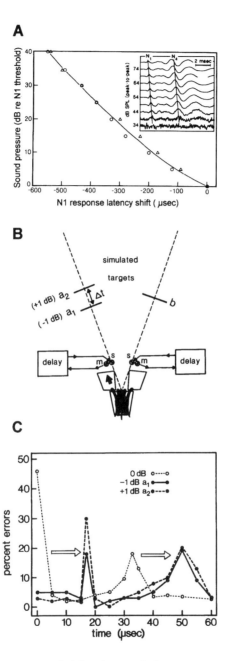

FIGURE 3.6 Caption on facing page.

actively stabilizing its sonar emissions over short durations as it probes the sonar targets. In this control study, Simmons et al. measured the bat's jitter performance using a digital playback system for delay changes of 5–60 μsec and found that the bat's performance dropped when the echoes of the jittering target differed by 1 dB and the echoes jittered in delay by about 17 μsec (Fig. 3.6C) These results suggest that amplitude-induced latency shifts can cancel the bat's perception of target jitter under experimentally controlled conditions. Spectral cues would not be affected by the amplitude manipulation in this experiment, indicating that the bat is using echo delay alone to discriminate target jitter in the microsecond range.

Simmons et al. (1990) also measured behavioral performance of jitter discrimination using time changes in playback echoes much smaller than those used by Moss and Schnitzler (1989) and Menne et al. (1989). In this study, they presented to the bat jitter values in the nanosecond range using two different methods, analog delay lines and coaxial cables differing in length to produce desired amounts of electronic delay. Transfer functions revealed no systematic delay-dependent spectral variations in the playback echoes, and they suggested that the only cue available to the bat for discriminating between jittering and nonjittering targets was echo arrival time. Using either the analog delay line or the cables of varying lengths, they found that the threshold for jitter discrimination in *Eptesicus* was about 10 nsec (range jitter < 0.0017 mm; see Fig. 3.5D). This report has been met with great skepticism (see, e.g., Pollak 1993), and indeed it is difficult to imagine how the nervous system might encode such small differences in echo arrival time, especially considering movements of the bat associated with its respiration and vocalization that must occur between echoes. The control experiment described earlier showing that amplitude-induced latency shifts can affect the bat's perception of target distance was only conducted with jitter in echo delay in the microsecond range. The technical difficulty of studying this phenomenon in the nanosecond range precluded the use of this control at around threshold, and confirmation of the 10-nsec jitter discrimination threshold estimates requires further study using a digital system. Moreover, the amplitude-latency trading data have implications for evaluating the threshold estimate of 10 nsec, and this is discussed further in Section 4, Receiver Models in Bat Sonar.

3.3 Range Resolution and the Discrimination of Two-Wavefront Targets

In the range difference and range jitter experiments just described here, the bat's task was to discriminate differences in the arrival time of echoes from simple targets with single reflecting surfaces. Thus, for each sonar emission produced by the bat, it received back a single echo, the delay of which conveyed information that guided its discrimination behavior. Such experiments are thus designed to measure the bat's accuracy of range estimation, but do not speak to the range resolution of its sonar receiver. Many natural targets of interest to the bat contain multiple reflecting surfaces whose separation may convey information about depth structure or range profile. Perception of the spatial separation of multiple reflecting surfaces might draw upon the range resolution of the bat's sonar receiver.

Most research on range resolution by sonar in bats has employed target stimuli containing (or simulating) two closely spaced reflecting surfaces (or wavefronts). In these experiments, the time delay corresponding to the range separation of the targets is small with respect to the duration of the bat's echolocation sounds. Thus, echoes from the two surfaces are largely overlapping. This results in a spectral interference pattern, with frequency regions of cancellation and reinforcement that correspond directly to the time separation between the sounds. For example, playback replicas of two FM sonar sounds separated by 100 μsec (1.74 cm) result in spectral notches at 10-kHz intervals, starting at 25 kHz in the echolocation signal. As the delay separation between overlapping echoes increases, the interval between spectral notches decreases (Fig. 3.7).

Three general questions have guided research on range resolution by echolocation: What is the minimum range separation between two closely spaced reflecting surfaces that the bat can resolve? What acoustic information in the echo does the bat use to perform in a range resolution task? How are targets with closely spaced reflecting surfaces represented in the bat sonar receiver? Each of these questions is considered next. Here, the discussion is organized around studies employing two different perceptual tasks, one in which the bat discriminates between two-wavefront targets that differ in the range/delay of their component echoes and another in which the bat discriminates between a one-wavefront target and a two-wavefront target that varies in the range/delay separation of its component echoes.

3.3.1 Discrimination Between Two-Wavefront Targets Differing in the Range/Delay Separation of Component Echoes

Two range resolution studies have been conducted on FM bats using real targets (Simmons et al. 1974; Habersetzer and Vogler 1983), and both show that the bat can discriminate range differences of less than 1 mm. The stimuli in these experiments consisted of Plexiglas plates with holes drilled to different depths. Simmons et al. (1974) trained *Eptesicus fuscus* in a 2-AFC procedure to discriminate between Plexiglas plates with holes drilled at a depth of 8 mm and a smaller depth, ranging between 6.5 and 7.6 mm, and reported that bats can discriminate differences as small as 0.6–0.9 mm in the depth of holes. The intensity of the echoes reflecting from the plates differed by less than 0.5 dB across all the pairs of stimuli presented to the bat, while the spectra of the echoes differed systematically with plate hole

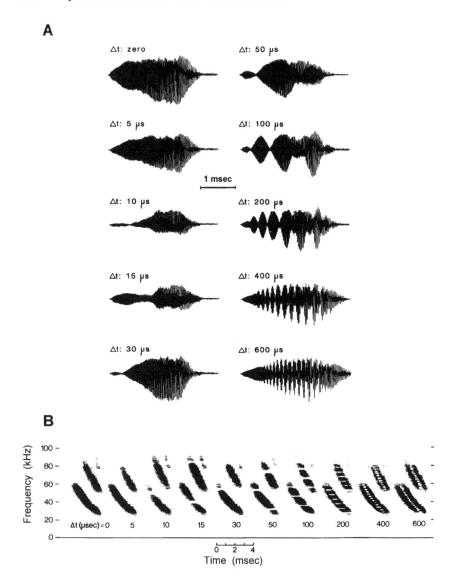

FIGURE 3.7A,B. (A) Time waveforms of two overlapping FM sounds (two-glint echoes) separated by delays ranging between 0 and 600 μsec. The signals are based on the sonar sound used by *Eptesicus fuscus* (from Simmons 1989). (B) Spectrograms of the same signals displayed in (A) (from Simmons 1992).

depth. The largest differences in the spectra appeared between 30 and 60 kHz, with notches occurring at lower frequencies for deeper holes. Simmons et al. suggested that *Eptesicus* can discriminate targets on the basis of information contained in the echo spectra. They also noted that the same signal-processing operation may be common to ranging (echo delay) and range resolution (echo spectrum) within the bat's auditory system, just as time-domain and frequency-domain displays of waveforms are mutually related through the Fourier transform. This might be interpreted to suggest that the bat experiences the overlapping echoes from the closely spaced surfaces as echoes arriving at different delays.

In the Habersetzer and Vogler study, *Myotis myotis* performed the discrimination task using two different reference targets, one with an 8-mm hole depth (as in Simmons et al. 1974) and the other with a 4-mm hole depth. They found that the bat's discrimination threshold was about 1 mm for the 8-mm reference target and about 0.8 mm for the 4-mm reference target. These data are in general agreement with those reported by Simmons et al. (1974) for *Eptesicus fuscus*.

Questions about the information the bat uses to discriminate echoes from closely spaced surfaces and the representation of such complex targets has been the subject of more recent research using target simulation methods. Target simulation methods permit control of the acoustic information arriving at the bat's ears and eliminate the resonance phenomena associated with real targets, such as the plates drilled with holes.

Schmidt (1988, 1992) studied the performance of *Megaderma lyra* trained to discriminate between virtual targets containing two replicas of the bat's sonar sounds at different delays (Fig. 3.8A). The delays were selected to simulate overlapping echoes from two closely spaced planar surfaces with different spatial offsets. In most of her experiments, the bat was rewarded in a 2-AFC task for selecting a reference delay offset of 7.77μsec, which simulated a distance difference between reflecting surfaces of 1.3 mm. Schmidt found that the smallest difference in temporal offset the bat could reliably discriminate was about 1 μsec (spatial separation of 0.17 mm), but discrimination performance was not a monotonic function of the difference in echo delay separation of the targets (see Fig. 3.8B). The reference target contained its first spectral notch at 64.4 kHz, and other echo delay offsets producing notches at this frequency (e.g., 23.3 μsec) caused a small drop in discrimination performance. From these data Schmidt hypothesized that the bat uses echo spectral information to discriminate target surface structure. She also used modeling to evaluate this hypothesis, which is discussed in Section 4 of this chapter.

Mogdans, Schnitzler, and Ostwald (1993) conducted a study on two-wavefront discrimination by *Eptesicus fuscus*, also using a 2-AFC behavioral paradigm. In one experiment, *Eptesicus* was first rewarded for selecting a two-wavefront target with an internal delay of 12 μsec, and its discrimination performance was tested against two-wavefront targets with internal delays of 28, 32, 36, 40, and 44 μsec. In a subsequent experiment, *Eptesicus* was rewarded for selecting a two-wavefront target with an internal delay of 36 μsec against others with internal delays of 28, 32, 36, 40, and 44 μsec. The bat successfully discriminated two-wavefront echoes with delay separations of either 12 or 36 μsec from those with delay separations of 28, 32, 40, or 44 μsec; however, discrimination between two-wavefront echoes with delay separations of 12 and 36 μsec fell to chance. This result

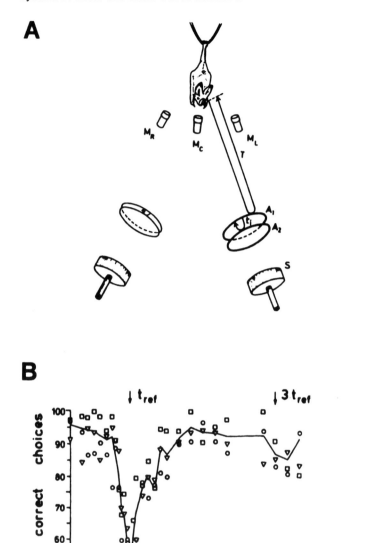

FIGURE 3.8A,B. (A) Schematic of behavioral setup for two-wavefront discrimination experiments with *Megaderma lyra*. M_R, right microphone; M_L, left microphone; M_C, center microphone. A_1 and A_2 represent the individual echo components of the two-wavefront target; S, playback speaker. (B) Behavioral data from three individual bats (*squares, circles, triangles*) trained to discriminate a two-wavefront target (t_{ref}) with an internal echo delay of 7.77 μsec from other two-

can be explained by a prominent spectral notch at about 42 kHz, which was common to the two-wavefront echoes with internal delays of 12 and 36 μsec (Fig. 3.9A). The authors noted, however, that these data do not establish that the bat uses a spectral representation of the signal to perform the task, because a temporal representation can also account for the data. Indeed, the energy density spectrum and cross-correlation function of a signal are mathematically related by the Fourier transform, an issue that is discussed further later in this chapter.

3.3.2 Discrimination Between One-Wavefront and Two-Wavefront Phantom Target Echoes

In another set of experiments, Mogdans, Schnitzler, and Ostwald (1993) trained *Eptesicus fuscus* to discriminate between a one-echo wavefront and a two-echo wavefront whose components differed in delay separation (see also Mogdans and Schnitzler 1990). The amplitude of the one-wavefront echo in this study was twice that of the individual components of the two-wavefront echo, and therefore the total energy of the one-wavefront echo was equal to that of the two-wavefront echo with an internal delay of 0 μsec. When the internal delay of the two-wavefront echo changed, so did its total energy and spectrum.

Mogdans, Schnitzler, and Ostwald (1993) reported that *Eptesicus* successfully discriminated between the targets, but only for particular temporal offsets between the echoes of the two-wavefront targets (see Fig. 3.9B). While the total energy of the two-wavefront target differed from that of the one-wavefront target, and the magnitude of this difference depended on the delay separation between the echoes of the two-wavefront target, the bat's pattern of performance indicates that it did *not* use total echo energy in the task. Instead, the data suggest that the spectrum of the playback echoes determined the bat's discrimination performance. When the spectral dissimilarity between the one- and two-wavefront targets was large, discrimination performance was well above chance. As mentioned earlier, these data do not demonstrate that a spectral model alone accounts for the bat's performance, because a time-domain representation can be derived from a frequency-domain representation (illustrated in Fig. 3.9B; see also Section 4, Receiver Models in Bat Sonar).

The basic pattern of results obtained from a one-wavefront versus

---

FIGURE 3.8. (*continued*) wavefront targets differing in internal delay. Percent correct responses plotted as a function of the temporal (t, μsec) and spatial (s, mm) offset of the two echoes of the test signal. Along the bottom abscissa is the frequency location of the first spectral notch of the two-wavefront test echo. The *solid line* shows the linear interpolation of the mean performance of three animals. (From Schmidt 1992.)

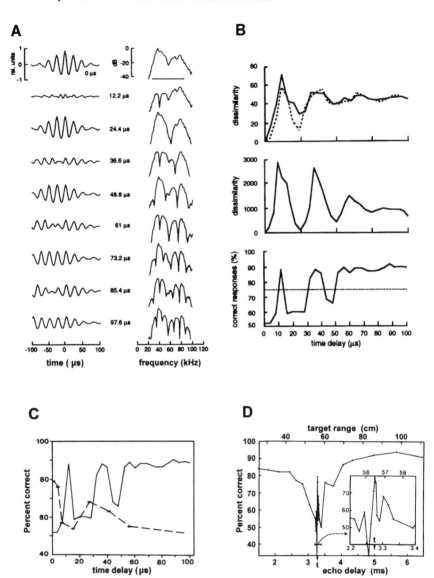

FIGURE 3.9A–D. (A) Cross-correlation functions of a representative echolocation sound used by *Eptesicus fuscus* and two-wavefront echoes at different internal time delays, ranging from 0 to 97.6 μsec (*left*). Amplitude spectra of the two-wavefront echoes at the same internal time delays (*right*). (B) Top panel: comparison of spectral dissimilarity between one-wavefront and two-wavefront echoes differing in internal delay (calculated after Schmidt 1988). Spectral similarity was based on the total frequency range of the sound (*solid line*) and the first harmonic *(stippled line)*. Middle panel: dissimilarity of the cross-correlation function (CCF) between a sonar sound and a one-wavefront echo and the CCF between a sound and a two-wavefront echo differing in internal delay. Bottom panel: mean performance

two-wavefront discrimination experiment can be strongly influenced by the initial task conditions. For example, in another experiment *Eptesicus fuscus* was trained in a 2-AFC procedure to detect a test echo at a fixed delay in the presence of clutter (Simmons et al. 1989). The clutter targets were single echoes, presented to the left and right of the bat's observing position. On one side, a clutter echo appeared alone, and on the other side, a test echo and a clutter echo appeared together where the test and clutter echoes were overlapping (i.e., one-wavefront versus two-wavefront). In this experiment, the amplitude of the individual echoes of the two-wavefront target (test plus clutter echoes) was the same as that of the one-wavefront target (clutter echo alone). Thus, for a 0-μsec internal delay of a two-wavefront target (no offset between test and clutter echoes), its energy was twice that of the one-wavefront target. For delay separations between the test echo and clutter echo of less than 100 μsec, the bat's performance fluctuated; however, the shape of the behavior curve differed markedly from that obtained by Mogdans, Schnitzler, and Ostwald (1993; see Fig. 3.9C).

In evaluating the data on two-wavefront discrimination, Mogdans, Schnitzler, and Ostwald (1993) also considered the representation of the complex target in the bat's sonar receiver. Based on the bat's near-chance performance for particular delay offsets, they conclude that it does not perceive the two-wavefront targets as containing distinct surfaces (or glints) along the range axis. This view contrasts with that presented by Simmons, Moss, and Ferragamo (1990), who proposed that a two-wavefront target is displayed in the bat's sonar receiver as discrete glints along the range axis.

In the study by Simmons, Moss, and Ferragamo (1990), the bat was trained to discriminate between a complex target containing two-echo wavefronts, each at 15 dB above threshold and offset by 100 μsec, and a simple target containing a one-echo wavefront. The delay (range) and amplitude of the complex target was fixed, while the delay and the

←
——

FIGURE 3.9   (*continued*) of six bats discriminating between a one-wavefront echo and two-wavefront echoes differing in internal delay. Note that the behavioral performance curve resembles the dissimilarity curves shown in top and middle panels of B (from Mogdans, Schnitzler, and Ostwald 1993). (C) Comparison of *Eptesicus*'s performance on a two-wavefront discrimination task (*solid line*; replotted from Fig. 3.9B, bottom panel) and on a detection task in which a clutter echo was presented at different delays with respect to the target echo (*stippled line*; replotted from Simmons et al. 1989, with an expanded time scale; see also Fig. 3.9D). Both curves plot behavioral performance as a function of the temporal separation between two echoes. Note that the shape of the two performance curves differs dramatically, reflecting differences in the test procedures in the two experiments (see text). (D) Detection of a test target echo at 3.27 msec in the presence of clutter (two-wavefront echo) as a function of the temporal separation between the test and clutter echoes (taken from Simmons et al. 1989). Mean performance of five individual bats. The inset shows a magnified segment of the curve between 3.2 and 3.4 msec.

amplitude of the simple target changed over the course of the experiment. The delay of the simple target varied in 25-μsec steps from 150 μsec before to 175 μsec after the first glint of the complex–echo pair; the amplitude of the simple target varied in 3-dB steps, from 6 dB below to 9 dB above the individual components of the complex target. In this experiment, the bat showed an increase in errors when the simple target was presented at the same delay and amplitude as the first or second echoes of the complex target, and they suggested that this rise in errors can serve as an index of the bat's perceived distance of the simple target with respect to the individual components of the complex target. From a series of experiments in which bats discriminated simple and complex target echoes, they concluded that the bat sonar receiver converts the spectral information contained in a two-wavefront echo (see Fig. 3.7) into a temporal representation, encoding the underlying delay or range separation of the component target echoes (see Simmons et al., this volume, Chapter 4).

In conclusion, research on range resolution suggests that the bat, at least in early stages of information processing, may use the spectrum of echoes to discriminate targets with two closely spaced reflecting surfaces; however, it is important to note that frequency-and time-domain representations of the echoes are mathematically equivalent. We would also like to emphasize that the shape of the bat's performance curve appears strongly dependent on the initial training procedures used in behavioral experiments, and for this reason it is often difficult to make clear comparisons across seemingly similar studies.

3.4 Summary

The behavioral studies summarized in this section contribute to our understanding of the processes important to the perceptual dimension of target range in echolocating bats and suggest that this perceptual dimension is not a single entity. Data on range difference discrimination, range jitter discrimination, and range resolution each appear to tap different perceptual processes in the bat sonar receiver. Range difference data describe the bat's accuracy at estimating the distance to a target and show that all bat species can discriminate distance with an accuracy of less than about 40 mm, and some species down to about 6–12 mm. Range jitter discrimination data reflect the bat's sensitivity to changes in target range, or perhaps to apparent target motion along the range axis. Data from range jitter experiments show that bats are sensitive to changes in echo delay/range of less than 0.5 μsec/0.1 mm. Studies on range resolution examine the bat's perception of target depth structure, and range resolution data show that bats are sensitive to very small spatial-temporal offsets of two-wavefront stimuli. FM bats can discriminate the separation of two closely spaced reflecting surfaces of about 1 mm or the difference in delay offset of two overlapping playback echoes of about 1 μsec. To discriminate a one-wavefront target from a two-wavefront target, the delay offset between the component echoes of the two-wavefront target must be about 12 μsec. The representation of complex targets in the sonar receiver of the bat is not well understood, and further research on the representation of depth-structured targets is particularly important for building our knowledge of spatial information processing by sonar.

4. Receiver Models in Bat Sonar

Since the modern-day discovery of echolocation in bats, engineers and biologists alike have explored the similarities between animal and man-made sonar/radar systems (e.g., Griffin 1944; McCue 1966). The comparison most commonly discussed is that between the bat and an ideal or matched filter receiver. An ideal receiver cross-correlates the outgoing sonar transmission and the incoming echo and reads a time-domain representation of the signal. If the echo reflects from a point target, the autocorrelation function of the sonar signal is a good approximation to the cross-correlation function of the sonar transmission and echo.

An ideal receiver may be either coherent or semicoherent, referring to the use of phase information in the signals. An ideal coherent receiver uses all the information contained in the cross-correlation of the sonar emission and echo to estimate target range. By contrast, an ideal semicoherent receiver does not make use of the phase information and uses the envelope of the correlation function for range estimation. The accuracy of time estimation depends on the form of the correlation function (which is determined by the signal type used) and internal and external noise. Wideband signals produce distinct central peaks with small sidelobes in the correlation function, whereas narrowband signals produce a less distinct central peak and prominent sidelobes (Simmons and Stein 1980).

The six panels in Figure 3.10 show different representations of the FM echolocation sound of *Eptesicus*. A time waveform representation of the sound is shown in Fig. 3.10A, the corresponding spectrogram in Fig 3.10B, and the amplitude spectrum in Fig. 3.10C. Figure 3.10D displays the autocorrelation function of this sound, which is also shown with an expanded time scale in Fig. 3.10E. These panels illustrate that the central peak of the autocorrelation function of a broadband signal is a good time marker. In Fig. 3.10F the bat's FM sound is cross-correlated with a time-reversed echo, and the prominent central peak (Figs. 3.10D and 3.10E) is absent.

4.1 Detection

A matched filter receiver concentrates the signal energy within a small time interval. This compression not only allows for an optimal estimate of echo

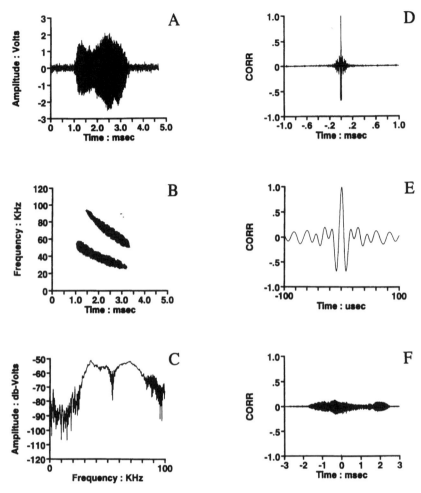

FIGURE 3.10A–F. Echolocation sound of *Eptesicus fuscus*. (A) Time waveform. (B) Spectrogram. (C) Spectrum. (D) Autocorrelation function. (E) Autocorrelation function on expanded time scale. (F) Cross-correlation function of signal and a time-reversed replica.

arrival time but also for optimal detection performance, because of an effective increase in S:N (see Skolnik 1980). If a bat uses a matched filter receiver for detection, the threshold should be very low and depend on the similarity between the echo and the receiver's signal template (Møhl 1988).

The very low threshold for echo detection calculated from Kick's (1982) data on *Eptesicus fuscus* can be interpreted as evidence in favor of coherent reception by bats (Møhl 1988). In another study, Moss and Simmons (1993) showed that the detection performance of *Eptesicus* depended on changes in the arrival time and the phase of phantom target playback signals. Bats were trained in a 2-AFC procedure to detect playback echoes that jittered in delay and in phase. The bat's detection performance for echoes that jittered in delay was improved compared to baseline (no jitter); however, for echo jitter of about 30 μsec, the bat's detection threshold was unchanged from baseline. When the phase of one echo from the jittered echo pair was inverted, detection performance increased compared to baseline, except for jitter values around 15 and 45 μsec. It appears that the temporal jitter in echo arrival time enhances detection performance, similar to a flickering target's enhancement of visual detection (Kulikowski and Tolhurst 1973). One possible interpretation of these data is that perceptual ambiguity of target jitter would arise from the sidelobes of the cross-correlation function of the bat's signal and returning echo at 30 μsec for phase-normal echoes, and at 15 and 45 μsec for phase-inverted echoes. The results of this study do not, however, demonstrate cross-correlation processing.

Arguments against the hypothesis of a matched filter for detection come from experiments using time-reversed playback signals and masking of echoes by white noise. In *Pipistrellus pipistrellus* (Møhl 1986) and *Eptesicus fuscus* (Masters and Jacobs 1989), species using broadband FM sounds that sweep in frequency from high to low, echo detection thresholds for time-reversed FM sweeps were unchanged from baseline measures with natural echolocation signals. (For a comparison of the cross-correlation functions of the baseline and time-reversed signals, see Fig. 3.10F.) Furthermore, Troest and Møhl (1986) reported that echo detection thresholds in noise were much higher than would be predicted of an ideal receiver. Together, these studies do not support the hypothesis that the bat uses a matched filter receiver for echo detection.

4.2 Ranging

Considerable debate has surrounded the question of whether the echolocating bat performs as an ideal receiver for range estimation (e.g., Simmons 1979; Schnitzler and Henson 1980; Schnitzler, Menne, and Hackbarth 1985; Hackbarth 1986; Menne and Hackbarth 1986; Simmons 1993). An ideal receiver achieves high ranging accuracy by cross-correlating the sonar emission and returning echo and using the time of the cross-correlation function peak as its best estimate of echo delay (target range). Here we consider the results of experiments on target ranging in bats and discuss their implications for assessing sonar receiver models.

4.2.1 Range Difference Discrimination

In 1973, Simmons reported a relationship between the width of the envelope of the autocorrelation function of the bat's echolocation sound and its ranging performance. Simmons compared the bat's range discrimination performance and that predicted from the envelope of the signal correlation function for four species, *Eptesicus fuscus, Phyllostomus hastus, Ptero-*

notus suapurensis, and *Rhinolophus ferrumequinum*, and found a correspondence between the empirical and predicted performance curves for
each of these species. The width of the correlation functions for each species
depends on the bandwidth of the animal's sonar sounds. Simmons reported
that species using broadband signals with narrow correlation function peaks
(sharper registration along the delay axis) showed lower range difference
discrimination thresholds than species using narrowband signals with broad
correlation function peaks. He used the general correspondence between
empirical performance curves for the four species and theoretical curves
derived from the envelope of the autocorrelation functions to argue that the
bat operates as an ideal semicoherent (phase-insensitive) receiver (see, for
example, empirical and theoretical curves shown in Figs. 3.4A and 3.4D).
However, Schnitzler, Menne, and Hackbarth (1985) pointed out that the
shape of the behavioral performance curve does not provide direct evidence
for cross-correlation processing.

Masters and Jacobs (1989) conducted a range difference discrimination
study with *Eptesicus fuscus* using normal and time-reversed FM playback
echoes. In this experiment, they found that the bat's range difference
discrimination threshold for time-reversed echoes was about 18 fold higher
than that measured in baseline experiments. While this result can be
explained by other receiver models (e.g., Menne 1988), it is consistent with
a matched filter receiver for range estimation.

Other experimental results are not consistent with the hypothesis that the
bat performs as an ideal receiver for the estimation of target distance. For
example, Simmons (1973) reported that the bat's range discrimination
performance was relatively stable at several test distances (30–240 cm; see
Fig. 3.4C), a finding that is not consistent with the predictions of matched
filtering. In this experiment, the S:N presumably decreased with increasing
range, and one would predict a rise in threshold with test distance; however,
because the S:N was not measured at these different test distances, this issue
is difficult to evaluate.

4.2.2 Range Jitter Discrimination

Using the bat's performance in the range jitter discrimination task, Simmons (1979) proposed that the bat operates as an ideal coherent (phase-
sensitive) receiver. This idea is based on the shape of the bat's performance
curve (i.e., percent correct performance as a function of echo jitter) in this
task rather than the absolute threshold for range jitter discrimination. In
particular, the bat showed a rise in errors for a jitter in echo delay of about
30 μsec relative to smaller and larger jitter delay values (Fig. 3.11). Simmons
reasoned that an animal performing the neural equivalent of matched
filtering would experience perceptual ambiguity for the presence of jitter at
30 μsec, because it would confuse the central peak and the first sidelobe of
the correlation function. Using a computer model to simulate a fully

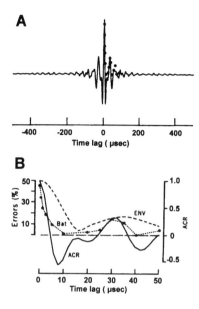

FIGURE 3.11A,B. Jitter discrimination performance compared with the autocorrelation function of the sonar sound used by *Eptesicus fuscus* (from Simmons 1979). (A) Percentage error data points plotted along the autocorrelation function of *Eptesicus*'s echolocation signal for time lag ± 400 μsec. (B) Percentage errors plotted for jitter discrimination performance for temporal jitter between 0 and 50 μsec. The *solid line* plots the autocorrelation function (ACR) of *Eptesicus*'s sonar sound; the *dashed line* shows the envelope (ENV) of the ACR.

coherent cross-correlation receiver, Menne and Hackbarth (1986) found no sidelobe (increase in errors) at 30 μsec in the predicted performance curves. A sidelobe in the performance curve, however, can be demonstrated using other receiver models, for example, a bank of filters with envelope processing (Hackbarth 1986).

In a later experiment, Simmons et al. (1990) showed that *Eptesicus* can discriminate a 180° phase alternation in playback echoes. The results of this study also show a drop in the bat's jitter discrimination performance for a combination of echo phase reversal and temporal jitter of 15 or 45 μsec. These findings are consistent with those predicted of an ideal coherent receiver (Altes 1981), and a computational model predicts this pattern of performance using cross-correlation processing (Saillant et al. 1993).

Additional evidence for an ideal coherent receiver in bat sonar comes from Simmons et al. (1990); they reported that *Eptesicus fuscus* can discriminate jitter in echo delay of 10 nsec with a S:N of about 30 dB, a value close to that predicted in a computer simulation (Menne and Hackbarth 1986). This corresponds to a change in target distance of about 0.0017 mm. Although appropriate calibration of the equipment was made and there were no obvious stimulus artifacts the bat might have used to perform in the task, this measurement demands replication using a digital system capable of producing changes in the delay of playback echoes in the nanosecond range.

The results of some jitter discrimination data are inconsistent with the predictions of a matched filter receiver. For example, Menne et al. (1989) studied the bat's performance with both time and phase jitter. They found that *Eptesicus fuscus* could discriminate changes in echo delay of 0.4 µsec (the smallest value tested) and 90° phase alternation with no added temporal jitter. While these data indicate that the bat is sensitive to very small changes in echo delay and phase, they do not support the hypothesis that the bat performs as an ideal receiver. For a coherent matched-filter receiver, an echo–phase alternation of 90° should cancel the bat's perception of jitter when combined with a 7-µsec alternation in echo delay. Thus, one would predict a drop in discrimination performance for this particular combination of phase and delay jitter. The closest jitter value they tested was 8 µsec, and for this value, the performance was not disturbed. While it is possible that a change in performance was not observed because 7 µsec was not tested, data from Simmons (1979; Simmons et al. 1990) suggest that errors associated with the sidelobe of the correlation function occur over a rather large time window. These data therefore suggest that the bat does not experience the perceptual ambiguity predicted for a matched filter receiver presented with a particular combination of phase and delay jitter.

The report by Simmons et al. (1990) that *Eptesicus* can discriminate jitter in echo delay of 10 nsec is not consistent with the idea that the bat's autocorrelation function can be used to predict its ranging performance (e.g., Simmons 1979; Simmons 1987; Simmons and Stein 1980), for this threshold jitter value is a factor of 10 smaller than the width of the central peak of the correlation function (see Fig. 3.10E). Furthermore, if the bat's range jitter discrimination threshold is 10 nsec, why would temporal ambiguity associated with the side peak of the correlation function contribute to the bat's performance over a time window that is several microseconds wide? And finally, it is difficult to imagine that the bat's head movements between sonar emissions and small changes in the sound level of returning echoes do not introduce variability in estimates of echo arrival time (see Fig. 3.6) that is greater than a few nanoseconds. Clearly more work is required in this area to resolve these issues.

4.2.3 Range Resolution

Receiver models have also been tested using behavioral data from two-wavefront discrimination tasks. For example, Schmidt (1988, 1992), proposed that a spectral correlation model best accounts for the performance of *Megderma lyra* in a task in which the bat is trained to discriminate between two-wavefront targets differing in internal delay. She contrasts the spectral correlation model with a highly simplistic temporal model, which does not yield predictions consistent with the empirical performance curves.

Mogdans, Schnitzler, and Ostwald (1993) used data from an experiment in which *Eptesicus fuscus* was trained to discriminate a one-wavefront target from a two-wavefront target to demonstrate that both spectral correlation and cross-correlation models predict the bat's range resolution performance (see Fig. 3.9B). They noted that the time-domain and frequency-domain models are mathematically related through the Fourier transform, and that it is not possible to differentiate between the two models using the bat's performance in the behavioral task.

4.3 Summary

The behavioral tests of receiver models in bat sonar indicate that detection and ranging are different processes. Many questions still surround the view that the bat performs as an ideal coherent receiver for the estimation of target range, and future research can directly address many of these questions through careful experimental design and stimulus control.

5. Horizontal and Vertical Sound Localization

To successfully maneuver in the environment and intercept prey, echolocating bats must continuously determine the azimuth, elevation, and range of objects in space. Bats use the time delay between sonar emission and returning echo to estimate target distance, and spatial perception along the range axis was discussed in detail in sections of this chapter. Here, we focus on behavioral studies of target localization in the horizontal and vertical planes.

5.1 Horizontal Localization

Mammals compare the arrival time, level, and spectrum of a sound at their two ears to determine azimuthal position (Yost and Gourevitch 1987; Wightman and Kistler 1993; Brown 1994). For a bat, the sound source is a target reflecting an echo of its sonar emission, and the directionality of its sonar emission will impart additional information about the spatial location of a target (e.g., Schnitzler and Grinnell 1977; Hartley and Suthers 1987, 1989).

The maximum arrival time difference occurs when a sound source is located at 80°–90° with respect to the midline of the head, and for bats with short interaural distances this time difference is small. For example, in *Eptesicus fuscus* whose head diameter is less than a centimeter, the maximum interaural time difference is approximately 40–50 μsec. Maximum interaural level differences measured with pure tones for the same species are about 25–30 dB (Jen and Chen 1988). Interaural spectral differences vary continuously with azimuth because of complex interactions between sound reflections off the pinna and tragus (see Shimozawa et al. 1974).

The threshold for horizontal sound localization has been estimated for the FM bat *Eptesicus fuscus*. Using a 2-AFC operant task, Simmons et al. (1983) found that the bat's threshold for discriminating the angular separation of thin rods was 1.5°. This angular separation creates interaural level differences of 0.3 to 1.5 dB, depending on the absolute direction of the targets with respect to the head aim of the bat. The interaural time difference associated with a 1.5° angular separation is about 1 μsec, a time separation that Simmons et al. (1983) suggested may be discriminated by *Eptesicus fuscus*.

5.2 Vertical Sound Localization

The cues associated with vertical sound localization in bats differ across species. Some species make large pinna movements that create interaural intensity differences, which can be used for vertical sound localization. For example, *Rhinolophus ferrumequinum* moves its pinna in coordination with its sonar emissions. Its highly mobile pinnae scan the sound field, moving in opposite directions. As one ear moves forward, the other ear moves backward, resulting in large differences in the echo levels at the two ears (Schneider and Möhres 1960).

Mogdans, Ostwald, and Schnitzler (1988) demonstrated the importance of ear movements for vertical localization in *Rhinolophus ferrumequinum*. Using a wire-avoidance task, they found that surgical immobilization of the pinnae selectively disrupted performance in vertical localization. In the baseline (preoperative) condition, bats successfully flew through vertical and horizontal arrays of wires separated by 15 cm. After the bat's pinnae were immobilized, it continued to successfully fly through the vertical array of wires (requiring horizontal localization) but showed many more collisions with the horizontal array of wires (requiring vertical localization). The bat's vertical localization was not, however, entirely impaired, suggesting that changes in the bat's head position during the task may have also provided information guiding flight through the horizontal wire array.

In other species, pinna movements are small by comparison with *Rhinolophus*, and the cues for vertical sound localization originate largely from the acoustic properties of the external ear. When sound enters the ear, it reflects off the pinna and tragus, creating spectral interference patterns from overlapping echoes. The interference patterns depend on the delay path between the reflecting surfaces of the external ear, which changes systematically with the vertical location of a sound source. In *Eptesicus fuscus*, sounds entering the ear produce primary and secondary reflections, separated by 45–60 μsec (Lawrence and Simmons 1982b). These reflections presumably represent echoes from the pinna and tragus, which mix to create notches in the echo spectrum that can be used by the bat to encode sound location in the vertical plane.

A psychophysical study of vertical sound localization has been conducted in *Eptesicus fuscus* using a 2-AFC procedure (Lawrence and Simmons 1982b). In this experiment, the bat was trained to discriminate between pairs of horizontal rods separated by different vertical angles. Threshold estimates for angular discrimination in the vertical plane was about 3° and depended on the intact tragus. When the bat's tragus was deflected, the bat's vertical discrimination performance deteriorated.

5.3 Summary

There are psychophysical data on horizontal and vertical sound localization thresholds for one species of FM bat, *Eptesicus fuscus*, but there are no comparable threshold data from short CF-FM or long CF-FM bats. In future work, it would be of interest to compare the sound localization capabilities of bats using different types of echolocation signals.

6. Fluttering Target Detection and Classification

Fluttering insects continuously change their reflecting surfaces, thus producing modulations in the echoes from the sonar signals of bats. When emitting short FM signals (below 5 msec), bats such as *Myotis lucifugus* and *Eptesicus fuscus* may use amplitude variations across echoes to recognize flying insects (Griffin 1958; Roeder 1962; Kober and Schnitzler 1990). Such FM sounds are too short to carry echo information during an insect's complete wingbeat cycle (Moss and Zagaeski 1994). By contrast, the rather long CF-FM signals (10–100 msec) that are characteristic of rhinolophids, hipposiderids, noctilionids, and of the mormoopid bat *Pteronotus parnellii* can carry much more acoustic information about a fluttering insect, often over more than one wingbeat cycle, thus increasing the probability of detecting fluttering insects (e.g., Goldman and Henson 1977).

6.1 Echoes from Fluttering Insects

Fluttering insects mounted in the acoustic beam of a CF signal reflect sounds that are rhythmically modulated in amplitude and frequency and in synchrony with the insect's wingbeat. These amplitude and frequency modulations not only reveal that there is a fluttering insect, they also contain species-specific information, such as wingbeat rate and movement patterns, that may be used by the bat to classify its prey. Furthermore, the acoustic parameters carried by fluttering insects may be used to encode the target's angular orientation with respect to the incident sound (Schnitzler and Henson 1980; Schnitzler and Ostwald 1983; Schnitzler et al. 1983; Schnitzler 1987; Ostwald, Schnitzler, and Schuller 1988; Kober and Schnitzler 1990).

The most prominent features of echoes from fluttering insects are very short and strong amplitude peaks or "acoustical glints," which are produced when the wings are perpendicular to the impinging sound waves. The wings act as "acoustic mirrors," causing amplitude peaks as much as 20 dB above the echo from the insect's body when positioned to create maximum echo reflection. The moving wings also produce spectral broadening in the echoes of CF signals from Doppler shifts. These spectral changes are largest when the wing velocity is highest, and their characteristics depend on the direction of the wing movement at the instant of glint production. For example, when the wings move down and backward, sound waves from ahead (0°) produce a negative Doppler shift and sound waves from the rear (180°) produce a positive Doppler shift (Fig. 3.12A). All other aspect angles also produce orientation-specific frequency deviations. Information about an insect species may be contained in the glint rate (encoding wingbeat rate), in the magnitude of the Doppler shift (encoding the size of the wings and the velocity of their motion), and in the temporal and spectral fine structure of the echo, related to other movements and morphological features of the insect (Kober and Schnitzler 1990).

6.2 Fluttering Target Detection and Discrimination by Bats Using CF Signals

Many species of bats that use CF-FM echolocation sounds are particularly well suited to make use of echo information from fluttering targets, as these bats compensate for Doppler shifts in echoes introduced by their own flight speed (Schnitzler 1968). For example, *Rhinolophus ferrumequinum* lowers the frequency of sonar emissions in an amount proportional to its flight velocity, receiving echoes from stationary objects at about 83 kHz, a frequency to which the bat shows high sensitivity (see Fig. 3.1). Thus, Doppler shift compensation behavior enables the bat to decouple echo Doppler shifts created by its own flight velocity from those created by moving targets in the environment (e.g., fluttering insect wings) and allows the bat to detect small shifts in echo frequency produced by the fluttering wings of insects.

Several observations indicate that bats with CF-FM signals use echo information produced by target flutter to recognize insect echoes even in dense clutter. In a laboratory study, *Rhinolophus ferrumequinum* caught fluttering moths not only when they were flying in the middle of the room but also on the walls or on the ground. Stationary moths were ignored (Schnitzler and Henson 1980; Trappe 1982). *Pteronotus parnellii* (Goldmann and Henson 1977) and *Hipposideros ruber* (Bell and Fenton 1984), and *Rhinolophus rouxi, Hipposideros bicolor*, and *Hipposideros speoris*

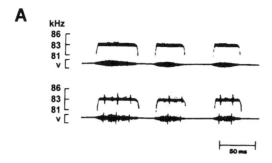

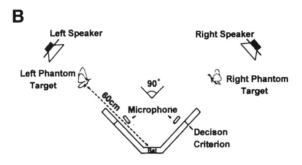

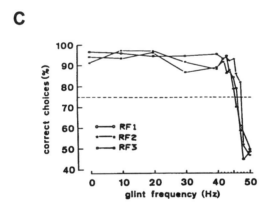

FIGURE 3.12A–C. (A) Sonar sounds used by *Rhinolophus ferrumequinum* (*top*) and playback echoes containing glints (spectral broadening and amplitude peaks) produced by fluttering insect wings (*bottom*). (B) Schematic of behavioral apparatus used to study fluttering target discrimination in *Rhinolophus ferrumequinum*. The bat was trained to make echolocation sounds into two microphones, positioned to the left and right of its start position. The signals picked up by the microphones were modified online and played back through two speakers, positioned behind the microphones. The playback sounds simulated echoes returning from fluttering insects (see echoes shown in A). (C) Fluttering target discrimination performance of three animals. Bats discriminated between an insect beating its wings at 50 Hz and at a slower, variable rate. Discrimination threshold (75% correct criterion) is about 4 Hz (adapted from von der Emde and Menne 1989).

(Link, Marimuthu, and Neuweiler 1986), also pursued fluttering insects whether flying or sitting and ignored the prey when they did not move their wings. Furthermore, hunting behavior in these bats was aborted when the insects stopped fluttering. A rotating propeller that produces an echo glint pattern similar to that of a fluttering insect (von der Emde and Schnitzler 1986) has been shown to attract *Pteronotus parnellii* (Goldmann and Henson 1977; Schnitzler et al. unpublished data) and *Rhinolophus ferrumequinum* (Schnitzler, unpublished data) in the laboratory. Such a fluttering target apparatus has also attracted the insect-hunting bats *Hipposideros ruber* (Bell and Fenton 1984) and *Rhinolophus ferrumequinum* and *Pteronotus parnellii* (Schnitzler et al. unpublished data), in the field. Together, these data suggest that bats using CF-FM signals respond selectively to echo changes produced by target movement and may use acoustic glint information to recognize insect echoes even in dense background clutter. Another type of echo modulation is used by the greater bulldog bat *Noctilio leporinus*, a species that pursues only jumping fish. Interference between echoes from the body of the fish and the water disturbances it creates by jumping produce a spectral pattern that the bat presumably uses to identify its prey (Schnitzler et al. 1994).

Laboratory studies also show that bats with long CF-FM signals are able to detect sinusoidal frequency modulations in their echoes. *Rhinolophus ferrumequinum* is much more sensitive than hipposiderid bats. In an experiment in which the bats had to discriminate an oscillating target from a similar motionless one, *Rhinolophus ferrumequinum* required a minimum oscillation frequency of 40 Hz (typical wingbeat frequency in many moths) at the threshold oscillation amplitude of 0.24 mm (Schnitzler and Flieger 1983). In *Hipposideros lankadiva*, the threshold amplitude of the oscillating target was about five fold higher than in *Rhinolophus ferrumequinum*, and *Hipposideros speoris* never learned the task (von der Emde and Schnitzler 1986). However, both hipposiderids reacted strongly to a glint-producing moving propeller.

The species differences in discrimination performance appear to correspond to differences in signal duration and duty cycle. *Rhinolophus ferrumequinum* produces very long signals (50–60 msec), whereas *Hippodideros lankadiva* produces signals that last about 10 msec and *Hippodideros speoris* signals of about 5 msec. While there are differences in the signal duty cycles among these species, it is interesting to note that they all increased sound duration or repetition rate when exposed to a fluttering target. These changes in the temporal features of the bat's sounds may indicate the bat's active effort to obtain acoustic information about target movement (Trappe and Schnitzler 1982; Schnitzler and Flieger 1983; von der Emde and Schnitzler 1986).

As acoustic information contained in echoes from fluttering targets differs across insect species, it has been asked whether bats can use echo flutter to classify their prey (Schnitzler and Ostwald 1983). In the laboratory, *Rhinolophus ferrumequinum* often selected one species of insect over another, presumably by using wingbeat rate as a cue (Trappe 1982). Studies of the food habits of this species also suggest selective prey acquisition (Ransome and Stebbings, personal communication). By contrast, there is no evidence for prey selection in *Hipposideros ruber* (Bell and Fenton 1984). A reason for this species difference could be the much shorter signal duration used by hipposidid bats, thus limiting the acoustic information about wingbeat signature in insect echoes.

The role of wingbeat rate as a possible cue for target characterization in bats with CF-FM signals has been studied in *Rhinolophus ferrumequinum* (von der Emde and Menne 1989), in *Rhinolophus rouxi* and *Hipposideros lankadiva* (Roverud, Nitsche, and Neuweiler 1991), and in *Pteronotus parnellii* (Schnitzler and Kaipf 1992). All experiments showed that bats are able to discriminate different wingbeat rates and that the performance of a given species is correlated with the duty cycle of its signals. The lowest flutter discrimination threshold was reported for the horseshoe bat, the species using the highest duty cycle. These bats were able to sense differences in wingbeat rate of 4%–9% (see Fig. 3.12C), whereas *Pteronotus parnellii* needed a difference of 20% and *Hipposideros lankadiva* a difference of 15%.

Many insects have rather similar wingbeat rates so that this cue alone would not be very suitable for the classification of prey. In discrimination experiments in which fluttering insects were simulated as phantom targets, *Rhinolophus ferrumequinum* and *Pteronotus parnelli* were able to discriminate different insect species exhibiting the same wingbeat rate (von der Emde and Schnitzler 1990; Schnitzler and Kaipf 1992). In a generalization experiment, bats were initially trained to select phantom target echoes of a particular insect species viewed from the side and later tested for recognition of the same species at novel aspect angles. The bat's mean performance for recognizing insects from new perspectives was mostly greater than 80%. Because the angular orientation of the prey strongly affects the modulation pattern in the echo (Kober and Schnitzler 1990), this kind of discrimination reveals astonishing cognitive abilities. This finding suggests that bats may develop a three-dimensional representation of a fluttering insect from acoustic information contained in echoes from a single view. The echo of a fluttering insect ensonified from a single orientation may thus be sufficient to construct a complete three-dimensional representation of the moving prey, and bats may use this representation for target classification (von der Emde and Schnitzler 1990).

The ability to use flutter information is not only helpful for target classification; it also allows CF-FM bats to hunt in areas where insect echoes are masked by strong background clutter (Neuweiler et al. 1988). Scattered field observations on feeding behavior in these bats show that they hunt for insects close to or in bushes and trees, along walls, and near the ground. Some species also hunt for insects from perches like a flycatcher (reviewed in Ostwald, Schnitzler, and Schuller 1988).

6.3 Fluttering Target Discrimination in Bats Using FM Signals

Fluttering target information may also be important to bats that hunt out in the open and produce rather long signals (up to 20 msec) of nearly constant frequency when searching for prey. As these bats never produce such signals in the laboratory, it is not possible to determine whether they use glint frequency in quasi-CF echoes to classify prey. However, it is evident that a glint in the echo of search pulses increases the likelihood of target detection and also perhaps indicates the presence of a fluttering insect (Moss and Zagaeski 1994).

In the laboratory, these bats emit short wideband FM sounds, and fluttering target discrimination has been studied with two species using these signals. In experiments in which a rotating propeller simulated a fluttering insect, *Pipistrellus stenopterus* (Sum and Menne 1988) and *Eptesicus fuscus* (Roverud, Nitsche, and Neuweiler 1991; Moss et al. 1992) were able to discriminate between different simulated wingbeat frequencies; however, the fluttering target discrimination threshold was higher in these FM bats than reported for *Rhinolophus* using long CF-FM signals.

What acoustic information does an FM bat use to discriminate between targets fluttering at different rates? In the behavioral discrimination experiments, the duration of the bat's echolocation sounds was too short and the duty cycle too low for glints in echoes to carry information about wingbeat period. It may be that the interference pattern between the Doppler-shifted echo from the moving blades of the propeller and the unmodulated echo from stationary parts of the wingbeat simulator contained the information that was used by the bats (Sum and Menne 1988). It is not clear whether this kind of information allows FM bats in the field to identify fluttering insects.

6.4 Summary

Flutter information is used by different species of bats in different ways. Bats with very long CF-FM signals (more than 20 msec), such as all rhinolophids, the mormoopid bat *Pteronotus parnellii*, and some hipposiderids (e.g., *Hipposideros lankadiva*), not only use flutter information to identify insect echoes and to discriminate them from unmodulated background clutter but may also use echo information from fluttering insects for target classification. In bats that use shorter CF-FM signals, such as most of the hipposiderids and the two species of noctilionids, fluttering target discrimination is not as well developed. This also appears to be the case for bats that hunt insects in the open, using quasi-CF pulses when searching for insects. For bats that use sounds of short duration and low duty cycle, echoes from fluttering insects may be most important in signaling the presence of prey and less useful for carrying detailed information about the wingbeat rate of the prey. In future research it would be of interest to examine the perceptual representations of fluttering insects in bat species using different echolocation signal design.

7. Conclusion

The computations performed by the auditory system of the bat support both passive listening and active echolocation processes. In studies of passive listening, absolute sensitivity data suggest that the bat's auditory system is well adapted to detect species-specific echolocation and communication signals. Basic psychoacoustic data on other aspects of hearing in bats are limited, but what exist suggests that the bat's auditory system may incorporate some specializations into the general mammalian plan.

The bulk of research on auditory information processing in bats has focused on the active echolocation system, requiring the animal to probe its environment by emitting sonar signals and listening to echoes from objects in the path of the sound beam. This chapter surveys research on the bat's active biosonar system for the detection, localization, and identification of targets. In each of these areas, we have examined data that reveal the limits of the bat's echolocation system for extracting information about the environment through acoustic channels. Moreover, when possible we have also considered data that speak to the representation of sonar targets in the bat's auditory system. In attempting to synthesize data from different experiments, we continuously encountered variations in the experimental design, the bat's task, and the acoustic environment that make cross-study comparisons difficult. In future studies on echolocation performance in bats, we propose that researchers pay careful attention to the design of each experiment and the details of the acoustic environment. Moreover, because bats have active control over the echo information returning to their ears, it is essential to frequently monitor the duration, bandwidth, and level of the sounds the bat produces under experimental conditions. We believe that future work in this field will yield exciting and important insights in comparative psychoacoustics.

Acknowledgments Preparation of this chapter was supported by a National Science Foundation Young Investigator Award to CFM and by the Deutsche Forschungsgemeinschaft SFB 307. We thank Richard Fay, Marc Hauser, Arthur Popper, and Doreen Valentine for comments on the manuscript and Ingrid Kaipf and Anne Grossetête for assistance with some of the figures.

References

Altes RA (1981) Echo phase perception in bat sonar? J Acoust Soc Am 69:505–509.

Airapetianz ESH, Konstantinov A I (1974) Echolocation in nature. Leningrad: Nauka. (English translation, Joint Publications Research Service, no. 63328, 1000 North Glebe Road, Arlington, VA 22201.)

Bell GP, Fenton MB (1984) The use of Doppler-shifted echoes as a clutter rejection system: the echolocation and feeding behavior of *Hipposideros ruber* (Chiroptera: Hipposideridae). Behav Ecol Sociobiol 15:109–114.

Brown CH (1994) Sound localization. In: Popper AN, Fay RR (eds) Comparative Hearing: Mammals. New York: Springer-Verlag, pp. 57–96.

Bruns V (1980) Structural adaptation in the cochlea of the horseshoe bat for the analysis of long CF-FM echolocating signals In: Busnel RG, Fish J F (eds) Animal Sonar Systems. New York: Plenum, pp. 867–869.

Burkard R, Moss C F (1994) The brainstem auditory evoked response (BAER) in the big brown bat (*Eptesicus fuscus*) to clicks and frequency modulated sweeps. J Acoust Soc Am 96:801–810.

Dalland JI (1965) Hearing sensitivity in bats. Science 150:1185–1186.

de Boer E (1985) Auditory time constants: A paradox? In: Michelsen A (ed) Time Resolution in Auditory Systems. Berlin: Springer, pp. 141–158.

Denzinger A, Schnitzler H-U (1994) Echo SPL influences the ranging performance of the big brown bat, *Eptesicus fuscus.* J Comp Physiol *175*:563–571.

Fay RR (1988) Hearing in Vertebrates. A Psychophysics Databook. Winnetka, IL: Hill-Fay.

Fletcher H (1940) Auditory patterns. Rev Mod Phys 12:47–65.

Fullard JH, Fenton M B (1977) Acoustic and behavioural analyses of the sounds produced by some species of Nearctic Arctiidae (Lepidoptera) Can J Zool 55:1213–1224.

Goldman LJ, Henson OW (1977) Prey recognition and selection by the constant frequency bat, *Pteronotus p. parnellii.* Behav Ecol Sociobiol 2:411–419.

Griffin D (1944) Echolocation by blind men, bats and radar. Science 100:589–590.

Griffin D (1953) Bat sounds under natural conditions, with evidence for the echolocation of insect prey. J Exp Zool 123:435–466.

Griffin D (1958) Listening in the Dark. New Haven: Yale University Press. (Reprinted by Cornell University Press, Ithaca, NY 1986.)

Griffin D, Webster FA, Michael CR (1960) The echolocation of flying insects by bats. Anim Behav 8:141–154.

Grinnell AD, Schnitzler H-U (1977) Directional sensitivity of echolocation in the horseshoe bat, *Rhinolophus ferrumequinum.* II. Behavioral directionality of hearing. J Comp Physiol A 116:63–76.

Habersetzer J, Vogler B (1983) Discrimination of surface-structured targets by the echolocating bat, *Myotis myotis,* during flight. J Comp Physiol A 152:275–282.

Hackbarth H (1986) Phase evaluation in hypothetical receivers simulating ranging in bats. Biol Cybern 54:281–287.

Hartley DJ (1989) The effect of atmospheric sound absorption on signal bandwidth and energy and some consequences for bat echolocation. J Acoust Soc Am 8:1338–1347.

Hartley DJ (1992a) Stabilization of perceived echo amplitudes in echolocating bats. I. Echo detection and automatic gain control in the big brown bat, *Eptesicus fuscus*, and the fishing bat, *Noctilio leporinus*. J Acoust Soc Am 91:1120–1132.

Hartley DJ (1992b) Stabilization of perceived echo amplitudes in echolocating bats. II. The acoustic behaviour of the big brown bat, *Eptesicus fuscus*, when tracking moving prey. J Acoust Soc Am 91:1133–1149.

Hartley DJ, Suthers RA (1987) The sound emission pattern and the acoustical role of the noseleaf in the echolocating bat, *Carollia perspicillata*. J Acoust Soc Am 82:1892–1900.

Hartley DJ, Suthers RA (1989) The sound emission pattern of the echolocating bat, *Eptesicus fuscus*. J Acoust Soc Am 85:1348–1351.

Hartridge H (1945) Acoustical control in the flight of bats. Nature 156:490–494; 692–693.

Henson OW Jr (1965) The activity and function of the middle ear muscles in echolocating bats. J Physiol (Lond) 180:871–887.

Jen PHS, Chen D (1988) Directionality of sound pressure transformation at the pinna of echolocating bats. Hear Res 34:101–118.

Kick SA (1982) Target detection by the echolocating bat, *Eptesicus fuscus*. J Comp Physiol A 145: 431–435.

Kick SA, Simmons JA (1984) Automatic gain control in the bat's sonar receiver and the neuroethology of echolocation. J Neurosci 4:2725–2737.

Kober R, Schnitzler H-U (1990) Information in sonar echoes of fluttering insects available for echolocating bats. J Acoust Soc Am 87:882–896.

Kulikowski JJ, Tolhurst DJ (1973) Psychophysical evidence for sustained and transient detectors in human vision. J Physiol (Lond) 232:149–162.

Lawrence BD, Simmons JA (1982a) Measurements of atmospheric attenuation at ultrasonic frequencies and the significance for echolocation by bats. J Acoust Soc Am 71:585–590.

Lawrence BD, Simmons JA (1982b) Echolocation in bats: the external ear and perception of the vertical positions of targets. Science 218:481–483.

Link A, Marimuthu G, Neuweiler G (1986) Movement as a specific stimulus for prey-catching behaviour in rhinolophid and hipposiderid bats. J Comp Physiol A 159:403–413.

Long G (1977) Masked auditory thresholds from the bat, *Rhinolophus ferrumequinum*. J Comp Physiol A 116:247–255.

Long G (1994) Psychoacoustics. In: Fay RR, Popper AN (eds) Comparative Hearing: Mammals. New York: Springer-Verlag pp 18–56.

Long G, Schnitzler H-U (1975) Behavioral audiograms form the bat *Rhinolophus ferrumequinum*. J Comp Physiol A 100:211–220.

Masters WM, Jacobs S C (1989) Target detection and range resolution by the big brown bat (*Eptesicus fuscus*) using normal and time-reversed model echoes. J Comp Physiol A 166:65–73.

McCue JJG (1966) Aural pulse compression in bats and humans. J Acoust Soc Am 40:545–548.

Menne D (1988) A matched filter bank for time delay estimation in bats. In: Nachtigall PE, Moore PWB (eds) Animal Sonar Processes and Performance. New York: Plenum, pp. 835–842.

Menne D, Hackbarth H (1986) Accuracy of distance measurement in the bat *Eptesicus fuscus*: theoretical aspects and computer simulations. J Acoust Soc Am 79:386–397.

Menne D, Kaipf I, Wagner I, Ostwald J, Schnitzler H-U (1989) Range estimation by echolocation in the bat *Eptesicus fuscus*: trading of phase versus time cues. J Acoust Soc Am 85:2642-2650.

Miller LA (1983) How insects detect and avoid bats. In: Huber F, Markl H (eds) Neuroethology and Behavioral Physiology. Berlin: Springer-Verlag, pp. 251-266.

Miller LA (1991) Arctiid moth clicks can degrade the accuracy of range difference discrimination in echolocating big brown bats, *Eptesicus fuscus*. J Comp Physiol A 168:571-579.

Mogdans J, Schnitzler H-U (1990) Range resolution and the possible use of spectral information in the echolocating bat, *Eptesicus fuscus*. J Acoust Soc Am 88:754-757.

Mogdans J, Schnitzler H-U, Ostwald J (1993) Discrimination of 2-wavefront echoes by the big brown bat, *Eptesicus fuscus*: behavioral experiments and receiver simulations. J Comp Physiol A 172:309-323.

Mogdans J, Ostwald J, and Schnitzler H-U (1988) The role of pinna movement for the localization of vertical and horizontal wire obstacles in the greater horseshoe bat, *Rhinolophus ferrumequinum*. J Acoust Soc Am 84:1676-1679.

Møhl B (1986) Detection by a pipistrellus bat of normal and reversed replica of its sonar pulses. Acustica 61:75-82.

Møhl B (1988) Target detection by echolocating bats. In: Nachtigall PE, Moore PWB (eds) Animal Sonar Processes and Performance. New York: Plenum Press, pp. 435-450.

Møhl B, Surlykke A (1989) Detection of sonar signals in the presence of pulses of masking noise by the echolocating bat, *Eptesicus fuscus*. J Comp Physiol A 165:119-124.

Moss CF, Schnitzler H-U (1989) Accuracy of target ranging in echolocating bats: acoustic information processing. J Comp Physiol A 165:383-393.

Moss CF, Simmons JA (1993) Acoustic image representation of a point target in the bat *Eptesicus fuscus:* evidence for sensitivity to echo phase in bat sonar. J Acoust Soc Am 93:1553-1562.

Moss CF, Zagaeski M (1994) Acoustic information available to bats using frequency-modulated echolocation sounds for the perception of insect prey. J Acoust Soc Am 95:2745-2756.

Moss CF, Gounden C, Booms J, Roach J (1992) Discrimination of target movement by the FM bat, *Eptesicus fuscus*. Abstracts of the 15th Midwinter Research Meeting of the Society for Research in Otolaryngology, p. 142.

Neuweiler G, Bruns V, Schuller G (1980) Ears adapted for the detection of motion, or how echolocating bats have exploited the capacities of the mammalian auditory system. J Acoust Soc Am 68:741-753.

Neuweiler G, Link A, Marimuthu G, Rübsamen R (1988) Detection of prey in echocluttering environments. In: Nachtigall PE, Moore PWB (eds) Animal Sonar Processes and Performance. New York: Plenum, pp. 613-618.

Ostwald J, Schnitzler H-U, Schuller G (1988) Target discrimination and target classification in echolocating bats. In: Nachtigall PE, Moore PWB (eds) Animal Sonar Processes and Performance. New York: Plenum, pp. 413-434.

Pollak GD (1993) Some comments on the proposed perception of phase and nanosecond time disparities by echolocating bats. J Comp Physiol A 172:523-531.

Poussin C, Simmons JA (1982) Low-frequency hearing sensitivity in the echolocating bat, *Eptesicus fuscus*. J Acoust Soc Am 72:340-342.

Reetz G, Schnitzler H-U (1992) Signal design in the bat *Eptesicus fuscus* when detecting targets differing in range and size. In: Elsner, N and Richter, D.W. (eds.) Stuttgart: Georg Thieme Verlag. Proceedings of the 20th Göttingen Neurobiology Conference, p. 213.

Roeder KD (1962) The behaviour of free-flying moths in the presence of artificial ultrasonic pulses. Anim Behav 10:300–304.

Roverud RC (1989a) Harmonic and frequency structure used for echolocation sound pattern recognition and distance information processing in the rufous horseshoe bat. J Comp Physiol A 166:251–255.

Roverud RC (1989b) A gating mechanism for sound pattern recognition is correlated with the temporal structure of echolocation sounds in the rufous horseshoe bat. J Comp Physiol A 166:243–249.

Roverud RC, Grinnell AD (1985a) Discrimination performance and echolocation signal integration requirements for target detection and distance determination in the CF/FM bat, *Noctilio albiventris*. J Comp Physiol A 156:447–456.

Roverud RC, Grinnell AD (1985b) Echolocation sound features processed to provide distance information in the CF/FM bat, *Noctilio albiventris*: evidence for a gated time window utilizing both CF and FM components. J Comp Physiol A 156:457–469.

Roverud RC, Nitsche V, Neuweiler G (1991) Discrimination of wingbeat motion by bats correlated with echolocation sound pattern. J Comp Physiol A 168:259–263.

Saillant PA, Simmons JA, Dear SP, McMullen T A (1993) A computational model of echo processing and acoustic imaging in frequency-modulated echolocating bats: The spectrogram correlation and transformation receiver. J Acoust Soc Am 94:2691–2712.

Schmidt S (1988) Evidence for a spectral basis of texture perception in bat sonar. Nature 331:617–619.

Schmidt S (1992) Perception of structured phantom targets in the echolocating bat, *Megaderma lyra*. J Acoust Soc Am 91:2203–2223.

Schmidt S (1993) Perspectives and problems of comparative psychoacoustics in echolocating bats. In: Abstracts of the Sixteenth Midwinter Research Meeting of the Association for Research in Otolaryngology, St. Petersburg Beach, FL, p. 145.

Schmidt S, Türke B, Vogler B (1983) Behavioural audiogram from the bat, *Megaderma lyra* (Geoffroy, 1810; Microchiroptera). Myotis 21/22:62–66.

Schneider H, Möhres F P (1960) Die Ohrbewegungen der Hufeisenfledermäse (Chiropetera, Rhinolophidae) und der Mechanismus des Bildhörens. Z Vgl Physiol 44:1–40.

Schnitzler H-U (1968) Die Ultraschall-Ortungslaute der Hufeisen-Fledermaüse (Chiroptera-Rhinolophidae) in verschiedenen Orientierungssituationen. Z Vgl Physiol 57:376–408.

Schnitzler H-U (1987) Echoes of fluttering insects: information for echolocating bats. In: Fenton MB, Racey P, Rayner JMV (eds) Recent Advances in the Study of Bats. Cambridge: Cambridge University Press, pp. 226–243.

Schnitzler H-U, Flieger E (1983) Detection of oscillating target movements by echolocation in the greater horseshoe bat. J Comp Physiol A 153:385–391.

Schnitzler H-U, Grinnell AD (1977) Directional sensitivity of echolocation in the horseshoe bat, *Rhinolophus ferrumequinum*. I. Directionality of sound emission. J Comp Physiol A 116:51–61.

Schnitzler H-U, Henson OW Jr (1980) Performance of airborne animal sonar systems: 1. Microchiroptera. In: Busnel RG, Fish JE (eds) Animal Sonar Systems. New York: Plenum, pp. 109–181.

Schnitzler H-U, Kaipf I (1992) Classification of insects by echolocation in the mustache bat, *Pteronotus parnellii*. Deustche Gesellshaft für Sägertierkunde, Vol. 66. Karlsruhe: Hauptversammlung.

Schnitzler H-U, Ostwald J (1983) Adaptation for the detection of fluttering insects by echolocation in horseshoe bats. In: Ewert JP, Capranica RR, Ingle DJ (eds) Advances in Vertebrate Neuroethology. New York: Plenum, pp. 801–827.

Schnitzler H-U, Menne D, Hackbarth H (1985) Range determination by measuring time delays in echolocating bats In: Michelsen A (ed) Time Resolution in Auditory Systems. Berlin: Springer-Verlag, pp. 180–204.

Schnitzler H-U, Kalko EKV, Kaipf I, Grinnell A (1994) Fishing and echolocation behavior of the greater bulldog bat, *Noctilio leporinus*, in the field. Behav Ecol Sociobiol 35:327–345.

Schnitzler H-U, Menne D, Kober R, Heblich K (1983) The acoustical image of fluttering insects in echolocating bats. In: Huber F, Markel H (eds) Neuro-ethology and Behavioral Physiology. Heidelberg: Springer-Verlag, pp. 235–249.

Shimozawa T, Suga N, Hendler P, Schuetze S (1974) Directional sensitivity of echolocation system in bats producing frequency modulated signals. J Exp Biol 60:53–69.

Simmons JA (1973) The resolution of target range by echolocating bats. J Acoust Soc Am 54:157–173.

Simmons J A (1979) Perception of echo phase information in bat sonar. Science 204:1336–1338.

Simmons JA (1987) Acoustic images of target range in the sonar of bats. Nav Res Rev 39:11–26.

Simmons JA (1989) A view of the world through the bat's ear: the formation of acoustic images in echolocation. Cognition 33:155–199.

Simmons JA (1992) Time-frequency transforms and images of targets in the sonar of bats. In: Bialek W (ed) Princeton Lectures on Biophysics–1991. Princeton: NEC Research Institute.

Simmons JA (1993) Evidence for perception of fine echo delay and phase by the FM bat, *Eptesicus fuscus*. J Comp Physiol A 172:533–547.

Simmons JA Stein RA (1980) Acoustic imaging in bat sonar: echolocation signals and the evolution of echolocation. J Comp Physiol A 135:61–84.

Simmons JA, Moffat AJM, Masters WM (1992) Sonar gain control and echo detection thresholds in the echolocating bat, *Eptesicus fuscus*. J Acoust Soc Am 91:1150–1163.

Simmons JA, Moss CF, Ferragamo M (1990) Convergence of temporal and spectral information into acoustic images of complex sonar targets perceived by the echolocation bat, *Eptesicus fuscus*. J Comp Physiol A 166:449–470.

Simmons JA, Ferragamo MJ, Dear SP, Haresign T, Fritz J (1995) Auditory computations for acoustic imaging in bat sonar. In: Hawkins H, McMullem T (eds) Springer Handbook of Auditory Research: Auditory Computations. New York: Springer-Verlag, (in press).

Simmons JA, Ferragamo M, Moss CF, Stevenson SB, Altes RA (1990) Discrimi-nation of jittered sonar echoes by the echolocating bat, *Eptesicus fuscus*: the shape of target images in echolocation. J Comp Physiol A 167:589–616.

Simmons JA, Freedman EG, Stevenson SB, Chen L, Wohlgenant TJ (1989) Clutter interference and the integration time of echoes in the echolocating bat, *Eptesicus fuscus*. J Acoust Soc Am 86:1318–1332.

Simmons JA, Kick SA, Moffat AJM, Masters WM, Kon D (1988) Clutter interference along the target range axis in the echolocating bat, *Eptesicus fuscus*. J Acoust Soc Am 84:551–559.

Simmons JA, Kick SA, Lawrence BD, Hale C, Bard C, Escudié B (1983) Acuity of horizontal angle discrimination by the echolocating bat, *Eptesicus fuscus*. J Comp Physiol A 153:321–330.

Simmons JA, Lavender WA, Lavender BA, Doroshow CA, Kiefer SW, Livingston R, Scallet AC, Crowley DE (1974) Target structure and echo spectral discrimination by echolocating bats. Science 186:1130–1132.

Skolnik MI (1980) Introduction to Radar Systems. New York: McGraw-Hill.

Sokolov BV (1972) Interaction of auditory perception and echolocation in bats (Rhinolophidae) during insect catching. Vestn Leningr Univ Ser Biol 27:96–104 (in Russian).

Stapells DR, Picton TW, Smith AD (1982) Normal hearing thresholds for clicks. J Acoust Soc Am 72:74–79.

Suga N, Jen P (1975) Peripheral control of acoustic signals in the auditory system of echolocating bats. J Exp Biol 62:277–311.

Suga N, Schlegel P (1972) Neural attenuation of responses to emitted sounds in echolocating bats. Science 177:82–84.

Suga N, Shimozawa T (1974) Site of neural attenuation of responses to self-vocalized sounds in echolocating bats. Science 183:1211–1213.

Sum YW, Menne D (1988) Discrimination of fluttering targets by the FM-bat *Pipistrellus stenopterus?* J Comp Physiol A 163:349–354.

Surlykke A (1992) Target ranging and the role of time-frequency structure of synthetic echoes in big brown bats, *Eptesicus fuscus*. J Comp Physiol A 170:83–92.

Surlykke A, Miller L A (1985) The influence of arctiid moth clicks on bat echolocation: jamming or warning? J Comp Physiol A 156:831–843.

Suthers RA, Summers C A (1980) Behavioral audiogram and masked thresholds of the megachiropteran echolocating bat, *Rousettus*. J Comp Physiol A 136:227–233.

Trappe M (1992) Verhalten und Echoorting der Grossen Hufeisemase beim Isektenfang. Ph.D. dissertation, University of Marburg.

Trappe M, Schnitzler H-U (1982) Doppler-shift compensation in insect-catching horseshoe bats. Naturwissenschaften 69:193–196.

Troest N, Møhl B (1986) The detection of phantom targets in noise by serotine bats; negative evidence for the coherent receiver. J Comp Physiol A 159:559–567.

von der Emde G, Schnitzler H-U (1986) Fluttering target detection in hipposiderid bats. J Comp Physiol A 14:43–55.

von der Emde G, Menne D (1989) Discrimination of insect wingbeat-frequencies by the bat *Rhinolophus ferrumequinum*. J Comp Physiol A 164:663–671.

von der Emde G, Schnitzler H-U (1990) Classification of insects by echolocating greater horseshoe bats. J Comp Physiol A 167:423–430.

Yost WA, Gourevitch G (eds) (1987) Directional Hearing. Berlin: Springer-Verlag.

Zwicker E, Flottorp G, Stevens SS (1957) Critical bandwidths in loudness summation. J Acoust Soc Am 29:548–557.

4

Auditory Dimensions of Acoustic Images in Echolocation

James A. Simmons, Michael J. Ferragamo,
Prestor A. Saillant, Tim Haresign, Janine M. Wotton,
Steven P. Dear, and David N. Lee

1. Introduction

1.1 Echolocation and Hearing

Echolocation in bats is one of the most demanding adaptations of hearing to be found in any animal. Transforming the information carried by sounds into perceptual images depicting the location and identity of objects rapidly enough to control the decisions and reactions of a swiftly flying bat is a prodigious task for the auditory system to accomplish. The exaggeration of aspects of auditory function to achieve spatial imaging reflects the vital role of hearing in the lives of bats — for finding prey and perceiving obstacles to flight (Neuweiler 1990). It also highlights the mechanisms behind these functions to make echolocation a useful model for studying how the auditory system processes information and creates auditory perceptions in the most extreme circumstances.

The auditory capacity most at a premium for echolocation with wideband signals (see Fenton, Chapter 2) is the ability to determine the arrival time of sonar echoes, both to perceive the delay of individual echoes and to separately perceive several closely spaced echoes at their slightly different delays. Behavioral evidence accumulated from a variety of recent experiments reveals that bats can determine echo delay with an accuracy and fine resolution in the range of at least 1–5 μsec, and under certain conditions with an accuracy of 10–40 nsec (Menne et al. 1989; Moss and Schnitzler 1989; Moss and Simmons 1993; Simmons 1979, 1989, 1993; Simmons et al. 1990; see Moss and Schnitzler, Chapter 3). However, this degree of temporal acuity has been seen as unattainable from neural responses typically recorded in the bat's auditory system (Pollak et al. 1977; Pollak and Casseday 1989; Schnitzler, Menne, and Hackbarth 1985). Nevertheless, the recent behavioral evidence is so compelling (Simmons 1993) that we have reexamined earlier, widely accepted experimental results and found that they, too, reveal an underlying echo–delay accuracy and resolution in the microsecond range. This chapter summarizes the findings of our reexamination.

1.2 Scope of This Chapter

Our chapter reviews acoustic and behavioral evidence for fine temporal precision and resolving power as the basis for acoustic imaging in echolocation. This evidence is drawn largely from experiments conducted with bats in naturalistic situations that have not previously been considered relevant to determining the detailed content of the bat's sonar images. The conceptual framework for our treatment of echo processing by bats has been introduced elsewhere (Simmons 1989, 1992; Simmons, Moss, and Ferragamo 1990), and the computational validity of this approach has been demonstrated in a quantitative model of echolocation (Saillant et al. 1993). A companion chapter in another forthcoming volume of the *Springer Handbook of Auditory Research* series describes this imaging process and how it is supported by neural responses in the bat's auditory system (see Simmons et al. in press).

In this chapter, we examine the content of the sonar images perceived by bats that broadcast frequency-modulated (FM) sonar sounds. We focus on determining what echolocating bats perceive along the axis of distance, or *target range*, from the arrival time of FM sonar echoes returning to their ears. While this chapter thus initially appears to have limited scope—concentrating on depth perception—our concern for the dimension of target range will expand to encompass perception of other aspects of targets such as shape and direction.

Many of the behavioral experiments considered here were carried out rather early in the history of research on echolocation; indeed, some of them were first used to demonstrate the existence of sonar in bats. While the basic issues they bring forth were recognized quite early from a physiological perspective (Grinnell 1967; Suga 1967), the full implications of their results for understanding the images perceived by bats have only now come to light. Our goal here is to discover what the results of these behavioral experiments reveal about the acoustic dimensionality of the images underlying echolocation by FM bats. A methodologically oriented review of the more recent, controversial experiments on FM echolocation that have drawn attention to the question of temporal acuity in bats is given in another chapter in this volume (see Moss and Schnitzler, Chapter 3). Several other chapters in this volume of the Handbook provide additional background for our chapter.

2. Echolocation by FM Bats

2.1 The Big Brown Bat, Eptesicus fuscus

The experiments discussed in this chapter have been carried out with the North American big brown bat, *Eptesicus fuscus* (family Vespertilionidae) or with several related species of FM bats in this same family. *Eptesicus* is an insectivorous bat that uses echolocation to find its prey (see Fenton, Chapter 2). Figure 4.1 shows the big brown bat in the act of capturing a mealworm suspended on a fine thread. The bat has located and approached the target by sonar and is just about to seize the insect in its tail membrane, which is a mode of capture employed by many species of bats (Webster and Griffin 1962). Details about the feeding habits and general biology of *Eptesicus* have been conveniently summarized by Kurta and Baker (1990).

2.2 Echolocation Signals of Eptesicus

Broadcasting sounds is a means for probing the environment in depth, and bats use FM echolocation sounds for this purpose (Simmons 1973; Simmons and Grinnell 1988). The transmitted signal illuminates or ensonifies objects located in the beam of sound to reflect or scatter back to the bat. Figure 4.2 illustrates spectrograms of echolocation sounds broadcast by the big brown bat, *Eptesicus fuscus*, during interception of prey. General information about the echolocation sounds of different species of bats is given in Chapter 2 of this volume; however, several details about the broadcast waveforms are relevant to our discussion and need to be summarized here. The signals in Figure 4.2 are FM sounds, with two

FIGURE 4.1. A big brown bat, *Eptesicus fuscus*, is about to capture a mealworm hanging on a fine filament. The bat uses its tail membrane to seize the target. (Photograph by S. P. Dear and P. A. Saillant.)

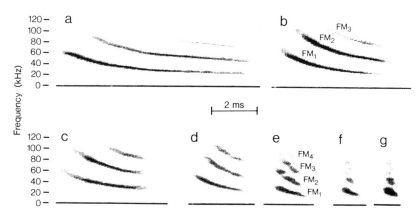

FIGURE 4.2. Spectrograms (*a–g*) of echolocation sounds broadcast by *Eptesicus* during flight toward prey. These frequency-modulated (FM) signals contain 3–4 harmonics (FM_1–FM_4) that collectively cover the frequency range of about 15–100 kHz. The bat progressively shortens its signals during the approach, with duration about equal to the two-way travel time of the sound, or echo delay.

prominent harmonics sweeping downward from about 55 kHz to about 20–25 kHz (the first harmonic, labeled FM_1 in Fig. 4.2*b*) and from about 100 kHz to 40–50 kHz (second harmonic, labeled FM_2 in Fig. 4.2*b*). Segments of the third harmonic (FM_3; Fig. 4.2*a–d*), and even the fourth harmonic (FM_4; Fig. 4.2*e–g*) are often present, too, at frequencies sweeping from 90 kHz to about 75 kHz (Griffin 1958; Simmons 1989; see also Fenton, Chapter 2). During the bat's approach to a target, the characteristics of the sounds being broadcast are adjusted from one transmission to the next according to the decreasing distance to the target, indicating that the bat monitors distance throughout its flight (Griffin 1958; Schnitzler and Henson 1980; Simmons 1989).

3. Acoustic Basis for the Bat's Sonar Images

3.1 The Echo Stream as Stimulus

When a bat transmits a sonar sound, the broadcast waveform travels outward into the environment to impinge on objects arrayed at different *distances* and then return as a series of echoes arrayed at different *times*. The image the bat perceives is a reconstruction of information about the *spatial* array of objects from the time series of echoes. Each reflecting point or surface returns a more-or-less complete replica of the incident sound back toward the bat's ears, with modifications caused by propagation through the air and also by the process of reflection itself. Echoes reach the bat's ears later than the transmission, at *delays* determined by target range, at *amplitudes* weaker than the transmission according to target size and distance, and with *spectral characteristics* determined by sound propagation in air, by the target's direction, and by its shape (Griffin 1958, 1967; Novick 1977; Pye 1980; Schnitzler and Henson 1980; Simmons 1989; Simmons and Kick 1984; Kober and Schnitzler 1990; Moss and Zagaeski 1994). The details of these acoustic effects constitute the physical basis for echolocation by defining the stimuli at the bat's ears.

The delay of echoes from targets at different ranges is 5.8 msec/m. An array of objects located at different distances from the bat yields a series of echoes at different arrival times, so that each of the bat's FM sonar transmissions is accompanied by a set of echoes that return within a specified window of time following the broadcast. For *Eptesicus*, with a maximum sonar operating range of about 5 m for insect-sized targets (Kick 1982), this time window is about 30 msec long. Physiological studies confirm that the auditory system of *Eptesicus* processes echo–delay information for arrival times to about 30–35 msec (Dear, Simmons, and Fritz 1993; Dear et al. 1993; see Simmons et al. in press). In the most general sense, the only information the bat has for reconstructing images of targets is the temporal sequence of echoes received by its two ears during this brief interval after each sound is broadcast. Consequently, the arrival time of echoes is the most important dimension for segregating individual reflections within the incoming stream of sound, which, in turn, makes the distance to objects the primitive perceptual quality underlying the images the bat perceives. To grasp the overriding significance of the arrival times of echoes in FM echolocation, the reader needs to know about some peculiarities of echoes formed by small objects in air.

3.2 The Inferential Nature of Echolocation

Echolocation is a classic example of an inverse problem: in sonar, targets are *inferred* from echoes using the properties of propagating sound as rules for reconstructing the locations and identifying features of scattering surfaces and points in space. The requirement that an inferential step formally intervene between the acoustic stimuli and the bat's images gives a sharp focus to our discussion of the perceptual mechanisms embodied in echolocation. For sonar in air (where all objects are significantly higher in acoustic impedance than the medium for propagation), a sonar target is a collection of one or more discrete points in space that reflect or scatter sounds back toward the sound source. In the process of reflection, the object becomes a source of sound itself; active sonar consists of inducing otherwise silent objects into returning echoes. Each reflecting point is called a glint, and in FM echolocation characterizing objects means locating and describing their constituent glints (Simmons 1989).

3.3 The Acoustic Nature of Targets

3.3.1 Targets as Groups of Glints

Contrary to the intuitive view of each target as a discrete object that reflects "an echo" that is straightforwardly separate from the echo reflected by another target, small objects such as flying insects behave as though they consist of two or more discrete reflecting points – that is, glints – which move to different locations within the target at different orientations as the target flutters and moves through the air (Simmons and Chen 1989; Kober and Schnitzler 1990; Moss and Zagaeski 1994). For such objects, there is no "target" apart from its constituent glints.

Figure 4.3 summarizes measurements of the echoes reflected by two types of targets that have been used as stimuli in important experiments with echolocating bats. One object is a mealworm, and the other object is a small plastic disk of roughly the same size as the mealworm (Simmons and Chen 1989). These targets are simplified versions of the more complex animated flying insects that bats encounter; they illustrate the principles of echo

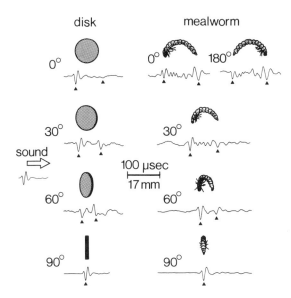

FIGURE 4.3. The acoustic nature of insect-sized targets in air (from Simmons and Chen 1989). A sonar sound (an impulse) is broadcast toward each target at different orientations (0°, 30°, 60°, 90°) in the incident sound field. The echo from the disk shows an impulse for the leading and trailing edges or glints at most orientations, and the echo from the mealworm shows an impulse from the head glint and the tail glint. As a first approximation, each target appears as a two-glint dipole whose most important feature is the separation of the principal glints along the axis of target range, or the time separation of the echo replicas from these two glints.

structure relevant to echolocation. Figure 4.3 shows the echo impulse responses for a mealworm and a disk at orientations of 0°, 30°, 60°, and 90° relative to the axis of the incident sound (an impulse containing the same frequencies as the bat's signals: labeled *sound* in Fig. 4.3). The disk yields a separate echo from its leading and trailing edges for all orientations except those very close to 0° or 90°, with a time separation between these echoes that corresponds to the range separation of the leading and trailing edges. The disk thus appears as an acoustic "dipole" target that reflects a two-glint echo. The target's shape consists chiefly of the momentary separation of its two glints along the range axis, with added secondary effects from more complex aspects of reflectivity (for disks, see Lhémery and Raillon 1994). The mealworm, too, returns two principal echoes at different delays from glints that correspond to its head and tail. At some orientations, weaker secondary echoes also are returned from other points along the mealworm's body, notably the legs and body segments. At most orientations, however, the mealworm is primarily a dipole target reflecting two-glint echoes. Consequently, the mealworm's most prominent shape feature is the separation of its head and tail along the range axis.

3.3.2 Flying Insects as Animated Glints

The echoes recorded from the mealworm and the disk (see Fig. 4.3) reveal a "quantal" character to the sounds reflected by these targets. The acoustic return from each object consists of several complete replicas of the impulse-like incident sonar sound arriving at slightly different delays. The leading and trailing edges of the target deliver the most prominent echo replicas, with finer structure in the target contributing a corresponding finer structure to the echoes between these principal reflections. It is evident that a complete description of these objects in terms of their echoes consists of the strength and temporal spacing of the individual reflected replicas of the incident sound. Echoes from flying insects contain added information, chiefly about the rhythmic motion of glints in relation to the beating of the wings at rates of about 10–100 Hz. Important information about the identity of prey can be obtained from wingbeat rate, which FM bats prove unexpectedly good at discriminating (Sum and Menne 1988; Roverud, Nitsche, and Neuweiler 1991; see Moss and Schnitzler, Chapter 3). Bats progressively increase the repetition rate of their sounds as they approach a target, from a rate of 5–10 Hz when at long range to as much as 150–200 Hz at close range (see Fenton, Chapter 2), and at some point in this progression the rate of emission momentarily matches the insect's wingbeat rate of 10–100 Hz. When this occurs, several successive echoes reveal the insect in a stroboscopically frozen posture, providing an opportunity to judge the target's shape independent of the wingbeat frequency itself (Feng, Condon, and White 1994; Moss and Zagaeski 1994; see Fig. 5D in Kick and Simmons 1984).

3.4 Structure of Echoes from Complex Targets

3.4.1 Overlap of Echoes from Glints

If the bat's sonar sounds were short-duration impulses such as the incident sound in Figure 4.3, the arrival of several echo replicas in a short span of time would be easy to recognize from their separate peaks in the received waveform. Instead, however, the bat's sonar sounds have relatively long durations of several milliseconds (see Fig. 4.2) compared to the short intervals of tens or hundreds of microseconds between echoes from different parts of the same target (see Fig. 4.3). Consequently, the waveform actually returned to the bat from the mealworm or the disk at different orientations is not as easily interpreted as the two primary "head" or "tail" impulses in the representative echoes in Figure 4.3 (Simmons 1989). The echoes reaching the bat's ears from these targets will indeed contain two primary reflections that are separated by a few tens of microseconds, but the individual reflected replicas will correspond to the long-duration FM sounds broadcast by the bat rather than short impulses. The waveform returning to the bat from a real object will contain several echo replicas that mix together to interfere (reinforce and cancel) at different frequencies according to the time separation of the replicas (Simmons et al. 1989; Mogdans and Schnitzler 1990; see Moss and Schnitzler, Chapter 3; also Fig. 21 in Dear et al. 1993).

3.4.2 Information on Time and Frequency Dimensions of Spectrograms

Figure 4.4 illustrates spectrograms of echoes composed of two reflected replicas of an incident *Eptesicus* sonar sound with a duration of 2.5 msec. These spectrograms display the principal information contained in two-glint echoes that have different time separations of the replica reflections from 0 to 600 μsec (for details about these waveforms, see Chapter 2; see also Altes 1980, 1984; Beuter 1980; Mogdans and Schnitzler 1990; Simmons 1992). These delay separations correspond to differences of 0–10 cm in the range spacing of glints, which covers the sizes of airborne objects encountered by bats, most insects having dimensions of a few centimeters at most. The broadcast signal for the echoes in Figure 4.4 contains two prominent harmonics sweeping from 55 to 25 kHz (FM_1) and from 90 to 50 kHz (FM_2), with a third, less prominent harmonic segment from 90 to 75 kHz (FM_3) (see 0-μsec echo at left in Fig. 4.4; this is the same as the incident sound).

In the two-glint echoes in Figure 4.4, the overlapping replicas of this incident sound interfere with each other so that the compound echo as a whole has a complicated spectrogram which is different from the spectrogram of the incident sound itself. In particular, although the two principal FM sweeps are visible in the echo spectrograms, they are streaked by horizontal bands at different spacings along the vertical frequency axis depending on the amount of two-glint echo–delay separation along the

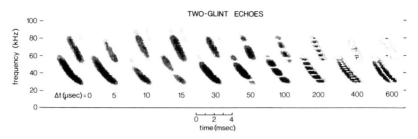

FIGURE 4.4. Spectrograms of two-glint echoes with different time separations of the reflected replicas of the incident sound, which is a 2.5-msec *Eptesicus* sonar signal (Simmons 1992). The incident sound is the same as the 0-μsec spectrogram (*left*). The integration time of these spectrograms is 350 μsec. Echo replicas that arrive separated by more than the integration time (400–600 μsec) are recognizable as separate FM signals, but echoes that arrive closer together (5–200 μsec) merge together; they interfere to form a single spectrogram with horizontal bands or spectral notches at frequencies determined by the reciprocal of the time separation. At short time separations, the only evidence that there are two echo components is this spectral structure.

horizontal time axis. In Figure 4.4, for delay separations of 400–600 μsec, there obviously are two echoes present. Each harmonic appears as a double FM signal with a short interval of time between the signals.

At these large echo–delay separations, the presence of two glints and their approximate spacing in time is immediately evident along the horizontal time axis of the spectrograms. In contrast, for two-glint echo–delay separations of 5, 10, 15, 30, 50, 100, and 200 μsec, the two separate echoes merge to form a single spectrogram. Each two-glint, compound echo contains just one FM sweep for each harmonic rather than a double sweep. Moreover, the appearance of these sweeps has been altered by the fact of overlap to form horizontal bands of interference at specific frequencies. These bands are notches in the spectrum of the echo, and their placement along the vertical frequency axis of the spectrograms is the only clue that the compound echoes really contain two overlapping components. For example, at a delay separation of 100 μsec these notches are spaced at frequency intervals of 10 kHz, while at a delay separation of 50 μsec the notches are 20 kHz apart ($\Delta f = 1 / \Delta t$).

The spectrogram for the two-glint echo with a delay separation of 400 μsec shows characteristics that are intermediate between the fully merged spectrograms of 5 - to 200-μsec echoes and the fully separated spectrograms of 400 - to 600-μsec echoes. At 400 μsec, information about the two glints is seen in both the time and frequency dimensions of the spectrogram, while at 400–600 μsec the information is entirely in the time separation of the sweeps and at 5–200 μsec it is entirely in the frequency spacing of the interference notches. Not only do double sweeps appear in this 400-μsec spectrogram, but a fine pattern of spectral notches also runs vertically along the frequency axis. This dual time–frequency quality of the spectrograms for two-glint echoes at time separations close to the integration time is also present in the bat's images (Simmons 1992), confirming the spectrogram-like time–frequency structure of the bat's auditory representation of echoes (see also Simmons, Moss, and Ferragamo 1990).

4. Reconstructing Target Scenes from Echo Streams

4.1 Implications of Spectrograms for Reconstruction of Glints

The crucial step for producing images in FM sonar is establishing an identity between each reflected version of the transmitted sound (each echo replica at different time separations in Fig. 4.4) and the corresponding glint that brought that echo replica into existence by reflecting or scattering the original broadcast sound back toward the receiver. Here, we consider an individual echo to be the echo replica reflected by a single glint. Because most real objects encountered by bats contain several glints, the target's "echo" in common parlance is really a "packet" of several discrete echo replicas of the incident sound that overlap each other to form a complex waveform (Beuter 1980; Altes 1984; Simmons 1989). The reason for making this distinction is not just that many targets consist of two or more glints.

Bats often encounter several targets together, located at about the same distance but in different directions. In this case, the bat would receive reflections from glints contained in different objects at about the same time. The waveform at the bat's ears still contains two or more overlapping echo replicas, as in Figure 4.4, but the "target" would not be a single two-glint object; it would be several objects. Because what would ordinarily be considered as "the echo" in fact contains several echo replicas from different glints located in different directions, the spatial scene composed of objects cannot be reconstructed from the acoustic stimulus composed of the whole echo packet without first decomposing it into its component echo replicas, one for each glint.

Accordingly, the fundamental role of the bat's auditory system for echolocation must be to recognize echo replicas as discrete components of sounds, even when they arrive so close together that their waveforms merge into a single spectrogram (see Fig. 4.4). Moreover, the fundamental limitation on echolocation performance must be the capacity to resolve two closely spaced replicas as separate reflections with distinct arrival times. However, the presence of two different kinds of information in echo spectrograms about the time separation of glints, separate spectrograms of each echo for long delay differences as opposed to bands of spectral notches for short delay differences (Fig. 4.4), means that inferences about glints must be carried out differently for long delay separations than for short delay separations (Simmons 1989). The location of the transition from the regime of representation that yields two separate spectrograms rather than a single, spectrally notched spectrogram evidently must be a fundamental characteristic of the bat's sonar receiver.

4.2 Auditory Integration Time and Segregation of Echoes

Bats frequently encounter echo replicas that overlap each other on reception and which must be separated to account for the bat's perceptions, that is, for behaviors demonstrably guided by perception of individual glints. The difficulty is, because echo replicas arrive at closely spaced delays, that an interval of time corresponding to the duration of a single sonar sound might have to be subdivided into a number of different estimates of arrival time for multiple, overlapping echo replicas. This might seem impossible; the duration of the sonar sound could be construed as the minimum interval of time for which a single delay value can be specified. This assumes, however, that the bat's ear records each echo in the manner of an oscilloscope, with one sweep of the screen covering the waveform of one transmission or one whole echo. This is equivalent to specifying the minimum integration time of echo reception as equal to the duration of the broadcast signal or one of its echoes (see discussion of integration-time measurements for bats in Chapter 3 by Moss and Schnitzler).

In fact, however, the bat's auditory system filters the wide span of frequencies covered in the FM sweep of each harmonic into numerous, narrower frequency segments (for details about tuning, see Chapter 6 by Covey and Casseday; for estimation of integration time, see Simmons et al. 1989; for discussion of temporal resolution, see Menne 1985; for auditory coding in relation to integration time, see Suga and Schlegel 1973). This auditory representation approximates a spectrogram, but with a frequency axis that is roughly hyperbolic instead of linear (see Simmons et al. in press).

As a result of auditory frequency tuning, an FM sweep that lasts several milliseconds is segmented into shorter pieces, each lasting for a fraction of the duration of the sound as a whole. The minimum integration time for echo reception corresponds to the duration of these segments (Beuter 1980; Altes 1984; Menne 1985; Simmons et al. 1989). This shortening of the effective duration of the FM sweeps is illustrated in Figure 4.4. The integration time of the spectrograms is the time axis width of the dark smear that traces each of the FM sweeps; it was set to be 350 μsec when the spectrograms were made. At a glint separation of 400 or 600 μsec in Figure 4.4, each of the echo replicas shows a distinct FM sweep for each harmonic, with no overlap of the segments at each frequency (even though the whole, unsegmented sweeps do overlap because the duration of the broadcast signal is 2.5 msec while the glint separation is only a fraction of a millisecond).

In contrast, at glint separations of 200 μsec or less, which are shorter than the width of the spectrogram smear, the two echo replicas merge to form a single, horizontally banded spectrogram. The effective duration of the bandpass-filtered segments of the sweep has been estimated from the sharpness of peripheral auditory tuning in *Eptesicus* and *Myotis* to be about 300–400 μsec (Simmons et al. 1989). Behavioral measurement of integration time in *Eptesicus* yields a value of about 350 μsec for double echoes to merge into a single echo (Simmons et al. 1989), which is in close agreement with neural integration time measured from recovery cycles to pairs of sounds or to amplitude-modulated sounds (Grinnell 1963; Suga 1964; Pollak and Casseday 1989; Covey and Casseday 1991; also see Covey and Casseday, Chapter 6).

In physiological terms, two separate sounds will evoke separate volleys of neural discharges as long as they are more than 300–400 μsec apart. Using integration time as a guide, bats should have no difficulty distinguishing two glints that are as close to each other as 5–7 cm in distance because their echo replicas will be 300–400 μsec apart, whereas glints that are closer together in range should be perceived as part of a single object because their echoes merge together into a single spectrogram (see Fig. 4.4). The difficulty with this simple interpretation of integration time is that a variety of behavioral studies unambiguously demonstrate that *Eptesicus* and other echolocating bats have little trouble separately perceiving the locations of glints that are closer together in range than 5–7 cm, which is the limit set by integration time alone.

4.3 Direct Psychophysical Measurements of Two-Glint Resolution

Several different experiments have been carried out specifically to assess the ability of *Eptesicus* to perceive echoes that arrive close together in time. Most of these experiments test the bat's ability to discriminate between two-glint echoes and single-glint echoes without determining whether the bat actually perceives the second glint as such (see Moss and Schnitzler, Chapter 3). However, one group of experiments (Simmons, Moss, and Ferragamo 1990; Simmons 1992, 1993) explicitly evaluates whether the bat can assign a discrete arrival-time value to each of two closely spaced echoes. Figure 4.5 shows the performance of two *Eptesicus* in an experiment that compares two-glint echoes (0-μsec, 10-μsec, 20-μsec, or 30-μsec echo-delay separations, corresponding to glint separations of 0 mm, 1.7 mm, 3.4 mm, or 5.2 mm, respectively) with single-glint echoes at different arrival times (using jittered echoes; see Simmons 1993; Simmons et al. 1990; see also Moss and Schnitzler, Chapter 3). The bats in Figure 4.5

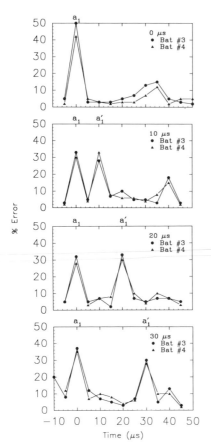

FIGURE 4.5. Performance of *Eptesicus* (bat #3 and #4) at perceiving the arrival times of two closely spaced echoes with delay separations of 0, 10, 20, or 30 μsec (from top to bottom) (Simmons 1993). In each case, the bats make a significant number of errors when the experimental probe echo (a_2) aligns with the delay of either the first glint (a_1) or the second glint (a_1') in the test echo. These error peaks indicate that the corresponding test glint has been encoded in terms of echo arrival time.

make significantly more errors in their discrimination when the arrival time of the single-glint echoes coincides with the arrival time of either the first-glint echo (a_1) or the second-glint echo (a_1'). Thus, the error peak at 0 μsec in Figure 4.5 corresponds to the bat's perception of the leading glint (a_1) at 3.2 msec (see legend), and the error peaks at 10, 20, or 30 μsec correspond to perception of the second glint (a_1') located slightly further away. The bats' errors indicate that both of the two-glint echoes were encoded in terms of their arrival times, even though the echo replicas for the individual glints arrive so close together in time that they merge into a single, spectrally complex spectrogram (see 10-μsec and 30-μsec spectrograms in Fig. 4.4). The results shown in Figure 4.5 demonstrate that the bat can translate the subtle changes in the spectrum of echoes caused by interference within the integration-time window to infer the value of the time separation of the echo replicas reflected by the two glints, so that both the first and the second replicas can have a known arrival time (see Menne 1985 for the concept of resolution and its enhancement; see Saillant et al. 1993 for deconvolution enhancement in echolocation).

The curves in Figure 4.5 show that *Eptesicus* has no difficulty perceiving the second of two glints (both error peaks are the same height) when the overlapping echoes arrive as close together as 10, 20, or 30 μsec. Evidently, the bat's limit of resolution is smaller than 10 μsec, or 1.7 mm in range. Figure 4.6 shows the performance of as many as three bats in a series of two-glint experiments with echo–delay separations of 10 μsec, 5 μsec, 3.8 μsec, 2.6 μsec, and 1.4 μsec (Saillant et al. 1993). The bats all show a pattern of errors similar to Figure 4.5, with a peak at 0 μsec (actual delay is

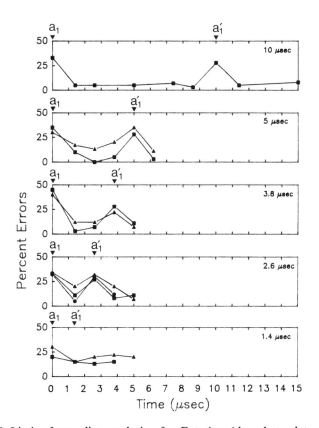

FIGURE 4.6. Limit of two-glint resolution for *Eptesicus* (three bats; data points are *squares, triangles, circles*). The bats perceive two arrival times for echo–delay separations of 10, 5, 3.8, and 2.6 μsec, (from top to bottom) but they perceive only one arrival time for a delay separation of 1.4 μsec (Saillant et al. 1993). The two-glint echo-resolution threshold is about 2 μsec, which corresponds to a separation in range of about 0.3 mm.

3.2 msec) for the first glint and a peak at 10, 5, 3.8, or 2.6 μsec for the second glint. At a separation of 1.4 μsec in Figure 4.6, however, the bats no longer register an error peak for the second glint, indicating that the second glint's echoes were not assigned a value of arrival time distinguishable from the arrival time of echoes for the first glint. From Figure 4.6, the limit of two-glint resolution in *Eptesicus* must be about 2 μsec (see also Simmons et al. 1989). This is an extraordinarily fine resolving power: first, because it is equivalent to only about 0.3 mm in range, and second, because it is shorter than the periods for any of the frequencies in the bat's sonar signals. *Eptesicus* broadcasts sounds containing frequencies approximately from 20 to 100 kHz (see Fig. 4.2), which have periods from 50 μsec down to 10 μsec. Resolution carried out within the time window of the shortest period in the sounds is within the Rayleigh limit for the signals themselves, which is a significant achievement that amounts to extrapolation of the broadcast spectrum to frequencies higher than those actually contained in the bat's sounds (Saillant et al. 1993).

4.4 Is There Other Evidence for Fine Temporal Acuity in Bats?

The ability of bats to separately perceive the arrival times of two echoes that reach the ears as close together as 2–30 μsec (Figs. 4.5 and 4.6) demonstrates that they are very good at segmenting the stream of incoming echoes following each broadcast into a series of discrete arrival-time values (see also Simmons 1992). The bat's 2-μsec limit of two-glint resolution is, of course, far shorter than the duration of its echolocation sounds, which is several milliseconds (Simmons et al. 1990; Simmons 1993), and it is shorter, too, than the integration-time window of 300–400 μsec. Consequently, the bat must be able to "read" spectrograms to determine time separations from information contained in their frequency axis as well as in their time axis.

The fine temporal acuity shown here, and the dual time-frequency process that must underlie it, have been considered physiologically improbable from the perspective of certain types of single-unit data (e.g., Pollak et al. 1977), leading to rejection of the types of behavioral experiments that generate such results (Pollak 1993). It therefore is important to know whether other types of behavioral experiments, whose results are widely accepted, might in fact conceal independent evidence for fine temporal acuity in echolocation. The following sections review the implications for sonar imaging of several different kinds of experiments that are considered classical in their impact on documenting the existence and natural history of echolocation, but that have not previously been brought to bear on determining the composition of the images themselves.

5. Performance of Echolocation in Natural Tasks

5.1 Interception of Targets in Simple and Complex Situations

5.1.1 Pursuit of Prey

In Figure 4.1, the bat's flight to the target, its detection of the target, and its identification of the object as potential prey are all guided by sonar (Schnitzler and Henson 1980). The behavior of bats during interception is surprisingly stereotyped, with a stable pattern of sonar emissions (see Fig. 4.2) and other responses occurring at regular points along the flight path (see Fenton, Chapter 2; also see Kick and Simmons 1984; Simmons 1989). The bat aims its mouth (the transmitter) and its external ears (the receiving antennas) at the target during the approach and will seize the target in its tail membrane at the moment of capture (Webster and Griffin 1962).

Figure 4.7 gives two examples of the bat's approach to a target in video motion-analysis studies of the interception process. These observations were initiated to evaluate the information the bat uses to control its approach flight (Lee et al. 1992), and they yield a wealth of data to reconstruct the stimuli reaching the bat's ears and the images it perceives each time a sound is broadcast (Saillant et al. 1993). In Figure 4.7, the bat is shown as a series of stick figures that trace its flight to the mealworm over the last 80 cm of the approach. Each stick figure depicts the bat's location and posture in a three-dimensional reconstruction of the bat's flight path taken with two synchronized video cameras at a frame rate of 60 Hz. In Figure 4.7A, the bat flies directly up to a single mealworm (mw) suspended on a fine thread and seizes the mealworm in its tail membrane before eating it. In Figure 4.7B, the bat is given a choice of two mealworms (mw 1, mw 2) hanging on threads; in this example the bat captures the target located on its right slightly farther away (mw 2). It approaches to seize this mealworm from its thread while at the same time withdrawing its right wing to avoid striking the thread holding the other mealworm (mw 1). These studies reveal no important distinctions in the acoustic behavior of bats approaching a single target as opposed to two targets.

5.1.2 Duration of Echolocation Signals for *Eptesicus*

For present purposes, the *duration* of the individual FM sonar sounds is their most critical feature for constraining subsequent processing of echoes to form images of targets. The duration of the signals shown in Figure 4.2 (*a–g*) progressively decreases from 8 msec to less than 0.5 msec as the bat approaches nearer to the target and the delay of echoes becomes shorter. At

A

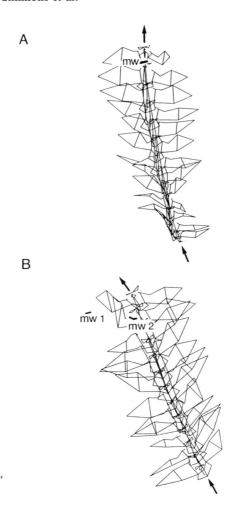

B

FIGURE 4.7A,B. Reconstruction of the flight path and posture of *Eptesicus* during captures of a mealworm hanging on a fine filament (see Fig. 4.1). The stick figures view the bat from the front based on 60-Hz video motion-analysis images as it flies up to the mealworm and seizes it in the tail membrane. In A, the bat captures a single mealworm (mw). In B, the bat captures one of two mealworms (mw 2) while simultaneously withdrawing its right wing to avoid striking the filament holding the other mealworm (mw 1). *Arrows* show direction of flight.

any point in the bat's approach to a target, the duration of the transmitted sound usually is only slightly shorter than the two-way travel time of the echo. Consequently, the pathlength to and from the target is nearly filled up with each sound (Hartley 1992; Saillant et al., in manuscript). Most experiments on echolocation focus on questions about what the bat perceives when it is at distances of about 0.5–1 m from targets, when the sonar transmissions typically are nearly 3–6 msec in duration. With such long durations for the incident sonar sounds, the individual echo replicas from two different parts of the same target (the mealworm's head and tail in Fig. 4.7A) or from different parts of two closely spaced targets (heads and tails of both mealworms in Fig. 4.7B) overlap each other for most of their duration.

5.1.3 Acoustic Stimuli During Capture of Tethered Mealworms

In Figure 4.7A the mealworm returns an echo packet composed of two overlapping echo replicas of roughly equal strength from its head and its tail (see Fig. 4.3). The maximum head-to-tail separation of a mealworm is about 17 mm, so the largest time separation of echo replicas from the head and tail glints will be only about 100 μsec. Thus, as the bat approaches the target in Figure 4.7A, it will receive echoes corresponding to examples of 5- to 100-μsec glint separations in Figure 4.4. In Figure 4.7B, the two mealworms collectively return an echo packet composed of four overlapping echo replicas of roughly equal strength from their heads and tails (see Fig. 4.3). The head-to-tail separation within each mealworm will be less than 17 mm, but the separation of the heads and tails between mealworms depends on the difference in distance from one mealworm to the other as well as on their orientations. In Figure 4.7B, the overall range difference is only about 2 cm, so the head of one mealworm could easily be at the same distance from the bat as the tail of the other.

Because each mealworm really is a dipole (Fig. 4.3), the spacing of the glints frequently will place two parts of different targets in closer range proximity than *two parts of the same target*, so that the objects cannot be identified along the range axis simply from which glints are nearer each other and which glints are farther apart. Consequently, the bat cannot select one of the two targets for capture on the basis of the echo packet alone; all the echo replicas from both targets arrive within a single integration-time window and will merge into a single, spectrally complex packet.

During its approach to the two mealworms in Figure 4.7B, the bat will receive echoes corresponding to the examples of 5-μsec to 100-μsec two-glint echoes (see Fig. 4.4) for *intra*target spacings mixed with examples of 5 μsec to about 300–400 μsec for *inter*target spacings. To perceive the mealworm in Figure 4.7A as an object composed of two principal glints, the bat has to decompose the pair of overlapping echo replicas, typically separated by 5 μsec to no more than 100 μsec, from the head and tail (Simmons and Chen 1989) and assign each echo replica a specific, and different, arrival time. To be sure, the bat does not actually have to perceive that there are two glints in the echo from the mealworm, only locate the mealworm well enough to intercept it.

In stark contrast, however, to perceive the two mealworms in Figure 4.7B as separate objects and guide its flight to intercept just one of them, the bat does have to decompose the series of four overlapping echo replicas from the heads and tails of both mealworms and assign each one a specific arrival time. In this case, perception of the glint structure of the target scene is necessary to locate one target in the presence of the other because the echo packet by itself portrays a nonexistent target in the wrong location. Moreover, the segregation of individual echo replicas in Figure 4.7B has to precede regrouping of replicas into pairs corresponding to the two mealworms because the bat has to avoid incorrectly associating the head of one mealworm with the tail of the other. Such a misassociation would lead the bat to attack a much lengthened "phantom mealworm" located halfway between the two real mealworms — an error of orientation that is not observed.

5.1.4 Interception of Targets in Clutter

Eptesicus and other FM bats routinely intercept prey in open spaces and clearings, with just the insect near the bat reflecting echoes that arrive as isolated events at the bat's ears (see Fenton, Chapter 2). Other objects such as vegetation or the walls of buildings are located at least a meter or two further away than the insect. However, once an insect is detected, the bats are often capable of following the target into the thick of vegetation, tracking it through an environment of leaves and branches located at similar distances all around (Webster and Brazier 1965; Webster 1967).

Figure 4.8 shows a multiple-exposure photograph of the little brown bat, *Myotis lucifugus* (a close relative of *Eptesicus*), pursuing a moth through branches. The bat is shown in numbered images (#1–8) at strobe-flash intervals of 100 msec, with the corresponding image of the moth connected to each image of the bat by a dashed line (#1–3). In image #1 the bat is about 50 cm from the moth, and by image #3 (200 msec later) it has closed to within 10–15 cm. At this point, the background vegetation has components that rival the moth in being at similarly close range, and the overall strength of the echoes from all the vegetation probably exceeds the strength of the echoes from the moth.

In Figure 4.8, the bat clearly has the moth in its mouth in image #6 and probably is already in contact with the moth in image #5. In spite of the numerous extraneous echoes from many branches that might confuse perception of echoes from the moth, the bat comes out of the clutter with the moth in its mouth (images #7–8). Not only does the bat intercept the moth, but it avoids collisions with the vegetation located all around its flight path. The clutter does not prevent the bat from tracking the moth to a successful capture in much the same manner as would have occurred out in the open, but the clutter probably constrains the bat's flight path while capture takes place (Webster 1967). Figure 4.8 documents the bat's ability to segregate the echo replicas created by the insect from what might seem to be an overwhelming number of interfering replicas from objects located at

FIGURE 4.8. Multiple-exposure photograph of *Myotis lucifugus* pursuing and capturing a moth that flies into vegetation during the bat's interception maneuver (Webster and Brazier 1965). Each numbered image of the bat is connected to the corresponding image of the moth with a *dashed line*.

such a wide range of different distances that there would always be some echo components from the background arriving at about the same time as the echo from the insect.

5.2 Avoidance of Obstacles

5.2.1 Wire-Avoidance Tests

Demonstrations of the ability of FM bats such as *Eptesicus fuscus* or *Myotis lucifugus* to perceive the locations of several targets at once are not confined to photographs or videorecordings of bats intercepting insects in clutter. Echolocation was first shown to exist by flying bats through arrays of obstacles and recording their sounds while they actively maneuvered past the obstacles, and most of the characteristics of echolocation (e.g., use of FM or constant-frequency [CF] sounds by different species; changes in the duration and repetition rate of sonar signals during approach) can be observed in this type of task as well as in pursuit maneuvers (Griffin 1958; Grinnell and Griffin 1958; for obstacle avoidance by *Eptesicus*, see Jen and Kamada 1982; for further reviews, see Novick 1977; Schnitzler and Henson 1980).

Figure 4.9 shows two examples of three-dimensional reconstructions of the flight of *Eptesicus* through an array of vertically stretched wires (0.7 mm diameter) with a spacing of 25 cm between wires (the bat's maximum wingspan is about 30 cm). In each case, the bat is viewed as approaching the observer, with about 80 cm of the bat's flight path shown by the stick figures. In Figure 4.9A, the bat flies up to the middle pair of wires (wires b and c) and passes more or less directly between them, tilting its right wing up and left wing down during the most critical part of its passage between the wires. At each point in its flight the bat is approximately equidistant from the wire on its left (c) and the wire on its right (b). Measurements made from the three-dimensional reconstruction of the flight in Figure 4.9A reveal that the difference in distance from the bat to the left and the right wires is no greater than 2 cm throughout the bat's approach.

In Figure 4.9B, the bat flies almost directly toward one of the wires (wire b) while keeping approximately equidistant from the next two wires (wires a and c). Just before it would have collided with the wire located to its front (b), the bat swerves sharply to its right while sharply tilting its right wing up and its left wing down to avoid striking the wire. The bat slides past the nearest wire (b) with a distance of about 2–3 cm to spare at its closest point. Significantly, the bat remains about the same distance from the wire located on its left (c) and the wire located on its right (a) even while coping with the presence of the wire located straight ahead (b), halfway between the two other wires (a, c). Until the bat swerves in flight, the difference in distance from the bat to these other wires (a and c) is only about 2–3 cm.

5.2.2 Acoustic Stimuli During Obstacle Avoidance

Each of the wires in an obstacle array returns an echo replica to the bat from the point along the wire where the wire is perpendicular to an imaginary line connecting the bat's head and ears to the wire itself. That is, each wire acts as though it consists primarily of a single glint located at this point of perpendicularity. Changes in the bat's height during flight merely move this principal glint along the wire to new locations. Because the wires are stretched to form parallel elements of an array, the array appears to the bat as a series of point reflectors located at distances and directions

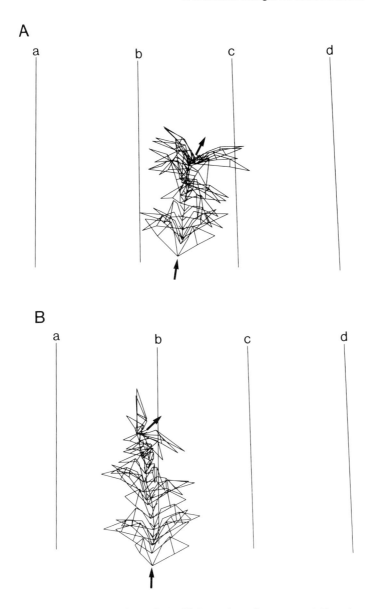

FIGURE 4.9A,B. Reconstruction of the flight path and posture of *Eptesicus* during obstacle-avoidance flights. The bat is viewed from the front as it approaches and avoids hitting the wires *(a-d)*. In A, the bat passes between two adjacent wires; in B, the bat heads straight for one wire but dodges around it at the last moment before colliding. *Arrows* show direction of flight.

corresponding to the positions of the wires in the horizontal plane relative to the bat.

From observations of its flight path during obstacle-avoidance tests, the bat must frequently have selected which wires to pass between by the time it has approached to a distance of 0.5–1 m. The angular separation of the adjacent wires thus is only about 15°–30° when the bat has made at least a preliminary evaluation of their locations. From recordings of the sounds during approach to the wires, the bat progressively shortens its sonar signals and increases their repetition rate, giving further evidence that it is monitoring the location of the obstacles as it flies nearer (see Griffin 1958; Novick 1977; Schnitzler and Henson 1980; for *Eptesicus*, see Jen and Kamada 1982). Crucially, the duration of each signal is far larger than the relatively short interval that separates echoes from the wire on the left and the wire on the right.

The chief factor determining the structure of the stimuli is the distance from the bat to each wire. The bat often makes a nearly perpendicular approach to the plane of the wires (Fig. 4.9A), and their angular separation of only about 15°–30° at distances of 0.5–1 m results in range differences being only a few centimeters. Echoes from two adjacent wires thus will overlap each other because they are at approximately the same distance from the bat, one on the right and one on the left. Consequently, during the bat's approach to the obstacle array, it will receive echoes resembling the examples shown in Figure 4.4. Just which of the two-glint examples actually corresponds to the echo received from the two nearest wires following a given sonar emission depends on how close the bat comes to flying along a path that keeps it equally far from each wire. From one sonar emission to the next, the difference in range from one wire to the other is no greater than several centimeters, so the difference in the arrival time of their echoes is nearly always less than several hundred microseconds and often only a few tens of microseconds.

In contrast to the small delay separation of echoes, however, the durations of the sounds themselves are several milliseconds, so the echoes overlap each other when they reach the bat's ears. Echoes from wires located further away on the left and right arrive later, too, and these will overlap with each other as well as overlap with the echoes from the nearest wires. The delay separation of echoes from the nearest pair of wires and the next nearest pair depends on how far the bat is from the array. For example, in the experiments with *Eptesicus* shown in Figure 4.9, when the bat is 1 m from the array as a whole, the nearest wires on the left and the right are about 101 cm away and the next nearest wires are about 107 cm away, for a difference in range of only 6 cm. Thus, the delay separation of the two sets of echoes is only about 350 μsec, and the delay separation of the echoes from the single wire on the left and right in each set will be smaller still.

5.2.3 Implications of Stimuli for Perception of Wires

As a consequence of the acoustic geometry of the parallel vertically stretched wires, following each broadcast the bat receives a packet of several echo replicas from the left and several echo replicas from the right, with the overlap times as small as a few tens of microseconds for corresponding pairs of wires (the nearest wires, the next nearest wires, etc.) located approximately equidistant from the bat on the left and the right. Because the echoes from each pair of wires will have a single spectrogram with interference notches along the frequency axis (see Fig. 4.4), an inference based on the undifferentiated echo packet from the wire on the left and the wire on the right would indicate the existence of only a single, nonexistent "target" located midway between the two real wires.

The bat cannot locate either of the real wires without decomposing the spectrogram into estimates of the arrival times of both echo replicas. The bat's ability to avoid the obstacles, by dodging between the adjacent wires in Figure 4.9A and by swerving to pass close to the wire located straight ahead in Figure 4.9B while also receiving echoes from the wires located further to the right and left, demonstrates that it can segregate the incoming stream of overlapping echoes which have spectrograms of the type shown in Figure 4.4 into estimates of the arrival time of each echo replica so that a distance and direction can be attributed to each wire.

5.3 Discrimination of Airborne Targets

5.3.1 Capture of Mealworms and Rejection of Spheres and Disks

There is yet another naturalistic task used with echolocating bats that gives evidence for the bat's ability to segment a stream of echoes into closely spaced echo replicas. In a series of very important experiments, flying bats (the FM bats *Myotis lucifugus* and *Eptesicus fuscus*) have been trained to discriminate between targets of different shapes thrown up into the air to tumble along trajectories that the bat can intercept (Griffin, Friend, and Webster 1965; Webster and Brazier 1965). One of the targets is an edible mealworm, which the bat is trained to catch, while the other targets are inedible plastic spheres or disks of different sizes, which the bat learns to reject. Figure 4.10 shows a series of drawings made from a multiple-exposure stroboscopic-flash photograph of a discrimination trial with *Myotis* intercepting a mealworm thrown into the air along with two small disks (Webster and Brazier 1965). The images of the bat (B_3, B_4, B_5), the mealworm (M_3, M_4, M_5), and the two disks (a_3, a_4, a_5; b_3, b_4, b_5) are numbered to correspond to each other at strobe-flash intervals of 100 msec. Each frame in the drawing contains a square marking the location in space that will be occupied by the bat's mouth just before capture; the trajectories of the targets can be judged from this fixed reference point, along with the

170   James A. Simmons et al.

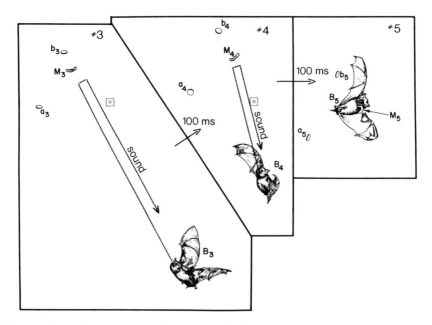

FIGURE 4.10. Drawings made from a multiple-exposure photograph of *Myotis* discriminating an airborne mealworm (M) from two disks (*a, b*). The original drawing (Webster and Brazier 1965) contained six strobe images 100 msec apart; here, the third, fourth, and fifth images have been separated into panels (#3, #4, #5) to show the image of the bat (B_3, B_4, B_5), the mealworm (M_3, M_4, M_5), and each disk (a_3, a_4, a_5; b_3, b_4, b_5) at their corresponding locations. A *small square* containing an *open circle* has been added to show the location that will be occupied by the bat's mouth in image #5. The approximate spatial extent of the bat's sonar sounds has been plotted as a *dashed line (sound)* based on durations of sounds recorded in these experiments.

bat's approach path. In tests of this kind, bats achieve 80%–90% correct captures of the mealworm.

5.3.2 Acoustic Stimuli During Airborne Discrimination Tests

Each frame in Figure 4.10 contains a solid line (labeled sound) going from the bat's mouth to the targets and back again. This line traces the approximate spatial extent of the bat's sonar signals from the average duration of sounds recorded at each distance from the mealworm. The bat's sounds have a duration of several milliseconds in time that spreads their waveforms in space over most of the travel path from the bat out to the target and back again. Note, however, that the mealworm and the disks are located at about the same distance from the bat and are separated by differences in distance that are small compared to the total distance to any one of these targets. Consequently, their echoes are separated by an interval of time smaller than the duration of the incident sonar sounds. The echoes from the mealworm and the disks thus arrive at the bat's ears in the form of one continuous, overlapping sound several milliseconds long. Nevertheless, as in the case of obstacle-avoidance experiments, the bat perceives the locations of several targets at once, but now the bat also perceives enough additional information to identify the mealworm and the disks even though their echoes do overlap.

The bat's sonar sounds return from the mealworm and the disks as discrete echo replicas at time separations for different target orientations given by the impulse reflections in Figure 4.3. The time separation of the echo replicas from different glints within a mealworm or a disk is in the region of 0–100 μsec, with typical separations of only a few tens of microseconds at most orientations. Consequently, the spectrograms of echoes with glint separations of 5–100 μsec in Figure 4.4 are typical of the stimuli returned by a single mealworm or disk. If a mealworm or a disk were presented alone, with no other targets near by, the echoes from the mealworm or the disk should differ enough in their interference spectra that the bat ought to perceive this difference after learning to compare the targets (Simmons and Chen 1989). The bat receives as many as a half-dozen echoes during its approach, while the target rotates as it soars along its trajectory to expose different orientations, and, taken together, these echoes provide enough information for the bat to determine whether the target is a mealworm or a disk on any particular trial (Griffin 1967).

The critical point is that for sequential presentation of an isolated mealworm on one flight and a disk on the next, the bat does not need to perceive the individual glints in the target, just the interference spectrum of the echoes from each target separately. That is, when the targets are considered one at a time, the echo interference spectrum is a plausible basis for the bat's discrimination of mealworms from disks (Griffin 1967; Simmons and Chen 1989). The problem is that the bat can perform the discrimination even when a mealworm and several disks are presented simultaneously, at roughly the same distances, on the same trial, as in Figure 4.10. In many trials a mealworm and one or more disks will follow trajectories that place them at about the same distance from the bat, so that the time separation of echo replicas from different glints between the targets will often be as small as the time separation of echo replicas within each target.

Figure 4.11 shows examples of the bat's approach to a mealworm and a lump of clay hanging on fine threads during six flights in discrimination tests for video motion studies. During its approach, the bat is capable of determining which target is the mealworm and capturing it even when the targets do not turn and tumble, as they do in airborne tests, and when the two targets are always within a few centimeters of the same distance from the bat (see also Fig. 4.7B). These examples serve to confirm that the bat can locate and identify one target from echoes that arrive virtually

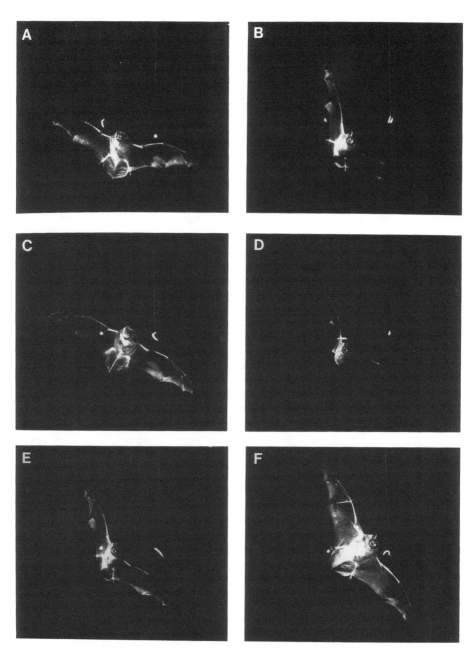

FIGURE 4.11A–F. The approach of *Eptesicus* to a mealworm and a small lump of clay from six flights in which the bat caught the mealworm. In each example, the bat's flight path kept it roughly equidistant from the two targets during the attack. (Photographs by S. P. Dear and P. A. Saillant.)

completely overlapped with echoes from another target located only a short distance away.

Figure 4.12 schematically illustrates the worse-case acoustic situation that often prevails in simultaneous discrimination trials with airborne targets. As the bat approaches a mealworm and a disk, which are thrown upward together and frequently remain close to each other throughout their brief flight, it receives echoes of about the same strength from two primary glints in both targets (a, b in mealworm; a, b in disk). Each target in Figure 4.12 consists of two discrete glints (a, b), each of which reflects a replica of the incident sound, so the whole echo packet contains two components from the mealworm and two from the disk all arriving at only slightly different delays. These four echo replicas overlap each other almost completely because the bat's sonar broadcasts are comparable in duration to the two-way path delay of echoes (sound lines in Fig. 4.10), which amount to several milliseconds, not to the much smaller delay separation between echoes from the two glints in each target (tens of microseconds between a and b in the mealworm or a and b in the disk) or the delay separation

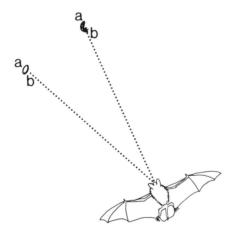

Figure 4.12. Diagram of worse-case situation in a bat's approach to a mealworm and a disk. The two targets are equidistant from the bat and are separated by an angle that keeps both targets within the directional beam of the sonar (see Fig. 4.13). The principal echo sources are the leading *(b)* and trailing *(a)* edges of the disk and the head *(b)* and tail *(a)* of the mealworm (see Fig. 4.3). Because the head of the mealworm and the leading edge of the disk are closer together in range than the tail is to the head or the trailing edge is to the leading edge, the bat will receive echo replicas that arrive closer together in time from glints *between* targets than *within* targets. To avoid misperceiving the two nearer glints as being part of the same target, the bat has to segregate all four echo replicas and perceive their locations in space before reassembling them into objects and choosing the one to attack.

between echoes from glints in different targets (tens or hundreds of microseconds between a in the mealworm and b in the disk, for instance).

The spectrograms in Figure 4.4 show examples of two-glint echo packets similar to the overall echo from a mealworm or a disk; the echo packet from a mealworm and a disk together is similar but more complex, with horizontal interference notches at spacings determined by the arrival time differences of all four principal glints from both targets. In situations such as shown in Figure 4.12, the echo packet taken as an undifferentiated whole has an arrival-time corresponding to a nonexistent "target" located about in the center of the volume of space collectively occupied by the four principal glints (a, b) in the mealworm and the disk. Nevertheless, the bat nearly always goes for the mealworm and approaches its location correctly, with no evidence for significant distortion of its perceived location as a consequence of the disk's presence close by. Moreover, the worse-case example in Figure 4.12 is a simplification because bats can successfully discriminate a mealworm from more than one disk presented on each trial (see Fig. 4.10). Figure 4.7B shows what might in fact be the most difficult situation of all — the bat locates and intercepts one mealworm in the presence of a second mealworm at about the same distance. Although each mealworm consists primarily of a head glint and a tail glint whose echoes all overlap when they reach the bat's ears at the about the same time, the bat nevertheless can separate the echo replicas into a pair for one mealworm and a pair for the other. The conclusion seems inescapable that the bat perceives the echo packet in terms of its constituent echo replicas, each with a discrete arrival time at each ear, to determine the location and identity of the mealworm.

6. Inferences About Images from the Bat's Performance in Naturalistic Tasks

6.1 Evidence for Decomposition of Targets into Glints

The performance of bats in several well-known, naturalistic tasks involving echolocation reveals that the bat can separately perceive the locations of multiple objects which are present together in the bat's sonar "field of view." To perceive the locations of different targets, the bat necessarily has to segregate the echoes returning from these objects so that each target can be located from its own specific reflections. Because the echoes from different objects often arrive within 300–400 μsec of each other, they overlap to form echo packets (see Fig. 4.4) that must be "unpacked" to reveal their composition in terms of echo replicas. This step is necessary because the characteristics of the echo packet betray no useful information about the individual objects, but instead about a "phantom" object located approximately at the "center of gravity" for the cluster of glints that collectively make up these objects.

The bat's ability to react to the location of one target when other targets are present at about the same distance means that the echo packets have been broken down into their constituent replicas so that the glints they represent can be separately perceived at their corresponding locations in space. However, information about the presence of different glints is distributed across more than one dimension of the spectrogram representation of echoes (see Fig. 4.4; Simmons 1989, 1992; see Moss and Schnitzler, Chapter 3). Some echo replicas can be segregated by the occurrence of discrete FM sweeps (glint separations greater than 300–400 μsec in Fig. 4.4), but others can only be segregated by recognizing that the notches in the spectrogram signify closely spaced glints (glint spacings of 5–200 μsec in Fig. 4.4). The boundary between separate registration of two echo replicas in time and joint spectral representation of overlapping replicas that are smeared together is fixed by the value of 300–400 μsec for the integration time of echo reception by *Eptesicus*, which also was used for the spectograms in Figure 4.4 (Beuter 1980; Altes 1984; Menne 1985).

If bats can perceive closely spaced glints, as obstacle-avoidance experiments and experiments on interception and discrimination of targets seem to demonstrate, they must be able to use both time axis registration and spectral interference notches to determine the time separation of overlapping echo replicas for reconstructing the spatial separation of the glints themselves in the final images (Simmons 1989). In particular, translation of the spectrum of overlapping echo replicas into arrival time estimates is essential for segregation of echoes from different objects if these objects are located at approximately the same distance from the bat (Saillant et al. 1993).

6.2 Echo–Delay Acuities Estimated from Performance

6.2.1 Minimal Requirements for Single-Target Interception

Most targets of interest to bats are small, with spatial extents (distances between glints) of no more than a few centimeters. Consequently, they reflect several echo replicas that arrive within an interval of no more than 100–200 μsec at the bat's ears. These replicas are blurred together into a single echo packet because the arrival times of the replicas differ by less than the integration time of echo reception, which is 300–400 μsec in *Eptesicus*. In Figure 4.4, the spectrograms for two-glint echoes with time separations of 5 μsec to 100–200 μsec are typical of the echoes encountered from single targets; actual echoes from flying insects have a similar appearance but are rhythmically animated (Kober and Schnitzler 1990; Moss and Zagaeski 1994).

If there is only one object near the bat to reflect echoes, the packet of sound it reflects probably is a good guide to the location of this target's "center of gravity," which is enough to approach the target and seize it in the tail membrane (see Fig. 4.1). With only one target present in the bat's "field of view," relatively uncomplicated echo-processing mechanisms are capable of tracking the target and guiding the bat's flight to a successful interception (Kuc 1994). Except for the fact that bats regulate the duration and repetition rate of their transmissions according to target range, determination of distance is not even formally required to accomplish interception (Kuc 1994), although, if echolocating bats can routinely intercept flying prey without accurate knowledge of target range, they must be "very lucky" (Grinnell 1967). From observations of the bat's reaching response at the moment of capture, an echo–delay accuracy of 50–100 μsec would be adequate to complete most interceptions (see Simmons 1989).

Estimates of the bat's best possible echo–delay acuity from the jitter observed in neural response latencies of single cells is also in the region of 50–100 μsec (Pollak et al. 1977; Bodenhamer and Pollak 1981; Covey and Casseday 1991). Furthermore, the bat does not need to perceive the separate glints in a single target merely to intercept it, so the ability to resolve two echo replicas as distinct could be limited to 300–400 μsec, the integration time of echo reception, without having any particularly deleterious effects on successful captures of isolated targets.

Although the target's spatial extent (separation of its glints) would not be perceived explicitly in spatial terms, the spectrum of the echo packet created by interference between the echo replicas from different parts of the target would be sufficient to distinguish one shape from another in many circumstances, especially for flying insects with different wingbeat rates (Simmons and Chen 1989; Kober and Schnitzler 1990; Neuweiler 1990; Schmidt 1992; Feng, Condon, and White 1994; Moss and Zagaeski 1994). This account of the physiological limitations on the content of the bat's images is widely accepted (Pollak et al. 1977; Pollak and Casseday 1989), but it nevertheless fails to account for obstacle avoidance or for interception in the presence of multiple targets, which are well-founded, classical observations about echolocation.

6.2.2 Minimal Requirements Implicit for Performance in Complex Conditions

To understand echolocation, the bat's performance cannot be characterized from observations made in the simplest cases; it has to be studied in realistically complicated situations with appropriately complex stimuli. Experiments have to be designed to engage whatever sophisticated mechanisms the bat might have for coping with complicated conditions, not just the minimal capabilities required to handle simple conditions. To account for the bat's ability to intercept one of two mealworms (see Fig. 4.7B), to capture an insect in clutter (Fig. 4.8), to avoid striking the adjacent wires in an obstacle-avoidance test (Fig. 4.9AB), to discriminate a mealworm from several disks (Fig. 4.10), or to choose a mealworm from a lump of clay when both are at the same distance (Fig. 4.12), it is not enough for the bat to perceive each packet of sound reaching its ears as an "echo" coming from a single "object." The bat has to segment the incoming stream of echoes into discrete estimates of the arrival time of each reflected replica of the broadcast signal (see Figs. 4.5 and 4.6). Without this explicit step, the bat cannot infer the locations of real objects in complicated environments, just "phantom" objects represented by the echo packets taken as whole units; the real objects are not accessible from the sounds at all except through the intermediary of their glints.

To account for successful obstacle avoidance on flights where the wires immediately on the left and right remain within roughly 5 cm of the same distance from the bat, the bat's ability to resolve two echo replicas as separate must be as good as 300 μsec, which is the lower limit of integration time and a feasible level of performance under the minimal assumptions given here that account for interception of single targets. It is also compatible with a straightforward interpretation of physiological recovery times for responses to the second of two sounds, which can occur for separations as short as 300–500 μsec (Grinnell 1963; Suga 1964; Pollak and Casseday 1989; Covey and Casseday 1991). However, if the bat keeps its flight path within 2 cm of the same distance to the wires, the bat's ability to resolve two echo replicas has to be be as good as 100 μsec, not 300 μsec. This figure is several times better than the minimum separation explainable from integration time alone. Furthermore, video tracking of bats during obstacle flights indicates that avoidance is successful even when the two nearest wires are as little as 1 cm apart in range, so the bat's ability to separate two overlapping echo replicas must be better than 50–60 μsec.

The earliest physiological studies obtained values of about 500 μsec for minimum recovery times in bats, but these experiments were interpreted with the knowledge from behavioral observations such as those described earlier that bats clearly must have true recovery times at least as small as 100 μsec, so a small population of neurons that could respond at time separations as short as 100 μsec was assumed to exist (Grinnell 1963, 1967). Although the fact has not received much attention more recently, the obstacle-avoidance results remain of direct concern for the bat's ability to perceive two echoes as separate: the value of 50–100 μsec for echo–delay resolution (an outside estimate from obstacle flights) is comparable in magnitude to the 50- to 100-μsec accuracy of delay determination for seizing a target in the tail membrane. However, resolution refers to segregation of two overlapping echoes, not accuracy for determining the delay of only one echo at a time (see Menne 1985; Schnitzler, Menne, and Hackbarth 1985). Because each echo replica must be assigned a specific arrival time along a scale of delay to locate its corresponding glint in range, the bat's ability to segregate two replicas only 50–100 μsec apart requires that the accuracy for specifying each of the arrival times by itself must be better than the difference between them (see Menne 1985; Simmons 1989). Figures 4.5 and 4.6 confirm that in well-regulated conditions the bat's ability to resolve two closely spaced echoes extends to separations as small as 2–30 μsec.

To account for interception of one mealworm in the presence of another object at about the same distance places even more severe demands on the bat's perceptual capacities because the glints in one object must be separated from the glints in the other object to locate the desired target, and that cannot occur unless all the glints first are separated out and assigned locations. In Figure 4.7B, the bat flies toward and seizes one mealworm in the presence of a second mealworm located close enough to the same target range that the echo replicas from the glints are completely intermingled; recognition of the individual glints in this case requires the ability to resolve replicas at least as close together as 10–20 μsec, and there are actually four primary glints to be sorted out, not two.

Consequently, the task shown in Figure 4.7B indicates that *Eptesicus* almost certainly can resolve echo replicas as close together as 5–10 μsec while in flight. This outside estimate of minimal echo–delay resolution is roughly 50 times smaller than the integration time of 300–400 μsec. Furthermore, because each echo replica has to be assigned a discrete arrival time, a resolution of 5–10 μsec implies that the accuracy for determining the arrival time of each replica by itself is better than 5–10 μsec as well.

6.2.3 Minimal Requirements Implicit for Target Localization

The estimates of required arrival-time acuity given here treat echo delay as a single imaginary axis from the bat's head to the target and back. The bat indeed broadcasts its sounds from a single site on the head, its open mouth, but it receives echoes at two sites, the left and right ears (see Fig. 4.1). Consequently, there is a separate echo–delay axis for each ear rather than a single axis for the bat's head as a whole. Segregation of echo replicas thus entails first of all separating them by their arrival times in the echo stream at each ear, so the bat really determines the distance to each glint at each ear. Beyond this, for avoiding obstacles or intercepting targets the bat perceives the directions of different objects, even when they are located at about the same distance; otherwise, it could not steer around wires or fly toward the mealworm. In the naturalistic tasks described earlier, there are implications of the stimulus regime for directional segregation and localization of targets as well as segregation along the range dimension.

It is widely assumed that bats use interaural differences in the amplitude and spectrum of echoes to determine the azimuth of a target because the bat's head is so small that interaural arrival-time differences are usually assumed be too small to be detected (Pollak 1988; Pollak and Casseday 1989; see also Schnitzler and Henson 1980; Schnitzler, Menne, and Hackbarth 1985). There are some observations of neural responses sensitive to small binaural time differences in FM bats (Harnischpfeger, Neuweiler, and Schlegel 1985), but these have been dismissed as too few in number (Pollak and Casseday 1989). However, the chief study that did not find such time sensitivity in FM bats (Pollak 1988) also did not use phase-controlled FM stimuli, whereas the study that did find time sensitivity did use phase-controlled stimuli (Harnischpfeger, Neuweiler, and Schlegel 1985), so the assumption that binaural time sensitivity is too poor in bats may be incorrect on physiological grounds.

In *Eptesicus*, interaural time differences amount to about 0.75 μsec per degree of azimuth, which is indeed small (Haresign et al., in manuscript). However, the overlap of echoes from multiple targets in the obstacle-avoidance task and during interception or discrimination of multiple targets renders interaural differences in the amplitude and spectrum of echoes useless for determining the azimuth of targets. This is a consequence of the integration time for echo reception; several echoes that arrive closer together than 300–400 μsec are assigned a single amplitude and a single spectrum over the whole interval, even though there are several glints actually responsible for the sounds. (Integration time *means* only one such estimate for the whole time window; allowing more than one amplitude estimate within this window is equivalent to reducing the length of the integration time itself.)

The single amplitude and single spectrum at each ear cannot be reduced to an estimate of the azimuth of one of the real glints, just a "phantom" glint at some intermediate location derived from binaural amplitude and spectral differences in the echo packets at the two ears. However, each echo packet has a spectrum that reflects primarily the time separation of the echo replicas, not the direction from which they come independent of this separation. Only by determining the arrival time of each replica separately at each ear and applying the time difference between ears to estimating azimuth can the bat perceive the correct location of a real target. Small as they are, interaural time differences are the only reliable cues for direction in obstacle-avoidance and airborne discrimination trials.

If the bat's accuracy for determining target azimuth is roughly 5° in obstacle-avoidance trials and during interception of mealworms, then its interaural echo–delay accuracy must be as good as 4 μsec (from an interaural time-difference ratio of 0.75 μsec/degree). If the bat's azimuth accuracy is 1.5° (a good approximation; see Simmons et al. 1983; Masters, Moffat, and Simmons 1985), then its interaural echo–delay accuracy must be about 1 μsec. These values refer to the separate registration of a single echo replica in both ears to determine the disparity in arrival time; because several echo replicas arrive in an interval of perhaps 100 μsec from different glints, the process of segregating echo replicas at each ear to determine the distance to each glint must really take place with an accuracy of no poorer than 1–4 μsec for all the glints. Otherwise, none of the real objects could be located in azimuth.

6.3 Reexamination of Conventional Localization Cues

The conclusion that FM bats must be able to use interaural arrival times to determine azimuth with an acuity of 1–4 μsec, and therefore that each delay estimate must have an associated accuracy of 1–4 μsec, may seem strong for being based on the performance of flying bats in tasks with continually changing bat–target relationships. In fact, however, during the bat's approach to vertically stretched wires or to airborne mealworms and disks, the critical overlap of echo replicas from several glints nearer each other in range than 5–7 cm is often present throughout the flight. Even though the bat flies progressively nearer the targets, so that the delay of echoes progressively shortens, it flies closer to each target at about the same rate. Consequently, the difference in range is relatively invariant on many trials. The bat's flight path toward several targets at once (see Figs. 4.7–4.12) keeps the separation of echo replicas smaller than the integration time of 300–400 μsec for the duration of entire flights. This hidden constancy in many of the obstacle-avoidance or airborne discrimination tests is documented by video motion-analysis studies, which track the distance from the bat to each part of several targets that are present in front of the bat, but it also can be concluded from descriptions of the original results from even the earliest experiments of these types (e.g., the bat's flight path is just about perpendicular to the plane of the wires [Griffin 1958; Grinnell and Griffin 1958; Jen and Kamada 1982]).

Second, the seemingly improbable conclusion from these naturalistic experiments that bats must have an inherent echo–delay acuity of at least 1–4 μsec becomes more plausible when the requirements for use of the conventional binaural amplitude and spectral cues in these same tasks are closely examined. The ruling constraint is the integration time for echo reception, which ensures that several echoes arriving at each ear within a 300 -to 400-μsec window will be assigned a single amplitude and spectrum at that ear (see Beuter 1980; Altes 1984; Menne 1985). Because all the objects whose echo replicas arrive within this window are lumped together for purposes of amplitude determination across frequencies (see 5-μsec to 200-μsec two-glint echoes in Fig. 4.4), there is no way to determine the binaural difference for each object, just for the echo packet as a whole. But there is no real object at the location represented by the binaural features of the entire packet. Moreover, each of the real objects is only accessible through the echo replicas it contributes to the packet; any information coming from other objects only complicates the problem by distorting the resulting estimate of that object's location from the packet alone. This distortion is nontrivial; in obstacle-avoidance flights, the packet would depict a phantom object located between two real wires, but the bat does not avoid this empty space, it flies through it (see Fig. 4.7A).

To make the binaural amplitude and spectral features of each echo replica available to determine each target's direction separately, the replicas first must be segregated out of the echo packet by their times of arrival. They cannot be separated by their amplitudes or their spectra because these dimensions have been combined across all the different replicas within the integration-time window. The spectrum of the packet can be used to determine the arrival-time separations of the replicas within the packet (Simmons 1989; Simmons et al. 1989; Saillant et al. 1993), but the separate spectra of the replicas cannot be assessed without first segregating the replicas by arrival time, and this requires temporal acuity much finer than the integration time itself.

Even disregarding the requirement that binaural time differences be used, the minimum two-glint resolving power implicit in the airborne interception results is 5–10 μsec. When the bat's ability to perceive the directions of different targets is considered, the minimum temporal resolution needed to segregate echo replicas and assign them their own binaural amplitude and spectral parameters is about as small as the temporal acuity required to locate the objects by interaural arrival-time differences alone. For these reasons, asserting that bats locate targets chiefly from binaural amplitude and spectral cues means that, in obstacle-avoidance and airborne discrimination tests, they must have a temporal resolving power of only a few microseconds to achieve this.

7. Directionality of Echolocation in *Eptesicus*

7.1 Directional Segregation of Targets?

It is sometimes assumed that the sonar of bats is sufficiently directional for echoes from two different objects to be isolated by aiming the sonar broadcasts and the ears toward one target rather than the other. For the directionality of the broadcast to be of practical use in segregating echoes from two different objects, the transmitted beam would have to be narrow enough to ensonify one object strongly while leaving the other object only weakly ensonified. Similarly, for the directionality of hearing to be useful, the receiving beam would have to be narrow enough to select echoes from one direction in preference to echoes from another direction. A commonly used index of target "separation" from directionality is reduction of echo strength from unwanted objects to half the amplitude (-6 dB) or half the energy (-3 dB) of echoes from the target of interest. This is only an arbitrary index of beamwidth, however; in practice, to effectively isolate echoes of one target from echoes of another, the unwanted echoes would have to be reduced sufficiently far that their presence does not impair determination of the delay, amplitude, and spectrum of the desired echoes. A more reasonable index of the minimal necessary separation would be nearer to a tenfold reduction of undesired echo amplitude (-20 dB) or power (-10 dB).

Observations of insect pursuit, obstacle avoidance, or target discrimination reveal that *Eptesicus* and other FM bats can perceive the locations of several objects simultaneously over a region of space that extends roughly from 30°–50° left to 30°–50° right (see Figs. 4.7–4.12). To isolate one object from another, the broadcasting or receiving beam would have to select a narrow window of azimuth from within this broad span of azimuths so that echoes from just the desired object become 10–20 dB stronger than echoes from unwanted objects located in the same overall region of space. In *Eptesicus*, neither the directionality of the sonar transmissions (Hartley and Suthers 1989) nor the directionality of hearing (Haresign et al. in manuscript; Jen and Chen 1988; Wotton 1994). is adequate to explain the bat's ability to perceive each of the adjacent wires in an obstacle-avoidance test or to locate one of two targets at about the same distance.

7.2 Broad Directionality of Transmissions

The open mouth of the bat in Figure 4.1 serves as the bat's broadcasting antenna, and the dimensions of the mouth in relation to the wavelengths of the sonar signals make this a directional antenna whose main axis is pointed to the front (Hartley and Suthers 1989). The equivalent acoustic diameter of the transmitting aperture is estimated to be about 9 mm, which is reasonably close to the actual dimensions of the bat's open mouth. Figure 4.13A illustrates the horizontal directionality of the bat's transmissions at two frequencies, 40 kHz and 60 kHz, that are representative of the sound as a whole (see Fig. 4.2). (Figure 4.13 shows a span of directions from 40° left to 40° right, covering the zone of frontal space within which the bat demonstrably can perceive several different objects at once.) At 40 kHz, the 6-dB width of the beam is ±30°, and at 60 kHz the width of the beam is ±25°. The equivalent 10-dB beamwidths are about ±50°–60°. These are surprisingly wide; in obstacle-avoidance or airborne discrimination flights, virtually all the objects that are present fall well within this subtended span of azimuths until the bat has approached quite close.

For the bat to ensonify one wire in an obstacle-avoidance flight with an incident sound that is as much as 10 dB stronger than the other wire, the bat would have to point its head directly at one wire when it has flown near enough for the other wire to be 50°–60° off to the left or right. For a wire array with a 25-cm distance between adjacent wires (Jen and Kamada 1982; see Fig. 4.7), the bat would have to be nearer than 25–30 cm to achieve 10-dB separation of the wires from the incident sounds alone. However, video studies of obstacle-avoidance flights (e.g., Fig. 4.7) show that the bat usually has settled on its flight path before reaching this distance, when it is as far away as 0.5–1 m and the angular separation of the adjacent wires is only about 15°–30°. At such distances, the largest possible amplitude separation of the broadcast sounds impinging on adjacent wires is only 3–6 dB (Fig. 4.13A).

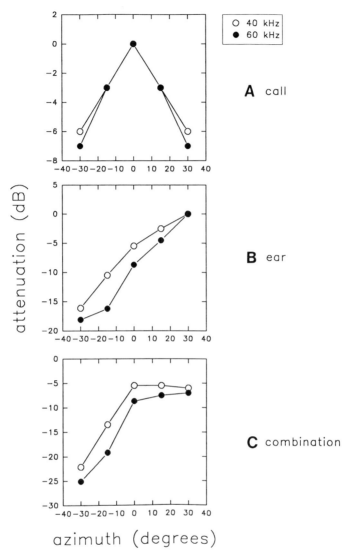

FIGURE 4.13A–C. Directional diagrams for the echolocation of *Eptesicus* (Wotton 1994). (A) The strength of the bat's sonar transmissions at azimuths from left (−) 30° to right (+) 30° at 40 kHz *(open circle)* and 60 kHz *(solid circle)* (Hartley and Suthers 1989); (B) the sensitivity of the right ear to 40 and 60 kHz frequency components in FM sounds from different directions; (C) the combined sensitivity of the sonar system at 40 and 60 kHz to a constant-strength target located in different directions, measured at the right ear. The broadcast is aimed to the front (0°), the right ear is aimed about 30°–40° to the right side at these frequencies, and the sonar system as a whole is omnidirectional in sensitivity to targets on the same side as the ear.

Moreover, the video motion studies confirm that the bat does not make large steering movements of its head to keep just one of the wires straight ahead in the middle of the broadcast beam, which is at 0° in Figure 4.13A. Thus, effective isolation of one wire from another beyond a few decibels clearly does not occur as a consequence of the directionality of the incident sound. During approaches to tethered mealworms (see Fig. 4.7B), the targets which the bat demonstrably can perceive as separate at distances of 0.5–0.75 m are located in a zone of azimuths anywhere from about 12° to 17° wide, so separation of echoes from the broadcast beam cannot exceed 3–4 dB. The bat still appears to have decided which target to capture before it has flown near enough that the unwanted target is as much as 50°–60° off to one side. Consequently, the crude directionality of transmissions is not adequate to segregate targets in well-documented tasks in which perceptual segregation in fact is achieved.

7.3 Broad Directionality of Echo Reception

The bat's ears act as receiving antennas for echoes, and their size and shape in relationship to the wavelengths of the ultrasonic signals make them about as directional for reception of sounds as the mouth is for emitting sounds. However, the bat's ears are not pointed toward the front but off to the side (see Fig. 4.1), and the shape of the receiving beam is not symmetrical (Haresign et al., in manuscript; Jen and Chen 1988; Wotton 1994). Figure 4.13B shows the receiving directionality for the right ear of *Eptesicus* at 40 and 60 kHz in the FM sweep of echolocation signals (Haresign et al., in manuscript; Wotton 1994; see also Jen and Chen 1988 for pure-tone directionality, which exaggerates the sharpness of the receiving beam). At both of these frequencies, which were presented in sweeps rather than as tones, the acoustic axis of the right ear is about 30°–40° off to the right side of the head, so the directional pattern in Figure 4.13B appears as a smooth decline in sensitivity for all directions to the left of this axis (left of +30° in Fig. 4.13B), with a shift to a steeper slope at directions toward the left side of the head (left of 0° in Fig. 4.13B).

Thus, for each ear, reception of echoes is progressively less sensitive at directions located toward the contralateral side of the acoustic axis. If the bat pointed its right-ear axis at the right-hand wire in the obstacle-avoidance array (Fig. 4.7A), the left-hand wire would have to be located 30°–45° contralateral for its echoes to be 10 dB weaker in the same right ear. To achieve this 10-dB separation of echoes in each ear, the bat would have to fly closer than 30–35 cm from the wire array. However, the bat frequently has determined its flight path at greater distances, so the crude directionality of hearing, like the crude directionality of the broadcasts, is apparently not the basis for perception of one target in the presence of others.

7.4 Directionality of the Sonar System

Both the bat's sounds and its hearing are directional, and the two directional patterns add together to determine the strength of echoes actually stimulating the bat from different azimuths. It would be possible for them to reinforce each other so that the directionality of the whole sonar system is sharper than either the broadcast or receiving component. However, the acoustic axis of the transmissions points straight ahead, while the axis of each ear points well off to the side. The combined directional patterns do not reinforce each other; instead, they yield a curious combination of omnidirectional sensitivity to targets on the same side as each ear and directionally graded rejection of echoes from targets on the opposite side.

Figure 4.13C shows this combined directionality for *Eptesicus* at frequencies of 40 and 60 kHz (Wotton 1994). For the right ear, a target located anywhere from straight ahead (0°) to at least 30° off to the right will return echoes of constant amplitude to the ear. A target located anywhere to the left will return echoes at progressively weaker strengths depending on how far to the left it is. For example, at 30° to the left, the echoes will be reduced by about 22-25 dB as a consequence of the combined directionality of transmissions and reception. This pattern of directional sensitivity is mirrored in the left ear. Thus, the bat's sensitivity to targets is omnidirectional but lateralized, with the ear pointing in the target's direction receiving a uniformly strong echo from the same target in many directions and the opposite ear receiving a weaker echo depending on how far in the contralateral direction the target is located.

The directionality of echolocation (see Fig. 4.13C) leads to an unanticipated pattern of stimulation during the bat's approach to obstacles. In Figure 4.7A, the nearest wire on the bat's left (*c*) returns a strong echo to the left ear, while the nearest wire on the right (*b*) returns a strong echo to the right ear. These same wires return substantially weaker echoes to the opposite ears. The arrival times of these echoes depend on the bat's flight path, but the stronger echo from the wire on the same side as the ear will typically be 1-50 μsec earlier than the weaker echo from the wire on the opposite side if the bat's head is exactly equidistant from the two wires. (The echo from the nearer wire, which is on the same side as the ear, arrives first.)

When the bat is at a distance of 0.5-1 m from the wires and the azimuthal separation of the wires is 12°-17°, the opposite ear from each wire receives an echo that is about 13-20 dB weaker than the ear on the same side. Thus, each ear receives a strong echo and a weak echo, at arrival times determined by the distance from each wire to each ear. However, the strong or weak echo in one ear comes from a different wire than the strong or weak echo in the other ear. Furthermore, as a consequence of the bat's nearly perpendicular approach to the wire array, all these echoes will typically fall within the 300 -to 400-μsec integration-time window at each ear. Interaural amplitude and spectral differences by themselves will be useless for localizing the wires because these binaural parameters will be dominated by the stronger echo in each ear, which comes from a different target on the left than on the right.

Because of a combination of the broad directional sensitivity pattern for targets (see Fig. 4.13A) and the nearly symmetric arrangement of the wires (see Fig. 4.7A), the obstacle-avoidance test actually provides a strong rejection of interaural amplitude and spectral hypotheses concerning horizontal localization of targets by *Eptesicus*. The integration time for echo reception so severely constrains this situation that the only way to determine the locations of both wires is to segregate the two echo replicas going to each ear from the spectral notches they create through interference (see Fig. 4.4) and then compare their arrival times between ears. (Behavioral experiments reveal that spectral information can be used to estimate echo–replica separations even when the sources of the overlapping replicas are as far apart as 40° in azimuth [Simmons, Moss, and Ferragamo 1990].) No other method can suppress the dominating effect of the strong echo from the nearest wire in each ear. Instead of being based on crude directionality, the bat's ability to separately perceive each target in an acoustic scene is derived from auditory computations that create images by segregating the glints making up the scene and then comparing the echo streams at the two ears in a glint-by-glint manner.

8. Summary

The bat's acuity for determining the arrival time of echo replicas and its acuity for resolving two overlapping echo replicas as separate are the most fundamental capacities underlying the formation of acoustic images in FM echolocation. In psychophysical experiments, the big brown bat *Eptesicus fuscus* can resolve echoes as separate even when they arrive as close together as 2–30 μsec. The bat's resolving power is independently estimated, from performance in obstacle-avoidance and airborne-interception tests, to be 5–10 μsec. The chief constraint on echo processing is the integration time of echo reception (300–400 μsec in *Eptesicus*), which combines all reflected replicas of the broadcast sound arriving within this window into a single echo "packet" and gives it a single amplitude and spectrum over this window.

Without segregation of closely spaced echo replicas by their arrival times, however, even within the integration-time window, the bat's images will not adequately depict the configuration of vertical wires or the location of one target (e.g., a mealworm) when other targets (e.g., a mealworm or several disks) are present also. Moreover, without very sharp registration of the delay of individual replicas at each ear, two closely spaced replicas cannot be segregated and then assigned individual arrival times, either to depict their distances or to determine their directions from binaural cues. Interaural arrival-time acuity is estimated, from psychophysical experiments, to be about 1 µsec, and independently from obstacle-avoidance and airborne-interception studies to be 1–4 µsec. Even the use of binaural amplitude and spectral cues requires prior temporal processing of echoes with an acuity and resolution of several microseconds to recognize which echoes come from which targets.

Acknowledgments This survey of evidence concerning the content of the bat's sonar images was supported by ONR Grant No. N00014-89-J-3055, by NIMH Research Scientist Development Award No. MH00521, by NIMH Training Grant No. MH19118, by NSF Grant No. BCS 9216718, by McDonnell-Pew Grant No. T89-01245-023, and by NIH Grant No. DC00511,and by DRF and NATO grants.

References

Altes RA (1980) Detection, estimation, and classification with spectrograms. J Acoust Soc Am 67:1232–1246.

Altes RA (1984) Texture analysis with spectrograms. IEEE Trans Sonics–Ultrasonics SU-31:407–417.

Beuter KJ (1980) A new concept of echo evaluation in the auditory system of bats. In: Busnel RG, Fish JF (eds) Animal Sonar Systems. New York: Plenum, pp. 747–761.

Bodenhamer RD, Pollak GD (1981) Time and frequency domain processing in the inferior colliculus of echolocating bats. Hear Res 5:317–355.

Covey E, Casseday JH (1991) The monaural nuclei of the lateral lemniscus in an echolocating bat: parallel pathways for analyzing temporal features of sound. J Neurosci 11: 3456–3470.

Dear SP, Simmons JA, Fritz J (1993) A possible neuronal basis for representation of acoustic scenes in auditory cortex of the big brown bat. Nature 364:620–623.

Dear SP, Fritz J, Haresign T, Ferragamo M, Simmons JA (1993) Tonotopic and functional organization in the auditory cortex of the big brown bat, *Eptesicus fuscus*. J Neurophysiol 70:1988–2009.

Feng AS, Condon CJ, White KR (1994) Stroboscopic hearing as a mechanism for prey discrimination in FM bats? J Acoust Soc Am 95:2736–2744.

Griffin DR (1958) Listening in the dark. New Haven: Yale University Press. (Reprinted by Cornell University Press, Ithaca, NY, 1986.)

Griffin DR (1967) Discriminative echolocation by bats. In: Busnel RG (ed) Animal Sonar Systems: Biology and Bionics. France: Jouy-en-Josas-78, Laboratoire de Physiologie Acoustique, pp. 273–300.

Griffin DR, Friend JH, Webster FA (1965) Target discrimination by the echolocation of bats. J Exp Zool 158:155–168.

Grinnell AD (1963) The neurophysiology of audition in bats: temporal parameters. J Physiol 167:67–96.

Grinnell AD (1967) Mechanisms of overcoming interference in echolocating animals. In: Busnel R-G (ed) Animal Sonar Systems: Biology and Bionics. France: Jouy-en-Josas-78, Laboratoire de Physiologie Acoustique, pp. 451–481.

Grinnell AD, Griffin DR (1958) The sensitivity of echolocation in bats. Biol Bull 114:10–22.

Harnischpfeger G, Neuweiler G, Schlegel P (1985) Interaural time and intensity coding in the superior olivary complex and inferior colliculus of the echolocating bat, *Molossus ater*. J Neurophysiol 53:89–109.

Hartley DJ (1992) Stabilization of perceived echo amplitudes in echolocating bats: II. The acoustic behavior of the big brown bat, *Eptesicus fuscus*, while tracking moving prey. J Acoust Soc Am 91:1133–1149.

Hartley DJ, Suthers RA (1989) The sound emission pattern of the echolocating bat, *Eptesicus fuscus*. J Acoust Soc Am 85:1348–1351.

Jen PH-S, Chen DM (1988) Directionality of sound pressure transformation at the pinna of echolocating bats. Hear Res 34:101–118

Jen PH-S, Kamada T (1982) Analysis of orientation signals emitted by the CF-FM bat *Pteronotus p. parnellii* and the FM bat *Eptesicus fuscus* during avoidance of moving and stationary obstacles. J Comp Physiol 148:389–398.

Kick SA (1982) Target detection by the echolocating bat, *Eptesicus fuscus*. J Comp Physiol 145:431–435.

Kick SA, Simmons JA (1984) Automatic gain control in the bat's sonar receiver and the neuroethology of echolocation. J Neurosci 4:2725–2737.

Kober R, Schnitzler H-U (1990) Information in sonar echoes of fluttering insects available for echolocating bats. J Acoust Soc Am 87:874–881.

Kuc R (1994) Sensorimotor model of bat echolocation and prey capture. J Acoust Soc Am 96: 1965–1978.

Kurta A, Baker RH (1990) *Eptesicus fuscus*. Mamm Species 356:1–10.

Lee DN, van der Weel FR, Hitchcock T, Matejowsky E, Pettigrew JD (1992) Common principle of guidance by echolocation and vision. J Comp Physiol A 171:563–571.

Lhémery A, Raillon R (1994) Impulse-response method to predict echo responses from targets of complex geometry: II. Computer implementation and experimental validation. J Acoust Soc Am 95:1790–1800.

Masters WM, Moffat AJM, Simmons JA (1985) Sonar tracking of horizontally moving targets by the big brown bat, *Eptesicus fuscus*. Science 228:1331–1333.

Menne D (1985) Theoretical limits of time resolution in narrow band neurons. In: Michelsen A (ed) Time Resolution in Auditory Systems. New York: Springer-Verlag, pp. 96–107.

Menne D, Kaipf I, Wagner I, Ostwald J, Schnitzler HU (1989) Range estimation by echolocation in the bat *Eptesicus fuscus*: trading of phase versus time cues. J Acoust Soc Am 85:2642–2650.

Mogdans J, Schnitzler H-U (1990) Range resolution and the possible use of spectral information in the echolocating bat, *Eptesicus fuscus*. J Acoust Soc Am 88:754–757.

Moss CF, Schnitzler H-U (1989) Accuracy of target ranging in echolocating bats: acoustic information processing. J Comp Physiol A 165:383–393.

Moss CF, Simmons JA (1993) Acoustic image representation of a point target in the bat, *Eptesicus fuscus*: evidence for sensitivity to echo phase in bat sonar. J Acoust Soc Am 93:1553–1562.

Moss CF, Zagaeski M (1994) Acoustic information available to bats using frequency-modulated sounds for the perception of insect prey. J Acoust Soc Am 95:2745-2756.

Neuweiler G (1990) Auditory adaptations for prey capture in echolocating bats. Physiol Rev 70:615-641.

Novick A (1977) Acoustic orientation. In: Wimsatt WA (ed) Biology of Bats, Vol. 3. New York: Academic Press, pp. 73-287.

Pollak GD (1988) Time is traded for intensity in the bat's auditory system. Hear Res 36:107-124.

Pollak GD (1993) Some comments on the proposed perception of phase and nanosecond time disparities by echolocating bats. J Comp Physiol A 172:523-531.

Pollak GD, Casseday JH (1989) The Neural Basis of Echolocation in Bats. New York: Springer-Verlag.

Pollak GD, Marsh DS, Bodenhamer R, Souther A (1977) Characteristics of phasic on neurons in inferior colliculus of unanesthetized bats with observations relating to mechanisms for echo ranging. J Neurophysiol 40:926-942.

Pye JD (1980) Echolocation signals and echoes in air. In: Busnel RG, Fish JF (eds) Animal Sonar Systems. New York: Plenum, pp. 309-353.

Roverud RC, Nitsche V, Neuweiler G (1991) Discrimination of wingbeat motion by bats, correlated with echolocation sound pattern. J Comp Physiol A 168:259-263.

Saillant PA, Simmons JA, Dear SP, McMullen TA (1993) A computational model of echo processing and acoustic imaging in frequency-modulated echolocating bats: the spectrogram correlation and transformation receiver. J Acoust Soc Am 94:2691-2712.

Schmidt S (1992) Perception of structured phantom targets in the echolocating bat, *Megaderma lyra*. J Acoust Soc Am 91:2203-2223.

Schnitzler H-U, Henson OW Jr (1980) Performance of airborne animal sonar systems: I. Microchiroptera. In: Busnel RG, Fish JF (eds) Animal Sonar systems. New York: Plenum, pp. 109-181.

Schnitzler H-U, Menne D, Hackbarth H (1985) Range determination by measuring time delay in echolocating bats. In: Michelsen A (ed) Time Resolution in Auditory Systems. New York: Springer-Verlag, pp. 180-204.

Simmons JA (1973) The resolution of target range by echolocating bats. J Acoust Soc Am 54:157-173.

Simmons JA (1979) Perception of echo phase information in bat sonar. Science 207:1336-1338.

Simmons JA (1989) A view of the world through the bat's ear: the formation of acoustic images in echolocation. Cognition 33:155-199.

Simmons JA (1992) Time-frequency transforms and images of targets in the sonar of bats. In: Bialek W (ed) Princeton Lectures on Biophysics. River Edge, NJ: World Scientific, pp.291-319.

Simmons JA (1993) Evidence for perception of fine echo delay and phase by the FM bat, *Eptesicus fuscus*. J Comp Physiol A 172:533-547.

Simmons JA, Chen L (1989) The acoustic basis for target discrimination by FM echolocating bats. J Acoust Soc Am 86:1333-1350.

Simmons JA, Grinnell AD (1988) The performance of echolocation: the acoustic images perceived by echolocating bats. In: Nachtigall P, Moore PWB (eds) Animal Sonar: Processes and Performance. New York: Plenum, pp. 353-385.

Simmons JA, Kick SA (1984) Physiological mechanisms for spatial filtering and image enhancement in the sonar of bats. Annu Rev Physiol 1984 46:599–614.

Simmons JA, Moss CF, Ferragamo M (1990) Convergence of temporal and spectral information into acoustic images of complex sonar targets perceived by the echolocating bat, *Eptesicus fuscus*. J Comp Physiol A 166:449–470.

Simmons JA, Ferragamo M, Moss CF, Stevenson SB, Altes RA (1990) Discrimination of jittered sonar echoes by the echolocating bat, *Eptesicus fuscus*: the shape of target images in echolocation. J Comp Physiol A 167:589–616.

Simmons JA, Freedman EG, Stevenson SB, Chen L, Wohlgenant TJ (1989) Clutter interference and the integration time of echoes in the echolocating bat, *Eptesicus fuscus*. J Acoust Soc Am 86:1318–1332.

Simmons JA, Kick SA, Lawrence BD, Hale C, Bard C, Escudié B (1983) Acuity of horizontal angle discrimination by the echolocating bat, *Eptesicus fuscus*. J Comp Physiol 153:321–330.

Simmons JA, Saillant PA, Ferragamo MJ, Haresign T, Dear SP, Fritz J, McMullen TA Auditory computations for biosonar target imaging in bats. In: Hawkins HL, McMullen TA, Popper AN, Fay RR (eds) Auditory computation. New York, Springer-Verlag. (in press)

Suga N (1964) Recovery cycles and responses to frequency modulated tone pulses in auditory neurons of echolocating bats. J Physiol 175:50–80.

Suga N (1967) Discussion (of presentation by O. W. Henson, Jr.). In: Busnel R-G (ed) Animal Sonar Systems: Biology and Bionics. Laboratoire de Physiologie Acoustique, France: Jouy-en-Josas-78, pp. 1004–1020.

Suga N, Schlegel P (1973) Coding and processing in the nervous system of FM signal producing bats. J Acoust Soc Am 84:174–190.

Sum YW, Menne D (1988) Discrimination of fluttering targets by the FM-bat *Pipistrellus stenopterus*? J Comp Physiol A 163:349–354.

Webster FA (1967) Performance of echolocating bats in the presence of interference. In: Busnel RG (ed) Animal Sonar Systems: Biology and Bionics. France: Jouy-en-Josas-78, Laboratoire de Physiologie Acoustique, pp. 673–713.

Webster FA, Brazier OG (1965) Experimental studies on target detection, evaluation, and interception by echolocating bats. TDR No. AMRL-TR-65-172, Aerospace Medical Division, USAF Systems Command, Tucson, AZ.

Webster FA, Griffin DR (1962) The role of the flight membrane in insect capture by bats. Anim Behav 10:332–340.

Wotton JM (1994) The basis for vertical sound localization of the FM bat, *Eptesicus fuscus*: acoustical cues and behavioral validation. Ph.D. dissertation, Brown University, Providence, RI.

5

Cochlear Structure and Function in Bats

Manfred Kössl and Marianne Vater

1. Introduction

The mammalian cochlea must extract loudness and frequency information about different and overlapping acoustic events from a single input channel. Unlike the visual system, where input from different spatial sources is separated early and distributed in parallel channels, all input to the cochlea converges in the middle ear to induce movement at the membrane of the oval window. The oval window acts as a point source of mechanical waves dissipating in the fluid-filled spaces of the cochlea. It is now left to the organ of Corti to filter relevant information from the total input. Because this is a difficult task, it is no wonder that the acquisition of highly developed cochleae that are able to analyze low-level signals at many different frequencies is a relatively recent step in animal evolution, one that is confined to higher vertebrates. The high-frequency hearing capabilities of reptiles and birds are restricted to frequencies below about 12 kHz (for review, see Manley 1990). As a specific mammalian adaptation, the middle ear and the cochlea have an extended sensitivity in the high-frequency range. For processing of high frequencies, the cochlea has developed macro- and micromechanical specializations and employs active processes that enhance frequency tuning and sensitivity. The hair cells are clearly distinguished as inner hair cells (IHCs) that are contacted by nearly all afferent nerve fibers to the brain and outer hair cells (OHCs) that function to a lesser degree as receptors but more as effectors. Outer hair cells are thought to be involved in amplifying and filtering the mechanical input in collaboration with additional structures such as the basilar and tectorial membranes (Dallos and Corey 1991). In the cochleae of bats, mechanisms to aquire sensitive and sharp tuning at high frequencies up to 160 kHz are fully exploited. The investigation of these cochleae may give clues to open questions about cochlear function in mammals, in particular about the frequency limitation of active processes, electromechanical feedback, and mechanical tuning mechanisms.

2. Evolutionary Adaptations of the Cochlea of Bats

A significant factor in the evolution of the cochlea in bats has doubtless been the use of echolocation for prey capture. In bats, the cochlea is not only involved in passive listening tasks but is an integral part of active orientation. This functional context is reflected in the sheer size of the cochleae of bats employing different echolocation strategies (Habersetzer and Storch 1992; Fig. 5.1). Most Macrochiroptera do not echolocate and instead rely on their eyes and olfactory system to find edible fruits. Their cochlea, consequently, is smaller relative to the skull size than that of echolocating Microchiroptera. According to the characteristics of their echolocation call, Microchiroptera can be divided into two groups (for a detailed discussion of the evolution of echolocation see Fenton, Chapter 2). The first group uses short broad-band calls that consist of different harmonics of a downward frequency-modulated (FM) component. The second group employs CF-FM calls, composed of a long, constant-frequency (CF) component followed by a short FM sweep, and displays characteristic Doppler-shift compensation behavior (Schnitzler 1968, 1970). CF-FM bats detect small frequency modulations in the CF component of the echo caused by the wingbeat of prey insects, even within a densely cluttered environment such as dense vegetation (see Neuweiler 1990). These capabilities crucially depend on a high cochlear frequency resolution.

Hydromechanical specializations in the cochlea create enhanced tuning and an acoustic fovea, an expanded representation of the biologically most important CF frequency range. As a consequence of the foveal frequency representation, typical CF-FM bats such as Rhinolophidae, the related Hipposideridae, and *Pteronotus parnellii* have a larger cochlea than less specialized FM bats such as Vespertilionidae (Fig. 5.1). Cochlear size depends more on functional requirements related to the echolocation signal than on the taxonomic relationships. This is clearly evident in the genus *Pteronotus*, New World bats of the mormoopid family (see Appendix in Fenton, Chapter 2, this volume). The species *Pteronotus personatus, P. davyi*, and *P. suapurensis*, which use FM calls, have much smaller cochleae than the species *P. parnellii*, whose different subspecies use CF-FM calls. This demonstrates the high degree of adaptability of the cochlear size to different environmental constraints (Habersetzer and Storch 1992). The size of the cochlea of extinct bats from the middle Eocene is at the lower margin of the vespertilionid family, indicating that the larger cochleae associated with CF call components may be a relatively recent development.

3. Physiology of the Cochlea

3.1 Auditory Threshold Curves

The absolute threshold of hearing in bats is obtained most reliably from data on single neurons of the eighth nerve or auditory brain stem and from

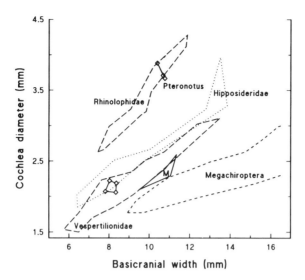

FIGURE 5.1. Relative size of the cochlea in different bat families, plotted as the cochlear diameter versus the basicranial width (after Habersetzer and Storch 1992). The cochleae of (CF-FM) bats (Rhinolophidae, Hipposideridae) are larger than in FM bats (Vespertilionidae), in Macrochiroptera, and in extinct bats from Messel (M). *Open symbols* mark the cochleae of different species of *Pteronotus* (see text).

behavioral audiograms. Cochlear microphonic (CM) potentials depend on the geometry of the cochlea in relation to the recording site and show a frequency-specific bias that makes them less useful for threshold determination. The N_1 evoked potential of the auditory nerve depends on the synchronicity of spike discharge in different nerve fibers. There are mechanical on/off processes in the cochlea of CF-FM bats at certain frequencies (Grinnell 1963, 1973), and the synchronization of onset spikes is influenced by cochlear resonance (Suga, Simmons, and Jen 1975). Therefore the N_1 potential does not unambiguously reflect the threshold of hearing.

Even in nonecholocating mammals, the behavioral threshold can extend up to about 100 kHz (Fig. 5.2A). However, auditory sensitivity steeply deteriorates for frequencies above 60 kHz and in the house mouse is at about 68 dB SPL (sound pressure level) at 100 kHz. In FM bats such as *Carollia perspicillata* or *Megaderma lyra* (Fig. 5.2A,B), the auditory thresholds of neurons measured in the inferior colliculus (Rübsamen, Neuweiler, and Sripathi 1988; Sterbing, Rübsamen, and Schmidt 1990) remain close to 0 dB SPL up to frequencies of about 100 kHz. Both bat species are extremely sensitive to sound frequencies between about 15 and 30 kHz with thresholds below −10 dB SPL. This frequency range is probably important for passive listening or communication because the range of the broad-band echolocation signals is higher in both species

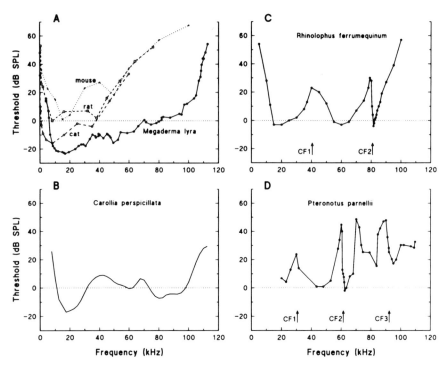

FIGURE 5.2A–D. Auditory threshold curves. (A) Behavioral thresholds in cat (from Neff and Hind 1955), rat (from Kelly and Masterton 1977), and the house mouse (from Ehret 1974) compared to a neuronal audiogram of *Megaderma lyra* (from Rübsamen, Neuweiler, and Sripathi 1988: inferior colliculus). (B) Neuronal audiogram in *Carollia perspicillata* (from Sterbing, Rübsamen, and Schmidt 1990: inferior colliculus). (C) Behavioral audiogram in *Rhinolophs ferrumequinum* (from Long and Schnitzler 1975). (D) Neuronal audiogram in *Pteronotus parnellii* (from Kössl and Vater 1990b: cochlear nucleus). CF_1, CF_2, and CF_3 indicate the first, second, and third harmonic of the constant frequency echolocation signal component. SPL, sound pressure level.

(40–100 kHz in *Megaderma*, 60–110 kHz in *Carollia*). In *Megaderma*, the high sensitivity around 20 kHz is partly caused by an amplifying effect of the large pinnae of the outer ear. The pinna gain is at a maximum of about 17 dB at 20 kHz (Obrist et al. 1993). In *Carollia*, a second minimum in the hearing threshold is evident within the frequency range of the echolocation signal (from 80 to 100 kHz).

In bats that use long CF-FM orientation calls, there are sharp maxima and minima in the threshold curves that are related to the CF frequencies. In *Rhinolophus ferrumequinum* (Fig. 5.2C), a narrow threshold minimum at about 83–85 kHz coincides with the range of the second-harmonic CF component (CF_2; Long and Schnitzler 1975). Between 79 and 82 kHz, there is a pronounced threshold maximum. A second threshold maximum is found close to the first CF component (CF_1) at about 41 kHz. DuringDoppler-shift compensation behavior, the frequency of the emitted CF_2 component is decreased by about 3 kHz (Schnitzler 1968; Schuller, Beuter, and Schnitzler 1974) and thus shifted into the region of acoustic insensitivity. The frequency of the Doppler-shifted CF_2 echo is now at the threshold minimum. Neuronal audiograms of *Rhinolophus rouxi* obtained in the inferior colliculus and lateral lemniscus (Schuller 1980; Metzner and Radtke-Schuller 1987) show a similar pattern; the CF_2 frequency of about 78 kHz is located at a steep slope between threshold maximum and minimum. In *Pteronotus parnellii*, the neuronal threshold curve shows multiple maxima and minima close to the CF_1, CF_2, and CF_3 frequencies (Fig. 5.2D). Again, the CF_2 frequency is located at a steeply sloping region of the audiogram.

Because the duration of the CF component ranges from 20 to 50 msec in both rhinolophids and *Pteronotus parnellii*, the outgoing call and the returning echo inevitably overlap in time (see Henson et al. 1987). The threshold maximum close to the call CF_2 component and the minimum at the echo frequency can be seen as an adaptation to minimize the cochlear response to the call and to maximize the responses to the echo (Henson et al. 1987). In hipposiderid CF-FM bats, the calls are much shorter (4–7 msec). Consequently, there is less overlap between call and echo, and the variations in the CF_2 frequency emitted by individual bats are larger (*Hipposideros speoris*, 0.5% –2%; *Hipposideros bicolor*, 0.75%) than in *Rhinolophus rouxi* (0.2%) (Schuller 1980; Habersetzer, Schuller, and Neuweiler 1984). Maxima and minima in the CF_2 ranges of the neuronal audiograms of both hipposiderid species are less pronounced than in long CF-FM bats or may even be absent (Schuller 1980; Rübsamen, Neuweiler, and Sripathi 1988). Hipposiderid bats may be seen as intermediate between FM and long CF-FM bats. This is also reflected in the absolute size of their cochleae (see Fig. 5.1).

3.2 Frequency Tuning

Data on the tuning properies of IHCs and OHCs are not yet available for bats. In the guinea pig, the tuning curves of IHCs are matched quite accurately by the activity of single auditory nerve fibers (Evans 1972; Robertson and Manley 1974; Russell and Sellick 1978). Therefore, the tuning measured in single units in the auditory nerve or the cochlear nucleus of different bat species can be taken as a close indicator of cochlear tuning. Figure 5.3 shows neuronal tuning curves from the auditory periphery of two CF-FM bats, *Rhinolophus ferrumequinum* and *Pteronotus parnellii*. The area above each curve corresponds to frequency and level combinations that lead to excitatory responses of the individual neuron. The frequency at which a neuron is most sensitive to pure tone bursts is called the best frequency (BF). The BFs vary between about 10 and 100 kHz in *Rhinolo-*

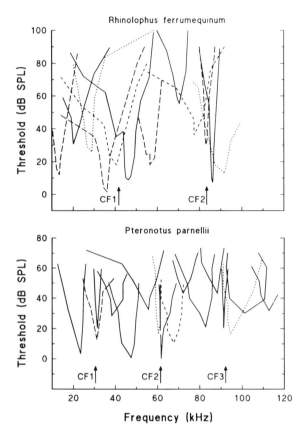

FIGURE 5.3. Tuning curves of single neurons in the cochlear nucleus of *Rhinolophus ferrumequinum* (*top*) (after Suga, Neuweiler, and Möller 1976) and *Pteronotus parnellii* (*bottom*) (after Kössl and Vater 1990b).

phus ferrumequinum and between about 10 and 115 kHz in *Pteronotus parnellii*. In neurons with BFs close to the CF_2 and the CF_3 frequency (only *Pteronotus*), the excitatory areas are narrower and the tips of the tuning curves are much sharper than for other BFs. The sharpness of tuning expressed as Q_{10dB} value (BF divided by the bandwidth of the tuning curve 10 dB above the threshold at the BF) varies from about 5–30 at non-CF frequencies to about 400 around the CF_2 frequency (Fig. 5.4) (Suga, Neuweiler, and Möller 1976). FM bats such as *Myotis lucifugus* have Q_{10dB} values below about 30 over their whole range of hearing (Suga 1964), and are comparable to nonecholocating mammals such as the cat and guinea pig in which maximum Q_{10dB} values of about 12 are found (Evans 1975: auditory nerve; Russell and Sellick 1978: inner hair cells). Hence, in CF-FM bats, cochlear processes that lead to sharp tuning go far beyond the scope seen in other mammals. The enhanced tuning around the CF frequencies guarantees the high-frequency resolution necessary to detect small fre-

FIGURE 5.4. Q_{10dB} values of single neurons in the cochlear nucleus of *Rhinolophus ferrumequinum* and *Myotis lucifugus* (after Suga 1964; Suga, Neuweiler, and Möller 1976).

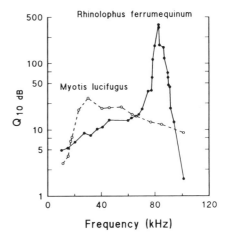

quency variations in the CF echoes caused by the wingbeat of prey insects, and is related to hydromechanical specializations in the cochlea (e.g., Neuweiler, Bruns, and Schuller 1980; Neuweiler 1990; see Section 5).

When stimulating CF-FM bats with tone bursts close to the CF ranges, pronounced neuronal on/off responses are observed (Grinnell 1973; Suga, Simmons, and Jen 1975; Suga, Neuweiler, and Möller 1976). The pure tone stimulus is only able to elicit a neuronal response at its onset or after offset (Fig. 5.5; Kössl and Vater 1990b). The on/off responses are not caused by neuronal inhibition and most probably are generated at the level of cochlear mechanics (Grinnell 1973; Suga, Simmons, and Jen 1975). They may be caused by the mechanisms that contribute to enhanced tuning and to the threshold insensitivity slightly below CF_2. Tuning curves of neurons with BFs just above the CF_2 frequency of *Pteronotus* not only show an on/off region on the low-frequency boundary but also show a shallow slope at the high-frequency side (Fig. 5.5). This is in contrast to normal mammalian tuning curves, and may arise from an increased longitudinal coupling of the basilar membrane in a cochlear region located basally to the cochlear place of the CF_2 frequency, where pronounced basilar membrane thickenings are present (see Sections 5.2 and 6.2.2). Similar tuning curves are also found in *Rhinolophus rouxi* (Metzner and Radtke-Schuller 1987).

3.3 Cochlear Emissions

The mammalian cochlea can emit sound at those frequencies at which the hearing sensitivity is high (Kemp 1978; Zwicker and Schloth 1984; review by Probst, Lonsbury-Martin, and Martin 1991). Spontaneous otoacoustic emissions (OAEs) are usually below the threshold of perception, but they can be measured as sharp spectral peaks with sensitive microphones placed in the outer ear. Evoked OAEs (EOAEs) appear as delayed oscillations

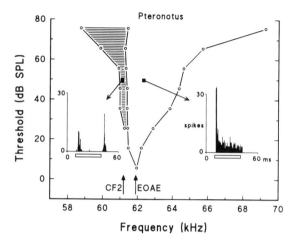

FIGURE 5.5. Tuning curve of a neuron in the cochlear nucleus of *Pteronotus parnellii* (from Kössl and Vater 1990b). Within the shaded area the neuron shows on/off temporal response properties. The arrows indicate the CF_2 frequency and the frequency of the evoked otoacoustic emission (EOAE).

coming from the ear after stimulation with click stimuli. Typically, each ear has multiple emissions at different frequencies. These emissions can also be measured as acoustic interference patterns in the frequency response obtained with pure tone sweeps. In humans, such "stimulus-frequency" OAEs saturate at stimulus levels of about 20–30 dB SPL. Active mechanical processes in the cochlea are thought to be involved in the generation of both spontaneous and evoked emissions.

In *Pteronotus*, each ear emits a single evoked OAE at about 62 kHz (Kössl and Vater 1985a). The stimulus-frequency OAE start to saturate at input levels above about 60 dB SPL (Fig. 5.6A). The maximum level of the evoked OAE is about 70 dB SPL, implying that the evoked emission is about 100 times stronger in this bat species than in other mammals. The evoked OAEs sometimes convert, for periods of a few days, to spontaneous OAEs (SOAEs) with a level up to 40 dB SPL (Fig. 5.6B). The spontaneous OAEs can be suppressed by additional sound stimuli of higher frequency (Fig. 5.6B). Prominent hydromechanical specializations found at and basal to the cochlear representation place of 62 kHz probably play a role in emission generation and enhanced tuning (see Sections 5.2 and 6.2.2). In *Rhinolophus rouxi*, another long CF-FM bat that was examined for otoacoustic emissions, only a few individuals produced weak stimulus-frequency OAEs. These were about 300 Hz above the CF_2 frequency of 78 kHz. Associated changes in the phase of the recorded signal were much broader than in *Pteronotus* (Kössl 1994), indicating a stronger damping of the underlying resonant mechanism (Henson, Schuller, and Vater 1985).

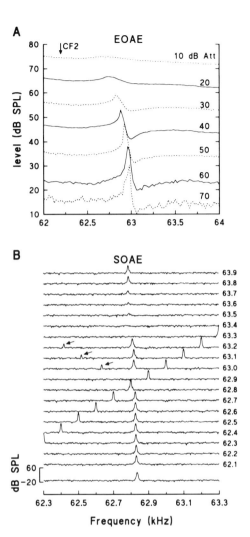

FIGURE 5.6A,B. (A) Evoked otacoustic emission (EOAE) measured with a micro-phone at the tympanum of *Pteronotus parnellii* (after Kössl 1994). A continuous pure tone is swept upward in frequency at different attenuations (indicated at each curve). At about 63 kHz, an outgoing emission interferes with the incoming stimulus, and maxima and minima are evident in the frequency response. (B) spontaneous otoacoustic emission (SOAE) in *Pteronotus* (after Kössl 1994). The lowest trace shows the SOAE without concomitant sound stimulation. The upper traces display the behavior of the SOAE during application of an additional pure tone stimulus of 40 dB SPL. The stimulus was moved across the range of the SOAE; its frequency is indicated to the left of each trace. *Arrows* indicate $2f_1$-f_2 distortion products produced by the SOAE and the stimulus. For stimulus frequencies between 63.3 and 63.7 kHz, the SOAE is suppressed.

3.4 Cochlear Resonance in Pteronotus parnellii

At the frequency of the evoked OAE, a pronounced maximum of the amplitude of cochlear microphonic potentials can be measured, and the CM threshold is at a minimum (Pollak, Henson, and Novick 1972; Henson, Schuller, and Vater 1985; Kössl and Vater 1985a). After stimulation with short tone bursts, the CM potentials show a strong long-lasting ringing that was attributed to a cochlear resonance (Suga, Simmons, and Jen 1975) and also could be measured as a delayed evoked OAE in the outer ear canal (Kössl and Vater 1985a). The cochlear resonance frequency (which equals the frequency of the OAE and the CM maximum) is about 400–900 Hz above the CF_2 frequency of Pteronotus (Kössl and Vater 1985a, 1990b; Kössl 1994). Neuronal thresholds in the cochlear nucleus reach minimum values at and slightly above the resonance frequency (Fig. 5.7). Maximum neuronal tuning sharpness, and hence highest Q_{10dB} values of about 400, are found for frequencies about 300 Hz below the resonance frequency (Fig. 5.7), that is, at the steep slope between maximum and minimum thresholds. This indicates that the cochlear mechanism that generates the evoked OAE is involved in creating sharp tuning close to the CF_2 frequency. It has to be pointed out, however, that sharply tuned neurons with Q_{10dB} values up to 300 are also found around 90 kHz, where there is no conspicuous OAE. Enhanced tuning is also found in the 30-kHz range where neuronal Q_{10dB} values can increase to about 70 (Suga and Jen 1977). In both cases, the

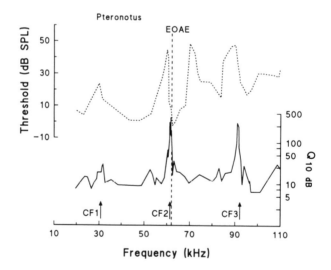

FIGURE 5.7. Lowest threshold *(dotted line)* and Q_{10dB} values *(solid line)* of single neurons in the cochlear nucleus of an individual of *Pteronotus* (from Kössl and Vater 1990b). The three harmonics of the constant-frequency component of the echolocation call are indicated by *arrows*. The frequency of the evoked otoacoustic emission (EOAE) is given by a *dashed line*.

maximum $Q_{10\,dB}$ values are located at the slopes between threshold maxima and minima.

The frequency and amplitude of the evoked OAE and corresponding CM potentials are affected by anaesthesia and changes in body temperature (Kössl and Vater 1985a; Henson et al. 1990). The largest variations in the resonance frequency are about 500 Hz, which is less than 1% of the OAE frequency. The magnitude of temperature-induced frequency shifts of the resonance frequency (39 Hz/°C) and of associated neuronal BFs (33 Hz/°C) (Huffman and Henson 1991, 1993a,b) is about 0.06%/°C. The frequency shifts are comparable to changes of spontaneous OAEs in the humans during menstrual or diurnal cycles (Wit 1985, maximally 0.12%/°C; Wilson 1985, 0.1%/°C or less). In *Pteronotus*, the small Q_{10} value (resonance frequency/resonance frequency measured at a 10°C lower temperature, of about 1.006 suggests that the cochlear resonance depends on complex mechanical interactions that are not greatly influenced in frequency by enzymatic or metabolic reactions (Q_{10} range, 2–3).

3.5 Cochlear Distortions

In the past few years, the measurement of acoustic two-tone distortions from the cochlea has gained increasing popularity. During stimulation with two tones of different frequency (f_1, f_2), pronounced cubic and quadratic distortion products can be measured acoustically in the outer ear canal (Probst, Lonsbury-Martin, and Martin 1991). The occurrence of distortion products results from nonlinear cochlear mechanics involved in the amplification of low-level sound stimuli. OHC motility is thought to contribute to the "cochlear amplifier". The distortion product otoacoustic emissions (DPOAEs) are a noninvasive indicator of hearing ability (Brown and Gaskill 1990; Gaskill and Brown 1990) and have the advantage that they appear over the whole range of hearing of an animal, in contrast to spontaneous and evoked OAEs, which are restricted in frequency. The DPOAEs are measured as distinct spectral peaks at both sides of the two primary stimuli (f_1, f_2) at the frequencies $(n + 1)f_1 - nf_2$ and $(n + 1)f_2 - nf_1$. The first lower sideband DPOAE at a frequency of $2f_1 - f_2$ is most prominent.

In FM and CF-FM bats, the DPOAEs are measurable from 5–100 kHz up to the frequency limit of the spectrum analyzer used, and they are affected by salicylate (Kössl 1992a,b), which is known to block OHC motility (Dieler, Shehata-Dieler, and Brownell 1991). Fig. 5.8 shows two examples of distortion products from *Pteronotus parnellii* for primary frequencies around 20 and 61 kHz. In both examples the frequency separation between the two stimuli was adjusted so that maximum $2f_1$-f_2 distortion level could be measured by giving an optimum overlap of the two corresponding traveling waves. The optimum frequency separation, Δf, between the two primary stimuli is taken as an indirect measure of the relative mechanical

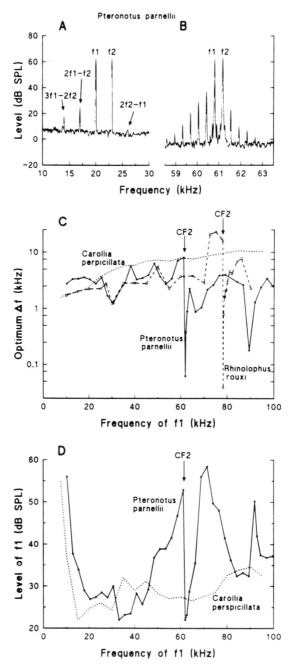

FIGURE 5.8A–D. Acoustic distortion products measured at the tympanum of different bat species. (A,B) The distortion products appear as distinct side peaks both on the low, and high-frequency side of the two stimuli (f_1, f_2; level of about 65 dB SPL). The stimulus frequencies were in the 20-kHz (A) and 60-kHz (B) range of

frequency separation in the cochlea (Kössl 1992b). In FM bats (Fig. 5.8C, *Carollia*), the optimum Δf continuously rises from about 2 to 10 kHz for f_1 frequencies between 10 and 90 kHz. In contrast, in the CF-FM bats *Pteronotus parnellii* and *Rhinolophus rouxi* there are sharp minima of the optimum Δf close to CF_2 and CF_3 (*Pteronotus* only; Fig. 5.8C). Minimum values of about 30–100 Hz are measured when f_1 exactly matched the frequency of the evoked OAE in *Pteronotus* (480 Hz above CF_2) and a frequency range about 300 Hz above CF_2 in *Rhinolophus*. The minima of the optimum Δf correlate well with neuronal tuning measurements (see Fig. 5.7). Additionally, the DPOAEs provide a quite accurate measure of the relative course of the hearing threshold. Figure 5.8D shows $2f_1$-f_2 threshold curves from *Carollia perspicillata* and *Pteronotus parnellii*. In both cases the curves run approximately parallel to the neuronal data (see Fig. 5.2) but are about 15–35 dB higher. In both *Pteronotus parnellii* and *Rhinolophus rouxi*, the CF_2 frequencies of the emitted call coincide with a maximum of the distortion threshold.

The distortion measurements indicate that in long CF-FM bats both the steep variations in threshold and the enhanced frequency tuning are implemented on the level of cochlear mechanics. The mechanical insensitivity at CF_2 may reduce the cochlear response to the emitted signal of the call to guarantee an undisturbed processing of the echoes during Doppler-effect compensation behavior.

4. Cochlear Frequency Maps

Single neuron recordings in the ascending auditory pathway of CF-FM bats show that a large proportion of the cells is tuned to a small frequency band around CF_2 (see Pollak and Park, Chapter 7). To investigate if this overrepresentation of CF_2 is caused by central neuronal connectivity or by cochlear specializations, it was necessary to obtain cochlear frequency maps.

An effective method to explore the frequency representation along the basoapical extension of the basilar membrane (BM) is the HRP frequency mapping technique. The neuronal tracer horseradish peroxidase (HRP) is injected intracellularly in auditory nerve fibers (Liberman 1982) or extra-

---

FIGURE 5.8A–D. (*continued*) *Pteronotus parnellii*. (C) The optimum frequency separation Δf between f_1 and f_2 to elicit maximum levels of the $2f_1$-f_2 distortion is plotted for different f_1 frequencies in *Carollia perspicillata, Pteronotus parnellii*, and *Rhinolophus rouxi*. (D) Threshold of the $2f_1$-f_2 distortion in *Carollia perspicillata* and *Pteronotus parnellii*. Plotted is the level of f_1 sufficient to produce a distortion of -10 dB SPL. For each f_1 frequency, the frequency of f_2 was adjusted to match the optimum Δf (adapted from Kössl 1992a,b, 1994).

cellularly close to the synaptic endings of the auditory nerve in the cochlear nucleus (Vater, Feng, and Betz 1985). The frequency tuning of the nerve fiber or the target neuron in the cochlear nucleus is determined before the injection. The tracer is transported toward the spiral ganglion cell body and also labels the afferent terminals at the IHCs. With multiple injections at different BFs, the frequency representation valid for IHCs can be obtained. HRP frequency maps have been published for *Rhinolophus rouxi* (Vater, Feng, and Betz 1985) and *Pteronotus parnellii* (Kössl and Vater 1985b; Zook and Leake 1989). In both CF-FM bat species, a narrow range around the CF_2 frequency is largely expanded in terms of BM length (Fig. 5.9A). The frequency ranges between about 76–83 kHz (*Rhinolophus*) and 59–66 kHz (*Pteronotus*) are expanded to about 30% of the BM length. In both species the maximum frequency expansion is found at CF_2 with about 40 mm BM/octave. In other parts of the cochlea, 2–3 mm of BM length are used for representation of one octave, which is similar to the frequency maps obtained in nonspecialized mammals such as the cat or rat (Fig. 5.9B). A frequency map of *Rhinolophus ferrumequinum*, obtained by analyzing morphological changes in OHCs after loud pure tone exposure (Bruns

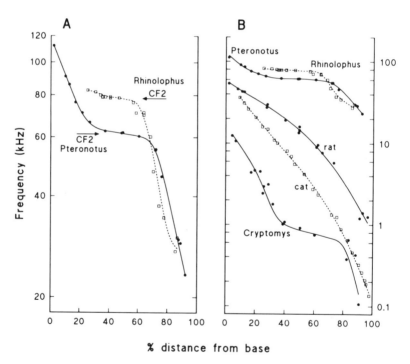

% distance from base

FIGURE 5.9A,B. Cochlear frequency maps of *Rhinolophus rouxi* (from Vater, Feng, and Betz 1985) and *Pteronotus parnellii* (from Kössl and Vater 1985b, 1990a) compared to maps from the rat (from Müller 1991), cat (from Liberman 1982), and mole rat *Cryptomys hottentottus* (from Müller et al. 1992).

1976b), shows a similar expansion of the CF_2 frequency range. However, this map is shifted toward the base (for discussion, see Vater 1988). The only other mammal with an expanded frequency representation of a narrow frequency range found so far is the African mole rat *Cryptomys hottentottus* (see Fig. 5.9B). In *Cryptomys*, frequencies between 0.6 and 1 kHz are expanded on about 50% of BM length (Müller et al. 1992), and the respective frequency correlates with a minimum in hearing threshold (Müller and Burda 1989). The sharp threshold minima of *Rhinolophus* and *Pteronotus* are within the range of the expanded cochlear region.

As a possible consequence of cochlear frequency expansion, high-frequency resolution and sharp tuning around the CF_2 frequency could emerge. However, the expanded frequency range extends a few kilohertz both apically and basally of CF_2 and reaches into frequency bands where neuronal tuning is poor. Enhanced cochlear tuning seems to be restricted to a small part of the expanded region. Moreover, in *Pteronotus* there is no conspicuous cochlear frequency expansion around 90 kHz, where also very sharp neuronal tuning and a small optimum frequency separation during distortion measurement are found. This leads to the conclusion that cochlear frequency expansion and enhanced tuning are not necessarily causally linked. Of course, an expanded cochlear region should be well suited to resolve sharp mechanical tuning provided from a different source.

5. Cochlear Anatomy

5.1 General Morphological Features

A schematic illustration of the gross anatomy of the bat cochlea (Fig. 5.10) serves as a reference for further detailed descriptions in later chapters. The basic structural composition of the cochlear duct is identical to the typical mammalian scheme (e.g., Lim 1986), but the relative dimensions of components and the organization of the anchoring system of the BM reflect adaptations for high-frequency hearing (Henson 1970; Firbas 1972; Bruns 1980; Vater, Lenoir, and Pujol 1992).

The receptor cells are organized into one row of IHCs and three rows of OHCs. The dimensions of OHC bodies and stereocilia are small as compared to mammals with good low-frequency hearing (guinea pig, cat, human; see Section 5.4.3.) The IHCs are situated medial to the pillar cells (PC) on the primary osseous spiral lamina (OSL) and therefore are not directly displaced by BM motion (e.g., Bruns 1980). The hair cell stereocilia are in close contact with the tectorial membrane (TM), which varies in shape in different regions of the cochlea (see also Section 5.3). The afferent innervation of hair cells is derived from the peripheral processes of spiral ganglion (SG) cells (see also Section 5.4.5). In addition to the sensory cells, the organ of Corti contains different types of well-developed supporting

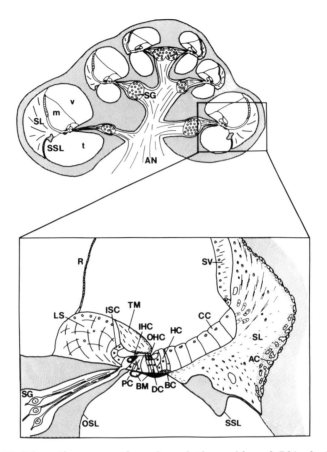

FIGURE 5.10. Schematic cross sections through the cochlea of *Rhinolophus rouxi* illustrate the composite structures of the organ of Corti in the second half-turn in further detail. Fluid spaces, *v*: scala vestibuli, *m*: scala media, *t*: Scala tympani are marked only in the first basal half-turn. AC, anchoring cells; AN, auditory nerve; BM, basilar membrane; BC, Boettcher cells; CC, Claudius cells; DC, Deiters cells; HC, Hensens cells; IHC, inner hair cell; ISC, inner sulcus cells; LS, spiral limbus; OC, otic capsule; OHC, outer hair cell; OSL, osseous spiral lamina; PC, pillar cells; R, Reissners membrane; SG, spiral ganglion; SL, spiral ligament; SSL, secondary spiral lamina; SV, stria vascularis; TM, tectorial membrane. (See text.)

cells. Sturdy pillar cells (PC) border the tunnel of Corti, and large Deiters cells (DC) provide the structural link between OHCs and the BM. Hensens cells (HC), Böttcher cells (BC), and the Claudius cells (CC) are situated lateral to the sensory epithelium (Henson, Jenkins, and Henson 1982, 1983).

The medial attachment of the BM is formed by the OSL and the lateral attachment by the spiral ligament (SL). Cross sections of the BM reveal two thickened portions (Figs. 5.10 and 5.11): pars arcuata (PA) situated beneath the pillar cells and pars pectinata (PP) lateral to the pillar cells. These thickenings are most prominent in the basal turn but can be found throughout most of the cochlear spiral except for the extreme apex. Such thickenings are also present in the basal turn of nonecholocating mammals with good high-frequency hearing and non-CF-FM bats, but the structure of the PP in the cochlear base of CF-FM bats is qualitatively different (see Section 5.2).

The spiral ligament is considerably enlarged in the basal turns and is composed of a matrix containing a network of extracellular fibers and

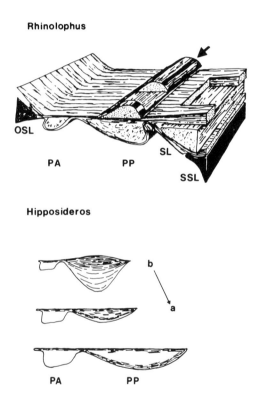

FIGURE 5.11. Structural organization of the basilar membrane in the basal turn of *Rhinolophus ferrumequinum* (*top*; adapted from Bruns 1980) and of *Hipposideros bicolor* (*bottom*; adapted from Dannhof and Bruns 1991). In *Rhinolophus*, the thickening of the scala tympani side of the pars pectinata (PP) contains radially directed fibers whereas the thickening at the scala vestibuli side is composed of longitudinally directed fibers. The thickening of the PP in Hipposideros (shown for different basoapical positions: 0.4 mm, 4 mm, 7 mm) only contain radially organized filaments. OSL, osseous spiral lamina; PA, pars arcuata; SL, spiral ligament; SSL, secondary spiral lamma.

several cell types (see Fig. 5.10). The stress fibers of the ligament and the anchoring cells (AC) along the bony otic capsule (OC) may be involved in creation of radial tension on the BM (Henson, Henson, and Jenkins 1984; Henson and Henson 1988). As a unique feature in *Pteronotus parnellii*, a local enlargement of the spiral ligament is found in the middle of the basal turn at the transition between the sparsely innervated and densely inner-vated cochlea region (Fig. 5.12) (Henson 1978; Kössl and Vater 1985b).

At the scala tympani side of the SL, a bony secondary spiral lamina (SSL) is present throughout most of the cochlea of bats whereas in other mammals it is confined to the basal turns (Firbas 1972). In some CF-FM bats (*Rhinolophus, Hipposideros*), the SSL has a specialized shape (Figs. 5.10, 5.11) that changes systematically along the cochlear spiral (Bruns 1976a; Dannhof and Bruns 1991).

The size of the perilymphatic spaces — scala vestibuli and scala tympani — decreases rather regularly and symmetrically from base to apex in the cochlea of *Rhinolophus* (Bruns 1976a) and *Myotis* (Ramprashad et al. 1979). In the mustached bat (*Pteronotus parnellii*) and the frog-eating bat (*Trachops cirrhosus*), however, deviations from this pattern were noted. In *Pteronotus parnellii*, the scala vestibuli of the basal turn is subdivided into two large chambers by a bony indentation that creates a focal narrowing (Henson, Henson, and Goldman 1977; Kössl and Vater 1985b). The volume of the scala tympani does not change in parallel; rather, it reaches its

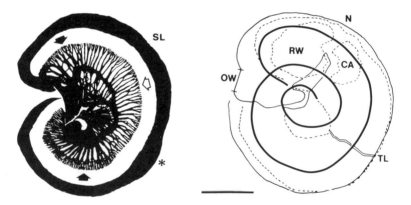

FIGURE 5.12. *Left*: Whole-mount preparations of the basal turn of the cochlea of *Pteronotus* illustrates local changes of nerve fiber densities (*black arrows*, location of maxima in innervation density; *white arrows*, sparsely innervated region) and in size of the spiral ligament (SL; *asterisk*). *Right*: Horizontal projection of the cochlea shows the course of the basilar membrane for all turns (*thick line*) and the size of fluid spaces of the basal turn (scala vestibuli; *thin lines*; scala tympani; *stippled lines*; N, narrowing of scala vestibuli), the location of the oval window (OW), the enlarged size of the round window (RW), and the cochlear aqueduct (CA) and the location of the apical border of the "thick lining" (TL). (Adapted from Henson and Henson 1991; Kössl and Vater 1985b.) Calibration bar, 1 mm.

maximal size beneath the narrowest point of scala vestibuli. The round window and the cochlear aquaeduct are located in this region and are much larger in size than in other species (Henson 1970). A further unique feature of the scala tympani in the basal turn of *Pteronotus* is the presence of a "thick lining" composed of cells associated with thick extracellular osmiophilic substance (Jenkins, Henson, and Henson 1983). The functions of these specializations are unknown. In the frog-eating bat *Trachops cirrhosus*, a species that appears to have exceptionally good low-frequency hearing (Bruns, Burda, and Ryan 1989), the volume of the scala tympani in the apical 50% of the cochlea is remarkably smaller than that of the scala vestibuli. Such asymmetries in scala volume are said to be typical for small mammals with good low-frequency hearing (Müller et al. 1992).

5.2 Basilar Membrane

The BM performs the first important steps in cochlear frequency analysis. Systematic increases in width and decreases in thickness from base to apex produce a stiffness gradient that creates a regular frequency representation along the membrane: high-frequency signals maximally displace basal cochlea regions where stiffness is high, and low frequencies are mapped progressively more apically following the decrease in stiffness (von Békésy 1960).

In general, the BM of Microchiroptera is significantly longer than that of nonecholocating mammals if relative body weight is taken into account (Table 5.1). The longest BMs are found in CF-FM bats. This feature is neither related to a widening of absolute hearing range nor typical for particular taxonomic groups, but results from the presence of an acoustic fovea (see also Sections 2 and 4). Furthermore, the BM of Microchiroptera is narrower and thicker than in most nonecholocating mammals, particularly those with good (cat, guinea pig) or predominantly (mole, rats: *Spalax, Cryptomys*) low-frequency hearing (Table 5.1). The highest values of BM thickenings are found in the basal cochlear regions of CF-FM bats. Although the thickening of the PP in the basal turn only contains radially directed filaments in *Hipposideros* (Dannhof and Bruns 1991) (see Fig. 5.11), the morphology of PP in bats emitting long CF-FM calls differs from the common mammalian pattern. In *Rhinolophus* (Bruns 1980) (see Fig. 5.11) and *Pteronotus* (Vater, unpublished data), the PP is composed of two parts: the part facing the scala tympani contains radially directed filaments, whereas the part facing the scala media is composed of longitudinally directed filaments. The latter feature suggests the presence of a longitudinal mechanical coupling within the membrane and has been suggested to create shallow high-frequency slopes in neurons with BFs slightly above CF_2 (Kössl and Vater 1990b; see also Figs. 5.3 and 5.5). Because the thickenings of PA and PP are not continuous but are linked via a thin segment, it is questionable if they can be viewed as structures simply increasing BM

TABLE 5.1. Basilar membrane (BM) dimensions in various mammalian species.

Species	Body weight	BM length (mm)	BM thickness (µm) Base	BM thickness (µm) Apex	BM width (µm) Base	BM width (µm) Apex	Reference
Rhinolophus ferrumequinum (horseshoe bat)	18 g	16	35	2	90	150	Bruns 1976a
Pteronotus parnellii (mustached bat)	12 g	14.3	22	2	50	110	Henson 1978; Kössl and Vater 1985b
Hipposideros fulvus	10 g	8.8	28	2	60	100	Kraus 1983
Hipposideros speoris	10 g	9.2	23	2	70	90	Dannhof and Bruns 1991
Molossus ater	37 g	14.6	10	1	60	130	Fiedler 1983
Rhinopoma hardwickii		11.2	14	2	70	150	Kraus 1983
Taphozous kachensis	50 g	14.4	12	2	60	140	Fiedler 1983
Myotis lucifugus	8 g	6.9	10	2	60	115	Ramprashad et al. 1979
Megaderma lyra (false vampire)	48 g	9.9	14	2	80	140	Fiedler 1983
Trachops cirrhosus (frog-eating bat)		14.9	19	2	60	160	Bruns, Burda, and Ryan 1989
Mus musculus (mouse)	40 g	6.8	15	1	100	170	Ehret and Frankenreiter 1977
Rattus norwegicus (rat)	400 g	10.4	19	8	140	240	Roth and Bruns 1992
Spalax ehrenbergi (mole rat)		13.7			110	200	Bruns et al. 1988
Cryptomys hottentottus (mole rat)	80 g	12.9	5	1.5	130	175	Müller et al. 1992
Cavia porcella (guinea pig)	400 g	18	7.4	1.3	100	245	Fernandez 1952
Felis cattus (cat)	3 kg	23.6	12	5	105	420	Cabuzedo 1978
Homo sapiens (human)	60 kg	32					Nadol 1988
Bos bovis (cattle)	500 kg	38			104	504	von Békésy 1960
Elephas max (elephant)	4000 kg	60					von Békésy 1960

stiffness. Rather, this arrangement might allow for relatively independent motion of inner and outer BM segments (Steele 1976; Ehret and Franken-reiter 1977; Bruns 1980).

The basoapical gradients in BM dimensions of different bat species are illustrated in Figure 5.13 together with measurements obtained in the cat, an auditory generalist with good low- and high-frequency hearing, and the African mole rat *Cryptomys*, specialized on low-frequency hearing. The gradients are clearly species specific, but common features are noted among those species sharing certain hearing characteristics.

In FM bats with unspecialized tuning properties (*Myotis lucifugus*, Ramprashad et al. 1979; *Megaderma lyra*, Fiedler 1983; see Fig. 5.13), both the width and the thickness of the basilar membrane change gradually toward the apex. Such regular patterns are also seen in most nonecholo-cating species (cat [Fig. 5.13], rat, guinea pig, mouse), but the absolute gradients in BM width can be considerably larger than in bats (see also Table 5.1). These features are expected to correlate with differences in absolute hearing range and the species-specific course of frequency maps (see also Fig. 5.9).

In *Trachops cirrhosus* (Fig. 5.13), the frog-eating bat, which emits FM calls but largely depends on passive listening to frequencies below 5 kHz for catching its prey, a conspicuous BM thickening in the basal turn is followed by a significant decrease in BM thickness to about 60% distance from base. From there up to the apex, thickness is less (about 2 μm). The BM width gradually increases from base to apex with the greatest basoapical differ-ences found in small mammals. Accordingly, *Trachops* is expected to have the largest frequency range among the species studied to date, and might have the best low-frequency hearing among bats (Bruns, Burda, and Ryan 1989).

The basoapical gradients in BM morphology of CF-FM bats deviate considerably from the patterns observed in other mammals. This feature appears related to specializations of the frequency map and tuning proper-ties of single units. Because the observed patterns differ among species of CF-FM bats, they are described individually.

Rhinolophus: In *Rhinolophus* (Fig. 5.13; Bruns 1976a), the BM in the basal turn is considerably thickened up to a point located at about 25% distance from the base, where the thickness abruptly drops from 35 μm to 10 μm within only a few hundred micrometers. From this point to about 55% distance from the base, BM thickness remains almost constant. The pattern of change in BM width found up to about 55% distance from the base is also highly unusual. Coinciding with the transition in BM thickness, the BM width decreases and then remains almost constant up to about 50% distance from the base. A systematic increase in BM width and a regular decrease in BM thickness, typical for a nonspecialized cochlea, is found only in more apical regions.

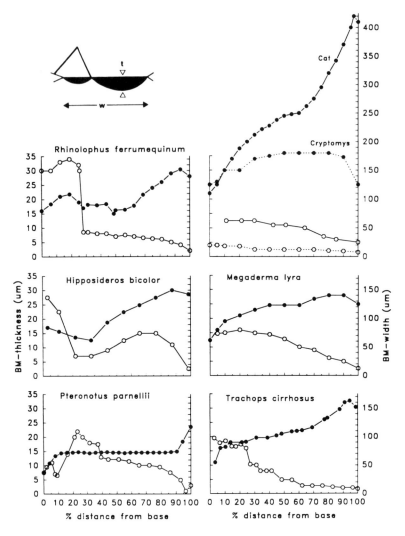

FIGURE 5.13. Basoapical gradients in BM width (*solid symbols*) and BM thickness (*open symbols*) represented on normalized BM length in different species. *Left column*: CF-FM bats (*Rhinolophus ferrumequinum*, after Bruns 1976a; *Hipposideros bicolor*, after Dannhof and Bruns 1991; *Pteronotus parnellii*, BM thickness after Kössl and Vater 1985b and BM width after Henson 1978). *Right column*: Nonecholocating mammals and FM bats (cat after Cabezudo 1978; *Cryptomys* after Müller et al 1992; *Megaderma lyra* after Fiedler 1983; *Trachops cirrhosus* after Bruns, Burda, and Ryan 1989).

Hipposideros: Some of the specialized features observed in *Rhinolophus* are also present in the cochlea of the related *Hipposideros* (Dannhof and Bruns 1991; see Fig. 5.13). The changes in BM width and thickness in the basal turn appear less abrupt than in *Rhinolophus*, but this feature probably results from a rather wide spacing of the measurement points. In contrast to other bats, *Hipposideros* possesses a second broad maximum in BM thickness located in the apical cochlea.

Pteronotus: In *Pteronotus* (Fig. 5.13; Kössl and Vater 1985b), low values of BM thickness are measured within the first 10% of cochlear length. These are followed by an apically directed increase in thickness between 10% and 23% of BM length. The area of increased BM thickness is apically terminated by an abrupt decrease of BM thickness at about 40% distance from the base. More apically, the decrease in BM thickness is very slight to about 80% distance from base. Contrasting with all other species, the BM width in *Pteronotus* appears to be constant throughout the cochlea, except for the very basal and apical portions (Henson 1978).

Although the exact gradients in BM morphology clearly differ among CF-FM bats, some common features are noted. First, in both *Rhinolophus* and *Pteronotus*, an abrupt decrease in BM thickness occurs just basal to the place of representation of the second-harmonic CF signal component. The functional role of these discontinuities is not fully understood, but they might play a role in enhancing tuning in a narrow frequency band (see Section 6.2.2). Second, areas of very little change in BM morphology, found just apical to the discontinuity, in both species include the expanded representation of the second harmonic CF component and coincide with a local maximum in innervation density (see also Section 5.4.5). In these regions, the stiffness gradient is probably very slight, thus leading to an expanded frequency mapping. Interestingly, the only other mammal in which expanded frequency mapping is found is the African mole rat (see Fig. 5.9). In the mole rat, long stretches of BM encompassing the acoustic fovea also have almost constant BM morphology (Fig. 5.13; Müller et al. 1992).

Thus, the presence of expanded mapping correlates with the presence of cochlear regions with almost constant BM stiffness. However, frequency expansion and enhanced tuning do not appear directly related. In *Pteronotus*, the representation places of CF_1 and CF_3 showing enhanced tuning do not coincide with BM specializations compatible to the CF_2 region. CF_1 is represented at about 90% distance from the base where a regular decrease in BM thickness occurs while CF_3 corresponds to a minimum in BM thickness at about 10% distance from base. The inverse pattern of BM thickness change within the most basal 20% of BM length in *Pteronotus* has been suggested to facilitate reverse traveling wave propagation, thus contributing to the exceptionally loud otoacoustic emissions in this species (see Section 6.2.2).

5.3 Tectorial Membrane

The receptor cell stereocilia are sheared by relative movement between the organ of Corti and the TM. The TM is either considered as a rigid beam providing stiffness and mass for shearing displacement or as a second resonator superimposed on the BM resonance (Zwislocki and Kletsky 1979; Zwislocki 1986). In most mammals, the TM increases in size from base to apex (e.g., Lim 1986). This pattern is also seen in *Hipposideros* (Dannhof and Bruns 1991) and *Trachops* (Bruns, Burda, and Ryan 1989) in which the cross-sectional area of the TM increases from base to apex with the largest changes observed in the upper half of the cochlea (Fig. 5.14). *Pteronotus* is most unusual because the cross-sectional area of the TM, and consequently its mass in the basal turn, is about 3.5-fold larger than in other species, and a distinct maximum of cross-sectional area is found at about 55% cochlear length. The basoapical changes in height of the spiral limbus reflect the pattern observed in TM dimensions. The peculiar TM morphology is thought to have substantial effects on hair cell stimulation in specific frequency bands (Henson and Henson 1991).

While it is generally accepted that OHC stereocilia are firmly embedded in the subsurface of the TM, the mode of linkage of IHC stereocilia is disputed: in the chinchilla, IHC stereocilia appear not embedded at all (Lim

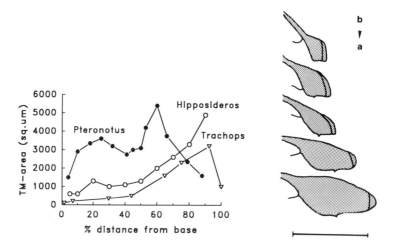

FIGURE 5.14. *Left*: Basoapical gradients in cross-sectional area of the TM in the CF-FM bats *Pteronotus parnellii* (after Henson and Henson 1991) and *Hipposideros bicolor* (after Dannhof and Bruns 1991), and the frog-eating bat, *Trachops cirrhosus* (after Bruns, Burda, and Ryan 1989). *Right*: Shape of the TM in cross-sections of the cochlea of *Hipposideros bicolor* obtained at different basoapical positions (from top to bottom at 0, 2, 4, 6, and 8 mm distance from base; after Dannhof and Bruns 1991). Note the prominent Hensen stripe at the subsurface of the TM. Calibration bar, 100 μm.

1986) whereas in the rat, imprints of IHC stereocilia in the subsurface of the TM are confined to the basal cochlea (Lenoir, Puel, and Pujol 1987). In the horseshoe bat *Rhinolophus rouxi*, imprints of IHC and OHC stereocilia have been observed throughout the cochlea (Vater and Lenoir 1992), and it was speculated that this feature is relevant for IHC stimulation in the high-frequency range of the mammalian audiogram. In all bat species studied so far, a prominent Hensen stripe is present on the subsurface of the TM above the region of the IHCs throughout most of the cochlea (Dannhof and Bruns 1991; Vater, Lenoir, and Pujol 1992). This structure may be relevant for IHC excitation at high frequencies, because in other mammals it is usually confined to basal cochlear regions (rat; Lenoir, Puel, and Pujol 1987).

5.4 Receptor Cells

5.4.1 Receptor Cell Arrangements

The structural organization of the receptor cells in the bat cochlea basically conforms to the patterns seen in most nonecholocating mammals: one single row of IHCs is located medial to three rows of OHCs. In the horseshoe bat and *Pteronotus*, the OHC arrangements are highly regular throughout the cochlea, and hair cell loss or additional hair cell rows typical for the rodent cochlea were not observed (Vater and Lenoir 1992; Vater, unpublished data). Such irregularities are, however, present in the apical cochlea of *Tadarida brasiliensis* and *Eptesicus fuscus* (Vater, unpublished data).

IHCs: Commonly, the elongated cuticular plates of IHCs form a tightly spaced single row in the longitudinal course of the organ of Corti (Lim 1986). A deviation from this bauplan was observed only in *Rhinolophus* (Bruns and Goldbach 1980; Vater and Lenoir 1992): in the lower basal turn, where the BM is considerably thickened, the spacing between neighboring IHCs is considerably enlarged because the size of cuticular plates and the number of stereocilia decrease. The transition from normal to specialized spacing coincides with the transition in BM dimensions (Vater and Lenoir 1992).

OHCs: Bat OHCs possess the typical W-shaped stereocilia bundles common to all mammals. As in most small mammals, three rows of stereocilia are present. The opening angle of the stereocilia bundles of bat OHCs is enlarged compared with other mammals (Vater and Lenoir 1992). This enlargement is paralleled by an increase in number of stereocilia per row and might produce a reinforced mechanical attachment of the TM relevant for high-frequency hearing.

Significantly, in *Rhinolophus* the transition in the morphology of IHC stereocilia bundles that parallels the change in BM morphology is not accompanied by a change in OHC organization.

5.4.2 Receptor Cell Ultrastructure

The ultrastructure of the receptor cell bodies of *Rhinolophus* is shown in Fig. 5.15 (Vater, Lenoir, and Pujol 1992). The salient cytological features closely conform to the typical mammalian pattern (Lim 1986). The IHC have a pearshaped body with an elongated neck portion carrying the cuticular plate. They are completely surrounded by supporting cells. An irregular arrangement of subsurface cisternae, specialized parts of the endoplasmic reticulum, is located along the lateral cell wall. The base of the IHC is contacted by numerous afferent endings of the dendrites of type I spiral ganglion cells. Each ending is opposed by a presynaptic bar or vesicle in the IHC. Efferent endings mainly synapse on the afferent dendrites.

The OHCs are cylindrical. Their cuticular plates are linked to the processes of supporting cells to form a continuous reticular lamina. The basal part of the OHC is attached to a specialized rigid cup formation of the Deiters cell. The lateral cell wall between the reticular lamina and the

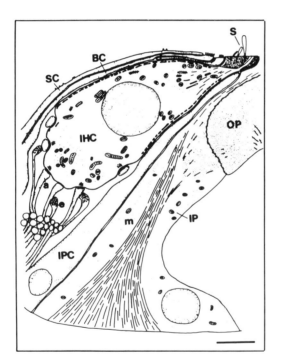
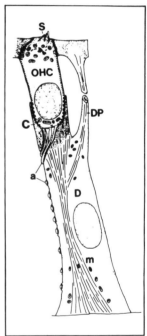

FIGURE 5.15. Ultrastructural organization of the organ of Corti of the horseshoe bat at the level of the IHC (left) and the level of the middle row of OHCs (right) (after Vater, Lenoir, and Pujol 1992). a, afferent endings; e; efferent endings; D; Deiters cell body; SC, inner sulcus cell; C, Deiters cup; DP, Deiters cell process; IHC, inner hair cell; IP, inner pillar cell; IPC, inner phalangeal cell; ISB; inner spiral bundle; m, microtubuli bundle; OP, head of outer pillar cell; S, stereocilia; ssc, subsurface cisternae. Calibration bar, 5 μm.

Deiters cup is surrounded by the fluid spaces of Nuel. A highly regular cytoskeleton is located along the lateral cell wall, composed of vesicle-shaped subsurface cisternae attached to the cell membrane via electron-dense pillars. Such a system is expected to play an integral role in mammalian OHC function: it forms a cytoskeletal spring maintaining the shape of the cylindrical cell, which is important for passive as well as active motion. Transmembrane motorproteins likely to underlie the fast force generation mechanism in OHCs of other mammals (Kalinec et al. 1992) remain to be demonstrated in bats. The innervation of OHCs in the horseshoe bat is unusual because it consists of afferent endings (dendrites of type II ganglion cells) only (Bruns and Schmieszek 1980; Bishop and Henson 1988; Vater, Lenoir, and Pujol 1992; and see also Section 5.4.5).

5.4.3 Receptor Cell Dimensions

In mammals, apical OHCs and their stereocilia are much longer than basal ones (Spoendlin 1966). The absolute dimensions and gradients in OHC length and stereocilia size are related to the frequency representation of the organ of Corti: the gradations of stereocilia length along the cochlea follow a tonotopic order (Strelioff, Flock, and Minser 1985), and there is evidence that sound-induced motility of the OHCs is tuned and correlated with cell length (Brundin, Flock, and Canlon 1989); the smaller the cell, the higher its BF.

The dimensions of the receptor cells and their stereocilia in bats, in particular those of the OHCs, are considerably smaller than in other mammals (Dannhof and Bruns 1991; Vater and Lenoir 1992; Vater, Lenoir, and Pujol 1992). Figure 5.16 shows that the maximal length of OHCs in the apical cochlea of *Hipposideros* (Dannhof and Bruns 1991) is lower than the minimal length measured in the guinea pig cochlea (Pujol et al. 1992). In fact, it appears to represent a basally directed continuation of the gradient seen in the other species. The short size of OHCs in the basal turn of *Hipposideros* correlates with the expansion of its high-frequency limit of hearing to values of at least 160 kHz as compared to about 40 kHz in the guinea pig (Heffner, Heffner, and Masterton 1971). Because OHC and stereocilia size (Fig. 5.17) in the apex of the bat cochlea are much smaller than those observed in other mammals, it might be expected that the low-frequency limit of hearing in the bat is located at higher frequencies.

The OHC stereocilia in bats show a clear trend toward miniaturization (see Fig. 5.17). The comparison with the human cochlea (Wright 1984) again shows that the gradients observed in bats appear as a basally directed continuation of the gradients seen in other species. *Rhinolophus* (Vater and Lenoir 1992), however, possesses smaller OHC stereocilia in the apical cochlea than *Hipposideros* (Dannhof and Bruns 1991). As physiological data are few, we can only speculate that the lower frequency limit of hearing and/or the representation of low frequencies might differ between the two bats.

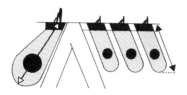

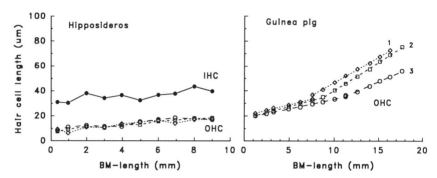

FIGURE 5.16. Basoapical gradients in the length of the cell bodies of IHCs and OHCs in *Hipposideros bicolor* (after Dannhof and Bruns 1991) and OHCs in the guinea pig (after Pujol et al. 1992). Different symbols and numbers denote measurements for innermost (first), middle (second), and outermost (third) row of OHCs.

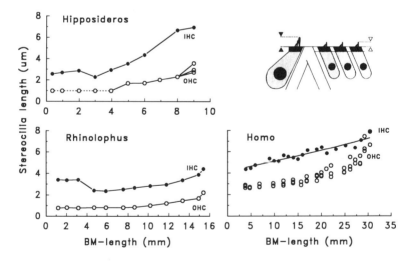

FIGURE 5.17. Basoapical gradients in sterocilia length for IHCs (*solid symbols*) and OHCs (*open symbols*) in the cochlea of CF-FM bats (*Hipposideros bicolor*, after Dannhof and Bruns 1991; *Rhinolophus rouxi*, after Vater and Lenoir 1992) and the human (after Wright 1984).

5.4.4 Receptor Cell Densities

The densities of OHCs in the cochlea of bats fall in the range reported for other mammals, but the density of IHCs appears slightly higher (Burda, Fiedler, and Bruns 1988). In particular, *Rhinolophus* has the highest density of IHC/mm of all bats studied (Bruns and Schmieszek 1980; Fig. 5.18[o]). Independent of the echolocation calls used, a trend for increasing OHC density towards the cochlear apex is observed while density of IHCs stays almost constant (Fig. 5.18).

5.4.5 Receptor Cell Innervation

Afferent innervation. The afferent innervation patterns of IHCs and OHCs have only been quantified in the horseshoe bat (Bruns and Schmieszek 1980) and are schematically illustrated in Figure 5.18. As in other mammals (Spoendlin 1973), the bulk of afferents contact the IHCs via highly convergent radially directed dendrites of type I spiral ganglion cells. Depending on cochlear location, between 80% and 90% of the afferent fibers terminate at the IHCs (Bruns and Schmieszek 1980). A small percentage of afferents contact the OHCs via outer spiral fibers that are probably derived from type II ganglion cells (Vater, Lenoir, and Pujol 1992; Zook and Leake 1989) coursing toward the cochlear base and making synaptic contacts with several sensory cells. The travel distance varies among turns (Bruns and Schmieszek 1980). As compared to other mammals in which only about 5% of the total afferents contact OHCs (Spoendlin 1973), the proportion of the afferent innervation of OHCs of the horseshoe bat is two- to fourfold higher.

Quantitative measurements of densities of spiral ganglion cells are available for several species (see Fig. 5.18), giving clues on variations of receptor cell innervation along the cochlea duct. However, only the study of Zook and Leake (1989) gives direct information on innervation densities of IHCs, because only type 1 ganglion cells were counted. In all species, distinct maxima and minima in innervation density are found. In *Pteronotus* (Henson 1973; Zook and Leake 1989), two maxima are located in the basal turn, separated by a sparsely innervated zone (see also Fig. 5.12). These innervation maxima correspond to the representation places of the second and third harmonic of the orientation call (Kössl and Vater 1985b; Zook and Leake 1989; also see Section 4). The sparsely innervated zone corresponds to the region of thickened BM. In *Rhinolophus* (Bruns and Schmieszek 1980), absolute densities of spiral ganglion neurons are much lower than in *Pteronotus*. The region of thickened BM is sparsely innervated, and the maximum of innervation density occurs in the upper basal turn, where according to Vater, Feng, and Betz (1985) the second-harmonic CF is represented. In *Taphozous* (Burda, Fiedler, and Bruns 1988), maxima and minima in innervation density are less pronounced than in CF-FM bats.

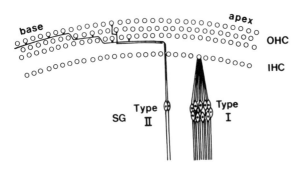

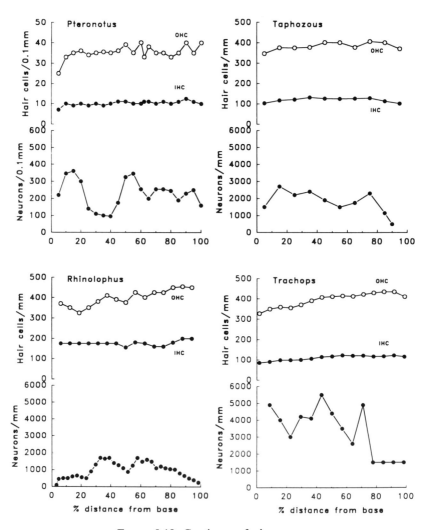

FIGURE 5.18. Caption on facing page.

The highest densities of spiral ganglion cells are found in the frog-eating bat *Trachops* (Bruns, Burda, and Ryan 1989).

The ratio of afferent fibers to IHCs in the region of innervation maxima amounts to 34–37:1 in *Pteronotus* (Zook and Leake 1989) and to 23.5:1 in *Rhinolophus* (Bruns and Schmieszek 1980). These values are in the range of maximal innervation densities reported for the cat at places in the cochlea representing frequencies above 2 kHz (Liberman, Dodds, and Pierce 1990). In the sparsely innervated zones of the cochlea of *Pteronotus* and *Rhinolophus*, only 8–12 afferent fibers contact IHCs. These values agree with the lowest values observed in the apex and extreme hook region of the cat cochlea (Liberman, Dodds, and Pierce 1990; Keithley and Schreiber 1986). In the FM bat *Myotis*, a ratio of 70:1 was calculated (Ramprashad et al. 1978), which is much higher than in CF-FM bats or other mammals. The exact numbers of afferent endings per OHC have not been quantified throughout the cochlear length in bats. Nevertheless, it is of interest to note that ratios of 4:1 in the basal turn of *Rhinolophus* (Vater and Lenoir 1992) and about 6:1 in the basal turn of *Pteronotus* (Bishop and Henson 1988) closely agree with values measured in basal cochlea regions of the cat (Liberman, Dodds, and Pierce 1990).

Efferent innervation. The efferent cochlear system in mammals is organized into two subsystems. The lateral olivocochlear system arises from small fusiform neurons located within or around the lateral superior olive (LSO) and synapses predominantly in the ipsilateral cochlea on afferent dendrites of type I ganglion cells contacting the IHCs. The medial olivocochlear system arises from large multipolar neurons located in the vicinity of the medial superior olive (MSO) and directly contacts OHC bodies of both cochleae with a distinct bias for the contralateral side (for review see: Warr 1992). Despite decades of intense research, the exact functions of the cochlear efferents remain unclear. It has been proposed that the lateral olivocochlear system exerts a tonic influence on primary auditory afferents (Liberman 1990). The medial olivocochlear efferents decrease cochlear sensitivity through a mechanism involving mechanical changes in the OHCs (Brown and Nuttal 1984) and might serve to protect the cochlea from sound-induced trauma (Rajan and Johnstone 1988), improve signal detectability in noise (Liberman 1988), or are involved in homeostasis (Johnstone, Patuzzi, and Yates 1986) by setting

←───

FIGURE 5.18. *Top*: Schematic representation of basic innervation pattern of the organ of Corti in the horseshoe bat (*Rhinolophus ferrumequinum*, after Bruns and Schmieszek 1980). *Bottom*: Basoapical gradients of hair cell densities (upper graphs) and corresponding densities of spiral ganglion cells (lower graphs) in different bat species: *Pteronotus parnellii* (after Zook and Leake 1989); *Rhinolophus ferrumequinum* (after Bruns and Schmieszek 1980); *Taphozous kachhensis* (after Burda, Fiedler, and Bruns 1988); and *Trachops cirrhosus* (after Bruns, Burda, and Ryan 1989).

the operating point of OHCs compensating for a bias in BM position introduced by fluctuations of endolymphatic pressure.

The origin of olivocochlear fibers has been studied in CF-FM bats (*Pteronotus*, Bishop and Henson 1987, 1988; *Rhinolophus*, Aschoff and Ostwald 1987) and non-CF-FM bats (*Rhinopoma, Tadarida, Phylostomus*; Aschoff and Ostwald 1987). In all species, the lateral efferent system is present and arises in nuclei located close to the LSO (*Pteronotus:* interstitial nucleus; *Rhinolophus,* nucleus olivocochlearis; Fig. 5.19) or in cells located within the main body of LSO (*Tadarida, Rhinopoma, Phylostomus*). It is remarkable that this system only contains the ipsilateral component. Putative neurotransmitters in the horseshoe bat include acetylcholine (Bruns and Schmieszek 1980), γ-aminobutyric acid (GABA) (Vater, Kössl, and Horn 1992), and enkephalins (Tachibana, Senuma, and Kumamoto 1992). Ultrastructural data (Vater, Lenoir, and Pujol 1992) show that in the horseshoe bat, as in other mammals, efferent endings predominantly contact dendrites of type I ganglion cells.

The nuclei of origin of the medial olivocochlear system are present in all

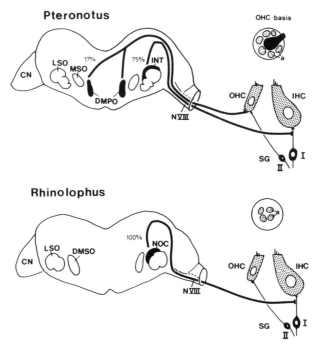

FIGURE 5.19. The origin and termination of the efferent innervation of the cochlea in *Pteronotus parnellii* (adapted from Bishop and Henson 1988) and *Rhinolophus* Spec. (adapted from Aschoff and Ostwald 1987; Vater, Lenoir, and Pujol 1992). CN, cochlear nucleus; DMPO, dorsomedial periolivary nucleus; DMSO, dorsomedial superior olive; INT, interstitial nucleus; LSO, lateral superior olive; MSO, medial superior olive; NOC, nucleus olivocochlearis; SG, spiral ganglion.

species (*Pteronotus:* dorsomedial periolivary nucleus, DMPO; Fig. 5.19) with the notable exception of *Rhinolophus* (Fig. 5.19). As in other mammals, the medial olivocochlear system of bats projects to the cochlea of both sides with the dominant component arising contralaterally. The lack of medial olivocochlear neurons in horseshoe bats agrees with reports that OHCs throughout the cochlea of horseshoe bats completely lack efferent synapses and thus considerably deviate from the typical mammalian scheme (Bruns and Schmieszek 1980; Bishop and Henson 1987, 1988; Vater, Lenoir, and Pujol 1992).

The medial olivocochlear system of *Pteronotus* has some specialized attributes. Each OHC only receives one efferent ending (Bishop and Henson 1988) centered at the OHC base and surrounded by afferent fibers (see Fig. 5.19), whereas in other mammals there are multiple efferent terminals on each OHC and their distribution varies along the cochlea (Liberman, Dodds, and Pierce 1990). The control of cochlear mechanics in *Pteronotus* appears very precise, because individual MOC fibers do not branch extensively and terminate on only one to as many as nine OHCs (Wilson, Henson, and Henson 1991), thus resembling the pattern in mice and deviating from the elaborate branching observed in the cat (Liberman, Dodds, and Pierce 1990). Acetylcholine esterase staining of efferent boutons in *Pteronotus* shows that they are particularly large within the auditory fovea (Xie et al. 1993).

The functional significance of the lack of the medial olivocochlear system in *Rhinolophus* and the presence of a uniform efferent system in *Pteronotus* is obscure as the two species are confronted with similar demands in analyzing echolocation signals. Explanations considering differences in cochlear mechanics, such as differences in overall damping (Bishop and Henson 1988), are intriguing for basal cochlear locations but of limited explanatory value because the lack of efferents is not confined to a particular cochlear place. Efferent terminals on OHCs are present in the cochleae of bats emitting FM calls (*Pteronotus suapurensis, Tadarida brasiliensis, Eptesicus fuscus*; Vater unpublished data). Interestingly, the only other mammal with no or considerably reduced efferent innervation of OHCs is the mole rat *Spalax ehrenbergi*, a species that is specialized to low-frequency hearing (Raphael et al. 1991). Again, no generally valid explanation for the presence or lack of this feature can be found.

6. Cochlear Mechanisms

6.1 Passive and Active Processes in the Cochlea of Mammals

The frequency-response characteristics of the mammalian cochlea are shaped to a large degree by cochlear macromechanics, that is, the more or less "passive" properties of the tectorial and basilar membranes and the supporting structures. In most nonspecialized mammals, a continuous basoapical decrease of BM thickness and an increase of BM width generate a gradient of decreasing membrane stiffness. Thus, sound energy of different frequencies can be dissipated along the BM at different locations enabling the frequency place transformation of the cochlea. However, to explain the sensitive threshold, the sharp tuning characteristics, and the occurrence of spontaneous otoacoustic emissions, cochlear models must include "active" components that introduce energy into the passive traveling wave, preferentially at cochlear locations located basal to the traveling wave maximum (de Boer 1983a,b; Neely and Kim 1983, 1986). The active components are likely to induce fast positive mechanical feedback, which results in negative damping or negative impedance of the BM, and a slower negative feedback system seems to be necessary to stabilize the traveling wave (Zweig 1991). The movement and sharp tuning of the BM also could be affected by different radial oscillation modes (Kolston et al. 1989; pars arcuata versus pars pectinata) or by radial differences in BM stiffness (Novoselova 1989). In addition, the TM could act as a second resonator and play a important role in a further filtering step to improve sharp tuning (Zwislocki and Kletsky 1979; Zwislocki 1986; Allen and Neely 1992).

The OHCs, which in vitro show motile responses to electrical and mechanical stimulation (e.g., Brownell et al. 1985; Zenner, Zimmermann, and Schmidt 1985; Ashmore 1987; Brundin, Flock, and Canlon 1989), seem to provide a cellular basis of the proposed active force generation in the cochlea. The extent to which movements of the OHC body are capable of following high-frequency sound stimuli and thus are able to act as mechanical amplifier on a cycle-by-cycle basis is still being investigated (Reuter et al. 1992; Santos-Sacchi 1992; He et al. 1993). Recent measurements of electrically induced traveling waves in the gerbil at about 30 kHz may indicate that fast movements of the OHC body are able to follow high frequencies (Xue, Mountain, and Hubbard 1993). In addition, slow contractions and elongations of OHCs that do not follow the frequency of the stimulus should significantly affect the geometry and interaction of the "passive" structures and may also contribute to a sharpening of the traveling wave.

6.2 Cochlear Structure and Function in Bats

To evaluate possible cochlear mechanisms from the viewpoint of a high-frequency echolocator, several features of structure and function in the cochlea of bats are discussed next.

6.2.1 Active Micromechanics

The strong spontaneous OAEs that occur in some individuals of *Pteronotus* at 60 kHz (up to 40 dB SPL) require a fast force generator which could be established by fast movements of the OHCs. In the foveal region of *Rhi-*

nolophus (Vater, Lenoir, and Pujol 1992), the OHCs are very short and are rigidly embedded in the Deiter cell cups and the reticular plate. This anatomical constellation, also valid for *Pteronotus,* should restrict the amplitude of fast movements of the OHC body. Indeed, in the FM bat *Carollia perspicillata,* the electrically induced movement in the isolated organ of Corti and of single OHCs of basal and apical cochlear turns is smaller than in the guinea pig and is below a noise level of 0.03 μm for frequencies above 1 kHz (Kössl et al. 1993). Therefore, the question is still open if the fast movements of the OHC body, which are likely to be very small at ultrasonic frequencies, can deliver enough energy to produce the large SOAEs. It has to be considered that the small OHC movements may feed into a nearly undamped resonator that depends on the passive macromechanics (see following).

One is not only looking for a mechanism to generate enhanced tuning and SOAEs at 60 kHz, but also for a more general process that ensures high sensitivity within a wide range of frequencies. The pronounced thickenings of the bat BM (pars pectinata, pars arcuata) should emphasize more or less passive mechanisms of sharp tuning that could be based on multiple oscillation modes in the radial direction (Kolston 1988; Kolston et al. 1989). Force generation in OHCs might influence the geometry of the organ of Corti and hence the possible oscillation modes.

It should be kept in mind that active force generation in mammalian OCHs also depends on transduction at the stereocilia. For low-level stimuli in the excitatory direction, the compliance of the mammalian hair bundle is increased (Russell, Kössl, and Richardson 1992), similar to the situation in the bullfrog sacculus (Howard and Hudspeth 1988). Such changes in stereocilia stiffness are a consequence of the asymmetric transduction mechanism at the tip links and may play a role in further increasing mechanical sensitivity. The large degree of tilt of the third-row stereocilia toward the second row in OHCs of *Rhinolophus* (Vater, Lenoir, and Pujol 1992) could enhance the excitatory versus inhibitory asymmetry of hair cells tuned to ultrasonic frequencies and also potentiate stiffness changes. The fact that in *Rhinolophus* the IHC stereocilia are embedded in the TM (Vater and Lenoir 1992) should enable a more direct communication between active processes in the body or stereocilia of the OHCs and the transduction site of the IHCs.

6.2.2 Macromechanics Exemplified in *Pteronotus*

In contrast to normal laboratory mammals, in *Pteronotus* and in other CF-FM bats the size and thickness of the BM and the TM vary considerably along the cochlear length.

i. At about 40% cochlear length the BM abruptly thickens toward the base (Fig. 5.13) which could provide a reflection zone for incoming waves. If the reflected waves reverberate between this discontinuity and the stapes, standing waves could occur and a more or less passive, but highly tuned resonator would be implemented (see also Kemp 1981; Duifhius and Vater 1985). The frequencies of the CF_2 range and of the OAEs are represented expandedly within a cochlear region just apical to the BM discontinuity. The OAEs are strongly suppressed by loud sound exposure with exposure frequencies between 62 and 70 kHz (Kössl and Vater 1985a, 1990b; see also Fig. 5.6). This range approximately matches the thickened BM region. The resonator would ensure the high sensitivity and sharp tuning just apically to the BM-thickness discontinuity.

ii. For frequencies at and just below CF_2, the cochlear resonance might, because of phase changes, produce an insensitivity. The neuronal on/off responses (Fig. 5.5) indicate that a delayed dampening or cancellation of the respective frequencies could be taking place. After switching off the tone burst stimulus, the resonator may oscillate on its own for a few cycles and produce the off response. Both OAEs and the ringing in the cochlear microphonic potentials can be evoked by slightly lower frequencies (Suga, Simmons, and Jen 1975; Kössl and Vater 1985a), that is, by stimuli within the on/off range. Therefore the on/off responses might arise from mechanical cancellations between the basilar membrane response to the stimulus and the concomitantly evoked resonance.

iii. The large size of the opening of the cochlear aqueduct and the "thick lining" of the scala tympani (see Section 5.1) could be adaptations to prevent damage from the reverberant oscillations. Apical to the 45% position, there is a local maximum of the TM area (Fig. 5.14; Henson and Henson 1991) whose functional significance is not yet understood. It could further dampen the 60-kHz response or act as a second resonator and enhance frequency tuning.

iv. At about 70 kHz there is a threshold maximum, the slopes of which do not coincide with enhanced tuning (see Fig. 5.7). Between 10% and 23% cochlear length a basoapical increase in BM thickness occurs (Fig. 5.13): This increase is just opposite to the normal mammalian gradient and may be responsible for the poor sensitivity. In addition, it could ensure that optimum reverse propagation of the OAEs does occur in the cochlea of *Pteronotus*, analogous to the "paradoxical waves" traveling toward the stapes in the mechanical model of von Békésy (1960), where the stiffness gradient was inverse.

v. It is still unclear whether the sharp neuronal tuning at about 90 and 30 kHz, which coincides with steep threshold slopes (Fig. 5.7), is generated by an harmonic effect of the 60-kHz resonator or by different mechanisms. After loud sound exposure at about 60 kHz, the 60-kHz threshold minimum of the N_1-off response from the auditory nerve is abolished but there are no significant effects on the thresholds at 30 and 90 kHz (Pollak, Henson, and Johnson 1979). This may suggest that independent cochlear mechanisms are acting at the three CF frequency ranges. The anatomical substrates for the specialized processing of CF_1 and CF_3 are largely unknown. There is a local decrease in BM thickness between about 6% and 9% of the cochlear length (Fig. 5.13) that coincides with the CF_3 area. It could be involved in producing another BM discontinuity responsible for 90-kHz reflections. However, OAEs have not yet been found in this frequency range.

7. Summary

The cochlea of bats is designed to yield high sensitivity at ultrasonic frequencies. For this purpose, the stiffness of the BM, as extrapolated from the thickness/width ratio, is increased. The size of the OHCs and their stereocilia is decreased, which may guarantee a high-speed micromechanical amplification of low-level signals by active OHC processes. In CF bats, the cochlear response is further shaped by macromechanical specializations of the BM, TM, and spiral ligament, and by the presence of an acoustic fovea, an expanded representation of the dominant CF frequency on the BM. A mechanical resonator that is evident in ringing of CM potentials and in otoacoustic emissions in the mustached bat is probably involved in creating extraordinarily sharp tuning to the CF frequency. In horseshoe bats, sharp tuning and high sensitivity in the CF range does not require the presence of the medial efferent system that is thought to affect OHC micromechanics.

References

Allen JB, Neely S T (1992) Micromechanical models of the cochlea. Physics Today, Vol. 45, July 1992, pp. 40–47.

Aschoff A, Ostwald J (1987) Different origins of cochlear efferents in some bat species, rats, and guinea pigs. J Comp Neurol 264:56–72.

Ashmore JF (1987) A fast motile response in guinea-pig outer hair cells: the cellular basis of the cochlear amplifier. J Physiol 288:323–347.

Bishop AL, Henson OW Jr (1987) The efferent cochlear projections of the superior olivary complex in the mustached bat. Hear Res 31:175–182.

Bishop AL, Henson OW Jr (1988) The efferent auditory system in Doppler-shift compensating bats. In: Nachtigall PE, Moore PWB (eds) Animal Sonar. New York: Plenum, pp. 307–311.

Brown AM, Gaskill SA (1990) Measurement of acoustic distortion reveals underlying similarities between human and rodent mechanical responses. J Acoust Soc Am 88:840–849.

Brown MC, Nuttall AL (1984) Efferent control of cochlear inner hair cell responses in the guinea pig. J Physiol 354:625–646.

Brownell WE, Bader CR, Bertrand D, Ribaupierre DEY (1985) Evoked mechanical responses of isolated cochlear outer hair cells. Science 227:194–196.

Brundin L, Flock A, Canlon B (1989) Sound induced motility of isolated cochlear outer hair cells is frequency selective. Nature 342:814–816.

Bruns V (1976a) Peripheral auditory tuning for fine frequency analysis by the CF-FM bat, *Rhinolophus ferrumequinum*. I. Mechanical specializations of the cochlea. J Comp Physiol 106:77–86.

Bruns V (1976b) Peripheral auditory tuning for fine frequency analysis by the CF-FM bat, *Rhinolophus ferrumequinum*. II. Frequency mapping in the cochlea.

J Comp Physiol 106:87–97.

Bruns V (1980) Basilar membrane and its anchoring system in the cochlea of the greater horseshoe bat. Anat Embryol 161:29–51.

Bruns V, Goldbach M (1980) Hair cells and tectorial membrane in the cochlea of the greater horseshoe bat. Anat Embryol 161:65–83.

Bruns V, Schmieszek E (1980) Cochlear innervation in the greater horseshoe bat; demonstration of an acoustic fovea. Hear Res 3:27–43.

Bruns V, Müller M, Hofer W, Heth G, Nevo E (1988) Inner ear structure and electrophysiological audiograms of the Subterranean mole rat, *Spalax ehrenbergi*. Hear Res 33:1–10.

Bruns V, Burda H, Ryan MJ (1989) Ear morphology of the frog-eating bat (*Trachops cirrhosus*, Family: Phyllostomidae): apparent specializations for low-frequency hearing. J Morphol 199:103–18.

Burda H, Fiedler J, Bruns V (1988) The receptor and neuron distribution in the cochlea of the bat, *Taphozous kachhensis*. Hear Res 32:131–136.

Cabezudo LM (1978) The ultrastructure of the basilar membrane in the cat. Acta Otolaryngol 86:160–175.

Dallos P, Corey M E (1991) The role of outer hair cell motility in cochlear tuning. Curr Opin Neurobiol 1:215–220.

Dannhof BJ, Bruns V (1991) The organ of Corti in the bat *Hipposideros bicolor*. Hear Res 53:253–268.

de Boer E (1983a) No sharpening? A challenge for cochlear mechanics. J Acoust Soc Am 73:567–573.

de Boer E (1983b) On active versus passive cochlear models — toward a generalized analysis. J Acoust Soc Am 73:574–576.

Dieler R, Shehata-Dieler WE, Brownell WE (1991) Concomitant salicylate-induced alterations of outer hair cell subsurface cisternae and electromotility. J Neurocytol 20:637–653.

Duifhuis H, Vater M (1985) On the mechanics of the horseshoe bat cochlea. In: Allen JB, Hall JL, Hubbard A, Neely ST, Tubis A (eds) Peripheral Auditory Mechanisms. Berlin: Springer, pp.89–96.

Ehret G (1974) Age-dependent hearing loss in normal hearing mice. Naturwissenschaften 11:506.

Ehret G, Frankenreiter M (1977) Quantitative analysis of cochlear structures in the house mouse in relation to mechanisms of acoustical information processing. J Comp Physiol 122:65–85.

Evans EF (1972) The frequency response and other properties of single fibres in the guinea pig cochlear nerve. J Physiol 226:263–287.

Evans EF (1975) Cochlear nerve and cochlear nucleus. In: Keidel WD, Neff WD (eds) Auditory Systems. (Handbook of Sensory Physiology, Vol.5/2.) Berlin: Springer, pp. 1–108.

Fernandez C (1952) Dimensions of the cochlea (guinea pig). J Acoust Soc Am 24:519–523.

Fiedler J (1983) Vergleichende Cochlea-Morphologie der Fledermausarten *Molossus ater, Taphozous nudiventris kachhensis* und *Megaderma lyra*. Ph.d. Thesis, University of Frankfurt, Germany.

Firbas W (1972) Über anatomische Anpassungen des Hörorgans an die Aufnahme höherer Frequenzen. Monatszeitschr Ohrenheilkd Laryngo-Rhinol 106:105–156.

Gaskill S A, Brown A M (1990) The behavior of the acoustic distortion product, f2$_1$-f$_2$, from the human ear and its relation to auditory sensitivity. J Acoust Soc Am 88:821-839.

Grinnell A D (1963) The neurophysiology of audition in bats: intensity and frequency parameters. J Physiol 167:38-66.

Grinnell A D (1973) Rebound excitation (off-responses) following non-neural suppression in the cochleas of echolocating bats. J Comp Physiol 82:179-194.

Habersetzer J, Storch G (1992) Cochlea size in extant chiroptera and middle eozene microchiropterans from Messel. Naturwissenschaften 79:462-466.

Habersetzer J, Schuller G, Neuweiler G (1984) Foraging behavior and Doppler shift compensation in echolocating bats, *Hipposideros bicolor* and *Hipposiideros speoris*. J Comp Physiol A 155:559-567.

He ZZ, Evans BN, Clark B, Sziklai I, Dallos P (1993) Voltage-dependent frequency response properties of isolated outer hair cells. Abstracts of the 30th Inner Ear Biology Workshop, Budapest, Hungary, p. 49.

Heffner R, Heffner H, Masterton RB (1971) Behavioural measurement of absolute and frequency-difference threshold in guinea pig. J Acoust Soc Am 49:1888-1895.

Henson MM (1973) Unusual nerve-fiber distribution in the cochlea of the bat *Pteronotus p. parnellii* (Gray). J Acoust Soc Am 53:1739-1740.

Henson MM (1978) The basilar membrane of the bat *Pteronotus p. parnellii*. Am J Anat 153:143-159.

Henson MM, Henson OW Jr (1988) Tension fibroblasts and the connective tissue matrix of the spiral ligament. Hear Res 35:237-258.

Henson MM, Henson OW Jr (1991) Specializations for sharp tuning in the mustached bat: the tectorial membrane and spiral limbus. Hear Res 56:122-132.

Henson MM, Henson OW Jr, Goldman LJ (1977) The perilymphatic spaces in the cochlea of the bat, *Pteronotus p. parnellii* (Gray). Anat Rec 187:767.

Henson MM, Henson OW Jr, Jenkins DB (1984) The attachment of the spiral ligament to the cochlear wall: anchoring cells and the creation of tension. Hear Res 16:231-242.

Henson MM, Jenkins DB, Henson OW Jr (1982) The cells of Boettcher in the bat, *Pteronotus parnellii*. Hear Res 7:91-103.

Henson MM, Jenkins DB, Henson OW Jr (1983) Sustentacular cells of the organ of Corti—the tectal cells of the outer tunnel. Hear Res 10:153-166.

Henson OW Jr (1970) The ear and audition. In: Wimsatt WA (ed) Biology of Bats. New York: Academic, Press pp. 181-256.

Henson OW Jr, Schuller G, Vater M (1985) A comparative study of the physiological properties of the inner ear in Doppler shift compensating bats (*Rhinolophus rouxi, Pteronotus parnellii*). J Comp Physiol 157:587-597.

Henson OW Jr, Bishop A, Keating A, Kobler J, Henson MM, Wilson B, Hansen R (1987) Biosonar imaging of insects by *Pteronotus parnellii*, the mustached bat. Natl Geogr Res 3:82-101.

Henson OW Jr, Koplas PA, Kating AW, Huffman RF, Henson MM (1990) Cochlear resonance in the mustached bat: behavioral adaptations. Hear Res 50:259-274.

Howard J, Hudspeth A J (1988) Compliance of hair bundle associated with gating of mechanoelectrical transduction channels in the bullfrog's saccular hair cell. Neuron 1:189-199.

Huffman RF, Henson OW Jr (1991) Cochlear and CNS tonotopy: normal physiological shifts in the mustached bat. Hear Res 56:79–85.

Huffman RF, Henson OW Jr (1993a) Labile cochlear tuning in the mustached bat I. Concomitant shifts in biosonar emission frequency. J Comp Physiol A 171:725–734.

Huffman RF, Henson OW Jr (1993b) Labile cochlear tuning in the mustached bat II. Concomitant shifts in neural tuning. J Comp Physiol A 171:735–748.

Jenkins DB, Henson MM, Henson OW Jr (1983) Ultrastructure of the lining of the scala tympani of the bat, *Pteronotus parnellii*. Hear Res 11:23–32.

Johnstone BM, Patuzzi R, Yates GK (1986) Basilar membrane measurements and the travelling wave. Hear Res 22:147–153.

Kalinec F, Holley CM, Iwasa K H, Lim DJ, Kachar B (1992) A membrane-based force generation mechanism in auditory sensory cells. Proc Natl Acad Sci USA 89:8671–8675.

Keithley EN, Schreiber RC (1986) Frequency map of the spiral ganglion in the cat. J Acoust Soc Am 81:1036–1042.

Kelly JB, Masterton B (1977) Auditory sensitivity of the albino rat. J Comp Physiol Psychol 91:930–936.

Kemp DT (1978) Stimulated acoustic emissions from within the human auditory system. J Acoust Soc Am 64:1386–1391.

Kemp DT (1981) Physiologically active cochlear micromechanics – one source of tinnitus. In: Evered D, Lawrencson G (eds) Tinnitus. (CIBA Foundation Symposium 85.) London: Pitman, pp. 54–81.

Kolston PJ (1988) Sharp mechanical tuning in a cochlear model without negative damping. J Acoust Soc Am 83:1481–1487.

Kolston PJ, Viergever MA, de Boer E, Diependaal RJ (1989) Realistic mechanical tuning in a micromechanical model. J Acoust Soc Am 86:133–140.

Kössl M (1992a) High frequency distortion products from the ears of two bat species, *Megaderma lyra* and *Carollia perspicillata*. Hear Res 60:156–164.

Kössl M (1992b) High frequency two-tone distortions from the cochlea of the mustached bat *Pteronotus parnellii* reflect enhanced cochlear tuning. Naturwissenschaften 79:425–427.

Kössl M (1994) Otoacoustic emissions from the cochlea of the 'constant frequency' bats, *Pteronotus parnellii* and *Rhinolophus rouxi*. Hear Res 72:59–72.

Kössl M, Vater M (1985a) Evoked acoustic emissions and cochlear microphonics in the mustache bat, *Pteronotus parnellii*. Hear Res 19:157–170.

Kössl M, Vater M (1985b) The cochlear frequency map of the mustache bat, *Pteronotus parnelli*. J Comp Physiol A 157:687–697.

Kössl M, Vater M (1990a) Tonotopic organization of the cochlear nucleus of the mustache bat, *Pteronotus parnelli*. J Comp Physiol A 166:695–709.

Kössl M, Vater M (1990b) Resonance phenomena in the cochlea of the mustache bat and their contribution to neuronal response characteristics in the cochlear nucleus. J Comp Physiol A 166:711–720.

Kössl M, Reuter G, Hemmert W, Preyer S, Zimmermann U, Zenner H-P (1993) Motility of outer hair cells in the organ of Corti of the bat, *Carollia perspicillata*. In: Elsner N, Heisenberg M (eds) Gene-Brain-Behaviour. (Proceedings of the 21th Göttingen Neuroscience Conference.) Stuttgart: Georg Thieme Verlag, p. 264.

Kraus H (1983) Vergleichende und funktionelle Cochleamorphologie der Fledermausarten *Rhinopoma hardwickii, Hipposideros speoris* und *Hipposideros fulvus* mit Hilfe einer Computergestützten Rekonstruktionsmethode. Ph.D. Thesis, University of Frankfurt, Germany.

Lenoir M, Puel J-L, Pujol R (1987) Stereocilia and tectorial membrane development in the rat cochlea. A SEM study. Anat Embryol 175:477–487.

Liberman MC (1982) The cochlear frequency map for the cat: labeling auditory-nerve fibers of known characteristic frequency. J Acoust Soc Am 72:1441–1449.

Liberman MC (1988) Response properties of cochlear efferent neurons: monaural vs. binaural stimulation and the effects of noise. J Neurophysiol (Bethesda) 60:1779–1798.

Liberman MC (1990) Effects of chronic cochlear de-efferentation on auditory-nerve response. Hear Res 49:209–224.

Liberman MC, Dodds LW, Pierce S (1990) Afferent and efferent innervation of the cat cochlea: quantitative analysis with light and electron microscopy. J Comp Neurol 301:443–460.

Lim DJ (1986) Functional structure of the organ of Corti: a review. Hear Res 22:117–146.

Long GR, Schnitzler H-U (1975) Behavioural audiograms from the bat, *Rinolophus ferrumequinum*. J Comp Physiol 100:211–219.

Manley GA (1990) Peripheral hearing mechanisms in reptiles and birds. Berlin: Springer-Verlag.

Metzner W, Radtke-Schuller S (1987) The nuclei of the lateral lemniscus in the rufous horseshoe bat, *Rhinolophus rouxi*. J Comp Physiol 160:395–411.

Müller M (1991) Frequency representation in the rat cochlea. Hear Res 51:247–254.

Müller M, Burda H (1989) Restricted hearing range in a subterranean rodent, *Cryptomys hottentotus* (Bathyergidae). Naturwissenschaften 76:134–135.

Müller M, Laube B, Burda H, Bruns V (1992) Structure and function of the cochlea in the African mole rat (*Cryptomys hottentottus*): evidence for a low frequency acoustic fovea. J Comp Physiol A 171:469–476.

Nadol JB Jr (1988) Comparative anatomy of the cochlea and auditory nerve in mammals. Hear Res 34:253–266.

Neely ST, Kim DO (1983) An active cochlear model showing sharp tuning and high sensitivity. Hear Res 9:123–130.

Neely ST, Kim DO (1986) A model for active elements in cochlear biomechanics. J Acoust Soc Am 79:1472–1480.

Neff W, Hind J (1955) Auditory thresholds of the cat. J Acoust Soc Am 27:480–483.

Neuweiler G (1990) Auditory adaptations for prey capture in echolocating bats. Physiol Rev 70:615–641.

Neuweiler G, Bruns V, Schuller G (1980) Ears adapted for the detection of motion, or how echolocating bats have exploited the capacities of the mammalian auditory system. J Acoust Soc Am 68:741–753.

Novoselova SM (1989) A possibility of sharp tuning in a linear transversally inhomogeneous cochlear model. Hear Res 41:125–136.

Obrist MK, Fenton MB, Eger JL, Schlegel PA (1993) What ears do for bats: a comparative study of pinna sound pressure transformation in chiroptera. J Exp Biol 180:119–152.

Pollak G, Henson OW Jr, Johnson R (1979) Multiple specializations in the peripheral auditory system of the CF-FM bat, *Pteronotus parnellii*. J Comp Physiol 131:255–266.

Pollak GD, Henson OW Jr, Novick A (1972) Cochlear microphonic audiograms in the 'pure tone' bat, *Chilonycteris parnellii parnellii*. Science 176:66–68.

Probst R, Lonsbury-Martin BL, Martin GK (1991) A review of otoacoustic emissions. J Acoust Soc Am 89:2027–2067.

Pujol R, Lenoir M, Ladrech S, Tribillac F, Rebillard G (1992) Correlations between the length of outer hair cells and the frequency coding of the cochlea. In: Cazals Y., Demany L., Horner K (eds) Advances in Biosciences, Auditory Physiology and Perception. Carcans: Pergamon.

Rajan R, Johnstone BM (1988) Electrical stimulation of cochlear efferents at the round window reduces auditory desensitization in guinea pigs. I. Dependence on electrical stimulus parameters. Hear Res 36:53–74.

Ramprashad F, Money KE, Landolt JP, Laufer J (1978) A neuroanatomical study of the little brown bat (*Myotis lucifugus*). J Comp Neurol 178:347–363.

Ramprashad F, Landolt JP, Money KE, Clark D, Laufer J (1979) A morphometric study of the cochlea of the little brown bat (*Myotis lucifugus*). J Morphol 160:345–358.

Raphael Y, Lenoir M, Wroblewski R, Pujol R (1991) The sensory epithelium and its innervation in the mole rat cochlea. J Comp Neurol 314:367–382.

Reuter G, Gitter AH, Thurm U, Zenner H-P (1992) High frequency radial movements of the reticular lamina induced by outer hair cell motility. Hear Res 60:236–246.

Robertson D, Manley GA (1974) Manipulation of frequency analysis in the cochlear ganglion of the guinea pig. J Comp Physiol 91:363–375.

Roth B, Bruns V (1992) Postnatal development of the rat organ of Corti. I. General morphology, basilar membrane, tectorial membrane and border cells. Anat Embryol 185:559–569.

Russell IJ, Sellick PM (1978) Intracellular studies of hair cells in the mammalian cochlea. J Physiol 284:261–290.

Russell IJ, Kössl M, Richardson GP (1992) Nonlinear mechanical responses of mouse cochlear hair bundles. Proc R Soc Lond B Biol Sci 250:217–227.

Rübsamen R, Neuweiler G, Sripathi K (1988) Comparative collicular tonotopy in two bat species adapted to movement detection, *Hipposideros speoris* and *Megaderma lyra*. J Comp Physiol A 163:271–285.

Santos-Sacchi J (1992) On the frequency limit and phase of outer hair cell motility: effects of the membrane filter. J Neurosci 12:1906–1919.

Schnitzler H-U (1968) Die Ultraschall-Ortungslaute der Hufeisen-Fledermäuse (*Chiroptera – Rhinolophidae*) in verschiedenen Orientierungssituationen. Z Vgl Physiol 57:376–408.

Schnitzler H-U (1970) Echoortung bei der Fledermaus *Chilonycteris rubiginosa*. Z Vgl Physiol 68:25–39.

Schuller G (1980) Hearing characteristics and Doppler shift compensation in South Indian CF-FM bats. J Comp Physiol 139:349–356.

Schuller G, Beuter K, Schnitzler H-U (1974) Response to frequency shifted artifical echoes in the bat *Rhinolophus ferrumequinum*. J Comp Physiol A 89:275–286.

Spoendlin H (1966) The organization of the cochlear receptor. In: Ruedi L (ed) Advances in Otorhinolaryngology, Vol. 13. Basel: Karger.

Spoendlin H (1973) The innervation of the cochlear receptor. In: Moller AR (ed) Basic Mechanisms in Hearing. New York: Academic, pp. 185–235.

Steele CR (1976) Cochlear mechanics. In: Keidel WD, Neff WD (eds) Handbook of Sensory Physiology, Vol. 513. Berlin: Springer-Verlag, pp. 443–478.

Sterbing S, Rübsamen R, Schmidt U (1990) Auditory midbrain frequency-place code and audio-vocal interaction during postnatal development in the phyllostomid bat, *Carollia perspicillata*. In: Elsner N, RothG (eds) Proceedings of the 18th Göttingen Neurobiology Conference. Stuttgart: Georg Thieme Verlag, p. 140.

Strelioff D, Flock A, Minser KE (1985) Role of inner and outer hair cells in mechanical frequency selectivity of the cochlea. Hear Res 18:169–175.

Suga N (1964) Single unit activity in cochlear nucleus and inferior colliculus of echolocating bats. J Physiol 172:449–474.

Suga N, Jen P-HS (1977) Further studies on the peripheral auditory system of CF-FM bats specialized for fine frequency analysis of Doppler shifted echoes. J Exp Biol 69:207–232.

Suga N, Neuweiler G, Möller J (1976) Peripheral auditory tuning for fine frequency analysis by the CF-FM bat, *Rhinolophus ferrumequinum*. J Comp Physiol 106:111–125.

Suga N, Simmons JA, Jen P-HS (1975) Peripheral specialization for fine analysis of Doppler shifted echoes in the auditory system of the CF-FM bat *Pteronotus parnellii*. J Exp Biol 63:161–192.

Tachibana M, Senuma H, Kumamoto K (1992) Enkephaline-like immunoreactivity in the cochlear efferents in the bat, *Rhinolophus ferrumequinum*. Hear Res 59:14–16.

Vater M (1988) Cochlear physiology and anatomy in bats. In: Nachtigall PE, Moore PWB (eds) Animal Sonar. New York: Plenum, pp. 225–242.

Vater M, Lenoir M (1992) Ultrastructure of the horseshoe bat's organ of Corti. I. Scanning electron microscopy. J Comp Neurol 318:367–379.

Vater M, Feng A S, Betz M (1985) An HRP-study of the frequency-place map of the horseshoe bat cochlea: morphological correlates of the sharp tuning to a narrow frequency band. J Comp Physiol A 157:671–686.

Vater M, Kössl M, Horn AKE (1992) GAD- and GABA-immunoreactivity in the ascending auditory pathway of horseshoe and mustached bats. J Comp Neurol 325:183–206.

Vater M, Lenoir M, Pujol R (1992) Ultrastructure of the horseshoe bat's organ of Corti. II. Transmission electron microscopy. J Comp Neurol 318:380–391.

von Békésy G (1960) Experiments in Hearing. New York: McGraw-Hill.

Warr WB (1992) Organization of olivocochlear efferent systems in mammals. In: Webster DB, Popper AN, Fay RR (eds) Mammalian Auditory Pathway: Neuroanatomy. New York: Springer-Verlag, pp. 410–448.

Wilson JP (1985) The influence of temperature on frequency-tuning mechanisms. In: Allen JB, Hall JL, Hubbard A, Neely ST, Tubis A (eds) Peripheral Auditory Mechanisms. New York: Springer, pp. 229–236.

Wilson JL, Henson MM, Henson OW Jr (1991) Course and distribution of efferent fibers in the cochlea of the mouse. Hear Res 55:98–108.

Wit HP (1985) Diurnal cycle for spontaneous oto-acoustic emission frequency. Hear Res 18:197–199.

Wright A (1984) Dimensions of the stereocilia in man and the guinea pig. Hear Res 13:89–98.

Xie DH, Henson MM, Bishop AL, Henson OW Jr (1993) Efferent terminals in the cochlea of the mustached bat: quantitative data. Hear Res 66:81-90.

Xue S, Mountain DC, Hubbard AE (1993) Direct measurement of electrically-evoked basilar membrane motion. In: Biophysics of Hair Cell Sensory Systems. Groningen: Academisch Ziekenhuis, pp. 303-309.

Zenner H-P, Zimmermann U, Schmidt U (1985) Reversible contraction of isolated cochlear hair cells. Hear Res 18:127-133.

Zook JM, Leake PA (1989) Connections and frequency representation in the auditory brainstem of the mustache bat, *Pteronotus parnellii*. J Comp Neurol 290:243-261.

Zweig G (1991) Finding the impedance of the organ of Corti. J Acoust Soc Am 89:1229-1254.

Zwicker E, Schloth E (1984) Interrelation of different oto-acoustic emission. J Acoust Soc Am 75:1148-1145.

Zwislocki JJ (1986) Analysis of cochlear mechanics. Hear Res 22:155-169.

Zwislocki JJ, Kletsky EJ (1979) Tectorial membrane: a possible effect on frequency analysis in the cochlea. Science 204:639-641.

6

The Lower Brainstem Auditory Pathways

Ellen Covey and John H. Casseday

1. Introduction

In the auditory system, more than any other sensory modality, extensive processing of incoming signals occurs in the brainstem. In all vertebrates, the auditory pathways below the inferior colliculus consist of a complex system of parallel pathways, each with its own centers for signal processing. The auditory structures of the lower brainstem act as filters to selectively enhance specific stimulus features and as computational centers to add, subtract, or compare signals in different channels. Some brainstem structures, such as the superior olive, have been studied extensively, and their function is at least partially understood. Others, such as the nuclei of the lateral lemniscus, have been largely ignored, and their functional roles are just beginning to be discovered.

In no vertebrate species are the brainstem auditory pathways larger or more highly differentiated than they are in echolocating bats. The large size and elegant organization of the lower brainstem auditory pathways in echolocating bats are not surprising when one considers the fact that these animals depend almost exclusively on their sense of hearing to navigate and capture prey in partial to complete darkness. It *is* surprising that the lower brainstem is the only part of the central auditory system that contains any clear anatomical specializations for echolocation (e.g., Baron 1974).

In this chapter, the structure and function of the auditory pathways of the lower brainstem in echolocating bats are compared and contrasted with those in other species of mammals; features peculiar to particular species of bats are discussed in terms of their possible role in echolocation behavior in each species.

2. Origins of Parallel Pathways in the Cochlear Nucleus

2.1 Structure of the Cochlear Nuclear Complex

In mammals, the auditory nerve branches as it enters the brainstem to innervate three major structures: the anteroventral cochlear nucleus

235

(AVCN), the posteroventral cochlear nucleus (PVCN), and the dorsal cochlear nucleus (DCN) (Ryugo 1992). This distribution of a common input to several distinct targets, each composed of a very different mixture of cell types (Cant 1992), forms the basis for the complex system of parallel pathways in the auditory brainstem.

The structure of each division of the cochlear nuclear complex has been described in several species of echolocating bats; these include the mustached bat, *Pteronotus parnellii* (Zook and Casseday 1982a; Kössl and Vater 1990), the greater horseshoe bat, *Rhinolophus ferrumequinum* (Poljak 1926; Schweizer 1981), and the rufous horseshoe bat, *Rhinolophus rouxi* (Feng and Vater 1985).

Figure 6.1 shows the cochlear nucleus of the mustached bat. In all species of echolocating bats that have been examined, the cochlear nucleus is extremely large relative to the size of the brainstem. This is true in species that have an expanded cochlear representation of certain spectral components of the echolocation call, such as the mustached and horseshoe bats, and also in those species that have a uniformly distributed frequency representation, such as the vespertilionids (Baron 1974).

In insectivorous echolocating bats, the AVCN is the largest division of the cochlear nuclear complex, comprising more than half of its entire volume (Baron 1974; Schweizer 1981; Zook and Casseday 1982a; Vater and Feng 1990). In addition to its large size, several cytoarchitectural features seem to be unique to and characteristic of the AVCN in all species of bats that have been studied. The first of these special features is the types of cells that are present in the AVCN and their distribution. In all mammals, the anterior subdivision of the AVCN contains mainly spherical cells, also known as bushy cells because they have a single thick dendrite that branches extensively at some distance from the cell body. In nonecholocating mammals, there are two populations of spherical bushy cells, distinguished by their size and location. Large spherical cells are found in the most rostral part of the AVCN, in an area where low frequencies are represented, whereas small spherical cells are distributed more caudally in areas responsive to high frequencies (see Cant 1992 for review). In echolocating bats, all the spherical cells in the AVCN are of the small type. In bats, small spherical cells are distributed throughout most of the rostrocaudal extent of the AVCN (Fig. 6.1A–C, G). These extend from the dorsal and medial part of the posterior subdivision, where they are intermixed with stellate cells, to the most rostral tip of the anterior subdivision, where they are the only cell type present (Zook and Casseday 1982a; Kössl and Vater 1990; Feng and Vater 1985). The absence of large spherical cells in bats, together with their rostral location in other mammals, suggests that large spherical cells may represent a special adaptation for processing low-frequency sounds. If so, this cell population in bats has either been lost or never developed during evolution. In spite of the difference in size, however, the spherical bushy cells of the bat have much in common with their counterparts in the cat. In both species, these cells receive end-bulbs of Held, large synaptic endings of

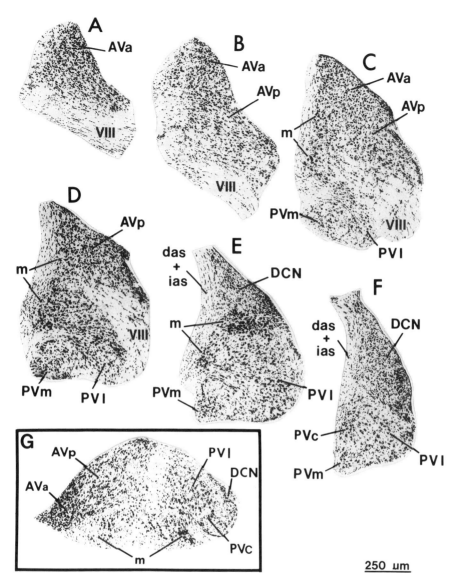

250 μm

Figure 6.1A–G. Cytoarchitecture of the cochlear nucleus in the mustached bat as seen in Nissl-stained frontal sections arranged from rostral (A) to caudal (F) and in a horizontal section (G). The sections illustrate several structural features characteristic of echolocating bats: (1) the large size of the anteroventral cochlear nucleus (AVCN) and posteroventral cochlear nucleus (PVCN) relative to the dorsal cochlear nucleus (DCN); (2) the presence of small spherical bushy cells throughout the AVCN, even in the rostral part; (3) the presence of the marginal cell area medial to the AVCN and separating DCN from PVCN; and (4) poor lamination in the DCN. AVa, anterior division of AVCN; AVp, posterior division of AVCN; AVm, medial division of AVCN; PVm, medial division of PVCN; PV, lateral division of PVCN; PVc, caudal division of PVCN; das + ias, dorsal acoustic stria and intermediate acoustic stria (From Zook and Casseday © 1982 J. Comp. Neurol. 207, reprinted by permission of Wiley-Liss, a division of John Wiley and Sons Inc.)

237

auditory nerve fibers that surround the cell bodies. This conservation of input morphology indicates that whatever the functional significance of the end-bulbs, it is not restricted to preserving phase locking to low-frequency sounds.

The caudal part of the AVCN in echolocating bats resembles that of other mammals in that a number of different cell types are present (see Fig. 6.1B–E). These include multipolar and globular cells in the ventral region and granule cells along the lateral edge and caudally at the border with the DCN.

A feature of the AVCN that appears to be peculiar to echolocating bats is the presence of a distinctive population of very large multipolar cells along the medial edge of the AVCN and posteroventrally at the border with the PVCN (Fig. 6.1C–E). The cells that make up this population are the largest in the cochlear nucleus. This cell group has been termed the "marginal" subdivision of the AVCN (Zook and Casseday 1982a; Kössl and Vater 1990) and has further been subdivided into a medial part at the medial border of the AVCN and PVCN and a lateral part that lies along the common border between AVCN, PVCN, and DCN (Kössl and Vater 1990).

The cytoarchitecture of the PVCN in bats is similar to that of other mammals (Schweizer 1981; Zook and Casseday 1982a; Vater and Feng 1990; Cant 1992) (see Fig. 6.1D, F–G). The PVCN has been separated into two subdivisions in some bat species and three in others. Throughout most of the PVCN there is a mixture of stellate, small multipolar, elongate, and small round cells. As in other mammals (Osen 1969a,b), the caudal part of the PVCN contains octopus cells, large neurons with two or more long thick dendrites that extend orthogonal to the fibers of the descending branch of the auditory nerve. In most bats the octopus cell area is prominent, and in some horseshoe bats the octopus cells are arranged in a particularly distinctive U-shaped formation (Poljak 1926; Schweizer 1981; Pollak and Casseday 1989).

Compared to the AVCN and PVCN, the DCN in most bat species is relatively small and poorly differentiated (Poljak 1926; Henson 1970; Baron 1974; Schweizer 1981; Zook and Casseday 1982a; Feng and Vater 1985) (see Fig. 6.1E–G). In many mammals, the chief characteristic of the DCN is the laminar arrangement of its cells (Cant 1992). Prominent features of the lamination are that the fusiform cells are surrounded by many granule cells, and further that the fusiform cells are aligned in a single layer with their long axes perpendicular to the surface of the nucleus so that one dendrite extends into the deep layer and the other into the molecular layer. In bats, the granule cells surrounding the fusiform cells are few in number, and the fusiform cells are not uniformly aligned perpendicular to the surface of the nucleus. In all species of bats that have been described, the lamination of the DCN is poorly developed, and the fusiform cells are not arranged in an orderly fashion. The near absence of lamination in DCN is a feature shared by primates, including humans (Moore 1987).

In the horseshoe bat, two divisions of DCN can be distinguished on the basis of the pattern of lamination. The part of the DCN that represents frequencies below the constant-frequency (CF) range is laminated, whereas the part of the DCN that represents frequencies at and above the CF is not. Because this division of DCN into laminated and unlaminated divisions is not seen in the mustached bat, it appears to be a feature peculiar to the horseshoe bat and not a specialization common to all bats that use a CF component in their echolocation calls.

2.2 Inputs to the Cochlear Nucleus

In bats, the pattern of connectivity of the cochlear nucleus is basically the same as in other mammals, with a few exceptions. The main input is via the afferent fibers of the auditory nerve. These fibers bifurcate to form an ascending branch that terminates in the AVCN and a descending branch that terminates in the PVCN and DCN. Anterograde labeling of auditory nerve fibers entering the cochlear nucleus shows that in both the rufous horseshoe bat (Feng and Vater 1985) and the mustached bat (Zook and Leake 1989) these fibers terminate in each division of the cochlear nucleus as thick bands or "slabs" which correspond to isofrequency contours. Structural and functional evidence of a slablike connectional organization of afferents has also been seen in nonecholocating mammals (Cajal 1909; Rose, Galambos, and Hughes 1959; Moskowitz and Liu 1972; Noda and Pirsig 1974; Bourke, Mielcarz, and Norris 1981; Ryan, Woolf, and Sharp 1982). Thus, the basic pattern of innervation of the cochlear nucleus in bats follows the standard mammalian pattern. Descending inputs to the cochlear nucleus originate mainly in the periolivary cell groups and project to all three divisions. Within the cochlear nucleus itself, there are intrinsic connections that link one division with another.

The connectional basis for tonotopy in each division of the cochlear nucleus and the pattern of innervation from the cochlea has been elegantly demonstrated in the mustached bat (Zook and Leake 1989). In these experiments, very small horseradish peroxidase (HRP) injections were placed in physiologically characterized sites in the cochlear nucleus. Retrograde transport from the injection sites labeled cells in the spiral ganglion, and thus identified the portion of the cochlea projecting to a given site in the cochlear nucleus. In addition, anterograde transport identified the target regions of each site within the superior olivary complex (Fig. 6.2). The distribution of labeled ganglion cells in the cochlea was always proportional to injection size, even within the expanded 60- to 63-kHz region. This finding supports and adds to physiological evidence for two ideas concerning frequency representation: first, the proportion of tissue devoted to processing any given frequency range is approximately the same at the level of the cochlear nucleus as it is at the cochlea, and second, the expanded

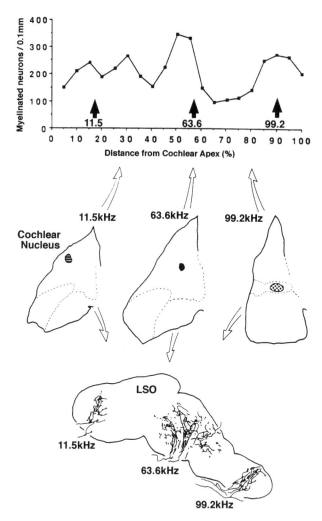

FIGURE 6.2. Patterns of connections of small regions in the AVCN as seen by anterograde and retrograde transport of horseradish peroxidase (HRP). In the cochlea, shown at the *top*, the location and extent of labeled ganglion cells varies with the frequency range and size of the injection sites, shown in *middle*. In the lateral superior olive (LSO), shown at the *bottom*, the region of terminal label is much larger for the injection at 63.6 kHz than for the injections at higher and lower frequencies, indicating a greater degree of divergence in this frequency range. (Adapted from Zook and Leake © 1988 J. Comp. Neurol. 290, reprinted by permission of Wiley-Liss, a division of John Wiley and Sons, Inc.)

representation of certain frequency ranges in the cochlea of CF-FM bats is maintained in a one-to-one projection to the cochlear nucleus.

2.3 Intrinsic Connections of the Cochlear Nucleus

Small HRP injections confined to subdivisions of the cochlear nucleus have been used to determine the patterns of branching and termination of auditory nerve fibers and patterns of intrinsic connections between cochlear nucleus subdivisions in the rufous horseshoe bat (Feng and Vater 1985) and in the mustached bat (Zook and Leake 1988; Kössl and Vater 1990). These results demonstrate frequency-specific reciprocal connections between the DCN and AVCN and between the DCN and PVCN. Like the afferent arborizations, the intrinsic connections between divisions of the cochlear nucleus are tonotopically organized and take the form of slabs. In the mustached bat, the intrinsic projection from the AVCN to the DCN molecular layer is especially large (Zook and Casseday 1985).

2.4 Physiology of the Cochlear Nucleus

The response properties of neurons in the cochlear nucleus have been studied in two FM species, the little brown bat (*Myotis lucifugus*) (Suga 1964) and the big brown bat (Haplea, Covey, and Casseday 1994), and in two CF-FM species, the mustached bat (Suga, Simmons, and Jen 1975) and the horseshoe bat (Neuweiler and Vater 1977; Feng and Vater 1985).

2.4.1 Tonotopy and Frequency Tuning

The tonotopic organization in the cochlear nucleus of bats is basically the same as that found in other mammals such as the cat (Bourke, Mielcarz, and Norris 1981). Thus, in bats, the tonotopic axis from high to low frequencies extends from dorsal to ventral in the DCN (Feng and Vater 1985), from dorsomedial to ventrolateral in the PVCN, and from caudal to rostral in the AVCN (Feng and Vater 1985; Haplea, Covey, and Casseday 1994).

The cochleas of CF-FM bats are specialized so that a very large proportion of the basilar membrane is tuned to the CF_2 frequency range, about 61 kHz (see Kössl and Vater, Chapter 5, this volume). As mentioned in Section 2.2, the cochlear frequency expansion is preserved in the tonotopic organization of each division of the cochlear nucleus. The result is a prominent central expansion of the same small frequency range that is expanded on the basilar membrane. Thus, more than half of all neurons in the cochlear nucleus are tuned to frequencies about 61 kHz in the mustached bat (Suga, Simmons, and Jen 1975), 81–88 kHz in the greater horseshoe bat (Neuweiler and Vater 1977), and 77–79 kHz in the rufous horseshoe bat (Feng and Vater 1985).

Figure 6.3 compares the distribution of best frequencies (BFs) and sharpness of tuning in the cochlear nucleus of a CF-FM bat with that of a bat that uses a purely frequency-modulated (FM) echolocation call. The cochleas of FM bats, so far as is known, have no specializations for fine frequency tuning nor do they show an expansion of any frequency range. Correspondingly, FM bats have a uniform representation of frequencies within the cochlear nucleus (Suga 1964; Suga, Simmons, and Jen 1975; Haplea, Covey, and Casseday 1994). These observations suggest that the frequency representations seen in the cochlear nucleus reflect the cochlear frequency representation in approximately a one-to-one relationship. In horseshoe bats, however, the expansion of the CF_2 frequency range is more pronounced in the AVCN and PVCN than in the DCN (Feng and Vater 1985). This finding suggests that although the cochlear frequency expansion is largely responsible for establishing frequency representation in the cochlear nucleus of CF-FM bats, this one-to-one correspondence may not be preserved at all levels. There is evidence that specific frequency ranges are progressively expanded in the central auditory system, especially in FM species which have no cochlear frequency expansion. Figure 6.4 shows that the BFs of neurons in the cochlear nucleus of the big brown bat are approximately evenly distributed across the entire audible range just as they are in the little brown bat (Fig. 6.3B); this distribution is probably a direct reflection of the cochlear frequency map. However, at the nuclei of the

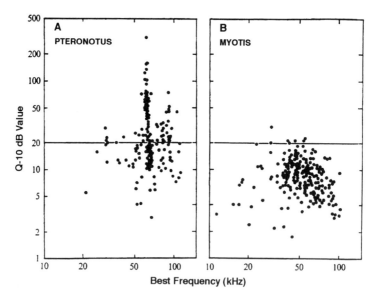

FIGURE 6.3A,B. Comparison of the best frequency (BF) distribution and sharpness of frequency tuning in the cochlear nucleus of (A) a constant frequency, frequency-modulated (CF-FM) bat (the mustached bat) and (B) an FM bat (the little brown bat). (From Suga, Simmons, and Jen 1975.)

FIGURE 6.4A–C. Comparison
of the BF distribution and
sharpness of frequency
tuning in an FM species, the
big brown bat, at the cochlear nucleus (C), nuclei of
the lateral lemniscus (B),
and inferior colliculus (A).
At the cochlear nucleus and
nuclei of the lateral lemniscus, the distribution of BFs
is approximately uniform
throughout the audible
range, and Q_{10dB}s are nearly
all below 20; at the inferior
colliculus, there is an expanded representation of the
frequency range between 20
and 30 kHz, and the Q_{10dB}s
within this range are high,
with a large proportion
greater than 20. (From Haplea, Covey, and Casseday
1994.)

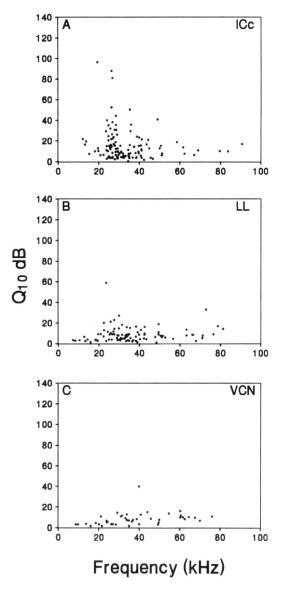

Frequency (kHz)

lateral lemniscus, the proportion of neurons with BFs in the frequency
range between 20 and 30 kHz is slightly increased. In the inferior colliculus,
there is a large expansion of this frequency range, so that about one-third
of all BFs are in the range between 20 and 30 kHz. Thus, it seems selective
convergence or divergence of projections results in portions of the frequency map being selectively contracted or expanded.

In CF-FM bats, the expanded representation of the CF_2 frequency range
is accompanied by specializations in frequency tuning. In the cochlear

nucleus of horseshoe bats and mustached bats, most neurons with BFs outside the CF_2 frequency range have ordinary V-shaped tuning curves similar to those in nonecholocating mammals; most units with BFs in the CF_2 range have extremely narrow tuning curves (Suga, Simmons, and Jen 1975; Neuweiler and Vater 1977; Feng and Vater 1985). For the narrowly tuned neurons in all these species, the quality factor at 10 dB above threshold (Q_{10dB}) is above 20, with some as high as 400.

Neurons with frequency selectivity comparable to that of the narrowly tuned units in CF-FM bats have not been found in the cochlear nucleus of FM bats. In the cochlear nucleus of the little brown bat and the big brown bat, neurons with BFs throughout the entire audible frequency range have V-shaped tuning curves comparable in breadth to those in cats or other nonecholocating mammals (Suga 1964; Haplea, Covey, and Casseday 1994).

Although it is likely that the fine frequency tuning in CF-FM bats is largely the result of cochlear mechanical specializations (see Kössl and Vater, Chapter 5, this volume), frequency tuning may be further modified in the central auditory system through neural inhibition. Inhibitory side-bands are associated with the tuning curves of some cochlear nucleus neurons in both CF-FM bats and FM bats, although they do not appear to be very common at this level (Suga 1964; Neuweiler and Vater 1977). Because inhibitory sidebands are found in both narrowly tuned and broadly tuned neurons, there is no reason to believe that the presence or absence of inhibitory sidebands is correlated with BF or narrow cochlear tuning. Suga (1964) proposed that the abrupt transition between excitation and inhibition in neurons with inhibition at only one side (Fig. 6.5) could play a role in creating directional selectivity for FM sweeps, a function that could be common to both FM and CF-FM bats because both use FM echolocation signals.

It is important to note that FM bats do have neurons at the level of the inferior colliculus (IC) that are narrowly tuned across a wide amplitude range, with Q_{10dB}, Q_{30dB} and Q_{40dB} values of 20 and higher. In the big brown bat, the BFs of all the narrowly tuned neurons fall within a range (20–30 kHz) that corresponds to the portion of the FM call that is lengthened to a "quasi-CF" during the search phase of echolocation (Casseday and Covey 1992). This finding suggests that extremely narrow frequency tuning is necessary at some level for any echolocating bat that must analyze echoes of a CF or quasi-CF call; furthermore, narrow tuning can be produced entirely by neural mechanisms. Thus, narrow tuning may have evolved independently through very different mechanisms in the two types of bats.

In the big brown bat, fine frequency tuning first appears at the IC and is created through neural inhibitory mechanisms (Casseday and Covey 1992; Covey et al. 1993; Haplea, Covey, and Casseday 1994). In the mustached bat, the already fine frequency tuning present at lower levels is sharpened at

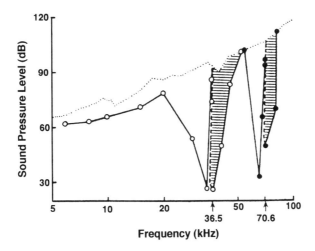

FIGURE 6.5. Turning curves of two spontaneously active neurons in the cochlear nucleus of the little brown bat have excitatory areas (*unshaded*) in which discharge increases above spontaneous rate, and inhibitory areas (*shaded*) in which discharge decreases below spontaneous rate. For both these neurons, the excitatory area occupies a range of frequencies below the inhibitory area. (From Suga 1964.)

the IC (Yang, Pollak, and Resler 1992). In either species it is unknown whether the inhibitory input responsible for modifying frequency tuning arises from interneurons in the IC or from projection neurons in lower brainstem pathways. The finding that narrow frequency tuning first appears at the level of the IC in FM bats suggests that even though fine frequency tuning is present throughout the brainstem auditory system of CF-FM bats, it may not be exploited until the level of the IC. The fact that the IC is an early stage of output to systems for motor control (Covey, Hall, and Kobler 1987; Schuller, Covey, and Casseday 1991) suggests that fine frequency tuning might be important for motor tasks such as Doppler-shift compensation (see Chapter 3, this volume).

2.4.2 Discharge Patterns and Timing Characteristics

The discharge patterns of cochlear nucleus units in both CF-FM and FM bats are basically similar to those in nonecholocating mammals such as the cat (Rose, Galambos, and Hughes 1959; Pfeiffer 1966). Transient responses of one or a few spikes may be correlated with either the onset or offset of sound. Sustained responses include primary-like, primary-like with notch, tonic (nonadapting), pauser, chopper, and buildup. In most species, including the little brown bat (Suga 1964), the big brown bat (Haplea, Covey, and Casseday 1994), the rufous horseshoe bat (Feng and Vater 1985), and the mustached bat (Suga, Simmons, and Jen 1975), sustained responses are reported to be the predominant type. In the greater horseshoe bat (Neuweiler and Vater 1977), the most common discharge pattern was reported to be transient, probably because in this study recording was limited to the PVCN and DCN. In addition, in this study a significant number of neurons were reported to have complex response patterns that changed depending on frequency or intensity of the stimulus. This phenomenon has also been seen in the DCN of nonecholocating mammals (e.g., Goldberg and Brownell 1973; Godfrey, Kiang, and Norris 1975; Adams 1976; Rhode and Kettner 1987).

Several lines of evidence suggest that some initial analysis of temporal information occurs in the cochlear nucleus. Calbindin-like immunoreactivity is thought to provide a marker for neural pathways that preserve timing information with great precision (Carr 1986; Takahashi et al. 1987). In the cochlear nucleus of the mustached bat, calbindin reactivity is prevalent; it is found in the end-bulbs of Held, auditory nerve root neurons, multipolar and globular cells in AVCN, multipolar and octopus cells in PVCN, and small- and medium-sized cells in DCN (Zettel, Carr, and O'Neill 1991). There is functional evidence that at least some of these cell types do play a role in transmitting precise information about the temporal structure of sounds.

In CF-FM bats, many cells that respond with a transient onset discharge also have a transient discharge correlated in time with stimulus offset. Usually the onset response is present throughout the excitatory frequency range, but the offset response appears only within a restricted frequency range. *On/off* responses are most commonly seen in cells with BFs around the CF_2 frequency (Suga, Simmons, and Jen 1975; Neuweiler and Vater 1977). In the mustached bat, the off responses at 60–61 kHz do not appear to be rebounds from neural inhibition because they are not preceded by suppression of spontaneous activity. Instead, the off discharges are correlated with cochlear microphonic off or "after" activity, suggesting that on/off responses in CF-FM bats are a result of cochlear mechanical specializations (Suga, Simmons, and Jen 1975; Suga and Jen 1977; Bruns 1976ab; Suga, Neuweiler, and Möller 1976).

It is not known whether on/off responses serve some purpose in neural processing of echolocation sounds. However, they persist at some higher levels of the auditory system such as the medial superior olive (MSO) (Covey, Vater, and Casseday 1991; Grothe et al. 1992) and IC (Pollak and Bodenhamer 1981; Lesser et al. 1990), but are rare or absent at other levels such as the lateral superior olive (LSO) (Covey, Vater, and Casseday 1991).

An unusual response pattern common to both FM and CF-FM bats is the so-called afterdischarge, a prolonged period of neural activity lasting up to 600 msec after the cessation of a 5- or 10-msec stimulus. Afterdischarges are especially common following sounds at high amplitudes (Suga 1964; Neuweiler and Vater 1977; Haplea, Covey, and Casseday 1994). Presentation of a second, identical tone during the afterdischarge causes brief inhibition of firing (Suga 1964). This finding suggests that the sequence of synaptic events leading to afterdischarge is inhibition followed by prolonged excitation, although cochlear mechanisms have not been ruled out as a possible cause of afterdischarge.

Some cochlear nucleus neurons in horseshoe bats are suppressed rather than excited for a period of several tens of milliseconds following their response to a sound (Neuweiler and Vater 1977). It seems likely that long periods of excitation or suppression following the presentation of a stimulus could provide a neural trace that marks the occurrence of a stimulus for a prolonged time period, thus facilitating or suppressing the response to subsequent sounds that occur during the period of the afterdischarge or suppression. Afterdischarges thus might play a role in echoranging or in the analysis of long sequences of sounds such as the rapid and continuous train of pulses and echoes that commonly occurs just before prey capture in many bat species.

Additional evidence that cells in the bat cochlear nucleus may be an initial stage in analyzing timing information comes from experiments in the greater horseshoe bat and big brown bat in which a population of neurons in the PVCN fire a single spike correlated precisely in time with stimulus onset. Virtually no spikes occur at any other time. These neurons maintain the same latency and probability of firing over a wide range of stimulus intensities (Neuweiler and Vater 1977; Haplea, Covey, and Casseday 1994). Neurons with similar timing characteristics have also been found in the marginal division of AVCN in the mustached bat (Kössl and Vater 1990).

In the mustached bat, most transient responses seen in the cochlear nucleus are recorded from neurons in the marginal division of AVCN. Most neurons in the marginal division have BFs within the frequency range of the first FM harmonic (FM_1), from 24 to 32 kHz (Kössl and Vater 1990). The finding of precisely timed transient responses in a population of neurons tuned to the FM_1 may be important for understanding how target range is determined by the mustached bat. The thalamic and cortical neurons that are thought to encode information about target range respond to the FM_2 or FM_3 only if it is preceded by the FM_1, and the maximal response occurs when the two sounds are separated by a specific time interval (O'Neill and Suga 1979, 1982; Olsen and Suga 1991a, b). It seems likely that FM_1-selective neurons in the marginal division of AVCN provide a sharp and precise timing marker for the occurrence of the FM_1, which could optimize time delay measurements at higher levels of the auditory system. The marginal division of AVCN in the mustached bat contains a high density of noradrenergic fibers that originate mainly in the locus coeruleus (Kössl, Vater, and Schweizer 1988). It has been shown that noradrenaline enhances the phasic nature and precise timing of discharges of cells in this region (Kössl and Vater 1989). This finding suggests that attention may play a role in regulating the responses of cochlear nucleus neurons, and that behavioral attention is especially important in the system that originates in the marginal division of the AVCN.

Besides providing timing information from which to derive target range, cochlear nucleus neurons are capable of providing highly accurate information about the fine timing characteristics of rapidly changing complex stimuli. Transient responders in both CF-FM and FM bats have recovery times as short as 0.3 msec, and this is reflected in their ability to follow amplitude modulations up to about 3000 Hz (Suga, Simmons, and Jen 1975; Neuweiler and Vater 1977).

Neurons in the cochlear nucleus of horseshoe bats respond with synchronized discharges to both sinusoidal amplitude modulations and sinusoidal frequency modulations. For amplitude-modulated sounds, the minimum modulation depth needed to produce synchronization appears to be independent of BF; in contrast, for frequency-modulated sounds, minimum modulation depth is related to BF range (Vater 1982). Although all neurons follow sinusoidal frequency modulations, those with BFs in the CF_2 range are much more sensitive to small frequency modulations than are any others. Neurons in the CF_2 range can synchronize their discharges to a modulation depth of ± 20 Hz applied to a carrier frequency of more than 80 kHz. This represents a modulation of only 0.00025%. This exquisite sensitivity of the CF_2 filter neurons to sinusoidal frequency modulations is a neural correlate of the behavioral observations that CF-FM bats can discriminate periodic Doppler shifts of the CF_2 frequency caused by the beating wings of an insect.

All these findings suggest that neurons in the cochlear nucleus of echolocating bats transmit information about the temporal pattern of sound in the form of discharges which are highly synchronized to sound onset, amplitude modulations, or frequency modulations. At higher brainstem levels, this synchronized pattern of neural activity undergoes various transformations such as bandpass or lowpass filtering at the MSO (Grothe 1990), further sharpening of time locking in the nuclei of the lateral lemniscus (Covey and Casseday 1991), or conversion to pattern selectivity at the inferior colliculus (Covey et al. 1993; Casseday, Ehrlich, and Covey 1994).

2.5 Central Connections of the Cochlear Nucleus

In all mammals, the three divisions of the cochlear nucleus are the origin of multiple parallel pathways (e.g., Warr 1982). In the bat, the pattern of ascending projections from the cochlear nucleus is similar to the basic mammalian plan, with some exceptions that may be specializations for localization of high-frequency sounds, adaptations for target range determination, or adaptations for identification of targets using echolocation. Figure 6.6 shows the basic pattern of outputs from each division of the cochlear nucleus in echolocating bats. Three major groups of targets receive projections from the cochlear nucleus: the superior olivary complex (SOC), the monaural nuclei of the lateral lemniscus, and the central nucleus of the IC.

In all bat species that have been studied, the AVCN supplies the predominant input to the principal nuclei of the SOC (Feng and Vater 1985;

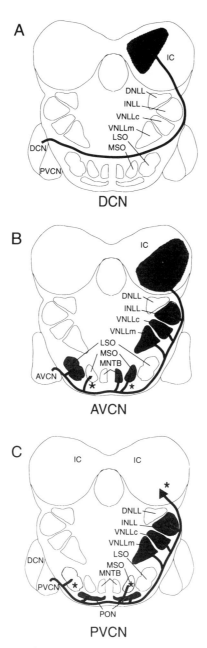

FIGURE 6.6A–C. Outputs of DCN (A), AVCN (B), and PVCN (C) shown in sche-
matized frontal sections from the brain of an echolocating bat. *Shaded areas* are
targets of the different divisions of the cochlear nucleus; *heavy stippling* indicates
major projections; *light stippling* indicates sparse projections; and *asterisks* indicate
pathways that have been described only in echolocating bats, or that differ significantly between bats and other mammals. DNLL, dorsal nucleus of lateral lemniscus;
INLL, intermediate nucleus of lateral lemniscus; VNLLc, columnar division of
ventral nucleus of lateral lemniscus; VNLLm, multipolar cell division; MSO, medial
superior olive; MNTB, medial nucleus of trapezoid body; PON, periolivary nuclei.

Zook and Casseday 1985, 1987; Covey and Casseday 1986; Vater and Feng 1990). However, in several species of bats, the PVCN supplies a significant projection to the LSO and MSO (Zook and Casseday 1985; Vater and Feng 1990). Because a projection from the PVCN to the SOC has not been described in other mammals, it may be restricted to echolocating bats. Surrounding the principal nuclei of the SOC are a number of different periolivary cell groups. The periolivary nuclei are highly variable among all mammalian species, especially among different species of bats. All divisions of the cochlear nucleus contribute projections to the periolivary cell groups, but the largest proportion of input comes from the PVCN, followed by the DCN (Zook and Casseday 1985).

The second major set of targets of the cochlear nucleus are the interme- diate and ventral nuclei of the lateral lemniscus (see Fig. 6.6B,C). In all mammals, most of the input to the intermediate and ventral nuclei of the lateral lemniscus arises from the contralateral cochlear nucleus, from both AVCN and PVCN (Glendenning et al. 1981; Zook and Casseday 1982b, 1985; Covey and Casseday 1986).

The third major target of direct output from the cochlear nucleus is the contralateral central nucleus of the inferior colliculus (ICc) (Fig. 6.6). The direct projections to the ICc originate from two main sources, the AVCN and DCN. However, in the species of echolocating bats that have been studied, projections to the ICc also originate in the PVCN (Zook and Casseday 1982b; Vater and Feng 1990). Because projections from the PVCN to the IC have not been described in other mammals, this pathway may be a feature peculiar to and characteristic of echolocating bats. Like the pathways to the SOC and nuclei of the lateral lemniscus, the direct projections from the cochlear nucleus to the ICc are tonotopically orga- nized (Zook and Casseday 1982b; Casseday and Covey 1992). In all mammals that have been studied, including several species of bats, the projections from the cochlear nucleus to the ICc terminate in a banded pattern (e.g., Zook and Casseday 1985, 1987; Casseday and Covey 1992; Oliver and Huerta 1992).

In the mustached bat and the big brown bat, small spherical cells of rostral AVCN project directly to the anterolateral, low-frequency area of the ICc; cells in the caudal, high-frequency part of AVCN project directly to the ventromedial, high-frequency part of the ICc, and small multipolar, globular, elongate, ovoid, and spherical cells at intermediate locations and frequencies project to corresponding intermediate positions in the ICc (Zook and Casseday 1982b, 1987). In nonecholocating mammals, the projection to the ICc originates mainly in the multipolar, or stellate, cell population of the AVCN (see Oliver and Huerta 1992). The fact that in the bat not only stellate cells but also spherical and globular cells project to the ICc may indicate that a different and larger subset of inputs are integrated in the IC of the echolocating bat than in nonecholocating mammals. The projection from the PVCN to the ICc does not originate in the octopus cell area but from other cell types (Zook and Casseday 1982b).

In the bat species that have been examined, the projection from the DCN to the ICc originates in fusiform and giant cells, just as it does in other mammals. This projection is tonotopically organized so that the dorsal DCN projects to the ventromedial, high-frequency area of the ICc while the ventral DCN projects to the anterolateral, low-frequency area of the ICc. The target of the DCN within the ICc extends more medially and dorsally than the AVCN target (Zook and Casseday 1987), suggesting that there is partial segregation of the two systems of projections.

3. Superior Olivary Complex

3.1 Structure of the Superior Olivary Complex

In all species of echolocating bats that have been studied, the SOC is large and well developed. It appears to contain all the same structures that are present in nonecholocating mammals, including the LSO, MSO, medial nucleus of the trapezoid body (MNTB), and various periolivary cell groups, although there is some controversy as to whether the large and specialized MSO of bats can be considered equivalent to MSO in nonecholocating animals. The SOC of the mustached bat is illustrated in Figure 6.7.

3.1.1 Principal Nuclei

In all bats that have been examined, the LSO and MNTB are unusually large relative to the size of the brainstem (Baron 1974). In some bat species, the LSO has more convolutions than does the LSO of most nonecholocating mammals. Nevertheless, in terms of their cytoarchitecture and patterns of connections, these nuclei in bats are virtually identical to the LSO and MNTB of other mammals (Schweizer 1981; Zook and Casseday 1982a; Schwartz 1992). The LSO is composed mainly of elongate cells oriented with their long axis perpendicular to the outer contour of the nucleus. Surrounding the LSO is a dense fiber plexus (Fig. 6.7B). The MNTB in bats and other mammals is a conspicuous group of densely packed cells, most of which are principal cells. These are round or oval, with an eccentric nucleus, and are contacted on a one to one basis by large calyces of Held that originate in the AVCN (Zook and Casseday 1982a; Kuwabara and Zook 1991; Kuwabara, DiCaprio, and Zook 1991).

In the mustached bat and other bats that have been examined, there is a large cell group in the position normally occupied by the MSO. This structure is cytoarchitecturally similar to MSO in other mammals in that a dense fiber plexus surrounds a column of cells (Fig. 6.7A,B) that are mainly elongate with their long axis oriented medial to lateral (Zook and Casseday 1982a). However, there is one structural aspect in which the bat MSO differs from that of most nonecholocating mammals. The MSO of a primate, for example, is many cells thick in the rostrocaudal and dorsoven-

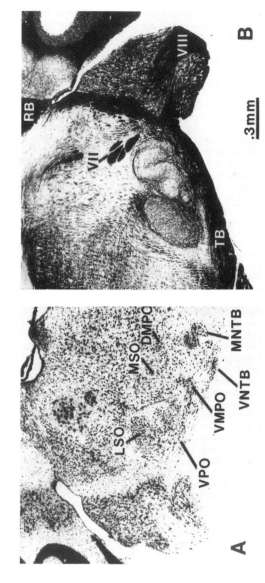

FIGURE 6.7A,B. Subdivisions of the superior olivary complex (SOC) in the mustached bat, *Pteronotus parnellii*. (A) Nissl-stained frontal section through the left side of the brainstem; (B) adjacent frontal section through the right side of the brainstem stained to show fibers. DMPO, dorsomedial periolivary nucleus; VPO, ventral periolivary nucleus; VMPO, ventromedial periolivary nucleus; VNTB, ventral nucleus of trapezoid body; TB, trapezoid body; RB, restiform body; VIII, eighth cranial nerve. (From Zook and Casseday © 1982a. J. Comp Neurol. 207, reprinted by permission of Wiley-Liss a division of John Wiley and Sons, Inc.)

tral dimensions, but only a few cells thick in the mediolateral dimension; in most echolocating bats, the MSO is many cells thick in all these dimensions. In the mustached bat, the MSO is convoluted to form a large dorsal limb and a small ventral limb. In horseshoe bats there are two structures, each of which is more like MSO in cytoarchitecture and connections than like any of the periolivary cell groups (Schweizer 1981). These have been termed the dorsal MSO and the ventral MSO (Casseday, Covey, and Vater 1988).

3.1.2 Periolivary Nuclei

The most detailed description of the cytoarchitecture and organization of the periolivary cell groups is in the mustached bat (Zook and Casseday 1982a), but some information is also available for a number of additional bat species (Aschoff and Ostwald 1987). Most bats have a prominent group of large multipolar neurons dorsomedial to the MSO. Although this cell group has been called the dorsomedial periolivary nucleus in the mustached bat (Zook and Casseday 1982a), it may be the same as the superior paraolivary nucleus (SPN) of rodents, which occupies a similar location and is made up of large multipolar neurons (e.g., Ollo and Schwartz 1979). Ventral to the principal nuclei of the SOC are the ventral nucleus of the trapezoid body and the ventral and ventromedial periolivary nuclei (Fig. 6.7A) (Zook and Casseday 1982a). In at least some bat species a large group of neurons is located lateral and anterolateral to the LSO. In the mustached bat and the horseshoe bat, these cells have been called the lateral nucleus of the trapezoid body (LNTB) (Zook and Casseday 1982a; Casseday, Covey, and Vater 1988; Kuwabara and Zook 1992), although it is not clear whether this is the same structure that is called LNTB in the cat. Rostral to the LSO is a group of very large multipolar neurons that are prominent in the mustached bat and in horseshoe bats, but are not well developed in vespertilionid bats. This cell group has been called the anterolateral periolivary nucleus (ALPO: Zook and Casseday 1982a; Covey, Hall, and Kobler 1987; Kobler, Isbey, and Casseday 1987) and more recently, the nucleus of the central acoustic tract (NCAT), because it is the source of the central acoustic tract, an extralemniscal pathway to the superior colliculus and thalamus (Casseday et al. 1989). It is highly likely that the NCAT in echolocating bats is a more robust homolog of a group of cells in the ventromedial lateral lemniscus of the cat, which has been shown to have sparse projections to the thalamus (Henkel 1983).

3.2 Variability of SOC Organization

Because of differences in the structure of MSO and the periolivary nuclei, the overall organization of the SOC varies considerably from one family of bats to another. Some of these differences are illustrated in Figure 6.8, in which the general structural features of the SOC are compared for the

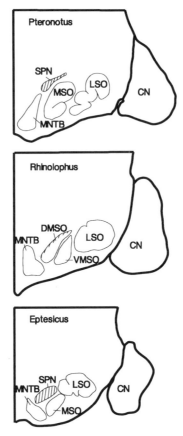

FIGURE 6.8. Structural features of the SOC in three different species of echolocating bats. The LSO and MNTB are similar in all three species, but the medial cell groups are variable. The *shaded area* (SPN) indicates a population of large multipolar neurons just dorsal to MSO. SPN is especially prominent in vespertilionid bats such as the big brown bat, intermediate in size in the mustached bat, and nearly absent in horseshoe bats, where there are only a few multipolar neurons intermingled with the cells along the dorsal border of the DMSO. The MSO in the mustached bat is large, prominent, and convoluted to form a dorsal and a ventral limb. In the big brown bat and other vespertilionids, the MSO is an ovoid structure of intermediate size. In horseshoe bats, the MSO consists of two distinct subdivisions, the dorsal MSO (DMSO) and the ventral MSO (VMSO).

mustached bat, a horseshoe bat, and a vespertilionid bat. It is not known whether these differences in organization of the SOC are systematically correlated with differences in echolocation behavior or hearing capabilities. The differences are systematically correlated with taxonomy, and they may provide as reliable a criterion for classification as do any of the external morphological features commonly used, such as tooth structure or forearm length.

Because of the controversial status of the MSO in bats, the MSO connectional pattern is described here in detail. In the best-studied example, the mustached bat, there are connectional data from both retrograde and anterograde transport of neural tracers as well as from single-fiber injections.

3.3 Inputs to the SOC

The pattern of input from the cochlear nucleus to the SOC in bats is in most respects similar to that in nonecholocating mammals. The LSO receives direct excitatory input from the ipsilateral AVCN (Casseday, Covey, and Vater 1988; Zook and DiCaprio 1988; Covey, Vater, and Casseday 1991; Kuwabara and Zook 1991), mainly from spherical bushy cells (Zook and Casseday 1985; Casseday, Covey, and Vater 1988; Kuwabara and Zook 1991). The input to the LSO from the contralateral AVCN is relayed via neurons in the MNTB that provide glycinergic inhibitory input. The input to MNTB is mainly from globular neurons in AVCN. The axons of the globular cells terminate in large calyces of Held, which synapse in a one-to-one manner on principal cells of the MNTB (Zook and DiCaprio 1988; Casseday, Covey, and Vater 1988; Zook and Leake 1989; Kuwabara and Zook 1991; Kuwabara, DiCaprio, and Zook 1991).

The one major respect in which the brainstem auditory system of echolocating bats appears to diverge from the general mammalian plan is in the pattern of inputs to the MSO. In nonecholocating animals, the MSO receives direct input from the AVCN of both sides, and the relative timing of these two inputs provides an important cue for sound localization (see Irvine 1992 for review). Although there is no unique input to the MSO in any bat that has been studied, bats do differ from nonecholocating mammals in the relative proportion of ipsilateral to contralateral inputs from the AVCN. In the horseshoe bat and mustached bat, for example, the contralateral input is robust while the ipsilateral input is sparse. Furthermore, the ipsilateral and contralateral projections to MSO seem to originate in different populations of cells in the AVCN (Casseday, Covey, and Vater 1988; Covey, Vater, and Casseday 1991; Vater, Casseday, and Covey 1995).

In all mammals, including bats, the projections to the principal nuclei of the SOC are tonotopically organized. Within this organization, the CF harmonic ranges, especially the CF_2, are greatly expanded in the horseshoe and mustached bats (Casseday, Covey, and Vater 1988; Covey, Vater, and Casseday 1991). Although these are the same frequency ranges that are expanded at the cochlea and cochlear nucleus, they appear to be even further expanded at the level of the SOC. Zook and Leake (1989) showed that in the mustached bat the extent of anterograde transport from the cochlear nucleus to the principal nuclei of the SOC is not proportional to the size of the injection site, but rather is differentially expanded in each target structure (see Fig. 6.2). Thus, the representations of certain frequency ranges, especially the CF_2 range from 60 to 63 kHz, are further expanded at the SOC. The relative amount by which each range is expanded differs in the different nuclei. This pattern of expanded CF harmonic frequency representation is reflected in the distribution of BFs in the SOC of the mustached bat (Covey, Vater, and Casseday 1991). In the LSO, approximately one-third of the total volume is devoted to representation of the frequency range from 60 to 63 kHz, and much of the remaining volume to the CF_1, CF_3, and CF_4 frequency ranges. In the MSO, an even larger proportion of the total volume, nearly one-half, is devoted to the CF_2; additionally, there is a large expansion of the CF_3 frequency range, which occupies most of the ventral limb of the MSO.

As would be expected from observations of most other mammals (Friauf and Ostwald 1988; Smith et al. 1991), the projections from the cochlear nucleus to the LSO are ipsilateral only, while those to the MNTB are exclusively contralateral (Kuwabara, DiCaprio, and Zook 1991). The projections to the MSO are unusual in two respects. First, instead of an equal distribution to the two sides, most of the projections are to the contralateral MSO (Casseday, Zook, and Kuwabara 1993; Vater, Casseday, and Covey 1995). Second, in most mammals, the projections from AVCN terminate in the half of MSO nearest their side of origin (Stotler 1953; Strominger and Strominger 1971 Warr 1982). In the mustached bat, there are only sparse projections to the ipsilateral MSO; these are confined to the lateral edge as would be expected. On the contralateral side, however, heavy projections are distributed throughout the entire mediolateral extent of MSO (Casseday, Zook, and Kuwabara 1993; Vater, Casseday, and Covey 1995).

Figure 6.9A shows the results of intraaxonal injection of a fluorescent dye into two fibers of the trapezoid body. The fiber that arises from a spherical bushy cell projects only to the contralateral MSO. The fiber that arises from a stellate cell projects bilaterally. These projection patterns are examples of a rule: most of the projections to MSO arise from spherical bushy cells in the contralateral AVCN, but a small proportion of the projections arise from stellate cells in the AVCN, mainly from the ipsilateral side (Fig. 6.10).

In nonecholocating mammals with good low-frequency hearing, the input to MSO originates mainly from the large spherical cells of the AVCN bilaterally (Cant and Casseday 1986; Casseday and Covey 1987; Schwartz 1992). However, large spherical cells are absent in the AVCN of echolocating bats, and the projection to the MSO originates mainly from the small spherical cells that occupy the rostral AVCN (Covey, Vater, and Casseday 1991; Casseday, Zook, and Kuwabara 1993). Despite these differences in projection pattern, the details of innervation of individual cells in bat MSO seem to be like those of other mammals (Clark 1969a,b; Schwartz 1980; Kiss and Majorossy 1983; Schwartz 1984; Zook and Leake 1989). The fibers from the AVCN form varicosities on the cell body and along one of the two main dendrites, usually the one nearest the origin of the fiber (Zook and Leake 1989).

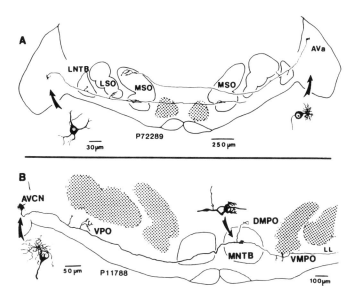

FIGURE 6.9A,B. Projections from intracellularly labeled axons of AVCN neurons in the mustached bat. (A) The axon of a spherical bushy cell (on *right*) projects to the ipsilateral LNTB and the contralateral MSO (*arrows*). The axon of a labeled stellate cell (on *left*) projects to the ipsilateral LNTB, ipsilateral LSO, ipsilateral MSO, and contralateral MSO. (B) The axon of a globular bushy cell projects to periolivary cell groups bilaterally and to the MNTB contralaterally (*large arrows*). In MNTB, the axon terminates in a large calyx of Held. (From Casseday, Zook, and Kuwabara 1993.)

The inputs to the bat MNTB are virtually identical to those in other mammals (Warr 1972; Friauf and Ostwald 1988). In the mustached bat (Kuwabara, DiCaprio, and Zook 1991), the calyces of Held in the MNTB arise from globular bushy cells in the AVCN (Fig. 6.9B). The cells that receive the calyces provide inhibitory projections to the LSO (Moore and Caspary 1983; Spangler, Warr, and Henkel 1985; Zook and DiCaprio 1988; Adams and Mugnaini 1990; Bledsoe et al. 1990; Kuwabara and Zook 1991). Recently it has been shown that MNTB cells also project to MSO (cat, Adams and Mugnaini 1990; bat, Kuwabara and Zook 1991) and that this projection follows the tonotopic arrangement just as it does in LSO (Kuwabara and Zook 1991). Thus, the MNTB is a major source of input to MSO from the contralateral AVCN. The input to MSO, like that to LSO, is inhibitory (Grothe 1990; Grothe et al. 1992).

Another recently recognized potential source of inhibitory input to both LSO and MSO is the LNTB. In bats and other mammals, LNTB cells are contacted by fibers from the ipsilateral cochlear nucleus (Warr 1992; Tolbert, Morest, and Yurgelun-Todd 1982; Roullier and Ryugo 1984; Spirou, Brownell, and Zidanic 1990; Smith et al. 1991; Kuwabara, DiCa

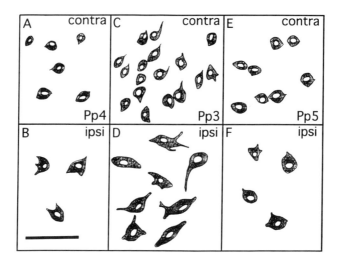

FIGURE 6.10 A–F. Labeled cells in AVCN of the mustached bat following three MSO injections that resulted in bilateral transport. In all three cases, most labeled cells on the contralateral side (*contra*) are small and oval or round in shape, probably spherical bushy cells (A,C,E). Most of the cells labeled on the ipsilateral side *(ipsi)* are larger and multipolar in shape (B,D,F). Calibration bar = 100 μm. (From Vater, Casseday, and Covey © 1995. J. Comp. Neurol. 351, reprinted by permission of Wiley-Liss a division of John Wiley and Sons, Inc.)

prio, and Zook 1991). Labeling of single axons of LNTB cells shows that they project ipsilaterally to both LSO and MSO. Because cells in the LNTB stain for glycine (Peyret, Geffard, and Aran 1986; Peyret et al. 1987; Wenthold et al. 1987; Helfert et al. 1989; Bledsloe et al. 1990), it seems likely that the LNTB cells transform excitatory input from the cochlear nucleus to ipsilateral inhibitory input to the LSO and MSO. At present there is no clue as to the function of an ipsilateral inhibitory input to LSO in bats or any other animal.

3.4 Physiology of the SOC

The functional properties of the principal nuclei of the SOC have been studied extensively in nonecholocating mammals. The resulting view is that information about differences in sound level at the two ears is first encoded in the LSO, whereas information about differences in binaural timing is first encoded in the MSO. The anatomical and physiological bases for binaural hearing in nonecholocating mammals have been reviewed elsewhere (Casseday and Covey 1987; Kuwada and Yin 1987; Irvine 1992). The responses of neurons in the SOC have been studied in an FM bat, *Molossus ater* (Harnischfeger, Neuweiler, and Schlegel 1985) and in two CF-FM bats, the rufous horseshoe bat (Casseday, Covey, and Vater 1988) and the mustached bat (Covey, Vater, and Casseday 1991).

3.4.1 Tonotopy and Frequency Tuning

In echolocating bats, just as in other mammals, the LSO and MSO are tonotopically organized, with a progressive increase in BF going from lateral to medial in the LSO and dorsal to ventral in the MSO. However, it seems clear, based on the observed connectional patterns and results of electrophysiological studies, that both the LSO and MSO in bats are adapted for the specific echolocation behavior of each species. CF-FM bats have clear specializations in the pattern of frequency representation in LSO and MSO (Covey, Vater, and Casseday 1991). Figure 6.11B shows that, in the mustached bat, the CF_2 range from about 60 to 63 kHz has a very large representation in both LSO and MSO. In the LSO, the CF_2 representation occupies the middle one-third of the nucleus, and in MSO, more than half of the nucleus. The representation of the third harmonic, CF_3, although only slightly expanded in the LSO, is greatly expanded in the MSO, where it includes nearly the entire ventral limb of the nucleus. In both LSO and MSO, only a very small proportion of volume is devoted to the range of frequencies in the FM portion of the echolocation signal. It appears that not only are the CF harmonic ranges expanded, but the FM ranges are contracted. The great expansion of the CF representation in CF-FM bats suggests that whatever the respective roles of the LSO and MSO in echolocation may be, both MSO and LSO are concerned mainly with processing the CF components of the call rather than the FM components.

In FM bats, the pattern of frequency representation in the SOC is similar to that in the cochlear nucleus in that no frequency range is expanded. Figure 6.11A shows that in the FM bat *Molossus*, BFs of neurons in MSO are rather uniformly distributed across frequency throughout the audible range (Harnischfeger, Neuweiler, and Schlegel 1985). This uniform frequency representation is not only very different from that in CF-FM bats, it is also very different from the expanded representation of low frequencies seen in the MSO of nonecholocating mammals with good low-frequency hearing (e.g. Guinan, Norris, and Guinan 1972). It is possible that the MSO in FM bats is in some sense less specialized than the MSO of animals that perform low-frequency sound localization, in which a major portion of the MSO contains an expanded low-frequency representation.

Thus, although the general tonotopic sequence in the LSO and MSO follows the standard mammalian plan, the relative representation of specific frequency ranges within the tonotopic sequence is highly variable and species dependent. The result is that the LSO and MSO of each species have an expanded representation of biologically important sounds.

3.4.2 Binaural Response Characteristics

Some clues as to how biologically important sounds may be processed can be seen in the responses of cells in the LSO and MSO to tests with binaural stimuli. In the bat species that have been studied, the responses of neurons in LSO are like those in other mammals. That is, nearly all LSO cells are

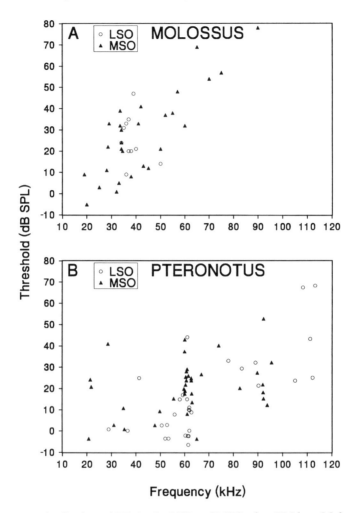

FIGURE 6.11. Distribution of BFs in the LSO and MSO of an FM bat, *Molossus ater* (A) compared with the distribution of BFs in the MSO of a CF-FM species, the mustached bat (*Pteronotus*) (B). (Data for the mustached bat are from Covey, Vater, and Casseday 1991; data for *Molossus* are from Harnischfeger, Neuweiler, and Schlegel 1985.)

excited by sounds at the ipsilateral ear, and this excitatory response is inhibited by sound at the contralateral ear. Thus, the LSO in echolocating bats presumably processes information about relative sound level at the two ears. In CF-FM species, this processing utilizes mainly the CF components of the call.

In contrast, the responses of neurons in MSO appear to vary considerably across bat species and, at least in CF-FM bats, to differ substantially from the responses of MSO cells in nonecholocating mammals. In the dog or cat

MSO, approximately two-thirds of neurons are excited by a sound at either ear, a small percentage are excited by sound at one ear and inhibited by a simultaneous sound at the other ear, and only a very few neurons are monaural (Goldberg and Brown 1968; Yin and Chan 1990). In the rat, MSO neurons are approximately evenly divided between those excited by sound at either ear and those excited by a contralateral sound but inhibited by a simultaneous ipsilateral sound (Inbody and Feng 1981). Functionally, the MSO of the FM bat *Molossus* more closely resembles that of the rat than that of the dog or cat. In *Molossus*, approximately one-fourth of MSO neurons are excited by sound at either ear, and about one-third are excited by a contralateral sound and inhibited by an ipsilateral sound. *Molossus* differs from the rat, however, in that monaural neurons are found in the MSO. If neurons excited by either ear are primarily responsible for encoding interaural phase differences of low-frequency sounds, perhaps their lower incidence in MSO of the rat and their even lower incidence in an FM bat reflects the decreased importance of interaural time or phase differences in sound localization for animals with poor low-frequency hearing, regardless of echolocation capability.

In CF-FM bats, the response properties of MSO neurons are even farther removed from those of dogs, cats, and rats than are those of FM bats. In horseshoe bats, the neurons in the dorsal MSO are approximately evenly divided into three groups: (1) those excited by sound at either ear, (2) those excited by a contralateral sound and inhibited by a simultaneous ipsilateral sound, and (3) monaural cells, responsive to sound at the contralateral ear only. In the ventral MSO, virtually all neurons are monaural, excited by the contralateral ear (Casseday, Covey, and Vater 1988). In the mustached bat, more than three-fourths of MSO neurons are monaural, responding only to a contralateral sound; the remainder are binaural, with the largest binaural class those excited by a contralateral sound and inhibited by a simultaneous ipsilateral sound. Only a very small percentage of neurons in the MSO of the mustached bat are excited by sound at either ear (Covey, Vater, and Casseday 1991; Grothe et al. 1992).

3.4.3 Discharge Patterns and Timing Characteristics

Temporal response properties reinforce the view that in bats LSO neurons are like those in other mammals whereas MSO neurons are specialized. In both FM and CF-FM bats, LSO neurons typically fire in a sustained, "fast chopper" pattern (Harnischfeger, Neuweiler, and Schlegel 1985; Casseday, Covey, and Vater 1988; Covey, Vater, and Casseday 1991). However, the discharge patterns of MSO neurons in both FM and CF-FM bats are somewhat different from those seen in mammals with low-frequency hearing. In the cat, nearly 90% of MSO units respond in a sustained pattern (Yin and Chan 1990). As in the case of binaural responses, the discharge patterns of MSO cells in FM bats more closely resemble those in nonecholocating species than do those of CF-FM bats. In the FM bat *Molossus*, about three-fourths of MSO units respond in a sustained pattern, but the remaining one-fourth respond transiently at the onset of a sound (Harnisch-feger, Neuweiler, and Schlegel 1985). In both species of CF-FM bat that have been studied, the percentage of transient responses in MSO is even higher, 40% in the horseshoe bat (Casseday, Covey, and Vater 1988) and about 70% in the mustached bat (Covey, Vater, and Casseday 1991). For many MSO neurons with transient responses in these species, the discharge switches from onset to offset with small changes in frequency. This change from on to off is seen most commonly in neurons with a BF near the CF_2 or CF_3.

The transient on and off responses in the MSO of the mustached bat have been shown to arise through the interaction of a sustained excitatory input with a sustained inhibitory input. Whether the response is to the onset or offset of the sound is determined by the relative timing of the two inputs (Grothe et al. 1992). The direct excitatory input is almost certainly from spherical bushy cells in the contralateral AVCN. The inhibitory input probably arises from the ipsilateral MNTB, which in turn receives its input from globular bushy cells in the contralateral AVCN. A similar type of interaction could account for the responses of MSO units that are excited by a contralateral sound and inhibited by an ipsilateral sound. Ipsilateral inhibitory input, relayed by LNTB cells, could reach MSO cells before or concurrent with the direct excitatory input from the contralateral cochlear nucleus.

3.5 Possible Function of the MSO in Echolocation

The MSO of the bat differs from that of other mammals in two important respects. Its inputs are altered to emphasize the contralateral ear, and its responses are altered to emphasize transients in the CF frequency range. Of what significance are these alterations for the bat? Although there is no single obvious answer, the data suggest several possibilities. The first involves localization in the vertical plane. Because of the aerial acrobatics that bats perform to pursue and capture flying prey, they must be as adept at localizing sound in the vertical plane as in the horizontal plane. The external ears of bats and other animals impose elevation-dependent changes in the intensity of specific spectral components of sounds, and it has been suggested that these systematic changes in the spectrum provide cues that can be analyzed in the central auditory system to compute the vertical location of a sound source (Simmons and Lawrence 1982; Fuzessery and Pollak 1985). Spectral changes might provide a particularly robust cue for elevation in echolocating bats, as the outgoing pulse could provide a reference for comparison.

In the IC of the mustached bat, the thresholds of all binaural neurons are lowest for sound at the midline in the horizontal dimension. This sensitivity pattern is independent of the neuron's frequency selectivity. In the vertical dimension, however, neurons tuned to the different harmonics of the echolocation call are differentially sensitive to a sound, depending on signal elevation. For example, neurons tuned to the CF_2 are most sensitive to about 0° elevation, whereas neurons tuned to the CF_2 are most sensitive to elevations of about $-40°$ (Fuzessery and Pollak 1985). In the MSO and LSO of the mustached bat, units tuned to the CF_2 frequency range, about 60 kHz, have the lowest thresholds and the widest range of thresholds; units tuned to the CF_3 frequency range, about 90 kHz, have higher thresholds (Covey, Vater, and Casseday 1991). As there is an increase in intensity for the frequencies about 90 kHz at lower elevations, the units in the LSO and MSO with BFs around 90 kHz would be more likely to be activated at lower elevations of the echo source. Thus, one function of MSO might be to transmit spectral cues that could be integrated at the IC with interaural intensity difference cues to derive the vertical and horizontal coordinates of an object in space.

A second possible function of the MSO in bats concerns localization of objects in the third dimension, that is, estimation of the distance to an object. This potential function for the MSO would by no means preclude the simultaneous transmission of spectral cues for elevation. When the bat flies toward a stationary or slowly moving object, the returning echoes are Doppler shifted to a frequency slightly higher than the emitted pulse. Through a reflex compensatory mechanism, the bat lowers the frequency of its own emitted call to keep the frequency of the echoes within the narrow excitatory range of filter neurons with BFs about 60 kHz. The response pattern of many MSO neurons in the mustached bat changes from on to off as sound frequency decreases. This on/off pattern is seen mainly in neurons tuned to the intense CF_2 harmonic of the echolocation call, about 60 kHz. Thus, on/off neurons in MSO could respond at the offset of the lower frequency emitted pulse and at the onset of the higher frequency Doppler-shifted echo; presumably, if the two overlapped, the response would be facilitated. Field studies have shown that the bat systematically decreases the duration of the CF component of its call as it approaches its prey (Novick and Vaisnys 1964). Thus, the facilitated response could provide a mechanism to signal the bat to adjust the duration of its echolocation signal in relationship to the distance from its target, thus acting as a sort of vocal yardstick (Casseday, Zook, and Kuwabara 1993).

A third possibility is that MSO neurons encode information about the fine temporal structure of the envelope of sounds. Nearly all MSO neurons in the mustached bat are capable of following amplitude modulations (AM) with on or off responses. Approximately one-third of MSO neurons exhibit bandpass selectivity for AM rate, while the remainder have low-pass filter characteristics with upper limits between 100 and 500 Hz, with most between 200 and 300 Hz (Grothe 1990, 1994). The upper AM filter limits for MSO neurons are considerably lower than those found at the level of the cochlear nucleus, where individual neurons show synchronized firing up to 1000 Hz (Vater 1982). Blocking glycinergic inhibitory input to MSO shows that the upper AM filter limit is created through an interplay of excitatory and inhibitory inputs, offset in time relative to one another. This mechanism for producing AM filter characteristics is almost certainly related to, or identical with, the mechanism by which frequency-specific on and off responses are produced in the bat MSO (Grothe et al. 1992; Grothe, 1994).

Behavioral studies have shown that CF-FM bats are capable of discriminating among different species of insects on the basis of the pattern of frequency or amplitude modulations imposed on the CF portion of the echo by an insect's wingbeats (Schnitzler et al. 1983; Von der Emde and Schnitzler 1986, 1990; Kober and Schnitzler 1990). An as-yet-untested hypothesis is that MSO neurons are selective for the wingbeat rates of insects that are attractive to the bat and that their activity could signal the presence of a prey worth pursuing.

A fourth possibility, given that bat MSO neurons exhibit phase locking to amplitude modulations of a high-frequency carrier tone, is that monaural phase information from the two MSOs could ultimately come together in some form at a higher level, the dorsal nucleus of the lateral lemniscus (DNLL) or the IC, for example, to provide localization cues based on interaural timing differences of low-frequency amplitude modulations of a high-frequency carrier.

Finally, echoes from three-dimensional objects are known to contain characteristic interference patterns or "glints" that can be seen as amplitude modulations in the envelope of the sound (Simmons 1989). Because each of the bat's ears is at a slightly different angle with respect to the object that produces the echo, the only conditions under which the echo would be the same at both ears would be for a perfectly symmetrical object located at the midline. Thus, another untested hypothesis about the function of MSO is that it could convey information about binaural disparities in the glint pattern of echoes, that could then be integrated at a higher level to derive information about an object's three-dimensional structure much as binocular disparity is used in the visual system to perceive depth.

3.6 Do Bats Really Have a Medial Superior Olive?

Animals with very small heads and high-frequency hearing are thought to have little or no usable range of binaural phase or envelope time differences (Masterton et al. 1975). This limitation is consistent with the idea, proposed several decades ago, that bats do not have an MSO, or at best have one that is extremely small and rudimentary (Harrison and Irving 1966; Irving and Harrison 1967). A different point of view, suggested by more recent studies of the superior olive in bats, is that the unusual requirements for localizing high-frequency sound reflected from flying prey in three-dimensional space have imposed an evolutionary pressure that has resulted in a highly developed and differentiated MSO homolog that is functionally different from the MSO in animals with good low-frequency hearing. This functional difference appears to result from a difference in the strength of projections from the ipsilateral cochlear nucleus. Otherwise, the cytoarchitecture and connections basically resemble those of MSO in mammals with low-frequency hearing.

If the MSO is defined as being a structure that receives direct excitatory input in equal measure from the two ears and compares interaural phase differences, then the bat "MSO" is clearly not an MSO. If, on the other hand, the MSO is defined as being a population of elongate cells between the LSO and the MNTB that receives direct or indirect input in equal or unequal measure from the two ears, and that provides tonotopically organized projections to the ipsilateral DNLL and IC, then the bat "MSO" clearly is an MSO. Unfortunately, no developmental data are available to answer the obvious question of whether the bat MSO is derived from the same embryonic precursor as the MSO in the cat or other large mammals. However, even if this question were resolved, the problem of the monaurality of MSO in the bat would remain. The fact that there seems to be a continuum of increasing monaurality from cat to rat to FM bat to CF-FM bat suggests that small changes in the relative strength of one input to a common precursor structure can result in very different functional properties. These properties customize the structure to fit the particular needs of each species. Just as the bat's head, pinnae, and hands have undergone evolutionary changes, the MSO has also undergone changes to adapt it to the special requirements of aerial navigation and hunting guided by echolocation.

3.7 Outputs of the SOC

The SOC is the source of three very different classes of outputs. First, and most thoroughly studied, are the pathways that ascend within the *lemniscal system*. The contribution of the SOC to the lemniscal system includes all the ascending projections that originate in the LSO and MSO and that terminate in the DNLL and the IC. The major lemniscal outputs of the SOC are summarized in Figure 6.12. Second is the *efferent system*, which includes all the projections that originate in the periolivary cell groups or principal nuclei of the SOC and terminate in the cochlea or cochlear nucleus. The third output of the SOC is a small but distinct *extralemniscal pathway* that originates in the NCAT and bypasses the IC to terminate in the superior colliculus and auditory thalamus.

3.7.1 SOC Lemniscal System: LSO, MSO, and DNLL

The lemniscal system as a whole includes all the pathways that ascend within the fiber tract of the lateral lemniscus. These pathways originate in the cochlear nucleus, the SOC, and the nuclei of the lateral lemniscus. All the lemniscal pathways terminate, directly or indirectly, in the IC. The main sources of input to the DNLL are the LSO and MSO; the DNLL in turn projects to the IC. Thus, the LSO, MSO, and DNLL together make up an integral system of direct and indirect pathways to the auditory midbrain, largely concerned with binaural processing. The MNTB does not project directly to the IC. However, in bats and other mammals, the MNTB provides dense input to the intermediate nucleus of the lateral lemniscus (INLL) and the ventral nucleus of the lateral lemniscus (VNLL) (Fig. 6.12C). This input is probably inhibitory, as is the input from the MNTB to the LSO and MSO.

As in all other mammals that have been studied, the projections from the LSO to the IC in echolocating bats are bilateral (Fig. 6.12A) (Schweizer 1981; Zook and Casseday 1982b, 1987; Casseday, Covey, and Vater 1988; Ross, Pollak, and Zook 1988; Ross and Pollak 1989; Vater, Casseday, and Covey 1995). The target of the projection from the LSO occupies roughly the ventral two-thirds of the IC, and the terminal fields take the form of bands or slabs oriented roughly parallel to the orientation of disk-shaped cells, and also roughly parallel to isofrequency contours (Zook and Casseday 1985, 1987; Casseday, Covey, and Vater 1988).

The ascending projections from the MSO in echolocating bats (see Fig. 6.12B) resemble those of other mammals in that they terminate only in the ipsilateral DNLL and IC (Adams 1979; Brunso-Bechtold, Henkel, and Linville 1990; Schweizer 1981; Zook and Casseday 1982b, 1987; Casseday, Covey, and Vater 1988; Ross, Pollak, and Zook 1988; Ross and Pollak 1989; Vater, Casseday, and Covey 1995). Like the LSO projections, the MSO projections extend throughout the ventral two-thirds of the IC and are tonotopically organized. The MSO projections, again like the LSO projections, appear to terminate in a pattern of bands or slabs. The monaural nature of MSO responses in the mustached bat appears to be preserved up through the level of the midbrain. Electrophysiological experiments combined with retrograde transport of HRP show that in this species the MSO is the main source of input to the monaural region of the enlarged 60-kHz contour of the inferior colliculus (Ross and Pollak 1989). Thus, although the overall target areas of MSO and LSO are similar, there still appears to be functional segregation according to monaural or binaural response properties. Nevertheless, there is considerable overlap between the targets of LSO and those of MSO, so it is highly likely that many individual cells in DNLL and IC receive input from both LSO and MSO (Vater, Casseday, and Covey 1995).

The DNLL: structure and function. Although published information about the DNLL in echolocating bats is not extensive, it is sufficient to indicate that its structure, function, and connections are essentially the same as they are in nonecholocating mammals. In both FM bats and CF-FM bats, the

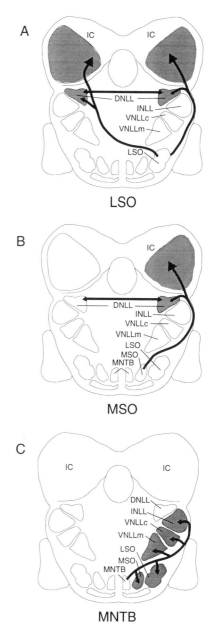

FIGURE 6.12 A–C. Outputs of the SOC. The *shaded areas* indicate the targets of LSO (A), MSO (B), and MNTB (C). These pathways are essentially the same in echolocating bats as they are in other mammals.

DNLL is a wedge-shaped structure composed mainly of medium to large fusiform cells arranged in clusters between the dense fascicles of ascending fibers of the lateral lemniscus (see Fig. 6.15, later in this chapter). The long axes of the cells are typically oriented perpendicular to the ascending fibers. The DNLL receives bilateral projections from the LSO and ipsilateral projections from the MSO. Each DNLL projects to the opposite DNLL via the commissure of Probst, and to the IC bilaterally, where its target region is approximately coextensive with that of the AVCN, LSO, and MSO (Zook and Casseday 1987; Casseday, Covey, and Vater 1988). In addition, in both bats and nonecholocating mammals, there is a projection from the rostral region of the DNLL and the adjacent auditory paralemniscal tegmentum to the deep layers of the superior colliculus (Casseday, Jones, and Diamond 1979; Kudo 1981; Henkel 1983; Tanaka et al. 1985; Covey, Hall, and Kobler 1987). As in other mammals, most DNLL neurons in the bat use γ-aminobutyric acid (GABA) as a neurotransmitter, indicating that the DNLL provides mainly inhibitory input to its targets (Adams and Mugnaini 1984; Covey 1993a).

As would be expected in a structure that receives bilateral input from the SOC, the responses of most DNLL neurons are binaural. This is true in the big brown bat (Covey 1993a), the mustached bat (Markovitz and Pollak 1994), and the cat (Aitkin, Anderson, and Brugge 1970; Brugge, Anderson, and Aitkin 1970), the three species in which binaural responses in DNLL have been studied. Most commonly, DNLL neurons are excited by sound at the contralateral ear and inhibited by sound at the ipsilateral ear. In the big brown bat, some DNLL neurons are excited by sound at either ear while others are facilitated by ipsilateral sound at some levels and inhibited at other levels (Covey 1993a). Because so few data have been obtained in nonecholocating mammals, it is impossible to say whether these response properties are common to the DNLL of all mammals or whether they are peculiar to FM bats. In either case, binaural neurons in the DNLL clearly play a role in shaping, through GABAergic inhibition, the binaural responses of neurons at the IC (Li and Kelly 1992); this must also be the case in echolocating bats.

In the cat, the DNLL contains a complete tonotopic representation that is clearly separate from the tonotopic representation in the more ventral parts of the lateral lemniscal nuclei (Aitkin, Anderson, and Brugge 1970; Brugge, Anderson, and Aitkin 1970). In horseshoe bats and big brown bats, the tonotopic representation in the DNLL is not as highly organized as in the cat (Metzner and Radtke-Schuller 1987; Covey and Casseday 1991). Nevertheless, even in these species, the tonotopy in the DNLL is distinct from that in the intermediate and ventral nuclei of the lateral lemniscus.

Although the role of DNLL in binaural hearing is almost certainly the same in echolocating bats as it is in nonecholocating mammals, there are several features of DNLL responses that may be specializations for echolocation. In the CF-FM species, the rufous horseshoe bat, some DNLL neurons have multiple BFs that are harmonically related. Because the BFs of these multiple-tuned neurons are not within the CF harmonic ranges, it has been suggested that they may play a role in range determination, which depends on the FM frequency ranges, or in social communication between individuals (Metzner and Radtke-Schuller 1987).

In the FM species, the big brown bat, some DNLL neurons are delay tuned in that they respond in a highly facilitated manner to the second of two identical sounds when it occurs at a specific time after the first. The best delays of these neurons are all within the range that would be relevant for echolocation behavior (Covey 1993a). Similar facilitated responses to the second of two identical sounds have also been described in the lateral lemniscus of another FM species, the little brown bat (Suga and Schlegel 1973). Although recording sites were not localized in the latter study, it seems likely, given the data in the big brown bat, that the facilitated neurons were in the DNLL. Delay-tuned neurons in the DNLL of FM bats may represent the first stage of neural selectivity for echoes that originate from objects at specific distances from the bat.

3.7.2 Efferent System

Although the ascending lemniscal pathways are the largest and most prominent outputs from cell groups in the SOC, there is also a system of decending pathways that are thought to provide feedback to modulate neural activity in the cochlea and cochlear nucleus. Although the organization of the efferent system differs greatly even among nonecholocating mammals, a few generalizations can be made. In most mammals, two distinct populations of neurons project to the cochlea. These are large olivocochlear neurons that terminate on outer hair cells and small olivocochlear neurons that terminate on inner hair cells (Warr and Guinan 1979; Guinan, Warr, and Norris 1983; Aschoff and Ostwald 1987; Warr 1992). The projections of olivocochlear neurons have been described in a number of different species of echolocating bats, including the greater horseshoe bat (Bruns and Schmieszek 1980), the mustached bat (Bishop and Henson 1987, 1988), the rufous horseshoe bat (Aschoff and Ostwald 1987; Bishop and Henson 1988), *Hipposideros lankadiva* (Bishop and Henson 1988), and *Rhinopoma*, *Tadarida*, and *Phyllostomus discolor* (Aschoff and Ostwald 1987). The pattern of distribution of large and small olivocochlear neurons in different species of bats is summarized in Figure 6.13. The reason for this seemingly inordinate amount of comparative data is the finding that certain species of bats lack efferent projections to the outer hair cells (see Kössl and Vater, Chapter 5, this volume). These species include the greater horseshoe bat, the rufous horseshoe bat, and *Hipposideros lankadiva*. All these species are horseshoe bats that use CF-FM calls. In the other unrelated CF-FM species, the mustached bat, each outer hair cell receives a single large efferent ending. Corresponding to the lack of outer hair cell innervation, horseshoe bats have only one population of olivocochlear neurons. These are small cells located in a region between the LSO and MSO, and they project only to the ipsilateral cochlea (Aschoff and Ostwald 1987).

In the mustached bat, two populations of neurons project to the cochlea. First, a population of small olivocochlear neurons located ipsilaterally in a region between the LSO and MSO probably innervate the inner hair cells.

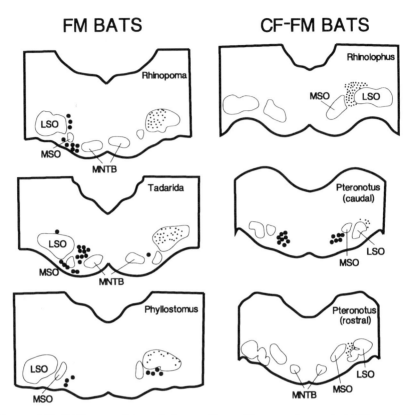

FIGURE 6.13. Different origins of olivocochlear efferent pathways in FM bats (*left*) and CF-FM bats (*right*). In FM bats, as in rodents, small olivocochlear neurons are located within the ipsilateral LSO. In CF-FM bats, they are ipsilateral, but outside the LSO, mainly in the region between the LSO and MSO. Large olivocochlear neurons are mostly located medial and ventral to the MSO. In some species they are bilateral, but in others they are contralateral only. In horseshoe bats, large olivocochlear neurons are absent. (Data for *Rhinopoma*, *Tadarida*, *Phyllostomus*, and the horseshoe bat are redrawn from Aschoff and Ostwald 1987; data for the mustached bat are redrawn from Bishop and Henson 1988.)

Second, a population of large olivocochlear neurons located bilaterally in a region caudal to the MSO probably innervate the outer hair cells (Bishop and Henson 1987, 1988). In the FM bats *Rhinopoma* and *Tadarida*, and in the whispering FM bat *Phyllostomus*, the organization of the olivocochlear efferents is very different. In all three species, small olivocochlear neurons are found inside the ipsilateral LSO, where they make up a significant proportion of the total cells. Large olivocochlear neurons are present bilaterally in the ventral and medial periolivary regions. Except for the lack of small olivocochlear neurons in the contralateral LSO, the organization in these bat species resembles that in rodents such as the guinea pig (Aschoff and Ostwald 1987). Thus, there is a clear difference between CF-FM bats and FM bats. The efferent systems of the FM bats are similar to those of rodents while those of the CF-FM bats are highly specialized and deviate considerably from what seems to be the general mammalian plan of organization.

Decending projections from the SOC terminate not only in the cochlea, but also in the cochlear nucleus. In the rufous horseshoe bat, efferents to all three divisions of the cochlear nucleus originate in the ventral periolivary cell groups bilaterally and in the LNTB ipsilaterally (Vater and Feng 1990). This pattern of efferent projections is essentially the same as that seen in the FM species, the big brown bat (Covey, unpublished data), suggesting that whereas the origins of efferents to the cochlea in CF-FM bats are very different from those in FM bats, the origins of efferents to the cochlear nucleus are similar.

3.7.3 The Central Acoustic Tract: An Extralemniscal Pathway to the Thalamus

In the lemniscal pathway, a variety of signals from the auditory brainstem reach the thalamus indirectly via one or more synapses in the inferior colliculus, which is often thought of as an obligatory relay (Goldberg and Moore 1967; Aitkin and Phillips 1984). However, the central acoustic tract, an extralemniscal pathway from the auditory medulla to the thalamus, was first described in nonecholocating mammals, including man (Papez 1929a,b). Perhaps because of the small size of this pathway relative to the lemniscal system, it has largely been ignored. However, in some echolocating bats, the central acoustic tract is large and robust. In the mustached bat it originates in a group of large multipolar neurons just rostral to the principal nuclei of the SOC, the nucleus of the central acoustic tract (NCAT) (Kobler, Isbey, and Casseday 1987; Casseday et al. 1989). The NCAT receives bilateral input from the cochlear nucleus and gives rise to a fiber bundle that courses medial to the lateral lemniscus and ventral to the inferior colliculus. It terminates in the deep layers of the superior colliculus and in the suprageniculate nucleus of the thalamus, a group of large cells just medial to the medial geniculate body. The suprageniculate nucleus, in turn, projects to the auditory cortex and frontal cortex. These connections are summarized in Figure 6.14. The fact that the central acoustic tract is especially highly developed in some species of echolocating bats suggests that it may play a role in specialized motor behaviors such as Doppler-shift compensation (Casseday et al. 1989).

3.7.4 The Paraleminscal Tegmentum and Auditory–Vocal Interactions

In echolocating bats, there is evidence for interaction between the auditory pathways of the lower brainstem and the motor systems for vocalization. In experiments on the gray bat, Suga and Schlegel (1972) and Suga and

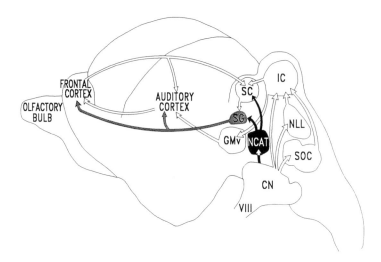

FIGURE 6.14. The central acoustic tract in the mustached bat and its relationship with the lemniscal auditory pathways and pathways to the cortex. The components of the central acoustic tract are shown by the *black arrows* and the lemniscal system by the *white arrows*. The *grey arrows* indicate further projections beyond the thalamic target of the central acoustic tract. NCAT, nucleus of central acoustic tract; IC, inferior colliculus; GMv, central division of medial geniculate body. (From Casseday et al. © 1989. J. Comp. Neurol. 287 reprinted by permission of Wiley-Liss, a division of John Wiley and Sons, Inc.)

Shimozawa (1974) measured evoked potentials in response to self-vocalized FM sounds and tape recordings of the same sounds. They found that in the region of the lateral lemniscus, the responses to self-vocalized sounds were considerably smaller than the responses to artificial sounds and suggested that the nuclei of the lateral lemniscus are the site of neural attenuation of vocalized sounds. Experiments on the rufous horseshoe bat (Metzner 1989; 1993) showed that the responses of many neurons in the paralemniscal tegmentum are affected by vocalization; about half of these neurons were inhibited when the bat vocalized. Thus, it is possible that the origin of the neural attenuation described in the early studies on the gray bat originated in the paralemniscal tegmentum.

Other auditory-vocal neurons in the paralemniscal tegmentum of the horseshoe bat show a variety of complex interactions between sound-evoked responses and vocal activity. Metzner (1989, 1993) has suggested that at least some of these neurons may play an active role in Doppler-shift compensation. Electrical stimulation of the paralemniscal tegmentum in several bat species elicits species-specific echolocation sounds (Suga et al. 1973; Schuller and Radtke-Schuller 1990). Thus it appears that the paralemniscal tegmentum acts as an interface between the sensory pathways of the auditory system and the motor pathways involved in vocalization.

4. Monaural Nuclei of the Lateral Lemniscus

In all mammals, groups of cell bodies located among the ascending fibers of the lateral lemniscus receive input from the cochlear nucleus and project to the IC. In echolocating bats these nuclei are hypertrophied and highly differentiated. Like the nuclei of the SOC, the nuclei of the lateral lemniscus represent a stage of specialized neural processing that is accomplished before the outputs of the different brainstem pathways converge at the IC. With the exception of the DNLL, which is clearly part of the binaural system, all the cell groups of the lateral lemniscus receive the bulk of their input from the contralateral cochlear nucleus and project densely and almost exclusively to the ipsilateral IC (for review, see Schwartz 1992). Thus, the monaural nuclei of the lateral lemniscus together make up a system of pathways that is organized in parallel to those from the SOC to the IC.

4.1 Structure

In nonecholocating animals such as cats and gerbils, the monaural nuclei of the lateral lemniscus are at least as large in size as the SOC, and their projections to the IC are at least equal in magnitude to those from the SOC (Adams 1979). In echolocating bats, the monaural nuclei of the lateral lemnsicus are extraordinarily large relative to the rest of the brainstem, and are exquisitely differentiated into morphologically distinct regions (Poljak 1926; Baron 1974; Zook and Casseday 1982a; Covey and Casseday 1986).

In all species of echolocating bats that have been examined, there are at least four separate cell groups embedded among the fibers of the lateral lemniscus: the dorsal nucleus (DNLL), the intermediate nucleus (INLL), and two parts of the ventral nucleus (VNLL), the columnar division (VNLLc) and the multipolar cell division (VNLLm). All of these cell groups are easily distinguished from one another on the basis of the arrangement and morphology of their neurons. Figure 6.15 shows the overall organization of the nuclei of the lateral lemniscus in the mustached bat. The INLL is a thick triangular wedge-shaped structure, which in many bats is so large that it forms a conspicuous protrusion on the side of the brainstem. It is bounded medially and laterally by thick bundles of ascending fibers. Fine fascicles of fibers course throughout the INLL, some parallel to the ascending fibers and others orthogonal to them. The principal cell type in the INLL is elongate, with dendrites that are oriented orthogonal to the ascending fibers of the lateral lemniscus. The VNLLc of echolocating bats is especially prominent because of the unusually high packing density and distribution of cells in columns separated by thin fiber bundles. Virtually every neuron in VNLLc is of the same type, small and round, with one thick dendrite that branches extensively at some distance from the cell

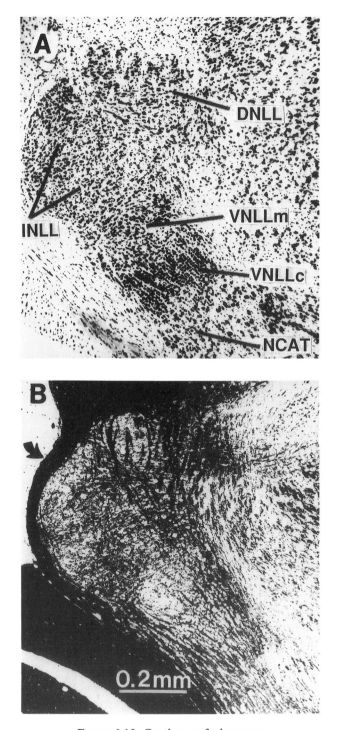

FIGURE 6.15. Caption on facing page.

body. They are very similar in appearance to spherical bushy cells in the AVCN. They appear to receive terminals mainly or exclusively on the cell body, some as large, calyx-like endings (Zook and Casseday 1985; Covey and Casseday 1986). As its name implies, the principal cell type in VNLLm is multipolar in shape. The dendrites of the multipolar cells do not appear to have any preferred orientation.

All echolocating bats have the same cell populations in INLL and VNLL, but the cells in the VNLL are distributed somewhat differently, according to species. Examples of these different patterns of organization are shown in Figure 16.6. In the mustached bat, the VNLLc is most ventral and the VNLLm lies between the INLL and the VNLLc. In the big brown bat and other vespertilionid bats, the locations of the two ventral nuclei are reversed, with the VNLLc located in the middle and the VNLLm in the ventralmost position. In horseshoe and hipposiderid bats, the VNLLc and VNLLm are side by side, with the VNLLc medial and the VNLLm lateral. Thus, just as in the case of SOC organization, there is a strong relationship between taxonomy and VNLL morphology.

In nonecholocating mammals there is no clear segregation of cell types within the INLL and VNLL, as there is in the bat, although the same basic cell types seem to be present (Adams 1979). Echolocating dolphins resemble bats in that the MNTB and monaural nuclei of the lateral lemniscus are hypertrophied and the VNLL is differentiated into columnar and multipolar cell regions (Zook et al. 1988). The similarity between bats and dolphins suggests that the neural processing performed by the monaural nuclei of the lateral lemniscus must be of crucial importance for echolocation. The variability in the location of the two cell populations of VNLL among different taxonomic groups of bats suggests that segregation of bushy cells from multipolar cells is somehow important for echolocation, but this segregation is accomplished in different ways in different species.

4.2 Inputs to the Monaural Nuclei of the Lateral Lemniscus

Anatomical studies in echolocating bats and nonecholocating mammals show that the INLL and VNLL are major targets of projections from the ventral cochlear nucleus and that this pathway is almost entirely contralateral (Warr 1966, 1969, 1982; Glendenning et al. 1981; Zook and

←───────────────────────────────────

FIGURE 6.15A,B. The nuclei of the lateral lemniscus in the mustached bat. (A) Nissl-stained frontal section through the left brainstem; (B) fiber-stained frontal section through the left brainstem. (From Zook and Casseday © 1982. J. Comp. Neurol, 207 reprinted by permission of Wiley-Liss, a division of John Wiley and Sons, Inc.)

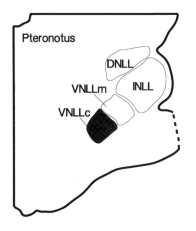

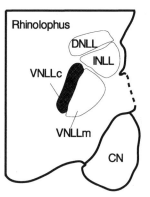

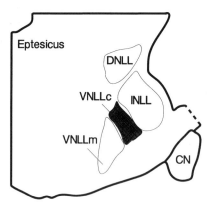

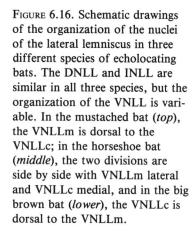

FIGURE 6.16. Schematic drawings of the organization of the nuclei of the lateral lemniscus in three different species of echolocating bats. The DNLL and INLL are similar in all three species, but the organization of the VNLL is variable. In the mustached bat (*top*), the VNLLm is dorsal to the VNLLc; in the horseshoe bat (*middle*), the two divisions are side by side with VNLLm lateral and VNLLc medial, and in the big brown bat (*lower*), the VNLLc is dorsal to the VNLLm.

Casseday 1985). In all mammals, the projections to the INLL and VNLL originate in both the AVCN and PVCN, probably from several different cell types with different response properties (Zook and Casseday 1985; Covey and Casseday 1986; Friauf and Ostwald 1988; Covey 1993b). Additionally, the INLL and VNLL in all mammals receive indirect projections from the contralateral cochlear nucleus via the MNTB and periolivary nuclei (Glendenning et al. 1981; Zook and Casseday 1985, 1987).

In the mustached bat, injections of retrograde tracers placed in the INLL label many cells in the AVCN, MNTB, and periolivary cell groups; only a few cells are labeled in the PVCN. Retrograde tracers placed in the VNLL label many cells in the AVCN and PVCN; only a few cells are labeled in MNTB and periolivary cell groups (Zook and Casseday 1985). A similar pattern of projections is seen in the cat (Glendenning et al. 1981; see Covey 1993b for review of connections in nonecholocating mammals). These results suggest that in all mammals including bats the AVCN projects densely throughout the entire region of the INLL and VNLL, but the MNTB and PVCN project with a dorsal-to-ventral density gradient. Thus, the most dense MNTB projection is to the INLL and the most dense PVCN projection is to the VNLLm.

Injections of anterograde tracers in the cochlear nucleus of both cats and bats result in multiple patches or bands of anterograde label in the region ventral to DNLL (Covey 1993b). Similarly, injection of HRP in axons of cells in the cochlear nucleus of the rat shows that each axon provides several collaterals that terminate in the region below the DNLL (Friauf and Ostwald 1988). Taken together, this evidence strongly suggests that the region below the DNLL contains multiple tonotopic representations; these multiple representations originate, at least partly, from collaterals of single axons.

4.3 Physiology of the Monaural Nuclei of the Lateral Lemniscus

Despite a convincing body of anatomical evidence to indicate that the monaural nuclei of the lateral lemniscus provide a major system of parallel pathways to the midbrain, there have to date been only four published studies of single unit response properties in the INLL and VNLL of any animal. Two of these are in the cat (Aitkin, Anderson, and Brugge 1970; Guinan, Norris, and Guinan 1972), and two are in the echolocating species, the rufous horseshoe bat (Metzner and Radtke-Schuller 1987) and the big brown bat (Covey and Casseday 1991). An abstract is also available on the mustached bat (O'Neill, Holt, and Gordon 1992). Despite a few inconsistencies among the results reported in these studies, the data present a reasonably coherent picture of the response properties of neurons in the mammalian INLL and VNLL, and provide functional evidence for the importance of these structures in echolocation.

4.3.1 Monaurality

As would be expected from the connections, the responses of most neurons in the INLL and VNLL are monaural, driven by sound at the contralateral ear. It was on the basis of monaural versus binaural responses in the cat lateral lemniscus that the distinction was first made between the DNLL, where responses are binaural, and the auditory region ventral to DNLL, where responses are nearly all monaural (Aitkin, Anderson, and Brugge 1970; Brugge, Anderson, and Aitkin 1970). In the big brown bat, all units in the VNLLc and VNLLm and nearly all units in the INLL are monaural, excited by sound at the contralateral ear. The few binaural units found in the INLL are at marginal locations, mainly near the border with the DNLL (Covey and Casseday 1991; Covey 1993a).

4.3.2 Tonotopy and frequency tuning.

In the CF-FM horseshoe bat as well as the FM species, the big brown bat, the INLL, VNLLc, and VNLLm each has a separate and distinct tonotopic representation. The INLL is organized so that low frequencies are represented laterally and high frequencies medially. In the VNLLc of both the big brown bat and the horseshoe bat, there is a well-defined tonotopy in which low frequencies are dorsal and high frequencies are ventral. However, there is one major difference between the two species. Whereas the VNLLc of the big brown bat contains a complete tonotopic sequence, in the horseshoe bat the CF_2 frequency range is absent (Metzner and Radtke-Schuller 1987). This finding is particularly striking in view of the fact that the CF_2 frequency range is greatly expanded in most other auditory nuclei in the horseshoe bat. It has also been reported that in the mustached bat, another CF-FM species, neurons in VNLLc do not respond to the CF harmonics of the echolocation call (O'Neill, Holt, and Gordon 1992). The absence of CF representation suggests that the VNLLc is specialized for processing the FM components of the echolocation call. This idea is supported by evidence that neurons in the VNLLc are the source of an extemely precise timing signal that provides the information necessary to determine the delay between the emission of the FM component of the echolocation signal and the return of the FM component of the echo. This delay information is essential for target range determination (Covey and Casseday 1991).

The VNLLm in the big brown bat has only a rough tonotopic organization. A central core of neurons tuned to high frequencies appears to be surrounded by concentric layers of neurons tuned to progressively lower frequencies. In the rufous horseshoe bat, the situation is much more-straightforward: all neurons in the VNLLm have BFs within the CF_2 range around 79 kHz. This highly specialized frequency representation indicates that the role of the VNLLm in CF-FM bats must be in processing the CF component of the echolocation call. The lack of a precise tonotopic map in either species suggests that a highly organized tonotopy is not important for the function of VNLLm.

Neurons in the INLL and VNLLm are not sharply tuned for sound frequency, and many neurons in the VNLLc in the big brown bat have extremely broad tuning curves. The frequency tuning of the majority of neurons in the INLL and VNLL basically reflects that of neurons in the cochlear nucleus (Neuweiler and Vater 1977; Metzner and Radtke-Schuller 1987; Covey and Casseday 1991; Haplea, Covey, and Casseday 1994). There is no evidence in any species to indicate that sharpness of frequency tuning increases in the nuclei of the lateral lemniscus; in the big brown bat, frequency tuning may actually become broader in the VNLLc. For example, in the big brown bat, frequency tuning has been compared at several brainstem levels (Fig. 6.17). The broadest tuning curves are found in the

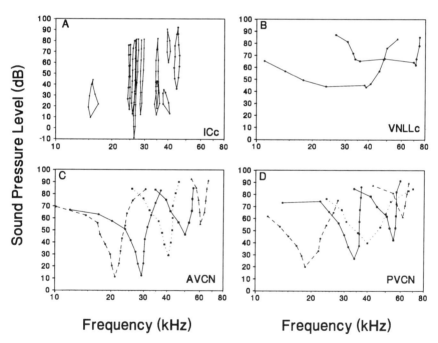

Frequency (kHz)   Frequency (kHz)

FIGURE 17.A–D Comparison of frequency tuning at three levels of the brainstem of the big brown bat. (A) The level-tolerant narrow-band tuning curves and closed tuning curves are found in the IC but not in the nuclei of the lateral lemniscus or cochlear nucleus. (B) Neurons in VNLLc are responsive to a very broad range of frequencies at only a few decibels above threshold. (C) Neurons in the AVCN have V-shaped tuning curves that resemble those found in AVCN of other mammals. (D) Likewise, neurons in PVCN have V-shaped tuning curves. (Modified from Haplea, Covey, and Casseday 1994).

VNLLc (Fig. 6.17B). These curves can be compared with the V-shaped tuning curves of neurons in the AVCN and PVCN (Fig. 6.17C,D) and with the "filter" and "closed" type tuning curves seen in the inferior colliculus (Fig. 6.17A). The broad frequency tuning of VNLLc neurons probably reflects integration across frequency to increase precision in the temporal domain (Covey and Casseday 1991; Covey 1993b).

Broad frequency tuning in the VNLLc may be a feature common to mammalian auditory systems rather than a specific adaptation for echolocation. In the cat, most units in the VNLL also have V-shaped tuning curves, but some "wide" tuning curves have been reported, possibly similar to those seen in VNLLc of the big brown bat (Guinan, Norris, and Guinan 1972). Thus, broad frequency tuning may be a correlate of the processing mechanisms common to the VNLL of all mammals. However, in echolocating bats, this processing mechanism has acquired an increased degree of importance.

4.3.3 Discharge Patterns and Timing Characteristics

Although neurons in the INLL, VNLLc, and VNLLm of the big brown bat exhibit a variety of response properties, they share certain features that make them ideally suited to transmit information about the timing of auditory events (Covey and Casseday 1991; Covey 1993b). Because neurons throughout these nuclei have little or no spontaneous activity, the occurrence of spikes that are synchronized to temporal features of the sound provides an unambiguous timing signal. Our main examples are from studies in the big brown bat, the species for which the most physiological information is available.

The discharge patterns of neurons in the INLL and VNLL include both sustained and transient responses. This is true in horseshoe bats and big brown bats, as well as in the cat (Aitkin, Anderson, and Brugge 1970; Metzner and Radtke-Schuller 1987; Covey and Casseday 1991). Based on regularity of firing, the sustained responses include choppers, which have regular interspike intervals, and nonchoppers, which have irregular interspike intervals. On the basis of the amount by which discharge rate decreases over the course of a response, sustained discharges can be classified as adapting or nonadapting. In the big brown bat, at least, there is some segregation of response types.

Transient responses in the INLL, VNLLc, and VNLLm typically consist of one or a few spikes. The most striking class of transient responding cells, found mainly in the VNLLc, are those with "phasic constant latency" responses (Covey and Casseday 1991). Virtually all neurons in the VNLLc of the big brown bat, and probably also in the mustached bat (O'Neill, Holt, and Gordon 1992), are phasic constant latency responders. These neurons discharge only one spike per stimulus, with a standard deviation in latency less than 0.1 msec. In addition, the latency of the first spike is virtually independent of stimulus level and frequency. Figure 6.18 shows a comparison of the timing properties of a neuron in VNLLc with those of neurons in AVCN and PVCN. A phasic constant latency neuron provides an extremely precise marker of the time of onset of a sound, either for the onset of a tone at any frequency within its range of sensitivity, or for the time when a frequency- or amplitude-modulated stimulus enters its range of sensitivity (Suga 1970; Pollak et al. 1977; Bodenhamer, Pollak, and Marsh 1979; Bodenhamer and Pollak 1981; Covey and Casseday 1991). Thus, phasic constant latency responses may be important in providing timing markers for specific portions of the bat's echolocation call and the returning echoes.

Neurons in the VNLLc are not the only ones that exhibit constant latency. These properties are also seen in the responses of some neurons in the INLL, VNLLm, and DNLL of the big brown bat, although they are not common in these areas (Covey 1993a). Not all transiently responding neurons are constant latency; in fact, the majority of transient responses in INLL and VNLLm have a variable first-spike latency that shifts with changes in sound amplitude and frequency much as in the auditory nerve and cochlear nucleus (see Fig. 6.18C–F).

In the INLL and VNLLm of the big brown bat, about two-thirds of neural responses are sustained. Most of the sustained responses in both INLL and VNLLm are the nonadapting type, which means that a neuron continues to fire at a fairly constant rate as long as the frequency and amplitude of a sound remain within its range of sensitivity. Sustained, nonadapting responses may be important in transmitting information about the duration of specific components of the echolocation call and the returning echo.

As in the cochlear nucleus, nearly all neurons in the monaural nuclei of the lateral lemniscus, whether transient or sustained responders, have very short integration times. These neurons respond robustly to stimuli a millisecond or less in duration, so that they should be capable of responding to transient auditory events such as an FM sound that sweeps rapidly through their range of frequency sensitivity, or to rapid frequency or amplitude modulations. In fact, neurons in the VNLLc of the mustached bat do synchronize their discharge to follow high rates of sinusoidal amplitude modulations, but they are unresponsive to modulation rates of less than about 600 Hz (O'Neill, Holt, and Gordon 1992). This selectivity of VNLLc neurons for high rates of amplitude modulation is in contrast to neurons in the cochlear nucleus, which have no lower limit for synchronizing to (SAM) (Vater 1982), and neurons in the MSO, which have an upper limit of about 200–300 Hz (Grothe 1990, 1994). This comparison further suggests that VNLLc neurons are specialized to respond to stimuli having very rapid changes in amplitude or frequency.

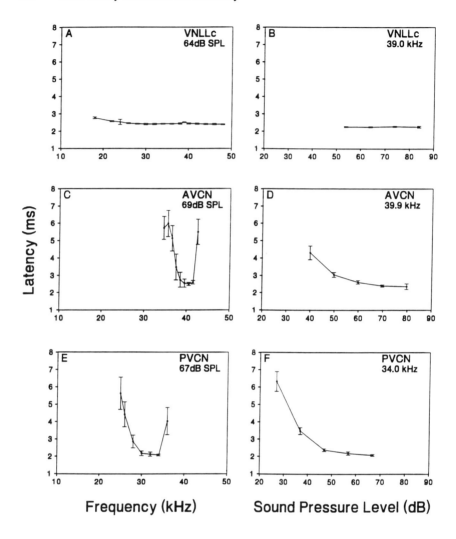

Frequency (kHz)   Sound Pressure Level (dB)

FIGURE 6.18A–F Comparison of the effects of changing sound frequency and amplitude on the latency variability of neurons in the VNLLc, AVCN, and PVCN of the big brown bat. In all graphs, mean first-spike latency is plotted as a function of frequency or amplitude; *vertical bars* represent standard deviations. (A) For neurons in the VNLLc, there is little change in mean latency across a wide range of sound frequencies; latency variability is uniformly low. (B) In the VNLLc, latency values and variability also remain constant across a wide range of sound amplitudes. (C) Responses of a neuron in AVCN to a sound 30 dB above threshold as frequency was varied. At BF, latency variability was as low as for the neuron in VNLLc. However, as frequency deviated from BF, mean first-spike latency of the AVCN neuron increased by several milliseconds and latency variability also increased. (D) Responses of the same neuron in AVCN to a sound at BF as amplitude was varied. Mean latency and latency variability progressively decreased as sound level was

4.4 Selection and Enhancement of Temporal Features of Sound: Two Streams of Processing

Both connectional and physiological evidence indicate that the INLL, VNLLc and VNLLm receive convergent parallel inputs from cell populations in the cochlear nucleus and other brainstem structures such as the MNTB and certain periolivary nuclei. These inputs are then transformed in a way that not only preserves the temporal pattern of sounds, but selects and enhances certain features of this pattern. Like auditory nerve fibers and certain neurons in the ventral cochlear nucleus, neurons throughout the INLL and VNLL respond robustly to transient stimuli of short duration. Therefore, they can follow very rapid frequency or amplitude modulations. It has been shown that low-BF globular cells in the cochlear nucleus are more precise phase lockers than are auditory nerve fibers (Smith et al. 1991); it therefore seems likely that the enhancement of temporal features is a multistage process that begins in the cochlear nucleus and culminates in the nuclei of the lateral lemniscus. In echolocating bats, the important temporal features may be rapid frequency or amplitude modulations or a rapid sequence of echolocation pulses and their echoes.

As summarized in Figure 6.19, the neurons in INLL and VNLL can be divided into two broad populations. The first population consists of the phasic constant latency neurons of the VNLLc, which signal very precisely the onset time of a sound. The second population includes the sustained nonadapting neurons of the INLL and VNLLm, which respond continuously throughout the time a sound is present and thus can transmit information about its intensity and duration. At least in the bat, these two streams of processing can clearly be distinguished from one another on the basis of their neuronal morphology, the type of terminals supplied by axons originating in the ventral cochlear nucleus and breadth of frequency tuning. In addition, they differ in their patterns of termination in the IC.

The population of onset encoders includes virtually all the neurons in the VNLLc. These cells have shorter response latencies than any others in the INLL, VNLL, or the nuclei of the SOC, averaging just a little over 3 msec. They appear to be glycinergic and provide dense and widespread projections to the IC, possibly organized so that a single sheet of cells in the VNLLc sends divergent projections to a broad frequency range within the IC

FIGURE 6.18A–F. (*continued*) increased. (E) Responses of a neuron in PVCN to a sound 40 dB above threshold as frequency was varied. At BF, latency variability was low, but as frequency deviated from BF, mean first-spike latency and latency variability increased. (F) Responses of the same neuron in PVCN to a sound at BF as amplitude was varied. Mean latency and latency variability progressively decreased as sound level was increased. (From Haplea, Covey and Casseday 1994)

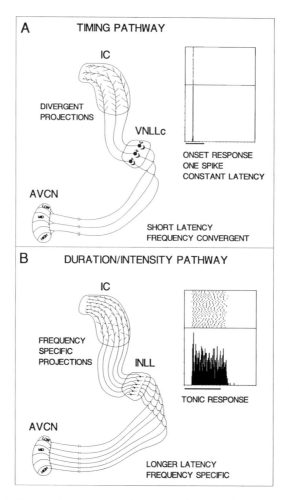

FIGURE 6.19A,B Schematic diagram of hypothesis that there are two different streams of processing within the monaural auditory pathways of the brainstem, each transmitting a different class of information about temporal features of sounds. (A) Constant latency neurons in VNLLc respond with one spike locked to the onset of a sound. (B) Neurons with sustained responses in INLL respond continuously throughout the duration of a sound. The insets show representative post-stimulus time histograms (PSTHs) (*lower parts*) and dot raster displays (*upper parts*) of spikes. The bar below each horizontal axis represents the stimulus duration.

(Covey and Casseday 1986). Thus, the output of the VNLLc is short latency, frequency tolerant and level tolerant, and is distributed widely in the inferior colliculus. The VNLLc provides the IC with very precise information about the onset of sound and about frequency and intensity transitions of sound. The VNLLc may be the source of transient inhibition that precedes excitatory responses in the IC. (Covey et al. 1993)

The population of intensity/duration encoders include the nonadapting sustained units in the INLL and VNLLm. These units probably vary somewhat in their morphology, but all have multiple large dendrites that are contacted by punctate terminals of axons originating in the ventral cochlear nucleus. Many of the sustained units have a relatively broad dynamic range over which firing rate changes in response to sound pressure level. Their responses would therefore provide a signal either of stimulus duration for a tone, or for an FM signal, the dwell time within their range of frequency-sensitivity. They respond with latencies that are 1 to several milliseconds longer than those of VNLLc neurons, averaging about 4–5 msec, and spanning a considerably wider range of latencies. Neurons in INLL and VNLLm appear to project in a tonotopic manner to the IC. The outputs of both timing pathways probably play an important role in shaping the responses of IC neurons to temporal features such as the duration of sounds (Casseday, Ehrlich, and Covey 1994).

4.5 Outputs of the Monaural Nuclei of the Lateral Lemniscus

The INLL, VNLLc, and VNLLm give rise to parallel pathways that terminate in the ipsilateral IC. Both anterograde and retrograde tracing experiments show that all these pathways terminate densely in the ventral two-thirds of the IC in a region that is largely if not entirely coextensive with the targets of the AVCN, LSO, MSO, and DNLL. In at least some species of echolocating bats, this region is also the target of projections from the PVCN (Schweizer 1981; Zook and Casseday 1982b, 1987; Covey and Casseday 1986).

5. Summary: Convergence and Integration of Parallel Pathways at the Inferior Colliculus

The auditory system described up to this point is really a set of very separate subsystems. Each subsystem has its own tonotopy, and each provides some separate and unique transformation of the auditory stimulus. However, the outputs of the subsystems converge at the inferior colliculus in a manner that reassembles the separate components back into one tonotopy. Because it is clear that a major reorganization of the input signals occurs at the IC, we conclude by reviewing the variety of these signals.

Inputs to the IC arrive directly from the cochlear nucleus and indirectly via one or more stages of processing. Inputs are both monaural and binaural. Monaural inputs arise from the three divisions of the cochlear nucleus, and from the INLL and VNLL. In addition, the MSO provides mainly monaural input in bats. Binaural inputs arise from the LSO and DNLL. Either monaural or binaural inputs may be excitatory or inhibitory.

For example, the DNLL provides almost exclusively inhibitory inputs, whereas the LSO provides excitatory and inhibitory inputs. The cochlear nuclei provide mainly excitatory inputs but may provide some inhibitory input. The INLL and VNLL contribute substantial inhibitory inputs.

When one considers the temporal characteristics of the responses in the different pathways, it becomes clear that much of the processing in the IC concerns the integration of inputs with different temporal patterns. First, different pathways have different latencies. The various inputs arrive at the IC at different times, depending on the latency of the pathway through which they are transmitted. This range of input latencies has the potential to create a prolonged and complex sequence of inhibitory and excitatory synaptic events at the IC, a sequence that is highly labile and can change depending on which systems are active and the order in which they are activated. Second, in a manner analogous to the XYZ system in visual pathways, the different auditory pathways in the lower brainstem have different temporal patterns of response. The temporal patterns range from a single spike locked to the onset or offset of a sound, to chopper or pauser patterns, to a continuous train of spikes that persists for the entire duration of a sound or even longer in the case of an afterdischarge. These differences in discharge pattern mean that a variety of excitatory and inhibitory inputs with activity distributed differentially over time could interact at the IC to produce complex sequences of postsynaptic excitation and inhibition (Casseday, Ehrlich, and Covey 1994). These sequences could last for tens of milliseconds or longer, thereby modulating the response of a neuron to subsequently occurring sounds. The resulting time-varying changes in the state of the neuron might be manifested in the simplest case as windows of facilitation or suppression, or in a more complex case as neural filters for biologically important temporal sequences of sounds.

Acknowledgments   The authors were supported by U.S National Institutes of Health (NIH) grants DC-00607 (E.C.) and DC-00287 (J.H.C.) during the preparation of this chapter. Special thanks to Boma Rosemond for help in preparing the illustrations.

References

Adams JC (1976) Single unit studies on the dorsal and intermediate acoustic striae. J Comp Neurol 170:97–106.

Adams JC (1979) Ascending projections to the inferior colliculus. J Comp Neurol 183:519–538.

Adams JC, Mugnaini E (1984) Dorsal nucleus of the lateral lemniscus: a nucleus of GABAergic projection neurons. Brain Res Bull 13:585–590.

Adams JC, Mugnaini E (1990) Immunocytochemical evidence for inhibitory and disinhibitory circuits in the superior olive. Hear Res 49:281–298.

Aitkin LM, Phillips SC (1984) Is the inferior colliculus an obligatory relay in the cat auditory system? Neurosci Lett 44:259–264.

Aitkin LM, Anderson DJ, Brugge JF (1970) Tonotopic organization and discharge characteristics of single neurons in nuclei of the lateral lemniscus of the cat. J Neurophysiol 33:421–440.

Aschoff A, Ostwald J (1987) Different origins of cochlear efferents in some bat species, rats, and guinea pigs. J Comp Neurol 264:56–72.

Baron G (1974) Differential phylogenetic development of the acoustic nuclei among chiroptera. Brain Behav Evol 9:7–40.

Bishop AL, Henson OW (1987) The efferent cochlear projections of the superior olivary complex in the mustached bat. Hear Res 31:175–182.

Bishop AL, Henson OW (1988) The efferent auditory system in Doppler-shift compensating bats. In: Nachtigall PE, Moore PWB (eds) Animal Sonar: Processes and Performance. New York: Plenum, pp. 307–310.

Bledsoe SC, Snead CR, Helfert RH, Prasad V, Wenthold RJ, Altschuler RA(1990) Immunocytochemical and lesion studies support the hypothesis that the projection from the medial nucleus of the trapezoid body to the lateral superior olive is glycinergic. Brain Res 517:189–194.

Bodenhamer RD, Pollak GD (1981) Time and frequency domain processing in the inferior colliculus of echolocating bats. Hear Res 5:317–335.

Bodenhamer RD, Pollak GD, Marsh DS (1979) Coding of fine frequency information by echoranging neurons in the inferior colliculus of the Mexican free-tailed bat. Brain Res 171:530–535.

Bourke TR, Mielcarz JP, Norris BE (1981) Tonotopic organization of the anteroventral cochlear nucleus of the cat. Hear Res 4:215–241.

Brugge JF, Anderson DJ, Aitkin LM (1970) Responses of neurons in the dorsal nucleus of the lateral lemniscus of cat to binaural tonal stimulation. J Neurophysiol (Bethesda) 33:441–458.

Bruns V (1976a) Peripheral auditory tuning for fine frequency analysis by the CF-FM bat, *Rhinolophus ferrumequinum*. I. Mechanical specializations of the cochlea. J Comp Physiol 106:77–86.

Bruns V (1976b) Peripheral auditory tuning for fine frequency analysis by the CF-FM bat, *Rhinolophus ferrumequinum*. J Comp Physiol 107:87–97.

Bruns V, Schmieszek ET (1980) Cochlear innervation in the greater horseshoe bat: demonstration of an acoustic fovea. Hear Res 3:27–43.

Brunso-Bechtold JK, Henkel CK, Linville C (1990) Synaptic organization in the adult ferret medial superior olive. J Comp Neurol 294:389–398.

Cajal Ramon SY (1909) Histologie du systeme nerveux de l'homme et des vertebres. Tome I. Madrid: Instituto Ramon y Cajal (1952), pp. 778–848.

Cant NB (1992) The cochlear nucleus: neuronal types and their synaptic organization. In: Popper AN, Fay RR (eds) The Mammalian Auditory Pathway: Neuroanatomy. New York: Springer-Verlag, pp. 66–116.

Cant NB, Casseday JH (1986) Projections from the anteroventral cochlear nucleus to the lateral and medial superior olivary nuclei. J Comp Neurol 247:457–476.

Carr CE (1986) Time coding in electric fish and barn owls. Brain Behav Evol 28:122–134.

Casseday JH, Covey E (1987) Central auditory pathways in directional hearing. In: Yost W, Gourevitch G (eds) Directional Hearing. New York: Springer-Verlag, pp. 109–145.

Casseday JH, Covey E (1992) Frequency tuning properties of neurons in the inferior colliculus of an FM bat. J Comp Neurol 319:34–50.

Casseday JH, Covey E, Vater M (1988) Connections of the superior olivary complex in the rufous horseshoe bat, *Rhinolophus rouxi*. J Comp Neurol 278:313–329.

Casseday JH, Ehrlich D, Covey E (1994) Neural tuning for sound duration: role of inhibitory mechanisms in the inferior colliculus. Science 264:847–850.

Casseday JH, Jones DR, Diamond IT (1979) Projections from cortex to tectum in the tree shrew, *Tupaia glis*. Comp Neurol 185:253–292.

Casseday JH, Zook JM, Kuwabara N (1993) Projections of cochlear nucleus to superior olivary complex in an echolocating bat: relation to function. In: Merchan MA, Juiz JM, Godfrey DA (eds) The Mammalian Cochlear Nuclei: Organization and Function. New York: Plenum, pp. 303–319.

Casseday JH, Kobler JB, Isbey SF, Covey E (1989) The central acoustic tract in an echolocating bat: an extralemniscal auditory pathway to the thalamus. J Comp Neurol 287:247–259.

Clark GM (1969a) The ultrastructure of nerve endings in the medial superior olive of the cat. Brain Res 14:298–305.

Clark GM (1969b) Vesicle shape versus type of synapse in the nerve endings of the cat medial superior olive. Brain Res 15:548–551.

Covey E (1993a) Response properties of single units in the dorsal nucleus of the lateral lemniscus and paralemniscal zone of an echolocating bat. J Neurophysiol (Bethesda) 69:842–859.

Covey E (1993b) The monaural nuclei of the lateral lemniscus: parallel pathways from cochlear nucleus to midbrain. In: Merchan MA, Juiz JM, Godfrey DA (eds) The Mammalian Cochlear Nuclei: Organization and Function. New York: Plenum, pp. 321–334.

Covey E, Casseday JH (1986) Connectional basis for frequency representation in the nuclei of the lateral lemniscus of the bat, *Eptesicus fuscus*. J Neurosci 6:2926–2940.

Covey E, Casseday JH (1991) The monaural nuclei of the lateral lemniscus in an echolocating bat: parallel pathways for analyzing temporal features of sound. J Neurosci 11:3456–3470.

Covey E, Hall WC, Kobler JB (1987) Subcortical connections of the superior colliculus in the mustache bat, *Pteronotus parnellii*. J Comp Neurol 263:179–197.

Covey E, Vater M, Casseday JH (1991) Binaural properties of single units in the superior olivary complex of the mustached bat. J Neurophysiol (Bethesda) 66:1080–1094.

Covey E, Johnson BR, Ehrlich D, Casseday JH (1993) Neural representation of the temporal features of sound undergoes transformation in the auditory midbrain: evidence from extracellular recording, application of pharmacological agents and *in vivo* whole cell patch clamp recording. Neurosci Abstr 19:535.

Feng AS, Vater M (1985) Functional organization of the cochlear nucleus of rufous horseshoe bats (*Rhinolophus rouxi*): frequencies and internal connections are arranged in slabs. J Comp Neurol 235:529–553.

Friauf E, Ostwald J (1988) Divergent projections of physiologically characterized rat ventral cochlear nucleus neurons as shown by intra-axonal injection of horseradish peroxidase. Exp Brain Res 73:263–284.

Fuzessery ZM, Pollak GD (1985) Determinants of sound location selectivity in bat inferior colliculus: A combined dichotic and free-field stimulation study. J Neurophysiol 54:757–781.

Glendenning KK, Brunso-Bechtold JK, Thompson GC, Masterton R B (1981) Ascending auditory afferents to the nuclei of the lateral lemniscus. J Comp Neurol 197:673–704.

Godfrey DA, Kiang NYS, Norris BA (1975) Single unit activity in the dorsal cochlear nucleus of the cat. J Comp Neurol 162:269–284.

Goldberg JM, Brown PB (1968) Functional organization of the dog superior olivary complex: an anatomical and electrophysiological study. J Neurophysiol (Bethesda) 31:639–656.

Goldberg JM, Brownell WE (1973) Discharge characteristics of neurons in antero-ventral and dorsal cochlear nuclei of cat. Brain Res 64:35–54.

Goldberg JM, Moore RY (1967) Ascending projections of the lateral lemniscus in the cat and monkey. J Comp Neurol 129:143–156.

Grothe B (1990) Versuch einer Definition des medialen Kernes des oberen Olivenkomplexes bei der Neuweltfledermaus *Pteronotus parnellii*. Ph.D. dissertation, Ludwig-Maximilians Universität, Munich, Germany.

Grothe B (1994) Interaction of excitation and inhibition in processing of pure tone and amplitude-modulated stimuli in the medial superior olive of the mustached bat. J Neurophysiol 71:706–721.

Grothe B, Vater M, Casseday JH, Covey E (1992) Monaural interaction of excitation and inhibition in the medial superior olive of the mustached bat: an adaptation for biosonar. Proc Natl Acad Sci USA 89:5108–5112.

Guinan JJ, Norris BE, Guinan SS (1972) Single auditory units in the superior olivary complex. II: Locations of unit categories and tonotopic organization. Int J Neurosci 4:147–166.

Guinan JJ, Warr WB, Norris BE (1983) Differential olivocochlear projections from lateral vs. medial zones of the superior olivary complex. J Comp Neurol 221:358–370.

Haplea S, Covey E, Casseday JH (1994) Frequency tuning and response latencies at three levels in the brainstem of the echolocating bat, *Eptesicus fuscus*. J Comp Physiol A 174:671–683.

Harnischfeger G, Neuweiler G, Schlegel P (1985) Interaural time and intensity coding in superior olivary complex and inferior colliculus of the echolocating bat, *Molossus ater*. J Neurophysiol 53:89–109.

Harrison JM, Irving R (1966) Visual and nonvisual auditory systems in mammals. Science 154:738–743.

Helfert RH, Bonneau JM, Wenthold RJ, Altschuler RA (1989) GABA and glycine immunoreactivity in the guinea pig superior olivary complex. Brain Res 501:269–286.

Henkel CK (1983) Evidence of sub-collicular projections to medial geniculate nucleus in the cat: an autoradiographic and horseradish peroxidase study. Brain Res 259:21–30.

Henson OW (1970) The central nervous system. In: Wimsatt WA (ed) Biology of Bats, Vol. 2. New York: Academic, pp. 57–152.

Inbody SB, Feng AS (1981) Binaural response characteristics of single neurons in the medial superior olivary nucleus of the albino rat. Brain Res 210:361–366.

Irvine DRF (1992) Physiology of the auditory brainstem. In: Popper AN, Fay RR (eds) The Mammalian Auditory Pathway: Physiology. New York: Springer-Verlag, pp. 153–231.

Irving R, Harrison JM (1967) Superior olivary complex and audition: A comparative study. J Comp Neurol 130:77–86.

Kiss A, Majorossy K (1983) Neuron morphology and synaptic architecture in the medial superior olivary nucleus. Exp Brain Res 52:15–327.

Kober R, Schnitzler H-U (1990) Information in sonar echoes of fluttering insects available for echolocating bats. J Acoust Soc Am 87:874–881.

Kobler JB, Isbey SF, Casseday JH (1987) Auditory pathways to the frontal cortex of the mustache bat, *Pteronotus parnellii*. Science 236:824–826.

Kössl M, Vater M (1989) Noradrenaline enhances temporal auditory contrast and neuronal timing precision in the cochlear nucleus of the mustached bat. J Neurosci 9:4169–4178.

Kössl M, Vater M (1990) Tonotopic organization of the cochlear nucleus of the mustache bat, *Pteronotus parnellii*. J Comp Physiol A 166:695–709.

Kössl M, Vater M, Schweizer H (1988) Distribution of catecholamine fibers in the cochlear nuclei of horseshoe bats and mustache bats. J Comp Neurol 269:523–535.

Kudo M (1981) Projections of the lateral lemniscus in the cat: an autoradiographic study. Brain Res 221:57–69.

Kuwabara N, Zook JM (1991) Classification of the principal cells of the medial nucleus of the trapezoid body. J Comp Neurol 314:707–720.

Kuwabara N, Zook JM (1992) Projections to the medial superior olive from the medial and lateral nuclei of the trapezoid body in rodents and bats. J Comp Neurol 324:522–538.

Kuwabara N, DiCaprio RA, Zook JM (1991) Afferents to the medial nucleus of the trapezoid body and their collateral projections. J Comp Neurol 314:684–706.

Kuwada S, Yin TCT (1987) Physiological studies of directional hearing. In: Yost WA, Gourevitch G (eds) Directional Hearing. New York: Springer-Verlag, pp. 146–176.

Lesser HD, O'Neill WE, Frisina RD, Emerson RC (1990) On-off units in the mustached bat inferior colliculus are selective for transients resembling "acoustic glint" from fluttering insect targets. Exp Brain Res 82:137–148.

Li L, Kelly JB (1992) Inhibitory influence of the dorsal nucleus of the lateral lemniscus on binaural responses in the rat's inferior colliculus. J Neurosci 12:4530–4539.

Markovitz NS, Pollak G.D. (1994) Binaural processing in the dorsal nucleus of the lateral lemniscus Hear. Res. 73:121–140.

Masterton RB, Thompson GC, Bechtold JK, RoBards MJ (1975) Neuroanatomical basis of binaural phase-difference analysis for sound localization: a comparative study. J Comp Physiol Psychol 89:379–386.

Metzner W (1989) A possible neuronal basis for Doppler-shift compensation in echo-locating horseshoe bats. Nature 341:529–532.

Metzner, W (1993) An audio-vocal interface in echolocating horseshoe bats. J. Neurosci. 13:1899–1915.

Metzner W, Radtke-Schuller S (1987) The nuclei of the lateral lemniscus in the rufous horseshoe bat, *Rhinolophus rouxi*. J Comp Physiol 160:395–411.

Moore JK (1987) The human auditory brain stem: a comparative view. Hear Res 29:1–32.

Moore MM, Caspary DM (1983) Strychnine blocks binaural inhibition in lateral superior olivary neurons. J Neurosci 3:237–242.

Moskowitz N, Liu J-C (1972) Central projections of the spiral ganglion of the squirrel monkey. J Comp Neurol 144:335–344.

Neuweiler G, Vater M (1977) Response patterns to pure tones of cochlear nucleus units in the CF-FM bat, *Rhinolophus ferrumequinum*. J Comp Physiol A 115:119–133.

Noda Y, Pirsig W (1974) Anatomical projection of the cochlea to the cochlear nuclei of the guinea pig. Arch Otolaryngol 208:107–120.

Novick A, Vaisnys JR (1964) Echolocation of flying insects by the bat *Chilonycteris parnellii*. Biol Bull 127:478–488.

Oliver DL, Huerta MF (1992) Inferior and superior colliculi. In: Popper AN, Fay RR (eds) The Mammalian Auditory Pathway: Neuroanatomy. New York: Springer-Verlag, pp. 168–221.

Ollo C, Schwartz I (1979) The superior olivary complex in C5BL/6 mice. Am J Anat 155:349–374.

Olsen JF, Suga N (1991a) Combination sensitive neurons in the medial geniculate body of the mustached bat: encoding of relative velocity information. J Neurophysiol (Bethesda) 65:1254–1274.

Olsen JF, Suga N (1991b) Combination sensitive neurons in the medial geniculate body of the mustached bat: encoding of target range information. J Neurophysiol 65:1275–1296.

O'Neill WE, Suga N (1979) Target range-sensitive neurons in the auditory cortex of the mustache bat. Science 203:69–73.

O'Neill WE, Suga N (1982) Encoding of target range and its representation in the auditory cortex of the mustached bat. J Neurosci 2:17–31.

O'Neill WE, Holt JR, Gordon M (1992) Responses of neurons in the intermediate and ventral nuclei of the lateral lemniscus of the mustached bat to sinusoidal and pseudorandom amplitude modulations. Assoc Res Otolaryngol Abstr 15:140.

Osen KK (1969a) Cytoarchitecture of the cochlear nuclei in the cat. J Comp Neurol 136:453–484.

Osen KK (1969b) The intrinsic organization of the cochlear nuclei in the cat. Acta Otolaryngol 67:352–359.

Papez JW (1929a) Central acoustic tract in cat and man. Anat Rec 42:60.

Papez JW (1929b) Comparative Neurology. New York: Crowell, pp. 270–293.

Peyret D, Geffard M, Aran J-M (1986) GABA immunoreactivity in the primary nuclei of the auditory central nervous system. Hear Res 23:115–121.

Peyret D, Campistron G, Geffard M, Aran J-M (1987) Glycine immunoreactivity in the brainstem auditory and vestibular nuclei of the guinea pig. Acta Otolaryngol 104:71–76.

Pfeiffer RR (1966) Classification of response patterns of spike discharges for units in the cochlear nucleus: tone-burst stimulation. Exp Brain Res 1:220–235.

Poljak S (1926) Untersuchungen am Oktavussystem der Säugetiere und an den mit diesem koordinierten motorischen Apparaten des Hirnstammes. J Psychol Neurol 32:170–231.

Pollak GD, Bodenhamer R (1981) Specialized characteristics of single units in the inferior colliculus of mustache bats: frequency representation, tuning and discharge patterns. J Neurophysiol 46:605–620.

Pollak GD, Casseday JH (1989) The Neural Basis of Echolocation in Bats. Berlin: Springer-Verlag.

Pollak GD, Marsh DS, Bodenhamer R, Souther A (1977) Echo-detecting characteristics of neurons in inferior colliculus of unanesthetized bats. Science 196:675–678.

Rhode WS, Kettner RE (1987) Physiological study of neurons in the dorsal and posteroventral cochlear nucleus of the unanesthetized cat. J Neurophysiol 57:414–442.

Rose JE, Galambos R, Hughes JR (1959) Microelectrode studies of the cochlear nuclei of the cat. Bull Johns Hopkins Hosp 104:211–251.

Ross LS, Pollak GD (1989) Differential ascending projections to aural regions in the 60-kHz contour of the mustache bat's inferior colliculus. J Neurosci 9:2819–2834.

Ross LS, Pollak GD, Zook JM (1988) Origin of ascending projections to an isofrequency region of the mustache bat's inferior colliculus. J Comp Neurol 270:488–505.

Rouiller EM, Ryugo DK (1984) Intracellular marking of physiologically characterized cells in the ventral cochlear nucleus of the cat. J Comp Neurol 225:167–186.

Ryan AF, Woolf NK, Sharp FR (1982) Tonotopic organization in the central auditory pathway of the mongolian gerbil: a 2-deoxyglucose study. J Comp Neurophysiol 207:369–380.

Ryugo DK (1992) The auditory nerve: peripheral innervation, cell body morphology, and central projections. In: Popper AN, Fay RR (eds) The Mammalian Auditory Pathway: Neuroatomy. New York: Springer-Verlag, pp. 23–65.

Schnitzler H-U, Menne D, Kober R, Heblich K (1983) The acoustical image of fluttering insects in echolocating bats. In: Huber F, Markl H (eds) Neuroethology and Behavioral Ethology: Roots and Growing Points. Berlin: Springer-Verlag, pp. 235–250.

Schuller G, Radtke-Schuller S (1990) Neural control of vocalization in bats: mapping of brainstem areas with electrical microstimulation eliciting species-specific echolocation calls in the rufous horseshoe bat. Exp. Brain Res. 79: 192–206.

Schuller G, Covey E, Casseday JH (1991) Auditory pontine grey: connections and response properties in the horseshoe bat. Eur J Neurosci 3:648–662.

Schwartz IR (1980) The differential distribution of synaptic terminal classes on marginal and central cells in the cat medial superior olivary nucleus. Am J Anat 159:25–31.

Schwartz IR (1984) Axonal organization in the cat medial superior olivary nucleus. In: Neff WD (ed) Contributions to Sensory Physiology, Vol. 8. New York: Academic, pp. 99–129.

Schwartz IR (1992) The superior olivary complex and lateral lemniscal nuclei. In: Popper AN, Fay RR (eds) The Mammalian Auditory Pathway: Neuroanatomy. New York: Springer-Verlag, pp. 117–167.

Schweizer H (1981) The connections of the inferior colliculus and the organization of the brainstem auditory system in the greater horseshoe bat (*Rhinolophus ferrumequinum*). J Comp Neurol 201:25–49.

Simmons JA (1989) A view of the world through the bat's ear: the formation of acoustic images in echolocation. Cognition 33:155–199.

Simmons JA, Lawrence BD (1982) Echolocation in bats: The external ear and perception of the vertical positions of targets. Science 218:481–483.

Smith PH, Joris PX, Carney LH, Yin TCT (1991) Projections of physiologically characterized globular bushy cell axons from the cochlear nucleus of the cat. J Comp Neurol 304:387–407.

Spangler KM, Warr RB, Henkel CK (1985) The projections of principal cells of the medial nucleus of the trapezoid body in the cat. J Comp Neurol 238:249–262.

Spirou GA, Brownell WE, Zidanic M (1990) Recordings from cat trapezoid body and HRP labeling of globular bushy cell axons. J Neurophysiol (Bethesda) 63:1169–1190.

Stotler WA (1953) An experimental study of the cells and connections of the superior olivary complex of the cat. J Comp Neurol 98:401–432.

Strominger NL, Strominger AI (1971) Ascending brain stem projections of the anteroventral cochlear nucleus in the rhesus monkey. J Comp Neurol 143:217–242.

Suga N (1964) Single unit activity in cochlear nucleus and inferior colliculus of echo-locating bats. J Physiol (Lond) 172:449–474.

Suga N (1970) Echo-ranging neurons in the inferior colliculus of bats. Science 170:449–452.

Suga N, Jen PHS (1977) Further studies on the peripheral auditory system of "CF-FM" bats specialized for fine frequency analysis of Doppler-shifted echoes. J Exp Biol 69:207–232.

Suga N, Schlegel P (1972) Neural attenuation of responses to emitted sounds in echolocating bats. Science 177:82–84.

Suga N, Schlegel P (1973) Coding and processing in the auditory systems of FM-signal-producing bats. J Acoust Soc Am 54:174–190.

Suga N, Neuweiler G, Möller J (1976) Peripheral auditory tuning for fine frequency analysis by the CF-FM bat *Rhinolophus ferrumequinum*. IV. Properties of peripheral auditory neurons. J Comp Physiol 106:111–125.

Suga N, Shimozawa T (1974) Site of neural attenuation of responses to self-vocalized sounds in echolocating bats Science 183:1211–1213.

Suga N, Simmons JA, Jen PHS (1975) Peripheral specializations for fine frequency analysis of Doppler-shifted echoes in the CF-FM bat, *Pteronotus parnellii*. J Exp Biol 63:161–192.

Takahashi TT, Carr CE, Brecha N, Konishi M (1987) Calcium binding protein-like immunoreactivity labels the terminal field of nucleus laminaris of the barn owl. J Neurosci 7:1843–1856.

Tanaka K, Otani K, Tokunaga A, Sugita S (1985) The organization of neurons in the nucleus of the lateral lemniscus projecting to the superior and inferior colliculi in the rat. Brain Res 341:252–260.

Tolbert LP, Morest DK, Yurgelun-Todd DA (1982) The neuronal architecture of the anteroventral cochlear nucleus in the cat in the region of the cochlear nerve root: horseradish peroxidase labeling of identified cell types. Neuroscience 7:3031–3052.

Vater M (1982) Single unit responses in cochlear nucleus of horseshoe bats to sinusoidal frequency and amplitude modulated signals. J Comp Physiol A 149:369–388.

Vater M, Feng AS (1990) Functional organization of ascending and descending connections of the cochlear nucleus of horseshoe bats. J Comp Neurol 292:373–395.

Vater M, Casseday JH, Covey E (1995) Convergence and divergence of ascending binaural and monaural pathways from the superior olives of the mustached bat. J Comp Neurol 351:632–646.

Von der Emde G, Schnitzler H-U (1986) Fluttering target detection in hipposiderid bats. J Comp Physiol 159:765–772.

Von der Emde G, Schnitzler H-U (1990) Classification of insects by echolocating greater horseshoe bats. J Comp Physiol A 167:423–430.

Warr WB (1966) Fiber degeneration following lesions in the anterior ventral cochlear nucleus of the cat. Exp Neurol 14:453–474.

Warr WB (1969) Fiber degeneration following lesions in the posteroventral cochlear nucleus of the cat. Exp Neurol 23:140–155.

Warr WB (1982) Parallel ascending pathways from the cochlear nucleus: neuroanatomical evidence of functional specialization. In: Neff WD (ed) Contributions to Sensory Physiology, Vol. 7. New York: Academic, pp. 1–38.

Warr WB (1992) Organization of olivocochlear efferent systems in mammals. In: Popper A N, Fay R R (eds) The Mammalian Auditory Pathway: Neuroanatomy. New York: Springer-Verlag, pp. 410–448.

Warr WB, Guinan JJ (1979) Efferent innervation of the organ of Corti: two separate systems. Brain Res 173:152–155.

Wenthold RJ, Huie D, Altschuler RA, Reeks KA (1987) Glycine immunoreactivity localized in the cochlear nucleus and superior olivary complex. Neuroscience 22:897–912.

Yang L, Pollak GD Resler C (1992) GABAergic circuits sharpen tuning curves and modify response properties in the mustache bat inferior colliculus. J Neurophysiol (Bethesda) 68:1760–1774.

Yin TCT, Chan JCK (1990) Interaural time sensitivity in medial superior olive of cat. J Neurophysiol (Bethesda) 64:465–488.

Zettel ML, Carr CE, O'Neill WE (1991) Calbindin-like immunoreactivity in the central auditory system of the mustached bat, *Pteronotus parnelli*. J Comp Neurol 313:1–16.

Zook JM, Casseday JH (1982a) Cytoarchitecture of auditory system in lower brainstem of the mustache bat, *Pteronotus parnellii*. J Comp Neurol 207:1–13.

Zook JM, Casseday JH (1982b) Origin of ascending projections to inferior colliculus in the mustache bat, *Pteronotus parnellii*. J Comp Neurol 207:14–28.

Zook JM, Casseday JH (1985) Projections from the cochlear nuclei in the mustache bat, *Pteronotus parnellii*. J Comp Neurol 237:307–324.

Zook JM, Casseday JH (1987) Convergence of ascending pathways at the inferior colliculus in the mustache bat, *Pteronotus parnellii*. J Comp Neurol 261:347–361.

Zook JM, DiCaprio RA (1988) Intracellular labeling of afferents to the lateral superior olive in the bat, *Eptesicus fuscus*. Hear Res 34:141–148.

Zook JM, Leake PA (1989) Connections and frequency representation in the auditory brainstem of the mustache bat, *Pteronotus parnellii*. J Comp Neurol 290:243–261.

Zook JM, Jacobs MS, Glezer I, Morgane PJ (1988) Some comparative aspects of auditory brainstem cytoarchitecture in echolocating mammals: speculations on the morphological basis of time-domain signal processing. In: Nachtigall PE, Moore PWB (eds) Animal Sonar: Processes and Performance. New York: Plenum, pp. 311–316.

Abbreviations

ALPO	anterolateral periolivary nucleus
AN	auditory nerve
AV, a, d, m, p	anteroventral cochlear nucleus, anterior division, dorsal division, medial division, posterior division
AVCN	anteroventral cochlear nucleus

BP	brachium pontis
CB, CER	cerebellum
CG	central grey
CP	cerebral peduncle
DCN, d, v	dorsal cochlear nucleus, dorsal, ventral
DMPO	dorsomedial periolivary nucleus
DMSO	medial superior olive, dorsal division
DNLL	dorsal nucleus of lateral lemniscus
DPO	dorsal periolivary nucleus
GM, v	medial geniculate body, ventral division
ICc	inferior colliculus, central nucleus
INLL	intermediate nucleus of the lateral lemniscus
LNTB	lateral nucleus of the trapezoid body
LSO	lateral superior olive
MNTB	medial nucleus of the trapezoid body
MSO	medial superior olive
NCAT	nucleus of the central acoustic tract
PV, a, c, l, m	posteroventral cochlear nucleus, anterior division, caudal division, lateral division, medial division
PVCN, d, v	posteroventral cochlear nucleus, dorsal division, ventral division
Pyr, Py	pyramidal tract
RB	restiform body
SC	superior colliculus
TB	trapezoid body
VII	seventh cranial nerve
VIII	eighth cranial nerve
VMPO	ventromedial periolivary nucleus
VMSO	medial superior olive, ventral division
VNLL, c, m	ventral nucleus of the lateral lemniscus, columnar division, multipolar cell division
VNTB	ventral nucleus of the trapezoid body
VPO	ventral periolivary nucleus

7

The Inferior Colliculus

GEORGE D. POLLAK AND THOMAS J. PARK

1. Introduction

The mammalian inferior colliculus sits as a protuberance on the dorsal surface of the midbrain (Fig. 7.1) and is composed of several subdivisions (see Oliver and Huerta 1992 for a discussion of various subdivisions). The largest division is the central nucleus of the inferior colliculus (ICc), which is the target of the ascending auditory projections of the lateral lemniscus (Adams 1979; Schweizer 1981; Zook and Casseday 1982; Aitkin 1986; Irvine 1986; Ross, Pollak, and Zook 1988; Ross and Pollak 1989; Frisina, O'Neill, and Zettel 1989; Vater and Feng 1990; Casseday and Covey 1992; Oliver and Huerta 1992). In most echolocating bats, the ICc is especially large while the other divisions of the inferior colliculus are considerably smaller than in other animals (e.g., Pollak and Casseday 1989). The ICc is truly a nexus in the mammalian auditory system where the ascending fibers from 10 or more of the lower auditory nuclei make an obligatory synaptic connection (Fig. 7.2). The ICc neurons, in turn, send a large projection to the medial geniculate (Olsen 1986; Frisina, O'Neill, and Zettel 1989; also see review by Winer 1992, and Wenstrup, Chapter 8) and lesser projections to other divisions of the inferior colliculus as well as to the superior colliculus (Covey, Hall, and Kobler 1987; Oliver and Huerta 1992). In addition, the ICc sends a large descending projection to numerous lower nuclei, including the lateral superior olive, cochlear nucleus, and pontine nuclei (Schuller, Covey, and Casseday 1991; also see reviews by Aitkin 1986; Oliver and Huerta 1992).

In this chapter the principal features that characterize the ICc are discussed. Particular attention is given to those features that are clearly related to the processing of information required for particular aspects of echolocation. However, the general theme is that the structure, immunocytochemistry, and physiology of the bat's inferior colliculus are strikingly similar to the inferior colliculus of other, less specialized mammals. Throughout the text, we point out how the particular features of the ICc of echolocating bats compare with those reported for other mammals.

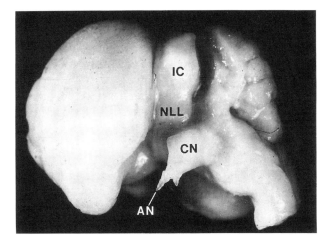

FIGURE 7.1. Photograph of brain of mustache bat. The cerebellum has been partially removed to more clearly reveal auditory structures. AN, auditory nerve; CN, cochlear nucleus; IC, inferior colliculus; NLL, nuclei of the lateral lemniscus.

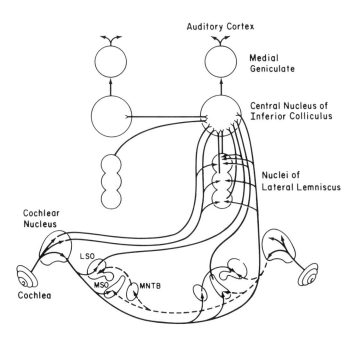

FIGURE 7.2. Wiring diagram shows principal connections of the mammalian auditory pathway. LSO, lateral superior olive; MNTB, medial nucleus of the trapezoid body; MSO, medial superior olive.

1.1 Features of the Projections from Lower Centers

The massive projections from most lower auditory nuclei terminate in the ICc in an orderly manner (Ross and Pollak 1989; also see Oliver and Huerta 1992). The magnitude of the projections, in combination with their orderly termination, underlie two other key characteristics of the inferior colliculus: (1) the remarkable amount of processing that transforms the properties of the converging inputs into new response properties expressed by ICc neurons (Faingold, Gelhbach, and Caspary 1989; Faingold, Boersma-Anderson, and Caspary 1991; Vater, Kössl, and Horn 1992; Yang, Pollak, and Resler 1992; Park and Pollak 1993a,b; Pollak and Park 1993); and (2) the orderly arrangement of those response properties in the ICc (Roth et al. 1978; Wenstrup, Ross, and Pollak 1985, 1986a; Schreiner and Langner 1988; Pollak and Casseday 1989; Park and Pollak 1993b).

The projections from the lower nuclei that terminate in the ICc are both excitatory and inhibitory. Excitatory projections originate from several lower nuclei. Among these are the cochlear nucleus (Semple and Aitkin 1980) and lateral superior olive (LSO) (Saint Marie et al. 1989; Glendenning et al. 1992). It is also likely that excitatory projections originate from the medial superior olive and the intermediate nucleus of the lateral lemniscus.

The inhibitory projections are as large as, if not larger than, the excitatory projections and are both glycinergic and γ-aminobutyric acid (GABA)ergic (Figs. 7.3 and 7.4) (Vater et al. 1992; Winer, Larue, and Pollak 1995). There are three prominent examples of nuclei that provide inhibitory inputs to their targets in the ICc. The first is the columnar division of the ventral nucleus of the lateral lemniscus (VNLLv), whose neurons are predominantly glycinergic (Fig. 7.3) (Wenthold and Hunter 1990; Pollak and Winer 1989; Larue et al. 1991; Winer, Larue, and Pollak 1995) and project heavily to the ICc (Schweizer 1981; Zook and Casseday 1982, 1987; Ross, Pollak, and Zook 1988; O'Neill, Frisina, and Gooler 1989; Vater and Feng 1990; Casseday and Covey 1992).

The second example is the LSO. The LSO projects bilaterally to the ICc (Beyerl 1978; Roth et al. 1978; Schweizer 1981; Zook and Casseday 1982; Shneiderman and Henkel 1987; Ross, Pollak, and Zook 1988) and contains a large number of glycinergic neurons (Saint Marie et al. 1989; Park et al. 1991; Glendenning et al. 1992; Winer, Larue, and Pollak 1995). Recent studies indicate the projection from the LSO to the ipsilateral ICc is largely, and possibly entirely, glycinergic (see Fig. 7.3), whereas the projection to the contralateral ICc is most likely excitatory (Saint Marie et al. 1989; Park et al. 1991; Glendenning et al. 1992; Park and Pollak 1993a).

The third example is the bilateral projections from the dorsal nucleus of the lateral lemniscus (DNLL). The neurons of the DNLL are predominantly

FIGURE 7.3. Diagram of auditory brainstem of the mustache bat shows ascending γ-aminobutyric acid (GABA)ergic (*solid lines*) and glycinergic (*dashed lines*) inhibitory projections to the inferior colliculus. DNLL, dorsal nucleus of the lateral lemniscus; DpD, dorsoposterior division of the inferior colliculus (the region containing neurons sharply tuned to 60 kHz); INLL, intermediate nucleus of the lateral lemniscus; LSO, lateral superior olive; MSO, medial superior olive; VNLLv; ventral or columnar division of the ventral nucleus of the lateral lemniscus; VNLLd, dorsal division of the ventral nucleus of the lateral lemniscus; VNTB; ventral nucleus of the trapezoid body.

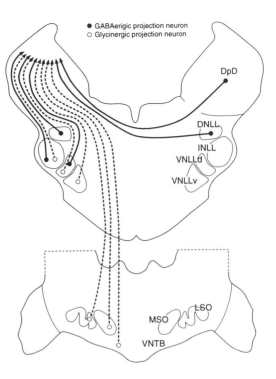

GABAergic (Adams and Mugniani 1984; Thompson, Cortez, and Lam 1985; Pollak et al. 1992; Vater et al. 1992; Winer, Larue, and Pollak 1995) and provide a substantial inhibitory input to the ICc (Fig. 7.3) (Roth et al. 1978; Brunso-Bechtold, Thompson, and Masterton 1981; Schweizer 1981; Zook and Casseday 1982, 1987; Covey and Casseday 1986; Ross, Pollak, and Zook 1988; Shneiderman, Oliver, and Henkel 1988; O'Neill, Frisina, and Gooler 1989; Ross and Pollak 1989; Shneiderman and Oliver 1989). In addition, many of the neurons in the ICc are GABAergic (Adams and Wenthold 1979; Mugniani and Oertel 1985; Roberts and Ribak 1987; Oliver, Nuding, and Beckius 1988; Pollak and Winer 1989; Vater et al. 1992; Winer, Larue, and Pollak 1995), and thus may act locally to complement the inhibition provided by the incoming fibers from lower centers (Oliver et al. 1991; also see Oliver and Huerta 1992). The inhibitory projections and intrinsic inhibitory neurons act to modify the response features of the excitatory projections to produce the coding properties of collicular cells (Faingold, Gelhbach, and Caspary 1989; Faingold, Boersma-Anderson, and Caspary 1991; Yang, Pollak, and Resler 1992; Vater, Kössl, and Horn 1992; Park and Pollak 1993a,b; Pollak and Park 1993). As we explore in the following section, the ordering of projection systems creates, in turn, an arrangement of response properties within the tonotopic framework of the colliculus.

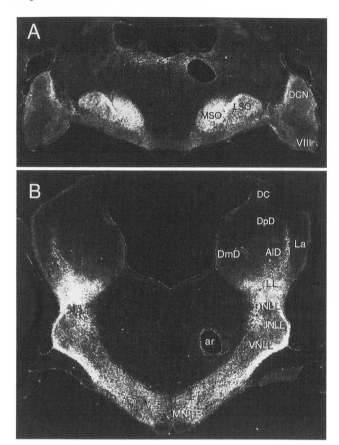

FIGURE 7.4A,B. Sections of the medullary (A) and pontine and midbrain auditory regions (B) of the mustache bat. Sections were reacted with antibodies against conjugated glycine and were photographed under darkfield illumination. Notice the massive projections of glycine-positive fibers in B that terminate in the inferior colliculus. Mark made by investigator during sectioning is indicated by *ar*. ALD, anterolateral division of the inferior colliculus (region representing frequencies below 60 kHz); DmD, medial division of the inferior colliculus (region representing frequencies above 60 kHz); DpD, dorsoposterior division of the inferior colliculus (region representing neurons sharply tuned to 60 kHz); DC, dorsal cortex of the inferior colliculus; DCN, dorsal cochlear nucleus; DNLL, dorsal nucleus of the lateral lemniscus; INLL, intermediate nucleus of the lateral lemniscus; LL, lateral lemniscus; LSO, lateral superior olive; MSO, medial superior olive; VNLL, ventral nucleus of the lateral lemniscus; VIII, auditory nerve. (Adapted from Winer, Larue, and Pollak, 1995.)

1.2 Organization of Response Properties

The results from a wide variety of studies during the past decade have shown that the projections to the ICc, as well as the response properties of its neurons, are remarkably well organized. This organization is expressed on different levels. The most fundamental and well-known is tonotopic organization, the remapping of each point of the cochlear surface on sheets of collicular cells (Fig. 7.5) (e.g., Aitkin 1976; Irvine 1986; Pollak and Casseday 1989). Thus, the cells that compose each sheet of cells in the colliculus are tuned to the same frequency as the cells in the lower nuclei from which they receive their innervation. Because all the cells in a sheet have the same best frequency, the sheet is said to be isofrequency. The sheets in turn are stacked one on top of another dorsoventrally, so that dorsal sheets represent low frequencies (the apical cochlea) and ventral sheets represent progressively higher frequencies and thus the more basal cochlear regions. This general form of tonotopy is seen in all common laboratory mammals, as well as in most of the bat species that have been studied.

The strict tonotopic organization of the lower auditory centers that is preserved in their projections to the ICc is conceptually significant. The importance of this feature is that it provides justification for thinking about the auditory system as having a modular organization. The concept of a modular organization derives from studies of sensory cortices, in which a module is an elementary unit of neuronal organization that encompasses the total processing from one segment of the sensory surface (Hubel and Wiesel 1977; Mountcastle 1978; Sur, Merzenich, and Kass 1980). Thus, the total processing from a given segment of the cochlear surface receives representation in the corresponding isofrequency contour of the ICc. An isofrequency contour is, then, the unit of neuronal organization, or the unit module, and like a cortical module each contour is, in principle, an iterative version of any other contour. In the following sections, we consider the next level of organization, the orderly arrangement of projections and arrangement of response properties within isofrequency contours.

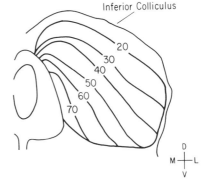

FIGURE 7.5. Drawing of transverse section through the inferior colliculus of the big brown bat, *Eptesicus fuscus*, showing its tonotopic arrangement. The orientation of isofrequency contours is shown by the *solid lines*, and the frequency in kilohertz represented in each contour is indicated. (Reproduced from Pollak and Casseday 1989.)

1.3 The 60-kHz Isofrequency Contour of the Mustache Bat's Inferior Colliculus Is a Model Contour

Both horseshoe bats (*Rhinolophus ferrumequinum* and *R. rouxi*) and mustache bats (*Pteronotus parnellii*) display a special variation of the general tonotopic pattern (Pollak and Schuller 1981; Zook et al. 1985; Rübsamen 1992). As described by Fenton in Chapter 2 and Schnitzler in Chapter 3, the dominant component of the mustache bat's orientation call is the 60-kHz constant-frequency component, and that of the horseshoe bat is the 83-kHz constant-frequency component. The dominant constant-frequency component of their call is the critical element that these bats use for Doppler-shift compensation and for the detection and recognition of insects, features discussed in the next section. As a consequence of their reliance on the dominant constant-frequency component, that frequency is greatly overrepresented throughout their auditory system, including the ICc. The overrepresentation, in turn, has distorted the general tonotopic arrangement in the ICc.

This distortion is well illustrated by the tonotopy of the mustache bat's ICc (Fig. 7.6). As described by Kössl and Vater in Chapter 5, the cochlear region representing 60 kHz is greatly expanded in the mustache bat and has adaptations that create exceptionally sharp tuning curves in the auditory

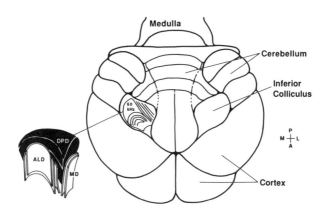

FIGURE 7.6. Schematic of dorsal view of the mustache bat's brain to show location of the 60-kHz region in the inferior colliculus. The hypertrophied inferior colliculi protrude between the cerebellum and cerebral cortex. In the left colliculus are shown the isofrequency contours, as determined in anatomical and physiological studies. A three-dimensional representation of the laminar arrangement is shown on the far left. Low-frequency contours, representing an orderly progression of frequencies from about 59 kHz to about 10 kHz, fill the anterolateral division (ALD). High-frequency contours, representing frequencies from about 64 to more than 120 kHz, occupy the medial division (MD). The 60-kHz region is the dorsoposterior division (DPD), and is the only representation of the filter units in the bat's colliculus. (Reproduced from Pollak and Casseday 1989.)

nerve fibers which innervate that region (Suga, Simmons, and Jen 1975; Suga and Jen 1977; Kössl and Vater 1985, 1990; Zook and Leake 1989). The auditory nerve fibers have $Q_{10\ dB}$ values that range from about 50 to more than 300 and are, on average, an order of magnitude sharper than the fibers tuned just a few hundred hertz higher or lower in frequency. This overrepresentation of sharply tuned 60-kHz neurons is manifest in each nucleus of the auditory pathway, from cochlear nucleus to cortex (e.g., Suga and Jen 1976; Schlegel 1977; Pollak and Bodenhamer 1981; Kössl and Vater 1990; Covey, Vater, and Casseday 1991), and is expressed in the ICc as an isofrequency contour composed of sharply tuned 60-kHz cells that occupy roughly a third of its volume (Figs. 7.6 and 7.7) (Zook et al. 1985; Wenstrup, Ross, and Pollak 1985, 1986a; Ross, Pollak, and Zook 1988). Indeed, the central nucleus of the mustache bat's inferior colliculus can be divided into three divisions (see Fig. 7.6): an anterolateral division, in which the sheets of isofrequency cells are stacked more or less vertically and create an orderly representation of frequencies from about 59 kHz to about 20 kHz; a medial division, where the isofrequency sheets are also stacked vertically and create an orderly representation of frequencies from about 64 kHz to about 120 kHz; and, wedged between these, the dorsoposterior division, in which all the units are sharply tuned to a small frequency band around 60 kHz. In short, the inferior colliculus of this animal has a greatly expanded isofrequency contour that is easily distinguished both by the uniformity of the best frequencies of its neuronal population and by their exceptionally sharp tuning curves. We use here the 60-kHz isofrequency contour of the mustache bat's colliculus as a model to illustrate the ways in which the ascending projections and response properties are organized within a collicular contour.

1.4 Projections to the 60-kHz Contour of the Mustache Bat's Inferior Colliculus

In the previous section we emphasized that the inferior colliculus possesses a tonotopic organization and receives projections from a large number of lower auditory nuclei. Each isofrequency region of the colliculus, then, presumably receives a complement of projections from corresponding isofrequency regions of the lower brainstem nuclei. This is also true of the 60-kHz contour in the mustache bat's inferior colliculus. Figure 7.8 shows the results of a study in which the afferent inputs to the 60-kHz contour were determined by initially mapping its boundaries physiologically, and then making injections of horseradish peroxidase (HRP) in the contour. The 60-kHz contour receives projections from the same set of lower auditory nuclei that project to the entire central nucleus of the inferior colliculus (Ross, Pollak, and Zook 1988). These include: (1) projections from the three divisions of the contralateral cochlear nucleus; (2) projec-

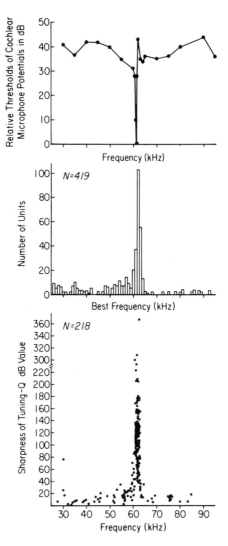

FIGURE 7.7. Distribution of best frequencies and Q_{10dB} values of neurons recorded from the inferior colliculus of the mustache bat. *Top panel*, a cochlear microphonic audiogram recorded from the inner ear of a mustache bat, shows that the cochlea is sharply tuned to 60 kHz. *Middle panel* shows that about a third of the neurons in the inferior colliculus have best frequencies around 60 kHz and correspond closely to the frequency to which the cochlea is tuned. *Lower graph* shows that the tuning curves of the 60-kHz neurons have large Q_{10dB} values, and thus their tuning curves are much sharper than are those of neurons tuned to higher or lower frequencies.

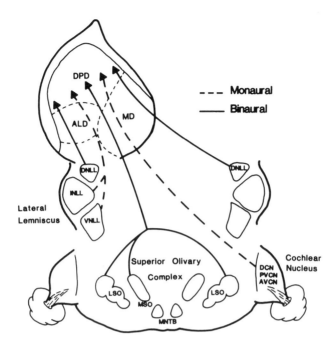

FIGURE 7.8. Ascending projections to the 60-kHz region (DPD) of the mustache bat's inferior colliculus. Projections from binaural auditory nuclei that receive innervation from the two ears are shown as *solid lines*; projections from monaural nuclei that receive innervation from only one ear are shown as *dashed lines*. Note that the 60-kHz region receives projections from all lower auditory nuclei, and thus receives the same types of inputs as the entire inferior colliculus. Although not shown, the neurons that project to the 60-kHz region in each lower nucleus are restricted to regions that presumably represent 60 kHz. (Drawing based on Ross, Pollak, and Zook 1988.)

tions from the ipsilateral medial superior olive; (3) bilateral projections from the LSO; (4) ipsilateral projections from the ventral and intermediate nuclei of the lateral lemniscus; (5) bilateral projections from the DNLL; and (6) projections from the contralateral inferior colliculus (not shown). The projections arise from discrete segments of the various projecting nuclei, each of which represents 60 kHz.

1.5 Aural Types Are Topographically Arranged in the 60-kHz Contour of the Inferior Colliculus

The inferior colliculi of all mammals have three types of neurons that are classified on the basis of aural preference (e.g., Semple and Aitkin 1979; Roth et al. 1978; Semple and Kitzes 1985; Fuzessery and Pollak 1985;

Wenstrup, Ross, and Pollak 1986a; Wenstrup, Fuzessery, and Pollak 1988a). These are monaural neurons (EO) and two types of binaural neurons. One binaural type receives excitatory inputs from both ears; these are called excitatory-excitatory (EE) neurons. The second type, called excitatory-inhibitory (EI) neurons, receives excitation from one ear and inhibition from the other ear. The noteworthy feature is that monaural and binaural neurons are arranged topographically within the 60-kHz contour of the inferior colliculus (Fig. 7.9) (Wenstrup, Ross, and Pollak 1985, 1986a; Ross and Pollak 1989). Monaural units are located along the dorsal and lateral portions of the contour. EE cells occur in two regions, one in the ventrolateral region and the other in the dorsomedial region. EI neurons also are restricted to two regions. The main population is in the ventromedial region, and a second population occurs along the very dorsolateral margin of the contour, which is most likely in the external nucleus of the inferior colliculus.

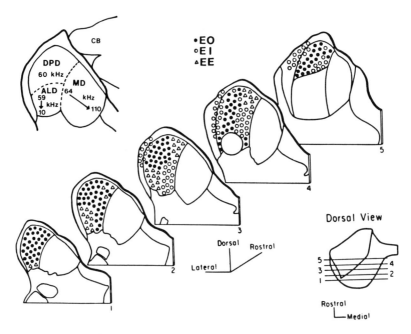

FIGURE 7.9. Schematic shows segregation of neurons with different aural preferences in the 60-kHz region (DPD) of the mustache bat's inferior colliculus. Transverse sections of the inferior colliculus are arranged in a caudal-to-rostral direction. Each section is separated by 180 μm, and the rostrocaudal position of each section is illustrated on the dorsal view of the inferior colliculus in the lower right. The medial division (MD), where high frequencies are represented, is the large area to the right of the DPD in each section. The anterolateral division (ALD), representing frequencies below 60 kHz, is shown below the DPD in sections 4 and 5. EO, monaural neurons; EI, excitatory-inhibitory neurons; EE, excitatory-excitatory neurons. (Adapted from Wenstrup, Ross, and Pollak 1986a.)

1.6 Each Aural Subregion of the 60-kHz Contour Receives Afferent Projections from a Unique Set of Lower Auditory Nuclei

The topographic arrangement of neurons with particular monaural and binaural properties in the 60-kHz contour suggests that the aural properties are a consequence of the projections that terminate in each subregion. By making small iontophoretic deposits of HRP within each of the physiologically defined aural regions, it was shown that each monaural and binaural region within the 60-kHz contour receives its chief inputs from a different subset of nuclei in the lower brainstem (Ross and Pollak 1989). The principal connections of the EI and both EE aural subregions are shown in Fig. 7.10. For most regions, the response properties reflect the subset of

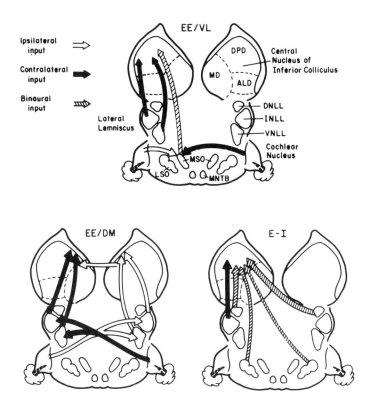

FIGURE 7.10. Schematic shows some of the major projections to the ventrolateral EE region *(top panel)*, dorsomedial EE region *(lower left panel)*, and ventromedial EI region *(lower right panel)* of the 60-kHz contour of the mustache bat's inferior colliculus. Notice that the EI region receives bilateral projections from the LSO and DNLL, and that those nuclei do not project to either of the EE regions. (Adapted from Ross and Pollak 1989.)

inputs. Of particular significance are the projections to the ventromedial EI region. The afferents to this area arise largely from binaural nuclei, especially the DNLL (bilaterally) and from the LSO (bilaterally). A major input also originates in the intermediate nucleus of the lateral lemniscus, which is a monaural nucleus. The robust inputs from the DNLL and LSO distinguish the EI region from all other aural regions of the 60-kHz contour, which receive few or no projections from either of these nuclei.

1.7 Rate-Intensity Functions and Temporal Discharge Patterns Are Other Characteristic Features of Auditory Neurons

Two additional properties that characterize all auditory neurons are the way in which spike counts change with increasing sound intensity, and the temporal discharge patterns that neurons display to tone bursts presented at their best frequencies. In response to increasing intensity, spike counts of auditory neurons change in one of two principal ways. In some neurons the spike counts increase as intensity is raised, and then either reach a plateau or continue to increase even with the highest intensities presented. Such neurons are said to have monotonic spike-count functions, that are either saturated or nonsaturated. In the majority of ICc neurons, however, spike counts at first increase with intensity and then decrease, sometimes to zero spikes, as intensity is increased further. These neurons are said to have nonmonotonic rate-intensity functions. In many neurons, the discharge rate falls to zero at higher intensities. These neurons not only have a lower threshold but an upper threshold as well. In the 60-kHz contour of the mustache bat, about 60% of the neurons have nonmonotonic rate-intensity functions, whereas 40% have monotonic functions (Pollak and Park 1993).

Another characteristic feature of auditory neurons is their temporal discharge pattern evoked by tone bursts at the neuron's best frequency. Discharge patterns can be complex, and in lower nuclei the number of different types is quite large (e.g., Rhode and Greenberg 1992). For purposes of simplicity, we classify collicular neurons into one of two types; those that fire with sustained discharges for the duration of the tone burst, and those that fire only to the onset, or to both the onset and offset of the signal. The discharge patterns of neurons in most lower nuclei are sustained (Goldberg and Brown 1968; Brugge, Anderson, and Aitkin 1970; Guinan, Guinan, and Norris 1972; Harnischfeger, Neuweiler, and Schlegel 1985; Covey, Vater, and Casseday 1991; Rhode and Greenberg 1992) and onset patterns are less common. In the ICc of bats, the population of onset neurons exceeds the population of sustained neurons (Pollak et al. 1978; Vater, Schlegel, and Zöller 1979; Jen and Suthers 1982; O'Neill 1985; Harnischfeger, Neuweiler, and Schlegel 1985; Park and Pollak 1993b). In the 60-kHz contour, for example, about 66% of the neurons have onset patterns and only about 33% have sustained patterns (Park and Pollak 1993b).

Neurons with monotonic and nonmonotonic rate-intensity functions, and neurons with phasic and sustained discharge patterns, occur throughout the dorsoventral extent of the 60-kHz contour, and thus at all levels of the EI, EE and monaural subregions (Park and Pollak 1993b).

1.8 Inhibitory Thresholds of EI Neurons are Topographically Organized in the Ventromedial Region of the 60-kHz Collicular Contour

The population of EI neurons is of particular interest because they differ in their sensitivities to interaural intensity disparities (IIDs). These neurons compare the sound intensity at one ear with the intensity at the other ear by subtracting the activity generated in one ear from that in the other, and thus play an important role in coding sound location. The way in which EI neurons code for sound location is discussed in detail in a later section. Suprathreshold sounds delivered to the excitatory (contralateral) ear evoke a certain discharge rate that is unaffected by low-intensity sounds delivered simultaneously to the inhibitory (ipsilateral) ear. However, when the ipsilateral intensity reaches a certain level, and thus generates a particular IID, the discharge rate declines sharply, and even small increases in the ipsilateral intensity will in most cases completely inhibit the cell (Fig. 7.11). Thus, each EI neuron has a steep IID function and reaches a criterion inhibition at a specified IID that remains relatively constant over a wide range of absolute intensities. We refer to the IID that produces a 50% reduction in discharge rate as the neuron's "inhibitory threshold" (Wenstrup, Ross, and Pollak 1986a; Wenstrup, Fuzessery, and Pollak 1988a,b). An inhibitory threshold is assigned a negative value if the criterion inhibition occurred when the ipsilateral signal was more intense than the contralateral signal, and was assigned a positive value if the intensity at the ipsilateral ear was lower than the contralateral ear when the criterion inhibition was achieved. The inhibitory threshold of the EI cell in Fig. 7.11 is 0 dB. Inhibitory thresholds among EI neurons vary from about +15 dB to about −25 dB, encompassing much of the range of IIDs that the bat would normally experience (Wenstrup, Fuzessery, and Pollak 1988a).

Particularly noteworthy is that the inhibitory thresholds of EI neurons are arranged in an orderly fashion in the EI subregion of the 60-kHz contour (Fig. 7.12). EI neurons with negative inhibitory thresholds (i.e., neurons requiring a louder ipsilateral stimulus than contralateral stimulus to produce 50% inhibition) are located in the dorsal EI region. Subsequent EI neurons display a progressive shift to more positive inhibitory thresholds. The most ventral EI neurons have the most positive inhibitory thresholds: they are suppressed by ipsilateral sounds equal or less intense than the contralateral sounds.

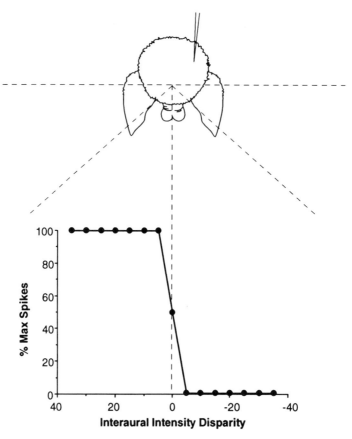

FIGURE 7.11. Schematic of interaural intensity function of an EI neuron. The bat's head is shown in the *upper portion* with a microelectrode that would record the activity of the EI neuron in the colliculus on the side shown. The IID function is obtained by presenting sound through earphones situated in the bat's ears (not shown). A 60-kHz sound is presented to the excitatory ear, the ear contralateral to the colliculus from which the recording is taken. The sound intensity is fixed at some intensity above threshold and evokes a discharge that is represented as 100% maximum spikes on the ordinate. A low-intensity sound is then presented simultaneously at the ipsilateral (inhibitory) ear, and has no effect on the discharge rate evoked by the sound presented to the excitatory ear. As the sound intensity at the ipsilateral ear is increased, indicated by the decreasing IIDs, the discharge rate remains unaffected until the IID reaches a certain value, in this case about 0 dB, at which point the spike count declines markedly. Further increases of the intensity at the ipsilateral ear (more negative IIDs) result in a complete inhibition of discharges. The bat's head is drawn to indicate how the IIDs translate into spatial locations and how the cell might respond to free-field sounds. The IID function suggests that discharges should be inhibited by 50% when the sound is directly in front of the bat, at 0° azimuth. The cell should not be inhibited by sounds in the hemifield contralateral to the recording electrode, but the cell should be completely inhibited by sounds from the hemifield ipsilateral to the electrode.

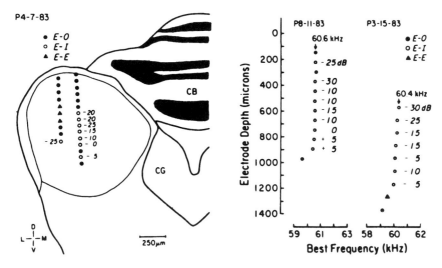

FIGURE 7.12. *Left:* Systematic shifts in the inhibitory thresholds (IIDs at which the discharge rate declines by 50%) of EI unit clusters shown in a transverse section of the mustache bat's inferior colliculus. Numbers indicate IIDs of the inhibitory thresholds recorded at those locations. All units were sharply tuned to frequencies around 60 kHz. *Right:* Systematic increase of inhibitory thresholds of EI neurons with depth in the dorsoposterior division of the mustache bat's inferior colliculus recorded in two dorsoventral penetrations from different bats. (Adapted from Wenstrup, Ross, and Pollak 1985).

1.9 Latencies Are Systematically Arranged Within the Aural Subregions of the 60-kHz Contour

Latency is another response feature that is arranged in an orderly fashion within isofrequency contours of the ICc (Schreiner and Langner 1988; Park and Pollak 1993b). Figure 7.13 shows plots of latency as a function of depth in the 60-kHz contour and reveals two salient features. The first is that neurons in the dorsal colliculus have, on the average, longer latencies than neurons located ventrally; the average latency in the more dorsal part of the colliculus is about 15.4 msec, which decreased to an average of about 9.8 msec in the more ventral region. The second feature is that the range of latencies changes markedly with depth. Dorsally there is a broad distribution of latencies while deeper regions have a much narrower range of latencies. Long-latency cells are found only dorsally whereas short-latency cells are found at all depths. The latencies of dorsal neurons, within 200–400 μm of the collicular surface, range from 30 to 6 msec, whereas the latencies at depths from 1000 to 1300 μm range from 5 to 10 msec. Thus, dorsal regions are characterized by a wide distribution of latencies and a relatively high average latency, while ventral neurons have a narrower range of relatively short average latencies.

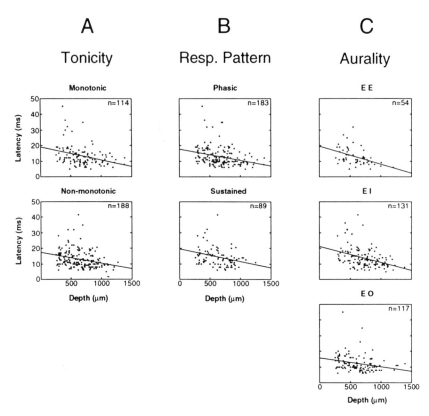

FIGURE 7.13A–C. Plots of latency as a function of depth for cells with different response types. (A) Latency distributions for cells with monotonic and nonmonotonic rate-intensity functions. (B) Latency distributions for cells with phasic and sustained response patterns. (C) Distributions for cells with different binaural and monaural properties. (Adapted from Park and Pollak 1993b.)

The foregoing features are also seen in each of the aural subregions. As shown in Fig. 7.13C, a wide distribution of latencies and a relatively high average latency are found dorsally in each of the aural subregions, where either monaural, EI, or EE units predominate. In more ventral portions of each aural subregion, the neuronal populations have a narrower range of latencies and shorter average latencies.

Finally, the neuronal populations that have monotonic and nonmonotonic rate-intensity functions, and those that have phasic and sustained discharge patterns, also express the same characteristic pattern of latency change with depth. This is illustrated in Figure 7.13A,B by the characteristic decrease in average latency and latency range with depth for neurons with either monotonic or nonmonotonic rate-intensity functions, and by neurons with either a sustained or phasic discharge pattern.

1.10 Overall View of the Organization of the Auditory Brainstem

The general picture of the auditory brainstem is that it is constructed from a hierarchical series of organizational principles. Once an organization is established at a certain level of the system, that organization is maintained throughout the system. At each level, a new organization is created and becomes embedded within the framework of the organization that was created at the lower level. Thus, the primary level of ordering, tonotopy, is initially established in the cochlea. The tonotopy is maintained throughout the brainstem auditory pathway, and results in the isofrequency contours of each nucleus. The projections from the isofrequency regions in each of the lower centers are further ordered and target regional areas of the same isofrequency contours in the inferior colliculus. The regional targeting of projection systems creates regions within each collicular contour dominated by monaural or one of the two types of binaural neurons (Roth et al. 1978; Wenstrup, Ross, and Pollak 1986a; Ross and Pollak 1989). We refer to these regions as aural subregions. Imposed on the neurons in each aural subregion are additional organizational features (Wenstrup, Ross, and Pollak 1986a; Schreiner and Langner 1988; Ross and Pollak 1989; Pollak and Casseday 1989; Park and Pollak 1993b).

Because sustained and phasic neurons occur throughout the contour, they can be considered to be arranged as parallel arrays that run dorsoventrally within each aural subregion of the 60-kHz contour. Furthermore, neurons that have a sustained discharge pattern and a monotonic rate-intensity function are also found throughout the contour, and may be arranged parallel to sustained neurons that have nonmonotonic rate-intensity functions. These populations, in turn, are arranged in parallel with neurons that have phasic, monotonic and phasic, nonmonotonic response types. Thus, aural type, temporal discharge pattern, and rate-intensity function are qualitative features that appear to be constant within a dorsoventral "column" or array of cells. Other features are then arranged orthogonal to the dorsoventral axis and change quantitatively along this spatial dimension; specifically, a characteristic range of latencies and a characteristic average latency are present at each dorsoventral level (Park and Pollak 1993b), and in the case of EI neurons, a characteristic interaural intensity disparity that causes the discharge rate to decline by 50% as well (inhibitory threshold) (Wenstrup, Ross, and Pollak 1985, 1986a). The values of both the average latency and range of latencies and the inhibitory thresholds (the interaural intensity disparities of the 50% cutoffs) change systematically with depth. Thus, an array of cells is conceptually analogous to a cortical column, and the complement of arrays may be analogous to a cortical hypercolumn (Hubel and Wiesel 1977).

In the following sections the way in which biologically relevant acoustic cues are encoded by collicular neurons are discussed. Where possible, we show how the organization of the response features within an isofrequency contour contributes to the representation of a particular cue, or to the establishment of new features at the next level. Finally, we will also show how, for certain features, the projection patterns from lower regions act to shape new response features, and how such transformations contribute to one or another of the different organizational features of the inferior colliculus.

2. Response Properties of Collicular Neurons That Underlie Doppler-Shift Compensation and the Coding of Target Features

One of the benefits of using echolocating bats for neurophysiological investigations of the auditory system is that the way in which they manipulate their orientation calls is known in considerable detail, as are the acoustic cues from which these animals derive information about objects in the external world (see Grinnell, Chapter 1; Fenton, Chapter 2; and Schnitzler, Chapter 3, this volume). This knowledge allows investigators to generate signals that mimic the types of signals that bats emit and receive during echolocation, as well as the sorts of acoustic cues which are contained in the echoes reflected from flying insects. In addition, it permits the investigator to make informed inferences about how the nervous system encodes and represents a particular cue from the discharge properties of the cells that are monitored. In this section, we briefly discuss one of the most remarkable behaviors exhibited by horseshoe and mustache bats, Doppler-shift compensation. We then consider the acoustic cues these bats use for the detection and recognition of insects, and why Doppler-shift compensation is important for the perception of those cues. Finally, we discuss some of the neural features that are important both for Doppler-shift compensation and for the coding of the acoustic cues these bats use for the identification of their targets.

2.1 Some Aspects of Doppler-Shift Compensation

Doppler-shift compensation is the expression of the bat's extreme sensitivity for motion. One way that relative motion is created is by the difference in flight speed between the bat and background objects in its environment. Because a flying bat approaches an object at a certain speed, the echo from that object will be Doppler shifted upward, and will therefore have a higher frequency than the emitted signal. Long constant-frequency, frequency-modulating (CF/FM) bats are exquisitely sensitive to upward Doppler shifts in the CF component of the echo. They compensate for the Doppler shifts by lowering the frequency of the next emitted CF component by an amount that is nearly equivalent to the Doppler shift in the preceding echo (Fig. 7.14) (Schnitzler 1970; Schuller, Beuter, and Schnitzler 1974; Simmons 1974; Henson et al. 1982; Trappe and Schnitzler 1982; Gaioni, Riquima-roux, and Suga 1990).

The flight of the bat, however, is not the only motion in the night sky, and thus is not the only source of Doppler shifts. Flying insects beat their wings to stay aloft, and the motion of the wings creates periodic Doppler shifts, which impose periodic frequency modulations on the CF component of the echo (Fig. 7.15) (Schnitzler and Ostwald 1983; Schuller 1984; Kober 1988; Ostwald, Schnitzler, and Schuller 1988). In addition, the wingmotion, or flutter, presents a reflective surface whose area alternates with wing position, thereby also creating periodic amplitude fluctuations (i.e., amplitude modulations) in the CF component of the echo. The modulations imposed upon the echo CF component by the fluttering wings of an insect are of critical importance to long CF/FM bats, because it is from these modulations that the bats discriminate flying insects from background objects and recognize the particular insect that has wandered into their acoustic space (Goldman and Henson 1977; Kober 1988; Ostwald, Schnitzler, and Schuller 1988; von der Emde 1988).

One of the key adaptations that underlies the perception of both Doppler shifts and modulation patterns is the large population of sharply tuned neurons. In the following sections we first consider some properties of these neurons and the mechanisms that create the sharp tuning curves. We then describe how the sharp tuning permits the processing of Doppler-shifted

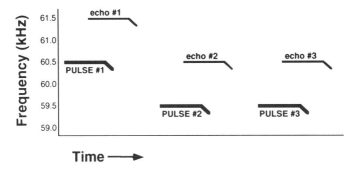

FIGURE 7.14. Schematic of Doppler-shift compensation in a flying mustache bat. The first pulse is emitted at about 60.5 kHz, and the Doppler-shifted echo returns with a higher frequency, at about 61.5 kHz. The mustache bat detects the difference in frequency between the pulse and echo, and lowers the frequency of its subsequent emitted pulses by an amount almost equal to the Doppler shift. Thus, the frequencies of the subsequent echoes are held constant and return at a frequency very close to that of the first emitted signal. Also notice that because of the length of the emitted CF component, the echoes return to the ear while the bat is still emitting the pulse, resulting in substantial periods of pulse–echo overlap.

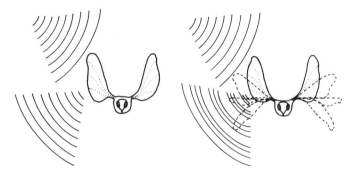

FIGURE 7.15. Drawings illustrate the echoes of tone bursts reflected from an insect in which the wings are stationary *(left)* and from an insect whose wings are in motion *(right)*. Echoes reflected from the insect whose wings are stationary will also be have the same frequency as the tone burst, while the echoes from the insect whose wings are beating will be frequency modulated, as indicated by the alternate spacing of the reflected waves. The frequency modulations are created by the Doppler effect caused by the velocity of wing movements. (Reproduced from Pollak and Casseday 1989.)

echoes and its significance for coding the variety of modulation patterns these bats receive from flying insects.

2.2 Long CF-FM Bats Have Large Populations of Neurons Sharply Tuned to the CF Components of Their Biosonar Signals

Perhaps the most dramatic neuronal response feature of the long CF/FM bats is the overrepresented population of neurons that have very sharp tuning curves and best frequencies that are close to the dominant CF component of the bat's orientation calls (see Fig. 7.7) (Suga, Simmons, and Jen 1975; Suga and Jen 1976, 1977; Schuller and Pollak 1979; Pollak and Bodenhamer 1981; Ostwald 1984; Vater, Feng, and Betz 1985; Kössl and Vater 1990). Because of the high degree of frequency selectivity, these neurons are also referred to as "filter neurons" (Neuweiler and Vater 1977).

The narrow tuning of filter neurons is a consequence of at least two processes. The first is the cochlear specializations that were mentioned earlier and which are considered in detail by Kössl and Vater in Chapter 5 (also see Suga, Simmons, and Jen 1975; Suga and Jen 1977; Kössl and Vater 1990). The second process is neural inhibition (Suga and Tsuzuki 1985; Yang, Pollak, and Resler 1992; Vater et al. 1992; Casseday and Covey 1992). In higher auditory nuclei, inhibitory circuits often modify the shape of tuning curves in at least two ways. The first way is to further sharpen the tuning curve with lateral inhibition; either the high- or low-frequency skirts of the tuning curve, or both, are trimmed by inhibitory inputs from neurons whose best frequencies are on the flanks of the excitatory tuning curve. In filter units, the inhibition most often affects the width of the tuning curve at intensities 20 dB or more above threshold, and has little or no effect on the width of the curve at 10 dB above threshold (Suga and Tsuzuki 1985; Vater et al. 1992; Yang, Pollak, and Resler 1992). Such sharpening is illustrated by the neuron in the top panel of Fig. 7.16. This figure shows the expansion of the tuning curve of a filter neuron in the mustache bat's inferior colliculus that occured when GABAergic inhibition was blocked by the drug bicuculline. The origin of the GABAergic inhibition is unknown, but it seems likely that the GABAergic inhibition that shapes tuning curves is caused by local GABAergic collicular neurons,

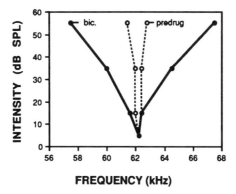

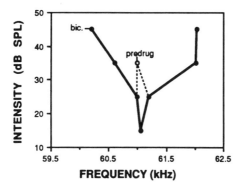

FIGURE 7.16. Two 60-kHz units from the mustache bat's inferior colliculus whose tuning curves broadened from the iontophoretic application of bicuculline. The unit in the *top panel* originally had a very sharp tuning curve (predrug), which opened substantially when GABAergic inhibition was blocked with bicuculline (bic). The cell in the *lower panel* originally had an upper-threshold tuning curve, which was transformed into a much wider, open tuning curve with bicuculline. (Adapted from Yang, Pollak, and Resler 1992.)

although the involvement of GABAergic projection neurons cannot be ruled out.

The second way that inhibition can modify a tuning curve is to suppress discharges at high intensities, thereby producing a closed tuning curve that has both a minimum threshold and an upper threshold (Vater et al. 1992; Yang, Pollak, and Resler 1992). The bottom panel in Figure 7.16 shows the closed tuning curve of a filter unit in the mustache bat's inferior colliculus that was changed to a more conventional "open" tuning curve when GABAergic inhibition was blocked with bicuculline. Although neurons with closed tuning curves are common in the ICc and at higher levels of the auditory system (Grinnell 1963; Suga 1964; Jen and Schlegel 1982; Jen and Suthers 1982; O'Neill 1985; Olsen and Suga 1991a,b; Suga and Tsuzuki 1985; Casseday and Covey 1992; Vater et al. 1992; Yang, Pollak, and Resler 1992), the significance of these tuning curves for echolocation is not entirely clear. One suggestion is that these neurons might be inhibited by the relatively loud emitted call, but would be left in a responsive condition to respond to the fainter echo that arrives shortly thereafter (Grinnell 1963; Casseday and Covey 1992). A mechanism of this sort could be important for processing the FM components of biosonar signals, but it apparently does not play the same role in the processing of the dominant CF components. The reason is that the CF components of the pulse and echo are processed by different neuronal populations, as explained next.

2.3 The Sharp Tuning of Filter Neurons Is Relevant for Doppler-Shift Compensation

The long CF/FM bats derive two major benefits by having sharply tuned filter units. The first benefit is that it allows the echo CF component to be perceived during periods of pulse–echo overlap, when the emitted and echo CF components are stimulating the ear at the same time. The second benefit is that the sharp tuning imparts exceptional sensitivity for encoding frequency modulation patterns that are important for detecting and char-acterizing insects (Suga and Jen 1977; Schuller 1979; Pollak and Schuller 1981; Schnitzler and Ostwald 1983; Bodenhamer and Pollak 1983; Suga, Niwa, and Taniguchi 1983; Ostwald 1988; von der Emde 1988; Lesser et al. 1990). We first consider the issue of pulse–echo overlap, and then turn to the detection and characterization of targets.

2.4 Sharp Tuning Allows Spatial Segregation of Activity Evoked by the Pulse and Echo During Periods of Pulse-Echo Overlap

Because the emitted CF components are as long as 30 msec in mustache bats (Novick and Vaisnys 1964) and up to 90 msec in horseshoe bats (Griffin and

Simmons 1974), the CF component of the echo always returns to the bat's ear while the animal is still emitting its pulse (see Fig. 7.14) (Novick 1971). This creates relatively long periods of pulse–echo overlap at the ear, a condition that seemingly would interfere with the bat's perception of the echo CF component. The sharp tuning of filter neurons is significant in this regard, because it segregates the neuronal populations that respond to the emitted CF component from the filter neurons that respond to the CF component of the Doppler-shifted echo (Schuller and Pollak 1979). Consider, for example, that the flight speed of mustache bats can create Doppler shifts up to about 1500 Hz in the echoes from stationary objects. When a mustache bat compensates for a 1500-Hz Doppler shift, its emitted CF component is about 58.5 kHz and the echo CF component is about 60 kHz. Under these conditions, the filter neurons tuned to 60 kHz would respond to the Doppler-shifted echo, but the extreme narrowness of their tuning curves would prevent them from discharging to the lower frequency of the emitted CF component. Especially important in this regard are the very sharp slopes on the low-frequency sides of the tuning curves, which are, in part, a consequence of lateral inhibition (Suga and Jen 1977; Suga and Tsuzuki 1985; Kössl and Vater 1990; Yang, Pollak, and Resler 1992). Conversely, because the high-frequency skirts of the tuning curves of units tuned to 58 and 59 kHz are steep (Suga and Jen 1977; O'Neill 1985; Kössl and Vater 1990), the majority of units in the 58-kHz and 59-kHz collicular contours would discharge to the emitted CF component, but would be largely unaffected by the higher frequency of the Doppler-shifted echo. Thus, the activity evoked by the emitted CF would be spatially segregated from the activity evoked by the Doppler-shifted echo. Such an arrangement effectively prevents the emitted pulse from masking the discharge activity in filter units evoked by the echo CF during periods of pulse–echo overlap, a feature of clear importance for the effective operation of a Doppler-based sonar system.

2.5 Advantage of Doppler-Shift Compensation

The adaptive advantage of Doppler-shift compensation is that it enhances the ability of long CF/FM bats to readily detect and recognize a fluttering insect in the presence of background echoes (Neuweiler 1983, 1984; Pollak and Casseday 1989). All bats that utilize this type of biosonar system hunt under the forest canopy (Neuweiler 1983, 1984; Bateman and Vaughan 1974). This habitat provides a rich source of food for which there is probably little other competition, but it is also an environment that requires the bat to be able to both detect and perceive prey in the midst of the echoes from background objects. Long CF/FM bats cope with the clutter by compensating for the Doppler shifts in the echoes from the background objects, and thereby clamp those echoes at a constant frequency that corresponds closely to the best frequencies of the overrepresented filter neurons. When a small insect crosses the path of a hunting horseshoe or mustache bat, the presence of modulations in the echo are clear and unambiguous cues that a flying insect has invaded the bat's acoustic space, and the modulation pattern provides the information for characterizing the insect (Goldman and Henson 1977; Ostwald, Schnitzler, and Schuller 1988; von der Emde 1988). Thus, by manipulating the frequencies of their emitted CF components, long CF/FM bats confine echoes to a very narrow frequency band, and this behavior ensures the bat that modulated echoes will be processed by an exceptionally large number of sharply tuned filter neurons. For reasons explained here, the sharp tuning curves are also important because they impart an exceptional sensitivity and effectiveness for encoding the patterns of the periodic frequency modulations.

2.6 Filter Neurons Are Exceptionally Sensitive to Modulated Signals

The periodic amplitude and frequency modulations created by the wingbeat pattern of a fluttering insect are complex (Schnitzler and Ostwald 1983: Schuller 1984; Kober 1988). The way in which filter units encode these patterns is usually studied with electronically generated signals that mimic the natural echoes (Suga and Jen 1977; Schuller 1979; Pollak and Schuller 1981; Vater 1982; Ostwald, Schnitzler, and Schuller 1988; Bodenhamer and Pollak 1983; Suga, Niwa, and Taniguchi 1983; Lesser et al. 1990). The response features of filter units evoked by simulated echoes have proven to be a good approximation to the way they respond to natural echoes from a fluttering insect (Schuller 1984; Ostwald 1988). The echoes are mimicked by modulating the frequency or amplitude of a carrier tone with low-frequency sinusoids of 50–500 Hz (Fig. 7.17). The low-frequency sinusoid is referred to as the modulating waveform to distinguish it from the carrier or center frequency. This arrangement creates either sinusoidally amplitude modulated (SAM) or sinusoidally frequency modulated (SFM) signals. With SFM signals, the depth of modulation, the amount by which the frequency varies around the carrier, mimics the degree of Doppler shift created by the wing motion of a flying insect. The other modulating parameter, the modulation rate, simulates the insect's wingbeat frequency.

When SFM or SAM signals are presented to the ear of any mammal, be it a bat (e.g., Schuller 1979), rat (M;daller 1972), or cat (Schreiner and Langner 1988), neurons in the mammalian auditory system respond with discharges that are tightly locked to the modulating waveform. This is also the case for both filter and nonfilter collicular neurons in long CF/FM bats. The feature that distinguishes filter neurons from all other neurons is that they are far more sensitive to very small periodic modulations than are more broadly tuned units (Schuller 1979; Bodenhamer and Pollak 1983; Vater 1982). The filter neurons fire only when the frequencies sweep into the

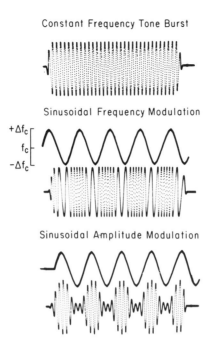

FIGURE 7.17. The fine structure of a tone burst *(upper record)*, a sinusoidally frequency-modulated (SFM) burst *(middle record)*, and a sinusoidally amplitude-modulated (SAM) burst *(lower record)*. The tone burst in the upper record is a shaped sine wave in which the frequency, or fine structure, of the signal remains constant. The fine structure of the SFM burst changes sinusoidally in frequency. F_c indicates the center or carrier frequency; the degree to which the frequency changes around the carrier is indicated by ΔF. The modulation waveform is shown above the record of the signal's fine structure. In the SAM burst, the fine structure of the signal is a constant frequency, but the amplitude varies with the sinusoidal modulating waveform. (Reproduced from Pollak and Casseday 1989.)

narrow confines of the tuning curve and remain silent when the signal is outside of the tuning curve (Suga and Jen 1977; Schuller 1979; Pollak and Schuller 1981; Bodenhamer and Pollak 1983; Ostwald 1988). The truly spectacular synchronization of discharges of filter neurons to the modulation waveform, even when the depth of modulation is as small as ±10 Hz, is illustrated in Figure 7.18.

2.7 Selectivity of Filter Neurons for Parameters of Modulated Signals

Although filter neurons having firing patterns tightly locked to the modulating waveform are commonly encountered in the colliculus, most are selective for some parameter of the modulated signal to which they will

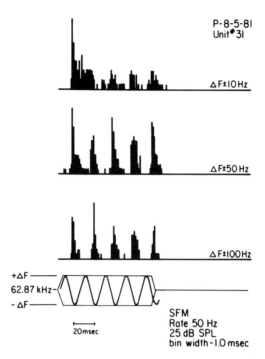

FIGURE 7.18. Peristimulus time histograms of a sharply tuned 60-kHz neuron in the mustache bat's inferior colliculus that phase locked to sinusoidally frequency-modulated (SFM) signals. The signal envelope and modulating waveform are shown below. The depth of modulation was varied in the three records around a carrier frequency of 62.87 kHz. This unit was unusually sensitive, and phase locked when the frequency swings (Δf) were as small as ±10 Hz around the 62.87-kHz center frequency. (Reproduced from Bodenhamer and Pollak 1983.)

synchronously discharge (Schuller 1979; Pollak 1980; Pollak and Schuller 1981; Bodenhamer and Pollak 1983; Ostwald 1988). This feature is well illustrated by considering how discharge synchrony is influenced by signal intensity, modulation depth and rate, and carrier frequency.

Many filter neurons respond with discharges that are about equally well synchronized to the modulating waveform at all intensities above threshold. Other filter neurons, however, discharge in registry with the modulation waveform only over a preferred range of intensities. Typically, the sharpest locking and most vigorous responses are evoked only at low or moderately low intensities, and at higher intensities the discharge registration either declines significantly or disappears completely. Four neurons exhibiting this feature are shown in (Fig. 7.19). It should also be pointed out that the preferred range of intensities of the four neurons in Figure 7.19 differs slightly from neuron to neuron. The population, then, forms a continuous gradient in which some neurons encode the modulation pattern at all intensities above threshold, whereas others are more selective and only

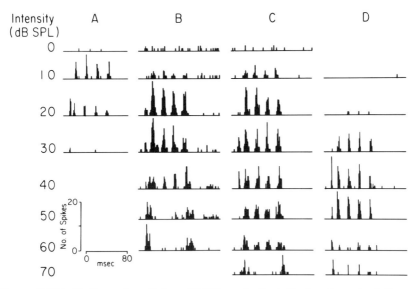

FIGURE 7.19A–D. Examples of four 60-kHz units that displayed a selectivity, or tuning, of their locked discharges for intensity. Each of the units phase locked to the SFM signals only over a narrow intensity range that was specific to each neuron. Modulation depths: ±100 Hz (A), ±50 Hz (B), ±200 Hz (C), and ±100 Hz (D). Modulation rate was 50 Hz for all units. Signal duration was 80 msec. (Reproduced from Bodenhamer and Pollak 1983.)

encode the modulation waveform if the echo falls within a narrow intensity slot.

Filter neurons also exhibit preferences for the range of modulation depths to which they will phase lock. As was the case for intensity, most filter neurons exhibit a continuous gradation in their abilities to encode modulation depth. This feature is illustrated by the three filter neurons in Figure 7.20. The neuron on the far left of Figure 7.20 locked best for modulations depths between ±100 Hz to ±400 Hz. In contrast, the neuron in the middle panel locked best to modulation depths that ranged from ±50 Hz to ±200 Hz, while the neuron in the far right panel locked best to ±50 Hz. Notice that the modulation *rate* presented to all three neurons was 50 Hz.

As is the case for intensity and modulation depth, filter neurons also display a selectivity for modulation rate, which simulates the various wingbeat frequencies of different insects. A few neurons accurately encode modulation rates from 25 Hz to 200–300 Hz, whereas most other units show a preference for a range of rates.

The effects of various SFM parameters are generally evaluated when the carrier frequency of the modulated signal is set at the neuron's best frequency. Under natural conditions, however, the center frequency of the echo will not correspond to the best frequencies of all the neurons excited by the signal. For example, consider a situation where a Doppler compen-

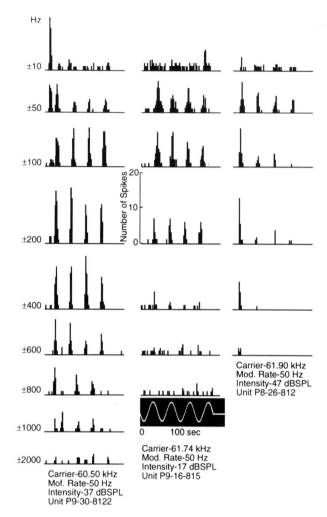

FIGURE 7.20. Three 60-kHz units from the mustache bat's inferior colliculus that had preferences for different SFM modulation depths. The modulation depth used to generate each histogram is shown at the far left. The modulation rate was 50 Hz and the signal duration was 80 msec for each unit. (Reproduced from Bodenhamer and Pollak 1983.)

sating mustache bat receives an echo in which the frequency modulations vary by ±500 Hz around a 60-kHz carrier. Under these conditions, units having a best frequency of 60 kHz will be excited, but so will other units having best frequencies slightly above or slightly below 60 kHz. Given these conditions, it becomes of some interest to assess how filter units encode modulation patterns for a variety of different carrier frequencies.

Carrier frequency affects synchronized discharges to SFM signals in two general ways. The first way is that many units lock in a symmetrical manner for carrier frequencies around the unit's best frequency (Fig. 7.21, left panel). Whether the carrier frequency is above or below the neuron's best frequency is unimportant to such units. Rather, the magnitude of the locked discharge rate depends largely on the extent to which the SFM signal encroaches upon the tuning curve. The second way is displayed by other units that respond in an asymmetric manner for different carrier frequencies. In some of these units the magnitude of the synchronized discharges is clearly greater when the carrier frequency is below the neuron's best frequency, or, in other units, when the carrier frequency is above the best frequency. An asymmetric unit in which the neuron favored carrier frequencies above the best frequency is shown in the right panel of Figure 7.21.

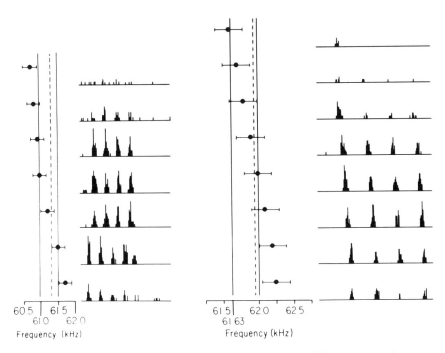

FIGURE 7.21. Phase-locked discharges in two filter units for different center frequencies. *Left:* Neuron that phase locked regardless of carrier frequency, so long as a portion of the SFM signal encroached upon the tuning curve. *Right:* Neuron displaying a marked preference, or asymmetry, for SFM carrier frequencies on the high-frequency side of its tuning curve. *Solid lines* to left of histograms show the limits of the unit's tuning curve at the same intensity as the SFM signal. *Dashed line* indicates the unit's best frequency. *Horizontal lines* indicate the position of the SFM signal. SFM signal in left panel was ±100 Hz at 20 dB SPL (sound pressure level), and SFM signal in right panel was ±200 Hz at 30 dB SPL. (Reproduced from Bodenhamer and Pollak 1983.)

2.8 Representation of an Insect's Signature in the Filter Region of the Inferior Colliculus

Insights into the spatial extent of activity evoked by an echo from a fluttering insect can be obtained from considerations of the sharp tuning curves and the way in which filter units respond to SFM signals. For example, on detecting a target, a mustache bat can determine if the target is inanimate or animate, for example, a falling leaf or a flying insect, on the basis of the temporal patterns of the responding neurons: an echo from a falling leaf will be encoded in a manner similar to a tone burst whereas a fluttering insect will elicit synchronized discharges locked to the modulation pattern of the echo. The active region of the colliculus will have boundaries sharply confined to the 60-kHz contour because of the narrow tuning curves of the filter units. The response characteristics evoked by SFM signals suggest that the discharge rate and synchrony should be maximal in units whose best frequencies coincide with the echo center frequency and which respond symmetrically to various SFM carrier frequencies. However, the activity should also be maximal in the asymmetric units whose best frequencies are slightly above or below the center or carrier frequency of the echo. In these neurons, the synchronized discharges are evoked only when the carrier frequency is below, or in other units, above, the neuron's best frequency.

Changes in position, orientation, and speed of either the bat or its target will cause subsequent echoes to differ more or less in carrier frequency, modulation pattern, and intensity from the previous echo(es). In principle, each echo will be encoded in a similar manner. However, the preferences of many filter units for selective ranges of intensity as well as modulation rate and depth suggest that the changes in echo parameters will cause some neural elements to drop out and new elements to be recruited, while others will simply change response vigor or firing registration to reflect the changes in echo characteristics. In short, the properties of filter neurons endow the system with the ability to encode the features of the echo CF component. Thus, the sum total of neural activity is a dynamic pattern that differs from echo to echo.

3. Sound Localization and the Processing of Localization Cues by the EI Circuit

The ability to localize a sound source in both azimuth and elevation is of obvious importance to a nocturnal predator that hunts flying insects in the night sky. The cues animals use to associate a sound with its position in space are interaural disparities in time or intensity (Erulkar 1972; Mills

1972; Gourevitch 1980). Animals that hear "high" frequencies, such as echolocating bats, rely on interaural intensity disparities (IIDs) to localize sounds composed of those frequencies (Erulkar 1972; Mills 1972).

The intensity disparities are generated by acoustic shadowing and the directional properties of the ear (Grinnell and Grinnell 1965; Blauert 1969/1970; Erulkar 1972; Flannery and Butler 1981; Musicant and Butler 1984; Fuzessery and Pollak 1984, 1985; Makous and O'Neill 1986; Jen and Sun 1984; Jen and Chen 1988; Pollak and Casseday 1989). Once produced, the intensity disparities are conveyed into the central nervous system where the information from the two ears is initially "compared" in the LSO (Stotler 1953; Boudreau and Tsuchitani 1968; Warr 1982; Cant and Casseday 1986; Harnischefeger, Neuweiler, and Schlegel 1985; Covey, Vater, and Casseday 1991; Finlayson and Caspary 1989). The comparison is a subtractive process, whereby signals from the ipsilateral ear excite and signals from the contralateral ear inhibit LSO cells (Boudreau and Tsuchitani 1968; Brownell, Manis, and Ritz 1979; Caird and Klinke 1983; Harnischefer, Neuweiler, and Schlegel 1985; Sanes and Rubel 1988; Covey, Vater, and Casseday 1991). These so-called EI cells are sensitive to intensity disparities and express the comparison of IIDs in their firing rates.

The comparisons of IIDs must be performed on a frequency-by-frequency basis, a requirement permitted by tonotopic organization. The frequency-by-frequency comparison is necessary because the directional properties of the ear amplify sounds emanating from certain positions in space in a frequency-dependent manner (Grinnell and Grinnell 1965; Blauert 1969/1970; Flannery and Butler 1981; Fuzessery and Pollak 1984, 1985; Makous and O'Neill 1986; Jen and Sun 1984; Jen and Chen 1988; Musicant and Butler 1984). Thus, a particular frequency emanating from a given region of space off the midline will generate an IID of a particular value, but a different frequency originating from the same location will create a different IID. This feature is illustrated in the top panel of Figure 7.22 for the IIDs generated at the mustache bat's ear by three harmonics of its orientation calls, at 30 kHz, 60 kHz, and 90 kHz.

Consistent with the requirement that binaural cues be processed on a frequency-by-frequency basis is the finding that LSO cells are binaurally innervated from comparable regions of the cochlear surface, and thereby compare the intensity of a given frequency from one ear with the intensity of the same frequency from the other ear (Boudreau and Tsuchitani 1968; Caird and Klinke 1983; Semple and Kitzes 1985; Sanes and Rubel 1988). Once the computation is made by the LSO, the encoded information is conveyed bilaterally to the inferior colliculus and DNLL (see Fig. 7.2). The DNLL, in turn, projects bilaterally to the inferior colliculus (Casseday, Covey, and Vater 1988; Glendenning et al. 1981; Irvine 1986; Roth et al. 1978; Ross, Pollak, and Zook 1988). Both the DNLL (Brugge, Anderson, and Aitkin 1970; Covey 1993; Markovitz and Pollak 1993) and inferior

Distribution of IIDS

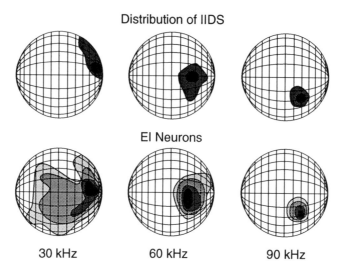

EI Neurons

30 kHz   60 kHz   90 kHz

FIGURE 7.22. Interaural intensity disparities (IIDs) generated by 30-, 60-, and 90-kHz sounds are shown in *top row*. Each sphere represents the bat's frontal sound field. The bat's head would lie in the center of each sphere. Each horizontal line represents a 20° increment in elevation, and each vertical line represents a 13° increment in azimuth. The *blackened areas* in each panel indicate the spatial locations of the largest IIDs (20 dB or greater), and the *grey areas* indicate those locations where the IIDs are at least 10 dB. The panels in the *lower row* show the spatial selectivity of three EI units, one tuned to 30 kHz, one to 60 kHz, and one to 90 kHz. The *blackened areas* indicate the spatial locations where the lowest thresholds were obtained. Isothreshold contours are drawn for threshold increments of 5 dB. Areas not contained within isothreshold contours indicate spatial locations from which sounds failed to evoke discharges. (Adapted from Fuzessery 1986.)

colliculus (Roth et al. 1978; Schlegel 1977; Semple and Aitkin 1979; Wenstrup, Fuzessery and Pollak 1988a,b; Irvine and Gago 1990) have large populations of EI cells.

Thus, the task of the auditory system is to encode each of the IIDs generated by the various frequencies of a complex sound. The location of a sound source, in both azimuth and elevation, must then be represented in the inferior colliculus by the pattern of activity among the population of EI neurons within each of the tonotopically organized isofrequency contours.

In the following sections, the 60-kHz contour of the mustache bat's inferior colliculus is used as a model to illustrate the manner in which IIDs are represented within an individual contour. Consideration of how the IIDs generated by other frequencies are represented in their contours leads to a hypothesis of how sound location, in both azimuth and elevation, is represented in the inferior colliculus. Finally, the role played by the inhibition from lower auditory nuclei for creating the representation of IIDs in the colliculus is discussed.

3.1 Directional Properties of the Ear and IIDs Generated by 60 kHz Along the Azimuth

Before describing the binaural and spatial properties of 60 kHz EI units, we consider how 60-kHz sounds are affected by the ear at different spatial locations. For simplicity, only spatial locations along the azimuth, at 0° elevation, are considered. The directional properties of the ear for 60 kHz are shown in the left panel of Figure 7.23. The noteworthy feature of the directional pattern is that the ear is most sensitive to sounds at about 26–40° from the midline, and that thresholds increase progressively as the source moves toward the midline and then into the sound field on the opposite side of the bat's head. The IIDs generated by 60 kHz were estimated by taking the difference in threshold between the two ears at mirror image locations in the ipsilateral and contralateral sound fields (right panel of Fig. 7.23). The largest IIDs originate at about 40° azimuth, the location at which the sound is most intense in one ear and least intense in the other. Notice that the IIDs change roughly linearly, by about 0.75 dB/degree, for locations ±40° around the midline. The importance of this linearity is that specifying the IID automatically specifies azimuthal location, but this is only true within ±40° around the midline and only when elevation is held constant. Beyond about 40°, the magnitude of the IIDs decline. In the following sections, we discuss the coding of IIDs within 40° of the midline. In later sections, we discuss the significance of the nonlinear IID changes that occur beyond 40° and the influence of elevation.

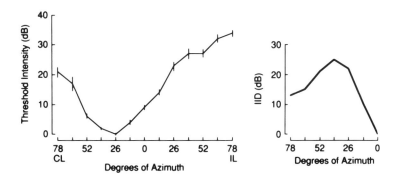

FIGURE 7.23. Measures of directional properties of the mustache bat's ear at 60 kHz *(left panel)* and the interaural intensity disparities (IIDs) generated by 60-kHz sounds *(right panel)*. Directional properties describe the changes in threshold of either the cochlear microphonic potential or monaural single units for a 60-kHz sound at different azimuthal positions in the acoustic hemifield. (Adapted from Wenstrup, Fuzessery, and Pollak 1988b.)

3.2 The Medial Border of Each 60-kHz EI Neuron is Determined by the Disparities Generated by the Ears and the Neuron's Inhibitory Threshold

When sounds are presented from various regions of space, EI neurons display a spatial selectivity, in that they discharge only when the sounds emanate from some regions of space and not from others. The regions of space from which discharges can be evoked constitute the neuron's receptive field (right panel of Fig. 7.24). The receptive fields of EI collicular units are characterized by a medial border in space. Sounds emanating from one side of the border, which correspond to more intense sound at the contralateral (excitatory) ear, evoke a more or less strong discharge rate depending on the signal intensity. Sounds emanating from the other side of the border, corresponding to locations closer to the ipsilateral (inhibitory) ear, evoke few or no discharges. The border is the location in space at which the discharge rate changes from one that is fairly vigorous to a much lower rate, or a complete inhibition.

To determine how binaural properties shape a neuron's receptive field, the binaural properties of 60-kHz EI cells were first determined with loudspeakers inserted into the ear canals, and subsequently the spatial properties of the same neurons were evaluated with free-field stimulation delivered from loudspeakers located around the bat's hemifield (Fuzessery and Pollak 1984, 1985; Wenstrup, Fuzessery, and Pollak 1988b; Fuzessery, Wenstrup, and Pollak 1990). By correcting for the directional properties of the ear at 60 kHz and the IIDs generated at each location, the quantitative aspects of the binaural responses could be associated with, and thus predict, the neuron's spatially selective properties (Wenstrup, Fuzessery, and Pollak 1988b). These studies showed that the receptive field of a 60-kHz EI neuron could be constructed from a family of IID functions. Each IID function was obtained when the intensity at the contralateral (excitatory) ear was held constant and the intensity at the ipsilateral (inhibitory) ear was progressively increased, as shown in the left panel of Figure 7.24. The neuron's receptive field was then constructed from a family of such IID functions, each obtained with a different sound intensity at the contralateral ear. The receptive field that was constructed from the IID functions was an accurate representation of the azimuthal receptive field of the same unit determined with free-field stimulation (Wenstrup, Fuzessery, and Pollak 1988b). These studies also showed: (1) that the medial borders and the interaural intensity disparities that are generated at the locations of those borders differ among EI cells; and (2) that the medial border and the corresponding IID correlate closely with, and thus can be predicted by, the neuron's inhibitory threshold (the IID at which the response declines by 50%) (Fuzessery and Pollak 1985; Wenstrup, Fuzessery, and Pollak 1988b).

That the medial border can be predicted by the cell's inhibitory threshold

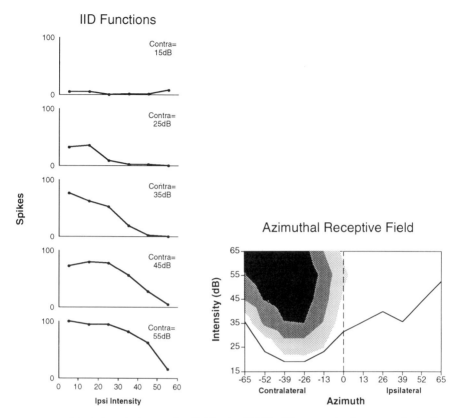

FIGURE 7.24. Azimuthal receptive field of a 60-kHz EI neuron in the mustache bat's inferior colliculus is shown on the *right*. *Blackened regions* show discharge rates ranging from 75% to 100% of the maximum rate. *Dark grey area* indicates discharge rates from 50% to 74% of maximum; *light grey areas*, regions of discharge rates from 25% to 49% of maximum. The *solid line* below the receptive field indicates the regions of space from which discharges could be evoked in a monaural neuron. The receptive field was constructed from the family of IID functions shown on the *left*. Each IID function was obtained by holding the intensity at the contralateral ear constant while changing the intensity at the ipsilateral ear. The algorithm described in Wenstrup, Fuzessery, and Pollak (1988b) was used to construct the receptive field.

is illustrated in Figure 7.25. This figure shows the receptive fields of three 60-kHz EI units, each having a different inhibitory threshold, and the receptive field of a monaural unit. As a starting point, consider the receptive field of the monaural neuron (Fig. 7.25A). Monaural neurons tuned to 60 kHz are always most sensitive at the same spatial position, about 26°–40° from the midline, which is a consequence of the directional properties of the ear for 60 kHz. Thus, a 60-kHz tone presented at about 40° in the contralateral sound field will have the lowest threshold, and a

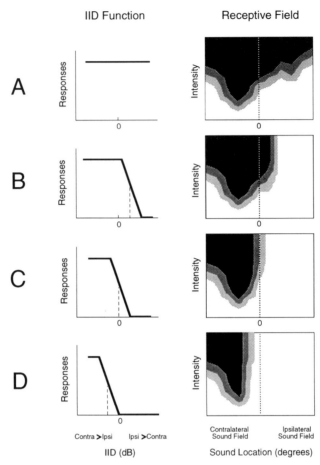

FIGURE 7.25A–D. Schematic shows influence of inhibitory thresholds of the IID functions on the azimuthal receptive fields of 60-kHz EI neurons. *Black area* in each receptive field shows locations and intensities that evoke discharge rates between 75% and 100% of maximum. *Dark grey* contours, firing rates between 50% and 74% of maximum; *light grey contour*, firing rates 25% to 49% of maximum. (A) The receptive field of a monaural neuron extends throughout the acoustic hemifield. The thresholds are lowest in the contralateral sound field and highest in the ipsilateral sound field because of the directional properties of the ear, together with the shadowing produced by the head and ears. (B) An EI neuron with a negative inhibitory threshold (sound at the ipsilateral ear has to be more intense than at the contralateral ear to produce a 50% reduction of discharge rate), The receptive field no longer extends throughout the ipsilateral sound field, because sounds presented from those regions are sufficiently intense at the ipsilateral ear to completely inhibit any discharges. (C) and (D) Two EI units with progressively more positive inhibitory thresholds have receptive fields that are correspondingly more limited to the contralateral sound field. (Based on model in Wenstrup, Fuzessery, and Pollak 1988b.)

low-intensity sound at that location will evoke a low discharge rate, as indicated by the lightly shaded area. Progressively louder sounds at that location will, in turn, elicit progressively higher discharge rates as indicated by the darker shading. If the sound source is moved to 40° in the ipsilateral sound field, a much more intense sound is needed to reach threshold because of the shadowing effect of the head and ears. To elicit the same range of discharge rates from the ipsilateral sound field as were evoked from the contralateral sound field, the sound in the ipsilateral field need only be made more intense. Thus, a monaural neuron can discharge over its entire dynamic range to sounds presented from any azimuthal location. However, the sound intensity required to evoke a given discharge rate varies with location because of the acoustic shadow cast by the head and ears and the directional properties of the ear. It is for these reasons that physical properties dictate the shape of the receptive field.

The receptive fields of EI neurons have the same general shape as monaural neurons, but with one difference; a portion of the receptive field in the ipsilateral sound field is cut off. The particular spatial location at which the cell's receptive field is cut off, that is, its medial border, is largely predicted by the cell's inhibitory threshold (Fig. 7.25B–D). The explanation is simply that the inhibitory projection that has been added from the other ear has a certain threshold, and when that threshold is exceeded, the inhibition begins to dominate. Thus, as the sound is moved around the acoustic hemifield, from the contralateral to the ipsilateral side, the sound initially drives only the excitatory ear, and thus evokes a discharge rate similar to the rate that would be evoked only by monaural stimulation. However, when the sound moves into the ipsilateral field, the increasing intensity at the ipsilateral ear will, at a certain location, exceed the threshold of the inhibitory neurons driven by sound at the ipsilateral ear. The activity evoked from the ipsilateral ear will then suppress the discharges evoked by sound at the contralateral ear. Thus, 60-kHz units with high inhibitory thresholds, such as the unit in Figure 7.25B, require a more intense stimulation of the inhibitory ear for complete inhibition, and therefore the medial borders of these units are in the ipsilateral sound field. In 60-kHz units with positive inhibitory thresholds, as in Figure 7.25 B,C, their medial borders occur along the midline or are even in the contralateral sound field. For these units, sounds presented ipsilateral to their borders are incapable of eliciting discharges, even with high-intensity stimulation. The important spatial feature of each EI unit receptive field is its medial border, and the particular azimuth of that border is determined principally by its inhibitory threshold.

3.3 The Systematic Arrangement of Inhibitory Thresholds Creates a Representation of IIDs

The finding that an EI neuron's inhibitory threshold determines where along the azimuth the cell's medial border is located, coupled with the orderly arrangement of inhibitory thresholds in the ventromedial EI region as presented previously, has implications for the representation of the azimuthal position of a sound in the mustache bat's inferior colliculus (Fuzessery and Pollak 1985; Wenstrup, Fuzessery, and Pollak 1988b; Pollak, Wenstrup, and Fuzessery 1986; Pollak and Casseday 1989). Specifically, the value of an IID is represented in the ventromedial 60-kHz contour as a border separating a region of discharging from a region of inhibited cells, as shown in Figure 7.26. Consider, for instance, the pattern of activity in the 60-kHz contour of one inferior colliculus generated by a 60-kHz sound that is 15 dB louder in the ipsilateral ear than in the contralateral ear. The IID in this case is −15 dB. Because neurons with positive inhibitory thresholds are situated ventrally, the high relative intensity in the ipsilateral (inhibitory) ear will inhibit all the EI neurons in the ventral portion of the contour. The same sound, however, will not be sufficiently intense at the ipsilateral ear to inhibit the EI neurons in the more dorsal portion of the contour, where neurons require an even more intense ipsilateral sound for inhibition. The topology of inhibitory thresholds and the steep IID functions of EI neurons, then, can create a border between excited and inhibited cells within the ventromedial region of the contour. The locus of the border, in turn, should shift with changing IID, and therefore should shift correspondingly with changing sound location (Fig. 7.26, lower panel).

3.4 Spatial Properties of Neurons Tuned to Other Frequencies

The binaural processing found in the 60-kHz contour appears to be representative of binaural processing within other isofrequency contours (Fuzessery and Pollak 1984, 1985). This feature is most readily appreciated by considering the spatial properties of EI units tuned to 90 kHz, the third harmonic of the mustache bat's orientation calls (see Fig. 7.22, lower panel

---→

FIGURE 7.26. *Top panel:* Schematic of systematic changes in the medial borders of the receptive fields of EI neurons in the 60-kHz contour of the inferior colliculus. Shifts in the medial borders result from the progressive shifts of the inhibitory thresholds of EI neurons along the dorsoventral axis of the 60-kHz contour. *Bottom panel*: Stylized illustration of relationship between the value of the interaural intensity disparity produced by a sound source at a given location (moths at the top of the panel) and the pattern of activity in the ventromedial EI region of the left colliculus, where IID sensitivities are topographically organized. The activity in this region, indicated by the *blackened area*, spreads ventrally as a sound source moves from the ipsilateral (−15 dB IID) to the contralateral sound field (+15 dB IID).

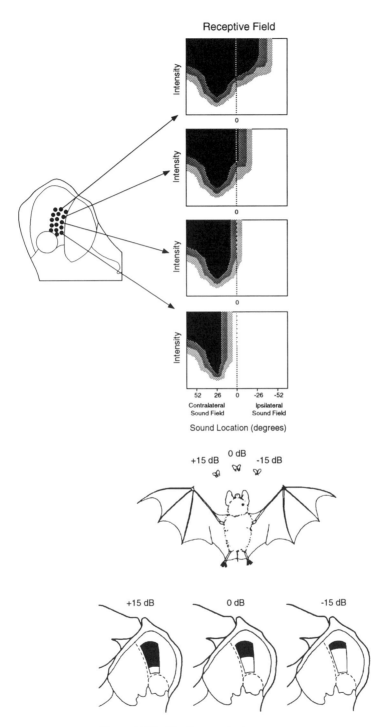

FIGURE 7.26. Caption on facing page.

on right). The significant feature is that the maximal interaural intensity disparities generated by 90-kHz tones occur in a region of space that is different from the region that generates the maximum disparity with 60-kHz tones. The largest disparities obtained for 90 kHz are at about 40° along the azimuth and − 40° in elevation (Fig. 7.22, top panel on right). The 90-kHz EI neurons, like those tuned to 60 kHz, are most sensitive to sounds presented from the same spatial location at which the maximal interaural intensity disparities are generated. Additionally, the inhibitory thresholds of these neurons determine the azimuthal border defining the region in space from which 90-kHz sounds can evoke discharges from the region where sounds are incapable of evoking discharges (Fuzessery and Pollak 1985). The population of 90-kHz EI neurons has a variety of inhibitory thresholds that appear to be topographically arranged within that contour (Wenstrup, Ross, and Pollak 1986b). Therefore, a particular interaural intensity disparity will be encoded by a population of 90-kHz EI cells, having a border separating the inhibited from the excited neurons, in a fashion similar to that shown for the 60-kHz contour. The same argument can be applied to the 30-kHz cells, but in this case the maximal interaural intensity disparity is generated from the very far lateral regions of space (see Fig. 7.22, top panel on left).

3.5 The Representation of Auditory Space in the Mustache Bat's Inferior Colliculus

We can now begin to see how the cues that define both the azimuth and elevation of a sound source are derived. The directional properties of the ears generate different interaural intensity disparities among frequencies, an interaural spectral difference, when a broadband sound emanates from a particular location. The interaural spectral difference will evoke a specific pattern of activity across each of the isofrequency contours in the bat's midbrain that are driven by the spectral differences received at the ears. Figure 7.27 shows a stylized illustration of the interaural intensity disparities generated by 30, 60, and 90 kHz within the mustache bat's hemifield, and below is shown the loci of borders in the EI region of each contour that would be generated by a biosonar signal containing the three harmonics emanating from different regions of space. Consider first a sound emanating from 40° along the azimuth and 0° elevation (Fig. 7.27, white circle in top panel). This position creates a maximal interaural intensity disparity at 60 kHz, a lesser interaural intensity disparity at 30 kHz, and yet a different interaural intensity disparity at 90 kHz. The borders created within each of the isofrequency contours by these interaural intensity disparities are shown in the bottom panel of Figure 7.27. Next, consider the interaural intensity disparities created by the same sound, but from a slightly different position in space, at about 60° azimuth and − 20°

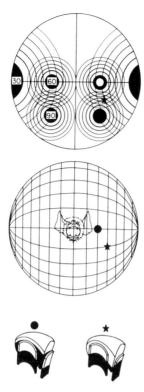

FIGURE 7.27. Loci of borders in the 30-, 60-, and 90-kHz EI regions generated by biosonar signals emanating from two regions of space. The *top panel* is a schematic representation of IIDs produced by 30-, 60-, and 90-kHz sounds that occur in the mustache bat's acoustic field. *Blackened areas* indicate the regions in space where the maximum IID is generated for each harmonic. The *circular lines* surrounding each blackened region indicate the spatial locations that produce iso-IIDs for that frequency. Each successive circle indicates a progressively smaller IID. The *middle panel* depicts the bat's head and the spatial positions of two sounds (*star* and *circle*), where each sound is composed of the three harmonics that have equal intensities. The *lower panel* shows the borders separating the regions of excited from regions of inhibited neurons in the 30-, 60-, and 90-kHz contours that would result from the IIDs generated by the sounds at the two locations. (Reproduced from Pollak and Casseday 1989.)

elevation (Fig. 7.27, star in top panel). In this case, there is a decline in the 60-kHz interaural intensity disparity, and an increase in the 90-kHz interaural intensity disparity, but the 30-kHz interaural intensity disparity will be the same as it was when the sound emanated from the previous position. The fact that the interaural intensity disparity at 30 kHz did not change although the location of the sound source changed is a crucial point. It illustrates that the interaural intensity disparity generated by a particular frequency is not uniquely associated with one position in space, but rather can be generated from a variety of positions. It is for this reason that sound localization with only one frequency is ambiguous (Blauert 1969/1970; Butler 1974; Musicant and Butler 1984; Fuzessery 1986). However, spatial location, *in both azimuth and elevation,* is rendered unambiguous by the simultaneous comparison of three interaural intensity disparities, because their values in combination are associated with a unique spatial location in the bat's acoustic hemifield, except for one special region of space.

The exception is the vertical midline, at 0° azimuth. The representation by EI regions becomes ineffective along the vertical midline because the interaural intensity disparities will be 0 dB at all frequencies and all elevations. The borders among the EI populations, therefore, will not change with elevation because the interaural intensity disparities remain constant. Several types of binaural cells may be important for the encoding of elevation along the midline (Fuzessery and Pollak 1985; Fuzessery 1986; Fuzessery, Wenstrup, and Pollak 1990). One type is the EI/f cell (Fig. 7.28). The distinguishing feature of EI/f cells is that as sound intensity increases at the ipsilateral (inhibitory) ear, the increase initially causes the neuron's firing rate to *increase* by at least 25% above the rate evoked by the sound at the contralateral (excitatory) ear alone (Fig. 7.28, left panel). Additional intensity increases at the ipsilateral ear then result in a marked decline in

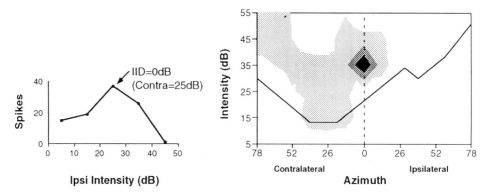

FIGURE 7.28. IID function *(left)* and azimuthal receptive field *(right)* of an 60-kHz EI/f unit recorded from the mustache bat's inferior colliculus. The IID function on the left was obtained when the contralateral intensity was fixed at 25 dB SPL and the intensity at the ipsilateral ear was varied from 5 to 45 dB. With low intensities at the ipsilateral ear (5 and 15 dB), the neuron's discharge rate was equal to that evoked when the 25-dB signal was presented only to the contralateral ear (not shown). When the ipsilateral intensity increased to 25 dB, generating an IID of 0 dB, the discharge rate increased markedly. However, additional intensity increments at the ipsilateral ear then resulted in a progressively greater inhibition. Note that the highest discharge rates could be evoked only from limited regions of space close to the midline. These regions generate the IIDs at which the facilitation is expressed in the IID functions. The receptive field was constructed from a family of IID functions (not shown), each generated with a different intensity at the contralateral ear.

response rate, as in regular EI neurons. The receptive fields of EI/f cells are variations of the fields of EI cells (Fuzessery and Pollak 1985; Fuzessery, Wenstrup, and Pollak 1990). Like EI cells, they respond poorly or not at all to sounds located in the ipsilateral acoustic field, and their borders are determined by the cell's inhibitory threshold.

Unlike EI cells, the discharge rates of EI/f cells are enhanced for sound locations that generate the IIDs at which they are facilitated. The IIDs producing facilitation are usually at, or close to, 0 dB and thus occur around the midline. EI/f cells, then, respond most vigorously to sounds located directly ahead as a consequence of the binaural facilitation. However, the elevation at which they discharge maximally depends on their best frequency. EI/f units tuned to different frequencies exhibit selectivities for different elevations; EI/f units tuned to 60 kHz are maximally sensitive at about 0° to −10° elevation, whereas units tuned to 90 kHz are most sensitive at about −40° (Fuzessery and Pollak 1985; Fuzessery, Wenstrup, and Pollak 1990). The elevation selectivity is a consequence of the directional properties of the ear for 30, 60, and 90 kHz. One can, therefore, visualize that as the elevation of a sound source along the vertical meridian shifts, the response magnitude also shifts among the EI/f units in different isofrequency contours. In combination then, these two binaural types could provide a neuronal representation of sound located anywhere within the bat's acoustic hemifield.

A striking feature of the foregoing scenario is its similarity, in principle, to the ideas about sound localization proposed previously by Pumphery (1948) and Grinnell and Grinnell (1965). What the recent studies provide are the details of how interaural disparities are encoded and how they are topologically represented in the acoustic midbrain.

3.6 A Wiring Diagram of the Excitatory and Inhibitory Projections to the Inferior Colliculus Can Be Constructed

The previous sections showed how the spatial properties of 60-kHz EI neurons are shaped by both the directional properties of the ear and the inhibition provided by the projections activated by stimulation of the ipsilateral ear. In the following sections, the nuclei that provide the inhibition evoked by the ipsilateral ear are identified, and the role that the inhibitory neurons in each of those nuclei could play in shaping the binaural properties and receptive fields of EI neurons in the inferior colliculus are discussed. This issue is addressed in two stages. In the first stage, the origins of the GABAergic and glycinergic inhibitory neurons that innervate the EI region of the colliculus are determined. In the second stage, the changes in binaural properties and receptive fields of EI neurons are described after inhibitory inputs are blocked by pharmacological antagonists applied iontophoretically.

The identification of inhibitory projections takes advantage of the fact that the majority of auditory nuclei have large numbers of projection neurons that are inhibitory (Adams and Mugniani 1984; Mugniani and Oertel 1985; Saint Marie et al. 1989; Park et al. 1991; Vater et al. 1992; Winer, Larue, and Pollak 1995). The prevalence of long-distance inhibitory projections is illustrated in Figure 7.4, which shows a section through the mustache bat's auditory brainstem that has been reacted with antibodies against the inhibitory neurotransmitter glycine. It is apparent from this photomicrograph that there is a massive glycinergic projection system ascending in the lateral lemniscus which innervates the inferior colliculus. As we pointed out previously, there are also strong GABAergic projections to the inferior colliculus (see Fig. 7.4). Determining which nuclei send glycinergic or GABAergic cells to the EI region of the colliculus was accomplished by injecting HRP in the 60-kHz isofrequency contour and then colocalizing the HRP reaction product from retrograde transport in sections that were stained alternately with antibodies against GABA and glycine (Larue et al. 1991; Park et al. 1991). Data of this sort show not only which cells project to the 60-kHz region, but also whether the cells are GABAergic, glycinergic, or neither, and thus are probably excitatory. If these findings are considered together with those from earlier studies, which revealed the particular subset of nuclei that project to the EI region, not only can the wiring diagram of the EI region be constructed but in addition an excitatory or inhibitory role can also be assigned to each projection. This neurotransmitter wiring diagram is shown in Figure 7.29.

Because the excitatory/inhibitory projection system involves both contralateral and ipsilateral structures, as well as crossed and uncrossed projections, we hereafter refer to structures ipsilateral to the left colliculus as being on the left, and structures contralateral to that colliculus as being on the right. Hence, EI cells in the left colliculus are excited by sound at the right (contralateral) ear and inhibited by sound at the left (ipsilateral) ear. The relevant features of Figure 7.29 concern the inhibitory projections that are activated by the (left) ipsilateral ear, because it is the inhibition evoked by stimulation of that ear which shapes the receptive fields of EI neurons in the left colliculus. Four nuclei are especially important: (1) the left (ipsilateral) LSO; (2) the right (contralateral) LSO; (3) the left (ipsilateral) DNLL; and (4) the right (contralateral) DNLL.

Turning first to the right (contralateral) LSO, this nucleus receives excitation from the right ear and inhibition from the left ear, which generates the IE properties of the cells in this nucleus. Indeed, LSO neurons in all mammals are overwhelming inhibitory/excitatory (Boudreau and Tsuchitani 1968; Caird and Klinke 1983; Sanes and Rubel 1988: Covey, Vater, and Casseday 1991). The right (contralateral) LSO, however, sends a putative excitatory projection to the contralateral (left) colliculus (Saint Marie et al. 1989; Park et al. 1991; Glendenning et al. 1992), and through this excitatory projection it can impose EI properties upon its targets in the

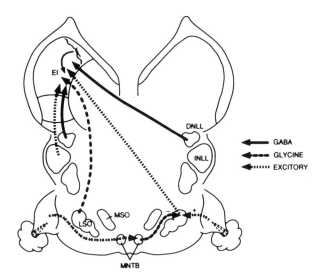

FIGURE 7.29. Wiring diagram shows sources of GABAergic, glycinergic, and excitatory projections to the 60-kHz EI region of the mustache bat's inferior colliculus. DNLL, dorsal nucleus of the lateral lemniscus; INLL, intermediate nucleus of the lateral lemniscus; LSO, lateral superior olive; MSO, medial superior olive; MNTB, medial nucleus of the trapezoid body.

left colliculus. The left (ipsilateral) LSO, in contrast, receives excitation from the left ear and inhibition from the right ear. However, the left LSO sends a glycinergic projection to the colliculus on the same side (Saint Marie et al. 1989; Park et al. 1991; Glendenning et al. 1992). It, therefore, can provide inhibition to the left colliculus when the left ear is stimulated. The DNLL, like the LSO, has mostly excitatory/inhibitory neurons (Brugge, Anderson, and Aitkin 1970; Covey 1993; Markovitz and Pollak 1993). Thus, stimulation of the left (ipsilateral) ear will provide an excitatory drive to neurons of the right DNLL. Because these neurons are GABAergic, they can, in turn, provide a potent inhibition to the left colliculus when the left (ipsilateral) ear is stimulated. This is in contrast to the left (ipsilateral) DNLL, which would be inhibited by sound presented to the left ear and excited by sound presented to the right ear. Thus, the left DNLL could provide inhibition to the left colliculus when sound was presented to the right ear. The effects of each of these complex connections are explained in greater detail in the following sections.

3.7 Effects of GABAergic and Glycinergic Innervation on Spatial Receptive Fields Can Be Evaluated

Armed with this information, we assessed the influences of inhibitory inputs on the binaural response properties of 60-kHz collicular EI neurons with microiontophoretic application of the GABA$_A$ receptor antagonist bicuculline and the glycine receptor antagonist strychnine. The rationale is that if EI properties are created in the colliculus by the convergence of excitatory and inhibitory inputs, then removing the influence of the inhibitory inputs should reduce or eliminate the acoustically evoked inhibition. On the other hand, if EI response properties are created in a lower nucleus and imposed upon the collicular cell via an excitatory projection, then the blockade of inhibitory inputs at the colliculus should have no effect on the expression of those response properties. We constructed the receptive field from the binaural properties obtained before blocking inhibition, and then compared it to the receptive field constructed from the binaural properties obtained when GABAergic inhibition was blocked by the application of bicuculline, or when glycinergic inhibition was blocked by the application of strychnine. The feature of interest is how inhibition evoked by the left (ipsilateral) ear limits the degree to which the receptive field extends into the left (ipsilateral) sound field. In this way, the effect of GABAergic inhibition on the spatial properties of an EI collicular neuron can be compared to the effects of glycinergic inhibition, and informed inferences can then be made as to which of the lower nuclei produced each of the effects.

3.8 Binaural Properties of EI Neurons Are Formed in at Least Five Ways

These studies show that a variety of EI properties are formed in the inferior colliculus by the mixing and matching of the various excitatory and inhibitory circuits that converge upon these binaural cells. The EI properties, and hence the features of receptive fields, are formed in at least five ways. The first way is through the complete generation of EI properties in the right (contralateral) LSO that are then imposed, without modification, on the left (ipsilateral) inferior colliculus via the crossed, excitatory projection from the LSO to the inferior colliculus. Because neither GABAergic nor glycinergic innervation of the inferior colliculus contributes to the binaural properties of the collicular cell, blocking either of these inhibitory transmitters should have no effect on the neuron's receptive field. An example of a neuron exhibiting these features, and the circuitry that could account for the absence of any changes in the receptive field of the cell, are shown in Figure 7.30.

The second way is more subtle, and involves a shift of the neuron's inhibitory threshold caused by GABAergic inhibition. Such shifts were seen in cells in which blocking GABAergic inhibition did not reduce the degree of inhibition evoked by the left (ipsilateral) ear, but rather the blockage caused the neuron's inhibitory threshold to shift to a more positive value (Park and Pollak 1993a). Figure 7.31 shows an example of a unit whose

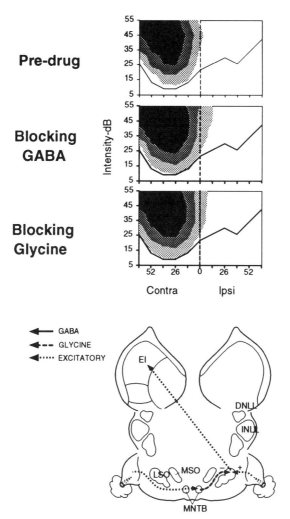

FIGURE 7.30. Example of a 60-kHz EI neuron whose receptive field was unaffected by either GABAergic or glycinergic inputs. *Top panel* shows the receptive field before drugs were applied; the *lower panels* show receptive fields obtained after blocking GABAergic and glycinergic inputs. The receptive field properties were most likely created in the lateral superior olive on the opposite side, and imposed upon the collicular neuron via the excitatory projection to the colliculus, as shown in the wiring diagram at the bottom of the figure. Thus, blocking either the GABAergic or glycinergic inputs had no effect on the cell's spatial receptive field.

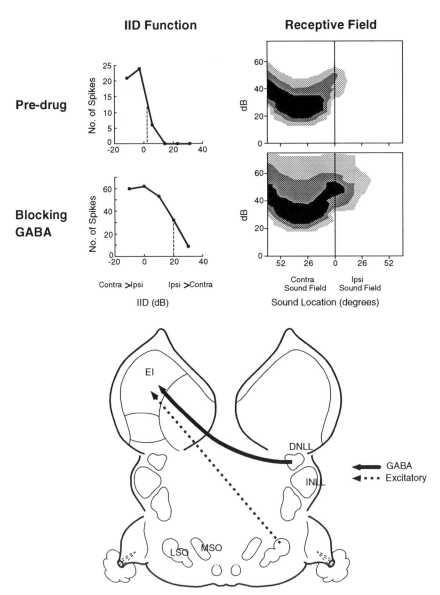

FIGURE 7.31. Example of a neuron in which blocking GABAergic inhibition had little effect on the magnitude of the ipsilaterally evoked inhibition, but caused a shift of 14 dB in its inhibitory threshold. The predrug IID function and the IID function obtained when GABAergic inhibition was blocked are shown in the *left panels*. Inhibitory thresholds are indicated by *dashed lines*. The changes in the azimuthal receptive fields are shown in the *right panels*. Note the expansion of the receptive field into the ipsilateral sound field by blockage of GABAergic inhibition. The *lower panel* shows a possible circuit that could generate these features. The simplest interpretation of such shifts is that in the normal condition there are at least two

inhibitory threshold changed by 14 dB because of the blockage of GA-BAergic inhibition and the effects of those changes on the unit's receptive field.

The simplest explanation for these effects is that the right (contralateral) LSO imparts its EI properties, and thus its inhibitory threshold, on the collicular cell, as was the case for the neuron in Figure 7.30. However, the inhibitory threshold of the LSO cell is shifted in the colliculus by the GA-BAergic inhibition evoked by stimulation of the left (ipsilateral) ear. The most likely source of that inhibition is the right (contralateral) DNLL, because its cells are excited by sound at the left (ipsilateral) ear and inhibited by sound at the right ear (Fig. 7.31, lower panel). This explanation is also supported by a recent study by Li and Kelly (1992a) in the rat. They blocked the DNLL pharmacologically while recording from EI cells in the colliculus and noted that the inhibitory thresholds (50% point IIDs) of many EI cells shifted because of the inactivation of the DNLL.

The third way is that the binaural properties are created entirely in the inferior colliculus by a monaural, excitatory projection from the right (contralateral) ear that is sculpted by GABAergic inhibition evoked from stimulation of the left (ipsilateral) ear (Faingold, Gelhbach, and Caspary 1989; Faingold, Boersma-Anderson, and Caspary 1991; Li and Kelly 1992a,b; Park and Pollak 1993a). These features are illustrated by the EI neuron in Figure 7.32. This neuron had a fairly sharply confined spatial receptive field before the application of any drug, but after blocking GABAergic inputs with bicuculline all the inhibitory inputs from the

---

FIGURE 7.31. (continued). projections that form the IID function in these collicular cells, one of which is the GABAergic projection from the contralateral DNLL. The other projection is from the contralateral LSO where the EI property is initially created. The LSO projection establishes both the strength of the inhibition and an inhibitory threshold is achieved when the intensity at the ipsilateral ear is substantially greater than the intensity at the contralateral ear. This projection is excitatory and therefore is not affected by iontophoresis of bicuculline. The second projection is GABAergic and innervates the same collicular cell as the LSO projection. The GABAergic circuit from the contralateral DNLL is driven only by stimulation of the ear ipsilateral to the IC, and has a lower absolute threshold than the inhibitory input to the LSO. Thus, the effect of the GABAergic circuit is to change the IID function at the colliculus, where it causes the discharge rate to decline with lower intensities at the ipsilateral ear than did the IID function created in the LSO. In short, the resultant IID function, caused by the summation of these circuits, has an inhibitory threshold that is shifted to a more positive IID (requires a less intense stimulus at the ipsilateral ear to inhibit the cell) but has the same maximum inhibition that was initially established in the LSO cell. The effect of bicuculline is to remove the influence of the GABAergic projection from the DNLL, thereby allowing the maximum inhibition and inhibitory threshold of the projection from the LSO to be expressed by the collicular cell.

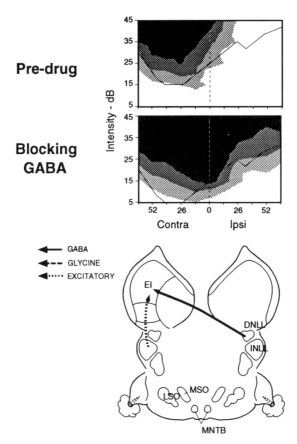

FIGURE 7.32. Changes in the receptive field of a 60-kHz EI neuron caused by blockage of GABAergic inputs. Notice that after blocking GABAergic inputs the receptive field expanded into the ipsilateral sound field and was similar to the receptive fields seen in monaural units. The wiring diagram (*below*) shows the circuit that could account for the expansion of the field. (See text for further explanation.)

ipsilateral ear were eliminated and the cell was rendered monaural. These charges arc sccn in the neuron's binaural properties (not shown) as well as its receptive field. Thus, the binaural properties of this cell were created entirely in the inferior colliculus by the convergence of excitatory projections from the right (contralateral) ear and GABAergic projections that most likely emanated from the right DNLL (Fig. 7.32). Because all inhibition evoked by stimulation of the ipsilateral ear was eliminated by bicuculline, no role for glycinergic projections needs to be proposed.

The EI neuron in Figure 7.33 illustrates the fourth way in which inhibitory circuits that converge on a collicular cell can shape its receptive field. In this case, the binaural properties and the receptive field properties were formed in the colliculus by both GABAergic and glycinergic inputs.

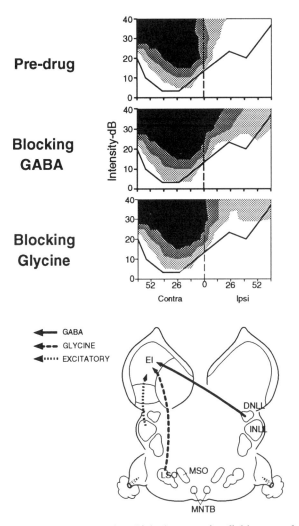

FIGURE 7.33. A 60-kHz EI neuron in which the receptive field was sculpted by both GABAergic and glycinergic inputs. (See text for further explanation.)

This neuron, like the previous one, had a receptive field sharply confined to the contralateral sound field. Blocking GABAergic inputs with bicuculline reduced the efficacy of the inhibition evoked by the ipsilateral ear, although it did not completely eliminate it. Because the inhibition evoked by the ipsilateral ear was reduced, the receptive field expanded into the ipsilateral acoustic hemifield. The unit, however, did not become completely monaural, because the highest discharge rates (shown in black) were still somewhat restricted to one side of the sound field. Effects similar to those produced by blocking GABA were observed when glycinergic inputs were blocked with strychnine. The receptive fields for both drug conditions show that discharges could be evoked from sound locations that were inhibitory in the predrug condition (Fig. 7.33). For neurons like this, GABAergic and glycinergic inputs appear to work in tandem to sculpt a receptive field from a monaural excitatory input from the right (contralateral) ear. The most likely source of the GABAergic inhibition is again the contralateral DNLL, and the source of the glycinergic inhibition is almost certainly the ipsilateral (left) LSO, because the cells in both these nuclei are excited by stimulation of the left (ipsilateral) ear and thus could provide the appropriate inhibition to their targets in the left colliculus. The circuit likely to produce the receptive field of this neuron is shown in the lower panel of Figure 7.33.

The fifth way that inhibition shapes receptive fields is one of the most interesting, and concerns the binaural facilitation of EI/f cells. Blocking GABA eliminated the binaural facilitation in many EI/f units independent of changes in maximum inhibition (Park and Pollak 1993a). Thus, binaural facilitation was reduced or lost in some cells in which the inhibition evoked by the left (ipsilateral) ear was also reduced or abolished after blocking GABA, whereas in other cells the ipsilaterally evoked inhibition was affected minimally or not at all. An example of an EI/f cell in which blocking GABAergic inhibition hardly affected the inhibition evoked by the left (ipsilateral) ear but abolished the facilitation is shown in Figure 7.34. For purposes of clarity, a circuit in which GABAergic inhibition could shape the facilitated response of an EI/f cell whose ipsilaterally evoked inhibition and inhibitory threshold are not affected by bicuculline is shown in Figure 7.35. The circuit has two components: a projection from the right LSO, which is responsible for generating the inhibition evoked by the left, ipsilateral ear and thus the cell's receptive field border, and a GABAergic cell in the *left* (ipsilateral) DNLL that generates the facilitation. One of the interesting features of this explanation is the inhibition is the proposed agent for producing the "facilitation." The explanation of how this circuit could produce binaural facilitation in an EI/f cell of the inferior colliculus is presented in Figure 7.35.

3.9 Summary

The studies reviewed here are beginning to reveal how neurons in the ICc integrate the signals from a number of parallel pathways to produce an output that is a synthesis of the information conveyed from those sources. The specific form of the neuron's output, in this case its binaural properties, is a reflection of the particular subset of inputs that converge upon that neuron. By mixing and matching different combinations of inputs, a variety of effects can be achieved. For example, the convergence of the inputs from the contralateral DNLL and contralateral LSO can adjust the expression of

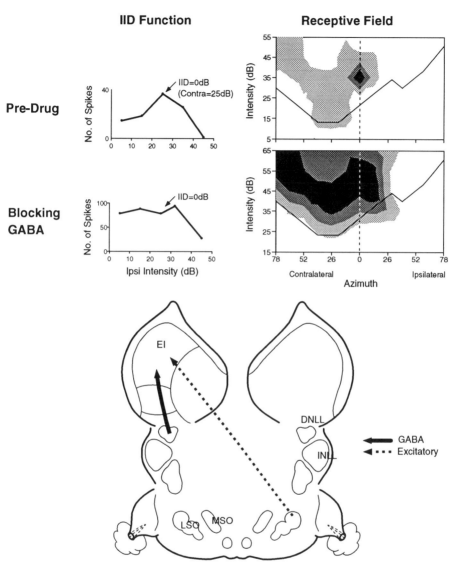

FIGURE 7.34. Change in the receptive field of an EI/f cell from blockage of GABAergic inhibition. The IID functions in the *left panels* show that blocking GABA eliminated the ipsilaterally evoked facilitation, but had only minimal effect on the magnitude of the ipsilaterally evoked inhibition. Thus, the effect of blocking GABAergic inhibition was to transform both the facilitated IID function and the focused activity of the spatial receptive field into those of a conventional EI cell. The simplest circuit that could generate these features would involve two projections: one from the contralateral LSO that generates the EI property, and another from the ipsilateral DNLL that generates the facilitation. A detailed description of how this could work is given in Figure 7.35.

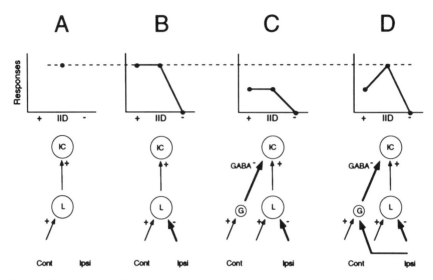

FIGURE 7.35. Construction of a circuit that could create binaural facilitation. (A) An excitatory input from the contra ear that makes a synaptic connection with cell L in a lower nucleus that drives the colliculus cell (IC) via an excitatory projection. The discharge rate of the IC cell is shown in the top graph. (B) An inhibitory input from the ipsi ear is added to L, making it EI. The IC cell thus also becomes EI (IID function in top panel). (C) A GABAergic input to the IC cell that originates from cell G is added next. G receives excitation from the contra ear. Sound at contra ear simultaneously evokes an excitation, via L, and an inhibition in the IC cell via G. Adding sound to the ipsi ear generates an IID function with a lower overall response rate than L, but has the same maximum inhibition. (D) Lastly, an inhibitory input evoked by the ipsi ear is added to G, making it EI. G has a *lower* 50% point than L. Sound at the contra ear alone at 10 dB above threshold, or with a subthreshold ipsi intensity, evokes the same reduced discharge rate as in panel C. When the intensity at the ipsi ear is increased, so that the inhibitory input is above threshold at G but is still below threshold for L, L still excites the IC cell. Since G is inhibited, and thus no longer imparts an inhibition at the IC, the discharge rate of the IC cell increases and expresses facilitation. Higher ipsi intensities now also suppress discharges in L. As the discharge rate of L falls to zero, so does the discharge rate of the IC cell. (Adapted from Park and Pollak 1993a.)

the inhibitory threshold of the LSO cell in its collicular target. An entirely different effect is achieved by the convergence of projections from the ipsilateral DNLL and contralateral LSO. Such a convergence does not shift the inhibitory threshold of the collicular target cell, but rather creates a new binaural property, a "facilitation" as expressed by the EI/f neurons. Other types of convergence create the binaural property in ICc neurons by combining inputs from a monaural, excitatory projection and an inhibitory projection from the contralateral DNLL. Such combinations are almost

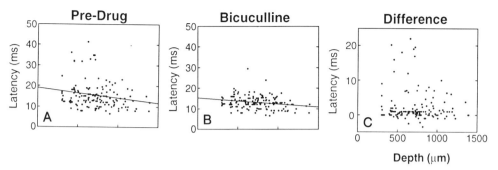

FIGURE 7.36A–C. Effects of blocking GABAergic inhibition with bicuculline on the pattern of latency with depth for 142 cells recorded from the 60-kHz region of the mustache bat inferior colliculus. (A) Plot of latency as a function of depth before bicuculline application. (B) Latencies of the same cells during application of bicuculline. (C) Difference in latency for each unit before and during blockage of GABAergic inhibition as a function of depth. (Adapted from Park and Pollak 1993b.)

certainly further affected by glycinergic input from the ipsilateral LSO, whose influence on binaural processing is not well understood at the present time.

4. Mechanisms That Create the Latency Distribution in the 60-kHz Contour and Some Functional Implications of That Distribution

In the first section of this chapter we showed that discharge latencies have a characteristic distribution in the 60-kHz contour of the inferior colliculus. Neurons in the dorsal colliculus have, on average, longer latencies than neurons located ventrally, and the range of latencies changes markedly with depth (Figs. 7.13 and 7.36) (Park and Pollak 1993b). Dorsally there is a broad distribution of latencies while deeper regions have a much narrower range of latencies. Long-latency cells are found only dorsally whereas short-latency cells are found at all depths. An orderly arrangement of latencies is apparently a general feature of collicular isofrequency contours, because it occurs not only in the mustache bat's inferior colliculus but in isofrequency contours of the cat's inferior colliculus as well (Schreiner and Langer 1988). The significance of such a distribution for information processing must be considerable, although at the present time it is not possible to exactly specify how different latencies affect binaural or other forms of processing. On the other hand, the ordering of latencies in collicular isofrequency contours could provide the topographical substrate for the generation of the combinatorial properties of neurons in the mustache bat's medial geniculate body. Before discussing this issue, we show that GABAergic inhibition is an important mechanism for producing the wide range of latencies expressed by collicular neurons, and that this inhibition apparently creates the particular form of the latency arrangement within the 60-kHz contour.

4.1 Effects of GABAeric Inhibition on Latency

GABAergic inhibition modifies the latencies of many, although not all, collicular neurons (Pollak and Park 1993; Park and Pollak 1993b). In neurons whose latencies are affected by inhibition, the effect is to increase latency thereby lengthening the time period between the presentation of a stimulus and the appearance of discharges. The change in latency is revealed when GABAergic inhibition is blocked by the application of bicuculline. In more than half of the neurons in the 60-kHz contour, blocking GABAergic inhibition with bicuculine caused latencies to shorten by 1–5 msec, and in about 20% of the population, latencies shortened by 5–30 msec. Examples of each type are shown in Figure 7.37.

Of significance is that the majority of neurons that had large (>8 msec) changes in latency with bicuculline were located in the dorsal region of the 60-kHz contour (Park and Pollak 1993b). Very few neurons in the more ventral regions of the colliculus had large bicuculline-induced latency changes. These features are illustrated by the four neurons in Figure 7.37 and by the three graphs in Figure 7.36. The top panel of Figure 7.36 shows the distribution of latencies as a function of depth before bicuculline was applied. The graph in the middle panel of Figure 7.36 shows the latency distribution during the application of bicuculline. The most apparent feature is the absence of neurons in the dorsal colliculus, between 200–600 μm from the surface, that had latencies longer than 20 msec during the application of bicuculline. There was also a reduction in the range of latencies at more ventral levels. At levels between 600 and 1000 μm, there was a smaller number of neurons with long latencies than in the predrug graph, although two neurons still had latencies well above 20 msec. In the most ventral colliculus, at depths of 1000–1400 μm, there was a reduction in the range of latencies, but the overall latency change from bicuculline was smaller than in the more dorsal regions.

The difference between the latencies before and during the application of bicuculline for neurons at the various levels of the colliculus is plotted graphically in the bottom panel of Figure 7.36. Here the largest changes in latency are clearly apparent for the neurons in the dorsal colliculus (200–600 μm from the surface). The smaller and moderate latency changes can also be seen for neurons at midlevels of the colliculus, while the neuronal population located ventrally, from about 1000–1400 μm, had, on average, the smallest latency change. It is also noteworthy that the latencies of large

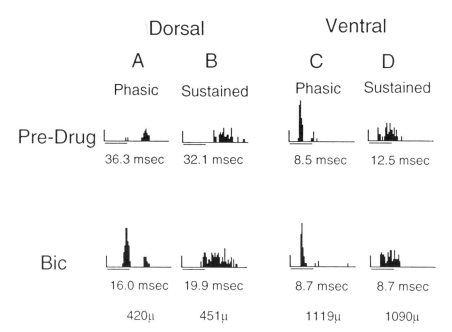

FIGURE 7.37A–D. Peristimulus time histograms of two neurons with sustained response patterns and two units with phasic discharge patterns from different dorsoventral locations in the 60-kHz region of the mustache bat's inferior colliculus. For each unit, the *top graph* shows responses before bicuculline (Bic) application; the *bottom graph* shows responses during the application of bicuculline. Numbers below the graphs are the median latency to the first spike in milliseconds and the depth of the unit (distance from the dorsal surface) in micrometers. (A) Phasic cell from dorsal colliculus. (B) Sustained cell from dorsal colliculus. (C) Phasic cell from ventral colliculus. (D) Sustained cell from ventral colliculus. Notice that the latencies of both the phasic and sustained neurons recorded from the dorsal colliculus (A and B) changed substantially with bicuculline, whereas the latencies of the phasic and sustained cells in the ventral colliculus changed very little. (Adapted from Park and Pollak 1993b.)

numbers of neurons at all levels of the colliculus were not changed by bicuculline.

The neurons in lower nuclei that send excitatory projections to the midbrain appear to have a much narrower range of latencies than the population of collicular neurons (Markovitz and Pollak 1993; Covey and Casseday 1991; Covey, Vater, and Casseday 1991 Haplea, Covey and Casseday 1994; Yang and Pollak 1994). The reason for the amplified range of collicular latencies is that some of the cells in the dorsal region of the contour, which receive fairly short latency excitatory innervation, are transformed into longer latency cells by GABAergic inhibition. Thus, the latencies of the cells receiving the appropriate inhibitory inputs are changed from the short-latency cells, seen after GABA is blocked, to the longer latencies which the cells normally express because of inhibitory innervation by their GABAergic neighbors.

4.2 Orderly Arrangement of Latencies Has Implications for the Creation of Combination-Sensitive Neurons in the Medial Geniculate

As discussed previously, the orderly arrangement of response properties seems to be a general feature of collicular isofrequency contours. The orderly arrangement of response features suggests that a neuron's location within a contour could specify its response properties. This arrangement may be important for the mustache bat because of the remarkable response properties that are generated in the medial and dorsal divisions of the medial geniculate body by the convergence of projections from at least two frequency contours (Olsen and Suga 1991a,b; Buttman 1992). The convergence results in the emergence of combination-sensitive neurons in the medial geniculate, a feature that is not present in the inferior colliculus (O'Neill 1985). If the organization of response properties in the other frequency contours is comparable to those in the 60-kHz contour, as we have argued they are, then all the response features required to generate combinatorial properties are already established and arranged by locus in the various frequency contours of the inferior colliculus. These considerations led us to suggest that the appropriate combinatorial properties could be created by the convergence of projections from a given locus in one isofrequency contour and the projections from a different locus in another isofrequency contour. We present here two hypothetical circuits that could generate the combinatorial properties in geniculate neurons.

The circuits can potentially account for the two general types of combination-sensitive neurons that have been found in the medial geniculate (Olsen and Suga 1991a,b). The characteristic features of these neurons are described in detail by Wenstrup in Chapter 8, and by O'Neill in Chapter 9. One type is selective for the frequencies in the constant frequency (CF) components of the bat's orientation calls, the so-called CF/CF neurons. The CF/CF neurons are characterized by their weak discharges to individual tone bursts and the vigorous firing when two tone bursts having different frequencies are presented together. These neurons are also distinguished by their relative insensitivity to the temporal interval separating the presentation of the two frequencies. CF/CF neurons are tuned to two of the harmonics of the CF components of the bat's orientation calls; thus, CF_1/CF_2 neurons are driven only when the fundamental and second harmonic of the CF components are combined, and CF_1/CF_3 neurons by the fundamental and third harmonic. The exact frequency requirements for optimally driving each type of CF/CF neuron are slightly different among the population, in that each neuron is tuned to a small deviation from an exact harmonic relationship. Thus, the population of CF_1/CF_2 neurons is optimally driven by the range of CF_1 frequencies the bat would emit during Doppler-shift compensation, combined with the CF_2 frequencies that correspond to the stabilized echoes it receives from Doppler-shift compensation. Similar precise frequency pairings are expressed by the population of CF_1/CF_3 neurons.

The other type, called FM/FM or delay-tuned neurons, are also most effectively driven by two harmonically related frequencies of the frequency-modulated (FM) components. Thus, there are FM_1/FM_2 neurons, as well as FM_1/FM_3 and FM_1/FM_4 neurons. For these neurons, the frequency specificity is not as strict as it is for the CF/CF neurons, but rather the neurons require that there be a specific delay between the two signals. For this reason, they are also referred to as delay-tuned neurons. The delay-tuned properties are a consequence of the latency differential with which the inputs evoked by each of the two frequencies arrive at the geniculate neuron (Olsen and Suga 1991b). We first propose a circuit for the generation of CF/CF neurons, and then suggest a slightly different circuitry that could lead to the formation of delay-tuned neurons.

The circuit for CF/CF neurons must account for the facilitated response from two specific frequencies and for the finding that the facilitation can be evoked over a wide range of temporal intervals separating the two tone bursts. To construct the appropriate circuit, we visualize first that a large number of 60-kHz cells project to a common neuron in the medial geniculate (Fig. 7.38, top panel). Furthermore, the projections arise from a dorsoventral column of cells in the contour. Thus the geniculate cell receives projections from a population of 60-kHz collicular cells in which each has a different latency. We next add projections from a population of topographically ordered cells in the 30-kHz contour, the fundamental frequency of the CF component, which then would also provide a wide distribution of latencies. The net result of this arrangement is that the presentation of a 30-kHz and a 60-kHz tone would evoke activity from the two frequency contours, and that the excitation evoked in the the geniculate neuron by the convergence of inputs from the two contours would be independent of the interval separating the two tones. If, for instance, the 30-kHz tone was presented 10 msec before the 60-kHz tone, the 30-kHz tone would evoke a response in the geniculate cell over a wide range of intervals, because each projection neuron would be activated sequentially. The 60-kHz tone, arriving 10 msec later, would also evoke activity distributed over time. The important feature is that because the population of 30-kHz cells and 60-kHz cells have distributed latencies, the discharges from at least some of the 30-kHz neurons will arrive simultaneously with the discharges from some 60-kHz cells and thereby evoke a facilitated response. Because the latency distribution of both the 30- and 60-kHz projection cells is large, there will be a convergence of excitation from some cells of both frequencies regardless of the time interval separating the two signals.

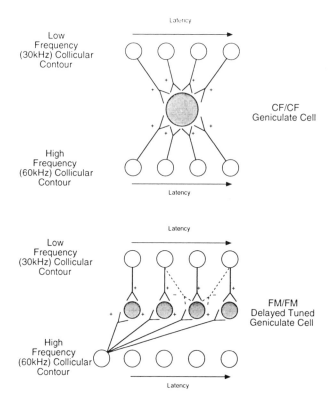

FIGURE 7.38. *Top*: Hypothetical circuit shows inputs from two isofrequency contours that could generate properties appropriate for CF/CF neurons in the mustache bat forebrain. *Bottom*: Hypothetical circuit shows inputs to an FM/FM neuron that could generate delay-tuned properties. (See text for discussion.) (Adapted from Park and Pollak 1993b.)

The projections from the two frequency contours that could generate the FM/FM delay-tuned neurons must be somewhat different. The circuit has to satisfy the additional criterion that the facilitated response occurs only for a small range of intervals between the two signals. Such properties could be achieved if projections from a region of the 30-kHz contour, where latencies are long, converged with the projections from the ventral 60-kHz contour, where latencies are relatively short (Fig. 7.38, lower panel). Under these conditions, the two tones presented simultaneously would arrive at their targets in the medial geniculate at different times and elicit a small response. However, if the 60-kHz signal was delayed relative to the 30-kHz signal by an interval equal to the latency differential evoked by the two frequencies in the colliculus, then the two inputs should converge simultaneously on the geniculate cell, resulting in a facilitated discharge.

An arrangement of the sort just described can be extended and thereby account for two additional features of delay-tuned neurons. The first feature is the distribution of delay-tuned values. This could be if a 60-kHz cell, having a particular latency, sent projections to several geniculate cells. These geniculate cells would also receive innervation from a population of topographically ordered 30-kHz cells, where the locus of each 30-kHz cell defined its latency. Each 30-kHz cell, however, would project to only one geniculate cell. Thus every geniculate cell in this array would be innervated with a common latency from the 60-kHz cell, but any given geniculate cell would be innervated with a different latency by the particular 30-kHz cell that projects to it. An arrangement of this sort would create a population of cells, each endowed with a different delay-tuned value. Another dividend derived from this arrangement is that it provides a simplified mechanism for generating a map of delay tuning. The ordered projection system described would at once generate a diversity of delay-tuned values and an orderly change in the delay-tuned values among the population of target cells in the medial geniculate. Assuming an orderly arrangment of latencies in other frequency contours, then this way of creating delay-tuned neurons can be generalized to any combination of two frequencies and does not have to include 60-kHz cells.

The second feature that requires explanation concerns the difference between neurons tuned to short best delays and those tuned to long best delays (Olsen and Suga 1991b). Medial geniculate neurons tuned to short delays are coincidence detectors and are thought to be generated only by the convergence of two excitatory inputs from the sort of circuit described here. Medial geniculate neurons tuned to long delays, however, are more complex and are created by interactions of excitation and inhibition. In these neurons, delays shorter or longer than the optimal delay inhibit the cell. Thus, the delay value to which the unit is tuned is sandwiched between periods of inhibition. The circuitry we proposed for generating delay-tuned neurons can easily incorporate inhibitory as well as excitatory projections (Fig. 7.38, lower panel). The mammalian inferior colliculus contains a large population of GABAergic principal neurons that presumably provide inhibitory inputs to the medial geniculate. One can postulate, for example, three topographically arranged projections from the 30-kHz contour; the first is an inhibitory projection with a short latency, the second is an excitatory projection with a slightly longer latency, and the third is another inhibitory projection with a latency somewhat longer than the second. The latency of the second neuron could generate an excitatory coincidence with the 60-kHz projection at a particular delay that would be sandwiched between the inhibitory periods produced by the shortest and longest latencies of the 30-kHz projections, thereby sharpening the neuron's delay tuning. These features satisfy many of the properties seen in long-delay geniculate neurons by Olsen and Suga (1991b).

The fact that an equivalent latency distribution occurs in monaural, as well as in the EE and EI subregions of the 60-kHz contour (see Fig. 7.13C), may also be important for creating combinatorial properties. An issue that has not previously been raised is that a mismatch in the aural properties of the convergent projections could confound the latency specificity of the delay-tuned projection system. The reason is that if the projection from one of the frequency contours were monaural and the projection from the other frequency were binaural, then interactions from the two ears could change the latency of the binaural projection. If this occurred, then the delay tuning of a given geniculate neuron would also vary with the interaural intensity disparity and hence with location of the sound source in space. The finding that the population of neurons in each of the 60-kHz aural subregions express a similar latency ordering seems significant in this regard. If a comparable arrangement exists in each of the other frequency contours, and we argue that it probably does, then a combinatorial projection system segregated according to aural type could be readily achieved. In such an arrangement, monaural projection neurons from the two frequency contours would converge on a common geniculate cell, and similarly for the other binaural types. If this proves to be correct, it would simplify what otherwise would appear to be a seriously confounded system.

In conclusion, the results indicate that the full range of delays, as well as other response properties required to form combinatorial neurons, are present in the inferior colliculus, and that the delays are arranged with a characteristic pattern in at least one frequency contour. We have also proposed a simplified model by which the convergence of projections from two frequency contours of the colliculus could theoretically generate the properties of the various types of combinatorial neurons in the medial geniculate. Whether the combinatorial properties are, in fact, created by such convergences remains for future studies to determine.

Abbreviations

ALD	anterolateral division of the inferior colliculus (region representing frequencies below 60 kHz)
AN	auditory nerve
CF	constant frequency
DC	dorsal cortex of the inferior colliculus
DCN	dorsal cochlear nucleus
DmD or MD	medial division of the inferior colliculus (region representing frequencies above 60 kHz)
DpD	dorsoposterior division of the inferior colliculus (the region containing neurons sharply tuned to 60 kHz)
EE	binaural neurons that are driven by acoustic stimulation presented to either ear
EI	binaural neurons that are driven by stimulation at one ear and inhibited by stimulation at the other ear

EO	monaural neurons that are are driven by stimulation at one ear and are unaffected by stimulation of the other ear
FM	frequency modulated
IC	inferior colliculus
ICc	central nucleus of the inferior colliculus
IID	interaural intensity disparity
INLL	intermediate nucleus of the lateral lemniscus
LL	lateral lemniscus
LSO	lateral superior olive
MNTB	medial nucleus of the trapezoid body
MSO	medial superior olive
NLL	nuclei of the lateral lemniscus
SAM	sinusoidal amplitude modulation
SFM	sinusoidal frequency modulation
SPL	sound pressure level
$VNLL_v$	ventral or columnar division of the ventral nucleus of the lateral lemniscus
VNLLd	dorsal division of the ventral nucleus of the lateral lemniscus
VNTB	ventral nucleus of the trapezoid body

References

Adams JC (1979) Ascending projections to the inferior colliculus. J Comp Neurol 183:519–538.

Adams JC, Mugniani E (1984) Dorsal nucleus of the lateral lemniscus: a nucleus of GABAergic projection neurons. Brain Res Bull 13:585–590.

Adams JC, Wenthold RJ (1979) Distribution of putative amino acid transmitters, choline acetyltransferase and glutamate decarboxylase in the inferior colliculus. Neuroscience 4:1947–1951.

Aitkin LM (1976) Tonotopic organization at higher levels of the auditory pathway. Intr Rev Physiol Neurophysiol 10:249–279.

Aitkin LM (1986) The Auditory Midbrain: Structure and Function in the Central Auditory Pathway. Clifton, NJ: Humana.

Bateman GC, Vaughan (1974) Nightly activities of mormoopid bats. J Mammal 55:45–65.

Beyerl BD (1978) Afferent projections to the central nucleus of the inferior colliculus in the rat. Brain Res 145:209–223.

Blauert J (1969/1970) Sound localization in the median plane. Acoustica 22:205–213.

Bodenhamer RD, Pollak GD (1983) Response characteristics of single units in the inferior colliculus of mustache bats to sinusoidally frequency modulated signals. J Comp Physiol A 153:67–79.

Boudreau JC, Tsuchitani C (1968) Binaural interaction in the cat superior olive S-segment. J Neurophysiol (Bethesda) 31:442–454.

Brownell WE, Manis PB, Ritz LA (1979) Ipsilateral inhibitory responses in the cat lateral superior olive. Brain Res 177:189–193.

Brugge JF, Anderson DJ, Aitkin LM (1970) Responses of neurons in the dorsalnucleus of the lateral lemniscus to binaural tonal stimuli. J Neurophysiol (Bethesda) 33:441–458.

Brunso-Bechtold JK, Thompson GC, Masterton RB (1981) HRP study of the organization of auditory afferents ascending to the central nucleus of the inferior colliculus in the cat. J Comp Neurol 97:705–722.

Butler RA (1974) Does tonotopy subserve the perceived elevation of a sound? Fed Proc 33:1920–1923.

Buttman JA (1992) Inhibitory and excitatory mechanisms of coincidence detection in delay-tuned neurons of the mustache bat. In: Proceedings of the 3rd International Congress of Neuroethology, p. 34.

Caird D, Klinke R (1983) Processing of binaural stimuli by cat superior olivary complex neurons. Exp Brain Res 52:385–399.

Cant NB, Casseday JH (1986) Projections from the anteroventral cochelar nucleus to the lateral and medial superior olivary nuclei. J Comp Neurol 247:457–476.

Casseday JH, Covey E (1992) Frequency tuning properties of neurons in the inferior colliculus of an FM bat. J Comp Neurol 319:34–50.

Casseday JH, Covey E, Vater M (1988) Connections of the superior olivary complex in the rufous horseshoe bat, *Rhinolophus rouxi*. J Comp Neurol 278:313–329.

Covey E (1993) Response properties of single units in the dorsal nucleus of the lateral lemniscus and paralemniscual zone of an echolocating bat. J Neurophysiol (Bethesda) 69:842–859.

Covey E, Casseday JH (1986) Connectional basis for frequency representation in the nuclei of the lateral lemniscus of the bat, *Eptesicus fuscus*. J Neurosci 6:2926–2940.

Covey E, Casseday JH (1991) The monaural nuclei of the lateral lemniscus in an echolocating bat: parallel pathways for analyzing temporal features of sound. J Neurosci 11:3456–3470.

Covey E, Hall WC, Kobler JB (1987) Subcortical connections of the superior colliclus in the mustache bat, *Pteronotus parnellii*. J Comp Neurol 263:179–197.

Covey E, Vater M, Casseday JH (1991) Binaural properties of single units in the superior olivary complex of the mustached bat. J Neurophysiol (Bethesda) 66:1080–1093.

Erulkar S (1972) Comparative aspects of sound localization. Physiol Rev 52:237–360.

Faingold CL, Boersma-Anderson CA, Caspary DM (1991) Involvment of GABA in acoustically evoked inhibition in inferior colliculus. Hear Res 52:201–216.

Faingold CL, Gehlbach G, Caspary DM (1989) On the role of GABA as an inhibitory neurotransmitter in inferior colliculus neurons: iontophoretic studies. Brain Res 500:302–312.

Finlayson PG, Caspary DM (1989) Synaptic potentials of chinchilla lateral superior olivary neurons. Hear Res 38:221–228.

Flannery R, Butler RA (1981) Spectral cues provided by the pinna for monaural localization in the horizontal plane. Percept Psychophys 29:438–444.

Frisina RD, O'Neill WE, Zettel ML (1989) Functional organization of mustached bat inferior colliculus: II. Connections of the FM_2 region. J Comp Neurol 284:85–107.

Fuzessery ZM (1986) Speculations on the role of frequency in sound localization. Brain Behav Evol 28:95–108.

Fuzessery ZM, Pollak GD (1984) Neural mechanisms of sound localization in an echolocating bat. Science 225:725–728.

Fuzessery ZM, Pollak GD (1985) Determinants of sound location selectivity in the bat inferior colliculus: a combined dichotic and free-field stimulation study. J Neurophysiol 54:757–781.

Fuzessery ZM, Wenstrup JJ, Pollak GD (1990) Determinants of horizontal sound location selectivity of binaurally excited neurons in an isofrequency region of the mustache bat inferior colliculus. J Neurophysiol (Bethesda) 63:1128–1147.

Gaioni SJ, Riquimaroux H, Suga N (1990) Biosonar behavior of mustached bats swung on a pendulum prior to cortical ablation. J Neurophysiol (Bethesda) 64:1801–1817.

Glendenning KK, Brunso-Bechtold JK, Thompson GC, Masterton RB (1981) Ascending auditory afferents to the nuclei of the lateral lemniscus. J Comp Neurol 197:673–703.

Glendenning KK, Baker BN, Hutson KA, Masterton RB (1992) Acoustic chiasm V: inhibition and excitation in the ipsilateral and contralateral projections of LSO. J Comp Neurol 319:100–122.

Goldberg JM, Brown PB (1968) Functional organization of the dog superior olivary complex: an anatomical and electrophysiological study. J Neurophysiol (Bethesda) 31:639–656.

Goldman LJ, Henson OW Jr (1977) Prey recognition and selection by the constant frequency bat, *Pteronotus parnellii*. Behav Ecol Sociobiol 2:411–419.

Gourevitch G (1980) Directional hearing in terrestrial mammals. In: Popper AN, Fay RR (eds) Comparative Studies of Hearing in Vertebrates. New York: Springer-Verlag, pp. 357–374.

Griffin DR, Simmons JA (1974) Echolocation of insects by horseshoe bats. Nature 250:731–732.

Grinnell AD (1963) The neurophysiology of audition in bats: intensity and frequency parameters. J Physiol 167:38–66.

Grinnell AD, Grinnell VS (1965) Neural correlates of vertical localization by echolocating bats. J Physiol 181:830–851.

Guinan JJ Jr, Guinan SS, Norris BE (1972) Single auditory units in the superior olivary complex. I. Responses to sounds and classifications based on physiological properties. Int J Neurosci 4:101–120.

Haplea S, Covey E, Casseday JH (1994) Frequency tuning and response latencies at three levels in the brainstem of the echolocating bat, *Eptesicus fuscus* J Comp Physiol 174:671–684.

Harnischfeger G, Neuweiler G, Schlegel P (1985) Interaural time and intensity coding in superior olivary complex and inferior colliculus of the echolocating bat, *Molossusater*. J Neurophysiol (Bethesda) 53:89–109.

Henson OW Jr, Pollak GD, Kobler JB, Henson MM, Goldman LJ (1982) Cochlear microphonics elicited by biosonar signals in flying bats. Hear Res 7:127–147.

Hubel DH, Wiesel TN (1977) Functional architecture of macaque monkey visual cortex (Ferrier lecture). Proc R Soc Lond 198:1–59.

Irvine DRF (1986) The auditory brainstem. In: Autrum H, Ottoson D (eds) Progress in Sensory Physiology, vol. 7. Berlin-Heidelberg: Springer-Verlag.

Irvine DRF, Gago G (1990) Binaural interaction in high-frequency neurons in inferior colliculus of the cat: effects of variations in sound pressure level on sensitivity to interaural intensity differences. J Neurophysiol (Bethesda) 63:570-591.

Jen PH-S, Chen D (1988) Directionality of sound pressure transformation at the pinna of echolocating bats. Hear Res 34:101-118.

Jen PH-S, Schlegel PA (1982) Auditory physiological properties of the neurons in the inferior colliculus of the big brown bat, *Eptesicus fuscus*. J Comp Physiol A 147:351-363.

Jen PH-S, Sun X (1984) Pinna orientation determines the maximal directional sensitivity of bat auditory neurons. Brain Res 301:157-161.

Jen PH-S, Suthers RA (1982) Responses of inferior collicular neurons to acoustic stimuli in certain FM and CF-FM, paleotropical bats. J Comp Physiol A 146:423-434.

Kober R (1988) Echoes of fluttering insects. In: Nachtigall PE, Moore PWB (eds) Animal Sonar: Processes and Performance. New York: Plenum, pp. 477-482.

Kössl M, Vater M (1985) The frequency place map of the bat, *Pteronotus parnellii*. J Comp Physiol 157:687-697.

Kössl M, Vater M (1990) Resonance phenomena in the cochlea of the mustache bat and their contribution to neuronal response characteristics in the cochlear nucleus. J Comp Physiol A 166:711-720.

Larue DT, Park TJ, Pollak GD, Winer JA (1991) Glycine and GABA immunostaining defines functional subregions of the lateral lemniscal nuclei in the mustache bat. Proc Soc Neurosci 17:300.

Lesser HD, O'Neill WE, Frisina RD, Emerson RC (1990) ON-OFF units in the mustached bat inferior colliculus are selective for transients resembling "acoustic glint" from fluttering insect targets. Exp Brain Res 82:137-148.

Li L, Kelly JB (1992a) Inhibitory influences of the dorsal nucleus of the lateral lemniscus on binaural responses in the rat's inferior colliculus. J Neurosci 12:4530-4539.

Li L, Kelly JB (1992b) Binaural responses in rat inferior colliculus following kainic acid lesions of the superior olive: interaural intensity difference functions. Hear Res 61:73-85.

Makous JC, O'Neill WE (1986) Directional sensitivity of the auditory midbrain in the mustached bat to free-field tones. Hear Res 24:73-88.

Markovitz NS, Pollak GD (1993) The dorsal nucleus of the lateral lemniscus in the mustache bat: Monaural properties. Hearing Res 71:51-63. Abstracts, 16th Annual Meeting of the Association for Research in Otolarynology, p. 110.

Mills AW (1972) Auditory localization. In: Tobias JV (ed) Foundations of Modern Auditory Theory, Vol II. New York: Academic, pp. 303-348.

M;daller AR (1972) Coding of amplitude and frequency modulated sounds in the cochlear nucleus of the rat. Acta Physiol Scand 86:223-238.

Mountcastle VB (1978) An organizing principle for cerebral function: the unit module and the distributed system. In: The Mindful Brain. Cambridge: MIT Press.

Mugniani E, Oertel WH (1985) An atlas of the distribution of GABAergic neurons and terminals in rat CNS as revealed by GAD immunocytochemistry. In: Bjorlund A, Hokfelt T (eds) Handbook of Chemical Neuroanatomy, Vol. 4: GABA and Neuropeptides in the CNS, Part I. Amsterdam: Elsevier, pp. 436-608.

Musicant AD, Butler RA (1984) The psychophysical basis of monaural localization. Hear Res 14:185-190.

Neuweiler G (1983) Echolocation and adaptivity to ecological constraints. In: Huber F, Markl H (eds) Neuroethology and Behavioral Physiology: Roots and Growing Pains. Berlin: Springer-Verlag, pp. 280–302.

Neuweiler G (1984) Foraging, echolocation and audition in bats. Naturwissenschaften 71:446–455.

Neuweiler G, Vater M (1977) Response patterns to pure tones of cochlear nucleus units in the CF-FM bat, *Rhinolophus ferrumequinum*. J Comp Physiol A 115:119–133.

Novick A (1971) Echolocation in bats: some aspects of pulse design. Am Sci 59:198–209.

Novick A, Vaisnys JR (1964) Echolocation of flying insects by the bat, *Chilonycteris parnellii*. Biol Bull 127:478–488.

Oliver DL, Huerta MF (1992) Inferior and superior colliculi. In: Webster DB, Popper AN, Fay RR (eds) The Mammalian Auditory Pathway: Neuroanatomy, Vol 1. New York: Springer-Verlag, pp. 168–221.

Oliver DL, Nuding SC, Beckius G (1988) Multiple cell types have GABA immunoreactivity in the inferior colliculus of the cat. Proc Soc Neurosci 14:490.

Oliver DL, Kuwada S, Yin TCT, Haberly LB, Henkel CK (1991) Dendritic and axonal morphology of HRP-injected neurons in the inferior colliculus of the cat. J Comp Neurol 303:75–100.

Olsen JF (1986) Processing of biosonar information by the medial geniculate body of the mustached bat, *Pteronotus parnellii*. Ph.D. dissertation, Washington University, St. Louis, MO.

Olsen JF, Suga N (1991a) Combination sensitive neurons in the medial geniculate body of the mustached bat: encoding of relative velocity information. J Neurophysiol (Bethesda) 65:1254–1274.

Olsen JF, Suga N (1991b) Combination sensitive neurons in the medial geniculate body of the mustached bat: encoding of target range information. J Neurophysiol (Bethesda) 65:1275–1296.

O'Neill WE (1985) Responses to pure tones and linear FM components of the CF/FM biosonar signals by single units in the inferior colliculus of the mustached bat. J Comp Physiol A 157:797–815.

O'Neill WE, Frisina RD, Gooler DM (1989) Functional organization of mustached bat inferior colliculus: I. Representation of FM frequency bands important for target ranging revealed by C-2-deoxyglucose autoradiography and single unit mapping. J Comp Neurol 284:60–84.

Ostwald J (1984) Tontopical organization and pure tone response characteristics of single units in the auditory cortex of the greater horseshoe bat. J Comp Physiol A 155:821–834.

Ostwald J (1988) Encoding of natural insect echoes and sinusoidally modulated stimuli by neurons in the auditory cortex of the greater horseshoe bat, *Rhinolophus ferrumequinum*. In: Nachtigall PE, Moore PWB (eds) Animal Sonar: Processes and Performance. New York: Plenum, pp. 483–487.

Ostwald J, Schnitzler H-U, Schuller G (1988) Target discrimination and target classification in echolocating bats. In: Nachtigall PE, Moore PWB (eds) Animal Sonar: Processes and Performance. New York: Plenum, pp. 413–434.

Park TJ, Pollak GD (1993a) GABA shapes sensitivity to interaural intensity disparities in the mustache bat's inferior colliculus: implications for encoding sound location. J Neurosci 13:2050–2067.

Park TJ, Pollak GD (1993b) GABA shapes a topographic organization of response latency in the mustache bat's inferior colliculus. J Neurosci 13:5172-5187.

Park TJ, Larue DT, Winer JA, Pollak GD (1991) Glycine and GABA in the superior olivary complex of the mustache bat: projections to the central nucleus of the inferior colliculus. Soc Neurosci Abstr 17:300.

Pollak GD (1980) Organizational and encoding features of single neurons in the inferior colliculus of bats. In: Busnel GR (ed) Animal Sonar Systems. New York: Plenum, pp. 549-587.

Pollak GD, Bodenhamer RD (1981) Specialized characteristics of single units in inferior colliculus of mustache bat: frequency representation, tuning, and discharge patterns. J Neurophysiol (Bethesda) 46:605-619.

Pollak GD, Casseday JH (1989) The Neural Basis of Echolocation in Bats. Berlin: Springer-Verlag.

Pollak GD, Park TJ (1993) The effects of GABAergic inhibition on monaural response properties of neurons in the mustache bat's inferior colliculus. Hear Res 65:99-117.

Pollak GD, Schuller G (1981) Tonotopic organization and encoding features of single units in inferior colliculus of horseshoe bats: functional implications for prey identification. J Neurophysiol (Bethesda) 45:208-226.

Pollak GD, Winer JA (1989) Glycinergic and GABAergic auditory brain stem neurons and axons in the mustache bat. Soc Neurosci Abstr 15:1115.

Pollak GD, Wenstrup JJ, Fuzessery ZM (1986) Auditory processing in the mustache bat's inferior colliculus. Trends Neurosci 9:556-561.

Pollak GD, Marsh DS, Bodenhamer R, Souther A (1978) A single-unit analysis of inferior colliculus in unanesthetized bats: response patterns and spike-count functions generated by constant frequency and frequency modulated sounds. J Neurophysiol (Bethesda) 41:677-691.

Pollak GD, Park TJ, Larue DT, Winer JA (1992) The role inhibitory circuits play in shaping receptive fields of neurons in the mustache bat's inferior colliculus. In: Singh RN (ed) Principles of Design and Function in Nervous Systems. New Delhi: Wiley, pp. 271-290.

Pumphery RJ (1948) The sense organs of birds. Ibis 90:171-190.

Rhode WS, Greenberg S (1992) Physiology of the cochlear nuclei. In: Webster DB, Popper AN, Fay RR (eds) The Mammalian Auditory Pathway: Neurophysiology, Vol. 2. New York: Springer-Verlag, pp. 94-152.

Roberts RC, Ribak CE (1987) GABAergic neurons and axon terminals in the brainstem auditory nuclei of the gerbil. J Comp Neurol 258:267-280.

Ross LS, Pollak GD (1989) Differential projections to aural regions in the 60-kHz isofrequency contour of the mustache bat's inferior colliculus. J Neurosci 9:2819-2834.

Ross LS, Pollak GD, Zook JM (1988) Origin of ascending projections to an isofrequency region of the mustache bat's inferior colliculus. J Comp Neurol 270:488-505.

Roth GL, Aitkin LM, Andersen RA, Merzenich MM (1978) Some features of the spatial organization of the central nucleus of the inferior colliculus of the cat. J Comp Neurol 182:661-680.

Rübsamen R (1992) Postnatal development of central auditory frequency maps. J Comp Physiol A 170:129-143.

Saint Marie RL, Ostapoff ME, Morest DK, Wenthold RJ (1989) Glycine-immunoreactive projection of the cat lateral superior olive: possible role in midbrain dominance. J Comp Neurol 279:382–396.

Sanes DH, Rubel EW (1988) The ontogeny of inhibition and excitation in the gerbil lateral superior olive J Neurosci 8:682–700.

Schlegel P (1977) Directional coding by binaural brainstem units of the CF-FM bat, *Rhinolophus ferrumequinum*. J Comp Physiol A 118:327–352.

Schnitzler H-U (1970) Comparison of echolocation behavior in *Rhinolophus ferrumequinum* and *Chilonycteris rubiginosa*. Bijdr Dierkd 40:77–80.

Schnitzler H-U, Ostwald J (1983) Adaptations for the detection of fluttering insects by echolocation in horseshoe bats. In: Ewert J-P, Capranica RR, Ingle DJ (eds) Advances in Vertebrate Neuroethology. New York: Plenum, pp. 801–828.

Schreiner CE, Langner G (1988) Periodicity coding in the inferior colliculus of the cat. II. Topographic organization. J Neurophysiol (Bethesda) 60:1823–1840.

Schuller G (1979) Coding of small sinusoidal frequency and amplitude modulations in the inferior colliculus of the CF-FM bat, *Rhinolophus ferrumequinum*. Exp Brain Res 34:117–132.

Schuller G (1984) Natural ultrasonic echoes from wing beating insects are coded by collicular neurons in the long CF-FM bat, *Rhinolophus ferrumequinum*. J Comp Physiol A 155:121–128.

Schuller G, Pollak GD (1979) Disproportionate frequency representation in the inferior colliculus of Doppler-compensating greater horseshoe bats: evidence for an acoustic fovea. J Comp Physiol A 132:47–54.

Schuller G, Beuter K, Schnitzler H-U (1974) Responses to frequency shifted artificial echoes in the bat, *Rhinolophus ferrumequinum*. J Comp Physiol A 89:275–286.

Schuller G, Covey E, Casseday JH (1991) Auditory pontine grey: connections and response properties in the horseshoe bat. Eur J Neurosci 3:648–662.

Schweizer H (1981) The connections of the inferior colliculus and organization of the brainstem auditory system in the greater horseshoe bat, *Rhinolophus ferrumequinum*. J Comp Neurol 201: 25–49.

Semple MN, Aitkin LM (1979) Representation of sound frequency and laterality by units in the central nucleus of the cat's inferior colliculus. J Neurophysiol (Bethesda) 42:1626–1639.

Semple MN, Aitkin LM (1980) Physiology of pathway from dorsal cochlear nucleus to inferior colliculus revealed by electrical and auditory stimulation. Exp Brain Res 41:19–28.

Semple MN, Kitzes LM (1985) Single-unit responses in the inferior colliculus: different consequences of contralateral and ipsilateral auditory stimulation. J Neurophysiol (Bethesda) 53:1467–1482.

Shneiderman A, Henkel CK (1987) Banding of lateral superior olivary nucleus afferents in the inferior colliculus: a possible substrate for sensory integration. J Comp Neurol 266:519–534.

Shneiderman A, Oliver DL (1989) EM autoradiographic study of the projections from the dorsal nucleus of the lateral lemniscus: a possible source of inhibitory inputs to the inferior colliculus. J Comp Neurol 286:28–47.

Shneiderman A, Oliver DL, Henkel CK (1988) The connections of the dorsal nucleus of the lateral lemniscus. An inhibitory parallel pathway in the ascending auditory system? J Comp Neurol 276:188–208.

Simmons JA (1974) Response of the Doppler echolocation system in the bat, *Rhinolophus ferrumequinum*. J Acoust Soc Am 56:672–682.

Stotler WA (1953) An experimental study of the cells and connections of the superior olivary complex of the cat. J Comp Neurol 98:401–432.

Suga N (1964) Single unit activity in the cochlear nucleus and inferior colliculus of echolocating bats. J Physiol 172:449–474.

Suga N, Jen PH-S (1976) Disproportionate tontopic representation for processing species specific CF-FM sonar signals in the mustache bat auditory cortex. Science 194:542–544.

Suga N, Jen PH-S (1977) Further studies on the peripheral auditory system of "CF-FM" bats specialized for the fine frequency analysis of Doppler-shifted echoes. J Exp Biol 69:207–232.

Suga N, Tsuzuki K (1985) Inhibition and level-tolerant frequency tuning in the auditory cortex of the mustached bat. J Neurophysiol (Bethesda) 53:1109–1145.

Suga N, Niwa H, Taniguchi I (1983) Representation of biosonar information in the auditory cortex of the mustached bat, with emphasis on representation of target velocity information. In: Ewert J-P, Capranica RR, Ingle DJ (eds) Advances in Vertebrate Neuroethology. New York: Plenum, pp. 829–870.

Suga N, Simmons JA, Jen PH-S (1975) Peripheral specializations for fine frequency analysis of Doppler-shifted echoes in the CF-FM bat, *Pteronotus parnellii*. J Exp Biol 63:161–192.

Sur M, Merzenich MM, Kass JH (1980) Magnification, receptive field area and "hypercolumn" size in areas 3b and 1 of somatosensory cortex in owl monkeys. J Neurophysiol (Bethesda) 44:295–311.

Thompson GC, Cortez AM, Lam DM-K (1985) Localization of GABA immunoreactivity in the auditory brainstem of guinea pigs. Brain Res 339:119–122.

Trappe M, Schnitzler H-U (1982) Doppler-shift compensation in insect-catching horseshoe bats. Naturwissenschaften 69:193–194.

Vater M (1982) Single unit responses in the cochlear nucleus of horseshoe bats to sinusoidal frequency and amplitude moduated signals. J Comp Physiol A 149:369–388.

Vater M, Feng AS (1990) The functional organization of ascending and descending connections of the cochlear nucleus of horseshoe bats. J Comp Neurol 292:373–395.

Vater M, Feng AS, Betz M (1985) An HRP study of the frequency-place map of the horseshoe bat cochlea: morphological correlates of the sharp tuning to a narrow frequency band. J Comp Physiol A 157:671–686.

Vater M, Kössl M, Horn AKE (1992) GAD- and GABA-immunoreactivity in the ascending auditory pathway of horseshoe and mustache bats. J Comp Neurol 325:183–206.

Vater M, Schlegel P, Zöller H (1979) Comparative auditory neurophysiology of the inferior colliculus of two molossid bats, *Molossus ater* and *Molossus molossus*. I. Gross evoked potentials and single unit responses to pure tones. J Comp Physiol A:131–137.

Vater M, Habbicht H, Kössl M, Grothe B (1992) The functional role of GABA and glycine in monaural and binaural processing in the inferior colliculus of horseshoe bats. J Comp Physiol A 171:541–553.

von der Emde G (1988) Greater horseshoe bats learn to discriminate simulated echoes of insects fluttering with different wingbeat patterns. In: Nachtigall PE, Moore PWB (eds) Animal Sonar: Processes and Performance. New York: Plenum, pp. 495–500.

Warr WB (1982) Parallel ascending pathways from the cochlear nucleus: neuro-anatomical evidence of functional specialization. In: Neff WD (ed) Contributions to Sensory Physiology. New York: Academic, pp. 1–38.

Wenstrup JJ, Ross LS, Pollak GD (1985) A functional organization of binaural responses in the inferior colliculus. Hear Res 17:191–195.

Wenstrup JJ, Ross LS, Pollak GD (1986a) Binaural response organization within a frequency-band representation of the inferior colliculus: implications for sound localization. J Neurosci 6:962–973.

Wenstrup JJ, Ross LS, Pollak GD (1986b) Organization of IID sensitivity in isofrequency representations of the mustache bat's inferior colliculus. In: IUPS Satellite Symposium on Hearing, University of California, San Francisco, CA (Abstr. 415).

Wenstrup JJ, Fuzessery ZM, Pollak GD (1988a) Binaural neurons in the mustache bat's inferior colliculus: I. Responses of 60 kHz E-I units to dichotic sound stimulation. J Neurophysiol (Bethesda) 60:1369–1383.

Wenstrup JJ, Fuzessery ZM, Pollak GD (1988b) Binaural neurons in the mustache bat's inferior colliculus: II. Determinants of spatial responses among 60 kHz E-I units. J Neurophysiol (Bethesda) 60:1384–1404.

Wenthold RJ, Hunter C (1990) Immunocytochemistry of glycine and glycine receptors in the central auditory system. In: Ottersen OP, Storm-Mathisen J (eds) Glycine Neurotransmission. Chichester: Wiley, pp. 391–417.

Winer JA (1992) The functional architecture of the medial geniculate body and the primary auditory cortex. In: Webster DB, Popper AN, Fay RR (eds) The Mammalian Auditory Pathway: Neuroanatomy, Vol. 1. New York: Springer-Verlag, pp. 222–409.

Winer JA, Larue DT, Pollak GD (1995) GABA and glycine in the central auditory system of the mustache bat: structural substrates for inhibitory neuronal organization. J Comp Neurol (in press).

Yang L, Pollak GD, Resler C (1992) GABAergic circuits sharpen tuning curves and modify response properties in the mustache bat inferior colliculus. J Neurophysiol (Bethesda) 68:1760–1774.

Yang L, Pollak GD (1994) GABA and glycine have different effects on monaural response properties in the dorsal nucleus of the lateral lemniscus of the mustache bat. J Neurophysiol (Bethesda) 71:2014–2024.

Zook JM, Casseday JH (1982) Origin of ascending projections to inferior colliculus in the mustache bat, *Pteronotus parnellii*. J Comp Neurol 207:14–28.

Zook JM, Casseday JH (1987) Convergence of ascending pathways at the inferior colliculus of the mustache bat, *Pteronotus parnellii*. J Comp Neurol 261:347–361.

Zook JM, Leake PA (1989) Connections and frequency representation in the auditory brainstem of the mustache bat, *Pteronotus parnellii*. J Comp Neurol 290:243–261.

Zook JM, Winer JA, Pollak GD, Bodenhamer RD (1985) Topology of the central nucleus of the mustache bat's inferior colliculus: correlation of single unit properties and neuronal architecture. J Comp Neurol 231:530–546.

8

The Auditory Thalamus in Bats

Jeffrey J. Wenstrup

Introduction

During the past several years, studies of the auditory cortex in bats have revealed striking examples of functional specializations in the analysis of biosonar pulses and echoes (see O'Neill, Chapter 9). By means of sharp timing or frequency selectivities, some bat cortical neurons encode particular features of sonar targets, and these features are mapped across the cortical surface. Target range is one such feature. Much of the cortical analysis of sonar signals may depend on neural interactions that occur in the medial geniculate body (MGB), the thalamic relay in the ascending pathway to the auditory cortex. For example, *combination-sensitive* neurons, selective for combinations of spectrally or temporally distinct signal elements in the sonar pulse and echo, have been well described in regions of the auditory cortex; physiological studies suggest that these responses may be created in the medial geniculate body (Olsen and Suga 1991a,b). Thus, the auditory thalamus in bats may provide new insights into the processing of complex sounds, as well as furthering our understanding of biosonar mechanisms in these animals.

Although the MGB in mammals is regarded primarily as the thalamic relay in the ascending, tonotopic pathway to the auditory cortex, its roles are multiple and diverse. It is composed of several structurally and functionally distinct nuclei, and these are believed to form the basis for functionally distinct, parallel ascending systems to the auditory cortex (Clarey, Barone, and Imig 1992; Winer 1992). The correspondence between medial geniculate nuclei in bats and their counterparts in other species is varied. How these nuclei differ among bats, and between bats and other mammals, may reveal much about their role both in the highly developed acoustic behavior of bats and in less specialized, nonchiropteran species.

Until recently, the *lateral geniculate nucleus*, the visual thalamic relay, had received more attention in bats than the MGB. Thus, despite its role in transmitting and modifying auditory information en route to the cortex, the auditory thalamus of bats is poorly understood. With a few exceptions, adequate descriptions of the structure, connections, chemical anatomy, or physiology are lacking. Most recent studies have concentrated on the mustached bat (*Pteronotus parnellii*), because its MGB has been implicated in creating novel responses for the processing of biosonar echoes. However, even the few current studies lead one to expect that other bats may differ substantially. Hence, a broader comparative approach is crucial to our understanding of the roles played by the MGB and auditory cortex in the acoustic behavior of bats.

This chapter presents studies on the mustached bat as a framework for discussion of bat medial geniculate structure and function generally. It first describes the basic organization, then examines the structural and functional properties underlying several circuits that involve different medial geniculate nuclei.

2. General Features of the Medial Geniculate Body

The MGB in bats is well developed, even among nonecholocating species (Baron 1974). However, in bats with poorly developed visual systems, such as the mustached bat (Covey, Hall, and Kobler 1987), it dominates the dorsal thalamus (Winer and Wenstrup 1994a) and forms the dorsolateral surface of the thalamus throughout the rostrocaudal extend of the MGB.

2.1 Architectonic Organization

Nuclei of the MGB are often grouped into three divisions — dorsal, ventral, and medial — although significant differences of opinion exist regarding the grouping of nuclei and their boundaries. This review follows an architectonic scheme previously identified in the cat and based on the dendritic and axonal architecture from Golgi-impregnated neurons as well as cytoarchitecture and myeloarchitecture (Morest 1964, 1965; Winer 1985). It has been extended to several other mammalian species, for example, the tree shrew (Oliver 1982), human (Winer 1984), opossum (Morest and Winer 1986; Winer, Morest, and Diamond 1988), rat (Clerici and Coleman 1990; Clerici et al. 1990), and recently the mustached bat (Winer and Wenstrup 1994a,b). Following is a brief description of some distinguishing features of each division to provide a basis for the ensuing discussion. See Winer (1992) for a more detailed treatment of the anatomical organization of the medial geniculate body in mammals.

2.1.1 The Ventral Division

The ventral division is the thalamic component of the ascending, tonotopic auditory pathway. Several features combine to distinguish it from other divisions, including (1) a tonotopic projection from the central nucleus of the

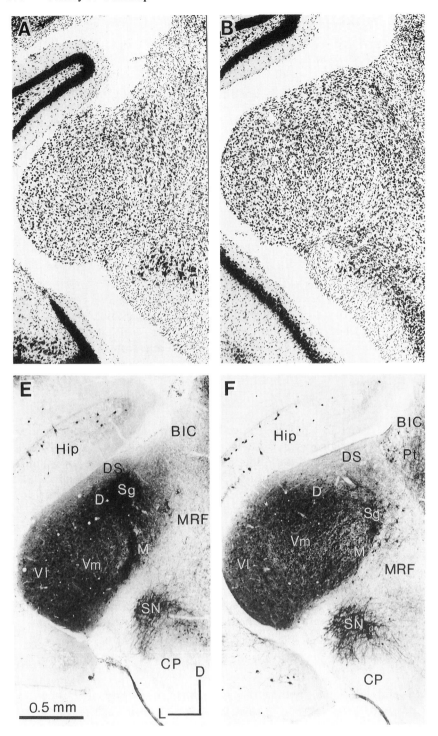

FIGURE 8.1A,B,E,F. Caption on page 372.

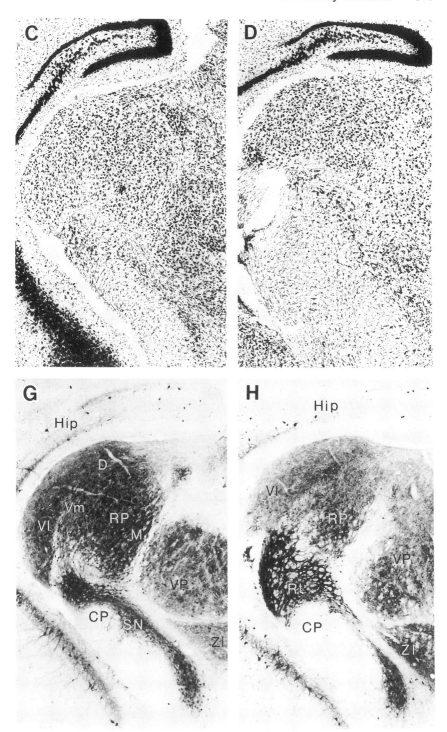

FIGURE 8.C,D,G,H. Caption on page 372.

inferior colliculus (ICC), (2) a laminar arrangement of principal cell dendrites, (3) sharply tuned, tonotopically organized responses to sound frequency, and (4) the major projection to the tonotopic, primary auditory cortical field. In many species, two major subnuclei, lateral and medial, have been identified. The lateral part (*Vl*; Figs. 8.1 and 8.2) contains the clearest laminar organization, in which the major dendrites of the principal cells lie parallel to the lateral surface of the MGB. In the medial part (*Vm*; Figs. 8.1 and 8.2), the laminar pattern may be distorted or less apparent than in the lateral part. The features of the mustached bat's ventral division generally conform to the pattern in other mammals (Olsen 1986; Wenstrup, Larue, and Winer 1994; Winer and Wenstrup 1994a,b).

2.1.2 The Dorsal Division

Several nuclei comprise the structurally and functionally diverse dorsal division. In other mammals these include superficial dorsal, dorsal, deep dorsal, suprageniculate, ventral lateral, and posterior limitans nuclei; more have been recognized in some species (Oliver 1982; Winer, Morest, and Diamond 1988). Dorsal division nuclei are united mainly in their differences from the ventral division (Winer 1992). Thus, dorsal division neurons do not have dendrites as strongly tufted as in the ventral division, and their sizes vary more among nuclei. Except for the deep dorsal nucleus, dorsal division nuclei do not receive strong ascending input from the ICC, but inputs originate instead from other inferior collicular nuclei, adjacent midbrain tegmental regions, and brainstem nuclei. The major targets of dorsal division nuclei are correspondingly varied, but all lie outside the primary tonotopic area of auditory cortex.

Most dorsal division nuclei have been identified in architectonic studies of the mustached bat (Figs 8.1 and 8.2) (Winer and Wenstrup 1994a,b). Suprageniculate neurons are among the largest in the MGB, and have radiating dendrites. Dorsal superficial neurons are bipolar or weakly tufted, while dorsal nucleus neurons are either weakly tufted or stellate. Additional

---

FIGURE 8.1A–H. Architecture of the medial geniculate body in the mustached bat, shown in Nissl-stained (A–D) and parvalbumin-immunostained (E–H) series. Each series is from a different animal, arranged in a caudal-to-rostral (left-to-right) sequence. Sections are located about 20%, 40%, 60%, and 80% through the caudal-to-rostral dimension. Both cells and neuropil are intensely parvalbumin immunoreactive in most medial geniculate subdivisions, with the exception of the superficial dorsal nucleus (DS; E,F). The rostral extreme of the medial geniculate body is only lightly or moderately labeled (H). Note also the immunostained neurons of the thalamic reticular nucleus (Rt; H). Protocol for parvalbumin immunocytochemistry: monoclonal mouse antibody (Sigma, St. Louis, MO); primary antibody dilution 1:2000; avidin-biotin-peroxidase method using heavy metal-intensified diaminobenzidine reaction.

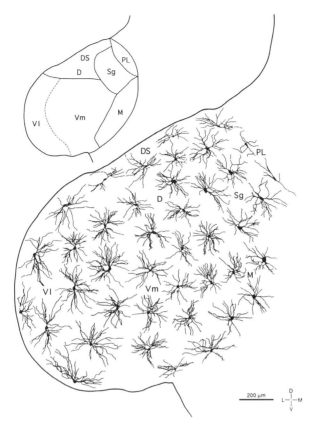

FIGURE 8.2. Illustration of Golgi preparations of the medial geniculate body (MGB) in the mustached bat. Section located near the border of the caudal and middle thirds of the MGB, corresponding roughly to Figure 8.1B. In both the lateral (Vl) and medial (Vm) parts of the ventral nucleus, principal neurons have bushy dendritic branches, but neurons in the medial part have more spherical dendritic fields. In the dorsal division, neurons with polarized dendritic fields are common in the superficial dorsal nucleus (DS), but dorsal nucleus (D) neurons have more radiate, spherical dendritic fields. Suprageniculate nucleus (Sg) neurons are larger, and their radiating branches distinguished them from the thin strip of sparsely branched, elongated neurons of the posterior limitans nucleus (PL). In the medial division (M), several neuronal varieties are impregnated, including the conspicuous magnocellular neurons. (Adapted from Winer and Wenstrup 1994b, copyright © 1994, reprinted by permission of Wiley-Liss, a division of John Wiley and Sons, Inc.)

subdivisions are also apparent, but their correspondence to other species is not always clear. For example, the rostral pole nucleus (see Fig. 8.1) was recognized by Winer and Wenstrup (1994a) as a distinctive region that has characteristics of both ventral and dorsal division neurons. However, its connections and physiology do not correspond closely to any currently known dorsal or ventral division nucleus in other mammals. Together with other parts of the dorsal division, it contains neuronal response properties that appear specialized for the analysis of sonar target features (see Section 5). These and other findings suggest that dorsal division nuclei in bats, more than the ventral division, may differ from their counterparts in other mammals and may have evolved to serve in species-specific auditory signal processing roles.

2.1.3 The Medial Division

The medial division is considered to be less closely associated with the ascending, frequency-specific auditory pathway than either the ventral or dorsal divisions (Winer and Morest 1983; but see also Rouiller et al. 1989). It is distinguished from these chiefly by its cellular population and connections. Characteristic are its large neurons (the largest in the medial geniculate) with radial dendrites, but other cell types also occur (Winer and Morest 1983; Winer 1992). The connections of the medial division reflect a broader, multisensory role, including auditory inputs from the central and external nuclei of the inferior colliculus, but also others, for example, from somatosensory and vestibular systems. Its output also differs; it projects broadly to primary and other auditory cortical regions. However, medial division projections to primary auditory cortex terminate more heavily in layer I or VI, rather than in layers III and IV (Sousa-Pinto 1973; Niimi and Naito 1974; Mitani, Itoh, and Mizuno 1987; Conley, Kupersmith, and Diamond 1991).

The medial division of the mustached bat shares a similar neuronal population with other mammals, but is much smaller in relative size than that of cats and primates; it comprises only 10% of the medial geniculate body (Figs. 8.1 and 8.2) (Winer and Wenstrup, 1994a,b). As in other mammals, the medial division in the mustached bat receives input from the ICC (see Figs. 8.4 and 8.6, later in this chapter) (Wenstrup, Larue, and Winer 1994), but its other connections are not known. In bats, functional properties of medial division neurons have not been described.

2.2 Neurochemistry

2.2.1 γ-Aminobutyric Acid (GABA)

The distribution of GABAergic neurons in the MGB differs strikingly between the two species described to date (Fig. 8.3). In the rufous horseshoe bat (*Rhinolophus rouxi*), GABAergic neurons are common in all subdivisions, but are most numerous in the ventral division (Vater, Kössl, and Horn 1992). Dendritic appendages of ventral and dorsal division GABAergic neurons are often complex and closely apposed to non-GABAergic neurons, suggesting their involvement in local circuits. GABAergic neurons are generally small, and may correspond to the small, Golgi type II cells described in other species (Morest 1975; Majorossy and Kiss 1976; Winer and Larue 1988). The mustached bat's MGB has a very different pattern (Vater,

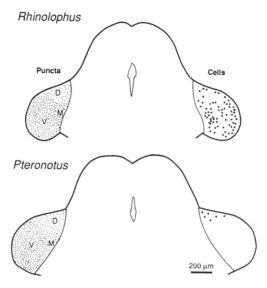

FIGURE 8.3. Schematic illustration of the distribution of GAD/GABA (glutamic acid decarboxylase/γ-aminobutyric acid)-immunoreactive puncta (*left*) and cells (*right*) in the MGB of the rufous horseshoe bat (*Rhinolophus*) and the mustached bat (*Pteronotus*). The distributions of immunopositive puncta are similar in the two species; there are fewer puncta in the dorsal division than in the other medial geniculate divisions. In contrast, the distributions of immunopositive cells are very different. In the horseshoe bat, GABAergic neurons are common in all three divisions, but are most numerous in the ventral division. GABAergic neurons in the mustached bat are rare, and most of these are located in the dorsal division. (Adapted from Vater, Kössl, and Horn, copyright © 1992, reprinted by permission of Wiley-Liss, a division of John Wiley and Sons, Inc.)

Kössl, and Horn 1992; Winer, Wenstrup, and Larue 1992). Only a very few GABAergic neurons have been found, probably less than 1% of medial geniculate neurons, and most of these were in the dorsal division.

In contrast, the form and distribution of GABAergic puncta (putative terminals) show close similarities between the two bats (see Fig. 8.3) (Vater, Kössl, and Horn 1992; Winer, Wenstrup, and Larue 1992). Thus, the dorsal and ventral divisions contain relatively fine puncta, while those in the medial division are much coarser. The dorsal division contains the fewest puncta. These patterns occurred despite differences in numbers of intrinsic GABAergic neurons, suggesting that the pattern of GABAergic terminals in the MGB may be imposed primarily by extrinsic neurons.

The dramatic difference between GABAergic neurons in the two species is very surprising, yet the results seem reliable. For example, each of the two studies (Vater, Kössl, and Horn 1992; Winer, Wenstrup, and Larue 1992) documented the scarcity of GABAergic neurons in the mustached bat using both GABA and glutamic acid decarboxylase (GAD) immunocytochemistry. These studies found few immunostained medial geniculate cells in the same histological sections where other neurons were intensely immuno-stained. Furthermore, the comparative study (Vater, Kössl, and Horn 1992), using both GABA- and GAD-immunocytochemistry, found that the numbers of GABAergic neurons differed between the two species but the distribution of puncta did not.

What is particularly surprising is that the different patterns are found in bat species that use similar biosonar signals and echo information, and that seem to use similar neuronal processing strategies. Unless some other inhibitory neurotransmitter in the mustached bat performs the same function as GABA, these results suggest major differences in the processing of ascending and descending input by medial geniculate neurons. Although MGB neurons in both species receive GABAergic input, only in the horseshoe bat do local, that is, intrageniculate, inhibitory mechanisms seem to occur. In the mustached bat, inhibitory interactions must result largely from external GABAergic input, perhaps from the thalamic reticular nucleus (Winer, Wenstrup, and Larue 1992; Wenstrup and Grose 1993), or from neurons utilizing other inhibitory neurotransmitters. However, glyci-nergic inhibition, although common in lower auditory centers, does not appear to play a role in medial geniculate processing by mammals (Aoki et al. 1988; Wenthold and Hunter 1990).

2.2.2 Calcium-Binding Proteins

Studies of the distribution of calcium-binding proteins in auditory systems have served to distinguish subsystems that process different types of information, as in the temporal processing pathways of the barn owl auditory system (Takahashi et al. 1993) or in the thalamocortical projec-tions of monkeys (Hashikawa et al. 1991). Zettel, Carr, and O'Neill (1991) examined immunoreactivity to calbindin throughout mustached bat audi-tory structures. In medial geniculate nuclei, calbindin-immunopositive neurons were abundant in dorsal and ventral divisions. Immunostaining of the neuropil was heaviest in the dorsal division, particularly the superficial dorsal nucleus. Virtually unlabeled were the suprageniculate nucleus and the posterior complex (corresponding to the caudal part of the medial division here), regions containing the largest medial geniculate neurons. In the horseshoe bat, calbindin-immunopositive cells are common in all divisions of the MGB, whereas calretinin-immunopositive cells occur only in the dorsal division and along the lateral and ventrolateral rims of the ventral division. Few axon terminals are calbindin- or calretinin-immuno-positive (Vater and Braun 1994).

Parvalbumin, another calcium-binding protein, is distributed differently in the medial geniculate body (see Fig. 8.1E–H). Both cells and neuropil are heavily labeled in several areas: the ventral division, medial division, and dorsal, rostral pole, and suprageniculate nuclei. Labeling in the most rostral part of the MGB was weaker, although present (see Fig. 8.1H). In contrast, the superficial dorsal nucleus contained virtually no immunopositive labeling (Fig. 8.1E,F). Generally, the parvalbumin-immunostained regions corresponded to those receiving input from the central nucleus of the inferior colliculus (Fig. 8.4) (Wenstrup, Larue, and Winer 1994). In the horseshoe bat, all divisions of the MGB contained parvalbumin-immunopositive cells and terminals (Vater and Braun 1994).

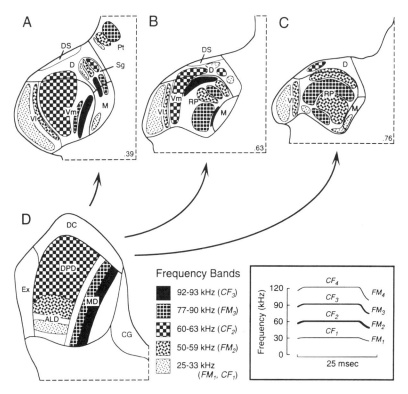

FIGURE 8.4A–D. Schematic summary of the distribution of inputs to the mustached bat's MGB (A–C) from five frequency band representations (D) in the central nucleus of the inferior colliculus (ICC) analyzing major elements of the bat's sonar signal. Three projection systems are described in the text. One system terminates in the lateral (Vl) and medial (Vm) parts of the ventral division and is tonotopically organized (A–C). A second terminates in the suprageniculate nucleus (Sg) and is also tonotopic (A). The third system was found principally in the rostral pole (RP) and dorsal (D) nuclei (B,C). These inputs may be organized according to functional biosonar components; CF_2 and CF_3 projections terminate at middle MGB levels (B), while FM_2 and FM_3 projections are extensive and terminate well into the most rostral part of MGB (B,C). Note also the projection to the pretectum from ICC regions representing frequency-modulated (FM) components of the sonar signal (A). *Inset at right*: sonogram of mustached bat biosonar pulse with signal components labeled. Thickness of lines in sonogram indicates relative intensity of harmonics. CF, constant frequency. (From Wenstrup, Larue, and Winer, copyright © 1994, reprinted by permission of Wiley-Liss, a division of John Wiley and Sons, Inc.)

Species comparisons raise questions about the functional implications of the distribution of calcium-binding proteins. For example, the distribution of parvalbumin in both the mustached bat and horseshoe bat agrees well with its distribution in monkeys (Hashikawa et al. 1991), but not in rats (Celio 1990). However, the distribution of calbindin agrees more with data in rats than monkeys (Celio 1990; Hashikawa et al. 1991). Furthermore, results differ depending on the antibody used; thus, in the mustached bat, monoclonal mouse anticalbindin labels the same medial geniculate regions (Wenstrup, unpublished data) as does polyclonal anticalbindin (Zettel, Carr, and O'Neill 1991), but it labels virtually none of the brainstem auditory regions labeled by the polyclonal antibody.

2.3 Connections of the Medial Geniculate Body

As in other mammals, the MGB in bats receives a major ascending input from the tonotopically organized central nucleus of the inferior colliculus (see Fig. 8.4). In all bats so far examined, including the mustached bat (Casseday et al. 1989; Frisina, O'Neill, and Zettel 1989; Wenstrup, Larue, and Winer 1994), the greater horseshoe bat (*Rhinolophus ferrumequinum*) (Schweizer 1981), the pallid bat (*Antrozous pallidus*) (Wenstrup and Fuzessery, unpublished data), and the big brown bat (*Eptesicus fuscus*) (Covey, unpublished data), the ICC projection is exclusively ipsilateral. In contrast, the contralateral ICC-MGB projection in the cat is significant, although smaller than the ipsilateral projection (Kudo and Niimi 1978; Rouiller and de Ribaupierre 1985).

Perhaps because bats have a special reliance on acoustic information, the MGB may receive broader input from the ICC than is the case in other mammals. This is clear in the mustached bat, where the ICC forms distinct projections to three parallel MGB systems (see Fig. 8.4), each probably serving different functional roles in acoustic orientation. These systems, considered next, include (1) the tonotopically organized system through the ventral division, (2) systems involving the suprageniculate nucleus, and (3) regions in the dorsal and rostral MGB containing specialized responses to biosonar signals. What is apparent is that these systems, and their affiliated MGB nuclei, differ considerably in the degree of their correspondence to other mammals.

3. Ventral Division: The Tonotopically Organized Medial Geniculate Body

3.1 Functional Architecture of the Ventral Division

The distinctive architectonic feature of the ventral division in the mustached bat, as in other mammals, is its laminar organization of principal cell dendrites (see Fig. 8.2) (Winer and Wenstrup 1994b). Dendrites of principal cells arise in tufts from the two somatic poles and fill a more or less planar or sheetlike expanse. The laminar arrangement is particularly evident in the lateral part, where the dendrites of tufted neurons run parallel to the lateral surface of the medial geniculate. In the medial part, the laminar organization is more variable. It is not apparent in the region, located caudally and dorsally, that receives input from the hypertrophied 60- to 63-Hz representation of the ICC. It is clearer more medially, however, where dendritic laminae lie along a ventrolateral-to-dorsomedial axis.

The laminar dendritic organization in the ventral division is matched by the arrangement of ICC afferents (Fig. 8.5A–C) (Wenstrup, Larue, and Winer 1994). Thus, in the lateral part, horseradish peroxidase-filled axons from low-frequency parts of the ICC terminate in sheets that parallel the lateral surface of the MGB (Fig. 8.5B). In the medial part, 60-kHz ICC axons do not terminate in a laminar pattern, even though the size of individual terminal fields is highly restricted (Fig. 8.5A). In contrast, ICC axons tuned above 63 kHz form terminal fields more medially, which extend along a ventrolateral-to-dorsomedial axis (Fig. 8.5C). The correspondence between dendritic arborization patterns and axon terminal fields is likely to preserve the segregation and topographic arrangement of frequencies that occurs in lower auditory nuclei. Thus, this organization closely corresponds to that in the mustached bat ICC (Zook et al. 1985), where three architectonic patterns characterize regions representing frequencies less than 60 kHz (the anterolateral division), 60–63 kHz (the dorsoposterior division), and greater than 63 kHz (the medial division). In the ICC as in the MGB the 60- to 63-kHz representation lacks a clear laminar pattern and distorts the overall laminar organization by its large size.

In other bats, the neuronal architecture of the ventral division is unknown. However, the organization of ICC input suggests somewhat different laminar patterns. Thus, in the big brown bat (Covey, unpublished data) and pallid bat (Wenstrup and Fuzessery, unpublished data), deposits in restricted ICC frequency bands result in ventral division label which extends in a dorsolateral-to-ventromedial pattern. In the little brown bat (*Myotis lucifugus*), restricted ICC deposits are reported to result in laminar patterns of labeling (Shannon and Wong 1987).

In cats, small stellate cells having only local axonal projections contribute significantly to the architecture of the ventral division (Morest 1975). Their dendrites receive synaptic contacts from tectal and cortical axons, and form axodendritic and dendrodendritic contacts with principal neurons. Many of these cells are probably GABAergic (Winer 1992), and may contribute a variety of inhibitory influences regulating the output of the larger principal cells. In the mustached bat's ventral division, small stellate cells are scarce in Golgi-impregnated material (Winer and Wenstrup 1994b), and there are very few GABAergic neurons (Winer, Wenstrup, and Larue 1992). Thus,

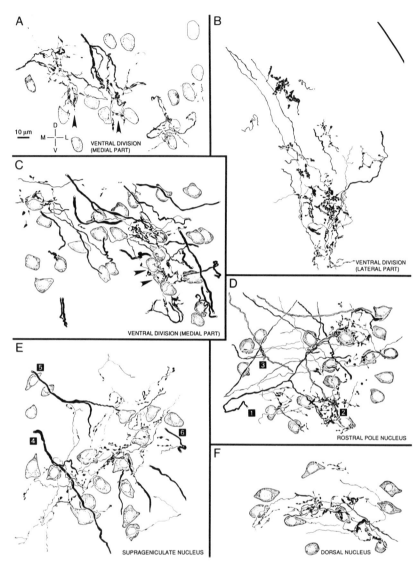

FIGURE 8.5A–F. Architecture of ICC axons in subdivisions of the medial geniculate body. (A) Labeling in the medial part of the ventral division after ICC deposits in regions tuned near 62 kHz. Boutons ended both in neuropil and near perikarya. (B) Labeling in the lateral part of the ventral division after deposits in regions tuned to 30–33 kHz. Axons and terminals formed an arc recapitulating the arrangement of principal cell dendrites in Golgi material (see Fig. 8.2). No perikarya are visible in this unstained section. *Heavy curved line*, dorsolateral surface of MGB. (C) Labeling in the medial part of the ventral division after deposits in regions tuned to 93 kHz. (D) Labeling in the rostral pole nucleus after 77- to 82-kHz deposits. Many axons (1–3) formed two branches separated by at least 100 μm. Two axons, (1) and (2), entered ventrolaterally, and each had two separate terminal plexuses that were

what is considered to be an important local inhibitory circuit in the cat's MGB apparently does not exist in the mustached bat. Whether some other circuit or some other inhibitory neurotransmitter serves the same function is unknown. As noted previously, this circuit may well exist in other bats, because horseshoe bats have many ventral division GABAergic neurons. These anatomical differences suggest that physiological properties, such as discharge patterns, tuning curves, or rate-level functions, may differ between ventral division neurons in the two bat species. Further effects may be revealed in neuropharmacological studies of ventral division neurons.

3.2 Connections of the Ventral Division

3.2.1 Inputs from the Inferior Colliculus

The ventral division receives a topographically organized projection from the central nucleus of the inferior colliculus. This projection has been studied most extensively in the mustached bat, where ICC tracer deposits were placed within regions representing each of the biosonar components of the first three harmonic elements (Frisina, O'Neill, and Zettel 1989; Wenstrup, Larue, and Winer 1994). There are three main topographic features of the ICC projection to the ventral division in this bat (see Fig. 8.4). First, low-frequency representations project laterally to the ventral division, while higher frequencies project to successively more medial loci. This topographic pattern is in general agreement with that observed in cats (Andersen et al. 1980; Kudo and Niimi 1980; Calford and Aitkin 1983). Second, 60-kHz input is expanded, in agreement with its representation in the ICC and throughout the ascending auditory system (Figs. 8.4 and 8.6).

The third feature is particularly noteworthy. Some frequency band representations in ICC appear to project only lightly to the ventral division. Thus, only modest labeling occurs within the ventral division after ICC deposits in the 50- to 59-kHz and 77- to 90-kHz frequency band representations, but much stronger labeling is found in the rostral MGB (Figs. 8.4 and 8.6). This is a significant departure from the pattern established in the cochlea and preserved within the ICC (Frisina, O'Neill, and Zettel 1989), showing that the tectothalamic projection can modify the organization within the primary auditory pathway. In this case, these frequency band

---

FIGURE 8.5A–F (*continued*) nearly congruent. (E) Labeling in the suprageniculate nucleus after 62-kHz deposits. Several axons branched in the suprageniculate nucleus (4–6). Some (4) sent collaterals into the suprageniculate while the main trunk proceeded to the ventral division. (F) Labeling in the dorsal nucleus after 62-kHz deposits. Boutons were similar to those of the ventral division in the same experiment (cf. A). (From Wenstrup, Larue, and Winer, copyright © 1994, reprinted by permission of Wiley-Liss, a division of John Wiley and Sons, Inc.)

382   Jeffrey J. Wenstrup

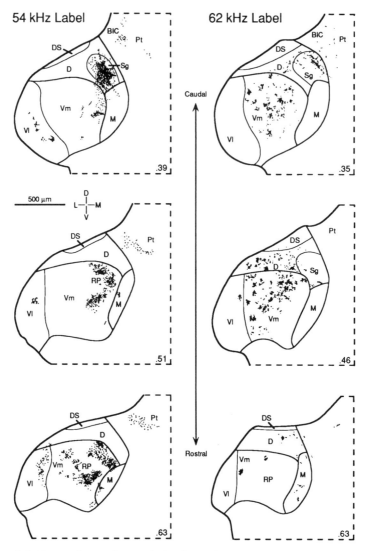

FIGURE 8.6. Comparison of tectothalamic projections from 54-kHz and 62-kHz representations of the ICC; these frequencies are contained within the FM_2 and CF_2 components, respectively, of the sonar signal. In the 54-kHz case, transport results from two large iontophoretic deposits of wheat germ agglutinin–horseradish peroxidase conjugate (WGA-HRP). In the 62-kHz case, six smaller, iontophoretic HRP deposits were placed. Note that 54-kHz regions project only lightly to the lateral part of the ventral division (Vl), but more heavily to the rostral pole nucleus (RP). In contrast, the 62-kHz ICC representation projects strongly to the medial part of the ventral division (Vm) and the adjacent dorsal division (D). Rostrally, the 62-kHz input is very weak. Both frequency representations project to the suprageniculate nucleus (Sg). (Adapted from Wenstrup, Larue, and Winer, copyright © 1994, reprinted by permission of Wiley-Liss, a division of John Wiley and Sons, Inc.)

representations correspond to higher harmonics of the frequency-modulated (FM) components of the sonar pulse, and their strong projection to the rostral MGB suggests their primary role is to carry information to specialized neurons in the rostral MGB that analyze target distance (Olsen and Suga 1991b). Furthermore, their reduced input to the ventral division suggests that they may play less of a role in other types of signal analysis, for example, elevational and azimuthal localization.

There is less information on the MGB of other species. In the pallid bat, tracer deposits in ICC frequency band representations result in terminal labeling arranged in sheets extending along a ventromedial-to-dorsolateral orientation; low-frequency label (~15 kHz) occurs ventrolaterally, while higher frequency label (~40 kHz) associated with this bat's sonar pulse occurs in more dorsal and medial sheets (Wenstrup and Fuzessery, unpublished data). A similar arrangement appears in the big brown bat (*Eptesicus fuscus*) (Covey, unpublished data). In view of the results in the mustached bat, the organization of tectothalamic input to the ventral division in other bats is of considerable interest. Does the ventral division receive more or less input from neurons tuned to frequencies within the biosonar sound? Do neurons tuned to some sonar frequencies project more heavily outside the lemniscal pathway? The pattern may be very different for bats using FM signals, because, unlike the mustached bat, they must utilize the same frequency bands to obtain several target characterizations.

3.2.2 Inputs from Other Sources

Additional inputs to the ventral division arise from the auditory cortex and the thalamic reticular nucleus (Olsen 1986; Wenstrup and Grose 1993, in press). Olsen (1986) found that deposits in AI, the tonotopically organized auditory cortex, resulted in congruent anterograde and retrograde labeling in the ventral division. Thus, corticothalamic projections follow the tonotopic organization in the ventral division. The ventral division also receives strong input from the thalamic reticular nucleus (Wenstrup and Grose 1993, in press). The auditory sector of this nucleus, located just ventral to the rostral MGB projects to most MGB regions, although its organization is not currently known.

3.2.3 Outputs

The major output of the ventral division is a topographically organized projection to tonotopically organized auditory cortex (Olsen 1986; Casseday and Pollak 1988). Thus, low frequencies in the lateral subdivision project caudally, while higher frequencies in the medial part project to more rostral cortical loci. These data agree well with the pattern of anterograde labeling in the ventral division following ICC deposits, as well as the tonotopic organization revealed in physiological studies (Olsen 1986).

3.2.4 Evidence of Intrinsic Connections

There is little evidence that the ventral division projects to or receives input from other MGB subdivisions. For example, tracer deposits in combination-sensitive regions of the dorsal division or rostral pole do not label ventral division regions outside the deposit site either retrogradely or anterogradely (Wenstrup and Grose 1993). Thus, combination-sensitive neurons do not receive low-frequency (24–30 kHz) input by way of the lateral part of the ventral division (see Section 5).

3.3 Physiological Properties of the Ventral Division

In all mammals, the ventral division is believed to maintain frequency-specific processing of acoustic information. Most ventral division neurons are sharply tuned and tonotopically organized, responding reliably and with temporal fidelity to tonal signals (Aitkin and Webster 1972; Calford 1983; Rodrigues-Dagaeff et al. 1989; Clarey, Barone, and Imig 1992). Although the evidence is limited, this generally appears to be the case for bats as well.

3.3.1 Frequency and Amplitude Tuning

Early physiological studies of MGB neurons in *Myotis oxygnathus* and the horseshoe bat (*Rhinolophus ferrumequinum*) showed that greater numbers of neurons were tuned to frequency bands within the biosonar signals of these species than to nonsonar frequencies (Vasil'ev and Andreeva 1972). In *M. oxygnathus*, a bat using broadband FM signals, nearly all neurons had moderately sharp tuning curves with Q_{10dB} values (defined as the best frequency of a neuron's tuning curve divided by the bandwidth 10 dB above threshold) less than 20; the distribution of Q_{10dB} values was similar to those in the cochlear nucleus and the auditory cortex. In the horseshoe bat, many MGB neurons were more sharply tuned, particularly at frequencies corresponding to the constant-frequency (CF) component of the bat's biosonar pulse. Ayrapet'yants and Konstantinov (1974) suggest that there is an increase in the sharpness of tuning in horseshoe bat auditory neurons from the cochlear nucleus to the MGB. In both species, some neurons were tuned in amplitude, having nonmonotonic rate-level functions and upper thresholds. Most neurons in these experiments were recorded from the "parvicellular part" of the MGB (Vasil'ev and Andreeva 1972), which includes both the ventral division and part of the dorsal division.

In the mustached bat's MGB, the best frequencies of neurons responding only to single tones ranged from 5 to 120 kHz, although very few were below 20 kHz or above 100 kHz (Olsen 1986; Olsen and Suga 1991a). Neurons tuned to frequencies in the CF sonar echoes of the second (60 kHz) and third (90 kHz) harmonic (CF_2 and CF_3, respectively) were more common than those tuned to other frequency bands in the sonar signal or to nonsonar frequencies. Many of these neurons probably lay within the ventral division, although their location by division was not reported.

Medial geniculate neurons tuned to the CF_2 and CF_3 sonar components were more sharply tuned in frequency than other MGB neurons (Fig. 8.7) (Olsen and Suga 1991a). Sharper tuning was evident near threshold (Fig. 8.7A) and well above threshold (Fig. 8.7B, C). Q_{10dB} values agree with studies on other auditory nuclei in the mustached bat (Suga, Simmons, and Jen 1975; Pollak and Bodenhamer 1981; Suga and Manabe 1982), and they reflect sharp cochlear tuning in these frequency bands, particularly at the frequencies of the CF_2 component (see Kössl and Vater, Chapter 5, this volume). However, neural inhibitory mechanisms also play a role. Thus, medial geniculate neurons display broadly tuned (i.e., broader than the excitatory tuning curve) inhibition that both sharpens the tuning curve at higher levels and restricts the range of levels to which the neuron responds (Olsen and Suga 1991a). These effects have also been documented among ventral division neurons in the cat (Aitkin and Webster 1972; Whitfield and Purser 1972; Rouiller et al. 1990). In the mustached bat, these inhibitory effects result in narrow, level-tolerant tuning curves that may contribute to the fine frequency analysis of sonar echoes (Suga and Tsuzuki 1985). This may have several functions, including (1) improved sensitivity to the magnitude of frequency (i.e., Doppler) shifts; (2) improved sensitivity to frequency-shifted CF echoes while reducing sensitivity to louder, temporally overlapping outgoing CF signals; and (3) increased sensitivity to small, periodic frequency modulations in echoes from fluttering insects.

Do circuits within the MGB contribute to inhibitory shaping of tuning curves? No studies to date have examined such effects via iontophoretic application of inhibitory transmitter antagonists. In the inferior colliculus, many of the inhibitory effects found among medial geniculate neurons have also been observed, some the result of GABAergic inhibition (Yang, Pollak, and Resler 1992; Pollak and Park, Chapter 7). While it is reasonable to suspect that inhibitory mechanisms in the cat MGB contribute to tuning sharpness (because there are many GABAergic neurons), the possibility is much less clear in the mustached bat with its lack of such neurons. If tuning sharpness is increased in the MGB, it must depend on other sources of GABAergic input or other neurotransmitters.

Some ventral division neurons may display more complex tuning than has been described thus far. In the primary auditory cortex, Fitzpatrick et al. (1993) found that many neurons tuned to the 60-kHz (CF_2) component are combination-sensitive; they are facilitated by a preceding signal tuned to frequencies in the fundamental FM biosonar component (FM_1). These have been recorded in the ventral division of the MGB (Wenstrup and Grose, in press). The origin of these responses within the ascending tonotopic system is unclear. Because preliminary studies (Mittmann and Wenstrup 1994) have reported that combination-sensitive neurons are common in the ICC, many ventral division neurons, including those responding to 60-kHz sounds,

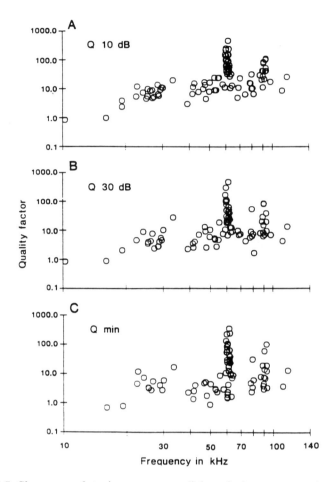

FIGURE 8.7. Sharpness of tuning among medial geniculate neurons that are not combination sensitive. Three measures of tuning were used. (A) Q_{10dB} and (B) Q_{30} dB values were calculated as the best frequency divided by the bandwidth of the tuning curve 10 dB and 30 dB above the neuron's threshold, respectively. (C) Q_{min} values were calculated as the best frequency divided by the bandwidth of the tuning curve at its widest point between threshold and 100 dB SPL (Sound Pressure Level). Although Q_{10dB} values probably reflect peripheral tuning mechanisms, Q_{30dB} and Q_{min} values also reflect the degree of neural sharpening of tuning curves. Neurons with best frequencies near 60 kHz and 90 kHz were most sharply tuned by all three measures. Data include singly tuned neurons from all parts of the medial geniculate body, not only the ventral division. (From Olsen and Suga 1991a, copyright © 1991, reprinted by permission of the American Physiological Society.)

may display sensitivity to multiple spectral elements in signals as the result of combination-sensitive ICC input.

3.3.2 Tonotopic Organization

Very few data exist on the frequency organization of the ventral division in bats. In the mustached bat, physiological mapping of best frequencies is generally consistent with data from anterograde and retrograde transport studies. Olsen (1986) recorded low-frequency responses in the lateral part of the ventral division, with higher frequencies more medially. There is a relatively large representation of the 60-kHz frequency band, corresponding to its large representation elsewhere. Also recorded are pure tone responses in the 48- to 59-kHz and 72- to 90-kHz bands, corresponding to the second and third harmonic FM components (FM_2 and FM_3, respectively) of the sonar signal (Olsen 1986). This supports connectional evidence that these frequency band representations do indeed project to the ventral division. However, their relative sizes appear to be reduced compared to the ICC. There are no published physiological studies of the frequency organization in other bats.

3.3.3 Modulation Sensitivity

Periodic modulations of frequency and amplitude have particular salience to insectivorous bats, since these modulations improve the detectability and identification of fluttering insects (see Schnitzler, Chapter 3; Pollak and Park, Chapter 7). In many other vertebrates, the ability of auditory neurons to code modulations by their temporal response pattern decreases at higher levels in the auditory pathway, and this also seems to be true in bats. Thus, cochlear nucleus/auditory nerve units (Suga and Jen 1977) lock to much higher modulation rates than do ICC units (Schuller 1979; Bodenhamer and Pollak 1983). Only one study has described modulation sensitivity in the MGB of bats. Andreeva and Lang (1977), recording evoked potentials from the horseshoe bat's MGB, were unable to demonstrate temporally locked responses to periodic amplitude modulation, whereas they obtained strong responses to such stimuli in the brainstem and inferior colliculus. Further studies with single units are needed to establish whether and in what form sensitivity to echo modulations exists among medial geniculate neurons, and also whether such sensitivity is limited to certain medial geniculate subdivisions.

3.3.4 Binaural Responses

The tonotopically organized pathway to the auditory cortex plays an important role in the analysis of sound localization cues by other mammals (Jenkins and Masterton 1982; Jenkins and Merzenich 1984; Kavanagh and Kelly 1987). In bats, there are no published accounts of the binaural responsiveness or spatial selectivity of medial geniculate neurons, and thus it is not known what role the MGB plays in analyzing and representing sound localization cues.

A preliminary report (Wenstrup 1992a) has examined projections to the MGB from aural response regions in the ICC, that is, regions characterized by a particular response to binaural sounds (Wenstrup, Ross, and Pollak 1986). Anterograde tracers were placed in parts of the 60-kHz ICC representation containing either monaural (contralaterally excited) neurons or excitatory-inhibitory (EI) neurons (excited by contralateral sound, inhibited by ipsilateral sound). Single deposits in EI regions generally labeled two target zones; one was located ventrally, in the medial part of the ventral division, while the second was located dorsally, spanning the border between the dorsal division and the medial part of the ventral division. The 60-kHz monaural region in the ICC projects mostly between the two EI inputs, to the dorsal part of the ventral division. These results suggest that aural response-specific regions may be preserved in the projection from ICC to the ventral division. Moreover, because EI neurons in ICC are sensitive to and topographically represent the interaural intensity difference (Wenstrup, Ross, and Pollak 1986), a sound localization cue, some ventral division regions may represent this target feature and project the information to primary auditory cortex.

Several questions need to be explored concerning the representation of sound location in the auditory tectothalamocortical pathway of bats, some at the cortical level but others within the MGB. Is the topographic representation of interaural intensity differences in the ICC maintained in the projection of EI neurons to the MGB? Are target elevation and azimuth analyzed by all frequency bands within sonar echoes, or are some frequencies excluded from this analysis, for example, those within the FM sweeps in the mustached bat that provide relatively little input to the ventral division? What medial geniculate subdivisions and their related cortical areas participate in analyses and representations of azimuthal and elevational spatial information?

4. The Suprageniculate Nucleus

The suprageniculate nucleus is considered to be part of the dorsal division, although it is clearly distinct from other dorsal division nuclei; others have placed it with the posterior group of thalamic nuclei (Jones and Powell 1971; Casseday et al. 1989). In the mustached bat, recent findings concerning the connections of this nucleus suggest an important role in its acoustic behavior.

The morphology of suprageniculate neurons in the mustached bat corresponds closely to that in other mammals (Casseday et al. 1989; Winer and Wenstrup 1994a,b). The principal cell is a distinctive, large, multipolar neuron with radial dendrites, about as large as the magnocellular neurons of the medial division (see Figs. 8.1 and 8.2). Medium-sized axons originating in the inferior colliculus course ventrolaterally through the suprageniculate nucleus, imparting a distinct architecture. Some ICC axons form collaterals en route to the ventral division, and these terminate within the suprageniculate nucleus (see Fig. 8.5E) (Wenstrup, Larue, and Winer 1994).

The connections of the suprageniculate nucleus are unusual in several respects. First, it receives strong input from the nucleus of the central acoustic tract (Casseday et al. 1989), a brainstem auditory nucleus that receives contralateral or possibly bilateral input from the ventral cochlear nuclei (Casseday et al. 1989) (Fig. 8.8). Physiological responses in this nucleus of the mustached bat have not been well studied, but in horseshoe bats these neurons were monaurally responsive to input from the contralateral ear (see Casseday and Covey, Chapter 5). Thus, the suprageniculate nucleus is the only thalamic auditory nucleus in the mustached bat known to receive direct auditory brainstem input. Similar, although weaker, inputs have been described in other species (Papez 1929; Morest 1965; Henkel 1983).

The suprageniculate nucleus in the mustached bat is also unusual in the input it receives from the tectum. As in other species, it receives input from the superior colliculus, but it also is the target of strong input from the central and external nuclei of the inferior colliculus (Casseday et al.

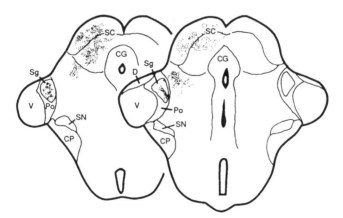

FIGURE 8.8. Anterograde labeling in the mustached bat's suprageniculate nucleus (Sg) and superior colliculus (SC) after a deposit of WGA-HRP near the nucleus of the central acoustic tract. This projection is unusual because it bypasses the inferior colliculus, proceeding to the medial geniculate body and superior colliculus by way of the central acoustic tract. (Modified from Casseday et al., copyright © 1989, reprinted by permission of Wiley-Liss, a division of John Wiley and Sons, Inc.)

1989; Wenstrup, Larue, and Winer 1994). In anterograde transport studies, Wenstrup, Larue, and Winer (1994) showed that all frequency band representations in ICC that received tracer deposits projected strongly to the suprageniculate nucleus in a frequency-specific pattern (see Figs. 8.4–8.6). Thus, lower frequency neurons (25–33 kHz; 50–59 kHz) terminate in the dorsolateral part of the nucleus, while 60-kHz inputs terminate dorsomedially (Fig. 8.6). Inputs from 80 and 90 kHz are placed more ventrally (Fig. 8.4). It is unclear how this topographic ICC projection is organized relative to inputs from the nucleus of the central acoustic tract. Moreover, because dendrites of suprageniculate neurons fill much of the nucleus, they may sample across a broad range of ICC inputs. Thus, it is unclear whether the ICC tonotopic projection will confer a physiological tonotopy onto this nucleus. The projection from the ICC is particularly interesting because it is not known to occur in any nonchiropteran species (Winer 1992).

The suprageniculate nucleus projects to broad regions of auditory cortex, but it also provides input to a limited region of frontal cortex containing robust auditory responses (Fig. 8.9) (Kobler, Isbey, and Casseday 1987). This frontal cortical region in turn projects to the superior colliculus (Kobler, Isbey, and Casseday 1987). Thus, the suprageniculate forms part of an extralemniscal (mostly) auditory subsystem involving the nucleus of the central acoustic tract, inferior collicular nuclei, the superior colliculus, and frontal cortex. Casseday et al. (1989) have suggested that the suprageniculate nucleus–frontal cortex–superior colliculus pathway serves a "priming" purpose for the activation of acousticomotor responses mediated by the superior colliculus. Inactivation of the frontal cortex during superior colliculus recordings could address possible roles of this suprageniculate nucleus–frontal cortical pathway.

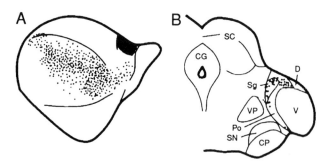

Figure 8.9A,B. Connection of the suprageniculate nucleus (Sg) with frontal cortex in the mustached bat. (A) WGA-HRP deposit site (*blackened region*) in frontal cortex and retrograde labeling in auditory cortical regions (*dots*). (B) Retrograde labeling in the suprageniculate nucleus and dorsal division (D). Suprageniculate neurons project to auditory cortex and to regions of frontal cortex. (Modified by permission from Kobler, Isbey, and Casseday, copyright © 1987 by the AAAS.)

5. The Dorsal and Rostral Medial Geniculate: Specialized Responses to Biosonar Signals

Dorsal and rostral areas of the MGB are perhaps the least understood of auditory thalamic regions. Their neurons often display longer latencies, broader tuning, and more variable responses to acoustic stimuli (Calford 1983), suggesting more complex analyses of sounds. In guinea pigs, dorsal division neurons display changes in frequency-receptive fields after the animals learned conditioned responses to tonal signals (Edeline and Weinberger 1991). In monkeys, dorsal division neurons may respond selectively to some social communication sounds (Olsen and Rauschecker 1992). In bats, they may participate in specialized analyses of biosonar echoes (see following) and social communication signals. Such findings suggest greater species differences in the structure and function of these areas than in the ventral division. Among bats, extensive studies have been reported only in the mustached bat. However, a preliminary study (Shannon and Wong 1987) suggested that the dorsal division in the little brown bat may also play a specialized role in biosonar information processing.

5.1 Architecture

The region considered here encompasses much of the dorsal and rostral MGB in the mustached bat (see Fig. 8.1C,D,G,H). In the present architectonic scheme, its major constituents are the dorsal nucleus and rostral pole nucleus, but not the lateral part of the ventral division. The architecture of the dorsal nucleus is similar to other mammals; it is distinguished from the ventral division by its population of stellate neurons and radiate neurons with weakly tufted dendrites (Winer and Wenstrup 1994a,b). The rostral pole nucleus, described by Winer and Wenstrup (1994a), constitutes a large part of the rostral MGB, replacing the medial part of the ventral division at levels into the rostral half (see Fig. 8.1). Although the architecture changes somewhat, neurons are similar to those in the ventral division, having relatively small somata ($\sim 8 \times 10$ μm). No somatic orientation is distinct. The neuronal architecture is less clear, because neurons in Golgi material are not well stained (Winer and Wenstrup 1994b).

Much of this region is considered by others (Olsen 1986; Olsen and Suga 1991a) to form the deep dorsal nucleus, and like the deep dorsal nucleus in the cat (Andersen et al. 1980; Calford and Aitkin 1983), it receives mainly high-frequency input from the ICC (Frisina, O'Neill, and Zettel 1989; Wenstrup 1992b; Wenstrup, Larue, and Winer 1994). However, the physiological properties and cortical targets appear quite different. Further studies are needed to determine the similarity of these regions to the deep dorsal nucleus of other mammals.

5.2 Connections of the Dorsal and Rostral Medial Geniculate Body

5.2.1 Inputs

Major inputs to the dorsal and rostral MGB are from the inferior colliculus, thalamic reticular nucleus, and auditory cortex. Input from the ICC is very strong, covering much of the rostral half of the medial geniculate body (see Fig. 8.4B,C) (Frisina, O'Neill, and Zettel 1989; Wenstrup, Larue, and Winer 1994). Axons from the ICC terminating caudally in the dorsal nucleus are similar to those in the underlying ventral division (see Fig. 8.5F), but the axons targeting the rostral pole nucleus and the rostral part of the dorsal nucleus are different, often diverging widely within these nuclei (see Fig. 8.5D) (Frisina, O'Neill, and Zettel 1989; Wenstrup, Larue, and Winer 1994). These axonal patterns may underlie the patchy organization of combination-sensitive responses observed in these regions (Olsen 1986).

The organization of ICC inputs differs from the ventral division (see Fig. 8.4B,C) (Wenstrup, Larue, and Winer 1994). Axons terminating in the rostral MGB originate in each of the ICC frequency band representations analyzing components in the first three harmonic elements. However, their organization does not follow a tonotopic pattern. Instead, they may be grouped according to their role in echolocation (and possibly social communication). Thus, ICC neurons tuned to CF_2 and CF_3 sonar components terminate in the dorsal and rostral pole nuclei of the middle third of the MGB. Input from frequency representations associated with FM_2 (48–60 kHz) and FM_3 (72–90 kHz) sonar components are very strong and cover much of the rostral third. Their projections appear to interdigitate, and it is these axons that bifurcate extensively. ICC neurons tuned to the FM_4 sonar component (96–120 kHz) terminate in a pattern similar to FM_2 and FM_3 inputs (Wenstrup, unpublished data). Responses to FM_4 frequencies are recorded in the same parts of the rostral MGB as are those to FM_2 and FM_3 frequencies (Olsen 1986; Olsen and Suga 1991a).

Although ICC neurons tuned to the fundamental of CF and FM sonar calls project to the rostral MGB, they terminate in different regions, generally located along the margins of the rostral MGB: the dorsal nucleus, the medial division, and the ventromedial extreme of the rostral pole nucleus (see Fig. 8.4). As described here, these results have significant implications for understanding the neuronal interactions that create combination-sensitive neurons.

The strong ICC input to the dorsal and rostral MGB differs from other mammals in both its organization and extent. In most mammals, only the ventral division receives such strong ICC input. However, some studies in other mammals have reported ICC input to the deep dorsal nucleus (Andersen et al. 1980; Calford and Aitkin 1983). Thus, to the extent that this region in the mustached bat corresponds to the deep dorsal nucleus, the connections are not entirely unconventional.

Inferior collicular input to combination-sensitive regions also originates in the external nucleus and the adjacent pericollicular tegmentum (Wenstrup and Grose 1993). These regions receive input from ICC frequency representations tuned to higher harmonics of the sonar signal (Wenstrup, Larue, and Winer 1994). Thus, combination-sensitive neurons in the MGB receive high-frequency input from several midbrain sources, including direct and indirect projections of the ICC.

Descending projections from the auditory cortex form a major input to combination-sensitive neurons of the rostral MGB. The strongest cortical input originates in layer VI of the dorsal auditory cortex (Wenstrup and Grose 1993, in press), an area examined in many physiological studies (see O'Neill, Chapter 9). Its combination-sensitive neurons are topographically arranged in separate CF/CF and FM-FM areas. The heaviest corticothalamic projection appears to connect MGB and cortical areas having similar functional properties. Thus, deposits of anterograde tracer placed in the dorsolateral extreme of auditory cortex, where neurons respond to FM-FM combinations (O'Neill and Suga 1982; Suga et al. 1983; Suga and Horikawa 1986), label the rostral half of the medial geniculate, where similar FM-FM neurons have been recorded (Olsen 1986). The converse is also true; deposits of retrograde tracer in FM-FM regions of the medial geniculate result in heavy labeling in the dorsolateral auditory cortex (Wenstrup and Grose 1993). A similar conclusion applies to corticothalamic input to CF/CF regions of the MGB.

The caudal part of the thalamic reticular nucleus provides strong input to the dorsal and rostral MGB, (Wenstrup and Grose 1993, in press). It is unclear whether specific thalamic reticular regions project to specific medial geniculate regions.

In the little brown bat (*Myotis lucifugus*), the dorsal division has been implicated in echo–delay-sensitive responses of auditory cortical neurons. Retrograde tracer, placed at cortical sites where delay-sensitive neurons were recorded, labeled neurons in the dorsal division of the medial geniculate body, outside the tonotopic axis of the ventral division (Shannon and Wong 1987). This finding reinforces the view that the dorsal division may be involved in the specialized processing of biosonar information in bats.

5.2.2 Outputs

The primary output is to restricted zones in the dorsolateral auditory cortex (Olsen 1986). These MGB regions appear to project to the appropriate CF/CF or FM-FM cortical combination-sensitive area. For example, tracer deposits at MGB sites showing CF/CF responses target a more lateral region than do deposits in MGB FM-FM regions.

5.3 Physiological Properties of Combination-Sensitive Neurons

One of the most distinctive features of the dorsal and rostral MGB in the mustached bat is the presence of large numbers of combination-sensitive neurons (Olsen and Suga 1991a,b). These neurons respond best when signals contain two spectral elements, generally corresponding to the fundamental component from the emitted biosonar pulse and a higher harmonic from the sonar echo (Fig. 8.10). (Recent evidence, however, suggests that some neurons also respond to social communication sounds [Ohlemiller, Kanwal, and Suga 1992]. There is a large body of data concerning the physiological properties of cortical combination-sensitive neurons in the mustached bat (see O'Neill, Chapter 9). This section focuses

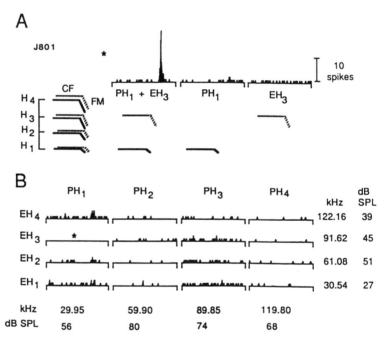

FIGURE 8.10A,B. Selectivity of an FM-FM neuron in the medial geniculate body to combinations of simulated pulse–echo pairs, shown in poststimulus time histograms. (A) For the maximum facilitated response, the neuron requires the fundamental of the pulse (PH_1) and the third harmonic of the echo (EH_3). The neuron responds poorly to either component presented alone. *Solid lines* indicate sonograms of pulse harmonics; *dashed lines*, sonograms of echo harmonics. (B) Response of same neuron to other pulse–echo pairs. The neuron did not respond well to any other combination. *Numbers at bottom* indicate constant frequency and intensity of test pulse harmonic, while *numbers at right* indicate constant frequency and intensity of the test echo harmonic. (From Olsen and Suga 1991b, copyright © 1991, reprinted by permission of the American Physiological Society.)

on the reports concerning thalamic combination-sensitive neurons and how they may differ from cortical neurons.

Two major types of combination-sensitive neurons are found in the rostral MGB and cortical areas outside the primary auditory cortex. CF/CF neurons respond best to combinations of the CF_1 component and a higher harmonic CF component (CF_n, near 60 or 90 kHz), with the frequencies of the two components being a critical stimulus feature. CF/CF neurons are believed to encode the velocity of sonar targets (Suga et al. 1983; Olsen and Suga 1991a). The second class, FM-FM neurons, responds best to a combination of spectral elements from the fundamental FM sweep and a higher harmonic FM sweep; the delay between the two components is an critical stimulus feature here. FM-FM neurons are believed to encode the distance of sonar targets (O'Neill and Suga 1982; Suga et al. 1983; Olsen and Suga 1991b). Muscimol-induced inactivation of cortical FM-FM regions decreases performance in temporal (i.e., delay) discrimination tasks by behaving bats (Riquimaroux, Gaioni, and Suga 1991).

5.3.1 CF/CF Neurons

Olsen (1986) and Olsen and Suga (1991a) examined the physiological properties of CF/CF neurons in the MGB. Only two CF combinations were recorded: CF_1/CF_2 (70%) and CF_1/CF_3 (30%). MGB CF/CF neurons typically show both a lower threshold and greater response magnitude when presented with the appropriate signal combination than in response to single tones (Fig. 8.11). The degree of facilitation varied among CF/CF neurons by 110%–5000% of the best single tone response. Reductions in threshold were correspondingly varied among units.

The facilitated response of CF/CF neurons was very sharply tuned in frequency, particularly to the higher harmonic. For the neurons shown in Figure 8.11, shifting the frequency of the CF_2 component by ± 1 kHz eliminated the response. The tuning of CF/CF neurons to CF_2 and CF_3 components was at least as sharp as singly tuned neurons responding only to these frequencies, in either the MGB or inferior colliculus. Sharpness of tuning to the CF_1 component was more variable; Olsen and Suga distinguished two populations (Fig. 8.11). Group I CF/CF neurons were relatively sharply tuned to the CF_1 (mean Q_{10dB} = 38). Their CF_1 response was always tuned within 0.7 kHz of the bat's resting frequency, and their CF_1 response area always included the resting frequency. Moreover, the best frequencies of facilitation for the CF_1 and CF_n components were in a near-exact harmonic relationship. Group II neurons differed in each of these features: they had much lower average Q_{10dB} values (9.6); they were tuned to frequencies greater than 0.7 kHz below the CF_1 resting frequency; their response areas never included the resting frequency; their best facilitation frequencies were not in an exact harmonic relationship. These two groups also differed in other ways; group I neurons did not respond

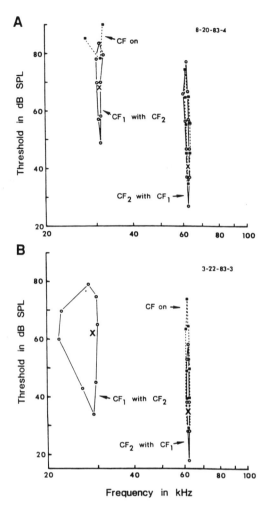

FIGURE 8.11A,B. Frequency tuning curves of CF_1/CF_2, combination-sensitive neurons, representing group I (A) and group II (B). Single tone tuning curves (*filled squares, dashed lines*); facilitative tuning curves (*open circles, solid lines*). Facilitative tuning curves were obtained by fixing the tone for one component at its best facilitative amplitude and frequency (**X**) and obtaining a tuning curve for the other component. (A) Tuning curves for group I CF/CF neuron. These neurons had sharp tuning to the CF_1 component, with the best facilitative frequency near the CF_1 resting frequency. (B) Tuning curves for group II CF/CF neuron. These neurons had broader tuning curves that excluded the CF_1 resting frequency. In both neurons, CF_2 tuning curves were very narrow. (From Olsen and Suga 1991a, copyright © 1991, reprinted by permission of the American Physiological Society.)

well when the two components were presented simultaneously, while group II neurons did.

Olsen and Suga (1991a) pointed out two significant functional aspects of the differing responses of group I and II neurons. First, each group is unresponsive to combinations of CF_1 and CF_n components within the emitted pulse or within echoes. Group I neurons are unresponsive because of their temporal sensitivity; they are inhibited when the two components are presented simultaneously. Group II neurons are unresponsive because of their spectral sensitivity; exact harmonic relationships in the pulse or in the echo do not elicit facilitation. The second feature regards the echolocation conditions under which these neurons are active; group I neurons respond to pulse–echo combinations when the bat is at rest, while group II neurons will respond when the flying bat receives Doppler-shifted CF_n echoes.

CF/CF neurons in the MGB are qualitatively similar to their counterparts in the auditory cortex, although cortical neurons may display a larger degree of facilitation (Olsen and Suga 1991a). It is unknown whether cortical neurons form populations similar to group I and II medial geniculate neurons. Nevertheless, anatomical and physiological evidence strongly suggests that medial geniculate CF/CF neurons are the source of CF/CF responses in the auditory cortex.

5.3.2 FM-FM Neurons

Medial geniculate FM-FM neurons display three response selectivities— frequency, amplitude, and delay tuning—that are well suited to encode target distance by the delay between an emitted FM_1 pulse and a returning higher harmonic FM echo (Olsen and Suga 1991b). Both the frequency and delay tuning distinguish medial geniculate FM-FM neurons from CF/CF neurons. Thus, although nearly all FM-FM neurons respond equally well to combinations of pure tone bursts as they do to FM sweeps, they are clearly tuned to frequencies within the range of the FM_1 and FM_n signal components. Moreover, they are considered to be more sharply tuned to the timing of two pulses than are CF/CF neurons.

Olsen and Suga (1991b) recorded facilitated responses to three FM combinations: FM_1-FM_2 (38%), FM_1-FM_3 (34%), and FM_1-FM_4 (28%). The frequency tuning of medial geniculate FM-FM neurons was much broader than that of CF/CF neurons, further distinguishing these two classes of medial geniculate combination-sensitive neurons (Fig. 8.12). The results of Olsen and Suga (1991b) suggest that FM-FM frequency tuning is level tolerant; facilitative tuning curves do not broaden significantly from about 10 to 50 dB or more above the threshold for facilitation. Above that level, the degree of facilitation often decreases, and many neurons have upper limits on their tuning curves. To elicit the facilitated response, FM

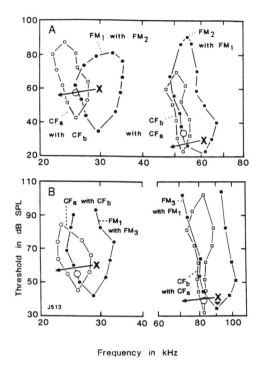

FIGURE 8.12A,B. Facilitative frequency tuning curves of FM_1-FM_2 (A) and FM_1-FM_3 (B) neurons, obtained using tone bursts and FM downsweeps. Facilitative tuning curves defined by tone bursts (*open symbols*) were obtained as described in Fig. 8.11. CF_a and CF_b refer to the test frequencies of tone bursts, where $CF_a <$ CF_b. *Large open circles* indicate best facilitative frequencies and amplitudes using tone burst stimuli. FM facilitative tuning curves (*filled symbols*) and best facilitative frequencies and amplitudes (**X**, *arrow*) are plotted using the initial frequency of the test FM sweep. To elicit a facilitated response, FM sweeps must pass into the tuning curves obtained with tone burst stimuli. Note the broader tuning of FM-FM neurons compared to CF/CF neurons (cf. Fig. 8.11). (From Olsen and Suga 1991b, copyright © 1991, reprinted by permission of the American Physiological Society.)

sweeps must pass through the facilitative tuning curves measured using combinations of tone bursts. The two facilitative tuning curves of FM-FM neurons are not in an exact harmonic relationship; generally, the higher harmonic FM curve is shifted upward. This suggests that FM-FM neurons are designed to respond to the combination of an emitted FM_1 signal and a Doppler-shifted higher harmonic FM echo. Moreover, their broader frequency tuning enables them to respond in a Doppler-tolerant fashion, that is, insensitive to the relative velocity between the bat and its target.

FM-FM neurons are also tuned in amplitude. On average, the facilitated response of FM-FM neurons was greatest about 20 dB above the threshold of their facilitated response. Significantly, both the thresholds and the best amplitudes of the facilitated response of the FM_1 component averaged 10–20 dB greater than those of the higher frequency components. Thus, the appropriate signal requires the combination of a strong FM_1 component with a weaker FM_n component. This suggests that the neurons are primarily responding to the bat's outgoing FM_1 component and the weaker FM_n component in the returning echo.

The sensitivity of FM-FM neurons to echo delay is one of their most significant features, and one closely related to the mustached bat's perception of target distance (Moss and Schnitzler, Chapter 3; Simmons et al., Chapter 4). These neurons typically display a peak response as the FM_n signal is delayed beyond the presentation of the FM_1 component (Fig. 8.13). Neurons can be characterized by their *best delay*, the delay eliciting the strongest facilitated response. In the MGB, FM-FM neurons have best delays ranging from 0 to 23 msec, although most are between 1 and 10 msec. Delay tuning curves are broader among neurons with longer best delays. Delay tuning remains relatively stable with changes in signal amplitude; significant shifts were noted mainly when amplitudes were changed near the lower and upper thresholds of the two signal components.

Olsen and Suga found several differences in stimulus locking and inhibition between medial geniculate FM-FM neurons having short best delays (<4 msec) versus long best delays (≥ 4 msec). For example, neurons with short best delays respond poorly to single tones or FM sweeps. The latency of their facilitated response to an FM_1-FM_n combination is more closely time locked to the FM_1 signal than to the delayed FM_n signal. Third, they show little evidence of inhibition immediately following the FM_1 pulse, either tested by the introduction of a second FM_1 pulse at variable delays,

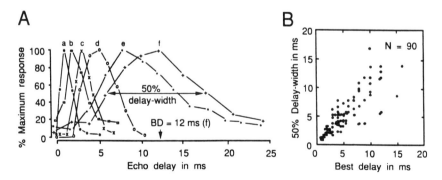

FIGURE 8.13A,B. Delay tuning of FM-FM neurons in the medial geniculate body of the mustached bat. (A) Rate versus delay functions for six representative FM-FM neurons, obtained with FM sweeps at the best facilitative amplitudes and frequencies. Best delays ranged from 0 to 23 msec. (B) Distribution of best delays and widths of delay tuning curves (50% delay width) among FM-FM neurons. Best delays were positively correlated with delay widths. (From Olsen and Suga 1991b, copyright © 1991, reprinted by permission of the American Physiological Society.)

or by the suppression of spontaneous activity. These results suggest that inhibitory mechanisms do not play a large role in the delay-sensitive, facilitated response of such neurons. One interpretation of these results is that excitatory FM_1 input initiates a brief, subthreshold facilitative period, during which the activation of an excitatory FM_n input will cause the discharge of action potentials.

FM-FM neurons with long best delays are significantly different in each of these features, suggesting fundamental differences in the mechanisms underlying their delay sensitivity. First, most such neurons respond to one or both signals presented singly, particularly to the FM_n signal (Fig. 8.14). Second, the latency of the facilitated response to FM-FM combinations is rigidly time locked to the delayed FM_n signal. Third, the FM_1 signal has an inhibitory effect. Thus, about half of these neurons show a reduction in spontaneous activity when presented with an FM_1 signal alone. In these, it appears that an early inhibitory period is triggered by the onset of the fundamental pulse, and is independent of its duration (Fig. 8.14). By examining the effect of a second FM_1 signal presented after the first, Olsen and Suga demonstrated that the second FM_1 signal could either suppress or reset the delay response. For long best-delay neurons, Olsen and Suga hypothesized that facilitation occurs when the excitation produced by a delayed FM_n signal outlasts the inhibitory period evoked by FM_1 onset, and coincides with the long-latency, FM_1-evoked excitation.

The shorter latencies of medial geniculate (compared to cortical) FM-FM neurons (Suga and Horikawa 1986; Olsen and Suga 1991b) and their input to FM-FM regions of the auditory cortex (Olsen 1986) indicate that the MGB neurons are the source of FM-FM neurons in the cortex. Olsen and Suga noted that medial geniculate neurons are very similar to cortical neurons in their delay tuning (i.e., best delays, width of delay tuning curves). However, other features differ. Nearly all medial geniculate neurons (96%) respond as well to combinations of pure tone bursts (within the appropriate frequency bands) as they do to FM sweeps, and most (83%) respond to one or both components when presented individually. In contrast, about 40% of cortical FM-FM neurons do not respond to combinations of pure tone bursts (Taniguchi et al. 1986), and most do not respond to components presented individually (Suga and Horikawa 1986). Furthermore, cortical neurons may show stronger and longer lasting facilitation. These comparisons suggest that additional processing may occur in the auditory cortex, but not with respect to delay tuning.

5.3.3 Organization of Combination-Sensitive Neurons

In the auditory cortex, distinct areas of FM-FM and CF/CF neurons occur. In each, two organizational features have been found. First, each region is topographically organized according to a particular response feature; best

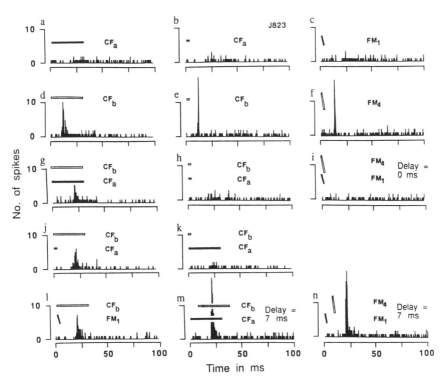

FIGURE 8.14. Inhibition in the response of an FM_1-FM_4 neuron having long best delay (7 msec). Peristimulus time histograms are shown in response to tone burst (*horizontal bars*) or FM (*oblique bars*) stimuli. *Dark bars* represent stimuli within the frequencies of the FM_1 component, while *open bars* are frequencies within the FM_4 component. CF_a and CF_b stimuli as in Fig. 8.12. (a–c) FM_1 and long or short CF_a stimuli each evoke early inhibition of spontaneous firing, followed by later increased excitability. (d–f) FM_4 and long or short CF_b stimuli evoke a short-latency, excitatory, phasic response. (g–l) When combinations of stimuli were presented at 0-msec delay, the neuron only responds if CF_b is sufficiently long to outlast a period of early inhibition evoked by FM_1 or CF_a signals (g, j, l). Moreover, the periods of inhibition and excitation are independent of the duration of FM_1 or CF_a stimuli (g, j, l). Note that the response under these conditions occurs at longer latency than the response to FM_4 or CF_b alone. The best response of this neuron is obtained when the FM_4 or CF_b signal is delayed by 7 msec. Olsen and Suga (1991b) concluded that, among long best-delay neurons, delay sensitivity is determined by a period of inhibition followed by period of increased excitability, produced by the onset of the fundamental component. (From Olsen and Suga 1991b, copyright © 1991, reprinted by permission of the American Physiological Society.)

delay for FM-FM neurons and frequency shift for CF/CF neurons. Second, the topographic organization occurs in parallel for each existing signal combination. For example, the best-delay axis of FM_1-FM_2 neurons in a cortical area is aligned with an adjacent group of FM_1-FM_4 neurons, which is in turn aligned with adjacent FM_1-FM_3 neurons (Suga et al. 1983; O'Neill, Chapter 9).

In the MGB, the organization of best-delay and frequency-shift response features is poorly understood. However, CF/CF and FM-FM neurons are segregated, with the former placed more caudally, dorsally, or laterally. Thus, Olsen and Suga (1991a) reported that CF/CF neurons were only recorded in the deep dorsal division of the central one-third of the MGB, whereas FM-FM neurons lay more rostrally in the deep dorsal division. In the architectonic scheme used here, these areas correspond to the dorsal nucleus and the adjacent rostral pole nucleus, both considered parts of the dorsal division.

This agrees well with the distribution of medial geniculate inputs from CF_n and FM_n ICC representations terminating in the dorsal and rostral MGB (Wenstrup 1992b; Wenstrup, Larue, and Winer 1994). As described previously, CF_2 and CF_3 inputs terminate dorsally in the middle one-third of the MGB, while FM_2 and FM_3 (and probably FM_4) inputs terminate more rostrally (see Fig. 8.4). Surprisingly, in neither FM-FM nor CF/CF regions of MGB is there a good correspondence between the distribution of these response properties and ICC input from the fundamental harmonic element (30–24 kHz). The implications of these results are discussed next.

5.4 Mechanisms for Constructing Combination-Sensitive Neurons

Mechanisms for constructing combination-sensitive neurons are of particular interest because these neurons represent one of the best examples of the convergence of information across frequency channels, a process that appears necessary for the analysis of spectrally complex sounds. Although facilitated CF/CF and FM-FM neurons have been recorded in the MGB and auditory cortex, they were not found in the ICC (O'Neill 1985). Thus, one possibility is that such responses arise in the MGB as the result of a convergence of fundamental and higher harmonic inputs from separate parts of the tonotopically organized ICC. To date, however, such convergence has not been supported by anatomical studies.

5.4.1 Connectional Evidence

Tracer studies have failed to demonstrate a significant direct projection from ICC representations of the fundamental sonar component to combination-sensitive areas of the MGB. Wenstrup, Larue, and Winer

(1994) examined the anterograde label resulting from deposits of tritiated leucine at ICC loci responding to frequencies in the FM_1 or CF_1 components; they found very little overlap with the anterograde label obtained in other experiments involving deposits in higher frequency, FM_n or CF_n representations. Wenstrup and Grose (Wenstrup, 1992b, in press), in dual anterograde tracer experiments, placed deposits of wheat germ agglutinin conjugated to horseradish peroxidase (WGA-HRP) into FM_1 or CF_1 representations and deposits of biocytin into FM_2, CF_2, or FM_3 representations. There was little overlap between the projections from the fundamental and the higher harmonic representations in the rostral MGB. In further experiments, retrograde tracers placed in medial geniculate combination-sensitive regions strongly labeled higher harmonic CF_n or FM_n representations in the ICC, but labeled few or no cells in representations tuned to the fundamental sonar component (Wenstrup and Grose 1993, in press). These retrograde experiments also showed little evidence that other parts of the MGB, including the ventral division and other MGB regions receiving low-frequency input, projected to combination-sensitive regions.

These studies raise doubts that ICC neurons tuned to the fundamental sonar component project directly or via another MGB subdivision onto combination-sensitive neurons in the rostral and dorsal MGB. Other possibilities must be investigated. For example, low-frequency input may arrive via an indirect pathway, possibly involving a combination of two or more of the following: the lateral part of the ventral division in the MGB, the thalamic reticular nucleus, or the auditory cortex. Alternatively, combination-sensitive responses may already exist among some neurons providing input to these MGB regions. For example, the external nucleus of the inferior colliculus and the pericollicular tegmentum both provide input to combination-sensitive MGB regions (Wenstrup and Grose 1993), but their physiological properties have not been explored in the mustached bat. Recently, Mittmann and Wenstrup (1994) reported that combination-sensitive neurons occur in the ICC. This finding may explain the surprising lack of CF_1 or FM_1 inputs from the ICC to combination-sensitive regions of the MGB, because the necessary frequency convergence may occur at auditory levels below the MGB. These results suggest that high frequency representations of the ICC supply combination-sensitive response properties to MGB neurons.

5.4.2 Pharmacological Evidence

Preliminary studies by Butman and Suga (1989, 1990; Butman 1992) examined receptor mechanisms involved in the response of FM-FM neurons. For some MGB neurons having long best delays, local application of bicuculline to block $GABA_A$ receptors had the effect of shifting the neuron's best delay from long to short. This suggests that both GABAergic and non-GABAergic FM_1 inputs synapse onto combination-sensitive medial geniculate neurons.

Butman and Suga (1990; Butman 1992) also examined the excitatory amino acid sensitivity of onset and later burst components in the response of delay-sensitive neurons. Local application of APV, an N-methyl-D-aspartate (NMDA) receptor channel antagonist, eliminated the burst response, but left the delay-tuned onset response unaffected. In contrast, CNQX, a non-NMDA glutamate receptor antagonist, eliminated the onset response but left the delay-tuned burst response unaffected. A nonspecific excitatory amino acid antagonist, kyenurenic acid, eliminated both features of the component. Thus, different excitatory amino acid receptor mechanisms mediate different components to the delay-tuned response. It is not clear whether these results apply specifically to short best-delay FM-FM neurons or to the entire population.

5.4.3 Origin of Delay Tuning in Combination-Sensitive Neurons

Delay-tuned FM-FM neurons in the MGB function as coincidence detectors (Suga, Olsen, and Butman 1990; Olsen and Suga 1991b). Facilitation occurs when excitatory FM_1 and FM_n influences overlap temporally. Because the emitted FM_1 signal precedes the echo FM_n signal by several milliseconds (5.8 msec per meter of bat-target distance), the FM_1 excitatory influence must be delayed neurally to coincide with the FM_n excitatory influence.

Several mechanisms have been proposed to explain aspects of the delayed FM_1 excitation. Kuwabara and Suga (1993) reported evidence suggesting that FM_1 neural delays are created at levels below the MGB. Recording from the brachium of the inferior colliculus, presumably among axons exiting the ICC, they found a broad distribution of FM_1 latencies (3.5–15.0 msec) and a restricted distribution of FM_n latencies (3.8–6.5 msec). They concluded that mechanisms below the MGB create FM_1 delays that may account for much of the delay tuning observed among MGB neurons. However, others have reported a broader distribution of latencies among ICC neurons responding to FM_n signals (Hattori and Suga 1989; Park and Pollak 1993; Pollak and Park, Chapter 7). These studies suggest that the distribution of latencies is broad among both FM_1 and FM_n neurons in the ICC, and that delay sensitivity results from the match of appropriately timed FM_1 and FM_n inputs at the level of the MGB.

Physiological differences between short and long best-delay neurons suggest different mechanisms in their creation. Short best-delay neurons show no evidence that inhibition contributes to their delay sensitivity (Olsen and Suga 1991b); these may be constructed by a facilitating convergence that matches FM_1 and FM_n inputs having the appropriate latencies. Because these latencies need to differ by no more than 4 msec for short best-delay neurons, variations in FM_1 and FM_n latencies observed among ICC neurons could easily account for the delay sensitivity.

FM-FM neurons having long best delays seem to require a different mechanism. In physiological (see Fig. 8.14) (Olsen and Suga 1991b) and pharmacological (Butman and Suga 1989) studies, these neurons clearly showed the effects of an FM_1-elicited inhibitory mechanism. The source of the inhibition is not clear. It is unlikely that an intrageniculate inhibitory pathway exists, because so few intrinsic GABAergic neurons can be found in the MGB (Vater, Kössl, and Horn 1992; Winer, Wenstrup, and Larue 1992), and because there is little evidence of a projection to the rostral and dorsal MGB from other low-frequency MGB regions (Wenstrup and Grose 1993). If this inhibition occurs in the MGB, one possibility is that it originates from the thalamic reticular nucleus, a nucleus that is GABAergic in the mustached bat (Winer, Wenstrup, and Larue 1992), contains neurons tuned between 24 and 30 kHz (Olsen, unpublished data; Wenstrup, unpublished data), and projects to FM-FM regions of the MGB (Olsen 1986; Wenstrup and Grose 1993). In these long best-delay neurons, delay sensitivity may depend on different FM_1 input latencies, on the kinetics of GABA-activated membrane channels, or on both.

The selectivity of CF/CF neurons for the delay between the CF_1 and CF_n components is considered to be relatively broad (Olsen and Suga 1991a), perhaps reflecting differences in the mechanisms which form CF/CF versus FM-FM neurons (Park and Pollak 1993). However, delay sensitivity of CF/CF neurons is generally tested using simulated biosonar signals with long CF components or relatively long (30 msec) tone bursts, not the brief tone bursts or FM sweeps used to evaluate delay sensitivity in FM-FM neurons. Such long CF stimuli, although biologically appropriate, probably broaden the apparent delay tuning of CF/CF neurons.

For each of the mechanisms thought to underlie delay tuning, the assumption has been that MGB neurons are the coincidence detectors. However, the lack of FM1 input to combination-sensitive MGB regions (Wenstrup and Grose 1993, in press) and the finding of combination-sensitive neurons in the inferior colliculus (Mittmann and Wenstrup 1994) suggest that integration may occur lower than the MGB, and that mechanisms that create delay lines may operate on brain stem inputs to the ICC and on the ICC itself. Thus, several questions remain concerning the mechanisms of the delay-tuned facilitation. These include the site(s) at which delay-tuned responses are constructed, the sources of FM_1 input, and the mechanism(s) by which FM_1 input is delayed relative to FM_n inputs.

6. The Thalamic Reticular Nucleus

The thalamic reticular nucleus in mammals is a sheet of neurons located along the lateral and rostral margins of the thalamus, just medial to the internal capsule. It receives major inputs from nuclei of the dorsal thalamus and from corresponding cortical regions, and its main output is to the nuclei of the dorsal thalamus (Scheibel and Scheibel 1966; Jones 1975). The thalamic reticular nucleus is composed of sectors related to specific dorsal thalamic nuclei (Jones 1975). All or nearly all the neurons are GABAergic (Houser et al. 1980), thus providing a negative feedback circuit to regulate the activity of specific thalamic relay neurons. It has been implicated in several aspects of dorsal thalamic activity (see review by Shosaku et al. 1989): (1) in the generation of electroencephalogram oscillations that appear in states of reduced alertness, (2) in setting excitation levels of thalamic relay neurons, and (3) in postexcitatory and surround inhibition. In cats, thalamic reticular neurons of the auditory sector display frequent bursts of spontaneous discharge. In response to auditory stimuli, they have longer latencies and broader and more complex frequency tuning than neurons in the medial geniculate body (Simm et al. 1990).

In the mustached bat, the thalamic reticular nucleus may provide the strongest GABAergic influence on neurons of the MGB, including combination-sensitive neurons. In view of its potential role in auditory thalamic signal processing in bats, this review considers its structure and possible functions.

6.1 Anatomy

The thalamic reticular nucleus in bats is well developed, although its location and shape differ somewhat from other species. Caudally, the thalamic reticular nucleus is located just ventral to the rostral MGB, lying directly over the internal capsule and rostral cerebral peduncle (see Fig. 8.1H). Its shape is unconventional at this point, being several cell layers thick, and the neurons are scattered among fascicles of the auditory radiation. More rostrally, the nucleus assumes its more conventional appearance, a laminated shell of neurons placed along the lateral and rostral borders of the thalamus (Fig. 8.15). In the mustached bat, thalamic reticular neurons are GABAergic (Fig. 8.15) (Winer, Wenstrup, and Larue 1992) and are intensely labeled by antibodies to parvalbumin (see Fig. 8.1H).

Connectional evidence suggests that the caudal, unconventionally shaped part of the thalamic reticular nucleus is the auditory sector (Wenstrup and Grose 1993, in press). Thus, the caudal region alone is labeled by medial geniculate deposits of retrograde tracer. Whether there is a finer organization of connections with the MGB is unclear, although preliminary evidence suggests that the ventral division of the MGB receives input from somewhat different parts than does the dorsal division (Wenstrup and Grose 1993).

In bats, inputs to the thalamic reticular nucleus have not been examined in detail, although Olsen (1986) reported a projection from the MGB. In other species, both the MGB and auditory cortical fields provide input to the thalamic reticular nucleus (Jones 1975; Conley, Kupersmith, and Diamond 1991). The connections are of particular interest in the mustached bat, because the nucleus is a possible source of low-frequency input to combination-sensitive neurons of the MGB.

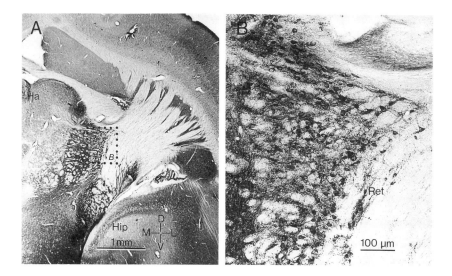

FIGURE 8.15A,B. GAD-immunolabeling of the thalamic reticular nucleus. (A) Overview of the rostral thalamus in transverse section. *Dotted square* frames the region shown at higher power in (B). (B) GAD-immunopositive neurons of the thalamic reticular nucleus (Ret) are evident. (From Winer, Wenstrup, and Larue, copyright © 1992, reprinted by permission of Wiley-Liss, a division of John Wiley and Sons, Inc.)

6.2 Physiological Responses

Very little is known about the physiology of neurons in the thalamic reticular nucleus of bats. Unpublished studies in the mustached bat (Olsen; Wenstrup) have recorded auditory responses, predominantly to frequencies in the 24- to 30-kHz band of the fundamental sonar component. These results, while preliminary, are consistent with the possible role of thalamic reticular neurons as a source of low-frequency input to combination-sensitive regions of the MGB. Such results suggest that the thalamic reticular nucleus may be involved directly in signal processing mechanisms of the MGB rather than in a modulatory role alone. Further studies clearly are necessary to address the role of this nucleus in signal processing by medial geniculate neurons in bats.

7. Summary and Conclusions

The mustached bat's medial geniculate body is composed of three major divisions distinguished by neuronal populations, connections, and physiological responses. The ventral division, part of the ascending tonotopic system, contains sharply tuned, tonotopically organized neurons. The dorsal division contains several subdivisions; most noteworthy are the dorsal and rostral parts whose combination-sensitive neurons are specialized for the analysis of species-specific, complex sounds. Suprageniculate neurons, also of the dorsal division, participate in separate pathways involving the frontal cortex and the superior colliculus. The medial division, although clearly different from the others, is not well understood. The studies in the mustached bat demonstrate in a striking way what different functional roles are played by the parallel subsystems that involve medial geniculate nuclei.

These subsystems—their mechanisms and their behavioral roles—need further study. For instance, the functional properties of the suprageniculate–frontal cortex system are poorly understood. What are the physiological responses within these areas, and how do they contribute to the acoustic behavior of the mustached bat? What processing of biosonar information occurs in the ventral division? Do frequencies in the echoes of FM components participate in the analysis of target elevation and azimuth? What mechanisms are responsible for the frequency convergence and temporal selectivity of combination-sensitive neurons? If combination-sensitive neurons are constructed at levels below the MGB, how are these responses modified in the MGB by inputs from the auditory cortex and thalamic reticular nucleus.

Necessarily, this review has focused on the mustached bat. A major gap is our lack of knowledge about medial geniculate processing in other bat species. Even the limited data from other species demonstrate sharp contrasts with the mustached bat, for example, in GABAergic inhibition and in responses encoding target distance. Basic studies of structure and function, as well as specific investigations of processing mechanisms in these other bats are likely to demonstrate different but equally interesting mechanisms of complex signal processing and neural correlates of echolocation.

Most medial geniculate subdivisions in bats correspond well to those described in other species, yet each shows modifications that may be related to the demanding acoustic behavior of bats. Even the tonotopic ventral division shows these effects in the mustached bat, whereby certain frequency bands within the biosonar signal may be underrepresented. Other parts, including the suprageniculate nucleus and rostral medial geniculate body may have undergone significantly greater modification, with different connections and response properties. These observations suggest how diverse are the roles of cell groups in the medial geniculate body of different species. However, much further work on bats is needed to understand whether these modifications are specific adaptations or part of broader phylogenetic trends.

Acknowledgments. The author thanks E. Covey for sharing unpublished data and Ms. Carol Grose for assistance in experiments and preparing the figures. This work was supported by the National Institute for Deafness and Other Communication Disorders (DC00937).

Abbreviations

Anatomical Names

ALD	anterolateral division of the inferior colliculus
BIC	brachium of the inferior colliculus
CG	central gray
CP	cerebral peduncle
D	dorsal nucleus *or* dorsal division of the medial geniculate body
DC	dorsal cortex of inferior colliculus
DNLL	dorsal nucleus of the lateral lemniscus
DPD	dorsoposterior division of the inferior colliculus
DS	superficial dorsal nucleus of the medial geniculate body
Ex	external nucleus of the inferior colliculus
Ha	habenula
Hip	hippocampus
ICC	central nucleus of the inferior colliculus
M	medial division of the medial geniculate body
MD	medial division of the inferior colliculus
MGB	medial geniculate body
MRF	mesencephalic reticular formation
PL	posterior limitans nucleus
Po	posterior thalamic nuclear group
Pt	pretectum
Pyr	pyramid
RP	rostral pole nucleus of the medial geniculate body
Rt, Ret	thalamic reticular nucleus
SC	superior colliculus
Sg	suprageniculate nucleus of the medial geniculate body
SN	substantia nigra
V	ventral division of the medial geniculate body
Vl	lateral part of the ventral division of the medial geniculate body
Vm	medial part of the ventral division of the medial geniculate body
VP	ventroposterior nucleus
ZI	zona incerta

Planes of Section

D	dorsal
L	lateral
M	medial
V	ventral

Other Abbreviations

CF	constant frequency
CF/CF	combination-sensitive neuron responding to CF components of the sonar signal
CF_a, CF_b	tone burst test frequencies used for FM-FM neurons
CF_n	nth harmonic of constant-frequency biosonar component
FM	frequency modulated
FM-FM	combination-sensitive neuron responding to FM components of the sonar signal
FM_n	nth harmonic of frequency modulated biosonar component
GABA	γ-aminobutyric acid
GAD	glutamic acid decarboxylase
HRP	horseradish peroxidase
$Q_{10\ dB}$	tuning sharpness expressed as best frequency divided by the bandwidth 10 decibels above threshold
$Q_{30\ dB}$	tuning sharpness expressed as best frequency divided by the bandwidth 30 decibels above threshold
Q_{min}	tuning sharpness expressed as best frequency divided by the maximum bandwidth of the tuning curve
WGA-HRP	wheat germ agglutinin conjugated to horseradish peroxidase

References

Aitkin LM, Webster WR (1972) Medial geniculate body of the cat: organization and responses to tonal stimuli of neurons in ventral division. J Neurophysiol (Bethesda) 35:365–380.

Andersen RA, Roth GL, Aitkin LM, Merzenich MM (1980) The efferent projections of the central nucleus and the pericentral nucleus of the inferior colliculus in the cat. J Comp Neurol 194:649–662.

Andreeva NG, Lang TT (1977) Evoked responses of the superior olive to amplitude-modulated signals. Neurosci Behav Physiol 8:306–310.

Aoki E, Semba R, Keino H, Kato K, Kashiwamata S (1988) Glycine-like immunoreactivity in the rat auditory pathway. Brain Res 442:63–71.

Ayrapet'yants ES, Konstantinov AI (1974) Echolocation in Nature. Arlington, VA: Joint Publications Research Service.

Baron G (1974) Differential phylogenetic development of the acoustic nuclei among chiroptera. Brain Behav Evol 9:7–40.

Bodenhamer RD, Pollak GD (1983) Response characteristics of single units in the inferior colliculus of mustache bats to sinusoidally frequency modulated signals. J Comp Physiol 153:67–79.

Butman JA (1992) Inhibitory and excitatory mechanisms of coincidence detection in delay tuned neurons of the mustache bats. In: Proceedings of the 3d International Congress in Neuroethology. McGill University, Montreal.

Butman JA, Suga N (1989) Bicuculline modifies the delay-tuning of FM-FM neurons in the mustached bat. Soc Neurosci Abstr 15:1293.

Butman JA, Suga N (1990) NMDA receptors are essential for delay-dependent facilitation in FM-FM neurons in the mustached bat. Soc Neurosci Abstr 16:795.

Calford MB (1983) The parcellation of the medial geniculate body of the cat defined by the auditory response properties of single units. J Neurosci 3:2350-2364.

Calford MB, Aitkin LM (1983) Ascending projections to the medial geniculate body of the cat: evidence for multiple, parallel auditory pathways through thalamus. J Neurosci 3:2365-2380.

Casseday JH, Pollak GD (1988) Parallel auditory pathways: I. Structure and connections. In: Nachtigall PE, Moore PWB (eds) Animal Sonar: Processes and Performance. New York: Plenum, pp. 169-196.

Casseday JH, Kobler JB, Isbey SF, Covey E (1989) Central acoustic tract in an echolocating bat: an extralemniscal auditory pathway to the thalamus. J Comp Neurol 287:247-259.

Celio MR (1990) Calbindin D-28k and parvalbumin in the rat nervous system. Neuroscience 35:375-475.

Clarey JC, Barone P, Imig TJ (1992) Physiology of thalamus and cortex. In: Popper AN, Fay RR (eds) Springer Handbook of Auditory Research, Vol. 2, The Mammalian Auditory Pathway: Neurophysiology. New York: Springer-Verlag, pp. 232-334.

Clerici WJ, Coleman JR (1990) Anatomy of the rat medial geniculate body: I. Cytoarchitecture, myeloarchitecture and neocortical connectivity. J Comp Neurol 297:14-31.

Clerici WJ, McDonald AJ, Thompson R, Coleman JR (1990) Anatomy of the rat medial geniculate body: II. Dendritic morphology. J Comp Neurol 297:32-54.

Conley M, Kupersmith AC, Diamond IT (1991) The organization of projections from subdivisions of the auditory cortex and thalamus to the auditory sector of the thalamic reticular nucleus in *Galago*. Eur J Neurosci 3:1089-1103.

Covey E, Hall WC, Kobler JB (1987) Subcortical connections of the superior colliculus in the mustache bat, *Pteronotus parnellii*. J Comp Neurol 263:179-197.

Edeline J-M, Weinberger NM (1991) Subcortical adaptive filtering in the auditory system: associative receptive field plasticity in the dorsal medial geniculate body. Behav Neurosci 105:154-175.

Fitzpatrick DC, Kanwal JS, Butman JA, Suga N (1993) Combination-sensitive neurons in the primary auditory cortex of the mustached bat. J Neurosci 13:931-940.

Frisina RD, O'Neill WE, Zettel ML (1989) Functional organization of mustached bat inferior colliculus: II. Connections of the FM_2 region. J Comp Neurol 284:85-107.

Hashikawa T, Rausell E, Molinari M, Jones EG (1991) Parvalbumin- and calbindin-containing neurons in the monkey medial geniculate complex: differential distribution and cortical layer specific projections. Brain Res 544:335-341.

Hattori T, Suga N (1989) Delay lines in the inferior colliculus of the mustached bat. Soc Neurosci Abstr 15:1293.

Henkel CK (1983) Evidence of sub-collicular auditory projections to the medial geniculate nucleus in the cat: an autoradiographic and horseradish peroxidase study. Brain Res 259:21-30.

Houser CR, Vaughn JE, Barber RP, Roberts E (1980) GABA neurons are the major cell type of the nucleus reticularis thalami. Brain Res 200:345-354.

Jenkins WM, Masterton RB (1982) Sound localization: effects of unilateral lesions in central auditory system. J Neurophysiol (Bethesda) 47:987–1016.

Jenkins WM, Merzenich MM (1984) Role of cat primary auditory cortex for sound-localization behavior. J Neurophysiol (Bethesda) 52:819–847.

Jones EG (1975) Some aspects of the organization of the thalamic reticular complex. J Comp Neurol 162:285–308.

Jones EG, Powell TPS (1971) An analysis of the posterior group of thalamic nuclei on the basis of its afferent connections. J Comp Neurol 143:185–216.

Kavanagh GL, Kelly JB (1987) Contribution of auditory cortex to sound localization by the ferret (*Mustela putorius*). J Neurophysiol (Bethesda) 57:1746–1766.

Kobler JB, Isbey SF, Casseday JH (1987) Auditory pathways to the frontal cortex of the mustache bat, *Pternontus parnellii*. Science 236:824–826.

Kudo M, Niimi K (1978) Ascending projections of the inferior colliculus onto the medial geniculate body in the cat studied by anterograde and retrograde tracing techniques. Brain Res 155:113–117.

Kudo M, Niimi K (1980) Ascending projections of the inferior colliculus in the cat: an autoradiographic study. J Comp Neurol 191:545–556.

Kuwabara N, Suga N (1993) Delay lines and amplitude selectivity are created in subthalamic auditory nuclei: the brachium of the inferior colliculus of the mustached bat. J Neurophysiol (Bethesda) 69:1713–1724.

Majorossy K, Kiss A (1976) Specific patterns of neuron arrangement and of synaptic articulation in the medial geniculate body. Exp Brain Res 26:1–17.

Mitani A, Itoh K, Mizuno N (1987) Distribution and size of thalamic neurons projecting to layer I of the auditory cortical fields of the cat compared to those projecting to layer IV. J Comp Neurol 257:105–121.

Mittmann DH, Wenstrup JJ (1994) Combination-sensitive neurons in the inferior colliculus of the mustached bat. In: Proceedings of the 17th Midwinter Meeting of the Association for Research in Otolaryngology, p 93. Association for Research in Otolaryngology, Des Moines.

Morest DK (1964) The neuronal architecture of the medial geniculate body of the cat. J Anat (Lond) 98:611–630.

Morest DK (1965) The laminar structure of the medial geniculate body of the cat. J Anat (Lond) 99:143–160.

Morest DK (1975) Synaptic relationships of golgi type II cells in the medial geniculate body of the cat. J Comp Neurol 162:157–193.

Morest DK, Winer JA (1986) The comparative anatomy of neurons: homologous neurons in the medial geniculate body of the opossum and the cat. Adv Anat Embryol Cell Biol 97:1–96.

Niimi K, Naito F (1974) Cortical projections of the medical geniculate body in the cat. Exp Brain Res 19:326–342.

Ohlemiller KK, Kanwal JS, Suga N (1992) Responses of cortical and thalamic FM-FM and CF/CF neurons of the mustached bat to species-specific communication sounds. Soc Neurosci Abstr 18:883.

Oliver DL (1982) A Golgi study of the medial geniculate body in the tree shrew (*Tupaia glis*). J Comp Neurol 209:1–16.

Olsen JF (1986) Processing of biosonar information by the medial geniculate body of the mustached bat, *Pteronotus parnellii*. Ph.D. dissertation, Washington University, St. Louis, MO.

Olsen JF, Rauschecker JP (1992) Medial geniculate neurons in the squirrel monkey sensitive to combinations of components in a species-specific vocalization. Soc Neurosci Abstr 18:883.

Olsen JF, Suga N (1991a) Combination-sensitive neurons in the medial geniculate body of the mustached bat: encoding of relative velocity information. J Neurophysiol (Bethesda) 65:1254–1274.

Olsen JF, Suga N (1991b) Combination-sensitive neurons in the medial geniculate body of the mustached bat: encoding of target range information. J Neurophysiol (Bethesda) 65:1275–1296.

O'Neill WE (1985) Responses to pure tones and linear FM components of the CF-FM biosonar signal by single units in the inferior colliculus of the mustached bat. J Comp Physiol A 157:797–815.

O'Neill WE, Suga N (1982) Encoding of target range and its representation in the auditory cortex of the mustached bat. J Neurosci 2:17–31.

Papez JW (1929) Central acoustic tract in cat and man. Anat Rec 42:60.

Park TJ, Pollak GD (1993) GABA shapes a topographic organization of response latency in the mustache bat's inferior colliculus. J Neurosci 13:5172–5187.

Pollak GD, Bodenhamer RD (1981) Specialized characteristics of single units in inferior colliculus of mustache bat: frequency representation, tuning, and discharge patterns. J Neurophysiol (Bethesda) 46:605–620.

Riquimaroux H, Gaioni SJ, Suga N (1991) Cortical computational maps control auditory perception. Science 251:565–568.

Rodrigues-Dagaeff C, Simm G, de Ribaupierre Y, Villa A, de Ribaupierre F, Rouiller EM (1989) Functional organization of the ventral division of the medial geniculate body of the cat: evidence for a rostro-caudal gradient of response properties and cortical projections. Hear Res 39:103–126.

Rouiller EM, de Ribaupierre F (1985) Origin of afferents to physiologically defined regions of the medial geniculate body of the cat: ventral and dorsal divisions. Hear Res 19:97–114.

Rouiller EM, Rodrigues-Dagaeff C, Simm G, DeRibaupierre Y, Villa A, DeRibaupierre F (1989) Functional organization of the medial division of the medial geniculate body of the cat: tonotopic organization, spatial distribution of response properties and cortical connections. Hear Res 39:127–142.

Rouiller EM, Capt M, Hornung JP, Streit P (1990) Correlation between regional changes in the distributions of GABA-containing neurons and unit response properties in the medial geniculate body of the cat. Hear Res 49:249–258.

Scheibel ME, Scheibel AB (1966) The organization of the nucleus reticularis thalami: a Golgi study. Brain Res 1:43–62.

Schuller G (1979) Coding of small sinusoidal frequency and amplitude modulations in the inferior colliculus of 'CF-FM' bat, *Rhinolophus ferrumequinum*. Exp Brain Res 34:117–132.

Schweizer H (1981) The connections of the inferior colliculus and the organization of the brainstem auditory system in the greater horseshoe bat (*Rhinolophus ferrumequinum*). J Comp Neurol 201:25–49.

Shannon S, Wong D (1987) Interconnections between the medial geniculate body and the auditory cortex in an FM bat. Soc Neurosci Abstr 13:1469.

Shosaku A, Kayama Y, Sumitomo I, Sugitani M, Iwama K (1989) Analysis of recurrent inhibitory circuit in rat thalamus: neurophysiology of the thalamic reticular nucleus. Prog Neurobiol 32:77–102.

Simm GM, de Ribaupierre F, de Ribaupierre Y, Rouiller EM (1990) Discharge properties of single units in auditory part of reticular nucleus of thalamus in cat. J Neurophysiol (Bethesda) 63:1010–1021.

Sousa-Pinto A (1973) Cortical projections of the medial geniculate body of the cat. Adv Anat Embryol Cell Biol 48:1–42.

Suga N, Horikawa J (1986) Multiple time axes for representation of echo delays in the auditory cortex of the mustached bat. J Neurophysiol (Bethesda) 55:776–805.

Suga N, Jen PH-S (1977) Further studies on the peripheral auditory system of the CF-FM bats specialized for fine frequency analysis of Doppler-shifted echoes. J Exp Biol 69:207–232.

Suga N, Manabe T (1982) Neural basis of amplitude-spectrum representation in auditory cortex of the mustached bat. J Neurophysiol (Bethesda) 47:225–254.

Suga N, Olsen JF, Butman JA (1990) Specialized subsystems for processing biologically important complex sounds: cross-correlation analysis for ranging in the bat's brain. Cold Spring Harbor Symp Quant Biol 55:585–597.

Suga N, Tsuzuki K (1985) Inhibition and level-tolerant frequency tuning in the auditory cortex of the mustached bat. J Neurophysiol (Bethesda) 53:1109–1145.

Suga N, Simmons JA, Jen PH-S (1975) Peripheral specialization for fine analysis of doppler-shifted echoes in the auditory system of the "CF-FM" bat *Pteronotus parnellii*. J Exp Biol 63:161–192.

Suga N, O'Neill WE, Kujirai K, Manabe T (1983) Specificity of combination-sensitive neurons for processing of complex biosonar signals in auditory cortex of the mustached bat. J Neurophysiol (Bethesda) 49:1573–1626.

Takahashi TT, Carr CE, Brecha N, Konishi M (1993) Calcium binding protein-like immunoreactivity labels the terminal field of nucleus laminaris of the barn owl. J Neurosci 7:1843–1856.

Taniguchi I, Niwa H, Wong D, Suga N (1986) Response properties of FM-FM combination-sensitive neurons in the auditory cortex of the mustached bat. J Comp Physiol A 159:331–337.

Vasil'ev AG, Andreeva NG (1972) Characteristics of electrical responses by the medial geniculate bodies in vespertilionidae and rhinolophidae to ultrasonic stimuli with different frequencies. Neurophysiology 3:104–109.

Vater M, Braun K (1994) Parvalbumin, calbindin D-28k, and calretinin immuno-reactivity in the ascending auditory pathway of horseshoe bats. J Comp Neurol 341:534–558.

Vater M, Kössl M, Horn AKE (1992) GAD- and GABA-immunoreactivity in the ascending auditory pathway of horseshoe and mustached bats. J Comp Neurol 325:183–206.

Wenstrup JJ (1992a) Monaural and binaural regions of the mustached bat's inferior colliculus project differently to targets in the medial geniculate body and pons. In: Proceedings of the 16th Midwinter Meeting of the Association for Research in Otolargnology, p. 77.

Wenstrup JJ (1992b) Inferior colliculus projections to the medial geniculate body: a study of the anatomical basis of combination-sensitive neurons in the mustached bat. Soc Neurosci Abstr 18:1039.

Wenstrup JJ, Grose CD (1993) Inputs to combination-sensitive neurons in the medial geniculate body of the mustached bat. Soc Neurosci Abstr 19:1426.

Wenstrup JJ, Grose CD Inputs to combination-sensitive neurons in the medial geniculate body of the mustached bat: the missing fundamental. J Neurosci (in press).

Wenstrup JJ, Ross LS, Pollak GD (1986) Binaural response organization within a frequency-band representation of the inferior colliculus: implications for sound localization. J Neurosci 6:962–973.

Wenstrup JJ, Larue DT, Winer JA (1994) Projections of physiologically defined subdivisions of the inferior colliculus in the mustached bat: targets in the medial geniculate body and extrathalamic nuclei. J Comp Neurol 346:207–236.

Wenthold RJ, Hunter C (1990) Immunocytochemistry of glycine and glycine receptors in the central auditory system. In: Ottersen OP, Storm-Mathisen J (eds) Glycine Neurotransmission. Chichester: Wiley, pp. 391–416.

Whitfield IC, Purser D (1972) Microelectrode study of the medial geniculate body in unanaesthetized free-moving cats. Brain Behav Evol 6:311–322.

Winer JA (1984) The human medial geniculate body. Hear Res 15:225–247.

Winer JA (1985) The medial geniculate body of the cat. Adv Anat Embryol Cell Biol 86:1–98.

Winer JA (1992) The functional architecture of the medial geniculate body and the primary auditory cortex. In: Webster DB, Popper AN, Fay RR (eds) Springer Handbook of Auditory Research, Vol. 1, The Mammalian Auditory Pathway: Neuroanatomy. New York: Springer-Verlag, pp. 222–409.

Winer JA, Larue DT (1988) Anatomy of glutamic acid decarboxylase (GAD) immunoreactive neurons and axons in the rat medial geniculate body. J Comp Neurol 278:47–68.

Winer JA, Morest DK (1983) The medial division of the medial geniculate body of the cat: implications for thalamic organization. J Neurosci 3:2629–2651.

Winer JA, Wenstrup JJ (1994a) Cytoarchitecture of the medial geniculate body in the mustached bat (Pteronotus parnellii). J Comp Neurol (346:161–182).

Winer JA, Wenstrup JJ (1994b) The neurons of the medial geniculate body in the mustached bat (Pteronotus parnellii). J Comp Neurol (346:183–206).

Winer JA, Morest DK, Diamond IT (1988) A cytoarchitectonic atlas of the medial geniculate body of the opossum, Didelphys virginiana, with a comment on the posterior intralaminar nuclei of the thalamus. J Comp Neurol 274:422–448.

Winer JA, Wenstrup JJ, Larue DT (1992) Patterns of GABAergic immunoreactivity define subdivisions of the mustached bat's medial geniculate body. J Comp Neurol 319:172–190.

Yang L, Pollak GD, Resler C (1992) GABAergic circuits sharpen tuning curves and modify response properties in the mustache bat inferior colliculus. J Neurophysiol (Bethesda) 68:1760–1774.

Zettel ML, Carr CE, O'Neill WE (1991) Calbindin-like immunoreactivity in the central auditory system of the mustached bat, Pteronotus parnellii. J Comp Neurol 313:1–16.

Zook JM, Winer JA, Pollak GD, Bodenhamer RD (1985) Topology of the central nucleus of the mustache bat's inferior colliculus: correlation of single unit response properties and neuronal architecture. J Comp Neurol 231:530–546.

9

The Bat Auditory Cortex

WILLIAM E. O'NEILL

1. Introduction

Situated as it is at the top of the hierarchy of nuclei comprising the auditory pathway, the auditory cortex should provide a wealth of information about higher order analysis of acoustic signals. The auditory cortex in bats is arguably the most intensively studied and best understood of all mammals. One species in particular, the mustached bat *Pteronotus parnelli*, has provided a wealth of detailed information about neuronal specialization and cortical organization. Despite these advances, many auditory neuroscientists tend to dismiss bat auditory neurobiology as "out of the mainstream" because they view bats as highly specialized mammals, with only one form of acoustic behavior, echolocation (but see Pollak, Winer, and O'Neill, Chapter 10). Some authors (e.g., Clarey, Barone, and Imig 1992) have questioned the applicability of the bat model of cortical organization because of the perception that bat cortex is highly specialized to analyze only the few stereotyped sounds used in echolocation. Bat researchers have inadvertently reinforced this bias by focusing their efforts only on echolocation, while ignoring the rich acoustic signal structures that bats use for communication. Only recently has there been any attention paid to the processing of the astonishingly rich variety of communication sounds used by colonial bats in social interactions (Kanwal, Ohlemiller and Suga 1993; Kanwal ct al. 1994; Ohlemiller, Kanwal, and Suga 1993).

Just as there is no one typical rodent, there is no one stereotypical bat. By taking advantage of the diversity of the chiropteran order, one might find clues to the fundamental "plan" of cortical organization, if there is such a thing. Indeed, this chapter shows that, while cortical neurons may have similar properties in different bats, these neurons are not organized into a single "bat-typical" cortical pattern. That cortical organization can differ markedly even within a single mammalian order sharing a specialized acoustic behavior is a sobering discovery. It implies that there may not be an archetypal model of mammalian auditory forebrain organization upon which we can base theories about the neural bases of auditory perception.

Unfortunately, only 5 of more than 900 known bat species representing only four genera have been studied at the single-unit electrophysiological level, although anatomical studies have been carried out on the cortices of some other species. All the species in which the cortex has been studied are aerial insectivores: there are so far no studies published on other microchiropteran bats with different feeding habits, let alone their nonecholocating megachiropteran cousins. Generalizations from this small sample of chiropteran life should be judged carefully, indeed.

This chapter by necessity focuses solely on these five species, with incidental references to a few others as needed. These species are divided bioacoustically into two groups, the "CF/FM" and "FM" bats, based on their biosonar signals (see list of abbreviations). These two groups express rather divergent adaptations for hunting flying insects in very different habitats (see Fenton, Chapter 2). In very general terms, FM bats are known to hunt in open areas, relatively free from the acoustic clutter of vegetation. By contrast, CF/FM bats generally hunt in wooded habitats near vegetation, and it is thought that their more complex sonar signals reflect acoustic adaptations to permit them to operate in high levels of clutter (Simmons, Fenton, and O'Farrell 1979).

Regarding the auditory cortex, the best studied CF/FM bat is the mustached bat, a common neotropical species belonging to the family Mormoopidae. The best-studied FM bat is the little brown bat, *Myotis lucifugus*, a common species found throughout most of temperate North America, belonging to the largest family of "microbats," the Vespertilionidae (see Fenton, Chapter 2). Studies of two other CF/FM bats, the horseshoe bats *Rhinolophus ferrumequinum* and *R. rouxi*, and most recently, the FM bat *Eptesicus fuscus*, have provided contrasting views of cortical organization that suggest that there is no single typical plan common to all bats.

2. Defining the Auditory Cortex

A thorough definition of the boundaries of any species' auditory cortex requires detailed anatomical study of cytoarchitecture and the pattern of thalamocortical connections, combined with electrophysiological mapping of cortical cell responses to acoustic stimulation. Unfortunately, no definitive cytoarchitectonic studies have yet been published on any of the five bats mentioned here, and there have been only a few attempts at defining cortex by the pattern of thalamocortical afferents (Shannon and Wong 1987; Wenstrup and Grose 1993; see Wenstrup, Chapter 8).

We are currently forced to rely mainly on neurophysiologically defined maps of the cortex. This technique has the limitation that only those cortical areas containing cells responsive to arbitrary search stimuli are discovered and characterized. Because such stimuli might not elicit responses from cells with complex response properties, one can never rule out the possibility that cortical boundaries are underestimated. Thus, it is not surprising to find a rough correlation between the number of studies published about a particular species and the total known area of its auditory cortex.

2.1 Cytoarchitecture

The cortex in all bats studied is predominantly lissencephalic, and it therefore has no pattern of gyri and sulci by which one can immediately recognize regions of neocortex. The temporal cortex in most bats is bounded dorsally only by a shallow groove along which blood sinuses are found (Henson 1970); only in the mustached bat does one see a prominent fissure homologous to the Sylvian sulcus (Fig. 9.1A–C). Studies of cortical cytoarchitecture have concluded that the chiropteran cortex is primitive (i.e., phylogenetically old), resembling that of insectivores (Sanides 1972; Sanides and Sanides 1974; Ferrer 1987; Fitzpatrick and Henson 1994). Like the neocortex in primitive mammals, the cortex in bats is not strongly laminated (Fig. 9.1B,D). In mustached bat, nonauditory areas in the temporal cortex range in thickness from 600 to 800 μm. However, the auditory cortex is about 900–1000 μm thick (Fig. 9.1B). In both *Rhinolophus* and *Pteronotus*, fibers within the auditory cortex are more heavily myelinated than that of surrounding cortex (Fig. 9.1C).

Recent studies employing the metabolic marker [^{3}H]2-deoxyglucose (2-DG) provide a functional view of auditory cortex in the mustached bat (Fig. 9.1D; Duncan and Henson in press; Duncan, Jiang, and Henson in manuscript). The results show that auditory cortex, bounded by the piriform cortex ventrally and the bottom of the temporal sulcus dorsally, is metabolically very active in echolocating, flying bats. This is in stark contrast to the lack of cortical 2-DG labeling in restrained, nonecholocating mustached bats passively exposed to sonar-like stimuli (O'Neill, Frisina, and Gooler 1989), and in resting bats (Duncan and Henson in press). Although quantitative analyses are not available, by inspection of the stained tissue sections it is clear that the auditory cortex is disproportionately large, occupying nearly the entire dorsoventral extent of the temporal cortex and from one-third to one-half the entire cortical hemisphere (Fig. 9.1D).

The cytoarchitecture of mustached bat primary auditory cortex is shown from Nissl-stained material in Figure 9.1E. Note the thick layer 1, thin and densely packed layers 2 and 4, and the low density of cells in layers 3 and 5. Ferrer (1987) has described the cortical cytoarchitecture of the echolocating bat *Miniopterus sthreibersi* from Nissl and Golgi material (Fig. 9.1F). In this species, there is a thick, cell-sparse molecular layer 1, bounded mainly by triangular (extraverted pyramidal) cells densely packed in a thin layer 2 (Fig. 9.1F). Unlike the occipital or central cortex in this species, the temporal cortex has a distinct granular layer 4 that in Nissl material is composed mainly of densely packed, small spherical cells lying about halfway down the depth of cortex (e.g., Fig. 9.1E). This layer in Golgi material contains spiny stellate cells (Fig. 9.1F). In *Miniopterus*, both layer 3, composed of a mixture of medium-sized triangular and pyramidal cells, and layer 5, containing large triangular and globular cells, are packed less densely than layers 2 and 4 (Fig. 9.1F). Layer 6 is more varied, including medium-sized triangular, polygonal, and globular cells.

2.2 Thalamocortical Connections

Traditionally, studies of auditory cortex have focused on its cochleotopic or tonotopic organization. All bats studied to date show some form of tonotopic organization in at least one region of cortex, and because the representation of frequencies follows the plan seen in the tonotopic primary area (AI) of other species, it has been assumed that this region is homologous. Although there is abundant neurophysiological evidence, there is scant hard anatomical evidence in favor of this homology. Olsen (1986), and more recently Wenstrup and Grose (1993), showed that the tonotopic region of mustached bat cortex is reciprocally connected to the ventral division of the medial geniculate body (MGB), a pattern typical of other mammalian species. Shannon and Wong (1987) also showed this to be true for *Myotis*. Wenstrup and Grose (1993) have demonstrated that the medial division of the mustached bat MGB is connected to all auditory cortical regions, including areas showing poor or no tonotopic organization surrounding the putative AI region (see Section 4). In some cases, the frequency tuning of neurons is poorly expressed in both the nontonotopic cortical fields and the divisions of MGB with which they are connected. Two regions dorsal to AI are connected to the unique "rostral pole" of the dorsal division of the MGB (Olsen 1986; Wenstrup and Grose 1993). As is discussed in Section 4.1, these two regions contain "combination-sensitive" neurons (cells tuned to more than one acoustic element in complex signals). Such cells first arise in the dorsal division of the MGB (Olsen 1986; see Wenstrup, Chapter 8).

The suprageniculate nucleus of the thalamus receives auditory input from the inferior colliculus (Wenstrup, Larue, and Winer 1994) and nucleus of the central auditory tract (anterolateral periolivary nucleus; Kobler, Isbey, and Casseday 1987). The suprageniculate connects broadly to the entire neocortex, auditory and nonauditory (Kobler, Isbey, and Casseday 1987).

3. Tonotopically Organized Cortex

The auditory cortex of all mammals studied so far is dominated by a tonotopic topographical organization, and bats are no exception to this rule. In fact, there are two general patterns of organization, tonotopic and

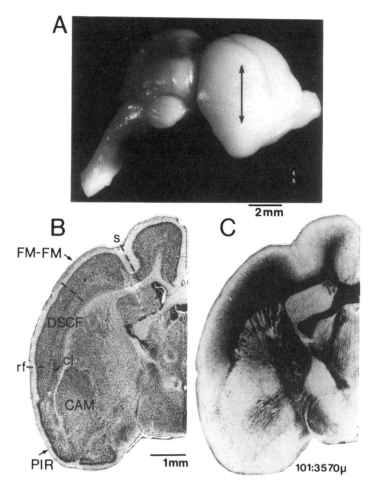

FIGURE 9.1A–E. Anatomy of auditory cortex in the mustached bat. (A) *Right side*, lateral view of forebrain. *Arrow* shows location of cross sections in B and C. Sulcus homologous to Sylvian fissure is visible (S. Radtke-Schuller, unpublished data). (B) Nissl-stained cross section through the temporal cortex at the level shown by *arrow* in A. Auditory area is bounded by Sylvian sulcus (s) dorsally and rhinal fissure (rf) ventrally. Locations of DSCF region of AI and FM-FM area are indicated. cl, claustrum; CAM, amygdala; PIR, piriform cortex (S. Radtke-Schuller, unpublished data). (C) Myelin-stained section at same level as B. Note heavy myelination in auditory cortex (S. Radtke-Schuller, unpublished). (D) Autoradiographs of serial cross sections through most of the auditory cortex from a flying bat, stained for [³H]2-deoxyglucose (2-DG) (from Duncan, Jiang, and Henson, in manuscript, reprinted by permission of the authors). *Arrows* indicate rostrocaudal direction.

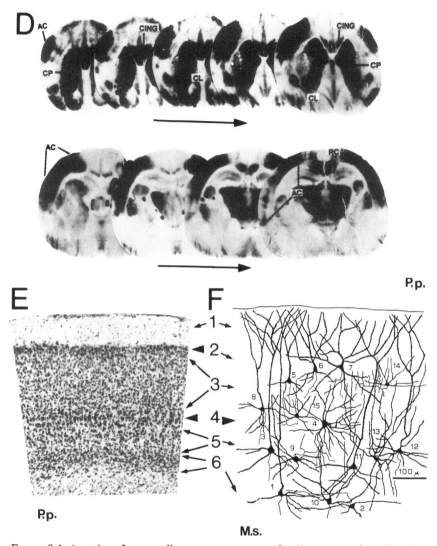

FIGURE 9.1. (*continued*) ac, auditory cortex; cp, caudate/putamen; cing, cingulate cortex; rc, retrosplenial cortex; cl, claustrum. (E) Nissl-stained cross section shows cortical layers in mustached bat AI (S. Radtke-Schuller, unpublished data). (F) Camera lucida drawings of representative Golgi-stained cortical neurons in *Miniopterus sthreibersi*, with approximate cortical layers indicated to *left* (from Ferrer 1987, copyright © 1987, J. Hirnforschung 28(2): 237–243, reprinted by permission of Academie Verlag GmbH). *1, 3, 5, 8, 11, 12*: pyramidal cells; *6,7*: extraverted pyramidal cells in layer 2; *4*: large multipolar cell with smooth dendrites; *13*: large multipolar neuron; *14*: chandelier cell; *15*: spiny stellate cell in layer 4.

nontonotopic. Primary auditory cortex is traditionally described as tono-topically organized, with neurons showing sharp tuning and relatively short latencies to tones at a single characteristic, or best, frequency (BF), arrayed by BF along the cortical surface (Woolsey 1960). The secondary auditory cortex (AII) is described as containing cells responding with longer latencies that are more broadly tuned and are often not well driven by tones (Schreiner and Cynader 1984). Some mammalian species express more than one tonotopic primary field (for review, see Clarey, Barone, and Imig 1992). Modern studies in the cat (Reale and Imig 1980) have shown that the primary auditory cortex is actually composed of four tonotopically orga-nized fields (AAF, AI, P, PES), flanked by three secondary areas (AII, DP, and V). In macaque monkey, AI is also surrounded by two other primary fields and four secondary fields, in a "core-belt" organization (Merzenich and Brugge 1973). By contrast, many species, including the ferret, only express a single primary field (Kelly, Judge, and Phillips 1986; Phillips, Judge, and Kelly 1988).

In CF/FM bats, there is typically one major tonotopic primary area, akin to AI, bordered by as many as seven secondary fields containing cells with complex response properties that typically do not respond well to single tones. Unlike neuronal responses in cat AII, which are often described as sluggish, habituating, broadly tuned, and otherwise enigmatic, cells in the nonprimary fields of CF/FM bats respond vigorously and have relatively similar latencies to neurons in AI, when presented with the appropriate complex stimulus. In mustached bat, one of these bordering fields (the CF/CF area) is actually tonotopically organized for a very narrow range of frequencies. However, it represents frequency along *two* axes, thereby forming a *bicoordinate* frequency representation (Suga et al. 1983; see Section 4.1.3).

In FM bats, cells with both simple and complex response properties are intermingled within one large tonotopic field, and there is usually one or two additional fields where tonotopy is absent or reversed in direction along the cortical surface (see Sections 3.3.2 and 3.4.2). The details of these arrangements are described next.

3.1 Mustached Bat

Auditory cortex in the mustached bat has been by far the most intensively studied. In fact, one could argue that it is the most studied of any single mammalian species, save perhaps the domestic cat. Nearly all the research on mustached bat cortex has been done in the lab of Nobuo Suga, and he and his colleagues have reviewed this work in numerous articles (Suga, Kujirai, and O'Neill 1981; Suga 1981, 1982, 1988a, b, c, 1990a, b; Suga, Niwa and Taniguchi 1983; Suga, Olsen, and Butman 1990). The mustached bat cortex is distinguished by a number of unusual (some might say "remarkable") features. The most important of these include an enormous overrepresentation of a very narrow frequency band centered around the second harmonic of the sonar pulse, the existence of two novel classes of neurons specialized to encode information related either to target ranging or velocity/motion, and a profusion of separate fields in temporal cortex where the different neurons are segregated by type, forming "processing areas" for specific types of acoustic analyses.

The terminology used to define different neuronal types and regions in mustached bat cortex is complex, but it has become somewhat of a standard for other species as well. Therefore, a short primer is necessary to aid the reader in deciphering abbreviations (see Table 9.1). Mustached bats emit signals consisting of a long constant-frequency (CF) component followed, and often preceded by, a short frequency-modulated (FM) sweep (Novick and Vaisnys 1964; Schnitzler 1970; Suga and Shimozawa 1974; Gooler and O'Neill 1987). Each cry contains up to four or five harmonics, designated H_1-H_4, and the second harmonic is usually dominant (although the bat can vary the energy in the harmonics; Gooler and O'Neill 1987). Each CF or FM component is addressed by a subscripted number (e.g., FM_1, CF_3, etc.). Cell types are assigned these labels if they are most sensitive to frequencies related to the sonar pulse. Thus, a "CF_2" neuron would be tuned to a tone at a frequency near the second harmonic CF component of its sonar pulse. "Combination-sensitive" cells (Section 4) require two or more components, such as FM_1-FM_3. Table 9.1 summarizes the properties and distribution of different cortical neurons in the mustached bat.

The duration of sonar pulses varies from as long as 30 msec during cruising or search phase to as short as 5–7 msec during terminal phase (Novick and Vaisnys 1964; Henson et al. 1987). Repetition rates increase from about 10/sec to about 100/sec as the bat progresses through search, approach, and terminal phases of target pursuit (see Fenton, Chapter 2; Moss and Schnitzler, Chapter 3; and Simmons et al., Chapter 4 for further details on sonar signals).

3.1.1 Response Properties in Tonotopic Cortex

Discharge patterns. Response properties of neurons in the tonotopically organized region of the mustached bat cortex are dependent on BF. The tonotopic region is noted for a large central region overrepresenting frequencies associated with the predominant CF_2 component of the sonar signal (i.e., the bat's "resting frequency") called the "Doppler-shifted CF" (DSCF) processing area (Suga and Jen 1976). Neurons in the DSCF have been studied most intensively. Cells in this area respond at a latency of 9–10 msec, and show two basic discharge patterns to tonal stimuli, phasic-on and sustained-on. However, discharge patterns vary widely dependent on stimulus amplitude and frequency (Suga and Manabe 1982). At BF within about 20 dB of minimum threshold (MT), most cells are transient-on

Table 9.1. Cortical units in *Pteronotus parnelli*

Type	Frequency tuning	Delay dependency	Key parameters	Field(s)	Organization	Information encoded
CF_2	Extremely sharp	None	Frequency, amplitude	DSCF ("fovea")	Bicoordinate: frequency vs. amplitude	Target flutter; velocity
FM_1-FM_2 FM_1-FM_3 FM_1-FM_4	Broad	Sharp	Pulse–echo time interval	FM-FM DF VF	Clustered by type Chronotopic (delay axis)	Target range
CF_1/CF_2 CF_1/CF_3	CF_1: broad; CF_2, CF_3: extremely sharp	Broad	Frequency of CF components	CF/CF	Clustered by type Bicoordinate: tonotopic	Relative target velocity (Doppler-shift magnitude)
FM_1-CF_2 (H_1-H_2)	FM_1: broad; CF_2: sharp	Broad	Pulse–echo delay; echo frequency/ amplitude	DSCF VA	As DSCF above	Target detection

CF, constant frequency; FM, frequency-modulated; DSCF, Doppler-shifted CF region of mustached bat; DF, dorsal fringe area of bat cortex; VA, area H_1-H_2) of mustached bat cortex; VF, ventral fringe of bat cortex.

responders. Sustained-on responders are comparatively rare. At frequencies below the bat's individual resting frequency, or amplitudes more than about 30 dB above threshold, transient on-off or pure off-discharge patterns occur. These periods of elevated activity at stimulus onset or offset are typically followed by periods of suppression, and rebound-off responses also occur in many cells up to 30–40 msec after cessation of the tone burst stimulus.

On-off discharge patterns are common in neurons tuned to the CF_2 frequencies throughout the mustached bat's auditory pathway. This pattern is presumably caused by prolonged "ringing" of the basilar membrane when stimulated near the bat's resting frequency (see Kössl and Vater, Chapter 5). The cochlear resonator is responsible for extremely sharp tuning of neurons tuned to this frequency band. Earlier work in the auditory periphery showed that off-discharges in auditory nerve fibers could be abolished by adding a downward sweeping FM to the end of the tone burst, mimicking the bat's sonar pulse (Suga, Simmons, and Jen 1975). However, in the inferior colliculus, about 14% of CF_2 cells show on-off responses not just at high amplitudes or lower frequencies, but at all frequencies and amplitudes within their response areas (Lesser 1987). Off-responses in these cells do not disappear with addition of an FM sweep at the end of a tone burst. Collicular on-off cells show interesting full-wave rectification properties that suggest specialization for detecting sudden but infrequent acoustic transients (Lesser et al. 1990). Whether these "pure" on-off cells occur in cortex remains to be seen.

Frequency tuning (spectral domain) properties. As is true for CF_2-tuned cells at lower levels (Suga and Jen 1977; Pollak and Bodenhamer 1981; O'Neill 1985; see Casseday and Covey, Chapter 6; Pollak and Park, Chapter 7), excitatory frequency response areas (tuning curves) are extremely sharply tuned in DSCF cells. Suga and Manabe (1982) reported Q_{10dB}* values ranging from 15 to 225, with outliers near 300. Q_{10}s this high are extraordinary, and have only been found in CF/FM bats. By contrast, Q_{10dB} of AI neurons in squirrel monkey do not exceed 20, and cells with Q_{10}s greater than 4 are considered "sharply tuned" (Shamma and Symmes 1985). Moreover, as Suga and Manabe (1982) demonstrated, central processing further sharpens frequency tuning beyond that seen at the auditory periphery. They compared tuning curve bandwidths of auditory nerve fibers and cortical cells not just near threshold with Q_{10}, but also at 30 dB and (where possible) 50 dB above threshold. They found that whereas Q_{10} values in cortical cells largely overlapped those of auditory nerve fibers, further sharpening of the response area was evident from Q_{30}

---

*Q_{10dB} equals the bandwidth of tuning curve at 10 dB above minimum threshold, divided by the best frequency. Q_{30} and Q_{50dB} are calculated from bandwidths at 30 and 50 dB above threshold, respectively.

and Q_{50dB} values at the cortical level. Therefore, cortical neurons show evidence of pronounced narrowing of the excitatory areas at moderate to high stimulus levels, and inhibition, both on- and off-BF, plays a critical role in shaping these response areas.

"Amplitude tuning" properties. Response areas can be thought of as neuronal activity maps plotted along two domains or axes, frequency and amplitude. Inhibition can sculpt the response area of a cortical neuron not only in the frequency domain, but also at both extremes of the amplitude domain. One can describe the amplitude domain behavior of a cell with an "intensity," or rate-level (input-output) function. When presented with a free-field BF stimulus (i.e., one exciting both ears equally), most *Pteronotus* cortical cells have bell-shaped, or nonmonotonic, intensity functions (Suga and Manabe 1982). This effectively "tunes" the cell to a particular "best amplitude" (BA). In a minority of this population, high stimulus levels may suppress activity to zero, producing an "upper threshold." At low stimulus levels, there is recent evidence to suggest that inhibition may also play a direct role determining a cell's threshold. Yang, Pollak, and Ressler (1993) showed in mustached bat inferior colliculus (IC) neurons that response areas, including thresholds, are shaped by γ-aminobutyric acid (GABA)-ergic inhibition (see Pollak and Park, Chapter 7). Whole-cell patch recordings in intact *Eptesicus* preparations of IC neurons with high thresholds have shown that inhibitory synaptic potentials can occur at subthreshold stimulus levels, and may therefore be important in setting threshold (Covey et al. 1993; Casseday and Covey, Chapter 6).

Inhibitory response areas. Suga and Manabe (1982) employed a modified forward masking paradigm to measure inhibition in cortical cells. Inhibitory areas were found to be extensive, in some cells covering more than an octave above and below the BF. In the extreme, cortical cells show completely circumscribed, spindle-shaped excitatory response areas. Suga and Manabe (1982) called such cells "level tolerant", that is, their frequency selectivity was essentially unchanged over the entire range of effective stimulus amplitudes. In the vast majority of DSCF neurons the discharge rate is maximized at particular values of both frequency and amplitude (i.e., at BF and BA). In this sense, DSCF neurons are tuned in both the frequency and intensity domains.

As in other mammals, the proportion of cells with nonmonotonic rate-level functions increases as one ascends from the cochlear nucleus to the cortex. At the cortical level, the net effect of the interaction of excitatory and inhibitory circuitry produces level-tolerant units functioning as frequency detectors. Such a mechanism might underlie the high acuity of frequency discrimination shown by human subjects when stimulated with moderate to high intensity signals (Hawkins and Stevens 1950; Scharf and Meiselman 1977).

3.1.2 Organization of Primary Tonotopic Area

The organization of the auditory cortex in the mustached bat is shown in
Figure 9.2. By making penetrations perpendicular to the cortical surface, it
was found that BFs, BAs, MTs, and response areas were roughly similar for
cells within cortical columns in AI (Suga and Jen 1976; Suga and Manabe
1982). Along the cortical plane, BFs of the columns are tonotopically
organized, high frequencies being represented rostrally and low frequencies
caudally, resembling cat primary auditory cortex. However, the tonotopic
sequence is interrupted by the DSCF area (Fig. 9.2A–C). The DSCF area
accounts for fully one-third of the tonotopic region, yet it represents only a
very narrow frequency band, from about 61 kHz to 63 kHz, encompassing
the bat's resting and reference frequencies. The mustached bat's auditory
system is especially sensitive within this frequency range, and just below this
band, between about 58 and 60 kHz, the bat's ear has a pronounced
insensitive region, or notch (Pollak, Henson, and Novick 1972; Suga,
Simmons, and Jen 1975; Pollak, Henson, and Johnson 1979; see Kössl and
Vater, Chapter 5). During Doppler-shift compensation, the bat lowers its
pulse frequency to stabilize the echo CF_2 within the narrow frequency band
represented within the DSCF area, and the emitted signal enters the
insensitive notch, thereby reducing the difference in sensation level between
the pulse and echo.

Schuller and Pollak (1979) dubbed the frequency band of extremely sharp
tuning the "acoustic fovea" (see also Section 3.2). At the cortical level, the
"fovea" is "personalized," in that the DSCF area represents frequencies
centered around the particular cochlear resonance of the individual bat's
ears. Moreover, there are sexual differences in the representation: males
emit lower sonar frequencies than females, and their ears are tuned lower
than females as well (Suga et al. 1987).

Within the DSCF area, Suga and Jen (1976) found that isofrequency
contours were decidedly circular, suggesting a tonotopic organization with
radial symmetry (Fig. 9.2B,C). Outside the DSCF area, isofrequency
contour lines are slightly curved, but are mainly dorsoventrally oriented,
similar to cat AI (Fig. 9.2A). Asanuma, Wong, and Suga (1983) showed
that rostral to the DSCF area, the CF_3 frequency band between about 89
and 95 kHz (especially 92–94 kHz) was also overrepresented (Fig. 9.2A).
Thus, the two strongest harmonics of the mustached bat's sonar cry are
both represented disproportionately in AI. By contrast, the dominant FM_2
(sweeping from 60 to 50 kHz) and FM_3 (89 to 74 kHz) components of the
species' biosonar signals were poorly represented along this strip of cortex.
These frequency bands are displaced dorsally into what is called the
"FM-FM" area, described in Section 4.1.1 (Suga and Jen 1976; Suga,
O'Neill, and Manabe 1978; O'Neill and Suga 1982). Similarly underrepresented are frequencies between the dominant harmonics of the sonar signal

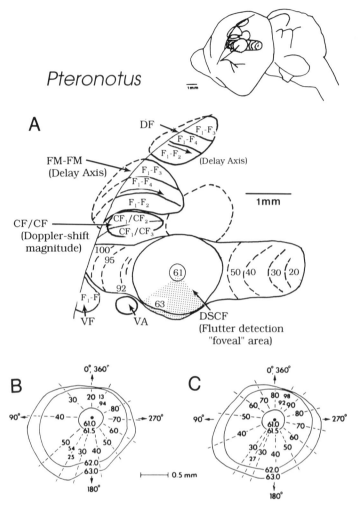

Pteronotus

FIGURE 9.2A–C. Functional organization of mustached bat (*Pteronotus*) auditory cortex. (A) Cortical fields include tonotopic zone with Doppler-shifted CF region (DSCF) area in center, three dorsal fields including the FM-FM, dorsal fringe (DF), and CF/CF areas, and two ventral fields including ventral fringe (VF) and ventral anterior area (VA). *Shaded area* in DSCF indicates region containing excitatory-excitatory (EE) cells; remaining part of DSCF is predominantly excitatory-inhibitory (EI). Cells in DSCF area are exquisitely sensitive to minor frequency modulations caused by fluttering targets (Suga 1990a, reprinted from Neural Networks 3, N. Suga, "Cortical computational maps for auditory imaging", pp. 3–21, © 1990, with kind permission of Elsevier Science Ltd.). (B,C) Organization of DSCF area. Overrepresentation of frequencies near second harmonic of sonar pulse is indicated by concentric isofrequency contours. Iso-best amplitude contours are radially organized, forming a bicoordinate frequency-amplitude map with two different patterns, N type (B), where the BA *increases* in the counterclockwise direction, and V type (C), where BA *decreases* dorsoventrally in each half of the area (from Suga 1982, © 1982, reprinted by permission of Humana Press).

(e.g., 35–50 kHz, 65–74 kHz). It is clear from these data that the mustached bat primary tonotopic cortex is heavily weighted toward the CF components in the biosonar signal, and the FM components are displaced into a completely different region dorsal to the tonotopically mapped area (see Section 4.1).

Most importantly, Suga (1977) showed that not only frequency, but also amplitude, was systematically represented within the DSCF area. Unit thresholds within the DSCF area were all quite similar, but BA varied systematically (Fig. 9.2B,C). Suga and Manabe (1982) showed a systematic "amplitopic" representation, with iso-BA contours arrayed roughly orthogonal to the iso-BF contours, much like spokes in a wheel (Fig. 9.2B, C). Frequency and amplitude are thereby distributed in the DSCF area in a bicoordinate map.

Tunturi (1952) found evidence of a systematic difference in threshold along isofrequency contours in the dog cortex, and recent evidence from cat suggests that there is a systematic dorsoventral variation in both threshold and BA along isofrequency contours (Schreiner, Mendelson, and Sutter 1992). Neither of these examples matches the level of systematic representation of these parameters found in the mustached bat. As pointed out by the latter authors, the implication of this type of representation is that signals with different intensities will activate spatially distinct subdivisions of the cortex, and the pattern of activation will shift with intensity changes in a fairly predictable manner. Patterns of activation might then be associated with stimuli with different amplitude spectra.

3.1.3 Binaural Properties of DSCF Neurons

Despite a wealth of data concerning binaural interactions in the *Pteronotus* midbrain (see Pollak and Park, Chapter 7), only two studies have investigated binaural properties in the auditory cortex (Manabe, Suga, and Ostwald 1978; Suga, Kawasaki, and Burkhard 1990). Using a quasi-dichotic stimulation technique, Manabe, Suga, and Ostwald (1978) studied binaural interactions in DSCF neurons. Three general classes of response were found. These included binaurally excited (EE), binaurally suppressed (EI), and monaural (EO) response types. Excitatory input came from the contralateral ear in the majority of EI neurons. They were found dorsally, and EE neurons ventrally, in the DSCF area (Fig. 9.2A). Thus EE neurons summating responses to the two ears are found in the region with the lowest BAs, and EI neurons sensitive to interaural intensity differences in the region with higher BAs. This pattern is analogous to the banded, patchy organization of EE and EI cells in cat cortex (Imig and Adrian 1977; Middlebrooks, Dykes, and Merzenich 1980). Manabe, Suga, and Ostwald (1978) asserted that EE neurons are best suited for target detection because they presumably have large receptive fields and low BAs. EI neurons, on the other hand, are better suited to target localization, as they have higher BAs and smaller receptive fields than EE cells.

Beyond this segregation of binaural types, little further information exists concerning the encoding of azimuth and elevation cues in the tonotopic part of auditory cortex. Suga, Kawasaki, and Burkhard (1990) examined the spatial receptive fields of neurons in the FM-FM area of mustached bat cortex, as noted in Section 4.1.1.4. However, a systematic study akin to those on IC cells by Fuzessery and Pollak (1985), in which binaural interactions and free-field spatial receptive fields are both investigated, has not yet been attempted in the mustached bat cortex.

3.1.4 Facilitation of DSCF Neurons by Tone or FM Pairs

Recent studies using pairs of tones and frequency-modulated signals, mimicking the mustached bat's sonar signal, have surprisingly revealed facilitation of activity to stimulus combinations in DSCF neurons (Fitzpatrick et al. 1993). Specifically, DSCF cells are facilitated by pairing an FM_1 with a CF_2 signal, and this facilitation is dependent on a time delay between the two signals. The properties of these cells are best understood after considering the response properties of cells in the FM-FM area of mustached bat cortex, and discussion is therefore deferred to Section 4.1.4.

3.2 Horseshoe Bat

Bats in the family Rhinolophidae occupy an ecological niche similar to the mustached bat, and show similar echolocation behavior, including Doppler-shift compensation (Schnitzler 1968). Yet horseshoe bats are not sympatric with the mustached bat, and phylogenetically they are not closely related (see Fenton, Chapter 2). The remarkable behavioral and physiological similarities have apparently evolved independently, converging presumably from similar selective pressures.

The search phase biosonar signals of *Rhinolophus* are much longer (about 60 msec) than those of mustached bats. However, the cries are less complex, having only a very weak fundamental (H_1 component) and dominant second harmonic (H_2 component). Unlike mustached bats, the sensitivity of the ear of *Rhinolophus* does not extend to frequencies much beyond that of the second harmonic of the sonar pulse spectrum (Neuweiler 1970; Bruns 1976; Bruns and Schmieszek 1980; see Kössl and Vater, Chapter 5).

Most of what is actually known about echolocation by long CF/FM bats comes from work with horseshoe bats. However, although there is much known about auditory processing at and below the level of the inferior colliculus, there are only two published studies on the auditory cortex, one from the European greater horseshoe bat *Rhinolophus ferrumequinum*, and the other from the rufous horseshoe bat *R. rouxi*, of southern India and Sri Lanka. The cortex in these bats provides revealing contrasts to the organization of the mustached bat auditory forebrain.

3.2.1 Response Properties of Neurons in Tonotopic Cortex

Ostwald (1984) reported that more than half the units recorded in the primary auditory cortex of *R. ferrumequinum* showed only transient-on responses to tonal stimuli at all excitatory frequencies and intensities above threshold. Eighteen percent showed sustained-on responses at all stimulus levels, and 4% showed on-off responses exclusively. The remaining 24% changed their discharge patterns with changes in either intensity, frequency, or both. Like the mustached bat, on-off patterns typically occurred at intensities 20–30 dB or more above threshold, or at frequencies usually below, but sometimes higher than, BF.

Table 9.2 summarizes the response properties of cortical cells in the horseshoe bat. Similar to the mustached bat, neurons in the central part of primary cortex have BFs near or just above the frequency of the CF_2 component of the bat's signal. Excitatory areas of of units within this region are typically V shaped and very sharply tuned (Q_{10} dB values from 20 to >400), whereas those elsewhere are relatively broad (Q_{10} lower than 20). Thresholds are also lowest within the CF_2 frequency range, and although very few units show upper thresholds, most have nonmonotonic intensity functions similar to the mustached bat.

3.2.2 Frequency Representation

The representation of the cochlear partition in the horseshoe bat cortex has gross similarities to that of the mustached bat. Ostwald (1980, 1984) found a tonotopic axis in AI (defined by units with short latency responses to tones) running roughly rostrocaudally (Fig. 9.3A,B, unshaded region). Rostral and ventral to the tonotopic area, units were difficult to excite with tones and had longer latencies, and there was no clear frequency organization. In the rostral half of the tonotopic area, the frequency map is distorted by a large "CF" region, where the CF_2 component of the individual bat's sonar signal (ranging between 81 and 85 kHz in *R. ferrumequinum*) is overrepresented (Fig. 9.3B). This region is therefore functionally analogous to the DSCF area of mustached bat cortex in that it encompasses the individual bat's resting and reference frequencies. For the sake of consistency in nomenclature, the CF region shall be hereafter referred to as the "CF_2" region, because it represents the second harmonic CF component of the horseshoe bat signal. Outside the CF_2 area, isofrequency contours are generally oriented dorsoventrally. Inside the CF_2 area, contours are strongly curved, but are not quite round or ellipsoid like those in the mustached bat DSCF area. The sequence of frequency representation is nearly reversed in the CF_2 area relative to the rest of AI, with lower frequencies ventral and rostral and higher frequencies dorsal and caudal. A sharp discontinuity occurs at the dorsal border of the CF_2 area, above which frequencies found mainly in the FM_2 component (70–83 kHz) of the sonar pulse are overrepresented. Ostwald's reconstructions of the FM_2 area show a rostro-

TABLE 9.2. Cortical Units in *Rhinolophus* spp.

Type	Frequency tuning	Delay dependency	Key parameters	Field(s)	Organization	Infomation encoded
CF_2	Very sharp $(Q_{10}:20 \sim 400)$	None	Frequency Amplitude	CF_2 area of AI	Tonotopic	Echo frequency Target flutter
FM_1-FM_2	Broad	Sharp	Echo delay, amplitude	FM-FM	Chronotopic	Target range
CF_1/CF_2	CF_1; broad CF_2; sharp	Broad	Doppler shift	CF/CF	Unknown	Doppler-shift magnitude
CFM_1-CFM_2	CF_1;broad CF_2; sharp FM_1; broad FM_2; broad	Sharp	Echo delay Doppler shift	Mixed	Unknown	Target range; Doppler-shift magnitude

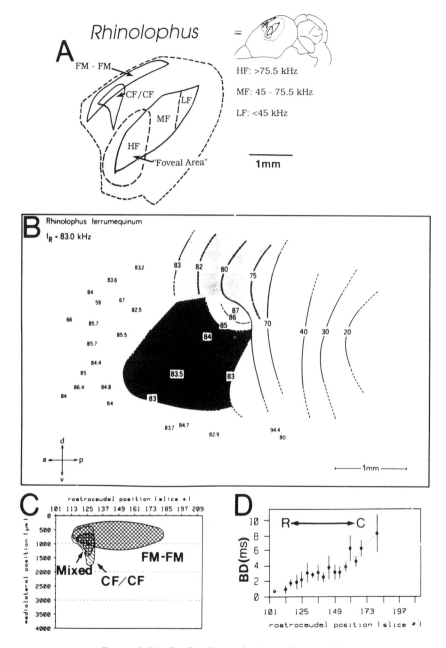

FIGURE 9.3A–D. Caption at bottom of page 434.

caudal, high-to-low frequency organization similar to that found caudal to the CF_2 area. Thus, as in *Pteronotus*, the FM_2 components of the *Rhinolophus* sonar pulse are displaced to a separate dorsal field (more detail is in Section 4.2). Ostwald (1984) showed that, relative to the innervation of receptors on the cochlear partition, there is a significant magnification of the CF_2 frequencies in the cortex.

3.3 Little Brown Bat

The cortical plan of two species whose echolocation behavior is typical of the majority of aerial insectivorous bats employing predominantly FM signals is now described. As with any classification scheme, it is easy to oversimplify the acoustical environment of the "FM bats" by ignoring the many circumstances in which these species use other types of signals (for a review, see Pye 1980; and Fenton, Chapter 2). For example, many species commonly included in this group (e.g., the Mexican free-tailed bat, *Tadarida brasiliensis*) use CF signals during search phase. Nevertheless, FM sounds are the predominant signals used for obstacle avoidance and prey capture, and research on the auditory cortex has accordingly been focused on the processing of FM sweeps.

The little brown bat, *Myotis lucifugus*, is a mosquito specialist often found foraging over water. The typical search-phase sonar pulse consists of a brief FM sweep descending over an octave from about 80 to 40 kHz in about 2–5 msec. These signals shorten to less than 1 msec, and a second harmonic appears as well, during the terminal phase.

3.3.1 Response Properties

Suga (1965a,b) recorded cortical unit responses in barbiturate anaesthetized *Myotis* to brief (4-msec) tones and FM sweeps. He described their discharge

---

FIGURE 9.3A–D. Functional organization of horseshoe bat (*Rhinolophus*) cortex. (A) Cortical fields (drawn to same scale as Figs. 9.2A, 9.4A, and 9.5A) include ventral tonotopic area and nontonotopic dorsal fields containing combination-sensitive cells. "*Foveal Area*" indicates overrepresented CF_2 region, homologous to DSCF area in mustached bat. *Inset:* Position of cortical fields in lateral view of brain (courtesy S. Radtke-Schuller). (B) Frequency representation within AI. CF_2 region shows expanded representation, reversal of frequency trend (from Ostwald 1984, © 1984, J. Comp. Physiol A 155:821–834, reprinted by permission of Springer-Verlag). (C) Functional organization of nontonotopic dorsal fields. FM-FM and CF/CF areas overlap rostrally, and CFM_1-CFM_2 ("mixed") cells are found in the region of overlap. (D) Systematic map of best delays is demonstrated within FM-FM area (C and D, adapted from Schuller, O'Neill, and Radtke-Schuller 1991, © 1991, Eur. J. Neurosci. 3:1165–1181, reprinted by permission of Oxford University Press).

patterns as "very phasic," although one might argue that the stimulus durations employed were too short to reveal more tonic response patterns. Four general types of response areas were found, including "narrow" (sharp V-shaped), "closed" (spindle-shaped), "double-peaked" (two regions of sensitivity), and "wide" (broad V- or U-shaped). In response to FM sweeps, both double-peaked and wide response area cells showed about the same thresholds regardless of sweep direction. In cells showing narrow or closed receptive fields, excitatory response areas to tones were often flanked by inhibitory areas on one or both sides. Cells with these response areas could show good responses to downward but poor responses to upward-sweeping FM, or vice versa. Smaller percentages of cells had upper thresholds for FM sweeps, showed no responses to FM sweeps ("FM insensitive" or "CF specialized"), or had thresholds to FM sweeps 10 dB or more lower than that to pure tones ("FM sensitive"). In the extreme, some cells only responded to FM sweeps, showing no response to tonal stimuli ("FM specialized"). Two-thirds of the cells exhibited nonmonotonic (peaked) rate-level functions.

Shannon-Hartman, Wong, and Maekawa (1992) have reexamined the FM sweep selectivity of *Myotis* cortical neurons. Nearly every neuron (96%) in their sample showed significantly stronger responses to FM sweeps compared to pure tones. Eighty-three percent, dubbed "type I FM-sensitive units," responded to both tones and FM signals that swept through the unit's pure tone response area. Similar to Suga's (1965a) earlier study, 13 % of the sample, termed "type II" cells, were specialized for FM sweeps, showing little or no response to tones. The remaining 4% showed little or no response to FM sweeps, only responding to tones (equivalent to Suga's "FM insensitive" units). The latter "CF specialized" cells were segregated in the cortex, as described later. The predominance of neurons preferring complex signals (i.e., FM) over simple pure tones is a hallmark of cortical organization in *Myotis*. Table 9.3 summarizes the response properties of *Myotis* cortical cells.

3.3.2 Organization of *Myotis* Cortex

By contrast to the CF/FM bats, the auditory cortex in *Myotis* is not as clearly subdivided into fields containing cells with different physiological properties. Two fields have been recognized. The larger field contains a single tonotopic map with the AI-typical rostrocaudal progression of high to low frequencies, formed by type I FM-sensitive cells (Fig. 9.4) (Suga 1965a; Wong and Shannon 1988). Type II cells are scattered about the tonotopic field. The much smaller rostral field contains CF-specialized neurons tuned to low (30–40 kHz) frequencies (Shannon-Hartman, Wong, and Maekawa 1992). Caudally, most FM-sensitive cells are delay tuned. Dear et al. (1993) have argued that the presence of delay-tuning properties within a tonotopic AI provides for an efficient integration of temporal and spectral domain processing (see Section 5).

TABLE 9.3 Cortical units in *Myotis lucifugus*

Type	Frequency tuning	Delay dependency	Key parameters	Field(s)	Organization	Infomation encoded
CF specialized	Sharp	Unknown	Pure tone	AI, AAF	Tonotopic	Stimulus spectrum
FM specialized	Broad	Unknown	Frequency modulation	AI only	Tonotopic (center frequency?)	FM sweeps Sonar signals
FM-FM P type E type	Broad	Sharp Moderately broad	Paired FM Echo delay Pulse and echo amplitude difference Repetition rate	AI only	Tonotopic No delay axis	Target range

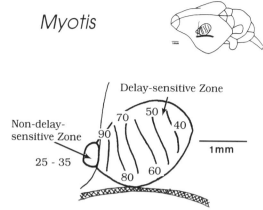

FIGURE 9.4. Organization of *Myotis* auditory cortex. By contrast to the CF/FM bats, there are only two cortical fields, both tonotopically organized. The tonotopic gradient in the larger caudal field is rostral to caudal, high to low frequency, and overrepresents the band of the sonar pulse from 80 to 40 kHz. The rostral field's tonotopic gradient is reversed, and only represents frequencies from 25 to 35 kHz. Cells in the caudal, but not the rostral, field are delay sensitive. Scale of drawings is same as in Figs. 9.2A, 9.3A, and 9.5A. (Courtesy of D. Wong.)

3.4 Big Brown Bat

The auditory behavior of this species is by far the most intensively studied, and there has been much research on the midbrain and auditory brainstem (see Covey and Casseday, Chapter 6). However, there is a paucity of data on the auditory cortex of big brown bats. The search/approach phase sonar pulse is a short FM sweep with two harmonics (FM_1 and FM_2). FM_1 sweeps downward over about an octave from about 50 kHz to 25 kHz. One or two additional harmonics appear, and the entire signal decreases in frequency, during late approach and terminal phase.

3.4.1 Response Areas and Intensity Functions

Jen, Sun, and Lin (1989) described the basic response properties of *Eptesicus* cortical neurons. They reported that all units discharged phasically at the onset of tonal stimuli. However, as in Suga's (1965a) study, the tonal stimulus they used was too short to draw conclusions about the relative abundance of phasic versus tonic response patterns. The lowest thresholds occurred in the 50- to 70-kHz range of BFs, and the highest thresholds were found in the 20- to 40-kHz BF range. Tuning curves were all V shaped, and Q_{10dB} values ranged from 2 to 30. However, despite the open shape of the tuning curves, nearly every unit tested had nonmonotonic rate-level functions.

3.4.2 Frequency Representation

Jen, Sun, and Lin (1989) described the representation of both frequency and space within the *Eptesicus* cortex. With regard to frequency, they reported columnar organization of BFs and a tonotopic axis in the cortical plane along which frequency decreased systematically from anterior to posterior. This tonotopic region spans the frequency range between 20 and 90 kHz, but there is an overrepresentation of frequencies between 30 and 75 kHz. Similarly, Dear et al. (1993) reported an overrepresentation of the 20- to 60-kHz range in the tonotopic region (their "area A"; Fig. 9.5A). Both laboratories describe a "variable" region anterior to the tonotopic zone ("area B") where frequency is not tonotopically represented either within or across animals. Dear et al. (1993) also recognized a third area "C" rostral to area B where a tonotopic pattern also occurs (Fig. 9.5A). In area C, the axis of tonotopic organization is reversed from that seen in area A. Dear et al.

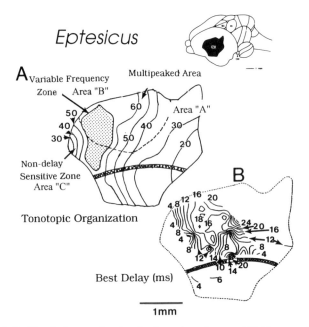

FIGURE 9.5A,B. Cortical organization in *Eptesicus fuscus*. (A) Tonotopic organization, showing three functional divisions. Tonotopic zone *(Area "A")* is caudal to the variable frequency zone *(Area "B", stippled),* and rostral tonotopic zone *(Area "C"),* which shows a frequency representation reversal. *"Multipeaked Area"* (bounded by *dashed line*) in dorsorostral part of area A contains cells tuned to two frequencies. (B) Iso-best delay (BD) contour map. Delay-tuned cells are found in both area A and especially, variable-frequency area B; they are absent in area C. In area B, there is a rostrocaudal, short-to-long BD progression, but in tonotopic area A, no clear delay map is visible. (From Dear et al. 1993, © 1993, J. Neurophysiol. 70:1988–2009, by permission of the American Physiological Society.)

reported that the dominant FM_1 frequencies of the *Eptesicus* sonar pulse (20–50 kHz) occupy about 74% of auditory cortex. As mentioned earlier, specialized delay-sensitive neurons are intermingled with unspecialized neurons in the tonotopic zone. As is discussed later, delay-tuned neurons have been found throughout areas A and B.

By contrast to Jen, Sun, and Lin (1989), Dear et al. (1993) found many instances where BFs did not show columnar organization in penetrations that were apparently perpendicular to the cortical surface. They have ruled out the possibility that this result is simply caused by improperly aligned penetration angles. One is forced to conclude that frequency representation in *Eptesicus* is not two dimensional within the cortical plane, as in other mammals, but includes a third depth dimension *along* cortical columns.

3.4.3 "Multipeaked" Neurons

Although the measurement of frequency response areas was not done systematically in their study, Dear et al. (1993) found that a number of neurons showed two threshold minima. They dubbed these cells "multipeaked neurons." BFs varied widely in multipeaked cells, and did not often coincide with frequencies of the sonar signals. The ratio of the two BFs was typically 1:3, and in most cells, the lower BF was below 20 kHz. Multipeaked cells were found in the anterior half of auditory cortex, mainly in the high-frequency part of the tonotopic area A and throughout area B. This is somewhat analogous to the distribution of delay-tuned cells (see Fig. 9.5B and Section 4.4.3).

Dear et al. (1993) pointed out that complex targets reflect multiwavefront echoes with spectral peaks and notches, the notches being caused by wavefront interference from the spatially separated surfaces (Beuter 1980; Habersetzer and Vogler 1983; Kober 1988; Kober and Schnitzler 1990; Simmons and Chen 1989; see Simmons et al., Chapter 4). Bats apparently use these spectral features to discriminate complex targets (Habersetzer and Vogler 1983; Mogdans and Schnitzler 1990; Schmidt 1988; Simmons et al. 1989). Dear et al. suggested that multipeaked neurons could decode spectral notches, as the notch frequencies are odd harmonics of each other (e.g., 1, 3, 5, etc.), similar to the 1:3 ratio of BFs in these cells. The problem with this hypothesis is that multipeaked cells respond best to spectral *peaks*, not notches. Dear et al. are content with the proposition that notched echoes would actually *decrease* the firing rate of multipeaked neurons. However, they provide no evidence that these cells have sufficient levels of background activity against which decreases in firing rate might be detected.

3.4.4 Auditory Space Representation

Jen, Lin and Sun (1989) determined the spatial receptive fields of cortical neurons using a roving free-field loudspeaker. All receptive field centers were located in the contralateral hemifield, and grew larger with amplitude. Receptive fields also depended on BF, such that higher BF units showed smaller receptive fields. The response centers of units with higher BFs tended to be located more toward the midline than those with lower BFs. Thus, the representation of auditory space systematically changes within the tonotopic axis of the cortex. Receptive field azimuths shift from frontal to lateral along the rostrocaudal axis of decreasing BF. By contrast, the receptive field elevations show no such dependency on location/BF. Contrary to the results of Dear et al. (1993), Jen, Lin and Sun (1989) found very little variation of BF, MT, tuning-curve shape, or azimuth of the spatial receptive field center within cortical columns.

The dependence of receptive field location on BF is predictable from the directionality of the pinna (Jen and Chen 1988). Pinna directionality improves at higher frequencies, and may be optimized at certain locations (Fuzessery, Hartley, and Wenstrup 1992). As frequency increases, the area of best sensitivity moves toward the midline, and often below the horizontal axis of the head, in every bat species studied so far (Grinnell and Grinnell 1965; Shimozawa et al. 1974; Grinnell and Schnitzler 1977; Fuzessery and Pollak 1984; Makous and O'Neill 1986).

3.5 Summary

In summary, each species of bat so far studied shows at least one tonotopically organized cortical field, representing frequency along a rostrocaudal gradient similar to AI in the cat. There is an overrepresentation of the biosonar frequencies in each species that is especially pronounced in CF/FM bats. However, in CF/FM bats the representation of the FM components of the sonar pulse is poor within AI, and is instead segregated into separate, nontonotopic fields. FM bats, on the other hand, have a tonotopically organized cortex composed of cells that are typically more responsive to FM stimuli than they are to pure tones. These fundamental distinctions are further explored next.

4. "Nontonotopic" Cortical Organization

The most intriguing feature of the auditory cortex in all bats so far studied is that new types of organization beyond tonotopy emerge from populations of cells with complex response properties. In this context, "complex" means that cells are specialized in some way for complex signals. "Specialization" can refer, for example, to a cell that responds to a complex, time-varying signal, such as an FM sweep, but does not respond to pure tones. As mentioned in Section 3, such cells are relatively common in the *Myotis* cortex (Suga 1965b; Shannon-Hartman, Wong, and Maekawa 1992). Specialization can also refer to a cell that shows facilitation or a markedly lower threshold to two or more signal elements ("combinations"), but is unresponsive to these signal elements presented alone, or in the wrong temporal order. In the microbat cortex, the most common type of facilitation occurs when pairs of FM signals are presented with particular time delays between the two sweeps. Combination-sensitive cells responsive to this stimulus are called "delay sensitive" or "delay tuned," and they occur in some form in the cortex of all bats studied so far. They also are known to occur in the MGB in at least one species, the mustached bat (Olsen 1986; Olsen and Suga 1991a,b; see Wenstrup, Chapter 8), and were actually first recorded in the midbrain intercollicular (pretectal) area of *Eptesicus* (Feng, Simmons, and Kick 1978). The properties of these cells are well suited for temporal and spectral pattern recognition. Moreover, their arrangement has revolutionized the traditional view of auditory cortex, emphasizing functional (task-oriented) rather than anatomical (cochleotopic) organizational principles.

4.1 Mustached Bat

Dorsal and ventral to the tonotopically organized area of mustached bat cortex are additional fields containing neurons with response properties that are much more complex than cells in AI (Table 9.2; see however Section 4.1.4 concerning the DSCF area). The first two of these fields to be discovered, and still by far the most intensively studied, are the FM-FM and CF/CF areas. It is within these regions that combination sensitive neurons were first found in cortex.[†] The typical combination-sensitive neuron is facilitated by two stimuli that alone do not elicit any response. As such, they are relatively "invisible" to tonal or wideband noise search stimuli routinely used to explore auditory centers. Such neurons combine two signal elements, and in the case of the mustached and horseshoe bats, these elements are always (and rather unexpectedly) derived from *different* harmonics of the bat's sonar signal.

4.1.1 FM-FM Area

Facilitation and delay tuning. The FM-FM area (Fig. 9.2B, Fig. 9.7) dorsal and rostral to the tonotopic area in the anterior auditory cortex of the mustached bat is notable because it contains a pure population of one type of combination-sensitive neuron. These neurons are specialized to respond to time-separated *pairs* of frequency-modulated sweeps (Suga, O'Neill, and Manabe 1978; Fig. 9.6A). The initial essential component of the pair

---

[†]Delay tuning and combination sensitivity were concurrently reported in both the auditory midbrain, and also, apparently, the auditory cortex in *Eptesicus* by Feng, Simmons, and Kick (1978). However, this paper only reported the response properties of midbrain cells. Details of *Eptesicus* cortical cells have only recently been described (see Section 4.4).

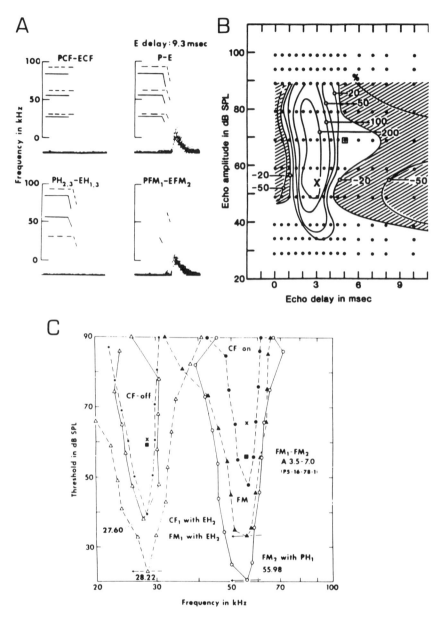

FIGURE 9.6A–C. Response properties of FM-FM cells in mustached bat. (A) Peristimulus time (PST) histograms and schematic spectra demonstrating selectivity of delay-sensitive FM-FM cells for combinations of a first-harmonic FM and (in this case) a second-harmonic FM stimulus. On *top right* is shown the strong facilitation produced by stimuli mimicking first three harmonics of a natural CF-FM sonar pulse *(solid lines)* and a 9.3-msec delayed echo *(dashed lines)*, compared to lack of response to the CF components alone *(top left)*. On the *bottom*, the facilitated

mimics the FM_1 (25–30 kHz). The delayed second sweep must mimic either the FM_2, FM_3, or FM_4 component (depending on the neuron), but never the FM_1 (Fig. 9.6C; Suga, O'Neill, and Manabe 1978). The majority are therefore classified as FM_1-FM_2, FM_1-FM_3, or FM_1-FM_4 cells. Each cell type is dorsoventrally segregated in the FM-FM area into three separate bands (O'Neill and Suga 1979, 1982). At the borders between the bands, cells facilitated by more than one FM component in the delayed sweep can be found (O'Neill and Suga 1982). These include FM_1-$FM_{2,3}$, FM_1-$FM_{3,4}$ and even FM_1-$FM_{2,3,4}$ neurons.

The most striking attribute of FM-FM neurons is that they are "delay tuned." Most cells respond best at a specific delay between the FM sweeps, called the "best delay" (BD). When presented with FM pairs at different repetition rates mimicking the different stages of target pursuit, the majority of cells show stable delay-tuning functions ("delay tuning curves"; Figs. 9.6B, 9.7A). Echo delay is the behavioral cue used to estimate target range (Simmons 1971, 1973). If one assumes that these cells are stimulated by the FM_1 component in the emitted pulse, and one (or more) FM components in returning echoes, then the probable role of delay-sensitive cells is to encode target range.

A small number of FM-FM neurons had best delays that were not stable, but rather became shorter at higher repetition rates. Therefore, Suga, O'Neill, and Manabe (1978) called cells with stable best delays *delay tuned*, and those with labile best delays *tracking* neurons. The delay-tuning curves of tracking neurons effectively "retune" themselves to shorter echo delays at higher repetition rates, such as when a bat increases its emission rate during

---

FIGURE 9.6A–C. (*continued*) response to a pair of FM_1 and delayed FM_2 components *(right)* contrasts to lack of discharges in response to multiharmonic stimulus lacking the first harmonic in the pulse and second harmonic in the delayed echo *(left)* (adapted from Suga et al. 1983, © 1983, J. Neurophysiol. 49:1573–1626, by permission of American Physiological Society). (B) Delay-tuning curve of area VF FM_1-FM_3 cell showing temporal inhibition. Contour values are expressed as a percentage relative to the sum of spike counts to FM_1 and FM_3 presented alone. Facilitation occurs when FM_1 is paired with FM_3 at delays and intensities indicated by *unshaded contours*. Inhibition is produced when a second FM_3 echo is presented with FM_1-FM_3 stimulus, indicated by *shaded areas* (FM_3 level and delay in the pair shown by *small square* at 6-msec delay, 70 dB sound pressure level [SPL]). Inhibition sharpens the delay-tuning curve, but does not alter BD (adapted from Edamatsu and Suga 1993, © 1993, J. Neurophysiol. 69:1700–1712, by permission of American Physiological Society). (C) Frequency-tuning curve of FM_1-FM_2 neuron in FM-FM area. Response areas for pure tones are shown as *filled circles, dashed lines*; for FM sweeps alone as *filled triangles* and *dashed lines*; and for various pairs (combinations) of CF and FM components, as labeled. Lowest thresholds were for FM_1-FM_2 pairs. *Arrows* indicate frequency range of FM sweeps (from Suga et al. 1983; © 1983, American Physiological Society).

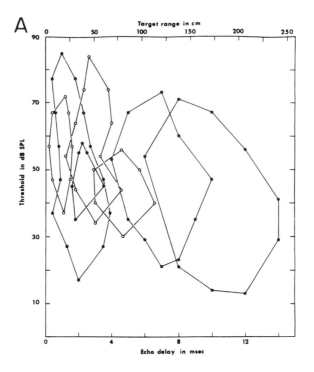

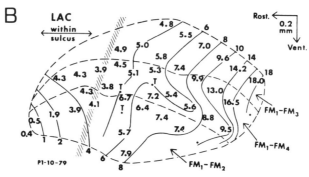

Best delay in msec (BR = BD × 17.2 cm)

FIGURE 9.7A,B. Organization of the FM-FM area in mustached bat. (A) Delay-tuning curves of delay-tuned neurons show increase in delay-tuning curve widths in cells with longer best delays. Different echo delays can be encoded by different neurons. (B) Iso-best delay map reconstructed from rostrocaudally oriented penetrations parallel to cortical plane in left hemisphere (LAC). Cells recorded within banks of Sylvian fissure are displayed unfolded to *left* of *shaded line*. Three bands containing FM_1-FM_2, FM_1-FM_3, and FM_1-FM_4 cells are indicated. (From O'Neill and Suga 1982; © 1982, J. Neurosci. 2:17–31, by permission of the Society for Neuroscience.)

target approach. By having a labile temporal window, tracking neurons would in effect "lock on" to targets throughout the attack, unlike delay-tuned cells that would only discharge when the target was in a specific range "window."

Edamatsu and Suga (1993) have found that when two echoes are presented after a pulse, one in the delay-sensitive period and the other at shorter or longer delay, the delay tuning curve becomes sharper (Fig. 9.6B, shaded areas). The function of these temporal inhibitory areas is analogous to that of lateral inhibition in the spectral domain, but in this instance it improves range acuity and echo-amplitude tolerance.

Suga and colleagues have dubbed facilitation involving different harmonic components "heteroharmonic facilitation" (Suga et al. 1983). To test the hypothesis that FM-FM neurons are responding to emitted pulses and returning echoes, Kawasaki, Margoliash, and Suga (1988) recorded from vocalizing bats under anechoic conditions (i.e., they faced the bats out a window!), and played back artificial echoes time locked to the vocalizations. They found that the self-stimulation of the bat's own emissions activated FM-FM cells when FM_n $(n=2,3, \text{ or } 4)$ echo stimuli were presented alone at the appropriate time delays; they were not responsive to the bat's emissions or the artificial echoes alone.

Why do FM-FM neurons require this seemingly odd set of spectral components to encode target range? One advantage of heteroharmonic facilitation that has been suggested (Suga and O'Neill 1979; O'Neill and Suga 1982; Suga et al. 1983) is that it reduces the likelihood of jamming of the target ranging system by the sonar signals of other bats. If the ranging system relied on high-amplitude signals in both the pulse and echo, the sonar cries of other bats might activate delay-tuned cells inappropriately. Because the emitted FM_1 signal is 30–40 dB weaker than the H_2 component, only the bat's own cries would be sufficiently intense to stimulate its FM-FM cells. This affords each bat its own "private line" for target ranging, and could perhaps explain the remarkable ability of these bats to echolocate successfully in the confined and crowded conditions typical of bat roosting sites.

Frequency and amplitude domain properties of FM-FM neurons. Two additional properties remain to be mentioned regarding FM-FM neurons. One is that they are rather broadly tuned in the frequency domain (Fig. 9.6C). This "frequency tolerance" enables these cells to respond to the broadband FM components of pulses and echoes regardless of the magnitude of Doppler compensation behavior or echo Doppler shift, without impairing the ability to encode target distance.

The second interesting property is that FM-FM cells are often sharply tuned in the amplitude domain. Echo amplitude is affected by both target size and distance from the bat (collectively, the subtended angle of the target). The facilitation of FM-FM cells is typically nonmonotonically related to the amplitude of the echo FM_n component, and most cells showed "best facilitation amplitudes" (BFA; Suga et al. 1983).

One question that arises is whether BFAs are matched to the stimulus levels impinging upon the bat's auditory system in natural conditions. The actual level of stimulation reaching the auditory forebrain is affected by such factors as the directionality of the ear and oral cavity, middle ear muscle contraction during vocalization, and neuronal suppression (Suga and Schlegel 1972; Suga and Shimozawa 1974). Kawasaki, Margoliash, and Suga (1988) estimated from cochlear microphonic recordings that the vocal self-stimulation by the FM_1 component is equivalent to about 70 dB sound pressure level (SPL), somewhat higher than the mean BFA for the FM_1 component in cortical cells (63 dB SPL). Middle ear muscle activation during vocalization (not accounted for by their calibration) would further attenuate the signal and might make up for the difference observed here.

For the echo FM components, the system is adapted to a fairly wide range of moderate amplitudes. However, there is behavioral evidence that mustached bats compensate echo amplitude by adjusting the intensity of their pulses, and thereby stabilize the stimulation of the ears at a nearly constant level regardless of target distance (Kobler et al. 1985). Estimates from *Eptesicus* by Kick and Simmons (1984) of the level of echo stimulation reaching the bat during approach phase range from 25 to 30 dB SPL, near the lower end of the range of BFAs for echo FM_2 components (56.7 ± 14.5 dB SPL; Suga et al. 1983).

Functional organization of FM-FM area. The best delays represented in the FM-FM area range from approximately 0.4 to 18 msec, but the majority of cells have BDs between 3 and 8 msec (Fig. 9.7A) (Suga and O'Neill 1979; O'Neill and Suga 1982). In terms of target range, these BDs span the biologically relevant time delays encoding distances between a few centimeters and about 3 m (Griffin 1958, 1971). Cells recorded in penetrations perpendicular to the cortical surface show similar delay tuning curves, suggesting columnar organization of this response property. Moreover, there is an orderly "chronotopic" representation of BD along the rostro-caudal axis of the FM-FM area, with an overrepresentation of echo delays of 3–8 msec (Fig. 9.7B; Suga and O'Neill 1979). Iso-best delay contours drawn on the cortical surface traverse the boundaries of the FM_1-FM_2, FM_1-FM_3, and FM_1-FM_4 cortical strips. Thus the FM-FM area contains a single map of echo delay, and thereby target range, that is congruent across echo harmonics. Because targets will reflect different echo spectra depending on their size, having delay-tuned cells tuned to all three upper harmonics enables the system to encode distance for a wide range of target sizes.

Suga and O'Neill (1979) made a rough calculation of target-range acuity based upon the distance along the cortical surface occupied by the range map, the range of BDs expressed along the range axis, and the estimated width of cortical columns. Assuming a column occupies about 20 μm, maximum range resolution in the linear region from 3 to 8 msec BD calculated to be about 2 cm, in agreement with range discrimination data for real (as opposed to jittered phantom) targets (Simmons 1971, 1973; see Chapter 4).

The discovery of these unusual cells, their clustering into separate fields, and their orderly arrangement along a temporal rather than a spectral axis, was the first clear demonstration in the mammalian auditory system of an organizational principle not based on the tonotopy of the cochlear partition. This place map representation for target distance followed closely the remarkable discovery of a "space map" in a division of the barn owl midbrain containing cells selective for both stimulus azimuth and elevation (Knudsen and Konishi 1978), and provided independent evidence that computational properties of the auditory pathways can create data structures in neural space that mimic spatial structures in extrapersonal space.

Directional selectivity of FM-FM neurons. Delay-sensitive FM-FM neurons are able to encode spatial information along one dimension related to target distance. But might they also be able to encode the location of a target in all three dimensions? Suga, Kawasaki, and Burkhard (1990) investigated the spatial tuning of range-tuned neurons in the FM-FM area to sound azimuth and elevation, to see whether they might encode target location in three-dimensional space. They found that the free-field spatial receptive fields determined from FM-FM best-delay stimuli were very large, more than 70° in both azimuth and elevation, even for amplitudes near threshold. Receptive fields in some cells ended at the midline, suggesting that they are suppressed by signals in the ipsilateral hemifield. In other cells, activity could be elicited in both hemifields, suggesting binaural excitation. Regardless of the type of cell, the receptive field centers were typically within the contralateral hemifield, but their centers differed according to the essential harmonic for the echo FM stimulus in a pair. Mean best azimuths were confined to the zone between midline and about 25° contralateral, nearly the same as more peripheral measures of ear directionality for these frequency bands (Fuzessery and Pollak 1984; Makous and O'Neill 1986). Best delays remained more or less constant within the receptive fields. Suga, Kawasaki, and Burkhard (1990) concluded that delay-tuned cortical cells are unsuited for target localization, but can provide reliable estimates of target range in a large region of frontal auditory space.

4.1.2 Dorsal Fringe, Ventral Fringe, and "H_1-H_2" Areas

Three other fields of mustached bat cortex contain delay-sensitive cells. The three regions are ipsilaterally connected "in series," and are also connected to the corresponding contralateral fields via the anterior commissure (Fritz et al. 1981). The dorsal fringe (DF) area lies just dorsal to the FM-FM area, rather close to the midline and straddling the terminus of the temporal sulcus (see Fig. 9.2A) (Suga and Horikawa 1986). Although smaller, this region is almost a carbon copy of the FM-FM area, in that there are the same three rostrocaudally elongated bands of cells tuned to the different echo harmonics, and there is a map of BD. However, the range of BDs found along the rostrocaudal axis extends out only to 9 msec (Suga and Horikawa 1986). Anterograde tracing studies have shown that DF receives a projection from the ipsi- and contralateral FM-FM areas (Fritz et al. 1981), but surprisingly there is no indication of a thalamocortical projection to this field. In this sense, the DF area could be considered secondary auditory cortex.

Another even smaller field containing FM-FM neurons is the ventral fringe (VF) area (Edamatsu, Kawasaki, and Suga 1989). The VF lies beneath the rostral end of the tonotopic AI region and receives a projection from the DF area (see Fig. 9.2A) (Fritz et al. 1981). The distribution of BDs in VF cells is truncated even more than that in the DF area, extending only to 5–6 msec. The small size of VF makes it difficult to determine whether there are discrete bands, clusters, or randomly intermingled FM-FM neurons tuned to different echo harmonics.

Smaller still is the H_1-H_2 region (also referred to as area VA), lying caudal to VF and just underlying the rostral border of the DSCF area. Area VA receives a weak intracortical projection from the FM-FM area (Suga 1984) and the CF/CF area (Olsen 1986). In this region, most neurons are facilitated by any combination of first and second harmonic CF and FM components: CF_1-FM_2, CF_1/CF_2, FM_1-FM_2, and FM_1/CF_2. They are somewhat similar to "CFM_1-CFM_2" neurons described recently in the horseshoe bat cortex (Schuller, O'Neill, and Radtke-Schuller 1991; see Section 4.2). In many respects, they are also similar to DSCF area cells recently found to be facilitated by FM_1-CF_2 combinations (Fitzpatrick et al. 1993; see Section 4.1.4). Because of this similarity in response properties, Fitzpatrick et al. (1993) have suggested that the H_1-H_2 region may in fact not constitute a separate field, but rather might be simply a previously unrecognized part of the DSCF area.

Differences in response properties exist between the FM-FM, DF, and VF areas, but the differences are small and do not clearly indicate a different function for each area. Edamatsu and Suga (1993) hypothesized that neurons in the three areas would show increasing specialization, based on the chainlike interconnection of the FM-FM → DF → VF pathway. In fact, what they found was that only rather subtle differences distinguish the FM-FM and VF fields. In both areas, for example, the delay-tuning curves of about 80% of the cells in both areas show temporal inhibition for pulse–echo delays both shorter and longer than the best delay (Edamatsu and Suga 1993). On the other hand, VF cells are strongly adapted by rapidly repeated FM stimulus pairs, and consequently can only function during search phase, whereas most FM-FM area cells can follow up to at least 100 pairs/sec (terminal-phase rates) with little change in either BD or delay-tuning curve shape. This result suggests that different cortical fields may take part in target range processing only during particular phases of echolocation. As is discussed in Section 4.3.3, an analogous situation occurs in the *Myotis* auditory cortex. To what end this strategy is applied by the brain is still unknown.

4.1.3 CF/CF Area

Response properties. A second type of combination-sensitive neuron discovered in the mustached bat is the CF/CF[‡] cell. CF/CF cells are facilitated by combinations of two constant-frequency tones (Fig. 9.8A) (Suga, O'Neill and Manabe 1979; Suga et al. 1983), and there are only two types, CF_1/CF_2 and CF_1/CF_3. As with FM-FM neurons, CF/CF cells are confined to a separate cortical field, overlying the rostral end of the tonotopic area, just ventral to the FM-FM area (see Fig. 9.2A). Another gross similarity to the organization of the FM-FM area is that CF_1/CF_2 and CF_1/CF_3 cells are segregated into two narrow strips running rostrocaudally (Fig. 9.8C). That there are no CF_1/CF_4 cells is not unexpected, as cells with BFs greater than about 112 kHz are rare in the mustached bat auditory system.

The similarities between CF/CF and FM-FM cells end at that point. The spectral and temporal tuning properties of CF/CF neurons are almost diametrically opposite to FM-FM cells, as summarized in Table 9.2. Whereas FM-FM cells are broadly frequency tuned and sharply delay tuned, CF/CF cells are sharply frequency tuned and broadly delay tuned.

The facilitation tuning curves for CF/CF cells (Fig. 9.8B) resemble a plot combining the tuning curves of CF_1-tuned cells with CF_2 or CF_3 tuned cells from the tonotopic field of auditory cortex (Suga, O'Neill, and Manabe 1979; Suga et al. 1983; Suga and Tsuzuki 1985) or lower centers like the IC (O'Neill 1985). The tuning for the CF_3 component in CF_1/CF_3 cells is especially sharp, with Q_{10dB} values of 65–70 versus 40–50 for CF_3-tuned cells in AI or the periphery (Suga and Jen 1977; Asanuma, Wong, and Suga 1983; Suga and Tsuzuki 1985). This is significant because it is the only example so far reported of central neurons showing better Q_{10} values than peripheral neurons. Suga and Tsuzuki (1985) showed that inhibitory sidebands make important contributions to the sharp tuning of these cells even at very high stimulus levels, rendering the cells level tolerant.

Suga et al. (1983) examined the relationship between the best facilitation frequencies (BFFs) for the essential components. Some CF/CF cells have harmonically related BFFs (i.e, the CF_n is an integer multiple of CF_1),

---

[‡]The use of the "/" in the abbreviations of the CF components facilitating these cells is deliberate, and indicates that there must be temporal *overlap* between the two components. This is in contrast to abbreviations for FM-FM cells, where "-" indicates that there must be temporal *separation* between the FM components of the stimulus.

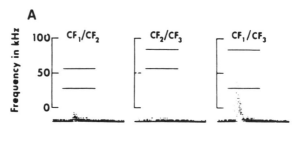

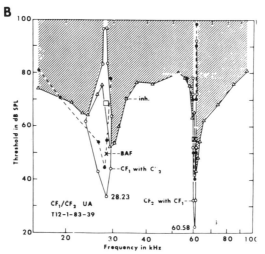

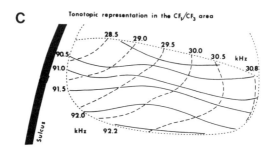

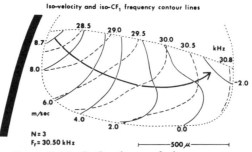

FIGURE 9.8A–C. Caption on facing page.

implying that they could respond to pulse or echo alone, although response to the latter is unlikely given the weak first harmonic. However, the majority of CF/CF cells are slightly "mistuned", that is, the best frequency for the CF_n component is not harmonically related to CF_1. Consequently, these cells cannot respond to harmonically related elements in pulses or echoes alone. Instead, they can respond only to signals with different fundamental frequencies, such as an emitted pulse and an overlapping Doppler-shifted echo. CF/CF cells could thereby encode the magnitude of Doppler shifts by comparing the fundamental frequency of the pulse to the higher harmonics of the echo. Because of their extraordinary tuning sharpness, each cell in effect represents a single Doppler-shift magnitude. Because echo Doppler shifts are caused by relative velocity differences between the bat and its surroundings, each CF/CF cell could represent a particular relative velocity.

By and large, CF/CF neurons are not delay tuned. Although some cells prefer a particular echo delay between the two CF components for maximum response, strong facilitation occurs at nearly any delay, as long as there is some temporal overlap. This temporal overlap can be as short as 1 msec and still produce noticeable facilitation. In a behavioral context, this broad "delay tolerance" enables CF/CF cells to encode relative velocity over a wide range of target distances.

Olsen (Olsen 1986; Olsen and Suga 1991a) found this type of neuron in the medial geniculate body of the mustached bat, but found that thalamic CF/CF cells responded much better to individual CF components presented alone than did cortical cells (see Wenstrup, Chapter 8, this volume).

Organization of the CF/CF area. The CF/CF area is divided such that CF_1/CF_2 cells are found dorsally and CF_1/CF_3 are found ventrally (see Fig. 9.2A) (Suga et al. 1983). Cells responding to either the CF_2 or the CF_3 component ($CF_1/CF_{2,3}$ cells) are located at the border between the two

←
___

FIGURE 9.8A–C Response properties of cells in CF/CF area of mustached bat cortex. (A) PST histograms and schematic spectra show responses to various pairs of CF signals mimicking sonar components. Cell shown is from the ventral half of the CF/CF area, selective for combinations of CF_1 and CF_3 (from Suga et al. 1983, © 1983, by permission of American Physiological Society). (B) Excitatory *(open)* and inhibitory *(shaded)* response areas of a CF_1/CF_3 neuron, measured with tones alone *(dashed lines)* or pairs of CF stimuli *(solid lines)*. Contrast these areas to those for the FM-FM cell in Figure 9.7. *x* indicates the BAF, best amplitude for facilitation (from Suga and Tsuzuki 1985, © 1985, J. Neurophysiol. 53:1109–1145, by permission of American Physiological Society). (C) Tonotopic representation *(top)* and isovelocity contours of Doppler-shift magnitude *(bottom)* in the CF_1/CF_3 region. CF_1 and CF_3 are represented roughly orthogonally, and the axis of increasing velocity is shown by the *arrow* (from Suga, Kujirai and O'Neill 1981; © 1981, in Syka J., Aitkin L. (eds), Neuronal Mechanisms of Hearing, pp. 197–219, by permission of Plenum Press.)

strips. CF/CF cells recorded in single cortical columns have essentially identical BFFs, that is, they are tuned to the same Doppler shift. Suga et al. (1983) showed that there is a bicoordinate representation of BFFs, with iso-BFF contours for the CF_1 component oriented more or less orthogonally to those for the CF_2 or CF_3 components (Fig. 9.8C, top). Although the map only covers a limited range of frequencies around the CF components of the sonar signal, this was the first demonstration of a *bicoordinate* tonotopic representation in the auditory system. This map encodes frequency differences ranging from 0 ventrocaudally to about 3 kHz rostrodorsally (Fig. 9.8C).

The relatively large and constant Doppler shift caused by a flying bat's motion relative to its surroundings has been termed the "DC" component of the Doppler-shifted echo, whereas the small, periodic frequency modulations caused by the local movement of insect targets can be considered the "AC" component (Suga and Manabe 1982). Because each CF/CF cell could precisely encode a specific Doppler shift, Suga et al. (1983) proposed that the CF/CF area systematically encodes flight velocity, that is, the DC component of the Doppler-shift, within a maplike representation on the cortical surface. In the CF/CF area, the represented Doppler shifts compute to velocities ranging from about -1 or 0 to about 9 m/sec, with an overrepresentation of the range from 0 to 5 m/sec (Fig. 9.8C, bottom).

However, Schuller and Pollak (1979) have argued that evolution has not selected for an exquisitely tuned ear and Doppler-shift compensation behavior simply for the purpose of velocity perception per se. They pointed out that velocity is readily available to all bats from the rate of change in echo delay across successive echo presentations. In their view, these extraordinary adaptations have evolved to enable the detection of small frequency and amplitude modulations (i.e., the echo AC component), caused by the beating of an insect's wings (Schnitzler et al. 1983). Indeed, Goldman and Henson (1977) have shown that mustached bats will only attack fluttering targets. Suga, Niwa, and Taniguchi (1983) have also shown that CF/CF cells show exquisite phase-locking to sinusoidal FM of the echo CF_n component, a stimulus that crudely approximates the AC component of the Doppler-shifted echo from a fluttering target. This suggests a possible role for the CF/CF area in local target motion analysis.

Gooler and O'Neill (1987, 1988) and O'Neill and Basham (1992) have proposed an additional possible function of the CF/CF area: encoding the magnitude of the Doppler shift not for measuring relative velocity, but rather for computation of the error signal needed by the vocal motor system to control Doppler-shift compensation behavior itself. For further discussion of vocal motor control systems and Doppler-compensation circuitry, see Schuller (1974, 1977), Suga, Simmons, and Shimozawa (1974), Gooler and O'Neill (1987), Schuller and Radtke-Schuller (1990), and Metzner (1993).

It seems possible that the CF/CF area could in fact represent both velocity and local target movement. Trappe and Schnitzler (1982) showed that flying horseshoe bats Doppler-shift compensate their flight velocity relative to their surroundings, and not to that of insect prey they were pursuing. Thus, echoes from the surroundings would stimulate a specific "iso-velocity" contour in the CF/CF area encoding the bat's flight velocity, whereas the insect echo, modulated by the movement of the wings, would stimulate other parts of the CF/CF area, within which cells would encode the echo modulation patterns by phase-locking their discharges.

4.1.4 Facilitation in the DSCF Area

Fitzpatrick et al. (1993) have found that about 75% of neurons in the DSCF area can also exhibit delay-dependent facilitation to paired stimuli mimicking biosonar components. These "FM_1-CF_2" cells constitute a third type of facilitation neuron, similar to those originally described in the H_1-H_2 area (Table 9.2; see Section 4.1.2). The best delays of DSCF cells are relatively long, from 5 to 30 msec, with a mean at about 21 msec corresponding to a target distance of 3.6 m. Thus, facilitation improves the sensitivity of DSCF neurons to faint echoes from distant targets. Fitzpatrick et al. (1993) pointed out that the target range at which facilitation first occurs in this population corresponds to the distance at which mustached bats initiate the pursuit of flying insects, and they suggest that facilitation might play a role in initiating approach-phase behavior. They point out that this result has important implications for the function of primary auditory cortex, namely that even AI neurons can exhibit complex response properties.

4.1.5 Role of Cortical Fields in Mustached Bat Echolocation Behavior

Recent experiments have examined the possible involvement of the DSCF and FM-FM areas in frequency and target range discrimination. Riquimaroux, Gaioni, and Suga (1991, 1992) attempted to inactivate the DSCF or the FM-FM area selectively with muscimol (a GABA receptor agonist), and then determined the bat's acuity for frequency and distance discriminations using a conditioned avoidance procedure. When muscimol was applied to the DSCF area, subject bats could discriminate large, but not small, frequency differences. By contrast, target distance discriminations remained unimpaired. When muscimol was instead applied to the FM-FM area, then fine, but not coarse, target range discrimination was impaired, and frequency discrimination remained normal. Thus, as predicted from the characteristics of cells and their functional organization, the FM-FM area is involved in the perception of distance, but has apparently little to do with frequency discrimination. The opposite is true for the DSCF area. Individual cortical fields appear to play pivotal roles in specific aspects of perception related to particular features of acoustic signals. In no other species so far investigated has there been as clear a demonstration of the function of specific cortical fields in perception.

4.2 Horseshoe Bat

Schuller and colleagues (Schuller, Radtke-Schuller, and O'Neill 1988; Schuller, O'Neill, and Radtke-Schuller 1991) carried out an extensive study of the FM region of the horseshoe bat cortex, and compared the results with data from the FM-FM, DF, and CF/CF areas in mustached bats gathered using identical experimental techniques in the same laboratory. In general, the cells in the FM region (as defined by Ostwald 1984) preferred hetero-harmonic combinations like those in mustached bat (cf. Tables 9.2 and 9.3). However, facilitation in horseshoe bats was less striking than in mustached bats, and many cells in the horseshoe bat showed good responses to CF or FM sounds presented alone. The fields containing combination-sensitive cells, although distinct, are also smaller in horseshoe bats than those in mustached bats, and so far there is no indication that more than one field of cortex contains neurons of each type (see Fig. 9.3A).

Consistent with the simpler biharmonic structure of the horseshoe bat sonar signal, only three types of combination-sensitive neurons were found in the horseshoe bat: FM_1-FM_2, CF_1/CF_2, and CFM_1-CFM_2. This latter class of "mixed" cells is facilitated by both CF_1/CF_2 and FM_1-FM_2 combinations, and is functionally equivalent to H_1-H_2 cells in area VA of the mustached bat (see Section 4.1.2 and 4.1.4). Mixed cells are also found in the FM-FM region of mustached bat cortex (Suga et al. 1983; Schuller, O'Neill, and Radtke-Schuller 1991), but they are not as prevalent, and have not been previously recognized as a separate class.

4.2.1 Response Areas of Combination-Sensitive Neurons

The tuning properties of horseshoe bat combination neurons resemble those in mustached bat, with some exceptions. As in mustached bat, horseshoe bat CF_1/CF_2 cells are typically broadly tuned for the CF_1 component and very sharply tuned for the CF_2 component. For FM_1-FM_2 cells, tuning for both components was broad, but there were cases in which the response area for the FM_1 component was narrower than that to the CF_1 component in a typical CF_1/CF_2 cell.

4.2.2 Delay Dependency

As in the mustached bat, horseshoe bat FM-FM cells are sharply delay tuned, whereas CF/CF cells were more often broadly tuned. Unlike the mustached bat, the majority of horseshoe bat FM-FM cells were tuned to short best delays between 1.5 and 4.5 msec (longest, 9.5 msec). However, there is a chronotopic representation of BD within a rostrocaudally elongated strip in the horseshoe bat cortex, distorted by an overrepresentation of BDs from 2 to 4 msec (see Fig. 9.3B,C). Within this strip, facilitated neurons are intermingled with nonfacilitated neurons, resembling FM bats (see following) but unlike the mustached bat. CF/CF and mixed cells are found in a mediolaterally elongated strip overlapping the rostral end of the FM-FM zone (see Fig. 9.3B).

Interestingly, facilitated cells were recorded only within cortical layer 5, at depths of 400–800 μm. This is similar to the preferred depth for recording such cells in the mustached bat (Schuller, O'Neill, and Radtke-Schuller 1991). Cells near the surface responded to FM_1 signals alone, whereas those below layer 5 often responded best to FM_2 signals alone. Thus, facilitation is restricted to layer 5 neurons sandwiched between nonfacilitated neurons in the superficial and deep layers tuned to the individual harmonics.

One of the more puzzling results to come from the study by Schuller, O'Neill, and Radtke-Schuller (1991) was that most horseshoe bat combination-sensitive neurons were tuned to frequencies that were appropriate for *negative* rather than positive Doppler shifts. In the case of CF/CF cells, rather than being tuned to frequencies equal to or higher than the second harmonic of the best CF_1 frequency for facilitation (i.e., positive Doppler shifts), the BFF for CF_2 was usually *lower* than twice the CF_1. The same was true for FM_1-FM_2 cells. Tuning for negative Doppler shifts also occurs in mustached bat, but to a lesser extent. Because only positive Doppler shifts can occur when the bat approaches a target, tuning to negative Doppler shifts would seem to be maladaptive. Of course, FM_1-FM_2 cells are broadly tuned to the lower FM component. However, CF/CF neurons are sharply tuned for the upper frequency for facilitation, making them less able to tolerate any mismatches between the stimulus and their response area. Thus, what exact role these cells play in perception is clouded by this rather puzzling finding.

One possible explanation comes from field observations of horseshoe bats (Link, Marimuthu, and Neuweiler 1986). These bats often adopt a strategy of ambushing prey from a hanging perch. In this instance, negative Doppler shifts would occur when the flying prey passed by the bat's position. This situation might elicit responses from CF/CF and mixed-type neurons. Unfortunately, whereas the echo frequency might be optimal when the prey has passed the bat's position, the pulse-echo time delay would likely be too long to excite the majority of FM-FM cells, whose BDs are 2–4 msec (34–68 cm target range).

Aside from the foregoing rather weak counterarguments, for the horseshoe bat it is difficult to reconcile the apparent selectivity for negative Doppler shifts with the hypothesis that combination-sensitive cells encode information about Doppler-shift magnitude or target range. Either the hypothesis needs to be revised, or something vital about the behavior of these bats is being overlooked (hardly surprising) that would reconcile the dilemma.

In general, then, despite many similarities in basic response properties and functional organization, it is clear that the horseshoe bat cortex is less differentiated than mustached bat cortex. Although the mustached bat has multiple large cortical fields and partial redundancy of representation of target range information, the horseshoe bat apparently makes due with only one relatively small ranging area, and the chronotopic axis heavily favors nearby targets. If one conceives of cortical maps as "data representations" that are involved in some way with perception, then one could argue that, because it has more maps, the mustached bat's perceptual world must be richer than that of the horseshoe bat, despite close similarities in echolocation behavior. Do redundant or even partially redundant maps in cortex subserve different percepts, or control different behaviors? This exciting and important question remains mostly unanswered, despite the relative accessibility for experimentation of these cortical fields. As noted in Section 4.1.5, Riquimaroux, Gaioni, and Suga (1991, 1992) have shown that one type of percept can be selectively impaired by inactivation of the appropriate cortical field, leaving other percepts intact. Does this imply that more widespread disruption of perception would occur in bats with less cortical differentiation in a similar experiment?

4.3 Little Brown Bat

4.3.1 Delay Tuning

Following closely on the discovery of delay-sensitive neurons in *Eptesicus* midbrain (Feng, Simmons, and Kick 1978) and *Pteronotus* cortex (Suga, O'Neill and Manabe 1978), Sullivan (1982a,b) investigated the response of cortical neurons in *Myotis* to paired FM stimuli. He found that fully 84% of the recorded cells were indeed facilitated by pairs of identical FM sweeps (*Myotis* sonar pulses typically have only one harmonic), and most of these were delay tuned as well (Sullivan 1982a). Unlike the CF/FM bats in which facilitation requires heteroharmonic combinations, delay-tuned facilitation in this species was expressed only when there was an *amplitude* difference between the two FM stimuli.

Delay-sensitive neurons fall into two categories, P-type and E-type (see Table 9.4). P-type (Fig. 9.9A, left) cells have short best echo delays, narrow delay tuning curves, and response latencies to FM pairs similar to that for loud (70–80 dB SPL) FM pulses delivered alone (Fig. 9.9A, top left). E-type cells (Fig. 9.9A, right), by contrast, have long best delays, broader delay tuning curves, and response latencies time locked to the echo FM (Fig. 9.9A, top right). That is, P-type cells discharge at a fixed latency to the pulse, and are facilitated only over a narrow range of echo delays. The latency of E-type units is locked to the echo, and their response is simply facilitated at particular delays by the presence of the earlier pulse FM. In this regard, P-type cells respond more like mustached and horseshoe bat FM-FM cells than do E-type cells.

On the basis of these observations, Sullivan (1982a) suggested that range might be encoded in *Myotis* by two distinct mechanisms operating at different target distances. Targets far away would excite E-type units, and distance might be encoded by the temporal pattern of facilitated discharge,

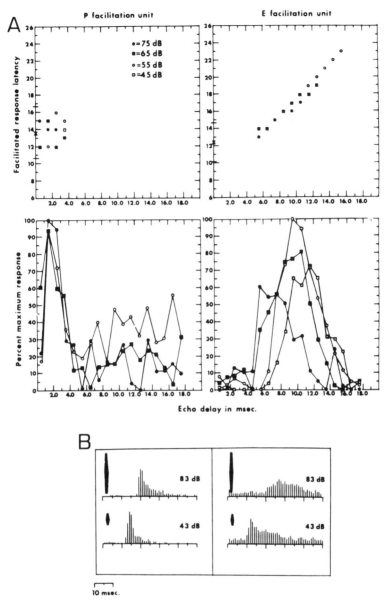

FIGURE 9.9A,B. Response properties of delay-sensitive cells in *Myotis*. (A) Response latency *(top)* and magnitude *(bottom)* as a function of echo delay for a P-type *(left)* and an E-type *(right)* cell. P-type cells were tuned to short-echo FM delays, and the response latencies were locked in time to the pulse FM component. E-type cells were tuned for longer echo delays, and the responses are locked to the echo component, rather than the pulse (from Sullivan 1982a; © 1982, J. Neurophysiol. 48:1011–1032, by permission of American Physiological Society). (B) Two examples of paradoxical latency shift. In each, the latency in response to a brief, high-amplitude FM stimulus *(top)* is longer than that to a low-amplitude stimulus *(bottom)*. Best delay is related to the *difference* in latency (from Sullivan 1982b; © 1982, J. Neurophysiol. 48:1033–1047, by permission of American Physiological Society).

457

rather than by a delay-tuned place mechanism. Targets nearby would excite P-type cells, and these might encode distance by a place mechanism (as in CF/FM bats), by virtue of their stable delay tuning.

However, Sullivan made an intriguing observation that suggested that these response properties might be somewhat interchangeable. He found that by lowering the amplitude of the pulse FM stimulus, an E-type cell could be made to respond like a P-type unit. He suggested that the bat might be able to control the type of range coding used by the system, simply by altering the amplitude of its sonar emissions.

4.3.2 Paradoxical Latency Shift

Sullivan (1982b) investigated further the stimulus parameters essential for facilitation of delay-sensitive cells. He found that in many units there was a "paradoxical" relationship between response latency and stimulus amplitude (Fig. 9.9B). That is, rather than latency decreasing with increased amplitude, he showed that latency *increased* in delay-sensitive cells. Units showing such a "paradoxical latency shift" had level-response functions with two peaks, one at low and the other at high amplitudes. Responses to stimuli in the low-amplitude peak had shorter latencies than those in the high-amplitude peak. Consequently, the latency to vocalized pulses would be longer than that for most echoes, because pulses are normally much more intense than echoes. In addition, Sullivan discovered that for P-type cells, the latency to high-amplitude FM sweeps was related to the best delay, and that best delay was related to the *difference* between the latencies for strong and weak FM sweeps presented singly. Olsen later found that a similar relationship existed for FM-FM delay-tuned cells in the mustached bat medial geniculate body (Olsen 1986; Olsen and Suga 1991b; see also Wenstrup, Chapter 8). Thus, in *Myotis* it appears that pulses and echoes are differentiated simply on the basis of amplitude, rather than frequency as found in CF/FM bats.

4.3.3 Effect of Repetition Rate and Duration on Delay Tuning

In a series of recent papers, Wong and colleagues have identified novel and important additional factors that affect target range coding by delay-sensitive neurons. Pinheiro, Wu, and Jen (1991) had already shown that IC neurons in the big brown bat (*Eptesicus*) were selective for repetition rate. Wong, Maekawa, and Tanaka (1992) and Teng and Wong (1993) found that *Myotis* delay-sensitive cells also preferred certain repetition rates between 5 and 50/sec. As in mustached bat, delay-tuned neurons showed little change in delay tuning, whereas tracking neurons changed their delay tuning, with increases in repetition rate.

Another parameter affecting delay sensitivity is pulse duration. *Myotis* biosonar signals range from about 2–4 msec in search phase to as short as 0.2 ms in terminal phase (Griffin, Webster, and Michael 1960; Sales and

Pye 1974). Tanaka, Wong, and Taniguchi (1992) found that shortening the stimulus either extended delay-dependent facilitation to higher repetition rates, decreased the range of repetition rates over which which a cell was delay sensitive, or abolished facilitation altogether.

The combined effects of repetition rate and duration of biosonar signals on facilitation can thereby shift the number and location of active range-tuned cells in the cortex. During each phase of target pursuit, different but overlapping sets of delay-sensitive neurons are activated. The largest number of cells would be recruited to encode delay at repetition rates between 10 and 20 Hz, that is, during early approach phase. Moreover, at these repetition rates virtually the entire range of best delays is represented in the active population. This means that during search phase neurons are available that can encode target distances from about 17 cm to about 300 cm, that is, just about the entire range physically perceivable by an echolocating bat (Griffin 1958, 1971). In addition, neurons preferring higher repetition rates, those presumably active during mid-approach and terminal phase, have shorter BDs.

It is somewhat puzzling that there is such a diminution of the active range-encoding population during the later phases of target approach, because intuitively it seems that precise range encoding would be needed at that time. Perhaps, as has recently been hypothesized by Dear, Simmons, and Fritz (1993), these bats develop a detailed acoustic "image" of the objects arrayed in front of them only during the search phase, before the start of target pursuit. In *Myotis* this image would seem to deteriorate during the approach phase, assuming that the number of range encoding neurons impacts on image "quality." However, it must be remembered that despite their fewer numbers, delay acuity is maximum among neurons tuned to short delays.

4.3.4 Effect of Frequency on Delay Tuning

There is considerable evidence that, in all bats so far studied, target distance information is not carried over one spectral channel. Instead, at least two spectrally distinct channels separately encode pulses and echoes. Berkowitz and Suga (1989) have shown that, unlike CF/FM bats, delay-sensitive neurons in *Myotis* do not rely on different harmonics of the FM sweep for facilitation, but this is not to say that different frequencies were preferred in the pulse and echo. Using pure tone pairs (CF-CF stimuli), they found that the BFFs for pulses were about 8 kHz *higher* than for echoes. Maekawa, Wong, and Paschal (1992) divided the idealized *Myotis* FM sweep into four spectral quartiles, and found that most delay-tuned neurons preferred different spectral quartiles in pulses and echoes. Most units required the lower quartile in pulse FM sweeps; in echoes, most preferred one or more of the three lower quartiles.

In summary, most neurons in *Myotis* auditory cortex are delay tuned and presumably contribute to the processing of target range information. The mechanism for creating delay tuning is quite different from that seen in CF/FM bats. Rather than relying on spectral features (different harmonics) to differentiate pulses from echoes, the *Myotis* system utilizes amplitude differences. Moreover, delay tuning is strongly dependent on the duration and repetition rate of sonar signals, and the availability of range-encoding neurons is heavily influenced by the vocal behavior of the bat. As such, the ranging mechanism is more labile and dependent on the behavioral situation than it is in CF/FM bats.

4.4 Big Brown Bat

Dear et al. (1993) have completed the first detailed investigation of facilitation and delay tuning in *Eptesicus* cortical cells (Table 9.4). They found that, like *Myotis*, there is no apparent segregation of delay-sensitive neurons outside the tonotopically organized primary field. However, by contrast to *Myotis* and the CF/FM bats, they reported that only a small proportion of cortical neurons (about one in six) were delay tuned. Like delay-tuned neurons in *Myotis*, those in *Eptesicus* showed facilitation to paired FM stimuli, and only weak or no response to FM stimuli presented singly. Facilitation occurred only over a delimited set of pulse–echo delays, measured in this study with a single (5/sec) stimulus repetition rate. Regarding CF/CF-type facilitation, although there are a significant number of multipeaked cortical cells (Section 3.4.3), Dear et al. did not test these cells with simultaneously presented CF stimuli.

4.4.1 Response Properties and Latency Variation in Delay-Tuned Cells

Dear et al. (1993) found that nearly all delay-tuned cells had brief onset discharge patterns (average 1 spike/stimulus pair). Pulse and echo amplitude were significant for facilitation of delay-sensitive cells, and there is a reasonable, if not perfect, match between level of self-stimulation and pulse BFA. Not unexpectedly, echo BFAs were lower than pulse BFAs. Both the average echo BFA and the lack of correlation between echo BFA and BD are properties consistent with observations in *Pteronotus* delay-tuned cells (Taniguchi et al. 1986).

Dear et al. (1993) measured pulse and echo facilitation latencies in delay-tuned cells. The range of pulse facilitation latencies measured was enormous, from 9 to 42 msec. Echo facilitation latencies also varied widely, from about 5 to 35 msec. The authors make the important point that for any given echo delay (i.e., target range), the neuronal discharges signaling the occurrence of a target at a particular distance would be temporally dispersed over many milliseconds within the subpopulation of cells tuned to that delay.

Dear, Simmons, and Fritz (1993) addressed the consequences of such a

TABLE 9.4 Cortical Units in *Eptesicus fuscus*

Type	Frequency tuning	Delay dependency	Key parameters	Field(s)	Organization	Information encoded
CF	Sharp	Unknown	Frequency	A,C B	Tonotopic Variable	Stimulus spectrum
FM-FM	Broad	Sharp	Paired FM; echo delay; pulse and echo amplitude difference	A,B only	Tonotopic; delay axis in area B?	Target range

temporal dispersion of responses in range-tuned neurons. The authors pointed out that in a complex environment, each pulse emitted by a bat would generate multiple echoes from targets located at different distances. This array of echoes constitutes the "acoustic scene" perceived by the bat at a given moment in time. Of course, the echoes from each object in the scene actually arrive at different times, and were they to be encoded in real time by delay-tuned neurons responding at a single, fixed latency, perceptual "binding" of the elements in the acoustic scene would have to be accomplished by linking neural responses distributed in time. However, because delay-tuned cells exhibit a wide array of response latencies for any given echo delay, targets located at various distances from the bat could elicit responses *simultaneously* from subpopulations of neurons with differing best delays (recall the single-spike, stimulus-evoked discharge pattern and lack of spontaneous discharge). In the words of the authors, this process " . . . transform(s) the sequential arrival times of echoes with different delays into a concurrent, accumulating neural representation of multiple objects at different ranges. . . ." (Dear, Simmons, and Fritz 1993).

Consider, for example, the simple case of two targets, one near and the other far from the bat. At any given moment following the arrival of both echoes, two subpopulations of delay-tuned cells will be *simultaneously* active: (1) a group of *short*-BD cells with *long* latencies, encoding the nearby target, and (2) a group of *long*-BD cells with *short* response latencies, encoding the distant target. At each successive moment in time, new subpopulations encoding each target are recruited in the order of their increasing response latencies. Ensembles of cells with different best delays discharging simultaneously can thus evoke "snapshots" of the entire acoustic scene. These "snapshots" are updated continuously, adding the images of progressively more distant objects until the next pulse is emitted. This process is analogous to viewing visual scenes with stroboscopic illumination. Thus, the array of latencies in delay-tuned cells could be considered as yet another example of *delay lines* by which behaviorally related events occurring at different times (representing in this case targets at different distances) can be encoded at the neuronal level by coincident activity in a population of cells.

An additional feature of the Dear, Simmons, and Fritz (1993) scene-analysis schema is that for at least some delay-tuned neurons, range acuity improves with time after vocalization. They showed that delay-tuning curve sharpness is better in cells with longer facilitation latencies, and claim that successive acuity improvements are consistent with a computational algorithm called "multiresolution decomposition," an image-analysis technique useful for such things as edge detection and object recognition. This view of cortical range processing is provocative and worthy of further consideration, because all bats studied so far show similar facilitation latency distributions for their delay-tuned populations.

4.4.2 "Amplitude-Shift" Delay-Tuned Neurons

About 13% of the delay-tuned neurons recorded by Dear et al. (1993) changed their BDs with changes in pulse amplitude. These cells were dubbed "amplitude-shift" neurons, and were nearly equally divided between cortical areas A and B. Most of these cells exhibited a fairly large decrease in BD with a decrease in pulse amplitude, but a couple showed an increase in BD under similar conditions. Small changes in the effective pulse amplitude (i.e., the amplitude stimulating the auditory system versus that actually impinging upon the ears) could thereby have drastic effects on the range tuning of this small population of cells.

As mentioned in Section 4.1.1 and 4.3.3, cells in *Pteronotus* and *Myotis* called "tracking neurons" also have labile delay-tuning properties (O'Neill and Suga 1979, 1982; Wong, Maekawa, and Tanaka 1992). Like amplitude-shift cells in *Eptesicus*, tracking neurons are few in number and are found scattered among delay-tuned neurons. While amplitude-shift cells in *Eptesicus* behave somewhat similarly, it is not clear that they are the equivalent of tracking neurons in *Pteronotus* and *Myotis*, because tracking neurons change their delay tuning not with changes in pulse *amplitude*, but rather with changes in pulse–echo *repetition rate*. Only one study of *Pteronotus* directly investigated whether delay tuning changed with pulse or echo amplitude (Taniguchi et al. 1986). They found that BD shortened with decreased amplitude in about one-third of their sample, but unlike amplitude-shift cells, the shift in BD was very small. Dear et al. (1993) also point out that while the delay tuning curves of tracking neurons become narrower with higher pulse repetition rates, those of amplitude-shift cells are unaffected by amplitude. Other disimilarities in the response properties of these cells bring into question the functional homology between amplitude-shift cells and tracking neurons, and this issue requires further study.

4.4.3 Organization of Delay-Tuned Neurons in *Eptesicus*

Similar to *Myotis*, delay-tuned neurons were found mainly in the anterior half of the tonotopic area A and throughout the nontonotopic area B of auditory cortex in *Eptesicus*, but none were found in the anterior tonotopic area C (Dear et al. 1993). Contour plots of BD averaged from eight bats (Fig. 9.5B) showed no single overall place representation of echo delay like that seen in CF/FM bats. However, it could be argued that BD is systematically represented in smaller patches of cortex, especially within the dorsorostral part of area B. Area B, in turn, is a region superficially homologous to the nontonotopic cortical fields in CF/FM bats where delay-tuned cells are found. Whether these small regions constitute a map of target range in *Eptesicus* is debatable. Not debatable, however, is the fact that delay-tuned cells in *Eptesicus* are intermingled with cells not sensitive to delay, as they are in both *Rhinolophus* and *Myotis*. They are also tonotopically organized in primary cortex, like *Myotis* but quite different from *Pteronotus* and *Rhinolophus*.

Based on the results from other species, one might expect to find an overrepresentation of certain BDs related to behaviorally significant target ranges. Unique to *Eptesicus*, Dear et al. (1993) found a bimodal distribution of BDs, forming two groups of cells with BDs less than 9 msec and more than 12 msec. Why there is this curious gap in BD representation is not immediately obvious. Does this mean that *Eptesicus* has little or no range acuity for objects between about 175 and 200 cm away? A behavioral test of this hypothesis would seem worthwhile. Dear et al. (1993) found that the BDs of the majority of cells in the long-BD group clustered between 12 and 22 msec, although there were a few outliers with BDs greater than 30 msec. This BD range corresponds to distances between 200 and 400 cm. The prominence of long-BD cells in the population is distinctly different from the situation in the CF/FM bats, where cells with BDs longer than 12 msec are very rare. Dear et al. also found that short-BD neurons were more typically found in the tonotopic area A, whereas long-BD neurons were clustered in the nontonotopic area B (see Fig. 9.5).

No studies have as yet systematically examined either the spectral response preferences or repetition rate sensitivity of delay-tuned neurons in *Eptesicus*. These factors are critical to whether cells even show delay sensitivity in *Myotis*. Until their effects are studied, it will remain unclear whether the representations of echo delay currently published accurately reflect the organization of *Eptesicus* cortex.

5. Circuit Models of Delay Tuning

The discovery of neurons able to encode time differences between temporally separated acoustic events has important implications for the auditory perception of complex signals. From his discovery of paradoxical latency shift and its relationship to delay sensitivity in *Myotis* cortical neurons, Sullivan (1982b) suggested a model that establishes delay tuning from the convergence of two afferent pathways,one processing high-amplitude and the other low-amplitude signals. This model can also be extended to apply to CF/FM bats, if one substitutes heteroharmonic facilitation for the paradoxical latency shift mechanism. One variant of the model involves delay lines, whereby the high-amplitude (FM bat)/low-frequency (CF/FM bat) pathway is delayed to produce longer latencies than the low-amplitude (high-frequency) pathway (Fig. 9.10A). The delay-tuned cell acts as a coincidence detector (logical AND gate), responding to the echo only when it is delayed acoustically by an amount equal to the difference in latencies of the two pathways. This delay-line model resembles that first put forth by Jeffress (1948) to explain detection of interaural time differences, for which

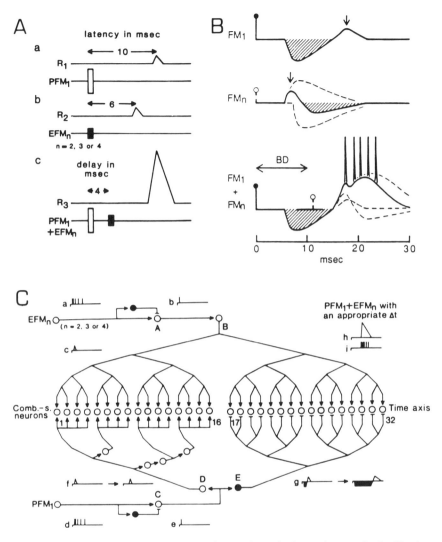

FIGURE 9.10A–C Circuit models for delay tuning via heteroharmonic facilitation mechanism, as used by CF/FM bats. (A) *Delay line model*, in which facilitation by a pulse FM and an echo FM at the best delay (c) is related to the difference between latencies to the pulse (a) and the echo (b) presented alone. In FM bats, amplitude differences between pulse and echo produce paradoxical latency shift having a similar effect (see Fig. 9.9B). This model is thought to be valid for delay-tuned cells with short BDs, and involves delay lines in the FM_1 channel produced by pathlength differences shown on the left half of the circuit model shown in (C) (from Suga 1990a; © 1990, with permission from Elsevier Science Ltd.). (B) *Neural inhibition* model for long BD cells. The response to the FM_1 alone *(top)* is inhibitory. At the end of the inhibitory period, there is a slight rebound in response that might give rise to a weak, long-latency response. The response to the echo FM component alone *(middle)* is excitatory, with a short latency, and may or may not be followed by

there is strong experimental support from work in both the avian (Konishi et al. 1988) and mammalian (Yin and Chan 1988) auditory system.

The other model suggested by Olsen (1986) and elaborated by Suga (1990a,b) involves similar coincidence detection, but the delay lines are created from the interplay of excitation and inhibition (Fig. 9.10B). In this model, the high-amplitude (low-frequency) afferent pathway is inhibitory, and the low-amplitude (high-frequency) pathway is excitatory, on the target delay-tuned cell. In this case, the best delay is related to the difference between the latency for recovery from inhibition (technically, the pulse-alone latency), which presumably varies across afferents to cells with different BDs, and the latency for the echo, which may be relatively similar across afferents. Olsen (1986) and Olsen and Suga (1991b) showed that, in delay-tuned neurons in the medial geniculate body, a delay-line mechanism might account for delay tuning for short delays (<4 msec), while the inhibition mechanism seems responsible for long best delays(>4 msec). The complete model (Suga 1990b) combining both delay-line mechanisms is shown in Figure 9.10C.

One difficulty facing the paradoxical latency shift theory is its vulnerability to disturbance by the many factors influencing pulse and echo amplitudes. There are many mechanisms, both behavioral and physiological, that reduce the amplitude differences between pulses and echoes. In FM bats, attenuation of self-stimulation during pulse emission amounts to about 35–40 dB, attributable to the combined effects of ear directionality (Grinnell and Grinnell 1965), vocal tract directionality (Hartley and Suthers 1989), middle ear muscle activation (Suga and Jen 1977), and central suppression at the level of the midbrain (Suga and Schlegel 1972; Suga and Shimozawa 1974). Moreover, the attenuation of the auditory system wanes with time following pulse emission, reducing the effective amplitude of echoes from nearby, but not distant, targets. This resembles an "automatic gain control," and its effect is to disassociate echo amplitude from target

---

Figure 9.10A–C. (*continued*) inhibition (tonic *[dashed line]* vs. phasic *[solid line]* response). When pulse is combined with an echo delayed by an amount equal to the latency difference between pulse and echo alone (i.e., at BD), the excitation in the echo afferent channel combines nonlinearly with the inhibitory rebound in the pulse afferent channel, resulting in strong facilitation and high discharge rates (*dashed lines* show underlying membrane responses to pulse and echo alone, and to pulse paired with echo) (from Suga, Olsen, and Butman 1990, © 1990, by permission of Cold Spring Harbor Laboratory Press). (C) Circuit model proposed by Suga (1990b) for delay tuning. *Circles* in the center represent delay-tuned neurons. Input from the echo (EFMn) afferent channels *(top)* arrives with little or no delay. Input from the pulse (PFM$_1$) afferent channel passes through delay lines established by pathlength differences (short best delay cells, *left half*) from pathway labeled "D", or inhibition (long best delay cells, *right half*) from pathway labeled "E" (from Suga 1990b, © 1990, Sci. Am. 262:60–68, by permission of Scientific American, Inc.).

range (Kick and Simmons 1984; Hartley and Suthers 1990). In CF/FM bats there is also evidence for "amplitude compensation" whereby the bat decreases pulse amplitude as it approaches obstacles, effectively stabilizing the amplitude difference between pulses and echoes (Kobler et al. 1985). The net effect of both these mechanisms is to attenuate the pulse amplitude and thereby reduce the amplitude differences between pulses and echoes. As Sullivan (1982b) noted, it is possible under certain conditions that echo amplitude might exceed vocal self-stimulation. How then can the *Myotis* auditory system discriminate pulses from echoes?

6. Discussion and Summary

It should be clear from the foregoing that, while there are many similarities in the response properties of cortical cells in different bats, there are also important species differences in the organization of those cells in the auditory cortex. These differences seem to be most pronounced between FM and CF/FM bats, and it is intriguing to speculate about which type of cortical organization is more representative. Common features include at least one tonotopically organized cortical field, overrepresentation of sonar signal frequencies, and specialized delay-sensitive neurons predominantly or exclusively responsive to paired FM sweep stimuli. In both *Myotis* and *Eptesicus*, the primary tonotopic field dominates the cortex, and columns of FM-specialized, delay-sensitive neurons are embedded among columns of unspecialized, frequency-tuned cells. In *Eptesicus*, neurons with multi-peaked tuning curves also lie within this matrix of columns. The intermingling of delay-tuned, multipeaked, and single-frequency tuned columns may well have some significance in terms of information processing. Dear et al. (1993) have put forth the idea that this type of cortical organization better subserves the processing of echo information, nearly all of which they feel is expressed in the time domain. In their opinion, it is logical for multipeaked cells that possibly encode spectral notches to be located near delay-tuned cells reporting overall target range. This is because spectral notches are caused by small-scale time differences in the multiple wave-fronts reflected from a complex target (Beuter 1980), and Simmons, Moss, and Ferragamo (1990) have provided evidence suggesting that bats perceive notches as if they were time-domain encoded. In this view, spectral and temporal information is encoded along a single temporal axis, and it would be logical to find cells encoding this information intermingled in the same field of auditory cortex. Whether in fact FM bats encode spatial information exclusively in the time domain is currently the topic of considerable debate (see Pollak 1993).

As discussed in Section 4.4.1, Dear, Simmons, and Fritz (1993) have pointed out that the delay-sensitive cells respond with a variety of latencies, and could provide the bat a type of "acoustic scene analysis" analogous to the way visual scenes are represented. While it has not yet been determined in *Eptesicus*, in *Myotis* many delay-tuned neurons only reveal delay sensitivity at certain stimulus rates, suggesting that different subsets of neurons are recruited into the analysis of range at different stages of target approach. This imparts an adaptive lability to the information processing in *Myotis* cortex. By contrast, if one ignores the small population of tracking neurons, there is little or no dependence of range tuning on vocal emission rate in the mustached bat. Mustached bat cortical organization is thereby more machine-like, providing a stable representation of echo information regardless of the behavioral state. It is not yet known whether other CF/FM bats such as *Rhinolophus* also show a stable cortical range representation under different conditions of stimulation.

The two CF/FM species have other links to the FM bats. In the mustached bat, multipeaked and delay-sensitive neurons have recently been found in the DSCF area of AI, similar to the organization of AI in FM bats (Fitzpatrick et al. 1993). However, delay-sensitive DSCF cells are not well suited for target ranging, in that their delay tuning is broad, variable, and favors very distant targets. The role played by these cells may be simply to amplify the probability of target detection at the physical limits of the bat's echo detection envelope. FM-FM cells with short BDs that are well suited for ranging are completely segregated in the mustached and horseshoe bats into distinct cortical fields. CF/CF cells, which seem especially suited to signal processing in species with long CF components in their sonar signals, also are segregated both from the tonotopic and the range-tuned areas in the CF/FM bats. However, in at least one respect, *Rhinolophus* may represent a "missing link" between FM bat and mustached bat cortical organization, because nearly half the cells recorded in the *Rhinolophus* combination-sensitive zones were tuned to tonal stimuli and were not combination sensitive. Like FM bats, combination-sensitive cells are intermingled in *Rhinolophus* cortex with cells tuned to simpler stimuli. However, in *Rhinolophus* this area is separate from AI and is not tonotopically organized.

Other differences between the two groups of species are more clearly definable. First, there is a major difference in the mechanism producing delay sensitivity. In CF/FM bats, the two channels bringing range information to the forebrain are tuned to different harmonic components that identify the pulse and the echo stimulus. O'Neill and Suga (1982) named this "heteroharmonic" facilitation. The hypothesis for this mechanism is that the pulse is encoded by first-harmonic-tuned cells, and routed through an array of delay lines, while the echo is encoded by higher harmonic-tuned cells and presumably routed through a fast pathway. In the FM bats, pulse and echo FM signals are apparently differentiated by amplitude: high-amplitude pulse information is captured by a yet-unknown mechanism and passed via an array of delay lines to the forebrain (paradoxical latency shift; Section 4.3.2), whereas low-amplitude echoes are routed through a fast pathway. Interestingly, even in the amplitude-based system of *Myotis*, the delay-tuned neurons are facilitated by different FM sweep parcels in the pulse and echo. It is interesting to speculate whether the heteroharmonic facilitation seen in CF/FM bats that use multiharmonic signals is simply an evolutionary development from the FM bat system that utilizes different parts of the broadband single-harmonic FM sweep in pulses and echoes.

The second big difference between CF/FM and FM bat cortex is in the expression of a neural map of target range, and the segregation of that map to a discrete cortical field. Maplike representations of sensory information are widespread and prominent in the visual, somatosensory, and auditory systems. Typically, these maps simply recapitulate the spatial arrangement of receptors in the sensory periphery, accompanied by overrepresentation of particular receptive fields. However, maps representing the spatial location of sound sources, including target range information in echolocating bats, must be computed by neural circuitry in the auditory system. The circuitry for computing target range clearly involves both midbrain (inferior colliculus and nuclei of the lateral lemniscus) and forebrain structures (auditory thalamus and cortex). In all bats so far studied, this circuitry culminates in delay-tuned forebrain neurons that can encode target range. However, the expression of a systematic target range map is highly developed only in the mustached bat, where it extends into at least three separate cortical fields, while such a map is distinct but less well developed in the horseshoe bat. By contrast, even though there are many range-encoding neurons in both the FM bats studied, there is no clearly defined map of echo delay (although it has been argued here that such an organization might occur on a small scale in *Eptesicus*).

This major difference in the organization of delay-tuned cells brings up the question of the role played by the mapped fields in CF/FM bats. No one would argue that target-range processing is any less sophisticated in FM bats because they lack a map of target range. Indeed, there is behavioral evidence that FM bats have better range acuity than CF/FM bats (Simmons 1973), as predicted from the advantage their broadband signal structure gives them in temporal resolving power (Simmons, Howell, and Suga 1975; Simmons and Stein 1980).

Might the answer to this question lie in the difference in habitats preferred by these species for hunting insects? CF/FM bats have evolved their elaborate sonar specializations to enable them to hunt in cluttered habitats (Neuweiler et al. 1987), where they must parse out many conflicting echoes vying for their attention with those from their intended prey. FM bats typically hunt over open ground or water, where there are presumably fewer objects cluttering their sonar "screen," and local target motion is not a critical identifying factor necessary to spot prey. When FM bats increase the pulse emission rate during a pursuit, they bring groups of range-tuned cells "on line," and the activation pattern in the cortex is complex and unsystematic.

CF/FM bats, on the other hand, put a premium on detecting and tracking a moving target in a background of echo clutter. Having a maplike representation of the acoustic scene that segregates the targets systematically along a range axis seems intuitively easier to interpret than having a more random excitation pattern like that in FM bats. The well-conceived experiments of Riquimaroux, Gaioni, and Suga (1991) showed that fine, but not coarse, range discrimination was disrupted by lesions of the FM-FM area in the mustached bat. Further, Altes (1989) has argued that having a maplike representation provides the bat with a data structure well suited for computation of target range on a finer scale than is possible just from considering the granularity of BD representation or delay-tuning-curve widths alone (sort of a range hyperacuity phenomenon). This may permit mustached and horseshoe bats to overcome some of the range acuity limitations inherent in their narrow-bandwidth sonar signals (Simmons 1973; Simmons, Howell, and Suga 1975), as well as provide a better neural substrate to assist in the spatial localization of small targets in highly cluttered habitats.

Despite the progress made in the last decade or so, much remains to be discovered about the auditory cortex in bats. For example, there have only been three, fairly small-scale, studies done on spatial sensitivity or binaural interaction in the bat cortex (Manabe, Suga, and Ostwald 1978; Jen, Sun, and Lin 1989; Suga, Kawasaki, and Burkhard 1990). Efforts need to be made to link bat cortical organization more clearly with that in other mammals. For example, it would be worthwhile to know whether such properties as tuning-curve sharpness, FM sweep rate selectivity, minimum threshold, or best amplitude of cells in the tonotopic fields are organized along isofrequency contours as they are in cat cortex (see Clarey, Barone, and Imig 1992 for review). Also, the current experiments of Ohlemiller, Kanwal, and Suga (1993), examining how cells tuned to range and Doppler-shift magnitude respond to the enormously rich variety of mustached bat communication sounds, should inspire further work on this important and long-neglected area of the bat acoustic behavior. The structural similarity of acoustic elements in these sounds to syllables in human speech is often striking (Kanwal et al. 1994), and understanding how these elements are processed by specialized cortical neurons in bats might give rise to useful models of speech sound analysis at higher levels of the auditory system.

Finally, it is hoped that the wealth of information already gleaned from study of the cortex in bats will inspire others to combine biologically relevant complex and simple stimuli in their attempts to understand the organization of auditory cortex in other species.

Acknowledgments.   I gratefully acknowledge the contribution of G. Schuller, S. Radtke-Schuller, G. E. Duncan, O. W. Henson and D. Wong for anatomical material on bat cortex, Martha Zettel for assistance with figures, Nobuo Suga, George Pollak, Donald Wong, and Jeff Wenstrup for valuable discussions about the auditory forebrain, and Willard Wilson, Art Popper, and Dick Fay for helpful comments on the manuscript. Support for the author was provided in part by the National Institute for Deafness and Communicative Disorders (R01-DC00267) and the National Institute on Aging (P01-AG09524).

Abbreviations

AI	primary auditory cortex
AII	secondary auditory cortex
BA	best amplitude
BD	best delay
BF	best frequency
BFA	best amplitude for facilitation
BFF	best frequency for facilitation
CF	constant frequency (signal component)
CF/CF	CF/CF area of mustached bat cortex; stimulus pair consisting of two CF components
CF/FM	sonar signal consisting of CF and FM components
CF_1/CF_2	combination-sensitive neuron facilitated by a first- and second-harmonic CF signal presented simultaneously
CF_1/CF_3	combination-sensitive neuron facilitated by a first- and third-harmonic CF signal presented simultaneously
CF_1, CF_2, CF_3, CF_4	CF components of the four harmonics of mustached bat sonar signal
CFM-CFM	combination-sensitive neuron facilitated by both CF and FM components ($= H_1 - H_2$ [below])
DF	area DF (dorsal fringe) of mustached bat cortex
DSCF	Doppler-shifted CF region of mustached bat AI
FM	frequency-modulated signal; a type of sonar signal.
FM-FM	region of mustached bat cortex containing delay-sensitive neurons tuned to pairs of FM signals
FM_1, FM_2, FM_3, FM_4	FM components of the four harmonics of mustached bat sonar signal
FM_1-CF_2	combination of signal elements that facilitates neurons in mustached bat DSCF area
FM_1-FM_2	delay-sensitive neuron facilitated by pairs of first- and second-harmonic FM signals
FM_1-FM_3	delay-sensitive neuron facilitated by pairs of first- and third-harmonic FM signals
FM_1-FM_4	delay-sensitive neuron facilitated by pairs of first- and fourth-harmonic FM signals
H_1-H_4	four harmonics of the mustached bat sonar signal (includes both CF and FM components)

H_1-H_2	delay-sensitive neuron facilitated by pairs of either CF or FM components (see "CFM-CFM" above); region of mustached bat cortex (= area VA)
IC	inferior colliculus
MGB	medial geniculate body
MT	minimum threshold
$Q_{10\ dB}$, $Q_{30\ dB}$, $Q_{50\ dB}$	best frequency divided by the tuning-curve bandwidth 10, 30, or 50 dB above MT
VA	area VA of mustached bat cortex (= area H_1-H_2)
VF	area VF (ventral fringe) of mustached bat cortex

References

Altes RA (1989) An interpretation of cortical maps in echolocating bats. J Acoust Soc Am 85:934–942.

Asanuma A, Wong D, Suga N (1983) Frequency and amplitude representations in anterior primary auditory cortex of the mustached bat. J Neurophysiol (Bethesda) 50:1182–1196.

Berkowitz A, Suga N (1989) Neural mechanisms of ranging are different in two species of bats. Hear Res 41:255–264.

Beuter KJ (1980) A new concept of echo evaluation in the auditory system of bats. In: Busnel RG, Fish JF (eds) Animal Sonar Systems. New York: Plenum, pp. 747–764.

Bruns V (1976) Peripheral auditory tuning for fine frequency analysis of the CF-FM bat, *Rhinolophus ferrumequinum*. II. Frequency mapping in the cochlea. J Comp Physiol A 106:87–97.

Bruns V, Schmieszek E (1980) Cochlear innervation in the greater horseshoe bat: demonstration of an acoustic fovea. Hear Res 3:27–43.

Clarey JC, Barone P, Imig TJ (1992) Physiology of thalamus and cortex. In: Popper AN, Fay RR (eds) Springer Handbook of Auditory Research, Vol 2, The Mammalian Auditory Pathway: Neurophysiology. New York: Springer-Verlag, pp. 232–334.

Covey E, Johnson BR, Ehrlich D, Casseday JH (1993) Neural representation of the temporal features of sound undergoes transformation in the auditory midbrain: evidence from extracellular recording, application of pharmacological agents, and in vivo whole cell patch clamp recording. Soc Neurosci Abstr 19:535.

Dear SP, Simmons JA, Fritz J (1993) A possible neuronal basis for representation of acoustic scenes in auditory cortex of the big brown bat. Nature 364:620–623.

Dear SP, Fritz J, Haresign T, Ferragamo M, Simmons JA (1993) Tonotopic and functional organization in the auditory cortex of the big brown bat, *Eptesicus fuscus*. J Neurophysiol (Bethesda) 70:1988–2009.

Duncan GE, Henson OW (1994) Brain activity patterns in flying, echolocating bats *(Pteronotus parnellii)*: assessment by high resolution autoradiographic imaging with ^{3}H-2-deoxyglucose. Neurosci 59:1051–1070.

Edamatsu H, Suga N (1993) Differences in response properties of neurons between two delay-tuned areas in the auditory cortex of the mustached bat. J Neurophysiol (Bethesda) 69:1700–1712.

Edamatsu H, Kawasaki M, Suga N (1989) Distribution of combination-sensitive neurons in the ventral fringe area of the auditory cortex of the mustached bat. J Neurophysiol (Bethesda) 61:202–207.

Feng AS, Simmons JA, Kick SA (1978) Echo detection and target-ranging neurons in the auditory system of the bat *Eptesicus fuscus*. Science 202:645–648.

Ferrer I (1987) The basic structure of the neocortex in insectivorous bats (*Miniopterus sthreibersi* and *Pipistrellus pipistrellus*). A Golgi study. J Hirnforsch 28:237–243.

Fitzpatrick DC, Henson OW (1994) Cell types in the mustached bat auditory cortex. Brain Behav Evol 43:79–91.

Fitzpatrick DC, Kanwal JS, Butman JA, Suga N (1993) Combination-sensitive neurons in the primary auditory cortex of the mustached bat. J Neurosci 13:931–940.

Fritz JB, Olsen J, Suga N, Jones EG (1981) Connectional differences between auditory fields in a CF-FM bat. Soc Neurosci Abstr 7:391.

Fuzessery ZM, Pollak GD (1984) Neural mechanisms of sound localization in an echolocating bat. Science 225:725–728.

Fuzessery ZM, Pollak GD (1985) Determinants of sound location selectivity in bat inferior colliculus: a combined dichotic and free-field stimulation study. J Neurophysiol (Bethesda) 54:757–781.

Fuzessery ZM, Hartley DJ, Wenstrup JJ (1992) Spatial processing within the mustache bat echolocation system: possible mechanisms for optimization. J Comp Physiol A 170:57–71.

Goldman LJ, Henson OW (1977) Prey recognition and selection by the constant frequency bat, *Pteronotus p. parnellii*. Behav Biol Sociobiol 2:411–419.

Gooler DM, O'Neill WE (1987) Topographic representation of vocal frequency demonstrated by microstimulation of anterior cingulate cortex in the echolocating bat, *Pteronotus parnelli parnelli*. J Comp Physiol A 161:283–294.

Gooler DM, O'Neill WE (1988) Central control of frequency in biosonar emissions of the mustached bat. In: Nachtigall PE, Moore PWB (eds) Animal Sonar: Processes and Performance. New York: Plenum, pp. 265–270.

Griffin DR (1958) Listening in the Dark. New Haven: Yale University Press.

Griffin DR (1971) The importance of atmospheric attenuation for the echolocation of bats. Anim Behav 19:55–61.

Griffin DR, Webster FA, Michael C (1960) The echolocation of flying insects by bats. Anim Behav 8:141–154.

Grinnell AD, Grinnell VS (1965) Neural correlates of vertical localization by echolocating bats. J Physiol 181:830–851.

Grinnell AD, Schnitzler H-U (1977) Directional sensitivity of echolocation in the horseshoe bat *Rhinolophus ferrumequinum*. II. Behavioral directionality of hearing. J Comp Physiol A 116:63–76.

Habersetzer J, Vogler B (1983) Discrimination of surface structured targets by the echolocating bat *Myotis myotis* during flight. J Comp Physiol A 152:275–282.

Hartley DJ, Suthers RA (1989) The sound emission pattern of the echolocating bat, *Eptesicus fuscus*. J Acoust Soc Am 85:1348–1351.

Hartley DJ, Suthers RA (1990) Sonar pulse radiation and filtering in the mustached bat, *Pteronotus parnellii rubiginosus*. J Acoust Soc Am 87:2756–2772.

Hawkins JE, Stevens SS (1950) The masking of pure tones and speech by white noise. J Acoust Soc Am 22:6–13.

Henson OW (1970) The central nervous system of Chiroptera. In: Wimsatt WA (ed) Biology of Bats. New York: Academic Press, pp. 57–152.

Henson OW, Bishop A, Keating A, Kobler J, Henson M, Wilson B, Hansen R

(1987) Biosonar imaging of insects by *Pteronotus p. parnellii*, the mustached bat. Natl Geogr Res 3:82–101.

Imig TJ, Adrian HO (1977) Binaural columns in the primary field (AI) of cat auditory cortex. Brain Res 138:241–257.

Jeffress (1948) A place theory of sound localization. J Comp Psychol 41:35–39.

Jen PHS, Chen D (1988) Directionality of sound pressure transformation at the pinna of echolocating bats. Hear Res 34:101–118.

Jen PHS, Sun X, Lin PJJ (1989) Frequency and space representation in the primary auditory cortex of the frequency modulating bat *Eptesicus fuscus*. J Comp Physiol A 165:1–14.

Kanwal JS, Ohlemiller KK, Suga N (1993) Communication sounds of the mustached bat: classification and multidimensional analyses of call structure. Assoc Res Otolaryngol Abstr 16:111.

Kanwal JS, Matsumura S, Ohlemiller KK, Suga N (1994) Analysis of acoustic elements and syntax in communication sounds emitted by mustached bats. J Acoust Soc Amer 96:1229–1254.

Kawasaki M, Margoliash D, Suga N (1988) Delay-tuned combination-sensitive neurons in the auditory cortex of the vocalizing mustached bat. J Neurophysiol (Bethesda) 59:623–635.

Kelly JB, Judge PW, Phillips DP (1986) Representation of the cochlea in primary auditory cortex of the ferret. Hear Res 24:111–115.

Kick SA, Simmons JA (1984) Automatic gain control in the bat's sonar receiver and the neuroethology of echolocation. J Neurosci 4:2725–2737.

Knudsen EI, Konishi M (1978) Space and frequency are represented seperately in auditory midbrain of the owl. J Neurophysiol (Bethesda) 41:870–884.

Kober R (1988) Echoes of fluttering insects. In: Nachtigall PE, Moore PW (eds) Animal Sonar: Processes and Performance. New York: Plenum, pp. 477–482.

Kober R, Schnitzler H-U (1990) Information in sonar echoes of fluttering insects available for echolocating bats. J Acoust Soc Am 87:882–895.

Kobler JB, Isbey SF, Casseday JH (1987) Auditory pathways to the frontal cortex of the mustache bat, *Pteronotus parnellii*. Science 236:824–826.

Kobler JB, Wilson BS, Henson OW Jr, Bishop AL (1985) Echo intensity compensation by echolocating bats. Hear Res 20:99–108.

Konishi M, Takahashi TT, Wagner H, Sullivan WE, Carr CE (1988) Neurophysiological and anatomical substrates of sound localization in the owl. In: Edelman GM, Gall WE, Cowan WM (eds) Auditory Function: Neurobiological Bases of Hearing. New York: Wiley, pp. 721–745.

Lesser HD (1987) Encoding of amplitude-modulated sounds by single units in the inferior colliculus of the mustached bat, *Pteronotus parnelli*. Ph.D. Thesis, University of Rochester, Rochester, NY.

Lesser HD, O'Neill WE, Frisina RD, Emerson RC (1990) ON-OFF units in the mustached bat inferior colliculus are selective for transients resembling "acoustic glint" from fluttering insect targets. Exp Brain Res 82:137–148.

Link A, Marimuthu G, Neuweiler G (1986) Movement as a specific stimulus for prey catching behavior in rhinolophid and hipposiderid bats. J Comp Physiol A 159:403–413.

Maekawa M, Wong D, Paschal WG (1992) Spectral selectivity of FM-FM neurons in the auditory cortex of the echolocating bat, *Myotis lucifugus*. J Comp Physiol A 171:513–522.

Makous JC, O'Neill WE (1986) Directional sensitivity of the auditory midbrain in the mustached bat to free-field tones. Hear Res 24:73–88..

Manabe T, Suga N, Ostwald J (1978) Aural representation in the Doppler-shifted-CF processing area of the auditory cortex of the mustache bat. Science 200:339–342.

Merzenich MM, Brugge JF (1973) Representation of the cochlear partition on the superior temporal plane of the macaque monkey. Brain Res 50;275–296.

Metzner W (1993) An audiovocal interface in echolocating horseshoe bats. J Neurosci 13:1862–1878.

Middlebrooks JC, Dykes RW, Merzenich MM (1980) Binaural response-specific bands in primary auditory cortex (AI) of the cat: Topographical organization orthogonal to isofrequency contours. Brain Res 181:31–48.

Mogdans J, Schnitzler H-U (1990) Range resolution and the possible use of spectral information in the echolocating bat, *Eptesicus fuscus*. J Acoust Soc Am 88:754–757.

Neuweiler G (1970) Neurophysiologische Untersuchungen zum Echoortungssystem der Grossen Hufeisennase *Rhinolophus ferrumequinum* Schreber. J Comp Physiol A 67:273–306.

Neuweiler G, Metzner W, Heilmann U, Rubsamen R, Eckrich M, Costa HH (1987) Foraging behavior and echolocation in the rufous horseshoe bat (*Rhinolophus rouxi*) of Sri Lanka. Behav Ecol Sociobiol 20:53–67.

Novick A, Vaisnys JR (1964) Echolocation of flying insects by the bat, *Chilonycteris parnellii*. Biol Bull 127:478–488.

Ohlemiller KK, Kanwal JS, Suga N (1993) Do cortical auditory neurons of the mustached bat have a dual function for processing biosonar signals and communication sounds? Assoc Res Otolaryngol Abstr 16:111.

Olsen JF (1986) Processing of biosonar information by the medial geniculate body of the mustached bat, *Pteronotus parnellii*. Ph.D. Thesis, Washington University, St. Louis, MO.

Olsen JF, Suga N (1991a) Combination-sensitive neurons in the medial geniculate body of the mustached bat: Encoding of relative velocity information. J Neurophysiol (Bethesda) 65:1254–1274.

Olsen JF, Suga N (1991b) Combination-sensitive neurons in the medial geniculate body of the mustached bat: encoding of target range information. J Neurophysiol (Bethesda) 65:1275–1296.

O'Neill WE (1985) Responses to pure tones and linear FM components of the CF-FM biosonar signal by single units in the inferior colliculus of the mustached bat. J Comp Physiol A 157:797–815.

O'Neill WE, Basham M (1992) Pulse-echo stimulus combinations can facilitate sonar signal vocalizations elicited by electrical stimulation of the anterior cingulate cortex in the mustache bat. In: Proceedings of Third International Congress of Neuroethology, Montreal, Quebec, CA, Aug 9–14, 1992. Soc for Neuroethology: p. 272.

O'Neill WE, Suga N (1979) Target-range sensitive neurons in the auditory cortex of the mustached bat. Science 203:69–73.

O'Neill WE, Suga N (1982) Encoding of target range and its representation in the auditory cortex of the mustached bat. J Neurosci 2:17–31.

O'Neill WE, Frisina RD, Gooler DM (1989) Functional organization of mustached bat inferior colliculus: I. Representation of FM frequency bands important for target ranging revealed by ^{14}C-2-deoxyglucose autoradiography and single unit mapping. J Comp Neurol 284:60–84.

Ostwald J (1980) The functional organization of the auditory cortex in the CF-FM bat *Rhinolophus ferrumequinum*. In: Busnel RG, Fish JF (eds) Animal Sonar Systems. New York: Plenum, pp. 953–956.

Ostwald J (1984) Tonotopical organization and pure tone response characteristics of single units in the auditory cortex of the greater horseshoe bat. J Comp Physiol A 155:821–834.

Phillips DP, Judge PW, Kelly JB (1988) Primary auditory cortex in the ferret (*Mustela putorius*): neural response properties and topographic organization. Brain Res 443:281–294.

Pinheiro AD, Wu M, Jen PH-S (1991) Encoding repetition rate and duration in the inferior colliculus of the big brown bat, *Eptesicus fuscus*. J Comp Physiol A 169:69–85.

Pollak GD (1993) Some comments on the proposed perception of phase and nanosecond time disparities by echolocating bats. J Comp Physiol A 172:523–531.

Pollak GD, Bodenhamer RD (1981) Specialized characteristics of single units in inferior colliculus of mustache bat: frequency representation, tuning, and discharge patterns. J Neurophysiol (Bethesda) 46:605–620.

Pollak GD, Henson OW Jr, Johnson R (1979) Multiple specializations in the peripheral auditory system of the CF-FM bat, *Pteronotus parnellii*. J Comp Physiol 131:255–266.

Pollak GD, Henson OW Jr, Novick A (1972) Cochlear microphonic audiograms in the pure tone bat *Chilonycteris parnellii parnellii*. Science 176:66–68.

Pye JD (1980) Echolocation signals and echoes in air. In: Busnel R-G, Fish JF (eds) Animal Sonar Systems. New York: Plenum, pp. 309–354.

Reale RA, Imig TJ (1980) Tonotopic organization of auditory cortex in the cat. J Comp Neurol 192:265–291.

Riquimaroux H, Gaioni SJ, Suga N (1991) Cortical computational maps control auditory perception. Science 251:565–568.

Riquimaroux H, Gaioni SJ, Suga N (1992) Inactivation of DSCF area of the auditory cortex with muscimol disrupts frequency discrimination in the mustached bat. J Neurophysiol (Bethesda) 68:1613–1623.

Sales G, Pye D (1974) Ultrasonic Communication by Animals. London: Chapman and Hall.

Sanides F (1972) Representation in the cerebral cortex and its areal lamination patterns. In: Bourne GH (ed) The Structure and Function of the Nervous System. New York: Academic Press, pp. 329–453.

Sanides D, Sanides F (1974) A comparative Golgi study of the neocortex in insectivores and rodents. Z Mikrosk Anat Forsch (Leipz) 88:957–977.

Scharf B, Meiselman CH (1977) Critical bandwidth at high intensities. In: Evans EF, Wilson JP (eds) Psychophysics and Physiology of Hearing. London: Academic Press, pp. 221–232.

Schmidt S (1988) Evidence for spectral basis of texture perception in bat sonar. Nature 331:617–619.

Schnitzler H-U (1968) Die Ultraschall-Ortungslaute der Hufeisen-Fledermause (Chiroptera-Rhinolophidae) in verschiedenen Orientierungssituationen. Z Vgl Physiol 57:376–408.

Schnitzler H-U (1970) Comparison of the echolocation behavior in *Rhinolophus ferrum-equinum* and *Chilonycteris rubiginosa*. Bijdr Dierkd 40:77–80.

Schnitzler H-U, Menne D, Kober R, Heblich K (1983) The acoustical image of fluttering insects in echolocating bats. In: Huber F, Markl H (eds) Neuroethology and Behavioral Physiology, Berlin: Springer-Verlag, pp. 235–250.

Schreiner CD, Cynader MS (1984) Basic functional organization of second auditory cortical field (AII) of the cat. J Neurophysiol (Bethesda) 51:1284–1305.

Schreiner CE, Mendelson JR, Sutter ML (1992) Functional topography of cat primary auditory cortex: representation of tone intensity. Exp Brain Res 92:105–122.

Schuller G (1974) The role of overlap of echo with outgoing echolocation sound in the bat *Rhinolophus ferrumequinum*. Naturwissenshaften 61:171–172.

Schuller G (1977) Echo delay and overlap with emitted orientation sounds and Doppler-shift compensation in the bat, *Rhinolophus ferrumequinum*. J Comp Physiol A 114:103–114.

Schuller G, Pollak GD (1979) Disproportionate frequency representation in the inferior colliculus of Doppler-compensating greater horseshoe bats: evidence for an acoustic fovea. J Comp Physiol 132:47–54.

Schuller G, Radtke-Schuller S (1990) Neural control of vocalization in bats: mapping of brainstem areas with electrical microstimulation eliciting species-specific echolocation calls in the rufous horseshoe bat. Exp Brain Res 79:192–206.

Schuller G, O'Neill WE, Radtke-Schuller S (1991) Facilitation and delay sensitivity of auditory cortex neurons in CF-FM bats, *Rhinolophus rouxi* and *Pteronotus p. parnellii*. Eur J Neurosci 3: 1165–1181.

Schuller G, Radtke-Schuller S, O'Neill WE (1988) Processing of paired biosonar signals in the cortices of *Rhinolophus rouxi* and *Pteronotus parnellii*. In: Nachtigall PE, Moore PWB (eds) Animal Sonar: Processes and Performance. New York: Plenum, pp. 259–264.

Shamma SA, Symmes D (1985) Patterns of inhibition in auditory cortical cells in awake squirrel monkeys. Hear Res 19:1–13.

Shannon S, Wong D (1987) Interconnections between the medial geniculate body and the auditory cortex in an FM bat. Soc Neurosci Abstr 13:1469

Shannon-Hartman S, Wong D, Maekawa M (1992) Processing of pure-tone and FM stimuli in the auditory cortex of the FM bat, *Myotis lucifugus*. Hear Res 61:179–188.

Shimozawa T, Suga N, Hendler P, Schuetze S (1974) Directional sensitivity of echolocation system in bats producing frequency-modulated signals. J Exp Biol 60:53–69.

Simmons JA (1971) Echolocation in bats: signal processing of echoes for target range. Science 171:925–928.

Simmons JA (1973) The resolution of target range by echolocating bats. J Acoust Soc Am 54:157–173.

Simmons JA, Chen L (1989) The acoustic basis for target discrimination by FM echolocating bats. J Acoust Soc Am 86:1333–1350.

Simmons JA, Stein RA (1980) Acoustic imaging in bat sonar: echolocation signals and the evolution of echolocation. J Comp Physiol A 135:61–84.

Simmons JA, Fenton MB, O'Farrell MJ (1979) Echolocation and pursuit of prey by bats. Science 203:16–21.

Simmons JA, Howell DJ, Suga N (1975) Information content of bat sonar echoes. Am Sci 63:204–215.

Simmons JA, Moss CF, Ferragamo M (1990) Convergence of temporal and spectral information into acoustic images of complex sonar targets perceived by the echolocating bat, *Eptesicus fuscus*. J Comp Physiol A 166:449–470.

Simmons JA, Freedman EG, Stevenson SB, Chen L, Wohlgenant TJ (1989) Clutter interference and the integration time of echoes in the echolocating bat, *Eptesicus fuscus*. J Acoust Soc Am 86:1318–1332.

Suga N (1965a) Functional properites of auditory neurones in the cortex of echolocating bats. J Physiol 181:671–700.

Suga N (1965b) Responses of cortical auditory neurones to frequency-modulated sounds in echo-locating bats. Nature 206:890–891.

Suga N (1977) Amplitude spectrum representation in the Doppler-shifted-CF processing area of the auditory cortex of the mustache bat. Science 196:64–67.

Suga N (1981) Neuroethology of the auditory system of echolocating bats. In: Katsuki Y, Norgren, Sato (eds) Brain Mechanisms of Sensation. New York: Wiley, pp. 45–60.

Suga N (1982) Functional organization of the auditory cortex: Representation beyond tonotopy in the bat. In: Woolsey CN (ed) Cortical Sensory Organization, Vol 3, Multiple Auditory Areas. Clifton, NJ: Humana, pp. 157–218.

Suga N (1984) The extent to which biosonar information is represented in the bat auditory cortex. In: Edelman GM, Gall WE, Cowan WM (eds) Dynamic Aspects of Neocortical Function. New York: Wiley, pp. 315–373.

Suga N (1988a) Auditory neuroethology and speech processing: Complex sound processing by combination-sensitive neurons. In: Edelman GM, Gall WE, Cowan WM (eds) Auditory Function: Neurobiological Bases of Hearing. New York: Wiley, pp. 679–719.

Suga N (1988b) What does single-unit analysis in the auditory cortex tell us about information processing in the auditory system? In: Rakic P, Singer W (eds) Neurobiology of Neocortex. New York: Wiley, pp. 331–349

Suga N (1988c) Parallel-hierarchical processing of biosonar information in the mustached bat. In: Nachtigal PE, Moore PWB (eds) Animal Sonar: Processes and Performance. New York: Plenum, pp. 149–159.

Suga N (1990a) Cortical computational maps for auditory imaging. Neural Networks 3:3–21.

Suga N (1990b) Biosonar and neural computation in bats. Sci Amer 262:60–66

Suga N, Horikawa J (1986) Multiple time axes for representation of echo delay in the auditory cortex of the mustached bat. J Neurophysiol (Bethesda) 55:776–805.

Suga N, Jen PH (1976) Disproportionate tonotopic representation for processing CF-FM sonar signals in the mustache bat auditory cortex. Science 194:542–544.

Suga N, Jen PH (1977) Further studies on the peripheral auditory system of 'CF-FM' bats specialized for fine frequency analysis of Doppler-shifted echoes. J Exp Biol 69:207–232.

Suga N, Manabe T (1982) Neural basis of amplitude-spectrum representation in auditory cortex of the mustached bat. J Neurophysiol (Bethesda) 47:225–255.

Suga N, O'Neill WE (1979) Neural axis representing target range in the auditory cortex of the mustached bat. Science 206:351–353.

Suga N, Schlegel P (1972) Neural attenuation of responses to emitted sounds in echolocating bats. Science 177:82–84.

Suga N, Shimozawa T (1974) Site of neural attenuation of responses to self-vocalized sounds in echolocating bats. Science 183:1211–1213.

Suga N, Tsuzuki K (1985) Inhibition and level-tolerant frequency tuning in the auditory cortex of the mustached bat. J Neurophysiol (Bethesda) 53:1109-1145.

Suga N, Kawasaki M, Burkard RF (1990) Delay-tuned neurons in auditory cortex of mustached bat are not suited for processing directional information. J Neurophysiol (Bethesda) 64:225-235.

Suga N, Kuzirai K, O'Neill WE (1981) How biosonar information is represented in the bat cerebral cortex. In: Syka J, Aitkin L (eds) Neuronal Mechanisms of Hearing. New York: Plenum, pp. 197-219.

Suga N, Niwa H, Taniguchi I (1983) Representation of biosonar information in the auditory cortex of the mustached bat, with emphasis on representation of target velocity information. In: Ewert J-P, Capranica RR, Ingle DJ (eds) Advances in Vertebrate Neuroethology. New York: Plenum, pp. 829-867.

Suga N, Olsen JF, Butman JA (1990) Specialized subsystems for processing biologically important complex sounds: Cross-correlation analysis for ranging in the bat's brain. Cold Spring Harbor Symp Quant Biol 55:585-597.

Suga N, O'Neill WE, Manabe T (1978) Cortical neurons sensitive to combinations of information-bearing elements of biosonar signals in the mustached bat. Science 200:778-781.

Suga N, O'Neill WE, Manabe T (1979) Harmonic-sensitive neurons in the auditory cortex of the mustache bat. Science 203:270-274.

Suga N, Simmons JA, Jen PH-S (1975) Peripheral specialization for fine analysis of Doppler-shifted echoes in the auditory system of the "CF-FM" bat, *Pteronotus parnellii*. J Exp Biol 69:207-232.

Suga N, Simmons JA, Shimozawa T (1974) Neurophysiological studies on echolocation systems in awake bats producing CF-FM orientation sounds. J Exp Biol 61:379-399.

Suga N, Niwa H, Taniguchi I, Margoliash D (1987) The personalized auditory cortex of the mustached bat: adaptation for echolocation. J Neurophysiol (Bethesda) 58:643-654.

Suga N, O'Neill WE, Kujirai K, Manabe T (1983) Specificity of "combination sensitive" neurons for processing complex biosonar signals in the auditory cortex of the mustached bat. J Neurophysiol (Bethesda) 49:1573-1626.

Sullivan WE (1982a) Neural representation of target distance in auditory cortex of the echolocating bat *Myotis lucifugus*. J Neurophysiol (Bethesda) 48:1011-1032.

Sullivan WE (1982b) Possible neural mechanisms of target distance coding in auditory system of the echolocating bat, *Myotis lucifugus*. J Neurophysiol (Bethesda) 48:1033-1047.

Tanaka H, Wong D, Taniguchi I (1992) The influence of stimulus duration on the delay tuning of cortical neurons in the FM bat, *Myotis lucifugus*. J Comp Physiol A 171:29-40.

Taniguchi I, Niwa H, Wong D, Suga N (1986) Response properties of FM - FM combination-sensitive neurons in the auditory cortex of the mustached bat. J Comp Physiol A 159:331-337.

Teng H, Wong D Temporal and amplitude tuning of delay-sensitive neurons in the auditory cortex of *Myotis lucifugus*. J Neurophysiol (Bethesda) (in press).

Trappe M, Schnitzler H-U (1982) Doppler-shift compensation in insect-catching horseshoe bats. Naturwissenshaften 69:193-194.

Tunturi AR (1952) A difference in the representation of auditory signals for the left and right ears in the iso-frequency contours of the right middle ectosylvian auditory cortex of the dog. Am J Physiol 168: 712–727.

Wenstrupp JJ, Grose CD (1993) Inputs to combination-sensitive neurons in the medial geniculate body of the mustached bat. Soc Neurosci Abstr 19:1426.

Wenstrup JJ, Larue DT, Winer JA (1994) Projections of physiologically defined subdivisions of the inferior colliculus in the mustached bat: targets in the medial geniculate body and extrathalamic nuclei. J Comp Neurol 346:207–236.

Wong D, Shannon SL (1988) Functional zones in the auditory cortex of the echolocating bat, *Myotis lucifugus*. Brain Res 453:349–352.

Wong D, Maekawa M, Tanaka H (1992) The effect of pulse repetition rate on the delay sensitivity of neurons in the auditory cortex of the FM bat, *Myotis lucifugus*. J Comp Physiol A 170:393–402.

Woolsey CN (1960) Organization of cortical auditory system: A review and a synthesis. In: Rasmussen G, Windle W (eds) Neural Mechanisms of the Auditory and Vestibular Systems. Springfield: Thomas, pp. 165–180.

Yang L, Pollak GD, Ressler C (1993) GABAergic circuits sharpen tuning curves and modify response properties in the mustache bat inferior colliculus. J Neurophysiol (Bethesda) 68:1760–1774.

Yin TCT, Chan JCK (1988) Neural mechanisms underlying interaural time sensitivity to tones and noise. In: Edelman GM, Gall WE, Cowan WM (eds) Auditory Function: Neurobiological Bases of Hearing. New York, Wiley: pp. 385–430.

10

Perspectives on the Functional Organization of the Mammalian Auditory System: Why Bats Are Good Models

GEORGE D. POLLAK, JEFFERY A. WINER, AND WILLIAM E. O'NEILL

1. Introduction

Echolocating bats, more than most other mammals, rely on their sense of hearing for obtaining information about their external world (Griffin 1958; also see Fenton, Chapter 2, and Moss and Schnitzler, Chapter 3). In keeping with their reliance on hearing, their auditory systems are not only well developed, but are also proportionately much larger than are the auditory systems of other mammals. Nevertheless, bats are rarely used as models to illustrate basic features of the mammalian auditory system. The reasons for this are partially historical. The cat has traditionally been employed in studies of the central auditory system, and thus the studies of other mammals are frequently overshadowed by the large number of reports on the cat. However, we believe there is also another reason. This reason stems from a notion that echolocation, the ability to "see" objects in the external world with ultrasonic echoes, required fundamental modifications of the auditory system. These modifications changed the nature of acoustic processing, and thus separated the bat auditory system from that of other mammals.

The purpose of this chapter is to frame the issue of the generality of the auditory systems of echolocating bats more explicitly by considering the functional organization and mechanisms of the brainstem and then the forebrain auditory systems of several species of bats, and contrasting their features with those acknowledged as common in other mammals. In the following sections we outline the reasons that led us to two main hypotheses. The first hypothesis is that the information conveyed from the cochlea to the central nervous system is processed in a similar manner with similar circuitry by the brainstem auditory system in most, if not all, mammals.

The second is that the forebrain auditory system differs considerably among mammals and expresses species-specific features that are of adaptive value to the particular species. Furthermore, we conclude by proposing that both the brainstem and forebrain auditory systems of bats are good models of the mammalian auditory system, but for very different reasons.

2. The Anatomy and Functional Organization of the Auditory Brainstem Is Conserved

Here we propose that the information that the cochlea conveys to the brain is processed in fundamentally the same way by similar mechanisms and structures in the brainstem auditory systems of mammals. The hypothesis of a common mammalian processing strategy is supported by a large number of studies which show that the anatomical and functional features of the brainstem auditory system, from cochlear nucleus to inferior colliculus, appear to be conserved throughout mammalian evolution. The homologies among the auditory systems of bats and other mammals have been discussed in a number of previous reports (e.g., Zook and Casseday 1982a,b, 1985, 1987; Zook et al. 1985, Zook and Leake 1989; Ross, Pollak, and Zook 1988; Pollak and Casseday 1989; Kuwabara and Zook 1991, 1992; Grothe et al. 1992, 1994; Pollak 1992; Winer, Larue, and Pollak 1995).

This does not imply that the brainstem auditory systems among mammals are identical, because they are not. The size and complexity of some cell groups, for example, are more highly elaborated in certain echolocating bats than they are either in other bats or in other mammals. A case in point is the marginal cell group in the mustache bat's anteroventral cochlear nucleus (Zook and Casseday 1982a). Other cell groups are more highly developed in all the echolocating bats that have been studied than they are in other mammals. The most prominent example is the intermediate nucleus of the lateral lemniscus (Zook and Casseday 1982a; Covey and Casseday 1991; Schwartz 1992). Such differences, however, are not limited only to bats. Considerable variations exist among brainstem structures in all mammals. These variations are especially evident in some periolivary nuclei (Casseday, Covey, and Vater 1988; Schofield 1991; Kuwabara and Zook 1992; Schofield and Cant 1992; Grothe et al. 1994), in the arrangement of the cell groups comprising the olivocochlear feedback system (Bishop and Henson 1987, 1988; Ostwald and Aschoff 1988; Warr 1992; Grothe et al. 1994), and in the variable expression of the medial superior olive (Harrison and Irving 1966; Casseday, Covey, and Vater 1988; Covey, Vater, and Casseday 1991; Grothe et al. 1992; Kuwabara and Zook 1992; Grothe et al. 1994). Nevertheless, the differences in the brainstem auditory nuclei among mammals appear to be largely in the degree to which one or another cell group is developed (Harrison and Irving 1966; Zook and Casseday 1982a; Pollak and Casseday 1989; Covey and Casseday 1991; Kuwabara and Zook 1992; Schwartz 1992), or in the relative position of a homologous cell group (Brown and Howlett 1972; Pollak and Casseday 1989; Schwartz 1992; Warr 1992), rather than in the emergence of new nuclei or novel processing strategies. The global view of the mammalian auditory brainstem, then, is one of striking similarity and fundamental continuity across many dimensions.

3. The Medial Geniculate Body Has Species-Specific Features

In contrast to the auditory brainstem, certain aspects of forebrain organization differ markedly among mammals. These differences are especially apparent in the auditory thalamus, where disparate patterns of γ-aminobutyric acid (GABA)ergic and calbindin-like immunoreactivity have been found in the medial geniculate body of the rat (Celio 1990), chinchilla (Kelley et al. 1992), bat (Zettel, Carr, and O'Neill 1991), and macaque monkey (Jones and Hendry 1989). In the rat (Winer and Larue 1988) and mustache bat (Vater, Kössl, and Horn 1992; Winer, Wenstrup, and Larue 1992; Winer, Larue, and Pollak 1995), for example, about 1% of the medial geniculate cells are GABAergic. In contrast, 25%-30% of the cells in the medial geniculate body of the cat (Rinvik, Ottersen, and Storm-Mathisen 1987) and squirrel monkey (Smith, Séguela, and Parent 1987) are GABAergic. Such immunocytochemical diversity occurs not only among orders of mammals, but also between species of bats. This is illustrated by the marked difference in the proportion of GABAergic medial geniculate cells in the mustache bat (Winer, Wenstrup, and Larue 1992; Winer, Larue, and Pollak 1995) compared to the horseshoe bat (Vater, Kössl, and Horn 1992): the horseshoe bat has an abundance of GABAergic cells, especially in the ventral division of the geniculate, whereas the ventral division of the mustache bat's geniculate is virtually devoid of GABAergic cells.

The disparate patterns of GABA and calbindin in the medial geniculate body imply that the neurochemical adaptations in the medial geniculate body are pivotal evolutionary features subserving some important facet of species-specific signal processing. This proposition is supported by recent neurophysiological studies of the medial geniculate in both the mustache bat and squirrel monkey. Turning first to the mustache bat, Olsen and Suga (1991a) found a prominent tonotopic organization in the ventral division, where cells are sharply tuned to only one frequency. The tonotopic organization and tuning features of neurons in the ventral division correspond closely to the patterns in other mammals. However, novel features emerge in portions of the medial and dorsal divisions of the medial geniculate body. In those divisions there is a convergence of projections from two or more isofrequency contours to produce combinatorial response properties that are not present in the mustache bat's inferior colliculus (O'Neill 1985). As explained next, the combinatorial neurons integrate information across frequency contours, corresponding to the harmonic frequencies of the mustache bat's orientation calls, to create emergent features that are of importance to the bat. More recently, Olsen and Rauschecker (1992) also found combination-sensitive neurons in the squirrel monkey medial geniculate. However, the combinatorial properties are for features of the communication calls of the squirrel monkey, and are therefore different from the combinatorial properties in the mustache bat geniculate.

3.1 The Medial Geniculate Body of the Mustache Bat Illustrates That Combinatorial Properties Are Species Specific

One of the primary factors that render combinatorial neurons species specific is the specificity with which the response properties of neurons from one frequency contour are matched with the response properties of neurons from another frequency contour. This is clearly illustrated by the specific acoustic features that are required to drive the various types of combinatorial neurons in the forebrain of the mustache bat. Because the relevance of those features are only apparent when related to the structure of the bat's biosonar signals, we first briefly consider the composition of the mustache bat's orientation calls and the types of information conveyed in components of the calls, and then turn to the acoustic signals that best drive combinatorial neurons.

The calls emitted by the mustache bat are characterized by an initial long constant-frequency (CF) component, and a terminal brief, downward sweeping frequency-modulated (FM) component (Fig. 10.1; also see Moss and Schnitzler, Chapter 3). Each call is emitted with a fundamental frequency and four harmonics, but the second harmonic always contains the most energy. This species relies to an inordinate extent on the second harmonic of the CF component. The importance of this CF component is reflected in Doppler-shift compensation, whereby a flying bat continually adjusts the frequency of its emitted CF component to compensate for the upward Doppler shifts that are imposed on the echo by the flight speed of the bat relative to background objects (Schnitzler 1970; Trappe and Schnitzler 1982). The CF and FM components convey different categories of information. The fluttering wings of a flying insect impose frequency and amplitude modulations on the echo CF component, and the modulation patterns are used for target recognition (see Moss and Schnitzler, Chapter 3). In addition, the difference in frequency between the emitted CF component and the Doppler-shifted echo CF component may be used to determine target flight speed (Suga 1988; Olsen and Suga 1991a; see O'Neill,

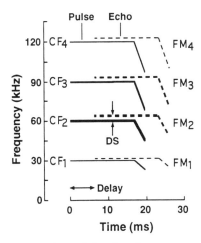

FIGURE 10.1. Schematic illustration of the time-frequency structure (sonogram) of the mustache bat's orientation call and its echo. Each call is emitted with four harmonics, and each harmonic has both a constant-frequency (CF) and frequency-modulated (FM) component. The second harmonic is always the most intense. The echo is higher in frequency than the emitted pulse because of Doppler shifts (DS). There is also a delay between the onset of the pulse and the onset of the echo due the distance between the bat and its target.

Chapter 9). In contrast, the FM component is used to evaluate target distance, which is conveyed by the time separating the emitted and echo FM components (O'Neill and Suga 1982; Olsen and Suga 1991b).

The combinatorial neurons are selective both for particular harmonic combinations and for one of the components of the bat's orientation call; one type of cell is selective for the frequencies in the CF components, the so-called CF/CF neurons, whereas other cells are selective for frequencies in the terminal FM components and are called FM-FM neurons (Olsen and Suga 1991a,b; also see Wenstrup, Chapter 8, and O'Neill, Chapter 9). The CF/CF neurons are characterized by their weak discharges to individual tone bursts and the vigorous firing when two tone bursts having different frequencies are presented together. They are also distinguished by their relative insensitivity to the temporal interval separating the presentation of the two frequencies. CF/CF neurons are tuned to two of the harmonics of the CF components of the bat's orientation calls; thus, CF_1-CF_2 neurons are driven only when the fundamental and second harmonic of the CF components are combined, and CF_1-CF_3 neurons by the fundamental and third harmonic. The exact frequency requirements for optimally driving each type of CF/CF neuron are slightly different among the population, in that each neuron is tuned to a small deviation from an exact harmonic relationship. Thus, the population of CF_1-CF_2 neurons are optimally driven by the range of CF_1 frequencies the bat would emit during Doppler-shift compensation, combined with the CF_2 frequencies that correspond to the stabilized echoes it receives from Doppler-shift compensation. Similar precise frequency pairings are expressed by the population of CF_1-CF_3.

FM-FM neurons are also most effectively driven by two harmonically related frequencies that correspond to the terminal FM components of the bat's orientation calls. Thus, there are FM_1-FM_2 neurons as well as FM_1-FM_3 and FM_1-FM_4 neurons. For these neurons the frequency specificity is not as strict as in the CF/CF neurons, but rather the neurons require that the two signals be presented in a specific temporal order, indicative of a target at a certain distance. Moreover, each neuron is tuned to a particular delay, its best delay, and the population expresses the range of pulse-echo delays that the mustache bat would normally receive during echolocation.

The foregoing features show the striking concordance between the highly specified spectral and temporal requirements of the signals that drive these neurons optimally and the spectral and temporal features of the biosonar signals that the mustache bat emits and receives. These combinatorial properties, then, are tailored to the mustache bat's orientation signals, and thus are unique to that animal.

4. The Functional Organization of the Mustache Bat's Auditory Cortex Is Species Specific

The various types of combinatorial and noncombinatorial neurons are elegantly arranged in the mustache bat cortex, and thereby illustrate the conceptually important point that the functional organization of the auditory cortex can be species specific. The auditory cortex of the mustache bat encompasses several cortical fields (Fig. 10.2, top left panel). One field has a pronounced tonotopic organization, and corresponds to the primary auditory cortices of other mammals. Within the tonotopically organized field, the 60-kHz contour, which Suga and Jen (1976) termed the "Doppler-shifted constant frequency" (DSCF) area, is greatly enlarged, as it is in all lower nuclei. Until recently, it was thought that the functional organization of the 60-kHz contour was similar to the frequency contours in the primary auditory cortices of other mammals (Suga and Jen 1976; Manabe, Suga, and Ostwald 1978). However, a recent study showed that combination sensitivity and delay tuning are also characteristics of neurons in the 60-kHz contour, but that these features are not present in the other frequency contours of the tonotopically organized field (Fitzpatrick et al. 1993). The properties of the combination sensitivity of neurons in the 60-kHz cortical contour also differ in a number of significant respects from the combinatorial properties of the CF/CF and FM-FM neurons described previously, and further illustrate the species-specificity of the mustache bat's auditory

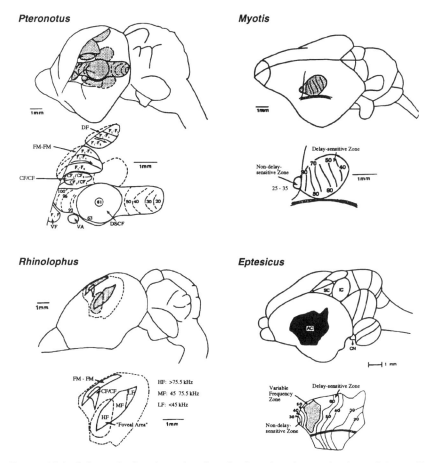

FIGURE 10.2. Schematic drawings showing the functional organization of the auditory corticies of four species of bats. On the *left* are shown two long CF/FM bats, the mustache bat (*Pteronotus parnellii, top*) and the horseshoe bat (*Rhinolophus* spp., *bottom*). On the *right* are shown two FM bats, the little brown bat (*Myotis lucifugus, top*) and the big brown bat (*Eptesicus fuscus, bottom*). For each species, the top drawing in each panel shows the location of the known auditory cortical fields in a side view of the brain and the bottom drawing showns an expanded view of the functional organization. Note that for the two CF/FM species, there is a segregation of combination-sensitive neurons into harmonically tuned (CF/CF) and delay-tuned (FM-FM) areas, distinct from the tonotopically organized zones of the primary auditory cortex. For the mustache bat, several additional combinatorial regions are shown: dorsal fringe area (DF), ventral fringe area (VF), and ventral area (VA). FM-FM areas contain a systematic map of best delays. By contrast, except for the extreme rostral part of the auditory cortex, delay-tuned cells are not completely segregated in FM bats: most cells within the tonotopically organized cortical field are also delay sensitive, and there is no clear map of best delays. CF/CF neurons are not found in FM bats, probably because they lack a long CF component in their biosonar signals. (Adapted from O'Neill, Chapter 9.)

cortex. Whether these features are formed in the cortex or are a consequence of processing in the geniculate has not been investigated.

Other cortical fields contain only CF/CF neurons or FM-FM neurons (Fig.10.2, top left panel) (Suga, O'Neill, and Manabe 1978, 1979; O'Neill and Suga 1982; Suga 1988; also see O'Neill, Chapter 9). Each of the combination-sensitive fields has additional levels of organization in which neurons that combine the first and second harmonics are arrayed in parallel with neurons that combine first and third, and first and fourth, harmonics. The CF/CF field is further characterized by an orderly change in the best frequencies of the neurons that is repeated in each of the harmonically related regions. The FM-FM region, on the other hand, is characterized by the orderly arrangement of neurons having different best delays, which creates a map of best delays that spans the harmonic regions. Finally, several fields that border the primary fields further elaborate the FM-FM delay-tuned features, and each field also expresses an orderly map of best delays (Suga and Horikawa 1986; Suga 1990).

In summary, both the response properties of neurons and their topographic arrangement in the mustache bat cortex are highly ordered. In addition, the functional organization corresponds to both the structure of the mustache bat's orientation calls and the types of information that the bat needs to extract from those signals. These features then appear to be tailored to the requirements of the mustache bat, which suggests that the functional organization of its cortex is species specific. Next, we show that in other bats the response properties of cortical neurons and their functional organization are different from that of the mustache bat. These data not only provide additional support for the hypothesis that mustache bat cortex is species specific, but suggest that the functional organization of the auditory cortices of other bats is also species specific.

5. Auditory Cortices of Other Bats Have Different Functional Organizations

The cortices of other bats also express combinatorial features, although no other species has been as thoroughly studied as the mustache bat. The general picture emerging from the studies of the auditory cortex of various species of bats is that the combinatorial requirements of their neurons, as well as their functional arrangement, are different, and sometimes markedly so, from that of the mustache bat.

Horseshoe bats, as one example, emit long CF/FM biosonar calls, similar to the orientation calls of the mustache bat. Horseshoe bats, like mustache bats, also have CF/CF and FM-FM cortical neurons that require harmonically related frequencies. The combinatorial neurons are topographically segregated from each other and from the tonotopically organized field (see Fig.10.2, lower left panel) (Schuller, O'Neill, and Ratke-Schuller 1991), features that are also comparable to the arrangement of the mustache bat cortex. There are, however, several notable differences between the horseshoe and mustache bats in the properties of the combinatorial cells and their arrangement in the cortex. One example is the properties of CF/CF neurons in the two species. The optimal frequency combinations in the horseshoe bat are not as strict as they are in the mustache bat, nor are the frequency combinations that best drive each of the cells arranged in an orderly manner, as they are in the mustache bat. Another example concerns the FM-FM neurons. Although the best delays of the neurons in the FM-FM field have an orderly arrangement, comparable to the arrangement in the mustache bat, in the horseshoe bat FM-FM neurons are represented in only one field. In the mustache bat, on the other hand, delay-tuned maps are present in several cortical fields.

The auditory cortices of both the big brown bat, *Eptesicus fuscus* (Jen, Sun, and Lin 1989; Dear et al. 1993) and the little brown bat, *Myotis lucifugus* (Sullivan 1982a,b; Wong and Shannon 1988; Berkowitz and Suga 1989; Maekawa, Wong, and Paschal 1992) have a very different functional organization from that of either the mustache or horseshoe bats (Fig. 10.2, right panels). While the cortices of the brown bats have delay-tuned neurons, the presence of CF/CF neurons has not been reported, and for reasons explained in a later section, it is unlikely that they would have such neurons. Delay-tuned neurons in the brown bats have some of the same combinatorial requirements as the delay-tuned neurons in the mustache bat forebrain. For instance, these cells respond best to different spectral components of the first and second signals when presented with a particular temporal delay (Sullivan 1982a,b; Berkowitz and Suga 1989; Maekawa, Wong, and Paschal 1992). However, the harmonic relationship is not required, as it is in the horseshoe and mustache bats, and the arrangement of neurons with different best delays are distributed in portions of the primary auditory cortex without apparent order (Wong and Shannon 1988; Berkowitz and Suga 1989; Dear et al.).

These studies show that although there are some shared properties of combinatorial neurons in the cortices of bats, there are also properties that are not shared and thus are different among species. In addition, these studies also show that the functional organization of the combinatorial regions is unique to each species.

6. Combinatorial Properties Reflect Adaptations for Some Ecological Niche and Are Selectively Advantageous

The studies reviewed here suggest that the combinatorial features expressed in the forebrain reflect adaptations for some ecological niche, and therefore may have evolved in response to selective pressures. This proposition is supported by the comparable combinatorial features in the forebrains of bats that exploit similar habitats. One example is the presence of CF/CF neurons in the forebrains of mustache and horseshoe bats. These bats, unlike most other bats, hunt for insects under the forest canopy (Bateman and Vaughan 1974; Neuweiler 1983, 1984). This habitat provides a rich source of food for which there is probably little other competition, but it is also an environment that requires the bat to detect and perceive prey in the midst of the echoes from background objects. To deal with the acoustic clutter, horseshoe and mustache bats evolved similar biosonar systems that emphasize a long constant-frequency component coupled with Doppler-shift compensation. Such a biosonar system greatly enhances the bat's ability to detect and recognize fluttering insects against a background of vegetation (Pollak and Casseday 1989; also see Moss and Schnitzler, Chapter 3, and Pollak and Park, Chapter 7). Concurrently, the bats also evolved numerous specialized features in their peripheral and central auditory systems for processing the CF component. Of interest in this regard is that horseshoe bats are found only in the Old World whereas mustache bats are confined to the tropics of the New World. Consequenly, the specialized constant-frequency sonar systems evolved independently in the two species. It follows, then, that there must also have been an independent genesis of CF/CF neurons in their forebrains, and that in both cases the genesis was driven by the selective advantages that accrue to a constant-frequency sonar system.

The proposition that combinatorial properties are adapted for an ecological niche is also supported by the studies of the forebrains of the brown bats. The brown bats employ a biosonar system that is very different from the long CF-FM bats and exploit a different ecological niche. These bats mainly use a brief, downward-sweeping FM chirp for echolocation, and generally hunt for insects in the open skies (Griffin 1958; Neuweiler 1983, 1984). While at times big brown bats add a short CF component to the FM chirp, the CF component is far less important to this bat and is used differently than it is by long CF-FM bats. This is evident because the CF component is only emitted when the bat is searching for prey and is eliminated once a target has been detected (Simmons, Fenton, and O'Farrell 1979). Furthermore, the brown bats do not compensate for Doppler shifts in the echo CF component. It is, therefore, not surprising that there have been no reports of CF/CF combinatorial neurons in the forebrains of the brown bats, and it is unlikely that these bats would have such neurons. Rather, the delay-tuned combinatorial neurons in their cortices emphasize the importance of extracting information from the FM chirp. Furthermore, the acoustic requirements of their delay-tuned combinatorial neurons, as well as the way they are functionally organized in the cortex, are different from the delay-tuned neurons of either the horseshoe bat or the mustache bat.

6.1 Other Mammals Should Also Have Combinatorial Neurons That Are Selectively Advantageous for the Particular Species

If portions of the auditory forebrains of bats evolved in response to selective pressures, then the forebrain of other mammals should also be adapted for processing specific classes of acoustic cues that confer selective advantages for the acoustic environment in which they evolved. This hypothesis gains additional credence from several recent reports concerned with the forebrain auditory systems of other mammals. One is the report by Olsen and Rauschecker (1992), mentioned previously, which showed that neurons in the medial geniculate of the squirrel monkey are sensitive to particular combinatorial features of the animal's communication calls. Others are recent studies of the auditory cortices of the cat and mouse, which have revealed specialized regions that may be analogous, if not homologous, to the combinatorial cortical regions in the long CF-FM bats. Sutter and Schreiner (1991), for example, reported that the dorsal region in the cat primary auditory cortex has populations of cells tuned to two, or in some cases three, frequencies. The response properties of these neurons are in some ways similar to those of CF/CF neurons, but in other ways differ considerably from the features that characterize those combinatorial neurons in the mustache bat. The authors point out that these neurons may be involved in complex sound processing, and suggest that the region of the cat cortex in which these neurons are found may be comparable to the CF/CF region in the cortex of the mustache bat. Similarly, Stiebler (1987) and Hoffstetter and Ehret (1992) reported that a separate cortical field in the mouse, adjacent to the primary auditory cortex, represents only high ultrasonic frequencies and lacks the prominent tonotopic organization characteristic of primary auditory cortex. These authors emphasize that this region is probably important for the analysis of ultrasonic signals related to mother–infant and sexual interactions, and also suggest that this region may be comparable to the combinatorial regions in the mustache bat's cortex.

The studies cited appear to support the general hypothesis that portions of the forebrain are adapted to extract features of acoustic signals, and that the particular features extracted may be species specific. However, we have little understanding of exactly what signal features are important to these animals, and what combinatorial features each of these animals needs to extract from those signals. The combinatorial features in the forebrain of bats relate to the information in the pulse–echo combinations they receive while echolocating. However, these features must surely be different from those that are important to mice or cats. Furthermore, relevant combinatorial features most likely differ among mammals as distantly related as rodents, carnivores, and primates.

7. Some Unifying Themes

The studies reviewed in the previous sections lead us to propose some fundamental differences between the functional organization of the auditory brainstem and auditory forebrain. Our scenario assumes that all mammals need to process similar types of acoustic cues in the lower stages of their central auditory systems, and that these cues are processed in the same way with similar mechanisms and structures in their auditory brainstems. Among the universal cues that must be extracted are binaural cues for spatial localization, temporal and spectral cues for the recognition of communication and other signals, and intensity cues for judgments of loudness. The cues in each spectral component of the signal are coded and represented in the frequency contours of the brainstem auditory nuclei. The coded cues are then partially recombined in the auditory forebrain, resulting in the emergence of new, combinatorial properties. The novel feature of the recombination is the integration of information from two or more frequency contours, thereby partially reconstituting the spectrum of the original signal. We are, therefore, led to the view that one of the functional distinctions between the brainstem and the forebrain is the role that each region plays in the decomposition of the spectrum of an acoustic signal, and subsequently in the partial reconstitution of the original spectrum. We envision the decomposition and reconstruction as a three-stage process that entails different strategies for signal processing in the cochlea, brainstem, and forebrain.

The first stage occurs in the cochlea where the spectral components of a sound are dissembled and the constituent frequencies are arranged as a place code along the cochlear partition. The second stage occurs in the brainstem, where the spectral decomposition is preserved in the various frequency channels of the auditory pathway. Thus, each frequency is processed in a series of parallel pathways that ultimately converge in the inferior colliculus. Some of the pathways merge the information from the two ears, and thus begin to code for the spatial location of the signal. Other pathways maintain the separation of the two ears, and thus process information that is not influenced by the signals received at the other ear (e.g., Pollak and Casseday 1989). The afferents to the lateral superior olive (LSO) and its ascending projections to the inferior colliculus constitute one of the binaural parallel pathways, while the projections from the dorsal cochlear nucleus to the inferior colliculus exemplify one of the monaural parallel pathways. The significant features of the parallel pathways are that processing is accomplished on a frequency-by-frequency basis in the isofrequency contours of each successive nucleus, and that the structures, connections, and mode of processing in each nucleus are fundamentally similar among mammals.

The third stage begins in the medial geniculate, which is the first site in the auditory system where information from two or more frequencies is integrated to create combination-sensitive neurons. By partially recombining the spectral features of the original signal, combinatorial neurons represent some aspect of the signal that cannot be represented adequately by neurons tuned to a single frequency. Thus, one of the features that distinguishes forebrain from brainstem processing is the partial reconstitution of the original spectrum by integrating information across frequency contours. The processing of combinatorial properties is then further elaborated in the various fields of the auditory cortex.

One common denominator of the mammalian forebrain, then, may be the presence of combinatorial properties that are initially created in the medial geniculate body and are functionally arranged in cortical fields adjacent to the primary auditory cortex. However, it appears that the medial geniculate possesses an evolutionary plasticity which allows for the construction of a diversity of combinatorial properties in different species. One indication of such a plasticity is the disparate patterns of neurochemical indicators in the medial geniculate body, which suggests that different forebrain strategies for processing information evolved among mammals. These considerations lead us to hypothesize that the medial geniculate does not construct a standard set of combinatorial response features that are present, with minor variation, in the forebrains of all mammals. Rather, what is constructed are species-specific combinatorial properties that are tailored to the needs of the species. The unique properties emerge from the ways in which various response properties of neurons from different frequency contours are integrated in the medial geniculate body. Furthermore, the disparate neurochemical patterns, particularly the differences in the population of GABAergic medial geniculate neurons, suggest that the means by which integration is achieved may also be different among species. The net result, however, is that the emergent combinatorial properties provide selective advantages to that animal for the perception of its communication signals or, in the case of echolocating bats, for the representation of objects in the external world.

8. The Brainstem and Forebrain Auditory Systems of Bats Are Both Good Models of the Mammalian Auditory System

We now return to the original topic of this chapter: whether the bat auditory system is an appropriate model for the mammalian auditory system. We propose that because the brainstem auditory system is highly conserved, the bat brainstem auditory system is a very good model of the mammalian auditory system. The contrary assertion, that the bat auditory brainstem possesses unique attributes that are fundamentally different from those of other, more generalized mammals, is inconsistent with the wealth of data showing a fundamental continuity in its physiological and anatomical arrangement.

Auditory forebrains, on the other hand, are more variable in their functional organization and apparently have species-specific features, which suggests that there is no prototypical model of the mammalian auditory forebrain. It follows that the functional organization of the forebrain is revealed only when probed with the "appropriate" stimuli. Furthermore, knowledge of what constitutes "appropriate" stimulation comes from an appreciation of the acoustic cues on which a particular species relies, and which features of the signals its auditory forebrain is designed to reconstruct. One advantage that accrues to using echolocating bats for neurophysiological investigations of the auditory system is that the way in which they manipulate their orientation calls is known in considerable detail, as are the acoustic cues from which these animals derive information about objects in the external world. This knowledge allows investigators to generate signals that mimic the types of signals that bats emit and receive during echolocation, as well as the sorts of acoustic cues that are contained in the echoes reflected from flying insects. The auditory forebrain of bats in general, and the mustache bat in particular, are excellent examples of species-specific adaptation in the mammalian auditory forebrain, in that they illustrate how a particular species solves the problem of constructing and uniquely organizing combinatorial properties of acoustic attributes that are of clear importance for that animal's perception of its external world. Deriving the rules that govern such processes in the forebrain will provide insights into the biological strategies of information processing that evolved in species with diverse acoustic behaviors.

Acknowledgments. We thank Evan Balaban, Mike Ferrari, David McAlpine, Lynn McAnelly, Tom Park, Carl Resler, Mike Ryan, Brett Schofield, Wesley Thompson, Harold Zakon, and John Zook for their helpful comments on this manuscript. Supported by NIH grant DC 20068.

References

Bateman GC, Vaughan TA (1974) Nightly activities of mormoopid bats. J Mammal 55:45–65.

Berkowitz A, Suga N (1989) Neural mechanisms of ranging are different in two species of bats. Hear Res 5:317–335.

Bishop AL, Henson OW Jr (1987) The efferent projections of the superior olivary complex in the mustached bat. Hear Res 31:175–182.

Bishop AL, Henson OW Jr (1988) The efferent auditory system in Doppler-shift compensating bats. In: Nachtigall PE, Moore PWB (eds) Animal Sonar: Processes and Performance. New York: Plenum, pp. 307–310.

Brown JC, Howlett B (1972) The olivo-cochlear tract in the rat and its bearing on the homologies of some constituent cell groups of the mammalian superior olivary complex: a thiochline study. Acta Anat 83:505–526.

Casseday JH, Covey E, Vater M (1988) Connections of the superior olivary complex in the rufous horseshoe bat, *Rhinolophus rouxi*. J Comp Neurol 278:313–329.

Celio MR (1990) Calbindin D- 28k and parvalbumin in the rat nervous system. Neuroscience 35:375–475.

Covey E, Casseday JH (1991) The monaural nuclei of the lateral lemniscus in an echolocating bat: parallel pathways for analyzing temporal features of sound. J Neurosci 11:3456–3470.

Covey E, Vater M, Casseday JH (1991) Binaural properties of single units in the superior olivary complex of the mustached bat. J Neurophysiol (Bethesda) 66:1080–1094.

Dear SP, Fritz J, Haresign T, Ferragamo M, Simmons JA (1993) Tontopic and functional organization in the auditory cortex of the big brown bat, *Eptesicus fuscus*. J Neurophysiol (Bethesda) 70:1988–2009.

Fitzpatrick DC, Kanwal J, Butman JA, Suga N (1993) Combination sensitive neurons in the primary auditory cortex of the mustached bat. J Neurosci 13:931–940.

Griffin DR (1958) Listening in the Dark. New Haven: Yale University Press.

Grothe B, Vater M, Casseday JH, Covey E (1992) Monaural interaction of excitation and inhibition in the medial superior olive of the mustached bat: an adapation for biosonar. Proc Natl Acad Sci USA 89:5108–5112.

Grothe B, Schweizer H, Pollak GD, Schuller G, Rosemann C. (1994) Anatomy and projection patterns of the superior olivary complex in the Mexican free-tailed bat, *Tadarida brasiliensis mexicana*. J Comp Neurol 343:630–646.

Harrison JM, Irving R (1966) Visual and nonvisual auditory systems in mammals. Science 154:738–742.

Hoffstetter KM, Ehret G (1992) The auditory cortex of the mouse: connections of the ultrasonic field. J Comp Neurol 323:370–386.

Jen PH-S, Sun X, Lin PJJ (1989) Frequency and space representation in the primary auditory cortex of the frequency modulating bat, *Eptesicus fuscus*. J Comp Physiol A 165:1–14.

Jones EG, Hendry SHC (1989) Differential calcium binding proten immunoreactivity distinguishes classes of relay neurons in monkey thalamic nuclei. Eur J Neuroci 1:222–246.

Kelly PE, Frisina RD, Zettel ML, Walton JP (1992) Differential calbindin-like immunoreactivity in the brain stem auditory system of the chinchilla. J Comp Neurol 320:196–212.

Kuwabara N, Zook JM (1991) Classifying the principal cells of the medial nucleus of the trapezoid body. J Comp Neurol 314:707–720.

Kuwabara N, Zook JM (1992) Projections to the medial superior olive from the medial and lateral nuclei of the trapezoid body in rodents and bats. J Comp Neurol 324:522–538.

Maekawa M, Wong D, Paschal WG (1992) Spectral sensitivity of FM-FM neurons in the auditory cortex of the echolocating bat, *Myotis lucifugus*. J Comp Physiol A 171:513–522.

Manabe T, Suga N, Ostwald J (1978) Aural representation in the Doppler-shifted-CF processing area of the primary auditory cortex of the mustached bat. Science 200:339–342.

Neuweiler G (1983) Echolocation and adaptivity to ecological constraints. In: Huber

F, Markl H (eds) Neuroethology and Behavioral Physiology: Roots and Growing Pains. Berlin Heidelberg: Springer-Verlag, pp. 280–302.

Neuweiler G (1984) Foraging, echolocation and audition in bats. Naturwissenschaften 71:446–455.

Olsen JF, Suga N (1991a) Combination sensitive neurons in the medial geniculate body of the mustached bat: encoding of relative velocity information. J Neurophysiol (Bethesda) 65:1254–1274.

Olsen JF, Suga N (1991b) Combination sensitive neurons in the medial geniculate body of the mustached bat: encoding of target range information. J Neurophysiol (Bethesda) 65:1275–1296.

Olsen JF, Rauschecker JP (1992) Medial geniculate neurons in the squirrel monkey sensitive to combinations of components in a species-specific vocalization. Soc Neurosci Abstr 18:883.

O'Neill WE (1985) Responses to pure tones and linear FM components of the CF/FM biosonar signals by single units in the inferior colliculus of the mustached bat. J Comp Physiol A 157:797–815.

O'Neill WE, Suga N (1982) Encoding of target-range information and its representation in the auditory cortex of the mustached bat. J Neurosci 47:225–255.

Ostwald J, Aschoff A (1988) Only one nucleus in the brainstem projects to the cochela in horseshoe bats: the nucleus olivo-cochlearis. In: Nachtigall PE, Moore PWB (eds) Animal Sonar: Processes and Performance. New York: Plenum, pp. 347–352.

Park TJ, Larue DT, Winer JA, Pollak GD (1991) Glycine and GABA in the superior olivary complex of the mustache bat: projections to the central nucleus of the inferior colliculus. Soc Neurosci Abstr 17(1):300.

Pollak GD (1992) Adaptations of basic structures and mechanisms in the cochlea and central auditory pathway of the mustache bat. In: Popper AN, Fay RR, Webster DB (eds) Evolutionary Biology of Hearing. New York: Springer- Verlag, pp. 751–778.

Pollak GD, Casseday JH (1989) The Neural Basis of Echolocation in Bats. Berlin Heidelberg New York: Springer-Verlag.

Rinvik E, Ottersen OP, Storm-Mathisen J (1987) Gamma-aminobutyrate-like immunoreactivity in the thalamus of the cat. Neuroscience 21:787–805.

Ross LS, Pollak GD, Zook JM (1988) Origin of ascending projections to an isofrequency region of the mustache bat's inferior colliculus. J Comp Neurol 270:488–505.

Schofield BR (1991) Superior paraolivary nucleus in the pigmented guinea pig: separate classes of neurons project to the inferior colliculus and the cochlear nuclues. J Comp Neurol 312:68–76.

Schofield BR, Cant NB (1992) Organization of the superior olivary complex in the guinea pig: II. Patterns of projections from the periolivary nuclei to the inferior colliculus. J Comp Neurol 317:438–455.

Schnitzler H-U (1970) Comparison of echolocation behavior in *Rhinolophus ferrumequinum* and *Chilonycteris rubiginosa*. Bijdr Dierkd 40:77–80.

Schuller G, O'Neill WE, Ratke-Schuller S (1991) Facilitation and delay sensitivity of auditory cortex neurons in CF-FM bats, *Rhinolophus rouxi* and *Pteronotus parnellii*. Eur J Neurosci 3:1165–1181.

Schwartz IR (1992) The superior olivary complex and lateral lemniscal nuclei. In: Webster DB, Popper AN, Fay RR (eds) The Mammalian Auditory Pathway: Neuroanatomy, Vol. 1. New York: Springer-Verlag, pp. 117–167.

Simmons JA, Fenton MB, O'Farrell M (1979) Echolocating and the pursuit of prey by bats. Science 203:16–21.

Smith Y, Séguela P, Parent A (1987) Distribution of GABA-immunoreactive neurons in the thalamus of the squirrel monkey (*Saimiri sciureus*). Neuroscience 2:579–591.

Stiebler I (1987) A distinct ultrasound-processing area in the auditory cortex of the mouse. Naturwissenschaften 74:96–97.

Suga N (1988) Auditory neuroethology and speech processing: complex-sound processing by combination sensitive neurons. In: Edelman GM, Gall EW, Cowan MW (eds) Auditory Function: Neurobiological Bases of Hearing. New York: Wiley, pp. 679–720.

Suga N (1990) Cortical computational maps for auditory imaging. Neural Networks 3:3–21.

Suga N, Horikawa J (1986) Multiple time axes for representation of echo delays in the auditory cortex of the mustached bat. J Neurophysiol 55:776–805.

Suga N, Jen PH-S (1976) Disproportionate tonotopic representation for processing species-specific CF-FM sonar signals in the mustache bat auditory cortex. Science 194:542–544.

Suga N, O'Neill WE, Manabe T (1978) Cortical neurons sensitive to particular combinations of information bearing elements of bio-sonar signals in the mustache bat. Science 200:778–781.

Suga N, O'Neill WE, Manabe T (1979) Harmonic sensitive neurons in the auditory cortex of the mustached bat. Science 203:270–274.

Sullivan WE (1982a) Neural representation of target distance in auditory cortex of the echolocating bat, *Myotis lucifugus*. J Neurophysiol (Bethesda) 48:1011–1031.

Sullivan WE (1982b) Possible neural mechanisms of target distance coding in the auditory system of the echolocating bat, *Myotis lucifugus*. J Neurophysiol (Bethesda) 48:1032–1047.

Sutter ML, Schreiner CE (1991) Physiology and topography of neurons with multipeaked tuning curves in cat primary auditory cortex. J Neurophyisol (Bethesda) 65:1207–1226.

Trappe M, Schnitzler H-U (1982) Doppler-shift compensation in insect-catching horseshoe bats. Naturwissenschaften 69:193–194.

Vater M, Kössl M, Horn AKE (1992) GAD- and GABA-immunoreactivity in the ascending auditory pathway of horseshoe and mustache bats. J Comp Neurol 325:183–206.

Warr WB (1992) Organization of olivocochlear efferent systems. In: Webster DB, Popper AN, Fay RR (eds) The Mammalian Auditory Pathway: Neuroanatomy, Vol.1. New York: Springer-Verlag, pp. 410–448.

Winer JA, Larue DT (1988) Anatomy of glutamic acid decarboxylase immunoreactive-reaction neurons and axons in the rat medial geniculate body. J Comp Neurol 278:47–68.

Winer JA, Wenstrup JJ, Larue DT (1992) Patterns of GAB Aergic immunoreactivity define subdivisions of the mustached bat's medial geniculate body. J Comp Neurol 319:172–190.

Winer JA, Larue DT, Pollak GD (1995) GABA and glycine in the central auditory system of the mustache bat: Structural substrates for inhibitory neuronal organization. J Comp Neurol (in press).

Wong D, Shannon SL (1988) Functional zones in the auditory cortex of the echolocating bat, *Myotis lucifugus*. Brain Res 453:349–352.

Zettel ML, Carr CE, O'Neill WE (1991) Calbindin-like immunoreactivity in the central auditory system of the mustached bat, *Pteronotus parnellii*. J Comp Neurol 313:1–16.

Zook JM, Casseday JH (1982a) Cytoarchitecture of auditory systems in lower brainstem of the mustache bat, *Pteronotus parnellii*. J Comp Neurol 207:1–13.

Zook JM, Casseday JH (1982b) Origin of ascending projections to inferior colliculus in the mustache bat, *Pteronotus parnellii*. J Comp Neurol 207:14–28.

Zook JM, Casseday JH (1985) Projections from the cochlear nuclues in the mustache bat, *Pteronotus parnellii*. J Comp Neurol 237:307–324.

Zook JM, Casseday JH (1987) Convergence of ascending pathways at the inferior colliculus of the mustache bat, *Pteronotus parnellii*. J Comp Neurol 261:347–361.

Zook JM, Leake PA (1989) Connections and frequency representation in the auditory brainstem of the mustache bat, *Pteronotus parnellii*. J Comp Neurol 290:243–261.

Zook JM, Winer JA, Pollak GD, Bodenhamer RD (1985) Topology of the central nucleus of the mustache bat's inferior colliculus: correlation of single unit properties and neuronal architecture. J Comp Neurol 231:530–546.

Index

Indexing is by scientific name of the various species. When the generic name is used alone this usually is because the reference is to several different species of the same genus. Most vertebrate species are referenced under the scientific names. The only exceptions are very common species such as mouse, rat, cat, chicken, dog, and human. A glossary of scientific and common names of most bat species used in this volume can be found in the appendix to chapter 2.

Sound localization is referred to in the index as sound source localization.